Science of Fullerenes and Carbon Nanotubes

Science of Fullerenes and Carbon Nanotubes

M. S. DRESSELHAUS
Department of Electrical Engineering and Computer Science and Department of Physics
Massachusetts Institute of Technology, Cambridge, Massachusetts

G. DRESSELHAUS
Francis Bitter National Magnet Laboratory
Massachusetts Institute of Technology, Cambridge, Massachusetts

P. C. EKLUND
Department of Physics and Astronomy and Center for Applied Energy Research
University of Kentucky, Lexington, Kentucky

ACADEMIC PRESS
San Diego Boston New York London Sydney Tokyo Toronto

This book is printed on acid-free paper. ♾

Academic Press, Inc.
A Division of Harcourt Brace & Company
525 B Street, Suite 1900, San Diego, California 92101-4495

United Kingdom Edition published by
Academic Press Limited
24-28 Oval Road, London NW1 7DX

Library of Congress Cataloging-in-Publication Data

Dresselhaus, M. S.
Science of fullerenes and carbon nanotubes / by M. S. Dresselhaus,
G. Dresselhaus, P. C. Eklund.
p. cm.
Includes bibliographical references and index.
ISBN 0-12-221820-5 (alk. paper)
1. Fullerenes. I. Dresselhaus, G. II. Eklund, P. C.
III. Title.
QD181.C1D74 1995
620.1'93--dc20 95-20529
CIP

PRINTED IN THE UNITED STATES OF AMERICA
95 96 97 98 99 00 MM 9 8 7 6 5 4 3 2 1

Contents

Preface . . . . *xvii*

CHAPTER 1
Historical Introduction **1**

1.1. Early History 1
1.2. Astronomical Observations 3
1.3. Carbon Cluster Studies 3
1.4. Recent History 6
1.5. Architectural Analogs 7
1.6. Biological and Geological Examples 9
1.7. Road Map 12
References 13

CHAPTER 2
Carbon Materials **15**

2.1. General Considerations 15
2.2. Graphite 18
2.3. Graphite Materials 20
2.4. Graphite Whiskers 21
2.5. Carbon Fibers 21
2.6. Glassy Carbon 24
2.7. Carbon Blacks 25
2.8. Carbon Coated Carbide Particles 29
2.9. Carbynes 31
2.10. Carbolites 32
2.11. Amorphous Carbon 33

2.12. Porous Carbons 34
2.13. Liquid Carbon 36
2.14. Graphite Intercalation Compounds 37
2.15. Diamond 39
2.16. Other Diamond Materials 41
2.16.1. Chemical Vapor Deposited Diamond Films 41
2.16.2. Diamond-like Carbon Films 43
2.17. Diamond-like and Cage Hydrocarbon Molecules 44
2.18. Synthesis of a Fully Unsaturated, All-Carbon Polymer 46
2.19. Metallo-Carbohedrenes (Met-Cars) 48
2.20. Carbon-Free Fullerenes 50
2.21. Metal-Coated Fullerenes 51
References 54

CHAPTER 3
Structure of Fullerenes 60

3.1. Structure of C_{60} and Euler's Theorem 60
3.2. Structure of C_{70} and Higher Fullerenes 66
3.3. The Projection Method for Specifying Fullerenes 74
References 77

CHAPTER 4
Symmetry Considerations of Fullerene Molecules 80

4.1. Icosahedral Symmetry Operations 80
4.2. Symmetry of Vibrational Modes 86
4.3. Symmetry for Electronic States 90
4.4. Going from Higher to Lower Symmetry 96
4.4.1. Symmetry Considerations for C_{70} 96
4.4.2. Symmetry Considerations for Higher-Mass Fullerenes 104
4.5. Symmetry Considerations for Isotopic Effects 104
References 109

CHAPTER 5
Synthesis, Extraction, and Purification of Fullerenes 110

5.1. Synthesis of Fullerenes 111
5.1.1. Historical Perspective 111
5.1.2. Synthesis Details 112
5.2. Fullerene Extraction 116
5.2.1. Solvent Methods 116
5.2.2. Sublimation Methods 118
5.2.3. Solubility of Fullerenes in Solvents 120
5.3. Fullerene Purification 121
5.3.1. Solvent Methods 121

5.3.2. Sublimation in a Temperature Gradient 125
5.3.3. Gas–Phase Separation and Purification 128
5.3.4. Vaporization Studies of C_{60} 129
5.4. Endohedral Fullerene Synthesis 131
5.5. Health and Safety Issues 138
References 138

CHAPTER 6
Fullerene Growth, Contraction, and Fragmentation 143

6.1. Fullerene Growth Models 143
6.1.1. Stone–Wales Model 144
6.1.2. Model for C_2 Absorption or Desorption 147
6.1.3. Fullerene Growth from a Corannulene Cluster 148
6.1.4. Transition from C_{60} to C_{70} 150
6.2. Mass Spectrometry Characterization 152
6.3. Stability Issues 153
6.4. Fullerene Contraction and Fragmentation 156
6.4.1. Photofragmentation 157
6.4.2. Collision of Fullerene Ion Projectiles 159
6.4.3. Collision of Fullerene Ions with Surfaces 163
6.4.4. Fragmentation of C_{60} by Energetic Ions 165
6.5. Molecular Dynamics Models 166
References 168

CHAPTER 7
Crystalline Structure of Fullerene Solids 171

7.1. Crystalline C_{60} 171
7.1.1. Ambient Structure 173
7.1.2. Group Theory for Crystalline Phases 175
7.1.3. Low-Temperature Phases 184
7.1.4. Merohedral Disorder 193
7.1.5. Model for Phase Transitions in C_{60} 194
7.2. Crystalline C_{70} and Higher-Mass Fullerenes 197
7.3. Effect of Pressure on Crystal Structure 203
7.4. Effect of Temperature on Crystal Structure 208
7.5. Polymerized Fullerenes 209
7.5.1. Photopolymerization of C_{60} 209
7.5.2. Electron Beam-Induced Polymerization of C_{60} 212
7.5.3. Pressure-Induced Polymerization of C_{60} 213
7.5.4. Plasma-Induced Polymerization of C_{60} 214
7.5.5. Photopolymerization of C_{70} Films 215
References 217

CHAPTER 8
Classification and Structure of Doped Fullerenes **224**

8.1. Classification of Types of Doping for Fullerenes 225
8.2. Endohedral Doping 228
8.3. Substitutional Doping 233
8.4. Exohedral Doping 234
8.4.1. Exohedral Charge Transfer Compounds 235
8.4.2. Exohedral Clathrate C_{60} Compounds 236
8.4.3. Positron Exohedral Trapping 238
8.5. Structure of Alkali Metal-Doped C_{60} Crystals 238
8.5.1. M_3C_{60} Alkali Metal Structures 241
8.5.2. M_1C_{60}, M_4C_{60}, and M_6C_{60} Structures (M = K, Rb, Cs) . . 247
8.5.3. Na_xC_{60} Structures 249
8.5.4. Structures for $M_2M'_1C_{60}$ and Related Compounds 251
8.5.5. Structure of Ammoniated M_3C_{60} Compounds 253
8.6. Structure of Alkaline Earth-Doped C_{60} 253
8.6.1. Ca_xC_{60} Structure 254
8.6.2. Sr_xC_{60} and Ba_xC_{60} Structures 254
8.6.3. Rare Earth R_xC_{60} Structures 256
8.7. Structure of Other Crystalline Fulleride Phases 256
8.7.1. Halogen-Doped Fullerenes 257
8.7.2. Fullerene Epoxides and Related Compounds 258
8.7.3. TDAE–C_{60} 260
8.7.4. Metal–C_{60} Multilayer Structures 261
8.8. The Doping of C_{70} and Higher-Mass Fullerenes 262
References 263

CHAPTER 9
Single-Crystal and Epitaxial Film Growth **271**

9.1. Single-Crystal Growth 272
9.1.1. Synthesis of Large C_{60} and C_{70} Crystals by Vapor Growth . 272
9.2. Thin-Film Synthesis 274
9.2.1. Epitaxial C_{60} Films 276
9.2.2. Free-Standing C_{60} Films, or Fullerene Membranes 281
9.3. Synthesis of Doped C_{60} Crystals and Films 283
9.3.1. Charge Transfer Compounds Based on Alkali Metals and Alkaline Earths 285
9.3.2. Fullerene-Based Clathrate Materials 288
References 289

CHAPTER 10
Fullerene Chemistry and Electrochemistry **292**

10.1. Practical Considerations in Fullerene Derivative Chemistry 293
10.2. General Characteristics of Fullerene Reactions 294

10.3. Reduction and Oxidation of C_{60} and C_{70} 297
10.3.1. Fullerene Reduction—General Remarks 298
10.3.2. Electrochemical Fullerene Reduction and Oxidation . . 300
10.3.3. Chemical Oxidation of C_{60} 303
10.4. Hydrogenation, Alkylation, and Amination 304
10.5. Halogenation Reactions 307
10.6. Bridging Reactions 307
10.7. Cycloaddition Reactions 312
10.8. Substitution Reactions 314
10.9. Reactions with Free Radicals 315
10.10. Host–Guest Complexes and Polymerization 316
10.10.1. Host–Guest Complexes 316
10.10.2. Polymerization 320
References 325

CHAPTER 11
Vibrational Modes 329

11.1. Overview of Mode Classifications 329
11.2. Experimental Techniques 331
11.3. C_{60} Intramolecular Modes 332
11.3.1. The Role of Symmetry and Theoretical Models 332
11.3.2. Theoretical Crystal Field Perturbation of the Intramolecular Modes of C_{60} 339
11.3.3. Isotope Effects in the Vibrational and Rotational Spectra of C_{60} Molecules 341
11.4. Intermolecular Modes 347
11.5. Experimental Results on C_{60} Solids and Films 355
11.5.1. Raman-Active Modes 355
11.5.2. Infrared-Active Modes 356
11.5.3. Higher-Order Raman Modes in C_{60} 358
11.5.4. Higher-Order Infrared Modes in C_{60} 361
11.5.5. Isotope Effects in the Raman Spectra 364
11.5.6. Silent Intramolecular Modes in C_{60} 364
11.5.7. Luminescence Studies of Vibrations in C_{60} 366
11.5.8. Neutron Scattering 367
11.5.9. HREELS Study of Vibrational Modes 369
11.5.10. Vibrational Modes as a Probe of Phase Transitions . . . 370
11.5.11. Vibrational States Associated with Excited Electronic States 375
11.6. Vibrational Modes in Doped Fullerene Solids 376
11.6.1. Doping Dependence of Raman-Active Modes 377
11.6.2. Doping Dependence of the F_{1u}-Derived Intramolecular Modes 382
11.6.3. Other Doping-Dependent Intramolecular Effects 388
11.6.4. Doping-Dependent Intermolecular Effects 388

11.7. Vibrational Spectra for C_{70} and Higher Fullerenes 390
11.7.1. Intramolecular Vibrational Spectra for C_{70} 391
11.7.2. Intermolecular Vibrational Spectra for C_{70} 392
11.7.3. Vibrational Modes in Doped C_{70} Solids 394
11.8. Vibrational Spectra for Phototransformed Fullerenes 397
11.9. Vibrational Spectra for C_{60} under Pressure 402
11.10. Vibrational Spectra of Other Fullerene-Related Materials 406
References 406

CHAPTER 12
Electronic Structure **413**

12.1. Electronic Levels for Free C_{60} Molecules 414
12.1.1. Models for Molecular Orbitals 414
12.1.2. Level Filling for Free Electron Model 416
12.2. Symmetry-Based Models 419
12.3. Many-Electron States for C_{60} and Other Icosahedral Fullerenes . 421
12.4. Multiplet States for Free Ions $C_{60}^{n\pm}$ 423
12.4.1. Ground States for Free Ions $C_{60}^{n\pm}$ 424
12.4.2. Excited States for Negative Molecular Ions C_{60}^{n-} 424
12.4.3. Positive Molecular Ions C_{60}^{n+} 428
12.5. Excitonic States for C_{60} 429
12.6. Molecular States for Higher-Mass Fullerenes 432
12.6.1. Molecular States for C_{70} 432
12.6.2. Metallofullerenes 435
12.7. Electronic Structure of Fullerenes in the Solid State 437
12.7.1. Overview of the Electronic Structure in the Solid State . 437
12.7.2. Band Calculations for Solid C_{60} 439
12.7.3. Band Calculations for Alkali Metal-Doped C_{60} 442
12.7.4. Electronic Structure of Alkaline Earth-Doped C_{60} . . . 447
12.7.5. Band Calculations for Other Doped Fullerenes 450
12.7.6. Band Calculations for Other Fullerenes 452
12.7.7. Many-Body Approach to Solid C_{60} 453
References 458

CHAPTER 13
Optical Properties **464**

13.1. Optical Response of Isolated C_{60} Molecules 464
13.1.1. Introduction to Molecular Photophysics 465
13.1.2. Dipole-Allowed Transitions 468
13.1.3. Optical Transitions between Low-Lying Herzberg–Teller Vibronic States 469
13.2. Optical Studies of C_{60} in Solution 476
13.2.1. Absorption of C_{60} in Solution 476
13.2.2. Photoluminescence of C_{60} in Solution 480

13.2.3. Dynamics of Excited States in Isolated C_{60} Molecules . . 483
13.2.4. Optically Detected Magnetic Resonance (ODMR) for Molecules 488
13.3. Optical Properties of Solid C_{60} 491
13.3.1. Overview 492
13.3.2. Optical Absorption in C_{60} Films 495
13.3.3. Low-Frequency Dielectric Properties 506
13.3.4. Optical Transitions in Phototransformed C_{60} 508
13.4. Optical Properties of Doped C_{60} 511
13.4.1. Optical Absorption of C_{60} Anions in Solution 512
13.4.2. Optical Properties of M_6C_{60} 517
13.4.3. Normal State Optical Properties of M_3C_{60} 523
13.4.4. Superconducting State Optical Properties 528
13.4.5. Optical Properties of M_1C_{60} Compounds 531
13.5. Optical Properties of C_{60}–Polymer Composites 533
13.6. Optical Properties of Higher-Mass Fullerenes 536
13.6.1. Optical Properties of C_{70} 537
13.7. Dynamic and Nonlinear Optical Properties of Fullerenes 540
13.7.1. Dynamical Properties 540
13.7.2. Nonlinear Absorption Effects 544
References 548

CHAPTER 14
Transport and Thermal Properties 556

14.1. Electrical Conductivity 557
14.1.1. Dependence on Stoichiometry 557
14.1.2. Temperature Dependence 564
14.1.3. Alkaline Earth-Doped C_{60} 566
14.1.4. Alkali Metal-Doped C_{70} 569
14.1.5. Transport in Metal–C_{60} Multilayer Structures 570
14.1.6. Lewis Acid-Doped C_{60} 572
14.2. Electron–Phonon Interaction 573
14.2.1. Special Properties of Fullerenes 573
14.2.2. Magnitudes of the Coupling Constant 575
14.2.3. Symmetry Considerations 578
14.2.4. Jahn–Teller Effects 579
14.3. Hall Coefficient 580
14.4. Magnetoresistance 581
14.5. Electronic Density of States 583
14.6. Pressure Effects 586
14.7. Photoconductivity 587
14.8. Specific Heat 594
14.8.1. Temperature Dependence 594
14.8.2. Specific Heat for C_{70} 598
14.8.3. Specific Heat for K_3C_{60} 598

14.9. Scanning Calorimetry Studies 600
14.10. Temperature Coefficient of Thermal Expansion 601
14.11. Thermal Conductivity 602
14.12. Thermopower 603
14.13. Internal Friction 608
References 609

CHAPTER 15
Superconductivity **616**

15.1. Experimental Observations of Superconductivity 616
15.2. Critical Temperature 624
15.3. Magnetic Field Effects 626
15.4. Temperature Dependence of the Superconducting Energy Gap . . 633
15.5. Isotope Effect 638
15.6. Pressure-Dependent Effects 641
15.7. Mechanism for Superconductivity 643
References 648

CHAPTER 16
Magnetic Resonance Studies **654**

16.1. Nuclear Magnetic Resonance 654
16.1.1. Structural Information about the Molecular Species . . . 655
16.1.2. Bond Lengths 656
16.1.3. Structural Information about Dopant Site Locations . . 656
16.1.4. Molecular Dynamics in C_{60} and C_{70} 657
16.1.5. Stoichiometric Characterization of M_xC_{60} Phases 662
16.1.6. Knight Shift and the Metallic Phase of M_3C_{60} 664
16.1.7. Density of States Determination by NMR 665
16.1.8. Determination of Superconducting Gap by NMR 665
16.1.9. NMR Studies of Magnetic Fullerenes 668
16.2. Electron Paramagnetic Resonance 669
16.2.1. EPR Studies of Endohedrally Doped Fullerenes 670
16.2.2. EPR Studies of Neutral C_{60} and Doped C_{60} 673
16.2.3. Time-Resolved EPR Study of Triplet State of Fullerenes . 678
16.2.4. EPR Studies of Strong Paramagnets 680
16.3. Muon Spin Resonance (μSR) 683
References 684

CHAPTER 17
Surface Science Studies Related to Fullerenes **689**

17.1. Photoemission and Inverse Photoemission 690
17.1.1. UV Spectra for C_{60} 691
17.1.2. XPS Spectra for C_{60} 694

17.1.3. PES and IPES Spectra for Doped Fullerenes 695
17.1.4. XPS Studies of Adsorbed Fullerenes on Substrates . . . 697
17.2. Electron Energy Loss Spectroscopy 697
17.2.1. Elastic Electron Scattering and Low-Energy Electron Diffraction Studies 699
17.2.2. Electronic Transitions near the Fermi Level 700
17.2.3. Core Level Studies 703
17.2.4. Plasmon Studies 704
17.2.5. HREELS and Vibrational Spectra 706
17.2.6. EELS Spectra for Alkali Metal-Doped Fullerenes 707
17.3. Auger Electron Spectroscopy 708
17.4. Scanning Tunneling Microscopy 712
17.4.1. STM Studies of the Fullerene–Surface Interaction . . . 712
17.4.2. Atomic Force Microscope Studies 714
17.5. Temperature-Programmed Desorption 715
17.6. Work Function 717
17.7. Surface-Enhanced Raman Scattering 717
17.8. Photoionization 719
17.9. Fullerene Interface Interactions with Substrates 719
17.9.1. C_{60} on Noble Metal Surfaces 721
17.9.2. C_{60} on Transition Metal Surfaces 725
17.9.3. C_{60} on Si Substrates 726
17.9.4. C_{60} on GaAs 730
17.9.5. Higher-Mass Fullerenes on Various Substrates 730
17.9.6. Metallofullerenes Adsorbed on Surfaces 733
References 733

CHAPTER 18
Magnetic Properties **739**

18.1. Diamagnetic Behavior 740
18.2. Magnetic Endohedral and Exohedral Dopants 742
18.3. Magnetic Properties of Fullerene Ions 744
18.4. Pauli Paramagnetism in Doped Fullerenes 745
18.5. *p*-level Magnetism 747
18.5.1. Observations in Alkali Metal–Doped C_{60} 747
18.5.2. Observations in TDAE–C_{60} 748
References 753

CHAPTER 19
C_{60}-Related Tubules and Spherules **756**

19.1. Relation between Tubules and Fullerenes 757
19.2. Experimental Observation of Carbon Nanotubes 761
19.2.1. Observation of Multiwall Carbon Nanotubes 761
19.2.2. Observation of Single-Wall Carbon Nanotubes 765

19.2.3. Tubule Caps and Chirality 769
19.2.4. Carbon Nanocones 777
19.2.5. Nanotube Synthesis 778
19.2.6. Alignment of Nanotubes 785
19.3. Growth Mechanism 785
19.4. Symmetry Properties of Carbon Nanotubes 791
19.4.1. Specification of Lattice Vectors in Real Space 791
19.4.2. Symmetry for Symmorphic Carbon Tubules 795
19.4.3. Symmetry for Nonsymmorphic Carbon Tubules 797
19.4.4. Reciprocal Lattice Vectors 800
19.5. Electronic Structure: Theoretical Predictions 802
19.5.1. Single-Wall Symmorphic Tubules 803
19.5.2. Single-Wall Nonsymmorphic Chiral Tubules 809
19.5.3. Multiwall Nanotubes and Arrays 814
19.5.4. 1D Electronic Structure in a Magnetic Field 818
19.6. Electronic Structure: Experimental Results 825
19.6.1. Scanning Tunneling Spectroscopy Studies 825
19.6.2. Transport Measurements 827
19.6.3. Magnetoresistance Studies 829
19.6.4. Magnetic Susceptibility Studies 833
19.6.5. Electron Energy Loss Spectroscopy Studies 838
19.7. Phonon Modes in Carbon Nanotubes 839
19.7.1. Phonon Dispersion Relations 840
19.7.2. Calculated Raman- and Infrared-Active Modes 845
19.7.3. Experiments on Vibrational Spectra of Carbon Nanotubes 850
19.8. Elastic Properties 854
19.9. Filled Nanotubes 858
19.10. Onion-Like Graphitic Particles 860
19.11. Possible Superconductivity in C_{60}-Related Tubules 863
References 864

CHAPTER 20
Applications of Carbon Nanostructures 870

20.1. Optical Applications 870
20.1.1. Optical Limiter 871
20.1.2. Photoexcited C_{60}–Polymer Composites 873
20.1.3. Photorefractivity in C_{60}–Polymer Composites 876
20.2. Electronics Applications 880
20.2.1. C_{60} Transistors 881
20.2.2. C_{60}-Based Heterojunction Diodes 884
20.2.3. C_{60}–Polymer Composite Heterojunction Rectifying Diode 885
20.2.4. C_{60}–Polymer Composite Heterojunction Photovoltaic Devices 886
20.2.5. Microelectronic Fabrication and Photoresists 888
20.2.6. Silicon Wafer Bonding 889

20.2.7. Passivation of Reactive Surfaces 891
20.2.8. Fullerenes Used for Uniform Electric Potential Surfaces . 891
20.3. Materials Applications 893
20.3.1. Enhanced Diamond Synthesis 893
20.3.2. Enhanced SiC Film Growth and Patterning 893
20.3.3. Catalytic Properties of C_{60} 894
20.3.4. Self-Assembled Monolayers 896
20.3.5. New Chemicals 897
20.4. Electrochemical Applications of C_{60} 898
20.4.1. Hydrogen Storage and Primary Batteries 899
20.4.2. C_{60} Electrodes for Secondary Batteries 899
20.5. Other Applications 901
20.5.1. Nanotechnology 901
20.5.2. Fullerene Coatings for STM Tips 904
20.5.3. Fullerene Membranes 905
20.5.4. Tribology—Lubricants 905
20.5.5. Separations 907
20.5.6. Sensors 907
20.6. Commercialization and Patents 908
References 911

Index **919**

Preface

Although the study of fullerenes and carbon nanotubes has recently undergone rapid development and still remains a fast moving research field, most of the fundamental concepts underlying this field are now becoming clearly defined. This is, therefore, an appropriate time to write a book providing a comprehensive review of the current status of this field. This review was initiated by a tutorial given for the fall 1992 meeting of the Materials Research Society. The tutorial proved to be so popular that it was repeated for the spring 1993 meeting, and resulted in several review articles. A special topics physics course was also taught at MIT in the fall of 1994 using the subject matter of this book.

Carbon is a remarkable element showing a variety of stable forms ranging from 3D semiconducting diamond to 2D semimetallic graphite to 1D conducting and semiconducting carbon nanotubes to 0D fullerenes, which show many interesting properties. Since much of the carbon phase diagram remains largely unexplored, it is expected that new forms of carbon remain to be discovered. To provide a context for the newly discovered fullerenes and carbon nanotubes, a brief review of carbon-based materials is provided at the beginning of this volume. This is followed by a comprehensive and pedagogical review of the field of fullerene and carbon nanotube research.

The structure and properties of fullerenes are reviewed, emphasizing their behavior as molecular solids. The structure and property modifications produced by doping are summarized, including modification to the electronic structure, lattice modes, transport, and optical properties. Particular emphasis is given to the alkali metal-doped fullerenes because they have been studied most extensively, due to their importance as superconductors. A review of the structure and properties of carbon nanotubes is also given, including a model for their one-dimensional electronic band structure. Potential applications for fullerene-based materials are suggested.

The authors acknowledge fruitful discussions with Professors M. A. Duncan, M. Endo, M. Fujita, J. P. Issi, R. A. Jishi, R. Saito, K. R. Subbaswamy, R. Taylor, D. R. M. Walton, J. H. Weaver, Drs. X. X. Bi, P. Brühwiler, J. C. Charlier, D. Eastwood, M. Golden, M. Grabow, A. V. Hamza, A. Hebard, J. Heremans, A. R. Kortan, R. Ochoa, B. Pevzner, and A. M. Rao. Ms. Kathie Sauer designed the cover of the book and contributed to the artwork. Ms. Laura Doughty has been indispensable in keeping the project organized and proceeding on schedule. The MIT authors gratefully acknowledge support from NSF grant DMR-92-01878 and from AFOSR grant F49620-93-1-0160. The work at UK was supported in part by NSF grant EH4-91-08764.

M. S. Dresselhaus, Cambridge, Massachusetts
G. Dresselhaus, Cambridge, Massachusetts
P. C. Eklund, Lexington, Kentucky

CHAPTER 1

Historical Introduction

The closed cage nearly spherical molecule C_{60} and related fullerene molecules have attracted a great deal of interest in recent years because of their unique structure and properties. For a variety of reasons, fullerenes are of broad-based interest to scientists in many fields, Some reasons for this interest are: to physicists for their relatively high T_c superconductivity (33 K), the fivefold local symmetry of the icosahedral fullerenes, and the quasi–one-dimensional behavior of fullerene-related nanotubules; to chemists for the molecular nature of the solid phase and the large family of new compounds that can be prepared from fullerenes; to earth scientists because of very old age of shungite, a carbon-rich mineral deposit which has been found to contain fullerenes; to materials scientists as representing a source of monodisperse nanostructures that can be assembled in film and crystal form and whose properties can be controlled by doping and intercalation; to device engineers because of the potential use in optical limiting and switching devices, in photoconductor applications for niche photoresist applications, among others. In this volume we review the structural and physical properties of fullerenes and materials, such as doped fullerenes or carbon nanotubes, related to fullerenes.

1.1. Early History

The discovery of C_{60} has a long and interesting history. The structure of the regular truncated icosahedron was already known to Leonardo da Vinci in about the year 1500 [1.1, 2]. A reproduction of Leonardo da Vinci's visualization of this highly symmetrical form is given in Fig. 1.1 showing the soccer-ball configuration. Another very early (also ~1500) rendition of the regular truncated icosahedron comes from a drawing of Albrecht Dürer

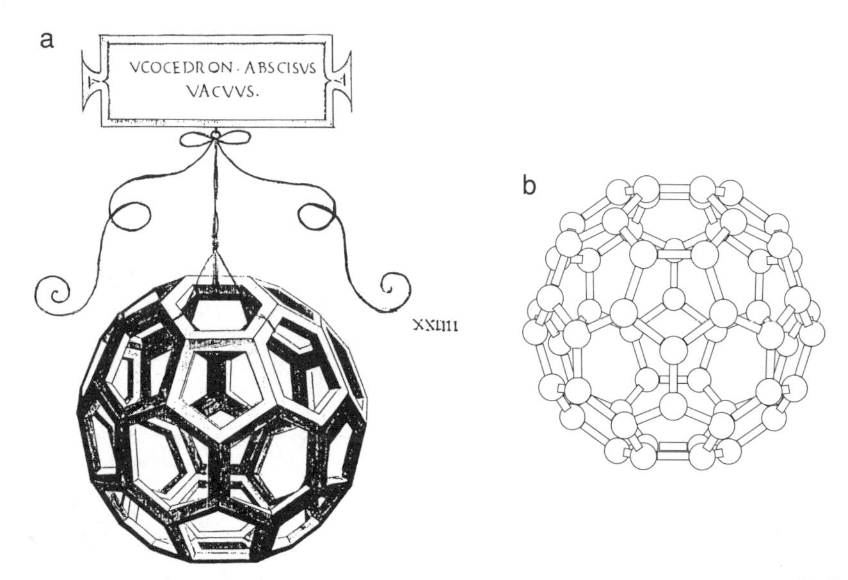

Fig. 1.1. (a) A regular truncated icosahedron from an Italian translation of the "Libellus" published in 1509 by the mathematician Luca Pacioli in his book "De Divina Proportione." The design is thought to be by Leonardo da Vinci [1.1, 2]. (b) In the C_{60} molecule each carbon atom is at an equivalent position on the corners of a regular truncated icosahedron.

[1.3] (cited by [1.4]) showing the construction of the icosahedron by folding up the 12 pentagonal and 20 hexagonal faces (see Fig. 1.2). In the present century a number of theoretical suggestions for icosahedral molecules (the highest possible molecular point group symmetry) predated their experimental discovery in the laboratory by several decades [1.5]. In very early work, Tisza (1933) [1.6] considered the point group symmetry for icosahedral molecules, and Osawa (1970) [1.7] suggested that an icosahedral C_{60} molecule [see Fig. 1.1(b)] might be stable chemically. Using Hückel

Fig. 1.2. Albrecht Dürer's drawing (~1500) of the construction of the *icosahedron truncum* by folding up a sheet of cardboard [1.3]. Notice the wedge-shaped gaps between the pentagons and the hexagons when these faces are made to lie in a plane. The pentagons are required in the structure of the icosahedron to generate the curvature of the fullerene shell.

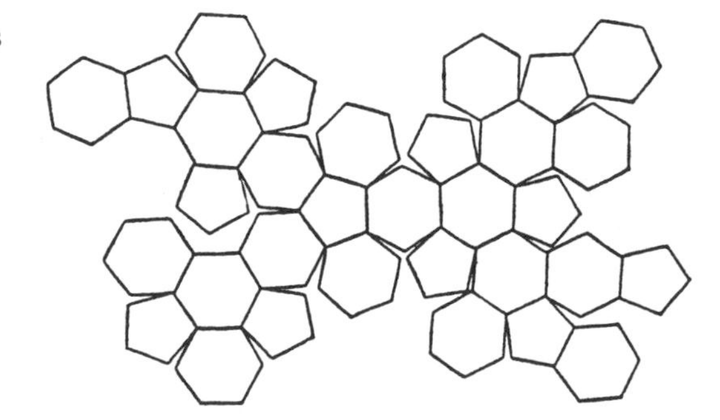

calculations, early Russian workers showed that C_{60} should have a large electronic gap between the highest occupied molecular orbital (HOMO) and the lowest unoccupied molecular orbital (LUMO) [1.8, 9]. These early theoretical suggestions for icosahedral C_{60} were not widely appreciated, and some of this literature was rediscovered only after the experimental work of Kroto, Smalley and co-workers in the middle 1980s [1.10] established the stability of the C_{60} molecule in the gas phase.

1.2. Astronomical Observations

The experimental identification of the C_{60} molecule as a regular truncated icosahedron occurred because of the coalescence of research activities in two seemingly independent areas. Astrophysicists were working in collaboration with spectroscopists [1.11] to identify some unusual infrared (IR) emission from large carbon clusters which had been shown to be streaming out of red giant carbon stars [1.12–14]. At the same time, the development of a laser vaporization cluster technique by Smalley and co-workers at Rice University for synthesizing clusters, in general, and carbon clusters in particular [1.15], suggested the possibility of creating unusual carbon-based molecules or clusters that would yield the same IR spectrum on earth as is seen in red giant carbon stars. This motivation led to a collaboration between Kroto and Smalley and co-workers to use the laser vaporization of a graphite target to synthesize and study cyanopolyynes [1.5, 16]. It was during these studies that a 60 carbon-atom cluster with unusually high stability was discovered. Shortly thereafter, this team of researchers proposed that the C_{60} cluster was indeed a molecule with icosahedral symmetry [1.10]. These advances in our understanding of carbon cluster formation on earth enhanced the understanding of astronomers regarding the role of large carbon clusters in the galactic carbon cycle [1.17]. Fullerene spectroscopy, from the infrared through the ultraviolet, remains a topic of broad interest to astrophysicists studying the properties of the intergalactic medium.

1.3. Carbon Cluster Studies

Early synthesis of fullerenes was done using laser ablation of graphite targets in He gas to create fullerenes in the gas phase [1.10]. An intense laser pulse (e.g., 10–100 mJ energy, 5 ns pulses, from a frequency-doubled Nd: YAG laser) was used to vaporize material from a graphite surface, thereby creating a hot carbon plasma (see Fig. 1.3). This hot plasma was quenched and entrained in a flowing inert gas (e.g., He) to form carbon

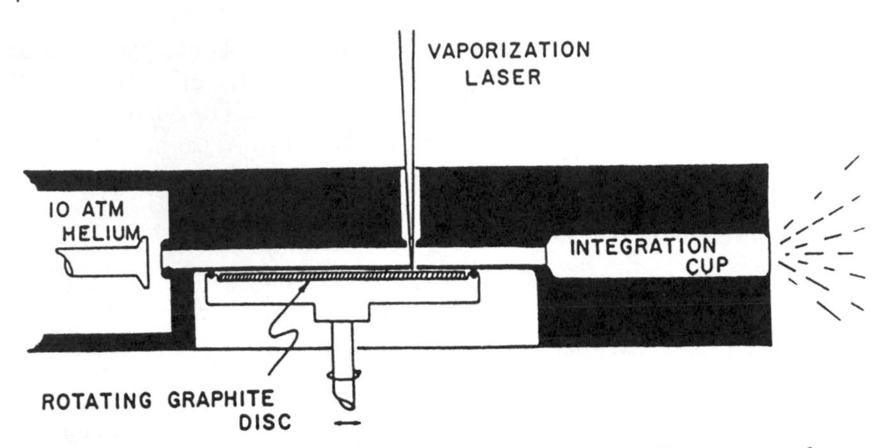

Fig. 1.3. Schematic view of a laser ablation source of fullerenes. The sequence of events is that a pulsed valve opens, flowing He gas over the surface. Near the middle of the He gas pulse, the laser pulse strikes the graphite surface, vaporizing the carbon material to atoms and ions. The carbon material is entrained in the He carrier gas and carbon clusters condense in the gas. After clustering the carbon material in the integration cup, the gas mixture undergoes supersonic expansion into a vacuum with resultant cooling [1.16].

clusters. The clustering was permitted to occur for 30–200μs, after which the gas stream was allowed to expand and cool, thereby terminating cluster growth. This laser ablation source was coupled to the front end of a time-of-flight mass spectrometer which allowed the mass spectrum of the clusters and molecules formed by the source to be measured [1.16, 18]. The oft-cited mass spectrum of Fig. 1.4 shows two main groupings of carbon clusters: (1) those with lower masses (<30 carbon atoms) containing predominantly odd numbers of carbon atoms and (2) the higher-mass clusters (above 36 carbon atoms) containing only even numbers of carbon atoms. Many years ago Pitzer and Clementi predicted that large carbon clusters could be formed from the vapor phase [1.19]. The spectrum of Fig. 1.4 was one of the first to show these high mass clusters experimentally. The mass spectrum above 40 carbon atoms shows a prominence for clusters of mass C_{60}. This preponderance of the C_{60} line in the mass spectra was recognized by Kroto and Smalley at an early stage. By allowing the plasma to react for a longer time, they found that the C_{60} peak could be enhanced relative to the other peaks [1.16].

Previous experimental and theoretical work had shown that the most stable form for carbon clusters up to about 10 atoms is a *linear chain* [1.18]. Furthermore, for clusters in the 10–30 carbon atom range, *rings* of carbon atoms are the most stable [1.20]. Certain ring sizes are more favorable energetically, as can be seen in Fig. 1.4, where local maxima in the peak

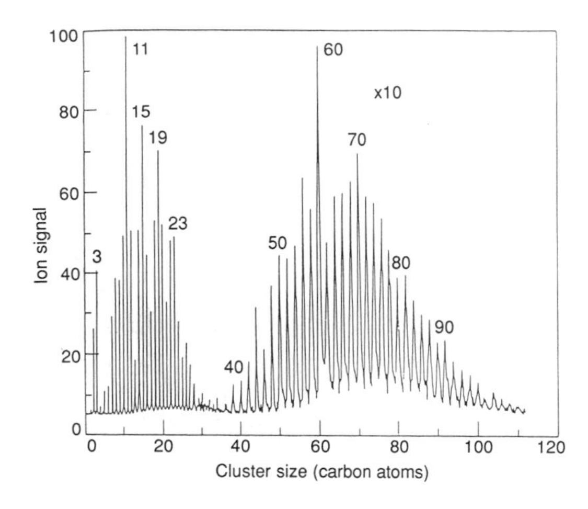

Fig. 1.4. Time-of-flight mass spectrum of carbon clusters produced in a supersonic beam by laser vaporization of a graphite target [1.18].

heights of the mass spectrum are found at $n_C = 11, 15, 19, 23$ carbon atoms. The high stability of molecules with certain ring sizes has been explained in terms of a Hückel model. These calculations showed that a neutral ring is stable at $4j+2$ carbon atoms, a negatively charged ring at $4j+1$ and a positively charged ring (as in Fig. 1.4) at $4j+3$ carbon atoms [1.21] for $j = 3, 4, 5$, and 6, while $10 < n_C < 30$. This figure, showing the mass distribution of the cluster emission from a graphite surface by laser vaporization, further indicates that the formation of C_{n_C} carbon clusters for $30 < n_C < 40$ is unlikely.

The realization that there was something special about C_{60} suggested to Kroto and Smalley that C_{60} and other clusters above $n_C = 40$ had a cage structure (thereby explaining the occurrence of peaks only for even values of n_C, as required by Euler's theorem (see §3.1)). Kroto *et al.* [1.10] were the first to infer from the mass spectrum that C_{60} had a cage structure formed by hexagons and pentagons, with carbon atoms at the 60 vertices of the regular truncated icosahedron. In this structure the carbon atoms have no dangling bonds, and the pentagons on the fullerene cage would be isolated from one another, thereby creating greater chemical and electronic stability. The stability of the C_{60} molecule, the largest stable molecule formed from a single elemental species, is noteworthy insofar as the C_{60} molecule can withstand collisions with species in the energy range of hundreds of eV and heating to above 1000°C. Kroto and Smalley recognized at an early time that carbon was a special element in the periodic table and that the sp^2 bonding of carbon fostered the formation of caged structures [1.22]. A stereographic projection of this unique molecule is shown in Fig. 1.5.

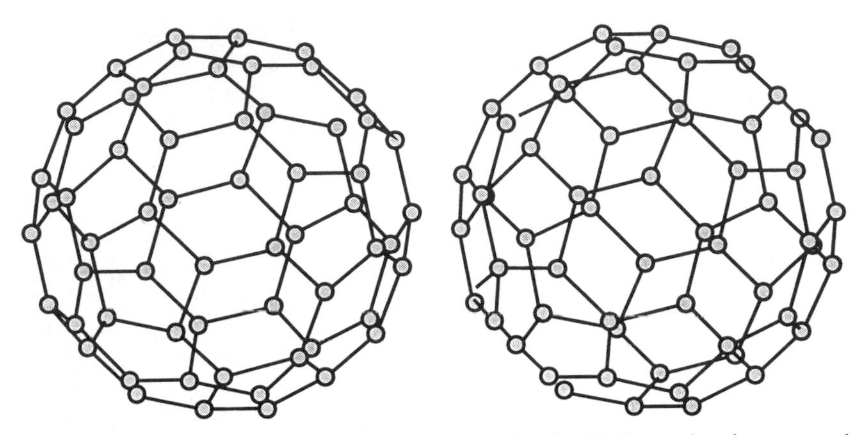

Fig. 1.5. Stereographic projection of C_{60}. To visualize the 3D image the picture must be held about 15 cm from the eyes and the eyes must focus at infinity.

1.4. Recent History

In the 1980s gas-phase work, in which clusters and molecular species were produced by the laser vaporization of a graphite target [1.10, 18], mass spectroscopy was used as the main characterization tool for fullerenes (see Fig. 1.4). Several important experiments were done at that time to support the cage-like structure of the C_{60} molecule, including chemical stability studies of C_{60} in the presence of reactive gas streams [1.23], endohedral doping experiments in which guest species are introduced inside the fullerenes [1.24, 25], and photofragmentation studies [1.25, 26], though none of these experiments directly confirmed the icosahedral symmetry of the C_{60} molecule [1.16]. Direct evidence that C_{60} exhibits icosahedral symmetry came somewhat later from nuclear magnetic resonance experiments [1.27–30] where a single resonance line was observed, consistent with only one type of chemical site for carbon atoms in this molecule (see Fig. 1.1). Likewise, the IR spectrum with four characteristic infrared-allowed lines provided further direct evidence for the icosahedral symmetry [1.31, 32].

In the fall of 1990 a new type of condensed carbon phase, based on C_{60}, was synthesized for the first time by Krätschmer, Huffman and co-workers [1.33]. The original motivation for the Krätschmer *et al.* discovery came from their astrophysical interests, similar to the early motivation of Kroto. The Krätschmer *et al.* experiments were performed initially in an attempt to identify the anomalous diffuse bands in the absorption spectra of interstellar dust; these bands had first been reported by astronomers in the 1920s [1.31]. Special attention was given to identifying features associated with the prominent peak in the absorption spectra at 217 nm (5.7 eV).

It had been conjectured that these spectral features were due to spherical carbon particles. To identify the origin of these absorption bands more directly, Krätschmer, Huffman and their colleagues embarked on a multiyear project to produce spherical carbon particles in the laboratory, to study the structure of these particles under different preparation methods, and to identify the optical absorption associated with each type of nanoparticle. In the course of these studies, Krätschmer and Huffman discovered a method for the preparation of a soot in the laboratory which reproduced some of the features in the astronomical spectra [1.31, 34]. Of particular importance to the subsequent development of the fullerene field is their discovery of a method (using resistive evaporation of graphite in a helium atmosphere) to produce large quantities of this soot, a method for the separation of C_{60} from the soot, and finally for making small C_{60} crystals by drying a benzene solution containing C_{60} molecules [1.34]. A convincing characterization for their C_{60} molecules was the unique infrared spectrum of this molecule (see §11.5.2). The major impact of the Krätschmer *et al.* discovery of a simple method for preparing gram quantities of C_{60}, which had previously been available only in trace quantities in the gas phase [1.10, 35], was to provide a great stimulus to this research field, allowing many workers to enter the field and allowing a wide variety of experiments to be carried out.

It was soon found by Haddon and co-workers [1.36] that the intercalation of alkali metals into C_{60} to a stoichiometry M_3C_{60} (where M = K, Rb, Cs) greatly modifies the electronic properties of the host fullerenes and yields relatively high transition temperature ($18 \leq T_c \leq 33$ K) superconductors. The discovery of superconductivity [1.36, 37] in these compounds further spurred research activity on the physical properties of C_{60}-related materials. The collection of reviews edited by Kroto *et al.* [1.38] gives a excellent summary of the state of understanding of fullerenes as of 1993.

1.5. Architectural Analogs

The name "fullerene" was given by Kroto and Smalley to the family of molecules observed in this gas-phase work [1.10], because of the resemblance of these molecules to the geodesic domes designed and built by R. Buckminster Fuller [1.39–44] as shown in Fig. 1.6. The name "buckminsterfullerene" or simply "buckyball" was given specifically to the C_{60} molecule.

The word "fullerenes" is now used to denote the whole class of closed cage molecules consisting of only carbon atoms. Euler's theorem (§3.1) can be used to show that a cage molecule with only hexagonal and pentagonal faces must contain exactly 12 pentagonal faces but could differ in the number of its hexagonal faces. For example, C_{60} has 20 hexagonal faces, while

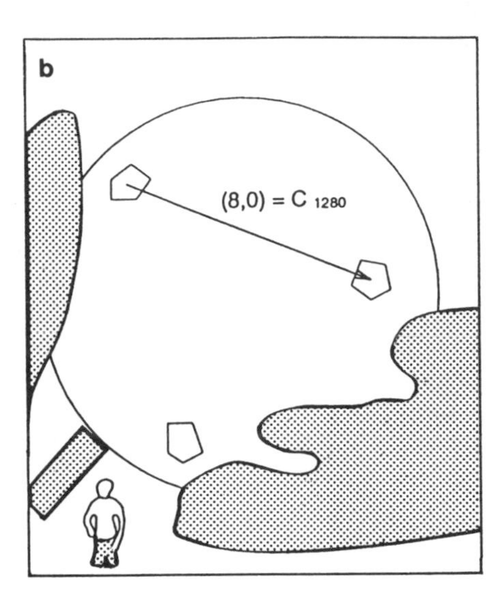

Fig. 1.6. (a) Photo of a geodesic dome located in Disneyworld, Orlando, Florida. (b) Schematic diagram identifying three of the pentagons on the geodesic dome. From the indicated distance, the size of the geodesic dome can be estimated.

C_{20} (if it were to exist as a carbon cage molecule) would have 12 pentagonal faces and no hexagons. Architects may use the nearest-neighbor distance between pentagonal faces on a regular truncated icosahedron, as illustrated in Fig. 1.6, to estimate the size of a geodesic dome, just as chemists use this approach to estimate the size of large fullerene molecules.

Dome-shaped structures are widely used worldwide in modern architecture, since domes have minimal surface area for maximum volume, and the structure itself is very strong. The basic unit in these structures is mainly a triangle, which is known as a truss structure. Five triangles sharing a vertex correspond to a pentagon and six triangles sharing a vertex correspond to a hexagon [see Fig. 1.6(a)]. The basic rule for closed polygons (known as Euler's theorem and discussed in §3.1) also holds in this case.

In Japan, fullerene-like cage structures are found in many arts and crafts objects based on cloth, china, and bamboo. In these cage structures, pentagons and hexagons are used as patterns, with pentagons representing the flower of a Japanese apricot and hexagons representing the back of turtles. Apricot blossoms and turtle backs are symbols of happiness as a precursor of spring and an animal with a long life. However, Japanese architecture seldom uses pentagons and hexagons. Japanese bamboo arts and crafts for making boxes, cages, and baskets were very popular before the

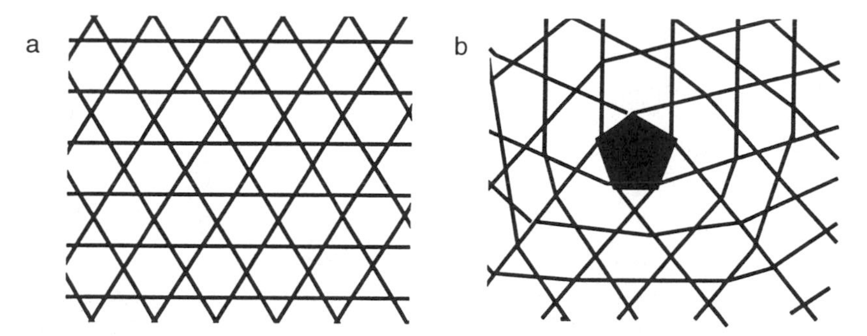

Fig. 1.7. Traditional Japanese kagome pattern, formed by the intersection of three bamboo reeds. The lines in (a) represent the pattern for a normal flat section, and (b) represents a pentagonal defect which is necessary to shape the concave basket.

developments of plastics. Crossing three bamboo strings with each other forms a stable but still porous basket. This pattern is generally called a "kagome" lattice structure (see Fig. 1.7) and consists of triangles and hexagons. The kagome lattice is clearly different from the fullerene structure, which consists of only pentagons and hexagons. However, the concept of fullerene and tubule curvature is well reproduced in Japanese arts and crafts, so that when a hexagon in a kagome lattice is replaced by a pentagon, a surface with positive curvature is formed. In the case where a heptagon replaces a hexagon, a negatively curved surface is obtained.

1.6. Biological and Geological Examples

In biology and biochemistry there are numerous examples of molecules or microstructures with fivefold symmetry axes, reminiscent of the symmetries occurring in fullerenes. Many of the constituents of living systems have been identified as nanometer-scale structures exhibiting fivefold symmetries (see Fig. 1.8), self-assembled in marvelous ways to achieve specific biological functions [1.45]. Looking into these self-assembly processes which have evolved over geological epochs is likely to provide us with many significant ideas for the self-assembly of nanometer-scale materials such as fullerenes.

The icosahedral structures of viruses have been studied by biologists for many years [1.46]. Recent computer simulations of the growth of icosahedral viruses [1.47] (see Fig. 1.9) have focused on how well ordered macroscopic icosahedral structures can be constructed using only a few algorithms regarding local order. The icosahedral symmetry of the bacteriophage ΦX174 can be seen in the reconstruction of the transmission electron microscope (TEM) image, showing the virus along a twofold axis

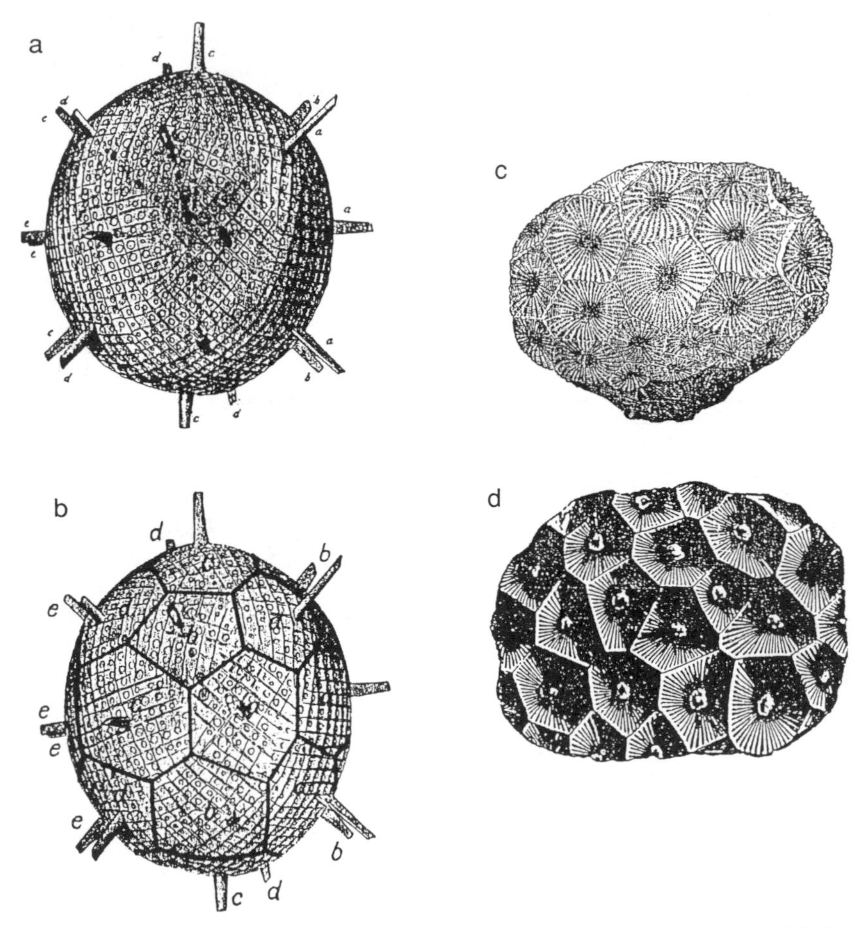

Fig. 1.8. Various biological systems showing pentagonal and hexagonal faces. (a) *Phatnaspis cristata Hkl*, (b) diagrammatic of (a), (c) *Cyathophyllum hexagonum*, and (d) *Arachnophyllum pentagonum* [1.45].

in Fig. 1.10(a) and along a fivefold axis in Fig. 1.10(b) [1.48]. The local symmetry algorithms [1.47] aim to explain the replication of these high symmetry cage biological molecules.

Subsequent to the discovery of fullerenes in 1985 and their intensive study in the solid phase, fullerenes have more recently been identified in geological samples [1.49, 50]. In the case of fulgurite specimens, the occurrence of fullerenes in these mineral deposits has been identified with lightning strikes, perhaps creating conditions resembling an arc discharge between carbon electrodes, which is the most common laboratory

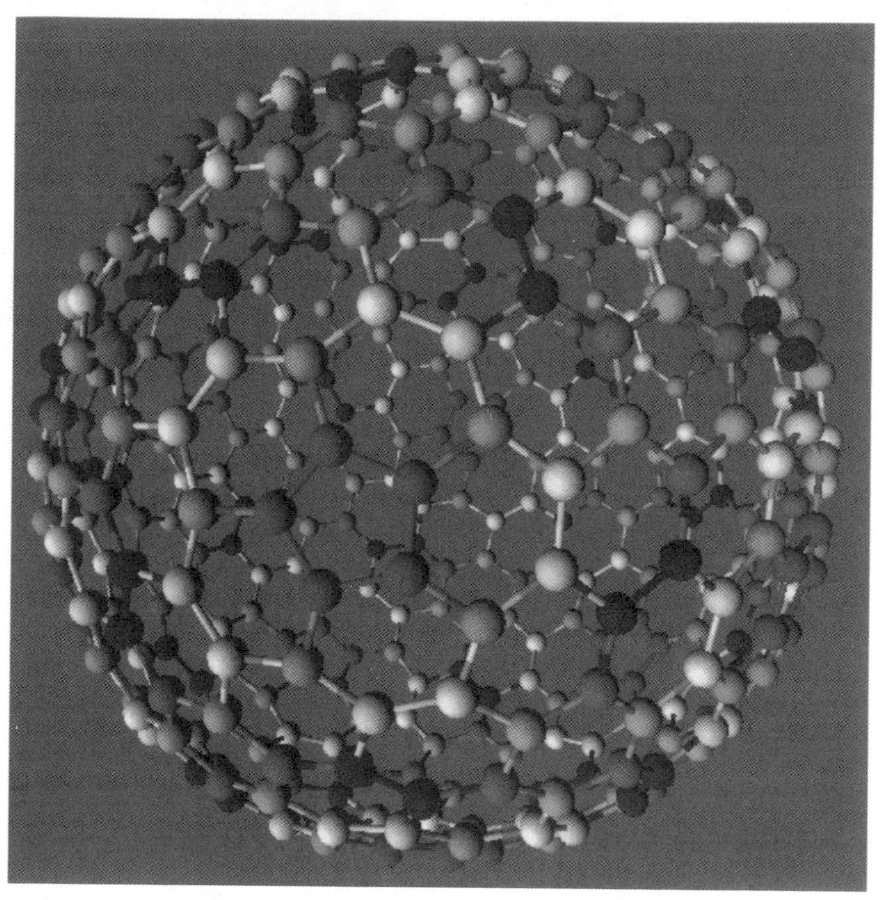

Fig. 1.9. Model for an icosahedral virus constructed using a local rule-based theory of virus shell assembly. In this model the shaded balls represent protein molecular units [1.47].

method for generating large amounts of C_{60} [1.49]. It is not believed, however, that the origin of fullerenes in other fullerene-containing minerals [1.50] is due to lightening strikes. For example, shungite is a very old geological fullerene-containing mineral deposit, which may predate living plants, and is of particular interest, regarding its origin. There is now much activity by geologists looking for fullerenes in other geological deposits.

Fullerenes are found in many places. For example, it has been said that dinner candles generate an estimated 2–40 fullerenes. The ability to produce fullerenes in flames has been explored in much detail [1.51].

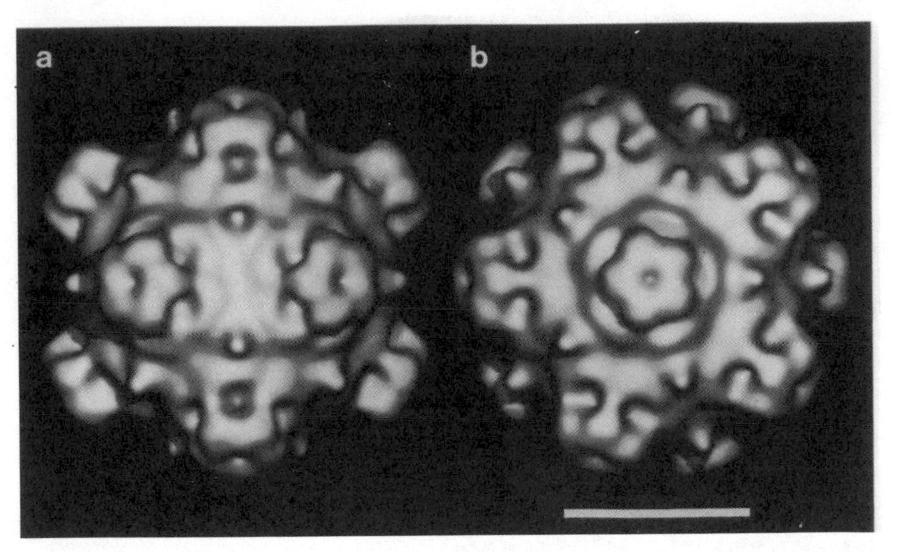

Fig. 1.10. Shaded surface representations of the bacteriophage ΦX174 reconstruction viewed along a twofold axis (a) and along a fivefold axis (b). The scale bar is 15 nm [1.48].

1.7. Road Map

The approach taken in this book is to initiate the discussion of each subject with a detailed treatment of the isolated fullerene molecules and then to consider the condensed state as a weakly interacting system. This book begins in Chapter 2 with a discussion of the fundamentals of carbon materials. Chapters 3–6 focus on molecular properties of fullerenes, with the structure of fullerene molecules reviewed in Chapter 3 and the group theoretical treatment of C_{60} and related fullerenes in Chapter 4. The synthesis, extraction, and purification of fullerenes are reviewed in Chapter 5, and Chapter 6 is devoted to the growth, contraction, and fragmentation of fullerenes. The crystalline structure of selected fullerene solids is reviewed in Chapter 7, and Chapter 8 treats the classification and structure of doped fullerenes. Chapter 9 reviews the synthesis of crystalline fullerene solids and thin films. Chemistry issues related to fullerenes are briefly discussed in Chapter 10. The spectra of vibrational modes in fullerenes are reviewed in Chapter 11; the electronic structure in Chapter 12; the optical properties in Chapter 13; the transport and thermal properties in Chapter 14; the superconducting properties of doped fullerenes in Chapter 15; nuclear magnetic resonance (NMR), electron spin resonance and magnetic susceptibility in Chapter 16; scanning tunneling microscopy (STM), surface science studies, and electron spectroscopies in Chapter 17; and the mag-

all materials, diamond along with graphite (in-plane) exhibits the highest thermal conductivity and the highest melting point [2.6, 7]. Diamond also has the highest atomic density of any solid (see Table 2.1). Graphite and diamond are two main allotropic forms of bulk carbon. Fullerenes and carbon nanotubes, as discussed in this volume, constitute other allotropic

Table 2.1

Properties of graphite and diamond

Property	Graphite[a]		Diamond
Lattice structure	Hexagonal		Cubic
Space group	$P6_3/mmc$ (D^4_{6h})		$Fd3m$ (O^7_h)
Lattice constant[b] (Å)	2.462	6.708	3.567
Atomic density (C atoms/cm^3)	1.14×10^{23}		1.77×10^{23}
Specific gravity (g/cm^3)	2.26		3.515
Specific heat (cal/g·K)	0.17		0.12
Thermal conductivity[b] (W/cm·K)[c]	30	0.06	~25
Binding energy (eV/C atom)	7.4		7.2
Debye temperature (K)	2500	950	1860
Bulk modulus (GPa)	286		42.2
Elastic moduli (GPa)	1060[d]	36.5[d]	107.6[e]
Compressibility (cm^2/dyn)	2.98×10^{-12}		2.26×10^{-13}
Mohs hardness[f]	0.5	9	10
Band gap (eV)	−0.04[g]		5.47
Carrier density (10^{18}/cm^3 at 4 K)	5		0
Electron mobility[b] (cm^2/Vsec)	20,000	100	1800
Hole mobility[b] (cm^2/Vsec)	15,000	90	1500
Resistivity (Ωcm)	50×10^{-6}	1	$\sim 10^{20}$
Dielectric constant[b] (low ω)	3.0	5.0	5.58
Breakdown field (V/cm)	0	0	10^7 (highest)
Magnetic susceptibility (10^{-6}cm^3/g)	−0.5	−21	—
Refractive index (visible)	—	—	2.4
Melting point (K)	4450		4500
Thermal expansion[b] (/K)	-1×10^{-6}	$+29 \times 10^{-6}$	$\sim 1 \times 10^{-6}$
Velocity of sound (cm/sec)	$\sim 2.63 \times 10^5$	$\sim 1 \times 10^5$	$\sim 1.96 \times 10^5$
Highest Raman mode (cm^{-1})	1582	—	1332

[a]For anisotropic properties, the in-plane (ab plane or a-axis) value is given on the left and the c-axis value on the right.

[b]Measurements at room temperature (300 K).

[c]Highest reported thermal conductivity values are listed.

[d]In-plane elastic constant is C_{11} and c-axis value is C_{33}. Other elastic constants for graphite are $C_{12} = 180$, $C_{13} = 15$, $C_{44} = 4.5$ GPa.

[e]For diamond, there are three elastic constants, $C_{11} = 1040$, $C_{12} = 170$, $C_{44} = 550$ GPa.

[f]A scale based on values from 0 to 10, where 10 is the hardest material (diamond) and 1 is talc [2.8].

[g]A negative band gap implies a band overlap, i.e., semimetallic behavior.

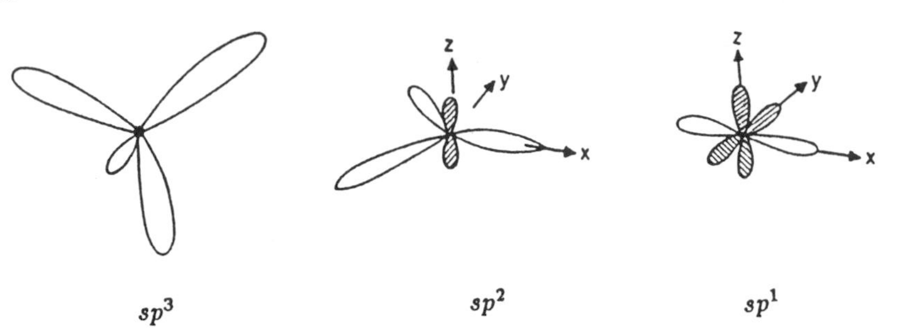

Fig. 2.3. Schematic presentation of sp^3, sp^2, and sp^1 bonding hybridizations [2.9]. The unshaded lobes denote strong bonds and the shaded lobes denote weak bonds.

forms. In addition, many solid-state variants of graphite and diamond have been developed, and these are also briefly reviewed in this chapter.

Carbon has an atomic number of 6 and is in a $1s^2 2s^2 2p^2$ electronic ground state configuration. In the graphite structure, strong in-plane bonds are formed between a carbon atom and its three nearest-neighbors from $2s$, $2p_x$, and $2p_y$ orbitals; this bonding arrangement is denoted by sp^2 (see Fig. 2.3). The remaining electron with a p_z orbital provides only weak interplanar bonding [2.10], but is responsible for the semimetallic electronic behavior in graphite [2.11]. In contrast, the carbon atoms in the diamond structure are tetrahedrally bonded to their four nearest-neighbors using linear combinations of $2s$, $2p_x$, $2p_y$, and $2p_z$ orbitals in an sp^3 configuration (see Fig. 2.3). The difference in the structural arrangement of these allotropic forms of carbon gives rise to the wide differences in their physical properties.

2.2. Graphite

The ideal crystal structure of graphite (see Fig. 2.2) consists of layers in which the carbon atoms are arranged in an open honeycomb network, such that the A and B atoms on consecutive layers are on top of one another, but the A′ atoms in one plane are over the unoccupied centers of the adjacent layers, and similarly for the B′ atoms [2.4]. This gives rise to an ABAB planar stacking arrangement shown in Fig. 2.2, with an in-plane nearest-neighbor distance a_{C-C} of 1.421 Å, an in-plane lattice constant a_0 of 2.462 Å, a c-axis lattice constant c_0 of 6.708 Å, and an interplanar distance of $c_0/2 = 3.3539$ Å (see Table 2.1). This structure is consistent with the $D^4_{6h}(P6_3/mmc)$ space group and has four atoms per unit cell (see Fig. 2.2).

Disorder tends to have little effect on the in-plane lattice constant, largely because the in-plane C-C bond is very strong and the nearest-neighbor

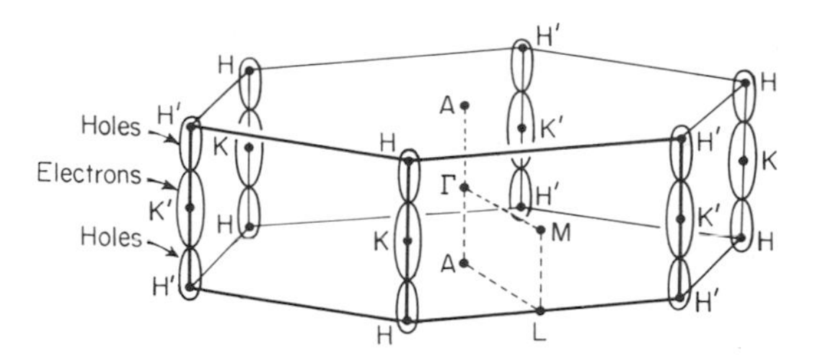

Fig. 2.4. Graphite Brillouin zone showing several high-symmetry points and a schematic version of the graphite electron and hole Fermi surfaces located along the HKH and $H'K'H'$ axes [2.12].

C-C spacing is very small. Interestingly, the intramolecular C-C spacing in fullerenes (see §3.1) is also close to $a_{\text{C-C}}$ for graphite. Disorder does, however, have a significant effect on the crystallite size in the basal plane and on the interplanar spacing because of the weak interplanar bonding. One consequence of the small value of $a_{\text{C-C}}$ is that impurity species are unlikely to enter the in-plane lattice sites substitutionally, but rather occupy some interstitial position between the layer planes. The weak interplanar bonding also allows entire planes of dopant atoms or molecules to be intercalated between the carbon layers, as discussed in more detail in §2.14.

Weak disorder results in stacking faults (meaning departures from the ABAB stacking order) giving rise to a small increase in the interlayer distance until a value of about 3.440 Å is reached, at which distance the stacking of the individual carbon layers (called graphene layers) becomes uncorrelated; the resulting two-dimensional (2D) honeycomb structure of uncorrelated graphene layers is called *turbostratic graphite* [2.12, 13]. The electronic structure of turbostratic graphite, a zero gap semiconductor, is qualitatively different from that of ideal graphite, a semimetal with a small band overlap (0.04 eV). Likewise, as the stacking disorder is increased, the interplanar spacing also increases, which further modifies the electronic properties.

To describe the electronic and phonon dispersion relations of 3D graphite, the Brillouin zone shown in Fig. 2.4 is used. Since the in-plane nearest-neighbor distance in real space is much larger than the interplanar separation, the Brillouin zone for graphite has a small length in reciprocal space along k_z. As the interplanar correlation becomes less and less important, as in the case of turbostratic 2D graphene layers, the Brillouin zone is reduced to a sheet. The special point K at the Brillouin zone corner in 2D graphite is the location of the Fermi level and the symmetry-imposed degeneracy of the conduction and valence bands [2.14]. The band structure model for graphite near E_F focuses on the electronic dispersion relations in the vicinity of the HKH and $H'K'H'$ edges of the Brillouin zone. Shown in the figure are electron and hole pockets which form along the zone edges [2.12].

In the following sections, a review of the most important forms of graphite-related materials is given, showing the wide variety of materials that have been prepared and investigated.

2.3. Graphite Materials

Many sources for crystalline graphite are available. For example, natural single-crystal graphite flakes are found in several locations around the world, especially in Madagascar, Russia, and the Ticonderoga area of New York State in the United States. These flakes can sometimes be as large as several millimeters in the basal plane and are typically much less than 0.1 mm in thickness. Natural graphite flakes usually contain defects in the form of twinning planes and screw dislocations and, in addition, contain chemical impurities such as Fe and other transition metals.

A synthetic single-crystal graphite, called "kish" graphite, is commonly used in scientific investigations. Kish graphite crystals form on the surface of high carbon content iron melts and are harvested as crystals from such solutions [2.15]. The kish graphite flakes are often larger than the natural graphite flakes, which makes kish graphite the material of choice when large single-crystal flakes are needed. The most perfect single-crystal flakes presently available are smaller size kish graphite flakes, ~1 mm in diameter.

In materials research laboratories, the most commonly used high-quality graphitic material today is *highly oriented pyrolytic graphite* (HOPG), which is prepared by the pyrolysis of hydrocarbons at temperatures above 2000°C and is subsequently heat treated to higher temperatures [2.16, 17]. When stress annealed above 3300°C, the HOPG exhibits electronic, transport, thermal, and mechanical properties close to those of single-crystal graphite, showing a very high degree of c-axis alignment. For the high temperature, stress-annealed HOPG, the crystalline order extends to about 1 μm within the basal plane and to about 0.1 μm along the c-direction. The HOPG basal planes are sufficiently parallel and ordered to be useful commercially for neutron and x-ray monochromator crystal applications. For all HOPG material however, there is no long-range in-plane a-axis alignment, and the a-axes of adjacent crystallites are randomly oriented [2.16, 17]. The degree of structural order and c-axis alignment of HOPG can be varied by control of the major processing parameters: heat treatment temperature T_{HT}, residence time at T_{HT}, and applied stress during heat treatment [2.16, 17]. Turbostratic pyrolytic graphites are obtained for $T_{HT} < 2300$°C, and higher T_{HT} values (typically above 2800°C) are needed to establish 3D ordering. Thus under normal x-ray diffraction measurements, HOPG appears to be polycrystalline but under selected area electron diffraction, HOPG can show a sharp single-crystal spot pattern. Large-area, ~5 mm thick plates of HOPG are used for x-ray and neutron spectrometers and are available commercially.

Recently, new precursor materials, such as polyimide (PI) and polyoxadiazole (POD) resins [2.18–22], have been used to synthesize graphite films, and, after suitable pyrolysis steps to drive off non-carbon constituents in the vapor phase and subsequent heat treatment steps, the resulting carbon films show a high degree of 3D structural ordering, especially when heat treated to $T_{HT} > 2800°C$. The quality of these thin-film graphite materials, especially those based on Kapton and Novax (polyimide) precursors, is rapidly improving and these graphite materials may soon become materials of choice for specific applications.

In addition to highly crystalline graphite, several less-ordered forms of graphite are available and are widely used for specific applications (e.g., high surface area or high strength applications).

2.4. Graphite Whiskers

A graphite whisker is a graphitic material formed by rolling a graphene sheet (an atomic layer of graphite) up into a scroll [2.23]. Except for the early work by Bacon [2.23], there is little literature about graphite whiskers. Graphite whiskers are formed in a dc discharge between carbon electrodes using 75–80 V and 70–76 A. In the arc apparatus, the diameter of the positive electrode is smaller than that of the negative electrode, and the discharge is carried out in an inert gas using a high gas pressure (92 atmospheres). As a result of this discharge, cylindrical boules with a hard shell were reported to form on the negative electrode. When these hard cylindrical boules were cracked open, scroll-like carbon whiskers up to ~3 cm long and 1–5 μm in diameter were found protruding from the fracture surfaces. The whiskers exhibited great crystalline perfection, high electrical conductivity, and high elastic modulus. Since their discovery [2.23], graphite whiskers have provided the benchmark against which the performance of carbon fibers is compared. The growth of graphite whiskers has many similarities to the growth of carbon nanotubes (see §19.2.5). It would be useful to verify once again the scroll structure of graphite whiskers.

2.5. Carbon Fibers

Carbon fibers represent an important class of graphite-related materials. Despite the many precursors that can be used to synthesize carbon fibers, each having different cross-sectional morphologies (see Fig. 2.5), the preferred orientation of the graphene planes is parallel to the fiber axis

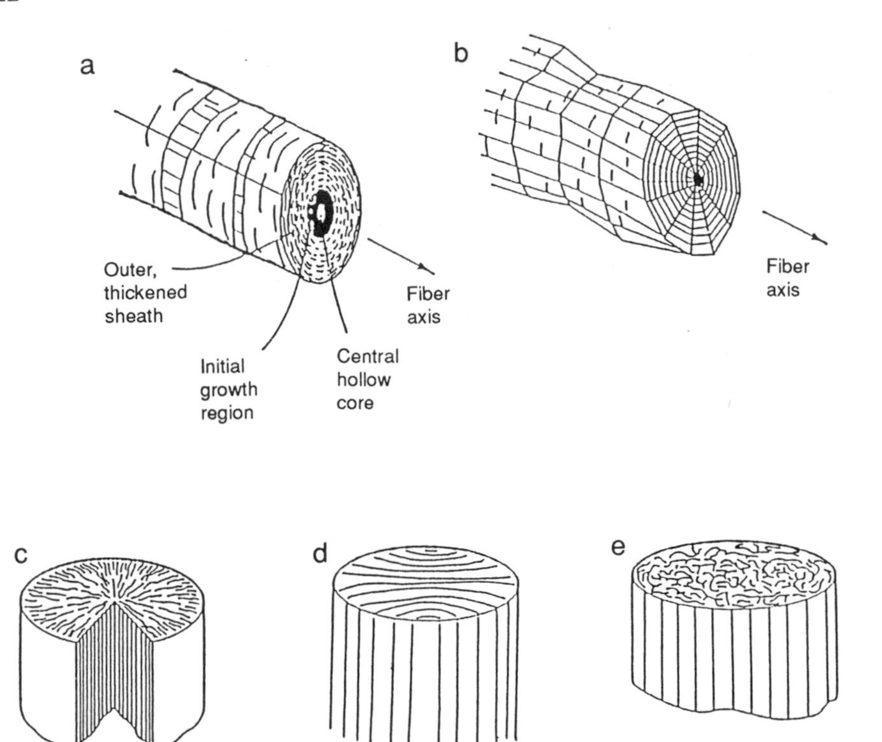

Fig. 2.5. Sketch illustrating the morphology of vapor-grown carbon fibers (VGCF): (a) as-deposited at 1100°C [2.12], (b) after heat treatment to 3000°C [2.12]. The morphologies for commercial mesophase-pitch fibers are shown in (c) for a "PAC-man" cross section with a radial arrangement of the straight graphene ribbons and a missing wedge and (d) for a PAN-AM cross-sectional arrangement of graphene planes. In (e) a PAN fiber is shown, with a circumferential arrangement of ribbons in the sheath region and a random structure in the core.

for all carbon fibers, thereby accounting for the high mechanical strength of carbon fibers [2.12]. Referring to the various morphologies in Fig. 2.5, the as-prepared vapor-grown fibers have an "onion skin" or "tree ring" morphology [Fig. 2.5(a)] and after heat treatment to about 3000°C, form facets [Fig. 2.5(b)]. Of all carbon fibers, these faceted fibers are closest to crystalline graphite in both crystal structure and properties. The commercially available mesophase pitch fibers, with either the radial morphology [Fig. 2.5(c)], the PAN-AM cross-sectional arrangement, or other morphologies (not shown), are exploited for their extremely high bulk modulus and high thermal conductivity, while the commercial PAN (polyacrylonitrile) fibers with circumferential texture [Fig. 2.5(e)] are widely used for their

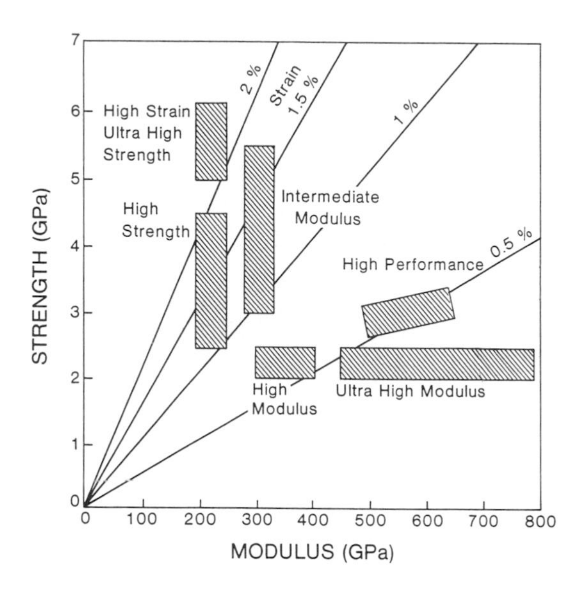

Fig. 2.6. The breaking strength of various types of carbon fibers plotted as a function of Young's modulus. Lines of constant strain are shown and can be used to estimate the breaking strains [2.12, 24, 25].

high strength [2.12]. The high modulus of the mesophase pitch fibers is related to the high degree of c-axis orientation of adjacent graphene layers, while the high strength of the PAN fibers is related to defects in the structure, which inhibit the slippage of adjacent planes relative to each other. Typical diameters for these individual commercial carbon fibers are $\sim$ 10 μm, and they can be very long. These fibers are woven into bundles called tows and are then wound up as a continuous yarn on a spool. The remarkable high strength and modulus of carbon fibers (see Fig. 2.6) are responsible for most of the commercial interest in these fibers. The superior mechanical properties of carbon fibers should be compared to steel, for which typical strengths and bulk modulus values are 1.4 and 207 GPa, respectively [2.12].

Vapor-grown carbon fibers can, however, be prepared over a wide range of diameters (from less than 1000 Å to more than 100 μm) and these fibers have hollow cores. The preparation of these fibers is based on the growth of a thin hollow tube of about 1000 Å diameter from a transition metal catalytic particle ($\sim$100 Å diameter) which has been supersaturated with carbon from a hydrocarbon gas present during growth at 1050°C. The thickening of the vapor-grown carbon fiber occurs through an epitaxial growth process whereby the hydrocarbon gas is dehydrogenated and sticks to the surface of the growing fiber. Subsequent heat treatment to $\sim$2500°C results in carbon fibers with a tree ring concentric cylinder morphology [2.26]. Vapor-grown carbon fibers with micrometer diameters and lengths

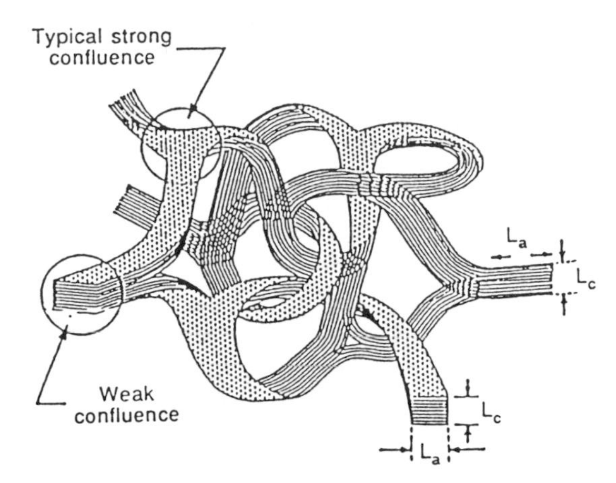

Fig. 2.7. Tangled structure within a grain proposed for many polymer-derived graphitic carbons including glassy carbon. The presence of voids accounts for the low mass density of glassy carbons, and the voids are likely sources of mechanical weaknesses [2.27].

of ~30 cm provide a close analogy to carbon nanotubes with diameters of nanometer dimensions (see Chapter 19).

2.6. Glassy Carbon

Glassy carbon (GC) is another common carbon material which is manufactured as a commercial product by slow, controlled degradation of certain polymers at temperatures typically on the order of 900–1000°C [2.27]. The name glassy carbon is thus given to a family of disordered carbon materials, which are glass-like and can be easily polished to attain a black, shiny appearance. Because they are prepared over a range of conditions, glassy carbons have a range of properties that depend somewhat on the precursor material and significantly on the processing conditions. Glassy carbons are granular, moderately hard, thermally conducting, impermeable, biocompatible, and stable at high temperatures. The apparent density of GC is in the range 1.46–1.50 g/cm^3 irrespective of heat treatment temperature, implying the existence of pores in the matrix.

According to one model [2.27], the microstructure of GC consists of a tangle of graphite-like ribbons or microfibrils, about 100 Å long and 30 Å in cross section (see Fig. 2.7), and resembles the polymer chain configuration from which the GC has been derived. Because of the tangled ribbon microstructure, Jenkins and Kawamura [2.27] have argued that glassy carbon does not fully graphitize, even when heat treated above 3000°C. X-ray diffraction studies of the radial distribution function show that within the graphite-like ribbons the carbon atoms are ordered in the honeycomb in-plane structure of the graphene layers, but that the 3D registry between the graphene layers is poor, so that the ribbons form a turbostratic structure, typical of hard carbons; a hard carbon denotes a carbon material

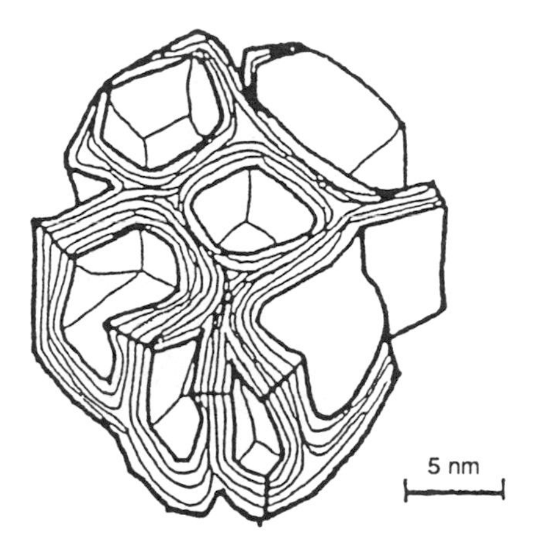

Fig. 2.8. Schematic diagram for the microstructure of the closed pore structure model for glassy carbon and other hard carbons [2.30, 31].

which does not fully graphitize, even under high-temperature (~3000°C) heat treatment [2.28, 29]. Detailed structural studies have indicated that glassy carbons have a network of closed pore structures, according to a competing structural model for glassy carbon proposed by Shiraishi (see Fig. 2.8) [2.30–32].

The temperature dependence of the electrical conductivity $\sigma(T)$ for glassy carbons prepared at various heat treatment temperatures T_{HT} is shown in Fig. 2.9, where T_{HT} in °C is indicated by the number following GC. This general behavior is seen in a variety of other disordered carbons. The temperature dependence of the conductivity of glassy carbon, which is due to hopping, exhibits an $\exp[-(T_0/T)^{1/4}]$ dependence, characteristic of 3D variable range hopping for carbon materials with $T_{HT} < 1000°C$. With increasing T_{HT}, it is shown in Fig. 2.9 that $\sigma(T)$ increases in magnitude and eventually becomes independent of temperature over a very wide temperature range. Glassy carbons have also been extensively characterized by the Raman effect, which is discussed in §2.7 in the context of carbon blacks.

2.7. Carbon Blacks

Classical carbon blacks represent many types of finely divided carbon particles that are produced by hydrocarbon dehydrogenation [2.33, 34] and are widely used in industry as a filler to modify the mechanical, electrical, and optical properties of the materials in which they are dispersed [2.34]. The various types of industrial carbon blacks are usually named following the processes by which they are produced. For example, thermal blacks are typically obtained by thermal decomposition of natural gas, channel blacks by

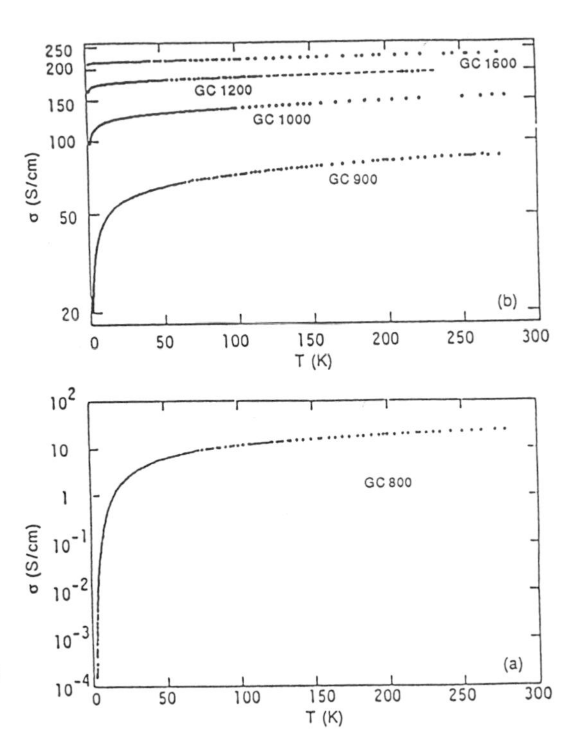

Fig. 2.9. Temperature dependence of the electrical conductivity for glassy carbon samples heat treated to various temperatures (indicated by numerals following GC) [2.31]. Note the many orders of magnitude change in σ for GC 800 in (a) as compared to glassy carbon samples with higher heat treatment temperatures shown in (b).

partial combustion of natural gas, acetylene blacks by exothermic decomposition of acetylene, furnace blacks by partial combustion of oil droplets, and plasma blacks by decomposition of ethylene in a plasma arc [2.33]. Other types of carbon blacks are synthesized on a laboratory scale by special processes, such as laser ablation of graphite [2.35]; CO_2 laser pyrolysis of acetylene (C_2H_2) [2.36, 37] and of ethylene (C_2H_4), which is catalytically assisted by small amount of $Fe(CO)_5$ [2.38]; and by the heat treatment of coal [2.33]. These synthesis routes have produced various types of carbons with different physical and chemical properties.

The microcrystalline structures of several types of carbon blacks (in sizes of 1000 Å and higher) in both their as-synthesized and heat-treated (up to 3000°C) forms have been established by study over many years, mainly using x-ray diffraction (XRD) [2.34, 39–46], high-resolution transmission electron microscope (TEM) lattice imaging [2.47–51], and Raman scattering techniques [2.38, 52, 53]. The earliest XRD studies on carbon blacks [2.39, 40] indicated that as-synthesized carbon blacks are composed of small graphite-like layers in which carbon atoms have the same relative atomic positions as in graphite. These studies led to a model in which the dimensions of the layers are described by two characteristic lengths, L_a and L_c, where L_a is the crystalline size in the plane of the layers and L_c is the size along the c-axis perpendicular to the planes. This simple model was frequently used in later studies to characterize the microcrystalline structures of carbon blacks including thermal blacks, channel blacks, acetylene blacks, plasma blacks, and furnace blacks [2.44].

Another characteristic signature associated with carbon blacks is a concentric organization of the graphite layers in each individual particle. This structural property was mainly established by studies involving high-resolution TEM lattice imaging [2.34, 43, 47–49]. These concentric graphene layers are found to be more pronounced in the region close to the particle surface than in the center. The development of the graphitic layers was also found to be correlated with carbon particle size, synthesis time and temperature, and microporosity of the particles [2.54, 55]. Furthermore, subsequent heat treatment of "as-synthesized" carbon blacks in Ar up to 3000°C was found to produce polygonized particles with an empty core and a well-graphitized carbon shell centered around the growth starting point [2.34, 38, 50, 51].

The morphology of carbon black particles changes as three-dimensional (3D) order is established, since 3D correlations can exist only over very short distances for curved concentric graphene layers. This same argument also applies to the nanometer scale carbon spherules (or onions) discussed in §19.10. Hence the transformation from 2D to 3D graphite requires graphene layer flattening through formation of faceted surfaces. Heat-treated carbon blacks do indeed display a high degree of faceting, as observed in transmission electron micrographs [2.56]. An idealized shape for partially graphitized carbon black particles was proposed [2.43] and is shown schematically in Fig. 2.10(a) as a faceted external surface in the shape of a polyhedron. As-prepared carbon blacks generally exhibit short layer segments which are roughly aligned with the particle surface, shown schematically in Fig. 2.10(b). As the particles are heat treated to higher temperatures, the segment lengths of the graphene layers increase and the graphene layers increase their parallelism with respect to the particle surface. The junctions between facets represent high-angle grain boundaries.

Raman scattering was first used by Tuinstra and Koenig [2.52] to characterize several forms of carbons including single-crystal graphite, pyrolytic graphite, activated charcoal, and carbon blacks. In their studies, all disordered graphite samples showed two prominent Raman features with peaks at 1580 and 1355 cm^{-1}. They assigned the dominant peak (1580 cm^{-1}) as due to the in-plane E_{2g} Raman-allowed mode vibration, which has the value of 1582 cm^{-1} for single crystal graphite, and the second peak (1355 cm^{-1})

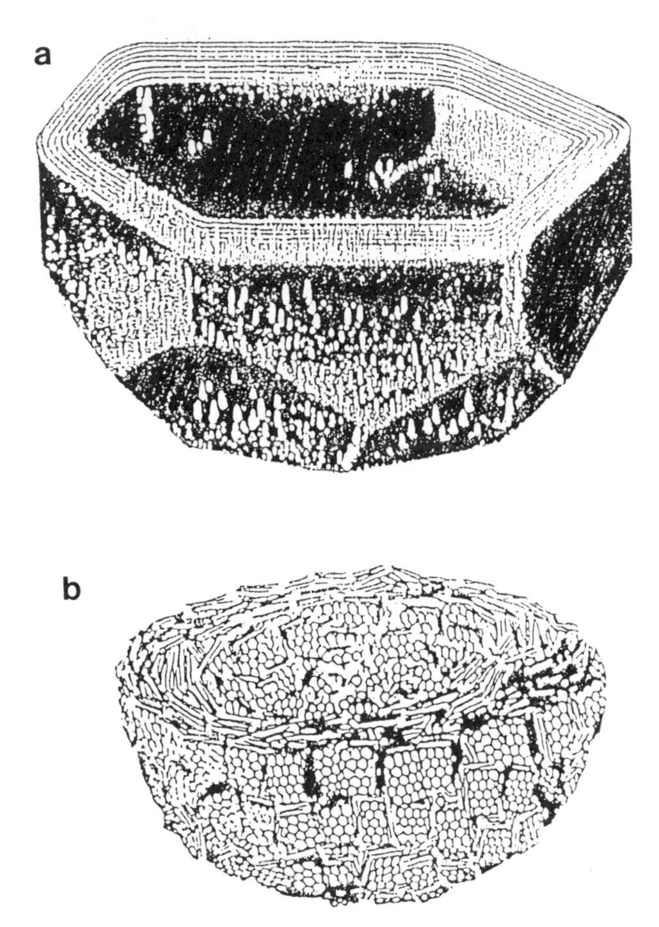

Fig. 2.10. (a) Idealized structure of a partially graphitized carbon black particle. (b) Another schematic view of a carbon black particle showing short graphitic segments [2.43, 55, 56].

was identified with in-plane disorder-induced scattering from graphitic modes throughout the Brillouin zone, but mostly emphasizing the zone edge, where the phonon density of states has a strong maximum. Of particular importance for characterizing the amount of disorder in the graphitic material is the close correlation between the in-plane crystalline size L_a as established by x-ray diffraction and the relative integrated intensity of the disorder-induced line (1355 cm^{-1}) to the first-order Raman-allowed line (1580 cm^{-1}) [2.38, 53, 57]. These two Raman lines are frequently used to monitor the graphitization process of carbon blacks upon annealing at elevated temperatures and for characterizing the internal microcrystalline size of nanoscale carbon black particles.

The synthesis of carbon blacks involving a CO_2 laser was carried out first by Yampolskii *et al.* [2.36] and later by Maleissye *et al.* [2.37]. In their studies, carbon blacks were obtained by decomposing acetylene (C_2H_2) gas using a CO_2 laser beam as a heating source. They reported that carbon blacks produced in this way are identical to those obtained by classic thermal pyrolysis of acetylene. However, no detailed studies were performed to characterize the physical properties of these carbon blacks in either the as-synthesized or heat-treated forms. Another approach producing nanoscale carbon blacks using a CO_2 laser as a heating source involves the pyrolysis of benzene (C_6H_6) vapor, assisted catalytically by a small amount of $Fe(CO)_5$ [2.38]. In this way, highly disordered and spherical carbon particles, 200–500 Å in diameter, have been synthesized. Subsequent heat treatment of these particles in argon gas to temperatures up to ~2800°C produces graphitized polygonal particles with central polygonal cavities. An important advantage of CO_2 laser pyrolysis for carbon black synthesis is that only carbon blacks are produced in the reaction zone defined by the intersection of the reactant gas stream and the CO_2 laser beam [2.38, 58, 59]. Since the reaction zone is away from the chamber wall, which remains at room temperature during the synthesis, no pyrolytic carbons are deposited on the hot chamber walls. The formation of pyrolytic carbons in heated furnaces [2.33] inhibited the development of carbon black formation theory because of the complex reaction products produced on the chamber walls. The CO_2 laser pyrolysis technique [2.60, 61] closely resembles the flame synthesis approach but does not require the presence of oxygen to initiate or sustain the reaction process, in contrast to the conventional flame techniquc in which oxygen must be used to provide heat to the reaction zone.

2.8. Carbon Coated Carbide Particles

Closely related to carbon blacks are graphitic structures (~100 Å in diameter) which form around nanoscale carbide particles such as cementite (Fe_3C) and are prepared by a CO_2 laser pyrolysis process [2.59] or by arc discharge methods [2.62]. These carbon-coated particles may have a significant bearing on the synthesis of single-wall carbon nanotubes [2.63–66] discussed in §19.2.2. The synthesis of cementite nanoparticles involves a gas-phase reaction of $Fe(CO)_5$ vapor with ethylene (C_2H_4) which strongly absorbs the CO_2 laser energy. A high degree of crystallinity has been observed for the graphitic structure that forms around the cementite particle, based on high-resolution transmission electron microscopy (TEM) measurements [2.59] of the lattice image from a single particle (diameter ~300 Å) taken from a batch of similarly prepared particles. X-ray diffraction (XRD) studies on these particles indicate that the particles are a nearly pure Fe_3C phase. Lattice fringe spacings were observed at $d \sim 6.75$ Å for $Fe_3C(001)$ and 3.5 Å for C(002) from both Fe_3C and the carbon coating material. The carbon coating was found to be thick (~60 Å), serving to passivate and stabilize the Fe_3C nanocrystalline particles. The carbon coating has also been characterized by Raman spectroscopy, where a broad doublet is observed with peaks centered at 1375 and 1580 cm^{-1}, and this doublet is identified with the well-known Raman spectrum for disordered pyrolytic carbons [2.12, 52] as discussed in §2.7. The formation mechanism of the carbon coatings has been identified with a catalytic decomposition of C_2H_4 at the particle surface. This decomposition occurs when the "hot" (~1000°C) carbide particle leaves the reaction zone defined by the intersection of the CO_2 laser beam and the reactant gas flow. The rapid cooling of the carbide particle at the reaction zone boundary "freezes" the carbons on the particle surface, rather than allowing these carbons to diffuse into the interior of the carbide particle. Furthermore, no carbon fiber growth has been observed on these cementite nanoparticles, presumably because of the rapid transfer of the Fe_3C particles to a cooler environment, which prevents carbon filament growth from occurring to any great extent. The same laser pyrolysis technique has also been used to produce nanoscale carbon blacks by pyrolyzing benzene (C_6H_6) vapor, catalytically assisted by a small amount of $Fe(CO)_5$ [2.38, 67] as discussed in §2.7. Transition metals or the transition metal carbides (Fe_3C, Ni_3C, Co_3C) form crystalline particles that are tightly wrapped by a thin layer of carbon. Carbon-encapsulated nanoscale Co particles have been produced similarly using the arc discharge method, in which one of the graphite electrodes was packed with Co powder (see §19.2.2).

The carbon encapsulation of pyrophoric, nanoscale lanthanide carbide single crystals was first observed by Ruoff *et al.* [2.62] and Saito *et al.* [2.68, 69] in a fullerene-related synthesis process using a dc arc discharge between two graphite electrodes. The positive electrode was impregnated in the center with La_2O_3 powder (or some other lanthanide oxide powder) with an La:C molar ratio of 0.02. Roughly 50% of nanoscale carbon particles produced by this method are observed to contain single crystalline LaC_2 crystals, and these carbon encapsulated metals are found to be stable in air, indicating that the otherwise pyrophoric LaC_2 metals are protected by the surrounding carbon layers. The carbons on the metal particle surface are found to be well graphitized, presumably due to the high temperature (~3000°C) in the arc discharge region. Because of the high temperature in the arc, the solidification starts on the surface, forming polyhedra of carbon with faceted faces inside which the lanthanide carbide forms a crystalline particle with composition La_2C. The carbon-coated lanthanide carbide particles have been called nanocapsules [2.68], and differ from the carbon wrapped transition metal carbide particles by having a small vacuum region formed during the cooling process from high temperature, whereas the transition metal carbides formed at lower temperatures are tightly wrapped. In both cases the carbon wrapping pacifies and stabilizes the carbides, which otherwise would be quite reactive [2.68].

2.9. Carbynes

Linear chains of carbon which have *sp* bonding have been the subject of research for many years [2.70]. Nevertheless, the present state of knowledge in this field is still fragmentary.

A polymeric form of carbon consisting of chains $[\cdots-C{\equiv}C-\cdots]_n$ for $n > 10$ has been reported in rapidly quenched carbons and is referred to as "carbynes." This carbon structure is stable at high temperature and pressure as indicated in the phase diagram of Fig. 2.1 as shock-quenched phases. Carbynes are silver-white in color and are found in meteoritic carbon deposits, where the carbynes are mixed with graphite particles. Synthetic carbynes have also been prepared by the sublimation of pyrolytic graphite [2.71, 72]. It has been reported that carbynes are formed during very rapid solidification of liquid carbon, near the surface of the solidified droplets formed upon solidification [2.73]. Some researchers [2.72–76] have reported evidence that these linearly bonded carbon phases are stable at temperatures in the range $2700 < T < 4500$ K.

Carbynes were first identified in samples found in the Ries crater in Bavaria [2.77] and were later synthesized by the dehydrogenation of acetylene [2.72, 78]. The carbynes have been characterized by x-ray diffraction, scanning electron microscopy (SEM), ion micromass analysis, and spectroscopic measurements which show some characteristic features that identify carbynes in general and specific carbyne polymorphs in particular. Since several different types of carbynes have been reported, and carbynes are often found mixed with other carbon species, it is often difficult to identify specific carbynes in actual samples. Carbynes have a characteristic negative ion mass spectrum with constituents up to C_{17}. This mass spectrum is different from that of graphite, which has negative ion emissions only up to C_4, with C_3 and C_4 being very weak. Thus ion micromass analysis has provided a sensitive characterization tool for carbynes.

The crystal structure of carbynes has been studied by x-ray diffraction through identification of the Bragg peaks with those of synthetic carbynes produced from sublimation of pyrolytic graphite [2.74, 78]. In fact, two polymorphs of carbynes (labeled α and β) have been identified, both being hexagonal and with lattice constants $a_\alpha = 8.94$ Å, $c_\alpha = 15.36$ Å; $a_\beta = 8.24$ Å, $c_\beta = 7.68$ Å [2.72]. Application of pressure converts the α phase into the β phase. The numbers of atoms per unit cell and the densities are, respectively, 144 and 2.68 g/cm^3 for the α phase and 72 and 3.13 g/cm^3 for the β phase [2.79]. These densities determined from x-ray data [2.72] are in rough agreement with prior estimates [2.80, 81]. It is expected that other less prevalent carbyne polymorphs should also exist.

In the solid form, these carbynes have a hardness intermediate between diamond and graphite. Because of the difficulty in isolating carbynes in general, and specific carbyne polymorphs in particular, little is known about their detailed physical properties, and some have even questioned the existence of carbynes.

2.10. Carbolites

In an effort to prepare larger quantities of carbynes for detailed measurements, a new crystalline form of carbon was synthesized. Because of its relatively low mass density ($\rho_m = 1.46$ g/cm^3) it was called carbolite [2.82]. The carbolite material was synthesized using a carbon arc powered by two automobile batteries ($I \sim 600$ A) to generate carbon vapor which was rapidly quenched onto a copper plate in an argon or argon-hydrogen atmosphere [2.82]. The carbolite material is a slightly orange-colored transparent crystal, suggesting nonconducting transport properties. Depending on whether or not hydrogen is contained with argon as the ambient gas, two crystal structures are obtained by analysis of x-ray diffraction patterns. For carbolites (type I) synthesized in argon gas, the crystal structure is hexagonal with $a_0 = 11.928$ Å, $c_0 = 10.62$ Å, and for carbolites (type II) formed using an argon-hydrogen gas mixture, the lattice constants are $a_0 = 11.66$ Å, $c_0 = 15.68$ Å, which could also be indexed to a rhombohedral structure with a unit cell length of 8.52 Å and a rhombohedral angle of 86.34°. From the structural analysis, the crystal model described in Fig. 2.11 was proposed, showing the stacking of four-atom carbon chains with a nearest-neighbor distance of 1.328 Å and 1.307 Å in a hexagonal lattice with 3.443 Å and 3.366 Å separation between the chains. In the type I structure, the four-atom chains have an AB stacking, while the type II structure shows ABC stacking. Infrared spectra suggest that the interchain bonding is of the –C≡C–C≡C– polyyne type. Evidence for modification of the electrical properties of carbolites by intercalation with K, Na, and I_2 was also reported [2.82]. Further structure/properties measurements on carbolites can be expected.

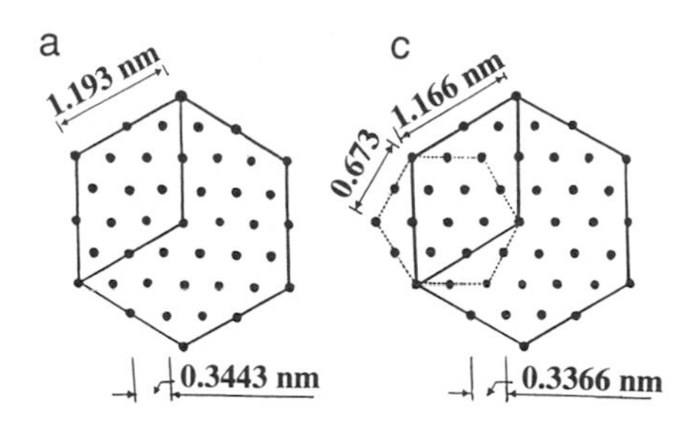

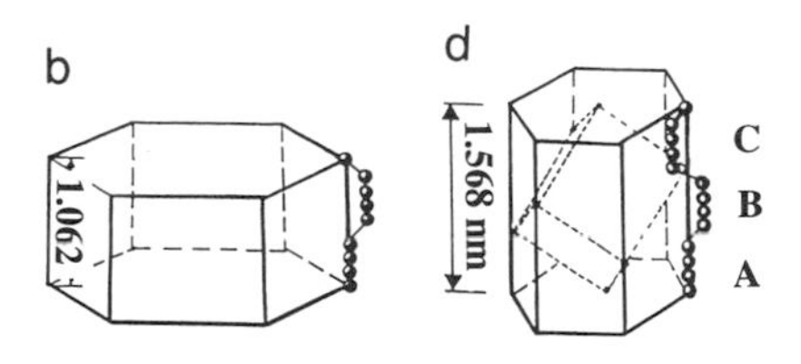

Fig. 2.11. Proposed structural model for the two forms of hexagonal carbolite. Hexagonal face of unit cell showing alignment of the chains for (a) type I carbolite and (c) for type II carbolite. View of the four-atom carbon chains (b) with AB stacking for type I, and (d) with ABC stacking for type II [2.82].

2.11. Amorphous Carbon

Amorphous carbon (a-C) refers to a highly disordered network of carbon atoms that have predominantly sp^2 bonding, with perhaps 10% sp^3 bonds and almost no sp^1 bonds (see Fig. 2.3) [2.9]. Although amorphous carbon has no long-range order, some short-range order is present. Since the nature of the short-range order varies significantly from one preparation method to another, the properties of amorphous carbon films likewise vary according to preparation methods [2.83]. Two parameters, the carbon bonding (expressed in terms of the ratio of sp^2/sp^3 bonds) and the hydrogen content, are most sensitive for characterizing the short-range order, which may exist on a length scale of ~10 Å. Thus the sp^2-bonded carbons of a-C may cluster into tiny warped layered regions, and likewise the sp^3-bonded carbons may also cluster and segregate, as may the hydrogen impurities, which are very effective in passivating the dangling bonds. Amorphous carbon is also commonly formed by neutron, electron, or ion beam irradiation of carbon-based materials which had greater structural order prior to irradiation [2.84].

A perfectly sp^2-ordered graphene sheet is a zero gap semiconductor. The introduction of disorder and sp^3 defects creates a semiconductor with localized states near the Fermi level and an effective band gap between mobile filled valence band states and empty conduction states. The greater the disorder and the greater the concentration of sp^3 bonds, the larger is the band gap. Amorphous carbon prepared by evaporation tends to have a higher room temperature conductivity $\sigma_{RT} \sim 10^3\ \Omega^{-1}\text{cm}^{-1}$ and smaller band gap (E_g ~0.4–0.7 eV) [2.83, 85–87] compared with a-C prepared by ion-beam deposition for which $\sigma_{RT} \sim 10^2\ \Omega^{-1}\text{cm}^{-1}$ and $E_g \sim 0.4 - 3.0$eV [2.88–90]. Increasing the sp^3 content of the films also tends to enhance the mechanical hardness of the films, to decrease σ_{RT}, and to increase E_g.

Since amorphous carbon has no long range order, it is customary to characterize a-C samples in terms of their radial distribution functions as measured in a diffraction experiment (electron, x-ray, or neutron). Amorphous carbon samples are also characterized in terms of their densities in comparison to graphite (ρ_m = 2.26 g/cm^3) to specify the packing density of the films. The ratio of the sp^2/sp^3 bonding in a-C samples is usually determined spectroscopically, for example by nuclear magnetic resonance (NMR), x-ray near-edge structure, and electron energy loss spectroscopy [2.9].

2.12. Porous Carbons

There are a number of carbons which form highly porous media, with very high surface areas and pores of nanometer dimensions similar to the dimensions of fullerenes. In this category of porous and high surface area carbons are included activated carbons, exfoliated graphite, and carbon aerogels. The nanopores may be in the form of cages or tunnels. Whereas the surfaces of the nanopores contain a high density of dangling bonds and surface states, the surfaces of fullerene molecules, which may be considered as nanoparticles, have no dangling bonds.

In preparing activated carbons (see Fig. 2.12), isotropic pitch and phenol are commonly used as precursor materials. Of the various activated carbon materials, activated carbon fibers have the narrowest distribution of pore sizes and nanopores. The dominance of the carbon nanopores in the properties of activated carbon fibers makes this fibrous material attractive for various applications [2.91] and of particular interest as a comparison material to fullerene carbon nanoparticles because of the similarity of the sizes of the pores and fullerenes. In the activation process, the precursor fiber is heated in O_2, H_2O, CO_2, or other oxidizing atmospheres in the temperature range 800–1200°C for pitch-based fibers and 1100–1400°C for phenolic-derived fibers. The main parameter that has been used to characterize activated carbon fibers is the specific surface area (SSA). High values for SSA are achieved (1000–3000 m^2/g) in activated carbon fibers by controlling the temperature and the time for activation, and the precursor materials. The SSA is usually measured using adsorption isotherms of N_2 gas at 78 K and CO_2 gas at 195 K, though the highest accuracy is achieved with helium gas near 4 K. Most of the surface area arises from a high density of nanopores with diameters < 1 nm. In the literature on porous carbon,

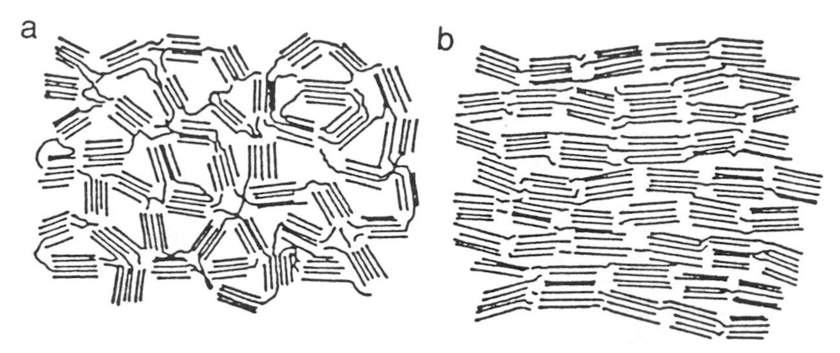

Fig. 2.12. A schematic model for the microstructure of activated carbon fibers: (a) for a high surface area fiber where the basic structural units are randomly arranged, (b) for a fiber after some heat treatment, showing partial alignment of the basic structural units [2.91].

the term macropore is used to describe pores with diameters greater than 100 nm, mesopores for diameters in the 2 to 100 nm range and micropores for diameters less than 2 nm [2.92].

Exfoliated graphite is another high surface area form of graphite which is prepared by heating a graphite intercalation compound (see §2.14) above some critical temperature above which a gigantic irreversible c-axis expansion occurs, with sample elongations of as much as a factor of 300 [2.93]. This gigantic elongation, called exfoliation, gives rise to spongy, foamy, low density, high surface area material (e.g., 85 m^2/g [2.94]). The commercial version of the exfoliated "wormy" material is called vermicular graphite and is used for high surface area applications. When pressed into sheets, the exfoliated graphite material is called grafoil and is widely used as a high temperature gasket or packing material because of its flexibility, chemical inertness, low transverse thermal conductivity, and ability to withstand high temperatures [2.95].

Carbon aerogels are a disordered form of sp^2-bonded carbon with an especially low bulk density and are made by a supercooling process. These materials are examples of a class of cluster-assembled low-density porous materials, consisting of interconnected carbon particles with diameters typically near 12 nm [2.57, 96]. Within each particle, a glassy carbon–like nanostructure is observed, consisting of an intertwined network of narrow graphitic ribbons of width $\sim$2.5 nm. The morphology is illustrated schematically in Fig. 2.13. This structure leads to high surface areas (600–800 m^2/g) with a significant fraction of the atoms covering the surfaces of the interconnected particles. For a given specific surface area, carbon aerogels tend to have larger size pores than the activated carbon fibers discussed above. Because of their large surface areas and consequently high density of dangling bonds, porous carbons tend to have

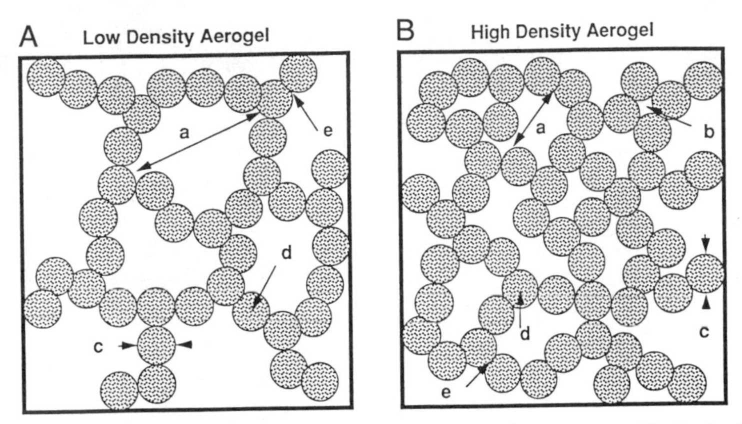

Fig. 2.13. Schematic diagram of the carbon aerogel microstructure. Each shaded circle represents an amorphous carbon particle. The microstructure is shown for (A) low (~0.1 g/cm^3) and (B) high (~0.6 g/cm^3) bulk density forms. The microstructure shows (a) mesopores that span the distance between chains of interconnected particles, (b) micropores sandwiched between particles, (c) individual particles (~12 nm diameter), (d) micropores within the particles, and (e) micropores between contiguous particles [2.57].

somewhat different electronic properties from those of other disordered carbons.

2.13. Liquid Carbon

Liquid carbon refers to the liquid phase of carbon resulting from the melting of pure carbon in a solid phase (graphite, diamond, or materials related to these pure carbon solids via disorder). Referring to the phase diagram for carbon (Fig. 2.1), we see that liquid carbon is stable at atmospheric pressure only at very high temperatures (the melting point of graphite T_m ~4450 K) [2.2]. Since carbon has the highest melting point of any elemental solid, to avoid contamination of the melt, the crucible in which the carbon is melted must itself be made of carbon, and sufficient heat must be focused on the sample volume to produce the necessary temperature rise to achieve melting [2.7, 97]. Liquid carbon has been produced in the laboratory by the laser melting of graphite, exploiting the poor interplanar thermal conductivity of the graphite [2.98], and by resistive heating [2.99], the technique used to establish the metallic nature of liquid carbon.

Although diamond and graphite may have different melting temperatures, it is believed that the same liquid carbon is obtained upon melting either solid phase. It is believed that liquid carbon is a monatomic form of carbon in contrast to a liquid crystal phase. It is also believed that the solidification of the melt under nonequilibrium conditions may yield some novel forms of carbon upon crystallization, as discussed above (see §2.9). Since the vaporization temperature for carbon (~4700 K) is only slightly higher than the melting point (~4450 K), the vapor pressure over liquid carbon is high. The high vapor pressure and the large carbon–carbon bonding energy make it energetically favorable for carbon clusters rather than independent atoms to be emitted from a molten carbon surface [2.6, 100]. Energetic considerations suggest that some of the species emitted from a molten carbon surface have masses comparable to those of fullerenes [2.97]. The emission of fullerenes from liquid carbon is consistent with the graphite laser ablation studies of Kroto, Smalley, and co-workers [2.101].

In considering the melting of fullerenes, several types of phase transitions must be taken into account. First the melting of the solid fullerene lattice into a liquid phase of freely moving fullerene molecules would be expected to occur, provided that the fullerene molecules do not evaporate into the gas phase or disintegrate before reaching this melting point. Even if a fullerene liquid were stable, it is unlikely that these molecules would be stable up to the temperature where finally the total dissociation of the fullerene occurs and liquid carbon forms (4450 K).

2.14. Graphite Intercalation Compounds

Because of the weak van der Waals interlayer forces associated with the sp^2 bonding in graphite, graphite intercalation compounds (GICs) may be formed by the insertion of layers of guest species between the layers of the graphite host material [2.102, 103], as shown schematically in Fig. 2.14. The guest species may be either atomic or molecular. In diamond, on the other hand, the isotropic and very strong sp^3 bonding (see Fig. 2.3) does not permit insertion of layers of guest species and therefore does not support intercalation. In the so-called donor GICs, electrons are transferred from the donor intercalate species (such as a layer of the alkali metal potassium) into the graphite layers, thereby raising the Fermi level E_F in the graphitic electronic states, and increasing the mobile electron concentration by two or three orders of magnitude, while leaving the intercalate layer positively charged with low mobility carriers. Conversely, for acceptor GICs, electrons are transferred to the intercalate species (which is usually molecular) from the graphite layers, thereby lowering the Fermi level E_F in the graphitic electronic states and creating an equal number of positively charged hole states in the graphitic π-band. Thus, electrical conduction in GICs (whether they are donors or acceptors) occurs predominantly in the graphene layers and as a result of the charge transfer between the intercalate and host layers. The electrical conductivity between adjacent graphene layers is very poor (especially in the acceptor compounds, where

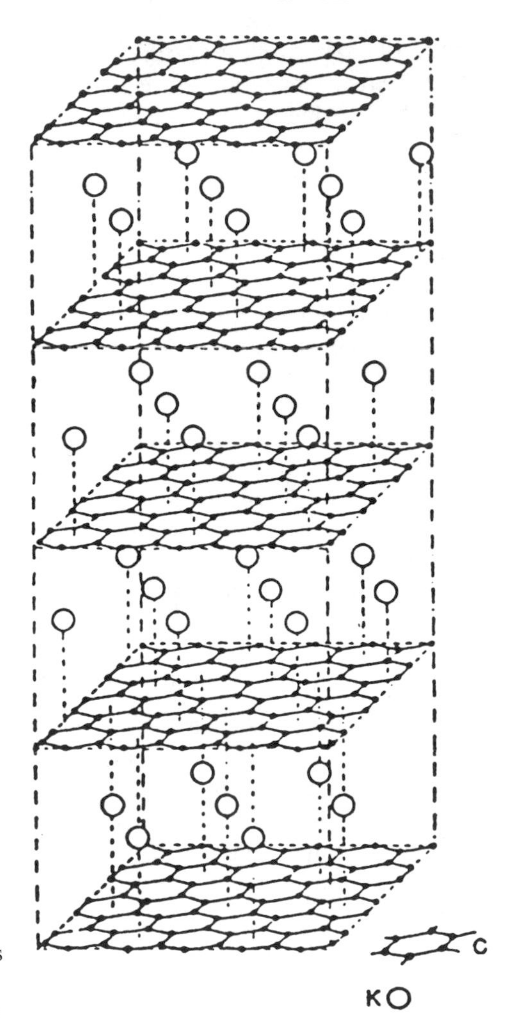

Fig. 2.14. Schematic model for a graphite intercalation compound showing the stacking of graphite layers (networks of hexagons on a sheet) and of intercalate (e.g., potassium) layers (networks of large hollow balls). For this stage 1 compound, each carbon layer is separated by an intercalate layer [2.104].

the graphene layer separation is generally larger). A similar charge transfer process is observed for alkali metals and alkaline earths, which dope fullerenes and transfer electrons forming fullerene anions. Also, the distances of the dopants to the nearest-neighbor carbon atoms are similar for doped fullerenes and for GICs. However, the conduction process in doped fullerenes is very different from that in GICs because of differences in electronic bandwidths and the dimensionality of the conduction process.

Because of the attractive electrostatic interaction between the intercalate and the adjacent graphene layers and the repulsive interplanar interaction resulting from the intercalation-induced lattice strain, the intercalate layers form an ordered superlattice structure, interleaved with the graphite layers, through a phenomenon called *staging* [2.102, 103]. A GIC of stage n has isolated intercalate layers separated from one another by n graphite layers, forming a unit cell along the c-axis normal to the layer planes of length

$$I_c = d_s + (n+1)c_0, \tag{2.1}$$

where for the potassium intercalate $d_s = 5.35$ Å is the thickness of a sandwich formed by two graphite layers between which the potassium layer is sandwiched, and $c_0 = 3.35$ Å is the interlayer distance between adjacent graphite layers. The distance d_s in K-GICs (5.35 Å) is close to twice the nearest-neighbor C–K distance for a K atom in a tetrahedral site in K_3C_{60} ($2 \times 2.62 = 5.24$ Å). For an acceptor GIC, such as prepared from the reaction of $SbCl_5$ and graphite, the intercalate unit cell consists of a Cl–Sb–Cl trilayer with $d_s = 9.42$ Å, so that the c-axis repeat distance is $I_c = 12.77$ Å for a stage 2 $SbCl_5$–GIC. As another example of a trilayer intercalate structure, the donor KHg–GIC has an intercalate structure containing a K–Hg–K trilayer with $d_s = 10.22$ Å.

Because of the high electronegativity of fullerenes, significant charge transfer in fullerenes occurs only for donor compounds, so that negatively charged molecules (or anions) are formed. Many similarities are found for the intercalation of the same donor intercalate species (e.g., an alkali metal) into graphite and into crystalline fullerene solids with regard to charge transfer and interatomic distances. There are, howcvcr, fewer similarities between acceptor GICs and doped fullerenes. Whereas charge transfer almost always occurs for acceptor GICs, fullerenes form clathrates which denote a mixture of neutral fullerene and dopant molecules; no significant charge transfer occurs during clathrate formation with fullerenes (see §8.4.2).

2.15. Diamond

The diamond structure is probably the most thoroughly investigated of all crystallographic structures of carbon. Silicon and germanium, the most commonly used elemental semiconductors, also exhibit the same "diamond" structure, and other important group III–V or II–VI compound semiconductors (such as GaAs or CdTe) crystallize in the closely related zincblende structure, the fundamental difference between the zincblende and diamond structures being that the two constituent atomic species of the zincblende structure occupy each of the distinct sites in the diamond structure [2.4].

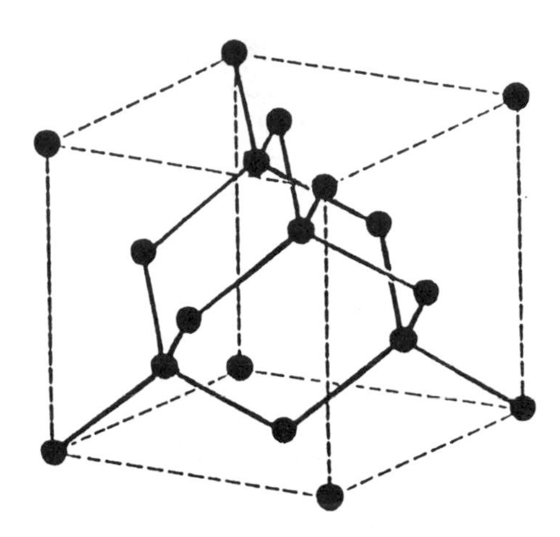

Fig. 2.15. The ideal diamond structure [2.4]

The ideal diamond structure shown in Fig. 2.15 has the characteristic property that every carbon atom is surrounded by four other carbon atoms at the corners of a regular tetrahedron with a cube edge length of $a_0 = 3.567$ Å, and this carbon atom is bonded to these neighbors by strong covalent sp^3 bonds. The diamond structure is therefore cubic and can be viewed (see Fig. 2.15) as two interpenetrating fcc structures displaced by $(1/4, 1/4, 1/4)a_0$ along the body diagonal. The nearest-neighbor carbon–carbon distance is 1.544 Å, nearly 10% larger than in graphite, yet the atomic density of diamond (1.77×10^{23} cm^{-3}) is 56% higher than in graphite, due to the high anisotropy of the graphite structure (see Table 2.1). The diamond crystal is highly symmetric with a cubic space group O_h^7 ($Fd3m$). Although cubic and nearly isotropic, diamond cleaves along {111} planes, while highly anisotropic graphite cleaves along {002} planes [2.3]. There also exists a hexagonal form of diamond with space group D_{6h}^4 ($P6_3/mmc$), the same as for graphite, but with different site locations. The packing in hexagonal diamond is similar to that of cubic diamond, except for a shift of one of the two carbon layers laterally along the {111} planes [2.105]. The c-axis unit vector for hexagonal diamond is $(a_0/\sqrt{3}) =$ 2.06 Å, considerably smaller than the interplanar separation of 3.35 Å for graphite.

Impurities in diamond are very important because of the changes they induce in the electrical and thermal properties of diamond; these modified properties find applications mainly in industrial processes. The best natural diamonds contain impurities with concentrations in the range ~1 part in 10^5. Only a few chemical species (e.g., B, N) can enter the diamond lattice substitutionally and, even when doping is possible, the concentration of such substitutional impurities is very low (less than 1 part in 10^4). This situation is similar to that for graphite, where substitutional impurities are limited to B. Synthetic diamonds grown under conditions of high temperature and pressure (see phase diagram in Fig. 2.1) have the same structure and defect types as natural diamonds. This is in contrast to diamond films discussed in §2.16.1, which have a higher concentration of impurities, especially hydrogen, non-diamond-bonded carbon defects, and numerous grain boundaries [2.106].

Diamonds have been historically classified according to their optical absorption properties, which are determined by impurities. The so-called type Ia diamonds exhibit strong absorption in the infrared. This is caused by fairly substantial amounts (up to 0.1%) of nitrogen, inhomogeneously distributed in the crystal and mainly concentrated in small agglomerates. Most natural diamonds belong to this group. Type Ib diamonds contain nitrogen as substitutional impurities. Although the type Ib diamonds are rarely formed in nature, most synthetic diamonds belong to this group. Diamonds with the highest purity are type IIa, and these exhibit the intrinsic semiconducting properties of diamond with a wide band gap of 5.47 eV. The best synthetic diamonds are type IIa, and these contain less N impurities than natural diamonds. The N defect in the best synthetic diamonds tends to be of the single interstitial type. The most perfect structurally of these synthetic diamonds are the best thermal conductors. Type IIb diamonds are naturally boron doped and show *p*-type conductivity due to an acceptor level introduced by the substitutional boron located ~0.35 eV above the valence band.

2.16. Other Diamond Materials

2.16.1. *Chemical Vapor Deposited Diamond Films*

In the last few years an important scientific and technological breakthrough occurred with the discovery that diamond thin films can be successfully grown by a large variety of chemical and physical vapor deposition techniques not requiring high temperatures and pressures [2.107–111]. The feature common to all these methods is that cracked hydrocarbon radicals impinge upon a hot (~ 900°C), usually prescratched, surface in the presence of atomic hydrogen. It should be emphasized that the essential catalytic agent necessary for the synthesis of diamond films is atomic hydrogen. Synthetic diamonds can be grown when the synthesis conditions are arranged to enhance the formation of sp^3 over sp^2 bonds. Diamond films have been prepared on a variety of substrates, including Si, quartz, Ni, and W. The films grown by these techniques are usually polycrystalline, consisting of randomly oriented, small diamond crystallites (several μm in size), and the films thus tend to have very rough surface morphologies, exhibiting columnar growth patterns. In addition, they tend to have a greater defect density and different kinds of defects than single crystal bulk diamonds, as noted in §2.15.

Raman scattering and temperature-dependent thermal conductivity measurements provide sensitive characterization tools for chemical vapor deposition (CVD) diamond films. Raman spectroscopy studies of diamond films usually show a sharp peak at 1332 cm^{-1}, typical for sp^3-bonded carbon (i.e., diamond), superimposed on a broad peak at about 1500 cm^{-1} which is due to graphitic sp^2 bonding. Figure 2.16 shows a typical Raman spectrum of a CVD diamond film. One accepted figure of merit for the evaluation of the quality of a diamond film is the ratio of the integrated intensity of the sp^3-related peak at 1332 cm^{-1} to that of the sp^2-related broad background at higher frequencies. Because of the higher (by a factor of ~50) Raman cross section for the sp^2-bonded material relative to the sp^3-bonded material, Raman spectroscopy provides a sensitive test for the quality of diamond films. The best CVD diamond films show no evidence for sp^2 bonding in the Raman spectra [2.113]. The physical properties of synthetic diamond films, to the extent they have been studied so far, remarkably resemble those of crystalline diamond, despite their polycrystallinity and their diver-

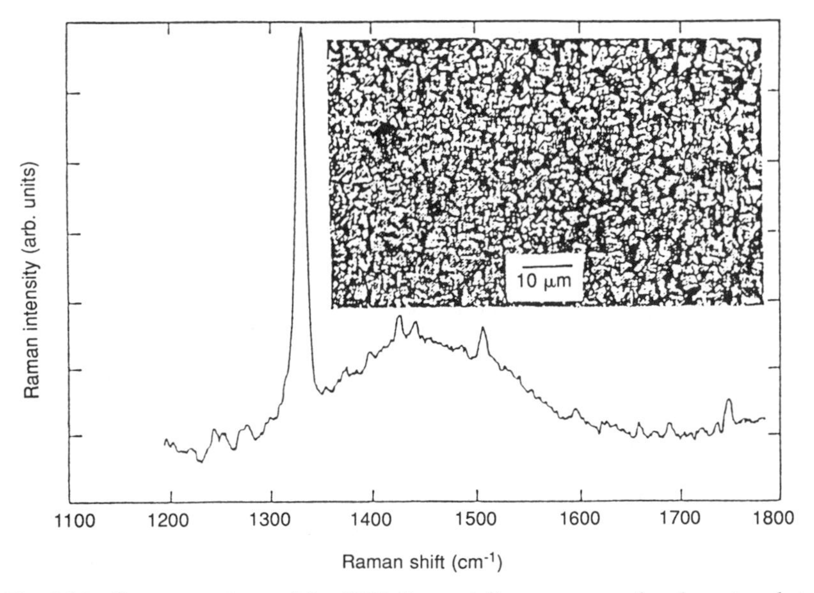

Fig. 2.16. Raman spectrum of the CVD diamond film grown on a fused quartz substrate. Inset: optical micrograph displaying the film morphology [2.112].

sity of defects (such as grain boundaries). The thermal conductivity $\kappa(T)$ of diamond films is an exception to the above, since $\kappa(T)$ is strongly affected by defects and grain boundaries. Thus $\kappa(T)$ for most synthetic diamond films shows a positive temperature coefficient in the range 300–700 K, in contrast to the negative slope typical of single-crystal diamond [2.114]. The best available CVD diamond films give $\kappa(T)$ comparable to that of single-crystal diamond, especially for the material far away from the substrate on which the films are grown. This is so because the diamond films have a columnar growth pattern and the grains tend to be small near the substrate interface and larger away from the interface [2.115]. The most sensitive test of the presence of non-sp^3 diamond-bonded carbon is the measurement of $\kappa(T)$ [2.113].

2.16.2. *Diamond-like Carbon Films*

In addition to crystalline diamond and CVD diamond films, there are so called diamond-like carbon (DLC) films which contain a mixture of both sp^3 and sp^2 bonding and have large concentrations of hydrogen impurities. DLC films are technologically important as hard, chemically inert, insulating coatings which are transparent in the infrared and are biologically compatible with human tissue [2.116, 117]. Diamond-like carbon, which should actually be called amorphous-hydrogenated carbon (a-C:H), is essentially a carbon-based material that bridges the gap between the sp^3-bonded diamond, the sp^2-bonded graphite, and hydrogen-containing organic and polymeric materials. It has been suggested [2.118, 119] that the structure of a-C:H consists of sp^2 carbon clusters, typically planar aromatic ring clusters, which are interconnected by randomly oriented tetrahedral sp^3 bonds to hydrogen. The hydrogen atoms in DLC films may be bonded on either tetrahedral sites, where they are required to reduce bond angle disorder, or on the edges of the ring structures, where they are needed for bond terminations. One major difference between DLC films and diamond (or graphite) is therefore the important role that hydrogen plays in stabilizing the DLC structure [2.120].

The DLC films are grown by a variety of deposition techniques in which a plasma is ignited in a hydrocarbon gas mixture and the ions and radicals in it are directed towards the substrate material. Unless a very H-rich atmosphere and a heated substrate are used (in which case diamond films can be grown), the resulting film has an amorphous structure, containing up to 40 atomic % hydrogen. DLC films are of great commercial interest because many of their physical and chemical properties are similar to those of diamond, including hardness, chemical inertness, electrical resistance, and some transparency in the visible and in the IR. Furthermore, the surface roughness of the as-grown DLC films is usually much less than that observed for CVD films. The properties and growth methods of a-C:H have been reviewed in a number of papers and conference proceedings [2.84, 120].

2.17. Diamond-like and Cage Hydrocarbon Molecules

Interest in diamond-like cage molecules dates back to the 1920s, when organic chemists first considered the possibility of obtaining diamondoid (i.e., sp^3-bonded) hydrocarbon molecules (for a review, see [2.121]). The first such molecule, tricyclodecane $C_{10}H_{16}$ or "adamantane" (see Fig. 2.17), was isolated from petroleum in Czechoslovakia by Landa in 1933 [2.122]. The structure of adamantane is shown schematically in Fig. 2.17(b), where the 10 C atoms in the figure are located at the vertices of the cage; the 16 hydrogen atoms required to satisfy the remaining C bonds are omitted from the figure for simplicity. In Fig. 2.18 the connection between adamantane and the crystal lattice of diamond is shown, and in this figure the dashed C–C bonds distinguish an adamantane C_{10} unit (dashed lines) as a structural repeating unit of the diamond lattice.

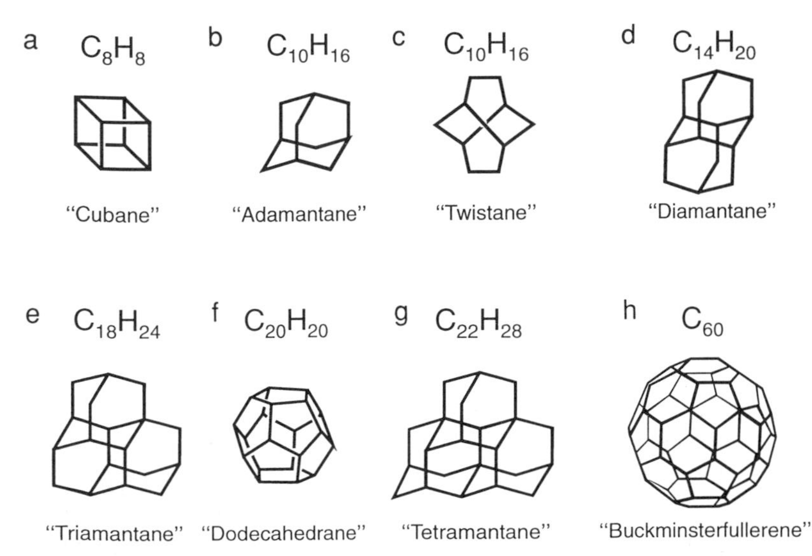

Fig. 2.17. Examples of cage hydrocarbons [2.123, 124]. (a) "Cubane" C_8H_8, (b) "adamantane" $C_{10}H_{16}$, (c) "twistane," an isomer of adamantane $C_{10}H_{16}$, (d) "diamantane" $C_{14}H_{20}$, in which two adamantane cages can be found, (e) "triamantane" $C_{18}H_{24}$ containing three adamantane cages, (f) "dodecahedrane" $C_{20}H_{20}$, (g) "tetramantane" $C_{22}H_{28}$ with four adamantane cages, and (h) "buckminsterfullerene" C_{60}. (a)–(g) Adapted from [2.123] and (h) from [2.101].

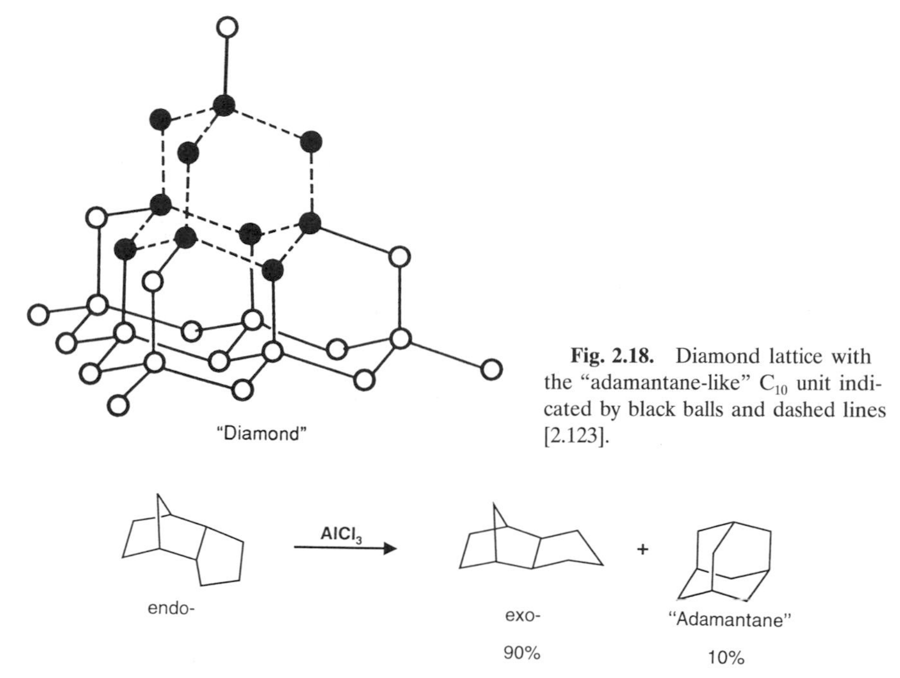

Fig. 2.18. Diamond lattice with the "adamantane-like" C_{10} unit indicated by black balls and dashed lines [2.123].

Fig. 2.19. Single-step rearrangement reaction using an $AlCl_3$ catalyst to produce adamantane [2.123]. The endo isomer of tetrahydrodicyclopentadiene is converted to the exo isomer with a small yield of adamantane (see also [2.121]).

The synthesis breakthrough for adamantane came in 1957, when Schleyer and co-workers [2.124] discovered a high-yield route to adamantane using an $AlCl_3$ catalyst (see Fig. 2.19). Progress in this field has typically been made by finding a suitable catalyst to induce a desired rearrangement reaction of the cage constituents. Many substantial improvements in the synthesis of adamantane followed [2.123], and as a result, adamantane is now a common commercially available chemical, either in pure form or in one of the many adamantane derivatives.

Adamantane and its derivatives also belong to a special class of molecules termed "cage hydrocarbons," and it is for this reason that we discuss adamantane here. Molecules in this class have three or more rings arranged topologically so as to enclose space in the center of the molecule (for a review, see [2.123]). The enclosed space is usually too small to envision the incorporation of additional atoms or ions. Stable examples of rigid, cage hydrocarbons are shown in Figs. 2.17(a–h), where the largest structure $C_{60}H_x(x = 1, \ldots, 30$ [2.125]) is, of course, hydrogenated buckminsterfullerene; $C_{60}H_{30}$ represents the highest hydrogen concentration

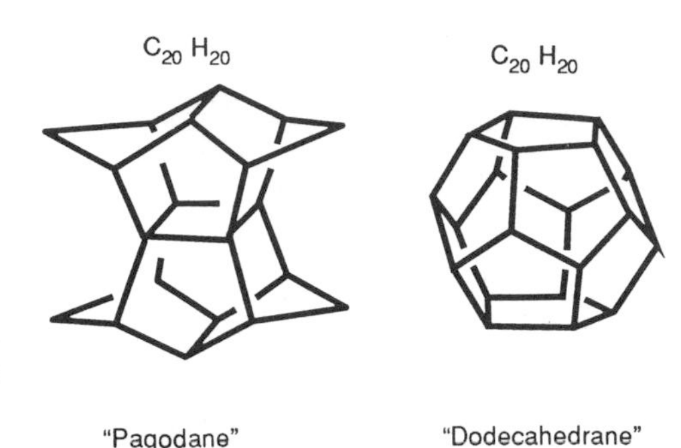

Fig. 2.20. A single-step rearrangement reaction of "pagodane" has been used to prepare "dodecahedrane", both having the same chemical composition $C_{20}H_{20}$ (from [2.123]).

achieved by this synthesis route [2.125]. Three molecules based on multiples of adamantane units appear in Fig. 2.17: "diamantane" or $C_{14}H_{20}$ (d) showing two such units, "triamantane" or $C_{18}H_{24}$ (e) showing three units, and "tetramantane" or $C_{22}H_{28}$ (g) showing four units; the anti- and the skew isomers of adamantane have also been synthesized and isolated [2.124], showing that interesting rearrangements of cage molecules can be prepared in a controlled way. Such rearrangements are analogous to the many possible isomers found for the higher-mass fullerenes (see §3.2). "Twistane," in Fig. 2.17(c), an isomer of adamantane, is synthesized by a rearrangement reaction of adamantane. "Cubane" C_8H_8, a cubic molecule, is available commercially and is being investigated for high density, carbon-based materials applications. All of the cage molecules, from "cubane" C_8H_8 [Fig. 2.17(a)] through "dodecahedrane" $C_{20}H_{20}$ [Fig. 2.17(f)], can be prepared via traditional organic chemistry methods. For example, dodecahedrane can be prepared in a one step rearrangement of "pagodane," shown schematically in Fig. 2.20 [2.126]. Dodecahedrane is a fully hydrogenated fullerene which exhibits 12 pentagonal (and no hexagonal) rings in its cage. The discovery of dodecahedrane in 1982 predates the proposal by Kroto and Smalley regarding the soccer-ball structure for C_{60}. Considering all the success that cage hydrocarbon chemistry has enjoyed since the pioneering work of Schleyer in the late 1950s [2.124], it will be interesting to see whether or not C_{60} might someday be synthesized by solution chemistry as well.

2.18. Synthesis of a Fully Unsaturated, All-Carbon Polymer

Recently, an all-carbon ladder polymer (i.e., no heteroatoms such as O or S are included in the polymer backbone) has been synthesized by Schluter *et al.* [2.127]. A precursor polymer was first obtained by a Diels–Alder polyaddition of a monomer [reaction (i), Fig. 2.21] and then converted (polymer analogous conversion) by dehydrogenation [reaction (ii),

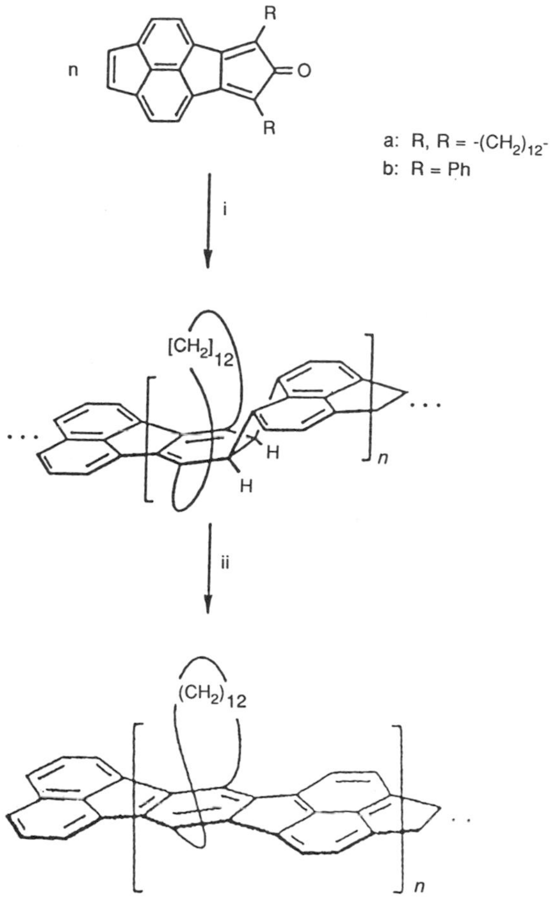

a: R, R = $-(CH_2)_{12}-$
b: R = Ph

Fig. 2.21. Synthesis of the precursor polymer (top) and its conversion into the unsaturated target polymer **a** (bottom). In the diagram *n* denotes a monomer unit, and only the reaction **a** is indicated. In reaction **b** a phenyl group (C_6H_5) attaches at each R site in reaction (i) [2.127].

Fig. 2.21] into the fully unsaturated ladder polymer, which is planar. The authors propose that this class of polymer might combine favorable electronic properties with processability. Possible uses of these polymer-derived carbons may be for electroluminescence and photovoltaic applications. Interestingly, the ladder polymer was found to be stable in oxygen. After exposure to air for several weeks, no apparent change in the ultraviolet–visible (UV-vis) absorption spectrum was noticed.

The ladder structure of the polymer is shown in Fig. 2.22. As can be seen, the polymer (case a) is defined by a repeating sequence of pyracylene units and carbon hexagons, and the pyracylene units are separated by a single hexagonal ring. For a hypothetical polymer (case **b**), the ladder structure is seen to be identical to the belt of an icosahedral C_{60} molecule

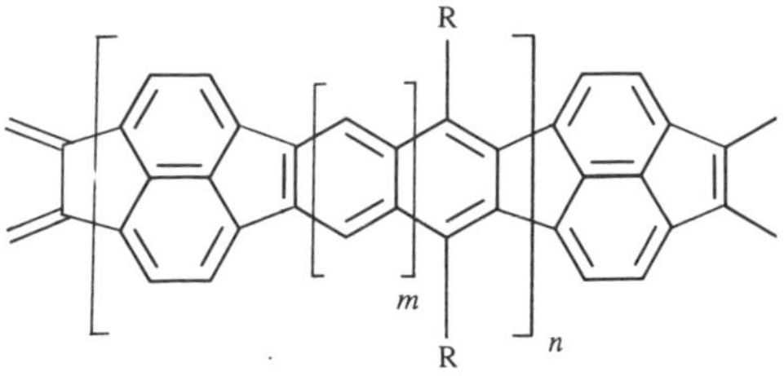

Fig. 2.22. Structures of the fully unsaturated target polymer (a) and a hypothetical, linearly extended fragment of C_{60} (b) containing an additional hexagon on the backbone. Here n denotes the monomer unit and R the side chain additions (see Fig. 2.21) [2.127].

normal to a twofold axis (i.e., the pyracylene units are separated by two hexagonal rings). The side chain groups R seen in Fig. 2.22 (case **a**) actually represent the same aliphatic loop $[CH_2]_{12}$, as verified by ^{13}C NMR [2.127]. Aliphatic side chains are often added to polymer precursors to improve their solubility and processability. In this case, the aliphatic loop was added to improve solubility of the precursor. Undoubtedly, the structural similarity of this planar, ladder polymer with the belt of C_{60} will induce many researchers to investigate the electronic and optical properties of this new carbon material.

2.19. Metallo-Carbohedrenes (Met-Cars)

Castleman and co-workers at the Pennsylvania State University have recently discovered small, stable clusters containing either $3d$, $4d$, or $5d$ transition metals and carbon [2.128–132]. These clusters have been named metallo-carbohedrenes, or "met-cars." A stoichiometry M_8C_{12}, where M = Ti, V, Sr, and Hf, was deduced for the simplest, lowest-mass met-cars from mass spectroscopy. Duncan and co-workers at the University of Georgia [2.133], using an apparatus similar to Castleman's, have confirmed the observation of peaks assignable to M_8C_{12} clusters by time-of-flight mass spectroscopy, where M = Ti, V. In addition, they have observed M_8C_{12} clusters for M = Cr, Fe, Mo, and Nb [2.134–137]. Although met-cars have been observed only in mass spectrometers, their unusual stability and their ability to complex with eight NH_3 molecules per cluster (presumably at the metal sites) have led Castleman to propose the pentagonal dodecahedral structure for the M_8C_{12} unit [Fig. 2.23(a)] and the symmetry shown in Fig. 2.23(b) for the corresponding $M_{14}C_{21}$ cluster. Note that the $M_{14}C_{21}$ cluster (or molecule) shares a pentagonal face common to two M_8C_{12} units.

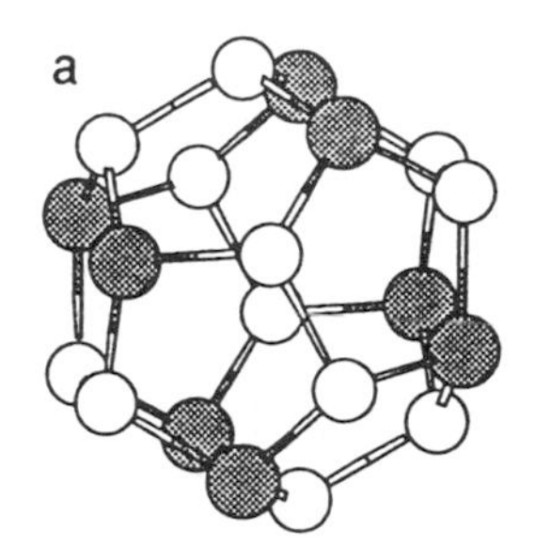

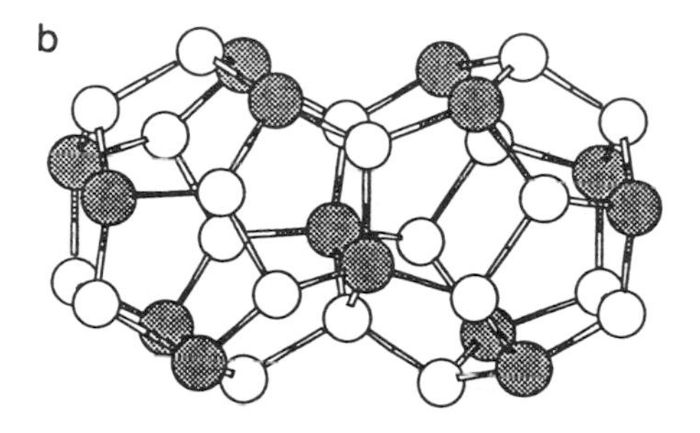

Fig. 2.23. Met-car clusters: (a) single-unit M_8C_{12} cluster and (b) double-unit $M_{14}C_{21}$ cluster where M is represented by dark balls and C by white balls. Because of their icosahedral symmetry, it is not possible to fill space using only units of this type, i.e., all units sharing pentagonal faces.

Identical to the proposed structure for a C_{20} cage molecule, the met-car shell consists of only pentagons, 12 in number, according to Euler's theorem (see §3.1). In each cage, each of the 12 faces contains two metal (M) atoms (darker balls) in symmetric locations and three C atoms (white balls) located at the remaining vertices of the dodecahedron. The observation of a small met-car cage molecule is in contrast to the noted instability of fullerene molecules with decreasing size and the absence of C_{20} in the fullerene mass spectra (see Fig. 1.4), although $C_{20}H_{20}$ is stable. Several theoretical investigations have also corroborated the stability of the met-car molecules [2.138–140].

Similar to the history of fullerenes, met-cars were first discovered using a laser ablation source coupled to a time-of-flight (TOF) mass spectrometer (see Fig. 1.3). A high-power pulsed laser is used to eject a plasma of metal atoms and ions from a metal target in the presence of low molecular weight hydrocarbon gases (e.g., CH_4, C_2H_2, C_2H_4, C_2H_6, and C_3H_6). Since metal atoms are known to induce rapid dehydrogenation of hydrocarbons, the met-cars probably form via the reaction of metal atoms and small C_n clusters, as proposed by the Castleman group [2.141]. A second synthesis route for met-cars, similar to the arc method that is used to generate gram quantities of fullerenes [2.125], has also been reported by the Castleman group [2.142]. This arc method, which involves either an ac or dc discharge in He between Ti:C electrodes (with stoichiometry 5:1), is estimated by them to provide a yield of ~1% met-cars in the Ti-C soot, and thus this technique shows promise for the production of macroscopic quantities of met-cars. To date, however, no reports of chemical separation of met-cars from the soot have been published, and the development of suitable separation techniques presents the next challenge toward carrying out detailed studies into the physical and chemical properties of these interesting molecules.

Finally, the Castleman group has recently reported a unique structural growth pattern for the met-cars in which multicage structures, e.g., Zr_mC_n, appear to be formed [2.132, 141]. In Fig. 2.23(b) we show a two-cage structure for $Zr_{14}C_{21}$, where the pentagonal face is shared between two Zr_8C_{12} cages. This structure was proposed to explain higher mass peaks in the mass spectra for met-cars. Unlike the fullerenes, where larger mass clusters are thought to occur in single cage molecules of larger diameters, this proposal for larger met-cars is for multicage structures with shared pentagonal faces.

Duncan *et al.* [2.133] have also observed higher-mass met-cars using lower laser fluence on the metal target, but these authors have observed clusters with the composition $M_{14}C_{13}$ (M = Ti, V). This met-car mass was assigned to a cubic nanocrystal, which is identical to the fcc crystal structure of TiC. The $M_{14}C_{13}$ nanocrystals were found to photodissociate into M_8C_{12} met-cars in addition to by-products. An M_8C_{13} species has also been identified and attributed to an M_8C_{12} met-car with an endohedral carbon (i.e., the additional carbon atom is proposed to be inside the cage) [2.136, 137]. In contrast to the multiple cage structure route to preparing higher met-car masses proposed by Castleman *et al.* [2.130], the cubic nanostructure is space filling. For experimental reasons, possibly identified with small differences in the cluster sources used by the two research groups, Duncan and co-workers have not yet observed the $M_{14}C_{21}$ mass cluster.

2.20. Carbon-Free Fullerenes

Examples of carbon-free molecules with the basic closed-cage fullerene structure are found in the family of solids $Na_{96}In_{97}Z_2$ (Z = Ni, Pd, Pt) [2.143]. These solids have a hexagonal structure in which large fullerene-like cages of In_{74} (having D_{3h} point group symmetry) share pentagonal faces with a C_{60}-like cage of $In_{48}Na_{12}$ (D_{3d} symmetry) leading to a double hcp structure (see Fig. 2.24). At the center of both of these larger polyhedra are $In_{10}Z$ clusters. Between these $In_{10}Z$ clusters and the In_{74} or $In_{48}Na_{12}$ clusters are shells of Na atoms, thus forming four-layered "onions" consisting of $Z@In_{10}@Na_{39}@In_{74}$ and $Z@In_{10}@Na_{32}@Na_{12}In_{48}$ for the two building block constituents, using an extension of the notation that has been adopted for endohedral fullerenes (see §8.2). The indium atoms on one pentagon

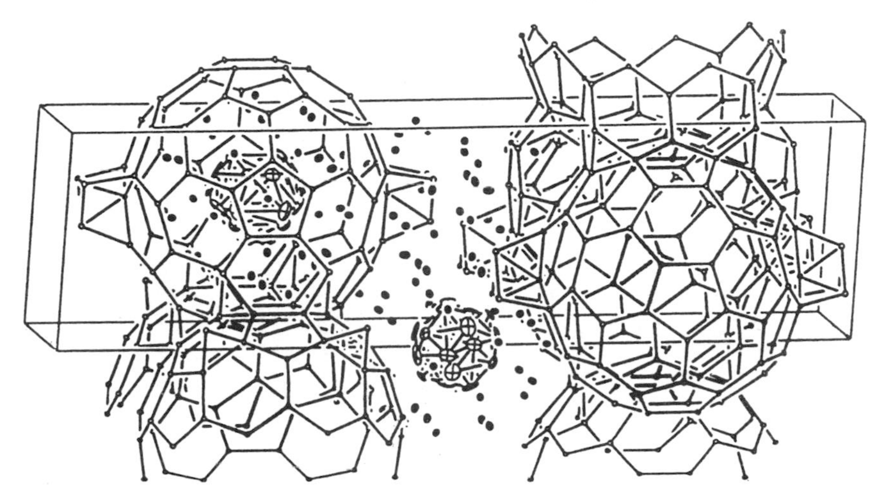

Fig. 2.24. A view of the proposed structure of $Na_{96}In_{97}Z_2$ (where Z = Ni, Pd, Pt) showing the layered nature of this material. Only two of the small $In_{10}Z$ units that center In_{74} and M_{60} are shown. The hexagonal cell that is shown has dimensions 16.0 × 47.4 Å^3. Indium atoms are found both in the network and in isolated clusters. The Na and Ni atoms are denoted by solid balls [2.143].

share their faces with adjoining cages to form layered structures, as shown in the figure. In addition to the two large constituents discussed above, there are small interstitial Z@In_{10} units which assist in the In–In bonding between In_{74} structures. In addition, Na atoms on the outside of the In_{74} structures are involved with intermolecular bonding. The proposed structure shown in Fig. 2.24 has been inferred by refinement of x-ray diffraction patterns for an $Na_{96}In_{97}Z_2$ sample. In this structure, the In–In distances on the In_{74} cage are 2.92 and 3.05 Å, and consequently the constituent balls in these compounds are much larger (diameter ~16 Å) than for C_{60} (diameter ~7.1 Å). A second family of indium-related compounds with cage-like structures is the orthorhombic compound $Na_{172}In_{197}Z_2$, where Z again may be Ni, Pd, or Pt. Structures for the larger indium clusters based on In_{70}, In_{74}, and In_{78} fullerene-like cages have been discussed. Electronically, these intermetallic systems are semiconductors, and because of electron localization in these materials, it is not likely that doping could produce a metallic state [2.143].

2.21. Metal-Coated Fullerenes

It has also been demonstrated that ordered metal-coated fullerene clusters can be synthesized [2.144] using a vapor synthesis method. For alkali metal–coated fullerenes, only one layer of metal atoms has thus far been placed

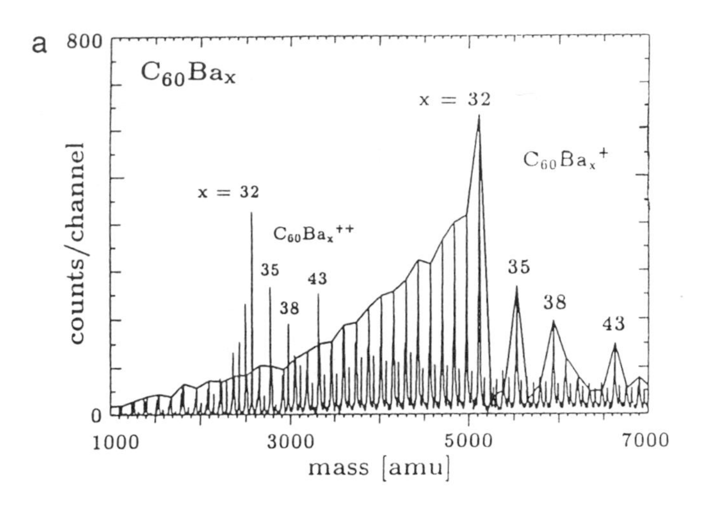

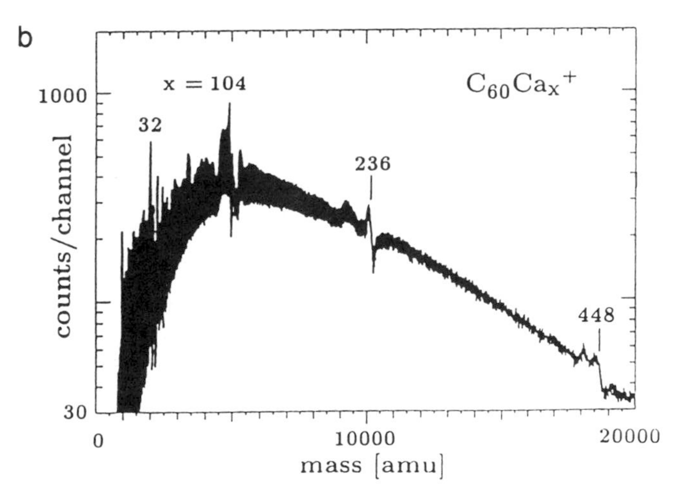

Fig. 2.25. Photoionization time of flight mass spectrometry measurements for: (a) the single and double ionization spectra of a single Ba_x overlayer on C_{60} denoted by $C_{60}Ba_x^+$ and $C_{60}Ba^{2+}$. (b) One, two, three, and four layers of metallic Ca_x overlaying a C_{60} molecule. The spectra show singly ionized $C_{60}Ca_x^+$, with peaks at $x = 32, 104, 236$ and $x = 448$ (see text) [2.144].

around a fullerene. For example, an Li atom has been proposed to reside over the centers of each pentagonal face of C_{60} [2.145, 146] to produce $Li_{12}C_{60}$.

A remarkable multilayer metal structure can also be grown over the C_{60} surface using small alkaline earth atoms such as Ca, Sr, and Ba. This is accomplished when the temperature and pressure conditions are appropriately regulated to favor C_{60}–M bonding or to favor M–M bonding. Photoionization time-of-flight mass spectrometry [see Fig. 2.25(a)] measurements show well-defined peaks in the singly charged and doubly charged spectra, corresponding to $Ba_{32}C_{60}$, consistent with placing one Ba over each of the 12 pentagonal faces and over each of the 20 hexagonal faces, as shown schematically in Fig. 2.26(a). The Ba atoms have just the right size (radius is 1.98 Å) to cover the C_{60} shell [2.144]. In support of this model is the observation of metal-coated $Ba_{37}C_{70}$ clusters based on C_{70}, consistent with five additional carbon hexagons in C_{70} relative to C_{60}. Even more remarkable is the formation of ordered structures with multiple Ca shells, as seen in the mass spectra of Fig. 2.25(b), where well-resolved peaks for Ca_xC_{60} are observed for $x = 32, 104, 236$, and 448, corresponding to 1, 2, 3, and 4 oriented layers of Ca atoms emanating from the C_{60} core, and

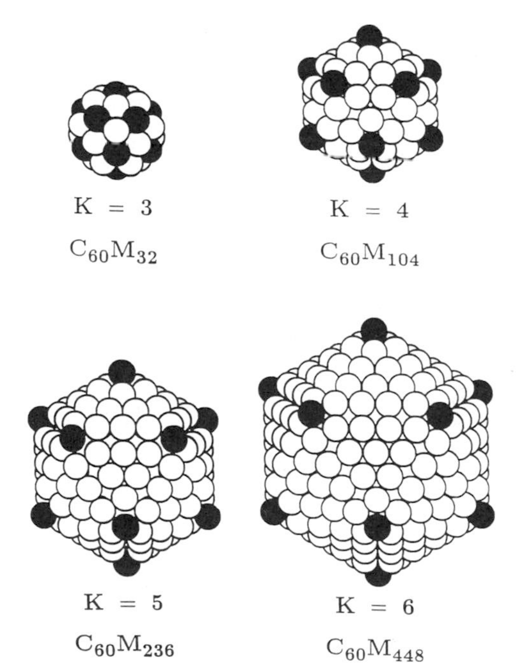

Fig. 2.26. Proposed arrangements in multilayer metal-covered fullerenes of the atoms M = Ca in the first four layers surrounding a C_{60} molecule. The M atoms over the icosahedral vertices, i.e., along the fivefold axes, of C_{60} are shaded. The number of metal layers is equal to K − 2, from Eq. 2.2. [2.144].

shown schematically in Fig. 2.26. In this figure the black balls denote the Ca atoms lying over the pentagonal faces of C_{60}, and for the $M_{32}C_{60}$ cluster, each white ball is over a hexagonal face. As the next layer is deposited, three white balls are closely packed relative to the lower layer, giving rise to 60 additional white balls and 12 additional black balls over the pentagonal faces and corresponding to the $M_{104}C_{60}$ stoichiometry $(32 + 12 + 3 \times 20 = 104)$. Likewise, the buildup of the white balls in the third and fourth layers is 6 and 10 per hexagonal face, respectively, thereby accounting for the mass spectroscopy peaks at $M_{236}C_{60}$ and $M_{448}C_{60}$, respectively. A formula giving the number $N(K)$ of metal atoms for $K - 2$ complete layers around the C_{60} molecule is

$$N(K) = \frac{1}{3}(10K^3 - 30K^2 + 56K - 72). \tag{2.2}$$

The layers can be described using the notation adopted for endohedral metallo-fullerenes, with $C_{60}@M_{32}@M_{72}@M_{132}@M_{212}$ describing the five concentric shells of the metal-coated cluster (see §8.2).

References

[2.1] F. P. Bundy. *J. Geophys. Res.*, **85**, 6930 (1980).

[2.2] F. P. Bundy. In S. Minomura (ed.), *Solid State Physics under Pressure: Recent Advance with Anvil Devices*, p. 1. D. Reidel, Dordrecht (1985).

[2.3] B. T. Kelly. *Physics of Graphite*. Applied Science, London (1981).

[2.4] R. W. G. Wyckoff. In *Crystal Structures*, vol. 1. Interscience, New York (1964).

[2.5] C. Kittel. *Introduction to Solid State Physics*. John Wiley and Sons, New York, 6th ed. (1986).

[2.6] J. Steinbeck, G. Braunstein, M. S. Dresselhaus, T. Venkatesan, and D. C. Jacobson. *J. Appl. Phys.*, **58**, 4374 (1985).

[2.7] M. S. Dresselhaus and J. Steinbeck. *Tanso*, **132**, 44 (1988). Journal of the Japanese Carbon Society.

[2.8] R. C. Weast. *CRC Handbook of Chemistry and Physics*. CRC Press, Inc., West Palm Beach, FL (1978). 59th edition.

[2.9] J. Robertson. *Advances in Physics*, **35**, 317 (1986).

[2.10] J. C. Charlier. *Carbon Nanotubes and Fullerenes*. Ph.D. thesis, Catholic University of Louvain (1994). Department of Physics of Materials.

[2.11] J. W. McClure. In D. L. Carter and R. T. Bate (eds.), *Proceedings of the International Conf. on Semimetals and Narrow Gap Semiconductors*, p. 127. Pergamon Press, New York (1971).

[2.12] M. S. Dresselhaus, G. Dresselhaus, K. Sugihara, I. L. Spain, and H. A. Goldberg. *Graphite Fibers and Filaments*, vol. 5 of *Springer Series in Materials Science*. Springer-Verlag, Berlin (1988).

[2.13] J. Maire and J. Méring. In *Proceedings of the First Conference of the Society of Chemical and Industrial Conference on Carbon and Graphite*, p. 204. London, (1958).

[2.14] P. R. Wallace. *Phys. Rev.*, **71**, 622 (1947).

[2.15] S. B. Austerman. In P. L. Walker, Jr. (ed.), *Chemistry and Physics of Carbon*, vol. 7, p. 137. Marcel Dekker, New York (1968).

[2.16] A. W. Moore. In P. L. Walker, Jr. and P. A. Thrower (eds.), *Chemistry and Physics of Carbon*, vol. 11, p. 69. Marcel Dekker, Inc., New York (1973).
[2.17] A. W. Moore. In P. L. Walker, Jr. and P. A. Thrower (eds.), *Chemistry and Physics of Carbon*, vol. 17, p. 233. Marcel Dekker, Inc., New York (1981).
[2.18] K. Matsubara, T. Tsuzuku, and M. Murakami. *Extended Abstracts of the 20th Biennial Conference on Carbon*, p. 556 (1991). Santa Barbara, CA.
[2.19] H. Yasujima, M. Murakami, and S. Yoshimura. *Appl. Phys. Lett.*, **49**, 499 (1986).
[2.20] Y. Hishiyama, S. Yasuda, and M. Inagaki. *J. Mat. Sci.*, **23**, 3722 (1988).
[2.21] M. Inagaki, S. Harada, T. Sato, T. Nakajima, Y. Horino, and K. Morita. *Carbon*, **27**, 253 (1988).
[2.22] Y. Hishiyama, Y. Kaburagi, and M. Inagaki. In P. A. Thrower (ed.), *Chemistry and Physics of Carbon*, vol. 23, p. 1. Marcel Dekker, New York (1991).
[2.23] R. Bacon. *J. Appl. Phys.*, **31**, 283 (1960).
[2.24] M. Endo, A. Katoh, T. Sugiura, and M. Shiraishi. In *Extended Abstracts of the 18th Biennial Conference on Carbon*, p. 151 (1987). Worcester Polytechnic Institute.
[2.25] M. Endo, T. Momose, H. Touhara, and N. Watanabe. *J. Power Sources*, **20**, 99 (1987).
[2.26] M. Endo. *CHEMTECH*, **18**, 568 (1988).
[2.27] G. M. Jenkins and K. Kawamura. In *Polymeric Carbons-Carbon Fibre, Glass and Char*. Cambridge University Press (1976).
[2.28] F. Rousseaux and D. Tchoubar. *Carbon*, **15**, 55, 63 (1977).
[2.29] A. Oberlin. In P. A. Thrower (ed.), *Chemistry and Physics of Carbon*, vol. 22, pp. 1–14. Marcel Dekker, New York (1989).
[2.30] Y. Shiraishi. In *Introduction to Carbon Materials*, pp. 29–40. Carbon Society of Japan, Tokyo (1984). Kaitei Tansozairyo Nyumon (in Japanese).
[2.31] Y. Kaburagi, S. Yasuda, and Y. Hishiyama. *Extended Abstracts of the 18th Biennial Conference on Carbon*, p. 476 (1987).
[2.32] A. Yoshida, Y. Kaburagi, and Y. Hishiyama. *Carbon*, **29**, 1107 (1991).
[2.33] J. Lahaye and G. Prado. In P. L. Walker, Jr (ed.), *Chemistry and Physics of Carbon*, vol. 16, p. 167. Marcel Dekker, New York (1978).
[2.34] J. B. Donnet, R. C. Bansal, and M. J. Wang. *Carbon Black*. Marcel Dekker, New York (1993).
[2.35] J. A. Howe. *J. Chem. Phys.*, **39**, 1362 (1963).
[2.36] T. P. Yampolskii, Y. V. Maximov, N. P. Novikov, and K. P. Lavrovskii. *Khim. Vys. Energ.*, **4**, 283 (1970).
[2.37] J. T. D. Maleissye, F. Lempereur, and C. Marsal. *C. R. Acad. Sci., Paris, Ser.*, **275**, 1153 (1972).
[2.38] X.-X. Bi, M. Jagtoyen, F. J. Derbyshire, P. C. Eklund, M. Endo, K. D. Chowdhury, and M. S. Dresselhaus. *J. Mat. Res.*, **10** (1995). (in press).
[2.39] B. E. Warren. *J. Chem. Phys.*, **2**, 551 (1934).
[2.40] B. E. Warren. *Phys. Rev.*, **59**, 693 (1941).
[2.41] R. E. Franklin. *Acta Cryst.*, **3**, 107 (1950).
[2.42] R. E. Franklin. *Acta Cryst.*, **4**, 253 (1951).
[2.43] R. D. Heidenreich, W. M. Hess, and L. L. Ban. *J. Appl. Crystallogr.*, **1**, 1 (1968).
[2.44] P. A. Marsh, A. Voet, T. J. Mullens, and L. D. Price. *Carbon*, **9**, 797 (1971).
[2.45] Y. Schwob. In J. Philip, L. Walker, and P. A. Thrower (eds.), *Physics and Chemistry of Carbon*, vol. 17, p. 109. Marcel Dekker, Inc., New York, NY (1979).
[2.46] X. Bourrat. *Carbon*, **31**, 287 (1993).
[2.47] C. E. Hall. *J. Appl. Phys.*, **297**, 315 (1948).
[2.48] H. P. Boehm. *Z. Anorg. U. Allgem. Chem.*, **297**, 315 (1958).

[2.49] V. I. Kasatotchkin, V. M. Loukianovitch, N. H. Popov, and K. V. Tchmoutov. *J. Chem. Phys.*, **37**, 822 (1960).
[2.50] L. L. Ban and W. M. Hess. *Extended Abstracts of the 10th Biennial Conference on Carbon*, p. 159 (1971). Defense Ceramic Information Center, Columbus, OH.
[2.51] L. L. Ban. *The Chem. Soc. London*, **1**, 54 (1972).
[2.52] F. Tuinstra and J. L. Koenig. *J. Chem. Phys.*, **53**, 1126 (1970).
[2.53] D. S. Knight and W. B. White. *J. Mater. Res.*, **4**, 385 (1989).
[2.54] M. Inagaki and T. Noda. *Bull. Chem. Soc. Japan*, **35**, 1652 (1962).
[2.55] J. S. Speck. *Structural correlations in graphite and its layer compounds*. Ph.D. thesis, Massachusetts Institute of Technology (1989). Department of Materials Science and Engineering.
[2.56] E. A. Kmetko. In S. Mrozowski and L. W. Phillips (eds.), *Proceedings of the First and Second Conference on Carbon*, p. 21. Waverly Press, Buffalo, NY (1956).
[2.57] A. W. P. Fung, Z. H. Wang, K. Lu, M. S. Dresselhaus, and R. W. Pekala. *J. Mat. Res.*, **8**, 1875 (1993).
[2.58] X. X. Bi, P. C. Eklund, J. M. Stencel, D. N. Taulbee, H. F. Ni, F. J. Derbyshire, and M. Endo. *Extended Abstracts of the 20th Biennial Conference on Carbon*, p. 518 (1991). Santa Barbara, CA.
[2.59] X. X. Bi, B. Ganguly, G. P. Huffman, F. E. Huggins, M. Endo, and P. C. Eklund. *J. Mater. Res.*, **8**, 1666 (1993).
[2.60] S. S. Abadzev, P. A. Tesner, and U. V. Smevchan. *Gazov, Prom.*, **14**, 36 (1969).
[2.61] B. N. Altshuler and P. A. Tesner. *Gazov, Prom.*, **6**, 41 (1969).
[2.62] R. S. Ruoff, D. C. Lorents, B. Chan, R. Malhotra, and S. Subramoney. *Science*, **259**, 346 (1993).
[2.63] S. Iijima and T. Ichihashi. *Nature (London)*, **363**, 603 (1993).
[2.64] D. S. Bethune, C. H. Kiang, M. S. de Vries, G. Gorman, R. Savoy, J. Vazquez, and R. Beyers. *Nature (London)*, **363**, 605 (1993).
[2.65] C. H. Kiang, W. A. Goddard III, R. Beyers, and D. S. Bethune. *Science* (1994).
[2.66] S. Seraphin and D. Zhou. *Appl. Phys. Lett.*, **64**, 2087 (1994).
[2.67] X. X. Bi, W. T. Lee, and P. C. Eklund. *Amer. Chem. Soc. (Fuel Chemistry Division)*, **38**, 444 (1993). Denver, CO.
[2.68] Y. Saito, T. Yoshikawa, M. Okuda, M. Ohkohchi, Y. Ando, A. Kasuya, and Y. Nishina. *Chem. Phys. Lett.*, **209**, 72 (1994).
[2.69] M. Tomita, Y. Saito, and T. Hayashi. *Jpn. J. Appl. Phys.*, **32**, L280 (1993).
[2.70] R. J. Lagow, J. J. Kampa, H. C. Wei, S. L. Battle, J. W. Genge, D. A. Laude, C. J. Harper, R. Bau, R. C. Stevens, J. F. Haw, and E. Munson. *Science*, **267**, 362 (1995).
[2.71] V. I. Kasatochkin, A. M. Sladkov, Y. P. Kudryavtsev, N. M. Popov, and V. V. Korshak. *Dokl. Chem.*, **177**, 1031 (1967).
[2.72] V. I. Kasatochkin, V. V. Korshak, Y. P. Kudryavtsev, A. M. Sladkov, and I. E. Sterenberg. *Carbon*, **11**, 70 (1973).
[2.73] A. G. Whittaker and P. L. Kintner. *Carbon*, **23**, 255 (1985).
[2.74] A. G. Whittaker and P. L. Kintner. *Science*, **165**, 589 (1969).
[2.75] A. G. Whittaker. *Science*, **200**, 763 (1978).
[2.76] A. G. Whittaker, E. J. Watts, R. S. Lewis, and E. Anders. *Science*, **209**, 1512 (1980).
[2.77] A. E. Gorsey and G. Donnary. *Science*, **363** (1968).
[2.78] V. I. Kasatochkin, M. E. Kasakov, V. A. Savransky, A. P. Nabatnikov, and N. P. Radimov. *Dokl. Akad. Nauk, USSR*, **201**, 1104 (1971).
[2.79] R. B. Heimann, J. Kleiman, and N. M. Slansky. *Nature (London)*, **306**, 164 (1983).
[2.80] A. G. Whittaker and G. M. Wolten. *Science*, **178**, 54 (1972).

[2.81] A. G. Whittaker, G. Donnay, and K. Lonsdale. *Carnegie Institution Year Book*, **69**, 311 (1971).
[2.82] S. Tanuma and A. Palnichenko. *J. Mater. Res.*, **10**, 1120 (1995).
[2.83] J. J. Hauser. *Solid State Commun.*, **17**, 1577 (1975).
[2.84] M. S. Dresselhaus and R. Kalish. *Ion Implantation in Diamond, Graphite and Related Materials*. Springer-Verlag; Springer Series in Materials Science, Berlin (1992). Volume 22.
[2.85] B. Dischler and G. Brandt. *Industrial Diamond Review*, **3**, 131 (1985).
[2.86] J. J. Hauser, J. R. Patel, and J. N. Rogers. *Appl. Phys. Lett.*, **30**, 129 (1977).
[2.87] J. Fink, T. Muller-Heinzerling, J. Pfluger, A. Bubenzer, P. Koidl, and G. Crecelius. *Solid State Commun.*, **47**, 687 (1983).
[2.88] N. Savvides. *J. Appl. Phys.*, **58**, 518 (1985).
[2.89] N. Savvides. *J. Appl. Phys.*, **59**, 4133 (1986).
[2.90] J. Zelez. *J. Vac. Sci. Technol.*, **1**, 306 (1983).
[2.91] M. S. Dresselhaus, A. W. P. Fung, A. M. Rao, S. L. di Vittorio, K. Kuriyama, G. Dresselhaus, and M. Endo. *Carbon*, **30**, 1065 (1992).
[2.92] A. W. P. Fung, K. Kuriyama, A. M. Rao, M. S. Dresselhaus, G. Dresselhaus, and M. Endo. *J. Mat. Sci.*, **8**, 489 (1993).
[2.93] M. Inagaki, K. Muramatsu, Y. Maeda, and K. Maekawa. *Synth. Met.*, **8**, 335 (1983).
[2.94] A. Thomy and X. Duval. *J. de Chimie Phys.*, **66**, 1966 (1969).
[2.95] W. H. Martin and J. E. Brocklehurst. *Carbon*, **1**, 133 (1964).
[2.96] R. W. Pekala and C. T. Alviso. In C. L. Renschler, J. J. Pouch, and D. M. Cox (eds.), *Novel Forms of Carbon*, vol. 270, pp. 3–14. Materials Research Society, Pittsburgh, PA (1992).
[2.97] J. Steinbeck, G. Dresselhaus, and M. S. Dresselhaus. *International Journal of Thermophysics*, **11**, 789 (1990).
[2.98] T. Venkatesan, J. Steinbeck, G. Braunstein, M. S. Dresselhaus, G. Dresselhaus, D. C. Jacobson, and B. S. Elman. In H. Kurz, G. L. Olson, and J. M. Poate (eds.), *Beam-Solid Interactions and Phase Transformations, MRS Symposia Proceedings, Boston*, vol. 51, p. 251, Pittsburgh, PA (1986). Materials Research Society.
[2.99] J. Heremans, C. H. Olk, G. L. Eesley, J. Steinbeck, and G. Dresselhaus. *Phys. Rev. Lett.*, **60**, 452 (1988).
[2.100] J. S. Speck, J. Steinbeck, G. Braunstein, M. S. Dresselhaus, and T. Venkatesan. In H. Kurz, G. L. Olson, and J. M. Poate (eds.), *Beam-Solid Interactions and Phase Transformations, MRS Symposia Proceedings, Boston*, vol. 51, p. 263, Pittsburgh, PA (1986). Materials Research Society Press.
[2.101] H. W. Kroto, J. R. Heath, S. C. O'Brien, R. F. Curl, and R. E. Smalley. *Nature (London)*, **318**, 162 (1985).
[2.102] M. S. Dresselhaus and G. Dresselhaus. *Advances in Phys.*, **30**, 139 (1981).
[2.103] In H. Zabel and S. A. Solin (eds.), *Graphite Intercalation Compounds I: Structure and Dynamics*, Springer Series in Materials Science, vol. 14. Springer-Verlag, Berlin (1990).
[2.104] W. Rüdorff and E. Shultze. *Z. Anorg. allg. Chem.*, **277**, 156 (1954).
[2.105] F. P. Bundy and J. S. Kasper. *J. Chem. Phys.*, **46**, 3437 (1967).
[2.106] D. Morelli. In P. A. Thrower (ed.), *Chemistry and Physics of Carbon*, vol. 24, p. 45. Marcel Dekker, NY (1994).
[2.107] R. C. DeVries. *Ann. Rev. Mater. Sci.*, **17**, 161 (1987).
[2.108] J. C. Angus and C. C. Hayman. *Science*, **241**, 877 (1988).
[2.109] J. C. Angus and C. C. Hayman. *Science*, **241**, 913 (1988).
[2.110] W. A. Yarbrough and R. Messier. *Science*, **247**, 688 (1990).

[2.111] B. V. Spitsyn, L. L. Bouitov, and B. V. Derjaguin. *J. Cryst. Growth*, **52**, 219 (1981).

[2.112] S. Prawer, A. Hoffman, and R. Kalish. *Appl. Phys. Lett.*, **57**, 2187 (1990).

[2.113] D. T. Morelli, T. M. Hartnett, and C. J. Robinson. *Appl. Phys. Lett.*, **59**, 2112 (1991).

[2.114] A. Feldman, H. P. R. Frederikse, and X. T. Ying. *Proc. SPIE. Int. Soc. Opt. Eng.*, **1146**, 78 (1990).

[2.115] J. E. Graebner, S. Jin, G. W. Kammlott, J. A. Herb, and C. F. Gardinier. *Nature (London)*, **359**, 401 (1992).

[2.116] In P. Koidl and P. Oelhafen (eds.), *Amorphous Hydrogenated Carbon Films*. European Materials Research Society (1987). Les Editions de Physique, France, Vol. XVII.

[2.117] J. T. A. Pollock and L. S. Wichinski. In J. S. Hanker and B. L. Giammara (eds.), *Biomedical Materials and Devices*, vol. 110. Materials Research Society, Pittsburgh, PA (1987).

[2.118] D. R. McKenzie, R. C. McPhedram, N. Savvides, and D. J. H. Cockayne. *Thin Solid Films*, **108**, 247 (1983).

[2.119] J. Robertson and E. P. O'Rielly. *Phys. Rev. B*, **35**, 2946 (1987).

[2.120] R. Kalish and E. Adel. In J. J. Pouch and S. A. Alterowitz (eds.), *Properties and Preparation of Amorphous Carbon Films*, vol. 52-53, p. 427. Trans. Tech. Pub., Switzerland (1990). Materials Science Forum.

[2.121] R. C. Fort, Jr. *Adamantane: the Chemistry of Diamond Molecules*. Marcel-Dekker, New York (1976).

[2.122] S. Landa, V. Machacek, and J. Mzourek. *Chemicke Listy*, **27**, 415 (1933).

[2.123] G. A. Olah. In G. A. Olah and P. von R. Schleyer (eds.), *Cage Hydrocarbons*, John Wiley & Sons, New York (1990).

[2.124] P. von R. Schleyer. In G. A. Olah and P. von R. Schleyer (eds.), *Cage Hydrocarbons*, J. Wiley & Sons, New York (1990). edition 1990.

[2.125] R. E. Haufler, J. J. Conceicao, L. P. F. Chibante, Y. Chai, N. E. Byrne, S. Flanagan, M. M. Haley, S. C. O'Brien, C. Pan, Z. Xiao, W. E. Billups, M. A. Ciufolini, R. H. Hauge, J. L. Margrave, L. J. Wilson, R. F. Curl, and R. E. Smalley. *J. Phys. Chem.*, **94**, 8634 (1990).

[2.126] R. J. Ternansky, D. E. Balogh, and L. A. Paquette. *J. Am. Chem. Soc.*, **104**, 4503 (1982).

[2.127] A. D. Schluter, M. Loffler, and V. Enkelmann. *Nature (London)*, **368**, 831 (1994).

[2.128] B. C. Guo, K. P. Kerns, and A. W. Castleman, Jr. *Science*, **255**, 1411 (1992).

[2.129] B. C. Guo, S. Wei, J. Purnell, S. Buzza, and A. W. Castleman, Jr. *Science*, **256**, 515 (1992).

[2.130] B. C. Guo, K. P. Kerns, and A. W. Castleman, Jr. *J. Am. Chem. Soc.*, **115**, 7415 (1993).

[2.131] S. Wei, B. C. Gao, J. Purnell, S. Buzza, and A. W. Castleman, Jr. *J. Phys. Chem.*, **96**, 4166 (1992).

[2.132] S. Wei and A. W. Castleman, Jr. *Chem. Phys. Lett.*, **227**, 305 (1994).

[2.133] M. A. Duncan (private communication).

[2.134] J. S. Pilgrim and M. A. Duncan. *J. Am. Chem. Soc.*, **115**, 495 (1993).

[2.135] J. S. Pilgrim and M. A. Duncan. *J. Am. Chem. Soc.*, **115**, 6958 (1993).

[2.136] J. S. Pilgrim, E. L. Stewart, and M. A. Duncan. *J. Am. Chem. Soc.*, **115**, 9724 (1993).

[2.137] J. S. Pilgrim and M. A. Duncan. *Intl. J. Mass Spectr. Ion Processes*, **138**, 283 (1994).

[2.138] B. V. Reddy, S. N. Khanna, and P. Jena. *Science*, **258**, 1640 (1992).

[2.139] R. W. Grimes and J. D. Gale. *J. Chem. Soc. Chem. Commun.*, **1992**, 1222 (1992).

[2.140] M. Methfessel, M. van Schilfgaarde, and M. Scheffler. *Phys. Rev. Lett.*, **70**, 29 (1993).

[2.141] S. Wei, B. C. Gao, J. Purnell, S. Buzza, and A. W. Castleman, Jr. *Science*, **256**, 818 (1992).

[2.142] S. F. Cartier, Z. Y. Chen, G. J. Walder, C. R. Sleppy, and A. W. Castleman, Jr. *Science*, **260**, 195 (1993).
[2.143] S. C. Sevov and J. D. Corbett. *Science*, **262**, 880 (1993).
[2.144] U. Zimmermann, N. Malinowski, U. Näher, S. Frank, and T. P. Martin. *Phys. Rev. Lett.*, **72**, 3542 (1994).
[2.145] T. P. Martin, N. Malinowski, U. Zimmermann, U. Näher, and H. Schaber. *J. Chem. Phys.*, **99**, 4210 (1993).
[2.146] J. Kohanoff, W. Andreoni, and M. Parrinello. *Chem. Phys. Lett.*, **198**, 472 (1992).

CHAPTER 3

Structure of Fullerenes

Many independent experiments show that the crystalline materials formed from fullerenes are molecular solids. Therefore the structure and properties of these solids are strongly dependent on the structure and properties of the constituent fullerene molecules. In this chapter we address the internal structure of the molecules, while in Chapter 7, the structures of the related molecular solids are reviewed.

3.1. Structure of C_{60} and Euler's Theorem

The 60 carbon atoms in C_{60} are now known to be located at the vertices of a truncated icosahedron where all carbon sites are equivalent [see Fig. 1.1(b)]. This is consistent with the observation of a single sharp line in the nuclear magnetic resonance (NMR) spectrum [3.1, 2] (see §16.1.1). Since (1) the average nearest-neighbor carbon–carbon (C–C) distance a_{C-C} in C_{60} (1.44 Å) is almost identical to that in graphite (1.42 Å), (2) each carbon atom in graphite and in C_{60} is trigonally bonded to three other carbon atoms in an sp^2-derived bonding configuration, and (3) most of the faces on the regular truncated icosahedron are hexagons, we can, to a first approximation, think of the C_{60} molecule as a "rolled-up" graphene sheet (a single layer of crystalline graphite). A regular truncated icosahedron has 90 edges of equal length, 60 equivalent vertices, 20 hexagonal faces, and 12 additional pentagonal faces to form a closed shell, consistent with Euler's theorem discussed below. Two single C–C bonds are located along a pentagonal edge at the fusion of a hexagon and a pentagon. In C_{60}, the bond lengths for the single bonds a_5 are 1.46 Å as measured by NMR [3.1–4] and 1.455 Å by neutron scattering [3.5] (see Fig. 3.1). The third bond is located at the fusion between two hexagons and is a double bond with a

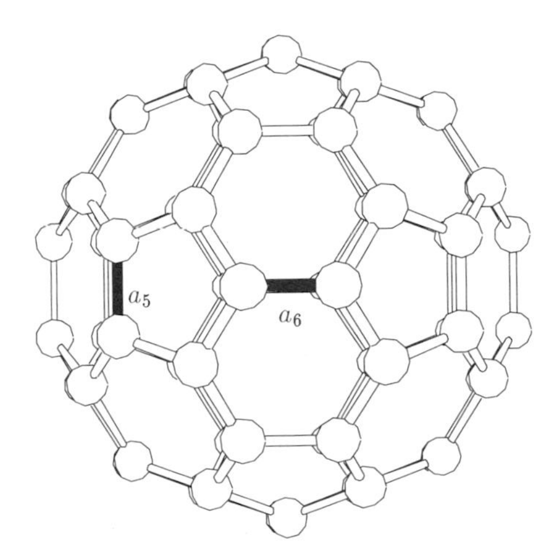

Fig. 3.1. The C_{60} molecule showing single bonds (a_5) and double bonds (a_6).

bond length a_6 which is measured to be 1.40 Å by NMR and 1.391 Å by neutron diffraction (see Fig. 3.1). Since the bond lengths in C_{60} are not exactly equal (i.e., $a_5 - a_6 \approx 0.06$ Å), the vertices of the C_{60} molecule form a truncated icosahedron but, strictly speaking, not a *regular* truncated icosahedron. In many descriptions of C_{60}, however, the small differences between the a_5 and a_6 bonds are neglected and C_{60} is often called a regular truncated icosahedron in the literature.

Since the bonding requirements of all the valence electrons in C_{60} are satisfied, it is expected that C_{60} has filled molecular levels. Because of the closed-shell properties of C_{60} (and also other fullerenes), the nominal sp^2 bonding between adjacent carbon atoms occurs on a curved surface, in contrast to the case of graphite where the sp^2 trigonal bonds are truly planar. This curvature of the trigonal bonds in C_{60} leads to some admixture of sp^3 bonding, characteristic of tetrahedrally bonded diamond, but absent in graphite.

Inspection of the C_{60} molecular structure [see Fig. 3.1] shows that every pentagon of C_{60} is surrounded by five hexagons; the pentagon, together with its five neighboring hexagons, has the form of the corannulene molecule [see Fig. 3.2(a)], where the curvature of the molecule is shown in Fig. 3.2(b). The double bonds in corannulene are in different positions relative to C_{60} because the edge carbons in corannulene are bonded to hydrogen atoms. Another molecular subunit on the C_{60} molecule is the pyraclene (also called pyracylene) subunit [see Fig. 3.2(c)], which consists of two pentagons and two hexagons in the arrangement shown. Again, the double bonds differ between the subunit of C_{60} and the pyracylene molecule because of the edge hydrogens in the molecular pyracylene form. The pyracylene molecule is of

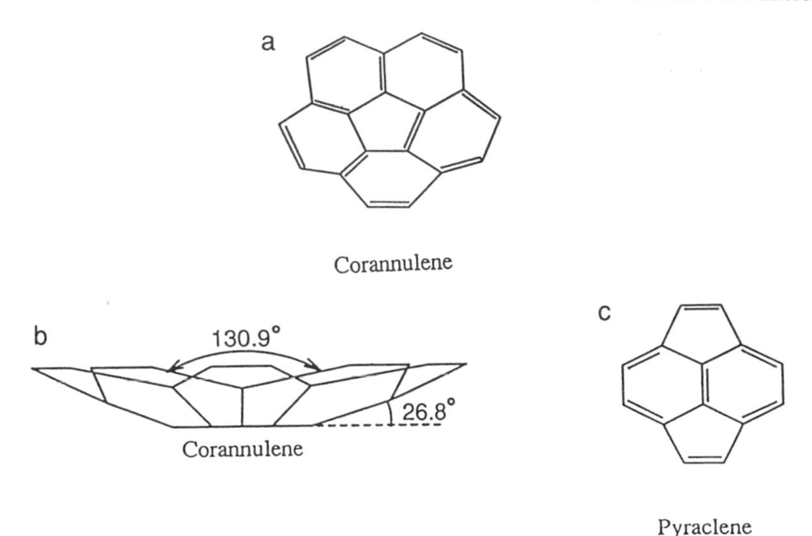

Fig. 3.2. (a) Top view and (b) side view of the nonplanar corannulene molecule. (c) A pyraclene or pyracylene molecule. The pyracylene and corannulene motifs are seen on the surface of C_{60} and other fullerenes.

particular interest in showing the separation between nearest pentagons. Adherence to the "isolated pentagon rule," whereby the distance between pentagons on the fullerene shell is maximized, reduces local curvature and stress and gives added stability to C_{60} [3.6, 7], as discussed below.

The symmetry operations of the truncated icosahedron consist of the identity operation, 6 fivefold axes through the centers of the 12 pentagonal faces, 10 threefold axes through the centers of the 20 hexagonal faces, and 15 twofold axes through centers of the 30 edges joining two hexagons. Each of the 60 rotational symmetry operations can be compounded with the inversion operation, resulting in 120 symmetry operations in the icosahedral point group I_h [3.8, 9] [see Fig. 3.1 and §4.1]. Molecules with I_h symmetry, of which C_{60} is the most prominent member, have the highest degree of symmetry of any known molecule. The diameter of 7.09 Å is calculated for the C_{60} molecule from geometric considerations treating the carbon atoms as points [3.10] and taking a_{C-C} = 1.40 Å for the bonds on the hexagons and 1.46 Å for the pentagonal bond lengths (see Table 3.1, where the physical constants for C_{60} molecules are gathered). Experimentally, the diameter for the C_{60} molecule is determined by NMR measurements to be 7.10±0.07 Å [3.4]. When taking account of the size of the π-electron cloud associated with the carbon atoms, the outer diameter of the C_{60} molecule can be estimated as 7.09+3.35=10.34 Å, where 3.35 Å is an estimate of

Table 3.1

Physical constants for C_{60} molecules.

Quantity	Value	Reference
Average C–C distance	1.44 Å	[3.5]
C–C bond length on a pentagon	1.46 Å	[3.2, 11]
C–C bond length on a hexagon	1.40 Å	[3.2, 11]
C_{60} mean ball diameter[a]	7.10 Å	[3.4]
C_{60} ball outer diameter[b]	10.34 Å	–
Moment of inertia I	1.0×10^{-43}kg m^2	[3.12]
Volume per C_{60}	1.87×10^{-22}/cm^3	–
Number of distinct C sites	1	–
Number of distinct C–C bonds	2	–
Binding energy per atom[c]	7.40 eV	[3.13]
Heat of formation (per g C atom)	10.16 kcal	[3.14]
Electron affinity	2.65±0.05 eV	[3.15]
Cohesive energy per C atom	1.4 eV/atom	[3.16]
Spin–orbit splitting of C($2p$)	0.00022 eV	[3.17]
First ionization potential	7.58 eV	[3.18]
Second ionization potential	11.5 eV	[3.19]
Optical absorption edge[d]	1.65 eV	[3.13]

[a]This value was obtained from NMR measurements. The calculated geometric value for the diameter is 7.09 Å (see text).

[b]This value for the outer diameter is found by assuming the thickness of the C_{60} shell to be 3.35 Å. In the solid, the C_{60}–C_{60} nearest-neighbor distance is 10.02 Å (see Table 7.1).

[c]The binding energy for C_{60} is believed to be ~0.7 eV/C atom less than for graphite, though literature values for both are given as 7.4 eV/C atom. The reason for the apparent inconsistency is attributed to differences in calculational techniques.

[d]Literature values for the optical absorption edge (see §13.2.1) for the free C_{60} molecules in solution range between 1.55 and 2.3 eV.

the thickness of the π-electron cloud surrounding the carbon atoms on the C_{60} shell. The estimate of 3.35 Å comes from the interplanar distance between graphite layers. The binding energy per carbon atom in graphite is 7.4 eV [3.20]. The corresponding energy has also been calculated by various authors for C_{60} [3.13, 21], showing that the binding energy for C_{60} per carbon atom is smaller than that for graphite by 0.4–0.7 eV, although the absolute values of the binding energies of C_{60} and graphite are not as well established. The high binding energy for C_{60} accounts for the high stability of the C_{60} molecule.

By definition, a fullerene is a closed cage molecule containing only hexagonal and pentagonal faces. This requires that there be exactly 12 pentagonal faces and an arbitrary number h of hexagonal faces. This result follows from Euler's theorem for polyhedra

$$f + v = e + 2 \tag{3.1}$$

where f, v, and e are, respectively, the numbers of faces, vertices, and edges of the polyhedra. If we consider polyhedra formed by h hexagonal faces and p pentagonal faces, then

$$\begin{aligned} f &= p + h, \\ 2e &= 5p + 6h, \\ 3v &= 5p + 6h. \end{aligned} \tag{3.2}$$

The three relations in Eq. (3.2) yield

$$6(f + v - e) = p = 12 \tag{3.3}$$

from which we conclude that all fullerenes with only hexagonal and pentagonal faces must have 12 pentagonal faces, and the number of hexagonal faces is arbitrary. Since the addition of each hexagonal face adds two carbon atoms to the total number n_C of carbon atoms in a fullerene, it follows that the number of hexagonal faces in a C_{n_C} fullerene can readily be determined. These arguments further show that the smallest possible fullerene is C_{20}, which would form a regular dodecahedron with 12 pentagonal faces and no hexagonal faces.

It is, however, energetically unfavorable for two pentagons to be adjacent to each other, since this would lead to higher local curvature on the fullerene ball, and hence more strain. The resulting tendency for pentagons not to be adjacent to one another is called the isolated pentagon rule [3.22, 23]. The smallest fullerene C_{n_C} to satisfy the isolated pentagon rule is C_{60} with $n_C = 60$. For this reason, fullerenes with many fewer than 60 carbon atoms are less likely (see Fig. 1.4), and, in fact, no fullerenes with fewer than 60 carbon atoms are found in the soot commonly used to extract fullerenes. Thus far, it has been possible to stabilize small fullerenes only by saturating their dangling bonds with hydrogen, and $C_{20}H_{20}$ is an example of such a molecule (see §2.17) [3.24]. While the smallest fullerene to satisfy the isolated pentagon rule is C_{60}, the next largest fullerene to do so is C_{70}, which is consistent with the absence of fullerenes between C_{60} and C_{70} in fullerene-containing soot. Furthermore, since the addition of a single hexagon adds two carbon atoms, all fullerenes must have an even number of carbon atoms, in agreement with the observed mass spectra for fullerenes as illustrated in Fig. 1.4 [3.25].

Although each carbon atom in C_{60} is equivalent to every other carbon atom, the three bonds emanating from each carbon atom are not equiva-

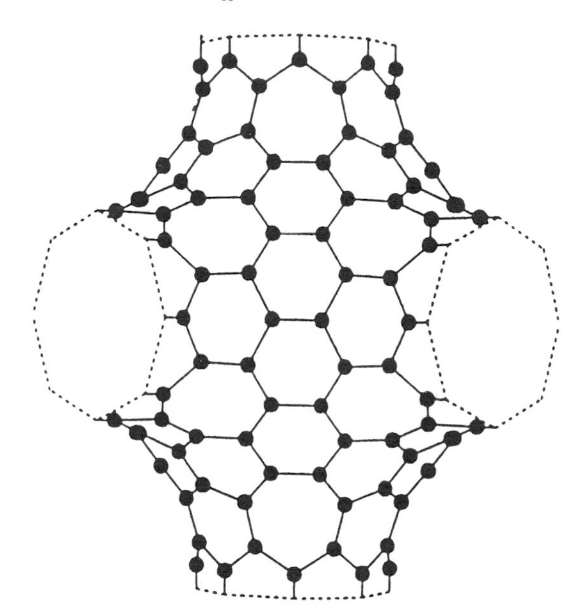

Fig. 3.3. Use of heptagons to yield a concave surface [3.28].

lent, two being electron-poor single bonds (on the pentagonal edges) and one being an electron-rich double bond (joining two hexagons). The alternating bond structure found around the hexagons of C_{60} (see Fig. 3.1) is called the Kekulé structure and stabilizes the C_{60} structure, consistent with x-ray and neutron diffraction evidence discussed above. We note that the icosahedral I_h symmetry is preserved for unequal values for a_5 and a_6. The single bonds that define the pentagonal faces are increased from the average bond length of 1.44 Å to 1.46 Å, and the double bonds are decreased to 1.40 Å [3.5, 10, 26, 27].

With regard to curvature for fullerene-related molecules, we note that pentagons are needed to produce closed (convex) surfaces, and hexagons by themselves lead to a planar surface. To produce a concave surface that spreads outward, heptagons need to be introduced. There are numerous references in the literature [3.28–30] to the use of heptagons to yield a concave surface (see Fig. 3.3). As mentioned in §19.2.3, heptagons are commonly observed in elbows and bends in carbon nanotubes.

Since the 60 carbon atoms in C_{60} form a cage with a very small value for the nearest-neighbor carbon–carbon distance a_{C-C}, it is expected that the C_{60} molecule is almost incompressible [3.31]. Of course, in the solid state the large van der Waals spacings between C_{60} molecules result in a solid that is quite compressible and "soft" [3.32] (see §7.3).

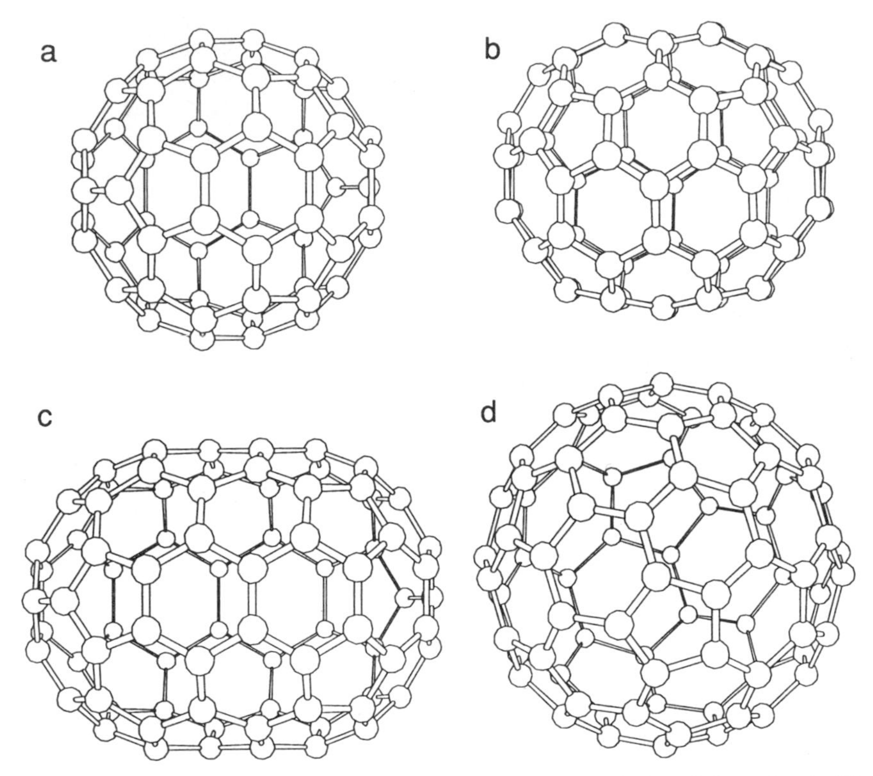

Fig. 3.4. (a) The icosahedral C_{60} molecule. (b) The rugby ball–shaped C_{70} molecule (D_{5h} symmetry). (c) The C_{80} isomer, which is an extended rugby ball (D_{5d} symmetry). (d) The C_{80} isomer, which is an icosahedron (I_h symmetry).

3.2. Structure of C_{70} and Higher Fullerenes

In the arc synthesis of C_{60} (see §5.1.2), larger molecular weight fullerenes C_{n_C} ($n_C > 60$) are also formed. By far the most abundant higher molecular weight fullerene present in the mass spectra for fullerenes (see Fig. 1.4) is the rugby ball–shaped C_{70} [Fig. 3.4(b)]. The high relative abundance of C_{70} is connected to its stability, which in part relates to its high binding energy, which is 7.42 eV/C atom or ~0.02 eV/C atom greater than that of C_{60} [3.13]. Although higher in binding energy, C_{70} is probably less abundant than C_{60} for reasons of kinetics rather than energetics. Even if the potential well for C_{70} may be deeper, the barriers that must be overcome to reach C_{70} upon formation are greater than for C_{60}, so that the synthesis rate is less favorable.

As the fullerene mass increases, the next fullerene beyond C_{70} to satisfy the isolated pentagon rule is C_{76}. For masses higher than C_{70}, the

As shown in Table 3.3, values calculated using tight binding molecular dynamics (TBMD), the local density approximation (LDA), modified neglect of differential overlap (MNDO), and Hartree–Fock (HF) methods for the bond lengths of C_{70} [3.49, 51, 53, 54] are in good agreement with experimental electron diffraction (ED), neutron inelastic scattering (NIS) and x-ray diffraction data [3.42, 52, 55]. Clearly, with increasing numbers of carbon atoms n_C in the fullerenes C_{n_C}, the number of distinct carbon sites and bond lengths increases, as does the complexity of the geometric structure.

Fullerenes often form isomers, since a given number n_C of carbon atoms C_{n_C} can correspond to molecules with different geometrical structures [3.21, 33, 49, 56, 57]. As mentioned above, not only does C_{70} have the next lowest n_C value to have an isomer obeying the isolated pentagon rule, but C_{70} also has only one isomer to do so. As another theoretical example, C_{80} might be formed in the shape of an elongated rugby ball by adding two rows of five hexagons normal to a fivefold axis of C_{60} at the equator [see Fig. 3.4(c)]. An icosahedral form of C_{80} can also be specified as shown in Fig. 3.4(d).

It is, in fact, easy to specify all possible *icosahedral* fullerenes using the projection method described in §3.3, where it is shown that an icosahedron consists of 20 equilateral triangles, each specified by a pair of integers (n, m) such that n_C in C_{n_C} is given by

$$n_C = 20(n^2 + nm + m^2) \tag{3.4}$$

and the diameter of the icosahedron d_i is given by

$$d_i = \frac{5\sqrt{3}a_{C-C}}{\pi}(n^2 + nm + m^2)^{1/2} \tag{3.5}$$

in which a_{C-C} is the nearest-neighbor carbon–carbon distance on the fullerene. In comparing the diameters of various fullerenes, it is useful to use an average value of a_{C-C}, which can be estimated from detailed measurements on C_{60}, discussed above (see Table 3.1). Using the average value of $a_{C-C} = 1.44$ Å, for C_{60} (see §3.1), the diameters of several icosahedral fullerenes can be found using Eq. (3.5), and the results are summarized in Table 3.4. For example, the smallest possible icosahedral fullerene $(n, m) = (1, 0)$ has only 20 carbon atoms and 12 pentagonal faces; this fullerene (C_{20}) has not yet been observed experimentally and is thought to be unstable from theoretical considerations. Increasing the diameter of the fullerene shell, the next icosahedral fullerene is C_{60} with $(n, m) = (1, 1)$, followed by C_{80} corresponding to $(n, m) = (2, 0)$. All icosahedral fullerenes with $n_C \geq 60$ obey the isolated pentagon rule. Listed in Table 3.4 are the various icosahedral fullerenes that are possible, together with their n_C values and diameters based on Eq. 3.5. Figure 3.6 shows fullerenes with

Table 3.4

Listing of icosahedral fullerenes, their symmetry and diameters d_i.

C_{n_C}	(n, m)	Symmetry	d_i (Å)[a]
C_{20}	$(1,0)$	I_h	3.97
C_{60}	$(1,1)$	I_h	6.88
C_{80}	$(2,0)$	I_h	7.94
C_{140}	$(2,1)$	I	10.50
C_{180}	$(3,0)$	I_h	11.91
C_{240}	$(2,2)$	I_h	13.75
C_{260}	$(3,1)$	I	14.31
C_{320}	$(4,0)$	I_h	15.88
C_{380}	$(3,2)$	I	17.30
C_{420}	$(4,1)$	I	18.19
C_{500}	$(5,0)$	I_h	19.85
C_{540}	$(3,3)$	I_h	20.63
C_{560}	$(4,2)$	I	21.00
C_{620}	$(5,1)$	I	22.10
C_{720}	$(6,0)$	I_h	23.82
C_{740}	$(4,3)$	I	24.15
C_{780}	$(5,2)$	I	24.79
C_{860}	$(6,1)$	I	26.03
C_{960}	$(4,4)$	I_h	27.50
C_{980}	$(5,3)$	I	27.79
C_{980}	$(7,0)$	I_h	27.79
⋮	⋮	⋮	⋮

[a] The average diameters given in this table for the icosahedral fullerenes are calculated from Eq. (3.5) and a_{C-C} is taken to be 1.44 Å. Other calculations, using different values for a_{C-C} and different definitions for d_i, yield slightly different values for d_i (see Table 3.1).

(n, m) = (2,1), (3,1), (1,3), and (4,4) corresponding to C_{140}, C_{260}, C_{260}, and C_{960}. Fullerenes with $n = m$ or with either $n = 0$ or $m = 0$ have full I_h symmetry (see Chapter 4) and were recognized at an early stage of fullerene research [3.59]. The icosahedral fullerenes for which $n \neq m$, and neither n nor m vanishes, have I symmetry but lack inversion symmetry. In this case, fullerenes with indices such as (1,3) and (3,1) are mirror images of each other, as shown in Fig. 3.6 [3.58].

For a given number of carbon atoms (such as $n_C = 80$), there may be more than one way to arrange the n_C carbon atoms into a fullerene. Each of these possibilities is called an isomer. In fact, for C_{80} there are seven

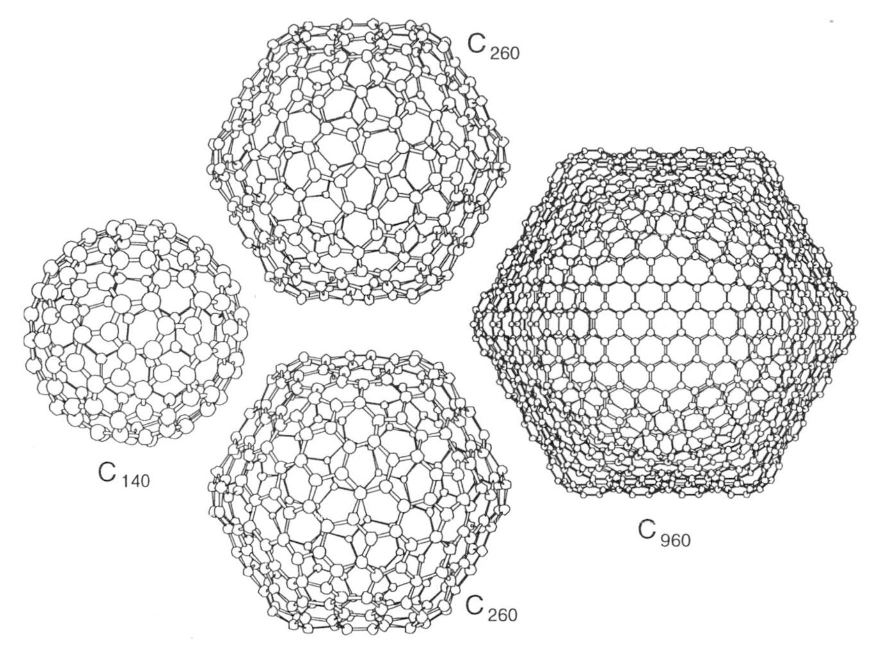

Fig. 3.6. Icosahedral fullerenes with $(n, m) = (2, 1)$, (3,1), (1,3), and (4,4) corresponding to C_{140}, C_{260}, C_{260}, and C_{960}. The (3,1) fullerene is below the (1,3) fullerene in the diagram [3.58].

distinct arrangements (two of which are shown in Fig. 3.4) of the 80 carbon atoms which conform to the isolated pentagon rule and have only pentagons and hexagons. In addition to the two isomers for C_{80} with I_h and D_{5d} point group symmetries that were discussed above (see §4.4), there are five other isomers for C_{80} which obey the isolated pentagon rule, each having a different symmetry group: D_2, C_{2v}, D_3, C_{2v}, D_{5d} [3.60]. The distinct bond lengths for each of these isomers and their symmetries have been calculated [3.49]. Listed in Table 3.5 is the number of isomers n_I for a given C_{n_C} obeying the isolated pentagon rule (out to $n_C = 88$). This table also lists the symmetries for the isomers following a notation used in the literature [3.60]. As another example of fullerene isomers, C_{78} has been shown to have five distinct isomers obeying the isolated pentagon rule, one with D_3 symmetry, two with C_{2v} symmetry, and two with D_{3h} symmetry [3.61] (see Table 3.5). Figure 3.7 shows examples of isomers for C_{84} with six different point group symmetries [3.21, 62]. The C_{84} molecule has a total of 24 isomers that obey the isolated pentagon rule.

Isomers giving rise to right-handed and left-handed optical activity are expected to occur in molecules belonging to group I (rather than I_h) and

Table 3.5

Isomers of common fullerene molecules [3.60].

C_{n_C}	n_I [a]	Symmetries
C_{60}	1	I_h
C_{70}	1	D_{5h}
C_{76}	1	D_2
C_{78}	5[b]	$D_3, C_{2v}, C_{2v}, D_{3h}, D_{3h}$
C_{80}	7[c]	$D_{5d}, D_2, C_{2v}, D_3, C_{2v}, D_{5h}, I_h$
C_{82}	9[d]	$C_2, C_s, C_2, C_s, C_2, C_s, C_{3v}, C_{3v}, C_{2v}$
C_{84}	24[e]	–
C_{86}	19[e]	–
C_{88}	35[e]	–
⋮	⋮	⋮

[a] n_I is the number of isomers obeying the isolated pentagon rule for given C_{n_C} fullerenes.

[b] The symmetries of the 5 isomers of C_{78} are: $C_{78}\mathbf{1}(D_3)$, $C_{78}\mathbf{2}(C_{2v})$, $C_{78}\mathbf{3}(C_{2v})$, $C_{78}\mathbf{4}(D_{3h})$, $C_{78}\mathbf{5}(D_{3h})$.

[c] The symmetries of the 7 isomers of C_{80} are: $C_{80}\mathbf{1}(D_{5d})$, $C_{80}\mathbf{2}(D_2)$, $C_{80}\mathbf{3}(C_{2v})$, $C_{80}\mathbf{4}(D_3)$, $C_{80}\mathbf{5}(C_{2v})$, $C_{80}\mathbf{6}(D_{5h})$, $C_{80}\mathbf{7}(I_h)$.

[d] The symmetries of the 9 isomers of C_{82} are: $C_{82}\mathbf{1}(C_2)$, $C_{82}\mathbf{2}(C_s)$, $C_{82}\mathbf{3}(C_2)$, $C_{82}\mathbf{4}(C_s)$, $C_{82}\mathbf{5}(C_2)$, $C_{82}\mathbf{6}(C_s)$, $C_{82}\mathbf{7}(C_{3v})$, $C_{82}\mathbf{8}(C_{3v})$, $C_{82}\mathbf{9}(C_{2v})$.

[e] The symmetries of the 19 isomers of C_{86} and the 35 isomers of C_{88} are listed in Table 4 of Reference [3.60].

not having inversion symmetry. Liquid chromatography can be used to separate various isomers of C_{n_C} according to their shapes, although this can be a more delicate distinction than separation according to mass. Experiments such as NMR and electron paramagnetic resonance (EPR) can in some cases identify the symmetry and shape of isomers of a given C_{n_C}. Such studies have been made on C_{78} and C_{82} [3.56, 57, 63], although no definitive agreement has yet been reached about the relative abundances of the various isomers, which may be sensitive to preparation methods.

It is now possible to obtain sufficient quantities of some of the higher fullerenes for characterization experiments using NMR techniques and various properties measurements as well, including EPR spectroscopy (§16.2), STM, and surface science studies (§17.9.5). Measurements of the electron affinity E_A have already been reported for C_{70} (2.72 ± 0.05 eV), C_{76} (2.88 ± 0.05 eV), C_{78} (3.1 ± 0.07 eV), and C_{84} (3.05 ± 0.08 eV), using Knud-

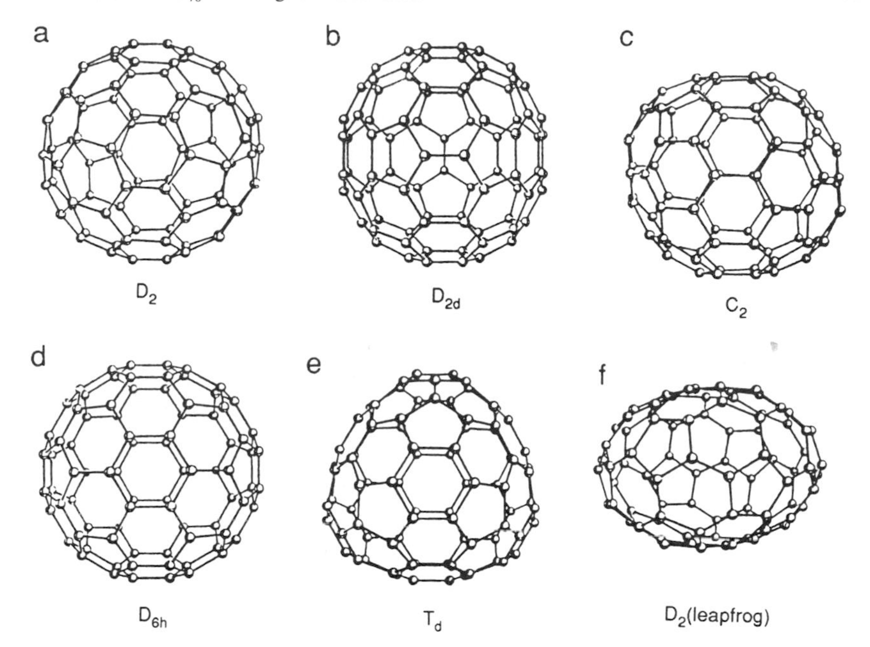

Fig. 3.7. Perspective views of several isomers of C_{84}: (a) structure with lowest energy D_2, (b) D_{2d}, (c) C_2, (d) D_{6h}, (c) T_d, and (f) leapfrog D_2 isomer of C_{84} [3.21].

sen cell mass spectrometry [3.45], showing a general increase in E_A with increasing n_C. The NMR technique, for example, is sensitive to the number and characteristics of the distinct carbon sites on the fullerene surface. The number of distinct NMR sites, the cohesive energy of the various isomers and their symmetry types, and the HOMO–LUMO band gap have been calculated for fullerenes in the range C_{60} to C_{94} [3.21]. Experimental NMR studies have already been initiated [3.56, 64] to study the isomers of C_{84}. To facilitate the analysis of such NMR spectra, calculations of the molecular electronic density of states have been carried out for a variety of isomers of C_{84} [3.35]. Density of states calculations for C_{76} have also been carried out [3.35] and have been used to compare with measured photoemission spectra [3.65]. The synthesis and isolation of these higher-mass fullerenes is now an active research field, and it is expected that as the availability of larger amounts of these materials becomes widespread, more extensive properties measurements of these higher-mass fullerenes will be carried out (see §17.9.5).

Whereas small-mass C_{n_C} fullerenes ($n_C < 100$) are likely to be single-walled, it is not known presently whether high-mass fullerenes ($n_C > 200$)

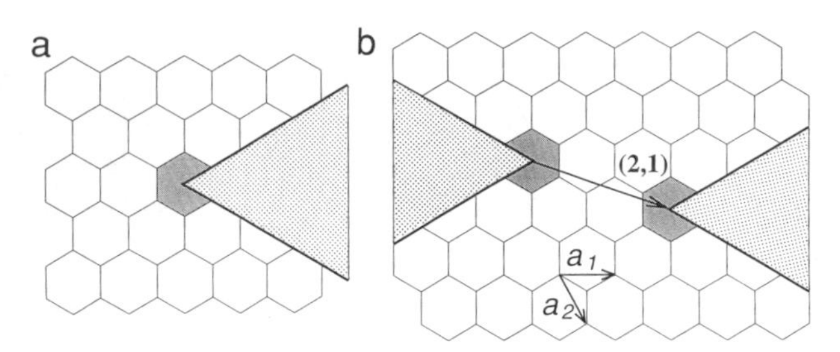

Fig. 3.8. (a) Formation of a pentagonal defect in a honeycomb lattice removes the area designated by the shaded outlined triangular wedge. (b) The vector (n, m) which connects two pentagonal defects fully specifies the icosahedral fullerene. The diagram is constructed for $(n, m) = (2, 1)$ and shows the related basis vectors of the unit cell.

are single-walled or perhaps may be multiwalled and described as a fullerene onion (see §19.10).

3.3. The Projection Method for Specifying Fullerenes

A useful construction for specifying the geometry of fullerene molecules is based on considering a curvature-producing pentagon as a defect in a planar network of hexagons. Shown in Fig. 3.8(a) is one such pentagonal defect. If a 60° wedge is introduced at each defect of this type, as indicated in Fig. 3.8(a), a five-sided regular polygon is formed. If the triangular wedge is cut out and the hexagons along the cut are joined together to each other, then a curved surface is produced. Two such pentagonal defects are shown in Fig. 3.8(b). The vector $\mathbf{d}_{nm}$ between the two pentagonal defects is

$$\mathbf{d}_{nm} = n\mathbf{a}_1 + m\mathbf{a}_2, \tag{3.6}$$

and the corresponding length is

$$d_{nm} = a_0(n^2 + m^2 + nm)^{1/2} \tag{3.7}$$

in which $\mathbf{a}_1$ and $\mathbf{a}_2$ are basis vectors of the honeycomb lattice and (n, m) are two integers which specify d_{nm}. The set of integers (n, m) further specify the fullerene. Using this concept, a planar projection of a fullerene on a honeycomb lattice can be made, as described below.

For the case of an icosahedral fullerene, only one vector $\mathbf{d}_{nm}$ needs be used to specify the fullerene, and the fullerene can be constructed from 20 equilateral triangles of length d_{mn} on a side, as shown in Fig. 3.9 [3.66]. Ten of these equilateral triangles form the belt section and five identical equilateral triangles form each of the two cap sections of the icosahedral fullerene. In this projection, the locations of the 12 pentagons needed to provide the curvature for the closed polyhedral structures are indicated by

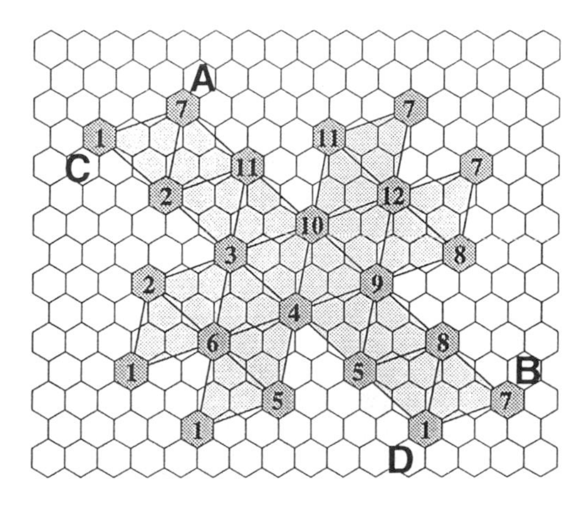

Fig. 3.9. Projection construction for an icosahedral fullerene. The size of each triangle is appropriate for C_{140} with $(n, m) = (2, 1)$, as defined in Fig. 3.8(b). The 12 pentagonal defects are indicated by 1–12. The C_{140} fullerene is obtained by superimposing pentagons with the same number to form a closed cage molecule. The parallelogram ABCD represents the "belt" region.

the numbers 1, ..., 12. By superimposing the numbered hexagons in Fig. 3.9 (which become pentagonal defects in the fullerene) with the same numbers wherever they appear, closed cage fullerenes are easily constructed. For illustrative purposes, the projections for the icosahedral C_{60}, C_{80}, and C_{140} molecules are shown in Fig. 3.10 [3.66].

If the lattice vector between two nearest-neighbor pentagonal defects is $n\mathbf{a}_1 + m\mathbf{a}_2$, where $\mathbf{a}_1$ and $\mathbf{a}_2$ are basis vectors of the honeycomb lattice, a_0 is the lattice constant of the 2D lattice ($a_0 = 2.46$ Å), and (n, m) are both integers, then a regular truncated icosahedron contains exactly 20 equilateral triangles. If the fullerene has an icosahedral shape, then the distance between nearest-neighbor pentagons is constant and is given by d_{nm}, as shown in Fig. 3.8(b). Using d_{nm} as the length of a side of an equilateral triangle, then the area of this equilateral triangle A_{nm} is

$$A_{nm} = \frac{\sqrt{3}a_0^2}{4}(n^2 + nm + m^2), \tag{3.8}$$

where a_0 is the lattice constant of the honeycomb lattice as indicated by vectors $\mathbf{a}_1$ and $\mathbf{a}_2$ in Fig. 3.8(b).

Every regular truncated icosahedron can be constructed from 20 equilateral triangles, each having an area A_{nm} given by Eq. (3.8). The number of carbon atoms per triangle is $(n^2 + nm + m^2)$, so that the number of carbon atoms in the fullerene n_C is given by Eq. (3.4). Figure 3.9 shows the generic format for constructing icosahedral fullerenes.

In Fig. 3.10 we show examples of projections of specific fullerenes, C_{60}, C_{80}, and C_{140} corresponding to $(n, m) = (1, 1)$, (2,0), and (2,1), respectively. The connection between Figs. 3.9 and 3.10 can be easily understood by comparing the two projections for C_{140} and drawing 10 equilateral triangles in

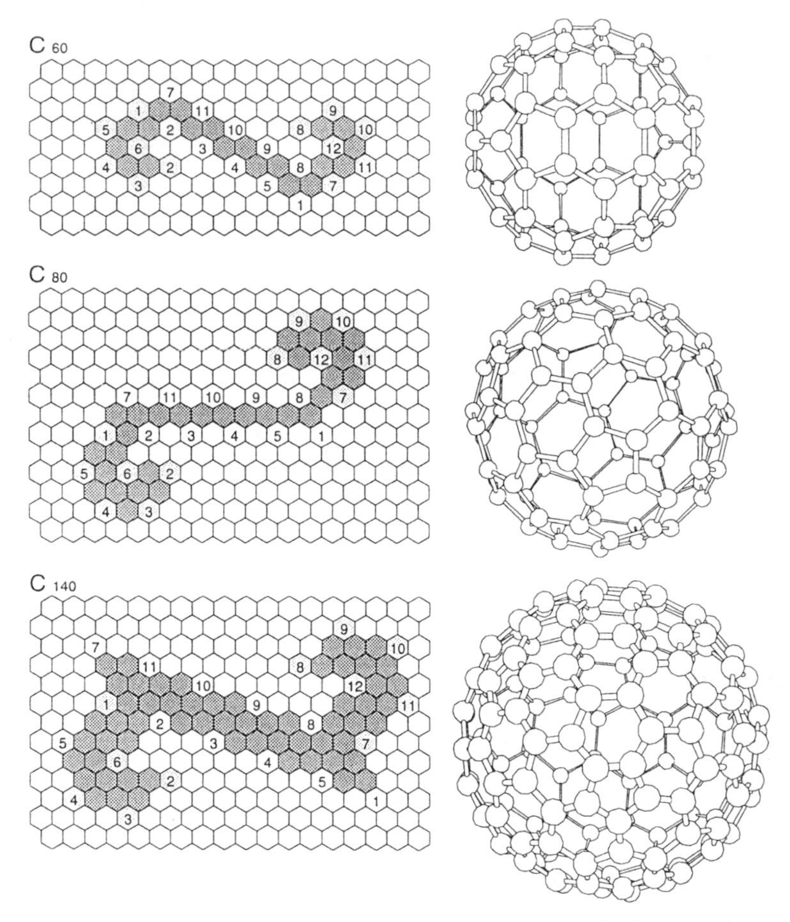

Fig. 3.10. The projection of three icosahedral fullerenes on a 2D honeycomb hexagonal lattice (a) for the C_{60} molecule, (b) the C_{80} molecule, and (c) the C_{140} molecule.

the belt area and 5 equilateral triangles in the cap regions of the projection in Fig. 3.10(c) [3.66]. Some rotations by 72° of the equilateral triangles in the cap regions are necessary to bring Fig. 3.9 into an equivalent form that looks identical to the projection shown in Fig. 3.10(c).

Although shown here only for icosahedral fullerenes, the projection method can also be used to construct polyhedra that do not have an icosahedral shape. The projection method is also valuable for specifying the structure of the end caps of carbon tubules (see §19.2.3). For the nanotubes, the tube wall is represented by many rows of hexagons, which are inserted parallel to *AB* and *CD* in Fig. 3.9, and the pentagons

'7' at points 'A' and 'B' are joined, as are pentagons '1' at points 'C' and 'D' to form a cylinder [3.66].

References

[3.1] R. Taylor, J. P. Hare, A. K. Abdul-Sada, and H. W. Kroto. *J. Chem. Soc. Chem. Commun.*, **20**, 1423 (1990).

[3.2] R. D. Johnson, G. Meijer, and D. S. Bethune. *J. Am. Chem. Soc.*, **112**, 8983 (1990).

[3.3] R. Tycko, R. C. Haddon, G. Dabbagh, S. H. Glarum, D. C. Douglass, and A. M. Mujsce. *J. Phys. Chem.*, **95**, 518 (1991).

[3.4] R. D. Johnson, D. S. Bethune, and C. S. Yannoni. *Accounts of Chem. Res.*, **25**, 169 (1992).

[3.5] W. I. F. David, R. M. Ibberson, J. C. Matthewman, K. Prassides, T. J. S. Dennis, J. P. Hare, H. W. Kroto, R. Taylor, and D. R. M. Walton. *Nature (London)*, **353**, 147 (1991).

[3.6] T. G. Schmalz, W. A. Seitz, D. J. Klein, and G. E. Hite. *Chem. Phys. Lett.*, **130**, 203 (1986).

[3.7] D. J. Klein, T. G. Schmalz, G. E. Hite, and W. A. Seitz. *J. Am. Chem. Soc.*, **108**, 1301 (1986).

[3.8] L. Tisza. *Zeitschrift für Physik*, **82**, 48 (1933).

[3.9] G. Dresselhaus, M. S. Dresselhaus, and P. C. Eklund. *Phys. Rev. B*, **45**, 6923 (1992).

[3.10] H. W. Kroto, J. R. Heath, S. C. O'Brien, R. F. Curl, and R. E. Smalley. *Nature (London)*, **318**, 162 (1985).

[3.11] R. D. Johnson, N. Yannoni, G. Meijer, and D. S. Bethune. In R. M. Fleming *et al.* (eds.), *Videotape of Late News Session on Buckyballs, MRS, Boston*. Materials Research Society Press, Pittsburgh, PA (1990).

[3.12] C. Christides, T. J. S. Dennis, K. Prassides, R. L. Cappelletti, D. A. Neumann, and J. R. D. Copley. *Phys. Rev. B*, **49**, 2897 (1994).

[3.13] S. Saito and A. Oshiyama. *Phys. Rev. B*, **44**, 11532 (1991).

[3.14] H. D. Beckhaus, C. Rüchart, M. Kas, F. Diederich, and C. S. Foote. *Angew. Chem. int. Edn. engl*, **31**, 63 (1992).

[3.15] D. L. Lichtenberger, K. W. Nebesny, C. D. Ray, D. R. Huffman, and L. D. Lamb. *Chem. Phys. Lett.*, **176**, 203 (1991).

[3.16] A. Tokmakoff, D. R. Haynes, and S. M. George. *Chem. Phys. Lett.*, **186**, 450 (1991).

[3.17] G. Dresselhaus, M. S. Dresselhaus, and J. G. Mavroides. *Carbon*, **4**, 433 (1966).

[3.18] J. de Vries, H. Steger, B. Kamke, C. Menzel, B. Weisser, W. Kamke, and I. V. Hertel. *Chem. Phys. Lett.*, **188**, 159 (1992).

[3.19] H. Steger, J. de Vries, B. Kamke, W. Kamke, and T. Drewello. *Chem. Phys. Lett.*, **194**, 452 (1992).

[3.20] C. Kittel. *Introduction to Solid State Physics*. John Wiley and Sons, New York, NY, 6th ed. (1986).

[3.21] B. L. Zhang, C. Z. Wang, and K. M. Ho. *Chem. Phys. Lett.*, **193**, 225 (1992).

[3.22] H. W. Kroto. *Nature (London)*, **329**, 529 (1987).

[3.23] T. G. Schmalz, W. A. Seitz, D. J. Klein, and G. E. Hite. *J. Am. Chem. Soc.*, **110**, 1113 (1988).

[3.24] L. A. Paquette, R. J. Ternansky, D. E. Balogh, and G. Kentgen. *J. Am. Chem. Soc.*, **105**, 5441 and 5446 (1983).

[3.25] E. A. Rohlfing, D. M. Cox, and A. Kaldor. *J. Chem. Phys.*, **81**, 3322 (1984).

[3.26] J. E. Fischer, P. A. Heiney, A. R. McGhie, W. J. Romanow, A. M. Denenstein, J. P. McCauley, Jr., and A. B. Smith III. *Science*, **252**, 1288 (1991).

[3.27] C. S. Yannoni, R. D. Johnson, G. Meijer, D. S. Bethune, and J. R. Salem. *J. Phys. Chem.*, **95**, 9 (1991).
[3.28] T. Lenosky, X. Gonze, M. Teter, and V. Elser. *Nature (London)*, **355**, 333 (1992).
[3.29] H. Terrones and A. L. MacKay. *Carbon*, **30**, 1251 (1992).
[3.30] S. Iijima. *Mater. Sci. and Eng.*, **B19**, 172 (1993).
[3.31] R. S. Ruoff and A. L. Ruoff. *Appl. Phys. Lett.*, **59**, 1553 (1991).
[3.32] S. J. Duclos, K. Brister, R. C. Haddon, A. R. Kortan, and F. A. Thiel. *Nature (London)*, **351**, 380 (1991).
[3.33] F. Diederich and R. L. Whetten. *Acc. Chem. Res.*, **25**, 119 (1992).
[3.34] H. Ajie, M. M. Alvarez, S. J. Anz, R. D. Beck, F. Diederich, K. Fostiropoulos, D. R. Huffman, W. Krätschmer, Y. Rubin, K. E. Schriver, D. Sensharma, and R. L. Whetten. *J. Phys. Chem.*, **94**, 8630 (1990).
[3.35] S. Saito, S. I. Sawada, N. Hamada, and A. Oshiyama. *Mater. Sci. Eng.*, **B19**, 105 (1993).
[3.36] R. E. Smalley. *Accounts of Chem. Research*, **25**, 98 (1992).
[3.37] P. W. Fowler and J. Woolrich. *Chem. Phys. Lett.*, **127**, 78 (1986).
[3.38] D. J. Klein, W. A. Seitz, and T. G. Schmalz. *Nature (London)*, **323**, 703 (1986).
[3.39] S. Maruyama, L. R. Anderson, and R. E. Smalley. *Rev. Sci. Instrum.*, **61**, 3686 (1990).
[3.40] Y. Maruyama, T. Inabe, H. Ogata, Y. Achiba, S. Suzuki, K. Kikuchi, and I. Ikemoto. *Chem. Lett.*, **10**, 1849 (1991). The Chemical Society of Japan.
[3.41] F. Negri, G. Orlandi, and F. Zerbetto. *J. Amer. Chem. Soc.*, **113**, 6037 (1991).
[3.42] A. V. Nilolaev, T. J. S. Dennis, K. Prassides, and A. K. Soper. *Chem. Phys. Lett.*, **223**, 143 (1994).
[3.43] H. D. Beckhaus, V. S, C. Ruchardt, F. N. Diederich, C. Thilgen, H. U. T. Meer, H. Mohn, and W. Muller. *Angew. Chem. int. Edn. engl*, **33**, 996 (1994).
[3.44] T. Kiyobayashi and M. Sakiyama. *Fullerene Sci. Tech.*, **1**, 269 (1993).
[3.45] O. V. Boltalina, L. N. Sidorov, A. Y. Borshchevsky, E. V. Seckhanova and E. V. Skokan. *Rapid Comm. Mass Spectrom.*, **7**, 1009 (1993).
[3.46] P. Wurz, K. R. Lykke, M. J. Pellin, and D. M. Gruen. *J. Appl. Phys.*, **70**, 6647 (1991).
[3.47] K. Harigaya. *Chem. Phys. Lett.*, **189**, 79 (1992).
[3.48] W. Andreoni, F. Gygi, and M. Parrinello. *Phys. Rev. Lett.*, **68**, 823 (1992).
[3.49] K. Nakao, N. Kurita, and M. Fujita. *Phys. Rev. B.*, **49**, 11415 (1994).
[3.50] G. Onida, W. Andreoni, J. Kohanoff, and M. Parrinello. *Chem. Phys. Lett.*, **219**, 1 (1994).
[3.51] W. Andreoni, F. Gygi, and M. Parrinello. *Chem. Phys. Lett.*, **189**, 241 (1992).
[3.52] D. R. McKenzie, C. A. Davis, D. J. H. Cockayne, D. A. Miller, and A. M. Vassallo. *Nature (London)*, **355**, 622 (1992).
[3.53] C. Z. Wang, C. T. Chan, and K. M. Ho. *Phys. Rev. B*, **46**, 9761 (1992).
[3.54] K. Raghavachari and C. M. Rohlfing. *J. Phys. Chem.*, **95**, 5768 (1991).
[3.55] S. van Smaalen, V. Petricek, J. L. de Boer, M. Dusek, M. A. Verheijen, and G. Meijer. *Chem. Phys. Lett.*, **223**, 323 (1994).
[3.56] K. Kikuchi, N. Nakahara, T. Wakabayashi, S. Suzuki, H. Shiramaru, Y. Miyake, K. Saito, I. Ikemoto, M. Kainosho, and Y. Achiba. *Nature (London)*, **357**, 142 (1992).
[3.57] K. Kikuchi, N. Nakahara, T. Wakabayashi, M. Honda, H. Matsumiya, T. Moriwaki, S. Suzuki, H. Shiromaru, K. Saito, K. Yamauchi, I. Ikemoto, and Y. Achiba. *Chem. Phys. Lett.*, **188**, 177 (1992).
[3.58] R. Saito (private communication).
[3.59] H. W. Kroto and K. G. McKay. *Nature (London)*, **331**, 328 (1988).
[3.60] P. W. Fowler, D. E. Manolopoulos, and R. P. Ryan. *Carbon*, **30**, 1235 (1992).
[3.61] D. E. Manolopoulos and P. W. Fowler. *Chem. Phys. Lett.*, **187**, 1 (1991).

[3.62] D. M. Poirier, J. H. Weaver, K. Kikuchi, and Y. Achiba. *Zeitschrift für Physik D: Atoms, Molecules and Clusters*, **26**, 79 (1993).

[3.63] F. Diederich, R. Ettl, Y. Rubin, R. L. Whetten, R. Beck, M. Alvarez, S. Anz, D. Sensharma, F. Wudl, K. C. Khemani, and A. Koch. *Science*, **252**, 548 (1991).

[3.64] U. Schneider, S. Richard, M. M. Kappes, and R. Ahlrichs. *Chem. Phys. Lett.*, **210**, 165 (1993).

[3.65] S. Hino, K. Matsumoto, S. Hasegawa, H. Inokuchi, T. Morikawa, T. Takahashi, K. Seki, K. Kikuchi, S. Suzuki, I. Ikemoto, and Y. Achiba. *Chem. Phys. Lett.*, **197**, 38 (1992).

[3.66] M. Fujita, R. Saito, G. Dresselhaus, and M. S. Dresselhaus. *Phys. Rev. B*, **45**, 13834 (1992).

CHAPTER 4

Symmetry Considerations of Fullerene Molecules

Many of the special properties that fullerenes exhibit are directly related to the very high symmetry of the C_{60} molecule, where the 60 equivalent carbon atoms are at the vertices of a truncated icosahedron. The regular truncated icosahedron is obtained from the regular icosahedron by passing planes normal to each of the 6 fivefold axes passing through the center of the icosahedron so that the edges of the pentagonal faces thus formed are equal in length to the edges of the hexagonal faces. Figure 1.1(a) shows this soccer-ball configuration for C_{60} as thought to have been constructed by Leonardo da Vinci in about 1500 [4.1, 2], and Fig. 1.1(b) shows the location of the carbon atoms at the vertices of the truncated icosahedron. The first application of the icosahedral group to molecules was by Tisza in 1933 [4.3].

In this chapter the group theory for the icosahedron is reviewed, and mathematical tables are given for a simple application of the icosahedral group symmetry to the vibrational and electronic states of the icosahedral fullerenes. The effect of lowering the icosahedral symmetry is discussed in terms of the vibrational and electronic states. Symmetry considerations related to the isotopic abundances of the ^{12}C and ^{13}C nuclei are also discussed. The space group symmetries appropriate to several crystalline phases of C_{60} are reviewed in §7.1.2. The symmetry properties of symmorphic and nonsymmorphic carbon nanotubes are discussed in §19.4.2 and §19.4.3.

4.1. Icosahedral Symmetry Operations

The truncated icosahedron (see Fig. 4.1) has 12 pentagonal faces, 20 hexagonal faces, 60 vertices, and 90 edges. The 120 symmetry operations

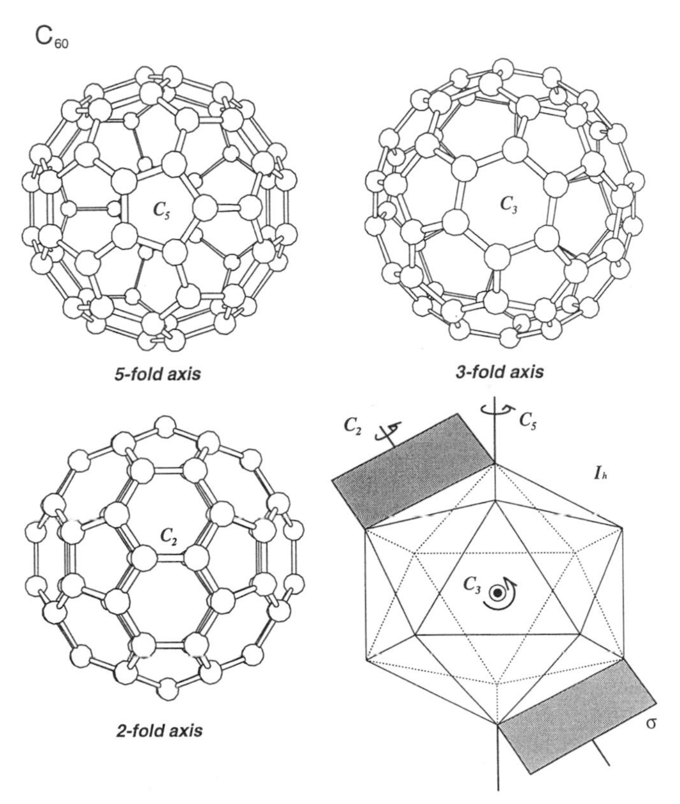

Fig. 4.1. Symmetry operations of the regular truncated icosahedron. (a) The fivefold axis, (b) the threefold axis, (c) the twofold axis, and (d) a composite of the symmetry operations of the point group I_h.

for the icosahedral point group are listed in the character table (Table 4.1) where they are grouped into 10 classes. These classes are the identity operator, which is in a class by itself, the 12 primary fivefold rotations (12 C_5 and 12 C_5^2) going through the centers of the pentagonal faces, the 20 secondary threefold rotations going through the centers of the 20 hexagonal faces, and the 30 secondary twofold rotations going through the 30 edges joining two adjacent hexagons. Each of these symmetry operations is compounded with the inversion operation. Also listed in the character table are the 10 irreducible representations of the point group I_h.

The C_{60} molecule has carbon atoms at the 60 equivalent vertices of a truncated icosahedron for which the lengths of the pentagonal edges are slightly longer ($a_5 = 1.46$ Å) than the bond lengths shared by two hexagons

Table 4.1

Character table[a,b,c] for I_h.

$\mathcal{R}$	E	$12C_5$	$12C_5^2$	$20C_3$	$15C_2$	i	$12S_{10}^3$	$12S_{10}$	$20S_3$	$15\sigma_v$
A_g	1	+1	+1	+1	+1	+1	+1	+1	+1	+1
F_{1g}	3	$+\tau^c$	$1-\tau$	0	−1	+3	$+\tau$	$1-\tau$	0	−1
F_{2g}	3	$1-\tau$	$+\tau$	0	−1	+3	$1-\tau$	$+\tau$	0	−1
G_g	4	−1	−1	+1	0	+4	−1	−1	+1	0
H_g	5	0	0	−1	+1	+5	0	0	−1	+1
A_u	1	+1	+1	+1	+1	−1	−1	−1	−1	−1
F_{1u}	3	$+\tau$	$1-\tau$	0	−1	−3	$-\tau$	$\tau-1$	0	+1
F_{2u}	3	$1-\tau$	$+\tau$	0	−1	−3	$\tau-1$	$-\tau$	0	+1
G_u	4	−1	−1	+1	0	−4	+1	+1	−1	0
H_u	5	0	0	−1	+1	−5	0	0	+1	−1

[a]Note: the symmetry operations about the fivefold axes are in two different classes, labeled $12C_5$ and $12C_5^2$ in the character table. Then $iC_5 = S_{10}^{-1}$ and $iC_5^{-1} = S_{10}$ are in the classes labeled $12S_{10}^3$ and $12S_{10}$, respectively. Also $iC_2 = \sigma_v$.

[b]See Table 4.2 for a complete listing of the basis functions for the I_h point group in terms of spherical harmonics.

[c]In this table $\tau \equiv (1+\sqrt{5})/2$.

(a_6 = 1.40 Å) (see §3.1). If we take this difference in bond length into account, then the C_{60} cage forms a truncated icosahedron but not a *regular* truncated icosahedron where all bond lengths would be equal.

Since the truncated icosahedron is close in shape to a sphere, it is suggestive to relate the basis functions of the icosahedron to those of the sphere, namely the spherical harmonics. The full set of basis functions for group I_h is listed in Table 4.2 in terms of spherical harmonics $Y_{\ell,m}$ with minimal ℓ values. Many physical problems dealing with fullerenes, such as the electronic states or vibrational modes, are treated in terms of spherical harmonics which are basis functions for the full rotational group. The spherical harmonics therefore form reducible representations of the I_h point group for $\ell > 2$ and irreducible representations for $\ell = 0, 1, 2$. Odd and even integers ℓ, respectively, correspond to odd and even representations of the group I_h. The basis functions for group I_h in an l-dimensional manifold are obtained by solving the eigenvalue problem for irreducible tensor operations. The decomposition of the spherical harmonics into irreducible representations of the point group I is given in Table 4.3 for both integral and half-integral values of the angular momentum J. The integral values of J are pertinent to the vibrational spectra (§4.2), while the half-integral J values are also needed to

Table 4.2

Basis functions for each of the irreducible representations $\mathscr{R}$ of point group I_h expressed as spherical harmonics $Y_{l,m}$.[a] For the multidimensional representations the basis functions for each partner are listed [4.4].

$\mathscr{R}$	Basis function
A_g	$Y_{0,0}$; and $\frac{\sqrt{7}}{5}(Y_{6,5} - Y_{6,-5}) + \frac{\sqrt{11}}{5} Y_{6,0}$
F_{1g}	$-\sqrt{3/5}Y_{6,-6} + \sqrt{66/10}Y_{6,-1} + \sqrt{22/10}Y_{6,4}$ $\sqrt{1/2}(Y_{6,5} + Y_{6,-5})$ $\sqrt{3/5}Y_{6,6} + \sqrt{66/10}Y_{6,1} - \sqrt{22/10}Y_{6,-4}$
F_{2g}	$-\sqrt{28/125}Y_{8,-8} + \sqrt{39/500}Y_{8,-3} + \sqrt{143/250}Y_{8,2} - \sqrt{63/500}Y_{8,7}$ $\sqrt{1/2}(Y_{8,5} + Y_{8,-5})$ $-\sqrt{28/125}Y_{8,8} - \sqrt{39/500}Y_{8,3} + \sqrt{143/250}Y_{8,-2} + \sqrt{63/500}Y_{8,-7}$
G_g	$\sqrt{8/15}Y_{4,-4} + \sqrt{7/15}Y_{4,1}$ $\sqrt{1/15}Y_{4,-3} + \sqrt{14/15}Y_{4,2}$ $-\sqrt{1/15}Y_{4,3} + \sqrt{14/15}Y_{4,-2}$ $\sqrt{8/15}Y_{4,4} - \sqrt{7/15}Y_{4,-1}$
H_g	$Y_{2,-2}$; $Y_{2,-1}$; $Y_{2,0}$; $Y_{2,1}$; $Y_{2,2}$
A_u	$-\frac{\sqrt{5 \cdot 7 \cdot 11 \cdot 13}}{250}(Y_{15,15} + Y_{15,-15}) - \frac{\sqrt{2 \cdot 3 \cdot 5 \cdot 11 \cdot 29}}{250}(Y_{15,10} - Y_{15,-10})$ $+ \frac{\sqrt{23 \cdot 29}}{50}(Y_{15,5} + Y_{15,-5})$
F_{1u}	$Y_{1,-1}$; $Y_{1,0}$; $Y_{1,1}$
F_{2u}	$-\sqrt{2/5}Y_{3,-3} + \sqrt{3/5}Y_{3,2})$ $Y_{3,0}$ $\sqrt{2/5}Y_{3,3} + \sqrt{3/5}Y_{3,-2})$
G_u	$\sqrt{3/5}Y_{3,-3} + \sqrt{2/5}Y_{3,2}$ $Y_{3,-1}$ $Y_{3,1}$ $-\sqrt{3/5}Y_{3,3} + \sqrt{2/5}Y_{3,-2}$
H_u	$\sqrt{3/10}Y_{5,-4} + \sqrt{7/10}Y_{5,1}$ $\sqrt{3/5}Y_{5,-3} - \sqrt{2/5}Y_{5,2}$ $\sqrt{1/2}(Y_{5,5} + Y_{5,-5})$ $\sqrt{3/5}Y_{5,3} + \sqrt{2/5}Y_{5,-2}$ $-\sqrt{3/10}Y_{5,4} + \sqrt{7/10}Y_{5,-1}$

[a] Ref. [4.5, 6].

Table 4.3

Decomposition of angular momentum basis functions in the full rotation group labeled by J into irreducible representations of the double group of I. Both integral and half-integral angular momentum basis functions are included.[a]

J	Γ_1 A	Γ_2 F_1	Γ_3 F_2	Γ_4 G	Γ_5 H	J	Γ_6	Γ_7	Γ_8	Γ_9
0 (S)	1					1/2	1			
1 (P)		1				3/2			1	
2 (D)					1	5/2				1
3 (F)			1	1		7/2		1		1
4 (G)				1	1	9/2			1	1
5 (H)		1	1		1	11/2	1		1	1
6 (I)	1	1		1	1	13/2	1	1	1	1
7 (K)		1	1	1	1	15/2			1	2
8 (L)			1	1	2	17/2		1	1	2
9 (M)		1	1	2	1	19/2	1	1	1	2
10 (N)	1	1	1	1	2	21/2	1		2	2
11 (O)		2	1	1	2	23/2	1	1	2	2
12 (Q)	1	1	1	2	2	25/2	1	1	1	3
13 (R)		1	2	2	2	27/2		1	2	3
14 (T)		1	1	2	3	29/2	1	1	2	3
15 (U)	1	2	2	2	2	31/2	2	1	2	3
16 (V)	1	2	1	2	3	33/2	1	1	3	3
17 (W)		2	2	2	3	35/2	1	1	2	4
18 (X)	1	1	2	3	3	37/2	1	2	2	4
19 (Y)		2	2	2	4	39/2	1	1	3	4
20 (Z)	1	2	2	2	4	41/2	2	1	3	4
21	1	3	2	3	3	43/2	2	2	3	4
22	1	2	2	3	4	45/2	1	1	3	5
23		2	3	3	4	47/2	1	2	3	5
24	1	2	2	4	4	49/2	2	2	3	5
25	1	3	3	3	4	51/2	2	1	4	5
26	1	3	2	3	5	53/2	2	2	4	5
27	1	3	3	4	4	55/2	2	2	3	6
28	1	2	3	4	5	57/2	1	2	4	6
29		3	3	4	5	59/2	2	2	4	6
30	2	3	3	4	5	61/2	3	2	4	6
31	1	4	3	4	5	63/2	2	2	5	6
32	1	3	3	4	6	65/2	2	2	4	7
33	1	3	4	4	5	67/2	2	3	4	7
34	1	3	3	4	6	69/2	2	2	5	7
35	1	4	4	4	6	71/2	3	2	5	7
36	2	4	3	5	6	72/2	3	3	5	7
⋮	⋮	⋮	⋮	⋮	⋮	⋮	⋮	⋮	⋮	⋮

[a] As an example in using the table, $\Gamma(J=5) = \Gamma_2(F_1) + \Gamma_3(F_2) + \Gamma_5(H)$.

describe the electronic states when electron spin is included in the wave function (§4.3).

Fivefold symmetry is not often found in solid-state physics, because it is impossible to construct a Bravais lattice based on fivefold symmetry. Thus fullerenes in the solid state crystallize into solids of lower point group symmetries, such as the fcc lattice (e.g., C_{60} at room temperature) or the hcp lattice (e.g., some phases of C_{70}). Nevertheless, the local point group symmetry of the individual fullerene molecules is very important because they crystallize into highly molecular solids in which the electronic and vibrational states are closely related to those of the free molecule. Therefore we summarize in this chapter the group theoretical considerations that are involved in finding the symmetries and degeneracies of the vibrational and electronic states of the C_{60} molecule, with some discussion also given to higher-mass fullerenes. Group theoretical applications of the space groups associated with the crystalline solids are made elsewhere in the book (see §7.1, §19.4).

To describe the symmetry properties of the vibrational modes and of the electronic levels, it is necessary to find the equivalence transformation for the carbon atoms in the molecule. The characters for the equivalence transformation $\chi^{\text{a.s.}}$ for the 60 equivalent carbon atom sites (a.s.) for the C_{60} molecule in icosahedral I_h symmetry are given in Table 4.4. Also listed in Table 4.4 are the number of elements for each class (top) and the characters for the equivalence transformation for the 12 fivefold axes, the 20 threefold axes, and the 30 twofold axes which form classes of the icosahedral I_h group. The decomposition of the reducible representations of Table 4.4 into their irreducible constituents is given in Table 4.5, which directly gives the number of π orbitals for each irreducible representation. For example, if guest species X are attached to each fivefold axis at an equal distance from the center of the icosahedron to yield a molecule $X_{12}C_{60}$, then the full icosahedral I_h symmetry is preserved. Table 4.4 also lists $\chi^{\text{a.s.}}$ for a few higher icosahedral fullerenes: C_{80} for which $(n, m) = (2, 0)$; C_{140} for which $(n, m) = (2, 1)$; and C_{240} for which $(n, m) = (2, 2)$ (see §3.3). We note that fullerenes with either $m = 0$, or $n = 0$, or those with $m = n$ have inversion symmetry and therefore are described by group I_h. Other icosahedral fullerenes with $m \neq n \neq 0$, lack inversion symmetry (e.g., C_{140}) and are described by the point group I. The decomposition of the equivalence transformation $\chi^{\text{a.s.}}$ which is a reducible representation of the group I_h (or I) into its irreducible constituents is given in Table 4.5 for every entry in Table 4.4, and the even and odd constituents are listed on separate lines.

Table 4.4

Characters $\chi^{\text{a.s.}}$ for the equivalence transformation of various atomic sites in icosahedral I_h symmetry. The corresponding irreducible representations of group I_h are listed in Table 4.5.

Cluster # elements	E 1	C_5 12	C_5^2 12	C_3 20	C_2 15	i 1	S_{10}^3 12	S_{10} 12	S_3 20	σ_v 15
X_{12}	12	2	2	0	0	0	0	0	0	4
C_{20}	20	0	0	2	0	0	0	0	0	4
X_{30}	30	0	0	0	2	0	0	0	0	4
C_{60}	60	0	0	0	0	0	0	0	0	4
C_{80}	80	0	0	2	0	0	0	0	0	8
C_{140}	140	0	0	2	0	—[a]	—[a]	—[a]	—[a]	—[a]
C_{180}	180	0	0	0	0	0	0	0	0	4
C_{240}	240	0	0	0	0	0	0	0	0	8

[a] Since C_{140} lacks inversion symmetry, entries in the table are made only for the classes that are pertinent to point group I.

4.2. Symmetry of Vibrational Modes

In this section we review the symmetries and degeneracies of the vibrational modes for the C_{60} molecule. There are 180 degrees of freedom (3×60) for each C_{60} molecule. Three of these degrees of freedom correspond to translations and three to rotation, leaving 174 vibrational degrees of freedom. Since icosahedral symmetry gives rise to a large number of degenerate modes, only 46 distinct mode frequencies are expected for the C_{60} molecule. The number of distinct modes N_ω for C_{60} and other icosahedral configurations is given in Table 4.6. The results given in Table 4.6 follow directly from group theoretical arguments, using the entries given in Table 4.4 for the characters for the equivalence transformation $\chi^{\text{a.s.}}(C_{60})$ for the 60 equivalent carbon atoms in C_{60}. Taking the direct product of $\chi^{\text{a.s.}}(C_{60})$ with the characters for the vector $\chi_{F_{1u}}$ (which transforms according to the irreducible representation F_{1u}), as given in Table 4.1, and subtracting off the irreducible representations for pure rotations (F_{1g}) and pure translations (F_{1u}) yields the irreducible representations for the vibrational modes of C_{60}. The symmetries of the resulting vibrational modes are listed in Table 4.6, where the multiplicities for each symmetry type are given. For example, X_{12} in I_h symmetry has $N_\omega = 8$ distinct modes, and the symmetry types that are

Table 4.5

Irreducible representations for various atomic sites in the icosahedral group I_h. The characters for the corresponding equivalence transformations are given in Table 4.4.

Cluster	$\chi^{\text{a.s.}}$
X_{12}	$A_g, H_g,$ F_{1u}, F_{2u}
C_{20}	$A_g, G_g, H_g,$ F_{1u}, F_{2u}, G_u
X_{30}	$A_g, G_g, 2H_g,$ F_{1u}, F_{2u}, G_u, H_u
C_{60}	$A_g, F_{1g}, F_{2g}, 2G_g, 3H_g,$ $2F_{1u}, 2F_{2u}, 2G_u, 2H_u$
C_{80}	$2A_g, F_{1g}, F_{2g}, 3G_g, 4H_g,$ $3F_{1u}, 3F_{2u}, 3G_u, 2H_u$
C_{140} [a]	$3A, 7F_1, 7F_2, 10G, 11H$
C_{180}	$2A_{1g}, 4F_{1g}, 4F_{2g}, 6G_g, 8H_g,$ $A_u, 5F_{1u}, 5F_{2u}, 6G_u, 7H_u$
C_{240}	$3A_g, 5F_{1g}, 5F_{2g}, 8G_g, 11H_g,$ $A_u, 7F_{1u}, 7F_{2u}, 8G_u, 9H_u$

[a] Since C_{140} lacks inversion symmetry, entries in the table are appropriate for group I.

Table 4.6

Symmetry properties of vibrational modes in molecules with icosahedral symmetry.

Cluster	N_ω [a]	A_g [b]	F_{1g}	F_{2g}	G_g	H_g [c]	A_u	F_{1u} [d]	F_{2u}	G_u	H_u
X_{12}	8	1			1	2		1	1	1	1
X_{20}	14	1		1	2	3		1	2	2	2
X_{30}	22	1	1	2	3	4		2	3	3	3
C_{60}	46	2	3	4	6	8	1	4	5	6	7
$X_{12}C_{60}$	56	3	4	4	7	10	1	6	6	7	8
C_{80}	62	3	6	5	8	10	1	5	6	8	10
C_{140} [e]	110	7	20	20	28	35	–	–	–	–	–
C_{180}	144	5	12	13	18	23	4	13	14	18	22
C_{240}	190	7	16	17	24	31	5	18	19	24	29

[a] Number of modes in icosahedral symmetry.

[b] Raman-active mode is seen only in $\|, \|$ polarization.

[c] Raman-active mode is seen in both $\|, \|$ and $\|, \perp$ polarizations.

[d] Infrared-active mode.

[e] I symmetry, thus no gerade or ungerade modes. In this case the F_1 modes are IR-active and the A and H modes are Raman-active.

found include $A_g + G_g + 2H_g + F_{1u} + F_{2u} + G_u + H_u$ where $2H_g$ indicates that there are two distinct mode frequencies corresponding to $(5 \times 2) = 10$ normal modes. We note that for the C_{60} molecule, every irreducible representation is contained at least once. In carrying out the direct product $\chi^{\text{a.s.}} \otimes \chi_{F_{1u}}$ the entries in Table 4.7 for the decomposition of direct products for the point group I are useful. We note that $I_h = I \otimes i$, indicating that the characters for the group I_h are obtained from those of the group I by taking the direct product of group I with the inversion group i.

The Raman-active modes have A_g and H_g symmetry (corresponding to the basis functions for all symmetrical quadratic forms, and the antisymmetric quadratic form transforming as F_{1g} does not contribute to the Raman scattering). The infrared-active modes have F_{1u} symmetry (the linear forms associated with a vector). One can see from the basis functions listed in Table 4.2 that the symmetry of the Raman tensor allows $\|, \|$ scattering for A_g modes and both $\|, \|$ and $\|, \perp$ scattering for H_g modes, where the directions $\|$ and $\perp$ refer to the polarization directions of the incident and scattered photon electric fields. For example, $\|, \perp$ implies that the polarizations for the incident and scattered beams are orthogonal. Table 4.6 shows that of the 46 distinct vibrational mode frequencies for C_{60}, only 4 are infrared-active with symmetry F_{1u} and only 10 are Raman-active (two with A_g symmetry and eight with H_g symmetry), while the remaining 32 modes are silent in the first-order infrared and Raman spectra. Many of these silent modes can, however, be observed by inelastic neutron scattering, electron energy loss spectroscopy, as vibronic side bands on the photoluminescence spectra, and most sensitively in the higher-order infrared and Raman spectra (see §11.5.3, §11.5.4, and §11.5.6), because of the different selection rules governing these higher-order processes.

Various possible attachments could be made to the C_{60} molecule without lowering its symmetry [e.g., by attaching 12 equivalent guest species (X) at each end of the 6 fivefold axes, or 20 guest species at each end of the 10 threefold axes, or 30 guest species at each end of the 15 twofold axes]. To preserve the I_h symmetry, all the equivalent sites must be occupied. However, if only some of these sites are occupied, the symmetry is lowered (see §4.4). Table 4.4 gives the characters $\chi^{\text{a.s.}}$ for the equivalence transformation for special arrangements of guest species that preserve the I_h or I symmetries, and these guest species may be attached through doping or a chemical reaction. The corresponding vibrational modes associated with such guest species are included in Table 4.6, both separately and in combination with the C_{60} molecule, as for example $X_{12}C_{60}$, where we might imagine a guest atom to be located at the center of each pentagonal face. Any detailed solution to the normal mode problem will involve solutions of a dynamical matrix in which tangential and radial modes having the same symmetry will mix because of the curvature of the molecule. Symmetry aspects of the vibrational modes of C_{60} are further treated in §11.3.1.

Vibrational modes that are silent in the first-order spectrum can, however, contribute to the second- and higher-order Raman [4.7] and infrared [4.8] spectra. Anharmonic terms in the potential couple the normal mode solutions of the harmonic potential approximation, giving rise to overtones ($n\omega_i$) and combination modes ($\omega_i \pm \omega_j$), many of which are observable in the second-order spectra. Group theory requires that the direct product of

Table 4.7

Decomposition of direct products[a] in the icosahedral point group I.

$\mathcal{R}$	Γ_1 (A)	Γ_2 (F_1)	Γ_3 (F_2)	Γ_4 (G)	Γ_5 (H)
Γ_1 (A)	Γ_1	Γ_2	Γ_3	Γ_4	Γ_5
Γ_2 (F_1)	Γ_2	Γ_1 Γ_2 Γ_5	Γ_4 Γ_5	Γ_3 Γ_4 Γ_5	Γ_2 Γ_3 Γ_4 Γ_5
Γ_3 (F_2)	Γ_3	Γ_4 Γ_5	Γ_1 Γ_3 Γ_5	Γ_2 Γ_4 Γ_5	Γ_2 Γ_3 Γ_4 Γ_5
Γ_4 (G)	Γ_4	Γ_3 Γ_4 Γ_5	Γ_2 Γ_4 Γ_5	Γ_1 Γ_2 Γ_3 Γ_4 Γ_5	Γ_2 Γ_3 Γ_4 $2\Gamma_5$
Γ_5 (H)	Γ_5	Γ_2 Γ_3 Γ_4 Γ_5	Γ_2 Γ_3 Γ_4 Γ_5	Γ_2 Γ_3 Γ_4 $2\Gamma_5$	Γ_1 Γ_2 Γ_3 $2\Gamma_4$ $2\Gamma_5$
Γ_6	Γ_6	Γ_6 Γ_8	Γ_9	Γ_7 Γ_9	Γ_8 Γ_9
Γ_7	Γ_7	Γ_9	Γ_6 Γ_8	Γ_6 Γ_9	Γ_8 Γ_9
Γ_8	Γ_8	Γ_6 Γ_8 Γ_9	Γ_7 Γ_8 Γ_9	Γ_8 $2\Gamma_9$	Γ_6 Γ_7 Γ_8 $2\Gamma_9$
Γ_9	Γ_9	Γ_7 Γ_8 $2\Gamma_9$	Γ_6 Γ_8 $2\Gamma_9$	Γ_6 Γ_7 $2\Gamma_8$ $2\Gamma_9$	Γ_6 Γ_7 $2\Gamma_8$ $3\Gamma_9$

[a] As an example of using this table, the direct product $\Gamma_4 \otimes \Gamma_2 = \Gamma_3 + \Gamma_4 + \Gamma_5$.

the second-order combination modes $\Gamma_i \otimes \Gamma_j$ (see Table 4.7) must contain the irreducible representations F_{1u} to be observable in the second-order infrared spectrum and A_g of H_g to be observable in the second-order Raman spectrum. By parity considerations alone, overtones (or harmonics) can be observed in the Raman spectrum but are not symmetry-allowed in the second-order infrared spectrum, since all second-order overtones have even parity. Because of the highly molecular nature of crystalline C_{60}, the second-order infrared and Raman spectra are especially strong in crystalline films (see §11.5.3 and §11.5.4, respectively). Whereas only about 10 strong features are seen experimentally in the first-order Raman spectrum, over 100 features are resolved and identified in the second-order Raman spectrum [4.7]. The observation of a multitude of sharp lines in the higher-order infrared and Raman spectra is a unique aspect of the spectroscopy of highly molecular solids. Usually, dispersion effects in the solid state broaden the higher-order spectra, so that detailed features commonly observed in the gas-phase spectra can no longer be resolved. However, for the case of C_{60}, analysis of the second-order infrared and Raman spectra provides a powerful method for the determination of the silent modes of C_{60}.

Although the icosahedral C_{80}, C_{140}, C_{180}, and C_{240} molecules have not yet been studied by Raman or infrared spectroscopy, the symmetry analysis for these molecules is included in Table 4.6. In contrast to C_{60}, the C_{80} molecule also has isomers of lower symmetry (see §3.2). The corresponding symmetry analysis for fullerene molecules of lower symmetry is given in §11.7. Symmetry effects in the vibrational spectra of fullerenes associated with higher-order spectra and with crystal field effects are discussed in §11.5. Symmetry effects associated with the vibrational spectra of carbon nanotubes are discussed in §19.7.

4.3. Symmetry for Electronic States

Symmetry considerations are also important for describing the electronic states of fullerenes and their related crystalline solids. In this section we consider a simple description of the electronic states of fullerene molecules. The basic concepts presented in this section are extended to the electronic states in crystalline solids in §12.7 and to carbon nanotubules in §19.4.2 and §19.4.3.

To treat the electronic energy levels of a fullerene molecule it is necessary to consider a many-electron system with the point group symmetry appropriate to the fullerene. From a group theoretical standpoint, treatment of the electronic states for the neutral C_{60} molecule or the charged $C_{60}^{\pm n}$ molecular ion requires consideration of both integral and half-integral angular momentum states. To describe the half-integral states, it is necessary to use the double group based on the point group I_h. The character table for the double group I is given in Table 4.8. The double group of I_h is found from that for I by taking the direct product group $I \otimes i$, and the irreducible representations and characters for the double group are obtained by taking appropriate direct products of the characters in Table 4.8 with those of the inversion group consisting of two elements (the identity and the inversion operator) and having two irreducible representations $(1, 1)$ and $(1, -1)$. Basis functions for each of the irreducible representations of the double group of I (Γ_i; $i = 6, 7, 8, 9$) are also listed in Table 4.9. Spin states enter both in the application of the Pauli principle and in considering the effect of the spin–orbit interaction. Although the spin–orbit interaction of carbon is small, its value has been determined in graphite (see Table 3.1) by means of detailed electron spin resonance (ESR) measurements [4.9]. Calculations of the g-factor in graphite [4.10–12] have made use of the spin–orbit splitting of the s^2p^2 configuration. In our discussion of the electronic states (§12.7) and the magnetism of fullerenes (§18.4), we will refer further to the double group of I_h (or I).

Each C_{60} molecule with icosahedral symmetry can be considered to have $60 \times 3 = 180$ σ electrons making bonds along the surface of the icosahedron and 60 π electrons with higher-lying energy levels for a given angular momentum state. On a graphene sheet, the bonding π states have nearest-neighbor orbitals parallel to one another, and the

Table 4.8

Character table[a,b] for the double point group I.

$\mathscr{R}$	E	R	C_5	$\overline{C}_5$	C_5^2	$\overline{C}_5^2$	C_3	$\overline{C}_3$	C_2
# elements	1	1	12	12	12	12	20	20	30
Γ_1 (A)	1	+1	+1	+1	+1	+1	+1	+1	+1
Γ_2 (F_1)	3	+3	$+\tau$	$+\tau$	$1-\tau$	$1-\tau$	0	0	−1
Γ_3 (F_2)	3	+3	$1-\tau$	$1-\tau$	$+\tau$	$+\tau$	0	0	−1
Γ_4 (G)	4	+4	−1	−1	−1	−1	+1	+1	0
Γ_5 (H)	5	+5	0	0	0	0	−1	−1	+1
Γ_6	2	−2	$+\tau$	$-\tau$	$-(1-\tau)$	$1-\tau$	+1	−1	0
Γ_7	2	−2	$1-\tau$	$-(1-\tau)$	$-\tau$	$+\tau$	+1	−1	0
Γ_8	4	−4	+1	−1	−1	+1	−1	+1	0
Γ_9	6	−6	−1	+1	+1	−1	0	0	0

[a]Note: C_5 and C_5^{-1} are in different classes, labeled $12C_5$ and $12C_5^2$ in the character table. The class R represents a rotation by 2π and classes $\overline{C}_i$ represent rotations by $2\pi/i$ between 2π and 4π. In this table $\tau = (1+\sqrt{5})/2$, $-1/\tau = 1-\tau$, $\tau^2 = 1+\tau$.

[b]The basis functions for $\Gamma_i; i = 1, \ldots 5$ are given in Table 4.2 and for $\Gamma_i; i = 6, \ldots 9$ are given in Table 4.9.

Table 4.9

Basis functions for the double group irreducible representations $\mathscr{R}$ of the point group I_h expressed as half-integer spherical harmonics ϕ_{j,m_j}.

$\mathscr{R}$	Basis function
Γ_6	$\phi_{1/2,-1/2}$; $\phi_{1/2,1/2}$
Γ_7	$(\sqrt{(7/10)}\phi_{7/2,-3/2} - \sqrt{(3/10)}\phi_{7/2,7/2})$ $(\sqrt{(7/10)}\phi_{7/2,3/2} + \sqrt{(3/10)}\phi_{7/2,-7/2})$
Γ_8	$\phi_{3/2,-3/2}$; $\phi_{3/2,-1/2}$; $\phi_{3/2,1/2}$; $\phi_{3/2,3/2}$
Γ_9	$\phi_{5/2,-5/2}$ $\phi_{5/2,-3/2}$ $\phi_{5/2,-1/2}$ $\phi_{5/2,1/2}$ $\phi_{5/2,3/2}$ $\phi_{5/2,5/2}$

antibonding states have antiparallel orbitals. More generally, for fullerenes with n_C carbon atoms, the molecular electronic problem involves n_C π electrons.

The electronic levels for the π electrons for a fullerene molecule can be found by starting with a spherical approximation where spherical harmonics can be used to specify the electronic wave functions according to their angular momentum quantum numbers. As stated above, for $\ell > 2$ these spherical harmonics form reducible representations for the icosahedral group symmetry. By lowering the symmetry from full rotational symmetry to icosahedral symmetry (see Table 4.3), the irreducible representations of the icosahedral group are found. In general, the bonding σ levels will lie well below the Fermi level in energy and are not as important for determining the electronic properties as the π electrons near E_F.

To obtain the symmetries for the 60 π electrons for C_{60} we focus our attention on the 60 bonding π electrons whose energies lie close to the Fermi level. Assigning angular momentum quantum numbers to this electron gas, we see from the Pauli principle that 60 π electrons will completely fill angular momentum states up through $\ell = 4$, leaving 10 electrons in the $\ell = 5$ level, which can accommodate a total of 22 electrons. In Table 4.10 we list the number of electrons that can be accommodated in each angular momentum state ℓ, as well as the splitting of the angular momentum states in the icosahedral field.

Table 4.10 thus shows that the $\ell = 4$ level is totally filled by $n_C = 50$. The filled states in icosahedral symmetry for $\ell = 4$ are labeled by the irre-

Table 4.10

Filled shell π-electron configurations for fullerene molecules.[a]

ℓ	electrons/state	n_C	HOMO in I_h symmetry
0	2	2	a_g^2
1	6	8	f_{1u}^6
2	10	18	h_g^{10}
3	14	24	f_{2u}^6
		26	g_u^8
		32	$(f_{2u}^6 g_u^8)$
4	18	40	g_g^8
		42	h_g^{10}
		50	$(g_g^8 h_g^{10})$
5	22	56	f_{1u}^6 or f_{2u}^6
		60	h_u^{10}
		62	$f_{1u}^6 f_{2u}^6$
		66	$(f_{1u}^6 h_u^{10})$ or $(f_{2u}^6 h_u^{10})$
		72	$(f_{1u}^6 f_{2u}^6 h_u^{10})$
6	26	74	a_g^2
		78	f_{1g}^6
		80	g_g^8 or $(a_g^2 f_{1g}^6)$
		82	h_g^{10} or $(a_g^2 g_g^8)$
		84	$(a_g^2 h_g^{10})$
		88	$(f_{1g}^6 g_g^8)$
		90	$(a_g^2 f_{1g}^6 g_g^8)$ or $(f_{1g}^6 h_g^{10})$
		92	$(a_g^2 g_g^8 h_g^{10})$
		96	$(f_{1g}^6 g_g^8 h_g^{10})$
		98	$(a_g^2 f_{1g}^6 g_g^8 h_g^{10})$
⋮	⋮	⋮	⋮

[a]The angular momentum for a spherical shell of π electrons is denoted by ℓ, while n_C denotes the number of π electrons for fullerenes with closed-shell $({}^1A_g)$ ground state configurations in icosahedral symmetry. The last column gives the symmetries of all the levels of the ℓ value corresponding to the highest occupied molecular orbital (HOMO); the superscript on the symmetry label indicates the total spin and orbital degeneracy of the level and all of the listed levels are assumed to be occupied.

ducible representations g_g^8 and h_g^{10} to accommodate a total of 18 electrons. On filling the $\ell = 4$ level, possible ground states occur when either the g_g level is filled with 8 electrons at $n_C = 40$ or the h_g level is filled with 10 electrons at $n_C = 42$, or when the complete shell $\ell = 4$ is filled (i.e., $g_g^8 h_g^{10}$) at $n_C = 50$. Following the same line of reasoning, the 22-fold degenerate $\ell = 5$ level in full rotational symmetry will be filled by C_{72}, which splits into the irreducible representations $H_u + F_{1u} + F_{2u}$ of the icosahedral group with filled shell occupations for these levels of 10, 6, and 6 electrons, respectively. Ten electrons in the $\ell = 5$ angular momentum states of C_{60} are sufficient to completely occupy the h_u level, leaving the f_{1u} and f_{2u} levels completely empty, so that the highest occupied molecular orbital (HOMO) corresponds to the h_u level and the lowest unoccupied molecular orbital (LUMO) corresponds to the f_{1u} level, in agreement with Hückel calculations for the one-electron molecular orbitals [4.13]. It should be noted that for Hückel calculations the next lowest unoccupied molecular orbital is not an f_{2u} level but rather an f_{1g} level, associated with the angular momentum state $\ell = 6$. The reason why an $\ell = 6$ derived level becomes lower than an $\ell = 5$ derived level is due to the form of the atomic potential. In fact, the C_{60} molecule has sufficiently large icosahedral splittings so that for the low-lying excited states, some of the $\ell = 6$ states become occupied before the $\ell = 5$ shell is completely filled. Such level crossings occur even closer to the HOMO level as n_C increases. The resulting electronic levels for the C_{60} molecule and its molecular ions are further discussed in §12.4.

Not only C_{60}, but also other higher-mass fullerenes, have icosahedral symmetry. As discussed in §3.2 and §3.3, all icosahedral (n, m) fullerenes can be specified by C_{n_C} where $n_C = 20(n^2 + nm + m^2)$. Using the same arguments as for C_{60}, the angular momentum states and electronic configurations for the n_C π electrons in these larger fullerenes can be found (see Table 4.11 for $n_C \leq 780$). In this table, the symmetry of each icosahedral fullerene is given. Fullerenes with (n, m) values such that $n = 0$, $m = 0$ or $m = n$ have I_h symmetry (including the inversion operation), while other entries have I symmetry, lacking inversion. Also listed in this table is ℓ_{max}, the maximum angular momentum state that is occupied, from which n_{tot}, the maximum number of electrons needed to fill a spherical shell can be calculated according to

$$n_{tot} = 2(\ell_{max} + 1)^2. \tag{4.1}$$

The number of valence electrons in the unfilled shell n_v is included in Table 4.11, which also lists the full electronic configuration of the π electrons, the J_{Hund} value for the ground state configuration according to Hund's rule and the icosahedral symmetries for the valence electron states. Then by decomposing these angular momentum states into irreducible representations

Table 4.11

Symmetries and configurations of the π electrons for icosahedral fullerenes

C_{n_C}	ℓ_{max} [a]	n_{tot} [b]	n_v [c]	Config. [d]	J_{Hund}	I_h (or I) symmetries [e]
C_{20}	3	32	2	$\ldots f^2$	4	G_g, H_g
C_{60}	5	72	10	$\ldots g^{18}h^{10}$	0	A_g
C_{80}	6	98	8	$\ldots h^{22}i^8$	16	$A_g, 2F_{1g}, F_{2g}, 2G_g, 3H_g$
C_{140}	8	162	12	$\ldots k^{30}l^{12}$	24	$A, 2F_1, 2F_2, 4G, 4H$
C_{180}	9	200	18	$\ldots l^{34}m^{18}$	0	A_g
C_{240}	10	242	40	$\ldots m^{38}n^{40}$	20	$A_g, 2F_{1g}, 2F_{2g}, 2G_g, 4H_g$
C_{260}	11	288	18	$\ldots n^{42}o^{18}$	36	$2A, 4F_1, 3F_2, 5G, 6H$
C_{320}	12	338	32	$\ldots o^{46}q^{32}$	72	...
C_{380}	13	392	42	$\ldots q^{50}r^{42}$	96	...
C_{420}	14	450	28	$\ldots r^{54}t^{28}$	0	A
C_{500}	15	512	50	$\ldots t^{58}u^{50}$	120	...
C_{540}	16	578	28	$\ldots u^{62}v^{28}$	56	...
C_{560}	16	578	48	$\ldots u^{62}v^{48}$	144	...
C_{620}	17	648	42	$\ldots v^{66}w^{42}$	112	...
C_{720}	18	722	72	$\ldots w^{70}x^{72}$	36	$2A_g, 4F_{1g}, 3F_{2g}, 5G_g, 6H_g$
C_{740}	19	800	18	$\ldots x^{74}y^{18}$	180	...
C_{780}	19	800	58	$\ldots x^{74}y^{58}$	200	...
C_{860}	20	882	60	$\ldots y^{78}z^{60}$	220	...
C_{960}	21	968	78	$\ldots z^{82}a^{78}$	144	...
C_{980}	22	1058	12	$\ldots a^{86}b^{12}$	192	...
⋮	⋮	⋮	⋮	⋮	⋮	⋮

[a] ℓ_{max} represents the maximum value of the angular momentum for the HOMO level.
[b] $n_{tot} = 2(\ell_{max} + 1)^2$ is the number of electrons needed for a filled shell.
[c] n_v represents the number of electrons in the HOMO level.
[d] The notation for labeling the angular momentum states is consistent with Table 4.3.
[e] The splittings of the Hund's rule ground state J_{Hund} in icosahedral symmetry.

of the icosahedral group (using Table 4.3 and the notation for the electronic configurations in this table), the symmetry designation for the ground state energy levels (according to Hund's rule) in icosahedral symmetry can be found (see Table 4.11) [4.4]. For example, the icosahedral C_{80} molecule has a sufficient number of π electrons to fill the $\ell = 5$ level, with eight electrons available for filling states in the $\ell = 6$ level. Hückel calculations for this molecule suggest that the f_{1u} and f_{1g} levels are completely filled, and the h_u level is partially filled with eight electrons [4.14].

There are, in general, many Pauli-allowed states that one can obtain from the spherical molecule configurations listed in Table 4.11. For example, the C_{20} molecule with the $s^2p^6d^{10}f^2$ or simply f^2 configuration has Pauli-allowed states with $S = 0$, $L = 0, 2, 4, 6$ and with $S = 1$, $L = 1, 3, 5$.

The Hund's rule ground state is the $J_{\text{Hund}} = 4$ state that comes from $S = 1$, $L = 5$. The symmetries of these Hund's rule ground states are listed in the column labeled J_{Hund} in Table 4.11 together with the decomposition of the states of the J_{Hund} reducible representation into the appropriate irreducible representations of the icosahedral group. If the perturbation to the spherical symmetry by the icosahedral potentials is small and Hund's rule applies, then the ground state will be as listed. If, however, the icosahedral perturbation is large compared with the electron correlation and exchange energies, then the icosahedral splitting must be considered first before the electrons are assigned to the spherical symmetry angular momentum states.

Since the ground and excited states of the fullerenes can be well understood only when actual numerical calculations are carried out, further discussion of the group theoretical aspects of the electronic structure is given in Chapter 12, both for the free fullerene molecules and the crystalline phases.

4.4. Going from Higher to Lower Symmetry

A lowering of the symmetry from full icosahedral symmetry occurs in a variety of fullerene-derived structure–property relations. One example of symmetry lowering results from elongation of the icosahedral shape of the fullerene molecules to a rugby-ball shape, as discussed below. Another example involves introduction of chirality into the fullerene molecule. A third example is found in many chemical or photochemical reactions which add side groups at various sites and with various symmetries (see Chapter 10). In these cases the symmetry-lowering effect is specific to the side groups that are added. A fourth example is the introduction of fullerenes into a crystal lattice. Since no Bravais lattice with fivefold symmetry is possible, symmetry lowering must occur in this case. Further discussion of this issue is presented in §7.1.2. As a fifth example, carbon nanotubes can be considered to be related to fullerenes through a symmetry-lowering process. A discussion of the group theory associated with carbon tubules, in general, and with chiral carbon tubules, in particular, is presented in §19.4.2 and §19.4.3.

4.4.1. *Symmetry Considerations for C_{70}*

The most common fullerene which has lower than icosahedral symmetry is C_{70}, and the structure and properties of this fullerene molecule have also been studied in some detail. As explained in §3.2, C_{70} can be constructed from C_{60} by appropriately bisecting the C_{60} molecule normal to a fivefold axis, rotating one hemisphere relative to the other by 36° (thereby losing inversion symmetry), then adding a ring of five hexagons around the equator (or belt), and finally reassembling these three constituents. The elongation of the icosahedral C_{60} in this way to yield C_{70} [see Fig. 3.4(b)] results in a lowering of the symmetry of the molecule from I_h to D_{5h}, but the point group D_{5h} does have a mirror plane normal to the fivefold axis. If a second ring of five hexagons is added around the equator [see Fig. 3.4(b)], we then obtain a C_{80} molecule with D_{5d} symmetry, which is symmetric under inversion but has no σ_h mirror plane. The character tables for the point groups D_{5h} and D_{5d} are given in Tables 4.12 and Tables 4.13, respectively, and the corresponding basis functions are given in Table 4.14. The corresponding compatibility relations for the irreducible representations of the point group I_h in going to lower symmetry groups (I, T_h, D_{5d}, D_5, and C_{1h}) are provided in Table 4.15. Since the group D_{5d} has inversion symmetry, the irreducible representations of I_h form reducible representations of D_{5d}, so that the compatibility relations between the two groups are easily written. For the group D_{5h}, which has a mirror plane but no inversion symmetry, one must use the lower symmetry icosahedral group I for relating the icosahedral irreducible representations to those in D_5, which is a subgroup of I. The compatibility relations for $I \rightarrow D_5$ are also included in Table 4.15, in addition to compatibility tables for groups $I_h \rightarrow T_h$ and $I \rightarrow C_{1h}$.

In treating the electronic levels and vibrational modes for the C_{70} molecule, we can either go from full rotational symmetry (see Table 4.3) to D_{5h} symmetry in analogy to §4.3, or we can first go from full rotational symmetry to I symmetry, and then treat D_5 as a subgroup of I, in going from I to D_5 in the sense of perturbation theory. Referring to Table 4.16, which shows the decomposition of the various angular momentum states ℓ into irreducible representations of point group I and then to group D_5 (or

Table 4.12

Character table for point group D_{5h}.

$\mathcal{R}$	E	$2C_5$	$2C_5^2$	$5C_2'$	σ_h	$2S_5$	$2S_5^3$	$5\sigma_v$
A_1'	+1	+1	+1	+1	+1	+1	+1	+1
A_2'	+1	+1	+1	−1	+1	+1	+1	−1
E_1'	+2	$\tau-1$	$-\tau$	0	+2	$\tau-1$	$-\tau$	0
E_2'	+2	$-\tau$	$\tau-1$	0	+2	$-\tau$	$\tau-1$	0
A_1''	+1	+1	+1	+1	−1	−1	−1	−1
A_2''	+1	+1	+1	−1	−1	−1	−1	+1
E_1''	+2	$\tau-1$	$-\tau$	0	−2	$1-\tau$	τ	0
E_2''	+2	$-\tau$	$\tau-1$	0	−2	τ	$1-\tau$	0

Table 4.13

Character table for point group D_{5d}.

$\mathcal{R}$	E	$2C_5$	$2C_5^2$	$5C_2'$	i	$2S_{10}^{-1}$ [a]	$2S_{10}$	$5\sigma_d$
A_{1g}	+1	+1	+1	+1	+1	+1	+1	+1
A_{2g}	+1	+1	+1	−1	+1	+1	+1	−1
E_{1g}	+2	$\tau-1$	$-\tau$	0	+2	$\tau-1$	$-\tau$	0
E_{2g}	+2	$-\tau$	$\tau-1$	0	+2	$-\tau$	$\tau-1$	0
A_{1u}	+1	+1	+1	+1	−1	−1	−1	−1
A_{2u}	+1	+1	+1	−1	−1	−1	−1	+1
E_{1u}	+2	$\tau-1$	$-\tau$	0	−2	$1-\tau$	$+\tau$	0
E_{2u}	+2	$-\tau$	$\tau-1$	0	−2	$+\tau$	$1-\tau$	0

[a]Note: $iC_5 = S_{10}^{-1}$ and $iC_5^2 = S_{10}$. Also $iC_2' = \sigma_d$

Table 4.14

Basis functions for the irreducible representations of groups D_{5h} and D_{5d}.

D_{5h}	D_{5d}	Basis functions
A_1'	A_{1g}	x^2+y^2, z^2
A_2'	A_{2g}	R_z
E_1'	E_{1u}	(x, y), (xz^2, yz^2), $[x(x^2+y^2), y(x^2+y^2)]$
E_2'	E_{2g}	(x^2-y^2, xy), $[y(3x^2-y^2), x(x^2-3y^2)]$
A_1''	A_{1u}	–
A_2''	A_{2u}	z, $z^3, z(x^2+y^2)$
E_1''	E_{1g}	(R_x, R_y), (xz, yz)
E_2''	E_{2u}	$[xyz, z(x^2-y^2)]$

directly from $\ell = 5$ to point group D_5), we obtain

$$\Gamma_{\ell=5} \rightarrow H + F_1 + F_2 \rightarrow A_1 + 2A_2 + 2E_1 + 2E_2 \tag{4.2}$$

where the irreducible representations of group I go into irreducible representations of group D_5

$$\begin{aligned} H &\rightarrow A_1 + E_1 + E_2 \\ F_1 &\rightarrow A_2 + E_1 \\ F_2 &\rightarrow A_2 + E_2 \end{aligned} \tag{4.3}$$

using the results of Table 4.15. Based on these symmetry considerations, the filling of the electronic levels for the rugby ball–shaped C_{70} molecule is discussed further in §12.1.2.

Symmetry considerations also play a major role in classifying the normal modes of the C_{70} molecule. To find the symmetries of the normal mode

Table 4.15

Compatibility relations between the icosahedral groups, I_h, I, and several point groups of lower symmetry.[a]

I_h	I	T_h	D_{5d}	D_5	C_{1h}
A_g	A	A_g	A_{1g}	A_1	A_1
F_{1g}	F_1	T_g	$A_{2g}+E_{1g}$	A_2+E_1	A_1+2A_2
F_{2g}	F_2	T_g	$A_{2g}+E_{2g}$	A_2+E_2	A_1+2A_2
G_g	G	A_g+T_g	$E_{1g}+E_{2g}$	E_1+E_2	$2A_1+2A_2$
H_g	H	E_g+T_g	$A_{1g}+E_{1g}+E_{2g}$	$A_1+E_1+E_2$	$3A_1+2A_2$
A_u	A	A_u	A_u	A_1	A_2
F_{1u}	F_1	T_u	$A_{2u}+E_{1u}$	A_2+E_1	$2A_1+A_2$
F_{2u}	F_2	T_u	$A_{2u}+E_{2u}$	A_2+E_2	$2A_1+A_2$
G_u	G	A_u+T_u	$E_{1u}+E_{2u}$	E_1+E_2	$2A_1+2A_2$
H_u	H	E_u+T_u	$A_{1u}+E_{1u}+E_{2u}$	$A_1+E_1+E_2$	$3A_1+2A_2$

[a]Symmetry-lowering can occur between $I_h \rightarrow T_h$; $I_h \rightarrow I$; $T_h \rightarrow D_{5d}$; $I \rightarrow D_5$; $T_h \rightarrow C_{1h}$; $D_{5d} \rightarrow C_{1h}$.

vibrations of C_{70}, we first find the symmetries for the transformation of the 70 carbon atoms denoted by $\chi^{\text{a.s.}}(C_{70})$ for the point group D_{5h} where "a.s." refers to atom sites (as in Table 4.5 for icosahedral symmetry). Because of the large number of degrees of freedom in fullerenes, it is advantageous to break up the 70 atoms in C_{70} into subunits which themselves transform as a subgroup of D_{5h}. This approach allows us to build up large fullerene molecules by summing over these building blocks. The equivalence transformation ($\chi^{\text{a.s.}}$) for each of the building blocks can be written down by inspection.

The characters for the equivalence transformation $\chi^{\text{a.s.}}$ for these subgroup building blocks, which are expressed in terms of sets of atoms normal to the fivefold axis, are listed in Table 4.17. The symmetry operations of the group transform the atoms within each of these subgroups into one another. The $\chi^{\text{a.s.}}$ entries in Table 4.17 under the various symmetry operations denote the number of carbon atoms that remain invariant under the various classes of symmetry operations. The irreducible representations of atomic sites $\chi^{\text{a.s.}}$ are given in Table 4.18. The set $C_{10}(\text{cap}^0)$ denotes the five carbon atoms around the two pentagons (10 atoms in total) through which the fivefold axis passes. Another 10 carbon atoms, that are nearest-neighbors to the 10 atoms on the axial pentagons, transform in the same way as the set $C_{10}(\text{cap}^0)$. The set $C_{10}(\text{belt})$ refers to the 10 equatorial atoms in the five hexagons on the equator that form a subgroup. There are also two sets of 20 carbon atoms on hexagon double bonds, labeled $C_{20}(\text{off-belt})$, that form another subgroup. The characters for the equivalence transformation

Table 4.16

Decomposition of spherical angular momentum states labeled by ℓ (for $\ell \leq 10$) into irreducible representations of lower symmetry groups.[a]

ℓ	T_h	D_{5d}	D_{5h}	C_{1h}
0	A_g	A_{1g}	A'_1	A_1
1	T_u	A_{2u}, E_{1u}	A''_2, E''_1	$2A_1$, A_2
2	E_g, T_g	A_{1g}, E_{1g}, E_{2g}	A'_1, E'_1, E'_2	$3A_1$, $2A_2$
3	A_u, $2T_u$	A_{2u}, E_{1u}, $2E_{2u}$	A''_2, E''_1, $2E''_2$	$4A_1$, $3A_2$
4	A_g, E_g, $2T_g$	A_{1g}, $2E_{1g}$, $2E_{2g}$	A'_1, $2E'_1$, $2E'_2$	$5A_1$, $4A_2$
5	E_u, $3T_u$	A_{1u}, $2A_{2u}$, $2E_{1u}$, $2E_{2u}$	A''_1, $2A''_2$, $2E''_1$, $2E''_2$	$6A_1$, $5A_2$
6	$2A_g$, E_g, $3T_g$	$2A_{1g}$, A_{2g}, $3E_{1g}$, $2E_{2g}$	$2A'_1$, A'_2, $3E'_1$, $2E'_2$	$7A_1$, $6A_2$
7	A_u, E_u, $4T_u$	A_{1u}, $2A_{2u}$, $3E_{1u}$, $3E_{2u}$	A''_1, $2A''_2$, $3E''_1$, $3E''_2$	$8A_1$, $7A_2$
8	A_g, $2E_g$, $4T_g$	$2A_{1g}$, A_{2g}, $3E_{1g}$, $4E_{2g}$	$2A'_1$, A'_2, $3E'_1$, $4E'_2$	$9A_1$, $8A_2$
9	$2A_u$, E_u, $5T_u$	A_{1u}, $2A_{2u}$, $4E_{1u}$, $4E_{2u}$	A''_1, $2A''_2$, $4E''_1$, $4E''_2$	$10A_1$, $9A_2$
10	$2A_g$, $2E_g$, $5T_g$	$3A_{1g}$, $2A_{2g}$, $4E_{1g}$, $4E_{2g}$	$3A'_1$, $2A'_2$, $4E'_1$, $4E'_2$	$11A_1$, $10A_2$

[a]Note that $D_{5d} = D_5 \otimes i$ and $D_{5h} = D_5 \otimes \sigma_h$.

Table 4.17

Characters of atomic sites $\chi^{\text{a.s.}}$ for D_{5h} of relevance to the C_{70} molecule[a,b] [4.15].

$\chi^{\text{a.s.}}(D_{5h})$	E	$2C_5$	$2C_5^2$	$5C_2'$	σ_h	$2S_5$	$2S_5^3$	$5\sigma_v$
$C_{10}(\text{cap}^0)$	10	0	0	0	0	0	0	2
C_{20}(off-belt)	20	0	0	0	0	0	0	0
C_{10}(belt)	10	0	0	0	10	0	0	0

[a]See text for a discussion of $C_{10}(\text{cap}^0)$, C_{20}(off-belt), and C_{10}(belt). The same building blocks listed in this table are found in C_{50}, C_{70}, C_{90}, etc.

[b]The irreducible representations for each $\chi^{\text{a.s.}}$ in this table are given in Table 4.18.

Table 4.18

Irreducible representations of atomic sites $\chi^{\text{a.s.}}$ for D_{5h} of relevance to the C_{70} molecule[a] [4.15].

$\chi^{\text{a.s.}}(D_{5h})$	Irreducible representations
$C_{10}(\text{cap}^0)$	$A_1' + A_2'' + E_1' + E_1'' + E_2' + E_2''$
C_{20}(off-belt)	$A_1' + A_2' + A_1'' + A_2'' + 2E_1' + 2E_1'' + 2E_2' + 2E_2''$
C_{10}(belt)	$A_1' + A_2' + 2E_1' + 2E_2'$

[a]The characters for the equivalence transformation for these sets of carbon atoms are given in Table 4.17.

for C_{70} are found by summing the contributions from the various layers appropriately:

$$\chi^{\text{a.s.}}(C_{70}) = 2\chi^{\text{a.s.}}[C_{10}(\text{cap}^0)] + 2\chi^{\text{a.s.}}[C_{20}(\text{off–belt})] + \chi^{\text{a.s.}}[C_{10}(\text{belt})]. \quad (4.4)$$

From Table 4.18 and Eq. (4.4), we then obtain the irreducible representations of D_{5h} contained in the equivalence transformation for C_{70} as a whole:

$$\chi^{\text{a.s.}}(C_{70}) = 5A_1' + 3A_2' + 2A_1'' + 4A_4'' + 8E_1' + 8E_2' + 6E_1'' + 6E_2''. \quad (4.5)$$

If instead of C_{70} we were to consider an isomer of C_{90} with D_{5h} symmetry, the same procedure as in Eqs. (4.4) and (4.5) would be used, except that an additional $\chi^{\text{a.s.}}$(off-belt) would be added to Eq. (4.4). The same building block approach could be used to describe C_{80} or C_{100} isomers with D_{5d} symmetry using Tables 4.19 and 4.20.

The symmetries of the molecular vibrations $\chi^{\text{m.v.}}$ (see Table 4.21) are then found using the relation

$$\chi^{\text{m.v.}}(C_{70}) = \chi^{\text{a.s.}}(C_{70}) \otimes \chi^{\text{vector}} - \chi^{\text{translations}} - \chi^{\text{rotations}} \quad (4.6)$$

Table 4.19

Characters of atomic sites $\chi^{\text{a.s.}}$ for D_{5d} of relevance to the D_{5d} isomer of the C_{80} molecule[a] [4.15].

$\chi^{\text{a.s.}}(D_{5d})$	E	$2C_5$	$2C_5^2$	$5C_2'$	i	$2S_{10}^{-1}$	$2S_{10}$	$5\sigma_d$
C_{10} (cap^0)	10	0	0	0	0	0	0	2
C_{20} (cap^1)	20	0	0	0	0	0	0	0
C_{20} (off–belt)	20	0	0	0	0	0	0	0
C_{20} (belt)	20	0	0	0	0	0	0	0

[a] For an explanation about the notation for the atom sites, see text. The building blocks in this table contribute to C_{80}, C_{100}, C_{120}, etc. with D_{5d} symmetry. Irreducible representations contained in $\chi^{\text{a.s.}}$ are given in Table 4.20.

Table 4.20

Irreducible representations of atomic sites $\chi^{\text{a.s.}}$ for D_{5d} of relevance to the D_{5d} isomer of the C_{80} molecule[a] [4.15].

$\chi^{\text{a.s.}}(D_{5d})$	Irreducible representations
C_{10} (cap^0)	$A_{1g} + A_{2u} + E_{1g} + E_{1u} + E_{2g} + E_{2u}$
C_{20} (cap^1), C_{20} (off–belt), C_{20} (belt)	$A_{1g} + A_{1u} + A_{2g} + A_{2u} + 2E_{1g} + 2E_{1u} + 2E_{2g} + 2E_{2u}$

[a] The characters for $\chi^{\text{a.s.}}$ in this table are given in Table 4.19.

in which the direct product is denoted by $\otimes$ and the irreducible representations for the χ^{vector}, $\chi^{\text{translation}}$, and χ^{rotation} for group D_{5h} are given by

$$\begin{aligned} \chi^{\text{vector}} &= A_2'' + E_1', \\ \chi^{\text{translation}} &= A_2'' + E_1', \\ \chi^{\text{rotation}} &= A_2' + E_1''. \end{aligned} \tag{4.7}$$

Table 4.22 lists the number of axial and radial molecular vibrations associated with each of the layers of carbon atoms of C_{70}. This division into axial and radial molecular modes is only approximate but often gives a good description of the physics of the molecular vibrations. The terms axial and transverse refer to modes associated with motions along and perpendicular to the fivefold axis, respectively. Also included in Table 4.22 are the total number of axial and radial modes for C_{70+20j}, which contains summaries of the mode symmetries for C_{70}, C_{90}, etc.

From Table 4.22 and Eq. (4.5), we see that the number of distinct mode frequencies for C_{70} is 122. It is sometimes useful to consider these modes as being approximately divided into cap modes and belt modes. Thus the 122

Table 4.21

The D_{5h} irreducible representations ($\mathcal{R}$) together with the number of distinct eigenvalues (N_ω) and the corresponding degeneracies g of the normal modes of the C_{70} molecule. The symbols N_ω^{belt} and N_ω^{cap} denote the number of distinct eigenvalues associated with the "belt" and "cap" modes, respectively, for each irreducible representation.

$\mathcal{R}$	N_ω^{belt}	N_ω^{cap}	N_ω	g
A_1'	2	10	12	1
A_2'	2	7	9	1
E_1'	4	17	21	2
E_2'	4	18	22	2
A_1''	1	8	9	1
A_2''	1	9	10	1
E_1''	2	17	19	2
E_2''	2	18	20	2

Table 4.22

Symmetries of molecular vibrational modes[a,b,c] for groups of carbon atoms with D_{5h} symmetry [4.15].

D_{5h}	A_1'	A_2'	E_1'	E_2'	A_1''	A_2''	E_1''	E_2''
C_{10}^{axial}(cap^0)	1	0	1	1	0	1	1	1
C_{10}^{radial}(cap^0)	2	2	2	1	2	2	2	1
C_{20}^{axial}(off-belt)	2	2	2	1	2	2	2	1
C_{20}^{radial}(off-belt)	2	2	2	1	2	2	2	1
C_{10}^{axial}(belt)	1	0	1	1	0	1	1	1
C_{10}^{radial}(belt)	2	2	2	1	2	2	2	1
$C_{70+20j}^{\text{axial}}$	4_c	1_c	6_d	6_d	3_c	4_c	8_d	8_d
$C_{70+20j}^{\text{radial}}$	8_d	8_d	15_e	16_e	6_d	6_d	11_e	12_e

[a]One A_2'' mode corresponding to translations of the center of mass of the free molecule along the fivefold axis and one A_2' mode corresponding to rotations of the free molecule about the fivefold axis have been subtracted.

[b]One E_1' mode corresponding to translations of the center of mass of the free molecule normal to the fivefold axis and one E_1'' mode corresponding to rotations of the free molecule about axes normal to the fivefold axis have been subtracted.

[c]The number of times an irreducible representation Γ_i occurs for $C_{70+20j}^{\text{axial}}$ and $C_{70+20j}^{\text{radial}}$ is given by symbols n_c, n_d, and n_e which, respectively, denote $n+j$, $n+2j$, and $n+4j$, where n is an integer. For example, 4_c, 6_d, and 15_e, respectively, denote $4+j$, $6+2j$ and $15+4j$.

modes for C_{70} are classified as 104 cap modes (corresponding to the 60 carbon atoms of the two hemispheres of C_{60}) and 18 belt modes. This division into cap and belt modes becomes more important in the limit of carbon nanotubes which are discussed in §19.7. The symmetries and degeneracies of the distinct mode frequencies for C_{70} are given in Table 4.22.

Among the modes given in Table 4.22, those that transform according to the A_1', E_2' or E_1'' irreducible representations are Raman-active, with the A_1' modes being observed only in the $(\|, \|)$ polarization geometry and the E_1'' mode observed in the $(\|, \perp)$ polarization. The E_2' symmetry mode is seen in both polarization geometries. The modes with A_2'' and E_1' symmetries are infrared-active.

4.4.2. Symmetry Considerations for Higher-Mass Fullerenes

Similar arguments can be made to classify the symmetries of the molecular vibrations of the rugby ball–shaped C_{80} which follows symmetry group D_{5d}. Table 4.19 lists the characters for the equivalence transformations for groups of carbon atoms comprising the C_{80} isomer with D_{5d} symmetry. Each of these equivalence transformations forms a reducible representation of D_{5d} and the decomposition of $\chi^{\text{a.s.}}$ into irreducible representations of D_{5d} is given in Table 4.20. The vibrations associated with groups D_{5h} and D_{5d} are found using a variant of Eq. (4.6), and the classification of the vibrational modes into irreducible representations of D_{5h} and D_{5d} can be obtained from Tables 4.22 and 4.23. Finally in Table 4.24, we give the number of distinct eigenfrequencies for the C_{80} isomer with D_{5d} symmetry, listed according to their symmetry type, and again distinguishing between the cap and belt modes. It should be noted that the building block approach using point group D_{5h} can be used to obtain $\chi^{\text{a.s.}}$ and $\chi^{\text{m.v.}}$ for C_{50}, C_{70}, C_{90}, etc., and using group D_{5d} the corresponding information can simply be found for D_{5d} isomers of C_{80}, C_{100}, etc. The building block approach provides a simple method for constructing the dynamical matrix of large fullerene molecules or for treating their electronic structure when using explicit potentials. Further discussion of the group theoretical aspects of the vibrational modes of the higher-mass fullerenes is given in §11.7. In §19.7, the symmetry of the molecular vibrations of carbon nanotubes is presented.

4.5. Symmetry Considerations for Isotopic Effects

Carbon has two stable isotopes: ^{12}C, which is 98.892% abundant and has a molecular weight of 12.011 and a zero nuclear spin, and ^{13}C with atomic weight 13.003, a natural abundance of 1.108%, and a nuclear spin of 1/2.

Table 4.23

Symmetries of molecular vibrational modes[a,b,c] for groups of carbon atoms with D_{5d} symmetry [4.15].

D_{5d}	A_{1g}	A_{2g}	E_{1g}	E_{2g}	A_{1u}	A_{2u}	E_{1u}	E_{2u}
C_{10}^{axial}(cap^0)	1	0	1	1	0	1	1	1
C_{10}^{radial}(cap^0)	2	2	2	1	2	2	2	1
C_{20}^{axial}(off-belt)	2	2	2	1	2	2	2	1
C_{20}^{radial}(off-belt)	2	2	2	1	2	2	2	1
C_{20}^{axial}(belt)	1	0	1	1	0	1	1	1
C_{20}^{radial}(belt)	2	2	2	1	2	2	2	1
C_{80+20j}^{axial}	5_c	2_c	8_d	8_d	3_c	4_c	8_d	8_d
C_{80+20j}^{radial}	8_d	8_d	15_e	16_e	8_d	8_d	15_e	16_e

[a]One A_{2u} mode corresponding to translations of the center of mass of the free molecule along the fivefold axis and one A_{2g} mode corresponding to rotations of the free molecule about the fivefold axis have been subtracted.

[b]One E_{1u} mode corresponding to translations of the center of mass of the free molecule normal to the fivefold axis and one E_{1g} mode corresponding to rotations of the free molecule about axes normal to the fivefold axis have been subtracted.

[c]The number of times an irreducible representation Γ_i occurs for C_{80+20j}^{axial} and C_{80+20j}^{radial} is given by symbols n_c, n_d, and n_e which, respectively, denote $n+j$, $n+2j$, and $n+4j$, where n is an integer. For example, 5_c, 8_d, and 15_e, respectively, denote $5+j$, $8+2j$, and $15+4j$.

Table 4.24

The D_{5d} irreducible representations ($\mathcal{R}$) together with the number of distinct eigenvalues (N_ω) and the corresponding degeneracies g of the normal modes of the C_{80} molecule. The symbols N_ω^{belt} and N_ω^{cap} denote the number of distinct eigenvalues associated with the "belt" and "cap" modes, respectively, for each irreducible representation.

$\mathcal{R}$	N_ω^{belt}	N_ω^{cap}	N_ω	g
A_{1g}	3	10	13	1
A_{2g}	2	8	10	1
E_{1g}	3	20	23	2
E_{2g}	2	22	24	2
A_{1u}	2	9	11	1
A_{2u}	3	9	12	1
E_{1u}	3	20	23	2
E_{2u}	2	22	24	2

It is the nuclear spin of the ^{13}C isotope that is exploited in the NMR experiments on fullerenes.

Although small in abundance, the ^{13}C isotope occurs on approximately half of the C_{60} molecules synthesized using the natural abundance of carbon isotopes as shown in Table 4.25. The probability p_m for m isotopic substitutions to occur on an n_C atom fullerene C_{n_C} is given by

$$p_m(C_{n_C}) = \binom{n_C}{m} x^m (1-x)^{n_C - m}, \tag{4.8}$$

where x is the fractional abundance of the isotope and the binomial coefficient appearing in Eq. (4.8) is given by

$$\binom{n_C}{m} = \frac{n_C!}{(n_C - m)!m!}. \tag{4.9}$$

For the natural abundance of carbon isotopes ($x = 0.01108$), we obtain the results for the C_{60} and C_{70} molecules listed in Table 4.25 and shown graphically in Fig. 4.2. Less than 1% of the C_{60} and C_{70} fullerenes have more than three ^{13}C isotopes per fullerene. Also included in the table are the corresponding results for an isotopic enrichment to 5% and 10% ^{13}C. The results in Table 4.25 show that as x increases (and also as n_C increases in C_{n_C}), the peak in the distribution moves to larger m values and the distribution gets broader. These conclusions follow from binomial statistics, where the average $\bar{m}$ and the standard deviation $\Delta m \equiv \langle (m - \bar{m})^2 \rangle^{1/2}$ give

$$\bar{m} = n_C x \tag{4.10}$$

and

$$\Delta m = \sqrt{n_C x(1-x)}, \tag{4.11}$$

respectively. For example, for a fullerene with $x = 0.05$, Eqs. (4.10) and (4.11) yield $\bar{m} = 3$ and $\Delta m = 1.7$. Thus as x increases, so do $\bar{m}$ and Δm, thereby accounting for the broader distribution with increasing x. In general, for isotopically enriched samples, the distribution $p_m(C_{n_C})$ is sufficiently shifted and broadened so that graphical displays are desirable, such as shown in Fig. 4.2.

The results of Table 4.25 have important consequences for both symmetry considerations and the rotational levels of fullerene molecules [4.16], and these topics are further discussed in §11.3.3. The molecules containing one or more ^{13}C atoms show much lower symmetry than that of the full I_h point group. In fact, the singly ^{13}C substituted molecule $^{13}C_1{}^{12}C_{59}$ has only one symmetry operation, a single reflection plane (point group C_{1h}, see Table 4.26); two or more substitutions generally show no symmetry,

Table 4.25

The probabilities of $p_m(C_{n_C})$ for ^{13}C among C_{60} and C_{70} fullerenes.[a]

x	m	$p_m(C_{60})$	$p_m(C_{70})$
0.0001	0	0.9940	0.9930
0.0005	0	0.9704	0.9656
0.001	0	0.9417	0.9324
0.005	0	0.7403	0.7041
0.01	0	0.5472	0.4948
0.01108	0	0.5125	0.4584
0.01108	1	0.3445	0.3595
0.01108	2	0.1139	0.1390
0.01108	3	0.0247	0.0353
0.01108	4	0.0039	0.0066
0.05	0	0.0461	0.0276
0.05	1	0.1455	0.1016
0.05	2	0.2259	0.1845
0.05	3	0.2298	0.2201
0.05	4	0.1724	0.1941
0.05	5	0.1016	0.1348
0.05	6	0.0490	0.0769
0.10	0	0.0018	0.00063
0.10	1	0.0120	0.00487
0.10	2	0.0393	0.01868
0.10	3	0.0844	0.04705
0.10	4	0.1336	0.08756
0.10	5	0.1662	0.12843
0.10	6	0.1693	0.15459
0.10	7	0.1451	0.15704

[a]Here x is the isotopic abundance, m is the number of ^{13}C per fullerene, and $p_m(C_{n_C})$ is the probability a fullerene C_{n_C} has m ^{13}C atoms. Although the table would normally be used for small concentrations of ^{13}C in C_{n_C}, the same probabilities as given in the table apply to 1.108% ^{12}C in 98.892% ^{13}C, or to 5% ^{12}C in 95% ^{13}C, or to 10% ^{12}C in 90% ^{13}C.

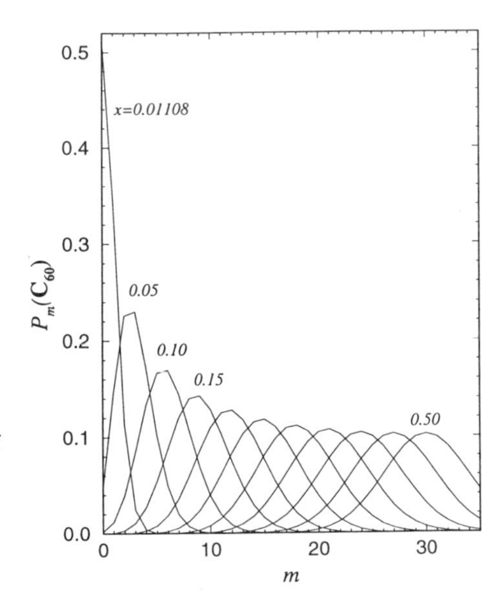

Fig. 4.2. Plot of the m dependence of $p_m(C_{60})$ the distribution of C_{60} molecules with m ^{13}C atoms for various concentrations of x from 5% to 50% in steps of 5%. A plot of $p_m(C_{60})$ for x equal to the natural abundance is also shown.

i.e., they belong to point group C_1 and have no symmetry-imposed degeneracies, which implies that all levels (electronic, vibrational, rotational, etc.) are nondegenerate and every mode is both IR and Raman-active. Group theory predicts this symmetry lowering, but the intensities of IR and Raman lines do not change much upon addition of ^{13}C isotopes. Optically-inactive modes before the isotopic symmetry-lowering effect remain mostly inactive, and the optically active modes still show strong intensity. The consequences of this symmetry lowering for the IR and Raman study of the normal mode vibrations are further discussed in §11.5.5.

The isotopic distribution has unique consequences with regard to the rotational levels of the C_{60} molecule and hence also with regard to the librational states of the corresponding solid. This is discussed further in §11.3.3. Other experiments sensitive to the isotopic abundance include NMR measurements (see §16.1) and studies of the isotope effect regarding the superconducting transition temperature T_c (see §15.5).

Table 4.26

Character table for point group C_{1h}.

$\mathcal{R}$	E	σ_h	Basis functions
A_1	1	1	1; x; y; x^2; etc.
A_2	1	−1	z; xz; yz; etc.

REFERENCES

[4.1] L. Saffaro. In C. Taliani, G. Ruani, and R. Zamboni (eds.), *Proc. of the First Italian Workshop on Fullerenes: Status and Perspectives*, vol. 2, p. 55, World Scientific, Singapore (1992).

[4.2] F. Chung and S. Sternberg. *American Scientist*, **81**, 56 (1993).

[4.3] L. Tisza. *Zeitschrift für Physik*, **82**, 48 (1933).

[4.4] R. Saito, G. Dresselhaus, and M. S. Dresselhaus. *Phys. Rev. B*, **46**, 9906 (1992).

[4.5] B. R. Judd. *Proc. Roy. Soc., London*, **A241**, 122 (1957).

[4.6] J. Raynal. *J. Math. Phys.*, **25**, 1187 (1984).

[4.7] Z. H. Dong, P. Zhou, J. M. Holden, P. C. Eklund, M. S. Dresselhaus, and G. Dresselhaus. *Phys. Rev. B*, **48**, 2862 (1993).

[4.8] K. A. Wang, A. M. Rao, P. C. Eklund, M. S. Dresselhaus, and G. Dresselhaus. *Phys. Rev. B*, **48**, 11375 (1993).

[4.9] G. Wagoner. *Phys. Rev.*, **118**, 647 (1960).

[4.10] G. Dresselhaus and M. S. Dresselhaus. *Phys. Rev.*, **140A**, 401 (1965).

[4.11] J. W. McClure. *Phys. Rev.*, **108**, 612 (1957).

[4.12] J. W. McClure and L. B. Smith. In S. Mrozowski, M. L. Studebaker, and P. L. Walker, Jr. (eds.), *Proc. Fifth Carbon Conf.*, vol. II, p. 3. Pergamon Press (1963).

[4.13] R. C. Haddon, L. E. Brus, and K. Raghavachari. *Chem. Phys. Lett.*, **125**, 459 (1986).

[4.14] R. C. Haddon. *Accounts of Chemical Research*, **25**, 127 (1992).

[4.15] M. S. Dresselhaus, G. Dresselhaus, and R. Saito. *Phys. Rev. B*, **45**, 6234 (1992).

[4.16] W. G. Harter and T. C. Reimer. *Chem. Phys. Lett.*, **194**, 230 (1992).

CHAPTER 5

Synthesis, Extraction, and Purification of Fullerenes

Fullerene molecules are formed in the laboratory from carbon-rich vapors which can be obtained in a variety of ways, e.g., resistive heating of carbon rods in a vacuum, ac or dc plasma discharge between carbon electrodes in He gas, laser ablation of carbon electrodes in He gas, and oxidative combustion of benzene/argon gas mixtures. Most methods for the production of large quantities of fullerenes simultaneously generate a mixture of stable fullerenes (C_{60}, C_{70}, ...), impurity molecules such as polyaromatic hydrocarbons, and carbon-rich soot. Therefore, the synthesis of fullerenes must be followed by procedures to extract and separate fullerenes from these impurities according to mass, and for the higher fullerenes, separation according to specific isomeric forms may also be required. In this chapter, we describe the laboratory methods commonly used to synthesize, extract, and purify fullerenes.

Fullerenes have also been reported to occur naturally in some forms of common carbon soot [5.1]; in shungite, a carbon-rich mineral [5.2]; and in fulgurite, which is formed by lightning strikes on carbon-containing rocks [5.3]. Ordinary materials such as gasoline [5.4] and coal [5.5] have also been reported as source materials for producing fullerenes. The field of fullerene synthesis, extraction, and purification methods is still developing rapidly. The most recent developments have also focused on the production and separation of higher-mass fullerenes and endohedral metallofullerenes, which are also discussed here. In this chapter we review the status of methods for the synthesis or generation of fullerenes (§5.1), extraction of fullerenes from fullerene-containing material (§5.2), and purification or separation of one fullerenes species from another (§5.3). A brief description of the preparation of endohedral fullerenes is reviewed in §5.4. A few comments on health and safety issues are given in §5.5.

5.1. Synthesis of Fullerenes

Fullerenes can be synthesized in the laboratory in a wide variety of ways, all involving the generation of a carbon-rich vapor or plasma. All current methods of fullerene synthesis produce primarily C_{60} and C_{70}, and these molecules are now routinely isolated in gram quantities and are commercially available. Higher-mass fullerenes and endohedral complexes can also be made and isolated, albeit in substantially reduced amounts. At present the most efficient method of producing fullerenes involves an electric discharge between graphite electrodes in ~200 torr of He gas. Fullerenes are embedded in the emitted carbon soot and must then be extracted and subsequently purified. A variation of the arc technique is used to synthesize graphene tubules (see §19.2.5). However, it appears that it will be difficult to extend the chemical methods now used to isolate particular fullerene isomers to separate the carbon tubules according to diameter and chiral angle (see §19.2.5).

5.1.1. Historical Perspective

Early approaches to fullerene synthesis used laser vaporization techniques, which produced only microscopic amounts of fullerenes, and these fullerenes in the gas phase were studied in an adjoining molecular beam apparatus (see §1.3). Using these techniques, early experiments by Kroto, Smalley, and co-workers in 1985 [5.6] indicated an especially high stability for C_{60} and C_{70} and stimulated a great deal of interest in the research community to find ways to produce and isolate macroscopic amounts of fullerenes. The breakthrough came in 1990 with the discovery by Krätschmer, Huffman, and co-workers [5.7] that carbon rods heated resistively in a helium atmosphere could generate gram quantities of fullerenes embedded in carbon soot, which was also produced in the process. This discovery, therefore, also necessitated research to discover efficient methods for the extraction of C_{60} and higher fullerenes from the carbon soot, as well as methods for the subsequent isolation of the extracted fullerenes according to mass and, in some cases, even according to their isomeric form (see §3.2).

An ac or dc arc discharge in ~200 torr He between graphite electrodes [5.8] has evolved as an efficient way to make gram quantities of fullerenes on the laboratory bench top. It is believed that fullerene formation requires a minimum gas pressure of ~25 torr [5.9]. It has also been reported that an increase in He gas pressure enhances the production of higher-mass fullerenes [5.10]. The heat generated in the discharge between the electrodes evaporates carbon to form soot and fullerenes, which both condense on the water-cooled walls of the reactor.

The arc discharge technique can also be used to make carbon nanotubes, if a dc, rather than an ac, voltage source is used to drive the discharge (see §19.2.5). Apparently, a variety of carbonaceous solids can be used as electrode materials in the arc process, and fullerenes have in fact been produced in this way from coal [5.5].

Laser vaporization is also used for fullerene production. In a typical apparatus a pulsed Nd:YAG laser operating at 532 nm and 250 mJ of power is used as the laser source and the graphite target is kept in a furnace (1200°C) [5.11–13]. Finally, it should be mentioned that fullerenes have also been produced in sooting flames involving, for example, the combustion of benzene and acetylene [5.14], although the yields are low.

5.1.2. Synthesis Details

The fact that fullerenes can be produced in electric arcs with temperatures well above 4000°C in inert atmospheres is testimony to their inherent stability. In experiments to produce fullerenes by high-frequency (500 kHz) inductive heating of carbon cylinders in an He atmosphere (150 kPa), it was noticed that although carbon begins to sublime from the cylinders at 2500°C, no fullerenes were produced until the cylinder temperature rose to 2700°C [5.15]. This temperature is noticeably higher than the flame temperature of ~1800°C used to produce fullerenes in premixed, low-pressure (~40 torr) benzene/oxygen/Ar flames [5.1], where about 3% of this gas mixture was converted to soot and about one third of the resulting soot was extractable as C_{60} and C_{70}. Subsequent searches for fullerenes in other carbon soots and powders (e.g., diesel exhaust, carbon black, activated charcoal, copy machine toner and an acetylene black produced with a torch without oxygen admixture) did not reveal any fullerenes [5.1]. Interestingly, the combustion of benzene/oxygen/Ar mixtures was reported [5.1] to yield a ratio for $(C_{70}/C_{60}) \sim 0.4$, about a factor of three times higher than obtained typically from the electric arc method. However, high current dc contact arc discharges between graphite electrodes in He gas with an estimated electrode tip temperature of ~4700°C have been reported to generate C_{70}/C_{60} ratios as high as ~ 0.5, and a total yield of fullerenes of 7–10% was obtained [5.16], significantly higher than that reported for flames.

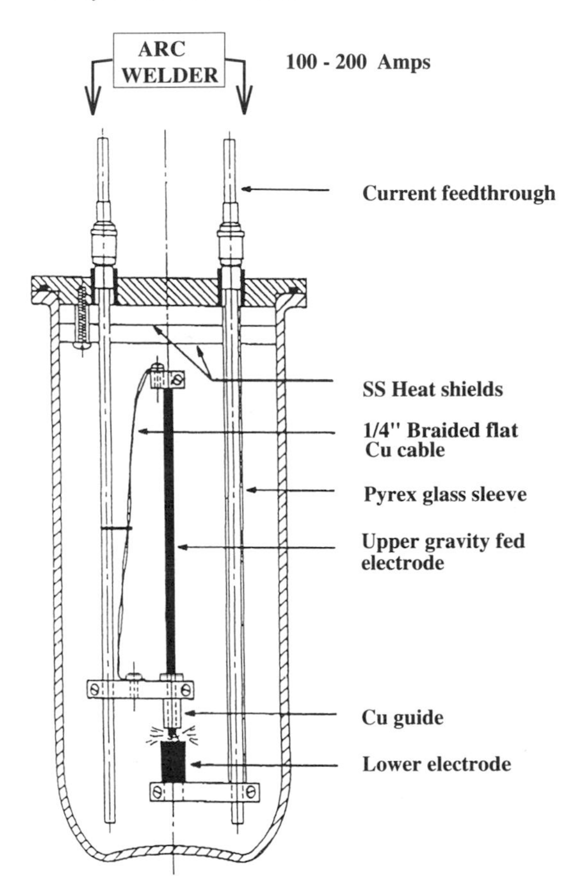

Fig. 5.1. Schematic diagram of a "contact-arc" apparatus used to produce fullerene-rich soot. Rather than incorporating a means of translating one or both of the carbon electrodes in the apparatus to maintain a fixed electrical discharge gap, the upper electrode in this design maintains contact with the lower electrode by the influence of gravity [5.17].

A simple, bench-top reactor used to produce fullerenes was reported by Wudl and co-workers [5.17], and yields about ~4% extractable C_{60}/C_{70} from the soot. As shown in Fig. 5.1, electrical feedthroughs are introduced through a flange at the top of the chamber, and an inexpensive ac arc welding power supply is used to initiate and maintain a contact arc between two graphite electrodes in ~200 torr of He gas. The inert gas lines and a vacuum line are not shown in Fig. 5.1. The apparatus is lowered into a tank of water to cool the cylindrical wall in contact with the cooling water. The initial design by Wudl and co-workers [5.17] used a glass cylinder for the vacuum shroud. Although this provides a convenient means for visual inspection of the electrodes through the wall, the Pyrex cylinder is easily broken, and it is advisable to replace it with a stainless steel cylinder containing a window. Since the arc discharge produces ultraviolet (UV) radiation harmful to the eyes and skin, a piece of filter glass from a welder's face shield is placed over the window if the arc process is to be observed.

A unique and useful feature of the design in Fig. 5.1 is that it does not require the electrodes to be translated together as the electrodes are consumed. Instead, the upper electrode is connected to the electrical feedthrough by a flexible braided cable, and gravity is used to maintain the contact between the electrodes automatically. Thus, this technique is termed the "contact arc method." Typically, about 1 g of 6-mm-diameter graphite electrodes is consumed in approximately 10 min at a current of 120 A (rms) [5.18]. Several electrodes 10–20 cm in length are successively mounted in the chamber and consumed before the soot is scraped from the walls and removed for the extraction step described in §5.2. Normally, inexpensive graphite electrodes with ~1.1% natural abundance of ^{13}C are used.

In Fig. 5.2 we show a scale diagram of a fullerene generator which uses a plasma arc in a fixed gap between horizontal carbon electrodes. A very high yield of 44% solvent extractables (26% by benzene and the rest by higher

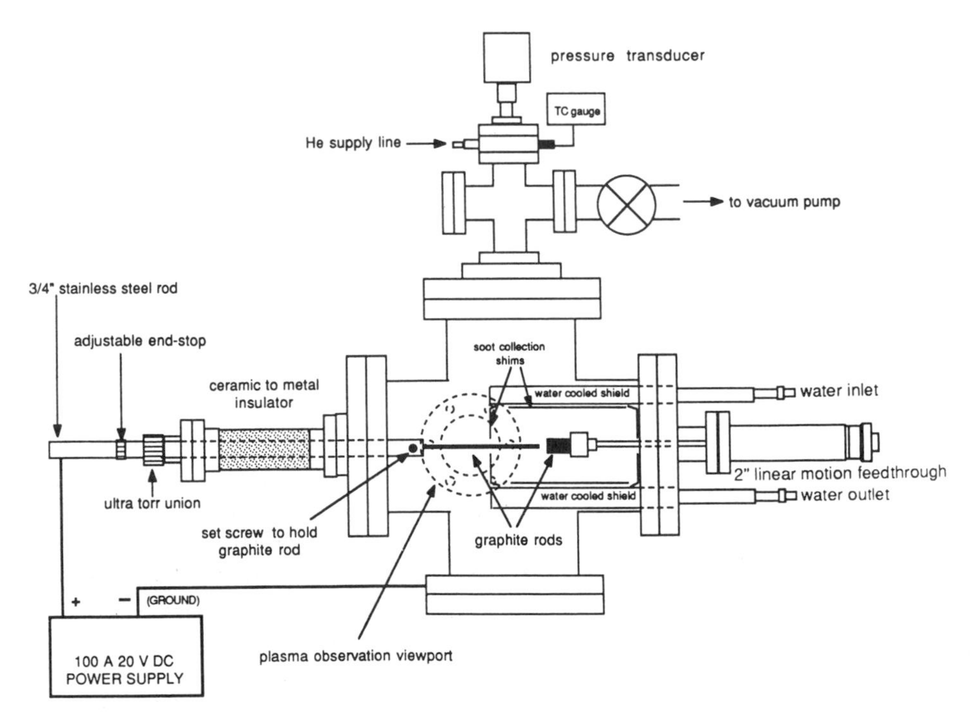

Fig. 5.2. Schematic diagram of a stainless steel plasma arc fullerene generator [5.19].

boiling point solvents) was reported for this apparatus [5.19]. The authors attributed the high yield to (1) the presence of the arc gap, as compared, for example, with resistive heating (1–10%) [5.7, 20, 21] or a contact arc (4%) [5.17], and (2) a static He atmosphere, in which no fullerenes leave the chamber in flowing He gas. The term "static He atmosphere" means that the inlet and outlet gas valves are closed during the arc synthesis. A dc power supply was used in Fig. 5.2 to maintain the arc discharge between a translatable 6-mm-diameter graphite electrode (30 cm maximum length) on the left and a fixed 12-mm-diameter electrode on the right [5.19]. Once the arc is initiated, a gap of $\sim$ 4 mm was maintained between electrodes, and this was judged to produce the maximum brightness in the arc. Six millimeters of electrode are consumed per minute in a static 200 torr He atmosphere for a constant current of 60 A; the arc gap determines the voltage developed across the arc (<20 V). The chamber is constructed from a conventional four-way 8-inch-diameter stainless steel cross with knife-edge seals, and the carbon soot generated in the arc is condensed onto a removable, water-cooled stainless steel cylinder. A good summary of the considerations pertinent to high-volume fullerene soot formation is given by Lamb and Huffman [5.9].

For the carbon arc discharge technique, several parameters are known to affect the conversion of the electrodes to fullerenes [5.9, 21]. Although fullerenes will form in a variety of inert gas atmospheres such as argon and molecular nitrogen, high-purity helium is usually used and is the preferred gas-quenching species. Most soot-generating chambers use a quenching-gas pressure in the 100–200 torr range. It is reported that the optimum pressure is highly sensitive to design specifications and should be determined for each soot generator. Although purity of the initial rod does not affect the soot production rate, higher purity rods are favored for producing soot with lower impurity content. Regarding the diameters, it is reported that smaller-diameter rods (e.g., 1/8 inch diameter) have higher yields of soot [5.9]. While it is believed that the dimensions, geometry, and convection of soot generators bear importantly on soot generation, these design parameters have not been optimized. Regarding contaminants in the generation chamber, small concentrations of hydrogen or H_2O are sufficient to suppress fullerene generation seriously, so that initial degassing of the system and constituents is highly desirable. The removal of impurities is also important for reducing the formation of polycyclic aromatic hydrocarbons [5.22, 23], which may be carcinogens. Other parameters affecting fullerene generation include the density, the binder present in the electrodes, the power dissipated in the arc, and the arc gap. Further research on the optimization of the arc parameters for the production of particular fullerenes, other than C_{60} and C_{70}, needs to be carried out.

5.2. Fullerene Extraction

In the process of fullerene formation from the carbon vapors generated from various carbonaceous materials or hydrocarbon gases, insoluble nanoscale carbon soot is generated together with soluble fullerenes and potentially soluble impurity molecules. Two distinct methods are employed to extract the fullerenes from the soot. In the most common method, the so-called solvent method, toluene, or some other appropriate solvent, is used to dissolve the fullerenes primarily, but the solvent also brings along other soluble hydrocarbon impurities. The soot and other insolubles are therefore easily separated from this solution by filtration, or simply by decanting the solution. In a second, or "sublimation method," the raw soot containing the fullerenes is heated in a quartz tube in He gas, or in vacuum ($\sim 10^{-5}$ torr), to sublime the fullerenes, which then condense in a cooler section of the tube, leaving the soot and other nonvolatiles behind in the hotter section of the tube. Both of these extraction methods also have a tendency to bring impurity molecules along with the most stable fullerenes (e.g., C_{60} and C_{70}). If a very pure fullerene solution or microcrystalline powder is desired, then a final chemical purification step must be carried out, which is discussed in the next section.

Regarding the separation of high-mass fullerenes, it is believed that as a by-product of the preparation of large amounts of C_{60} and C_{70} commercially, there will be available larger amounts of high-mass fullerenes. Improved separation methods hold promise for yielding isolated higher-mass (M) fullerene samples in a purity range $M \pm 2\%$ [5.9].

5.2.1. Solvent Methods

"Soxhlet" extraction in a hot solvent was used in early research to remove fullerenes from the carbon soot generated in the synthesis step (see §5.1.2). More recent solvent techniques discussed below are more applicable to large batch processing. The Soxhlet technique is used in a variety of extractions where the molecular species to be removed from the solid phase are soluble in organic solvents, e.g., polyaromatic hydrocarbons (PAHs) from coal. A schematic view of a Soxhlet extraction unit is shown in Fig. 5.3. The solid sample containing fullerenes, soot, and other materials is loaded into a thimble in the center section of the apparatus. The solvent is boiled in the flask at the bottom, and as a result solvent vapors rise through the side arm on the left and condense near the bottom of the condenser unit, dripping hot, distilled solvent into the thimble, or series of concentric thimbles, which are permeable to the solution. The solution containing the extracted molecules returns to the boiling flask via the side arm on the right. This closed-loop system is normally operated for several hours, during which

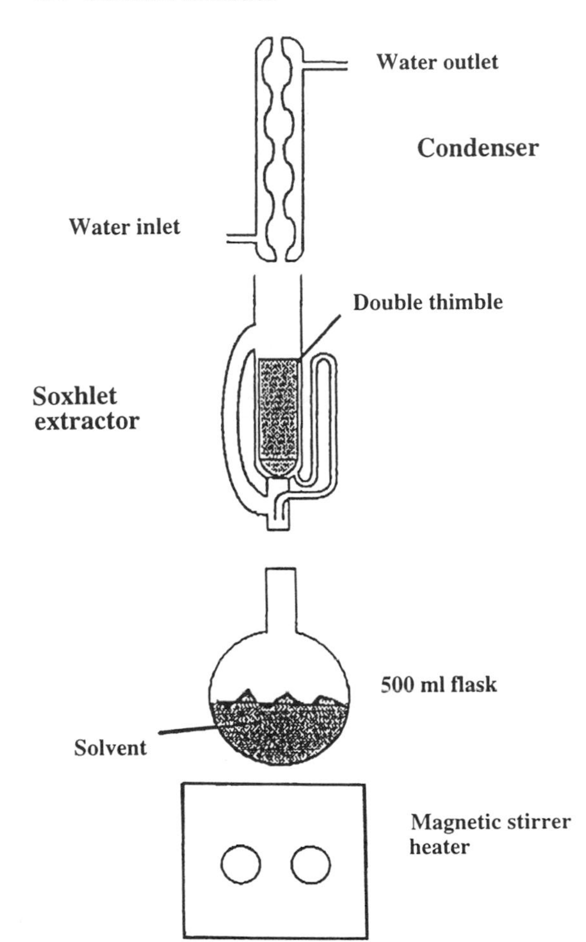

Fig. 5.3. Schematic view of a Soxhlet extraction unit used to separate fullerenes from soot.

time the fullerenes that are soluble in the solvent collect in the flask, and the portion of the solid sample that is not soluble in the solvent remains in the thimble. A reportedly efficient alternative to this method has been proposed [5.24] in which the carbon soot is simply dispersed in tetrahydrofuran (THF) (10 ml/g) at room temperature and sonicated in an ultrasonic bath for 20 min. After filtration to remove the insoluble matter, the THF is removed from the fullerene extract in a rotary evaporator at 50°C, leaving a powder containing fullerenes and soluble impurities in the bottom of the evaporator flask. Some authors have advised that potential PAH impurities should then be eliminated by first washing the fullerene extract in diethyl ether prior to the subsequent liquid chromatography purification step, which is discussed below.

The success of every efficient method for the extraction and purification of fullerenes is tied to the solubility of the fullerene molecules in solvents (see Table 5.1 and §5.2.3). More polar and higher boiling point solvents have been reported to be selective in extracting higher-mass fullerenes. A sequence of Soxhlet extractions was employed [5.19], first using benzene to remove primarily C_{60}/C_{70} (26%), then pyridine (4%), and finally 1,2,3,5-tetramethylbenzene or tetralin (14%), to achieve a total extracted yield of 44% from the carbon soot produced in the arc discharge apparatus shown in Fig. 5.2. Time-of-flight (TOF) mass spectroscopy was used to analyze the extracted material in the extraction sequence. It was found that the higher boiling point solvent, tetralin, isolates the higher-mass fullerenes, and it was proposed on the basis of mass spectroscopy data that fullerenes with masses up to C_{200} constitute over half of the soluble material extracted from the soot.

5.2.2. *Sublimation Methods*

Microcrystalline C_{60} and C_{70} powders are known to sublime in vacuum at relatively low temperatures, i.e., $T_s \sim 350°C$ (C_{60}) [5.25] and $T_s = 460°C$ (C_{70}) [5.26], and this fact can be used directly to separate C_{60} and C_{70} from arc-generated carbon soot without introducing solvents such as hexane, benzene, carbon disulfide, or toluene (see §5.2.1). For some experiments which are particularly sensitive to solvent contamination in the samples, this technique might provide a useful alternative to solvent extraction. In this solvent-free approach [5.19, 27], the soot is placed in one end of an evacuated and sealed (or dynamically pumped) quartz tube and the whole apparatus is then placed in a furnace with a temperature gradient. Dynamical pumping (i.e., the pump is left attached to the quartz tube during the sublimation process) is preferred, since it is possible that the soot may also contain polyaromatic hydrocarbons or other volatile impurities. The tube end containing the raw arc soot is maintained at the highest temperature $T \sim 600$–$700°C$, so that C_{60}, C_{70}, and possibly higher-mass fullerenes will sublime from the soot and drift in the temperature gradient toward the colder regions of the tube, condensing there on the walls. C_{70} and the relatively small amounts of higher fullerenes (e.g., C_{76} and C_{84}) will condense closer to the soot, since their sublimation temperatures are higher than that of C_{60}.

The difference in sublimation temperatures between C_{60} and C_{70} has been used to produce a C_{60} molecular beam from a microcrystalline mixture of C_{60} and C_{70} [5.26]. In this case, the microcrystalline mixture of C_{60} and C_{70} is first heated in a dynamic vacuum to ~ 250–$300°C$ for an extended time (> 4–6 h) to drive off most of the residual solvent, and then the

Table 5.1

Solubility of C_{60} in various solvents [5.29, 30].

Solvent	Parameter δ^a $[(cal/cm^3)^{1/2}]$	Solubility [mg/ml]
Alkanes		
isooctane[b]	6.8	0.026
n-pentane	7.0	0.005
n-hexane	7.3	0.043
octane[b]	7.4	0.025
dioxane[b]	9.9	0.041
docane[b]	7.6	0.070
decalin	8.8	4.6
dodecane[b]	7.8	0.091
tetradecane[b]	7.8	0.126
cyclohexane	8.2	0.036
Haloalkanes		
carbon tetrachloride	8.6	0.32
trichloroethylene	9.2	1.4
tetrachloroethylene	9.3	1.2
1,1,2,2-tetrachloroethane	9.7	5.3
methylene chloride[b]	9.7	0.254
Polar Solvents		
methanol	14.5	0.0
ethanol	12.7	0.001
n-methyl-2-pyrrolidone	11.3	0.89
Benzenes		
benzene	9.2	1.7
toluene	8.9	2.8
xylenes	8.8	5.2
mesitylene	8.8	1.5
tetralin	9.0	16.0
bromobenzene	9.5	3.3
anisole	9.5	5.6
chlorobenzene	9.2	7.0
1,2-dichlorobenzene	10.0	27.0
1,2,4-trichlorobenzene	9.3	8.5
Naphthalenes		
1-methylnaphthalene	9.9	33.0
dimethylnaphthalenes	9.9	36.0
1-phenylnaphthalene	10.0	50.0
1-chloronaphthalene	9.8	51.0
Miscellaneous		
carbon disulfide	10.0	7.9
tetrahydrofuran	9.1	0.0
2-methylthiophene	9.6	6.8
pyridine	10.7	0.89

[a] See Eq. (5.1) for the definition of δ. Unless noted otherwise, the data refer to values at 295 K [5.30].

[b] These data refer to values at 303 K [5.29].

powder is heated above the sublimation temperature of C_{60}. At $T \sim 400^{\circ}$C, the sublimation rate for C_{60} in vacuum is favored by a factor of 20 over that of C_{70} [5.26]. Therefore, since C_{70} is normally a factor of ~ 7 less abundant in arc soot than C_{60}, a reasonably pure molecular beam of C_{60} can be obtained. For example, if a Knudsen cell [5.28] containing this C_{60}/C_{70} mixture is heated to 400°C, one would expect to emit a molecular beam of C_{60} molecules with less than 1% C_{70} impurity.

5.2.3. *Solubility of Fullerenes in Solvents*

The solubility of C_{60} in organic solvents has been studied and correlated with pertinent solubility parameters [5.29, 30]. In Table 5.1 we display the experimental results for the solubility of C_{60} near room temperature (303 K) in a variety of organic and inorganic solvents [5.30], measured by high-performance liquid chromatography, discussed below in §5.3.1. In general, the solubility of a solute in a solvent is enhanced when the polarizability, polarity, molecular size, and cohesive energy density of the solvent are equal to the corresponding parameters for the solute. In this discussion of solubility, chemists define the polarizability as $(n^2-1)/(n^2+2)$ where n is the index of refraction at a reference optical wavelength, such as 5890 Å; the polarity is defined as $(\epsilon-1)/(\epsilon+2)$ where ϵ is the static (dc) dielectric constant; the molecular volume parameter V is defined as the ratio of the molecular weight to the density at a reference temperature, such as 298 K; the cohesive energy is normally related to the Hildebrand solubility parameter δ [5.31] which is defined below. From a detailed study of 47 solvents for C_{60} [5.30], some trends were found governing the solubility of C_{60} for each of these four parameters, but no single parameter could by itself be used to predict the solubility of C_{60} in a given solvent [5.30].

The following general trends were found to govern the solubility of C_{60} in various solvents. Increasing the polarizability parameter was found to yield higher solubilities, and a similar trend was found for the polarity, but with a lesser degree of correlation. Also, increasing the molecular weight of the solvent generally increases the solubility of C_{60}. Correlations were also found with the Hildebrand solubility parameter δ, which is found from the relation

$$\delta = [(\Delta H - RT)/V]^{1/2} = [\Delta E/V]^{1/2} \tag{5.1}$$

where ΔH is the heat of vaporization of the solvent, T is the temperature, V is the molar volume, R is the gas constant $[R = 8.31 \times 10^7$ erg/(K mol)$]$ and ΔE is a measure of the cohesive energy and is defined as the energy needed to convert one mole of liquid at 298 K to one mole of noninteracting gas [5.32]. Values of δ for the various solvents are listed in Table 5.1

[5.30]. Whereas for conventional solutes and solvents ΔH in Eq. (5.1) is interpreted as the molar heat of vaporization, a better correlation for the C_{60} solute was obtained using the molar heat of sublimation for C_{60}. Values of these four parameters (the polarizability, polarity, molecular volume V, and the Hildebrand solubility parameter δ) for C_{60} are found using $n = 1.96$, $\epsilon = 3.61$, $V = 4.29$ cm^3/mol and $\delta = 9.8$ cal$^{1/2}$ cm$^{-3/2}$ (where 41.4 kcal/mol is used for ΔH). Thus Table 5.1 shows that solvents with δ values close to that for C_{60} (i.e., $\delta \sim 10$ cal$^{1/2}$ cm$^{-3/2}$) tend to have higher solubilities for C_{60} than solvents with other values of δ.

Since the solubility of solutes in solvents tends to be significantly temperature dependent, it is important to specify the temperature at which solubility measurements are made (see Table 5.1). This temperature dependence can also be exploited to enhance or suppress the solubility of a particular fullerene by appropriate choice of solvent and temperature in the extraction and purification process.

5.3. Fullerene Purification

In this section we describe fullerene purification using both solvent methods based on liquid chromatography (§5.3.1) and sublimation methods based on temperature gradients (§5.3.2). Gas-phase separation is discussed in §5.3.3 and vaporization studies of C_{60} are reviewed in §5.3.4. By purification, we mean the separation of the fullerenes in the fullerene extract into C_{60}, C_{70}, C_{84}, etc. To verify the effectiveness of the purification process, the fullerenes are characterized by sensitive tools such as mass spectrometry, nuclear magnetic resonance (NMR), liquid chromatography, and infrared and optical absorption spectroscopy.

5.3.1. *Solvent Methods*

Liquid chromatography (LC) is the main technique used for fullerene purification. Briefly, LC is a wet chemistry technique in which a solution (termed the "mobile phase") containing a molecular mixture is forced to pass through a column packed with a high surface area solid (termed the "stationary phase") [5.33]. The identity of the separated fractions from the LC column is verified qualitatively by color (magenta or purple for C_{60} in toluene and reddish-orange for C_{70} in toluene) and more quantitatively, by the comparison of the observed infrared vibrational spectra (see §11.5.2), optical spectra (see §13.2 and §13.6.1), and NMR data (see §16.1) with published results [5.34]. Liquid chromatography generally allows separation of the fullerenes according to their molecular weights, but this method can also be used to isolate a single distinct chiral allotrope such as C_{76}, or to separate isomers with the same molecular weight but having different molecular shapes, e.g., separating C_{78} with C_{2v} symmetry from C_{78} with D_3 symmetry [5.10].

The principle of the liquid chromatography process is as follows. By any one, or several, physical or chemical mechanisms, a particular molecule in the mobile phase is differentiated by being forced to experience an interaction with the stationary phase. This interaction increases (or decreases) the retention time for that molecule in the column or, equivalently, decreases (or increases) the rate of migration for that molecular species through the column. Thus, as a function of time, separated molecular components from the mixture emerge (or "elute") in the order of decreasing interaction with the stationary phase. The least retarded molecular species elutes first, and so on. Separation of molecular species is obtained when a sufficient difference in the retention time in the column can be achieved. For example, impurity PAHs might be eluted in the mobile phase first, C_{60} might be eluted second, followed next by C_{70}, and then followed by the higher-mass fullerenes. Significant physical or chemical differences of the molecular species (e.g., mass, shape, surface adsorption) are required to achieve a clear chromatographic separation.

Historically, the earliest purification of C_{60}, C_{70}, and the higher fullerenes involved "flash" (or rapid) liquid chromatography of the crude fullerene extract in a column packed with neutral alumina (stationary phase) and using hexane/toluene (95/5 volume %) as the mobile phase [5.20, 21, 35]. This procedure was found to be reasonably effective but very labor intensive and consumed large quantities of solvent which was difficult to recycle. One of the first significant improvements to this approach was the development of an automated high-performance liquid chromatography (HPLC) separation process involving pure toluene as the mobile phase and styragel as the stationary phase [5.36–38]. This approach, which employs sophisticated LC equipment, successfully separates gram quantities of fullerenes automatically. Several other HPLC protocols using various stationary and mobile phases have also been reported [5.24, 39].

Several groups have published reports of fullerene purification procedures which combine the extraction and purification steps into one apparatus using pure hexane. An example of such a "modified Soxhlet chromatography" apparatus is shown in Fig. 5.4, which is reported to be capable of extracting a gram of C_{60} and $\sim$0.1 g of C_{70} in a day [5.40]. As can be seen, the apparatus in the figure is similar to a Soxhlet apparatus discussed above for the extraction of fullerenes from soot (Fig. 5.3), except that the Soxhlet thimble has been replaced with a liquid chromatography column. To carry out the purification step, fullerenes are first loaded by adsorption onto the top of the column, which has been packed with neutral alumina.

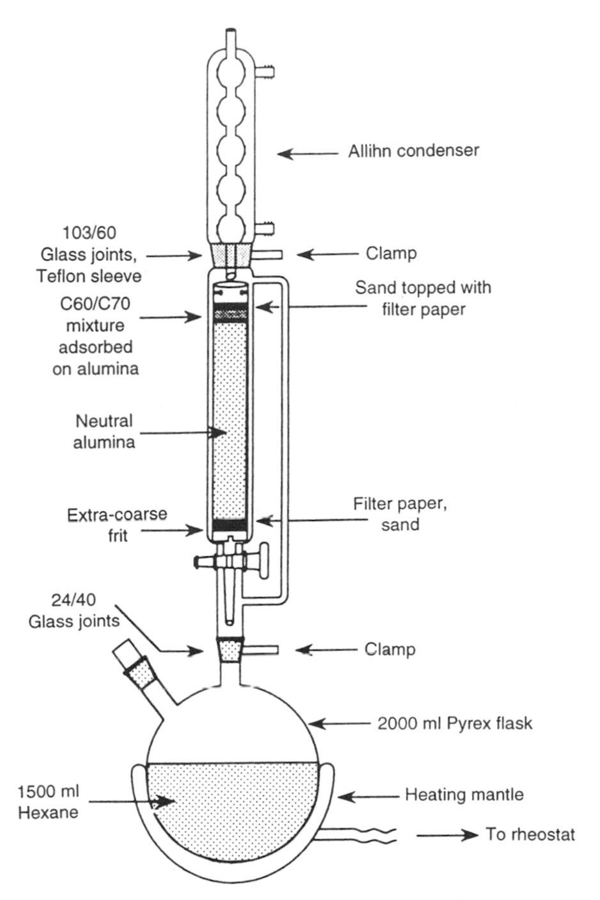

Fig. 5.4. Schematic representation of the modified "Soxhlet chromatography" apparatus used to combine the extraction and purification steps and used to separate fullerenes according to mass and molecular shape [5.40].

In operation, the distilled solvent (pure hexane, 1.5 L) is then condensed at the top of the column and passes through the column, returning to the distillation flask. In this way, the hexane is recirculated continuously as the C_{60} molecules migrate in a purple or magenta band down the column. This process proceeds for 20–30 h until all the C_{60} has passed slowly through the column to the distillation flask. The valve is then closed and the first flask containing C_{60} in hexane is removed. A second flask of hexane is then attached to the apparatus and the C_{70} remaining in the column is collected in the same way.

Recently, considerable improvements in the facile extraction of pure C_{60} from soot have been achieved. A new inexpensive and relatively simple method has been developed, which involves a simple filtration of toluene-extracted fullerenes through a short plug of charcoal/silica gel using toluene as the elutant [5.39]. The method is inexpensive and quite easy to carry out. Unfortunately, however, the extracted C_{70} and higher fullerenes are trapped in the charcoal/silica plug and are difficult to recover at a later time. This filtration separation of C_{60} [5.39] represents an improvement on the flash chromatography method reported previously [5.41], which yielded excellent separation of C_{60} from higher fullerenes and impurities using a solid phase consisting of silica gel/carbon Norit-A (2:1) and pure toluene as the mobile phase [5.41] and achieved a separation of $\sim$2 g of fullerene extract in 1 h with a yield of 63% out of a possible 75%. The two improvements to the previous method were the replacement of the Norit-A carbon by activated, acid-washed Darco G60 charcoal (Fluka) and the elimination of the flash chromatography step, thus replacing the column by a simple plug-filtration flask. A concentrated solution of extracted fullerenes in toluene was loaded in a fritted funnel and eluted by the application of a slight vacuum at the side of the filter flask. Only 15 min was needed to elute 1.5 g of C_{60} starting from 2.5 g of fullerene extract! Furthermore, HPLC analysis of the material found a contamination of less than 0.05% C_{70} or other fullerenes [5.39].

For C_{60} or C_{70} powders obtained by liquid chromatography, heat treatment in a dynamic vacuum of flowing inert gas (300°C for $\frac{1}{2}$ to 1 day) is used to reduce the solvent impurity. Furthermore, solid C_{60} has been shown to intercalate or accept oxygen quite easily at 1 atm, particularly in the presence of visible or UV light [5.42]. Thus, degassed powders should be stored in vacuum or in an inert gas (preferably in the dark).

Many organic molecules have been found to have a structural, mirror-image twin or enantiomer, which is derived from the Greek word enantios, meaning "against." These two structural types are also referred to as "left-handed" and "right-handed," and a common example is left- and right-handed glucose. When polarized light is made to pass through solutions containing molecules of one structural type (i.e., the left-handed form), the plane of polarization is observed to be rotated, while the other structural type also causes optical rotation, but in the opposite sense. If the polarization rotation is significant, these molecules can be used in optical applications. Because of its D_2 symmetry, the C_{76} molecule has right- and left-handedness (see §3.2). Thus C_{76} is found in two mirror-image, chiral forms (see Fig. 5.5) that can be extracted in equal concentrations as a minority constituent from carbon arc soot [5.44]. One solution to the puzzle of how to effectively separate these two C_{76} enantiomers, with potential for more general application to the separation of other chiral fullerenes [5.43],

Fig. 5.5. Schematic view of the two C_{76} enantiomers. A computer graphics rendition of these enantiomers is shown in [5.43]. The two molecules are mirror images of one another.

isolates the C_{76} molecules using liquid chromatography, first involving neutral alumina columns and then using a 50% toluene in hexane solution as the mobile phase in a "Bucky Clutcher I" column (Regis Chemical, Inc.). In the important next step, a combination of a plant alkaloid and an osmium-containing chemical is used to promote a rapid osmylation of one enantiomer over the other [5.43, 45–48]. Using liquid chromatography involving a silica gel column, the osmylated C_{76} enantiomer is then separated from the unreacted twin. In a final step, the osmium adduct is removed to obtain the other C_{76} enantiomer as well, via a reaction with $SnCl_2$ in pyridine. The two C_{76} enantiomers recovered by this process were found to exhibit a maximum specific optical rotation of 4000±400 °/decimeter g/100 ml at the sodium D line. (The definition of the specific optical rotation is in terms of $\alpha/\ell c$ where α is the angle of rotation in degrees, ℓ is the optical path length in 10^{-1} m, and c is the concentration of the species in g/100 ml.) The osmylation of C_{60} is discussed in §10.6.

5.3.2. *Sublimation in a Temperature Gradient*

Separation and purification of fullerenes by sublimation in a temperature gradient (STG) [5.49] does not involve solvents and therefore avoids any possible contamination from the solvent used in common solvent-based extraction and purification procedures. Furthermore, since higher fullerenes and endohedral fullerenes are, as a rule, much less soluble in organic solvents than C_{60} and C_{70}, it is a natural concern that a significant quantity of these less abundant fullerenes might remain behind in the soot when solvent extraction methods are used.

In the STG process, raw soot (containing fullerenes) is obtained directly from the arc apparatus and placed in one end of an evacuated quartz tube (diameter ~15 mm, length ~30 cm). The tube is evacuated, sealed off with a torch, and placed in a resistively heated oven maintained at a temperature in the range 900–1000°C at the center of the oven. The end of the tube containing the soot is placed at the center of the oven (hottest point) and the other end of the tube protrudes out of the oven and into the ambient environment. The natural temperature gradient of this arrangement allows fullerenes to sublime from the soot at the hot end of the tube and diffuse down toward the colder end. A specific fullerene molecule eventually sticks to the wall of the tube at a location dependent on its particular sublimation temperature. Thus, fullerenes with the higher sublimation temperatures (high-masses) are located on the walls in the hotter regions, and vice versa. A low pressure of inert buffer gas (e.g., argon) as well as internal baffles can be introduced into the system to ensure a slower diffusion of molecules down the tube, thereby increasing the chance for local thermal equilibrium with the tube wall [5.50].

A modification of the STG apparatus, to allow easy use of laser desorption mass spectrographic (LDMS) methods to study the spatial variation of the fullerene deposits along the rod, is the use of a ~3-mm-diameter quartz rod inside the sealed quartz tube as shown in Fig. 5.6(a) [5.51]. In the sublimation setup, 400 mg of fullerenes or metallofullerenes are placed at the closed end of the quartz tube and the fullerenes deposit on the quartz rod held at the center of the tube by a Teflon ring. The tube is in a resistively heated and temperature-controlled furnace held at 1050°C with the soot at the hot end and the Teflon ring kept 4 inches outside. In the LDMS experiments the quartz rod is placed directly in a TOF mass spectrometer for analysis, Fig. 5.6(b). Fullerenes are desorbed by a short laser

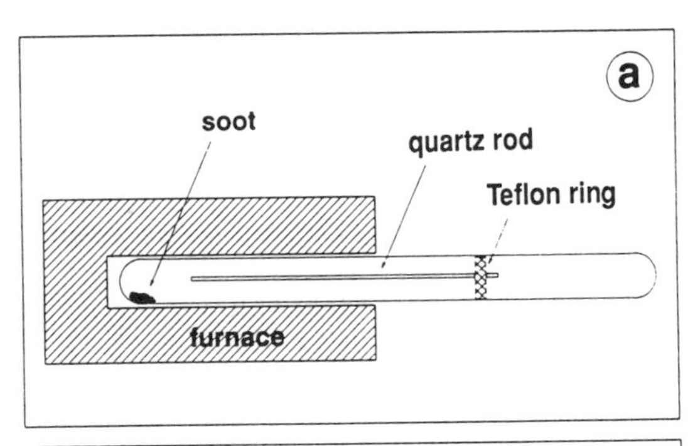

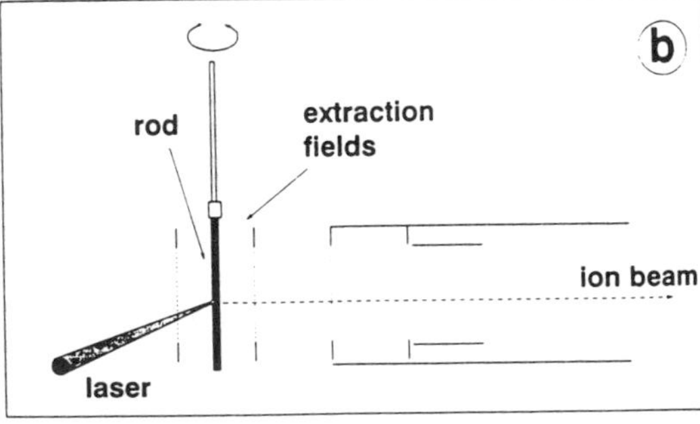

Fig. 5.6. Schematic diagrams of the sublimation and laser desorption mass spectrometer setups for separation (a) and purification (b) of fullerenes. The position-sensitive LDMS setup (b) shows the quartz rod of (a), coated with the sublimed fullerenes, hanging freely within the first extraction field of a two-stage Wiley–MacLaren type time-of-flight mass spectrometer for analysis of the mass distribution of the sample [5.51].

pulse (ArF excimer laser, 193 nm, 6.4 eV) directly into the high vacuum (5×10^{-6} torr). After 30–40 μs, the desorbed species are extracted by a pulsed field and mass-analyzed according to their time of flight within the mass spectrometer instrument. The enhanced concentration of C_{84} relative to C_{60} is shown in Fig. 5.7, where the LDMS instrument is operating in the positive ion mode, using an ArF excimer laser at 193 nm, and ~1.5 mJ/cm^2 power level for the desorption of the molecular species. In the hotter region, the concentration of higher-mass fullerenes on the quartz rod is seen to be greater.

As another example of separation by sublimation in a temperature gradient, the enhanced concentration of the metallofullerenes La@C_{74} and La@C_{82} relative to C_{70} is shown in Fig. 5.8, where LDMS data are shown for a 10-mm-long ribbon of fullerenes deposited along the temperature gradient. During the first 20 mm (not shown), a gradual increase

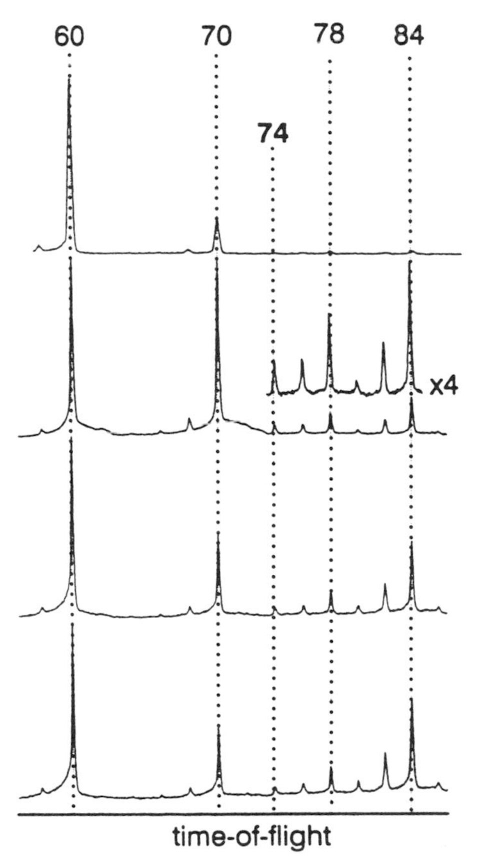

Fig. 5.7. Position-sensitive LDMS profile of the ~30-mm-long sublimed deposit along the temperature gradient of the fullerene-containing raw soot generated in the system in Fig. 5.6. Going from the top to the bottom spectrum, we see changes in the concentration of the higher-mass constituents of the deposit in going from colder to hotter regions of the temperature gradient [5.51].

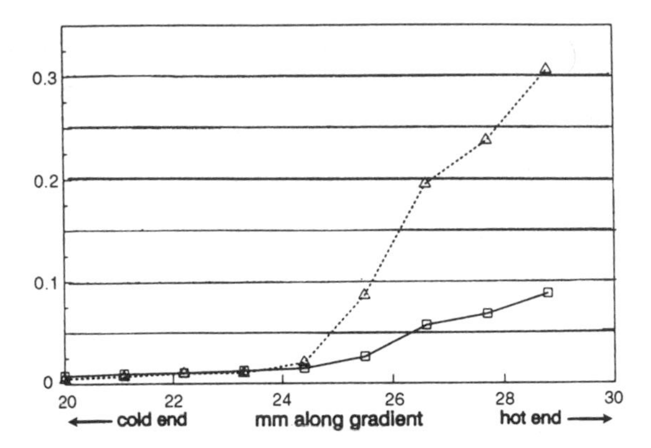

Fig. 5.8. Comparison of the intensities of the LDMS signal for La@C_{74} (open squares) and La@C_{82} (open triangles) relative to C_{70} as a function of the position along the temperature gradient [5.51].

of the overall intensities of the higher-mass peaks relative to C_{60} and C_{70} is observed, without any major intensity variations among the two metallofullerene species (see §5.4 and §8.2). However, in the last 10 mm, at the hot end of the gradient, an overall steep increase of the higher-mass fullerenes is observed with interesting variations of the relative intensity distribution. Most notable is the pronounced increase of La@C_{82} at the hot end of the ribbon [5.51].

The STG method has not been used to produce large quantities of purified fullerenes. Its potential application appears, for the moment, to lie more in the isolation and purification of particularly insoluble fullerenes, or perhaps as a method to preconcentrate certain higher fullerenes or endohedral fullerenes (see §5.4 and §8.2) in advance of their purification by a final liquid chromatographic step.

5.3.3. Gas-Phase Separation and Purification

Gas-phase purification utilizing the difference in vapor pressure for the various species is an attractive approach for the purification and isolation of species that are either highly reactive, not soluble, or suffer irreversible retention on the stationary phase during conventional chromatography [5.52, 53]. On this basis, some good candidates for gas-phase separation are the endohedral fullerenes La@C_{60} and U@C_{28} [5.54–56], which have not been successfully isolated by chromatography.

Gas-phase purification utilizes the difference in vapor pressure between C_{60} and the higher fullerenes. This method has been used successfully to produce ultrahigh purity (99.97%) C_{60} [5.57]. Raw fullerene soot or fullerene extract is introduced into one end of a distillation column lined with a series of evenly spaced baffles with circular perforations. The fullerene starting material is heated under high vacuum to ~1000 K, while a linear temperature gradient is simultaneously established along the length of the column. As the mixed fullerene vapor traverses to the cooler region of the column by effusion through the perforated baffles, it becomes enriched in the more volatile species. Because of the repeated evaporations and condensations necessary to move down the column, virtually all volatile impurities are removed and pumped away. Since the perforations in the baffles are small compared to the mean free path of the fullerenes, effusion is the dominant mass transport mechanism in the column. Using the Knudsen effusion equation [5.58], the theoretical rate of effusion dW_n/dt through an aperture for each fullerene component (in g/s) is written as

$$\frac{dW_n}{dt} = kp_n A\sqrt{\frac{M_n}{2\pi RT}}, \tag{5.2}$$

where p_n is the equilibrium vapor pressure of the nth component in dyn/cm^2, T is the temperature in degrees Kelvin, M_n is the molecular weight of the nth component, A is the effective aperture area in cm^2, k is a geometrical factor used to account for the thickness of the plate, and R is the gas constant. The apparatus consists of a meter-long fractional distillation column made from stainless steel tubing, which is lined with a removable quartz tube containing the separation baffles, consisting of many stainless steel disks with circular perforations [5.57].

The mass flow down the distillation column is initiated by subliming the starting material, and the fullerenes now in the vapor phase experience a pressure gradient induced by the temperature gradient. Starting with a higher-purity fullerene source material gives a higher-purity product and a significantly higher yield of ultrahigh purity material. It is believed that the gas-phase separation approach can also be used to purify higher-mass fullerenes and endohedral fullerenes.

5.3.4. Vaporization Studies of C_{60}

Separation of fullerenes by utilizing differences in their vapor pressure has been discussed as one approach to the purification of fullerenes (see §5.3.3). This separation method depends on knowledge of the temperature dependence of the vapor pressure of fullerenes and the associated heat of sublimation [5.26, 57, 59, 60]. These quantities have been measured for a C_{60}/C_{70} mixture obtained from an unpurified extract from arc soot [5.26] and for chromatographically purified C_{60} [5.59].

We summarize here the temperature dependence of the vapor pressure obtained from a purified C_{60} sample [5.59] employing a Knudsen cell as a source (see Fig. 5.9). A thermocouple which touched the base of the

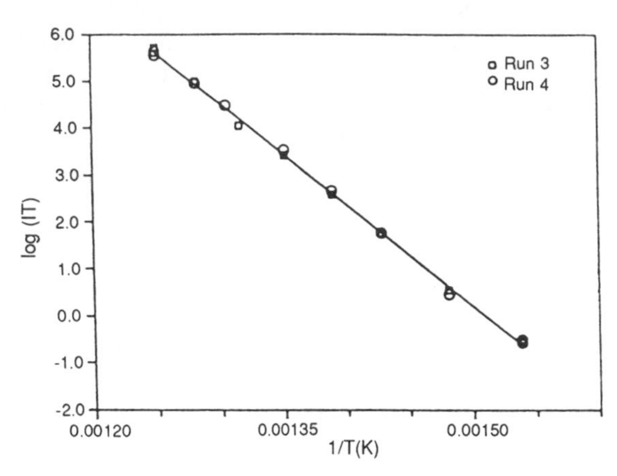

Fig. 5.9. Temperature dependence of the vapor pressure measured in terms of IT [see Eq. (5.3)] where I is the number of ionized C_{60}^+ ions measured in a mass spectrometer. The solid line is a least squares fit to the experimental points, obtained from a purified C_{60} sample [5.59].

alumina crucible in the Knudsen cell was calibrated against the melting point of silver. The molecular beam effusing from the Knudsen cell was ionized by 38 eV electrons and passed into a mass spectrometer which served both to identify the particular molecular species and to determine the vapor pressure of C_{60} above the pure solid. The vapor pressure p of the neutral C_{60} molecules is related to the ion intensity I (number of C_{60}^+ ions detected per second in the mass spectrometer) and the temperature T by the relation

$$p = cIT, \tag{5.3}$$

where c is a temperature-independent constant which is sensitive to instrumental factors and the ionization cross section for C_{60}. The constant c in Eq. (5.3) is determined using the time-integrated weight loss from the Knudsen cell for a standard cell. The following empirical equation is used to relate the vapor pressure (in units of Pa) to the temperature (in K):

$$\log p = c_1/T + c_2, \tag{5.4}$$

where the constants c_1 and c_2 are determined by a least squares fit to the measurements (see Fig. 5.9), yielding $c_1 = -9777 \pm 138$ K and $c_2 = 11.582 \pm 0.126$ for C_{60} [5.59] and $c_1 = -10219 \pm 78$ K and $c_2 = 11.596 \pm 0.126$ for C_{70} [5.61].

The entropy (ΔS) and enthalpy (ΔH) of sublimation are related to the vapor pressure p [5.62] by the relation

$$p = \exp\left[-\left(\frac{\Delta H}{RT}\right)+\left(\frac{\Delta S}{R}\right)\right], \tag{5.5}$$

where R is the gas constant. From Eqs. (5.3–5.5), it is apparent that the slope of a plot of log p *vs.* $1/T$ yields the heat of sublimation ΔH, and the intercept yields ΔS. The value obtained for the enthalpy of sublimation of C_{60} in this way is $\Delta H = 181.4 \pm 2.3$ kJ/mol [5.59], and for C_{70} a value of $\Delta H = 195.7 \pm 1.1$ kJ/mol is obtained [5.61]. These values can be compared to an earlier value reported for a C_{60}/C_{70} mixture, $\Delta H = 167.8 \pm 5.4$ kJ/mol [5.26].

The partial pressures of C_{60} and C_{70} in the C_{60}/C_{70} binary system have been measured as a function of temperature in the range 600–800 K at several compositions ranging from pure C_{60} to pure C_{70} [5.63], and the results of gas-phase study were compared with partial pressures over the solid phase. From the pressure–temperature–composition data, it was concluded that C_{60} and C_{70} are soluble in each other in the solid state up to 30% with a miscibility gap for intermediate compositions [5.63].

5.4. Endohedral Fullerene Synthesis

The initial discovery in 1985 that La atoms might be trapped inside a C_{60} molecule to form an "endohedral" fullerene stemmed from the observation of La-related peaks in the mass spectra of material ejected by laser vaporization of an La-impregnated graphite target [5.64]. Both La@C_{60} and La_2@C_{60} were discovered in these pioneering experiments (see §8.2). However, only minute quantities of these molecules could be prepared in this way. The endohedral doping configuration is denoted by La@C_{60} for one endohedral lanthanum in C_{60}, or Y_2@C_{82} for two Y atoms inside a C_{82} fullerene [5.65]. For the larger fullerenes, the endohedral addition of up to four metal ions has been demonstrated by mass spectroscopy techniques (see §6.2) [5.66, 67].

Widespread study of the properties of these and other endohedral fullerenes began shortly after the discoveries in 1991–1992 that mg quantities of metal-containing endohedral fullerenes or metallofullerenes could be obtained by laser [5.68] and arc [5.69] vaporization of metal–graphite composites in a rare gas (He or Ar) atmosphere (see §8.2). The endohedral fullerene molecules that have been made thus far involve mostly rare earth and transition metal atoms, although some alkali metals and other species have been introduced. Some of the more commonly known endohedral fullerenes are listed in Tables 5.2 and 5.3. Also, He or Ne can

Table 5.2

Table of endohedral fullerenes for cages $C_{n_C} \leq 72$.

Cluster	Synthesis method	References
$Hf@C_{28}$	arc discharge using graphite/HfO_2 electrode	[5.55]
$Ti@C_{28}$	arc discharge using graphite/TiO_2 electrode	[5.55]
$U@C_{28}$	photocontraction from higher $U@C_{n_C}$ species	[5.55]
$U_2@C_{28}$	photocontraction from higher $U@C_{n_C}$ species	[5.55]
$Zr@C_{28}$	arc discharge using graphite/ZrO_2 electrode	[5.55]
$U@C_{36}$	photocontraction from higher $U@C_{n_C}$ species	[5.55]
$K@C_{44}$	laser vaporization	[5.70]
$La@C_{44}$	laser vaporization of La-impregnated graphite	[5.64, 68, 70]
$U@C_{44}$	photocontraction from higher $U@C_{n_C}$ species	[5.55]
$Cs@C_{48}$	laser vaporization	[5.70]
$U@C_{50}$	laser vaporization of graphite/UO_2	[5.55]
$U_2@C_{58}$	laser vaporization of graphite/UO_2	[5.55]
$Ca@C_{60}$	laser vaporization	[5.71, 72]
$CCl_2@C_{60}$	only theory–no experiment	[5.73, 74]
$Co@C_{60}$	arc vaporization; no extraction	[5.54]
$Cs@C_{60}$	laser vaporization and photofragmentation	[5.75]
$Fe@C_{60}$	carbon arc with He–$Fe(CO)_5$ atmosphere	[5.76]
$He@C_{60}$	trace amounts in standard arc growth	[5.77–80]
$K@C_{60}$	photofragmentation	[5.75]
$La@C_{60}$	laser vaporization of La-impregnated graphite	[5.64, 68, 81]
$La_2@C_{60}$	laser vaporization of La-impregnated graphite	[5.64, 81]
$Mu@C_{60}$	muonium stopping in solid C_{60}	[5.82]
$Ne@C_{60}$	trace amounts in arc method growth	[5.80]
$Rb@C_{60}$	photofragmentation	[5.75]
$U@C_{60}$	laser vaporization of graphite/UO_2	[5.55]
$U_2@C_{60}$	laser vaporization	[5.55]
$Y_2@C_{60}$	laser vaporization of graphite/Y_2O_3	[5.65]
$La@C_{70}$	laser vaporization of La-impregnated graphite	[5.68]
$U@C_{70}$	laser vaporization of graphite/UO_2	[5.55]
$U@C_{72}$	laser vaporization of graphite/UO_2	[5.55]

be incorporated inside the C_{60} shell through arc vaporization of carbon electrodes in the presence of these gases, thereby forming small quantities of $He@C_{60}$ and $Ne@C_{60}$ endohedral complexes [5.80] (see §8.2). The main characterization techniques that have been used for endohedral fullerenes are mass spectra (see Fig. 5.10) and electron paramagnetic resonance (EPR; see §16.2.1).

Endohedral fullerenes for properties measurements are generally prepared in an arc generator [5.90, 91] where the positive electrode is a composite containing a high-purity oxide of the metal species (e.g., Sc_2O_3 to prepare $Sc@C_{82}$), graphite powder, and a high-strength pitch binder. Pro-

Table 5.3

Table of endohedral fullerenes for cages $C_{n_C} > 72$.

Cluster	Synthesis method	References
$La@C_{74}$	laser vaporization of La-impregnated graphite	[5.68]
$Sc@C_{74}$	arc discharge	[5.83]
$Sc_2@C_{74}$	arc discharge using Sc_2O_3 packed electrode	[5.83]
$La@C_{76}$	laser and arc vaporization of La-impregnated graphite	[5.64, 83]
$La_2@C_{76}$	arc vaporization of LaO_2/graphite electrode	[5.83]
$La_2@C_{80}$	arc vaporization of LaO_2/graphite electrode	[5.84, 85]
$Er@C_{82}$	arc vaporization with Er-oxides	[5.86]
$La@C_{82}$	laser vaporization of La-impregnated graphite	[5.68, 87–89]
$La@C_{82}$	carbon arc using La_2O_3 packed electrode	[5.83]
$La_2@C_{82}$	carbon arc using La_2O_3 packed electrode	[5.83]
$Sc@C_{82}$	carbon arc using Sc_2O_3 packed electrode; Sc metal chips and graphite	[5.90–92]
$Sc_2@C_{82}$	carbon arc using Sc_2O_3 packed electrode; Sc metal chips and graphite	[5.90–92]
$Sc_3@C_{82}$	carbon arc using Sc_2O_3 packed electrode; Sc metal chips and graphite	[5.92, 93]
$Y@C_{82}$	arc vaporization of graphite/Y_2O_3	[5.94]
$Y_2@C_{82}$	laser vaporization of graphite/Y_2O_3	[5.65, 94, 95]
$La@C_{84}$	arc vaporization of LaO_2/graphite electrode	[5.83]
$La_2@C_{84}$	arc vaporization of LaO_2/graphite electrode	[5.83]
$Sc_2@C_{84}$	carbon arc using Sc_2O_3 packed electrode	[5.69, 83, 87, 90, 92]
$Sc_3@C_{84}$	arc vaporization	[5.69, 87, 90, 92, 94]

duction efficiency has been reported to increase in some cases if metal carbides, rather than oxides, are used to introduce the metal component into the vapor phase [5.96]. The Sc-graphite electrodes are cured and carbonized at 1000°C for 5 h in vacuum (10^{-3} torr), and the rods are then further baked at 1200°C for 5 h at 10^{-5} torr. The composite rods are used as positive electrodes in a dc arc operating in an He atmosphere (50 torr). For quantitative studies, an anaerobic (air-free) production, sampling and solvent extraction method is used, whereby the carbon soot containing the endohedral metallofullerenes is produced in an He gas flow atmosphere (for example, 43 torr/l s at 100 torr) [5.96]. A current density of $\sim$2.4 A/mm^2 is used between a pure graphite rod (negative electrode) and the composite rod (positive electrode). Solvent vapor (e.g., toluene) is introduced into the He gas stream that leaves the arc chamber. This allows the entrained fullerenes and the endohedrally doped fullerenes to be frozen in a downstream liquid nitrogen cold trap. After collection of the soot in the frozen solvent, the sample is kept in a dry nitrogen atmosphere. All sample handling for characterization measurements is done in a dry nitrogen atmo-

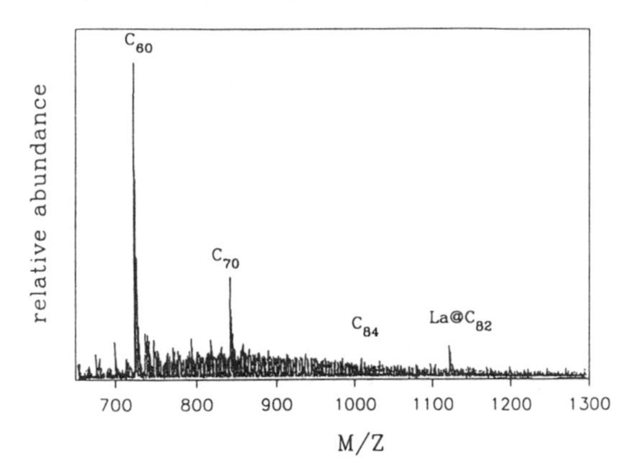

Fig. 5.10. Fast atom bombardment mass spectrum of the pyridine-soluble extract of the soot produced by arc vaporization of lanthanum-containing graphite electrodes. Evidence for La@C_{82} is clearly seen in this mass spectrum [5.11].

sphere [5.83], since many $M_p@C_{n_C}$ metallofullerenes (where M denotes a metal species) may be sensitive to oxygen/air [5.83, 97, 98]. With this technique, a sufficient quantity of material has been isolated for EPR, structural, and physical characterization measurements on selected endohedral fullerenes.

To produce macroscopic amounts of endohedral fullerenes by laser vaporization, the target for the laser vaporization is placed in an oven at 1200°C [5.68]. A slow flow of Ar or He gas is passed over the hot La-containing target, and the inert gas then flows down a 50 cm long × 2.5 cm diameter quartz tube centered in a resistively heated furnace. A frequency-doubled YAG laser (green light) operating at 300 mJ and 10 pulses per second was focused through a window on the end of the tube onto a rotating, composite target to vaporize both metal and carbon atoms. Fullerenes and metallofullerenes produced in the process at one end of the tube were entrained in the inert gas, then flowed down the tube, and finally condensed on the quartz tube wall near the opposite end of the furnace. Various mixtures of carbon, carbon-based binder and metal or metal-oxide powders have been consolidated into mechanically stable targets for laser vaporization, similar to the material used in the anode of the arc generator.

Although much progress has been made, no efficient method has yet been found for the production of metallofullerenes ($M_p@C_{n_C}$) either by laser ablation of M-impregnated targets or by arc vaporization of M-impregnated rods. The weight fraction of $M_p@C_{n_C}$ endohedral fullerenes in the carbon soot produced by arc vaporization is generally found to be no more than $\sim 10^{-3}$ (or 0.1%). For low concentrations of metal in the M-impregnated rod, the monometal species is favored, but with increasing metal concentration, the dimetal species become more prevalent [5.99].

The extraction of metallofullerenes is also made difficult by their low solubility in solvents for many of the $M_p@C_{n_C}$ species (i.e., the solubility is comparable to that for C_{n_C} where $n_C > 70$). For example, the following solubility sequence has been reported [5.84] for hexane (room temperature), toluene (room temperature or boiling), or CS_2 (room temperature): C_{60}, $C_{70} > C_{76}$, C_{78}, C_{84}, $La@C_{82}$, $La_2@C_{80} > C_{86-96} \gg C_{98-200}$.

The separation and isolation of $M_p@C_{n_C}$ species remain a challenging aspect of fullerene synthesis research, although steady progress is being made. The first such separation has now been achieved for $Sc_2@C_{n_C}$ (for $n_C = 74, 82, 84$) [5.83, 98] and for $La@C_{82}$ [5.88]. Soot from vaporizing composite rods (M/C $\sim$1 atomic %) was treated in CS_2 to extract the soluble fullerenes and $M_p@C_{n_C}$ species. The extraction is followed by a two-step chromatography process [5.88, 98]. In step I the relative concentration of endohedral fullerenes is enhanced by eliminating C_{60} and C_{70}, the most common fullerene species. This step involves preparative-scale HPLC, introducing the crude extract from the arc soot into a polystyrene-filled column and using CS_2 as the elutant. For example, in the separation of $La@C_{82}$, four fractions, I, II, III, IV [Fig. 5.11(a)], were observed to elute

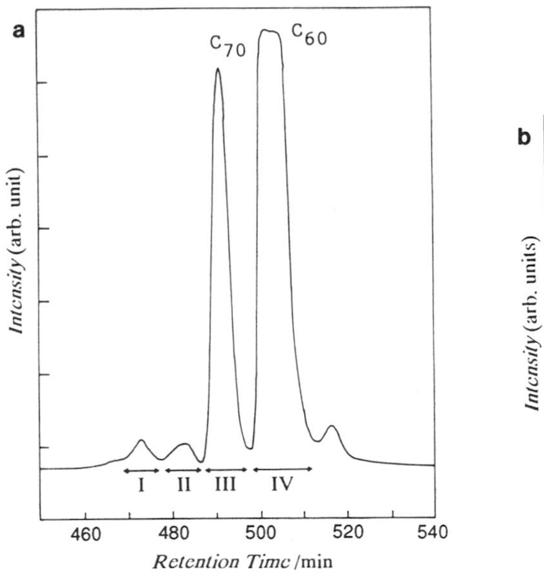

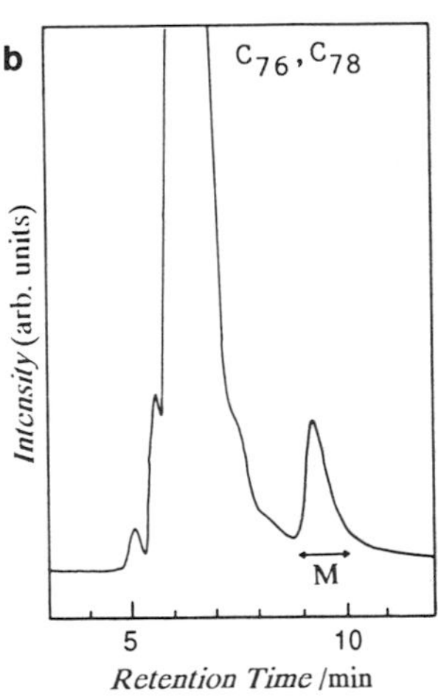

Fig. 5.11. High-performance liquid chromatography (HPLC) time profiles taken for (a) the crude extract from carbon soot containing $La@C_{82}$ and (b) fraction II, which contains mostly C_{76} and C_{78} in addition to the small peak labeled M, which is mostly associated with metallofullerenes [5.88].

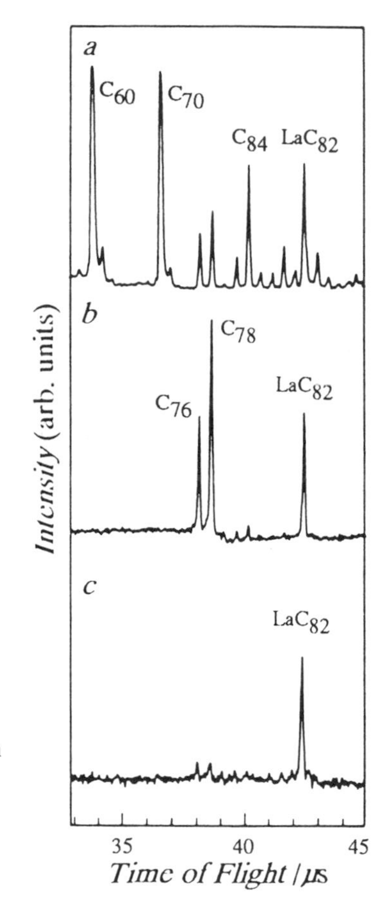

Fig. 5.12. Time-of-flight mass spectra taken from (a) the crude extract shown in Fig. 5.11(a), (b) fraction II in Fig. 5.11(a), and (c) fraction M in Fig. 5.11(b). Negative ions directly produced by 355 nm laser irradiation were detected as a function of flight time [5.88].

from the polystyrene column. Fraction II was found to contain C_{76}, C_{78}, and La@C_{82}, as shown in the time-of-flight mass spectrum in Fig. 5.12(b), which should be compared with the mass spectrum of the crude extract shown in Fig. 5.12(a). Step II of the separation process involves placing fraction II into the Bucky Clutcher I column (Regis Chemical Co.) in which toluene was used as the elutant. The HPLC time profile for the Bucky Clutcher I column is shown in Fig. 5.11(b), where the fraction M was found to contain primarily La@C_{82}, as shown in the mass spectrum of this fraction [Fig. 5.12(c)].

The HPLC separation of a typical $M_p@C_{n_C}$ endohedral fullerene is a very time-consuming process. It appears that the principal area requiring improvement would be the development of a more time-efficient preconcentration step for the metallofullerenes. Automated high-performance liq-

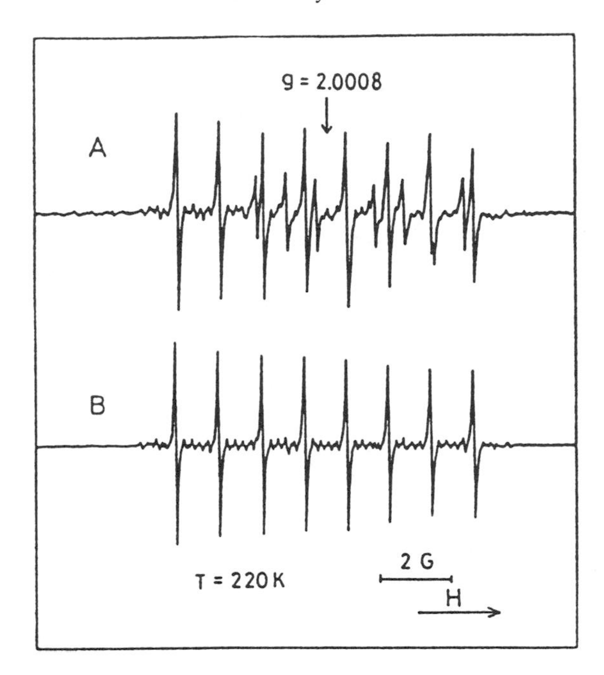

Fig. 5.13. Electron paramagnetic resonance (EPR) spectra of La@C_{82}: (A) normal aerobically prepared La@C_{82} sample dissolved in toluene at 220 K, and (B) an anaerobically prepared La@C_{82} sample showing the octet EPR pattern clearly. The chromatography was performed using an ethanol-deactivated silica gel column with a 100% toluene elutant. The deactivation was done using an ethanol–hexane (1:1) mixed solution [5.83].

uid chromatography using two polystyrene chromatographic columns in series has been shown to lead to more efficient separation of $Sc_m@C_{n_C}$ and $Y_m@C_{n_C}$ metallofullerenes ($m = 1, 2, 3$) [5.100].

To monitor the synthesis process, two characterization techniques have normally been used. The first is mass spectroscopy (see Figs. 5.10 and 5.12 and §5.3.2), which provides information on the various species present in the sample and their relative abundances (see §6.2). The second method is electron paramagnetic resonance, which measures the hyperfine interaction ($\sim \mathbf{I} \cdot \mathbf{J}$) and therefore provides a sensitive measure of the angular momentum J and charge state of the dopant ion. Strong evidence for a charge transfer of three electrons from La to these higher-mass fullerene shells is provided by the octet EPR hyperfine spectrum shown in trace B of Fig. 5.13 [5.83] (see §16.2.1). In the EPR spectrum for La@C_{82}, the spin of ^{139}La is 7/2, and the electron g-value is 2.0010 for an unpaired electron associated with the LUMO level (see §12.1). The spectra shown in Fig. 5.13 demonstrate the importance of anaerobic sample preparation and chromatography (see §5.2.1). In trace A of Fig. 5.13 the spectrum is for an La@C_{82} sample handled in air and then dissolved in toluene, while the spectrum in trace B is for an anaerobically prepared and chromatographed sample. An ethanol-deactivated silica gel column with 100% toluene elutant was used for the chromatography (see §5.2.1). EPR detection has also been used to provide feedback to greatly enhance the isolation capability of EPR-active endohedral metallofullerenes (such as $Y@C_{82}$ and $Sc_3@C_{82}$) using a chromatographic system with on-line EPR detection [5.100, 101].

5.5. Health and Safety Issues

There are as yet no reports of adverse health affects resulting from exposure to fullerenes or fullerene-containing material. Nevertheless, the health effects of fullerenes have not yet been extensively studied. Early work [5.102] suggests the absence of genotoxicity of C_{60} as demonstrated by somatic mutation and recombination tests.

Even if fullerenes and fullerene-derived materials turn out to be benign, a number of safety issues concerning the handling of fullerenes are identified below [5.11]. Since the inhalation of airborne fullerene soot particles may pose a potential health problem for lungs, the use of fume hoods is recommended. Appropriate gloves, masks, and clothing changes are recommended to avoid skin contact with fullerene soot. Of course, proper laboratory procedures should be used to contain the soot in appropriate vessels, minimizing the possibility of airborne soot in the workspace. Adequate design protection should be provided in the event of the failure of a glass reaction vessel used for soot generation. Suitable protection should be used to avoid eye damage and potential skin cancer when exposed to radiation from the carbon arc. Impurities in the arc rods may lead to polyaromatic hydrocarbon (PAH) generation; these chemicals are known carcinogens. The usual laboratory and safety precautions regarding solvents should be exercised in their use, storage, and disposal, since the solvents can be toxic, inflammable, or explosive. Because of the limited present knowledge of fullerenes and fullerene-containing materials, they should be treated with the respect given to chemical unknowns, and care should be taken to avoid ingestion, inhalation, and skin contact [5.9].

References

[5.1] J. T. McKinnon, W. L. Bell, and R. M. Barkley. *Combustion and Flame*, **88**, 102 (1992).
[5.2] P. R. Buseck, S. J. Tsipursky, and R. Hettich. *Science*, **257**, 215 (1992); ibid., p. 167.
[5.3] T. K. Daly, P. R. Buseck, P. Williams, and C. F. Lewis. *Science*, **259**, 1599 (1993).
[5.4] Y. Yosida. *Jpn. J. Appl. Phys.*, **31**, L505 (1992).
[5.5] L. S. K. Pang, A. M. Vassallo, and M. A. Wilson. *Nature (London)*, **352**, 480 (1991).
[5.6] H. W. Kroto, J. R. Heath, S. C. O'Brien, R. F. Curl, and R. E. Smalley. *Nature (London)*, **318**, 162 (1985).
[5.7] W. Krätschmer, L. D. Lamb, K. Fostiropoulos, and D. R. Huffman. *Nature (London)*, **347**, 354 (1990).

[5.8] R. E. Haufler, J. J. Conceicao, L. P. F. Chibante, Y. Chai, N. E. Byrne, S. Flanagan, M. M. Haley, S. C. O'Brien, C. Pan, Z. Xiao, W. E. Billups, M. A. Ciufolini, R. H. Hauge, J. L. Margrave, L. J. Wilson, R. F. Curl, and R. E. Smalley. *J. Phys. Chem.*, **94**, 8634 (1990).

[5.9] L. D. Lamb and D. R. Huffman. *J. Phys. Chem. Solids*, **54**, 1635 (1993).

[5.10] F. Diederich, R. Ettl, Y. Rubin, R. L. Whetten, R. Beck, M. Alvarez, S. Anz, D. Sensharma, F. Wudl, K. C. Khemani, and A. Koch. *Science*, **252**, 548 (1991).

[5.11] C. M. Lieber and C. C. Chen. In H. Ehrenreich and F. Spaepen (eds.), *Solid State Physics*, vol. 48, p. 109, Academic Press, New York (1994).

[5.12] H. W. Kroto. *Science*, **242**, 1139 (1988).

[5.13] J. T. Cheung and H. Sankur. *CRC Critical Rev. in Solid State and Mat. Sci.*, **15**, 63 (1988).

[5.14] J. B. Howard, A. L. Lafleur, Y. Makarovsky, S. Mitra, C. J. Pope, and T. K. Yaday. *Carbon*, **30**, 1183 (1992).

[5.15] G. Peters and M. Jansen. *Angew. Chem. Int. (Ed. English)*, **31**, 223 (1992).

[5.16] A. Mittlebach, W. Hönle, H. G. van Schnering, J. Carlsen, R. Janiak, and H. Quast. *Angew. Chem. Int. (Ed. English)*, **31**, 1640 (1992).

[5.17] A. S. Koch, K. C. Khemani, and F. C. Wudl. *J. Org. Chem.*, **56**, 4543 (1991).

[5.18] P. C. Eklund. Private communication.

[5.19] D. H. Parker, P. Wurz, K. Chatterjee, K. R. Lykke, J. E. Hunt, M. J. Pellin, J. C. Hemminger, D. M. Gruen, and L. M. Stock. *J. Am. Chem. Soc.*, **113**, 7499 (1991).

[5.20] R. Taylor, J. P. Hare, A. K. Abdul-Sada, and H. W. Kroto. *J. Chem. Soc. Chem. Commun.*, **20**, 1423 (1990).

[5.21] H. Ajie, M. M. Alvarez, S. J. Anz, R. D. Beck, F. Diederich, K. Fostiropoulos, D. R. Huffman, W. Krätschmer, Y. Rubin, K. E. Schriver, D. Sensharma, and R. L. Whetten. *J. Phys. Chem.*, **94**, 8630 (1990).

[5.22] M. S. de Vries, H. R. Wendt, H. Hunziker, E. Peterson, and S. Chang. In *Proc. of the XXII Lunar and Planetary Sciences Conference*, p. 315 (1991).

[5.23] M. A. Wilson, L. S. K. Pang, R. A. Quezada, K. J. Fischer, I. G. Dance, and G. D. Willett. *Carbon*, **31**, 393 (1993).

[5.24] M. Diack, R. L. Hettich, R. N. Compton, and G. Guiochon. *Anal. Chem.*, **64**, 2143 (1992).

[5.25] J. Abrefah, D. R. Olander, M. Balooch, and W. J. Siekhaus. *Appl. Phys. Lett.*, **60**, 1313 (1992).

[5.26] C. Pan, M. P. Sampson, Y. Chai, R. H. Hauge, and J. L. Margrave. *J. Phys. Chem.*, **95**, 2944 (1991).

[5.27] G. B. Vaughan, P. A. Heiney, J. E. Fischer, D. E. Luzzi, D. A. Ricketts-Foot, A. R. McGhie, Y. W. Hui, A. L. Smith, D. E. Cox, W. J. Romanow, B. H. Allen, N. Coustel, J. P. McCauley, Jr., and A. B. Smith III. *Science*, **254**, 1350 (1991).

[5.28] D. Schmicker, S. Schmidt, J. G. Skoffronick, J. P. Toennies, and R. Vollmer. *Phys. Rev. B*, **44**, 10995 (1991).

[5.29] N. Sivaraman, R. Dhamodaran, I. Kaliappan, T. G. Srinivasan, P. R. V. Rao, and C. K. Mathews. *J. Org. Chem.*, **57**, 6077 (1992).

[5.30] R. S. Ruoff, D. S. Tse, R. Malhotra, and D. C. Lorents. *J. Phys. Chem.*, **97**, 3379 (1993).

[5.31] J. Hildebrand, J. M. Prausnitz, and R. L. Scott. *Regular and Related Solutions: The Solubility of Gases, Liquids and Solids*. Van Nostrand Reinhold, New York (1970).

[5.32] J. Hildebrand and R. L. Scott. *The Solubility of Nonelectrolytes*. Van Nostrand Reinhold, New York (1950).

[5.33] H. H. Willard, L. L. Merritt, Jr., J. A. Dean, and F. A. Settle, Jr. *Instrumental Methods of Analysis*, 7th ed., p. 513. Wadsworth Publishing Co., Belmont, CA (1988).

[5.34] F. Diederich and R. L. Whetten. *Accounts of Chemical Research*, **25**, 119 (1992).

[5.35] P.-M. Allemand, A. Koch, F. Wudl, Y. Rubin, F. Diederich, M. M. Alvarez, S. J. Anz, and R. L. Whetten. *J. Am. Chem. Soc.*, **113**, 1050 (1991).

[5.36] M. S. Meier and J. P. Selegue. *J. Org. Chem.*, **57**, 1925 (1992).

[5.37] A. Gügel, M. Becker, D. Hammel, L. Mindach, J. Rader, T. Simon, M. Wagner, and K. Mullen. *Angew. Chem.*, **104**, 666 (1992).

[5.38] A. Gügel, M. Becker, D. Hammel, L. Mindach, J. Rader, T. Simon, M. Wagner, and K. Mullen. *Angew. Chem. Int. (Ed. English)*, **31**, 644 (1992).

[5.39] L. Isaacs, A. Wehrsig, and F. Diederich. Preprint (1993).

[5.40] K. C. Khemani, M. Prato, and F. Wudl. *J. Org. Chem.*, **57**, 3254 (1992).

[5.41] W. A. Scrivens, P. V. Bedworth, and J. M. Tour. *J. Am. Chem. Soc.*, **114**, 7917 (1992).

[5.42] A. M. Rao, K.-A. Wang, J. M. Holden, Y. Wang, P. Zhou, P. C. Eklund, C. C. Eloi, and J. D. Robertson. *J. Mater. Sci.*, **8**, 2277 (1993).

[5.43] J. M. Hawkins and A. Meyer. *Science*, **260**, 1918 (1993). See also Science News, vol. 144, July 3, 1993, p. 7.

[5.44] R. Ettl, I. Chao, F. Diederich, and R. L. Whetten. *Nature (London)*, **353**, 443 (1991).

[5.45] J. M. Hawkins, S. Loren, A. Meyer, and R. Nunlist. *J. Am. Chem. Soc.*, **113**, 7770 (1991).

[5.46] J. M. Hawkins. *Accounts of Chemical Research*, **25**, 150 (1992).

[5.47] J. M. Hawkins, A. Meyer, T. Lewis, and S. Loren. In G. S. Hammond and V. J. Kuck (eds.), *Fullerenes*, p. 91. American Chemical Society, Washington, DC (1992). ACS Symposium Series, 481, Chapter 6.

[5.48] G. A. Olah, I. Bucsi, R. Aniszfeld, and G. K. S. Prakash. *Carbon*, **30**, 1203 (1992).

[5.49] D. M. Cox, R. D. Sherwood, P. Tindall, K. M. Kreegan, W. Anderson, and D. J. Martella. In G. S. Hammond and V. J. Kuck (eds.), *Fullerenes: Synthesis, Properties and Chemistry of Large Carbon Clusters.*, p. 117, American Chemical Society, Washington, DC (1992). ACS Symposium Series, 481.

[5.50] N. J. Halas. Private communication.

[5.51] C. Yeretzian, J. B. Wiley, K. Holczer, T. Su, S. Nguyen, R. B. Kaner, and R. L. Whetten. *J. Phys. Chem.*, **97**, 10097 (1993).

[5.52] D. M. Cox, S. Behal, M. Disko, S. M. Gorun, M. Greaney, C. S. Hsu, E. B. Kollin, J. M. Millar, J. Robbins, W. Robbins, R. D. Sherwood, and P. Tindall. *J. Am. Chem. Soc.*, **113**, 2940 (1991).

[5.53] C. Yeretzian, K. Hansen, R. D. Beck, and R. L. Whetten. *J. Chem. Phys.*, **98**, 7480 (1993).

[5.54] D. S. Bethune, C. H. Kiang, M. S. de Vries, G. Gorman, R. Savoy, J. Vazquez, and R. Beyers. *Nature (London)*, **363**, 605 (1993).

[5.55] T. Guo, M. D. Diener, Y. Chai, J. M. Alford, R. E. Haufler, S. M. McClure, T. R. Ohno, J. H. Weaver, G. E. Scuseria, and R. E. Smalley. *Science*, **257**, 1661 (1992).

[5.56] L. Moro, R. S. Ruoff, C. H. Becker, D. C. Lorentz, and R. J. Malhotra. *J. Phys. Chem.*, **97**, 6801 (1993).

[5.57] R. D. Averitt, J. M. Alford, and N. J. Halas. *Appl. Phys. Lett.*, **65**, 374 (1994).

[5.58] G. Burrows. *Molecular Distillation*. Clarendon Press, Oxford (1960).

[5.59] C. K. Mathews, M. S. Baba, T. S. L. Narasimhan, R. Balasubramanian, N. Sivaraman, T. G. Srinivasan, and P. R. V. Rao. *J. Phys. Chem.*, **96**, 3566 (1992).

[5.60] C. K. Mathews, P. R. V. Rao, T. G. Srinivasan, V. Ganesan, N. Sivaraman, T. S. L. Narasimham, I. Kaliappan, K. Chandran, and R. Dhamodaran. *Current Sci. (India)*, **61**, 834 (1991).

[5.61] C. K. Mathews, M. S. Baba, T. S. L. Narasimhan, R. Balasubramanian, N. Sivaraman, T. G. Srinivasan, and P. R. V. Rao. *Fullerene Sci. Technol.*, **1**, 101 (1993).
[5.62] C. Kittel. *Elementary Statistical Physics*, p. 141. J. Wiley & Sons, New York (1958).
[5.63] M. S. Baba, T. S. L. Narasimhan, R. Balasubramanian, N. Sivaraman and C. K. Mathews. *J. Phys. Chem.*, **98**, 1333 (1994).
[5.64] J. R. Heath, S. C. O'Brien, Q. Zhang, Y. Liu, R. F. Curl, H. W. Kroto, F. K. Tittel, and R. E. Smalley. *J. Am. Chem. Soc.*, **107**, 7779 (1985).
[5.65] R. E. Smalley. *Accounts of Chem. Research*, **25**, 98 (1992).
[5.66] R. E. Smalley. In G. S. Hammond and V. J. Kuck (eds.), *Large Carbon Clusters*, p. 199, American Chemical Society, Washington, DC (1991).
[5.67] R. E. Smalley. *Materials Science and Engineering*, **B19**, 1 (1993).
[5.68] Y. Chai, T. Guo, C. M. Jin, R. E. Haufler, L. P. Felipe Chibante, J. Fure, L. H. Wang, J. M. Alford, and R. E. Smalley. *J. Phys. Chem.*, **95**, 7564 (1991).
[5.69] R. D. Johnson, D. S. Bethune, and C. S. Yannoni. *Accounts of Chem. Res.*, **25**, 169 (1992).
[5.70] F. D. Weiss, S. C. O'Brien, J. L. Eklund, R. F. Curl, and R. E. Smalley. *J. Am. Chem. Soc.*, **110**, 4464 (1988).
[5.71] L. S. Wang, J. M. Alford, Y. Chai, M. Diener, J. Zhang, S. M. McClure, T. Guo, G. E. Scuseria, and R. E. Smalley. *Chem. Phys. Lett.*, **207**, 354 (1993).
[5.72] L. S. Wang, J. M. Alford, Y. Chai, M. Diener, and R. E. Smalley. *Z. Phys. D (suppl. issue)*, **26**, 297 (1993).
[5.73] S. Saito. In R. S. Averback, J. Bernholc, and D. L. Nelson (eds.), *Clusters and Cluster-Assembled Materials*, MRS Symposia Proceedings, Boston, vol. 206, p. 115. Materials Research Society Press, Pittsburgh, PA (1991).
[5.74] A. Rosén. *Z. Phys. D*, **25**, 175 (1993).
[5.75] R. F. Curl. *Carbon*, **30**, 1149 (1992).
[5.76] T. Pradeep, G. U. Kulkarni, K. R. Kannan, T. N. G. Row, and C. N. Rao. *J. Am. Chem. Soc.*, **114**, 2272 (1992).
[5.77] T. Weiske, J. Hrusak, D. K. Bohme, and H. Schwarz. *Chem. Phys. Lett.*, **186**, 459 (1991).
[5.78] T. Weiske, D. K. Bohme, and H. Schwarz. *J. Phys. Chem.*, **95**, 8451 (1991).
[5.79] K. A. Caldwell, D. E. Giblin, C. C. Hsu, D. Cox, and M. L. Gross. *J. Am. Chem. Soc.*, **113**, 8519 (1991).
[5.80] M. Saunders, H. A. Jimenez-Vazquez, R. J. Cross, and R. J. Poreda. *Science*, **259**, 1428 (1993).
[5.81] R. Huang, H. Li, W. Lu, and S. Yang. *Chem. Phys. Lett.*, **228**, 111 (1994).
[5.82] R. F. Kiefl, T. L. Duty, J. W. Schneider, A. MacFarlane, K. Chow, J. W. Elzey, P. Mendels, G. D. Morris, J. H. Brawer, E. J. Ansaldo, C. Niedermayer, D. R. Nokes, C. E. Stronach, B. Hitti, and J. E. Fischer. *Phys. Rev. Lett.*, **69**, 2005 (1992).
[5.83] H. Shinohara, H. Yamaguchi, N. Hayashi, H. Sato, M. Inagaki, Y. Saito, S. Bandow, H. Kitagawa, T. Mitani, and H. Inokuchi. *Materials Science and Engineering*, **B19**, 25 (1993).
[5.84] M. M. Alvarez, E. G. Gillan, K. Holczer, R. B. Kaner, K. S. Min, and R. L. Whetten. *J. Phys. Chem.*, **95**, 10561 (1991).
[5.85] M. M. Ross, H. J. Nelson, J. H. Callahan, and S. W. McElvany. *J. Phys. Chem.*, **96**, 5231 (1992).
[5.86] M. E. J. Boonman, P. H. M. van Loosdrecht, D. S. Bethune, I. Holleman, G. J. M. Meijer, and P. J. M. van Bentum. *Physica B*, **211**, 323 (1995).
[5.87] R. D. Johnson, M. S. de Vries, J. Salem, D. S. Bethune, and C. S. Yannoni. *Nature (London)*, **355**, 239 (1992).

[5.88] K. Kikuchi, S. Suzuki, Y. Nakao, N. Nakahara, T. Wakabayashi, H. Shiromaru, I. Ikemoto, and Y. Achiba. *Chem. Phys. Lett.*, **216**, 67 (1993).

[5.89] Y. Achiba, T. Wakabayashi, T. Moriwaki, S. Suzuki, and H. Shiromaru. *Materials Science and Engineering*, **B19**, 14 (1993).

[5.90] H. Shinohara, H. Sato, Y. Saito, A. Izuoka, T. Sugawara, , H. Ito, T. Sakurai, and M. T. *J. Phys. Chem.*, **96**, 3571 (1992).

[5.91] H. Shinohara, H. Sato, Y. Saito, M. Ohkohchi, and Y. Ando. *Rapid Commun. Mass Spectrum*, **6**, 413 (1992).

[5.92] C. S. Yannoni, M. Hoinkis, M. S. de Vries, D. S. Bethune, J. R. Salem, M. S. Crowder, and R. D. Johnson. *Science*, **256**, 1191 (1992).

[5.93] H. Shinohara, H. Sato, M. Ohkohchi, Y. Ando, T. Kodama, T. Shida, T. Kato, and Y. Saito. *Nature (London)*, **357**, 52 (1992).

[5.94] A. Bartl, L. Dunsch, J. Froehner, and U. Kirbach. *Chem. Phys. Lett.*, **229**, 115 (1994).

[5.95] J. H. Weaver, Y. Chai, G. H. Kroll, C. Jin, T. R. Ohno, R. E. Haufler, T. Guo, J. M. Alford, J. Conceicao, L. P. Chibante, A. Jain, G. Palmer, and R. E. Smalley. *Chem. Phys. Lett.*, **190**, 460 (1992).

[5.96] D. S. Bethune, R. D. Johnson, J. R. Salem, M. S. de Vries, and C. S. Yannoni. *Nature (London)*, **366**, 123 (1993).

[5.97] S. Bandow, H. Kitagawa, T. Mitani, H. Inokuchi, Y. Saito, H. Yamaguchi, N. Hayashi, H. Sato, and H. Shinohara. *J. Phys. Chem.*, **96**, 9609 (1992).

[5.98] H. Shinohara, H. Yamaguchi, N. Hayashi, H. Sato, M. Ohkohchi, Y. Ando, and Y. Saito. *J. Phys. Chem.*, **97**, 4259 (1993).

[5.99] P. H. M. van Loosdrecht, R. D. Johnson, R. Beyers, J. R. Salem, M. S. de Vries, D. S. Berhune, P. Burbank, J. Haynes, T. Glass, S. Stevenson, H. C. Dorn, M. Boonman, P. J. M. van Bentum, and G. Meijer. In K. M. Kadish and R. S. Ruoff (eds.), *Recent Advances in the Chemistry and Physics of Fullerenes and Related Materials: Electrochemical Society Symposia Proceedings, San Francisco, May 1994*, vol. 94–24, p. 1320, Electrochemical Society, Pennington, NJ (1994).

[5.100] S. Stevenson, H. C. Dorn, P. Burbank, K. Harich, J. Haynes, Jr, C. H. Kiang, J. R. Salem, M. S. DeVries, P. H. M. van Loosdrecht, R. D. Johnson, C. S. Yannoni, and D. S. Bethune. *Annal. Chem.*, **66**, 2675 (1994).

[5.101] S. Stevenson, H. C. Dorn, P. Burbank, K. Harich, Z. Sun, C. H. Kiang, J. R. Salem, M. S. DeVries, P. H. M. van Loosdrecht, R. D. Johnson, C. S. Yannoni, and D. S. Bethune. *Annal. Chem.*, **66**, 2680 (1994).

[5.102] L. P. Zakharenko, I. K. Zakharov, S. N. Lunegov, and A. A. Nikiforov. *Doklady Akademii Nauk*, **335**, 261 (1994).

CHAPTER 6

Fullerene Growth, Contraction, and Fragmentation

This chapter reviews models for the growth, contraction, and fragmentation of fullerenes. The models for the growth process that have been proposed for fullerenes are discussed in §6.1. Models for the addition and subtraction of a single hexagon C_2 to and from a fullerene C_{n_C} are considered. Because of the great importance of mass spectrometry to growth, contraction, and fragmentation studies of fullerenes, a brief review of mass spectrometry techniques is given in §6.2. Special attention is given to the experimentally observed stability of C_{60} and C_{70} in §6.3. Experiments showing the accretion and contraction of fullerenes using photofragmentation and ion collision techniques are reviewed in §6.4. Finally, some discussion is given in §6.5 of molecular dynamics models for fullerene growth.

6.1. Fullerene Growth Models

Referring to the mass spectra for fullerenes (see Fig. 1.4), we see that the C_{60} and C_{70} species have the highest relative abundances and therefore are observed to have the greatest relative stability of the fullerenes. Other fullerenes with relatively high stability are C_{76}, C_{78} and C_{84} [6.1–3], but C_{78} and C_{84} have many isomers that obey the isolated pentagon rule (see §3.2). On the basis of Euler's theorem, a fullerene shell has 12 pentagons and an arbitrary number of hexagons (see §3.1). The addition of one extra hexagon to a fullerene requires two additional carbon atoms. Thus, all closed-shell fullerenes have an even number of carbon atoms (see §3.1), and therefore it would seem logical to go from one fullerene to another by the addition of a C_2 unit. However, to go from one highly stable fullerene, such as C_{60}, to the next highly stable fullerene, such as C_{70} by adding C_2 clusters, it is necessary to overcome a large potential barrier [6.4]. Therefore it has been suggested that, during synthesis at high temperature, stable fullerenes form from the coalescence of larger units such as C_{18} or C_{24} or C_{30}, perhaps in the form of rings, which then add or emit a few C_2 clusters to reach the highly stable fullerenes [6.5, 6].

To achieve the most stable molecular configurations for a given number of carbon atoms n_C, fullerenes must rearrange the locations of their carbon atoms to separate the pentagons from one another in order to minimize the local curvature of the fullerenes [6.7]. In §6.1.1 we discuss first the most common mechanism that has been proposed for the rearrangement of carbon atoms on a fullerene molecule, the Stone–Wales model [6.8], followed by a generalization of the Stone–Wales mechanism for the rearrangement of pentagons (5 edges), hexagons (6 edges), and heptagons (7 edges) and for the formation and movement of pentagon–heptagon (5,7) pairs on the surface of fullerenes. Models for the growth mechanism for fullerenes are then discussed.

6.1.1. Stone–Wales Model

In describing the growth of fullerenes, the formation of various fullerene isomers (see §3.2), or the growth of carbon nanotubes or nanospheres, the Stone–Wales transformation is often discussed [6.8, 9]. According to this model, the pyracylene motif [see Figs. 3.2(c) and 6.1] is, for example, transformed from a vertical orientation to a horizontal orientation, by breaking the bond connecting two hexagons and converting the two vertices on the horizontal line to two vertices on the vertical line, without modifying the bonds on any but the four polygons of the pyracylene structure, as illustrated in Fig. 6.1 [6.10].

The Stone–Wales transformation has been used by many workers as a method for moving pentagonal faces around the surface of fullerene structures, thereby providing a mechanism for achieving the maximum separations between all pentagonal faces and stabilizing the structure. The most stable structure is one for which the polygon arrangement best satisfies the isolated pentagon rule (see §3.1), or more precisely, the arrangement with minimal energy. The Stone–Wales model was first used to maximize the separation between the pentagonal faces in C_{28} [6.10]. Although the initial and final states for the Stone–Wales transformation may be relatively stable, corresponding to local energy minima, the transition between these states involves a large potential barrier [6.10] estimated to be ~7 eV, on the basis of cutting the two σ bonds [6.11] required to move a pentagon. This value

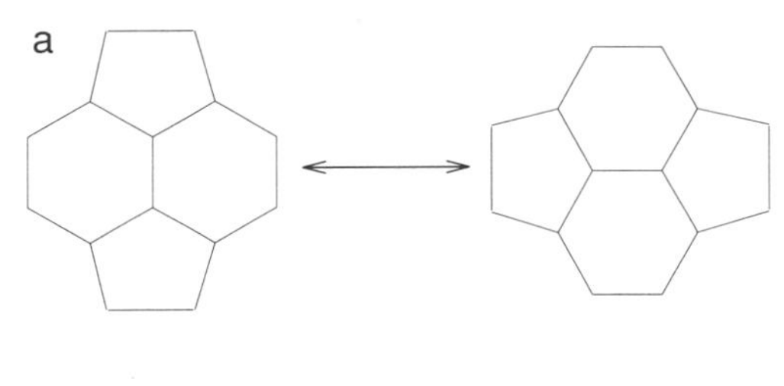

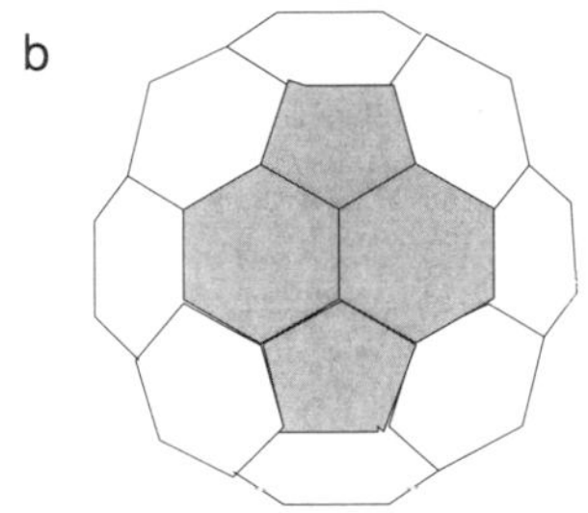

Fig. 6.1. (a) Stone–Wales transformation of the pyracylene motif between the vertical and horizontal orientations. (b) The embedding of the pyracylene motif (shaded) into a fullerene shell [6.10].

for the potential barrier is considerably higher than what a physical system can surmount under actual synthesis conditions (3000–3500°C or ~0.3 eV).

Even though the Stone–Wales mechanism may not be the actual physical mechanism that conveniently interchanges pentagons and hexagons on fullerene surfaces, the model is useful geometrically to generate all fullerene isomers obeying the isolated pentagon rule for fixed n_C [6.10, 12]. The Stone–Wales mechanism thus provides a mechanism for moving specific polygons around a fullerene surface, as illustrated below in the discussion of growth mechanisms. Figure 6.2 shows an extension of the Stone–Wales model for the basic interchange of pentagons, hexagons, and heptagons among four polygons without changing the number of carbon atoms. Although fullerenes contain only hexagons and pentagons, as discussed in connection with Euler's theorem, heptagons are observed experimentally in connection with the formation of a concave surface (see §3.1) and in the formation of bill-like structures for carbon nanotubes (see §19.2.3). In the growth process, heptagons can be used to move pentagons around the surface of fullerenes, by forming a (7,5) heptagon–pentagon pair, moving the pentagon, and then annihilating the (7,5) pair.

Each pentagon introduces long-range positive convex curvature to the fullerene shell corresponding to an angle of $\pi/6$, while a heptagon introduces a similar long-range concave curvature. On the other hand, a (7,5) pair produces only local curvatures, since the long-range effects of the pentagon and heptagon cancel. Thus, the energy to create a (7,5) pair should be relatively low. Furthermore, the introduction of the (7,5) pair does not

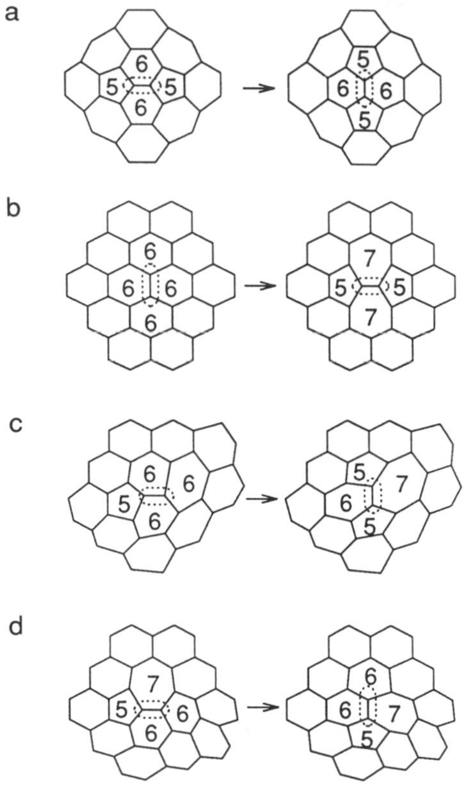

a b c d

Fig. 6.2. Diagrams describing basic steps for the Stone–Wales transformation and showing interchanges between polygons within a four-polygon motif that includes pentagons (5), hexagons (6), and heptagons (7), where the number in parentheses indicates the number of edges of a polygon. (a) Two pentagon movements as in the standard Stone–Wales transformation, (5,6,5,6)→(6,5,6,5) without creation of a (7,5) pair. (b) Creation of two (7,5) pairs according to the (6,6,6,6)→(7,5,7,5) transformation. (c) Creation of a single (7,5) pair, based on (5,6,6,6)→(6,5,7,5). (d) Motion of a (7,5) pair, based on (5,7,6,6)→(6,6,7,5) [6.13].

violate Euler's theorem, so that a fullerene C_{n_C} containing f faces including a (7,5) pair would in addition have 12 pentagonal faces and $f - 14$ hexagonal faces. Figure 6.2(a) shows the conventional Stone–Wales transformation for the rearrangement of the pentagons in the pyracylene motif [see Fig. 3.2(c)] [6.13]. Figure 6.2(b) shows the creation of two (7,5) pairs from four hexagons in the arrangement shown in the figure, or equivalently the annihilation of two (7,5) pairs into four hexagons. Figure 6.2(c) shows the creation (or annihilation) of a (7,5) pair in the presence of a pentagon by the extended Stone–Wales mechanism introduced in Fig. 6.2(b). Correspondingly, Fig. 6.2(d) shows a method for moving a (7,5) pair around the surface of a fullerene, thereby also moving a pentagon around the surface. The (7,5) pair mechanisms shown in Figs. 6.2(c) and (d) assume no dangling bond formation [6.13]. Although there is no experimental evidence for heptagons on fullerene surfaces, examples of heptagons have been reported to be responsible for introducing bends and diameter changes in

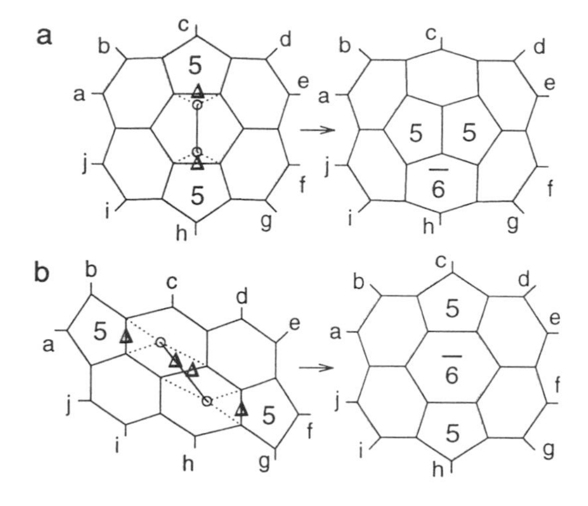

Fig. 6.3. Model for the absorption of a C_2 cluster near two pentagons without generating any new pentagons or heptagons. The new hexagon is denoted by $\overline{6}$. The relative distance between the two pentagons, expressed as a linear combination of unit vectors of the honeycomb lattice, becomes closer after C_2 absorption in (a) (2,0)→(1,0) and (b) (2,1)→(2,0), respectively. Dotted lines indicate new bonds formed to accommodate the C_2 unit, and Δ denotes bonds that are broken [6.13].

carbon nanotubes, as observed in transmission electron microscopy (TEM) experiments (see §19.2.3).

6.1.2. Model for C_2 Absorption or Desorption

One model for the growth of fullerenes is based on the absorption of a C_2 cluster, thereby adding one hexagon in accordance with Euler's theorem (see §3.1). Two possible methods for absorbing one C_2 cluster into a carbon shell, restricted to contain only pentagonal and hexagonal faces, are shown in Fig. 6.3 [6.13]. The absorption takes place in the vicinity of a pentagon, where bonds are broken (indicated on the figure by Δ) and new bonds are formed (dotted lines). In order to preserve the three-bond connectivity at each carbon atom, the numbers of new bonds and deleted bonds for each vertex must be equal. The chemical bonds denoted by $a, b, c, \ldots$ (which connect carbon atoms within the cluster to carbon atoms outside the cluster) do not change during the C_2 absorption (desorption) and thus only local bonds are changed by the C_2 absorption process.

The pathway for the absorption of a C_2 dimer by a C_{n_C} fullerene shown in Fig. 6.3(a) leads to two adjacent pentagons on the fullerene surface. This arrangement is energetically unfavorable because it does not satisfy the isolated pentagon rule. However, the absorption pathway shown in Fig. 6.3(b) results in a configuration C_{n_C+2} that is consistent with the isolated pentagon rule and therefore could be used more effectively to model the growth of fullerenes. In Fig. 6.4, we show a fullerene hemisphere with six pentagons, with the pentagon labeled 6 in a central position. The relative separation between pentagons 2 and 3 in Fig. 6.4(a) corresponds to the $(n, m) = (2, 1)$ distance (see §3.2), which is active for absorbing a C_2 unit in accordance

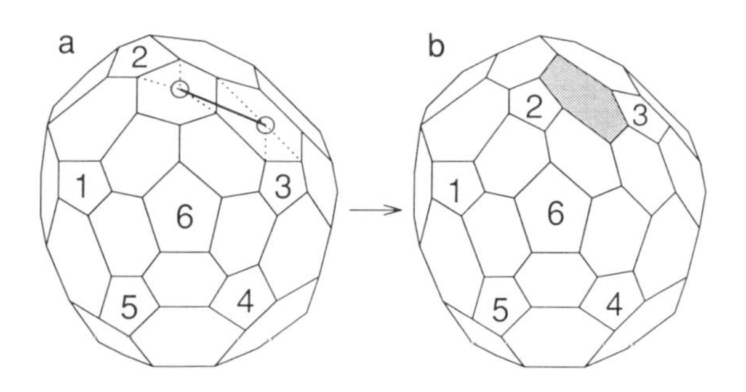

Fig. 6.4. A C_2 cluster is added to two hexagons, each of which is bounded by a pentagon. The process adds one hexagonal face (dark shading), two carbon vertices, and three edges, thereby satisfying Euler's theorem and yielding the fullerene shown in (b) [6.13]. The dashed lines represent the six new bonds that form to replace the three broken bonds [6.14].

with the mechanism shown in Fig. 6.3(b). When we introduce a C_2 unit between two pentagons according to the process of Fig. 6.3(b), a new hexagon is introduced [dark shading in Fig. 6.4(b)]. If we rotate the picture in Fig. 6.4(b) by $\approx 72^\circ$ around the axis of the hemisphere (i.e., an axis normal to pentagon 6), it is seen that the resulting shape of the hemispherical cap is identical to that of Fig. 6.4(a), except for the symmetric addition of five hexagons, symmetrically placed with respect to pentagon 6, indicative of a screw-type crystal growth. Repeating the process shown in Fig. 6.4 five times, we return to Fig. 6.4(a), but with the addition of five carbon hexagons to the fullerene. The application of this growth mechanism to carbon tubules is further discussed in §19.3.

Energetically, the absorption of a C_2 unit is exothermic because each carbon atom in the C_2 cluster, having two *sp*-hybridized σ bonds and two *sp*-hybridized π bonds, has a relatively high energy before C_2 absorption. However, after C_2 absorption, each of the two carbon atoms in C_2 has three σ bonds and one π bond, representing a decrease in energy of $\sim$4 eV per carbon atom [6.15]. In terms of the initial and final states, the absorption of a C_2 dimer is therefore expected to be energetically favorable, as shown in Fig. 6.5, although the transition between C_{n_C} and C_{n_C+2} requires surmounting a high ($\sim$7 eV) potential barrier [6.10] (not shown in Fig. 6.5), corresponding physically to the rearrangements of bonds that are required to make the transition, such as the bonds indicated schematically by Fig. 6.3(b).

6.1.3. Fullerene Growth from a Corannulene Cluster

It has been suggested that a corannulene-like carbon cluster (i.e., the same carbon atom organization as for $C_{20}H_{10}$ shown in Fig. 6.3) could be the nucleation site for the formation of a fullerene in a carbon plasma [6.17, 18].

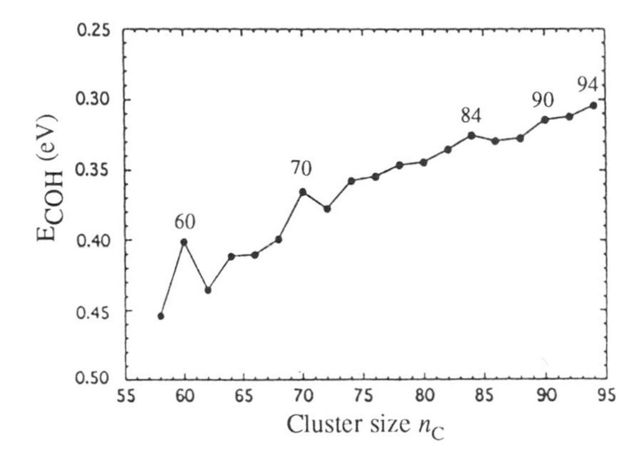

Fig. 6.5. The cohesive energy (eV/C atom) for fullerenes, plotted as a function of cluster size, where E_{COH} denotes the cohesive energy relative to bulk graphite [6.16]. Not shown in this figure is the large potential barrier between stable C_{n_C} fullerene, where n_C, the number of carbon atoms, is even.

In a hot carbon plasma (operating above ~3500°C), carbon monomers, dimers, clusters and rings are expected to be present, and the larger carbon structures that form at these high temperatures are rather flexible [6.19]. Each corannulene cluster contains five V-shaped edge sites (see Fig. 3.2(a)) and five flat edges with one carbon atom at each end of the edges. We can argue that the V-shaped sites are the more active sites because a single C_2 dimer can form a pentagon at that site (see Fig. 6.6). Once a new row of pentagons is formed at the five adjacent V-shaped sites, the addition of a single C_2 dimer is sufficient to form a new hexagonal face. After five C_2 dimers are absorbed, five hexagons are added, thus completing the second ring of polygons about the corannulene cluster and forming a hemisphere of the C_{60} molecule. The next ring of hexagonal faces can then be formed by subsequent addition of C_2 dimers following the approach just described. The following ring of pentagons can be formed by the addition of a single C monomer per new pentagon. The final completion of the whole fullerene can then be accomplished simply by five C_2 additions, resulting in five new hexagons (see Fig. 6.6).

Some authors have suggested [6.20] that fullerenes are assembled from ring networks in the plasma. The construction suggested above can also be carried out with ring networks of appropriate size, i.e., rings of five pentagons and rings of five hexagons. This is called the "ring road" to fullerene formation, in contrast to the "pentagon road" discussed above. In a sense, the correlated addition of multiple C_2 dimers and isolated C atoms is strongly related to the addition of rings of carbon atoms, as is further discussed in §6.1.4. A totally different method for the synthesis of fullerenes by a purely chemical route has been suggested by R. Taylor [6.21] and G. H. Taylor *et al.* [6.22].

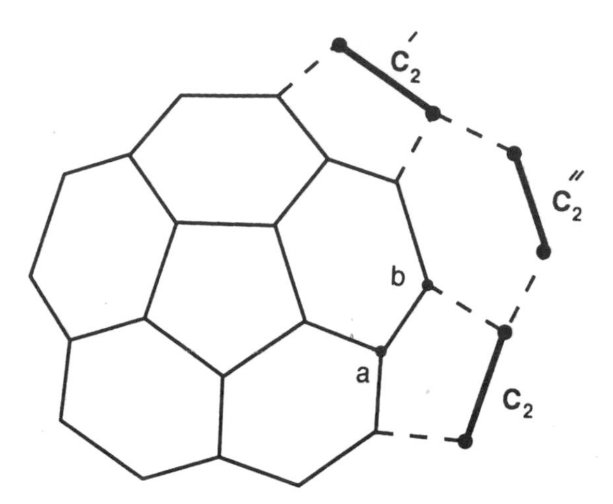

Fig. 6.6. A corannulene cluster grows with the addition of C_2 clusters. For the initial corannulene cluster, there are 15 carbon atoms at the periphery. Five of these (a sites) are at V-shaped sites and ten are at edge sites denoted by (b). The addition of two C_2 units (denoted by C_2 and C_2') forms pentagons, while the subsequent addition of C_2'' forms an added hexagon.

Closely related to the C_2 dimer absorption discussed above is the carbon dimer (C_2), tetramer (C_4), and hexamer (C_6) emission shown in Fig. 6.7 [6.4]. Kroto and Iijima *et al.* [6.23, 24] have suggested that the formation of fullerene balls on the surface of a carbon tubule (see §19.2.5) is due to the curling up of a fragment of a graphene sheet near a defect in the surface layer of the tubule.

6.1.4. *Transition from C_{60} to C_{70}*

As mentioned above, there is some difficulty in accounting for a reaction pathway between C_{60} and C_{70}, since both C_{60} and C_{70} have relative potential minima and both appear as relative maxima in the cohesive energy plot of Fig. 6.5. This cohesive energy plot implies that C_{60} and C_{70} are especially stable and the intermediate fullerenes $C_{62}, \ldots, C_{68}$ are less stable. In Fig. 6.5 we see that C_{60} has the most pronounced local maximum in the cohesive energy and C_{70} has the second most pronounced local maximum [6.16]. The general upward background in Fig. 6.5 reflects an increase in binding energy of ~11 meV per C_2 dimer addition. The limit $n_C \to \infty$ corresponds to the binding energy of graphite. It should be noted that the binding energy for C_2 is 3.6 eV/C atom. According to Fig. 6.5 the cohesive energy of C_{70} is about 35 meV higher than that for C_{60}. However, for each of the less stable fullerenes, as well as for the stable fullerenes, the sum of the cohesive energies for C_{n_C} and for C_2 is less than that for C_{n_C+2}, thus leading to the C_2 dimer absorption process. Thus, while the addition of one C_2 dimer from C_{60} is relatively unfavorable by the C_2 absorption pathway, the probability for the addition of a second C_2 dimer at the opposite side of the fullerene is

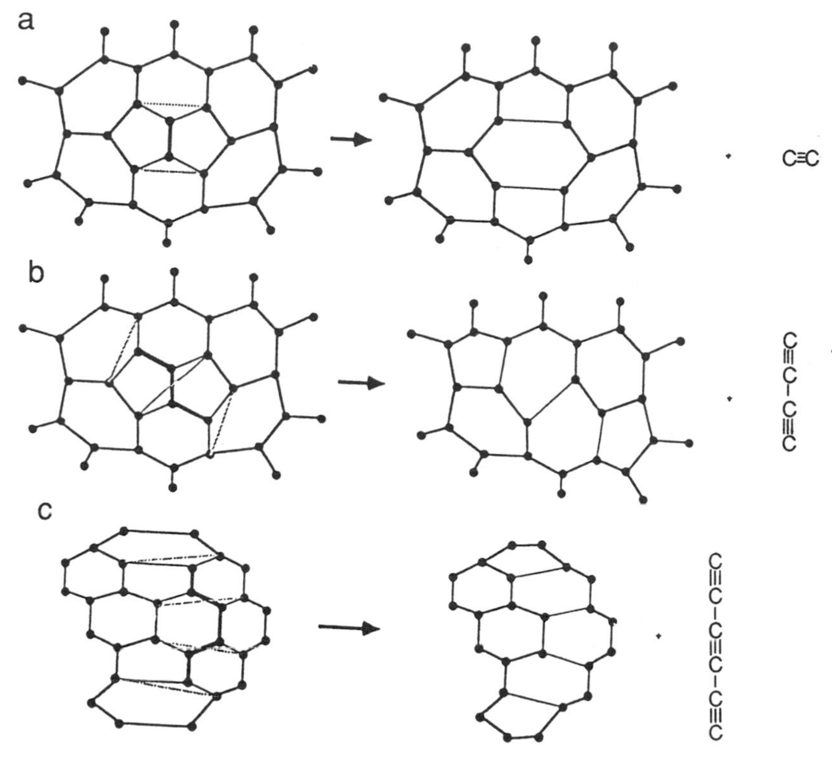

Fig. 6.7. Possible mechanisms for the fragmentation of a spheroidal carbon shell to lose (a) C_2, (b) C_4, or (c) C_6 clusters, while maintaining the integrity of the fullerene cage [6.4]. The new bonds are shown by dashed lines on the left.

enhanced before the first C_2 dimer is desorbed. Physically, the absorption of two or more C_2 dimers following the absorption of the first dimer is favorable since the C_2 absorption partially relieves the strain introduced by the absorption of the first dimer. If two dimers are thus absorbed, then the probability of quickly adding three additional C_2 dimers is enhanced, because the addition of five C_2 dimers completes the ring of five additional hexagons needed to form C_{70}.

The correlated addition of one or more C_2 dimers provides an intuitive mechanism for relieving local strain. Thus, the correlated sequential addition of five C_2 dimers is closely related to the addition of a ring of five hexagons, so that the sequential correlated C_2 dimer addition process provides a mechanism for the catalytic self-assembly of rings at preferred locations on fullerene balls. Either the correlated absorption of five C_2 dimers or the absorption of a whole ring of five hexagons provides a pathway for going from C_{60} to C_{70}. This concept can be generalized to a growth process from one relative potential minimum to another on the cohesive energy E_{COH} *vs.* n_C plot in Fig. 6.5. The same type of correlated self-assembly process can also be envisaged for the decay of fullerenes in going between C_{70} and C_{60}, under an emission process that could, for example, be initiated by ion or laser irradiation.

6.2. Mass Spectrometry Characterization

Mass spectrometry has provided an extremely valuable technique for the characterization of fullerene specimens regarding their mass distribution (§6.3), the detection of their various stable charge states (§6.3) and of metallofullerenes (§5.4 and §8.2), and the identification of the various oligomers in photopolymerized fullerenes (see §7.5.1). Mass resolution of $m/\Delta m > 10^6$ can be achieved with presently available mass spectrometers, and signals from as few as 10 ions can be detected [6.25]. The high-mass resolution is exploited for the unambiguous identification of high-charge states of fullerenes, and high-resolution techniques have been extensively utilized in photofragmentation studies of fullerenes to detect isotopes for a given molecular species. The high sensitivity of mass spectrometry instruments provides quantitative information about very small samples, which is of great use in studying high-mass fullerenes (such as C_{84}) and metallofullerenes [such as La@C_{82} (see §8.2)].

A variety of mass spectrometry instruments are used for the characterization of fullerenes, including quadrupole time-of-flight instruments and Fourier transform mass spectrometers. The specific instrument selection depends somewhat on the type of measurement to be performed and the sensitivity that is needed. In time-of-flight instruments, the mass m is proportional to the square of the time of flight t.

To steer, trap, manipulate, and detect the various molecules and clusters in a mass spectrometer, the fullerenes are normally charged either positively or negatively. Lasers are commonly used both to ionize fullerenes and to desorb fullerenes from the surfaces. Ionization capabilities to form positive ions are often provided in a mass spectrometer, although negative ions are also used. A variety of lasers are used for ionization purposes, such as Nd:YAG operating either on its fundamental frequency or on one of its harmonics (1064, 532, 355, 266 nm). Laser desorption is used to remove trace quantities of charged ions from a sample for mass spectral characterization. Single-photon ionization, however, requires high photon energies ($\hbar\omega > 7.58$ eV for C_{60} and 7.3 eV for C_{70} [6.26]). Other common modes of ionization include ^{252}Cf plasma desorption, fast atom bombardment, and electrospray ionization [6.25, 27].

Mass spectrometers are sensitive to the ratio M/q, rather than to M itself, where M and q denote the mass and number of charges per molecule or cluster. Thus high-resolution analysis is needed for measuring the expected ^{13}C contributions to each peak, since isotopic distributions are often used to identify the particular charge state of a fullerene (e.g., $C_{60}^{n\pm}$). The ability to charge the C_{60} beams both positively and negatively, without causing structural damage to the fullerenes, has been very important for the detection, characterization and properties measurements of C_{60} beams by mass spectroscopy techniques.

6.3. Stability Issues

The stability of fullerenes is in part due to their high binding energies per carbon atom. The experimental value for the binding energy per carbon atom in graphite is 7.4 eV [6.11]. Total energy calculations using a local density approximation (LDA) and density functional theory [6.28] show that the binding energy of C_{60} per carbon atom is only reduced by about 9% relative to graphite. The corresponding calculation for C_{70} [6.28] gives a binding energy per carbon atom that is slightly higher (by about 20 meV) than for C_{60}. These LDA calculations of the binding energy [6.28] predict trends similar to those calculated for the cohesive energy in Fig. 6.5 [6.16], although the quantitative predictions differ in detail.

A very impressive experimental confirmation of the stability of fullerenes comes from the production of fullerenes by laser ablation from graphite [6.29] or carbon-based polymer substrates [6.30, 31]. From the time-of-flight mass spectrum of polyimide shown in Fig. 6.8 we see three characteristic regions: the low-mass region shown in (a), which is characteristic of the substrate; the low- to intermediate-mass spectra in (b) showing evidence for fullerene formation, but with masses less than C_{60}; the intermediate-mass range (c) showing the highest intensity of fullerene products; and finally in (d) the high-mass region going out to high mass values and showing only a slow fall-off in intensity with increasing mass. Proof that the mass peaks are associated with pure carbon species comes from doing an analysis of each mass peak for its isotopic abundance composition, and the results provide good fits to theoretical predictions for the natural isotopic abundance (see §4.5) [6.26]. Carbon clusters are formed by laser ablation at high temperatures (>3000 K), and the energetic clusters are believed to cool by C_2 emission (reduction of a fullerene by one hexagon face), consistent with the requirements of Euler's theorem. The release of C_2 rather than individual carbon atoms is also favored by the high binding energy of C_2 (3.6 eV). From the peak intensities of the mass spectra of Fig. 6.8, it is possible to

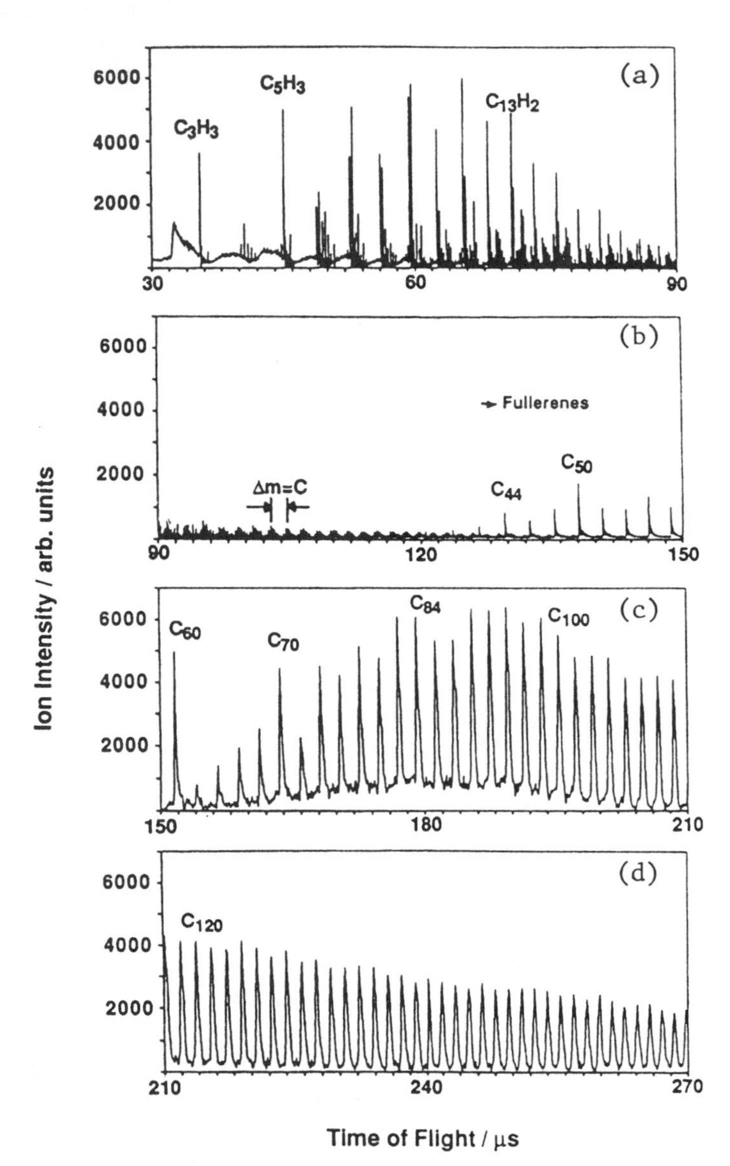

Fig. 6.8. Positive ion mass spectrum produced from 308 nm laser ablation of polyimide (see text) [6.26].

obtain rate constants for the fragmentation process, and the rate constant is more than seven times lower for C_{60} than for C_{62} [6.26], indicative of the high stability of C_{60} relative to C_{62}. The presence of such a large number of fullerenes in the mass spectra (Fig. 6.8) supports the stability of fullerenes generally. The large intensities for certain fullerene masses indicate a specially high stability for certain fullerenes, such as C_{60}, C_{70}, and C_{84}.

Returning to Fig. 6.8, the formation of fullerenes up to C_{84} is consistent with the data in the figure, as is also the C_2 emission process for masses beyond approximately C_{100}. Fullerenes between C_{70} and C_{60} are expected to arise primarily from the process of C_2 emission by higher-mass fullerenes (see §6.4).

Since positively and negatively charged fullerenes are important for many gas-phase experiments, the stability of these charged species is of interest. Positively charged C_{60}^{n+} ions up to $n = 5$ have been produced by electron impact ionization [6.32]. These highly charged fullerene cations have been unambiguously identified by high-resolution mass spectroscopy making use of the $^{12}C_{59}{}^{13}C$ and $^{12}C_{58}{}^{13}C_2$ isotopic species for calibration purposes, in accordance with the natural abundance of ^{13}C (see §4.5) [6.32]. For the highly charged C_{60}^{n+} fullerenes ($n = 4, 5$), contributions from C_{15}^{+} and C_{12}^{+} clusters due to fragmentation of C_{60} are also seen in the spectra of Fig. 6.9, but these lower-mass contributions can be subtracted from the C_{60}^{4+} and C_{60}^{5+} contributions, using the isotopic calibration procedure summarized in the figure. (One atomic mass unit in Fig. 6.9 is denoted by one dalton [6.32].) The data in Fig. 6.9 also document the stability of positively charged fullerene cations up to C_{60}^{5+} in the gas-phase.

Evidence for the stability of C_{60} anions comes from many experiments, including electrochemical studies where C_{60}^{n-} ions have been observed up through $n = 6$ in solution in the presence of an electric field (see §10.3.2). Laser ablation and carbon arc generation have both been successfully employed for the generation of negatively charged species. Specifically, negatively charged C_{60}^{-} and C_{60}^{2-} have been clearly identified through time-of-flight mass spectroscopy studies [6.33] and also using the isotopic composition of a mass peak to distinguish between C_{60}^{-} and C_{60}^{2-} (see Fig. 6.10); such data also provide information about stability issues. Just as for the positively charged species, the negatively charged species $C_{n_C}^{n-}$ appear only for even n_C in the mass spectra, and with very long lifetimes ($> 10^{-3}$ sec) [6.33]. As shown in Fig. 6.10, the intensity ratios C_{60}^{2-}/C_{60}^{-} and C_{70}^{2-}/C_{70}^{-} are unexpectedly high (0.02–0.20) since C_{60}^{-} is believed to be rather stable. Calculations show that the doubly charged C_{60}^{2-} anion has only a small binding energy [6.33]. To account for the relatively long lifetime and the relatively large abundance of C_{60}^{2-}, a model has been proposed for the formation of the negative ions. A binding energy of $\sim$2.5 eV (the electron affinity) has

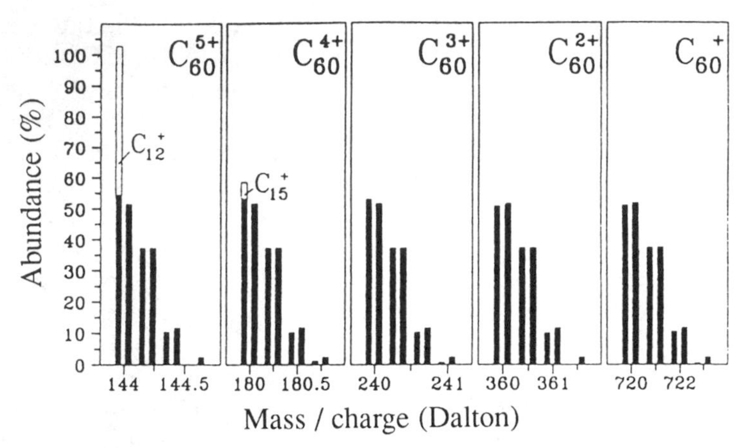

Fig. 6.9. Comparison of normalized measured isotopic distributions for different charge states of C_{60} and calculated distributions (filled bars) based on the natural abundance of ^{13}C and the intensity of the $^{12}C_{59}{}^{13}C$ peak in the mass spectra. The differences between the measured and calculated first peaks for pure ^{12}C for the C_{60}^{4+} and C_{60}^{5+} fullerene cations are due to contributions from singly charged fragments of C_{60}, i.e., C_{15}^{+} and C_{12}^{+} (open bars on top of the filled bars). The mass spectra were taken at an electron energy of 200 eV and 300 μA electron current [6.32].

been used to estimate the energy necessary to add one electron to C_{60}^{-}. Negative surface ionization is proposed as the mechanism for forming the observed high concentration of C_{60}^{2-} whereby the ratio of C_{60}^{2-} to C_{60}^{-} has been fit to the relation

$$\frac{C_{60}^{2-}}{C_{60}^{-}} = \frac{g_{2-}}{g_{-}} \exp\left(\frac{-\Delta E}{kT}\right) \tag{6.1}$$

where $(g_{2-}/g_{-}) = 4$, due to the possibility of singlet and triplet states for the doubly ionized species, and $\Delta E = 2.2$ eV has been attributed to the high thermal energies achieved in the carbon arc excitation method (~3000 K) [6.33].

6.4. Fullerene Contraction and Fragmentation

Closely related to studies of the growth and formation of fullerenes are studies of their contraction and fragmentation. Three main techniques are used for fragmentation studies: (1) photofragmentation, whereby incident photons excite fullerenes and thus promote subsequent fullerene fragmentation; (2) collision of charged C_{60} ions, which are used as projectiles impinging on surfaces; (3) energetic electron irradiation, which, similarly to

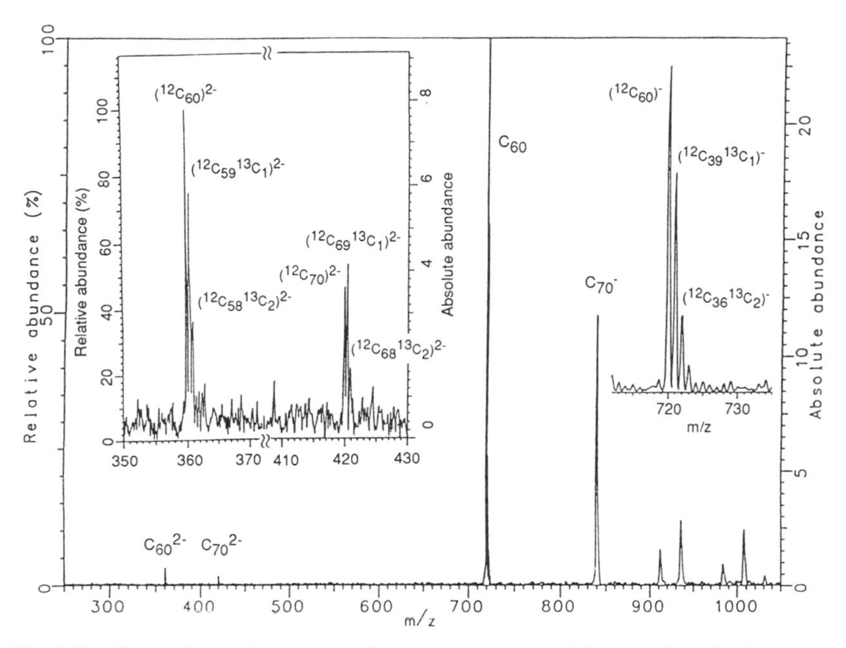

Fig. 6.10. Laser desorption negative ion mass spectrum of C_{60} and C_{70}. The insets show fine structure in the mass spectra for the C_{60}^{-}, C_{60}^{2-}, and C_{70}^{2-} peaks associated with ^{13}C isotopes, in good agreement with their natural abundances [6.33].

(1), excites the fullerene target and promotes fragmentation. Thermal fragmentation can also occur at elevated temperatures. The thermal dissociation of C_{60} into C_{58} and C_2 has been observed above 900 K on both SiO_2 and highly oriented pyrolytic graphite (HOPG) surfaces using molecular beam reaction spectroscopy techniques [6.34], showing that the activation energy for this thermal dissociation is reduced from 5–12 eV in the gas-phase [6.35–38] to ~2.5 eV on these two surfaces.

6.4.1. *Photofragmentation*

An experiment of historical significance is the study of photofragmentation of C_{60} and of endohedrally doped K@C_{60} and Cs@C_{60} [6.4]. The photofragmentation process tends to be a relatively gentle process and initially results in fullerene contraction, thereby shrinking the size of the fullerenes, without destroying them. In the photofragmentation process C_2 units are expelled (as discussed in §6.3) and the mass of the fullerene drops until the fullerene bursts because of its high strain energy. The disintegration of C_{60} occurs at about C_{32}, while with an endohedral dopant some reports claim

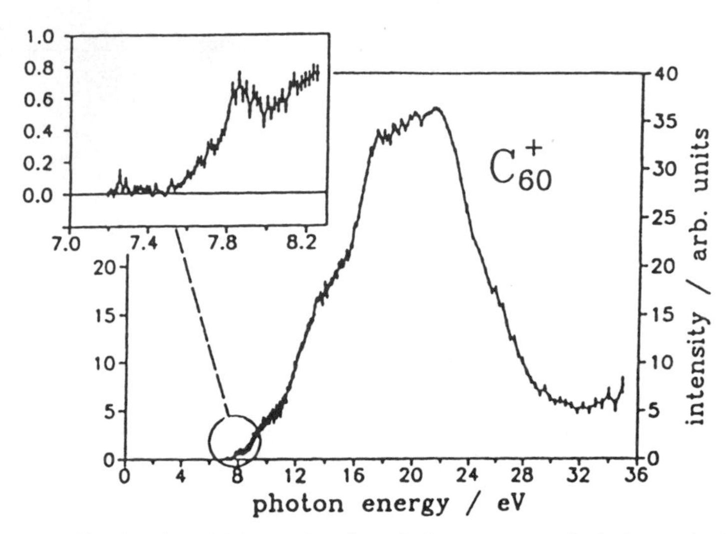

Fig. 6.11. C_{60}^+ photoion yield as a function of photon energy, displaying excitation of a giant plasmon resonance. Inset: The threshold region is shown on a magnified scale [6.41].

that the fullerenes burst at larger mass values, such as K@C_{44} and Cs@C_{48} [6.39], although other reports indicate that smaller metallofullerenes such as U@C_{28} are stable [6.40]. Early photofragmentation experiments provided strong evidence for a cage-like structure for C_{60} and focused strongly on study of the formation and disintegration mechanisms for fullerenes [6.4].

Single-photon (UV) excitation experiments have been carried out using synchrotron radiation to provide an intense photon source above the ionization energy of 7.58 eV for C_{60} and above 7.3 eV for C_{70} [6.26]. The photoion yield *vs.* $\hbar\omega$ for the single-photon excitation is shown in Fig. 6.11, where a strong and broad plasmon resonance is seen. This strong feature is attributed to collective motion of the valence electrons of C_{60}^+ [6.41], and the spectral line shape is in agreement with theoretical predictions [6.42].

Laser desorption of C_2 species provides an important technique for fullerene contraction studies. Laser desorption of C_2 can occur either at low photon energies, whereby several photons are absorbed by a single fullerene, or at high photon energies (above the ionization energy of 7.58 eV) where a single photon can produce electron ionization directly. For detection and measurement purposes, it is desirable to charge the C_{60} beam, which is normally emitted as a neutral beam from a C_{60} surface by laser ablation. The neutral beam can become charged by applying a second UV laser (~4 eV photon energy) to the incident neutral beam or by a variety of other means, while the beam is within a Fourier transform mass spectrometer. Neutral C_{60} molecules are often required to absorb several visible photons before they acquire enough energy to become ionized. The absorption of a photon normally results in strong vibrational excitations which can be quite energetic (~0.05–0.2 eV) and eventually may lead to thermionic emission or direct ionization [6.26], as discussed below.

For low photon intensities, the main effect of the laser beam is to desorb fullerenes from the surface without causing much fragmentation [6.43], as shown in Fig. 6.12(a), where time-of-flight mass spectra (see §6.2) are shown for 20 mJ/cm^2 photon irradiation at 248 nm. In this spectrum the peaks for the positive cations C_{60}^+ and C_{70}^+ are well separated, and the inset showing the $^{12}C_{59}{}^{13}C_1$ and $^{12}C_{58}{}^{13}C_2$ mass peaks confirms the identification of the spectral line with C_{60}^+ (see §4.5). When the photon intensity increases to 30 mJ/cm^2, as shown in Fig. 6.12(b), the intensity of the C_{84}^+ line increases dramatically. The intensities for all the fullerenes between C_{70}^+ and C_{84}^+ and also the intensities of the mass peaks for C_{58}^+ and C_{56}^+ increase with increasing photon intensity. However, increasing the photon intensity to 90 mJ/cm^2 [Fig. 6.12(c)] shows evidence for the formation of many higher-mass fullerenes and for their contraction by C_2 emission. The relative intensities of the peaks between C_{60}^+ and C_{70}^+ suggest that these species are formed by the emission of C_2 clusters from the C_{70}^+ ion, and likewise the peaks below C_{60}^+ are attributed to a similar contraction process by the C_{60}^+ ion. In contrast, the presence of relatively large amounts of high-mass fullerenes indicates the photoinduced absorption of C_2 units by the stable C_{60}^+ and C_{70}^+ fullerenes [6.43].

Delayed photoionization effects have also been reported for fullerenes. In this phenomenon, thermionic emission of electrons occurs from neutral fullerenes on a time scale of microseconds [6.44]. This effect has been explained by a statistical model by Klots [6.42], based on the following physical mechanism. When a photon of energy less than the ionization energy (7.58 eV) is absorbed by C_{60}, the photon energy is rapidly distributed among the vibrational degrees of freedom of the molecule. If several photons are consecutively absorbed, the temperature of the molecule rises, so that the molecule eventually has sufficient thermal energy for thermionic emission to occur. In such experiments a high-intensity laser of photon energy less than the ionization energy is used to energize the molecules, while a second laser with a different frequency is used for desorption to monitor the molecular ionization by the thermionic emission process [6.26].

6.4.2. *Collision of Fullerene Ion Projectiles*

In carrying out collision-induced fragmentation experiments, fullerenes are often used as the projectiles, and they are ionized so that the fullerene

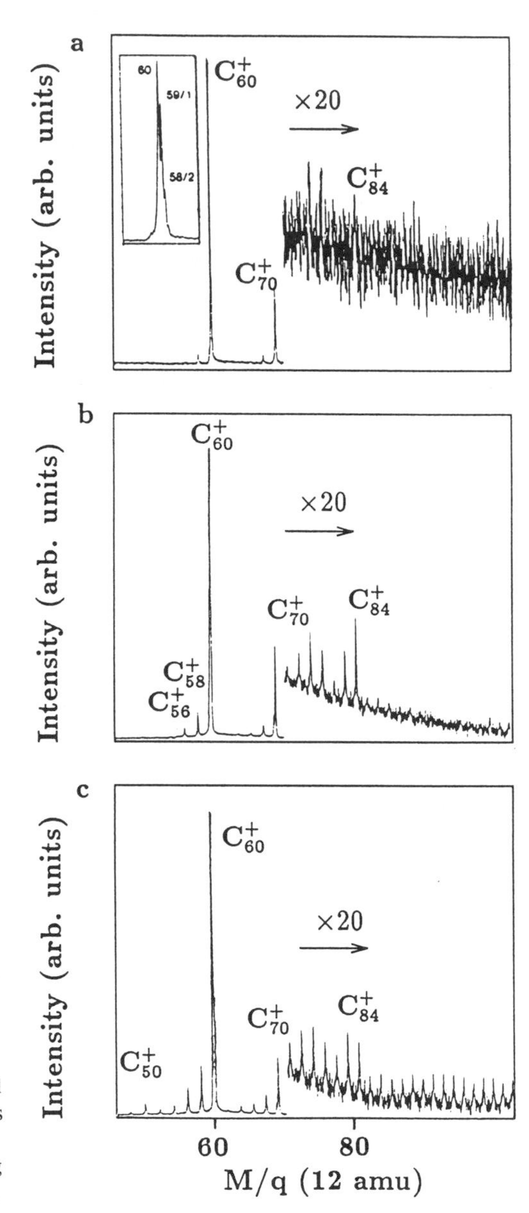

Fig. 6.12. Ions produced directly from desorption of a C_{60}/C_{70} extract from a silicon substrate using 248 nm photons and a laser fluence of (a) 20 mJ/cm^2, (b) 30 mJ/cm^2, and (c) 90 mJ/cm^2. The inset in (a) shows the isotope distribution of the C_{60}^+ peak, with the numbers x/y referring to the isotopic ratio $(^{12}C,^{13}C)^+$ [6.43].

ion beam can be separated in a magnetic field according to charge, mass, and energy. Since fullerenes with charge states between −2 and +5 can be prepared, a wide variety of fullerene ions can be used as projectiles. As a starting material, C_{60} powder heated to 300–600°C can be used as a supply for individual C_{60} molecules in the vapor phase.

One conclusion from these collision-induced fragmentation studies [6.45, 46] is that collision-induced fragmentation occurs by the loss of an even number of carbon atoms, as shown in Fig. 6.13, where the emission from the collision of 200-keV C_{60}^+ ions on a hydrogen target is plotted in terms of the yield for the various emission species which are selected by their mass-to-charge ratio. In this figure, even mass emission products ranging from C_{36}^+ to C_{58}^+ are shown. To explain the various experimental observations described below, it is assumed that in the center-of-mass frame, a gas atom colliding with a C_{60} ion will experience a number of collisions with individual carbon atoms *within* the C_{60} molecule. On a statistical basis, a rift will be opened in some of these charged fullerenes, because of the large local stresses and strains, thus resulting in the sequential and independent emission of m carbon atom pairs. Two characteristic times are introduced to describe the fragmentation process: t_1, the time to emit a carbon pair, and t_2, the time to close the rift. Figure 6.14 shows a semilog plot of the yield of fullerene fragments *vs.* m, the number of carbon pairs emitted. The results of Fig. 6.14 imply a p^m dependence for the yield, where p is defined below [6.46]. Further analysis shows that the probability

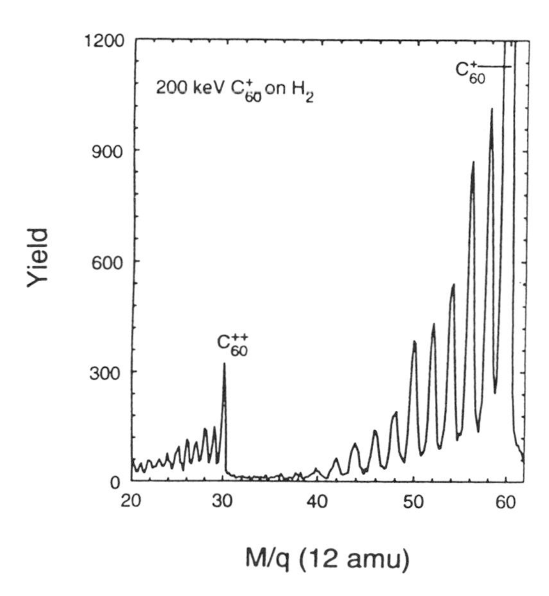

Fig. 6.13. M/q (mass divided by charge) spectrum resulting from the fragmentation of 200-keV C_{60} projectiles colliding with H_2 molecules in the gas-phase. The right- and left-hand parts of the spectrum reflect singly and doubly charged fragments, respectively, which appear to be replicas of one another [6.46].

$P_t(m)$ for the evaporation of m carbon pairs in a time interval t after a collision is

$$P_t(m) = \frac{1}{m!}\left(\frac{t}{t_1}\right)^m \exp(-t/t_1) \tag{6.2}$$

where t_1 is the evaporation lifetime, while the total probability for the evaporation of m carbon pairs integrated over time is

$$P(m) = \frac{t_1}{t_2}\left(\frac{t_2}{t_1 + t_2}\right)^{m+1} = \frac{t_1}{t_2} p^{m+1} \tag{6.3}$$

where $p = t_2/(t_1 + t_2)$, in which t_2 is the characteristic lifetime of the rift in the C_{60} molecule. The results of Fig. 6.14 show that for collisions of C_{60}^+ with a hydrogen molecule H_2, the time t_2 that the C_{60}^+ molecular ion is open is about twice the characteristic time t_1 for emission of the m carbon pairs (and $t_2 \simeq 3t_1$ for the He target), independent of the incident energy of the C_{60}^+ ion, although the average m value increases rapidly with increasing ion energy and mass of the target species [6.46]. Referring to Fig. 6.14, we see that the yield for $m = 5$ (corresponding to C_{50}^+) is exceptionally large, indicative of the special stability of the C_{50}^+ species. The special stability of C_{50} is related to its structure, which is like that of C_{60}, but with five fewer hexagons around the belt of the molecule, just as C_{70} has five additional hexagons around its belt.

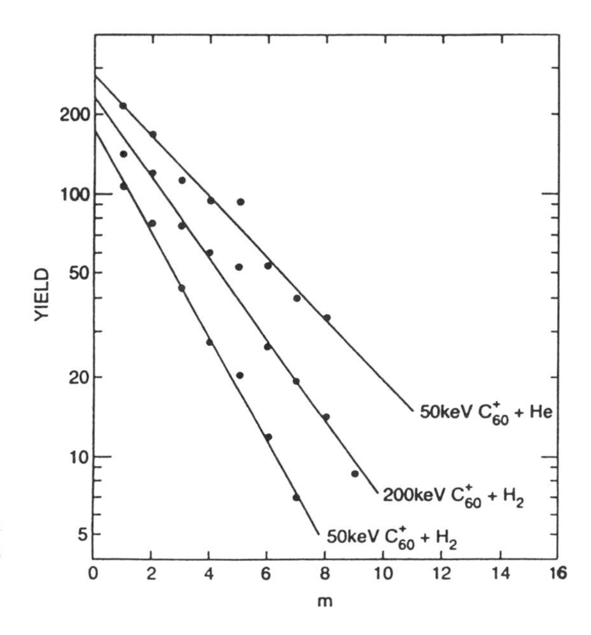

Fig. 6.14. Fragment-peak intensities as a function of the number m of pairs of carbon atoms (C_2) that are lost from an energetic C_{60}^+ ion upon collision with hydrogen and helium gas molecules [6.46].

The products observed by mass spectra in collision-induced fragmentation are also dependent on the charge of the fullerene projectile and the target molecules. For example, in the case of the H_2 target, the various features in the yield spectrum for 300-keV C_{60}^{+2} are similar to those in the 200-keV C_{60}^{+} spectrum (see Fig. 6.13), indicating electron absorption by C_{60}^{+2} to yield C_{60}^{+}, upon collision of the C_{60}^{+2} ion with the H_2 target.

By measuring the relative yields of the fragmentation process in terms of C_2 emission, going from C_{58}^{+} to C_{56}^{+} and from C_{56}^{+} to C_{54}^{+}, we can obtain the yields for the daughters in the decay chain. Decay lifetime measurements of the C_{58}^{+} and C_{56}^{+} molecular ions have also been carried out, assuming a simple decay formula $I(t) = I_0 \exp(-t/\tau)$. Results for τ of 337 μs for C_{58}^{+}, and 66 μs for C_{56}^{+} have been obtained by several groups [6.47, 48]. After decay to a neutral species, the lifetime for the neutral ions is much greater ($\sim$10 s), thus allowing better opportunities for experiments to be carried out with these neutral molecular beams.

In collisions of C_{60} positive ions, such as C_{60}^{3+}, charge can be transferred to the target. For a triply charged C_{60}^{3+} ion, such charge transfer occurs, for example, when the third ionization potential I_{III} for C_{60}^{3+} exceeds the ionization potential for the target species. Charge transfer from C_{60}^{2+} and C_{60}^{3+} ions can occur when these ions collide with a gas target, if the ionization potentials of the target are less than the second and third ionization potentials of carbon, 9.37 eV and 10.75 eV, respectively [6.49].

6.4.3. Collision of Fullerene Ions with Surfaces

Interesting results have been reported for the collision of C_{60}^{+} ions and various surfaces. Especially popular have been graphite surfaces, where the projectile and target are both carbon atoms. Collisions between energetic C_{60}^{+} ions (with energies up to 200 eV) and graphite surfaces result in large energy transfers to the surface, leaving the ions with final energies of only $\sim$15 eV, almost independent of the incident energy [6.36, 37, 50]. In this collision process, the C_{60}^{+} ion is momentarily heated to high temperatures, experiences major distortions, but transfers most of its initial kinetic energy into heating the surface. Only the elastic energy involved in the molecular deformation is retained by the C_{60}^{+}, and the retained energy is only weakly dependent on the incident energy. For higher incident energies (above 200 eV), the C_{60}^{+} ions fragment with the loss of C_2 units [6.26, 36]. Similar behavior is observed for C_{60}^{+} ions colliding with diamond surfaces, regarding the velocity distribution of the scattered C_{60}^{+} ions and the conditions for their fragmentation [6.36].

Similar experiments have been done by studying the collision of C_{60}^{+} with an Si (100) surface for energies $\leq$ 250 eV [6.51, 52]. The scattering of the

C_{60}^+ ion by the surface first neutralizes the C_{60}, as it picks up an electron from the surface. The energy of the scattered neutrals is determined by reionization of the C_{60} neutrals to C_{60}^+ using a pulsed ArF excimer laser (~15 ns, 193 nm, 6.4 eV). Just as for the graphite and diamond surfaces, almost all of the energy of the C_{60}^+ ion incident on silicon goes into heating the surface, with ~15 eV going into internal vibrations of the rebounding fullerene, consistent with molecular dynamics calculations [6.53], showing that the C_{60} projectile on impact with the surface is compressed to two thirds of its diameter (from ~7 Å to ~4.5 Å). This intense deformation couples strongly to the vibrational modes, with a dissipation of ~10 eV per C_{60} into vibrational energy and ~5 eV into its rebound energy (rebound velocity ~1200 m/s).

As another example, collisions of 275-eV C_{60}^+ were studied with a crystalline (111) fullerite film grown on mica [6.54]. Many of the effects that are observed for the collision of C_{60}^+ with graphite, diamond, and silicon surfaces, as described above, are also found for collisions of C_{60}^+ ions with a C_{60} film surface. One important new feature of the impact of C_{60}^+ ions on a C_{60} film is the tendency to form higher-mass fullerenes, especially near C_{120}, as shown in Fig. 6.15, with some intensity also found near C_{70}. Of particular interest are the many fullerenes observed between C_{110} and C_{130} [6.54]. Apparently, during the collision between the projectile and target fullerenes, much of the kinetic energy of the projectile is consumed in forming the larger fullerene, which is released from the surface. In contrast, Fig. 6.15 shows that the impact of C_{60}^+ ions on a graphite surface does not give rise to higher-mass fullerenes.

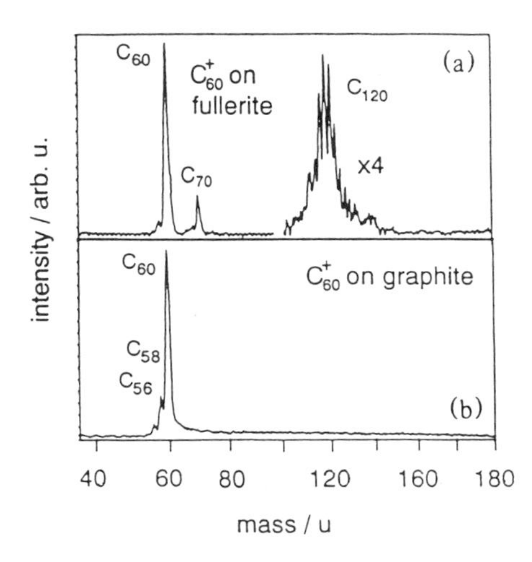

Fig. 6.15. Mass spectra of fullerene ions released from a surface after impact of C_{60}^+ ions (a) on a crystalline fullerite film target (E_{in} = 275 eV, α = 25°, v = 2000 m/s) and (b) on oriented graphite (HOPG). Note that the intensity scale for the heavy mass fullerenes has been expanded by a factor of 4 [6.54].

6.4.4. *Fragmentation of C_{60} by Energetic Ions*

We now describe a very different fragmentation study, where C_{60} is the target rather than the projectile. Under the influence of heavy-ion irradiation, it has been shown that C_{60} breaks up into individual carbon atoms [6.55]. In these experiments C_{60} films were exposed to 320-keV Xe ion irradiation for which the mean projected ion penetration depth R_p and the half-width at half-maximum intensity ΔR_p are $R_p \pm \Delta R_p = 120 \pm 18$ nm, and the Xe ion implantation was carried out at both room temperature and 200°C for ion doses in the range 10^{12} to 10^{16} Xe/cm^2. Monitoring the effect of the implantation by resistance measurements on the C_{60} film [see Fig. 6.16(a)], it was found that the onset for conduction occurred at a dose of 2.5×10^{12} Xe/cm^2, several orders of magnitude less than the onset for fused quartz, irradiated with 100-keV carbon atoms [6.55, 56], although in both cases the temperature dependence of the resistance of the irradiated sample followed a $\log R$ *vs.* $(1/T)^{1/4}$ law, characteristic of the 3D variable-range hopping transport mechanism [6.57]. Writing the functional dependence of the electrical resistance R for this regime as

$$R = 1/\sigma \sim \exp(\gamma N^{-1/3} \xi^{-1}) \tag{6.4}$$

where γ is a numerical factor, N is the concentration of hopping centers, and ξ is the average size of a hopping center, and noting that the hopping center density is proportional to the dose D, then N can be written as

$$N = \beta D/W, \tag{6.5}$$

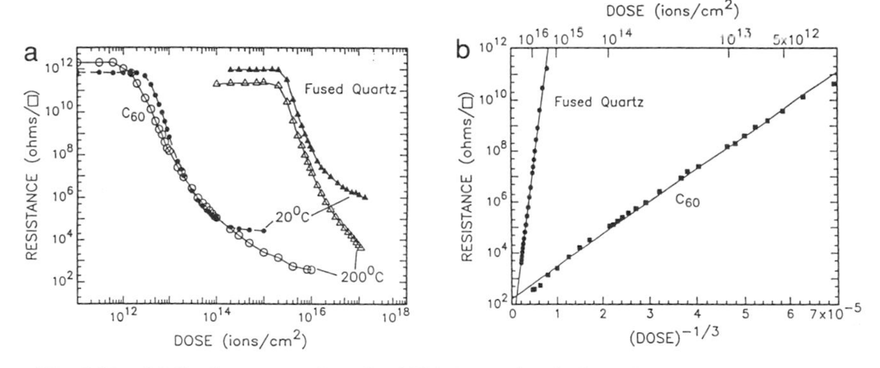

Fig. 6.16. (a) Resistance *vs.* dose for 320-keV Xe irradiation of C_{60} at room temperature (filled circles) and at 200°C (open circles) and for 100-keV C irradiation of fused quartz at room temperature (filled triangles) and at 200°C (open triangles). In (b) the data of (a) are replotted as $\log_{10}$(resistance) *vs.* (dose)$^{-1/3}$ for 100-keV C implanted fused quartz (circles) and 320-keV Xe implanted C_{60} (squares). Both implantations are at 200°C. The straight lines are least squares linear fits to the data [6.55].

where W is the thickness of the implanted region, yielding a prediction that $\log R \sim D^{-1/3}$. We see from Fig. 6.16(b) that this hypothesis is well supported by the experiments for 320-keV Xe implantation into C_{60} and for 100-keV C implantation into fused quartz. From the slopes of the lines in Fig. 6.16(b), the magnitude of ξ can be determined. For both the C_{60} and the fused quartz, the hopping center is found to be a single carbon atom of size ~1 Å [6.55]. The observation that the irradiation dose onset for conduction is about 60 times lower for C_{60} than for ion-implanted diamond [6.56, 58] suggests that the fragmentation of C_{60} by heavy Xe ions breaks up the C_{60} molecule into 60 individual carbon atoms. This hypothesis is supported by the magnitude of the thermal spike associated with the ion implantation [6.56], yielding a kinetic energy of 16 eV per carbon atom, far greater than the binding energy of carbon in the solid state (7.4 eV/C atom in graphite) [6.55]. The results obtained for Xe irradiation are consistent with those for proton irradiation of C_{60} films where disintegration of the C_{60} by protons of 100 keV was reported [6.59].

The tendency of C_{60} to fragment upon ion irradiation suggests that ion implantation is not an attractive method for the doping of C_{60}. These results also indicate that the abundance of fullerenes in the interstellar space may be low because of their tendency to disintegrate upon exposure to heavily damaging energetic particles in space [6.55].

6.5. Molecular Dynamics Models

Molecular dynamics simulations combined with tight-binding total energy calculations have been used to simulate the fragmentation process [6.60–62] of C_{60} and C_{70} by calculating bond lengths, the number of nearest-neighbors per carbon atom, the heat capacity, and the binding energy as a function of temperature. Bond breaking (or melting) is observed in these simulations between 3000 and 4000 K for both C_{60} and C_{70}, below that for diamond (4100 K). The bond breaking of fullerenes is already anticipated by $T \sim$ 3000 K because of large bond distortions shown in the molecular dynamics simulations. A decrease in the average number of nearest-neighbors (below 3) appears by 4000 K. Once a single bond is broken, stress accumulates near the broken bond, leading rapidly to further bond breaking at neighboring bonds. The bond breaking occurs preferentially at single bond sites. At melting, a latent heat of 1.65 eV and 1.29 eV is found for C_{60} and C_{70}, respectively [6.60], which is somewhat higher than for graphite (1.085 eV).

Not only the fragmentation but also the growth of fullerenes has been modeled using molecular dynamic simulations. The *ab initio* molecular dynamics method used in these simulations is called the Car–Parrinello method [6.63], in which the electronic interactions are treated in the density functional formalism, while the atoms follow Newtonian dynamics with forces derived directly from the local density equations. The electronic and ionic subsystems evolve simultaneously, allowing study of time-dependent simulations at finite temperature and of geometry optimization based on simulated annealing.

These molecular dynamics simulations have provided a number of insights into the growth process for fullerenes [6.60–62] and for tubules [6.64]. Simulations of a C_{20} unit showed the dodecahedron cage structure to have the lowest energy, with the corannulene C_{20} cluster (found in the C_{60} molecule) as discussed in §3.1 to be higher in energy by 0.75 eV, and the ring structure to have an energy higher by 2.65 eV than the cage structure [6.61]. However, when entropy is taken into account, the free energy at elevated temperatures (>1500 K) shows the opposite order, with the C_{20} ring structure having the lowest free energy, the corannulene C_{20} the second lowest free energy, and the C_{20} dodecahedron cage molecule having the highest free energy of the three configurations. Such calculations thus provide insight into reasons why mass spectra for $n_C \sim 20$ are interpreted in terms of the ring configuration [6.61].

Molecular dynamic simulations have been informative about the formation of defective C_{60} structures and the identification of the most likely defects (two adjacent pentagons) [6.5, 10, 65–68]. Simulations of the growth and annealing of C_{60} from the gas-phase are impeded by the inability of today's computers to simulate the long cooling times needed to reach equilibrium. Some molecular dynamic simulations [6.5] show that as a plasma of 60 carbon atoms is cooled from 7000 K, dimers form first, then chains, and finally polycyclic rings are formed. Only when the system is cooled to ~3000 K is a cage structure formed, and this structure is predominantly made of hexagonal and pentagonal faces, although the cage structure according to this model is a highly defective form of C_{60} [6.5, 6, 69, 70].

Molecular dynamics simulations have also been informative for studying the local stability and reactivity of fullerenes. For example, scenarios for the absorption of a C_2 cluster by C_{60} have been identified, although no route could be found for the absorption of a C_3 cluster [6.61], as might be expected from Euler's theorem. As another example, molecular dynamic simulations have been used to compute the heats of reactions of fullerenes with small C_2 or C_3 clusters, showing for example that energy is required for C_2 emission in the reaction $C_{60} + 13.6\ \text{eV} \rightarrow C_{58} + C_2$ or, as another example, $C_{58} + C_{62} \rightarrow 2C_{60} + 4.6\ \text{eV}$, reaffirming the high stability of C_{60} [6.61].

The energies for fragmentation of C_{60} and C_{70} by C_2 emission were also calculated [6.71], yielding C_{60}+ 11.8 eV → C_{58}+ C_2 and C_{70}+ 11.5 eV → C_{68}+ C_2, respectively, in good agreement with the work of others [6.61, 72].

In addition to these examples, molecular dynamics simulations have been widely used for the interpretation of specific experiments, such as nuclear magnetic resonance (NMR) dynamical studies [6.73], optimization of bond lengths for fullerenes [6.74, 75] and carbon nanotubes, temperature dependence of bond lengths for fullerenes [6.61], the energy partition between C_{60}^{+} and a diamond (111) surface upon collision [6.76], and vibrational modes of fullerenes [6.77, 78], to mention a few explicit examples.

References

[6.1] J. R. Heath, S. C. O'Brien, Q. Zhang, Y. Liu, R. F. Curl, H. W. Kroto, F. K. Tittel, and R. E. Smalley. *J. Am. Chem. Soc.*, **107**, 7779 (1985).
[6.2] D. R. McKenzie, C. A. Davis, D. J. H. Cockayne, D. A. Miller, and A. M. Vassallo. *Nature (London)*, **355**, 622 (1992).
[6.3] B. I. Dunlap. *Phys. Rev. B*, **47**, 4018 (1993).
[6.4] R. F. Curl. *Carbon*, **30**, 1149 (1992).
[6.5] J. R. Chelikowsky. *Phys. Rev. Lett.*, **67**, 2970 (1991).
[6.6] X. Jing and J. R. Chelikowsky. *Phys. Rev. B*, **46**, 5028 (1992).
[6.7] R. E. Smalley. *Accounts of Chem. Research*, **25**, 98 (1992).
[6.8] A. J. Stone and D. J. Wales. *Chem. Phys. Lett.*, **128**, 501 (1986).
[6.9] F. Diederich, R. Ettl, Y. Rubin, R. L. Whetten, R. Beck, M. Alvarez, S. Anz, D. Sensharma, F. Wudl, K. C. Khemani, and A. Koch. *Science*, **252**, 548 (1991).
[6.10] P. W. Fowler, D. E. Manolopoulos, and R. P. Ryan. *Carbon*, **30**, 1235 (1992).
[6.11] C. Kittel. *Introduction to Solid State Physics*. John Wiley and Sons, New York, 6th ed. (1986).
[6.12] T. Wakabayashi, H. Shiromaru, K. Kikuchi, and Y. Achiba. *Chem. Phys. Lett.*, **201**, 470 (1993).
[6.13] R. Saito, G. Dresselhaus, and M. S. Dresselhaus. *Chem. Phys. Lett.*, **195**, 537 (1992).
[6.14] M. S. Dresselhaus, G. Dresselhaus, and R. Saito. *Phys. Rev. B*, **45**, 6234 (1992).
[6.15] R. Saito. Private communication.
[6.16] B. L. Zhang, C. Z. Wang, and K. M. Ho. *Chem. Phys. Lett.*, **193**, 225 (1992).
[6.17] R. F. Curl and R. E. Smalley. *Scientific American*, **265**, 54 (1991).
[6.18] R. E. Smalley. *The Sciences*, **31**, 22 (1991).
[6.19] R. E. Smalley. *Materials Science and Engineering*, **B19**, 1 (1993).
[6.20] Y. Achiba, T. Wakabayashi, T. Moriwaki, S. Suzuki, and H. Shiromaru. *Materials Science and Engineering*, **B19**, 14 (1993).
[6.21] R. Taylor. Unpublished (1994).
[6.22] G. H. Taylor, J. D. Fitz Gerald, L. Pang, and M. A. Wilson. *Journal of Crystal Growth*, **135**, 157 (1994).
[6.23] H. Kroto. *Nature (London)*, **359**, 670 (1992).
[6.24] S. Iijima, T. Ichihashi, and Y. Ando. *Nature (London)*, **356**, 776 (1992).
[6.25] M. V. Buchanan and R. L. Hettich. *Analytical Chemistry*, **65**, 245A (1993).
[6.26] E. E. B. Campbell and I. V. Hertel. *Carbon*, **30**, 1157 (1992).
[6.27] S. R. Wilson and Y. H. Wu. *Organic Mass Spectrometry*, **29**, 186 (1994).
[6.28] S. Saito and A. Oshiyama. *Phys. Rev. B*, **44**, 11532 (1991).
[6.29] E. A. Rohlfing, D. M. Cox, and A. Kaldor. *J. Chem. Phys.*, **81**, 3322 (1984).
[6.30] E. E. B. Campbell, G. Ulmer, B. Hasselberger, and I. V. Hertel. *Appl. Surf. Sci.*, **43**, 346 (1989).

[6.31] G. Ulmer, B. Hasselberger, H. G. Busmann, and E. E. B. Campbell. *Appl. Surf. Sci.*, **46**, 272 (1990).
[6.32] P. Scheier, R. Robl, B. Schiestl, and T. D. Märk. *Chem. Phys. Lett.*, **220**, 141 (1994).
[6.33] R. L. Hettich, R. N. Compton, and R. H. Ritchie. *Phys. Rev. Lett.*, **67**, 1242 (1991).
[6.34] A. V. Hamza, M. Balooch, M. Moalem, and D. R. Olander. *Chem. Phys. Lett.*, **228**, 117 (1994).
[6.35] P. Sandler, C. Lifshitz, and C. E. Klots. *Chem. Phys. Lett.*, **200**, 445 (1992).
[6.36] H. G. Busmann, T. Lill, B. Reif, and I. V. Hertel. *Surface Science*, **272**, 146 (1992).
[6.37] H. G. Busmann, T. Lill, B. Reif, I. V. Hertel, and H. G. Maguire. *J. Chem. Phys.*, **98**, 7574 (1993).
[6.38] K. R. Lykke and P. Wurz. *J. Phys. Chem*, **96**, 3191 (1992).
[6.39] F. D. Weiss, S. C. O'Brien, J. L. Eklund, R. F. Curl, and R. E. Smalley. *J. Am. Chem. Soc.*, **110**, 4464 (1988).
[6.40] T. Guo, M. D. Diener, Y. Chai, J. M. Alford, R. E. Haufler, S. M. McClure, T. R. Ohno, J. H. Weaver, G. E. Scuseria, and R. E. Smalley. *Science*, **257**, 1661 (1992).
[6.41] I. V. Hertel, H. Steger, J. de Vries, B. Weisser, C. Menzel, B. Kamke, and W. Kamke. *Phys. Rev. Lett.*, **68**, 784 (1992).
[6.42] C. E. Klots. *Chem. Phys. Lett.*, **186**, 73 (1991).
[6.43] G. Ulmer, E. E. B. Campbell, R. Kühnle, H. G. Busmann, and I. V. Hertel. *Chem. Phys. Lett.*, **182**, 114 (1991).
[6.44] E. E. B. Campbell, G. Ulmer, and I. V. Hertel. *Phys. Rev. Lett.*, **67**, 1986 (1991).
[6.45] R. J. Doyle and M. M. Ross. *J. Phys. Chem.*, **95**, 4954 (1991).
[6.46] P. Hvelplund, L. H. Andersen, H. K. Haugen, J. Lindhard, D. C. Lorents, R. Malhotra, and R. Ruoff. *Phys. Rev. Lett.*, **69**, 1915 (1992).
[6.47] P. Radii, M. T. Hsu, M. E. Rincon, P. R. Kemper, and M. T. Bowers. *Chem. Phys. Lett.*, **174**, 223 (1990).
[6.48] E. E. B. Campbell, G. Ulmer, H. G. Bugmann, and I. V. Hertel. *Chem. Phys. Lett.*, **175**, 505 (1990).
[6.49] P. Hvelplund. In T. Andersen, B. Fastrup, F. Folkmann, and H. Knudsen (eds.), *Electronic and Atomic Collisions*, AIP Conference Proceedings, New York, vol. 295, pp. 709–718 (1993). AIP Conference Proc. Invited talk, Aarhus.
[6.50] T. Lill, H. G. Busmann, B. Reif, and I. V. Hertel. *Appl. Phys. A*, **55**, 461 (1992).
[6.51] C. Yeretzian, K. Hansen, R. D. Beck, and R. L. Whetten. *J. Chem. Phys.*, **98**, 7480 (1993).
[6.52] R. D. Beck, P. M. S. John, M. M. Alvarez, F. Diederich, and R. L. Whetten. *J. Phys. Chem.*, **95**, 8402 (1991).
[6.53] D. W. Brenner, J. A. Harrison, C. T. White, and R. J. Colton. *Thin Solid Films*, **206**, 220 (1991).
[6.54] T. Lill, F. Lacher, H. G. Busmann, and I. V. Hertel. *Phys. Rev. Lett.*, **71**, 3383 (1993).
[6.55] R. Kalish, A. Samoiloff, A. Hoffman, C. Uzan-Saguy, D. McCulloch, and S. Prawer. *Phys. Rev. B*, **48**, 18235 (1993).
[6.56] M. S. Dresselhaus and R. Kalish. *Ion Implantation in Diamond, Graphite and Related Materials*, Springer Series in Materials Science, vol. 22, Springer-Verlag, Berlin (1992).
[6.57] N. F. Mott and E. Davis. *Electronic Processes in Noncrystalline Materials*. Clarendon Press, Oxford (1979).
[6.58] S. Prawer, A. Hoffman, and R. Kalish. *Appl. Phys. Lett.*, **57**, 2187 (1990).
[6.59] R. G. Musket, R. A. Hawley-Fedder, and W. L. Bell. *Radiat. Eff.*, **118**, 225 (1991).
[6.60] E. Kim, Y. H. Lee, and J. Y. Lee. *Phys. Rev. B*, **48**, 18230 (1993).
[6.61] J. Bernholc, J. Y. Yi, Q. M. Zhang, C. J. Brabec, E. B. Anderson, B. N. Davidson, and S. A. Kajihara. *Zeitschrift für Physik D: Atoms, Molecules and Clusters*, **26**, 74 (1993).

[6.62] C. H. Xu and G. E. Scuseria. *Phys. Rev. Lett.*, **72**, 669 (1994).
[6.63] R. Car and M. Parrinello. *Phys. Rev. Lett.*, **55**, 2471 (1985).
[6.64] J. Y. Yi and J. Bernholc. *Phys. Rev. B*, **47**, 1708 (1993).
[6.65] C. Z. Wang, C. H. Xu, C. T. Chan, and K. M. Ho. *J. Phys. Chem.*, **96**, 3563 (1992).
[6.66] B. Ballone and P. Milani. *Phys. Rev. B*, **42**, 3201 (1990).
[6.67] J. Y. Yi and J. Bernholc. *J. Chem. Phys.*, **96**, 8634 (1992).
[6.68] K. Raghavachari and C. M. Rohlfing. *J. Phys. Chem.*, **96**, 2463 (1992).
[6.69] J. R. Chelikowsky. *Phys. Rev. B*, **45**, 12062 (1992).
[6.70] X. Jing and J. R. Chelikowsky. *Phys. Rev. B*, **46**, 15503 (1992).
[6.71] W. C. Eckhoff and G. E. Scuseria. *Chem. Phys. Lett.*, **216**, 399 (1993).
[6.72] J. Y. Li and J. Bernholc. *Phys. Rev. B*, **48**, 5724 (1993).
[6.73] Q. M. Zhang, J. Y. Yi, and J. Bernholc. *Phys. Rev. Lett.*, **66**, 2633 (1991).
[6.74] M. Menon, K. R. Subbaswamy, and M. Sawtarie. *Phys. Rev. B*, **48**, 8398 (1993).
[6.75] M. Menon, K. R. Subbaswamy, and M. Sawtarie. *Phys. Rev.*, **49**, 13966 (1994).
[6.76] P. Blaudeck, T. Frauenheim, H. G. Busmann, and T. Lill. *Phys. Rev. B*, **49**, 11409 (1994).
[6.77] G. B. Adams, J. B. Page, O. F. Sankey, K. Sinha, J. Menendez, and D. R. Huffman. *Phys. Rev. B*, **44**, 4052 (1991).
[6.78] B. P. Feuston, W. Andreoni, M. Parrinello, and E. Clementi. *Phys. Rev. B*, **44**, 4056 (1991).

CHAPTER 7

Crystalline Structure of Fullerene Solids

In this chapter we discuss the crystalline structure of C_{60} (§7.1), and the various phases that have been identified as a function of pressure (§7.3) and temperature (§7.4). Also included is a review of the polymeric phase formed by photopolymerization and the conditions under which this phase is formed (§7.5). Although less is known about the crystalline structure of C_{70}, an approximate picture of the temperature dependence of the various structural phases has already emerged (§7.2). A summary of the status of knowledge about the even less known crystal structure of higher fullerenes is also given (§7.2). The crystalline forms of fullerenes are often called fullerites in the literature. Knowledge of the crystal structure of any fullerite is, of course, necessary to gain a microscopic understanding of their solid-state physical properties that are discussed in subsequent chapters.

7.1. Crystalline C_{60}

In this section the structure of crystalline C_{60} is reviewed, including the structure at room temperature, phase transitions occurring as a function of temperature, group theoretical considerations for the crystalline phases, and merohedral disorder.

In the solid state, the C_{60} molecules crystallize into a cubic structure with a lattice constant of 14.17 Å, a nearest-neighbor C_{60}–C_{60} distance of 10.02 Å [7.1] and a density of 1.72 g/cm^3 (corresponding to 1.44 $\times 10^{21}$ C_{60} molecules/cm^3) (see Table 7.1). Whereas the C_{60} molecules themselves are almost incompressible, the molecular solid is quite compressible. Thus under hydrostatic pressure, it is expected that the mass density of solid

Table 7.1

Physical constants for crystalline C_{60} in the solid state.

Quantity	Value	Reference
fcc Lattice constant	14.17 Å	[7.2]
C_{60}–C_{60} distance	10.02Å	[7.1]
C_{60}–C_{60} cohesive energy	1.6 eV	[7.3]
Tetrahedral interstitial site radius	1.12 Å	[7.1]
Octahedral interstitial site radius	2.07 Å	[7.1]
Mass density	1.72 g/cm^3	[7.1]
Molecular density	1.44×10^{21}/cm^3	[7.1]
Compressibility ($-d\ln V/dP$)	6.9×10^{-12} cm^2/dyn	[7.4]
Bulk modulus[a]	6.8, 8.8 GPa	[7.5, 6]
Young's modulus[b]	15.9 GPa	[7.7]
Transition temperature (T_{01})[c]	261 K	[7.8]
dT_{01}/dp[c]	11 K/kbar	[7.9, 10]
Vol. coeff. of thermal expansion	6.1×10^{-5}/K	[7.11]
Optical absorption edge	1.7 eV	[7.12]
Work function	4.7 ± 0.1 eV	[7.13]
Velocity of sound v_t	2.1×10^5 cm/s	[7.7]
Velocity of sound v_l	$3.6 - 4.3 \times 10^5$ cm/s	[7.7, 14]
Debye temperature	185 K	[7.15, 16]
Thermal conductivity (300 K)	0.4 W/mK	[7.17]
Electrical conductivity (300 K)	1.7×10^{-7} S/cm	[7.18]
Phonon mean free path	50 Å	[7.17]
Static dielectric constant	4.0–4.5	[7.19, 20]
Melting temperature	1180°C	[7.21]
Sublimation temperature	434°C	[7.22]
Heat of sublimation	40.1 kcal/mol	[7.22]
Latent heat	1.65 eV/C_{60}	[7.23]

[a]The bulk modulus is 6.8 GPa at room temperature in the fcc phase and 8.8 GPa in the sc phase below T_{01} [7.5]. Earlier reported values for the bulk modulus were in the range 14–18 GPa [7.4, 24–26]. The conversion of units for pressure measurements is 1 GPa $= 10^9$ N/m^2 $= 0.99$ kbar $= 7.52 \times 10^6$ torr $= 10^{10}$ dyn/cm^2. Also, 1 $atm = 760$ torr $= 1.01 \times 10^5$ Pa.

[b]For solvated C_{60} crystals measured by vibrating reed method [7.7]. Measurements on film samples suggest a higher Young's modulus [7.27].

[c]Structural phase transition discussed in §7.1.2.

C_{60} should increase significantly, as discussed in §7.3. A proposed phase diagram for C_{60} [7.28] is shown in Fig. 7.1 [7.9, 28–31]. As the temperature and pressure are both increased, two distinct dense phases of C_{60} have been identified (see §7.3). In this section, we first describe the crystal structure of C_{60} at room temperature and then review the temperature dependence of the crystal structure.

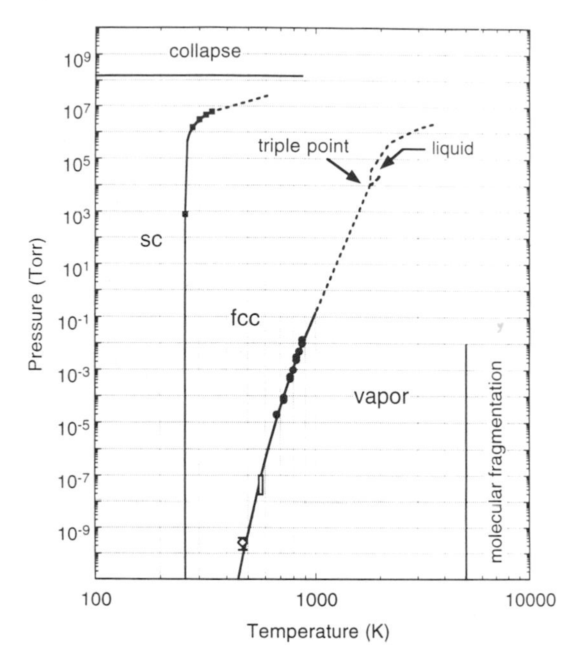

Fig. 7.1. Phase diagram for C_{60}. Solid lines are from experiment and dashed lines are theoretical or reflect extrapolations from experiment. The pressure dependence of the transition from the simple cubic (sc) to the face-centered cubic (fcc) phases is from Ref. [7.9] and the delineation between the fcc and vapor phase is based on the results of Refs. [7.28–30]. The data at high temperature and pressure reflect calculated boundaries between the liquid, solid, and vapor phases from Ref. [7.31]. The existence of a liquid phase remains to be proven. Molecular collapse and fragmentation are indicated at high pressure or high temperature [7.28].

7.1.1. *Ambient Structure*

At room temperature, the C_{60} molecules in the pristine solid have been shown by nuclear magnetic resonance (NMR) [7.32–36] to be rotating rapidly with three degrees of rotational freedom. The molecular centers themselves are arranged on a face-centered cubic (fcc) lattice (see Fig. 7.1) with one C_{60} molecule per primitive fcc unit cell, or four molecules per simple cubic unit cell (see Fig. 7.2). Since the molecules are spinning rapidly about their lattice positions, there is no orientational order, and all molecules are equivalent in measurements which require a time longer than the average rotational period ($t \gg 10^{-11}$ s). Thus the pertinent space group for the structure where the molecules are rotating rapidly is O_h^5 or $Fm\bar{3}m$, using the Schoenflies and International Crystallographic nomenclature, respectively. The symmetry of the space group for C_{60} at room temperature has been established directly by x-ray and neutron diffraction

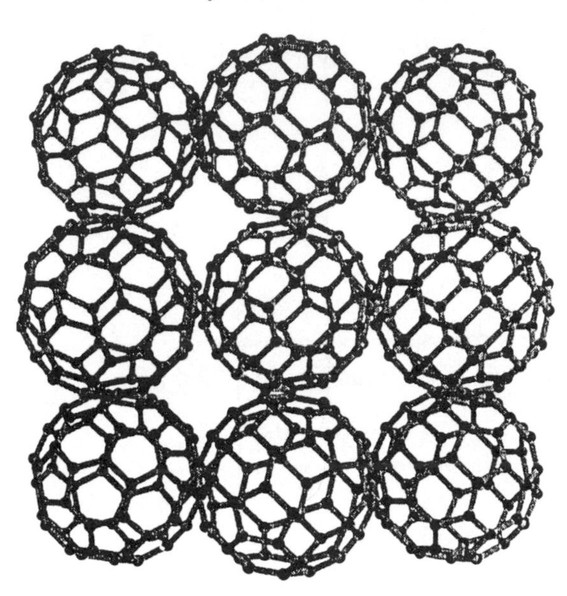

Fig. 7.2. Basal plane projection of the low-temperature crystal structure of solid C_{60} (space group $Pa\bar{3}$) [7.38].

[7.11, 39, 40]. The crystal structure pertinent to the room temperature ambient phase persists down to a temperature T_{01} where a transition to a simple cubic (sc) phase occurs (see Fig. 7.1 and §7.1.3).

The C–C bond lengths in the solid phase are essentially the same as in the gas-phase (see §3.1). An accurate measurement of the bond length difference between the single (a_5) and double bonds (a_6) has been carried out by measuring the neutron scattering intensity over a large momentum transfer range ($6.5 < Q < 20$ Å^{-1}), over which only the intramolecular structure factor contributes to the scattering [7.41]. In modeling the scattering results, the C_{60} molecule is considered as a truncated icosahedron with $(60)(59)/2 = 1770$ pairs of carbon atoms and 23 distinct interatomic distances. The results show a difference in bond lengths $a_5 - a_6 = 0.062$ Å at 295 K [7.41].

More detailed analysis of single crystal x-ray diffraction data shows that there are some orientational correlations between adjacent C_{60} molecules even at room temperature [7.42, 43], consistent with a model for rapid ratcheting motion, rather than constant angular momentum rotations. The rotational reorientation time at room temperature is 9–12 ps based on NMR [7.32, 33, 35, 36, 44], and μ meson spin resonance (μSR) [7.45] measurements (see §16.1.4 and §16.3). It has also been suggested that low-frequency elastic relaxation (ratcheting motion) may originate from the appearance and growth of small ordered clusters of a low-symmetry phase (i.e., order parameter fluctuations) [7.46].

Relative to the other allotropic forms of carbon, solid C_{60} is relatively compressible, with an isothermal volume compressibility (see Table 7.1) of 6.9×10^{-12} cm^2/dyn [7.4], approximately three times larger than that of graphite, because the van der Waals charge cloud around the C_{60} molecules can be compressed easily in three dimensions, rather than in one dimension as in the case of graphite.

It should be noted that the cubic structure described in this chapter is pertinent to single crystals grown from the vapor sublimation of C_{60} molecules. For single crystals prepared from a solution involving organic solvents, other crystal structures may be stabilized by solvent molecules incorporated into the structure. For example, C_{60} crystals grown from CS_2 solution crystallize in a black crystal with an orthorhombic structure (space group *Pbnm*) and lattice constants, $a = 25.012$ Å, $b = 25.583$ Å, and $c = 10.003$ Å, and with a unit cell volume of 6401 Å^3 [7.47]. On heating these orthorhombic single crystals above 100°C in vacuum, a transformation is made to the fcc structure obtained from vapor sublimation growth. Two different high-pressure phases have also been identified and are discussed on §7.3, but neither of these phases is included in the phase diagram of Fig. 7.1.

7.1.2. *Group Theory for Crystalline Phases*

In this section we consider the symmetry aspects of solid-state effects associated with the high-temperature and low-temperature phases of crystalline C_{60}. In discussing the symmetry properties of crystalline C_{60}, we must bear in mind that some properties are most sensitive to the symmetry of the constituent molecules (such as the vibrational spectra), while other properties are sensitive to crystalline properties (such as transport phenomena). The major problem to be addressed is the determination of the symmetry elements common to both the icosahedral symmetry of C_{60} and the crystal field symmetry of the crystal lattice.

Regarding the high-temperature fcc phase in Fig. 7.1, there are three possible cases to consider. In one case, the crystal symmetry is emphasized relative to the molecular symmetry, and in the other two cases, the molecular symmetry is emphasized relative to the crystal symmetry. The first case would be appropriate for phenomena where the molecular reorientation is so rapid that the physical property under consideration senses only spherical balls located at fcc lattice sites ($Fm\bar{3}m$ space group or O_h point group symmetry). This first case is seldom considered in the literature, because of the dominance of the molecular vibrational and electronic states in most of the observed physical properties of fullerenes in the crystalline phase. Thus, most treatments of crystalline C_{60} consider an arrangement of icosahedral C_{60} molecules relative to the crystal field coordinates, in one of two cases, as outlined below.

The simplest case to consider is the one where all the molecules are similarly oriented with respect to the same crystalline axes, leading to a space group *Fm*3 (or T_h^3) with the maximum point group symmetry T_h and four identical C_{60} molecules per fcc cubic unit cell. The other case is one where the molecules are merohedrally disordered with respect to common crystalline axes, as discussed in §7.1.4 and §8.5.1. If the C_{60} molecules are randomly oriented, there is no special site symmetry. It should be mentioned that $Im\bar{3}$ is the space group that maximizes the common symmetry operations between the icosahedral molecule (I_h) and the body-centered cubic (bcc) crystal lattice, as occurs for K_6C_{60}, Rb_6C_{60}, and Cs_6C_{60}. In the case of Cs_6C_{60} all the C_{60} anions have the same orientation of their twofold axes (see §8.5.2).

We first discuss the symmetry lowering due to the placement of icosahedral molecules in the *Fm*3 fcc lattice, preserving all the symmetry elements common between the icosahedrons and the fcc lattice; this is represented by the space group *Fm*3 (see Table 7.2) and the point group T_h ($m3$). This is an fcc lattice with four ordered molecules per cubic unit cell. Various other space group symmetries with the twofold axis of the C_{60} aligned with respect to the (100) crystalline axes are discussed by Harris and Sachi-

Table 7.2

Symmetry sites for the face centered cubic space group T_h^3 (or *Fm*3) [7.49].

No.	Site	Sym.	Site coordinates[a] $(0,0,0;\ 0,\frac{1}{2},\frac{1}{2};\ \frac{1}{2},0,\frac{1}{2};\ \frac{1}{2},\frac{1}{2},0)$ + listed sites
4	*a*	$m3$	$0,0,0$
4	*b*	$m3$	$\frac{1}{2},\frac{1}{2},\frac{1}{2}$
8	*c*	23	$\frac{1}{4},\frac{1}{4},\frac{1}{4};\ \frac{3}{4},\frac{3}{4},\frac{3}{4}$
24	*d*	$2/m$	$0,\frac{1}{4},\frac{1}{4};\ \frac{1}{4},0,\frac{1}{4};\ \frac{1}{4},\frac{1}{4},0;\ \frac{1}{2},\frac{1}{4},\frac{1}{4};\ \frac{1}{4},\frac{1}{2},\frac{1}{4};\ \frac{1}{4},\frac{1}{4},\frac{1}{2}$
24	*e*	mm	$x,0,0;\ 0,x,0;\ 0,0,x;\ \bar{x},0,0;\ 0,\bar{x},0;\ 0,0,\bar{x}$
32	*f*	3	$x,x,x;\quad x,\bar{x},\bar{x};\quad \bar{x},x,\bar{x};\quad \bar{x},\bar{x},x$ $\bar{x},\bar{x},\bar{x};\quad \bar{x},x,x;\quad x,\bar{x},x;\quad x,x,\bar{x}$
48	*g*	2	$x,\frac{1}{4},\frac{1}{4};\ \frac{1}{4},x,\frac{1}{4};\ \frac{1}{4},\frac{1}{4},x;\ \frac{1}{2}+x,\frac{1}{4},\frac{1}{4};\ \frac{1}{4},\frac{1}{2}+x,\frac{1}{4};\ \frac{1}{4},\frac{1}{4},\frac{1}{2}+x$ $\bar{x},\frac{1}{4},\frac{1}{4};\ \frac{1}{4},\bar{x},\frac{1}{4};\ \frac{1}{4},\frac{1}{4},\bar{x};\ \frac{1}{2}-x,\frac{1}{4},\frac{1}{4};\ \frac{1}{4},\frac{1}{2}-x,\frac{1}{4};\ \frac{1}{4},\frac{1}{4},\frac{1}{2}-x$
48	*h*	m	$0,y,z;\quad z,0,y;\quad y,z,0;\quad 0,\bar{y},\bar{z};\quad \bar{z},0,\bar{y};\quad \bar{y},\bar{z},0$ $0,y,\bar{z};\quad \bar{z},0,y;\quad y,\bar{z},0;\quad 0,\bar{y},z;\quad z,0,\bar{y};\quad \bar{y},z,0$
96	*i*	1	$x,y,z;\quad z,x,y;\quad y,z,x;\quad \bar{x},\bar{y},\bar{z};\quad \bar{z},\bar{x},\bar{y};\quad \bar{y},\bar{z},\bar{x}$ $x,\bar{y},\bar{z};\quad z,\bar{x},\bar{y};\quad y,\bar{z},\bar{x};\quad \bar{x},y,z;\quad \bar{z},x,y;\quad \bar{y},z,x$ $\bar{x},y,\bar{z};\quad \bar{z},x,\bar{y};\quad \bar{y},z,\bar{x};\quad x,\bar{y},z;\quad z,\bar{x},y;\quad y,\bar{z},x$ $\bar{x},\bar{y},z;\quad \bar{z},\bar{x},y;\quad \bar{y},\bar{z},x;\quad x,y,\bar{z};\quad z,x,\bar{y};\quad y,z,\bar{x}$

[a] Origin at center ($m3$).

Table 7.3

Character table for the point group T_h.

$\mathcal{R}$	E	$3C_2$	$4C_3$	$4C_3'$	i	3σ	$4S_3$	$4S_3'$
A_g	1	1	1	1	1	1	1	1
E_g	1	1	ω	ω^2	1	1	ω	ω^2
	1	1	ω^2	ω	1	1	ω^2	ω
T_g	3	−1	0	0	3	−1	0	0
A_u	1	1	1	1	−1	−1	−1	−1
E_u	1	1	ω	ω^2	−1	−1	$-\omega$	$-\omega^2$
	1	1	ω^2	ω	−1	−1	$-\omega^2$	$-\omega$
T_u	3	−1	0	0	−3	1	0	0

danandam [7.48]. The site symmetries for the $Fm3$ group are given in Table 7.2, where it is seen that the C_{60} molecules are located on a sites having $m3$ or T_h symmetry. The character table for the T_h point group, showing the symmetry classes and irreducible representations, is given in Table 7.3 [7.50, 51], while the corresponding basis functions are given in Table 7.4. The transformation properties of the various sites in Table 7.2 are given in Table 7.5 in terms of the irreducible representations of group T_h that are contained in the equivalence transformation for the various atom sites $(a, b, \ldots, i)$.

The compatibility relations between the I_h, T_h, and T groups are given in Table 4.15. These compatibility relations are used to obtain the crystal field splittings associated with the symmetry lowering that occurs when an icosahedral C_{60} molecule is placed in a cubic field. To have the full T_h point group symmetry, all the C_{60} molecules must be similarly oriented in the cubic field with their x, y, z molecular twofold axes along the same x, y, z

Table 7.4

Basis functions for the point group T_h.

$\mathcal{R}$	Basis functions
A_g	$x^2 + y^2 + z^2$
E_g	$x^2 + \omega y^2 + \omega^2 z^2$
	$x^2 + \omega^2 y^2 + \omega z^2$
T_g	$(R_x, R_y, R_z), (yz, zx, xy)$
A_u	xyz
E_u	$x^3 + \omega y^3 + \omega^2 z^3$
	$x^3 + \omega^2 y^3 + \omega z^3$
T_u	(x, y, z)

Table 7.5

Irreducible representations contained in the equivalence transformation $\chi^{\text{a.s.}}$ for the various symmetry sites for the simple cubic space group T_h^3 (or $Fm3$) [7.49].

No.	Site	Sym.	$\chi^{\text{a.s.}}$
4	a	$m3$	$A_g + T_g$
4	b	$m3$	$A_g + T_g$
8	c	23	$A_g + T_g + A_u + T_u$
24	d	$2/m$	$\begin{cases} A_g + E_g + 3T_g \\ A_u + E_u + 3T_u \end{cases}$
24	e	mm	$\begin{cases} A_g + E_g + 3T_g \\ A_u + E_u + 3T_u \end{cases}$
32	f	3	$\begin{cases} 2A_g + E_g + 4T_g \\ 2A_u + E_u + 4T_u \end{cases}$
48	g	2	$\begin{cases} 2A_g + 2E_g + 6T_g \\ 2A_u + 2E_u + 6T_u \end{cases}$
48	h	m	$\begin{cases} 2A_g + 2E_g + 6T_g \\ 2A_u + 2E_u + 6T_u \end{cases}$
96	i	1	$\begin{cases} 4A_g + 4E_g + 12T_g \\ 4A_u + 4E_u + 12T_u \end{cases}$

twofold axes of the crystalline solid. It is of significance that the group T_h has no fourfold symmetry axes, necessitated by the absence of fourfold axes in group I_h. In the T_h point group, the vector transforms as the T_u irreducible representation, which is used to establish the lattice modes of solid C_{60}, once the equivalence transformation for the atom sites within the unit cell is specified. Crystal field splitting effects, together with the zone folding for the lattice modes of crystalline C_{60}, are further discussed in §11.3.2.

When the C_{60} molecules are in $m3$ crystal field sites, there are three inequivalent carbon atoms on each molecule. These are shown in Fig. 7.3 from the perspective of a threefold axis and a twofold axis. In Fig. 7.3(b) the 4 threefold axes and one of the twofold axes of the T_h point group are shown [7.52].

According to Table 4.15, only the fourfold and fivefold vibrational modes in the molecular I_h symmetry will be split in an fcc crystal field, and the parity of the modes in the $I_h \rightarrow T_h$ symmetry-lowering transformation is preserved. Thus suppressing the g and u parity subscripts, we can express the mode splitting in the $I_h \rightarrow T_h$ symmetry-lowering transformation by

$$\begin{aligned} [G]_{I_h} &\rightarrow [A+T]_{T_h}, \\ [H]_{I_h} &\rightarrow [E+T]_{T_h} \end{aligned} \tag{7.1}$$

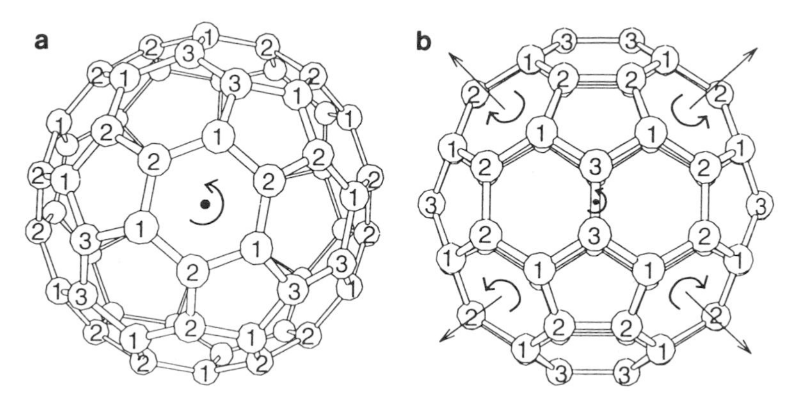

Fig. 7.3. Inequivalent sites for carbon atoms on the C_{60} molecule when placed in the $Fm3$ crystal structure on an $m3$ site: (a) view from a threefold axis and (b) view from a twofold axis [7.52].

where the subscripts are used to denote the pertinent point symmetry groups. A summary of the relations between the molecular modes in I_h symmetry and the modes in various lower symmetries is given in Table 7.6. As is discussed in §11.3.2, the magnitudes of the crystal field splittings in crystalline C_{60} are small and are not easily resolved over the observed linewidths of the Raman and infrared spectral features.

The basis functions for the irreducible representations of the point group T_h (see Table 7.4) imply that the A_g, E_g, and T_g modes are Raman active, with the A_g and E_g modes appearing only for the $\|, \|$ geometry and the

Table 7.6

Compatibility relations of the high symmetry group I_h with lower symmetry groups $\bar{3}$, 23, $m3$ and for the Γ point of the zone-folded T_h^6 space group.

I_h	$\bar{3}$	23 (T)	$m3$ (T_h)	Γ point in T_h^6
A_g	A_g	A	A_g	$A_g + T_g$
F_{1g}	$A_g + E_g$	T	T_g	$A_g + E_g + 3T_g$
F_{2g}	$A_g + E_g$	T	T_g	$A_g + E_g + 3T_g$
G_g	$2A_g + E_g$	$A + T$	$A_g + T_g$	$2A_g + E_g + 4T_g$
H_g	$A_g + 2E_g$	$E + T$	$E_g + T_g$	$A_g + 2E_g + 5T_g$
A_u	A_u	A	A_u	$A_u + T_u$
F_{1u}	$A_u + E_u$	T	T_u	$A_u + E_u + 3T_u$
F_{2u}	$A_u + E_u$	T	T_u	$A_u + E_u + 3T_u$
G_u	$2A_u + E_u$	$A + T$	$A_u + T_u$	$2A_u + E_u + 4T_u$
H_u	$A_u + 2E_u$	$E + T$	$E_u + T_u$	$A_u + 2E_u + 5T_u$

T_g modes only for the $\parallel, \perp$ geometry, where the two indicated orientations refer to the polarizations of the incident and scattered light.

At low temperature (as discussed in §7.1.3), the appropriate space group crystal structure is T_h^6 or $Pa\bar{3}$, as has been determined by x-ray, neutron, and electron diffraction experiments. In the low-temperature phase, there are four distinct C_{60} molecules per cubic unit cell. The T_h point group symmetry is maintained for the Γ point in the Brillouin zone by virtue of the four molecules interchanging places under the point group symmetry operations (see Table 7.6).

Table 7.7 gives the site symmetries for the low-temperature T_h^6 space group. The four C_{60} molecules per unit cell are located on the a sites listed in Table 7.7, which has $\bar{3}$ site symmetry. As a result of the lower site symmetry of the C_{60} molecules in $Pa\bar{3}$ relative to the $Fm3$ space group, the number of inequivalent sites for the carbon atoms of the C_{60} molecule in $Pa\bar{3}$ increases to 10, as shown in Fig. 7.4. It is shown in Table 7.8 that dopants can occupy all the b sites, or all the c sites, or all the d sites of space group $Pa\bar{3}$ without lowering the space group symmetry. The sites for common dopants in fullerene solids are discussed in §8.6 and §8.7.

The four fullerene molecules per unit cell in the $Pa\bar{3}$ structure transform into one another according to the irreducible representations of the equivalence transformation (A_g+T_g) as shown in Table 7.8 [7.51]. The irreducible representations for the equivalence transformation $\chi^{\text{a.s.}}$ for dopants on other sites of the space group T_h^6 $(Pa\bar{3})$ are given in Table 7.8. The character table for the point group $S_6(\bar{3})$ is given in Table 7.9. The irreducible

Table 7.7

Site symmetries for the simple cubic space group T_h^6 (or $Pa\bar{3}$) [7.49].

No.	Site	Sym.	Site coordinates[a]			
4	a	$\bar{3}$	$0,0,0$; $0,\frac{1}{2},\frac{1}{2}$; $\frac{1}{2},0,\frac{1}{2}$; $\frac{1}{2},\frac{1}{2},0$			
4	b	$\bar{3}$	$\frac{1}{2},\frac{1}{2},\frac{1}{2}$; $\frac{1}{2},0,0$; $0,\frac{1}{2},0$; $0,0,\frac{1}{2}$			
8	c	3	x,x,x;	$\frac{1}{2}+x,\frac{1}{2}-x,\bar{x}$;	$\bar{x},\frac{1}{2}+x,\frac{1}{2}-x$;	$\frac{1}{2}-x,\bar{x},\frac{1}{2}+x$
			$\bar{x},\bar{x},\bar{x}$;	$\frac{1}{2}-x,\frac{1}{2}+x,x$;	$x,\frac{1}{2}-x,\frac{1}{2}+x$;	$\frac{1}{2}+x,x,\frac{1}{2}-x$
24	d	1	x,y,z;	$\frac{1}{2}+x,\frac{1}{2}-y,\bar{z}$;	$\bar{x},\frac{1}{2}+y,\frac{1}{2}-z$;	$\frac{1}{2}-x,\bar{y},\frac{1}{2}+z$
			z,x,y;	$\frac{1}{2}+z,\frac{1}{2}-x,\bar{y}$;	$\bar{z},\frac{1}{2}+x,\frac{1}{2}-y$;	$\frac{1}{2}-z,\bar{x},\frac{1}{2}+y$
			y,z,x;	$\frac{1}{2}+y,\frac{1}{2}-z,\bar{x}$;	$\bar{y},\frac{1}{2}+z,\frac{1}{2}-x$;	$\frac{1}{2}-y,\bar{z},\frac{1}{2}+x$
			$\bar{x},\bar{y},\bar{z}$;	$\frac{1}{2}-x,\frac{1}{2}+y,z$;	$x,\frac{1}{2}-y,\frac{1}{2}+z$;	$\frac{1}{2}+x,y,\frac{1}{2}-z$
			$\bar{z},\bar{x},\bar{y}$;	$\frac{1}{2}-z,\frac{1}{2}+x,y$;	$z,\frac{1}{2}-x,\frac{1}{2}+y$;	$\frac{1}{2}+z,x,\frac{1}{2}-y$
			$\bar{y},\bar{z},\bar{x}$;	$\frac{1}{2}-y,\frac{1}{2}+z,x$;	$y,\frac{1}{2}-z,\frac{1}{2}+x$;	$\frac{1}{2}+y,z,\frac{1}{2}-x$

[a] Origin at center ($\bar{3}$).

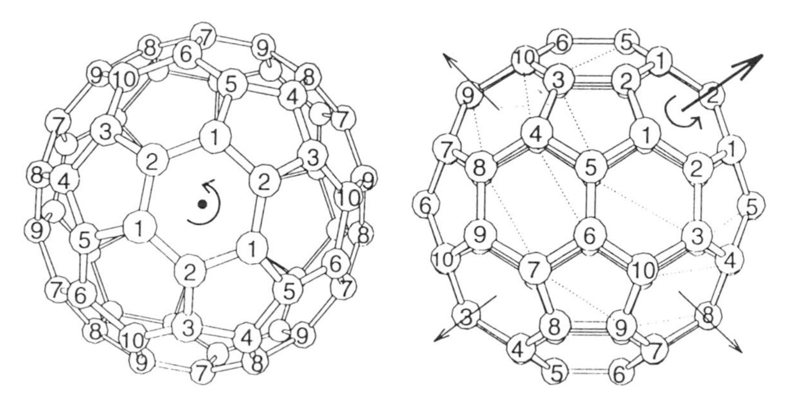

Fig. 7.4. Two views of C_{60} in $Pa\bar{3}$ symmetry showing the 10 inequivalent carbon sites. The dotted lines on the right indicate cross-sectional cuts normal to the threefold crystal symmetry axis, which is shown as the dark arrow in the figure. By adding mirror planes appropriate to the $m3$ site symmetry, it is easy to see how the 10 inequivalent sites in this figure go into the 3 sites shown in Fig. 7.3[7.52].

representations contained in $\chi^{\text{a.s.}}$ are used to obtain symmetry properties for the electronic and vibrational dispersion relations in the solid state.

The presence of four atoms per unit cell affects the optical lattice mode symmetries. Using $\chi^{\text{a.s.}}$ for the equivalence transformation (A_g+T_g) for the four C_{60} molecules per unit cell in the $Pa\bar{3}$ phase, the zone-folded vibrations for a mode with symmetry Γ at the zone center transforms as $\Gamma \otimes (A_g+T_g)$. We use Table 4.15 to relate the normal modes Γ of the I_h icosahedral group to groups of lower symmetry (such as T_h). We use Table 4.16 to obtain the irreducible representations after taking the indicated direct products. In this way the direct product $\Gamma \otimes (A_g + T_g)$ is evaluated and the results are given in Table 7.6 for various symmetries at the zone center, including $S_6(\bar{3})$, $T_h(m3)$, and the space group T_h^6 with four C_{60} molecules per unit cell. For example, the optically silent fourfold degenerate G_g modes in the molecule

Table 7.8

$\chi^{\text{a.s.}}$ for symmetry sites for the simple cubic space group T_h^6 (or $Pa\bar{3}$) [7.49].

No.	Site	Sym.	$\chi^{\text{a.s.}}$
4	a	$\bar{3}$	$A_g + T_g$
4	b	$\bar{3}$	$A_g + T_g$
8	c	3	$A_g + T_g + A_u + T_u$
24	d	1	$\{ A_g + E_g + 3T_g$ $A_u + E_u + 3T_u$

Table 7.9

Character table for the point group $S_6(\bar{3})$.

$\mathscr{R}$		E	$4C_3$	$4C_3'$	i	$4S_3$	$4S_3'$
A_g		1	1	1	1	1	1
E_g	{	1	ω^a	ω^2	1	ω	ω^2
		1	ω^2	ω	1	ω^2	ω
A_u		1	1	1	-1	-1	-1
E_u	{	1	ω	ω^2	-1	$-\omega$	$-\omega^2$
		1	ω^2	ω	-1	$-\omega^2$	$-\omega$

[a] $\omega = \exp(2\pi i/3)$.

(I_h symmetry) would split under T_h point group symmetry according to Eq. (7.1), $[G]_{I_h} \to [A+T]_{T_h}$. Now, taking into account the four molecules per unit cell for the low-temperature cubic phase ($\chi^{\text{a.s.}} = A_g + T_g$), we obtain seven distinct mode frequencies for the splitting of each G_g mode, in accordance with

$$(A_g + T_g) \otimes (A_g + T_g) = 2A_g + E_g + 4T_g, \tag{7.2}$$

using Tables 4.15 and 4.16. Thus the 6 fourfold degenerate G_g modes for the C_{60} molecule in icosahedral I_h symmetry (see §11.3.1) are transformed into 24 × 4 Raman-active modes, associated with 42 distinct frequencies (lines) which would be symmetry allowed in the low-temperature simple cubic symmetry phase. Since their Raman activity would arise only through the crystal field interaction, the G_g-derived Raman lines would be expected to give rise to weakly Raman-active spectral features.

Another way to obtain the splitting of a given mode Γ is to consider the $\bar{3}$ site symmetry for each of the molecules, as given in Table 7.6. Using the $\bar{3}$ site symmetry, the mode splitting for each of the intramolecular modes Γ can be found in analogy with the symmetry lowering discussed above for the case of the $m3$ site symmetry involving the T_h group. The mode splittings for all possible mode symmetries Γ in icosahedral symmetry are given in the last column of Table 7.6 for the case of T_h point group symmetry. If each of the four molecules is properly aligned in $Pa\bar{3}$ symmetry, the direct product for the transformation of the four molecules $\chi^{\text{a.s.}}$ in the unit cell $(A_g + T_g)$ with the mode symmetries in the column labeled $m3$ (T_h) in Table 7.6 gives the splittings listed in the right-hand column (labeled Γ point in T_h^6). A comparison of the entry for the G_g mode in Table 7.6 yields the same result as is given in Eq. (7.2). If merohedral disorder is present in the lattice (see §7.1.4), and the four molecules per unit cell become uncorrelated, but the $\bar{3}$ site symmetry is preserved, then the splitting of the G_g line would yield 12 lines with symmetries $4(2A_g + E_g)$ in accordance with Table 7.6.

Using crystal field and zone-folding arguments such as illustrated in Table 7.6, we see that the total number of Raman-allowed modes for C_{60} in the low-temperature T_h^6 structure becomes very large. Table 7.10 lists the symmetries of the normal modes for C_{60} in the crystal field for the $Fm\bar{3}m$ and $Pa\bar{3}$ structures above and below T_{01}, the latter including many Raman-active modes: 29 one-dimensional A_g modes, 29 two-dimensional E_g modes, and 87 three-dimensional T_g modes. Table 7.10 assumes that every molecule is properly oriented in the crystal lattice and that no disorder exists. The one-electron band calculations typically make this assumption, although the molecules in actual crystals are not ordered in this way (see §7.1.4).

Group theoretical considerations must also be taken into account in treating the intermolecular vibrational modes (see §11.4). For the case of one molecule per unit cell, the intermolecular rotational and translational motions, respectively, give rise to modes with $T_g + T_u$ symmetries, one of which (T_u) is the acoustic mode. For the T_h^6 space group ($Pa\bar{3}$) with four C_{60} molecules per unit cell, the number of intermolecular modes associated

Table 7.10

Symmetries of Brillouin zone-center intramolecular vibrational modes for carbon atoms[a] in C_{60}, first with icosahedral symmetry (as in solution), then in T_h point group symmetry (the fcc structure above T_{01}), and finally in space group T_h^6 symmetry (for $T < T_{01}$).

I_h	T_h	T_h^6 (Γ point)
$2A_g$	$2A_g$	$2A_g + 2T_g$
$3F_{1g}$	$3T_g$	$3T_g + 3(A_g + E_g + 2T_g)$
$4F_{2g}$	$4T_g$	$4T_g + 4(A_g + E_g + 2T_g)$
$6G_g$	$6A_g + 6T_g$	$6(A_g + T_g) + 6(A_g + E_g + 3T_g)$
$8H_g$	$8E_g + 8T_g$	$8(E_g + T_g) + 8(A_g + E_g + 4T_g)$
$1A_u$	$1A_u$	$A_u + T_u$
$4F_{1u}$	$4T_u$	$4T_u + 4(A_u + E_u + 2T_u)$
$5F_{2u}$	$5T_u$	$5T_u + 5(A_u + E_u + 2T_u)$
$6G_u$	$6A_u + 6T_u$	$6(A_u + T_u) + 6(A_u + E_u + 3T_u)$
$7H_u$	$7E_u + 7T_u$	$7(E_u + T_u) + 7(A_u + E_u + 4T_u)$

[a] The modes associated with the dopant atoms in doped solid C_{60} are not included in this listing and must be accounted for separately. The modes associated with translations and rotations (librations) of the center of mass of the C_{60} molecules also are not included in this table.

with translations and rotations of the center of mass is found by taking the direct product of the equivalence transformation $(A_g + T_g)$ with $(T_g + T_u)$ for the rotations and translations, to yield

$$(A_g + T_g) \otimes (T_g + T_u) \rightarrow A_g + E_g + 3T_g + A_u + E_u + 3T_u \qquad (7.3)$$

in which one of the T_u modes is the acoustic mode. These symmetries are used in §11.4 to guide the interpretation of the intermolecular vibrational mode spectra in crystalline C_{60}.

7.1.3. Low-Temperature Phases

Below a characteristic temperature of $T_{01} = 261$ K, the C_{60} molecules in solid C_{60} lose two of their three degrees of rotational freedom. Thus the rotational motion below T_{01} occurs around the four $\langle 111 \rangle$ axes and is a hindered rotation whereby adjacent C_{60} molecules develop strongly correlated orientations [7.11, 40, 53, 54]. In the ordered phase below T_{01}, the orientation of the C_{60} molecules relative to the cubic crystalline axes becomes important. In this section we consider the structural properties of crystalline C_{60} as a function of temperature with particular emphasis given to structural changes associated with phase transitions. The group theoretical issues associated with these phases are reviewed in §7.1.2.

Whereas the structure of solid C_{60} above T_{01} is the fcc structure $Fm\bar{3}m$ discussed in §7.1.1, the structure below ~261 K is a simple cubic structure (space group T_h^6 or $Pa\bar{3}$) with a lattice constant $a_0 = 14.17$ Å and four C_{60} molecules per unit cell, since the four molecules within the fcc structure (see Fig. 7.2) become inequivalent below T_{01}, as molecular orientation effects become important [7.11, 40]. From a physical standpoint, the lowering of the crystal symmetry as the temperature is reduced below T_{01} is caused by the assignment of a specific $\langle 111 \rangle$ direction (vector) to each of the four molecules within a unit cell as shown in Fig. 7.5. In this figure the four distinct molecules in the unit cell are shown, as well as the rotation axis for each of the molecules [7.55]. Thus the degrees of rotational freedom are greatly reduced from essentially free rotation above T_{01} to rotations about a specific threefold axis below T_{01}. This reduction in the degrees of freedom is the cause of the phase transition at T_{01} and affects many of the physical properties of C_{60} in the crystalline state.

Supporting evidence for a phase transition at T_{01} from the fcc structure (above T_{01}) to the simple cubic structure below T_{01} is provided by anomalies in many property measurements, such as x-ray diffraction [7.56], differential scanning calorimetry [7.57], specific heat [7.15, 58, 59], electrical resistivity [7.60], NMR [7.32, 33, 35, 57, 61–64], inelastic neutron scattering

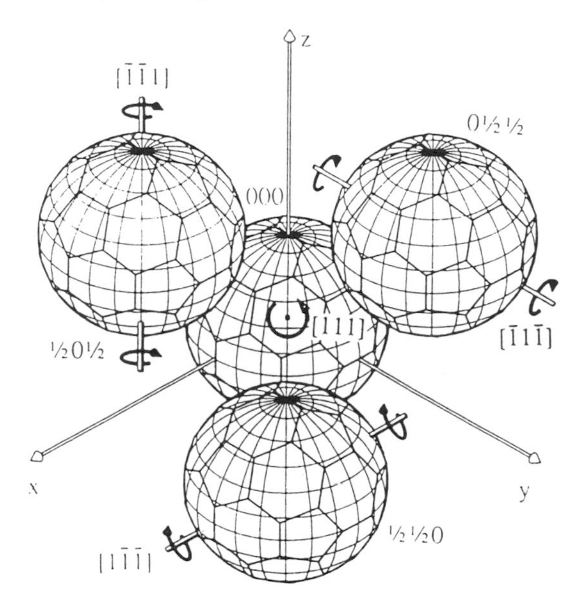

Fig. 7.5. The four molecules in the unit cell of C_{60} showing the same standard orientation, with twofold axes aligned parallel to the cube edges. Starting from this orientation, molecules at (0,0,0), $(\frac{1}{2},\frac{1}{2},0)$, $(\frac{1}{2},0,\frac{1}{2})$, $(0,\frac{1}{2},\frac{1}{2})$ are rotated by the same angle about local axes $\langle 111\rangle$, $\langle 1\bar{1}\bar{1}\rangle$, $\langle \bar{1}\bar{1}1\rangle$, and $\langle \bar{1}1\bar{1}\rangle$, respectively. The sense of rotation about each $\{111\}$ axis is indicated in the figure [7.55].

[7.65–68], electron diffraction [7.69], sound velocity and ultrasonic attenuation [7.7, 70], elastic properties [7.46], Raman spectroscopy [7.71–74], thermal conductivity [7.17], and the thermal coefficient of lattice expansion [7.11, 75].

As T is lowered below T_{01}, orientational alignment begins to take place, although the structural alignment occurs over a relatively large temperature range. The crystal structure of the ordered phase below temperature T_{01} can be understood by referring to Fig. 7.6(a), where one standard orientation of a C_{60} molecule is shown with respect to a cubic coordinate system [7.48]. In this standard orientation of the C_{60} molecule, the maximum point group symmetry (T_h) common to the icosahedral point group I_h and the fcc space group $Fm\bar{3}m$ is maintained. If all the C_{60} molecules in the lattice were to maintain the same standard orientation shown in Fig. 7.6(a), then the space group would be $Fm3$, as discussed in §7.1.2. Here all the $\{100\}$ axes pass through three mutually orthogonal twofold molecular axes (the centers of the electron-rich double-bond hexagon–hexagon edges), and four $\langle 111\rangle$ axes pass through diagonally opposite corners of the cube shown in Fig. 7.6. A second alternate standard orientation is shown in Fig. 7.6(b), which is obtained from Fig. 7.6(a) through rotation by $\pi/2$ about the [001] axis. In the case of M_6C_{60} (M = K, Rb, Cs), all the C_{60} molecules are similarly oriented with respect to a single set of x, y, z axes. However, this is not the case in crystalline C_{60}, where the molecules order with nearly equal probability in the two standard orientations. *Merohedral disorder* refers to

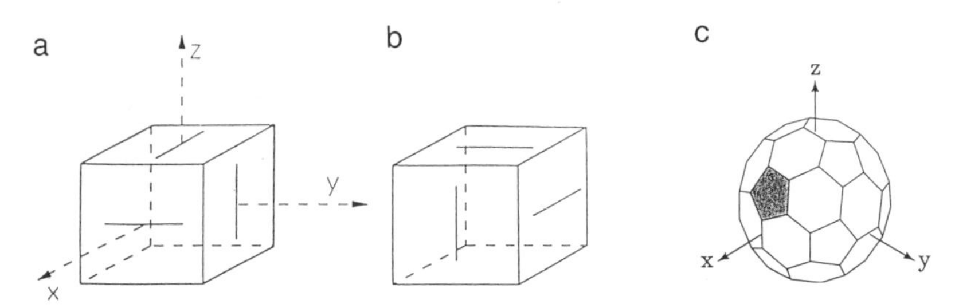

Fig. 7.6. (a) One "standard" orientation for the Cartesian axes in a cubic crystal so that these axes pass through three orthogonal twofold axes. (b) The other "standard" orientation, related to that in (a) by a 90° rotation about any of the Cartesian axes. (c) A view of the (b) "standard" orientation, showing twofold and threefold molecular axes aligned with the cubic axes of the C_{60} crystal. Placing the inscribed icosahedron (c) into (a) or (b) has mirror planes perpendicular to the three $\langle 100 \rangle$ directions. In this orientation, six of the icosahedral edges lie in planes parallel to the mirror planes, so that when the icosahedron is circumscribed by a minimal cube, these edges lie in the cube faces, as shown. Each $\langle 111 \rangle$ direction is a threefold axis [7.21, 48].

the random choice between one of the two possible standard orientations shown in Fig. 7.6(a) and (b) for each C_{60} molecule in the solid (see §7.1.4).

In the low-temperature ordered phase, the centers of four C_{60} molecules have coordinate locations (0,0,0), $(\frac{1}{2}, \frac{1}{2}, 0)$, $(\frac{1}{2}, 0, \frac{1}{2})$, and $(0, \frac{1}{2}, \frac{1}{2})$, and each molecule, respectively, is rotated by 22°–26° from the standard orientation shown in Fig. 7.6(a) about the threefold $\langle 1,1,1 \rangle$, $\langle 1,\bar{1},\bar{1} \rangle$, $\langle \bar{1},\bar{1},1 \rangle$, and $\langle \bar{1},1,\bar{1} \rangle$ axes [7.11]. These rotations of the {111} axes are shown in Fig. 7.5 [7.55]. This rotation angle of 22°–26° is determined experimentally and is governed by the intermolecular interactions between C_{60} molecules, as described below. Since the rotation angle is not fixed by symmetry [7.8], it must be measured experimentally.

In the idealized ordered structure, the relative orientation of adjacent molecules is stabilized by aligning an electron-rich double bond on one molecule opposite the electron-poor *pentagonal* face of an adjacent C_{60} molecule [see Fig. 7.7(a) and (c)]. This orientation corresponds to the 22°–26° rotation angle mentioned above. The preferred orientation of the C_{60} molecules is measured through an angular dependence of the Bragg peaks, using x-ray and neutron diffraction techniques [7.76]. As the temperature is lowered, the molecules align preferentially along these favored directions. A second "phase" transition at a lower temperature T_{02} is associated with this alignment in the sense that many physical properties also show an anomaly in their temperature dependence at this lower temperature T_{02}. However, the value of T_{02} depends on the property being measured, and especially the time required for the excitation necessary to make the measurement, as discussed below.

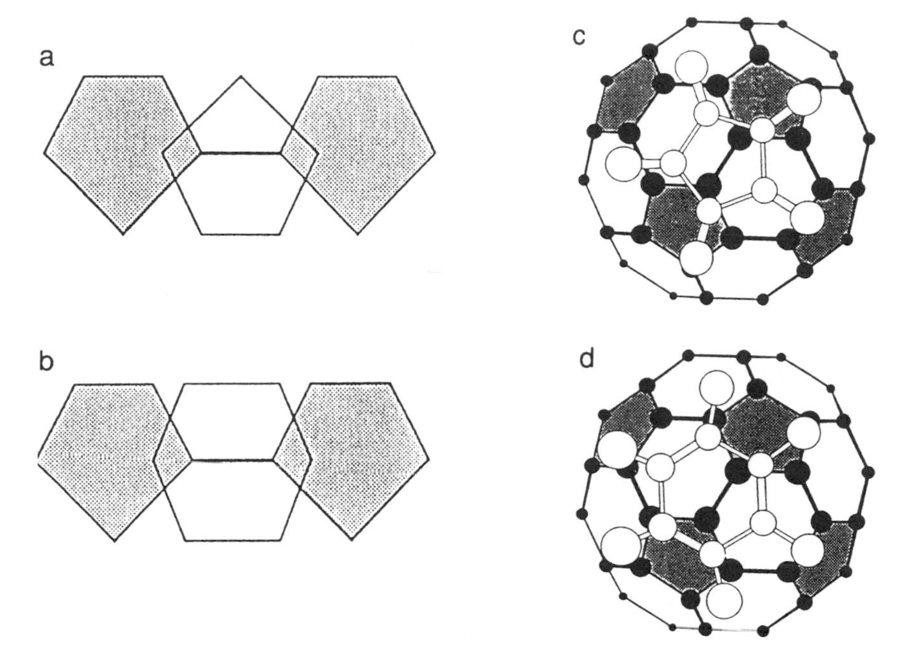

Fig. 7.7. Electron-rich double bond on one C_{60} molecule opposite an electron-poor (a) pentagonal face, (b) hexagonal face on the adjacent molecule [7.48]. (c) is a view along an axis joining centers of nearest-neighbor C_{60} molecules. (d) shows a rotation of (c) by 60° about a threefold axis to bring the hexagonal face of one molecule adjacent to the twofold axis of a second molecule [7.76]. In (c) and (d), the black circles refer to the molecule in back and the open circles refer to the molecule in front.

Another structure (called the "defect structure"), with only slightly higher energy, places the electron-rich double bond of one C_{60} molecule opposite an electron-poor *hexagonal* face [see Figs. 7.7(b) and (d)]. This orientation can be achieved from the lower-energy orientation described in Fig. 7.5 in one of two ways, as shown in Fig. 7.8: by rotation of the C_{60} molecule by 60° around a $\langle 111 \rangle$ threefold axis or the rotation of the molecule by $\sim 42°$ about any one of the three twofold axes normal to the body diagonal [111] direction in the lattice [7.53, 54]. The lower energy configuration is believed to lie only $\sim$ 11.4 meV below the upper energy configuration, and the two are separated by a potential barrier of $\sim$ 290 meV [7.46, 53, 54, 78, 79]. In Fig. 7.8, we see the dark rods denoting the six {110} directions where the double bonds of the indicated atom are adjacent to pentagonal faces of the nearest-neighbor C_{60} molecules. The light rods denote three of the six {110} directions for which the pentagons of the indicated C_{60} molecule are adjacent to double bonds on the nearest-neighbor molecules. For the

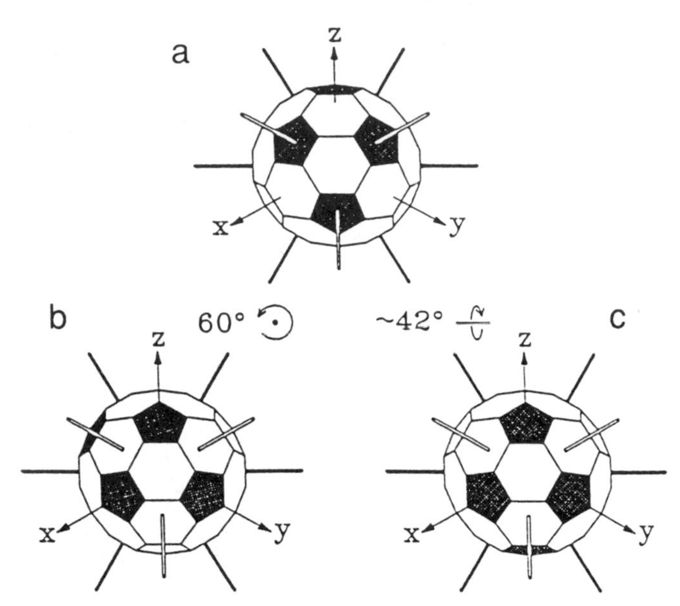

Fig. 7.8. Three possible orientations of a C_{60} molecule with respect to a fixed set of axes. The plane of each drawing is normal to a [111] direction. Thin shaded (unshaded) rods represent [110] directions that are normal (inclined at 35.26°) to the ⟨111⟩ direction. The 12 nearest-neighbors of the molecule are in the twelve {110} directions. At (a), which represents the majority orientation, pentagons face neighbors in the "inclined" ⟨1$\bar{1}$0⟩ directions, and double bonds face neighbors in the "normal" ⟨110⟩ directions. At (b) and (c), which represent the same minority orientation, hexagons have replaced pentagons in the "inclined" ⟨110⟩ directions. The transformation from (a) to (b) involves a 60° rotation about the [111] direction, and from (a) to (c) involves a ~ 42° rotation about the ⟨1$\bar{1}$0⟩ direction [7.77].

defect orientation, the double bonds are opposite a hexagonal face as in Figs. 7.7(b) and (d) [7.77].

As the temperature T is lowered below $T_{01} = 261$ K, the probability of occupying the lower energy configuration increases [7.53, 54]. This model for two orientational configurations (see Figs. 7.7 and 7.8) with nearly the same energy is called the David model [7.53, 54]. On the basis of studies of the temperature dependence of the ^{13}C-NMR spin correlation time τ, which is obtained from measurement of the NMR spin–lattice relaxation time T_1, a ratchet-type motion was proposed between these correlated orientations below T_{01}. These ratchet-type rotations are described by an activation-type behavior in which the diffusion constant D is given by $D = D_0 \exp(-T_A/T)$ [7.36]. Here the activation temperature T_A for ratchet motion below T_{01} is $T_A^{\text{rat}} = 2100 \pm 600$ K. As discussed in §7.1.1, some ratchet-type motion is also observed above T_{01}, and for this regime, the activation temperature is much lower, $T_A^{\text{rat}} = 695 \pm 45$ K [7.36], indicative of a motion much closer to free rotation of the molecule. The corresponding physical picture is that of a molecule that jumps via thermal excitation from one local potential minimum to another; these jumps are made quicker at high temperature and slower at low temperature, as discussed in more detail in §7.1.5.

Evidence for residual structural disorder in the low-temperature molecular alignment is found from neutron scattering experiments [7.53, 54], specific heat [7.15, 58, 59, 80], NMR motional narrowing studies [7.32, 33, 35, 57, 63, 64], thermal conductivity measurements [7.17, 81], and others. The mechanism by which the orientational alignment is enhanced is by the ratcheting motion of the C_{60} molecules around the $\langle 111 \rangle$ axes as the molecules execute a hindered rotational motion. Very rapid ratcheting motion has been observed from about 300 K (room temperature) down to the $T_{01} =$ 261 K phase transition. Slower ratcheting motion is the dominant molecular motion below T_{01} and this motion continues down to low temperatures. Below T_{01}, the rotational reorientation time τ is much slower than above T_{01} (e.g., $\tau \sim 2$ ns at 250 K and 5 ps at 340 K) [7.35, 36, 45, 63]. In the solid phase, librational motion takes place below T_{01} and also plays an important role in the molecular alignment (see §11.4).

The reorientation of C_{60} molecules is governed by an orientational potential $\bar{V}(\theta)$

$$\bar{V}(\theta) = \frac{\bar{V}_A}{2}\left[1 - \cos\left(2\pi \frac{\theta}{\theta_{\text{hop}}}\right)\right] \tag{7.4}$$

where θ_{hop} is the angle through which the molecule librates. The energy $\bar{V}_A$ is related to the librational energy E_{lib}, the moment of inertia I of the C_{60} molecule, and θ_{hop} by

$$\bar{V}_A = E_{\text{lib}}^2 \left(\frac{\theta_{\text{hop}}}{2\pi}\right) \frac{2I}{\hbar^2}, \tag{7.5}$$

showing that librations enable the molecules to overcome the potential barrier and hop to an equivalent potential minimum. Using values of $E_{\text{lib}} =$ 2.5 meV (see §11.4) and $I = 1.0 \times 10^{-43}$ kg m^2 (see Table 3.1), Fig. 7.9 is obtained showing $\bar{V}(\theta)$ plotted for ratcheting motion by $\theta_{\text{hop}} = 60°$ about the $\langle 111 \rangle$ axes (solid curves) and by $\theta_{\text{hop}} = 42°$ about the $\langle 1\bar{1}0 \rangle$ axes (dashed lines) [7.76]. The experimental measurements favor rotations of the molecules about $\langle 1\bar{1}0 \rangle$ axes with regard to the barrier height ($\sim$250 meV) and the presence of a second local minimum at $\theta = 42°$ corresponding to the orientation of the double bond opposite a hexagonal face (the defect orientation), with an energy $\sim$11 meV higher than $\bar{V}(0)$ [7.82]. Model calculations of the single-molecule orientational potential have been carried

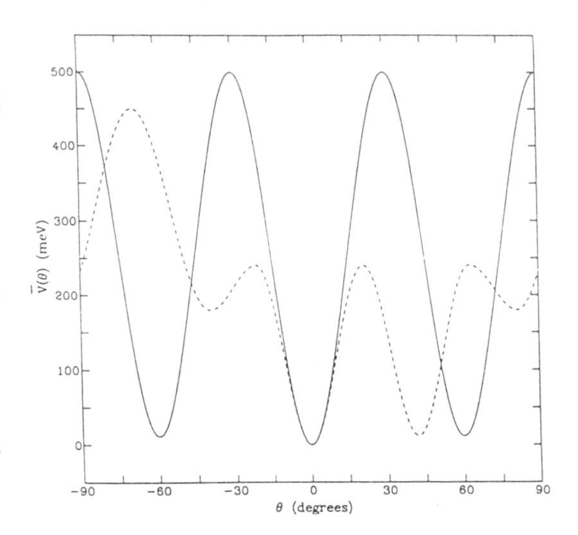

Fig. 7.9. Schematic diagram of the single-particle potential $\bar{V}(\theta)$ in the low-temperature $Pa\bar{3}$ phase. Here $\theta = 0$ corresponds to the equilibrium position of the C_{60} molecule, the solid line represents the potential for rotations of the molecule about the threefold axis aligned along a ⟨111⟩ direction, and the dashed line is the potential for rotations about the ⟨1$\bar{1}$0⟩ direction, which corresponds to a twofold axis of the molecule. The "defect" orientation can be reached by a 60° rotation about [111] or a ~42° rotation about [1$\bar{1}$0] as shown in Fig. 7.8 [7.76].

out, giving good agreement with the general results of Fig. 7.9, but with some differences in the barrier heights [7.43, 83–85].

Evidence for another "phase" transition at lower temperatures (90 K to 165 K) has been provided by a number of experimental techniques, including velocity of sound and ultrasonic attenuation studies [7.7, 70], specific heat measurements [7.16, 59], thermal conductivity [7.17, 81], elasticity [7.7, 46, 70], high-resolution capacitance dilatometry [7.75], dielectric relaxation studies [7.78], electron microscopy observations [7.86], neutron scattering measurements [7.53, 54] and Raman scattering studies [7.71, 74]. A wide range of transition temperatures T_{02} has been reported in the literature for this lower-temperature phase transition. A detailed study shows that the temperature T_{02} obtained for this transition depends on the time scale of the measurement [7.46]. Thus the temperature dependence of the relaxation time is given by the relation $\tau_t = \tau_{t0} \exp(E_{At}/k_B T)$, where E_{At} is the activation energy for the transition and τ_{t0} is a characteristic relaxation time with values of $\tau_{t0} = 4 \pm 2 \times 10^{-14}$ s and $E_{At} = 300 \pm 10$ meV [7.46]. The dependence of the transition temperature on the time scale of the perturbation is shown in Fig. 7.10, where the temperature dependence of the real part of the elastic constant C'_{eff} is plotted for various ac frequencies ω [7.46]. A comprehensive plot covering 11 orders of magnitude in ω is shown in Fig. 7.11, where the frequencies of many experimental probes are related to their respective transition temperatures T_{02} shown in Fig. 7.10 [7.46]. The plots in Figs. 7.10 and 7.11 show that the microscopic origins of the phase transition over the temperature range $80 < T < 182$ K have the same physical origin. The very low frequency points in this plot

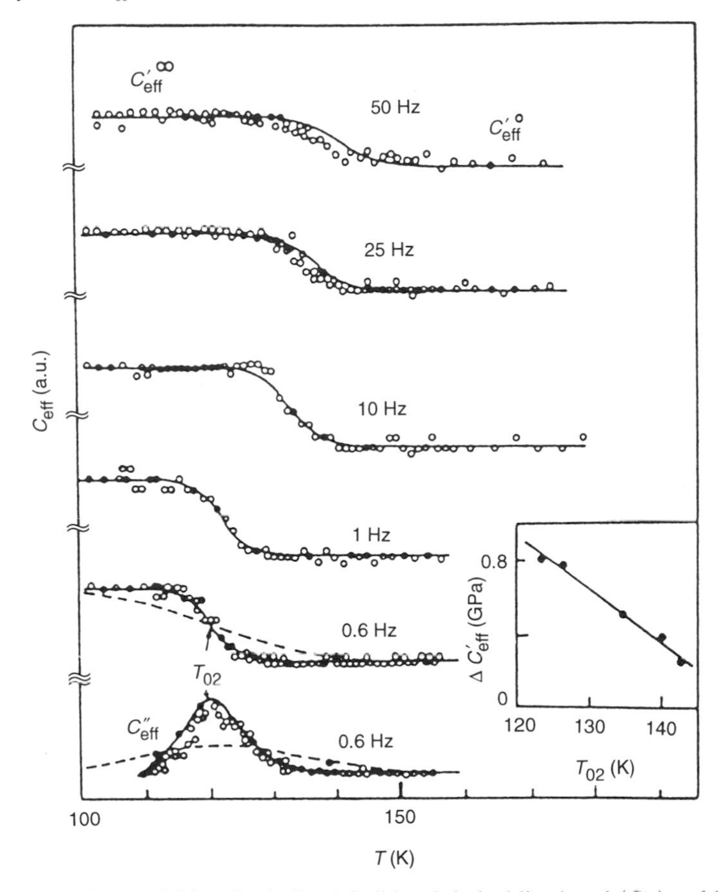

Fig. 7.10. Measured (○) and calculated (solid and dashed lines) real ($C'_{\rm eff}$) and imaginary ($C''_{\rm eff}$) parts of the complex elastic constant $C_{\rm eff}(T)$ *vs.* temperature at five frequencies between 0.6 and 50 Hz. The corresponding plot for the loss component $C''_{\rm eff}(T)$ is shown at 0.6 Hz. The peak in $C''_{\rm eff}$ determines the transition temperature T_{02}. Different scales for $C_{\rm eff}$ (arbitrary units) are used at different frequencies for clarity. The solid lines represent a least squares fit of a Debye relaxational response function with a single thermally-activated relaxation time, yielding good agreement with the experimental data. The dashed lines at 0.6 Hz were calculated with a Gaussian distribution of activation energies and a width of 40 meV; this calculation does not fit the data points. Inset: $\Delta C'_{\rm eff} \equiv C'_{\rm eff}(T \ll T_{02}) - C'_{\rm eff}(T \gg T_{02})$ as a function of the transition temperature T_{02}, which follows a straight line [7.46].

are determined from slow cooling rate experiments, where the sample is in equilibrium throughout the measurements [7.46]. For high-frequency measurement probes, the molecules cannot reorient quickly enough to follow the probe frequency, so that the molecules become frozen into positions corresponding to the two energy minima separated by 11.4 meV and de-

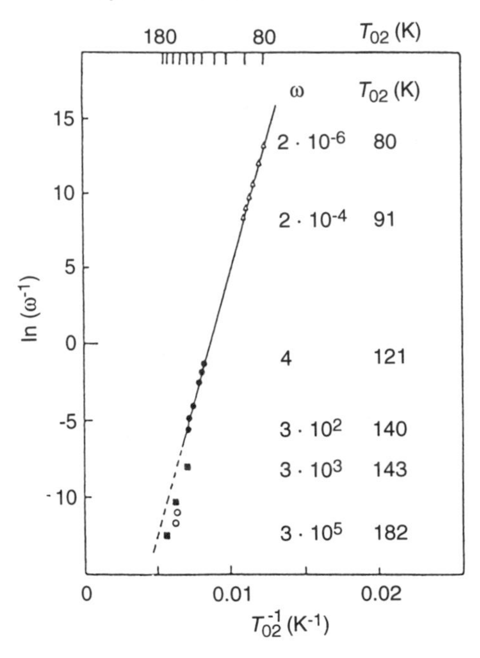

Fig. 7.11. Plot of $\ln(\omega^{-1})$ *vs.* T_{02}^{-1} for different measurement frequencies ω. The points, triangles, full squares, and open circles are taken from elastic constant [7.46], thermal expansion [7.75], dielectric [7.78], and sound velocity [7.7] measurements, respectively. The two columns on the right relate representative probe frequencies to their T_{02} values [7.46].

scribed in Fig. 7.7. Since there is no significant distribution of relaxation times over the wide temperature range of Fig. 7.11, it is concluded that the lower-temperature transition arises from a quenched disorder rather than the establishment of a glass phase [7.46].

Internal friction experiments which are sensitive to the vibrational damping were carried out with a mechanical oscillator [7.87] on a sample consisting of 85% C_{60} and 15% C_{70} with a 100 Å grain size. The fullerene sample was placed on a quartz piezoelectric oscillator and the vibrational damping was measured. The results showed that the magnitude of the internal friction is too low to support a low T-glassy phase. Internal friction measurements on films with larger grain size gave smaller internal friction values at low T. The observed resonant peak in the temperature dependence of the internal friction observed at high T is not sensitive to grain size and relates to measurements of one of the elastic constants [7.87].

When equilibrium is reached, the ratcheting motion described above allows a greater fraction of the C_{60} molecules to align, so that more of the electron-rich double bonds of one molecule are opposite electron-poor pentagons of adjacent C_{60} molecules as the temperature is decreased. As the temperature is decreased to $T_{02} = 90$ K, ~83% of the intermolecular alignments are reported to be in the lower energy state, leaving ~17% in the higher energy state, where the double bond of one molecule is opposite a hexagonal face on the adjacent molecule [7.53, 54]. Below $T_{02}(= 90$ K), some ratcheting motion continues down to about 50 K, but in this regime the same fractions (~83% and ~17%) of the relative intermolecular orientations are maintained and the ratcheting motion is from one orientation to an equivalent orientation. The random occurrence of the higher-energy alignment state gives rise to a disordered two-state tunneling system which has been studied by both specific heat [7.80] and thermal conductivity [7.17] measurements, as discussed in §7.1.4.

Some authors have also identified a low-temperature phase with a superlattice structure and structural unit cell of $2a_0$ [7.86] rather than the lattice constant a_0 for the simple cubic structure that sets in below ~261 K. To model such a phase, further neighbor interactions must also be considered.

7.1.4. Merohedral Disorder

Merohedral disorder in crystalline C_{60} arises because the C_{60} molecule has no fourfold symmetry, even though the molecules, if considered as points, are on face centered cubic (fcc) sites. Since the C_{60} molecule has no fourfold symmetry, each molecule in solid C_{60} chooses approximately randomly between one of the two distinct "standard" orientations of the twofold axes shown in Fig. 7.6, below the ordering temperature T_{01}. If all the icosahedral C_{60} molecules in the lattice had the same orientation relative to the same x, y, z axes as shown in Fig. 7.6(c), then the appropriate space group for the crystal would be $Fm3$. Since the two standard orientations of Fig. 7.6 occur approximately randomly among the C_{60} molecules, it is said that below T_{01} the crystal has *merohedral disorder*.

If the C_{60} molecules assume completely random orientations (without regard to the twofold axes), so that a time-averaged C_{60} molecule would appear as a spherical shell in a diffraction experiment, then the appropriate space group would be $Fm\bar{3}m$, which corresponds to the high-temperature phase [7.8]. This spherical shell model has been successfully used to explain the x-ray powder diffraction patterns above T_{01} which show no even-order ($h00$) peaks [7.4, 88].

Orientational disorder greatly affects physical properties dependent on the relative orientations of adjacent C_{60} molecules. Therefore merohedral disorder plays a major role in the transport properties of fullerene solids. On the other hand, other properties, such as the vibrational spectra, which are mainly determined by the molecular icosahedral symmetry, are much less sensitive to crystal field effects and merohedral disorder.

7.1.5. *Model for Phase Transitions in C_{60}*

Using an intermolecular model potential for C_{60}, several theoretical calculations of the molecular dynamics of the ratchet motion have been performed in C_{60} [7.43, 89]. Calculations have also been reported for the librational and intermolecular vibrational dispersion relations [7.84, 90, 91], and for a Landau theory for the phase transition in C_{60} at T_{01} [7.48, 84, 92]. Model calculations for the intermolecular C_{60}–C_{60} interactions yield a first-order transition at T_{01} and a low-temperature freezing transition broadly occurring in the 90–130 K range [7.85, 90].

One simple model that has had some success in explaining most of the characteristics identified with the first-order phase transition in solid C_{60} at T_{01} is described below [7.92, 93]. This model considers molecules in three states: (a) orientationally frozen state, (b) undergoing ratchet motion about a threefold axis with three positions per molecule for rotation, and (c) undergoing ratchet motion not only about the threefold axis but also about a fivefold (C_5) axis, thus yielding a total of 90 rotational positions.

In a C_{60} molecule, there are 10 C_3 axes and 6 C_5 axes as discussed in §4.1. If a molecule is restricted to rotate about only one $\langle 1,1,1 \rangle$ direction in the low-temperature simple cubic phase (see §7.1.4), then the only possible ratchet motion is about one of the C_3 axis, ratcheting from one potential minimum to another one of the three possible potential minima. When we also consider 30 symmetry operations about six C_5 axes at high temperature, all C_3 axes permute with each other, and thus we will have 90 potential minima. For each C_3 axis, we have three absolute potential minima and six local minima [7.84]. This simple model does not consider rotations about the 20 C_2 axes. According to this simple model, the motions which are most relevant to the phase transition are the hopping motion between local potential minima below and above the phase transition. The activation energies used to describe the hopping between potential minima for the C_3 and C_5 axes are taken from nuclear magnetic resonance experiments [7.36], which show two different activation energies of 2100 K (181 meV) and 700 K (60 meV) in the temperature-dependent spin correlation time τ for the lower- and higher-temperature phases, respectively.

In states (b) and (c), there are 3 and 90 equivalent configurations, respectively, which correspond to the potential minima for the ratchet motion. It is assumed that at a given temperature, N_1, N_2 and N_3 molecules of a total of N molecules are in (a), (b), and (c) states, respectively. The total number of states, W, is then given by

$$W = \frac{N!}{N_1!N_2!N_3!} 3^{N_2} 90^{N_3}. \tag{7.6}$$

Using the Stirling formula $\log N! \sim N \log N - N$, the entropy, S, is then given by

$$\begin{aligned} S &= k_B \log W \\ &= -Nk_B(X_1 \log X_1 + X_2 \log X_2 + X_3 \log X_3) \\ &\quad +Nk_B(X_2 \log 3 + X_3 \log 90), \end{aligned} \tag{7.7}$$

where

$$N = \sum_{i=1}^{3} N_i, \qquad X_i = N_i/N, \qquad \text{and} \qquad \sum_{i=1}^{3} X_i = 1. \tag{7.8}$$

The internal energy E is written as a sum of two terms

$$E = -NJ_a X_1 - \frac{NzJ_b}{2}(X_1 + X_2)^2, \tag{7.9}$$

in which J_a (> 0) stands for the energy gained from freezing the motion for the (a) states, J_b (> 0) is the intermolecular attractive interaction, and $z = 12$ is the number of the nearest-neighbor molecules. It is assumed in this simple model that the attractive interaction acts only when the two molecules in the (a) or (b) states are nearest-neighbors.

In thermal equilibrium the free energy, F,

$$F = E - TS \tag{7.10}$$

is minimized with respect to X_1, X_2 while keeping the temperature fixed and imposing the condition that $X_1 + X_2 + X_3 = 1$. This minimization of F results in three simultaneous self-consistent equations which yield the temperature dependence of N_i ($i = 1, 2, 3$) and the number of molecules in states (a), (b), and (c), respectively.

The temperature dependences of X_1, X_2 and X_3 thus obtained are plotted in Fig. 7.12 as dotted, dashed, and solid lines, respectively, as a function of T/T_b with a fixed ratio of $T_a/T_b = 1/3$, where $T_a = J_a/k_B$ and $T_b = zJ_b/k_B$, k_B being the Boltzmann constant. From Fig. 7.12, it follows first that X_1, the probability that a molecule is frozen, decreases monotonically with increasing T in the range $0.05 < T/T_b < 0.21$, and second a discontinuous change is found in many properties at $T/T_b = 0.21$, characteristic of a first-order transition. The characteristic temperature at $T/T_b = 0.21$ is identified with the phase transition temperature $T = T_{01} = 261$ K, taking $T_b = J_b/k_B = 1243$ K. The observed first-order transition results from the competition between the entropy gain by rotation and the energy gain by intermolecular attraction.

The model shows that X_1 decreases from a value of unity, and X_2 increases from zero over a wider range of temperature than the phase transition at T_{01} (see Fig. 7.12). The onset value for the lower-temperature

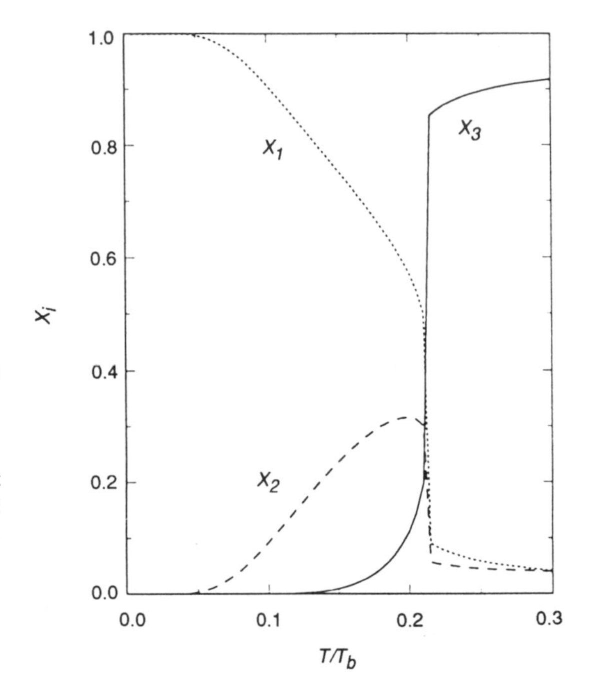

Fig. 7.12. Temperature dependence of the fractional occupation of states (a), (b), and (c) denoted by X_1 (dotted line), X_2 (dashed line), and X_3 (solid line), respectively, plotted as a function of T/T_b (see text). The large changes at $T/T_b = 0.21$ are identified with the first-order phase transition at T_{01}. The onset temperature at $T/T_b \sim 0.05$ is identified with the onset of quenched disorder at low temperatures [7.92].

phase transition T_{02} is operationally defined by the temperature where X_1 becomes 0.99. Figure 7.12 shows that the onset value thus defined occurs at $T/T_b = 0.05$. Using this criterion, it is found that T_{02} is proportional to T_a.

Whereas T_{01} does not depend on T_a/T_b for $T_a/T_b \leq 0.3$, a linear dependence of X_1 on T_a/T_b is found for $T_a/T_b \geq 0.3$. A discontinuous change of X_1 at T_{01} (denoted by ΔX_1) is observed for $T_a/T_b \leq 0.6$ with a maximum in ΔX_1 found at $T_a/T_b \sim 0.25$. For $T_a/T_b \geq 0.6$, a continuous change occurs at T_{01} (i.e., ΔX_1 is very small). Thus, the model is expected to be applicable for a wide range of T_a/T_b from 0.1 to 0.6. The ratio $T_a/T_b = 1/3$ gives the best results for this simple model in comparison to experimental data for solid C_{60}.

From the calculated temperature dependence of the order parameters X_1, X_2, and X_3, the free energy, and the internal energy, a number of thermodynamic quantities are calculated, including ΔH and ΔS, the discontinuity in the enthalpy and entropy, respectively at T_{01}, and the jump in the specific heat ΔC_v associated with the transition at lower temperature T_{02}. The calculated values for these thermodynamic quantities are listed in Table 7.11 and are also compared with experimental values.

Table 7.11

Thermodynamic quantities of the phase transition of solid C_{60} [7.92].

Quantity	Model	Experiment	
$\Delta H(\mathrm{Jg}^{-1})$	6.7^a	5.9^b	at $T = 261$ K
$\Delta S(\mathrm{JK}^{-1}\mathrm{mol}^{-1})$	19.1	30.0^c	at $T = 261$ K
$\Delta C_v(\mathrm{Jg}^{-1})$	8.3	7^c	at $T = 70$–90 K
ΔX_1	0.41	0.4–0.5^d	at $T = 261$ K

[a] For $T_b = 1243$ K.
[b] Reference [7.56].
[c] Reference [7.59].
[d] Intensity of fcc forbidden x-ray peaks (4,5,0) and (4,5,1) [7.56, 88].

The temperature dependence of the order parameter X_1 may be compared with that of x-ray diffraction spots such as (4,5,1) [7.88] and (4,5,0) [7.56], which are forbidden in the fcc structure. The (4,5,1) [or (4,5,0)] peak intensity decreases from its value at very low temperature and experiences a discontinuous jump of about half of its intensity at $T = 261$ K (see Table 7.11) [7.88]. The T dependence of this x-ray peak is similar to the T dependence for X_1 shown in Fig. 7.12. A discontinuity in the order parameter is observed in the low-energy region of the Raman spectrum ($\sim$30 cm^{-1}) [7.94], and it is noted that the observed temperature dependences of the libron intensity and the uncorrelated Lorentzian scattering intensity are similar to the temperature dependences of X_2 and X_3, respectively.

In accordance with the discontinuous change from ratchet motion to relatively free molecular motion, the lattice should expand as a result of surmounting the very weak van der Waals interaction, as the weak Coulomb interaction between electron-rich double-bond regions on one molecule with electron-poor pentagon face regions on the adjacent molecule (see Fig. 7.7) become less effective when the rapid spinning motion commences. Weak hysteresis effects near T_{01} are characteristic of a first-order transition in which there is a local maximum in the free energy between two local minima as a function of the order parameter. As stated above, the basic driving force behind the phase transition at T_{01} is the competition between entropy gain by rotation and energy gain by intermolecular attraction [7.92].

7.2. Crystalline C_{70} and Higher-Mass Fullerenes

The crystal structure of C_{70} is more complex than that of C_{60} [7.86, 95–98], evolving through several distinct crystal structures as a function of temperature (see Fig. 7.13), including the high temperature regime (above

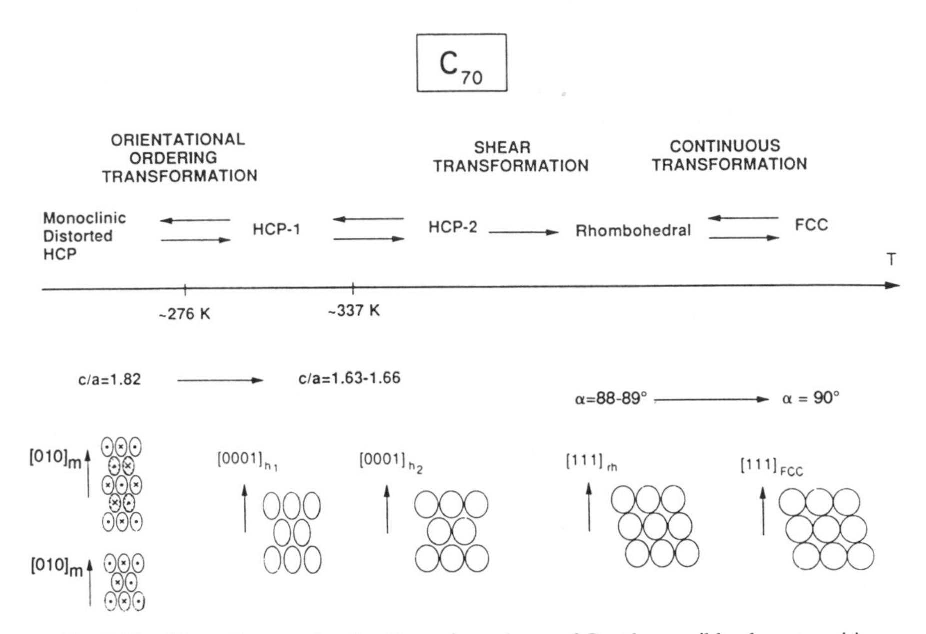

Fig. 7.13. Phase diagram showing the various phases of C_{70}, the possible phase transitions, and the stacking of the C_{70} molecules [7.96].

$T_{01} = 340$ K), the intermediate temperature regime ($275 < T < 340$ K), and the low temperature regime (below $T_{02} = 275$ K). We describe below the crystal structures that have been reported for each of these regimes. Although there is no agreement about the details of the structure in the three regimes, there is general agreement that there are three regimes, that the structure is isotropic above T_{01} and becomes more anisotropic as T is decreased. A compilation of physical parameters associated with solid C_{70} is shown in Table 7.12.

At high temperature ($T \gg T_{01}$) the close-packed fcc phase ($a_0 = 15.01$ Å) with freely rotating molecules is the most stable phase (see Table 7.12), but since the ideal hexagonal close packed (hcp) phase with $c/a_0 = 1.63$ is almost equally stable, fcc crystals of C_{70} tend to be severely twinned and show many stacking faults [7.107]. In this connection we note that the stacking sequences of the close-packed planes follow an ABCABC sequence in the fcc structure and an ABAB sequence for the hcp structure. Although there is some disagreement among the various research groups about the crystal structure of the high-temperature phase (or phases), there is agreement that the structure shows little anisotropy, since the C_{70}–C_{70} distances for the fcc and ideal hcp are essentially the same, except for the planar stacking along the $\langle 111 \rangle$ direction. NMR magic angle spinning experiments

Table 7.12

Physical constants for C_{70} solid materials.

Quantity	Value	Reference
fcc lattice constant	15.01 Å	[7.95, 96]
C_{70}–C_{70} distance	10.61 Å	[7.95, 96]
hcp lattice constant	$a = b = 10.56$Å $c = 17.18$Å	[7.95, 96, 99]
Bulk modulus (rhombohedral)	11 GPa	[7.100]
Structural transitions (T_{01}, T_{02})	361 K, 282 K	[7.97]
Heat of transition (270–380 K)	10.4 J/g	[7.97]
dT_{01}/dp	300 K/GPa	[7.100]
Optical absorption edge	1.6 eV	[7.101, 102]
Calculated HOMO–LUMO gap	1.7 ± 0.5 eV	[7.103, 104]
Calculated binding energy/C	7.42 eV	[7.103]
Ionization potential (1st)	7.61 eV	[7.105]
Ionization potential (2nd)	16.0 ± 0.5 eV	[7.106]
Ionization potential (3rd)	33.0 ± 1.0 eV	[7.106]
Static dielectric constant	3.75	[7.102]
Sublimation temperature	466 K	[7.22]
Heat of sublimation	43.0 kcal/mol	[7.22]
Latent heat	1.29 eV/C_{60}	[7.23]

show rapid molecular reorientation in the high temperature phase with little anisotropy [7.108, 109]. Some authors have reported that as the temperature is lowered, the fcc structure is continuously transformed by deformation into a rhombohedral structure with the long diagonal threefold axis aligned parallel to the $\langle 111 \rangle$ direction of the fcc structure (see Fig. 7.13) [7.96]. These authors further report at yet lower temperatures, but still above T_{01}, a stable hexagonal phase (space group $P6_3/mmc$) with lattice constants $a = b = 10.56$ Å and $c = 17.18$ Å and a nearly ideal c/a ratio of 1.63 (see Fig. 7.13). The observed crystal structures above 337 K tend to be strongly influenced by crystal quality, thermal history, amount of supercooling, and the amount of C_{60} impurity in the crystal [7.97]. Differences in transition temperatures at T_{01} are observed on heating and cooling. It is believed that the most reliable and reproducible results regarding phase transitions are obtained upon warming [7.21]. Some authors believe that the fcc phase is the lowest equilibrium energy phase from high temperatures down to $T_{01} = 345$ K and do not report phase transitions above T_{01} [7.21]. Differential scanning calorimetry studies on very high purity C_{70} solids show a sharp feature at 361 K and a second smaller feature at 343 K, associated with the onset of quasi-isotropic tumbling above T_{01} [7.97].

In the intermediate range between ~280 K and ~340 K, anisotropy in the crystal structure becomes established, as the long axis of the C_{70} rugby balls begin to align, and the C_{70} molecules tumble about rhombohedral axes. Many groups see evidence for a first-order structural phase transition near T_{01} ~340 K, below which anisotropy begins to appear. The phase transition near T_{01} has been studied by NMR [7.108–111], μSR [7.112], neutron scattering [7.113], x-ray diffraction [7.95, 114, 115], dilatometry [7.115], and thermal conductivity [7.81]. The value of the transition temperature T_{01} has been shown to be sensitive to the thermal history and the amount of solvent in the sample. Some authors identify the intermediate temperature phase with rhombohedral symmetry (space group $R\bar{3}m$), which is derived from the high temperature fcc structure by the elongation of the fcc structure along a unique $\langle 111 \rangle$ direction and is driven by the freezing out of the molecular rotation about an axis perpendicular to the long axis [7.8]. Some C_{70} samples show a transition to another hcp lattice below ~337 K, but with $a = b = 10.11$ Å and a c/a ratio of 1.82, larger than that above T_{01} [7.96]. This larger c/a ratio is associated with the orientation of the C_{70} molecules along their long axis, as the free molecular rotation (full rotational symmetry), that is prevalent in the higher temperature phase above T_{01}, freezes into rotations about only the fivefold axis of the C_{70} molecule (see Fig. 7.13) [7.96]. In this intermediate temperature phase (which encompasses room temperature), all research groups report crystal anisotropy. It is also observed that a ratcheting motion around the long axis of the C_{70} molecules is dominant, with the anisotropy of the molecular reorientation increasing with decreasing temperature [7.108–110]. The molecular orientational ordering of the C_{70} molecules in this phase is confirmed by μSR experiments [7.112].

As the temperature is further lowered to T_{02} ≈280 K, the free rotation about the c-axis also becomes frozen. The crystal phase below T_{02} is a monoclinic structure (see Fig. 7.13) with the unique axis along the c-axis of the intermediate temperature hcp or rhombohedral structures, and the monoclinic angle β is close to 120° [7.96]. Differential scanning calorimetry studies on very pure C_{70} crystals show T_{02} = 282 K and another small feature at 298 K has been reported [7.97]. The total heat of transformation from 270 K to 380 K is 10.4 J/g. The supercooling effect for the phase transition near T_{02} is large (with a width of 15–50 K) while the supercooling effect near the T_{01} transition is much smaller (2–3 K) [7.97]. The difference between the experimental values for T_{01} and T_{02} from one research group to another may be due to large supercooling effects, the presence of minority C_{70} phases, and to the presence of C_{60} and solvent impurities (which suppress T_{01}). The molar volume decrease at both the T_{01} and T_{02} phase transitions of C_{70} is ~3%, which is considerably lower than the corresponding decrease in solid C_{60} at T_{01}. Evidence for molecular alignment is also provided by the qualitative change in the intermolecular pair correlation function that is found between 200 K and room temperature [7.116].

The space group for the low-temperature phase has not yet been clearly identified, with C_m, C_2, P_2, and P_m being the most likely candidates, and special preference is currently being give to C_2 and P_2 [7.21]. It has also been suggested that the low-temperature phase may be orthorhombic *Pbcm* or *Pbnm* rather than monoclinic, based on a model for the low-temperature orientational ordering of solid C_{70} [7.117]. Detailed x-ray studies indicate that the low-temperature phase of a sample grown by sublimation as an hcp crystal above T_{01} is of the orthohexagonal *Pbnm* space group [7.107].

One suggestion for the stacking arrangement of the C_{70} molecules in the monoclinic phase is shown in Fig. 7.14, where it is noted that one of the

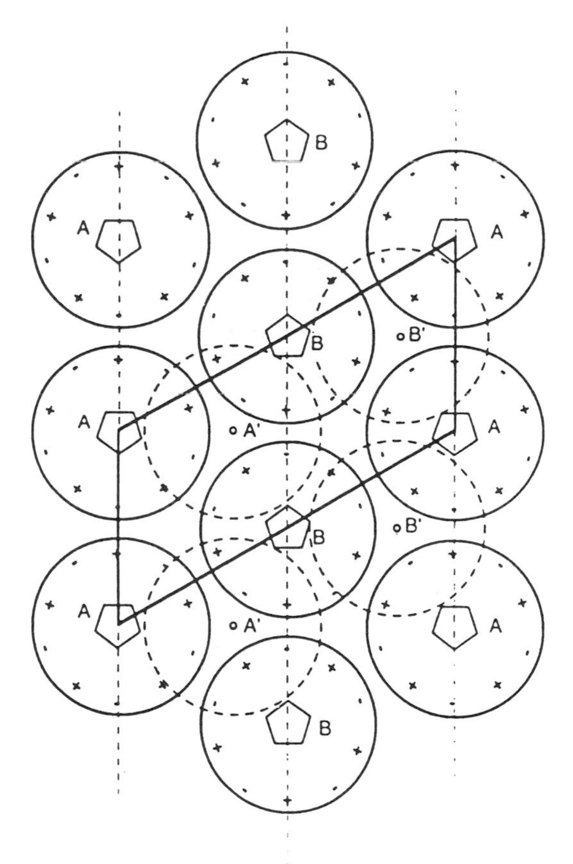

Fig. 7.14. One model for the stacking of orientationally ordered C_{70} molecules in close-packed planes in the monoclinic phase. *A* and *B* represent the two different orientations of molecules within a close-packed layer. The molecules *A′* and *B′* are positioned in the adjacent layer. Electron-rich and electron-poor regions are indicated by + and −, respectively [7.96].

former hexagonal a-axes of the hcp phase is doubled [7.96]. The former hexagonal plane in the low-temperature phase consists of alternating rows of fixed C_{70} molecules which are oriented with their main symmetry axes along the unique axis of the monoclinic structure and the cross sections are arranged within rows as shown in Fig. 7.14, so that the high electron density hexagon–hexagon edges are opposite low electron density hexagonal face centers [see Fig. 7.7(b)]. The stacking perpendicular to this plane can be arranged in either of two ways: one where the A and B rows of the C_{70} molecules are directly over one another, and the other where the C_{70} molecules in the next layer are shifted to locations $b_m/2 + c_m/2$, in which the monoclinic c_m axis is related to the hexagonal a_h axis by $c_m = 2a_h$ (see Fig. 7.14). Transmission electron microscopy (TEM) diffraction patterns for both stacking arrangements have been reported [7.86].

The molecular reorientation times τ_r along the long axis drop rapidly as T is lowered. Some typical values are $\tau_r = 5$ ps at 340 K, $\tau_r = 2$ ns at 250 K, and $\tau_r \sim 20$ μs at 100 K. Also reported are reorientations of the molecular axes at a temperature near T_{01}, with values of $\tau_r \sim 20$ μs reported for rotation of the long axis of the molecule above 300 K [7.108–110]. Ratcheting motions of the molecules along the long axis have been reported even at low T (down to 73 K [7.118]).

Because of the small sample sizes currently available for the higher fullerenes C_{n_C}, structural studies by conventional methods (such as x-ray diffraction and neutron scattering) are difficult. Therefore, only a few-single crystal structural reports have, thus far, been published regarding the higher fullerenes. The most direct structural measurements thus far have come from scanning tunneling microscopy (STM) studies [7.119, 120] and selected area electron diffraction studies using electron energy loss spectroscopy (EELS) [7.121].

The STM measurements have been done on C_{76}, C_{80}, C_{82}, and C_{84} fullerenes adsorbed on Si (100) 2×1 and GaAs (110) surfaces [7.119, 120] (see §17.9.5). The STM studies provide detailed information on the local environment of the higher-mass fullerenes in the crystal lattice. These STM studies show that all of these higher fullerenes crystallize in an fcc structure with $C_{n_C} - C_{n_C}$ nearest-neighbor distances of 11.3 Å, 11.0 Å, 11.74 Å, 12.1 Å for C_{76}, C_{78}, C_{82}, and C_{84}, respectively, and fcc lattice constants are obtained from these distances by multiplication by $\sqrt{2}$. Since higher fullerenes generally have multiple isomers (see §3.2), the fullerenes on the fcc lattice sites are not expected to be identical, and therefore more lattice disorder is expected for higher-mass fullerene crystals. Also the more massive fullerenes are less mobile rotationally, so that the rotational ordering phase transition at T_{01} is driven to higher temperatures, if a phase transition occurs at all.

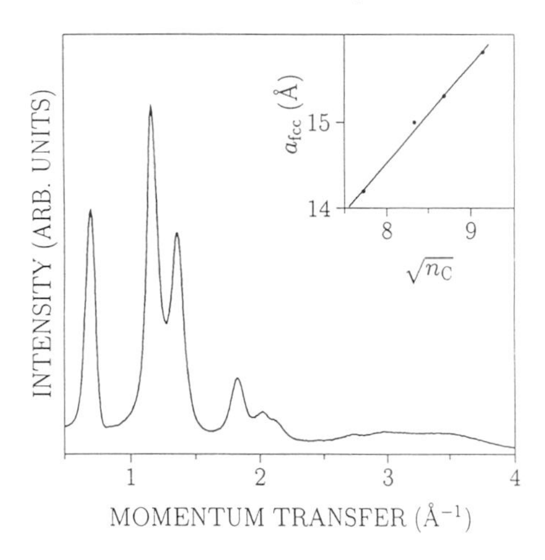

Fig. 7.15. Electron diffraction profile of a thin film of C_{76} recorded in an EELS spectrometer with a momentum transfer resolution of 0.06 Å [7.121]. The inset shows the lattice constant for the higher-mass fullerenes C_{n_C} as a function of $\sqrt{n_C}$, where $60 \leq n_C \leq 84$.

As mentioned above, electron diffraction measurements provide detailed results on lattice constants and grain sizes of very small crystalline samples of the higher fullerenes. Such measurements were made on thin films of higher fullerenes prepare by sublimation of a purified powder onto an NaCl substrate. The NaCl substrate was then dissolved to yield a free-standing fullerene film with grains of ~1000 Å in size [7.121]. A trace showing the diffracted beam intensities from elastic scattering within an electron energy loss apparatus is shown in Fig. 7.15 for C_{76}. By indexing the diffraction peaks observed for C_{60}, C_{70}, C_{76}, and C_{84} to an fcc lattice, the lattice constants a_{fcc} shown in the inset to Fig. 7.15 are obtained and plotted as a function of $\sqrt{n_C}$, where n_C is the number of carbon atoms in C_{n_C}. Results for the lattice constant are in good agreement with the STM results given above and in §17.9.5.

7.3. Effect of Pressure on Crystal Structure

The effect of pressure on the structure of fullerenes has been studied for crystalline C_{60} and C_{70}. Because of the weak van der Waals forces between the fullerene molecules in the lattice, C_{60} and C_{70} are expected to be highly compressible at low pressures where the intermolecular distance is reduced, but nearly incompressible at high pressures because of the very low compressibility of the molecules themselves [7.122]. Experiments show, however, that crystalline C_{60} and C_{70} undergo a phase transition before the incompressibility effect is expected to become important (at ~50 GPa) [7.122].

Experimental studies of the pressure dependence of the structures of crystalline C_{60} and C_{70} have focused on both the pressure dependence of the structural parameters for the stable phases at atmospheric pressure and the structures of the pressure-induced phases.

As stated above, solid C_{60} under pressure is rather compressible. Accurate measurements on high-quality single-crystal samples yield values for the bulk modulus $\beta = -Vdp/dV$ of 6.8 GPa for the fcc phase above T_{01} and 8.8 GPa for the sc phase below T_{01} (see Table 7.1) [7.5, 6], in contrast to early reports of values for β between 14 and 18 GPa [7.4, 24, 25]. These lower values of β imply a uniaxial compressibility of 2.3×10^{-12} cm^2/dyn, which is about the same as the interlayer compressibility $[-(1/c_0)(dc_0/dp)]$ of graphite 2.98×10^{-12} cm^2/dyn [7.123]. Regarding the structure of solid C_{60}, pressures of up to 15 GPa (keeping the temperature constant) have little effect on the crystal structure, except that the transition temperature T_{01} increases approximately linearly with pressure (see Fig. 7.16) at a rate of ~105–164 K/GPa depending on the pressure environment [7.5, 9, 10] (see §7.1.3). Model calculations for the intermolecular C_{60}–C_{60} interactions yield a value of $dT_{01}/dp = 115$ K/GPa [7.85, 90], in good agreement with bulk modulus experiments and also differential scanning calorimetry experiments (163 K/GPa) [7.5]. At room temperature, the phase transition starts at 135 MPa, but the transition is smeared out over the pressure range 135–500 MPa. The phase transition appears to be first order only in the limit $p \to 0$. A linear pressure dependence

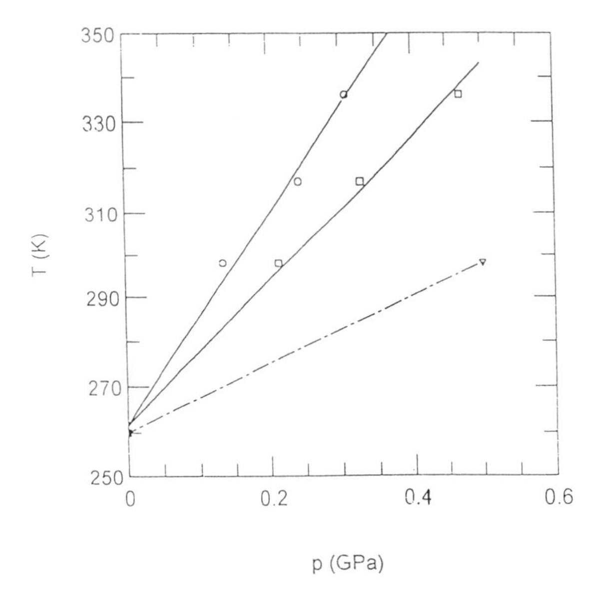

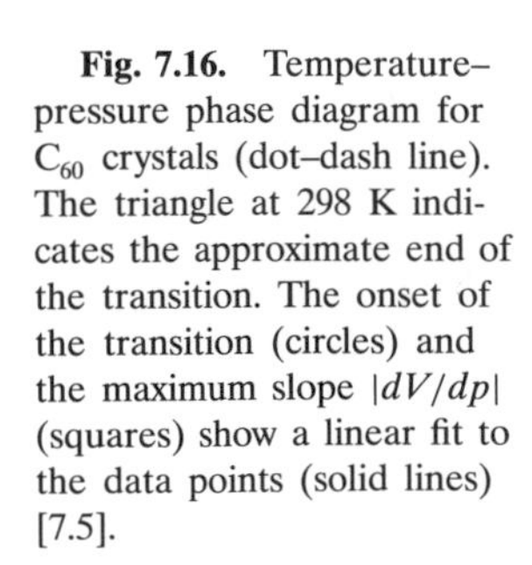
Fig. 7.16. Temperature–pressure phase diagram for C_{60} crystals (dot–dash line). The triangle at 298 K indicates the approximate end of the transition. The onset of the transition (circles) and the maximum slope $|dV/dp|$ (squares) show a linear fit to the data points (solid lines) [7.5].

of the bulk modulus β (see Fig. 7.17) was found in both the fcc and sc phases

$$\beta = \beta_0 + \beta' p \tag{7.11}$$

where the coefficients β_0 and β' were found to be temperature dependent in the sc phase (β_0 = 12.78 GPa at p = 0 and β' = 17.9 at 152 K), but less temperature dependent in the fcc phase (β_0 = 6.77 GPa at p = 0 and β' = 21.5 at 336 K). Thus upon transforming from the fcc to the sc phase, crystalline C_{60} becomes softer and the bulk modulus becomes less pressure dependent. This is in contrast to what is expected from the decrease in C_{60}–C_{60} distance which occurs during this orientational phase transition, and would suggest increased intermolecular interaction in the sc phase. The decrease in β_0 at the fcc–sc transition is explained by the alignment of the double bond of one C_{60} molecule near the center of the pentagon (or hexagon) of the neighboring C_{60} molecule (see Fig. 7.7), thereby minimizing p-orbital interactions. The smaller pressure dependence of β in the sc phase is attributed to the lower energy required to compress one orbital within a current ring in comparison to compressing orbitals on adjacent spheres [7.5]. As the pressure increases, the C_{60}–C_{60} distance decreases, until eventually the intramolecular and intermolecular C-C distances become comparable. When these distances become comparable, a phase transition occurs to another nonconducting transparent phase, which has not yet been

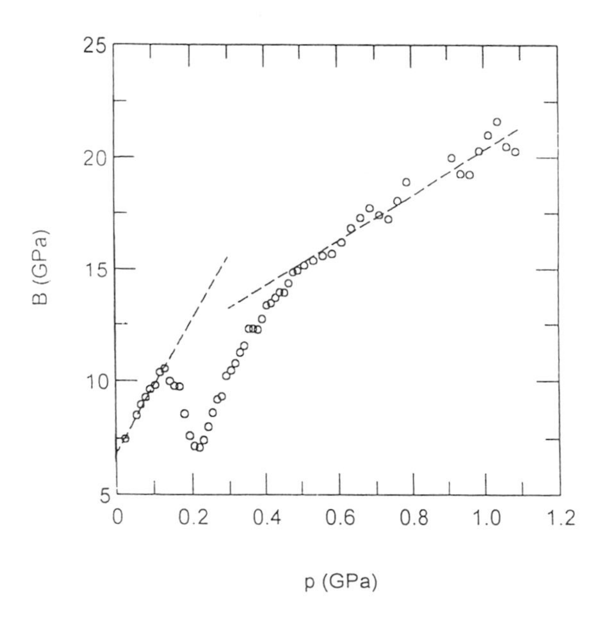

Fig. 7.17. Pressure dependence of the bulk modulus for C_{60} crystals, calculated between consecutive experimental points at 298 K. The fitted lines show the linear behavior in the fcc (low pressure) and sc (high pressure) crystalline phases, respectively [7.5].

clearly identified [7.124–127]. Placement of this phase on the phase diagram for carbon (Fig. 2.1) suggests that pressures of ~20 GPa may transform C_{60} into diamond [7.125].

However, if both the pressure and temperature are simultaneously increased, then two metastable forms of solid C_{60} can be stabilized by quenching from high temperature [7.128]. One phase, denoted by fcc(pC_{60}) where "p" refers to a pressure-induced phase, is prepared at 5 GPa using quasi-hydrostatic pressure from an anvil pressure apparatus and $300 \leq T \leq 400°C$ [7.128]. This phase has been described as a pressure-induced fcc structure with a lattice constant of 13.6 Å [about 4% smaller than that for the conventional fcc structure (see §7.1.1)]. A second pressure-induced rhombohedral $R\bar{3}m$ phase, denoted by rh(pC_{60}), is also stabilized at a pressure of 5 GPa but in a higher temperature range, $500 \leq T \leq 800°C$. This second structure can be indexed by a rhombohedral unit cell with $a_0 = 9.77$ Å and $\alpha = 56.3°$ or, equivalently, by a hexagonal unit cell with lattice parameters $a_0 = 9.22$ Å and $c_0 = 24.6$ Å and containing three C_{60} molecules at (0,0,0), $(\frac{1}{3}, \frac{2}{3}, \frac{1}{3})$, and $(\frac{2}{3}, \frac{1}{3}, \frac{2}{3})$. The $R\bar{3}m$ structure describes an fcc unit cell elongated along the $\langle 111 \rangle$ direction [7.128]. Values for the lattice constants for the various high-pressure phases are summarized in Table 7.13, in comparison with those for the fcc phase that is stable at ambient pressure.

For the two pressure-induced phases, the Bragg x-ray diffraction peaks are very broad, indicative of short-range crystalline order (~40 Å). The phase transition to fcc(pC_{60}) reduces the nearest-neighbor C_{60}–C_{60} distance from 10.02 Å to 9.62 Å, while in the rh(pC_{60}) phase, the C_{60}–C_{60} distance for the six nearest-neighbors normal to the rhombohedral axis is 9.22 Å, and along the rhombohedral axis the C_{60}–C_{60} distance is 9.76 Å. Table 7.13 also shows that the volume per molecule decreases from 711 Å^3 for the ambient fcc phase to 629 Å^3 for the fcc(pC_{60}) phase, and finally to 603 Å^3 for the rh(pC_{60}) phase [7.128].

Table 7.13

Lattice constants of various high-pressure phases of C_{60} in comparison to those for the ambient phase [7.128].

Phase	Cubic cell a_0 (Å)	Hexagonal cell a_0 (Å)	c_0 (Å)	Vol/C_{60} (Å^3)
Ambient fcc C_{60}	14.17	10.02	24.54	711
fcc(pC_{60})	13.6	9.62	23.6	629
rh(pC_{60})	—	9.22	24.6	603

Infrared and Raman spectroscopies show many more lines for the pressure-induced C_{60} phases than for conventional C_{60} [7.128], consistent with the loss of icosahedral I_h symmetry for the C_{60} molecule in both of these high-pressure phases. A structural model shows that the C_{60}–C_{60} distance in the pressure-induced phases becomes close enough to form a chemical bond for the fcc(pC_{60}) and rh(pC_{60}) phases, consistent with the reduced solubility of the pressure-induced phases in toluene and 1,2-dichlorobenzene, which are traditional solvents for conventional crystalline C_{60} (see §5.2.3) [7.128]. Infrared, Raman, and NMR spectra all give strong evidence for a drastic reduction in symmetry for the C_{60} molecules in the pressure-induced phases (see §11.9), consistent with the cross-linking of adjacent C_{60} molecules (see §7.5). The bonding between the C_{60} molecules in the fcc(pC_{60}) phase is increased relative to ambient fcc C_{60} and the increased intermolecular bonding is even greater for rh(pC_{60}). The NMR ^{13}C spectra for the pressure-induced C_{60} phases show several broad lines, indicative of the presence of several inequivalent carbon sites [7.128].

Regarding the stability of the fcc(pC_{60}) and rh(pC_{60}) phases, pressurizing them at 1000°C and 5 GPa results in the fracture of the C_{60} cage and the formation of amorphous carbon [7.128]. Furthermore, heating an rh(pC_{60}) sample above ~270°C shows an irreversible endothermic peak in a differential scanning calorimetry (DSC) trace with an enthalpy change of 23 J/g upon heating. Upon cooling fcc(pC_{60}) and rh(p C_{60}) samples from 300°C to room temperature, each reverts to the conventional fcc C_{60} structure, showing the metastability of these high pressure phases [7.128].

The pressure dependence of the crystalline structure of C_{70} has also been investigated [7.100] showing the fcc to be stable at high temperatures and pressures (see Fig. 7.18). At lower temperatures the dominant rhombohedral phase is often found with some admixture of fcc phase, especially near the phase boundary. The phase boundary between these phases is steep $(dT_{01}/dp) \sim 300 - 400$ K/GPa in comparison with that for C_{60} (104 K/GPa) [7.9], while the bulk modulus for C_{70} is found to be comparable to that for C_{60} (see Table 7.12). The C_{70} molar volume decreases by only ~3% at both the T_{01} and T_{02} phase transitions [7.98]. The crystalline diffraction pattern was found to disappear for $T > 1170$ K at 0.9 GPa [7.100].

Measurements of the Raman spectra of C_{70} at room temperature with pressures up to 17 GPa have shown pressure-induced shifts in frequency and increases in linewidth as well as anomalies at 2.0 ± 0.2 GPa and at $\sim 5.5 \pm 0.5$ GPa [7.129]. These effects have not yet been correlated with the structural phase diagram shown in Fig. 7.18.

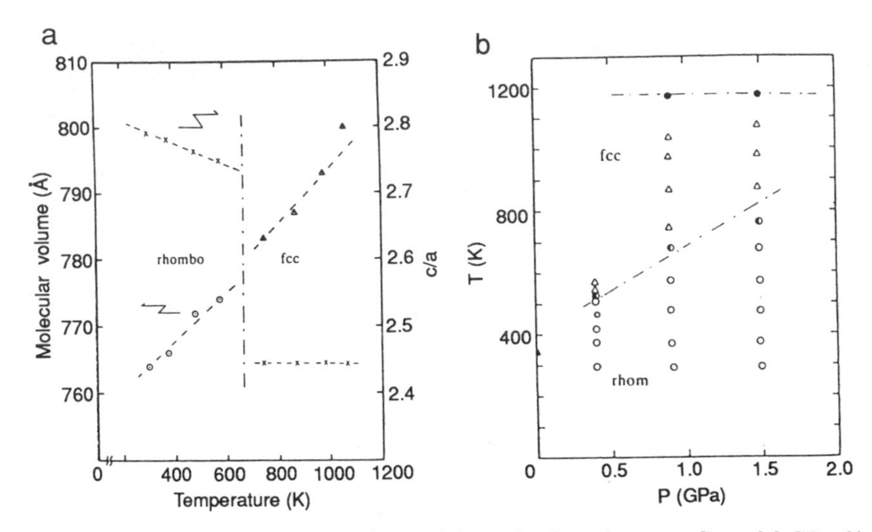

Fig. 7.18. (a) Temperature dependence of the molecular volume per C_{70} at 0.9 GPa. Circles and triangles denote the volumes (left scale) of the rhombohedral and fcc phases, respectively. The c/a ratios (right scale) are given by crosses [7.100]. (b) Pressure–temperature phase diagram of solid C_{70}. Circles and triangles denote the rhombohedral and fcc phases, respectively. Half-filled circles represent the coexistence of the rhombohedral and fcc phases. The dot–dashed curve gives the phase boundary between the rhombohedral and fcc phases, and closed circles indicate points where the diffraction peaks from solid C_{70} disappear [7.100].

7.4. Effect of Temperature on Crystal Structure

As a function of temperature, the most dramatic structural effects that are observed are the first-order phase transitions at T_{01} and the Schottky-type transition at lower temperatures discussed in §7.1.3. A discontinuous change of 0.044 Å occurs at temperature T_{01} in the lattice constant a_0 for C_{60} upon heating (see Fig. 14.21), with the larger lattice constant associated with the fcc phase [7.11] (see §14.10). A sharp change in the slope of the average isobaric volume thermal expansion coefficient α_v occurs at both T_{01} and ~90 K [7.11, 53, 54, 75].

In addition to the discontinuous changes in lattice constant associated with these phase transitions, continuous changes in lattice constant occur within the various phases. Below room temperature, the differences in the bond lengths $a_5 - a_6$ have been measured over the temperature range 4–295 K. Over this temperature range, the overall volume of the fullerene molecule remains almost constant, although $a_5 - a_6$ varies from 0.079 Å at 4 K to 0.062 at 295 K [7.41]. One suggested explanation for the large temperature dependence of the $a_5 - a_6$ bond length difference is that the electron distribution in a bond is somewhat dependent on interactions between neighboring carbon atoms on adjacent C_{60} molecules, and these interactions depend on the relative intermolecular orientations, which are temperature dependent [7.41]. This explanation is consistent with the finding that the a_5 value of a bond is larger if the pentagon on one molecule faces a hexagon, rather than any other configuration [7.130]. Nevertheless, no discontinuity in bond length difference $a_5 - a_6$ was noted as T crosses T_{01} [7.41].

At high temperature in the range 295–1180 K, $\alpha_v \cong 4.7 \times 10^{-5}$ /K [7.21], somewhat lower than the value near room temperature (6.1×10^{-5} /K) [7.98]. From these structural measurements on solid C_{60}, it is concluded that the melting point for solid C_{60} is above 1180 K. Estimates based on the Lindemann criterion [7.84] and molecular dynamics simulations [7.31] suggest a triple point temperature for C_{60} in the 1750 K range, which is to be compared to that of graphite, which is ~4200 K [7.131, 132]. These high-temperature experiments provide strong support for the stability of C_{60}. Further discussion on the temperature dependence of the C_{60} structure is given in §14.10.

7.5. Polymerized Fullerenes

Polymerization has been induced in C_{60} films through excitation by photons (§7.5.1) [7.12, 133, 134], electrons (§7.5.2) [7.135], in a plasma discharge (§7.5.4), and as a result of hydrostatic pressure (§7.5.3) [7.136]. Of the various processes, the phototransformation process has been studied most extensively (§7.5.1) and has also been studied in C_{70} (§7.5.5) [7.137]. The electron beam–induced polymerization (§7.5.2) was observed for both a 3 eV electron beam in an STM instrument and a 1500 eV external electron beam [7.135]. Pressure-induced polymerization of C_{60} (§7.5.3) has been reported using a diamond anvil cell [7.136]. In this section, we review the present status of polymerization studies on fullerenes. Not discussed in this section is the polymerization of C_{60} that has also been achieved by chemical means (see §10.10.2) such as in $(\text{-}C_{60}Pd\text{-})_n$ chains or by intercalation to yield $(\text{-}K_1C_{60}\text{-})_n$ "pearl necklaces" (see §8.5.2).

7.5.1. *Photopolymerization of C_{60}*

In the absence of diatomic oxygen, solid C_{60} has been found to undergo a photochemical transformation when exposed to visible or ultraviolet light with photon energies greater than the optical absorption edge (~1.7 eV) [7.12, 138]. Similar to observations for the pressure-induced phases discussed in §7.3, phototransformation renders the C_{60} films insoluble in toluene, a property which may be exploitable in photoresist applications. Because the optical penetration depth in pristine C_{60} in this region of the spectrum is on the order of ~0.1–1 μm, small particle-size powders or thin films can undergo a significant fractional volume transformation. The photochemical process is fairly efficient; a loss of 63% of the signature of monomeric units (i.e., van der Waals–bonded C_{60} molecules) is observed in ~20 min by Raman spectroscopy characterization using a modest light flux (20 mW/mm^2). Heating the transformed films to ~100°C has been reported to return the system to the pristine state [7.139].

On the basis of this and other experimental evidence (e.g., vibrational modes, luminescence, optical absorption), it has been proposed [7.12] that in the phototransformation process, covalent C–C bonds are formed between C_{60} molecules to produce a polymeric solid. Direct evidence for these covalent intermolecular bonds was obtained, for example, from laser desorption mass spectroscopy (LDMS) measurements [7.12, 140], as shown in Fig. 7.19 for a phototransformed C_{60} film. Using a pulsed nitrogen laser (337 nm wavelength, ~8 ns pulse, and 10 mJ/pulse energy) to desorb molecules from the film, mass peaks were observed at the C_{60} mass value [720 atomic mass units (amu)] and all integer multiples of this mass out to 20×720 amu, consistent with the desorption of a distribution of covalently linked clusters of C_{60} molecules from the film surface. The measurements show that the pulsed nitrogen laser itself produces some weak cross-linking of the fullerenes.

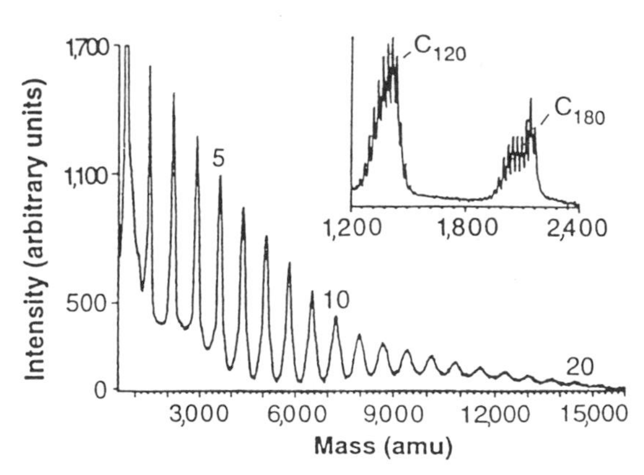

Fig. 7.19. Laser desorption mass spectrum (LDMS) of a C_{60} film (thickness ~2000 Å) was phototransformed by exposure to UV–visible radiation from a 300 W Hg arc lamp for 12 h. The material was desorbed from the film with a pulsed nitrogen laser (337 nm wavelength, ~8 ns pulse, and 10 mJ/pulse energy) under vacuum (~ 10^{-7} torr). A series of 20 peaks have been identified with N cross-linked fullerenes of mass ~ 720 N atomic mass units (amu), $N = 1, \ldots 20$. Inset: LDMS spectrum indicating C_2 loss or gain in the desorption process [7.12].

It has been shown [7.141] that the rate of transformation to the polymeric state at $T = 300$ K is proportional to the incident light intensity; i.e., the process involves a one-photon excitation of the molecule (see §13.3.4). As mentioned above, the presence of dioxygen in the film appears to prevent (or considerably retard) the photopolymerization process [7.12, 142]. This observation has been suggested as evidence that the lowest excited triplet state (see §12.5 and §13.1.1) is involved in the photochemical reaction, since O_2 is known to quench the population of the triplet state effectively. Furthermore, oxygen can be inadvertently introduced into the sample during the photopolymerization process itself, if the sample is illuminated in air, or in a significant background of O_2. For example, O_2 at room temperature diffuses rapidly into C_{60} films under the application of visible or ultraviolet light [7.143, 144], resulting in a saturation stoichiometry of approximately $C_{60}O_2$ in 2000 Å C_{60} films on a time scale of $\sim$ 30 min. Therefore, photopolymerization of C_{60} must be carried out in an inert gas or in vacuum.

The photopolymerization is most easily detected by Raman scattering using 488 or 514 nm excitation, as evidenced by a decrease in the intensity of the polarized $A_g(2)$ Raman line at 1469 cm^{-1} (intrinsic pentagonal pinch mode, see §11.3), and the concomitant emergence of a broader, unpolarized peak (with shoulders) at 1458 cm^{-1} identified with photochemically produced dimers and higher oligomers. The 1458 cm^{-1} line continues to grow at the expense of the 1469 cm^{-1} line as the phototransformation proceeds to completion (see §11.8).

The photopolymerization process has been found to occur only at temperatures $T > T_{01}$, where $T_{01} \simeq 260$ K is the orientational ordering temperature [7.142] discussed in §7.1.3. It has been proposed that molecular rotations occurring above this transition temperature allow adjacent C_{60} molecules to meet the general topochemical requirements for the formation of a four-membered ring linking adjacent molecules. This topochemistry requires that parallel double bonds (C=C) on adjacent molecules be separated by less than 4.2 Å [7.142, 145]. When these topochemical conditions are met, the photochemical formation of a four-membered ring between molecules occurs readily (see §10.7), and this ring formation is referred to as "2+2 cycloaddition" (see Fig. 7.20). Finally, the photopolymerization of C_{60} seems to be most effectively carried out near room temperature. At elevated temperatures ($\sim$100°C), the thermal scission rate of the covalent bonds between the molecules is high enough so that the film reverts to the pristine monomer form, as observed by Raman spectroscopy [7.139] (see §11.8). Thus the stability range of the phototransformed C_{60} is between T_{01} ($\sim$260 K) and T_p ($\sim$330 K).

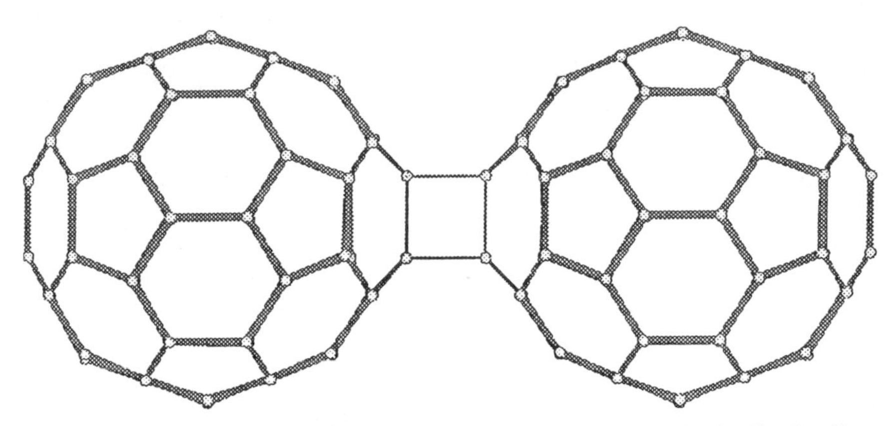

Fig. 7.20. Structure of a photodimerized C_{60} molecule (forming a dumbbell). The dimer structure shows a four-membered ring bridging adjacent fullerenes [7.12].

Phototransformation effects have also been observed in Brillouin scattering studies carried out on a single crystal of C_{60} in the presence of oxygen. The use of sufficient intensity of 541.5 nm laser excitation from an argon ion laser to observe Brillouin scattering at room temperature produced an irreversible phototransformation effect which was identified with fullerene cage opening and the formation of a spongy material, which from the Brillouin spectra was identified with an amorphous carbon or a carbon aerogel-like material (see §2.12), [7.14].

7.5.2. Electron Beam–Induced Polymerization of C_{60}

Electron beam–stimulated polymerization of C_{60} has been reported both for ~3 eV electrons from a scanning tunneling microscope (STM) tip at nanometer dimensions and for 1500 eV electrons from an electron gun on a larger spatial scale [7.135]. The STM-induced modifications to solid C_{60} were carried out both for monolayer coverage of C_{60} on a GaAs(110) substrate and for a 10-monolayer C_{60} film for 3 eV electron irradiation. Typical conditions used to induce polymerization in the C_{60} film were a flux of $\sim 1 \times 10^8$ electrons per second per C_{60} molecule [7.135]. The STM technique, operating at lower voltages (< 2 eV), was used to probe the effect of the electron irradiation, which was carried out by repeated scanning (for approximately 30 min) of a small area of the image (e.g., 80×80 Å^2) at a higher voltage (~3 eV). The electron irradiation-induced modification to the solid C_{60} was identified as the appearance of a modified "speckled" region in the STM scan. Annealing to 470 K restored the original ordered surface, as it appeared prior to electron irradiation [7.135]. These results also imply operating ranges where STM techniques can be used for the investigation of C_{60} surfaces, without introducing surface damage due to electron irradiation.

Electron irradiation–induced polymerization on a larger spatial scale was demonstrated using 1500 eV electrons from a 100 μA beam incident on a 2×3 mm^2 C_{60} surface, corresponding to ~100 primary electrons/s per C_{60} molecule, but also including many secondary electrons associated with electron scattering processes. Electron irradiation effects were observed as "speckled" patches in STM patterns [7.135], indicating that the nucleation of a polymerized event enhanced the probability for further polymerization or growth of the polymerized region. Such correlated behavior may be due to bond length changes and their associated strain effects. Beam-induced heating was found to provide a competing mechanism for breaking the cross-linking bond associated with the polymerization [7.135]. The observation that annealing at 470 K restores a polymerized film to its original monomer form supports a model for the C_{60} polymerization and is in qualitative agreement with the phototransformation studies discussed in §7.5.1, except for the value of the transition temperature T_p.

Electron beams can be used as probes without significant modification of the C_{60} surfaces under investigation. Low energy electron diffraction (LEED) experiments carried out on C_{60} films with electrons in the ~9–20 eV range with electron currents of ~1 mA and a beam size of ~0.1 mm^2 showed no polymerization effects, nor did inverse photoemission spectroscopy studies carried out on similar surfaces exposed to a 20 eV electron beam for 5 h with an electron current of ~2 μA and a beam size of ~5 mm^2, or photoemission studies on a 1 mm^2 area and a flux of ~ 10^9/s [7.135]. Electron beams can, however, be used to write patterns on C_{60} surfaces, either very fine lines on a nanometer scale using an STM tip operating at ~3 eV or thicker lines using an electron gun operating at ~1500 eV [7.135].

7.5.3. *Pressure-Induced Polymerization of C_{60}*

Using infrared (IR) spectroscopy studies, it has been shown that hydrostatic pressure can induce polymerization of solid C_{60} at room temperature following the procedure described below [7.136]. This result is consistent with the pressure-dependent studies discussed in §7.3 [7.128]. In these experiments a flake of solid C_{60} (~20 μm thick) was loaded into a diamond anvil cell for infrared transmission studies. Xe (or Kr) was used as the pressure-transmitting medium, generating a quasi-hydrostatic environment. In experiments up to 10 GPa, a continuous decrease in the intensity of the four F_{1u} IR-active lines was observed. After compression to 6 GPa, an ambient pressure spectrum revealed additional IR lines, many of which are in good agreement with frequencies reported for photopolymerized C_{60} [7.12]. Thus, these IR results were interpreted [7.136] as evidence for a pressure-induced polymerization of C_{60}. Although not explicitly discussed in their work, pressure-induced polymerization may have occurred in some of the pressure-dependent structural studies that have been carried out on C_{60} and other fullerenes (see §7.3).

7.5.4. *Plasma-Induced Polymerization of C_{60}*

The polymerization of mixed C_{60}/C_{70} films induced by the interaction of the film with a radio frequency (rf) Ar ion plasma has been reported [7.146]. The films were deposited on glass substrates by vacuum sublimation of microcrystalline C_{60} + 10% C_{70} powder in 10^{-2} torr Ar. An rf-driven plasma (13.6 MHz) was present during film deposition.

The primary evidence for polymerization of such films [7.146] is from laser desorption mass spectroscopy (see §6.2). A pulsed nitrogen laser is used to desorb material from the film and a series of six close peaks in the mass spectrum is observed at masses $m = Nm_0$ where $N = 1, 2, \ldots 6$, is the number of monomers in the oligomer and $m_0 = 720$ amu is the mass of a single C_{60} molecule. However, this conclusion neglects the possibility that the polymerization might have been caused by the nitrogen desorption laser itself. In fact, this effect has been shown to occur [7.12, 140], and six peaks in an LDMS spectrum of C_{60} were reported using a pulsed N_2 laser for C_{60} desorption. Under the same LDMS experimental conditions a series of 20 peaks ($N = 1, 2, \ldots 20$) were observed for a C_{60} film which had first been photopolymerized [7.12, 140].

We rule out the possibility that the N_2 laser was responsible for the polymerization of the film prepared in the presence of the rf plasma on the basis of the optical absorption spectrum in the UV–visible region, where it is found that the spectrum for the plasma-transformed C_{60} film is markedly different from that of a pristine C_{60} film. Whereas four prominent absorption bands are observed for pristine C_{60} in the UV–visible region (see §13.3.2), these absorption bands have broadened into one broad band in the case of the plasma-transformed film [7.146]. This observation is in qualitative agreement with the corresponding results on phototransformed films, where increasing the exposure of a C_{60} film to 488 nm radiation was also shown to broaden absorption bands in the UV–visible region [7.147].

The properties of the plasma-induced polymerized films were characterized by various physical property measurements including temperature-dependent conductivity (10^{-7} S/cm under ambient conditions), molecular weight distribution using LDMS, surface morphology using scanning electron microscopy (SEM), surface wetting properties using contact angle measurements, electron paramagnetic resonance (EPR) measurements to determine the unpaired spin concentrations ($10^{17}/cm^3$), Fourier transform infrared (FTIR), and UV–visible spectroscopy measurements to document polymerization-induced spectral changes, and x-ray scattering to detect structural order but found none [7.146].

To interpret plasma-induced polymerization data for C_{60}, the heats of formation were calculated for 1–2 cycloaddition of two C_{60} molecules to form a dimer as well as two isomeric versions of a 1–4 cycloaddition (one with C_{2v} symmetry, the other with C_{2h} symmetry) [7.146]. These results showed that the 1–2 cycloaddition was the most stable, and the authors confirmed the implications of this finding by calculating the expected infrared spectra for a C_{60} dimer joined by 1–2 cycloaddition and obtained agreement with the experimentally observed spectra [7.146].

7.5.5. *Photopolymerization of C_{70} Films*

Experimental evidence for the UV–visible photopolymerization of C_{70} films in the absence of dioxygen has also been reported [7.137, 148]. Results from laser desorption mass spectroscopy, vibrational spectroscopy, and the insolubility of thin irradiated films in toluene all have been interpreted [7.137] as evidence for the phototransformation of C_{70} to a polymeric solid. However, the cross section for the phototransformation in C_{70} is found to be considerably smaller than that observed for C_{60} [7.12], and this reduced cross section was attributed to a reduction in the number of reactive double bonds from 30 in C_{60} to only 5 in each polar cap of a C_{70} molecule.

The tendency of C_{70} to dimerize via the formation of a four-membered ring between C_{70} shells was also shown theoretically [7.137], using a generalized, tight-binding molecular dynamics (GTBMD) simulation [7.148–150]. These molecular dynamics simulations find that the lowest-energy dimer configuration contains a four-membered ring between monomers, the same dimeric bonding unit as is found theoretically for the C_{60} dimer [7.151] or for the "bucky dumbbell" using the GTBMD technique.

Laser desorption mass spectroscopy results for C_{70} films ($\sim$2000 Å thick) in vacuum are shown in Fig. 7.21(a) for a phototransformed film using a 300 W Hg arc lamp and in Fig. 7.21(b) for a pristine film not previously irradiated [7.137]. Desorption of material from the film surface (and deeper into the bulk) was accomplished using a pulsed N_2 laser (337 nm, 8 ns, 10 mJ per pulse), with parameter values similar to those used for LDMS studies of photopolymerized C_{60} [7.12, 140], as described in §7.5.1. As can be seen in Fig. 7.21, the LDMS data exhibit a strong mass peak

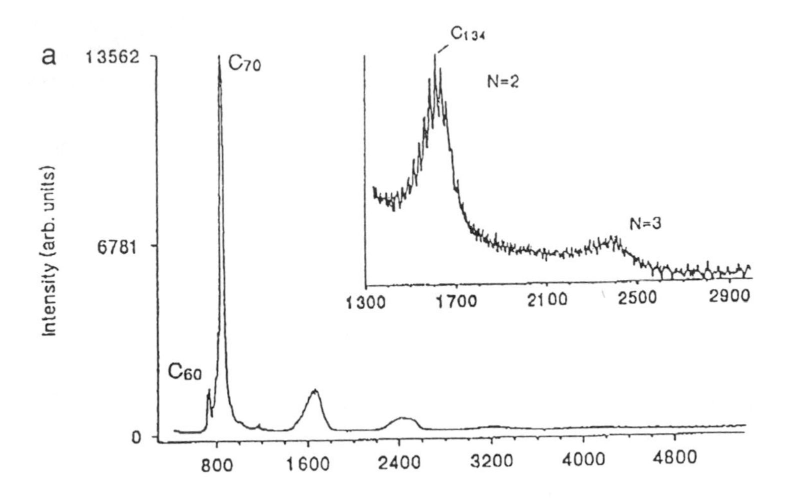

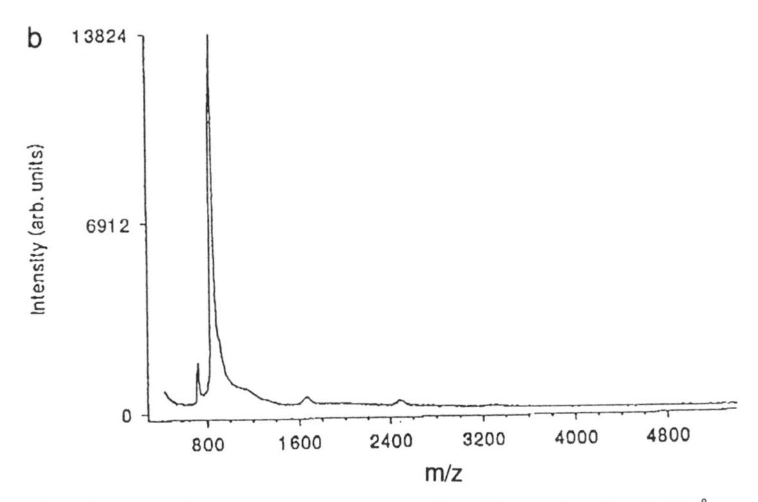

Fig. 7.21. (a) Laser desorption mass spectrum (LDMS) of a C_{70} film (2000 Å thickness) phototransformed with UV–visible radiation from a 300 W Hg arc lamp for 10 days. The mass peaks following the strong C_{70} peak at $m/z = 840$ amu are identified with 2, 3, and 4 cross-linked C_{70} molecules. Inset: LDMS spectrum for the same C_{70} film shown on an expanded scale in the region of the dimer (C_{140}) and trimer (C_{210}) mass values showing C_2 loss or gain in the desorption process. (b) LDMS spectrum of a pristine C_{70} film not exposed to light from a 300 W Hg arc lamp. The data were taken under conditions similar to those in panel (a). The weak peaks in the figure at $m/z > 840$ amu are identified with cross-linked C_{70} produced by the desorption laser itself. In both panels, the presence of a mass peak at $m/z = 720$ amu is associated with the presence of C_{60} in the sample [7.137].

at $m/z = (70 \times 12) = 840$ amu, along with a series of weaker peaks at $m/z = N \times 840$ amu, where $N = 2, 3$, and 4. The $N = 2, 3, 4$ peaks are identified with C_{70} dimers, trimers, etc. The presence of the weak peak at $m/z = 720$ amu is, of course, associated with the presence of C_{60} in the sample. A higher-resolution mass spectrum [see inset to Fig. 7.21(a)] for the C_{70} dimer and trimer peaks reveals fine structure at C_2 intervals. For comparison (see Fig. 7.19), similar studies on photopolymerized solid C_{60} [7.12, 140], show a series of mass peaks $N \times 720$ amu, for N as high as 20. Figure 7.21(b) shows the mass spectrum for a pristine (nonirradiated) C_{70} film deposited from the same batch as the sample featured in Fig. 7.21(a), and taken under identical conditions. Similar to Fig. 7.21(a), the spectrum in Fig. 7.21(b) also shows a series of peaks spaced by 840 amu. However, the intensity of the $N > 1$ peaks is substantially reduced relative to those seen in Fig. 7.21(a). A comparison of Figs. 7.21(a) and (b) indicates: (1) irradiation of C_{70} films with the 300 W Hg arc lamp increases the number of C_{70} dimers and trimers observed in the LDMS spectrum, and (2) the N_2 laser used in the LDMS measurements can also produce photopolymerization in the mass spectrometer, similar to observations in solid C_{60} films [7.12, 140]. However, comparison of Figs. 7.19 and 7.21 clearly show that the phototransformation of C_{70} has a much lower cross section than that observed for solid C_{60}. In contrast to photopolymerized C_{60} [7.12, 140], no clear signature for the formation of intermolecular bonds between C_{70} shells was observed in either the Raman spectrum or the UV–visible electronic absorption spectrum [7.137]. However, some evidence for photoinduced changes in the vibrational mode spectra of the C_{70} films was detected using FTIR transmission. A new, broad band (possibly a vibrational mode continuum) in the mid-IR region at $\sim$1100 cm^{-1} was observed in the phototransformed C_{70} [7.137]. This broad band is not due to oxygen contamination in the sample, as confirmed by FTIR studies of C_{70} films, intentionally exposed to oxygen under illumination from an Hg arc.

References

[7.1] P. W. Stephens, L. Mihaly, P. L. Lee, R. L. Whetten, S. M. Huang, R. Kaner, F. Diederich, and K. Holczer. *Nature (London)*, **351**, 632 (1991).

[7.2] A. R. Kortan, N. Kopylov, S. H. Glarum, E. M. Gyorgy, A. P. Ramirez, R. M. Fleming, F. A. Thiel, and R. C. Haddon. *Nature (London)*, **355**, 529 (1992).

[7.3] M. B. Jost, N. Troullier, D. M. Poirier, J. L. Martins, J. H. Weaver, L. P. F. Chibante, and R. E. Smalley. *Phys. Rev. B*, **44**, 1966 (1991).

[7.4] J. E. Fischer, P. A. Heiney, A. R. McGhie, W. J. Romanow, A. M. Denenstein, J. P. McCauley, Jr., and A. B. Smith III. *Science*, **252**, 1288 (1991).

[7.5] A. Lundin and B. Sundqvist. *Europhys. Lett.*, **27**, 463 (1994).

[7.6] A. Lundin, B. Sundqvist, P. Skoglund, A. Fransson, and S. Pettersson. *Solid State Commun.*, **84**, 879 (1992).
[7.7] X. D. Shi, A. R. Kortan, J. M. Williams, A. M. Kini, B. M. Saval, and P. M. Chaikin. *Phys. Rev. Lett.*, **68**, 827 (1992).
[7.8] J. E. Fischer. *Materials Science and Engineering*, **B19**, 90 (1993).
[7.9] G. A. Samara, J. E. Schirber, B. Morosin, L. V. Hansen, D. Loy, and A. P. Sylwester. *Phys. Rev. Lett.*, **67**, 3136 (1991).
[7.10] G. Kriza, J. C. Ameline, J. Jerome, A. Dworkin, H. Szwarc, C. Fabre, D. Schutz, A. Rassat, and P. Bernier. *J. Phys. I, France*, **1**, 1361 (1991).
[7.11] P. A. Heiney, G. B. M. Vaughan, J. E. Fischer, N. Coustel, D. E. Cox, J. R. D. Copley, D. A. Neumann, W. A. Kamitakahara, K. M. Creegan, D. M. Cox, J. P. McCauley, Jr., and A. B. Smith III. *Phys. Rev. B*, **45**, 4544 (1992).
[7.12] A. M. Rao, P. Zhou, K.-A. Wang, G. T. Hager, J. M. Holden, Y. Wang, W. T. Lee, X.-X. Bi, P. C. Eklund, D. S. Cornett, M. A. Duncan, and I. J. Amster. *Science*, **259**, 955 (1993).
[7.13] G. Gensterblum, K. Hevesi, B. Y. Han, L. M. Yu, J. J. Pireaux, P. A. Thiry, D. Bernaerts, S. Amelinckx, G. V. Tendeloo, G. Bendele, T. Buslaps, R. L. Johnson, M. Foss, R. Feidenhans'l, and G. LeLay. *Phys. Rev. B*, **50**, 11981 (1994).
[7.14] M. Manfredini, C. E. Bottani, and P. Milani. *Chem. Phys. Lett.*, **226**, 600 (1994).
[7.15] E. Grivei, B. Nysten, M. Cassart, A. Demain, and J. P. Issi. *Solid State Commun.*, **85**, 73 (1993).
[7.16] E. Grivei, M. Cassart, J. P. Issi, L. Langer, B. Nysten, J. P. Michenaud, C. Fabre, and A. Rassat. *Phys. Rev. B*, **48**, 8514 (1993).
[7.17] R. C. Yu, N. Tea, M. B. Salamon, D. Lorens, and R. Malhotra. *Phys. Rev. Lett.*, **68**, 2050 (1992).
[7.18] C. Wen, J. Li, K. Kitazawa, T. Aida, I. Honma, H. Komiyama, and K. Yamada. *Appl. Phys. Lett.*, **61**, 2162 (1992).
[7.19] S. L. Ren, Y. Wang, A. M. Rao, E. McRae, G. T. Hager, K. A. Wang, W. T. Lee, H. F. Ni, J. Selegue, and P. C. Eklund. *Appl. Phys. Lett.*, **59**, 2678 (1991).
[7.20] Y. Wang, J. M. Holden, A. M. Rao, W.-T. Lee, G. T. Hager, X. X. Bi, S. L. Ren, G. W. Lehman, G. T. Hager, and P. C. Eklund. *Phys. Rev. B*, **45**, 14396 (1992).
[7.21] J. E. Fischer and P. A. Heiney. *J. Phys. Chem. Solids*, **54**, 1725 (1993).
[7.22] C. Pan, M. P. Sampson, Y. Chai, R. H. Hauge, and J. L. Margrave. *J. Phys. Chem.*, **95**, 2944 (1991).
[7.23] E. Kim, Y. H. Lee, and J. Y. Lee. *Phys. Rev. B*, **48**, 18230 (1993).
[7.24] S. J. Woo, S. H. Lee, E. Kim, K. H. Lee, Y. H. Lee, S. Y. Hwang, and I. C. Jeon. *Phys. Lett. A*, **162**, 501 (1992).
[7.25] S. J. Duclos, K. Brister, R. C. Haddon, A. R. Kortan, and F. A. Thiel. *Nature (London)*, **351**, 380 (1991).
[7.26] Y. Wang, D. Tománek, and G. F. Bertsch. *Phys. Rev. B*, **44**, 6562 (1992).
[7.27] C. B. Eom, A. F. Hebard, L. E. Trimble, G. K. Celler, and R. C. Haddon. *Science*, **259**, 1887 (1993).
[7.28] D. M. Poirier, D. W. Owens, and J. H. Weaver. *Phys. Rev. B*, **51**, 1830 (1995).
[7.29] J. Abrefah, D. R. Olander, M. Balooch, and W. J. Siekhaus. *Appl. Phys. Lett.*, **60**, 1313 (1992).
[7.30] A. F. Hebard, R. C. Haddon, R. M. Fleming, and A. R. Kortan. *Appl. Phys. Lett*, **59**, 2109 (1991).
[7.31] A. Cheng, M. L. Klein, and C. Caccamo. *Phys. Rev. Lett.*, **71**, 1200 (1993).
[7.32] C. S. Yannoni, R. D. Johnson, G. Meijer, D. S. Bethune, and J. R. Salem. *J. Phys. Chem.*, **95**, 9 (1991).

[7.33] R. Tycko, R. C. Haddon, G. Dabbagh, S. H. Glarum, D. C. Douglass, and A. M. Mujsce. *J. Phys. Chem.*, **95**, 518 (1991).

[7.34] Q. M. Zhang, J. Y. Yi, and J. Bernholc. *Phys. Rev. Lett.*, **66**, 2633 (1991).

[7.35] R. Tycko, G. Dabbagh, R. M. Fleming, R. C. Haddon, A. V. Makhija, and S. M. Zahurak. *Phys. Rev. Lett.*, **67**, 1886 (1991).

[7.36] R. D. Johnson, C. S. Yannoni, H. C. Dorn, J. R. Salem, and D. S. Bethune. *Science*, **255**, 1235 (1992).

[7.37] R. M. Fleming, T. Siegrist, P. M. Marsh, B. Hessen, A. R. Kortan, D. W. Murphy, R. C. Haddon, R. Tycko, G. Dabbagh, A. M. Mujsce, M. L. Kaplan, and S. M. Zahurak. In R. S. Averback, J. Bernholc, and D. L. Nelson (eds.), *Clusters and Cluster-Assembled Materials*, MRS Symposia Proceedings, Boston, vol. 206, p. 691, Materials Research Society Press, Pittsburgh, PA (1991).

[7.38] K. Prassides, H. W. Kroto, R. Taylor, D. R. M. Walton, W. I. F. David, J. Tomkinson, R. C. Haddon, M. J. Rosseinsky, and D. W. Murphy. *Carbon*, **30**, 1277 (1992).

[7.39] R. M. Fleming, A. P. Ramirez, M. J. Rosseinsky, D. W. Murphy, R. C. Haddon, S. M. Zahurak, and A. V. Makhija. *Nature (London)*, **352**, 787 (1991).

[7.40] P. A. Heiney, J. E. Fischer, A. R. McGhie, W. J. Romanow, A. M. Denenstein, J. P. McCauley, Jr., A. B. Smith III, and D. E. Cox. *Phys. Rev. Lett.*, **67**, 1468 (1991).

[7.41] P. Damay and F. Leclercq. *Phys. Rev. B*, **49**, 7790 (1994).

[7.42] R. Moret, S. Ravey, and J. M. Godard. *J. Phys. I (France)*, **2**, 1699 (1992).

[7.43] M. Sprik, A. Cheng, and M. L. Klein. *J. Phys. Chem.*, **96**, 2027 (1992).

[7.44] C. S. Yannoni, P. P. Bernier, D. S. Bethune, G. Meijer, and J. R. Salem. *J. Am. Chem. Soc.*, **113**, 3190 (1991).

[7.45] R. F. Kicfl, J. W. Schncidcr, A. MacFarlanc, K. Chow, T. L. Duty, S. R. Kreitzman, T. L. Estle, B. Hitti, R. L. Lichti, E. J. Ansaldo, C. Schwab, P. W. Percival, G. Wei, S. Wlodek, K. Kojima, W. J. Romanow, J. P. McCauley, Jr., N. Coustel, J. E. Fischer, and A. B. Smith, III. *Phys. Rev. Lett.*, **68**, 1347 (1992). ibid. **69**, 2708 (1992).

[7.46] W. Schranz, A. Fuith, P. Dolinar, H. Warhanek, M. Haluska, and H. Kuzmany. *Phys. Rev. Lett.*, **71**, 1561 (1993).

[7.47] K. Kikuchi, S. Suzuki, K. Saito, H. Shiramaru, I. Ikemoto, Y. Achiba, A. A. Zakhidov, A. Ugawa, K. Imaeda, H. Inokuchi, and K. Yakushi. *Physica C*, **185–189**, 415 (1991).

[7.48] A. B. Harris and R. Sachidanandam. *Phys. Rev. B*, **46**, 4944 (1992).

[7.49] *The International Tables for X-ray Crystallography*, vol. 1. The Kynoch Press, Birmingham, England (1952).

[7.50] M. Tinkham. *Group Theory and Quantum Mechanics*. McGraw-Hill, New York (1964).

[7.51] G. Dresselhaus, M. S. Dresselhaus, and P. C. Eklund. *Phys. Rev. B*, **45**, 6923 (1992).

[7.52] R. Saito. Private communication.

[7.53] W. I. F. David, R. M. Ibberson, T. J. S. Dennis, J. P. Hare, and K. Prassides. *Europhys. Lett.*, **18**, 219 (1992).

[7.54] W. I. F. David, R. M. Ibberson, T. J. S. Dennis, J. P. Hare, and K. Prassides. *Europhys. Lett.*, **18**, 735 (1992).

[7.55] J. R. D. Copley, W. I. F. David, and D. A. Neumann. *Neutron News*, **4**, 20 (1993).

[7.56] P. A. Heiney. *J. Phys. Chem. Solids*, **53**, 1333 (1992).

[7.57] A. Dworkin, H. Szwarc, S. Leach, J. P. Hare, T. J. Dennis, H. W. Kroto, R. Taylor, and D. R. M. Walton. *C.R. Acad. Sci., Paris Serie II*, **312**, 979 (1991).

[7.58] T. Atake, T. Tanaka, R. Kawaji, K. Kikuchi, K. Saito, S. Suzuki, I. Ikemoto, and Y. Achiba. *Physica C*, **185–189**, 427 (1991).

[7.59] T. Matsuo, H. Suga, W. I. F. David, R. M. Ibberson, P. Bernier, A. Zahab, C. Fabre, A. Rassat, and A. Dworkin. *Solid State Commun.*, **83**, 711 (1992).

[7.60] Y. Maruyama, T. Inabe, H. Ogata, Y. Achiba, S. Suzuki, K. Kikuchi, and I. Ikemoto. *Chem. Lett.*, **10**, 1849 (1991). The Chemical Society of Japan.

[7.61] R. Taylor, J. P. Hare, A. K. Abdul-Sada, and H. W. Kroto. *J. Chem. Soc. Chem. Commun.*, **20**, 1423 (1990).

[7.62] R. D. Johnson, G. Meijer, and D. S. Bethune. *J. Am. Chem. Soc.*, **112**, 8983 (1990).

[7.63] R. D. Johnson, D. S. Bethune, and C. S. Yannoni. *Accounts of Chem. Res.*, **25**, 169 (1992).

[7.64] A. Dworkin, C. Fabre, D. Schutz, G. Kriza, R. Ceolin, H. Szwarc, P. Bernier, D. Jerome, S. Leach, A. Rassat, J. P. Hare, T. J. Dennis, H. W. Kroto, R. Taylor, and D. R. M. Walton. *C.R. Acad. Sci., Paris Serie II*, **313**, 1017 (1991).

[7.65] W. I. F. David, R. M. Ibberson, J. C. Matthewman, K. Prassides, T. J. S. Dennis, J. P. Hare, H. W. Kroto, R. Taylor, and D. R. M. Walton. *Nature (London)*, **353**, 147 (1991).

[7.66] D. A. Neumann, J. R. D. Copley, R. L. Cappelletti, W. A. Kamitakahara, R. M. Lindstrom, K. M. Creegan, D. M. Cox, W. J. Romanow, N. Coustel, J. P. McCauley, Jr., N. C. Maloszewskyi, J. E. Fischer, and A. B. S. III. *Phys. Rev. Lett.*, **67**, 3808 (1991).

[7.67] L. Pintschovius, B. Renker, F. Gompf, R. Heid, S. L. Chaplot, M. Haluška, and H. Kuzmany. *Phys. Rev. Lett.*, **69**, 2662 (1992).

[7.68] D. A. Neumann, J. R. D. Copley, W. A. Kamitakahara, J. J. Rush, R. L. Cappelletti, N. Coustel, J. E. Fischer, J. P. McCauley, Jr., A. B. Smith III, K. M. Creegan, and D. M. Cox. *J. Chem. Phys.*, **96**, 8631 (1992).

[7.69] G. Van Tendeloo, S. Amelinckx, S. Muto, M. A. Verheijen, P. H. M. van Loosdrecht, and G. Meijer. *Ultramicroscopy*, **51**, 168 (1993).

[7.70] S. Hoen, N. G. Chopra, X. D. Xiang, R. Mostovoy, J. Hou, W. A. Vareka, and A. Zettl. *Phys. Rev. B*, **46**, 12737 (1992).

[7.71] P. H. M. van Loosdrecht, P. J. M. van Bentum, and G. Meijer. *Phys. Rev. Lett.*, **68**, 1176 (1992).

[7.72] P. H. M. van Loosdrecht, P. J. M. van Bentum, M. A. Verheijen, and G. Meijer. *Chem. Phys. Lett.*, **198**, 587 (1992).

[7.73] M. Matus and H. Kuzmany. *Appl. Phys. A*, **56**, 241 (1993).

[7.74] S. P. Love, D. M. aand M. I. Salkola, N. V. Coppa, J. M. Robinson, B. I. Swanson, and A. R. Bishop. *Chem. Phys. Lett.*, **225**, 170 (1994).

[7.75] F. Gugenberger, R. Heid, C. Meingast, P. Adelmann, M. Braun, H. Wühl, M. Haluska, and H. Kuzmany. *Phys. Rev. Lett.*, **60**, 3774 (1992).

[7.76] J. D. Axe, S. C. Moss, and D. A. Neumann. In H. Ehrenreich and F. Spaepen (eds.), *Solid State Physics: Advances in Research and Applications*, vol. 48, pp. 149–224, Academic Press, New York (1994).

[7.77] J. R. D. Copley, D. A. Neumann, R. L. Cappelletti, and W. A. Kamitakahara. *J. Phys. Chem. Solids*, **53**, 1353 (1992).

[7.78] G. B. Alers, B. Golding, A. R. Kortan, R. C. Haddon, and F. A. Thiel. *Science*, **257**, 511 (1992).

[7.79] D. A. Neumann, J. R. D. Copley, W. A. Kamitakahara, J. J. Rush, R. L. Paul, and R. M. Lindstrom. *J. Phys. Chem. Solids*, **54**, 1699 (1993).

[7.80] W. P. Beyermann, M. F. Hundley, J. D. Thompson, F. N. Diederich, and G. Grüner. *Phys. Rev. Lett.*, **68**, 2046 (1992).

[7.81] N. H. Tea, R. C. Yu, M. B. Salamon, D. C. Lorents, R. Malhotra, and R. S. Ruoff. *Appl. Phys.*, **A56**, 219 (1993).

[7.82] P. C. Chow, X. Jiang, G. Reiter, P. Wochner, S. C. Moss, J. D. Axe, J. C. Hanson, R. K. McMullan, R. L. Meng, and C. W. Chu. *Phys. Rev. Lett.*, **69**, 2943 (1992); ibid.; 3591(E).

[7.83] T. K. Cheng, H. J. Zeiger, J. Vidal, G. Dresselhaus, M. S. Dresselhaus, and E. P. Ippen. *QELS Digest*, **11**, 60 (1991). Technical Digest of Conference on Quantum Electronics Laser Science, May 1991.
[7.84] T. Yildirim and A. B. Harris. *Phys. Rev. B*, **46**, 7878 (1992).
[7.85] J. P. Lu, X. P. Li, and R. M. Martin. *Phys. Rev. Lett.*, **68**, 1551 (1992).
[7.86] G. Van Tendeloo, S. Amelinckx, M. A. Verheijen, P. H. M. van Loosdrecht, and G. Meijer. *Phys. Rev. Lett.*, **69**, 1065 (1992).
[7.87] B. E. White, J. E. Freund, K. A. Topp, and R. O. Pohl. In P. Bernier, T. W. Ebbeson, D. S. Bethune, R. M. Metzger, L. Y. Chiang, and J. W. Mintmire (eds.), *Science and Technology of Fullerene Materials*, vol. 359, 411, MRS Symposium Proceedings (1995).
[7.88] P. A. Heiney, J. E. Fischer, A. R. McGhie, W. J. Romanow, A. M. Denenstein, J. P. McCauley, Jr., A. B. Smith III, and D. E. Cox. *Phys. Rev. Lett.*, **66**, 2911 (1991). See also Comment by R. Sachidanandam and A. B. Harris, Phys. Rev. Lett. **67**, 1467 (1991).
[7.89] A. Cheng and M. L. Klein. *Phys. Rev. B*, **45**, 1889 (1992).
[7.90] X. P. Li, J. P. Lu, and R. M. Martin. *Phys. Rev. B*, **46**, 4301 (1992).
[7.91] W. Que and M. B. Walker. *Phys. Rev. B*, **48**, 13104 (1993).
[7.92] R. Saito, G. Dresselhaus, and M. S. Dresselhaus. *Phys. Rev. B*, **49**, 2143 (1994).
[7.93] R. Kubo, H. Ichimura, T. Usui, and N. Hashizume. *Statistical Mechanics*. North-Holland Publishing Co., Interscience Publishers, New York (1965). Translation of part of the book Netsugaku, tokei-rikigaku.
[7.94] P. J. Horoyski and M. L. Thewalt. *Phys. Rev. B*, **48**, 11446 (1993).
[7.95] G. B. Vaughan, P. A. Heiney, J. E. Fischer, D. E. Luzzi, D. A. Ricketts-Foot, A. R. McGhie, Y. W. Hui, A. L. Smith, D. E. Cox, W. J. Romanow, B. H. Allen, N. Coustel, J. P. McCauley, Jr., and A. B. Smith III. *Science*, **254**, 1350 (1991).
[7.96] M. A. Verheijen, H. Meekes, G. Meijer, P. Bennema, J. L. de Boer, S. van Smaalen, G. V. Tendeloo, S. Amelinckx, S. Muto, and J. van Landuyt. *Chem. Phys.*, **166**, 287 (1992).
[7.97] A. R. McGhie, J. E. Fischer, P. W. Stephens, R. L. Cappelletti, D. A. Neumann, W. H. Mueller, H. Mohn, and H. U. ter Meer. *Phys. Rev. B*, **49**, 12614 (1994).
[7.98] G. B. M. Vaughan, P. A. Heiney, D. E. Cox, J. E. Fischer, A. R. McGhie, A. L. Smith, R. M. Strongin, M. A. Cichy, and A. B. Smith III. *Chem. Phys.*, **178**, 599 (1993).
[7.99] V. P. Dravid, S. Liu, and M. M. Kappes. *Chem. Phys. Lett.*, **185**, 75 (1991).
[7.100] H. Kawamura, Y. Akahama, M. Kobayashi, H. Shinohara, H. Sato, Y. Saito, T. Kikegawa, O. Shimomura, and K. Aoki. *J. Phys. Chem. Solids*, **54**, 1675 (1993).
[7.101] R. E. Haufler, L.-S. Wang, L. P. F. Chibante, C.-M. Jin, J. J. Conceicao, Y. Chai, and R. E. Smalley. *Chem. Phys. Lett.*, **179**, 449 (1991).
[7.102] S. L. Ren, K. A. Wang, P. Zhou, Y. Wang, A. M. Rao, M. S. Meier, J. Selegue, and P. C. Eklund. *Appl. Phys. Lett.*, **61**, 124 (1992).
[7.103] S. Saito and A. Oshiyama. *Phys. Rev. B*, **44**, 11532 (1991).
[7.104] W. Andreoni, F. Gygi, and M. Parrinello. *Chem. Phys. Lett.*, **189**, 241 (1992).
[7.105] P. Wurz, K. R. Lykke, M. J. Pellin, and D. M. Gruen. *J. Appl. Phys.*, **70**, 6647 (1991).
[7.106] M. S. Baba, T. S. L. Narasimhan, R. Balasubramanian, and C. K. Mathews. *Rapid Commun. Mass Spectrom.*, **7**, 1141 (1993).
[7.107] S. van Smaalen, V. Petricek, J. L. de Boer, M. Dusek, M. A. Verheijen, and G. Meijer. *Chem. Phys. Lett.*, **223**, 323 (1994).
[7.108] R. Tycko, G. Dabbagh, G. B. M. Vaughan, P. A. Heiney, R. M. Strongin, M. A. Cichy, and A. B. Smith, III. *J. Chem. Phys.*, **99**, 7554 (1993).
[7.109] R. Tycko, G. Dabbagh, D. W. Murphy, Q. Zhu, and J. E. Fischer. *Phys. Rev. B*, **48**, 9097 (1993).

[7.110] Y. Maniwa, A. Ohi, K. Mizoguchi, K. Kume, K. Kikuchi, K. Saito, I. Ikemoto, S. Suzuki, and Y. Achiba. *J. Phys. Soc. Jpn.*, **62**, 1131 (1993).
[7.111] R. Blinc, J. Dolinsek, J. Seliger, and D. Arcon. *Solid State Commun.*, **88**, 9 (1993).
[7.112] K. Prassides, T. J. S. Dennis, C. Christides, E. Roduner, H. W. Kroto, R. Taylor, and D. R. M. Walton. *J. Phys. Chem.*, **96**, 10600 (1992).
[7.113] C. Christides, T. J. S. Dennis, K. Prassides, R. L. Cappelletti, D. A. Neumann, and J. R. D. Copley. *Phys. Rev. B*, **49**, 2897 (1994).
[7.114] G. van Tenderloo, M. Op de Beeck, S. Amerlinkx, J. Bohr, and W. Krätschmer. *Europhys. Lett.*, **15**, 295 (1991).
[7.115] C. Meingast, F. Gugenberger, M. Haluska, H. Kuzmany, and G. Roth. *Appl. Phys. A*, **56**, 227 (1993).
[7.116] A. V. Nilolaev, T. J. S. Dennis, K. Prassides, and A. K. Soper. *Chem. Phys. Lett.*, **223**, 143 (1994).
[7.117] D. F. Agterberg and M. B. Walker. *Phys. Rev., B*, **48**, 5630 (1993).
[7.118] T. J. S. Dennis, K. Prassides, E. Roduner, L. Christofolini, and R. DeRenzi. *J. Phys. Chem.*, **97**, 8553 (1993).
[7.119] X.-D. Wang, T. Hashizume, H. Shinohara, Y. Saito, Y. Nishina, and T. Sakurai. *Phys. Rev. B*, **47**, 15923 (1993).
[7.120] Y. Z. Li, J. C. Patrin, M. Chander, J. H. Weaver, K. Kikuchi, and Y. Achiba. *Phys. Rev. B*, **47**, 10867 (1993).
[7.121] J. F. Armbruster, H. A. Romberg, P. Schweiss, P. Adelmann, M. Knupfer, J. Fink, R. H. Michel, J. Rockenberger, F. Hennrich, H. Schreiber, and M. M. Kappes. *Z. Phys. B*, **95**, 469 (1994).
[7.122] R. S. Ruoff and A. L. Ruoff. *Appl. Phys. Lett.*, **59**, 1553 (1991).
[7.123] P. Bridgeman. *Proc. Amer. Acad. Art. Sci.*, **76**, 9 (1945). *Ibid.*, **76**, 55 (1948).
[7.124] M. N. Regueiro, P. Monceau, A. Rassat, P. Bernier, and A. Zahab. *Nature (London)*, **354**, 289 (1991).
[7.125] M. N. Regueiro, P. Monceau, and J.-L. Hodeau. *Nature (London)*, **355**, 237 (1992).
[7.126] F. Moshary, N. H. Chen, I. F. Silvera, C. A. Brown, H. C. Dorn, M. S. de Vries, and D. S. Bethune. *Phys. Rev. Lett.*, **69**, 466 (1992).
[7.127] L. Zeger and E. Kaxiras. *Phys. Rev. Lett.*, **70**, 2920 (1993).
[7.128] Y. Iwasa, T. Arima, R. M. Fleming, T. Siegrist, O. Zhou, R. C. Haddon, L. J. Rothberg, K. B. Lyons, H. L. Carter, Jr., A. F. Hebard, R. Tycko, G. Dabbagh, J. J. Krajewski, G. A. Thomas, and T. Yagi. *Science*, **264**, 1570 (1994).
[7.129] A. A. Maksimov, K. P. Meletov, and Y. A. Osipyan. *JETP Lett.*, **57**, 816 (1993).
[7.130] W. I. F. David, R. M. Ibberson, T. J. S. Dennis, J. P. Hare, and K. Prassides. *Physica B*, **180–181**, 567 (1992).
[7.131] J. Steinbeck, G. Dresselhaus, and M. S. Dresselhaus. *International Journal of Thermophysics*, **11**, 789 (1990).
[7.132] A. Cezairliyan and A. P. Miller. *Int. J. Thermophysics*, **11**, 643 (1990).
[7.133] C. Yeretzian, K. Hansen, F. N. Diederich, and R. L. Whetten. *Nature (London)*, **359**, 44 (1992).
[7.134] M. Matus, J. Winter, and H. Kuzmany. In H. Kuzmany, J. Fink, M. Mehring, and S. Roth (eds)., *Electronic Properties of Fullerenes*, vol. 117, p. 253, Springer-Verlag, Berlin (1993). Springer Series in Solid-State Sciences.
[7.135] Y. B. Zhao, D. M. Poirier, R. J. Pechman, and J. H. Weaver. *Appl. Phys. Lett.*, **64**, 577 (1994).
[7.136] H. Yamawaki, M. Yoshida, Y. Kakudate, S. Usuba, H. Yokoi, S. Fujiwara, K. Aoki, R. Ruoff, R. Malhotra, and D. C. Lorents. *J. Phys. Chem.*, **97**, 11161 (1993).

[7.137] A. M. Rao, M. Menon, K. A. Wang, P. C. Eklund, K. R. Subbaswamy, D. S. Cornett, M. A. Duncan, and I. J. Amster. *Chem. Phys. Lett.*, **224**, 106 (1994).
[7.138] P. Zhou, A. M. Rao, K. A. Wang, J. D. Robertson, C. Eloi, M. S. Meier, S. L. Ren, X. X. Bi, P. C. Eklund, and M. S. Dresselhaus. *Appl. Phys. Lett.*, **60**, 2871 (1992).
[7.139] Y. Wang, J. M. Holden, X. X. Bi, and P. C. Eklund. *Chem. Phys. Lett.*, **217**, 413 (1994).
[7.140] D. S. Cornett, I. J. Amster, M. A. Duncan, A. M. Rao, and P. C. Eklund. *J. Phys. Chem.*, **97**, 5036 (1993).
[7.141] Y. Wang, J. M. Holden, Z. H. Dong, X. X. Bi, and P. C. Eklund. *Chem. Phys. Lett.*, **211**, 341 (1993).
[7.142] P. Zhou, Z. H. Dong, A. M. Rao, and P. C. Eklund. *Chem. Phys. Lett.*, **211**, 337 (1993).
[7.143] A. M. Rao, K.-A. Wang, J. M. Holden, Y. Wang, P. Zhou, P. C. Eklund, C. C. Eloi, and J. D. Robertson. *J. Mater. Sci.*, **8**, 2277 (1993).
[7.144] C. Eloi, J. D. Robertson, A. M. Rao, P. Zhou, K. A. Wang, and P. C. Eklund. *J. Mater. Res.*, **8**, 3085 (1993).
[7.145] K. Venkatesan and V. Ramamurthy. In V. Ramamurthy (ed.), *Bimolecular Photoreactions in Crystals*, p. 133, VCH Press, Weinheim (1991).
[7.146] N. Takahashi, H. Dock, N. Matsuzawa, and M. Ata. *J. Appl. Phys.*, **74**, 5790 (1993).
[7.147] Y. Wang, J. M. Holden, A. M. Rao, P. C. Eklund, U. Venkateswaran, D. Eastwood, R. L. Lidberg, G. Dresselhaus, and M. S. Dresselhaus. *Phys. Rev. B*, **51**, 4547 (1995).
[7.148] M. Menon, A. M. Rao, K. R. Subbaswamy, and P. C. Eklund. *Phys. Rev. B*, **51**, 800 (1995).
[7.149] M. Menon and K. R. Subbaswamy. *Phys. Rev. Lett.*, **67**, 3487 (1991).
[7.150] M. Menon, K. R. Subbaswamy, and M. Sawtarie. *Phys. Rev. B*, **48**, 8398 (1993).
[7.151] M. Menon, K. R. Subbaswamy, and M. Sawtarie. *Phys. Rev.*, **49**, 13966 (1994).

CHAPTER 8

Classification and Structure of Doped Fullerenes

Fullerene solids are unique insofar as they can be doped in several different ways, including endohedral doping (where the dopant is inside the fullerene shell), substitutional doping (where the dopant is included in the fullerene shell), and the most commonly practiced exohedral doping (where the dopant is outside or between fullerene shells). Doped fullerenes in the crystalline phase are often called fullerides, in contrast to the term fullerite, which refers to the crystalline phase prior to doping. In this chapter, the various approaches to doping are classified and reviewed (§8.1). In many instances, the doping of fullerenes with guest species leads to charge transfer between the guest species and the host, while in clathrate materials, such charge transfer does not occur. Examples of both charge transfer and clathrate fullerene-based compounds are given in this chapter (§8.4). Studies of the structure and properties of the doped molecules and of the materials synthesized from doped fullerenes are expected to be active research fields for at least the near future.

Several stable crystalline phases for exohedrally doped (or intercalated) C_{60} have been identified. At present there is only scanty structural information available for crystalline phases based on endohedrally doped fullerenes (§8.2) or for substitutionally doped fullerenes (§8.3), such as the fullerene BC_{59}. Most widely studied are the crystalline phases formed by the intercalation of alkali metals (§8.5), although some structural reports have been given for fullerene-derived crystals with alkaline earth (§8.6) [8.1–3], iodine [8.4, 5], and other intercalants (§8.7). In §8.8, we review the relatively sparse information on the doping of C_{70} and higher-mass fullerenes.

The practical aspects of the preparation of doped fullerene solids are reviewed in Chapter 9 on “Single Crystal and Epitaxial Film Growth.” The preparation, extraction, purification, and separation of endohedral metallofullerenes are discussed in §5.4. The electrochemistry of fullerenes and fullerene-derived compounds is reviewed in Chapter 10 on “Fullerene Chemistry.”

8.1. Classification of Types of Doping for Fullerenes

In this section the various types of doping are classified from both a structural standpoint [8.6] and an electrical point of view. Regarding the structure, we distinguish the doping according to the location of the dopant. The principal methods of doping include endohedral doping (whereby the dopant goes into the hollow core of the fullerene), substitutional doping (whereby the dopant replaces one or more of the carbon atoms on the shell of the molecule), and exohedral doping (whereby the dopant enters the host crystal structure in interstitial positions of the lattice; this method of doping is often called intercalation). Each of these types of doping is discussed in more depth in the following sections. The dopant can be further classified according to whether charge is transferred upon doping. Since charge transfer can modify the properties of fullerenes in scientifically interesting and practically important ways, there is a large literature on the synthesis, structure, and properties of doped fullerides, especially for charge transfer dopants. Dopants for which there is no charge transfer form clathrate compounds. In a clathrate structure, the fullerenes appear on a sublattice, and other molecules reside either on other sublattices or at random lattice sites.

Each carbon atom in a C_{60} molecule is in an identical environment, has four valence electrons, and bonds to each of the three nearest-neighbor carbon atoms on the shell of the molecule, as discussed in §3.1. Since all the intramolecular bonding requirements of the carbon atoms are satisfied, it is expected that C_{60} is a van der Waals insulator (semiconductor) with an energy gap between the occupied and unoccupied states, consistent with the observed electronic structure (see Chapter 12). To make C_{60} (and also other fullerites) conducting, doping is necessary to provide the charge transfer to move the Fermi level into a band of conducting states.

As mentioned above, crystalline compounds based on C_{60} can be subdivided into two classes: charge transfer compounds and clathrate compounds. In the first class, or “C_{60} charge transfer compounds,” foreign atoms, e.g., alkali metals or alkaline earths, are diffused into solid C_{60}, donating electrons to the filled shell C_{60} to form C_{60}^{n-} molecular anions on which the transferred charge is mainly delocalized over the molecular shell.

The resulting dopant cations, required for charge neutrality, reside in the interstitial voids of the C_{60} sublattice. Upon doping, Fig. 8.1 shows that the solid either may retain the fcc structure of the pristine crystal, or may transform it into a different structure [e.g., body-centered tetragonal (bct) or body-centered cubic (bcc)], because of steric strains introduced by the dopant [8.8].

Considerable research activity on fullerene-based materials has been expended on the study of the M_3C_{60} or $M_{3-x}M'_xC_{60}$ alkali metal compounds since the discovery in 1991 of moderately high temperature superconductivity (18–33 K) in these compounds [8.9]. The AT&T group has also demonstrated charge transfer in ammoniated alkali metal compounds, but in this case, the NH_3 molecules are not thought to contribute directly to the charge

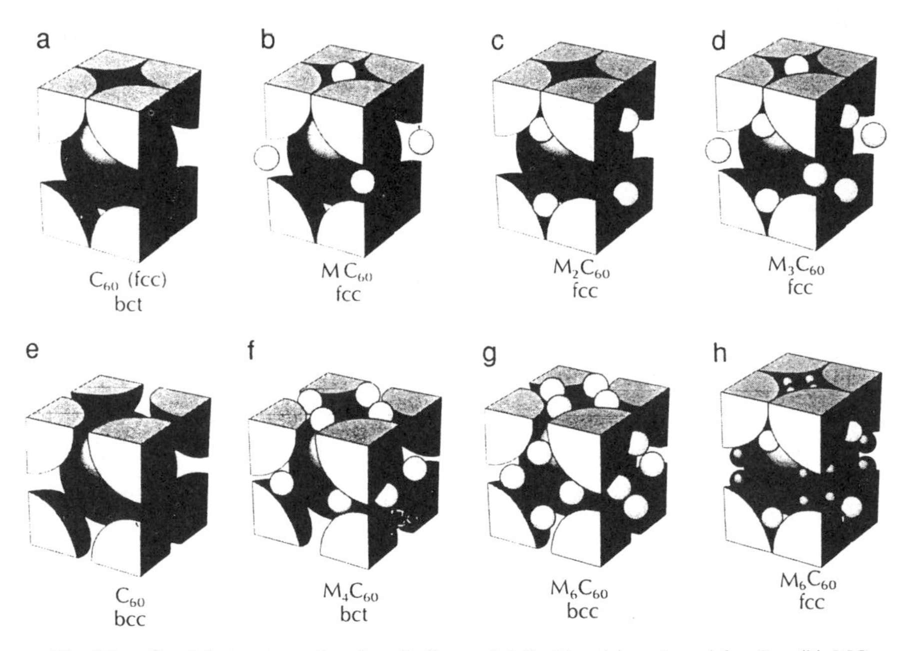

Fig. 8.1. Crystal structures for the alkali metal fullerides (a) undoped fcc C_{60}, (b) MC_{60}, (c) M_2C_{60}, (d) M_3C_{60}, (e) undoped hypothetical bcc C_{60}, (f) M_4C_{60}, and two structures for M_6C_{60}, (g) M_6C_{60} (bcc) for M = K, Rb, Cs, and (h) M_6C_{60} (fcc), which is appropriate for M = Na [8.7]. The large balls denote C_{60} molecules and the small balls are alkali metal ions. For fcc M_3C_{60}, which has four C_{60} molecules per cubic unit cell, the M atoms can be on either octahedral or tetrahedral sites. Undoped solid C_{60} also exhibits the fcc crystal structure at room temperature, but in this case all tetrahedral and octahedral sites are unoccupied. For (f) bct M_4C_{60} and (g) bcc M_6C_{60}, all the M atoms are on distorted tetrahedral sites. For (h) we see that four Na ions can occupy an octahedral site of this fcc lattice. The M_1C_{60} compounds (for M = Na, K, Rb, Cs) crystallize in the rock-salt structure shown in (b).

transfer process. These ammoniated compounds have attracted much attention through their enhancement of the superconducting transition temperature T_c. This enhancement of T_c has been explained by the diffusion of NH_3 into $M_{3-x}M'_xC_{60}$ (M, M′ = alkali metal dopants) to increase the lattice constant of the crystalline solid [8.9, 10] (see §8.5.5 and §15.2).

In the second class of C_{60}-based materials, or "C_{60} clathrate solids," the structure is stabilized by the van der Waals interaction between C_{60} molecules and other molecular species, such as a solvent molecule (e.g., hexane, CS_2), O_2 [8.11] or S_8 [8.12], or mixtures of a solvent molecule and a third species (e.g., $C_{60}S_8CS_2$ [8.13]). Since, by definition, no charge is transferred between the C_{60} molecules and the other molecular species in the structure, the basic physical properties of the clathrate compounds should be predictable from the properties of the constituent molecular species. One can also include in the clathrate class molecular mixtures of C_{60} adducts (see Chapter 10) and other molecular species. Several fullerene clathrates have been found to exhibit structural order in both the C_{60} and solvent sublattice, although the fullerene shells display orientational disorder [8.14–17].

Returning to the charge transfer dopants, the electrons transferred to the fullerenes are delocalized on the shell of the fullerene anions, and the distinction between double and single bonds becomes less important as charge transfer proceeds. This effect is supported by both experimental observations and theoretical calculations [8.18, 19] which show that the bond lengths for the single bonds along the pentagon edges decrease from a_5 = 1.46 Å upon charge transfer to form a C_{60}^{n-} anion, while the bond lengths for the double bonds between adjacent hexagons increase from a_6 = 1.40 Å for neutral C_{60}, as shown in Fig. 8.2. The high degree of charge transfer observed for some doped C_{60} compounds (such as those

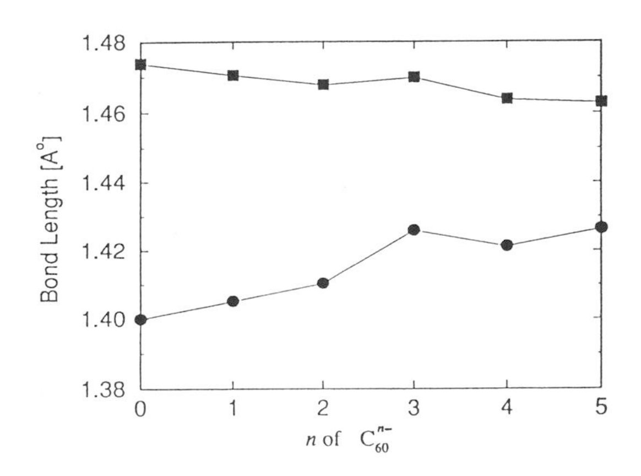

Fig. 8.2. Dependence of bond length on the number of electrons transferred to C_{60} for the pentagonal bonds a_5 (filled squares) and for the bond between two hexagons a_6 (filled circles) to form C_{60}^{n-} [8.18, 20].

formed by alkali metal doping) indicates that these compounds are highly ionic.

While doped C_{60} has been investigated in some depth, many gaps in our knowledge remain. For example, very little is presently known about the doping of higher-mass fullerenes. For these reasons, the study of the synthesis, structure, and properties of doped fullerenes is likely to remain an active field of study for some time. One method for monitoring the charge transfer is by cyclovoltammetry measurements (see §10.3.2). These electrochemical techniques have clearly shown that up to six electrons can be transferred from the alkali metal dopants to the C_{60} shell [8.21].

8.2. Endohedral Doping

One doping method is the addition of an atom or an ion into the interior hollow core of the fullerene molecule to form an endohedrally doped molecular unit, also called a metallofullerene or endofullerene [8.22]. In principle, many different atomic species can be inserted within fullerenes, and many have been (see Tables 5.2 and 5.3). The insertion of one, two, and three metal species inside a fullerene cage is common, and up to four metal atoms (ions) have thus far been introduced [8.23]. Slow progress with the isolation and purification of sufficient quantities of metallofullerenes has, however, delayed study of the structure and properties of metallofullerenes in the solid state.

The practical aspects of the synthesis, extraction, and purification of endohedral fullerenes are reviewed in §5.4. Examples of endohedrally doped rare earth, alkaline earth, and alkali metal fullerenes are given in Tables 5.2 and 5.3. Upon preparation of endohedrally doped fullerenes, the M@C_{60} (see Fig. 8.3(a)) species is thought to be the most common in the soot, but after sublimation of the soot, the most common form is M@C_{82} (see Fig. 8.3(b) and (c)), with lesser amounts of M@C_{74} and M@C_{80} [8.24]. Studies of La@C_{82} confirm the relatively high stability of this molecule [8.25].

From an historical perspective, soon after it was realized that C_{60} formed a closed-cage molecule, efforts were made to insert ions within the cage [8.26, 27]. Mass spectra suggesting endohedral doping of C_{60} with a single La ion, but not with two La ions, were observed soon after fullerenes were first identified, and this observation played an important role in supporting the hypothesis that C_{60} was a closed-cage molecule [8.28, 29]. In this early work the metallofullerenes La@C_{60}, La@C_{44}, and La@C_{76} were identified. It was also found at this early time that laser photodissociation could be used to remove C_2 units from the endohedral shell (see §6.4.1), thereby yielding smaller endofullerenes in a process which "shrink wrapped" the carbon cage around the endohedral core, eventually leading to $La@C_{44}$, the smallest member of this metallofullerene series [8.29]. The "shrink wrapped" experiments and related chemical reactivity experiments confirm the structural and chemical stability of endohedrally doped fullerene complexes [8.30]. Furthermore, infrared and Raman spectra for $La@C_{82}$ indicate that the cage structure is not much altered by La introduction [8.31, 32].

Early theoretical work [8.33] indicated that the introduction of La into C_{60} did not affect the electronic structure of the filled highest occupied molecular orbital (HOMO) level (see §12.6.2), but resulted in hybridization of the partially filled lowest unoccupied molecular orbital (LUMO) level between carbon $2p$ and La $5d$ orbitals. In calculating the electronic structure for endohedral fullerenes (see §12.6.2), it is necessary to consider electron correlation effects and the relaxation of the carbon shell atoms by the introduction of the endohedral species. Since this relaxation is dependent on the specific endohedral species as well as the isomer of the carbon shell, structural differences are expected for the various endohedral dopants, and, in fact, different isomers may have greater stability as the endohedral dopant is varied. For example, C_{82} has nine isolated-pentagon isomers [8.34]. Four of these with symmetries C_2, C_{2v}, C_{3v}, and C_2 have been identified by liquid chromatography from the soot [8.25, 35] and two isomers of $La@C_{82}$ (with C_2 and C_{3v} symmetries) have been identified in the electron paramagnetic resonance (EPR) spectra [8.36, 37].

Metallofullerenes are characterized by many experimental techniques. EPR, via lineshape and hyperfine interaction studies, provides a very sensitive method for characterizing the charge transfer for the various endohedral dopants [8.36–40] (see §5.4 and §16.2.1). While EPR is a powerful technique for characterizing endohedrally doped fullerenes, not all such species show EPR spectra; for example, $La_2@C_{80}$ does not give rise to an EPR spectrum [8.38], since, in this case, the two nuclear spins for each of the La ions pair to yield a total $I = 0$, leading to a vanishing hyperfine interaction. Trimetal species such as La_3C_{82} or Sc_3C_{82} do, however, give rise to EPR spectra [8.41] and give strong evidence that the three metal species are at corners of an equilateral triangle within the fullerene [see Fig. 8.3(d)]. Other important characterization techniques for metallofullerenes include scanning tunneling microscopy [8.42], x-ray absorption fine structure (EXAFS) [8.43], x-ray diffraction (XRD) and transmission microscopy (TEM) [8.44, 45], photoelectron spectroscopy, and Mössbauer spectroscopy studies. Raman and infrared spectroscopy [8.25], electron energy loss spectroscopy [8.46], and UV–visible spectroscopy [8.25, 47] have been used to study the solid phase [8.22].

It has been shown by a variety of techniques that when La is inside the fullerene molecule, it has a valence of +3 [8.6, 31, 38], indicating the pres-

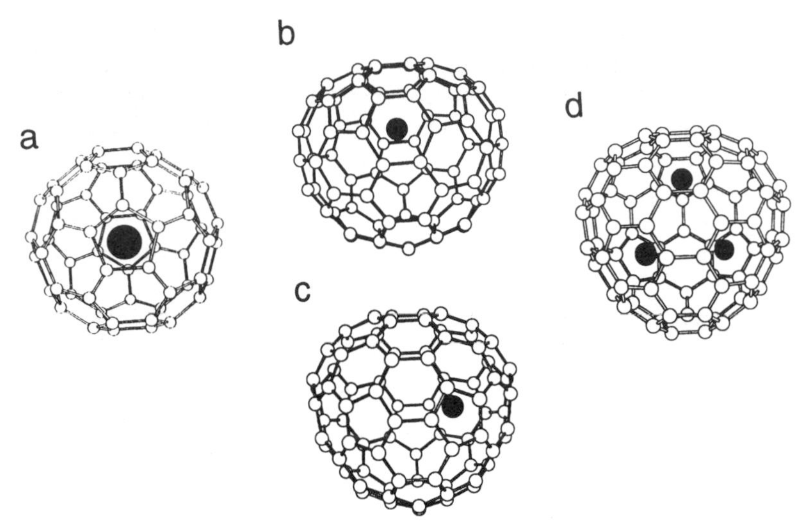

Fig. 8.3. Structural models for various endofullerenes. (a) One structural model for La@C_{60}, with La at the center of the C_{60} cage [8.52]. (b, c) Another structural model for La@C_{82}, with the La at two different off-center positions within the C_{82} cage [8.50]. (d) Structural model for Sc_3@C_{82} (assuming C_{3v} symmetry of the C_{82} cage), where the large black balls represent the three equivalent Sc^{3+} ions, which rapidly reorient within the C_{82} cage [8.31].

ence of three delocalized electrons on the C_{60} cage in antibonding states. These electrons could be available for conduction in a solid comprised of endohedrally doped C_{60} molecules, if such a solid could be formed. Isoelectronic Y^{3+} and Sc^{3+} have also been introduced as endohedral dopants in C_{60} [8.38]. Interestingly, La^{3+}, Y^{3+}, and Sc^{3+} are all filled-shell ions, corresponding to angular momentum $J = 0$, but endohedral ions with $J \neq 0$ could have magnetic moments (see §16.2.2). The most extensive characterization of the charge transfer of three electrons from La to the fullerene is available for La@C_{82}, where EPR measurements [8.39] of the ^{13}C hyperfine structure [8.36, 38] and x-ray photoemission studies of the $3d$ core levels [8.40] have all been carried out.

Although there is no universal acceptance, it is now widely believed that the dopants are inside the cage, although not necessarily at the central position. For example, early EXAFS studies [8.30] suggested that the endohedral dopant atoms may not be inside the C_{82} cage but rather bridge two adjacent fullerenes. Subsequently, EXAFS studies on Y@C_{82} and Y_2@C_{82} have provided strong evidence that the metal species are indeed inside the fullerene cage [8.43]. The same conclusion was reached on the basis of scanning tunneling microscopy (STM) studies [8.48], which provided strong evidence for a nearly spherical shape for Sc@C_{74} and Sc_2@C_{74} on an Si

(100) 2×1 surface and also showed that the metal species is within the fullerene cage.

One simple model for the structure of La@C_{82} places the La^{3+} ion centrally. Other structural model calculations for minimizing the energy of La@C_{82} show the La to be significantly displaced (by 1.49 Å) from the center of the nearly spherical carbon cage [see Fig. 8.3(b) and (c)], with the smallest La–C distances (to six nearest neighbors) at 2.53–2.56 Å and the second nearest-neighbor La–C distances (to six next-nearest neighbors) at 2.95–3.03 Å [8.31], consistent with EXAFS measurements [8.43]. Molecular dynamics simulations for La@C_{82} favor an off-center position for the La ion to stabilize its +3 charge state [8.22, 32, 49, 50].

As a result of the off-center position of the La in La@C_{82}, a dipole moment for La@C_{82} is expected. This dipole moment may be large (2–4 Debye) [8.31] and may strongly affect the solubility of La@C_{82} in various solvents such as CS_2 and toluene (see Table 5.1) [8.31]. Differences in solubility might be exploited in isolating and purifying endofullerenes (see §5.4).

Ab initio calculations of the structural symmetry for La_2@C_{82} have also been reported, indicating separation of the La ions within the cage [8.32]. A model for the locations of three La ions in C_{82} has also been developed and the results are presented schematically in Fig. 8.3(d) [8.38, 51].

Alkali metal dopants are of special interest because they can serve as either endohedral or exohedral dopants. Calculations of the ionization energies for K_2@C_{60} [8.53] were used to determine the location of the alkali metal atoms (ions) within the C_{60} cage. Calculations of the charge transfer for many endohedral dopants into C_{60} have already been carried out, including the donors K, Ca, Mn, Sr, Ba, La, Eu, and U and the presumed acceptors O and F [8.54].

Metallofullerenes based on uranium, which is believed to be tetravalent (U^{4+}) as an endohedral dopant, are especially interesting insofar as small-mass metallofullerenes can be formed with uranium in the cage [8.55]. Starting with U@C_{70}, U@C_{60}, and U@C_{50} in the gas-phase, the uranium species can be "shrink wrapped" by laser irradiation, which ejects a succession of C_2 units, giving rise to U@C_{44}, U@C_{38}, and eventually U@C_{28} (see §6.4.1). The cage of U@C_{28} is smaller than that reported for the smallest stable empty fullerene (C_{32}). Furthermore, mass spectroscopy evidence has also been given for endohedral fullerenes with other column IV-B ions, such as Ti@C_{28}, Zr@C_{28}, and Hf@C_{28}. The cross section for the production of metallofullerenes decreases as the atomic number of the dopant decreases (U–Hf–Zr–Ti) [8.55]. Endohedral doping of Ca into a C_{60} molecule has also been observed [8.55, 56]. In this case, ultraviolet photoelectron spectroscopy has shown that the electron affinity for Ca@C_{60} is 3.0 eV (~0.5 eV higher than for C_{60}) and that the valence for Ca in this case is +2. Rare earth metallofullerenes have also been observed, with $Eu_m@C_{n_C}$ being found for many n_C values, and the Eu ion always having a charge of +3 [8.57]. Issues that molecular dynamics calculations may address concern the motion of the metal dopant inside the fullerene cage, the extent of the electron–phonon coupling of the metallofullerene clusters, and the magnitudes and orientations of the dipole moments.

An interesting recent discovery is the endohedral doping of C_{60} with the rare gas atoms of He and Ne [8.58]. So far, only low concentrations ($\sim 10^{-6}$ per fullerene) of endohedral doping of He and Ne have been demonstrated. The He- and Ne-doped C_{60} compounds are stable at room temperature and release the rare gas upon heating. From the temperature dependence of the gas release, a potential barrier of ~80 kcal/mol is found for the binding of the rare gas atom inside the potential well of the fullerene. In this case, there is no charge transfer. A mechanism has been proposed for endohedral gas uptake, based on a reversible bond breaking (bond/reforming) of the fullerene cage to "open a window" temporarily for the entry of the gas molecule. A number of workers have shown that it is possible to insert rare gas atoms into the endohedral cavity by high energy ion collisions with C_{60} [8.59–61]. Studies of ion collisions as a function of collision energy between 0 and 150 eV indicate that elements with ionization potentials less than 7 eV could form endofullerenes [8.62]. One exception to this rule of thumb is the reported formation of $Fe@C_{60}$ [8.63], where Fe has an ionization potential of 7.87 eV.

Another important and interesting example of an endohedral system is found for the case where muonium is introduced endohedrally into the C_{60} trianions, denoted by $Mu@C_{60}^{3-}$. These trianions were available in Rb_3C_{60} powder samples which exhibited a superconducting transition temperature of $T_c = 29.4$ K [8.64, 65]. In this case, the conduction electrons on the C_{60}^{3-} anions screen the positive charge on the muon, so that no local dipole moment is present on the muon. The species $Mu@C_{60}^{3-}$ has provided an important probe in μ magnetic resonance (μMR) experiments which are important for determining the superconducting energy gap for fullerenes (see §15.4 and §16.3).

Thus far, only very small quantities of endohedrally doped fullerenes have been prepared, and most studies of metallofullerenes have been limited to the gas-phase. Isolation of sufficient quantities of $Sc@C_{74}$ and $Sc_2@C_{74}$ has allowed vacuum deposition of a monolayer of these species onto a clean Si (100) surface. These monolayer films were then characterized by scanning tunneling microscopy, yielding a diameter of ~9.5 Å for these endohedral fullerenes and indicating the presence of some charge delocalization over the cage [8.66].

Detailed structural information based on electron diffraction and high-resolution TEM is available thus far for only a few metallofullerene crystals such as $Sc_2@C_{84}$, which was prepared in crystalline form by dissolving the metallofullerene in CS_2 [8.45], and $Sc@C_{74}$, which was deposited on an Si surface [8.48]. Electron diffraction patterns showed $Sc_2@C_{84}$ to crystallize in the hexagonal close-packed (hcp) $P6_3/mmc$ structure with $a = 11.2$ Å, $c = 18.3$ Å, so that $c/a = 1.63$, which is the ideal value for c/a in the hcp structure. A detailed examination of the diffraction pattern and the TEM lattice images provides strong evidence that both Sc metal ions are contained within the fullerene cage [8.45]. It is interesting to note that the C_{84}–C_{84} nearest-neighbor distance is also 11.2 Å for crystalline face-centered cubic (fcc) C_{84} [8.46, 67], and the nearest-neighbor C–C distance between C_{84} molecules and between their endohedral counterparts remains at ~2.8 Å, which is close to the corresponding intermolecular C–C distance of 2.87 Å in crystalline C_{60}. This is in contrast to the report for $Sc@C_{74}$ and $Sc_2@C_{74}$, where the metallofullerene diameters were reported to be larger than for C_{74} itself [8.48].

As synthesis techniques improve and larger samples of isolated and purified metallofullerenes become available, it is expected that both the structure and properties of metallofullerenes in the solid state will be clarified.

8.3. Substitutional Doping

A second doping method involves the substitution of an impurity atom with a different valence state for a carbon atom on the surface of a C_{60} molecule. Substitutional doping is common for group IV solids, including silicon, germanium, diamond, and graphite. Since a carbon atom is very small, and since the average nearest neighbor C–C distance, denoted by a_{C-C}, is only 1.421 Å in graphite, the only likely substitutional dopant in graphite is boron [8.68]. However, for diamond, which has a larger nearest-neighbor distance of $a_{C-C} = 1.544$ Å, both boron and nitrogen can enter the diamond lattice substitutionally [8.69]. Since the average a_{C-C} on the C_{60} surface is only 1.44 Å [8.7, 26, 70–73], quite close to that of graphite, the only species that is expected to substitute for a carbon atom on the C_{60} shell is boron. However, since a_{C-C} on the C_{60} molecule is slightly larger than in graphite, and the force constants are somewhat weakened by the curvature of the C_{60} surface, it may be possible also to dope the C_{60} shell with nitrogen, although this has not yet been demonstrated experimentally. Electronic structure calculations [8.74, 75] indicate that electrons are transferred from the C_{60} shell to the substitutional boron, leaving hole states in the HOMO level of BC_{59}. Smalley and co-workers have demonstrated experimentally that it is also possible to replace two carbon atoms by two boron atoms on a given molecule [8.6]. It has also been reported that it is possible to place a potassium atom endohedrally inside the C_{60} molecule while at the same time substituting a boron for a carbon atom on the surface of the molecule [8.6]. ^{13}C substitution for ^{12}C occurs commonly in fullerenes and involves no charge transfer (see §4.5).

8.4. Exohedral Doping

New solid state compounds can be made from fullerenes in which the fullerene molecules form a sublattice and dopant atoms or molecules fill the interstitial voids in this sublattice. As mentioned previously (§8.1), this is referred to as "exohedral" doping. We do not include in this class of fullerene compounds materials for which chemical modification to the fullerene shell has occurred via covalent attachment of various chemical adducts; such compounds are discussed in Chapter 10.

An empirical criterion based on electronic energy arguments has been introduced [8.76] to determine whether or not $E_{\text{intc}} > 0$, which is a criterion that determines whether a specific metal species will intercalate into crystalline C_{60}

$$E_{\text{intc}} = E_{\text{LUMO}} - e\phi - E_{\text{COH}}, \tag{8.1}$$

where ϕ is the work function, E_{LUMO} is the energy of the lowest unoccupied orbital, and E_{COH} is the cohesive energy of the metal–fullerene complex. Application of Eq. (8.1) shows that alkali metal and alkaline earth metals should intercalate into crystalline C_{60} [8.76]. Because of their high cohesive energies, transition metals, noble metals, and column III metals yield negative values for E_{intc}. In contrast, the low E_{COH} for alkali metals and relatively low ϕ for alkaline earth metals yield positive E_{intc}, and consequently, lead to intercalation, in agreement with experiment [8.76]. Negative values for E_{intc} were reported for Zn, Cd, Al, Ga, In, Tl, Sn, Pb, Cr, Mo, W, Cu, Ag, and Au [8.76], implying that none of these metals would be attractive candidates for intercalation. However, mercury was found to have a positive value for E_{intc}, suggesting that Hg may be a possible exohedral dopant for C_{60} [8.76]. Other candidates for intercalation (with $E_{\text{intc}} > 0$) are certain rare earth elements, and, in fact, a recent success with the intercalation of Yb into C_{60} has been reported [8.77].

The discussion in §8.4.1 focuses on exohedral doping, where the dopant transfers charge, while §8.4.2 addresses the opposite limit where no charge transfer occurs. The use of a positron probe at the octahedral sites in C_{60} is reviewed in §8.4.3. Details of the doping synthesis are given in §9.3, and the crystal structure of alkali metal doped C_{60} compounds is given in §8.5.

8.4.1. Exohedral Charge Transfer Compounds

The most common method of doping full erene solids is exohedral donor doping (also called intercalation), whereby the dopant (e.g., an alkali metal or an alkaline earth, M) is introduced into the interstitial positions between adjacent molecules (exohedral locations) in a solid structure. Because of the large size of the fullerene molecules relative to the size of the typical dopant atoms or ions, the interstitial cavities between the molecules can also be large enough to accommodate various guest species in these cavities. Charge transfer can take place between the M atoms and the molecules, so that the M atoms become positively charged ions and the C_{60} molecules become negatively charged with predominately delocalized electrons. This method of doping forms exohedral fullerene solids (also called fullerides) and closely parallels the process of intercalating alkali metal ions into graphite. In this case, layers of guest atoms, molecules, or ions are sandwiched between the graphene layers, as discussed in §2.14, to form graphite intercalation compounds [8.78].

Considerable effort has been expended to explore the physical properties of the alkali metal (M)–doped C_{60} compounds and to a lesser extent M-doped C_{70} compounds. As the dopant is introduced into the fullerite lattice, electrons are donated from the alkali metal atoms to the fullerite host sublattice. For the case of the binary M_xC_{60} compounds, a wide range of x can be obtained ($0 < x \leq 6$) with well-defined crystal structures at $x = 1, 3, 4, 6$ for M = K, Rb, and Cs [8.7, 79–81]. For the small alkali metal species Na, stable compounds with x values as low as 1 and as high as 11 have been reported for Na_xC_{60} compounds [8.82].

M_xC_{60} compounds can be prepared by exposing solid C_{60} in the form of polycrystalline or crystalline films, microcrystalline powder, or single crystals to the vapors of the respective alkali metal, and these vapors diffuse into the host at elevated temperatures (100–200°C). Thus exohedral doping is sometimes referred to in the fullerene literature as "intercalation," although strictly speaking, this term was applied originally to the introduction of dopant atoms between the existing atomic layers in *layered* materials such as graphite or MoS_2 (see §2.14). All the alkali metals, Li, Na, K, Rb, and Cs [8.83], have been used as exohedral dopants for C_{60}, as well as intermetallic mixtures of these alkali metals such as Na_2K [8.84, 85], Rb_xCs_{3-x} [8.86], and more generally $M_xM'_{3-x}C_{60}$ [8.83, 86–91]. The interest in the $M_xM'_{3-x}C_{60}$ compounds relates to their relatively high conductivity and their high probability for superconductivity. Since the alkali metal–doped fullerenes are highly reactive under ambient conditions, they must be prepared and studied under strict exclusion of air and water. Since the

Li in Li_xC_{60} compounds reacts with glass, relatively little work has thus far been done on these compounds.

Some success has also been achieved with the intercalation of alkaline earth dopants, such as Ca, Ba, and Sr [8.1, 81, 92, 93], into C_{60}. The intercalation of alkaline earth dopants generally requires higher temperatures for intercalation and a longer intercalation time than for alkali metals [8.3]. In the case of the alkaline earth dopants, two electrons per metal atom M are transferred to the C_{60} molecules for low concentrations of metal atoms, i.e., until the t_{1u}-derived (C_{60} LUMO) band is filled. Because of hybridization between the metal bands and the next higher lying t_{1g}-derived C_{60} band, the charge transfer is less than two electrons per alkaline earth ion for high concentrations of metal atoms (see §12.7.4). In general, the alkaline earth ion is smaller than the corresponding alkali metal ion in the same row of the periodic table. For this reason the crystal structures formed with alkaline earth doping are often different from those for the alkali metal dopants (see §8.5 and §8.6).

Doping fullerenes with acceptors (i.e., with dopants which accept electrons from the C_{60} molecules to yield C_{60}^{n+} cations as discussed in §8.7) has been much more difficult than with donors because of the high electron affinity of C_{60} (see Table 3.1). A stable crystal structure has nevertheless been identified upon doping C_{60} with iodine, where the iodine is reported to form a sheet between layers of C_{60} molecules (see §8.7.1); no measurable conductivity has, however, been observed in the case of iodine doping of fullerenes [8.4, 94] and therefore it is concluded that charge transfer does not occur in this case. Thus $C_{60}I_x$ may then be considered a clathrate (i.e., a mixture of neutral C_{60} and I_2 molecules in which the iodine occupies interstitial sites in the C_{60} lattice).

8.4.2. Exohedral Clathrate C_{60} Compounds

Clathrate C_{60} solids contain a mixture of C_{60} and a second molecular species such as O_2, S_8, or *n*-pentane. Clathrate materials may exhibit well-ordered crystalline phases or may also be highly disordered or they may be partially disordered. The molecules interact via the weak van der Waals force as no charge transfer between the dopant molecules and the fullerenes takes place. In some cases, the molecular species are small enough (e.g., O_2, H_2) to fit comfortably into the interstitial octahedral voids in the pristine C_{60} fcc lattice, but not into the tetrahedral voids. In other cases, where larger dopant molecules are involved, substantial rearrangements of the fullerene molecules must occur, if clathrate crystals are to form. Further discussion of the actual synthesis of exohedral clathrate fullerene compounds is given in §9.3.2.

The crystal structure of $C_{60}(S_8)CS_2$ is shown in Fig. 8.4 as a projection of the structure along the *b*- and *c*-axes [8.12]. The orthorhombic structure consists of a three-dimensional framework of oriented C_{60} molecules with one-dimensional channels in the *a* direction, containing the S_8 and CS_2 molecules (see Fig. 8.4) [8.13]. The lattice constants of the rhombohedral structure are $a = 9.97$ Å, $b = 13.36$ Å, and $c = 28.65$ Å, and the space group is *Pnma* [8.12]. The structure consists of six-membered rings of C_{60} molecules in the *b*–*c* plane which share common edges and are stacked in the *a* direction. In this structure the C_{60} molecules are orientationally disordered, while the S_8 and CS_2 are well-ordered (see Fig. 8.4). The C–C bond lengths within a C_{60} molecule have been derived from the x-ray diffraction data (1.449 Å and 1.341 Å for the long and short bonds, respectively). The C–C bond length difference in this clathrate crystal is thus found to be significantly larger [8.95] than is reported for pure C_{60} (see §3.1), and showing an opposite trend relative to charge transfer compounds (see Fig. 8.2).

The crystal structure of several other clathrate compounds has been investigated, such as C_{60}–*n*-pentane. Crystals of C_{60}–*n*-pentane are grown from a saturated solution [8.96]. The room temperature structure of C_{60}–*n*-pentane is a centered orthorhombic phase [8.97] with $a = 10.101$ Å, $b = 10.163$ Å, and $c = 31.707$ Å and shows a phase transition to a low-temperature monoclinic phase which involves a distortion of the

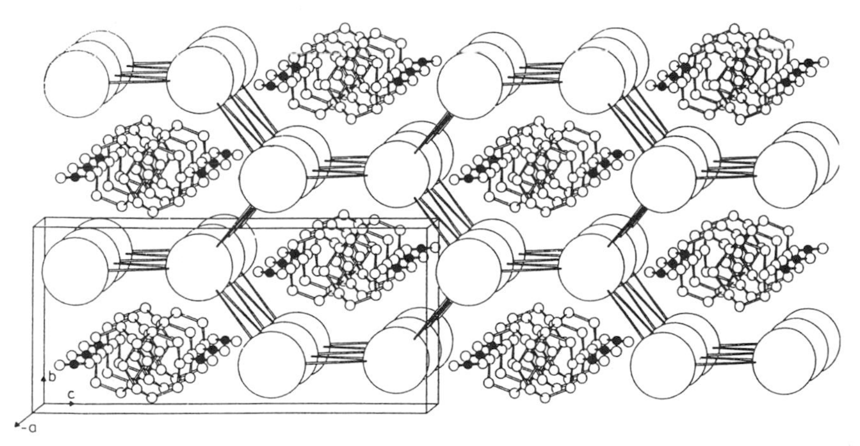

Fig. 8.4. Crystal structure of the compound $C_{60}(S_8)_2CS_2$ projected normal to the *a*-axis. Large circles denote C_{60}, small circles denote sulfur, black balls denote carbon. In this structure, the C_{60}–C_{60} distance is nearly 11 Å, and the diameter of the C_{60} molecule has been reduced relative to the other atoms for clarity [8.13].

basal plane. The phase transition occurs in the range 175 to 195 K, whereby the free rotation of the C_{60} molecules stops and the *n*-pentane molecules become frozen into quasi-equilibrium positions in the lattice [8.96]. This system has been used as one of the model C_{60}-based clathrate systems and has been studied by x-ray diffraction [8.96, 97], differential scanning calorimetry [8.98, 99], and by a molecular dynamics simulation [8.96].

8.4.3. Positron Exohedral Trapping

Both calculations [8.100] and experiments under pressure [8.101] show that positrons in C_{60} tend to stabilize in octahedral sites of the C_{60} fcc lattice. Positrons (e^+) provide a probe of the local electron charge density because their probability for recombination with an electron (e^-) to form two gamma rays is proportional to the local electron charge density. The angular correlation of the two gamma rays gives a measure of the wave vector of the electron. The positron lifetime τ_{e^+}, which is the mean time for electron–positron recombination, is 395 ps in fcc C_{60} at 280 K, and τ_{e^+} increases by $\sim 3 \times 10^{-2}$ ps/K due to the decrease in electron density associated with normal lattice expansion below T_{01} (261 K), but increases by 4.5 ± 0.5 ps at T_{01}, the simple cubic (sc)-to-fcc phase transition. The magnitude of this jump in the positron lifetime τ_{e^+} can be fully explained by the volume expansion of the lattice ($\Delta V/V \simeq 9.3 \times 10^{-3}$) [8.102] and the consequent lowered cross section for electron–positron recombination [8.103]. Whereas the introduction of oxygen by itself has little effect on τ_{e^+}, the combined effect of oxygen in the presence of light is to lower τ_{e^+} significantly, consistent with the photo-enhanced diffusion of O_2 into the C_{60} lattice (see §7.5.1) [8.11].

8.5. Structure of Alkali Metal–Doped C_{60} Crystals

When C_{60} is doped with the alkali metals (M = K, Rb), stable crystalline phases are formed for the compositions M_1C_{60}, M_3C_{60}, M_4C_{60}, and M_6C_{60} (see Fig. 8.5) [8.7, 104–106], and these stable phases are illustrated in the hatched areas of the binary phase diagram shown in Fig. 8.5 [8.106]. Less is known about the Cs_xC_{60} compounds, which have some differences in behavior because of the large size of the Cs^+ ion relative to K^+ and Rb^+ and relative to the cavity at the dopant lattice site (see Table 8.1). Also, the phase diagrams for the smaller ion alkali metal compounds Na_xC_{60} and Li_xC_{60} are substantially different from Fig. 8.5 because the ionic radius for the Na^+ ion is much smaller than that for K^+ and Rb^+ and is smaller than the tetrahedral cavity radius, as discussed in §8.5.3.

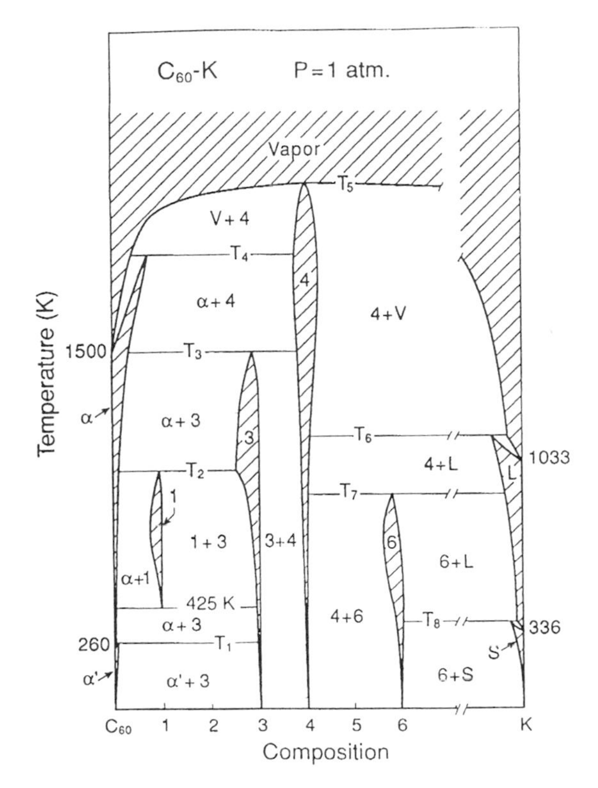

Fig. 8.5. Provisional binary phase diagram for K_xC_{60} at 1 atm pressure. Experimentally determined temperatures are indicated, and unknown temperatures of predicted transformations are labeled $T_1, \ldots, T_8$. [8.106].

The experimental temperatures given for the phase boundaries in Fig. 8.5 are for K-doped C_{60}, but the phase diagram is believed to be approximately applicable to the Rb_xC_{60} compounds also [8.79]. The phases corresponding to K_4C_{60} and K_6C_{60} are also observed for the Cs_xC_{60} compounds. Some underlying assumptions for the phase diagram in Fig. 8.5 are that there is no liquid phase and that dilute amounts of K in C_{60} will raise the solid-to-vapor transformation temperature, producing a peritectic-like transformation at T_4. The phase fields of K_1C_{60}, K_3C_{60}, K_4C_{60}, and K_6C_{60} are drawn schematically, since the widths remain to be determined. Areas of two-phase coexistence for equilibrium samples are also shown in Fig. 8.5. It has been reported that the stability of K_4C_{60} exceeds that of the other fullerides, based on decomposition experiments [8.106] and consistent with Fig. 8.5. The effect of C_{60} addition to a solid-state solution with potassium K(S) or in solution with liquid potassium K(L) is assumed to raise the S–L and L–V transformation temperatures relative to pure K. In the diagram, V, L, and S refer to the vapor, liquid, and solid phases of potassium. Between 425 K $< T < T_2$, Fig. 8.5 shows phase separation into a C_{60} and a K_1C_{60}

Table 8.1

Radii for various metal ions.[a]

Alkali metal	r_i (Å)	Alkaline earth	r_i (Å)
Li^+	0.76	Be^{+2}	0.30
Na^+	1.02	Mg^{+2}	0.65
K^+	1.38	Ca^{+2}	1.00
Rb^+	1.52	Sr^{+2}	1.13
Cs^+	1.67	Ba^{+2}	1.34
Fr^+	1.75	Ra^{+2}	1.37

[a]The radii of the cavities (or voids) available to tetrahedral and octahedral sites (see Fig. 8.1) are 1.12 Å and 2.07 Å, respectively.

phase for compositions $x < 1$ and into K_1C_{60} and K_3C_{60} coexistence for $1 < x < 3$. The diagram indicates that for $T < 425$ K and $0 < x < 3$, phase separation into the phases C_{60} and K_3C_{60} occurs [8.7, 106]. The diagram also distinguishes between the two crystal structures for C_{60} below and above $T_{01} = T_1 = 260$ K. Between $3 < x < 4$, phase separation into the fcc K_3C_{60} phase is shown coexisting with the body-centered tetragonal (bct) phase (stable at K_4C_{60}). Finally, for $4 < x < 6$, the bct and bcc phases coexist with the bcc phase stable at M_6C_{60}. The stability regions for each of the phases are shown schematically in Fig. 8.5.

In alkali metal doping, charge transfer of one electron per dopant atom to the C_{60} molecule occurs, resulting in dopant ions at the tetrahedral and/or octahedral interstices of the cubic C_{60} host structure (see Fig. 8.1). Values for the ionic radii of various alkali metal ions which serve as donor dopants for C_{60} are given in Table 8.1, and the relation of these radii to the tetrahedral and octahedral voids or cavities is also given in the table. For the tetrahedral sites, the radius of the cavity available to an ion is 1.12 Å (slightly smaller than the ionic radius, r_i, of K^+), while for an octahedral site, the cavity radius is 2.07 Å (considerably larger than r_i for K^+) [8.107]. The K^+ ionic radius is 1.38 Å in solid K_3C_{60} [8.107], compared with $r_i = 1.33$ Å for the K^+ ion in ionic salts, and compared with 1.03 Å, which is half the thickness of the potassium layer in the stage 1 graphite intercalation compound C_8K [8.78].

An important factor affecting the metal ion uptake in doping fullerene crystals is the size of the metal ion (see Table 8.1), which plays a major role in determining the lattice constant. In some cases the crystal structure of the doped fullerides is also dependent on the alkali metal species, as, for example, the Na_xC_{60} compounds [8.108] (see §8.5.3). In particular, the ionic radius of the species occupying the tetrahedral (rather than octahedral) sites most sensitively determines the lattice constant of the doped fulleride [8.109]. The shortest distance between a K^+ ion at a tetrahedral site and the adjacent C_{60} anion is 2.66 Å [8.107]. It is of interest to note that the Cs^+ ion is too large to form a stable Cs_3C_{60} compound with the fcc structure unless pressure is applied [8.90]. If no pressure is applied, Cs_3C_{60} crystallizes in a bcc structure, which is not superconducting (see §15.1).

Discussion of the doping of crystalline C_{60} with alkaline earth dopants is given in §8.6. Since the cavities in the C_{60} crystal structure are already large enough to accommodate alkali metal or alkaline earth metals, only a small lattice expansion occurs due to the intercalation of alkali metal species into fullerene crystals, in contrast to the behavior observed in graphite intercalation compounds (see §2.14) or in transition metal dichalcogenide intercalation compounds [8.110], where the lattice expansion is large (factors of 2 or 3 typically).

The effect of the addition of alkali metal dopants into the C_{60} lattice does expand the lattice somewhat, as can be seen in Tables 8.2 and 8.3. The larger the ionic radius, the larger is the lattice expansion, although in some cases, lattice contraction is seen because of the electrostatic attraction between the charged alkali metal cations and the C_{60} anions. Since the fcc lattice is compact and close-packed, higher doping levels or larger dopant radii favor the more open bcc lattice. Table 8.2 lists lattice constants for mostly nonsuperconducting alkali metal–doped fullerides but includes the alkaline earth and rare earth superconductors (see §8.6). Listed separately in Table 8.3 are the lattice constants for the alkali metal M_3C_{60}-related compounds, most of which form superconductors.

8.5.1. M_3C_{60} Alkali Metal Structures

Because of the discovery of relatively high T_c superconductivity in the M_3C_{60} (M = K, Rb) compounds, the M_3C_{60} system has become the most widely studied of the doped C_{60} compounds, and therefore more is known about the M_3C_{60} phase than the other phases shown in Fig. 8.5. For the composition M_3C_{60}, the metallic crystals for K_3C_{60} and Rb_3C_{60} have the basic fcc structure shown in Fig. 8.1(d). Within this structure, the alkali metal ions can sit on either tetrahedral $(\frac{1}{4}, \frac{1}{4}, \frac{1}{4})$ sites, where the metal ion is surrounded by four neighboring C_{60} molecules, or octahedral $(\frac{1}{2}, 0, 0)$ sites, where the metal ion is coordinated to six nearest-neighbor C_{60} molecules. The tetrahedral sites are twice as numerous as the octahedral sites. Surrounding each C_{60}^{3-} anion in M_3C_{60} compounds are 14 dopant ions, 8 of which are in tetrahedral sites, and 6 are in octahedral sites. As shown in Fig. 8.1(d), alkali metal atoms can be placed only on tetrahedral (c) sites giving the stoichiometry M_2C_{60}, or only on octahedral (b) sites giving the

Table 8.2

Lattice constants for stable C_{60}-related crystalline materials [8.7, 8].[a]

M_xC_{60}	Phase	Lattice constants (Å)	References
C_{60}	fcc ($Fm\bar{3}m$)	14.161	[8.90]
K_1C_{60}	fcc (rock salt)	14.07	[8.105]
Rb_1C_{60}	fcc (rock salt)	14.08	[8.105]
Cs_1C_{60}	fcc (rock salt)	14.12	[8.105]
Na_2C_{60}	sc ($Pa\bar{3}$)	14.189	[8.83]
Na_6C_{60}	fcc ($Fm\bar{3}m$)	14.380	[8.83]
$Na_{11-x}C_{60}$	fcc ($Fm\bar{3}m$)	14.59	[8.82]
K_4C_{60}	bct[b]	11.886(a), 10.774(c)	[8.90]
Rb_4C_{60}	bct[b]	11.962(a), 11.022(c)	[8.90]
Cs_4C_{60}	bct[b]	12.057(a), 11.443(c)	[8.90]
Ca_5C_{60}[d]	fcc ($Fm\bar{3}m$)	14.01	[8.1]
Ba_3C_{60}	bcc ($Pm\bar{3}n$)	11.34[c]	[8.111]
Ba_6C_{60}[d]	bcc ($Im\bar{3}$)	11.171[c]	[8.111]
K_6C_{60}	bcc ($Im\bar{3}$)	11.385[c]	[8.112]
Rb_6C_{60}	bcc ($Im\bar{3}$)	11.548[c]	[8.8]
Cs_6C_{60}	bcc ($Im\bar{3}$)	11.790[c]	[8.8, 112]
Sr_3C_{60}	bcc ($Pm\bar{3}n$)	11.140[c]	[8.93]
Sr_6C_{60}[d]	bcc ($Im\bar{3}$)	10.975[c]	[8.93]
Yb_3C_{60}[d]	sc ($Pa\bar{3}$)	13.954	[8.77, 113]

[a] See also Table 8.3, where the lattice constants, symmetries, and superconducting transition temperatures T_c for M_3C_{60} and related crystals are listed.

[b] For the (bct) body-centered tetragonal structure ($I4/mmm$), the in-plane and c-axis lattice constants are denoted by a and c, respectively.

[c] To compare $a_{\rm fcc}$ to $a_{\rm bcc}$, use the relation $a_{\rm bcc} = a_{\rm fcc}/\sqrt{2}$.

[d] The alkaline earth compounds Ca_5C_{60}, Ba_6C_{60}, Sr_6C_{60} and the rare earth Yb_3C_{60} are superconducting (see §15.1), with T_c values of 8.0 K, 6.2 K, 4.0 K, and 4.0 K, respectively.

stoichiometry M_1C_{60} (see Table 7.2 for site symmetries). If both tetrahedral and octahedral sites are fully occupied, then the composition M_3C_{60} is achieved [8.107]. Referring to Table 7.2, we see that the tetrahedral sites (c) and the octahedral sites (b) do not lower the T_h^3 symmetry of the space group (as discussed in §7.1.2) associated with C_{60} anions, provided that all the tetrahedral and all the octahedral sites are occupied [8.70, 121, 122]. The space group describing the crystal structure for K_3C_{60} and Rb_3C_{60} has been given as $Fm\bar{3}m$, the same structure as for C_{60} above T_{01} (see §7.1.1) [8.123]. The $Fm\bar{3}m$ accurately describes the structure for K_3C_{60} and Rb_3C_{60} if orientational effects of the C_{60} molecules are ignored and the molecules are considered as spheres.

Table 8.3

Lattice constants and superconducting T_c values for alkali metal M_3C_{60}, C_{60}, and related compounds.[a]

M_3C_{60}	Dopant site sym.[b]	a (Å)	T_c (K)	References
C_{60}	—	14.161	—	[8.90]
Li_2RbC_{60}	Li(T)Li(T)Rb(O)	13.896	<2.0	[8.91]
Li_2CsC_{60}	Li(T)Li(T)Cs(O)	14.120	12.0	[8.114]
Na_2KC_{60}	Na(T)Na(T)K(O)	14.025	—	[8.115]
Na_2RbC_{60}	Na(T)Na(T)Rb(O)	14.028	3.5	[8.115]
Na_2CsC_{60}	Na(T)Na(T)Cs(O)	14.134	12.5	[8.115]
KRb_2C_{60}	K(T)Rb(T)Rb(O)	14.243	23.0	[8.91]
$K_{1.5}Rb_{1.5}C_{60}$	O and T	14.253	22.2	[8.90]
K_2RbC_{60}	K(T)K(T)Rb(O)	14.243	23.0	[8.91, 116]
K_2CsC_{60}	K(T)K(T)Cs(O)	14.292	24.0	[8.91]
K_3C_{60}	O and T	14.240	19.3	[8.91]
$RbCs_2C_{60}$	Rb(T)Cs(T)Cs(O)	14.555	33.0	[8.91]
Rb_2KC_{60}	O and T	14.323	27	[8.91]
Rb_2CsC_{60}	Rb(T)Rb(T)Cs(O)	14.431	31.3	[8.90]
Rb_3C_{60}	O and T	14.384	29	[8.91]
Cs_3C_{60}[c]	O and T	—[c]	40.0	[8.117]
$Na_2Cs(NH_3)_4C_{60}$	O and T	14.473	29.6	[8.118, 119]
$K_3(NH_3)C_{60}$[d]	O and T	14.520	—	[8.120]

[a]All superconducting compounds in this table crystallize in the fcc structure (except for Cs_3C_{60}), neglecting the orientation of molecules with respect to crystal axes (see text). However, the compounds LiK_2C_{60}, Li_2KC_{60}, and NaK_2C_{60} are multiphase, and KCs_2C_{60} is unstable [8.115].

[b]The octahedral and tetrahedral site symmetries are denoted by O and T, respectively.

[c]The T_c for Cs_3C_{60} increases under application of pressure, with a maximum T_c of ~40 K achieved at ~12 kbar [8.117]. However, the crystal structure of the superconducting phase for Cs_3C_{60} is not believed to be fcc [8.117].

[d]Orthorhombic space group $P2221$, which is distorted from the fcc structure common to M_3C_{60} compounds. The value of 14.52 Å given in the table is an average of the a, b, c lattice constants of the orthorhombic structure, $a = 14.971$ Å, $b = 14.895$ Å, and $c = 13.687$ Å [8.120].

As discussed in §7.1.3, various interactions determine the orientation of the C_{60} anions in crystal structures, the most important being an orientational potential consisting of a van der Waals attractive force between the C_{60} anions and a short-range Lennard–Jones repulsive interaction, as shown in Fig. 8.6(a), where a potential minimum is seen at ~23°, corresponding to the orientation of C_{60} molecules in the $Pa\bar{3}$ structure (see §7.1.3). For alkali metal–doped compounds, two additional alkali metal–C_{60} anion interactions must be considered. Shown in Fig. 8.6(c) is the contribution to the orientational potential from the attractive electrostatic alkali metal–C_{60} anion force, which tends to orient the C_{60} anions to lie close to an

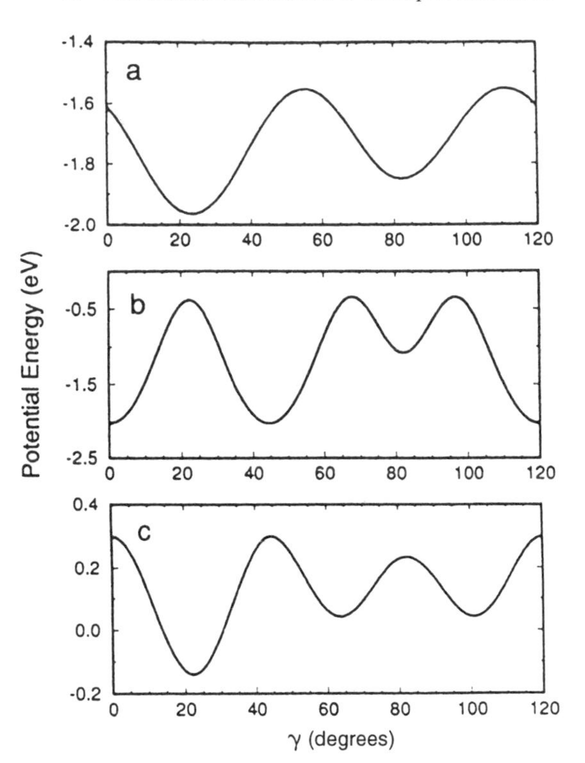

Fig. 8.6. Contribution to the orientational potential in cubic close-packed C_{60} and alkali metal M_xC_{60} intercalation compounds. The orientational dependence is with respect to rotation of a single C_{60} anion through an angle γ about a [111] axis. (a) Intermolecular C_{60}–C_{60} contribution including an atom-centered Lennard–Jones repulsive potential plus a Coulomb potential which is necessary to reproduce the orientational alignment in C_{60} itself. (b) Repulsive M–C_{60} interaction due to core orbital overlap with parameters chosen for K_3C_{60}. (c) Attractive M–C_{60} interaction due to charge transfer with parameters chosen for Na_2C_{60} [8.108, 123, 124].

electron-rich C=C double bond. This effect contributes to orienting the C_{60} molecules and also restrains their rotational degrees of freedom [8.7]. Thus K_3C_{60} and Rb_3C_{60} do not have rapidly rotating molecular anions at room temperature, as do the undoped C_{60} molecules. Finally, a repulsive interaction between the alkali metal cations and the C_{60} anions due to the size of the metal cations [see Fig. 8.6(b)] is important, especially for the larger ions (K^+ and Rb^+), tending to orient the C_{60} anions so that the hexagonal faces are adjacent to the alkali metal ions. Because the repulsive metal-C_{60} interaction is large for K_3C_{60} and Rb_3C_{60}, their crystal structures differ from that for the Na_xC_{60} compounds, where this contribution to the orientational potential is much less important (see §8.5.3).

The crystal structure resulting from consideration of orientational alignment effects in the case where the repulsive interaction shown in Fig. 8.6(b) is important is illustrated in Fig. 8.7. In this figure the C_{60} anions are seen to align so that twofold axes of the C_{60} anions are along x, y, and z axes. As discussed in §7.1.2, the point group T_h is a subgroup of I_h so that the alignment for each molecule as shown in Fig. 8.7 is consistent with the highest-symmetry alignment of the C_{60} molecules in the fcc basic crystal

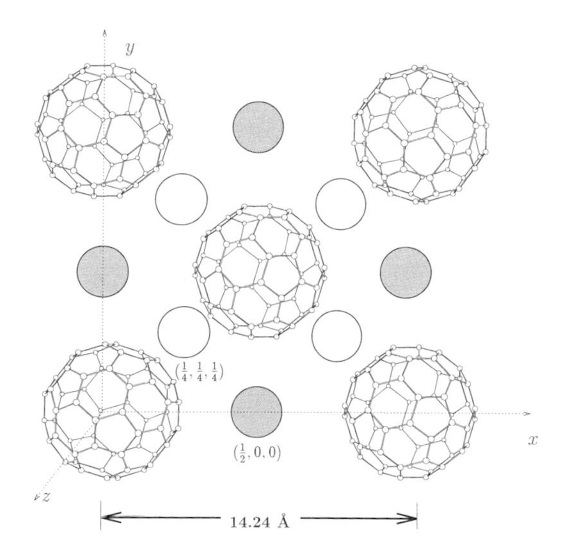

Fig. 8.7. Crystal structure of the K_3C_{60} and Rb_3C_{60} compounds. The open and dark spheres represent the alkali atom at tetrahedral and octahedral sites, respectively. Note the two possible orientations of the C_{60} molecules. Site locations for tetrahedral and octahedral sites relative to the origin (center of lower left C_{60} anion) and the cubic lattice constant for K_3C_{60} are given [8.107].

structure. However, as also pointed out in §7.1.2, the point group T_h has no fourfold axes, so that there are two ways to align the twofold axes of the C_{60} anions, and these two orientations are shown more explicitly in Fig. 8.8. Keeping the same orientation for the z-axis, we see that the x- and y-axes can have two different orientations: either (a) or (b, c) as shown in Fig. 8.8, where (b, c) are equivalent. To arrive at orientation (b, c) from orientation (a) in Fig. 8.8, we can rotate the C_{60} anion by $\pi/2$ about the z-axis to obtain (b), or by $\sim 44.5^\circ$ about the [111] direction to obtain (c). The shaded pentagon in Fig. 8.8 helps to follow the rotations in going from (a) to (b) or from (a) to (c). Although the symmetry of the structure in Fig. 8.7 is lower than $Fm\bar{3}m$, the literature often uses the $Fm\bar{3}m$ space group designation for the K_3C_{60} and Rb_3C_{60} structures, based on the argument that both orientations (a) and (b, c) occur with equal probability.

Because of the equivalence, the equal probability, and the random occurrence of each of the two standard orientations (usually called A and B) for the twofold axes in crystalline K_3C_{60} and Rb_3C_{60}, we say that these M_3C_{60} compounds exhibit merohedral disorder [8.124], in analogy with the merohedral disorder of C_{60} (see §7.1.4). Because of the merohedral disorder, the transfer integral for charge transfer between the C_{60}^{3-} anions has a random component, which can significantly increase the carrier scattering in the intermolecular hopping process. This merohedral disorder leads to relatively high values for the residual resistivity of the K_3C_{60} and Rb_3C_{60} compounds at low temperature. Explicit calculations of this residual temperature-independent contribution to the low-temperature re-

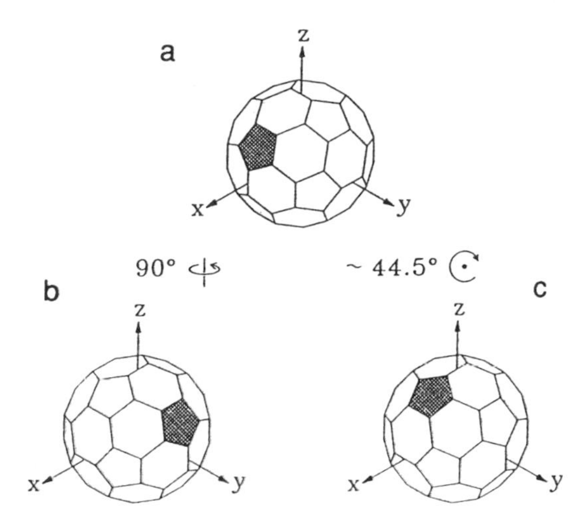

Fig. 8.8. Two "standard" orientations for the C_{60} anions (a) and (b, c) in M_3C_{60} (M = K or Rb), with respect to a fixed set of crystallographic axes. The plane of each drawing is normal to the [111] direction. The transformation from (a) to (b) involves a 90° rotation about the [001] direction. The transformation from (a) to (c) involves a ~44.5° rotation about the [111] direction. One of the pentagons is shaded to illustrate these rotations. Orientations shown in (b) and (c) are equivalent [8.127].

sistivity, due to the merohedral disorder mechanism [8.125], yield good agreement with experimental values of the residual resistivity for K_3C_{60} and Rb_3C_{60} (see §14.1.2). In diffraction experiments, diffraction patterns consistent with a fourfold axis have been reported [8.107] due to the rapid molecular reorientations and the merohedral disorder. Neutron diffraction experiments show that merohedral disorder persists down to very low temperatures (12 K) [8.126].

Using A and B to denote the two equivalent "standard" orientations of the C_{60} molecules in the K_3C_{60} and Rb_3C_{60} structures (Figs. 8.7 and 8.8), it has been proposed [8.128–130] that some correlations should exist between the molecular orientations of the C_{60} anions on nearest-neighbor sites, but that no long-range orientational correlation should be present. Some experimental evidence for such short-range correlations has been reported, based on detailed analysis of neutron diffraction profiles and the pair distribution function of Rb_3C_{60} looking for deviations from strict merohedral disorder, for which there are no nearest-neighbor correlations between molecules in the standard orientations A and B (see Fig. 7.6). The two standard orientations differ by a $\pi/2$ rotation about the $\langle 001 \rangle$ axis [8.131].

In connection with the merohedral disorder, Stephens pointed out that the site ordering of the dopants would introduce three inequivalent sites for the hexagon rings of the C_{60}^{3-} anions (see Fig. 7.2) [8.107]. These three inequivalent sites have been experimentally corroborated in Rb_3C_{60}, based on ^{87}Rb-NMR studies of the chemical effect [8.132], and in K_3C_{60} based on NMR studies of the jump rotation of the ^{13}C nuclei [8.133]. It is concluded from these experiments that the repulsive alkali metal–C_{60} interaction is important in K_3C_{60} and Rb_3C_{60} and plays a major role in determining the orientation of the molecule in the crystal lattice.

8.5.2. M_1C_{60}, M_4C_{60}, and M_6C_{60} Structures (M = K, Rb, Cs)

Three other stable phases are observed for M_xC_{60} (namely, MC_{60}, M_4C_{60} and M_6C_{60} for M = K, Rb, and Cs), and each has a different range of stability and crystalline structure (see Fig. 8.1), in accordance with the phase diagram of Fig. 8.5. The crystal structures of these phases are discussed below.

The MC_{60} alkali metal phase is stable at elevated temperatures only for a limited temperature range (410–460 K). This phase has been observed with M = Na, K, Rb, and Cs, where the M ion is in an octahedral site, thereby forming a rock-salt (NaCl) crystal structure [8.105, 108, 123, 134–136]. The reported lattice constants for $K_{1.4}C_{60}$, $Rb_{0.9}C_{60}$, CsC_{60} are 14.07 Å, 14.08 Å, and 14.12 Å, respectively, somewhat smaller than that for undoped C_{60} (14.16 Å) [8.105]. When cooled below ~100°C, the rock-salt structure becomes distorted to a pseudo–body-centered orthorhombic phase, with lattice constants for Rb_1C_{60} $a = 9.138$ Å, $b = 10.107$ Å, and $c = 14.233$ Å. The distance between the C_{60} molecules is very short along the a direction, as shown in Fig. 8.9, and a polymeric cross-linking between the C_{60} molecules has been reported for both K_1C_{60} and Rb_1C_{60} [8.137]. The differences in

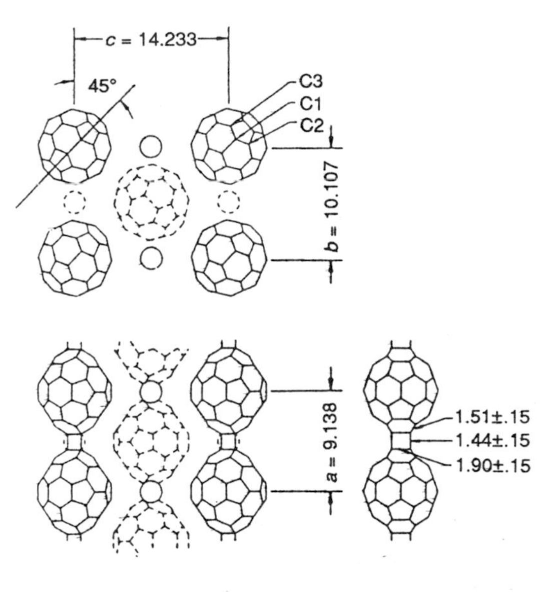

Fig. 8.9. Atomic structure of the polymeric Rb_1C_{60} phase determined from x-ray diffraction patterns [8.137]. Solid (broken) lines show fullerenes and cations centered at $x = 0$ ($x = 1/2$) in the b–c plane at the top, and at $y = 0$ ($y = 1/2$) in the a–c plane at the bottom. To the right is a projection of one chain perpendicular to the molecular mirror plane.

bond lengths in the region of the cross-linkage relative to the ideal C_{60} molecule are shown in Fig. 8.9 in the lower right corner [8.137]. Presently, there is no concensus about whether the M_1C_{60} phase is metallic or semiconducting. In the slow-cooled Rb_1C_{60} phase, infrared transmission studies have been interpreted in terms of enhanced conduction along the *a* direction where C_{60} cages are coupled by [2+2] cycloaddition bonds [8.137]. One feature of practical importance regarding the polymerized phase shown in Fig. 8.9 is the observation that this doped Rb_1C_{60} phase is stable in air, unlike all other reported alkali metal–doped C_{60} phases, which are highly reactive under ambient conditions. Regarding polymerization of M_1C_{60}, it is believed that slow cooling to room temperature (~25°C) results in a chain polymer which retains inversion symmetry, whereas rapid quenching to temperatures below room temperature gives rise to dimers and no inversion symmetry in the lattice [8.137–139], as evidenced by Raman and infrared spectra [8.140] (see §11.6.1). This orthorhombic crystal structure tends to be highly strained, has a very low heat of transformation, and transforms reversibly at 200°C to the rock-salt phase [8.137]. Presently available samples in the rock-salt phase show a high vacancy concentration at the metal dopant sites. The large volume available to the M ion in the rock-salt structure suggests that the C_{60} anions should have rotational freedom similar to that for undoped C_{60}. A stable NaC_{60} phase has been reported [8.108], which shows a phase transition to a simple cubic structure below 320 K, similar to the undoped C_{60} crystal phases above and below T_{01}.

The M_4C_{60} phase [see Fig. 8.1(f)], which appears to be difficult to prepare, has a body-centered tetragonal (bct) structure with a space group $I4/mmm$ [8.7, 112]. Some values for the structural parameters for K_4C_{60} and Rb_4C_{60} are given in Table 8.4. For Rb_4C_{60}, the nearest-neighbor C_{60}–C_{60} distance is 10.10 Å, and the closest approach between a C_{60} anion and an alkali metal cation is 3.13 Å, yielding a volume per unit cell of

Table 8.4

Unit cell dimensions and bond distances for K_4C_{60} and Rb_4C_{60} in the $I4/mmm$ structure.[a]

Compound	a (Å)	c (Å)	C_{60}–C_{60} (Å)	M–C_{60} (Å)	V (Å^3)[b]	Reference
K_4C_{60}	11.84	10.75	9.95	3.33	753	[8.112]
Rb_4C_{60}	11.96	11.04	10.10	3.13	789	[8.141]

[a] The location of the K ion in K_4C_{60} is at (0.21, 0.5, 0), close to the ideal location (0.25, 0.5, 0).

[b] The primitive unit cell volume V is found by $V = a^2c/2$, and the C_{60}–C_{60} distance by $[2(a/2)^2 + (c/2)^2]^{1/2}$.

$V = 7.89 \times 10^{-22}$ cm^3 [8.141]. In the lower-symmetry body-centered tetragonal phase [8.90], the C_{60} molecules are believed to occupy sites with C_{2v} symmetry. Orientational disorder is expected for the C_{60} anions relative to the alkali metal ions residing in distorted tetrahedral sites, and attempts have been made to account for this disorder on the basis of merohedral disorder [8.142]. Orientational rearrangement of the C_{60} anions in the M_4C_{60} alkali metal compounds is expected to occur rather easily [8.143].

The alkali metal–saturated compound M_6C_{60} for M = K, Rb, Cs has a body-centered cubic (bcc) structure, as shown in Fig. 8.1(g), and corresponds to the space group T_h^5 or $Im\bar{3}$. For this semiconductor, six electrons are transferred from the M ions to the C_{60} molecule to produce a C_{60}^{6-} anion with a filled t_{1u}-derived band [8.84] (see §12.7.3). In the M_6C_{60} structure, the alkali metal ions all occupy equivalent distorted tetrahedral sites, which are surrounded by four C_{60} anions, two with adjacent pentagonal faces and two with adjacent hexagonal faces. In contrast to other crystalline M_xC_{60} compounds, the C_{60} anions in the bcc M_6C_{60} alkali metal compounds are orientationally ordered as shown in Fig. 8.10 [8.112, 142], and this orientational ordering is maintained up to at least 520 K. The point group for the C_{60}^{6-} anion site in this structure is T_h (see §7.1.2). In this structure each C_{60} anion is surrounded by 24 M cations. Values for some of the structural parameters for the K_6C_{60}, Rb_6C_{60}, and Cs_6C_{60} crystal phases obtained from Rietveld refinement of x-ray diffraction data are given in Table 8.5. The C–M distances in Table 8.5 are all larger than the sum of the van der Waals carbon radii and the alkali metal radii r_i in Table 8.1. Also the M–M nearest-neighbor distances are all significantly larger than their respective ionic radii, as shown in Table 8.5. For the case of Rb_6C_{60}, the nearest-neighbor C_{60}–C_{60} distance is 9.99 Å, which is less than 10.03 Å for undoped C_{60}, while the closest C_{60}^{3-} approach to a Rb^+ cation is 3.42 Å, which can be contrasted to the C–M distance of 3.25 Å in Table 8.5; the volume per primitive unit cell for Rb_6C_{60} is 7.68×10^{-22} cm^3 which can be contrasted with the corresponding value of 7.89×10^{-22} cm^3 in Rb_4C_{60}, given in Table 8.4 [8.141].

8.5.3. *Na_xC_{60} Structures*

Since the ionic radii of the alkali metal ions Na^+ and Li^+ are both much smaller than the K^+, Rb^+, and Cs^+ ions, the repulsive term in the orientational potential [Fig. 8.6(b)] is much smaller for the small Na^+ and Li^+ ions, and consequently, a number of unusual structural phenomena might be expected and are indeed observed. First, a stable compound Na_2C_{60} is formed [see Fig. 8.1(b)], where the dopants fill the tetrahedral sites, leaving the octahedral sites unoccupied [8.83]. Second, it is possible to fit more

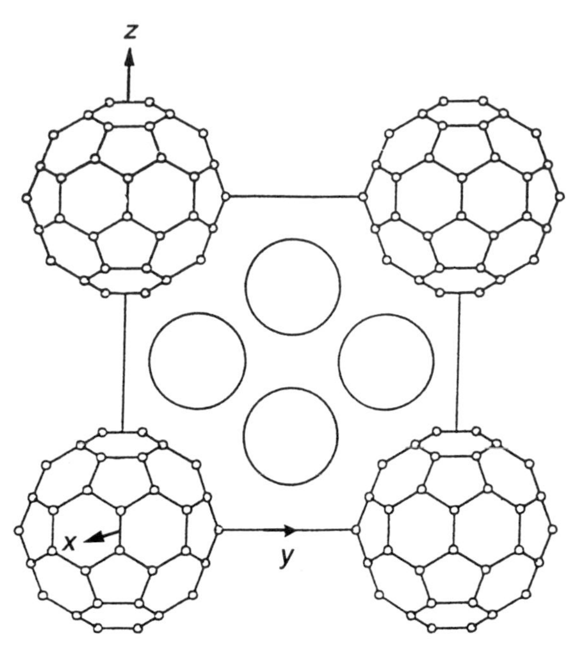

Fig. 8.10. Part of the Cs_6C_{60} structure projected on the cube face normal to the [100] direction. Large open circles represent Cs^+ cations on the same scale as the cube edge. Small circles denote C atoms, not to scale. In this bcc structure, an equivalent C_{60} molecule is centered at $(\frac{1}{2}, \frac{1}{2}, \frac{1}{2})$ (not shown). The Cs coordinates (clockwise from the top) obtained from x-ray data are $(0, \frac{1}{2}, 1-\delta)$, $(0, \frac{1}{2}+\delta, \frac{1}{2})$, $(0, \frac{1}{2}, \delta)$, and $(0, \frac{1}{2}-\delta, \frac{1}{2})$ where $\delta = 0.28$ with respect to an origin at the center of the C_{60}^{6-} anion at the bottom left corner. Faces normal to the y or z axes may be visualized by appropriately rotating the diagram by $\pm 90°$ [8.112].

than one Na^+ in an octahedral site [8.83, 114]. For example, the fcc structure can be preserved upon adding six Na atoms, with two Na ions going into tetrahedral sites and four Na atoms (not ions) into octahedral sites [see Fig. 8.1(h)] to form Na_6C_{60} as a stable fcc phase. This is in contrast to M_6C_{60} (M = K, Rb, Cs), which crystallizes in a bcc structure [see Fig. 8.1(g)] and where complete charge transfer is thought to lead to the formation of the hexanions C_{60}^{6-}. For Na_6C_{60} only about two electrons are donated by the total of six Na atoms, because of the large electrostatic energy that would be needed to stabilize four Na^+ ions in close proximity at a single octahedral site. The maximum stability for four Na atoms on an octahedral site is at the corners of a regular tetrahedron with respect to the site center [8.83, 144]. Only scant evidence is available for a stable Na_3C_{60} phase [8.109]. Most authors who have prepared materials with this stoichiometry report that Na_3C_{60} tends to phase separate (disproportionate) into Na_2C_{60} [Fig. 8.1(c)] and Na_6C_{60} [Fig. 8.1(h)] [8.83, 109, 144]. The small size of the Na atom also

Table 8.5

Some structural parameters for the bcc M_6C_{60} phases obtained from Rietveld refinement of x-ray data. C–M and M–M denote the near-neighbor carbon–metal and metal–metal distances, respectively. Values of ionic radii are taken from Table 8.1, ρ_m denotes the mass density, and V is the volume of the primitive unit cell [8.142].

M	a (Å)	C_{60}–C_{60} (Å)	C–M (Å)	M–M (Å)	r_i (Å)	ρ_m (g cm^{-3})	V (Å^3)
K	11.38	9.86	3.20	4.06	1.33	2.21	737
Rb	11.54	9.99	3.25	4.11	1.48	2.66	768
Cs	11.79	10.21	3.38	4.19	1.67	3.08	819

allows the intercalation of Na to high concentrations. For example, a stoichiometry as high as $Na_{11}C_{60}$ has been reported, for which nine Na atoms are expected to go into an octahedral site, as shown in Fig. 8.11 [8.145].

The Na_xC_{60} crystal structures also differ from those of the corresponding heavier alkali metal dopants with regard to the orientational alignment of the C_{60} anions. As mentioned above, the smaller size of the dopant species results in different relative magnitudes of the contributions to the orientational potential than for the case of the heavier alkali metals. For the Na_xC_{60} compounds the repulsive interaction between the alkali cations and the C_{60} anions is very much smaller than for the larger cations, thereby reducing the tendency for the alkali metal ions to lie close to the large hollows of the C_{60} shell (at the hexagonal faces). For example, the repulsive interaction between Na cations and C_{60} anions [shown in Fig. 8.6(b)] almost cancels the attractive Coulomb interaction [shown in Fig. 8.6(c)], so that the orientational alignment potential $V_A(\theta)$ for the Na_xC_{60} compounds is almost the same as for C_{60} itself (see §7.1.3). X-ray and neutron diffraction experiments are in agreement with this prediction, so that the Na_xC_{60} compounds tend to crystallize in the $Pa\bar{3}$ structure at low temperature, with an orientational phase transition occurring at T_{01} in analogy with the crystal phases of C_{60} [8.123].

8.5.4. *Structures for $M_2M'_1C_{60}$ and Related Compounds*

The $M_2M'C_{60}$ compounds [8.91, 109], where M = Li or Na is a small alkali metal ion and M′ = Rb or Cs is a large ion, show an important site ordering effect, placing the smaller M ions in tetrahedral sites and the larger M′ ions in octahedral sites. The orientational alignment is more sensitive to the dopant occupying the tetrahedral sites, because of their smaller cavity size. If the tetrahedral sites are occupied by dopants of small size (such as for Na_2RbC_{60}), then the repulsive interaction shown in Fig. 8.6(b) is less

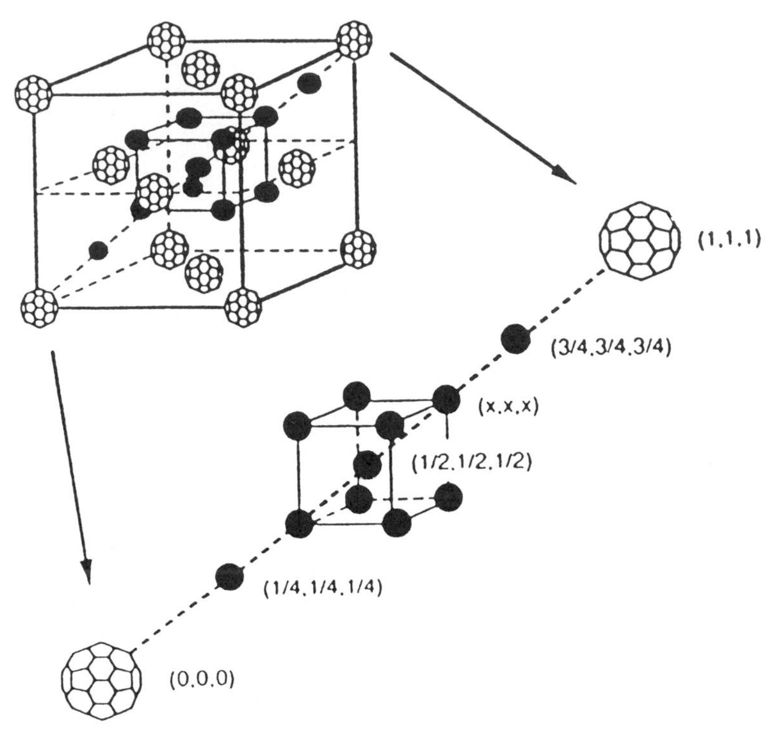

Fig. 8.11. Schematic representation of a proposed x-ray–derived structure of $Na_{11}C_{60}$ [8.144]. For clarity, isolated Na atoms on tetrahedral sites are shown along only one of the body diagonals, and only the octahedral cluster of Na atoms centered at $(\frac{1}{2}, \frac{1}{2}, \frac{1}{2})$ is shown. Note that the C_{60} molecules (shown on a reduced scale) are for simplicity orientationally ordered in this schematic diagram [8.145].

important, and for the case of Na ions the attractive and repulsive alkali metal–C_{60} anion interaction terms largely cancel and molecular alignment close to that for C_{60} itself results. Experimental evidence for such behavior in Na_2RbC_{60} has been reported in neutron scattering studies [8.146]. This effect is also similar to the behavior described in §8.5.3 for the Na_xC_{60} compounds [8.91, 147, 148]. Examples of site selectivity for the smaller and larger ions have been reported for the compound Rb_2CsC_{60} but not for $RbCs_2C_{60}$ [8.88, 149], and for K_2RbC_{60} but not for KRb_2C_{60} [8.116, 132], consistent with the above discussion. Of the various $M_2M'C_{60}$ compounds that become superconducting at low temperature, the Na_2RbC_{60} compound is the one having the best-established site selectivity (i.e., few Na ions occupy octahedral sites) [8.123]. In contrast, the Li_2CsC_{60} compound has recently been reported [8.150] to show a large amount of orientational disorder.

One unusual effect is that alkali metal doping with the alloy dopants Na_2Cs, Na_2Rb, Na_2K, and Li_2Cs results in a crystal with a closer C_{60}–C_{60} separation than is present in undoped C_{60} (see Table 8.2), presumably due to the attractive interaction between the metal ions and the fullerene molecules [8.141]. This interaction reduces the lattice constant of the $M_2M'C_{60}$ fcc phase relative to the value obtained for the M_3C_{60} (M = K, Rb, Cs) fcc binary phase.

8.5.5. *Structure of Ammoniated M_3C_{60} Compounds*

The x-ray structural analysis of two ammoniated alkali metal fulleride compounds $(NH_3)_4Na_2CsC_{60}$ [8.119] and $(NH_3)K_3C_{60}$ [8.120] has been reported. These compounds were initially prepared with the goal of increasing the lattice constant of the alkali metal $M_{3-x}M'_xC_{60}$ compounds and thereby increasing their superconducting transition temperatures (see §15.1). Since no significant change in charge transfer is anticipated upon adding the (NH_3) groups, it is appropriate to relate these compounds to the alkali metal–doped C_{60} compounds.

The structure of the $(NH_3)_4Na_2CsC_{60}$ compound has been identified with the $Fm\bar{3}$ space group and a lattice constant of a = 14.473 Å has been reported [8.119], with the Cs and half of the Na alkali metal dopants located on tetrahedral sites, and the remaining Na ions are located near the center of a tetrahedron of the (NH_3) molecules on an octahedral site. The nearest-neighbor distances are N–Na(O)=2.5 Å, N–N=2.89 Å, N–Cs(T)=3.76 Å, N–C=2.81 Å, N–H=1.02 Å, Cs(T)–C=3.37 Å, and H–C=2.29 Å [8.119], where (T) and (O), respectively, refer to tetrahedral and octahedral sites (see Table 8.3).

In contrast, the crystal structure of $(NH_3)K_3C_{60}$ has been described by an orthorhombic distortion of an fcc structure [8.120], with one K^+ and one NH_3 per distorted octahedral site, and with lattice constants $a = 14.971$ Å, $b = 14.895$ Å, $c = 13.687$ Å. Some of the bond lengths for this compound are K–C(3) = 3.20 Å, K–N=2.68 Å, K–C(2)= 3.05 Å, K–C(1)= 3.48 Å, where C(i), $i = 1, 2, 3$, denote distinct carbon sites in the lattice (see Fig. 7.3). The departure of $(NH_3)K_3C_{60}$ from a cubic crystal structure may be responsible for the absence of superconductivity in this compound [8.120], although it may be speculated that external pressure might restore the fcc structure and perhaps also restore superconductivity.

8.6. Structure of Alkaline Earth–Doped C_{60}

The structures of the alkaline earth–doped C_{60} compounds have been studied both experimentally and theoretically, because of their superconducting properties, as well as their interesting normal state properties. A variety of crystal structures and phases for M_xC_{60} (M = Ca, Sr, Ba) have been observed. The divalence of these ions in the tetrahedral and octahedral sites leads to a partial filling of the broadened (LUMO+1) t_{1g}-derived electronic energy band of C_{60} and leads to semimetallic behavior in the Sr_6C_{60} and Ba_6C_{60} compounds [8.143] (see §12.7.4).

8.6.1. Ca_xC_{60} Structure

The structure of the alkaline earth metal compound Ca_xC_{60} (for $x \leq 5$) follows the same space group $Pa\bar{3}$ as C_{60} below T_{01}. No phase transition to the $Fm\bar{3}m$ structure is seen all the way up to 400°C. In the $Pa\bar{3}$ structure, the Ca ions (nominally doubly charged) occupy both tetrahedral and octahedral sites. Because of the smaller size of the calcium ion (as also noted above for the Na^+ ion [8.83]), the octahedral sites can accommodate multiple Ca ions, and it is believed that for Ca_5C_{60} up to four Ca^{2+} cations are accommodated in a single octahedral site, despite the Coulomb repulsion between the Ca cations [8.1]. Since the charge transfer for the five Ca dopants/C_{60} is only about nine electrons (sufficient to fill all of the t_{1u}-derived band and approximately half of the t_{1g}-derived band), the average charge transfer is somewhat less than two electrons per Ca dopant (see §12.7.4). Also for Ca_5C_{60} the lattice constant $a_0 = 14.01$ Å is smaller than that for C_{60} prior to doping (see Table 8.2), again this must be associated with the attraction between the C_{60} anions and the charged Ca ions.

8.6.2. Sr_xC_{60} and Ba_xC_{60} Structures

The structures of strontium and barium intercalated fullerides Sr_xC_{60} and Ba_xC_{60} are more complicated than that of Ca_xC_{60}. The Sr_xC_{60} and Ba_xC_{60} systems are the only fulleride systems, thus far reported, where fcc and bcc phases compete in the same compositional range [8.93]. Near $x = 3$, a bcc A15 ($Pm\bar{3}n$) phase coexists with an fcc ($Fm\bar{3}$) phase, with lattice constants 11.140 Å and 14.144 Å for the bcc and fcc phases, respectively. Most of the Sr^{2+} ions in the fcc phase are on tetrahedral sites, and the ions on the octahedral sites are displaced from the central position and are off-centered along the {111} axes [8.93]. In the bcc ($Pm\bar{3}n$) phase for Sr_3C_{60}, the C_{60} anions are orientationally disordered. Studies as a function of x indicate that Sr_xC_{60} starts to transform from an fcc structure at small x to a bcc structure for $x < 3$. It is believed that the A15 phase is stabilized for the divalent alkaline earth cations because this structure allows the divalent cations to be located preferentially adjacent to the electron-deficient five-membered (pentagonal) rings [8.93].

Increasing the Sr dopant stoichiometry above $x = 3$ suppresses the fcc and A15 bcc structures and leads to the appearance of the bcc $Im\bar{3}$ phase, with $a_0 = 10.975$ Å for Sr_6C_{60}. This $Im\bar{3}$ phase is also found for Ba_6C_{60} and certain M_6C_{60} (M = K, Rb, Cs) alkali metal compounds (see §8.5.2). Figure 8.12 shows that the lattice constant for the $Im\bar{3}$ structure increases with increasing size of the dopant species for both the alkali metal and alkaline earth systems. This figure also shows that the lattice constant depends on the amount of charge transfer, which is physically reasonable, because of the greater attractive Coulomb interaction between the doubly charged alkaline earth ions and the C_{60} anions, bringing these anions even closer together. The C_{60}–C_{60} distance of 9.53Å for Sr_6C_{60} is the shortest C_{60}–C_{60} distance thus far reported for unpolymerized C_{60} systems [8.93]. The structural and electronic properties of the Sr_xC_{60} fullerides are intermediate between those of the Ca_xC_{60} and Ba_xC_{60} intercalated fullerides [8.151], consistent with size considerations for the dopant ions.

The structure of Ba_3C_{60} has also been investigated by x-ray powder diffraction techniques, yielding the A15 structure with a space group $Pm\bar{3}n$, typical of several BCS superconductors (such as Nb_3Sn) with T_c values above 10 K [8.111]. The lattice constant for the bcc A15 Ba_3C_{60} structure is 11.34 Å with the Ba ions separated from the three nearest-neighbor inequivalent carbons by 2.98 Å, 3.14 Å, and 3.39 Å, respectively. The nearest-neighbor carbon–carbon distance in this crystal structure is 1.47 Å, as determined by a detailed x-ray analysis [8.111]. In the A15

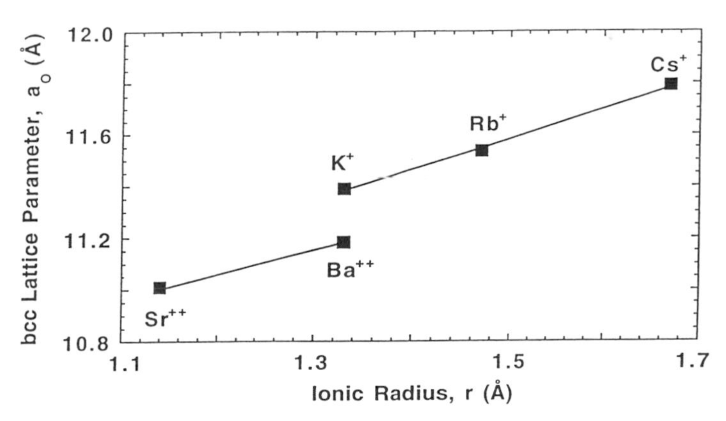

Fig. 8.12. The bcc lattice constant dependence on cation size is compared for alkali metal and alkaline earth fullerides. The discontinuity in the nearly monotonic dependence is probably associated with both the small size of Sr^{2+} and Ba^{2+} ions and the increased electrostatic attraction between these 2+ cations and the C_{60} anion sublattice giving rise to reduced lattice constants for the alkaline earth fullerides [8.93].

phase, C_{60} molecules rotationally orient themselves such that all cation sites of the lattice are surrounded by five-membered rings of the neighboring molecules. In contrast to the A15 crystal structure for Ba_3C_{60}, which is an insulator (see §14.1.3), the superconductor Ba_6C_{60} crystallizes in the bcc $Im\bar{3}$ space group with a lattice constant 11.171 Å. Unlike alkali metal–doped C_{60}, where each M atom transfers one electron to C_{60}, the alkaline earth metal dopants, as stated above, transfer two electrons per metal dopant for low metal concentrations, but lower charge transfer values are found for higher metal concentrations [8.1], as discussed in §12.7.4. The charge state of the metal ions in these intercalation compounds as a function of metal concentration is not as well established as for the alkali metals.

8.6.3. Rare Earth R_xC_{60} Structures

Closely related to the alkaline earth fullerides is the compound Yb_xC_{60} (see Table 8.2), which is also synthesized at relatively high temperatures (800–1000 K). The Yb cations are similar in size to Ca anions and occupy tetrahedral and octahedral sites in the C_{60} fcc structure. The lattice constant for the Yb_xC_{60} crystal structure is 13.92 Å, which is relatively small, reflecting the strong coupling between the dopant cations and the C_{60} anions [8.77, 113]. There is good reason to expect that other rare earth R_xC_{60} compounds will be prepared in the future.

8.7. Structure of Other Crystalline Fulleride Phases

The high electronegativity of pristine C_{60} hinders acceptor doping [8.143]. Consequently, relatively few C_{60}-based compounds have been synthesized with potential for hole charge transfer to the fullerene constituents. Thus, few structural studies have been made until now on such compounds, and the charge transfer is not well documented. Although it is expected that a number of crystalline fullerene-derived materials will be prepared with possible acceptors in time, it is less likely that charge transfer resulting in C_{60} cations will be found in such compounds. Two such fullerides have already been studied in some detail—$C_{60}I_4$ and $C_{60}O$ compounds—and for both materials, structural studies have been carried out, as reviewed below. Of the various compounds thus far prepared, the compounds with significant potential for acceptor charge transfer are those based on the reaction of $SbCl_5$, AsF_5, and SbF_5 with C_{60} [8.152, 153], although no structural studies of these fulleride compounds have yet been carried out.

8.7.1. *Halogen-Doped Fullerenes*

Solid-state ordered structures have been reported for C_{60} and iodine with a composition $C_{60}I_4$, for which the iodine molecules lie in planes between layers of C_{60} molecules [8.4, 5, 154], as shown in Fig. 8.13. A simple hexagonal structure (Fig. 8.13) with full site occupancy has been reported [8.4]. In this structure the C_{60} host material exhibits a sequence of (AAA...) layers, interleaved with guest iodine layers between the host layers, as in a graphite intercalation compound (see §2.14). The {111} planes of the fcc C_{60} become the {001} planes in the compound by relative shear motions which transform the layer stacking sequence from ABCABC... in fcc C_{60} to A/A/A/..., where / denotes an iodine guest layer in the $C_{60}I_4$ compound. The shortest I–I (in-plane) distance [see Fig. 8.13(b)] is 2.53 Å, which is close to the value of 2.72 Å for the molecular solid I_2 [8.155]. Mössbauer spectroscopy shows two distinct isomer shift patterns, which are consistent with the two different iodine site geometries within the unit cell as shown in Fig. 8.13(b) [8.145, 156]. Photoemission, x-ray absorption, and NMR results confirm the insulating (clathrate) nature of $C_{60}I_4$, i.e., the absence of guest–host charge transfer [8.5, 157–159].

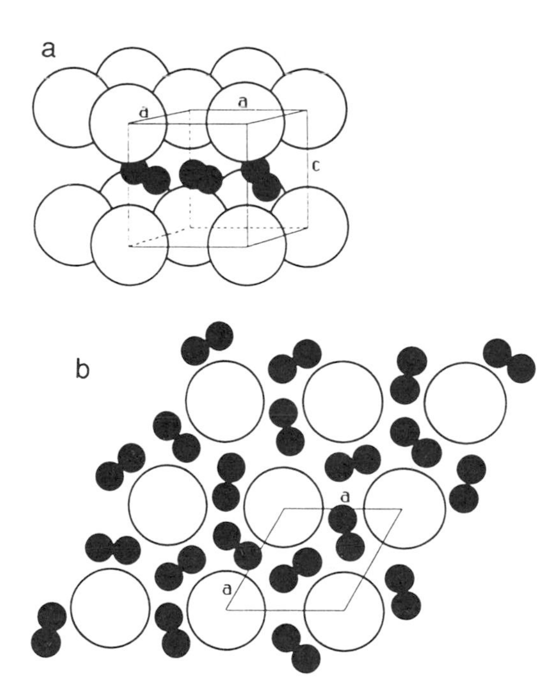

Fig. 8.13. Schematic diagram of the structure for $C_{60}I_4$ derived from analysis of x-ray diffraction patterns [8.4]. (a) A three-dimensional perspective view showing the C_{60} molecules as large white balls with a radius of ~ 3.4 Å, and I_2 molecules denoted by a pair of dark small circles, the radius of each small ball being ~ 1 Å. The drawing is scaled to a lattice constant of ~ 10 Å. (b) Basal plane projection of the $C_{60}I_4$ structure, using the same scaling. There is no long-range order in the orientation of the I_2 molecules on their sites, and the bonding between the I_2 and the C_{60} molecules is believed to be due to a van der Waals interaction, with no charge transfer [8.145].

The crystal structure of another iodine-doped layered phase C_{60}–I_2 [as in Fig. 8.13(a)] has also been reported [8.154]. In this case, the crystal structure is base-centered orthorhombic, with lattice constants $a = 17.33$ Å, $b = 9.99$ Å, and $c = 9.97$ Å, with an a/b ratio of approximately $\sqrt{3}$. The electrical resistivity of the compound under ambient conditions yielded a high value of 10^7 Ω-cm which could be decreased by about four orders of magnitude by applying a pressure of 20 GPa [8.154]. In addition, a C_{70}–I_2 phase has been identified [8.154], and has a bct structure with lattice constants $a = b = 10.13$ Å and $c = 17.00$ Å. The nearest-neighbor C_{70}–C_{70} distances are 10.13 Å, substantially smaller than for the corresponding distance in pristine C_{70} (10.58 Å) and the C_{70} molecule is oriented with the long axis in the c-direction.

Several bromine-doped fullerene phases have also been prepared and characterized [8.160] and are further discussed in §10.5. Calculations of the electronic structure of other proposed halogenated C_{60} systems [8.74, 143] suggest that several halogenated phases may be thermodynamically stable, but no evidence has been found for charge transfer in these compounds.

Recently the intercalation of C_{60} with the Lewis acids $SbCl_5$, SbF_5, and AsF_5 has been reported [8.152]. Although no structural measurements were made, nor are band calculations available for these systems, the experimental observation of greatly enhanced electrical conduction (see §14.1.6) raises the question that charge transfer in acceptor compounds of C_{60} might be possible. The C_{60} samples doped with these Lewis acids were characterized by mass spectrometry, infrared spectroscopy, and temperature-dependent resistivity measurements [8.152].

8.7.2. Fullerene Epoxides and Related Compounds

Another compound that has thus far been studied in considerable detail is $C_{60}O$, where the oxygen is covalently bonded to C_{60} [8.161]. NMR studies provide a sensitive technique for monitoring the formation of $C_{60}O$ through the appearance of additional NMR lines, since inequivalent carbon sites appear near the oxygen bridge bond to the C_{60} molecule (see §10.6). Figure 8.14 shows the epoxide structure for the $C_{60}O$ molecule with five inequivalent bond lengths. Values for these bonds for the low-temperature phase are 1.45, 1.39, 1.45, 1.55, and 1.43 Å for $B1, B2, B3, B4$, and $B5$, respectively [8.161].

Structural studies show that $C_{60}O$ crystallizes in the same crystal structure as C_{60} and that $C_{60}O$ undergoes a first-order phase transition at $T_{01} \sim 278$ K. In the high temperature fcc phase, $C_{60}O$ has four equivalent C_{60} molecules per cubic unit cell and $a_0 = 14.185$ Å, which is only slightly greater than a_0 for C_{60} (14.17 Å). For the high-temperature phase, the

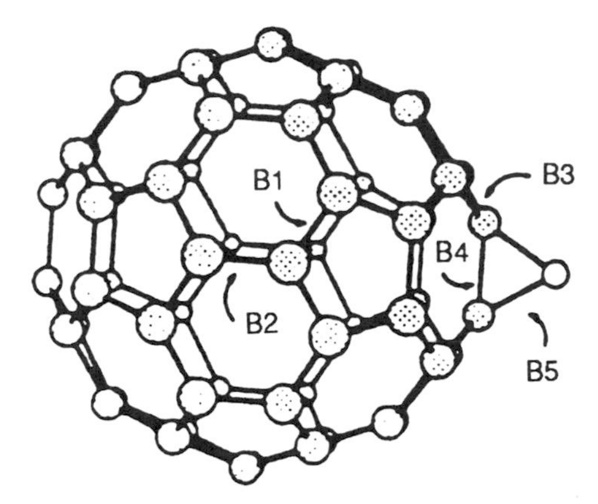

Fig. 8.14. Structure of the epoxide molecule $C_{60}O$. Labels indicate the inequivalent bonds used in the analysis of the low-temperature structure [8.161].

molecules have rather free rotational motion, while the low-T phase has only one degree of rotational freedom, corresponding to ratcheting motion about the main symmetry axis, and the nearest-neighbor C_{60}–C_{60} distance is 10.03 Å [8.161]. In the low-temperature phase the crystal structure is simple cubic with space group $Pa\bar{3}$, the same as for C_{60} itself. The transition temperature T_{01} is about 20 K higher in $C_{60}O$ than for a similar phase transition in C_{60} (see §7.1.3); the higher T_{01} value is attributed to the additional hindered rotational motion arising from the bridge bonding of the oxygen to nearby carbons (see §10.6) [8.161]. At the phase transition, the lattice constant decreases by 0.04 Å upon cooling, comparable to the contraction of C_{60} itself (see §14.10) [8.162]. At low temperature (19 K), the oxygen site attachment is statistically disordered with the six bonds directed toward the center of the octahedral voids having the highest probabilities for oxygen occupation. The oxygen tends to occupy the largest available spaces in the lattice. At room temperature approximately two thirds of the oxygen atoms point along ⟨100⟩ axes and one third along ⟨111⟩ axes (see Fig. 8.14), and the C_{60} molecules are disordered with respect to the axis of the oxygen atom [8.161]. The average isobaric volume expansion coefficients for $C_{60}O$ are 7.3×10^{-5}/K below T_{01} and 10.9×10^{-5}/K above T_{01}, in contrast to C_{60}, which has a smaller value of 6.1×10^{-5}/K (see Table 7.1). The larger volume expansion for $C_{60}O$ can be understood in terms of the oxygen seeking the octahedral void at low temperature, but this preferred orientation occurs to a lesser degree as T is increased.

The 6,5-annulene isomer of $C_{61}H_2$ shows similar phase transitions to the epoxide $C_{60}O$, with the orientationally disordered fcc phase above 290 K and the ordered simple cubic ($Pa\bar{3}$) phase below T_{01} [8.163, 164]. The mechanism driving this phase transition is believed to be the same as for the

C_{60} and $C_{60}O$ structures. A second $C_{61}H_2$ phase has also been identified, whereby the methylene CH_2 group bridges a 6:6 bond of C_{60}.

8.7.3. TDAE–C_{60}

One example of a different type of intercalation compound containing a relatively large organic molecule is tetrakis-dimethylamino-ethylene (TDAE), which has the formula $C_2N_4(CH_3)_8$ and forms a compound with a 1:1 ratio of TDAE to C_{60} (see Fig. 8.15) [8.165]. The structure of the TDAE molecule can be considered to be similar to that of the ethylene molecule, C_2H_4, where the two carbon atoms of this planar molecule are joined by double bonds and the remaining two bonds for each carbon atom are satisfied by two hydrogens. In the TDAE molecule, each of these four hydrogens is replaced by a nitrogen to which two methyl groups are attached (see inset to Fig. 8.15).

The TDAE–C_{60} compound is ferromagnetic with a Curie temperature of 16.1 K (see §18.5.2), the highest value known for an organic molecular ferromagnet [8.14]. In a synchrotron x-ray powder study (see Fig. 8.15), the structure for TDAE–C_{60} was determined. The diffraction data at 11 K were interpreted in terms of a centered monoclinic unit cell, $a = 15.807$ Å, $b = 12.785$ Å, $c = 9.859$ Å, and $\beta = 94.02°$, and two TDAE formula units per unit cell [8.165]. The diffraction pattern shown in Fig. 8.15 also contains

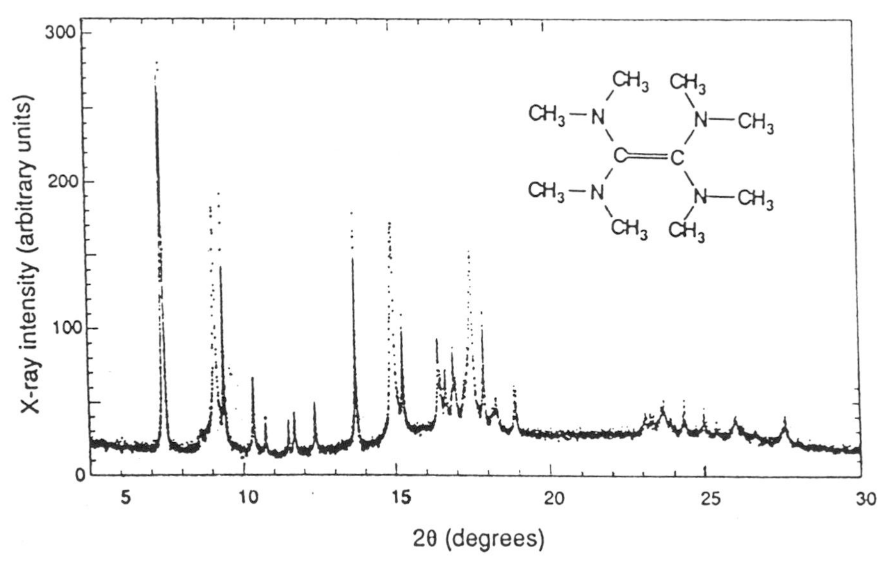

Fig. 8.15. A powder x-ray diffractogram of the TDAE–C_{60} compound. The constituents of the TDAE molecule, which has the basic ethylene structure, are shown in the inset [8.165].

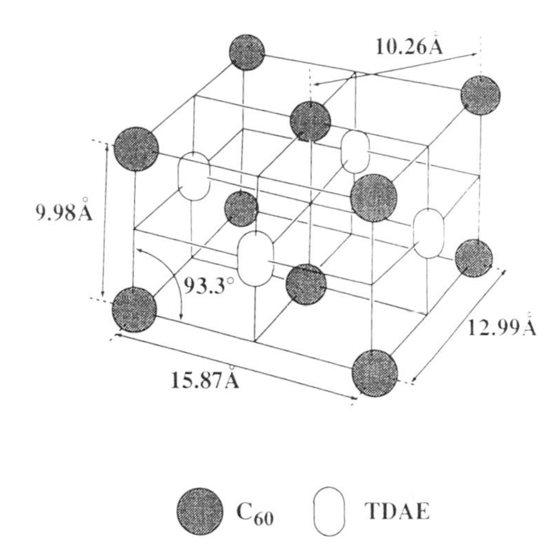

Fig. 8.16. Crystal structure of TDAE–C_{60} at room temperature determined by x-ray diffraction, showing the location of the C_{60} molecules and the orientation of the anisotropic TDAE molecules in the unit cell [8.165].

lines for unreacted fcc C_{60} ($a_0 = 14.18$ Å). Since the TDAE molecule lacks inversion symmetry, the logical choice of the space group is *P*2, with the C_{60} molecules centered at the 2*a* sites of the space group. In this structure shown in Fig. 8.16, the molecules in the monoclinic unit cell are located at (0,0,0) and ($\frac{1}{2}$,$\frac{1}{2}$,0), and the organic molecules are in the large interstices centered around the 2*d* sites at ($\frac{1}{2}$,0,$\frac{1}{2}$) and (0,$\frac{1}{2}$,$\frac{1}{2}$). The long axis of the molecule (a = 15.87 Å) is believed to be along the *c*-axis, as indicated in Fig. 8.16. The C–C bond lengths given for the C_{60} constituents in this compound are 1.37 Å and 1.45 Å, with one of the twofold axes of C_{60} along the *b*-axis of the crystal. The nearest-neighbor separation between C_{60} molecules in TDAE is 9.98 Å along the *c*-axis and 10.26 Å within the *a*–*b* plane, thereby giving further evidence for anisotropy in the crystal structure. If the van der Waals radius for C_{60} is taken as 5.0 Å, then 47% of the volume of the unit cell is available to accommodate the guest TDAE molecule. The nearest approach between the C_{60} and TDAE molecules in the solid phase is 2.58 Å [8.165]. The anisotropic crystal properties are expected to give rise to anisotropic magnetic properties as discussed in §18.5.2.

8.7.4. Metal–C_{60} Multilayer Structures

Reports on the structure and properties of synthetic metal-C_{60} multilayer structures are now becoming available [8.166]. In these synthetic materials, multilayers of alternating metal/C_{60} sequences are deposited under high-vacuum conditions on a substrate such as quartz. For example, up to 20 multilayers of Al/C_{60} on an insulating substrate [8.166] and C_{60} overlayers on thin films of Sn, Ba, and Ga have been studied [8.167].

Based on characterization of the Al/C_{60} multilayers using x-ray diffraction, *in situ* resistivity, and Raman scattering measurements, it is concluded that when depositing Al on C_{60} the aluminum penetrates the C_{60} layer because of the different sizes of the Al atoms and C_{60} molecules, so that a conducting Al_xC_{60} phase forms at the interface. When C_{60} is now deposited on the Al layer, a doped monolayer of C_{60} is formed by charge transfer from the Al to the C_{60} of up to six electrons per C_{60} [8.166] (see §14.1.5). The strong binding of C_{60} to metal substrates and the charge transfer that occurs at such interfaces have been studied in some detail (see §17.9).

8.8. The Doping of C_{70} and Higher-Mass Fullerenes

To date, most studies of doped fullerenes have been confined to C_{60}, although a number of studies of doped C_{70} have also been carried out (see §11.7.3, §12.7.5, and §14.1.4) and a very few have been made of the higher fullerenes (see §16.2 and §18.5). Almost all of these studies have involved alkali metal exohedral dopants and a few have involved the organic compound TDAE–C_{n_C}. The doping techniques that have been employed for C_{70} are similar to those for C_{60}, and the charge transfer effect is believed to be similar to that for C_{60}, or basically one electron transferred per alkali metal dopant, at least for low dopant concentrations. Because of the lower symmetry of C_{70}, the number of electrons required to fill electronic levels in the case of C_{70} is different from that in C_{60} (see §12.6.1). Correspondingly, half-filled and completely filled electronic bands occur for higher-mass fullerenes at different numbers of electrons relative to doped C_{60}.

Structural information is now available for M_xC_{70} (M = K, Rb, Cs), and stable phases have been reported for $x = 0, 1, 3, 4, 6, 9$ [8.168–170]. The various phases and lattice constants that have been reported are listed in

Table 8.6

Structure and lattice constants for the M_xC_{70} alkali metal compounds[a] [8.168–171].

M	$x = 1$ fcc	$x = 3$ fcc	$x = 4$ bct	$x = 6$ bcc	$x = 9$ fcc
K	–[b]	$a = 14.86$Å[c]	$a = 12.65$Å $c = 10.98$Å	$a = 12.02$Å	$a = 15.69$Å
Rb	$a = 14.71$Å		$a = 12.67$Å $c = 11.21$Å	$a = 12.11$Å	$a = 15.97$Å
Cs	$a \sim 14.7$Å		$a = 12.93$Å $c = 11.51$Å	$a \sim 12.3$Å	$a = 16.42$Å

[a] Pristine C_{70} has an fcc structure with $a = 14.92$ Å.
[b] This phase has not been confirmed.
[c] A K_3C_{70} fcc phase has been identified [8.169].

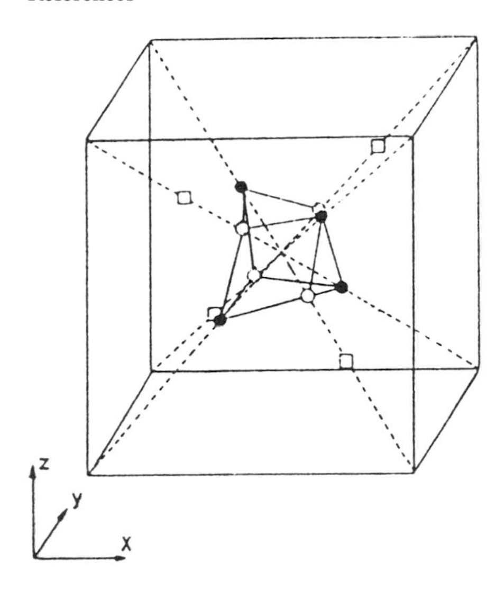

Fig. 8.17. Schematic drawing of the structure for Rb_9C_{70}. The large cube represents the fcc lattice of the C_{70} molecule. Solid and open circles represent Rb atoms in octahedral sites. Small open rectangles represent Rb atoms in tetrahedral sites [8.168].

Table 8.6. The $x = 0, 1, 4$, and 6 phases for M_xC_{70} correspond to the filling of the C_{70} a_2'' electronic level, half-filling of the a_1'' level, half-filling of the e_1'' level, and complete filling of the e_1'' level, respectively (see §12.6.1 and Fig. 12.3). The charge transfer associated with the $x = 9$ phase has not yet been determined, nor has the occupation of the energy levels. A model has, however, been proposed showing 12 possible sites for the locations of the nine alkali metal atoms (ions) within the fcc unit cell (see Fig. 8.17).

REFERENCES

[8.1] A. R. Kortan, N. Kopylov, S. H. Glarum, E. M. Gyorgy, A. P. Ramirez, R. M. Fleming, F. A. Thiel, and R. C. Haddon. *Nature (London)*, **355**, 529 (1992).

[8.2] A. R. Kortan, N. Kopylov, and F. A. Thiel. *J. Phys. Chem. Solids*, **53**, 1683 (1992).

[8.3] A. R. Kortan, N. Kopylov, S. H. Glarum, E. M. Gyorgy, A. P. Ramirez, R. M. Fleming, O. Zhou, F. A. Thiel, P. L. Trevor, and R. C. Haddon. *Nature (London)*, **360**, 566 (1992).

[8.4] Q. Zhu, D. E. Cox, J. E. Fischer, K. Kniaź, A. R. McGhie, and O. Zhou. *Nature (London)*, **355**, 712 (1992).

[8.5] G. Wortmann, Y. S. Grushko, A. Bolotov, E. A. Bychkov, W. Bensch, H. Werner, and R. Schlögl. *Mol. Cryst. Liq. Cryst.*, **245**, 313 (1994). 7th International Symposium on Intercalation Compounds.

[8.6] R. E. Smalley. In G. S. Hammond and V. J. Kuck (eds.), *Large Carbon Clusters*, p. 199, American Chemical Society, Washington, DC (1991).

[8.7] R. M. Fleming, T. Siegrist, P. M. Marsh, B. Hessen, A. R. Kortan, D. W. Murphy, R. C. Haddon, R. Tycko, G. Dabbagh, A. M. Mujsce, M. L. Kaplan, and S. M. Zahurak. In R. S. Averback, J. Bernholc, and D. L. Nelson (eds.), *Clusters and Cluster-Assembled Materials, MRS Symposia Proceedings, Boston*, vol. 206, pp. 691–696, Materials Research Society Press, Pittsburgh, PA (1991).

[8.8] D. W. Murphy, M. J. Rosseinsky, R. M. Fleming, R. Tycko, A. P. Ramirez, R. C. Haddon, T. Siegrist, G. Dabbagh, J. C. Tully, and R. E. Walstedt. *J. Phys. Chem. Solids*, **53**, 1321 (1992).

[8.9] R. C. Haddon. *Accounts of Chemical Research*, **25**, 127 (1992).

[8.10] O. Zhou, T. T. M. Palstra, Y. Iwasa, R. M. Fleming, A. F. Hebard, P. E. Sulewski, D. W. Murphy, and B. R. Zegarski. *Phys. Rev. B*, **52**, 483 (1995).

[8.11] A. M. Rao, P. Zhou, K.-A. Wang, G. T. Hager, J. M. Holden, Y. Wang, W. T. Lee, X.-X. Bi, P. C. Eklund, D. S. Cornett, M. A. Duncan, and I. J. Amster. *Science*, **259**, 955 (1993).

[8.12] G. Roth and P. Adelmann. *Appl. Phys. A*, **56**, 169 (1993).

[8.13] G. Roth, P. Adelmann, and R. Knitter. *Materials Lett.*, **16**, 357 (1993).

[8.14] P.-M. Allemand, K. C. Khemani, A. Koch, F. Wudl, K. Holczer, S. Donovan, G. Grüner, and J. D. Thompson. *Science*, **253**, 301 (1991).

[8.15] P. J. Fagan, J. C. Calabrese, and B. Malone. *Science*, **252**, 1160 (1991).

[8.16] P. J. Fagan, J. C. Calabrese, and B. Malone. *Accounts Chem. Res.*, **25**, 134 (1992).

[8.17] J. M. Hawkins, S. Loren, A. Meyer, and R. Nunlist. *J. Am. Chem. Soc.*, **113**, 7770 (1991).

[8.18] R. Saito, G. Dresselhaus, and M. S. Dresselhaus. *Chem. Phys. Lett.*, **210**, 159 (1993).

[8.19] W. Andreoni, F. Gygi, and M. Parrinello. *Phys. Rev. Lett.*, **68**, 823 (1992).

[8.20] R. Saito, G. Dresselhaus, and M. S. Dresselhaus. *Proc. of the 4th C_{60} Symposium* (1993). Japanese Chemical Society, Toyohashi, Japan, January 26–27, 1993.

[8.21] Q. Xie, E. Perez-Cordero, and L. Echegoyen. *J. Amer. Chem. Soc.*, **114**, 3978 (1992).

[8.22] D. S. Bethune, R. D. Johnson, J. R. Salem, M. S. de Vries, and C. S. Yannoni. *Nature (London)*, **366**, 123 (1993).

[8.23] R. E. Smalley. *Materials Science and Engineering*, **B19**, 1 (1993).

[8.24] C. S. Yannoni, H. R. Wendt, M. S. de Vries, R. L. Siemens, J. R. Salem, J. Lyerla, R. D. Johnson, M. Hoinkis, M. S. Crowder, C. A. Brown, D. S. Bethune, L. Taylor, D. Nguyen, P. Jedrzejewski, and H. C. Dorn. *Synthetic Metals*, **59**, 279 (1993).

[8.25] K. Kikuchi, S. Suzuki, Y. Nakao, N. Nakahara, T. Wakabayashi, H. Shiromaru, I. Ikemoto, and Y. Achiba. *Chem. Phys. Lett.*, **216**, 67 (1993).

[8.26] H. W. Kroto, J. R. Heath, S. C. O'Brien, R. F. Curl, and R. E. Smalley. *Nature (London)*, **318**, 162 (1985).

[8.27] J. R. Heath, S. C. O'Brien, Q. Zhang, Y. Liu, R. F. Curl, H. W. Kroto, F. K. Tittel, and R. E. Smalley. *J. Am. Chem. Soc.*, **107**, 7779 (1985).

[8.28] S. C. O'Brien, J. R. Heath, R. F. Curl, and R. E. Smalley. *J. Chem. Phys.*, **88**, 220 (1988).

[8.29] F. D. Weiss, S. C. O'Brien, J. L. Eklund, R. F. Curl, and R. E. Smalley. *J. Am. Chem. Soc.*, **110**, 4464 (1988).

[8.30] L. Soderholm, P. Wurz, K. R. Lykke, and D. H. Parker. *J. Phys. Chem.*, **96**, 7153 (1992).

[8.31] D. M. Poirier, M. Knupfer, J. H. Weaver, W. Andreoni, K. Laasonen, M. Parrinello, D. S. Bethune, K. Kikuchi, and Y. Achiba. *Phys. Rev. B*, **49**, 17403 (1994).

[8.32] S. Nagase, K. Kobayashi, T. Kato, and Y. Achiba. *Chem. Phys. Lett.*, **201**, 475 (1993).

[8.33] A. Rosén and B. Wästberg. *J. Am. Chem. Soc.*, **110**, 8701 (1988).

[8.34] P. W. Fowler and D. E. Manolopoulos. *Nature (London)*, **355**, 428 (1992).

[8.35] K. Kikuchi, N. Nakahara, T. Wakabayashi, S. Suzuki, H. Shiromaru, Y. Miyake, K. Saito, I. Ikemoto, M. Kainosho, and Y. Achiba. *Nature (London)*, **357**, 142 (1993).

[8.36] S. Suzuki, S. Kawata, H. Shiromaru, K. Yamauchi, K. Kikuchi, T. Kato, and Y. Achiba. *J. Phys. Chem.*, **96**, 7159 (1992).

[8.37] M. Hoinkis, C. S. Yannoni, D. S. Bethune, J. R. Salem, R. D. Johnson, M. S. Crowder, and M. S. de Vries. *Phys. Chem. Lett.*, **198**, 461 (1992).

[8.38] H. Shinohara, H. Yamaguchi, N. Hayashi, H. Sato, M. Inagaki, Y. Saito, S. Bandow, H. Kitagawa, T. Mitani, and H. Inokuchi. *Materials Science and Engineering*, **B19**, 25 (1993).

[8.39] R. D. Johnson, M. S. de Vries, J. Salem, D. S. Bethune, and C. S. Yannoni. *Nature (London)*, **355**, 239 (1992).

[8.40] J. H. Weaver, Y. Chai, G. H. Kroll, C. Jin, T. R. Ohno, R. E. Haufler, T. Guo, J. M. Alford, J. Conceicao, L. P. Chibante, A. Jain, G. Palmer, and R. E. Smalley. *Chem. Phys. Lett.*, **190**, 460 (1992).

[8.41] C. S. Yannoni, M. Hoinkis, M. S. de Vries, D. S. Bethune, J. R. Salem, M. S. Crowder, and R. D. Johnson. *Science*, **256**, 1191 (1992).

[8.42] X. D. Wang, Q. K. Xue, T. Hashizume, H. Shinohara, Y. Nishina, and T. Sakurai. *Phys. Rev. B*, **48**, 15492 (1993).

[8.43] C. H. Park, B. O. Wells, J. DiCarlo, Z. X. Shen, J. R. Salem, D. S. Bethune, C. S. Yannoni, R. D. Johnson, M. S. de Vries, C. Booth, F. Bridges, and P. Pianetta. *Chem. Phys. Lett.*, **213**, 196 (1993).

[8.44] P. H. M. van Loosdrecht, R. D. Johnson, R. Beyers, J. R. Salem, M. S. de Vries, D. S. Bethune, P. Burbank, J. Haynes, T. Glass, S. Stevenson, H. C. Dorn, M. Boonman, P. J. M. van Bentum, and G. Meijer. In K. M. Kadish and R. S. Ruoff (eds.), *Recent Advances in the Chemistry and Physics of Fullerenes and Related Materials: Electrochemical Society Symposia Proceedings, San Francisco, May 1994*, vol. 94-24, p. 1320, Electrochemical Society, Pennington, NJ (1994).

[8.45] R. Beyers, C. H. Kiang, R. D. Johnson, J. R. Salem, M. S. de Vries, C. S. Yannoni, D. S. Bethune, H. C. Dorn, P. Burbank, K. Harich, and S. Stevenson. *Nature (London)*, **370**, 196 (1994).

[8.46] J. F. Armbruster, M. Roth, H. A. Romberg, M. Sing, M. Schmitt, M. S. Golden, P. Schweiss, P. Adelmann, J. Fink, R. H. Michel, J. Rockenberger, F. Hennrich, and M. M. Kappes. *Phys. Rev. B*, **51**, 4933 (1995).

[8.47] H. Shinohara, H. Yamaguchi, N. Hayashi, H. Sato, M. Ohkohchi, Y. Ando, and Y. Saito. *J. Phys. Chem.*, **97**, 4259 (1993).

[8.48] X.-D. Wang, T. Hashizume, Q. Xue, H. Shinohara, Y. Saito, Y. Nishina, and T. Sakurai. *Jpn. J. Appl. Phys.*, **B32**, L866 (1993).

[8.49] K. Laasonen, W. Andreoni, and M. Parrinello. *Science*, **258**, 1916 (1992).

[8.50] S. Nagase and K. Kobayashi. *Chem. Phys. Lett.*, **228**, 106 (1994).

[8.51] P. H. M. van Loosdrecht, R. D. Johnson, M. S. de Vries, C. H. Kiang, D. S. Bethune, H. C. Dorn, P. Burbank, and S. Stevenson. *Phys. Rev. Lett.*, **73**, 3415 (1994).

[8.52] S. Saito, S. Sawada, and N. Hamada. *Phys. Rev. B*, **45**, 13845 (1992).

[8.53] B. Wästberg and A. Rosén. *Physica Scripta*, **44**, 276 (1991).

[8.54] A. H. H. Chang, W. C. Ermler, and R. M. Pitzer. *J. Chem. Phys.*, **94**, 5004 (1991).

[8.55] T. Guo, M. D. Diener, Y. Chai, J. M. Alford, R. E. Haufler, S. M. McClure, T. R. Ohno, J. H. Weaver, G. E. Scuseria, and R. E. Smalley. *Science*, **257**, 1661 (1992).

[8.56] L. S. Wang, J. M. Alford, Y. Chai, M. Diener, J. Zhang, S. M. McClure, T. Guo, G. E. Scuseria, and R. E. Smalley. *Chem. Phys. Lett.*, **207**, 354 (1993).

[8.57] E. G. Su *et al.* In F. Wudl (ed.), *Endohedral Rare-Earth Fullerene Complexes*, Pergamon, Oxford (1993). Abstract.

[8.58] M. Saunders, H. A. Jimenez-Vazquez, R. J. Cross, and R. J. Poreda. *Science*, **259**, 1428 (1993).

[8.59] M. M. Ross and J. H. Callahan. *J. Phys. Chem.*, **95**, 5720 (1991).

[8.60] K. A. Caldwell, D. E. Giblin, C. C. Hsu, D. Cox, and M. L. Gross. *J. Am. Chem. Soc.*, **113**, 8519 (1991).

[8.61] E. E. B. Campbell, R. Ehlich, A. Hielscher, J. M. A. Frazao, and I. V. Hertel. *Z. Phys. D*, **23**, 1 (1992).
[8.62] Z. Wan, J. F. Christian, and S. L. Anderson. *Phys. Rev. Lett.*, **69**, 1352 (1992).
[8.63] T. Pradeep, G. U. Kulkarni, K. R. Kannan, T. N. G. Row, and C. N. Rao. *J. Am. Chem. Soc.*, **114**, 2272 (1992).
[8.64] R. F. Kiefl, T. L. Duty, J. W. Schneider, A. MacFarlane, K. Chow, J. W. Elzey, P. Mendels, G. D. Morris, J. H. Brawer, E. J. Ansaldo, C. Niedermayer, D. R. Nokes, C. E. Stronach, B. Hitti, and J. E. Fischer. *Phys. Rev. Lett.*, **69**, 2005 (1992).
[8.65] R. F. Kiefl, W. A. MacFarlane, K. H. Chow, S. Dunsiger, T. L. Duty, T. M. S. Johnston, J. W. Schneider, J. Sonier, L. Brard, R. M. Strongin, J. E. Fischer, and A. B. Smith III. *Phys. Rev. Lett.*, **70**, 3987 (1993).
[8.66] X.-D. Wang, T. Hashizume, Q. Xue, H. Shinohara, Y. Saito, Y. Nishina, and T. Sakurai. *J. Appl. Phys.*, **32**, L147 (1993).
[8.67] Y. Saito, T. Yoshikawa, N. Fujimoto, and H. Shinohara. *Phys. Rev. B*, **48**, 9182 (1993).
[8.68] B. T. Kelly. *Physics of Graphite*. Applied Science, London (1981).
[8.69] J. E. Field. In J. E. Field (ed.), *Properties of Diamond*, p. 281. Academic Press, London (1979).
[8.70] W. I. F. David, R. M. Ibberson, J. C. Matthewman, K. Prassides, T. J. S. Dennis, J. P. Hare, H. W. Kroto, R. Taylor, and D. R. M. Walton. *Nature (London)*, **353**, 147 (1991).
[8.71] J. E. Fischer, P. A. Heiney, A. R. McGhie, W. J. Romanow, A. M. Denenstein, J. P. McCauley, Jr., and A. B. Smith III. *Science*, **252**, 1288 (1991).
[8.72] W. I. F. David, R. M. Ibberson, T. J. S. Dennis, J. P. Hare, and K. Prassides. *Europhys. Lett.*, **18**, 219 (1992).
[8.73] W. I. F. David, R. M. Ibberson, T. J. S. Dennis, J. P. Hare, and K. Prassides. *Europhys. Lett.*, **18**, 735 (1992).
[8.74] Y. Miyamoto, A. Oshiyama, and S. Saito. *Solid State Commun.*, **82**, 437 (1992).
[8.75] Y. Miyamoto, N. Hamada, A. Oshiyama, and S. Saito. *Phys. Rev. B*, **46**, 1749 (1992).
[8.76] G. K. Wertheim and D. N. E. Buchanan. *Solid State Commun.*, **88**, 97 (1993).
[8.77] E. Ozdas, A. R. Kortan, N. Kopylov, S. H. Glarum, T. Siegrist, and A. P. Ramirez. *Bulletin Amer. Phys. Soc.*, **39**, 330 (1994).
[8.78] M. S. Dresselhaus and G. Dresselhaus. *Advances in Phys.*, **30**, 139 (1981).
[8.79] Q. Zhu, O. Zhou, N. Coustel, G. B. M. Vaughan, J. P. McCauley, Jr., W. J. Romanow, J. E. Fischer, and A. B. Smith, III. *Science*, **254**, 545 (1992).
[8.80] P. W. Stephens, L. Mihaly, J. B. Wiley, S. M. Huang, R. B. Kaner, F. Diederich, R. L. Whetten, and K. Holczer. *Phys. Rev. B*, **45**, 543 (1992).
[8.81] D. M. Poirier. *Appl. Phys. Lett.*, **64**, 1356 (1993).
[8.82] T. Yildirim, O. Zhou, J. E. Fischer, N. Bykovetz, R. A. Strongin, M. A. Cichy, A. B. Smith III, C. L. Lin, and R. Jelinek. *Nature (London)*, **360**, 568 (1992).
[8.83] M. J. Rosseinsky, D. W. Murphy, R. M. Fleming, R. Tycko, A. P. Ramirez, T. Siegrist, G. Dabbagh, and S. E. Barrett. *Nature (London)*, **356**, 416 (1992).
[8.84] C. Gu, F. Stepniak, D. M. Poirier, M. B. Jost, P. J. Benning, Y. Chen, T. R. Ohno, J. L. Martins, J. H. Weaver, J. Fure, and R. E. Smalley. *Phys. Rev. B*, **45**, 6348 (1992).
[8.85] P. J. Benning, D. M. Poirier, T. R. Ohno, Y. Chen, M. B. Jost, F. Stepniak, G. H. Kroll, and J. H. Weaver. *Phys. Rev. B*, **45**, 6899 (1992).
[8.86] K. Tanigaki, T. W. Ebbesen, S. Saito, J. Mizuki, J. S. Tsai, Y. Kubo, and S. Kuroshima. *Nature (London)*, **352**, 222 (1991).
[8.87] K. Tanigaki. *Chem. Phys. Lett.*, **185**, 189 (1991).
[8.88] I. Hirosawa, K. Tanigaki, J. Mizuki, T. W. Ebbesen, Y. Shimakawa, Y. Kubo, and S. Kuroshima. *Solid State Commun.*, **82**, 979 (1992).

[8.89] I. Hirosawa, J. Mizuki, K. Tanigaki, T. W. Ebbesen, J. S. Tsai, and S. Kuroshima. *Solid State Commun.*, **87**, 945 (1993).

[8.90] R. M. Fleming, A. P. Ramirez, M. J. Rosseinsky, D. W. Murphy, R. C. Haddon, S. M. Zahurak, and A. V. Makhija. *Nature (London)*, **352**, 787 (1991).

[8.91] K. Tanigaki, S. Kuroshima, J. Fujita, and T. W. Ebbesen. *Appl. Phys. Lett.*, **63**, 2351 (1993).

[8.92] Y. Chen, F. Stepniak, J. H. Weaver, L. P. F. Chibante, and R. E. Smalley. *Phys. Rev. B*, **45**, 8845 (1992).

[8.93] A. R. Kortan, N. Kopylov, E. Özdas, A. P. Ramirez, R. M. Fleming, and R. C. Haddon. *Chem. Phys. Lett.* **223**, 501 (1994).

[8.94] T. R. Ohno, G. H. Kroll, J. H. Weaver, L. P. F. Chibante, and R. E. Smalley. *Nature (London)*, **350**, 401 (1992).

[8.95] S. Z. Liu, Y. J. Lu, M. M. Kappes, and J. A. Ibers. *Science*, **254**, 408 (1991).

[8.96] G. Faigel, G. Bortel, G. Oszànyi, S. Pekker, M. Tegze, P. W. Stephens, and D. Liu. *Phys. Rev. B*, **49**, 9186 (1994).

[8.97] S. Pekker, G. Faigel, K. Fodor-Csorba, L. Gránásy, E. Jakab, and M. Tegze. *Solid State Commun.*, **83**, 423 (1992).

[8.98] S. Pekker, G. Faigel, G. Oszlanyi, M. Tegze, T. Kemeny, and E. Jakab. *Synth. Met.*, **55**, 3014 (1993).

[8.99] G. Faigel, M. Tegze, S. Pekker, and T. Kemeny. *J. Mod. Phys.*, **23–24**, 3859 (1992).

[8.100] Y. Lou, X. Lu, G. H. Dai, W. Y. Ching, Y. N. Xu, M. Z. Huang, P. K. Tseng, Y. C. Jean, R. L. Meng, P. H. Hor, and C. W. Chu. *Phys. Rev. B*, **46**, 2644 (1992).

[8.101] Y. C. Jean, X. Lu, Y. Lou, A. Bharathi, C. S. Sundar, Y. Lyu, P. H. Hor, and C. W. Chu. *Phys. Rev. B*, **45**, 12126 (1992).

[8.102] P. A. Heiney, G. B. M. Vaughan, J. E. Fischer, N. Coustel, D. E. Cox, J. R. D. Copley, D. A. Neumann, W. A. Kamitakahara, K. M. Creegan, D. M. Cox, J. P. McCauley, Jr., and A. B. Smith, III. *Phys. Rev. B*, **45**, 4544 (1992).

[8.103] J. W. Dykes, W. D. Mosley, P. A. Sterne, J. Z. Liu, R. N. Shelton, and R. H. Howell. *Chem Phys. Lett.*, **232**, 22 (1995).

[8.104] Y. Murakami, T. Arai, H. Suematsu, K. Kikuchi, N. Nakahara, Y. Achiba, and I. Ikemoto. *Fullerene Sci. and Tech.* **1**, 351 (1993).

[8.105] Q. Zhu, O. Zhou, J. E. Fischer, A. R. McGhie, W. J. Romanow, R. M. Strongin, M. A. Cichy, and A. B. Smith, III. *Phys. Rev. B*, **47**, 13948 (1993).

[8.106] D. M. Poirier, D. W. Owens, and J. H. Weaver. *Phys. Rev. B*, **51**, 1830 (1995).

[8.107] P. W. Stephens, L. Mihaly, P. L. Lee, R. L. Whetten, S. M. Huang, R. Kaner, F. Diederich, and K. Holczer. *Nature (London)*, **351**, 632 (1991).

[8.108] T. Yildirim, J. E. Fischer, A. B. Harris, P. W. Stephens, D. Liu, L. Brard, R. A. Strongin, and A. B. Smith III. *Phys. Rev. Lett.*, **71**, 1383 (1993).

[8.109] K. Tanigaki, I. Hirosawa, T. W. Ebbesen, J. I. Mizuki, and J. S. Tsai. *J. Phys. Chem. Solids*, **54**, 1645 (1993).

[8.110] M. S. Dresselhaus. In M. S. Dresselhaus (ed.), *Intercalation in Layered Materials*, p. 1, Plenum Press, New York (1987).

[8.111] A. R. Kortan, N. Kopylov, R. M. Fleming, O. Zhou, F. A. Thiel, R. C. Haddon, and K. M. Rabe. *Phys. Rev. B*, **47**, 13070 (1993).

[8.112] O. Zhou, J. E. Fischer, N. Coustel, S. Kycia, Q. Zhu, A. R. McGhie, W. J. Romanow, J. P. McCauley, Jr., A. B. Smith III, and D. E. Cox. *Nature (London)*, **351**, 462 (1991).

[8.113] E. Ozdas, A. R. Kortan, N. Kopylov, A. P. Ramirez, T. Siegrist, K. M. Rabe, H. E. Bair, S. Schuppler, and P. H. Citrin. *Nature (London)*, **375**, 126 (1995).

[8.114] K. Tanigaki, I. Hirosawa, T. W. Ebbesen, J. Mizuki, Y. Shimakawa, Y. Kubo, J. S. Tsai, and S. Kuroshima. *Nature (London)*, **356**, 419 (1992).

[8.115] K. Tanigaki, I. Hirosawa, T. W. Ebbesen, J. Mizuki, and S. Kuroshima. *Chem. Phys. Lett.*, **203**, 33 (1993).

[8.116] I. Hirosawa, J. Mizuki, K. Tanigaki, and H. Kumura. *Solid State Commun.*, **89**, 55 (1994).

[8.117] T. T. M. Palstra, O. Zhou, Y. Iwasa, P. Sulewski, R. Fleming, and B. Zegarski. *Solid State Commun.*, **93**, 327 (1995).

[8.118] O. Zhou, R. M. Fleming, D. W. Murphy, M. J. Rosseinsky, A. P. Ramirez, R. B. van Dover, and R. C. Haddon. *Nature (London)*, **362**, 433 (1993).

[8.119] P. Zhou, Z. H. Dong, A. M. Rao, and P. C. Eklund. *Chem. Phys. Lett.*, **211**, 337 (1993).

[8.120] M. J. Rosseinsky, D. W. Murphy, R. M. Fleming, and O. Zhou. *Nature (London)*, **364**, 425 (1993).

[8.121] P. A. Heiney, J. E. Fischer, A. R. McGhie, W. J. Romanow, A. M. Denenstein, J. P. McCauley, Jr., A. B. Smith, III, and D. E. Cox. *Phys. Rev. Lett.*, **66**, 2911 (1991). See also Comment by R. Sachidanandam and A. B. Harris, Phys. Rev. Lett. **67**, 1467 (1991).

[8.122] P. A. Heiney, J. E. Fischer, A. R. McGhie, W. J. Romanow, A. M. Denenstein, J. P. McCauley, Jr., A. B. Smith, III, and D. E. Cox. *Phys. Rev. Lett.*, **67**, 1468 (1991).

[8.123] J. E. Fischer and P. A. Heiney. *J. Phys. Chem. Solids*, **54**, 1725 (1993).

[8.124] T. Yildirim, S. Hong, A. B. Harris, and E. J. Mele. *Phys. Rev. B*, **48**, 12262 (1993).

[8.125] M. P. Gelfand and J. P. Lu. *Appl. Phys. A*, **56**, 215 (1993).

[8.126] C. Christides, D. A. Neumann, K. Prassides, J. R. D. Copley, J. J. Rush, M. J. Rosseinsky, D. W. Murphy, and R. C. Haddon. *Phys. Rev. B*, **46**, 12088 (1992).

[8.127] D. A. Neumann, J. R. D. Copley, W. A. Kamitakahara, J. J. Rush, R. L. Paul, and R. M. Lindstrom. *J. Phys. Chem. Solids*, **54**, 1699 (1993).

[8.128] O. Gunnarsson, S. Satpathy, O. Jepsen, and O. K. Andersen. *Phys. Rev. Lett.*, **67**, 3002 (1991).

[8.129] S. Satpathy, V. P. Antropov, O. K. Andersen, O. Jepson, O. Gunnarsson, and A. I. Liechtenstein. *Phys. Rev. B*, **46**, 1773 (1992).

[8.130] I. I. Mazin, A. I. Liechtenstein, O. Gunnarsson, O. K. Andersen, and V. P. Antropov. *Phys. Rev. Lett.*, **70**, 1773 (1993).

[8.131] S. Teslic, T. Egami, and J. E. Fischer. *Phys. Rev. B*, **51**, 5973 (1995).

[8.132] R. E. Walstedt, D. W. Murphy, and M. Rosseinsky. *Nature (London)*, **362**, 611 (1993).

[8.133] M. Mehring, R. Rachdi, and E. Zimmer. *Phil. Mag.* **B70**, 787 (1994).

[8.134] J. Winter and H. Kuzmany. *Solid State Commun.*, **83**, 1321 (1992).

[8.135] D. M. Poirier and J. H. Weaver. *Phys. Rev. B*, **47**, 10959 (1993).

[8.136] R. Tycko, G. Dabbagh, D. W. Murphy, Q. Zhu, and J. E. Fischer. *Phys. Rev. B*, **48**, 9097 (1993).

[8.137] P. W. Stephens, G. Bortel, G. Falgel, M. Tegze, A. Jánossy, S. Pekker, G. Oszànyi, and L. Forró. *Nature (London)*, **370**, 636 (1994).

[8.138] A. Jànossy, O. Chauvet, S. Pekker, J. R. Cooper, and L. Forró. *Phys. Rev. Lett.*, **71**, 1091 (1993).

[8.139] S. Pekker, A. Jánossy, L. Mihaly, O. Chauver, M. Carrard, and L. Forró. *Science*, **265**, 1077 (1994).

[8.140] M. C. Martin, D. Koller, A. Rosenberg, C. Kendziora, and L. Mihaly. *Phys. Rev. B*, **51**, 3210 (1995).

[8.141] P. W. Stephens. *Nature (London)*, **356**, 383 (1992).

[8.142] O. Zhou and D. E. Cox. *J. Phys. Chem. Solids*, **53**, 1373 (1992).

[8.143] A. Oshiyama, S. Saito, N. Hamada, and Y. Miyamoto. *J. Phys. Chem. Solids*, **53**, 1457 (1992).
[8.144] T. Yildirim, O. Zhou, J. E. Fischer, R. A. Strongin, M. A. Cichy, A. B. Smith III, C. L. Lin, and R. Jelinek. *Nature (London)*, **360**, 568 (1992).
[8.145] J. E. Fischer. *Materials Science and Engineering*, **B19**, 90 (1993).
[8.146] C. Christides, I. M. Thomas, T. J. S. Dennis, and K. Prassides. *Europhys. Lett.*, **22**, 611 (1993).
[8.147] K. Kniaź, J. E. Fischer, Q. Zhu, M. J. Rosseinsky, O. Zhou, and D. W. Murphy. *Solid State Commun.*, **88**, 47 (1993).
[8.148] K. Prassides, C. Christides, I. M. Thomas, J. Mizuki, K. Tanigaki, I. Hirosawa, and T. W. Ebbesen. *Science*, **263**, 950 (1994).
[8.149] Y. Maniwa, K. Mizoguchi, K. Kume, K. Tanigaki, T. W. Ebbesen, S. Saito, J. Mizuki, J. S. Tsai, and Y. Kubo. *Solid State Commun.*, **82**, 783 (1992).
[8.150] I. Hirosawa, K. Prassides, J. Mizuki, K. Tanigaki, M. Gevaert, A. Lappus, and J. K. Cockcroft. *Science*, **264**, 1294 (1994).
[8.151] S. Saito and A. Oshiyama. *Jpn. J. Appl. Phys.* **32**, 1438 (1993).
[8.152] W. R. Datars, P. K. Ummat, T. Olech, and R. K.Nkum. *Solid. State Commun.*, **86**, 579 (1993).
[8.153] W. R. Datars, T. R. Chien, R. K. Nkum, and P. K. Ummat. *Phys. Rev. B*, **50**, 4937 (1994).
[8.154] M. Kobayashi, Y. Akahama, H. Kawamura, H. Shinohara, H. Sato, and Y. Saito. *Mater. Sci. Eng.*, **B19**, 100 (1993).
[8.155] F. van Bolhuis, P. B. Koster, and T. Migchelsen. *Acta Crystall.*, **23**, 90 (1967).
[8.156] Y. S. Grushko, G. Wortmann, M. F. Kovalev, L. I. Molkanov, Y. V. Ganzha, Y. A. Ossipyan, and O. V. Zharikov. *Solid State Commun.*, **84**, 505 (1992).
[8.157] G. Wortmann, H. Werner, and R. Schlögl. In H. Kuzmany, M. Mehring, and J. Fink (eds.), *Springer Series Solid State Sci.*, **113**, 492 (1994).
[8.158] Y. Maniwa, T. Shibata, K. Mizoguchi, K. Kume, K. Kikuchi, I. Ikemoto, S. Suzuki, and Y. Achiba. *J. Phys. Soc. Jpn.*, **61**, 2212 (1992).
[8.159] H. Werner, M. Wesemann, and R. Schlögl. *Europhys. Lett.* **20**, 107 (1992).
[8.160] F. N. Tebbe, R. L. Harlow, D. B. Chase, D. L. Thorn, G. C. Campbell, Jr., J. C. Calabrese, N. Herron, R. J. Young, Jr., and E. Wasserman. *Science*, **256**, 822 (1992).
[8.161] G. B. M. Vaughan, P. A. Heiney, D. E. Cox, A. R. McGhie, D. R. Jones, R. M. Strongin, M. A. Cichy, and A. B. Smith III. *Chem. Phys.*, **168**, 185 (1992).
[8.162] F. Gugenberger, R. Heid, C. Meingast, P. Adelmann, M. Braun, H. Wühl, M. Haluska, and H. Kuzmany. *Phys. Rev. Lett.*, **60**, 3774 (1992).
[8.163] A. N. Lommen, P. A. Heiney, G. B. M. Vaughan, P. W. Stephens, D. Liu, D. Li, A. L. Smith, A. R. McGhie, R. M. Strongin, L. Brard, and A. B. Smith, III. *Phys. Rev. B*, **49**, 12572 (1994).
[8.164] T. Suzuki, Q. Li, K. C. Khemani, and F. Wudl. *J. Am. Chem. Soc.*, **114**, 7301 (1992).
[8.165] P. W. Stephens, D. E. Cox, J. W. Lauher, L. Mihaly, J. B. Wiley, P. Allemand, A. Hirsch, K. Holczer, Q. Li, J. D. Thompson, and F. Wudl. *Nature (London)*, **355**, 331 (1992).
[8.166] A. F. Hebard, C. B. Eom, Y. Iwasa, K. B. Lyons, G. A. Thomas, D. H. Rapkine, R. M. Fleming, R. C. Haddon, J. M. Phillips, J. H. Marshall, and R. H. Eick. *Phys. Rev. B*, **50** 17740 (1995).
[8.167] W. Zhao, K. Luo, J. Cheng, C. Li, D. Yin, Z. Gu, X. Zhou, and Z. Jin. *J. Phys. Condens. Matter.*, **4**, L513 (1992).
[8.168] M. Kobayashi, M. Fukuda, Y. Akahama, H. Kawamura, Y. Saito, and H. Shinohara. In P. Bernier, T. W. Ebbesen, D. S. Bethune, R. Metzger, L. Y. Chiang, and

J. W. Mintmire (eds.), *Science and Technology of Fullerene Materials: MRS Symposia Proceedings, Boston, Fall 1994*, vol. 359, p. 313, Materials Research Society Press, Pittsburgh, PA (1995).

[8.169] M. Kobayashi, Y. Akahama, H. Kawamura, H. Shinohara, H. Sato, and Y. Saito. *Phys. Rev. B*, **48**, 16877 (1993).

[8.170] M. Kobayashi, Y. Akahama, H. Kawamura, H. Shinohara, H. Sato, and Y. Saito. *Fullerene Sci. and Tech.*, **1**, 449 (1993).

[8.171] M. Kobayashi. (private communication). Unpublished.

CHAPTER 9

Single-Crystal and Epitaxial Film Growth

In the first part of this chapter we discuss the methods used to grow single crystals and thin solid films of van der Waals-bonded C_{60} and C_{70}. In principle, the methods described here can also be used for higher-mass fullerenes, if they can be isolated in sufficient quantity. Although crystalline fullerene material can be obtained by solution growth in saturated solvents, this method for growing pristine (undoped) fullerene films or crystals often leads to trapped solvent in the lattice, and therefore this growth method is not emphasized here. However, the low sublimation temperatures of C_{60} and C_{70} allow vapor growth methods to be applied to both film and single crystal growth. Careful studies of C_{60} film growth have been made, and several substrates have been found which promote epitaxy in monolayer and thicker films. A more detailed treatment of the interaction between C_{60} and substrates is presented in §17.9 on surface science. The photochemical transformation of solid C_{60} into what appears to be a polymeric solid in which the C_{60} shells remain intact but are tightly coupled to neighboring C_{60} molecules by intermolecular C–C bonds is reviewed in §7.5.

The second part of this chapter deals with methods for introducing foreign atoms or molecules into solid C_{60} to produce crystalline derivatives, e.g., alkaline earths, alkali metals, NH_3, and sulfur. These doped fullerene-based materials are classified into two groups: (1) charge transfer materials, in which electrons are exchanged between the host fullerite sublattice and the dopants, and (2) clathrate materials, in which neutral molecules are incorporated into the lattice and no charge exchange occurs between these dopant molecules and the fullerenes. Synthesis methods for the production of carbon nanotubes are given in §19.2.5, and electrochemical synthesis is reviewed in §10.3.2. The preparation of endohedrally doped fullerenes is discussed in §5.4.

9.1. Single-Crystal Growth

9.1.1. Synthesis of Large C_{60} and C_{70} Crystals by Vapor Growth

The low sublimation temperatures of C_{60} and C_{70} can be exploited to obtain millimeter-size single crystals of these fullerites with prominent facets by vapor growth [9.1–6]. The discovery of the sublimation growth method for C_{60} [9.1, 2] is of particular historical importance, because up to that point, C_{60} had been crystallized primarily by growth from saturated solvents (e.g., toluene, hexane) [9.7]. Unfortunately, the solvent molecules were also incorporated into the lattice, leading to various crystal structures [9.3, 8–11] and crystals with many shapes, including whiskers [9.12] and fibers with 10-fold symmetry having 10 facets along the column axis [9.13]. It nas, however, been shown subsequently [9.14] that $\sim(0.5 \times 0.3 \times 0.3\ \text{mm}^3)$ face-centered cubic (fcc) single crystals of C_{60} could be obtained from C_{60}-saturated boiling benzene; these crystals were reported to be free of solvent.

The vapor growth technique for synthesizing single-crystal fullerites begins in a clean, dynamically pumped, quartz tube ($\sim$ 50 cm long by 1 cm in diameter) shown schematically in Fig. 9.1. C_{60} powder is first extracted with toluene from carbon soot via standard liquid chromatography techniques or by passing the toluene extract over activated carbon (see §5.2.1 and §5.3.1). The powder is placed in section I of the tube and then degassed by baking at 250°C under dynamic vacuum for 6–24 h [9.4–6]. Next a series of sublimations is carried out to purify the initial charge [9.4, 5]. Section I is heated above the C_{60} sublimation temperature to ~600°C and the C_{60} condenses on the colder walls of section II. Section I is then sealed off at A, with a torch. This process is continued two more times until purified C_{60} has been vapor transported into section IV, which is then sealed off at C and D. It is

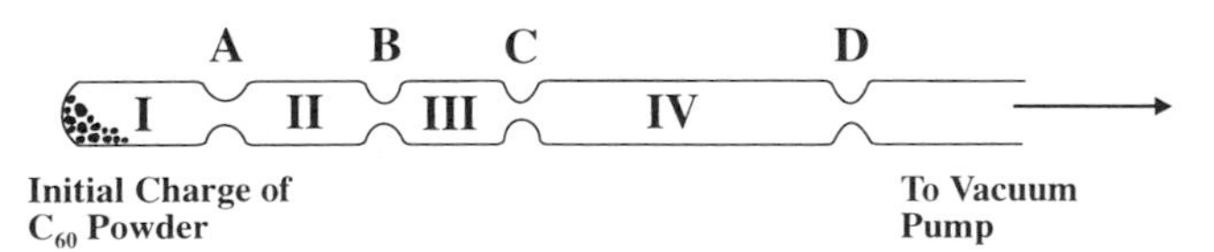

Fig. 9.1. Schematic view of a quartz tube (50 cm ×1 cm OD) used to degas, purify and finally grow small single crystals by vapor transport. A, B, C, D indicate constrictions in the tube used to facilitate sealing off sections with a torch. Sections I, II, and III are used to purify successively the starting material by sublimation. Tube section IV is used to grow the crystals by vapor transport (see text).

well known that a considerable amount of the initial C_{60} charge [9.4] cannot be sublimed, although the reason(s) for this is still not clear. Raman spectroscopy indicates that the major portion of the unsublimed residual is indeed C_{60} [9.5]. The long quartz tube (IV) containing the purified C_{60} in one end is then placed in a tube furnace for the final sublimation, vapor transport, and crystal growth steps. In the simplest case, the tube is loaded into the furnace initially at 500°C, and then the furnace is cooled to room temperature in 4 h [9.6]. Other researchers have chosen to place the sealed quartz tube containing C_{60} into a furnace with a temperature gradient [9.1]. Vapor transport of the C_{60} occurs from the hot end of the tube (560–620°C) to the cold end (480–540°C). In this gradient method, crystals of size visible to the unaided eye form in a period of hours. It is found that C_{60} has the highest vapor pressure among fullerenes, and this feature has been useful for separation applications (see §5.3.2) [9.13].

Untwinned single C_{60} crystals with 2 mm × 2 mm facets and C_{70} crystals with ~0.2 mm × 0.2 mm facets have been obtained [9.5] by a vapor growth technique, employing a double-temperature-gradient furnace. The gradient and heat treatment history are shown schematically in Fig. 9.2. For the case of C_{60}, a sealed quartz ampoule (20 cm long × 1 cm in diameter) containing the purified C_{60} in one end of the tube is placed in an oven whose temperature profile is shown in Fig. 9.2. Both ends of the tube are at temperature T_2 (~ 590°C), and a minimum temperature T_1 (~ 550°C) is obtained near the center of the tube, as shown. A sharp concave minimum in the temperature profile at an early time in the growth process was reported to yield one large fcc single crystal near the position of minimum temperature. Examples of C_{60} and C_{70} crystals obtained by this method are shown in Fig. 9.3, together with calculated faceted crystal forms [9.15] for an fcc material using a strong-faceting model. The configuration of the facets can be used to identify twinning in the crystals (d, i). An example of

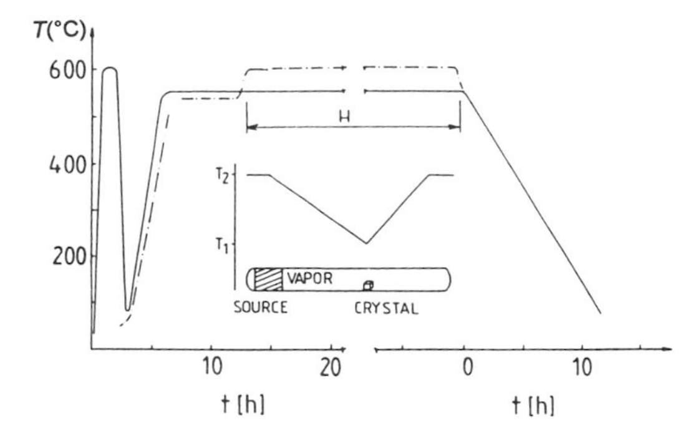

Fig. 9.2. The time dependence of the temperature and the temperature profile (inset) for the crystal growth process by the double-gradient technique. The region denoted by H is the time the crystal grows under steady-state conditions, and the time scales on the left and right, respectively, refer to times before and after the steady state period [9.5].

an untwinned crystal is shown in (a), and the theoretical fcc crystal shape is shown in (c), which exhibits (100) and (111) natural facets. The crystals and facets shown in Fig. 9.3 are large enough to carry out a wide variety of solid-state properties studies.

Another set of single-crystal growth procedures using the sublimation method consist of sealing purified C_{60} powder in a quartz tube containing a few hundred torr of argon gas [9.16]. The sealed tubes are placed in a gradient furnace with the powder held at 650°C. Single crystals are reported to form in the region of the tube that is at $\sim$ 450°C [9.16].

9.2. Thin-Film Synthesis

Thin films of fullerenes can be prepared by vapor growth and are useful for many applications, such as optical studies. Thin-film synthesis is normally carried out by vacuum sublimation of microcrystalline powder. The powder should first be degassed ($\sim$ 300°C, 3–6 h, $\sim 10^{-5}$ torr) to remove residual solvent, following standard extraction and purification steps. The powder is then placed in either a conventional evaporation boat or a Knudsen cell for vacuum sublimation. Following conventional thin-film growth practice [9.17, 18], a large substrate-to-source distance is used to prepare high-quality, epitaxial films. However, for the more difficult to obtain higher-mass fullerenes (e.g., C_{70}, C_{76}, C_{84}), much shorter source distances have been used to increase the amount of higher-mass fullerene material deposited on the substrates.

Several substrates have been reported to achieve epitaxial growth for fcc C_{60} films (e.g., GaAs (110) [9.19], mica (001) [9.20–23], GeS (001) [9.24], GaSe (001) [9.24], MoS_2 [9.25], and Sb [9.26]). Several of these substrates are discussed below for illustrative purposes. Other substrates (e.g., Pyrex, Suprasil, NaCl, KBr, sapphire, Si) have been reported to generate nanocrystalline C_{60} films with typical grain sizes of 10–20 nm, as determined by x-ray diffraction peak widths [9.27]. Several of these are also discussed below as examples of substrates that do not lead to epitaxial growth. Pristine C_{60} films can be deposited onto a substrate such as a clean silicon (100) surface by sublimation of the C_{60} powder in a vacuum of $\sim 10^{-6}$ torr. Ellipsometry [9.28] or dilatometry can be used to measure the thickness of the C_{60} films. By variation of the substrate material and crystalline orientation, and by variation of the film growth temperature, the crystallite size and microstructure of the film can be varied.

Typical of common practice in thin-film deposition, heated substrates lead to larger grain size in nanocrystalline films and better epitaxy in epitaxial films. However, because C_{60} sublimes at relatively low temperatures, only moderate substrate temperatures (150–200°C) can be used. For exam-

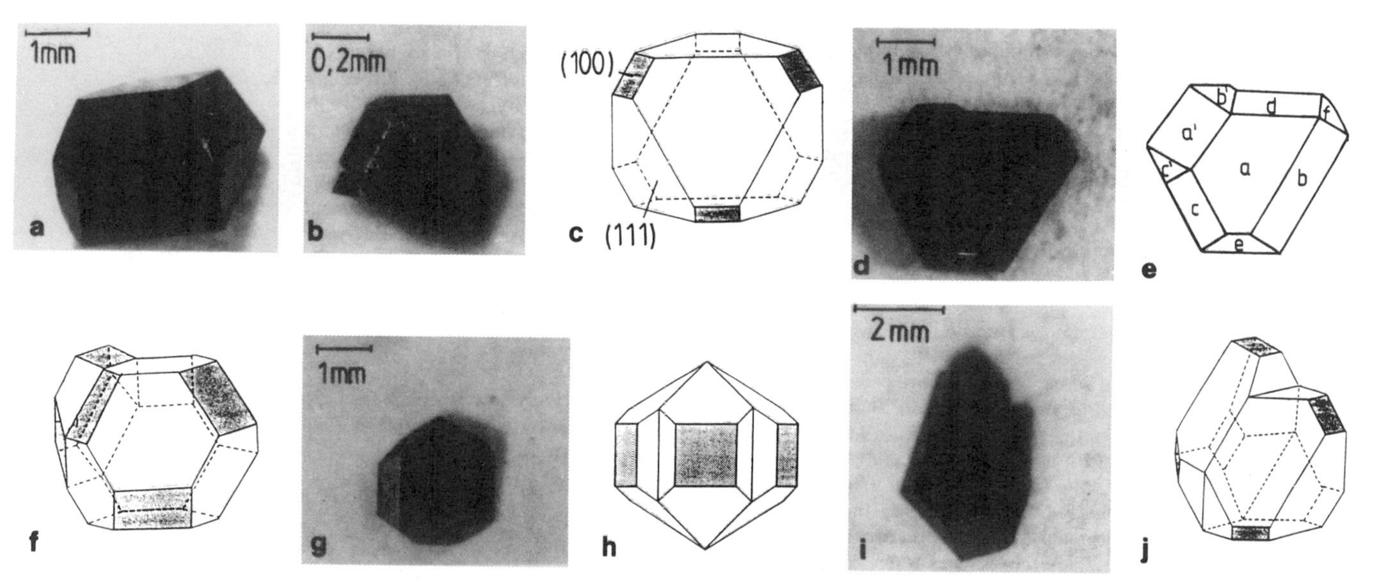

Fig. 9.3. Fullerite crystals prepared by the double-gradient technique [9.5] and compared to forms calculated by Marks [9.15] for fcc material using a "strong faceting" model. C_{60} (a); C_{70} (b); the shape of a theoretical equilibrium crystal (c); an asymmetric twin (d), with a detailed description of its faces a, e, f, b', c' octahedral {111}, b, c, d, a' hexahedral {100} (e); and its calculated form (f); multiply twinned particles (g, h); a symmetric twin (i, j) [9.5].

ple, at substrate temperatures $T = 300°C$, it was found that C_{60} sublimes from the film faster than it is being deposited (0.25 Å/s) [9.29].

In general, the C_{60}–C_{60} bonding is through weak van der Waals forces. For substrates where the C_{60}–substrate interaction is even weaker than the C_{60}–C_{60} interaction (such as mica), crystalline C_{60} films will grow in the (111) orientation normal to the substrate. Crystalline C_{60} film growth also occurs when the substrate is well lattice-matched to the lattice planes in fcc C_{60}. An example of such crystalline film growth occurs for C_{60} on GeS (001) substrates as discussed in §9.2.1.

9.2.1. *Epitaxial C_{60} Films*

Several studies have been made to determine the nature of the earliest stages of film growth for C_{60} which reveal equilibrium and nonequilibrium structures. Schmicker *et al.* [9.20], in He atom scattering studies of C_{60} monolayer (ML) growth on a clean (001) mica substrate at 300 K in high vacuum (10^{-10} mbar), found that the growth of the first C_{60} layer is hexagonal and in registry with the underlying hexagonal mica substrate. Mica was selected as a substrate material because of its weak bonding to C_{60}. Nevertheless, the film was found to grow with a 10.4 Å separation between C_{60} molecules, double the (001) mica spacing and considerably larger than the 10.18 Å spacing between molecules in the fcc solid lattice, indicating a significant C_{60}–substrate interaction. The lattice mismatch is about 3.4% between the muscovite mica and fcc C_{60}. The registry between the (001) mica surface and the C_{60} overlayer is shown schematically in Fig. 9.4, where the SiO_4 tetrahedra in the (001) mica substrate are shown together with the hexagonal C_{60} overlayer; the C_{60} molecules are represented by 10 Å diameter circles. He atom diffraction scans (Fig. 9.5) for 0.7 ML (b) and 1.7 ML (d) coverages of C_{60} on clean (100) mica are shown. The diffraction data were taken along the (110) and (100) directions as indicated, and the lines due to the hexagonal C_{60} overlayer can be identified by comparing the data in Fig. 9.5(b) and (d) to data for the clean substrate, (a) and (c). He atom scattering was also used to probe the vibrational properties of the 1.7 ML C_{60} film, and two dispersionless surface vibrational modes were observed. A mode at 1.5 meV (12 cm^{-1}), assigned to C_{60} vibrations normal to the substrate surface, and its overtone were reported [9.20].

Transmission electron microscopy (TEM) studies of C_{60} film growth on (001) mica were carried out at higher substrate temperatures [9.29], and a similar expansion of the in-plane lattice constant of single monolayer C_{60} films was found [9.29, 30]. Thicker films (2500 Å) of C_{60} grown on (100) mica at 200°C were found to grow epitaxially with the anticipated (111) orientation. The bright-field image shown in Fig. 9.6 demonstrates

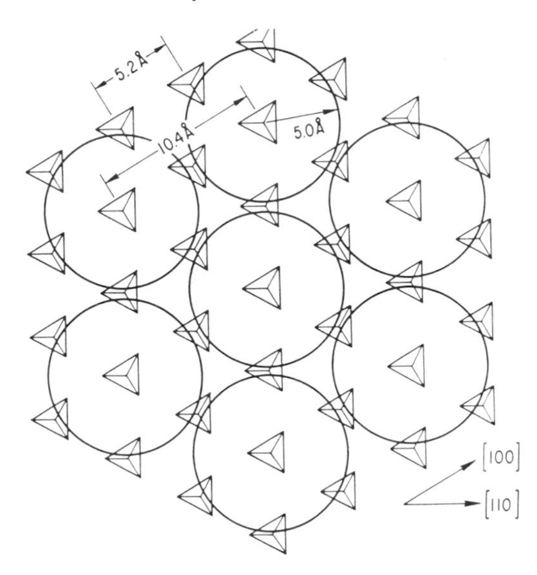

Fig. 9.4. Schematic view of a mica layer which consists of a hexagonal array of SiO_4 tetrahedra, upon which a C_{60} monolayer is then deposited. A silicon atom is located at the center of each tetrahedron and an oxygen at each corner. C_{60} molecules are depicted by the large circles [9.29].

the crystallinity of the film. It is inferred [9.22] that the interfacial strain between the mica and the C_{60} film is most likely removed in a few C_{60} monolayers, since C_{60} is a van der Waals solid. It was further noted [9.22] that the 3.7% lattice mismatch between the mica and C_{60} is much larger than that observed typically in strained layer epitaxy.

C_{60} film growth on NaCl (001) surfaces has also been studied [9.29] using TEM, and it was found that at lower substrate temperatures ($T <$ 100°C), the films (thickness 150–500 Å) were amorphous. For deposition onto 150°C (001) NaCl substrates, the films were observed to be polycrystalline, with a tendency toward a (110) orientation. Finally, at the still higher temperatures of 200 and 261°C, the films became more randomly oriented, and no tendency toward epitaxy was observed.

In contrast to the conventional deposition geometry, pseudoepitaxial (111) C_{60} films have been grown on (001) mica by a "hot-wall" method [9.22], such as used for the growth of device-grade epitaxial compound semiconductors. In this method, the substrates and C_{60} powder were loaded in opposite ends of a horizontal Pyrex tube (3 cm diameter bore by 30 cm long) located in an oven as shown schematically in Fig. 9.7. As the oven is heated, vapor transport of the C_{60} occurs toward the cooler end of the tube containing the substrates. Synthetic (Muscovite-1M) mica substrates were placed both horizontally and vertically near the open end of the hot-wall tube, which was pumped dynamically during the film growth process. Films of 1000–2000 Å were grown over a 12 h period. The C_{60} source and hot-wall temperatures were both approximately 400°C, and film growth

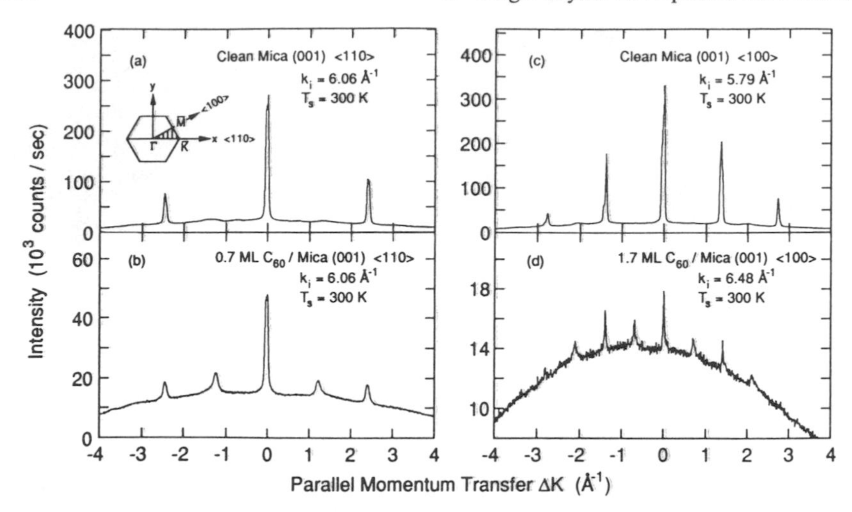

Fig. 9.5. He atom diffusion intensity as a function of parallel momentum transfer for the clean mica (001) surface at 300 K (a) in the ⟨110⟩ direction, (b) after deposition of a 0.7 monolayer of C_{60}, (c) in the ⟨100⟩ direction, and (d) after deposition of 1.7 monolayers of C_{60}. Both C_{60} depositions were done at the mica substrate temperature T_s of 300 K. The extra diffraction peaks are caused by the hexagonal crystal structure of the deposited C_{60}. The k_i values are for the incident wave vectors of the He atomic beam [9.20].

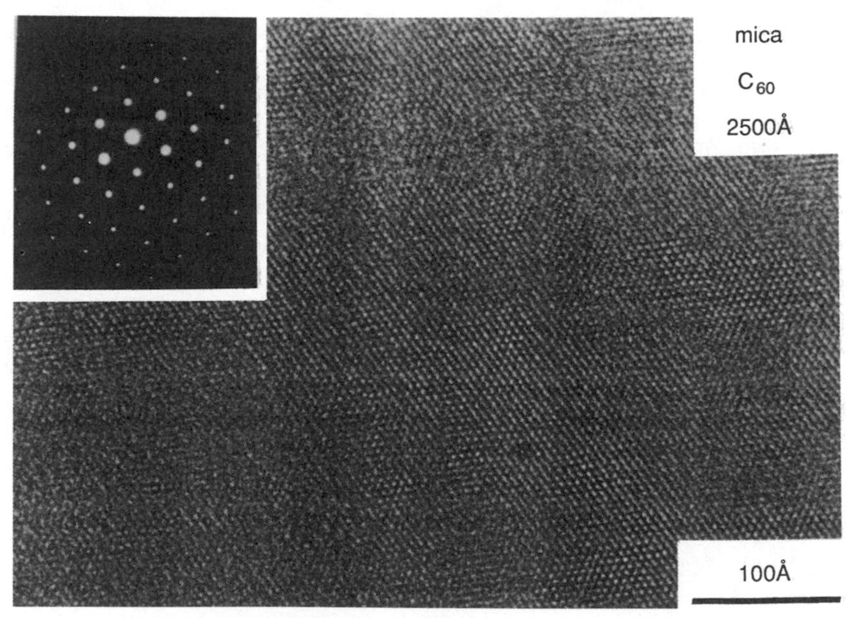

Fig. 9.6. Bright-field image of a (111) 2500 Å thick C_{60} film grown on mica at 200°C, with a magnification of 300,000 ×. The inset shows the electron diffraction pattern for this film [9.29].

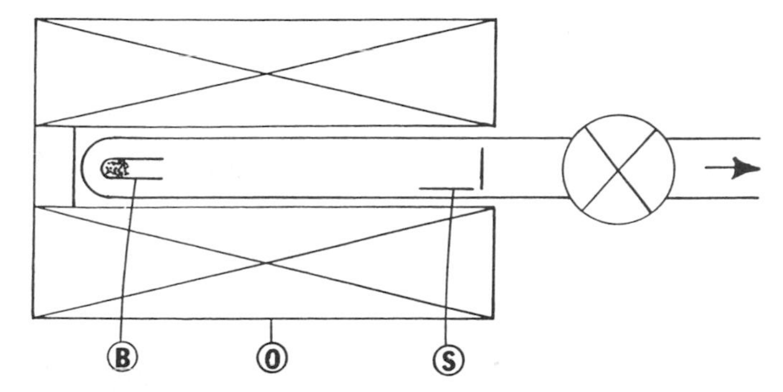

Fig. 9.7. Schematic diagram of a hot-wall oven for the sublimation of C_{60} films. Source material is contained in a small "bullet" (B) near the closed end of the hot-wall Pyrex tube, which in turn is connected to a vacuum system by a stopcock and O-ring fitting. Substrates (S) are supported horizontally or vertically near the open end of the oven (O) [9.22].

was studied for substrate temperatures between 20 and 100°C, somewhat lower than the value of 150°C reported to first yield crystalline (111) C_{60} film growth on (100) mica [9.29]. Using x-ray diffraction, it was shown that films grown by the hot-wall method were pseudoepitaxial with a 0.9° crystallite orientational spread, and with 850 Å and 450 Å out-of-plane and in-plane correlation lengths, respectively [9.22]. Subsequent work showed that with a specially designed effusion cell, a structural crystallite size of ~5000 Å could be obtained for C_{60} growth on mica at 185°C [9.23].

A comparative study of C_{60} thin-film growth in ultrahigh vacuum has been carried out on Si (100) and on the lamellar substrates GaSe (0001) and GeS (001) [9.30]. A Knudsen cell containing purified C_{60} in either a graphite or BN crucible was used for the source, and the films were characterized using low-energy electron diffraction (LEED) and high-resolution electron energy loss spectroscopy (HREELS) to study their structure and the vibrational modes. For the case of Si (100) substrates, where the C_{60}–substrate interaction is strong, C_{60} films of ~5 ML thickness were deposited on a clean surface maintained *near room temperature*. Although a LEED pattern associated with the film could not be detected, the off-specular HREELS beam was used to estimate a grain size of ~ 2.5 nm for the polycrystalline film. Because of the loss of longer-range order, the dipole selection rule in HREELS (see §11.5.9) was found to be relaxed and numerous intramolecular vibrational modes were observed as a result (see §11.3).

C_{60} film growth on the layered substrates GaSe (0001) and GeS (001) produced epitaxial films at 150–200°C [9.30]. At low coverage (~ 3 ML), epitaxial growth of fcc C_{60} films was observed on (0001) GaSe at a substrate temperature of ~150°C. The film exhibited a diffraction pattern of

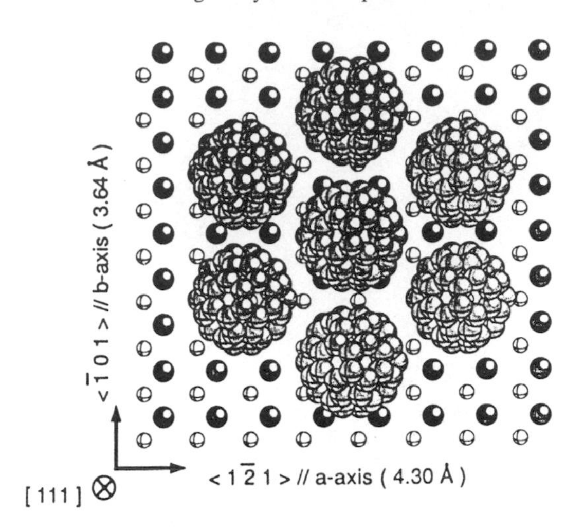

Fig. 9.8. Top view of the model structure of a C_{60} layer grown on a GeS (001) substrate. Small (very small) spheres represent the surface germanium (sulfur) atoms of GeS. The C_{60} $\langle\bar{1}01\rangle$($\langle 1\bar{2}1\rangle$) direction is parallel to the GeS *b*-axis (*a*-axis), so that the C_{60} (111) plane is parallel to the surface [9.30].

superimposed C_{60} (111) domains rotated by 23° with respect to each other. However, at thicker coverages, the LEED pattern was lost, indicating that thicker films grow in a polycrystalline form. The lattice mismatch between fcc C_{60} and GaSe (0001) is fairly large (12%) when compared to the mismatch of C_{60} to mica (4%) and MoS_2 (5%) [9.30], which probably explains why these latter substrates have been reported to grow thick epitaxial films.

Highly epitaxial, thick (~1000 Å) C_{60} films with a (111) fcc orientation were obtained in studies of C_{60} film growth on a (100) GeS substrate at ~200°C [9.20], for which only a ~0.75% lattice mismatch exists. A model structure based on a lattice constant with twice the *a*-axis of GeS ($2 \times 4.30 = 8.60$ Å), closely matching the separation between rows of C_{60} molecules along the $\langle\bar{1}01\rangle$ direction (8.68 Å), was used to model a C_{60} overlayer on a (001) GeS surface, as shown schematically in Fig. 9.8. It was reported that the epitaxy was achieved over millimeter square areas, despite the absence of a hexagonal surface symmetry in the substrate [9.30]. The authors attributed the good epitaxy they observed to deep corrugations of the GeS (001) surface (see Fig. 9.9). In another interesting experiment on C_{60} film growth on GeS (001), a film of several ML thickness was deposited onto a room temperature substrate and an amorphous film was obtained; subsequent annealing at ~200°C resulted in the desorption of the upper layers of the film, leaving an epitaxial (111) single ML film still attached to the substrate, indicating clearly that the substrate interaction with C_{60} molecules is stronger than the intermolecular C_{60}–C_{60} interaction.

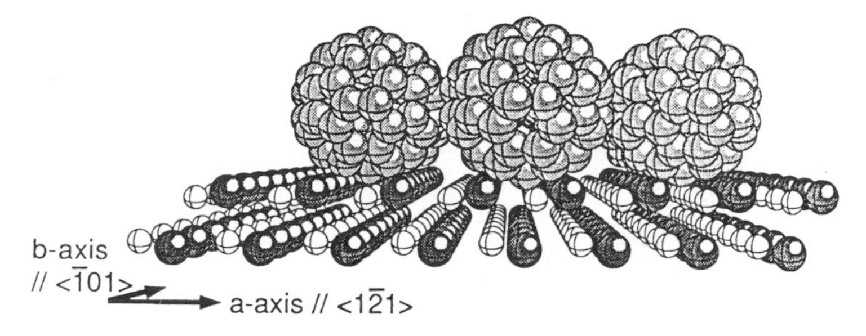

Fig. 9.9. Side view of the model structure of a C_{60} (111) monolayer on a GeS (001) substrate, showing a large corrugation in the *a*-direction of the basal plane and the systematic skipping of one interchain groove by the big C_{60} molecules [9.30].

Such phenomena have been observed on a number of surfaces with strong bonding to C_{60} (see §17.9).

The case of crystalline C_{60} growth on Si substrates is of particular interest for technological reasons. It is well established that the high density of dangling bonds on Si (111), Si (100), and Si (110) surfaces results in strong C_{60}–Si bonding, and because of the lattice mismatch between C_{60} and the various Si faces, the C_{60} film growth usually has poor crystallinity (crystallite size of only a few nm) [9.30]. It was, however, found that by passivation of the Si surface (through hydrogenation) to tie up the dangling bonds, the bonding of C_{60} to the passivated Si (100) and Si (111) surfaces could be greatly reduced, so that crystalline C_{60} films could be prepared [9.31]. Shown in Fig. 9.10 is a comparison of an x-ray scan taken from a C_{60} film on an Si (100) surface without passivation (the lower trace labeled hydrophilic) and with passivation (the upper trace labeled hydrophobic). The passivation or hydrogenation of the dangling bonds was done with a 5% HF solution for the Si (100) face, and with the same 5% HF solution followed by a 40% NH_4F solution for the Si (111) face. It was further found that the crystallinity of the C_{60} film improved with increased thickness of the film, as shown in Fig. 9.11 [9.31].

9.2.2. *Free-Standing C_{60} Films, or Fullerene Membranes*

Free-standing membranes of C_{60} films 2000–6000 Å thick with areas as large as 6×6 mm^2 have been fabricated successfully [9.32]. These membranes, supported on a Si frame, were found to be robust, despite the weak van der Waals bonding between C_{60} molecules. These membranes have been used for optical spectroscopy and mechanical properties measurements (e.g., Young's modulus) as well as for potential applications (e.g., micropore filters, x-ray windows, TEM sample supports). Figure 9.12 shows

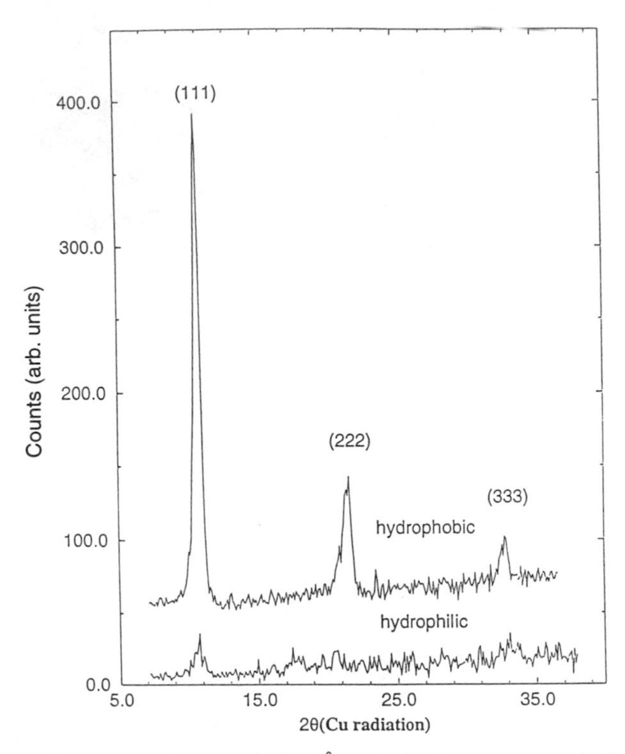

Fig. 9.10. Cu K_α x-ray $\theta - 2\theta$ scan of a 500 Å thick C_{60} film grown on a hydrophobic (top) and hydrophilic (bottom) Si (100) substrate. The data were collected against a fixed monitor count [9.31].

the sequence of deposition and etching steps used to fabricate free-standing C_{60} membranes [9.32]. First, low-stress, 1000 Å Si_3N_4 films are deposited on both the top and bottom of the central region of the 100-μm-thick Si (100) wafer. After a photolithography step, which allows the removal of the Si_3N_4 film from the central region of the top side of the substrate, an anisotropic KOH etching solution is used to remove the Si in the central region of the wafer, leaving the lower Si_3N_4 film intact. A thin (2000–6000 Å) C_{60} film is then deposited over the lower Si_3N_4 film by conventional sublimation techniques, and the Si_3N_4 film backing of the C_{60} film is then removed by CF_4 reactive plasma etching. What remains after this process is a C_{60} membrane supported by the surrounding Si wafer which functions as a frame. The C_{60} membrane was further thinned by continuing the last CF_4 reactive plasma etch without apparent damage to the film, as demonstrated in Fig. 9.13, which shows the evolution of the ultraviolet (UV)–visible optical spectrum of a membrane with increased exposure to the CF_4 plasma etch for 8, 14,

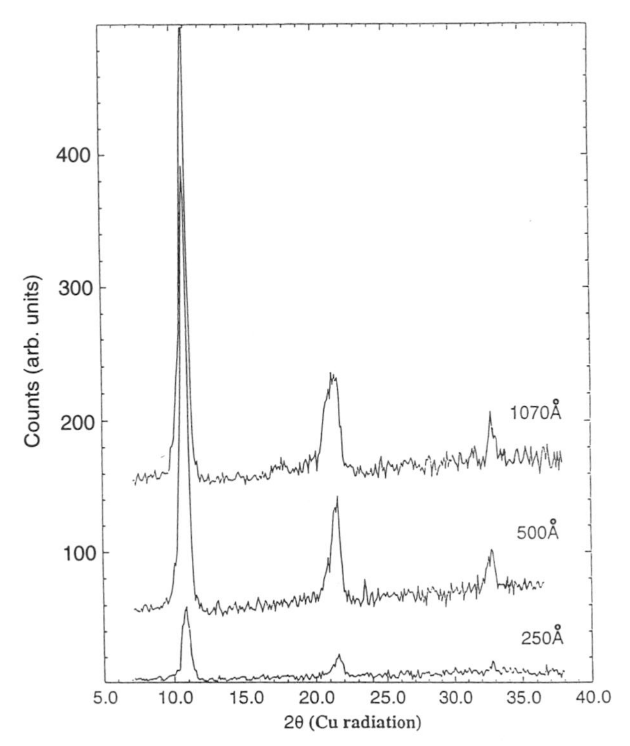

Fig. 9.11. X-ray diffraction patterns of three C_{60} films with indicated thickness grown on a hydrophobic Si (100) substrates. Data were collected in the $\theta - 2\theta$ scan mode and at a fixed monitor count. For covenience in plotting, the three traces are displaced from one another vertically [9.31].

and 18 min. The spectra are shown in comparison to that prior to etching (labeled 0 min). The spectra exhibit an optical absorption edge in the red, as well as the prominent, dipole-allowed absorption bands involving transitions between molecular states near the Fermi energy (see §13.3.2). As the membrane is thinned, the optical absorption decreases, as expected, and the absorption bands and absorption edge retain their shape, indicating that the C_{60} molecular shells, as well as the fcc crystal structure, remain intact.

9.3. Synthesis of Doped C_{60} Crystals and Films

In this section, we review general experimental techniques for the preparation of doped fullerene-based solids. Details concerning the preparation and purification of the fullerene starting material appear elsewhere (see

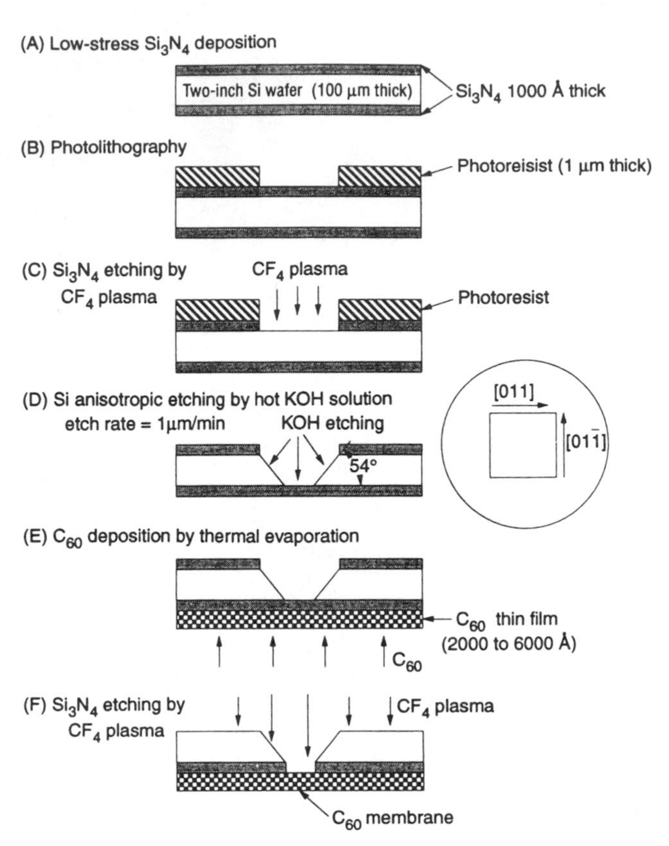

Fig. 9.12. Sequence of deposition and etching steps used to fabricate C_{60} membranes on (100) oriented Si [9.32].

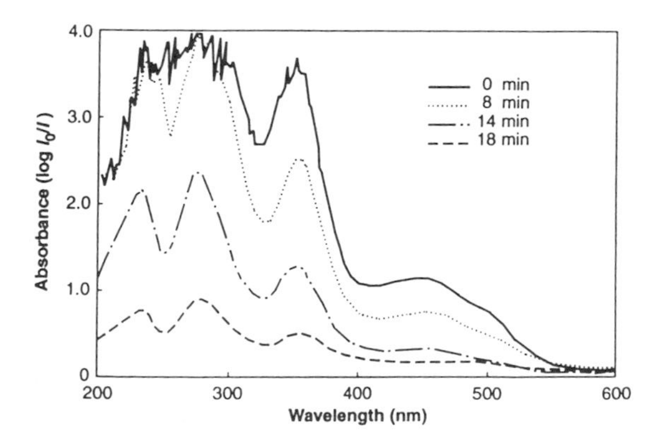

Fig. 9.13. Variation of the ultraviolet–visible optical absorbance of a C_{60} film initially 3200 Å thick deposited on a quartz substrate. The film was plasma etched with CF_4 for the time durations indicated in the legend, thereby thinning the film and reducing its absorbance, while retaining the spectral features [9.32].

§5.1, §5.2, §5.3, §9.1, §9.2), and techniques for the endohedral doping of fullerenes are reviewed in §5.4.

9.3.1. *Charge Transfer Compounds Based on Alkali Metals and Alkaline Earths*

One general approach to alkali metal (M) doping is to introduce a near-stoichiometric amount of M to *oxygen-free* C_{60} films, single crystals, or microcrystalline powder placed in a sealed glass tube. The tube is then evacuated and raised in temperature to 200–300°C to promote intercalation. Depending on the grain size and the longest diffusion length of the C_{60} starting material and the intercalation temperature, several hours to several days may typically be required to achieve homogeneous doping. Several studies of M_xC_{60} compounds have been carried out during intercalation as x passes from zero to its maximum value. In these types of experiments it is important to exercise caution to obtain homogeneous doping and a single phase region over the depth of the physical probe used for the sample characterization. Several phases (e.g., $x = 0, 1, 3, 4, 6$) may coexist, which complicates (or may even invalidate) the interpretation of experimental characterization data. Alkali metal alloy compounds, e.g., $Rb_xCs_yC_{60}$, have also been synthesized using a similar technique, primarily aimed at expanding the lattice to enhance the superconducting transition temperature [9.33] (see §15.1).

Doped charge transfer materials are prepared by exposing fcc C_{60} to the vapors of alkali metals or alkaline earths. The form of the starting C_{60} material, microcrystalline powder, single crystals or thin films is, of course, largely dictated by the nature of the experiments to be performed. Microcrystalline powders are easiest to work with, because it is relatively easy to obtain large quantities of fullerenes and fullerene derivatives in powder form. Synthesis of the doped compounds is also simplified since a significant mass of C_{60} powder can be placed in the ampoule, and therefore a stoichiometric amount of M dopants (or M and M′ dopants in the case of an alloy such as $Rb_{3-x}Cs_x$) can be added easily to the ampoule by weight. In the case of reactive alkali metals (M,M′), it is often convenient to siphon molten alkali metal into the bore of a glass capillary and to break off the desired length of the capillary to introduce a stoichiometric amount of alkali metal into the reaction chamber. The ampoule containing the mixture of C_{60} and alkali metal is loaded into a furnace and reacted for a time range from hours to days in an attempt to achieve the desired single-phase reaction product [9.13].

The doping of C_{60} with alkali metals can also be achieved in a two-temperature oven, similar to the apparatus used to prepare alkali metal graphite intercalation compounds [9.34]. This approach works well with powdered samples and has also been used to intercalate single crystals. The stoichiometry is controlled by the number of moles of alkali metal introduced into the sealed ampoule (quartz or Pyrex) relative to the number of moles of C_{60}, the temperatures of the reagents (usually less than the temperature of the C_{60}), and the duration of the reactions.

Alkali metal doping of thin (~1000 Å) C_{60} films can be carried out similarly, using a sealed quartz or Pyrex tube (with approximate dimensions 25 cm long × 1 cm diameter) which contains the C_{60} film/substrate sample and the alkali metal at opposite ends. The films are maintained at a higher temperature (200°C) than the alkali metal (M) (80–150°C) to avoid condensation of the alkali metal on the film surface. The *in situ* progress of the reactions can be monitored qualitatively by the color change observed in the films and more quantitatively by the doping-induced downshift of the strongest C_{60} Raman line (see Fig. 11.24) or by changes in the electrical conductivity (see §14.1.1).

The doping reaction in C_{60} single crystals has been studied during the preparation of doped single crystals for electrical transport measurements as a function of temperature [9.16]. Electrical contacts to the samples are made prior to doping, by first evaporating silver onto the crystal surfaces and then attaching gold wires with conducting silver paint. The mounted samples are then sealed together with fresh potassium metal in a Pyrex glass apparatus with tungsten feedthrough leads. Uniform doping can be accomplished using a repetitive high-temperature doping–anneal cycle. First, both the sample and dopant are heated uniformly in a furnace, while the sample resistance is continuously monitored. The temperature is raised from room temperature to about 200°C at a rate of ~6°C/min. At about 150°C, the resistance of the sample drops to a measurable range (~20 MΩ). As T is raised further, the resistivity of the sample drops continually to a few hundred mΩcm within a few min. The tube should be maintained at about 200°C for approximately one-half hour until the resistance of the sample reaches a minimum. The potassium end of the tube is now cooled to room temperature and the sample alone is annealed at about 200°C to 250°C overnight. Following this overnight anneal, the potassium end of the ampoule is reheated to ~200°C and the sample is further doped, until a lower resistivity is reached. The sample is then annealed again for several h. This doping and annealing process is repeated until the resistance reaches an equilibrium minimum resistivity value. For transport measurements, the sample cell is injected with helium exchange gas to ensure good thermal conduction, and the sample temperature is determined by a diode temperature sensor which is mounted in the cell adjacent to the crystal.

The lattice constant of M_xC_{60}-related compounds has in some cases been increased by the subsequent reaction with NH_3 gas [9.35]. Similar reactions were tried previously with alkali metal–doped graphite intercalation compounds [9.36]. The conceptual idea behind the ammoniation of the M-doped fullerene compounds is to have the ammonia complex and the alkali metal ion both occupy the same octahedral interstitial site, thereby expanding the fcc lattice. Under similar experimental conditions, $T = 100°C$ and 0.5 atm of dry NH_3 gas distilled from liquid ammonia, a microcrystalline $M_xM'_{3-x}C_{60}$ powder was reacted with the ammonia for ~1–10 days in a 9-cm sealed Pyrex tube, previously evacuated to 10^{-5} torr [9.35]. The following results were obtained: Rb_3C_{60} did not react with the NH_3, while K_3C_{60} yielded $NH_3K_3C_{60}$ [9.37, 38], and Na_2MC_{60} (M = Rb, Cs) resulted in $(NH_3)_4Na_2MC_{60}$ [9.39], where the NH_3 stoichiometry was determined by weight uptake measurements. Prior to this annealing step, the powder was first exposed to 0.5 atm NH_3 gas at room temperature for 1–2 days, and a substantial pressure drop was noticed during the first few min of exposure, indicating immediate ammonia uptake. The NH_3 reaction was found to be thermally reversible; *mild* heating of the ammoniated compounds in vacuum returned the lattice parameters and T_c to their original values (see §8.5.5 and §15.1). This observation suggests that amides are not being formed and that neutral NH_3 molecules are intercalating into the structure and solvating the alkali metal ions.

The preparation of the alkaline earth intercalation compounds represents more of a challenge than that faced in the synthesis of the alkali metals because of the higher reactivity of the alkaline earths with quartz and the higher reaction temperatures required to prepare the alkaline earth compounds. To prepare Ca_xC_{60} [9.40], high-purity (99.99%) dendritically solidified calcium powdered metal is weighed and mixed with purified C_{60} powder, in high-purity tantalum tubes, using various Ca:C_{60} ratios to obtain the desired stoichiometries for the end products. The tantalum tubes containing the Ca and the C_{60} are then loaded inside quartz tubes. Since quartz reacts with calcium, tantalum tubes are needed to avoid sample contamination by unwanted reaction products. All sample preparation is done in a controlled-atmosphere glove box where the oxygen or water vapor concentrations are maintained below a few parts per million. After the quartz tubes are loaded as described above, an ultrahigh-vacuum valve is connected to the open end of the quartz tube. The quartz tube is then taken out of the dry box and sealed under a vacuum of 10^{-6} torr. Heat treatments are done in tube furnaces at 550°C for 20 h, although the reaction is nearly complete after only 1 h at this temperature. Following heat treatment, the samples are removed from the tantalum tubes in the glove box and are loaded under vacuum into appropriate sample cells for characterization (e.g., x-ray diffraction studies) and properties measurements.

A similar preparation method is used for the preparation of strontium and barium fulleride compounds. Strontium metal is intercalated at temperatures 450–600°C and for periods ranging from minutes to weeks. The best samples showing the highest superconducting volume fraction are prepared by anneals at 550°C for a few days. Ba_3C_{60} and Ba_6C_{60} compounds are prepared [9.41, 42] at higher heat treatment temperatures (550–800°C) and for longer time periods (from hours to weeks) than the corresponding Sr_xC_{60} compounds [9.41, 42].

The reaction of C_{60} with the Lewis acids $SbCl_5$, AsF_5, and SbF_5 were carried out in a two-arm tube furnace (~150°C for the $SbCl_5$ reaction with C_{60}). After removing the sample from the furnace, the arm containing the reaction products was kept at 50°C and the second arm was placed in liquid nitrogen to distill the $SbCl_5$. A second method used for these reactions was a solution method based on $SbCl_5$ dissolved in CCl_4 [9.43]. It is inferred that the reaction of C_{60} with these Lewis acids results in the transfer of electrons from C_{60} to the Lewis acids [9.43].

9.3.2. Fullerene-Based Clathrate Materials

The methods for preparing clathrate materials depend to a large extent on the nature and size of the dopant molecule. If the dopant is a small molecule, an overpressure of the molecular gas above fcc solid C_{60} is sufficient to induce room temperature intercalation [9.44] at pressures of 1 kbar and 0.14 kbar to introduce O_2 and H_2, respectively, into a substantial fraction of the large, octahedral sites in the fcc C_{60} lattice. Exposure to 1 kbar O_2 for 116 h resulted in the occupancy of 25% of these interstitial spaces. By using visible light and 1 atm of O_2, it is possible to fill nearly all the octahedral sites in a thin film (~2000 Å) in a much shorter time (~ 1 h) [9.45, 46]. Through study of the isothermal adsorption or intercalation of N_2 and O_2 into microcrystalline C_{60} powder and from analysis of the N_2 isotherm at 77 K, it was concluded that the internal nanopore surface area associated with the octahedral sites was ~24 m^2/g [9.47]. It was also found [9.44, 47] that O_2 adsorption (and desorption) was reversible at 303 K, but the saturation concentration of O_2 was found to be higher than for N_2, namely one O_2 molecule per two C_{60} molecules, or 50% of complete saturation of the octahedral sites [9.47]. Presumably, for a large uptake of O_2 and N_2, there must be enough defects in the C_{60} crystalline lattice to link many octahedral sites effectively, otherwise the diffusion rate would be very slow. It was pointed out that lattice defects of more than 0.2 per unit cell can produce a percolating micropore structure [9.47]. Most recently, studies of the isothermal adsorption of He into microcrystalline C_{60} (80%)/C_{70} (20%) powders [9.48] showed that ^{4}He intercalation into these powders takes place more readily as the temperature is decreased, in contrast to previous studies of C_{60} films [9.49, 50], where ^{4}He was found not to intercalate C_{60} films at low temperatures.

Larger molecules, such as organic solvent molecules and ring sulfur (S_8) clusters, have been introduced into crystalline clathrate structures containing either C_{60} or C_{70}. For example, the crystal structures which result from heating mixtures of sulfur and microcrystalline C_{60} and C_{70} powders have been investigated [9.51] and small clathrate crystals containing oriented C_{60} or C_{70} molecules and crown-shaped S_8 rings were obtained.

To prepare sulfur–C_{60} clathrate compounds (see §8.4.2), the C_{60} (C_{70}) and excess sulfur were heated in an evacuated glass ampoule to a temperature slightly above the melting point of sulfur [9.51]. At this temperature, the S_8 rings are the dominant species in the melt. For the case of C_{60} and sulfur, the mixture was then cooled slowly (1°C/h) down to room temperature to produce intergrown clathrate crystals with composition $C_{60}(S_8)_2$. These crystals can also be grown at room temperature from solution, using CCl_4 and 10% CS_2 (volume) as the solvent [9.51]. Using higher concentrations of CS_2 in CCl_4, the compound $C_{60}(S_8)_2CS_2$ was formed by slow evaporation of the solvent (CS_2 evaporates preferentially). It was further shown that isolated isometric crystals of $C_{60}(S_8)_2$ with linear dimensions up to 200 μm can be grown (i.e., two S_8 rings per C_{60} unit) [9.52].

Noncubic clathrate crystals containing C_{60} and solvent molecules were discovered in 1991 by several groups [9.3, 11, 53]. More recently, C_{60} clathrates formed with aliphatic molecules [e.g., *n*-pentane (C_5H_{12}), diethyl ether ($C_4H_{10}O$), and 1,3 dibromopropane ($C_3H_6Br_2$)] have been prepared in single-crystal form [9.54]. These clathrate crystals, containing organic solvent molecules, are grown from solutions, either containing the pure solvent to be incorporated into the structure or in a solution containing a second solvent species. For example, aliphatic molecules were incorporated into C_{60} clathrate crystals from a toluene solution by precipitation methods [9.55]. These C_{60}–aliphatic clathrate molecules all exhibited an identical 0.1 eV blue shift of the optical absorption edge, indicating a nonspecific, weak interaction between C_{60} and the aliphatic molecules [9.54], consistent with clathrate behavior.

References

[9.1] R. L. Meng, D. Ramirez, X. Jiang, P. C. Chow, C. Diaz, K. Matsuishi, S. C. Moss, P. H. Hor, and C. W. Chu. *Appl. Phys. Lett.*, **59**, 3402 (1991).

[9.2] R. M. Fleming, T. Siegrist, P. M. Marsh, B. Hessen, A. R. Kortan, D. W. Murphy, R. C. Haddon, R. Tycko, G. Dabbagh, A. M. Mujsce, M. L. Kaplan, and S. M. Zahurak. In

R. S. Averback, J. Bernholc, and D. L. Nelson (eds.), *Clusters and Cluster-Assembled Materials, MRS Symposia Proceedings, Boston*, vol. 206, pp. 691–696, Materials Research Society Press, Pittsburgh, PA (1991).

[9.3] R. M. Fleming, A. R. Kortan, B. Hessen, T. Siegrist, F. A. Thiel, P. M. Marsh, R. C. Haddon, R. Tycko, G. Dabbagh, M. L. Kaplan, and A. M. Mujsce. *Phys. Rev. B*, **44**, 888 (1991).

[9.4] M. A. Verheijen, H. Meekes, G. Meijer, P. Bennema, J. L. de Boer, S. van Smaalen, G. V. Tendeloo, S. Amelinckx, S. Muto, and J. van Landuyt. *Chem. Phys.*, **166**, 287 (1992).

[9.5] M. Haluška, H. Kuzmany, M. Vybornov, P. Rogl, and P. Fejdi. *Appl. Phys. A*, **56**, 161 (1993).

[9.6] M. A. Verheijen, H. Meekes, G. Meijer, E. Raas, and P. Bennema. *Chem. Phys. Lett.*, **191**, 339 (1992).

[9.7] W. Krätschmer, L. D. Lamb, K. Fostiropoulos, and D. R. Huffman. *Nature (London)*, **347**, 354 (1990).

[9.8] J. M. Hawkins, S. Loren, A. Meyer, and R. Nunlist. *J. Am. Chem. Soc.*, **113**, 7770 (1991).

[9.9] J. M. Hawkins, T. A. Lewis, S. Loren, A. Meyer, R. J. Saykally, and F. J. Hollander. *J. Chem. Soc. Chem. Commun.*, p. 775 (1991).

[9.10] S. M. Gorun, K. M. Creegan, R. D. Sherwood, D. M. Cox, V. W. Day, C. S. Day, R. M. Upton, and C. E. Briant. *J. Chem. Soc. Chem. Commun.*, p. 1556 (1991).

[9.11] K. Kikuchi, S. Suzuki, K. Saito, H. Shiramaru, I. Ikemoto, Y. Achiba, A. A. Zakhidov, A. Ugawa, K. Imaeda, H. Inokuchi, and K. Yakushi. *Physica C*, **185–189**, 415 (1991).

[9.12] Y. Yosida. *Jpn. J. Appl. Phys.*, **31**, L505 (1992).

[9.13] A. R. Kortan, N. Kopylov, and F. A. Thiel. *J. Phys. Chem. Solids*, **53**, 1683 (1992).

[9.14] Y. Yosida, T. Arai, and H. Suematsu. *Appl. Phys. Lett.*, **61**, 1043 (1992).

[9.15] L. D. Marks. *J. Crystal Growth*, **61**, 556 (1983).

[9.16] X.-D. Xiang, J. G. Hou, G. Briceño, W. A. Vareka, R. Mostovoy, A. Zettl, V. H. Crespi, and M. L. Cohen. *Science*, **256**, 1190 (1992).

[9.17] K. K. Schuegraf. *Handbook of Thin-Film Deposition Processes and Techniques*. Noyes Publications, Park Ridge, NJ (1988).

[9.18] F. J. Hollander. In J. Russel and J. Hill (eds.), *Physical Vapor Deposition*, BOC Group, Inc., Berkeley, CA (1986).

[9.19] Y. Z. Li, J. C. Patrin, M. Chander, J. H. Weaver, L. P. F. Chibante, and R. E. Smalley. *Science*, **252**, 547 (1991).

[9.20] D. Schmicker, S. Schmidt, J. G. Skoffronick, J. P. Toennies, and R. Vollmer. *Phys. Rev. B*, **44**, 10995 (1991).

[9.21] W. Krakow, N. M. Rivera, R. A. Roy, R. S. Ruoff, and J. J. Cuomo. *J. Mater. Sci.*, **7**, 784 (1992).

[9.22] J. E. Fischer, E. Werwa, and P. A. Heiney. *Appl. Phys. A*, **56**, 193 (1993).

[9.23] A. Fartash. *Appl. Phys. Lett.*, **64**, 1877 (1994).

[9.24] G. Gensterblum, L.-M. Yu, J. J. Pireaux, P. A. Thiry, R. Caudano, P. Lambin, A. A. Lucas, W. Krätschmer, and J. E. Fischer. *J. Phys. Chem. Solids*, **53**, 1427 (1992).

[9.25] M. Sakurai, H. Tada, K. Saiki, and A. Koma. *Jpn. J. Appl. Phys.*, **30**, L1892 (1991).

[9.26] J. A. Dura, P. M. Pippenger, N. J. Halas, X. Z. Xiong, P. C. Chow, and S. C. Moss. *Appl. Phys. Lett.*, **63**, 3443 (1993).

[9.27] A. F. Hebard, R. C. Haddon, R. M. Fleming, and A. R. Kortan. *Appl. Phys. Lett*, **59**, 2109 (1991).

[9.28] S. L. Ren, Y. Wang, A. M. Rao, E. McRae, G. T. Hager, K. A. Wang, W. T. Lee, H. F. Ni, J. Selegue, and P. C. Eklund. *Appl. Phys. Lett.*, **59**, 2678 (1991).

[9.29] W. Krakow, N. M. Rivera, R. A. Roy, R. S. Ruoff, and J. J. Cuomo. *Appl. Phys. A*, **56**, 185 (1993).
[9.30] G. Gensterblum, L. M. Yu, J. J. Pireaux, P. A. Thiry, R. Caudano, J. M. Themlin, S. Bouzidi, F. Coletti, and J. M. Debever. *Appl. Phys. A*, **56**, 175 (1993).
[9.31] A. F. Hebard, O. Zhou, Q. Zhong, R. M. Fleming, and R. C. Haddon. *Thin Solid Films*, **257**, 147 (1995).
[9.32] C. B. Eom, A. F. Hebard, L. E. Trimble, G. K. Celler, and R. C. Haddon. *Science*, **259**, 1887 (1993).
[9.33] K. Tanigaki, S. Kuroshima, J. Fujita, and T. W. Ebbesen. *Appl. Phys. Lett.*, **63**, 2351 (1993).
[9.34] A. Hérold. In F. Lévy (ed.), *Physics and Chemistry of Materials with Layered Structures*, vol. 6, p. 323. Reidel Dordrecht, New York (1979).
[9.35] P. Zhou, Z. H. Dong, A. M. Rao, and P. C. Eklund. *Chem. Phys. Lett.*, **211**, 337 (1993).
[9.36] R. Setton. In H. Zabel and S. A. Solin (eds.), *Graphite Intercalation Compounds I: Structure and Dynamics*. Springer-Verlag, Berlin (1990). Springer Series in Materials Science, Vol. 14, p. 305.
[9.37] M. J. Rosseinsky, D. W. Murphy, R. M. Fleming, and O. Zhou. *Nature (London)*, **364**, 425 (1993).
[9.38] O. Zhou, T. T. M. Palstra, Y. Iwasa, R. M. Fleming, A. F. Hebard, P. E. Sulewski, D. W. Murphy, and B. R. Zegarski. *Phys. Rev. B*. **52**, 483 (1995).
[9.39] O. Zhou, R. M. Fleming, D. W. Murphy, M. J. Rosseinsky, A. P. Ramirez, R. B. van Dover, and R. C. Haddon. *Nature (London)*, **362**, 433 (1993).
[9.40] A. R. Kortan, N. Kopylov, S. H. Glarum, E. M. Gyorgy, A. P. Ramirez, R. M. Fleming, F. A. Thiel, and R. C. Haddon. *Nature (London)*, **355**, 529 (1992).
[9.41] A. R. Kortan, N. Kopylov, S. H. Glarum, E. M. Gyorgy, A. P. Ramirez, R. M. Fleming, O. Zhou, F. A. Thiel, P. L. Trevor, and R. C. Haddon. *Nature (London)*, **360**, 566 (1992).
[9.42] A. R. Kortan, N. Kopylov, R. M. Fleming, O. Zhou, F. A. Thiel, R. C. Haddon, and K. M. Rabe. *Phys. Rev. B*, **47**, 13070 (1993).
[9.43] W. R. Datars, P. K. Ummat, T. Olech, and R. K.Nkum. *Solid. State Commun.*, **86**, 579 (1993).
[9.44] R. A. Assink, J. E. Schirber, D. A. Loy, B. Morosin, and G. A. Carlson. *J. Mater. Res.*, **7**, 2136 (1992).
[9.45] P. Zhou, A. M. Rao, K. A. Wang, J. D. Robertson, C. Eloi, M. S. Meier, S. L. Ren, X. X. Bi, P. C. Eklund, and M. S. Dresselhaus. *Appl. Phys. Lett.*, **60**, 2871 (1992).
[9.46] C. Eloi, J. D. Robertson, A. M. Rao, P. Zhou, K. A. Wang, and P. C. Eklund. *J. Mater. Res.*, **8**, 3085 (1993).
[9.47] K. Kaneko, C. Ishii, T. Arai, and H. Suematsu. *J. Phys. Chem.*, **97**, 6764 (1993).
[9.48] C. P. Chen, S. Mehta, L. P. Fu, A. Petrou, F. M. Gasparini, and A. F. Hebard. *Phys. Rev. Lett.*, **71**, 739 (1993).
[9.49] K. S. Ketola, S. Wang, R. B. Hallock, and A. F. Hebard. *J. Low. Temp. Phys.*, **89**, 609 (1992).
[9.50] P. W. Adams, V. Pont, and A. F. Hebard. *J. Phys. C*, **4**, 9525 (1992).
[9.51] G. Roth and P. Adelmann. *Appl. Phys. A*, **56**, 169 (1993).
[9.52] G. Roth, P. Adelmann, and R. Knitter. *Materials Lett.*, **16**, 357 (1993).
[9.53] B. Morosin, P. P. Newcomer, R. J. Baughman, E. L. Venturini, D. Loy, and J. E. Schirber. *Physica C*, **184**, 21 (1991).
[9.54] K. Kamarás, A. Breitschwerdt, S. Pekker, K. Fodor-Csorba, G. Faigel, and M. Tegze. *Appl. Phys. A*, **56**, 231 (1993).
[9.55] S. Pekker, G. Faigel, K. Fodor-Csorba, L. Gránásy, E. Jakab, and M. Tegze. *Solid State Commun.*, **83**, 423 (1992).

CHAPTER 10

Fullerene Chemistry and Electrochemistry

Fullerene chemistry has become a very active research field, largely because of the uniqueness of the C_{60} molecule [10.1] and the variety of fullerene derivatives that appear to be possible. The synthesis of crystalline M_3C_{60} (M = K, Rb, Cs) compounds by the chemical reduction of C_{60} with alkali metals led to the discovery in 1991 of moderately high temperature ($T_c \sim 20$ K) superconductivity in these compounds [10.2]. Since that date, chemists have learned how to generate a diverse group of fullerene derivatives, where molecular fragments are bonded to the C_{60} cage, leaving the cage essentially intact, although the C–C bond lengths in the vicinity of the attachment are perturbed [10.3–7]. Many of these chemical reactions have taken advantage of the electrophilic properties of fullerenes, that is, their tendency to attract electrons.

Because all the carbon bonds in the fullerene molecule are satisfied within the molecular shell, substitution reactions involving the exchange of one chemical group for another, as is common in organic chemistry [10.8, 9], are not possible, although substitution of boron for carbon in the shell has been observed in molecular beam experiments [10.1]. Furthermore, catastrophic decomposition of the carbon shell occurs in oxygen above 400°C [10.10, 11].

Chemical groups have been attached to the fullerene molecule through bonds between carbon atoms in the fullerene shell and transition metal, nitrogen, oxygen, or carbon atoms in the adduct. The reactive site in the fullerene is, in most cases, the C=C (double) bond bridging pentagons [or in other words, the double bond which is located at the fusion of two

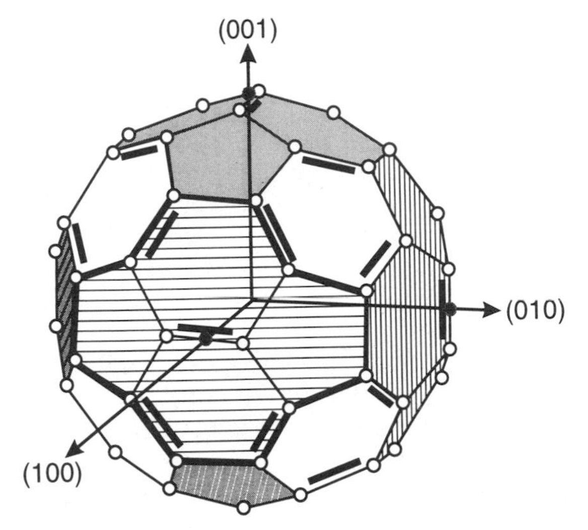

Fig. 10.1. Schematic view of a C_{60} molecule emphasizing the octahedral arrangement of pyracylene units. A pyracylene unit contains two pentagons and two hexagons and a central (reactive) double bond. The pyracylene unit centered on the (100) axis is highlighted, and all rings associated with these units are shaded in the figure. Double bonds are located at the fusion of hexagonal rings.

hexagonal rings (see Fig. 10.1)] in a pyracylene unit (shaded). The chemical behavior of C_{60} has been reported to be similar to that of electron-deficient polyalkenes, -(RC=CR)$_n$- where R represents an alkyl, hydrogen, or aromatic group [10.4, 7].

The limited availability of gram quantities of fullerenes, other than C_{60} and C_{70}, has largely confined fullerene chemistry to the study of chemical reactions based on C_{60} and C_{70}, with most of the chemical studies reported for C_{60} and some smaller effort given to chemical reactions based on C_{70}. Almost no chemical synthesis studies involving higher-mass fullerenes have thus far been published.

In this chapter, we first review some of the general characteristics of chemical reactions involving fullerenes, classify these reactions into major categories (reduction, bridging, addition, polymerization, host–guest complexing, etc.), then give a few examples of chemical reactions in several of these categories. Intercalation reactions in the solid state are discussed in Chapter 8 and in §9.3, while the synthesis, extraction, and purification of fullerenes are discussed in §5.1, §5.2, and §5.3, including the use of chromatography for separations. Also related to fullerene chemistry is fullerene surface science, discussed in §17.9.

10.1. Practical Considerations in Fullerene Derivative Chemistry

Since many of the fullerene reactions seem to be reversible and the amount of the fullerene material available to most chemists is relatively small, fullerene chemistry is one of the most challenging research areas at the cutting edge of the chemical sciences. To add to the challenge, chemical analysis of fullerene derivatives using a variety of spectroscopic techniques can be difficult, because of the poor solubility of these species in water or common organic solvents. Regarding solid-state chemistry, the crystallization of many of the fullerene-derived products has been found to be difficult, so that definitive structural determinations are complicated by the unavailability of single-crystal specimens of adequate quality and size. Fullerenes have also been found to form solvates with many organic solvents, so that attempts to grow crystals from solution instead yield crystals of the solvated species.

The chemical reactivity of C_{60} can be strongly photosensitive and oxygen sensitive. For example, C_{60} can be polymerized by ultraviolet (UV)–visible light (see §7.5) and the chemical reactivity of polymerized C_{60} differs from that of the individual C_{60} monomers. Furthermore, the characteristics of the polymerization process are highly sensitive to the presence of oxygen [10.12, 13]. Thus C_{60} must be stored and handled with appropriate care to avoid unwanted polymerization and photoinduced intercalation of O_2 prior to use, or while carrying out chemical reactions, or while characterizing the reaction products. In some cases, oxygen in the presence of light can be used as a cage-opening process [10.14–16] to promote specific chemical reactions.

10.2. General Characteristics of Fullerene Reactions

Since C_{60} is a unique molecule in a number of ways, it can be expected that the chemistry governing C_{60} will also have some unique characteristics. It is further believed that chemical reactions involving any fullerene have some common characteristics. In this section we discuss some of these characteristics.

The bonding in C_{60} can be described approximately by an sp^2 configuration. However, curvature of the fullerene cage leads to a small admixture of sp^3 character. All fullerene molecules share the closed cage structure (see the discussion of Euler's theorem in §3.1). As the number of carbon atoms in the cage increases, the curvature is reduced, and the chemical behavior should approach that of graphite without the dangling bond edge sites.

Another feature of importance to chemical reactions with C_{60} arises from its unique structure. The regular truncated icosahedron is the only one out of over 12,000 possible structures that can be formed from 60 carbon atoms [10.17] which has pentagons separated from each other (the isolated pentagon rule). It is believed that after chemical reactions have occurred, there will be little change to the C_{60} shell, and the isolated pentagon rule will still be valid [10.4], thus preventing double bonds and the attendant strain from occurring in the pentagonal rings [10.18]. Referring to Fig. 10.2, where the three possible placements of two pentagons relative to a hexagon in a fullerene shell are presented, it is seen that only the configuration shown in Fig. 10.2(a) satisfies the requirement that all five bonds of the pentagon are single bonds, and Fig. 10.2(a) is indeed the configuration found in the C_{60} molecule and most of its derivatives. Another way to reduce strain is to introduce aromaticity in the pentagonal ring to reduce charge accumulation, as discussed below.

For purposes of understanding the chemical reactivity of C_{60}, researchers view this molecule as a nonoverlapping, octahedral arrangement of six pyracylene units with one reactive double bond located in the center of each unit. This arrangement may be visualized from inspection of Fig. 10.1, where the pyracylene units are shaded for clarity. Four pyracylene units are arranged around the belt of the C_{60} molecule, and the remaining two units are centered on the top and bottom of the molecule. The reactive double bonds are also indicated symbolically in Fig. 10.1. Finally, four of the eight hexagons (unshaded) that are isolated from one another by the pyracylene units are evident in the figure.

Because C_{60} also contains double bonds, chemists often classify the C_{60} molecule (and other fullerenes as well) as an alkene. Most chemists do not generally classify fullerenes as aromatic molecules, which are ring molecules with alternating single and double bonds that resonate with one another [10.19]. The presence of the pentagons in C_{60}, with their single bonds, localizes the double bonds at specific sites, unlike the situation typical of aromatic molecules such as benzene, in which the π electrons in the double bond are delocalized around the hexagonal ring. Double bonds are thought to be absent in the pentagonal rings because of the attendant strain energy [10.4]. Among the class of alkenes, C_{60} has an unusually large number of equivalent reaction sites (30, corresponding to the number of double bonds per molecule), leading to the possibility of a large number of different re-

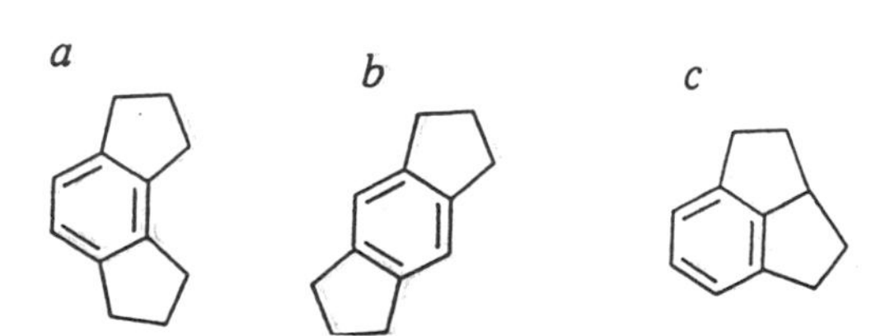

Fig. 10.2. The three possible dispositions of two pentagonal rings adjacent to a hexagonal ring: configurations (b) and (c) are inconsistent with having only single bonds on the pentagonal edges, while configuration (a) has only single bonds on the pentagonal edges and is the favored disposition for C_{60} and its derivatives [10.4].

action products resulting from reactions with a single reagent. Thus, most chemical reactions with fullerenes are not selective.

Many of the addition reactions of C_{60} and C_{70} involve the scission of the central double bond in the pyracylene unit. Several examples of fullerene derivatives obtained in this way are given below. As various chemical groups bond to the two carbons at the double bond location, the carbon atoms are pulled out of their equilibrium positions, the local structure of the fullerene is modified, and the high degree of symmetry of C_{60} is reduced, thereby activating many new (primarily C atom motion) Raman and infrared vibrational modes (see §11.3.1).

Whereas the C_{60} molecule rotates rapidly [10^{10} s^{-1} as determined from nuclear magnetic resonance (NMR) measurements [10.20, 21] see §16.1.4] in the lattice at room temperature, the formation of adducts in a chemical reaction serves to restrict the C_{60} rotations. Thus chemical reactions affect many physical properties of fullerenes, such as those relating to phase transitions associated with this rotational motion (see §7.1.3).

In general, the fullerenes are electron attracting, which chemists call "electrophilic" [10.4, 7]. Thus chemical reactions with electron-donating species (called nucleophiles by chemists) are preferred. It follows that C_{60} and C_{70} are easily reduced (i.e., they accept electrons from the nucleophile). Consequently, these fullerenes are considered to be (mild) oxidizing agents. Chemical reactions can also be used to "functionalize" fullerenes, which refers to the attachment of new chemical groups for the purpose of altering specific chemical or physical properties. The new surface group may also allow a subsequent reaction to occur, producing a new fullerene derivative. For example, the addition of ethylenediamine [$C_2H_4(NH_2)_2$] to C_{60} can be used to produce a "hairy ball" that is water soluble [10.22]. Reactions involving the scission of the π bond in the central C=C double bond of the pyracylene unit are further discussed in §10.6 and §10.7.

In Fig. 10.3 we show some general categories of chemical reactions known to occur for C_{60} (and to some extent also for C_{70}) [10.4]. In the following sections we comment on a number of these categories for chemical reactions and give a few examples for each case. Note that in Fig. 10.3, which is adapted from Ref. [10.4], we have rearranged the figure, grouping the various fullerene derivatives according to their structural similarity. Three main groups have been identified. Approximately half the products in the figure are derived from hydrogenation, alkylation, and halogenation, resulting in a radial covalent bond between the C_{60} and the adduct. In some cases, the species or groups of the same type that are attached to a single ball can be quite numerous [e.g., Br(24), F(60), H(60), CH_3(24), and the amine group RNH(12), where the number in parentheses is the reported maximum value of that species], and the maximum number of a

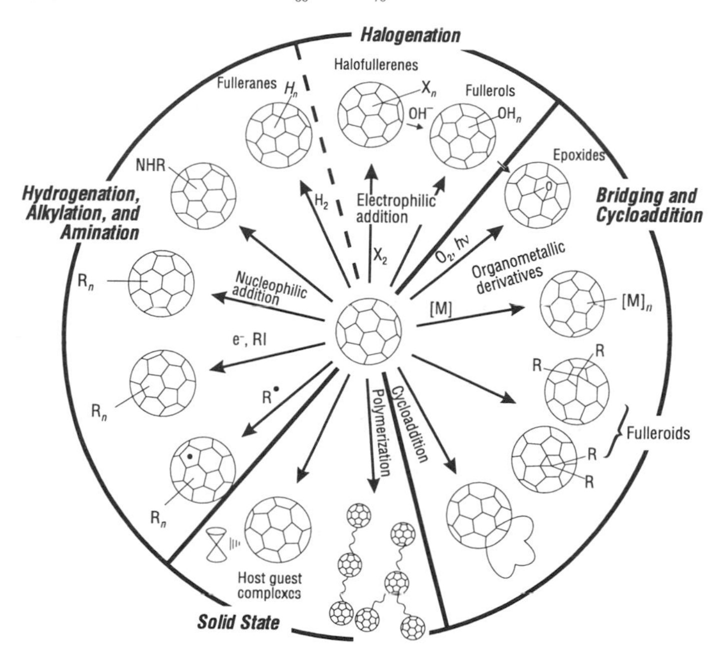

Fig. 10.3. Some general categories of reactions known to occur with C_{60} (and to a lesser extent with C_{70}), based on a figure of Ref. [10.4]. Here R denotes a functional group.

given adduct may be determined by steric hindrance. Another group in the figure corresponds to derivatives where the adduct is attached by a bridge whose ends are bonded across the reactive double bond in a pyracylene unit. Multiple additions of identical larger adducts, attached in this way, are also possible. Thus, similar to the case of the first group, one reagent can generate a variety of fullerene derivatives which differ simply by the number of additions. This multiplicity of products leads to an interesting diversity of derivatives, with different chemical and physical properties. The third (and last) group in Fig. 10.3 contains derivatives which one normally associates with the solid state, i.e., polymers and host–guest solids or clathrates and intercalation compounds (e.g., K_xC_{60}).

10.3. Reduction and Oxidation of C_{60} and C_{70}

The propensity of C_{60} and C_{70} to take on extra electrons is an important driving force in many fullerene-based chemical reactions and is the focal point of this section. In §10.3.1, we make general remarks about the relationship between the electrophilicity of the pyracylene units within the C_{60} and C_{70} molecules and the reducibility of C_{60}. In §10.3.2, we discuss the electrochemical reduction of C_{60} and C_{70} in solution, as well as the more difficult electrochemical oxidation of C_{60}. Battery applications of electrochemical redox reactions are presented in §20.4. In §10.3.3, we present limited evidence for the chemical oxidation of C_{60}.

10.3.1. Fullerene Reduction—General Remarks

The high electron affinity of C_{60} and C_{70} strongly favors their reduction relative to their oxidation. In a chemical reduction reaction, electronic charge is transferred to the fullerene, leading to anion formation C_{60}^{n-}, or more generally to a higher density of electron charge on the fullerene. Cyclic voltammetry studies of C_{60} have shown (see §10.3.2) that reversible reduction can be carried out to yield fullerene anions C_{60}^{n-} [10.23–25] and C_{70}^{n-} [10.26–28] for $1 \leq n \leq 6$ [10.23]. The addition of 6 electrons to C_{60} is sufficient to completely fill the lowest unoccupied molecular orbital (LUMO) t_{1u} (or f_{1u}) level (see §12.1); both the t_{1u} and f_{1u} notations appear in the literature.

In Fig. 10.4, the effect of addition of one electron to a pyracylene (shaded) unit of C_{60} is shown. It is proposed [10.4, 7] that the addition of a sixth π electron to the pentagonal ring (introducing a net charge $-e$ on the ring) is favored because it forms the aromatic cyclopentadienyl radical within the pyracylene unit. The five dots, located where the double bonds once existed (see Fig. 10.4), represent the unpaired electrons left after the dissociation of these bonds.

When considering reduction reactions from a chemical and structural point of view, the addition of six electrons to C_{60} presents an especially symmetric configuration. Since one pyracylene unit is involved with each

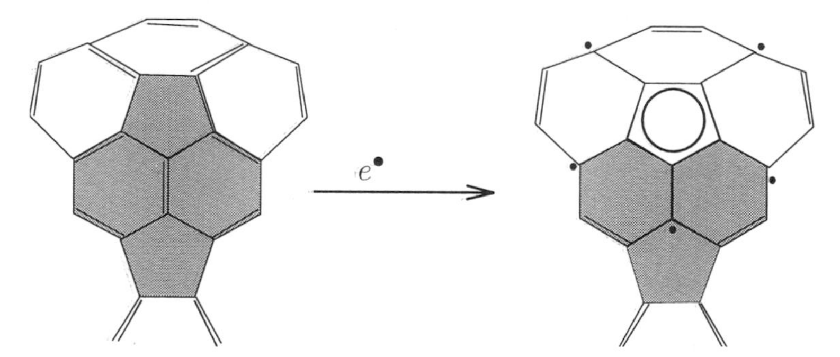

Fig. 10.4. Addition of an electron to a pyracylene unit (shaded) of a fullerene to give an aromatic pentagonal ring [10.4]. The pentagonal ring then contains an additional delocalized π electron, forming an aromatic 6π cyclopentadienyl radical. The five dots on the figure represent unpaired electrons created by forming the cyclopentadienyl radical.

electron addition of the type shown in Fig. 10.4, the addition of six electrons would symmetrically involve all six pyracylene constituents of the C_{60} molecule (see Fig. 10.1). Furthermore, the addition of six nucleophilic (electron-donating) groups, one per pyracylene unit, is commonly found in fullerene chemistry [10.4].

In discussing sites for chemical reactions, chemists often make use of projection diagrams of fullerenes, such as the Schlegel projection of C_{60} and C_{70} shown in Fig. 10.5 [10.4]. These diagrams show bonding relations for all bonding sites for these fullerenes on a planar projection. In each of these diagrams, three of the six pyracylene units are easily identified, since the two pentagons of these pyracylene units are outlined in boldface in the figure.

Many of the reduction reactions of C_{60} and C_{70} of importance to physical measurements in the solid state have been made with alkali metal and alkaline earth species. In many cases, the reaction is deliberately controlled so that exactly three electrons are added to each C_{60} molecule, as, for example, for superconductivity studies in M_3C_{60} compounds (M = K, Rb and appropriate alkali metal alloy compounds such as $M_{1-x}M'_xC_{60}$) (see §15.1). Another interesting example of a reduction reaction is the addition of the charge transfer complex TDAE (tetrakis-dimethylamino-ethylene) to form an unusual magnetic compound (see §8.7.3 and §18.5.2). The reducibility of C_{60} and C_{70} allows these molecules to be hydrogenated [10.29] or alkylated [10.30] as, for example, by CH_3 addition. In these reactions, the C_{60}^{n-} anion has been proposed as an intermediate species in the synthesis route [10.30].

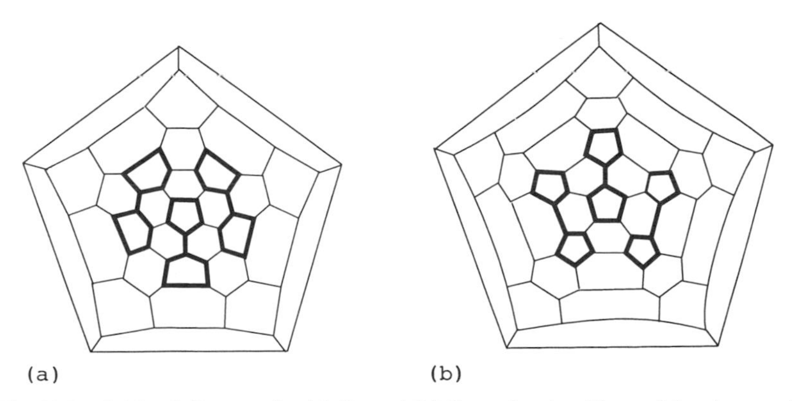

Fig. 10.5. Schlegel diagrams for (a) C_{60} and (b) C_{70} molecules. Three of the six pyracylene units for each molecule are highlighted. The three other pyracylene units in each molecule are located symmetrically with respect to those highlighted by bold lines [10.4].

10.3.2. *Electrochemical Fullerene Reduction and Oxidation*

Electrochemistry of the free molecule is important for identifying the various ionization states that can be achieved. Historically, cyclic voltammetry studies showed at an early time that C_{60} and C_{70} could easily be reduced [10.23, 25, 29]. In these experiments, a solution containing C_{60} or C_{70} is placed in an electrochemical cell containing two chemically inert (redox) electrodes and a third reference or "standard" electrode. A potential is established between the inert electrodes. The choice of the solvent in the cell is critical in keeping the electrochemically generated C_{60}^{n-} anions in solution. The solvent also affects, to some degree, the values of the reduction potentials at which specific anions are first formed. An electrolyte is also added to raise the cell conductance and should be chosen so that charge is exchanged between the fullerenes and the inert electrodes, and not with the electrolyte. Furthermore, the charge state of the electrolyte should remain unchanged during the electrochemical reaction. The cell potential is measured between one of the inert electrodes and the reference electrode.

For the case of C_{60}, a mixture of acetonitrile and toluene usually provides a suitable solvent for study of the charging of the free C_{60} molecules. In the electrochemical reduction process, C_{60} molecules or C_{60}^{n-} anions near the negative electrode are reduced, picking up one electron. Figures 10.6(a)–(d) show the cyclic voltammograms for the reversible reduction of (a, b) C_{60} and (c, d) C_{70} in a toluene/acetonitrile solution at $-10°C$ [10.27]. The electrolyte, in this case, is a phosphorus hexafluoride salt. In both Figs. 10.6(a) and (c), the upper curves are the voltammograms and the lower curves (b, d) are the differential pulsed polarograms (DPP), which provide a more accurate means of determining the cell potential required to form a specific fullerene anion. At the various voltage peaks shown in Fig. 10.6(a) and (c), the C_{60}^{n-} and C_{70}^{n-} molecular ions pick up one additional electron. The experimental results [10.27] show the possibility of adding six electrons to both C_{60} and C_{70}. These six electrons completely fill the t_{1u} LUMO level of the icosahedral C_{60}^{6-} molecular ion. The stability of the C_{60}^{n-} and C_{70}^{n-} anions suggests that they should engage in reversible electron transfer with many electron donors (nucleophiles), including metals and metal complexes. Electrochemistry can thus provide an important route for the synthesis of C_{60}–metal compounds and charge transfer salts.

Thus far, no report has been given for the *reversible* electrolytic oxidation of C_{60}, i.e., removing electrons from the highest occupied molecular orbital (HOMO) level in an electrolytic cell. All electrochemical attempts to oxidize C_{60} have led to the oxidation of the electrolyte instead [10.31, 32]. However, the *irreversible* formation of C_{60}^{+} has been reported [10.25].

Cyclic voltammetry studies, such as shown in Fig. 10.6, have been frequently used by chemists for the characterization of fullerene deriva-

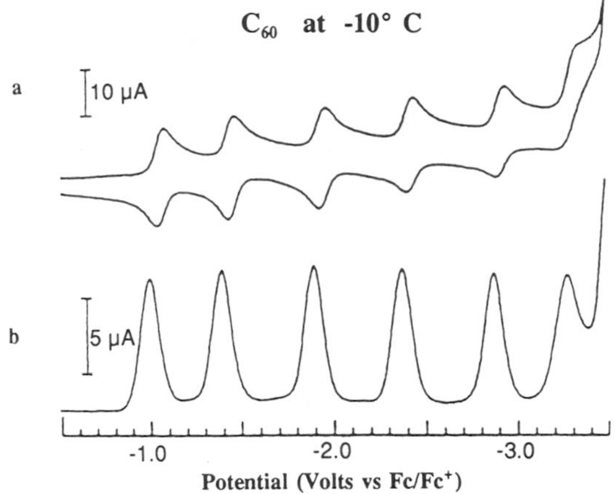

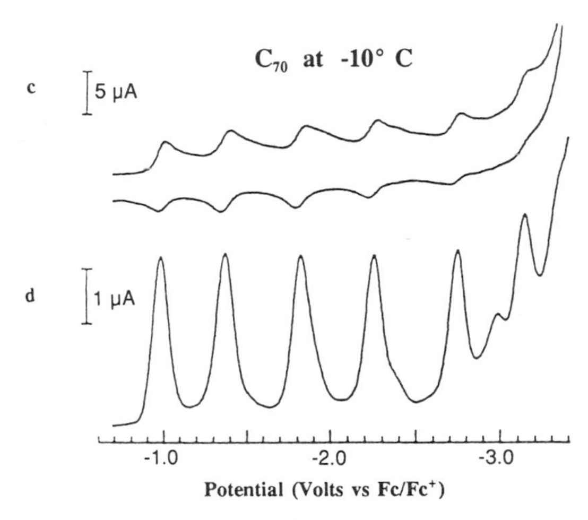

Fig. 10.6. (a) Cyclic voltammogram for C_{60} in a toluene/acetonitrile solvent using a phosphorus–hexafluoride salt electrolyte. The voltammogram shows six reversible reductions of C_{60} at −0.98, −1.37, −1.87, −2.35, −2.85, and −3.26 V. Shown in (b) is the differential pulse polarogram of the same solution, which is used to identify the reduction peak voltages more accurately. (c and d) The same diagrams as in (a and b), but for C_{70}, where the six reversible reductions occur at −0.97, −1.34, −1.78, −2.21, −2.70, and −3.07 V [10.27].

tives. Such studies can be useful in classifying categories of fullerene derivatives. For example, the addition of two phenyl groups through a carbon bridge $C(C_6H_5)_2$ to C_{60} yields the same cyclic voltammetry wave pattern (see Fig. 10.6 for such a pattern) as the addition of isobenzofuran $[(C_6H_4)(C_4H_3O)]$ to C_{60} [10.7] (see §10.6). This result shows that the addition of an adduct, such as isobenzofuran or $C(C_6H_5)_2$ to C_{60} does not significantly alter the electrochemistry of the fullerene.

Electrochemistry provides a powerful tool for the intercalation of guest species into a host material, as, for example, in layered materials such as graphite and transition metal dichalcogenides. This method has also been applied with significant success to the case of intercalation of guest species into solid C_{60}. Because of extensive prior experience with the electrochemical intercalation of Li, Na, K, and Rb into layered host materials, and in particular the intercalation of Li into Li ion battery electrodes, the electrochemical intercalation of Li into the interstitial voids of solid C_{60} has received particular attention [10.33] (see also §20.4.2). The positive electrode (or working electrode) is made from a mixture (composite) containing 60% purified C_{60} and 40% solid electrolyte (polyethylene oxide and $LiClO_4$), presumably for better contact of the reagents with the electrolyte; metallic Li is used for the negative electrode. Furthermore, additional electrolyte material of the same composition in the form of a film was wrapped about the working electrode. Electrochemical measurements were made by sequentially applying a step potential increase of 10 mV/h to the cell, and the current was monitored as a function of cell potential. The resulting voltammogram for the first reduction cycle from 3.0→0.2 V is shown as the negative current trace in Fig. 10.7(a), and the subsequent oxidation cycle

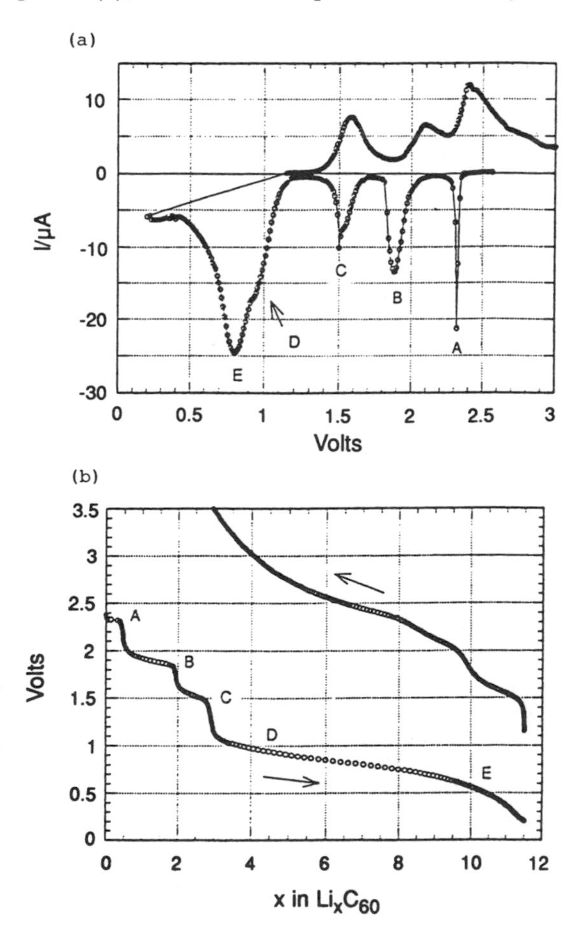

Fig. 10.7. (a) Voltammogram of the first reduction/oxidation cycle of pure C_{60} (3.16 mg). Dots are the current averaged over the 1 h duration of each 10 mV step. The open circuit voltage relaxed to 1.2 V after reducing to 0.2 V. (b) Voltage *vs* composition obtained by integration of voltammogram (a). The error in x is $\pm 5\%$. The structure associated with reduction occurs at 2.3, 1.9, 1.5, and 1.0 V [10.33].

is shown as the positive current trace [10.33]. By integration of the current that flows from the start of the cycle to each point on the reduction voltammogram, the amount of Li delivered to the working electrode is monitored. Assuming that one Li atom is inserted per electron passing through the external circuit, a plot is then made of the working electrode voltage *vs* the stoichiometry of the working electrode, as shown in Fig. 10.7(b). The well-defined features A, B, and C in Fig. 10.7(a) correspond to well-defined steps in Fig. 10.7(b), where A, B, and C correspond to $x = 0.5$, 2.0, and 3.0 in Li_xC_{60}. The weak feature at D corresponds to $x = 4$.

It is interesting to note that the roughly constant 0.5 V separation between A, B, C, and D in Fig. 10.7(a) corresponds approximately to that observed for electron reduction of C_{60}^{n-} up to $n = 4$ in solution. One should conclude, therefore, as do the authors of Ref. [10.33], that $(Li^+)_4C_{60}^{4-}$ has formed at D in Fig. 10.7(b). The passage of further charge leads to the insertion of neutral Li, and for $x > 4$ we presume that $Li_{x-4}(Li^+)_4C_{60}^{4-}$ is formed up to $x \simeq 11$. It remains an open question where the four Li^+ ions reside and how they are distributed between tetrahedral and possible multiply occupied octahedral sites.

For the oxidation cycle, the separation between each peak is ~0.5 V as seen in Fig. 10.7(a). The constant value of this peak-to-peak separation would be interpreted as implying small changes in energy between molecular ion states separated by a single electronic charge [10.33]. The magnitude and repetition rate of the potential steps in Fig. 10.7(b) are chosen to be small enough to give good resolution and large enough to monitor the time response of the system [10.33]. By taking large voltage steps, a general overview of the electrochemistry of the system is obtained over a wide voltage range, while study of the response for small steps provides detailed information about the equilibrium conditions pertaining to each feature in the voltammogram. A detailed analysis of the reduction peak "A" shows this process to be an unusual first-order transition, corresponding to a difficult initiation of the Li intercalation at 2.34 V. Peak "B" corresponds to a more usual first-order process, while peak "C" behaves like a solid solution process with a voltage overlap between the reduction and oxidation cycles [10.33]. Some examples for which electrochemical intercalation into C_{60} has been successfully demonstrated include Li [10.33] and Na [10.32].

10.3.3. Chemical Oxidation of C_{60}

By the term chemical oxidation of C_{60} we refer to a chemical reaction which converts neutral C_{60} to a cation C_{60}^+. This is to be distinguished from the formation of the covalently bonded $C_{60}O$ epoxide compound (see §10.6), or the physisorption of molecular O_2 within the crystalline C_{60} structure to form a clathrate (see §8.7.2 and §9.3.2), or the photochemical intercalation of dioxygen into C_{60} in the solid state (see §10.10).

As stated above, the reduction of C_{60} is much easier to carry out than its oxidation, although oxidation reactions have been observed. For example, a radical cation form of C_{60} has been reported based on the oxidation of C_{60} by XeF_2 [10.34] or the reaction with a strong superacid. In the latter case, C_{60} was oxidized by a mixture of fluorosulfuric acid and antimony pentafluoride $FSO_3H{:}SbF_5$ [10.35], which was shown to yield largely stable C_{60}^{+}/superacid mixtures. The C_{60}^{+} cation reacts readily with different types of nucleophiles (electron donors) to produce a variety of addition products, probably through a chain reaction mechanism in which carbocations (a group of carbons that possess a positive charge localized on one of the carbon atoms) act as intermediates, similar to the nucleophilic addition to carbon–carbon double bonds in alkenes [10.9]. In these reactions, multiple attachment of the adduct is carried out through the continuous transfer of the positive charge from one carbon atom of one double bond to another carbon atom in a different double bond, or to another species in the reaction medium. For example, C_{60}^{+} can produce symmetrically substituted alkoxylated [such as $C_{60}(OCH_3)_n$, $n = 2, 4$ or 6] and arylated [such as $C_{60}(C_6H_5)_{22}$] fullerene derivatives by reacting with alcohols such as methanol and butanol [10.35].

10.4. Hydrogenation, Alkylation, and Amination

The fullerene derivatives obtained by hydrogenation and alkylation attach functional groups to the fullerene shell by radical single bonds. Hydrogenated derivatives of C_{60} and C_{70} have been synthesized by chemical, electrochemical, and catalytic methods. Thus, $C_{60}H_{18}$ and $C_{60}H_{36}$ have been prepared by a Birch reduction [10.29], which refers to a chemical process in which Li metal in the presence of *t*-butanol is used to reduce the C_{60} (or C_{70}) to the monoanion C_{60}^{-} (or C_{70}^{-}), which then leads to hydrogen attachment. High levels of hydrogen attachment ($C_{60}H_{36}$ and $C_{70}H_{46}$) have also been obtained by gas-phase catalytic reactions [10.36]. Even further hydrogenation has been accomplished by high-pressure ($\sim$175 psi of H_2) reactions of C_{60} in an electrochemical cell, leading to $C_{60}H_{60}$ [10.36]. The working electrode in this case was a composite of C_{60} with $\sim$17% Ag, and these cells contained a 30% KOH solution. These results were interpreted as encouraging evidence for the use of $C_{60}H_x$ in battery applications [10.36] (see §20.4.2). Although hydrogenation of C_{60} is relatively easy to accomplish, the hydrogenated reaction products are unstable, change their solubility with time, and show signs of oxygen uptake [10.4].

Neutral C_{60} molecules do not readily react with an alkyl (C_nH_{2n+1}) group in an alkylation reaction. More generally, C_{60} does not readily react with electrophilic (electron-attracting) species, as already mentioned in §10.3.3. To promote reactions with electrophilic species such as alkyls, C_{60} is first reduced to an anion, as, for example, by reaction with an alkali metal species, similar to the Birch-type reaction mentioned above, and this is followed by reaction with an alkyl halide, such as CH_3I, to yield methylated C_{60} and KI as a by-product. Compounds such as $C_{60}(CH_3)_n$ with stoichiometries $n = 6, 8$, and 24 have been prepared by such a substitution reaction [10.30]. The stoichiometry determination of these compounds is usually based on proton and ^{13}C NMR characterization and field ionization mass spectrometry. In $C_{60}(CH_3)_n$, the methyl (CH_3) groups are connected to the fullerene shell by a single C–C bond (see Fig. 10.8), as is the case also for a hydrogen or a halogen attachment (see §10.5). The maximum number of alkyl groups that can be added to C_{60} without any groups being connected to adjacent C atoms on the shell is 24.

Alkylation of C_{60} and C_{70} may also be accomplished by nucleophilic (electron-donating) addition and substitution reactions of C_{60} and its derivatives. For example, addition products have been obtained from the reaction of an aromatic moiety with protonated C_{60}^+ (see §10.3.3). The formation of mixed derivatives [$C_{60}(CH_3)_{10}Ph_{10}$, $C_{60}(CH_3)t$-C_4H_9] and monoalkylated derivatives $C_{60}H(C_2H_5)$ have been produced by reactions with charged nucleophiles following standard synthetic routes for alkylation reactions [10.22]. In the above chemical formula, Ph represents a phenyl group (C_6H_5), and "*t*-" refers to a carbon linked to three groups. The tertiary carbon in this *t*-C_4H_9 group attaches to one of the carbons participating in a double bond prior to the reaction; the other CH_3 group in this fullerene derivative may attach to the other carbon in the same double bond or to a carbon at another double bond site.

Addition of an amino group to C_{60} is also possible. It has been reported that C_{60}, via nucleophilic addition reactions, may undergo multiple attachments of primary amines (RNH_2), where R is an alkyl or aryl (aromatic) group terminated by an amine group (-NH_2). The first step in the addition is the formation of a single C–N bond linking the amine group to one of the carbons at a reactive C=C bond site of the fullerene. The adjacent C atom at this double bond site attracts a hydrogen atom from the terminating amine group, forming a C–H bond, as shown in Fig. 10.9. For example, C_{60} has been functionalized by the nucleophilic (electron-donating) attachment of ethylenediamine ($NH_2CH_2CH_2NH_2$). Multiple attachments of this type to a single fullerene molecule have also been reported [10.22].

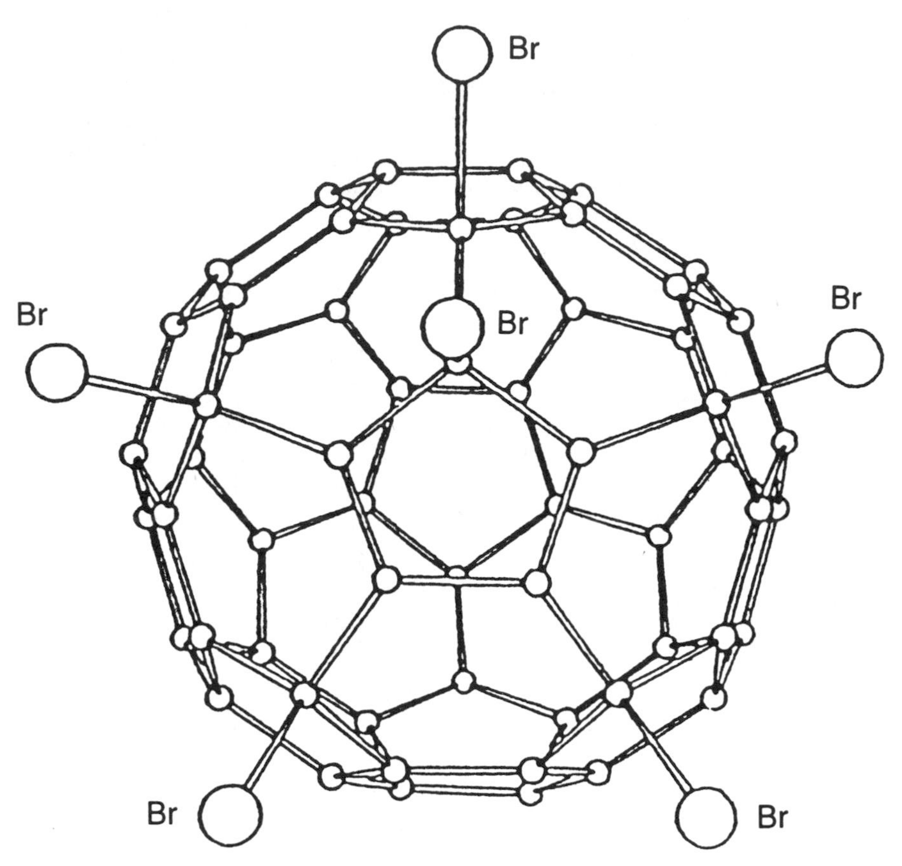

Fig. 10.8. Possible structure of $C_{60}Br_6$ in which six isolated Br atoms are attached to one of the carbon atoms on a pentagonal ring on the C_{60} shell, with one Br per pyracylene unit [10.4].

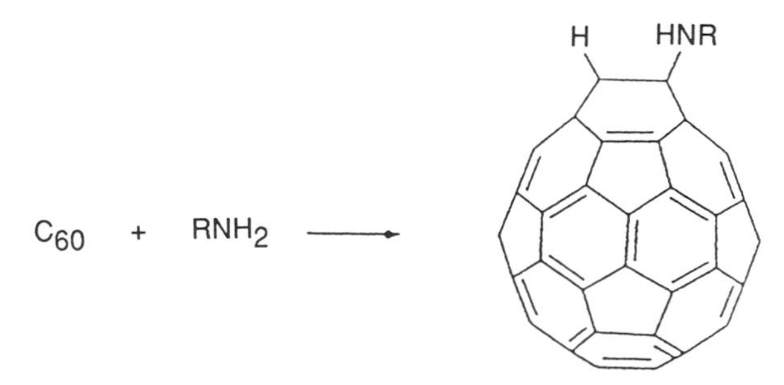

Fig. 10.9. Functionalization of C_{60} by the attachment of a primary amine group [10.4].

10.5. Halogenation Reactions

Similar to the hydrogenation reactions of C_{60} discussed in §10.4, halogenation reactions of C_{60} forming $C_{60}X_n$ (X = Br, Cl, F) have been reported for a wide variety of n values. In these reactions the halogen is attached to a carbon atom on the fullerene shell through a single radial covalent bond, formed by using one electron provided by the fullerene and the other by the halogen. The greatest reactivity is with fluorine and the species $C_{60}F_{36}$ has been reported by several groups [10.29, 37]. Complete fluorination to $C_{60}F_{60}$ has also been reported [10.38], based on a single ^{19}F line in the NMR spectrum. Polychlorofullerenes are produced by passing Cl_2 gas through C_{60} at elevated temperatures (300–400°C), and up to 24 Cl atoms have been added to a C_{60} molecule [10.39]. Bromination of C_{60} has also been reported, for chemical reactions carried out in the 20–55°C range. X-ray crystallographic evidence has been provided for $C_{60}Br_6$, $C_{60}Br_8$, and $C_{60}Br_{24}$ [10.40, 41]. These x-ray studies of $C_{60}Br_x$ ($x = 6, 8, 24$) showed that the attachment of Br is accomplished by placing Br at symmetrically bonding vertices of the C_{60} molecule (see Fig. 10.8). Other possibilities might be found for the placement of Br atoms around the C_{60} shell to obtain other $C_{60}Br_n$ compounds.

Ultraviolet irradiation of fullerene/chlorine mixtures in chlorinated solvents permits fast chlorination under very mild conditions. A very high chlorine content has been found in a photochlorinated fullerene sample, which by elemental analysis was found to fit the average formula $C_{60}Cl_{40}$ [10.42], although it was clear that a distribution of chlorinated products $C_{60}Cl_n$ had been obtained.

Halogenated fullerenes such as $C_{60}Cl_n$ can be used in substitution reactions to attach aromatic groups sequentially to the fullerene shell. In this case, the aromatic molecule reacts with the halogenated C_{60} in the presence of a Lewis acid catalyst. For example, starting with $C_{60}Cl_n$ in the presence of benzene and an $AlCl_3$ catalyst, phenylation of C_{60} with up to 22 phenyl (C_6H_5) groups to form $C_{60}(C_6H_5)_n$ (with $n \leq 22$) has been observed with mass spectroscopy [10.39].

10.6. Bridging Reactions

Addition of functional groups to C_{60} and C_{70} can be accomplished through the formation of a bridge across the reactive C=C double bonds. Nitrogen, carbon, oxygen, and transition metal atoms such as Ni, Pt, or Pd [10.3–7] have been found to bridge these C=C bonds.

The simplest example of a bridging fullerene derivative is the covalently bonded epoxide $C_{60}O$ [see Fig. 10.10(a)], which is an important building

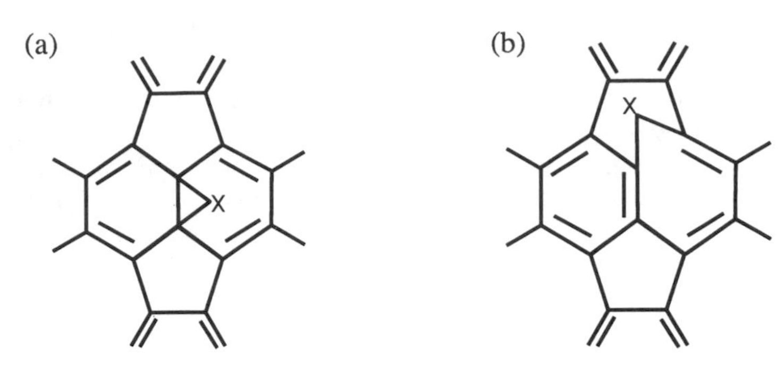

Fig. 10.10. Two possible sites for bridge attachments: (a) the symmetric attachment of X across the double bond of a C_{60} pyracylene unit to form $C_{60}X$, or (b) into the adjacent pentagonal ring bond of C_{60}. The structure in (a) is favored for C_{60} in forming $C_{60}O$ and the structure (b) is expected to occur for C_{70} in forming $C_{70}O$ [10.4].

block for more complex synthesis paths, whereby other species attach to the oxygen, or the oxygen participates in a substitution reaction (see §10.3.3). With dimethyl dioxirane $[(CH_3)_2CO_2]$, C_{60} either removes the oxygen to form the epoxide or forms $[C_{60}C(CH_3)_2O_2]$, a dioxirane derivative [10.43]. Use of ^{13}C NMR [10.44] and x-ray diffraction techniques [10.45] established the structure of $C_{60}O$ to be a bridge attachment with elongated bonds of a single oxygen at the double-bond position of a pyracylene unit, as shown in Fig. 10.10(a) [10.46–48]. It is interesting to note that the crystal structure for $C_{60}O$ is the same (fcc) as for C_{60} itself and with a similar lattice constant [10.45], except that the presence of the oxygen weakly inhibits the rotation of the fullerene in the lattice, thereby increasing the temperature T_{01} for the structural phase transition (see §8.7.2). Whereas $C_{60}O$ follows the bonding scheme shown in Fig. 10.10(a), the bonding of oxygen to C_{70} is expected to form an oxido-annulene type bridge structure, shown in Fig. 10.10(b), where the annulene structure is defined as $[\text{-CH=CH-}]_n$. Furthermore, in the oxido-annulene form, the oxygen was proposed to attach itself at the waist of the C_{70} molecule, where there is a high density of hexagons [10.49]. On the other hand, the waist hexagons are more graphitic and, therefore, less reactive, suggesting that the oxygen might instead attach at the polar caps, in analogy with C_{60}. The bridge arrangement shown in Fig. 10.10(b) has also been proposed for the addition of methylene (CH_2) to C_{60}, where the carbon atom is at the bridge site.

We now give a few examples of carbon bridge reactions. In one example, X in Fig. 10.10(a) is $X \equiv CAr_2$, where Ar denotes an aromatic group, and a carbon atom forms the bridge between the fullerene and each of the Ar groups. A specific example of a $C(Ar)_2$ attachment is the fullerene derivative $C_{61}(C_6H_5)_2$, which is illustrated in Fig. 10.11. Here the attachment of

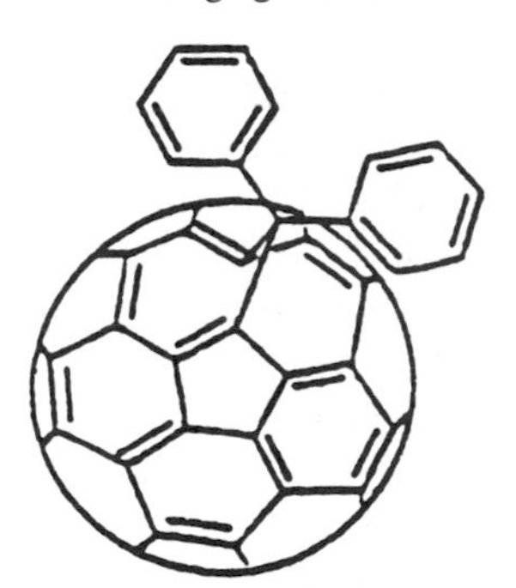

Fig. 10.11. The diphenylfulleroid $C_{61}(C_6H_5)_2$ is formed by the addition of two phenyl groups through a carbon CAr_2 bridge [10.7].

two aromatic (C_6H_5) phenyl groups forms a diphenylfulleroid $C_{61}(C_6H_5)_2$, where the term *fulleroid* refers to C_{61}, denoting a C_{60} cage to which a carbon atom is added at a double-bond site in a pyracylene unit [10.3], consistent with Fig. 10.10(a). In this process the double bond becomes saturated and the adjoined carbon atom has elongated bridge bonds to the C_{60} cage (see Fig. 10.11). The number of π electrons associated with the fulleroid (C_{61}) is two less than for the neutral C_{61}, with two dangling bonds available for bonding as shown in Fig. 10.11 [10.7]. As a result of the phenylation reaction, the authors place a double bond in a pentagonal ring, in contrast with the view [10.4] that the pentagons on the fullerene shell remain unstrained. However, it is possible that the stress created by the double bond in the pentagon is relieved by an increase in the bond length between the bridging carbon and the carbon from the C_{60} molecule.

Another example of a carbon bridge reaction of the type shown in Fig. 10.10(a) is the reaction yielding a fulleroid $(p\text{-BrC}_6H_4)_2C_{61}$ (see Fig. 10.12 where X $\equiv$ Br), which forms hexagonal crystals, and the molecular unit is closely related to Fig. 10.11. In the crystalline form, the structure of this 4-4′ dibromodiphenyl fulleroid has been studied by x-ray diffrac-

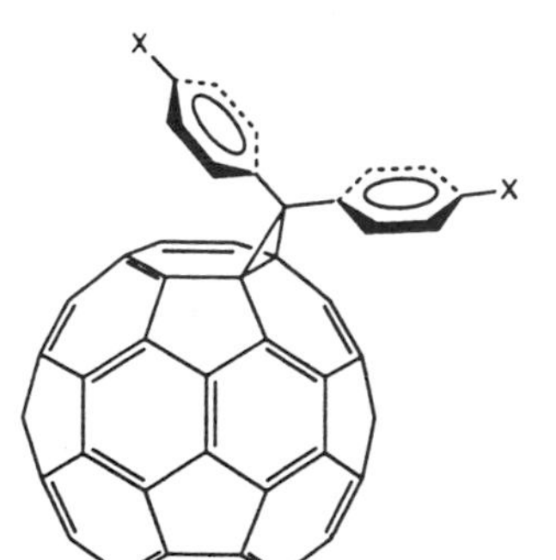

Fig. 10.12. A water-soluble fulleroid of the form $C_{60}C(C_6H_4X)_2$. One simple example of this molecular derivative is X $\equiv$ Br and another more complex example may have the potential to inhibit the key viral enzyme HIV-1 protease, where X = $HOC(O)(CH_2)_2C(O)NH(CH_2)_2-$. Connecting the species X to the fulleroid is a C_6H_4 group shown in the diagram [10.50–52].

tion [10.7]. It is of interest to note that, in the CAr_2 addition shown in Fig. 10.12, all the pentagons retain their single bonds, in contrast to the CAr_2 addition in Fig. 10.11. From a physicist's standpoint, the addition of a CAr_2 group in both Figs. 10.11 and 10.12 causes a charge redistribution on the rings surrounding the carbon bridge, which is not easily described by the simple schematic diagrams shown in either of these figures.

Fullerene derivatives of the CAr_2 carbon bridge type have been discussed for practical applications to control the binding of drugs. The fulleroid shown in Fig. 10.12 also serves as a schematic version of a water-soluble C_{60} derivative which has the potential for inhibiting growth of the human immunodeficiency virus (HIV) by blocking the active site of the viral enzyme HIV-1 protease. In this case the Ar group is a substituted phenyl group C_6H_4X, where X = $HOC(O)(CH_2)_2C(O)NH(CH_2)_2-$ [10.50–52]. The X group might also be a drug which is weakly bound to the phenyl group, so that selective transfer of the drug group X can take place in a controlled way.

In the preceding sections, we have referred to several fullerene derivatives that have involved the attachment of functional groups by bridging to the fullerene shell, through one of the oxygen and carbon bridges of the type shown in Figs. 10.10(a) and (b). We now discuss the formation of fullerene derivatives based on bridging via metal atoms [10.6, 53]. An early example of this type of fullerene derivative is shown in Fig. 10.13, where a direct bridge to C_{70} is made through the Ir atom to two triphenylphosphine groups, in terms of the bridge type shown in Fig. 10.10(a). As can be seen in Fig. 10.13, each of the phosphine groups donates two electrons. Two of these donated electrons bond to each of the CO and Cl and the remaining two electrons bond the Ir to the C_{70}. This example also points to the multiplicity of compounds that can be formed by attachments to lower-

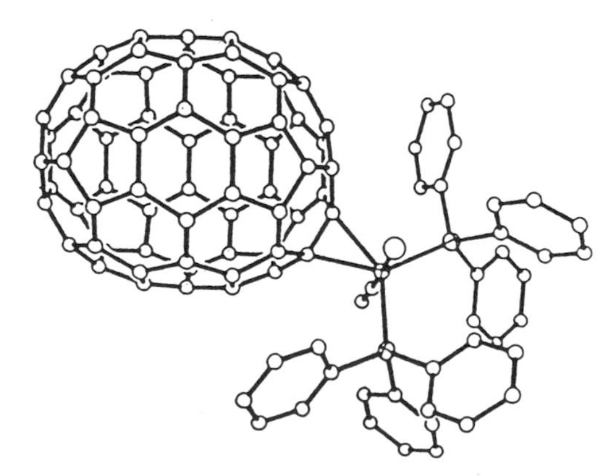

Fig. 10.13. The addition of $Ir(CO)Cl(PPh_3)_2$ to C_{70} as an example of a metal bridge reaction [10.54]. As shown, the Ir metal atom bonds to the two (PPh_3) groups, to the CO and Cl, and to the C_{70} shell.

symmetry fullerenes of higher-mass. For example, three crystallographically different pyracylene bridge site attachments can be made to C_{70}, leading to 15 different isomers of $Ir(CO)Cl(PPh_3)_2C_{70}$.

The metal Pt has been used to attach triethylphosphine groups by reacting C_{60} with $[(C_2H_5)_3P]_4Pt$ in benzene [10.5]. In this case, two triethylphosphide groups are bonded as ligands to each Pt atom through the donation of two electrons from each phosphorus atom to the Pt, forming a coordination bond which is covalent in character. In turn, the platinum atom shares two of its valence electrons with two adjacent carbons of the C_{60} molecule, thus forming two bridging bonds. The same authors have also used Pt and Pd metal bridges to synthesize symmetric structures (see Fig. 10.14), such as $[(Et_3P)_2Pt]_6C_{60}$ and $[(Et_3P)_2Pd]_6C_{60}$, where ethyl is denoted by $Et \equiv C_2H_5$, and the metal atom is attached at the double-bond site [see Fig. 10.10(a)], one metal atom per pyracylene unit [10.5]. Structural studies using x-ray diffraction have been applied to these Pt-bridged fullerene derivatives [10.5]. In preparing many of the metal bridge compounds, specific functional groups are added for special chemical reasons, while other

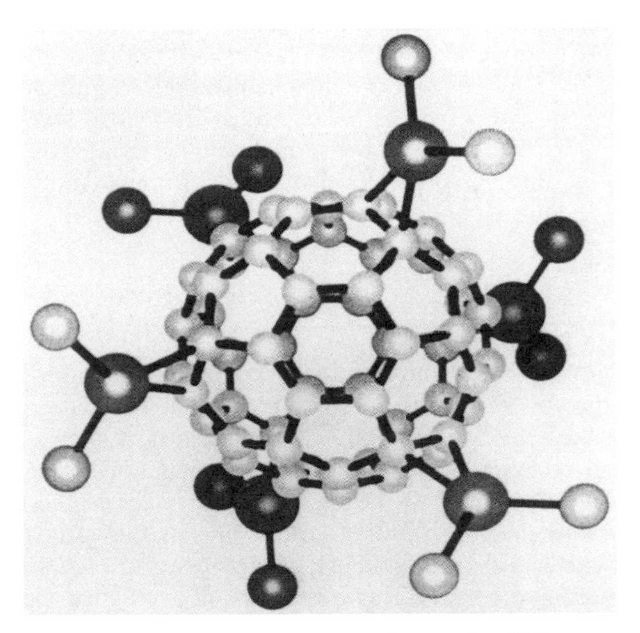

Fig. 10.14. Structure of $[(Et_3P)_2Pt]_6C_{60}$ looking down a threefold symmetry axis under idealized T_h point group symmetry. The Pt forms a direct metal bridge bond to the fullerene shell and also bonds to two triethylphosphine (Et_3P) groups. The $Et \equiv C_2H_5$ terminations are denoted by a single white ball [10.5].

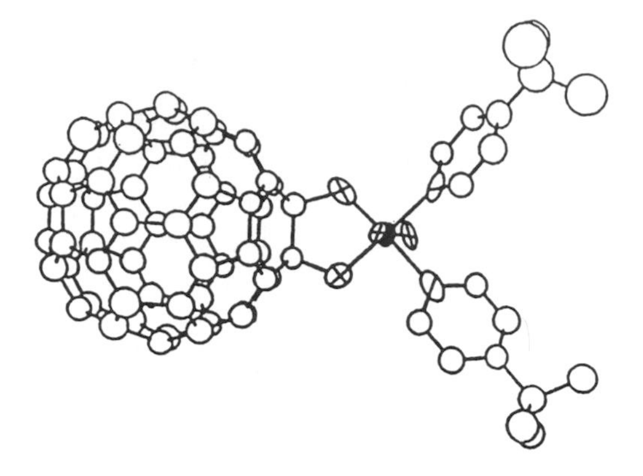

Fig. 10.15. The osmylation of C_{60} by OsO_4 in *t*-butylpyridine to form a large metal bridge derivative [10.53]. In the figure the small black ball denotes the Os and the four balls marked with an X denote the oxygen. The oval with a slash denotes the nitrogen of the pyridine and the four carbons of the *t*-butyl groups are shown.

chemical species might be selected to enhance the chemical stability of the compound.

Figure 10.15 shows a metal bridge reaction involving cycloaddition, where the metal complex is the osmylated C_{60} compound [10.53]. In this reaction a metal bridge is made by the Os atom through two oxygens to the fullerene forming a pentagonal ring, and for this reason this metal bridge reaction is also classified as a "cycloaddition" reaction, which is further discussed in §10.7. The OsO_4 group in Fig. 10.15 attaches as a bridge of the type shown in Fig. 10.10(a). Four oxygens are attached to the osmium, accounting for six electrons, since two oxygen bonds attach the entire complex back to the C_{60} cage (see Fig. 10.15). Two *t*-butylpyridine groups are coordinated to the Os atom through single covalent bonds, formed from two electrons provided by the nitrogen atom in the *t*-butylpyridine group.

10.7. Cycloaddition Reactions

Cycloaddition reactions are also possible with fullerenes. As a result of a cycloaddition reaction, a four-, five-, or six-membered ring is fused to the outside of the fullerene shell in such a way that one side of the ring is also part of the cage. An example of cycloaddition is shown in Fig. 10.16 in which a six-membered ring is fused to a C_{60} shell with the evolution of CO. Here the attachment to the C_{60} molecule is across the double bond connecting two pentagons of a pyracylene unit as shown in Fig. 10.10(a). It is readily checked that the bonding requirements are satisfied for all the atoms in the aromatic ring structures shown in the figure. Because of the bulkiness of many of the ring additions, the number of rings that can be attached to a single fullerene is limited by steric considerations.

C_{60} can undergo "4+2" and "2+2" cycloaddition reactions with other organic molecules. In these reactions, C=C double bonds are broken on both

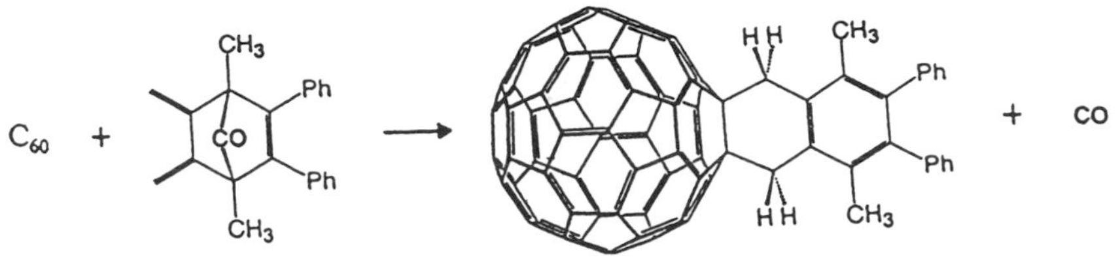

Fig. 10.16. An example of a Diels–Alder reaction yielding the cycloaddition product of a 5,6-dimethylene-1,4-dimethyl-2,3-diphenyl group to C_{60} to form a fullerene derivative that is unable to undergo the reverse reaction because of the release of the CO by-product. The chemical notation of the reaction product focuses on the benzene ring on the right, where two phenyl (C_6H_5) groups attach at sites 2 and 3, while methyl (CH_3) groups attach at sites 1 and 4, and finally at the two remaining sites, 5 and 6, methylene groups (CH_2) attach [10.55, 56].

molecules and new bonds form, depending on the number of π electrons involved in these bonds. A "4+2" addition implies the rearrangement of four π electrons from a conjugated diene unit (C=C–C=C) on the organic reactant molecule and two π electrons from a double bond in the fullerene. Similarly, a "2+2" cycloaddition to C_{60} implies a bond rearrangement in which two pairs of π electrons, one pair from C_{60} and the other pair from the reacting organic molecule, are redistributed to form new single bonds joining the molecules. The latter has been proposed for the formation of the C_{60} dimer in the solid state with photoassistance, in which two C_{60} shells are joined by a four-membered ring (see Fig. 7.20) [10.12].

An example of a "4+2" cycloaddition reaction in solution is the multiple addition of cyclopentadiene (C_5H_6, with two double bonds in the ring) to C_{60}. Up to six cyclopentadiene addends have been attached to the fullerene, each across the reactive double bond of the pyracylene unit [see Fig. 10.10(a)]. One such addition is shown in Fig. 10.17 [10.57]. The addition of each cyclopentadiene occurs by breaking the double bond in the pyracylene unit and forming two new single bonds to the cyclopentadiene, as shown in the figure. Since this addition reaction is reversible, analysis of the products has been facilitated by stabilization of the "cycloadded" C_{60} through saturation of the remaining double bond in the cyclopentene ring with hydrogen [10.57].

Gas-phase cycloaddition of cyclopentadiene (C_5H_6) and cyclohexadiene (C_6H_8) to C_{60} and C_{70} has also been accomplished. In this case, the ring molecule was added to a C_{60} and C_{70} radical cation ($C_{60}^{\bullet+}$, $C_{70}^{\bullet+}$), formed by 50 eV electron impact [10.58]. The addition has been proposed to occur at the site of the positive charge, which is localized on the C–C bond between the pentagonal rings of the pyracylene unit. The resulting adduct is structurally similar to that of Fig. 10.17, with the difference that a positive charge is delocalized on the C_{60} surface.

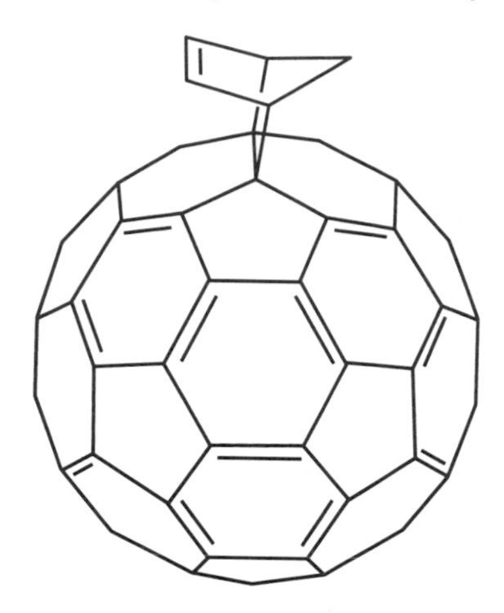

Fig. 10.17. Model for the structure of the C_{60}-cyclopentadiene adduct as an example of "4+2" cycloaddition. As an isolated molecule, cyclopentadiene (C_5H_6) has two C=C bonds, 120° apart, in the pentagonal ring [10.4].

An example of a "2+2" cycloaddition reaction in solution is given by the attachment of up to six benzyne (C_6H_4) groups to the C_{60} shell [10.59]. Analysis of these reaction products by ^{1}H NMR and ^{13}C NMR analysis indicates that the benzyne is connected to the fullerene shell by two single carbon bonds, thus forming a four-membered ring.

Cycloaddition reactions can be used not only to form derivatives of C_{60} but also to introduce functional groups to a C_{60} adduct. For example, photocatalytic addition of a substituted cyclohexenone (C_5H_7RCO, R=H, CH_3) to C_{60}, produces a "2+2" cycloaddition product in which the ketone [an organic molecule that contains the group –(C=O)] is linked to the shell by single C–C bonds forming a four-membered ring [10.60]. For other examples of "2+2" cycloaddition relevant to polymer chain formation, see §10.10.2 and §7.5.1.

10.8. Substitution Reactions

Since C_{60} contains only strongly bonded carbon atoms (and no hydrogen atoms), conventional substitution reactions do not occur for C_{60} or other higher-mass fullerenes. However, substitution reactions do occur on *derivatives* of C_{60} containing additional groups. For example, K_6C_{60} reacts with CH_3I to yield $(CH_3)_6C_{60}$ and KI, where methyl groups are substituted for potassium atoms. Numerous C_{60} derivatives can be prepared from nucleophilic (electron-donating) substitution of substituted C_{60}, especially in the case of halogenated compounds. Thus, fluorinated C_{60} reacts vigorously with strong nucleophiles like amines as well as with weak nucleophiles like acetic acid to produce C_{60} adducts containing oxygen and nitrogen groups [10.61]. Alcohol derivatives of C_{60} [e.g., $C_{60}(OH)_n$], called "fullerols," can also be prepared by nucleophilic substitution of a nitrosated [$C_{60}(NO)^+$] derivative [10.62].

Another example in which C_{60} participates in a substitution reaction is in the formation of alkoxylated and arylated derivatives obtained by electrophilic substitution of halogenated compounds in the presence of an acid catalyst, as explained in §10.5.

10.9. Reactions with Free Radicals

A free radical, denoted by a superscript dot, is a chemical group with a dangling bond. Figure 10.3 shows schematically the reaction of a radical group with C_{60} to yield a fullerene derivative [i.e., $C_{60} + nR^\bullet \rightarrow (C_{60}R_n)^\bullet$], where the attachment of the functional group is at one site and the unpaired electron is at another site. The electrophilicity of C_{60} makes it highly reactive toward the addition of free radicals. Reactions involving free radicals lead to the formation of addition products, as in the case of hydrogenated and alkylated C_{60} (see §10.4), or to the formation of highly stable radicals, as is the case for thc addition of benzyl radicals for C_{60} [10.61].

Some of the radicals which react readily with C_{60} include $Me^\bullet$, $Ph^\bullet$, $PhS^\bullet$, $PhCH_2^\bullet$, $CBr_3^\bullet$, $CCl_3^\bullet$, $CF_3^\bullet$, and $Me_3CO^\bullet$, where Me and Ph denote methyl and phenyl groups, respectively [10.4]. Models for the attachment of free radical groups to C_{60} are shown in Fig. 10.18 and lead to a free radical derivative of C_{60}. In Fig. 10.18(a), three $R^\bullet$ groups [where $R^\bullet$ could, for example, be a benzyl group (C_6H_5–$CH_2^\bullet$)] are attached at the sites indicated by R and the unpaired electron is localized within the pentagon ring, forming an "allylic" radical. However, when five groups are attached to all five available sites in a cyclopentadienyl configuration, then the unpaired electron becomes delocalized within the pentagon ring, forming a "cyclopentadienyl" radical, as shown in Fig. 10.18(b).

Thus, the addition of methyl radicals ($Me^\bullet$) to obtain $(CH_3)_nC_{60}$ for $1 \leq n \leq 34$, or benzyl radicals $C_6H_5CH_2^\bullet$ to obtain $(C_6H_5CH_2)_nC_{60}$ for $1 \leq n \leq 15$, has been demonstrated [10.61]. In the case of the addition of

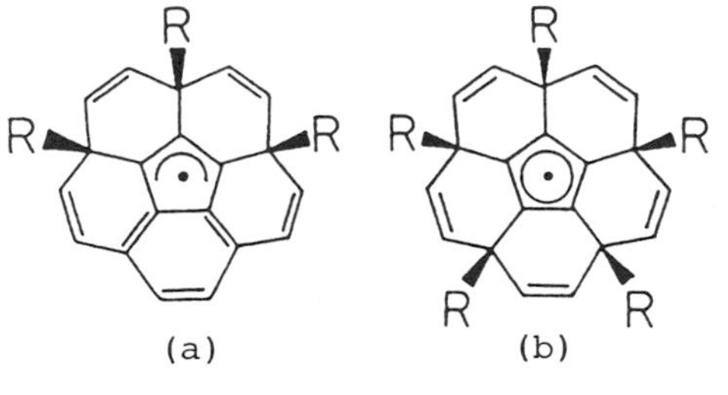

Fig. 10.18. Structure of free radical derivatives of C_{60}, where the unpaired spin is (a) localized forming an "allylic" free radical and (b) delocalized in the symmetric "cyclopentadienyl" configuration [10.4]. Note that only the region of the C_{60} molecule around a single pentagon is shown.

benzyl radicals, as was described in the previous paragraph, the C_{60} radical resulting from the addition of three or five benzyl radicals is specially stable above 50°C. The stability of these radicals has been attributed to the steric protection of the surface radical sites by the surrounding benzyl substituents. Finally, as described in §10.10.2, radical formation by fullerene molecules plays an important role in polymerization reactions.

10.10. Host–Guest Complexes and Polymerization

10.10.1. Host–Guest Complexes

The term "host–guest" complex refers to a cluster of molecules or atoms (the host structure), which defines a cavity, containing a "guest" molecule, atom, or ion. Implicit in the definition is the condition that the guest is bound to the host structure through either van der Waals or ionic interactions and not by any covalent bonds. In many host–guest complexes, the geometry of the host structure is such that the host cannot form first, followed by the insertion of the guest species. Rather the formation of the host–guest complex requires the simultaneous presence of the guest and the constituent units of the host structure. If the complex is stabilized by the van der Waals interaction, either a "clathrate" or an "inclusion" complex is formed, depending on whether the cavity is a tube or channel (inclusion complexes) or the cavity completely surrounds the guest species (clathrate complex). Complexes can bond together to form larger complexes leading to a solid-state structure. For example, we have discussed C_{60} clathrate compounds in previous chapters, in which C_{60} is the guest species located inside cages defined by organic solvent molecules. If the complex is stabilized by charge transfer, the complex is referred to as a "charge transfer" complex. Examples of clathrate and charge transfer complexes are given below.

Before discussing these complexes, however, we should remind the reader that C_{60} can also participate in a host–guest compound as the host. Two well-known examples discussed previously are the M_xC_{60} (M = alkali metal) compounds and $C_{60}(O_2)_x$. The former is a charge transfer compound, since the lattice is stabilized by the electron transfer from the M atoms to form C_{60} anions, while the latter does not involve charge transfer between the C_{60} lattice and the O_2 physisorbed in the octahedral interstices of the fcc lattice. We have also used the term "intercalation" compounds for the M_xC_{60} and $C_{60}(O_2)_x$ compounds, because the intercalant is observed to diffuse or "intercalate" into the host structure. In the case of the M_xC_{60} compounds, a structural rearrangement of the host occurs for $x = 1, 4, 6$. Strictly speaking, the term intercalation applies only to two-dimensional host materials and refers to the introduction of guest species into layers between the host lattice planes (see §2.14).

Several examples of C_{60} host–guest complexes have been reported where C_{60} is the guest molecule, e.g., C_{60}–hydroquinone [10.63], C_{60} γ-cyclodextrin [10.64], and C_{60} calixarenes [10.65]. For example, a 3:1 clathrate complex is formed from C_{60} and hydroquinone. Hydroquinone ($C_6H_6O_2$) is a benzenoid ring with OH groups on two opposing vertices of the ring, and these OH groups participate in intermolecular hydrogen bonding (i.e., an intermolecular H-O-H link). In this complex, each C_{60} molecule is in contact with a total of 18 hydroquinone molecules. According to a simple model, twelve of the 18 hydroquinone molecules are located in the region of the north and south poles (six on each pole) of the C_{60} molecules and their planar rings are parallel to the fullerene surface. The remaining six hydroquinone molecules are centered on a ring around the equator of the C_{60} and contact the fullerene "edge on." It has been suggested that the short distance between the C_{60} and the hydroquinone molecules near the north and south poles of the enclosed C_{60} molecules is caused by a charge transfer interaction between the hydroquinone and the C_{60}, resulting in a very compact structure. Classic past examples of hydroquinone host–guest complexes [10.63] involve 3:1 complexes with methanol (but not ethanol), SO_2, CO_2 and argon (but not neon) as the guest species, where the 3:1 refers to the relative concentrations of hydroquinone to guest species.

Two further examples of C_{60} clathrate complexes involve interesting "cup-shaped" molecules. In the case of cyclodextrin (CD) hosts, a 2:1 γ-cyclodextrin:C_{60} complex is formed from an aqueous solution of C_{60} and γ-cyclodextrin [10.64]. The C_{60} becomes trapped between two molecular "cups" of γ-cyclodextrin. The shape and size of the cups, or truncated cones, for α-, β-, and γ-cyclodextrin, which contain six, seven, and eight glucose units, respectively, are shown in Fig. 10.19, while in Fig. 10.20 we show, respectively, a schematic top view of β-cyclodextrin and a perspective view of α-cyclodextrin. As can be seen in Fig. 10.19, the cavity in the α-, β-, and γ-cyclodextrin (CD) is too small to accommodate a C_{60} molecule deep within the cup. Two γ-cyclodextrins, however, have been shown to complex around a C_{60} guest, and a side view and top view of the 2:1 complex are shown in Fig. 10.21 [10.64]. Cyclodextrins are nontoxic, and as a result they find applications in drug and food encapsulation [10.9].

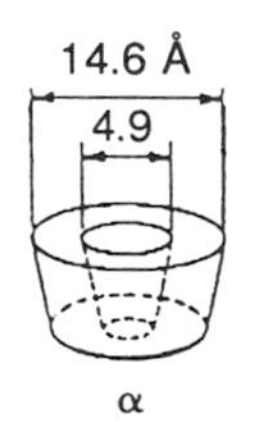

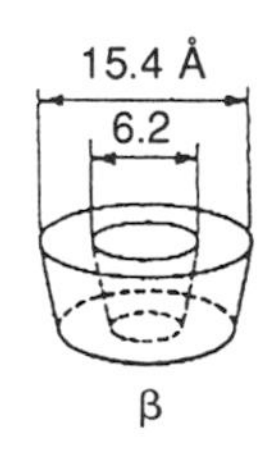

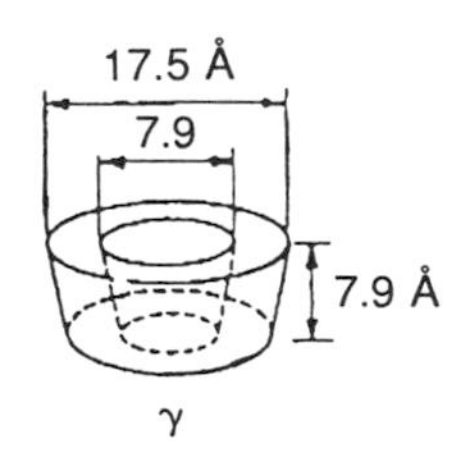

Fig. 10.19. Shape and dimensions of the α-, β-, and γ-cyclodextrin molecules [10.9].

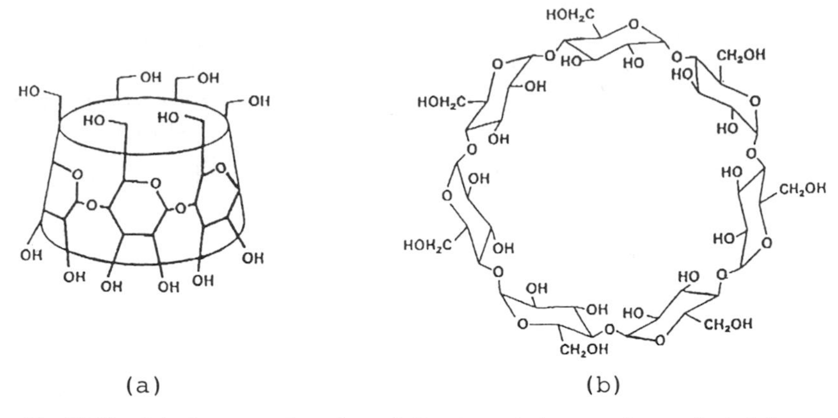

Fig. 10.20. (a) A perspective view of the truncated cone shape adopted by an α-cyclodextrin molecule. The OH groups situated at the rims of the cone make this molecule soluble in aqueous solutions. (b) The chemical structure of a β-cyclodextrin molecule consisting of seven glucose units [10.9].

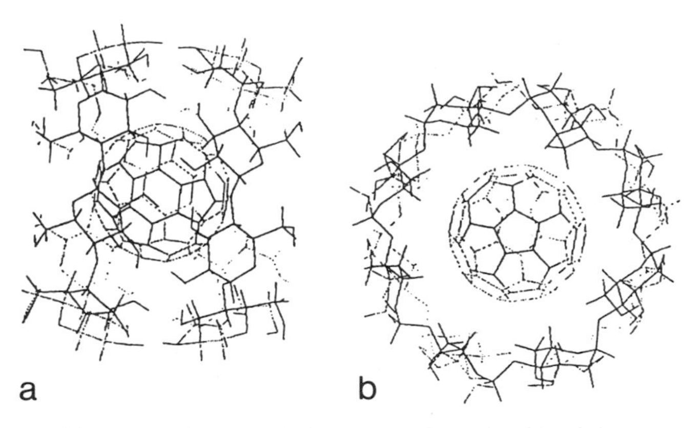

Fig. 10.21. (a) A 2:1 γ-CD:C_{60} complex as seen from the side of the two cyclodextrins (CDs). (b) An end view showing the CD cavity filled with C_{60} [10.64].

The calixarene molecule is another cup-shaped molecule containing phenol units which, interestingly, forms a 1:1 (rather than a 2:1) complex with C_{60} as shown in Fig. 10.22 [10.65], where R is a functional group and *p* denotes "*para*." Top views of the *p*-R-calix[8]arene and the *p*-R-calix[6]arene molecules are shown in Figs. 10.22(a) and (b), respectively [10.64]. The schematic "ball and socket" nanostructure for calix[8]arene-C_{60} is shown in Fig. 10.22(c), where the term nanostructure refers to ball and socket structures on the size scale of a few nm or less. In Fig. 10.22 it is seen that C_{60} can be suspended by a van der Waals interaction over the truncated cone of the

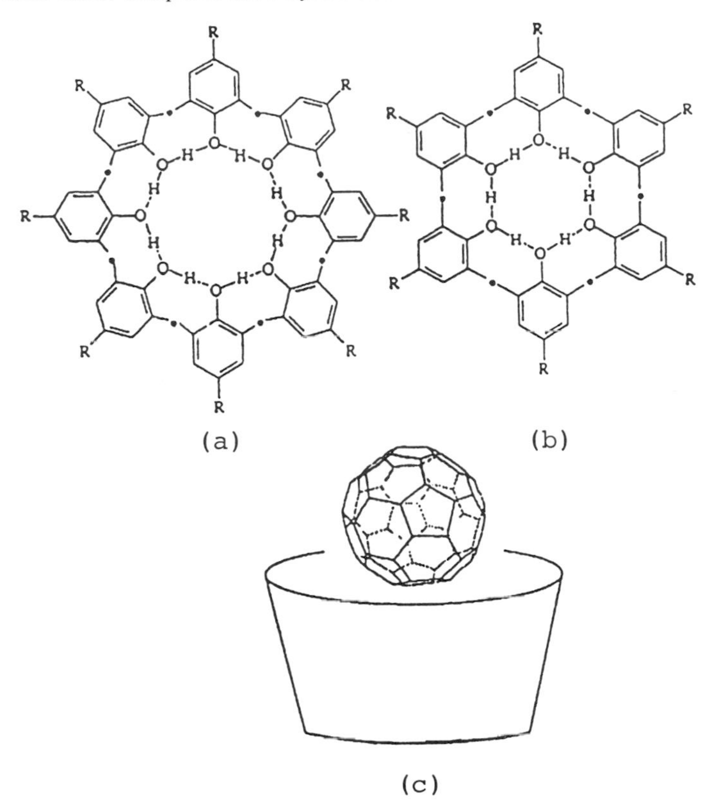

Fig. 10.22. Planar projections of the chemical structure for the molecules: (a) *p*-R-calix-[8]arene and (b) *p*-R-calix[6]arene. (c) Diagrammatic representation of a "ball and socket" nanostructure where the ball is C_{60} and the socket is a calixarene molecule [10.65].

p-R-calix[8]arene molecule. Furthermore, a *p*-*t*-butyl-calix[8]arene complex with C_{70} has also been reported [10.65], where *p* stands for para and denotes a location for the complex opposite the oxygen atom and "*t*-butyl" denotes $(CH_3)_3C$. Complexation of *p*-*t*-butyl-calix[8]arene with a mixture of a toluene extract of crude fullerene soot, followed by a series of recrystallizations, yielded >99.5% pure C_{60} [10.65], showing a use of complexation reactions for the separation of C_{60} from soot.

Another example of structural stabilization through weak intermolecular interactions is the formation of a 2:1 complex between C_{60} and bis(ethylenedithio)tetrathiafulvalene (ET). Evidence of charge transfer between C_{60} and the ET molecule was given by UV spectroscopy and x-ray crystallography [10.66]. The authors suggest that charge transfer occurs between a *p* orbital of the sulfur atoms in the ET molecule and a valence π orbital of

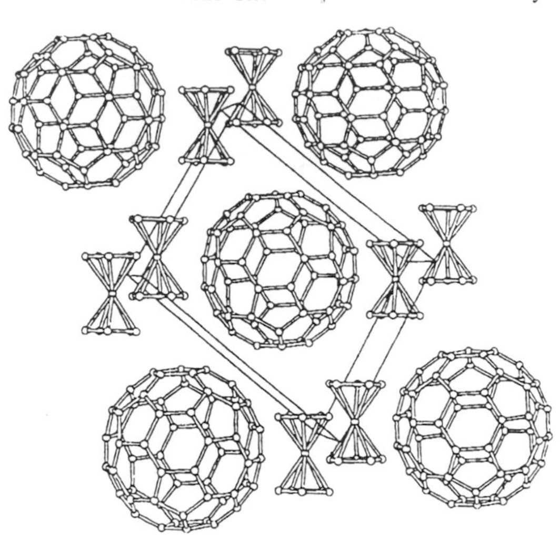

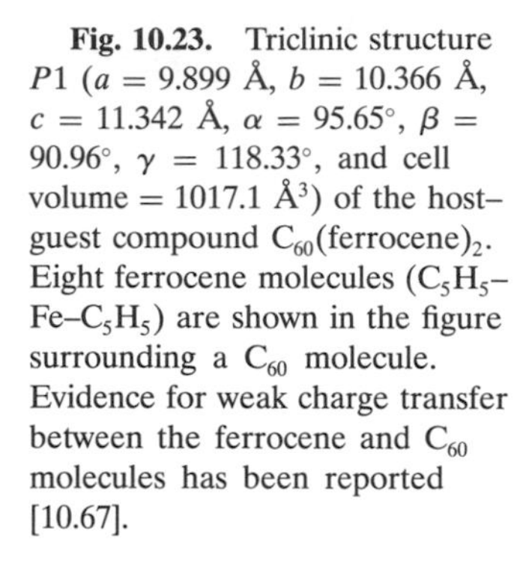

Fig. 10.23. Triclinic structure $P1$ ($a = 9.899$ Å, $b = 10.366$ Å, $c = 11.342$ Å, $\alpha = 95.65°$, $\beta = 90.96°$, $\gamma = 118.33°$, and cell volume = 1017.1 Å^3) of the host–guest compound C_{60}(ferrocene)$_2$. Eight ferrocene molecules (C_5H_5–Fe–C_5H_5) are shown in the figure surrounding a C_{60} molecule. Evidence for weak charge transfer between the ferrocene and C_{60} molecules has been reported [10.67].

the C_{60} [10.66]. Another example of stabilization through intermolecular interaction without formal charge transfer is provided by C_{60} and ferrocene complexes [10.67] (C_5H_5–Fe–C_5H_5, i.e., two parallel cyclopentane rings with a central Fe atom located midway between the rings). The weak ferrocene–C_{60} charge transfer complex is shown schematically in Fig. 10.23.

10.10.2. Polymerization

Polymeric materials contain macromolecules that are built from smaller molecular units, called monomers, which are linked together in a regular array by covalent bonds. The physical properties of these polymeric materials depend both on the properties of an average macromolecule and on the way in which these macromolecules bind together in the mixture, albeit by a van der Waals interaction or via covalent cross-linking. Examples of polymers involving C_{60} in the polymer chain and in polymer side chains have been reported and are discussed below after a few preliminary remarks on polymer terminology and structure.

Two common forms of polymers are the "homopolymers" and "copolymers" (see Fig. 10.24). The homopolymers are formed from a single monomer (X), and in the simplest case they bond end to end to form long-chain macromolecules (-X–X–X–X-). Copolymerization occurs when a mixture of two or more monomer units (X and Y) bond together (polymerize) so that both units appear in the polymer chain. Random bonding seldom occurs; i.e., a tendency toward an ordered structure is observed, in which X and Y monomeric units usually alternate in the chain (-X–Y–X–Y-). Ordered "graft" and "block" copolymers are also common.

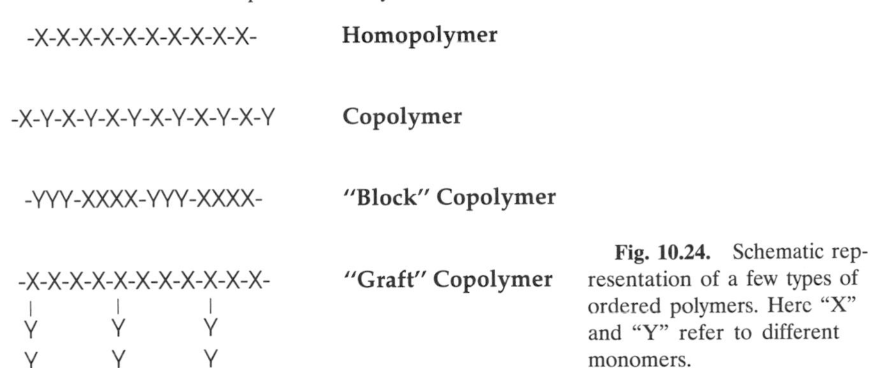

Fig. 10.24. Schematic representation of a few types of ordered polymers. Here "X" and "Y" refer to different monomers.

Block copolymers consist of interconnected homopolymer chains, such as $(\text{-}X_4Y_3\text{-})_n$ shown in Fig. 10.24. Graft copolymers, on the other hand, often exhibit homopolymer side chains.

The proclivity for C_{60} to form free radicals and react with various reagents, as already described throughout this chapter, indicates that the incorporation of C_{60} into a polymer structure should be possible. This incorporation might be accomplished in one of two general ways: "in-chain" addition, as shown in Fig. 10.25(a), and "side-chain" addition, as shown in Fig. 10.25(b). These C_{60} polymers are sometimes referred to as "pearl necklace" (in-chain) and "pendant" (side-chain) C_{60} polymers. Below we give a few examples of C_{60} pendant and pearl necklace polymers, which form either as ordered copolymers or as block copolymers.

$(\text{-}C_{60}Pd\text{-})_n$ has been proposed as an example of a fullerene-derived pearl necklace or an in-chain polymer structure [10.68]. This polymer was synthesized by the reaction of C_{60} with a palladium complex, $Pd_2(dba)CHCl_3$, where dba denotes dibenzylidene-acetone or $(C_6H_5CH{=}CH)_2C{=}O$. The authors proposed that the first step in the reaction produces $C_{60}Pd_n$, where the metal Pd atoms bridge the C_{60} molecules in the form of the pearl necklace structure shown in Fig. 10.25(a). This polymer is reported to be air stable and insoluble in common organic solvents. Further reaction of the polymer with Pd was proposed to yield interchain cross-links in the form of Pd bridges (see Fig. 10.26). The incorporation of Pd into the polymer, up to the ratio 3:1 for Pd:C_{60}, is viewed as completing the Pd bridging between chains. For Pd:C_{60} ratios greater than 3, excess Pd was then found to deposit on the polymer surface, as indicated schematically in the figure.

Single crystals of K_1C_{60} prepared by coevaporation of K and C_{60} have been reported [10.69] to contain linear polyanion chains whose molecular structure is shown schematically in Fig. 10.27 [10.70] (see §13.4.5). The authors have interpreted their electron spin resonance (ESR) measurements to indicate that their K_1C_{60} material is a metal, consistent with the dona-

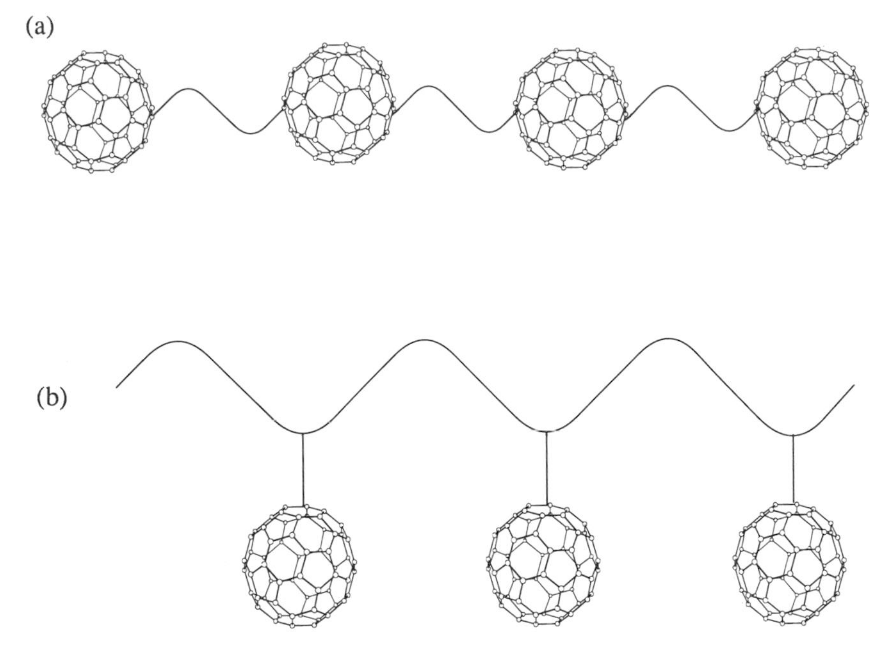

Fig. 10.25. Schematic representation of (a) "pearl necklace" and (b) "pendant chain" fullerene-derived polymers [10.4].

tion of electrons from the K atoms to the polymer chains. Treatment of the K_1C_{60} crystals in (oxygen-containing) toluene leads to the production of fiber bundles, interpreted as evidence of the linear polymer structure. The four-membered ring between C_{60} shells shown in the figure is the same intermolecular link that was previously proposed [10.12] to occur in polymeric C_{60} as a result of a photochemical "2+2" cycloaddition reaction (see §7.5.1). In contrast to the pearl necklace K_1C_{60} polymer, the C_{60}-polymer produced optically is thought to be extensively cross-linked.

Two reports of pendant C_{60} copolymers formed by the attachment of C_{60} to amino polymers have appeared [10.71, 72]. The coupling of C_{60} to the amino polymer is the same as for amines [10.7, 22] discussed in §10.4, as shown in the addition reaction of Fig. 10.9, where the amine (RNH_2) attaches to C_{60} at one of its double bonds. In one study [10.71], the reaction occurs between C_{60} dissolved in toluene and the amino polymer precursors (a, b, c) shown schematically in Fig. 10.28. The amino copolymer precursors had an average molecular weight consistent with ~250 monomeric units in (a), and ~100 monomeric units in (b). The C_{60} was found to add at ~20:1 (monomer:C_{60}) loading for both (a) and (b). Interestingly, the pendant C_{60} polymer made from the precursor (b) was reported to be soluble in toluene

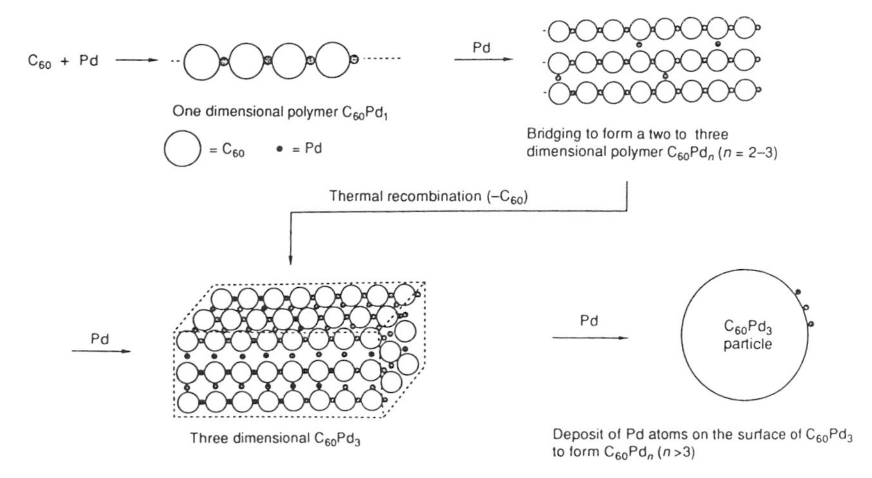

Fig. 10.26. Proposed mechanism for the formation of $C_{60}Pd_n$ showing pearl necklace formation at low Pd concentrations, interchain cross-links up to $n = 3$, and the deposition of Pd on the C_{60} surface for $n > 3$ [10.68].

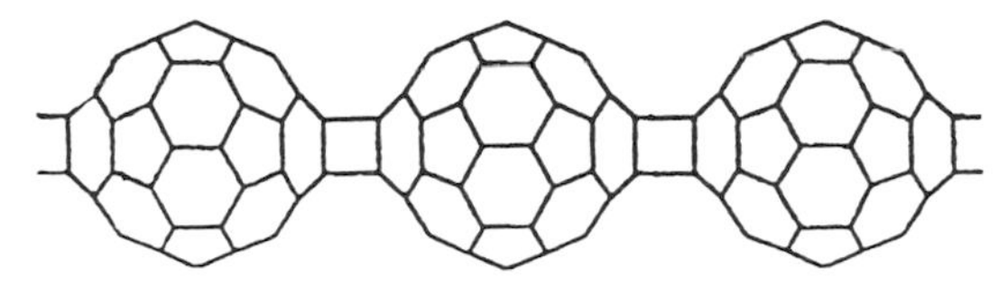

Fig. 10.27. Molecular structure of the "2+2" cycloadduct $(C_{60})_n$ polyanion in K_1C_{60}. The K^+ ions (not shown) are near the bridge sites between adjacent fullerenes [10.69].

and CS_2. In the second report [10.72], the amino polymer precursor was ethylene propylene terpolymer (EPDM-amine) (c). Copolymerization with (c) was inferred from viscosity measurements. Similar to the copolymer in Fig. 10.28(b), the copolymer in Fig. 10.28(c) was found to be soluble in common solvents.

The synthesis of a block copolymer of C_{60} through free radical intermediates has been reported using the diradical xylylene $\bullet CH_2$–C_6H_4–$CH_2\bullet$. The resulting block copolymer $(\text{-}[C_{60}]_p\ [\text{xylylene}]_q\text{-})_n$ has a ratio of $q/p = 3.4$ [10.73]. The structure of this copolymer is not known, although it is expected to be a side chain or pendant C_{60} polymer. ^{13}C NMR data were interpreted to indicate that the C_{60} attachment to the xylylene units was through benzylated C atoms on the C_{60}. The xylylene monomer was prepared by the flash thermolysis of paracyclophane $(C_{16}H_{20})$ at 650°C. This monomer can react with itself to form the polymer poly(*p*-xylylene) (Fig. 10.29). In the synthesis of the copolymer, the xylylene was swept from the furnace in which it was produced into a cooled solution of C_{60} in toluene. Upon warming to room temperature, a brown precipitate (polymer) formed, which

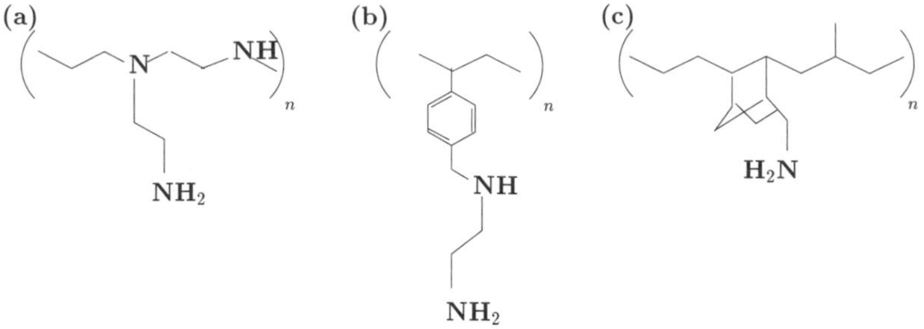

Fig. 10.28. Precursor polymers used in the preparation of C_{60} amino polymers. (a) poly(ethyleneimine), (b) poly{4-[(2-aminoethyl)imino]methyl}styrene, (c) ethylene propylene terpolymer (EPDM-amine). Each of these precursors can be substituted for R in Fig. 10.9 [10.71, 72]. The straight line segments represent C−C single bonds and double bonds are indicated in the aromatic ring, with C atoms at each unmarked vertex.

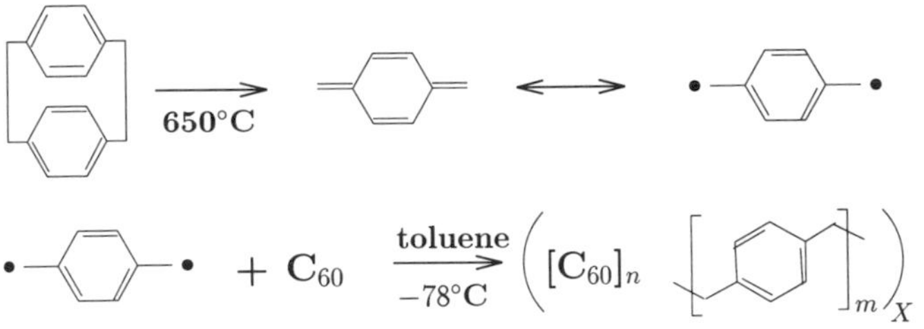

Fig. 10.29. Mechanism for formation of the xylylene diradical at 650°C, followed by its copolymerization with C_{60} [10.73].

was found to be unstable in air. C_{60} polymers, in which direct linkages between fullerenes are formed, may have also been produced. This proposal stemmed from a series of studies involving the phototransformation of thin solid films of C_{60} with visible–UV radiation [10.74].

The photopolymerization of C_{60} has been suggested to proceed by a photochemical "2+2" cycloaddition process (see §10.7) in which double bonds on adjacent C_{60} molecules are broken to form a four-membered ring between molecules (see §7.5.1) [10.12, 13]. The phototransformation reaction takes place in C_{60} films only above ~260 K, where the C_{60} molecules rotate relatively freely and can position themselves so that a double bond on one C_{60} molecule is parallel to a double bond on an adjacent C_{60} molecule [10.75], satisfying a topological requirement for "2+2" cycloaddition. The stability of this four-membered ring is measured by the thermal barrier ($E_b \sim 1.25$ eV) which must be overcome to break the bond in the four-membered bridge ring. At higher temperatures (above ~400 K), a rapid decrease in the population of oligomers is observed by Raman scattering

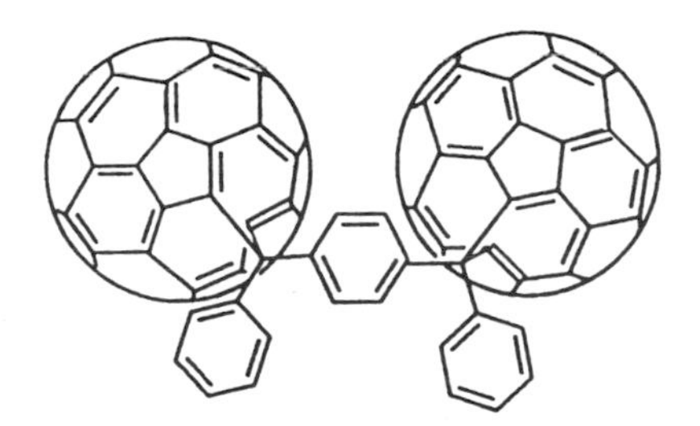

Fig. 10.30. Two fulleroids (C_{61}) joined by a phenyl group to form a bifulleroid. This bifulleroid dimer shows the pearl necklace conformation [10.7].

measurements [10.76]. Another route for preparing C_{60} polymers is by extending the synthetic route for the preparation of fulleroids of the type shown in Fig. 10.11.

Bifulleroid formation has also been demonstrated chemically. For example, Fig. 10.30 shows the formation of such a bifulleroid dimer through the bridging of two fulleroids through a phenyl group, suggesting the variety of polymeric structures that might be formed [10.7]. The coupling between the two fulleroids is confirmed by cyclic voltammetry studies (§10.3.2) showing two close doublet peaks for each wave on a diagram such as in Fig. 10.6 [10.27, 77–80]. Of course, bifulleroids can be formed by other aromatic hydrocarbon groups. The synthesis of higher polymers has been hindered by poor solubility of the bifulleroid molecules. Nevertheless, this difficulty has been overcome by the attachment of solubilizing groups on the phenylene moiety.

REFERENCES

[10.1] Q. L. Zhang, S. C. O'Brien, J. R. Heath, Y. Liu, R. F. Curl, H. W. Kroto, and R. E. Smalley. *J. Phys. Chem.*, **90**, 525 (1986).

[10.2] R. C. Haddon, A. F. Hebard, M. J. Rosseinsky, D. W. Murphy, S. J. Duclos, K. B. Lyons, B. Miller, J. M. Rosamilia, R. M. Fleming, A. R. Kortan, S. H. Glarum, A. V. Makhija, A. J. Muller, R. H. Eick, S. M. Zahurak, R. Tycko, G. Dabbagh, and F. A. Thiel. *Nature (London)*, **350**, 320 (1991).

[10.3] G. A. Olah, I. Bucsi, R. Aniszfeld, and G. K. S. Prakash. *Carbon*, **30**, 1203 (1992).

[10.4] R. Taylor and D. R. M. Walton. *Nature (London)*, **363**, 685 (1993).

[10.5] P. J. Fagan, J. C. Calabrese, and B. Malone. *Acc. Chem. Res.*, **25**, 134 (1992).

[10.6] J. M. Hawkins. *Acc. Chem. Res.*, **25**, 150 (1992).

[10.7] F. Wudl. *Acc. Chem. Res.*, **25**, 157 (1992).

[10.8] R. T. Morrison and R. N. Boyd. *Organic Chemistry*, 6th ed. Prentice Hall, Englewood Cliffs, NJ (1992).

[10.9] J. March. *Advanced Organic Chemistry: Reactions, Mechanisms and Structure*, 4th ed. Wiley, New York (1992).

[10.10] M. Gevaert and P. V. Kamat. *J. Chem. Soc. Chem. Commun.*, pp. 1470–1472 (1992).

[10.11] H. S. Chen, A. R. Kortan, R. C. Haddon, and R. A. Fleming. *J. Phys. Chem.*, **96**, 1016 (1992).

[10.12] A. M. Rao, P. Zhou, K.-A. Wang, G. T. Hager, J. M. Holden, Y. Wang, W. T. Lee, X.-X. Bi, P. C. Eklund, D. S. Cornett, M. A. Duncan, and I. J. Amster. *Science*, **259**, 955 (1993).

[10.13] P. Zhou, A. M. Rao, K. A. Wang, J. D. Robertson, C. Eloi, M. S. Meier, S. L. Ren, X. X. Bi, and P. C. Eklund. *Appl. Phys. Lett.*, **60**, 2871 (1992).

[10.14] R. Taylor, J. P. Parsons, A. G. Avent, S. P. Rannard, T. J. Dennis, J. P. Hare, H. W. Kroto, and D. R. M. Walton. *Nature (London)*, **351**, 277 (1991).

[10.15] S. Kawata, K. Yamauchi, S. Suzuki, K. Kikuchi, H. Shiromura, M. Katada, K. Saito, I. Ikemoto, and Y. Achiba. *Chem. Lett.*, pp. 1659–1662 (1992).

[10.16] H. Werner, D. Bublak, U. Göbel, B. Henschke, W. Bensch, and R. Schlögl. *Angew. Chem. Int. (Ed. English)*, **31**, 868 (1992).

[10.17] D. J. Klein, T. G. Schmalz, G. E. Hite, and W. A. Seitz. *J. Am. Chem. Soc.*, **108**, 1301 (1986).

[10.18] R. E. Smalley. *Acc. Chem. Res.*, **25**, 98 (1992).

[10.19] R. C. Haddon. *Science*, **261**, 1545 (1993).

[10.20] R. Tycko, R. C. Haddon, G. Dabbagh, S. H. Glarum, D. C. Douglass, and A. M. Mujsce. *J. Phys. Chem.*, **95**, 518 (1991).

[10.21] C. S. Yannoni, R. D. Johnson, G. Meijer, D. S. Bethune, and J. R. Salem. *J. Phys. Chem.*, **95**, 9 (1991).

[10.22] F. Wudl, A. Hirsch, K. C. Khemani, T. Suzuki, P. M. Allemand, A. Koch, H. Eckert, G. Srdanov, and H. M. Webb. In G. S. Hammond and V. J. Kuck (eds.), *Fullerenes: Synthesis, Properties and Chemistry of Large Carbon Clusters.*, p. 161, American Chemical Society, Washington, DC (1992). ACS Symposium Series, 481.

[10.23] P.-M. Allemand, A. Koch, F. Wudl, Y. Rubin, F. Diederich, M. M. Alvarez, S. J. Anz, and R. L. Whetten. *J. Am. Chem. Soc.*, **113**, 1050 (1991).

[10.24] D. DuBois, K. Kadish, S. Flanagan, R. E. Haufler, L. P. F. Chibante, and L. J. Wilson. *J. Am. Chem. Soc.*, **113**, 4364 (1991).

[10.25] D. DuBois, K. Kadish, S. Flanagan, and L. J. Wilson. *J. Am. Chem. Soc.*, **113**, 7773 (1991).

[10.26] Y. Ohsawa and T. Saji. *J. Chem. Soc., Chem. Commun.*, pp. 781–782 (1992).

[10.27] Q. Xie, E. Perez-Cordero, and L. Echegoyen. *J. Amer. Chem. Soc.*, **114**, 3978 (1992).

[10.28] F. Zhou, C. Jehoulet, and A. S. Bard. *J. Am. Chem. Soc.*, **114**, 11004 (1992).

[10.29] R. E. Haufler, J. J. Conceicao, L. P. F. Chibante, Y. Chai, N. E. Byrne, S. Flanagan, M. M. Haley, S. C. O'Brien, C. Pan, Z. Xiao, W. E. Billups, M. A. Ciufolini, R. H. Hauge, J. L. Margrave, L. J. Wilson, R. F. Curl, and R. E. Smalley. *J. Phys. Chem.*, **94**, 8634 (1990).

[10.30] J. W. Bausch, G. K. S. Prakash, G. A. Olah, D. S. Tse, D. C. Lorents, Y. K. Bae, and R. Malhotra. *J. Am. Chem. Soc.*, **113**, 3205 (1991).

[10.31] Y. Chabre. *NATO, ASI Series B: Physics*, **305**, 181 (1993). Chemical Physics of Intercalation II, Plenum, New York (1993) T. P. Verso and P. Bernier (eds).

[10.32] Y. Chabre, D. Djurado, and M. Barral. *Mol. Cryst. Liq. Cryst.*, **245**, 307 (1994). 7th International Symposium on Intercalation Compounds.

[10.33] Y. Chabre, D. Djurado, M. Armand, W. R. Romanow, N. Coustel, J. P. McCauley, Jr., J. E. Fischer, and A. B. Smith, III. *J. Am. Chem. Soc.*, **114**, 764 (1992).

[10.34] K. O. Christe and W. W. Wilson. *Fluorine Division* (1992). Paper #66, presented at 203rd American Chemical Society Meeting, San Francisco, CA.

[10.35] G. P. Miller, C. S. Hsu, L. Y. Chieng, H. Thomann, and M. Bernardo. *Chem. & Eng. News*, **69**, 17 (1991). December 16, issue: unpublished results from the Exxon Laboratory which were reported at the 1991 MRS meeting in Boston.

[10.36] D. Koruga, S. Hameroff, J. Withers, R. Loutfy, and M. Sundareshan. *Fullerene C_{60}: History, Physics, Nanobiology, Nanotechnology*. North-Holland, Amsterdam (1993). reported on p. 318.

[10.37] H. Selig, C. Lifshitz, T. Peres, J. E. Fischer, A. R. McGhie, W. J. Romanow, J. P. McCauley, Jr., and A. B. Smith, III. *J. Am. Chem. Soc.*, **113**, 5475 (1991).

[10.38] J. H. Holloway, E. G. Hope, R. Taylor, G. J. Langley, A. G. Avent, T. J. Dennis, J. P. Hare, H. W. Kroto, and D. R. M. Walton. *J. Chem. Soc., Chem. Commun.*, p. 966 (1991).

[10.39] G. A. Olah, I. Bucsi, C. Lambert, R. Aniszfeld, N. J. Trivedi, D. K. Sensharma, and G. K. S. Prakash. *J. Am. Chem. Soc.*, **113**, 9385 (1991).

[10.40] P. R. Birkett, P. B. Hitchcock, H. W. Kroto, R. Taylor, and D. R. M. Walton. *Nature (London)*, **357**, 479 (1992).

[10.41] F. N. Tebbe, R. L. Harlow, D. B. Chase, D. L. Thorn, G. C. Campbell, Jr., J. C. Calabrese, N. Herron, R. J. Young, Jr., and E. Wasserman. *Science*, **256**, 822 (1992).

[10.42] F. Cataldo. *Carbon*, **32**, 437 (1994).

[10.43] Y. Elemes, S. Silverman, C. Shen, M. Kao, S. Christopher, M. Alvarez, and R. Whetten. *Angew. Chem. Int. Edn. Engl.*, **31**, 351 (1992).

[10.44] K. M. Creegan, J. L. Robbins, W. K. Robbins, J. M. Millar, R. D. Sherwood, P. J. Tindal, D. M. Cox, A. B. Smith, III, J. P. McCauley, Jr., D. R. Jones, and R. T. Gallagher. *J. Am. Chem. Soc.*, **114**, 1103 (1992).

[10.45] G. B. M. Vaughan, P. A. Heiney, D. E. Cox, A. R. McGhie, D. R. Jones, R. M. Strongin, M. A. Cichy, and A. B. Smith III. *Chem. Phys.*, **168**, 185 (1992).

[10.46] F. Diederich, R. Ettl, Y. Rubin, R. L. Whetten, R. Beck, M. Alvarez, S. Anz, D. Sensharma, F. Wudl, K. C. Khemani, and A. Koch. *Science*, **252**, 548 (1991).

[10.47] R. Taylor, J. H. Holloway, E. G. Hope, A. G. Avent, G. J. Langley, T. J. Dennis, J. P. Hare, H. W. Kroto, and D. R. M. Walton. *J. Chem. Soc. Chem. Commun.*, pp. 667–668 (1992).

[10.48] A. A. Tuinman, P. Mukherjee, J. L. Adcock, R. L. Hettich, and R. N. Compton. *J. Phys. Chem.*, **96**, 7584 (1992).

[10.49] K. Raghavachari and C. M. Rohlfing. *Chem. Phys. Lett.*, **197**, 495 (1992).

[10.50] R. F. Schinazi, R. Sijbesma, G. Srdanov, C. L. Hill, and F. Wudl. *Antimicrobial Agents and Chemotherapy*, **37**, 1707 (1993).

[10.51] S. H. Friedman, D. L. DeCamp, R. P. Sijbesma, F. Wudl, and G. L. Kenyon. *J. Am. Chem. Soc.*, **115**, 6506 (1993).

[10.52] R. Sijbesma, G. Srdanov, F. Wudl, J. A. Castoro, C. L. Wilkins, S. H. Friedman, D. L. DeCamp, and G. L. Kenyon. *J. Am. Chem. Soc.*, **115**, 6510 (1993).

[10.53] J. M. Hawkins, A. Meyer, T. Lewis, and S. Loren. In G. S. Hammond and V. J. Kuck (eds.), *Fullerenes*, p. 91. American Chemical Society, Washington, DC (1992). ACS Symposium Series, 481, Chapter 6.

[10.54] A. L. Balch, V. J. Catalano, J. W. Lee, M. M. Olmstead, and S. R. Parkin. *J. Am. Chem. Soc.*, **113**, 8953 (1991).

[10.55] Y. Rubin, S. Khan, D. I. Freedberg, and C. Yeretzian. *J. Am. Chem. Soc.*, **115**, 344 (1993).

[10.56] P. Belik, A. Gügel, M. Spickerman, and K. Müllen. *Angew. Chem. Int. Edn. Engl.*, **32**, 78 (1993).

[10.57] M. F. Meidine, R. Roers, G. J. Langley, A. G. Avent, A. D. Darwish, S. Firth, H. W. Kroto, R. Taylor, and D. R. M. Walton. *J. Chem. Soc. Chem. Commun.*, p. 1342 (1993).

[10.58] H. Becker, G. Javahery, S. Petrie, and D. K. Bohme. *J. Phys. Chem.*, **98**, 5591 (1994).

[10.59] S. H. Hoke, J. Molstad, D. Dilettato, M. J. Jay, D. Carlson, B. Kahr, and R. G. Cooks. *J. Org. Chem.*, **57**, 5069 (1992).
[10.60] S. R. Wilson, N. Kaprinidis, Y. Wu, and D. I. Schuster. *J. Am. Chem. Soc.*, **115**, 8495 (1993).
[10.61] P. J. Krusic, E. Wasserman, P. N. Keizer, J. R. Morton, and K. F. Preston. *Science*, **254**, 1183 (1991).
[10.62] L. Y. Chiang, L. Upasani, and J. W. Swirczewski. *J. Am. Chem. Soc.*, **114**, 10154 (1992).
[10.63] O. Emmer and C. Robke. *J. Am. Chem. Soc.*, **115**, 10077 (1993).
[10.64] T. Andersson, K. Nilsson, M. Sundahl, G. Westman, and O. Wennerstrom. *J. Chem. Soc. Chem. Commun.*, p. 604 (1992).
[10.65] J. L. Atwood, G. A. Koutsantonis, and C. L. Raston. *Nature*, **368**, 229 (1994).
[10.66] A. Izuoka, T. Tachikawa, T. Sugawara, Y. Saito, and H. Shinohara. *Chem. Lett.*, pp. 1049–1052 (1992). Number 6.
[10.67] J. D. Crane, P. W. Hitchcock, H. W. Kroto, R. Taylor, and D. R. M. Walton. *J. Chem. Soc. Chem. Commun.*, pp. 1764–1765 (1992).
[10.68] H. Nagashima, A. Nakaoka, Y. Saito, M. Kato, T. Kawanishi, and K. Itoh. *J. Chem. Soc. Chem. Commun.*, pp. 377–380 (1992).
[10.69] S. Pekker, A. Jánossy, L. Mihaly, O. Chauver, M. Carrard, and L. Forró. *Science*, **265**, 1077 (1994).
[10.70] P. W. Stephens, G. Bortel, G. Falgel, M. Tegze, A. Jánossy, S. Pekker, G. Oszànyi, and L. Forró. *Nature (London)*, **370**, 636 (1994).
[10.71] K. E. Geckeler and A. Hirsch. *J. Am. Chem. Soc.*, **115**, 3850 (1993).
[10.72] A. O. Patil, G. W. Schriver, B. Cartensen, and R. D. Lundberg. *Polymer Bulletin*, **30**, 187 (1993).
[10.73] D. A. Loy and R. A. Assink. *J. Am. Chem. Soc.*, **114**, 3977 (1992).
[10.74] P. C. Eklund, A. M. Rao, Y. Wang, and P. Zhou. In Z. H. Kafafi (ed.), *Proceedings of the International Society for Optical Engineering (SPIE)*, vol. 2284, pp. 2–4, Bellingham, WA (1994). SPIE Optical Engineering Press. San Diego, CA, July 24–29, 1994.
[10.75] P. Zhou, Z. H. Dong, A. M. Rao, and P. C. Eklund. *Chem. Phys. Lett.*, **211**, 337 (1993).
[10.76] Y. Wang, J. M. Holden, X. X. Bi, and P. C. Eklund. *Chem. Phys. Lett.*, **217**, 413 (1994).
[10.77] T. Suzuki, Q. Li, K. C. Khemani, and F. Wudl. *J. Am. Chem. Soc.*, **114**, 7301 (1992).
[10.78] T. Suzuki, Y. Maruyama, T. Akasaka, W. Ando, K. Kobayashi, and S. Nagase. *J. Am. Chem. Soc.*, **116**, 1359 (1994).
[10.79] T. Suzuki, Q. Li, K. C. Khemani, and F. Wudl. *J. Am. Chem. Soc.*, **114**, 7300 (1992).
[10.80] J. Chlistunoff, D. Cliffel, and A. J. Bard. *Thin Solid Films*, **257**, 166 (1995).

CHAPTER 11

Vibrational Modes

In this chapter we review the lattice mode structure for the isolated fullerene molecules and for the corresponding molecular solid. Explicit results are given for C_{60}, C_{70}, and higher fullerenes. The effects of doping, photopolymerization, and pressure on the vibrational spectra are reviewed.

11.1. Overview of Mode Classifications

Because solid C_{60} is very nearly an ideal molecular solid, its vibrational modes can be subdivided naturally into two classes: intermolecular vibrations (or lattice modes) and intramolecular vibrations (or simply "molecular" modes). The intermolecular modes can be further divided into three subclasses: acoustic, optical, and librational modes. All solids have three *acoustic modes*, whose frequencies vanish at $q = 0$. In addition, *optic "lattice" modes* are found in all atomic solids with two or more atoms per primitive unit cell; in the case of solid C_{60}, the intermolecular optical lattice modes are associated with two or more C_{60} molecules per unit cell. For molecular solids, a third subclass of modes is identified with *librational modes*, which stem from the moment of inertia of the C_{60} molecule and involve a hindered rotation (rocking) of the C_{60} molecules about their equilibrium lattice positions. The frequencies of the librational modes are quite low because the molecules are only weakly coupled and also because the moment of inertia of the fullerene molecules is large. These distinctions are traditionally made for C_{60} (as shown in Fig. 11.1) [11.1] but can also be made for higher fullerenes and for doped fullerenes. A number of authors have discussed the vibrational spectra of molecules with icosahedral symmetry both experimentally [11.2–4] and theoretically [11.5–9].

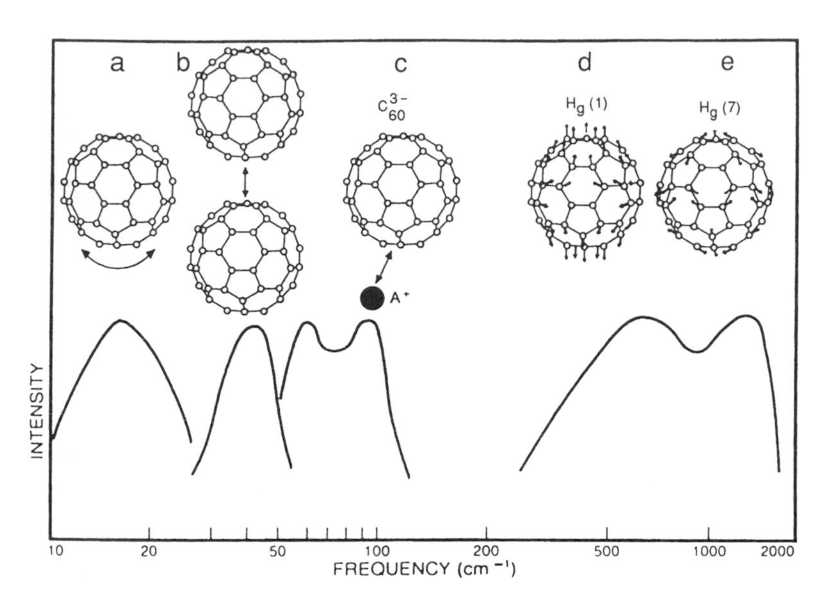

Fig. 11.1. Schematic view of the various classes of vibrations in M_3C_{60} compounds (see text). At low frequencies, the compounds exhibit librational modes of individual C_{60} molecules (a), intermolecular optic modes (b), and IR-active optic modes (c). At higher frequencies (above ~ 200 cm^{-1}), the intramolecular modes are dominant and follow the schematic vibrational density of states shown in the figure. These "intramolecular" modes have predominantly radial character (d) at lower frequencies ($\omega < 700$ cm^{-1}) [for example the $H_g(1)$ mode], and at higher frequencies ($\omega > 700$ cm^{-1}) the mode displacements have predominantly a tangential character (e) [for example, the $H_g(7)$ mode] [11.1].

The introduction of alkaline earth or alkali metal atoms (M) into the fullerene lattice leaves the molecules intact but may rearrange and/or reorient the fullerene molecules with respect to each other (see §8.5 and §8.6). Despite these rearrangements and the concomitant charge transfer of electrons from these metal dopant atoms to create C_{60} molecular anions and metal cations, the vibrational mode picture described above for solid C_{60} remains largely intact. However, a new subclass of intermolecular modes, involving the relative motion of the cation (M^+) with respect to the molecular anion sublattices, must now be included.

In Fig. 11.1 we show schematically the natural separation of the vibrational modes of a particular crystalline compound, M_3C_{60} (M = alkali metal) solid [11.1], which is superconducting for M = potassium (T_c = 18 K) or M = rubidium (T_c = 29 K). The general nature of the schematic vibrational mode structure shown in Fig. 11.1 should be independent of the extent of the alkali metal doping (i.e., $1 \leq x \leq 6$ in M_xC_{60}). Note that for clarity the acoustic lattice modes have been omitted from the figure and that the density of vibrational states has been plotted against a logarithmic frequency scale covering more than two decades in frequency. The characteristic atomic or molecular motion associated with each of the four mode types is also indicated above the respective band of frequencies. The librational (a) modes lie lowest in frequency (10–30 cm^{-1}), followed by the intermolecular optic (b) modes involving relative motion between neutral C_{60} molecules or C_{60}^{3-} anions (30–60 cm^{-1}), then the infrared-active optic modes (c) involving relative motion between the C_{60}^{3-} anions and the M^+ cations (50–120 cm^{-1}), and finally the C_{60} intramolecular or "molecular" (d, e) modes (270–1700 cm^{-1}), which involve relative C atom displacements on a single molecule or anion. (The conversion between units commonly used for the vibrational mode frequencies is: 1 meV is 8.0668 cm^{-1} = 11.606 K = 0.24184 THz, and 1 THz = 33.356 cm^{-1}.) In the limit that the molecule is treated as a "point," the molecular moment of inertia I approaches zero, and the librational modes are lost from the spectrum. The intermolecular optic modes (b) are generally treated in the molecular limit [11.10, 11], as is discussed in §11.4.

In general, the low-frequency ($\omega < 700$ cm^{-1}) and high-frequency ($\omega >$ 700 cm^{-1}) intramolecular modes tend to exhibit approximately radial (d) and tangential (e) displacements, respectively, as indicated for the specific H_g symmetry modes shown in the figure. Some of the molecular modes have vibrational quanta as large as 0.1 to 0.2 eV, which is significant when compared to the electronic bandwidths (~0.5 eV) for energy bands located near the Fermi level and has implications for electrical transport in metallic fullerene solids (see §14.1).

In treating the vibrational spectra for other fullerene molecules or crystalline solids, it is necessary to consider the shape, mass, and moment of inertia for each fullerene species and isomer (C_{70}, C_{76}, C_{82}, C_{84}, etc.). For each of these heavier fullerenes, the four types of vibrational modes, shown schematically in Fig. 11.1, are also expected, as are the schematic phonon density of states curves for the fullerene solid.

11.2. Experimental Techniques

The major experimental techniques for studying the vibrational spectra include Raman and infrared spectroscopy, inelastic neutron scattering, and electron energy loss spectroscopy. Of these techniques, the most precise values for vibrational frequencies are obtained from Raman and infrared spectroscopies. In most crystalline solids, Raman and infrared spectroscopy measurements focus primarily on first-order spectra which are confined to zone-center ($q = 0$) phonons. Because of the molecular nature of fullerene solids, higher-order Raman and infrared spectra also give sharp spectral features, as shown in §11.5.3 and §11.5.4. Analysis of the second-order Raman and infrared spectra for C_{60} provides a good determination of the 32 silent mode frequencies and their symmetries. If large single crystals are available, then the most general technique for studying solid-state phonon dispersion relations is inelastic neutron scattering. Unfortunately, large single crystals are not yet available even for C_{60}, so that almost all the reported inelastic neutron scattering measurements have been done on polycrystalline samples. Much of the emphasis of the inelastic neutron scattering studies thus far has been on the lowest-energy intermolecular dispersion relations, which have been studied on both single-crystal and polycrystalline samples [11.12, 13]. Electron energy loss spectroscopy (EELS), although generally having less resolution than the other techniques, is especially useful for the study of silent modes because of differences in selection rules. An unexpected rich source of information about molecular C_{60} modes comes from singlet oxygen photoluminescence side bands, which have been especially useful for studying the intramolecular vibrational spectrum of C_{60} [11.14], since the usual IR or Raman selection rules do not apply to these spectra.

11.3. C_{60} Intramolecular Modes

In this section we consider theoretical issues affecting the vibrational intramolecular spectra. The room temperature experimental Raman spectra [11.3, 4, 15, 16] suggest that solid-state effects due to the T_h^6 (or $Pa\bar{3}$) space group are very weak or give rise to very small unresolved splittings of the 10 main Raman-allowed modes in the isolated C_{60} molecule. For this reason it is appropriate to start the discussion of the vibrational spectra for the intramolecular modes with those for the free molecule. However, additional lines are observed in Raman and infrared spectroscopy, and the theoretical explanations for such effects are discussed in this section.

11.3.1. The Role of Symmetry and Theoretical Models

The vibrational modes of an isolated C_{60} molecule may be classified according to their symmetry using standard group theoretical methods (see §4.2 and Table 4.6) [11.17–19], by which it is found that the 46 distinct intramolecular mode frequencies correspond to the following symmetries:

$$\Gamma_{\rm mol} = 2A_g+3F_{1g}+4F_{2g}+6G_g+8H_g+A_u+4F_{1u}+5F_{2u}+6G_u+7H_u, \quad (11.1)$$

where the subscripts g (gerade or even) and u (ungerade or odd) refer to the symmetry of the eigenvector under the action of the inversion operator, and the symmetry labels (e.g., F_1, H) refer to irreducible representations of the icosahedral symmetry group (I_h) (see §4.2). (Note that some authors use the notation T_1 and T_2 in place of the F_1 and F_2 employed here for the three-dimensional representations of the icosahedral group, and for this reason, we use both notations.) The degeneracy for each mode symmetry (given in parentheses) also follows from group theory: $A_g(1)$, $A_u(1)$; $F_{1g}(3)$, $F_{2g}(3)$, $F_{1u}(3)$, $F_{2u}(3)$; $G_g(4)$, $G_u(4)$; and $H_g(5)$, $H_u(5)$ (see §4.1). Thus Eq. (11.1) enumerates a total of 174 normal mode eigenvectors corresponding to 46 distinct mode frequencies, such that two distinct eigenfrequencies have A_g symmetry, three have F_{1g} symmetry, etc. Starting with $60\times3=180$ total degrees of freedom for an isolated C_{60} molecule, and subtracting the six degrees of freedom corresponding to three translations and three rotations, results in 174 vibrational degrees of freedom. Group theory, furthermore, indicates that 10 of the 46 mode frequencies are Raman-active ($2A_g + 8H_g$) in first order, 4 are infrared (IR)-active ($4F_{1u}$) in first order, and the remaining 32 are optically silent. Experimental values for all of the 46 mode frequencies, determined by first- and second-order Raman [11.20] and IR [11.21] spectroscopic features, are displayed in Table 11.1 [11.21], where the mode frequencies are identified with their appropriate symmetries and are listed in order of increasing vibrational frequency for each symmetry type.

Despite the high symmetry of the C_{60} molecule, the eigenvectors are, in general, somewhat difficult to visualize. In Fig. 11.2 we display eigenvectors for the 10 Raman-active and the 4 infrared-active modes [11.27]. The nondegenerate A_g modes, because of the higher symmetry of their eigenvectors, are easier to visualize. As shown in Fig. 11.3, the $A_g(1)$ "breathing" mode (492 cm^{-1}) involves identical radial displacements for all 60 carbon atoms, whereas the higher-frequency $A_g(2)$ mode, or "pentagonal pinch" mode (1469 cm^{-1}), involves primarily tangential displacements, with a contraction of the pentagonal rings and an expansion of the hexagonal rings. Because these modes have the same symmetry, any linear combination of these modes also has A_g symmetry. Physically, the curvature of the C_{60} cage gives rise to a small admixture of the radial and tangential modes. As can be seen in Fig. 11.2, the 8 fivefold degenerate H_g modes are much more complex and span the frequency range from 273 [$H_g(1)$] to 1578 cm^{-1} [$H_g(8)$]. The tendency of the lower-frequency molecular modes to have displacements that are more radial in character than those at higher frequency can be seen in Fig. 11.2. It should be recalled that the eigenvectors shown for the H_g modes are not unique, as is the case for degenerate modes, generally. Each H_g mode is fivefold degenerate with five partners, so that equally valid sets of eigenvectors for a common frequency can be constructed by forming orthonormal linear combinations of these five partners.

Table 11.1

Intramolecular vibrational frequencies of the C_{60} molecule and their symmetries: experiment and models.[a] The Raman-active modes have A_g and H_g symmetry and the IR-active modes hare F_{1u} symmetries. The remaining 32 modes are silent modes.

Even-parity					Odd-parity				
$\omega_i(\mathscr{R})$	Frequency (cm^{-1})				$\omega_i(\mathscr{R})$	Frequency (cm^{-1})			
	Expt.[a]	J[b]	Q[c]	F[d]		Expt.[a]	J[b]	Q[c]	F[d]
$\omega_1(A_g)$	497.5	492	478	483	$\omega_1(A_u)$	1143.0	1142	850	1012
$\omega_2(A_g)$	1470.0	1468	1499	1470					
					$\omega_1(F_{1u})$	526.5	505	547	547
$\omega_1(F_{1g})$	502.0	501	580	584	$\omega_2(F_{1u})$	575.8	589	570	578
$\omega_2(F_{1g})$	975.5	981	788	879	$\omega_3(F_{1u})$	1182.9	1208	1176	1208
$\omega_3(F_{1g})$	1357.5	1346	1252	1297	$\omega_4(F_{1u})$	1429.2	1450	1461	1445
$\omega_1(F_{2g})$	566.5	541	547	573	$\omega_1(F_{2u})$	355.5	367	342	377
$\omega_2(F_{2g})$	865.0	847	610	888	$\omega_2(F_{2u})$	680.0	677	738	705
$\omega_3(F_{2g})$	914.0	931	770	957	$\omega_3(F_{2u})$	1026.0	1025	962	1014
$\omega_4(F_{2g})$	1360.0	1351	1316	1433	$\omega_4(F_{2u})$	1201.0	1212	1185	1274
					$\omega_5(F_{2u})$	1576.5	1575	1539	1564
$\omega_1(G_g)$	486.0	498	486	449					
$\omega_2(G_g)$	621.0	626	571	612	$\omega_1(G_u)$	399.5	385	356	346
$\omega_3(G_g)$	806.0	805	759	840	$\omega_2(G_u)$	760.0	789	683	829
$\omega_4(G_g)$	1075.5	1056	1087	1153	$\omega_3(G_u)$	924.0	929	742	931
$\omega_5(G_g)$	1356.0	1375	1296	1396	$\omega_4(G_u)$	970.0	961	957	994
$\omega_6(G_g)$	1524.5[e]	1521	1505	1534	$\omega_5(G_u)$	1310.0	1327	1298	1425
					$\omega_6(G_u)$	1446.0	1413	1440	1451
$\omega_1(H_g)$	273.0	269	258	268					
$\omega_2(H_g)$	432.5	439	439	438	$\omega_1(H_u)$	342.5	361	404	387
$\omega_3(H_g)$	711.0	708	727	692	$\omega_2(H_u)$	563.0	543	539	521
$\omega_4(H_g)$	775.0	788	767	782	$\omega_3(H_u)$	696.0	700	657	667
$\omega_5(H_g)$	1101.0	1102	1093	1094	$\omega_4(H_u)$	801.0	801	737	814
$\omega_6(H_g)$	1251.0	1217	1244	1226	$\omega_5(H_u)$	1117.0	1129	1205	1141
$\omega_7(H_g)$	1426.5	1401	1443	1431	$\omega_6(H_u)$	1385.0	1385	1320	1358
$\omega_8(H_g)$	1577.5	1575	1575	1568	$\omega_7(H_u)$	1559.0	1552	1565	1558

[a] All experimental mode frequencies in the table are derived from a fit to first-order and higher-order Raman [11.20] and IR spectra [11.21]. Mode frequencies are reported in the literature in units of cm^{-1} and of meV (1 meV $\equiv$ 8.0668 cm^{-1}). Notations of ω and ν are commonly used to denote the same numerical values of the mode frequencies.

[b] Calculated mode frequencies of Ref. [11.22].

[c] Calculated mode frequencies of Ref. [11.23].

[d] Calculated mode frequencies of Ref. [11.24].

[e] Interpretation of an isotopically induced line in the Raman spectra implies this mode should be at 1490 cm^{-1} [11.25].

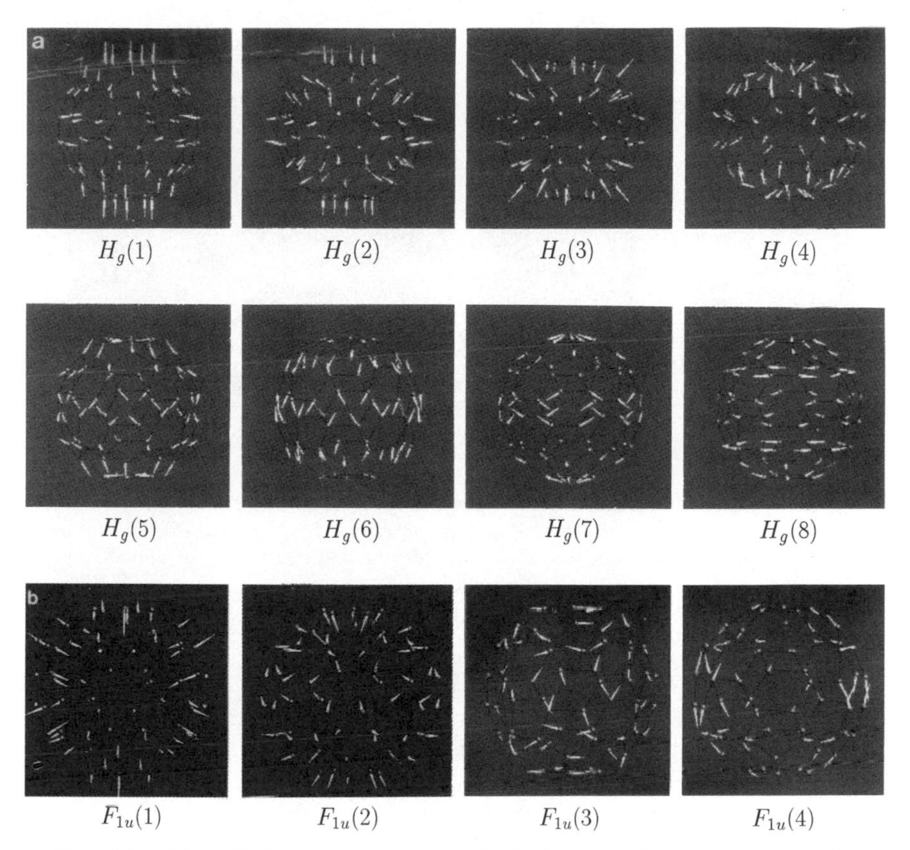

Fig. 11.2. Schematic diagram of normal mode displacements for a consistent set of partners of each of the (a) eight H_g Raman-active and (b) four F_u infrared-active modes in C_{60}. The eigenvectors are a function of the model and force constants, but the general trend is that modes with primarily radial displacements have low vibrational frequencies, and those modes with predominantly tangential displacements correspond to higher frequencies. The plots are a courtesy of M. Grabow at AT&T Bell Laboratories [11.26].

For the actual C_{60} molecule, the eigenvalues and eigenvectors for the intramolecular vibrational problem depend on the solutions to the 180×180 dynamical matrix. All 180 degrees of freedom must be considered in the dynamical matrix, since the translational and vibrational degrees of freedom result in modes with finite frequencies away from the zone center. The high symmetry of the molecule allows us to bring the dynamical matrix into block diagonal form at the high-symmetry points in the Brillouin zone using a unitary transformation which brings the coordinates into the correct symmetrized form, each block corresponding to a particular sym-

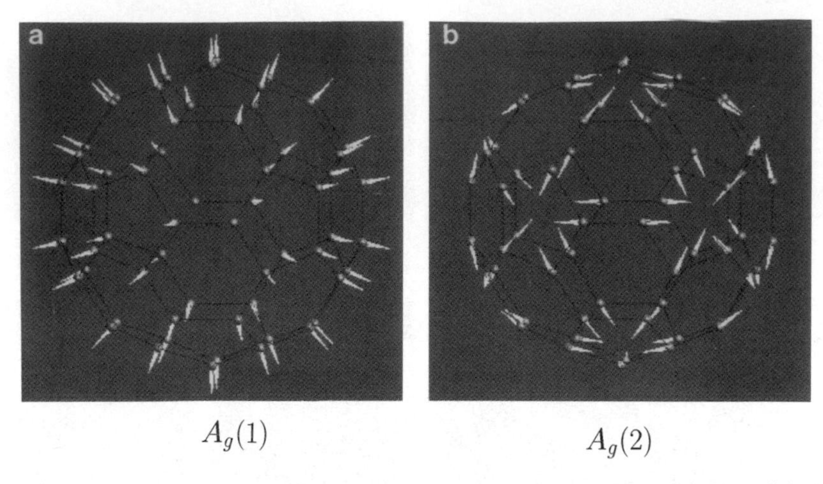

Fig. 11.3. Schematic diagram for the displacements corresponding to (a) the radial $A_g(1)$ breathing mode where all displacements are in the radial direction, and (b) the tangential $A_g(2)$ "pentagonal pinch" mode where all displacements are tangential. For both A_g modes all displacements are of equal magnitude and phase.

metry type, and only modes of the same symmetry are coupled. Thus the only eigenvector that can be uniquely specified is the one associated with the single A_u mode at the Brillouin zone center.

Just as it is instructive to consider the electronic states of a C_{60} molecule in a simple approximation where the 360 electrons (i.e., six electrons per carbon atom) are confined to a sphere of radius 7 Å (see §12.1.2), it is similarly instructive to consider the vibrational modes of a C_{60} molecule as elastic deformations of a thin spherical shell. This calculation has been carried out for the axisymmetric modes of C_{60} [11.28] and the calculation applies the classical results of Lamb for an elastic spherical shell published in 1883 [11.29]. For this shell model, there are two distinct families of axially symmetric mode frequencies displayed *vs.* mode number n in Fig. 11.4, an upper branch denoted by a_n and a lower branch denoted by b_n. The frequencies of these modes can be computed analytically in terms of a single parameter A which can be shown to depend on Young's modulus, the Poisson ratio γ (taken as 1/3), the shell radius, and the mass density [11.28]. To fit the continuum shell model to experiment, a value for A is obtained by setting the theoretical expression for the radial breathing mode frequency $\omega_1(A_g)$ equal to the experimental value of 497 cm^{-1}. In Fig. 11.4, the agreement between experimental Raman and IR frequencies (filled circles) [11.6, 30] and the model (open circles) is reasonably good, considering the simplicity of the model. In Fig. 11.5, the classical normal modes of the de-

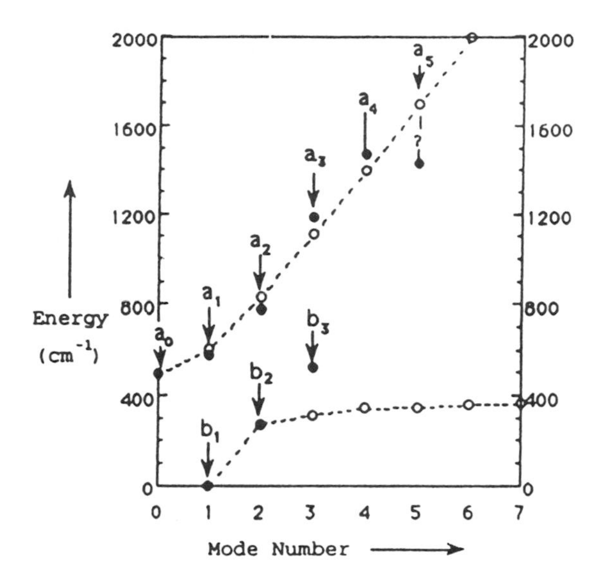

Fig. 11.4. Comparison of the continuum shell model predictions for axisymmetric frequencies of vibrations (open circles) and the experimental, infrared and Raman scattering results [11.30, 31] (filled circles). The dashed lines trace the two branches of the basic dispersion curve of the spherical shell model [11.28]. The higher mode numbers physically correspond to multiphonon processes.

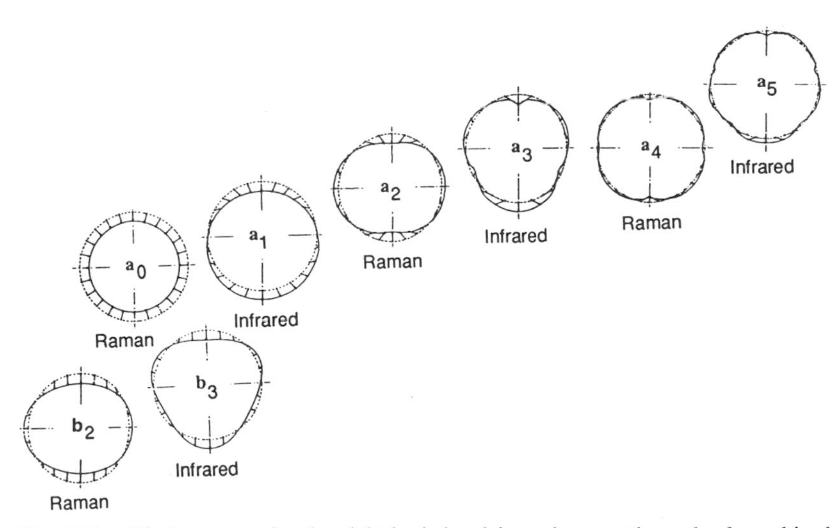

Fig. 11.5. Displacements for the eight basic low-lying axisymmetric modes for a thin-shell model of C_{60} are shown in schematic form, six on the upper (a_0–a_5) branch, and two on the lower branch (b_2, b_3), with b_1 denoting the acoustic mode. The "breathing" mode a_0 is considered to be the fundamental, while the b_3 mode departs the most from the initial spherical symmetry [11.28].

formable shell are displayed. Symmetry labels and experimental frequencies are provided to aid in the identification of these shell modes with actual C_{60} modes.

Several normal mode calculations for the C_{60} molecule involving discrete equilibrium positions for the C atoms have been carried out and can be compared with experiment. These calculations fall into two categories: classical force constant treatments involving bond-bending and bond-stretching displacements [11.22, 24, 32–35] and *ab initio* or semiempirical calculations [11.5, 36–40]. The minimal set of force constants for C_{60} includes two bond-stretching constants (for bonds along the pentagonal and hexagonal edges) and the two corresponding bond-bending constants. Bonding to more distant neighbors has been considered in the actual calculations. For example, a five-force-constant model for the C_{60} vibrational problem has been carried out [11.33, 34] with values for the force constants determined from other aromatic hydrocarbon molecules, while another empirical model [11.22], also using a bond-bending and bond-stretching force constant model, includes pairwise interactions through third-nearest-neighbor interactions. By adjusting the force constants to match the 10 observed first-order Raman-active mode frequencies, very good agreement was obtained between the calculated and experimental mode frequencies determined by Raman and infrared spectroscopy, inelastic neutron scattering, and electron energy loss spectroscopy. Also available are *ab initio* calculations for the molecular modes of C_{60} [11.6], which are also in reasonably good agreement with experiment.

Although these vibrational calculations are all made for an isolated molecule, the calculated results are usually compared to experimental data on C_{60} molecular vibrations in the solid state. The theoretical justification for this is the weak van der Waals bonds between C_{60} molecules in the face-centered cubic (fcc) solid. The direct experimental evidence in support of this procedure comes from comparison between the observed vibrational spectra of C_{60} in the gas-phase [11.41], the solid phase [11.19, 42], and in solution [11.42], the mode frequencies showing only small differences from one phase to another. Gas-phase values (527, 570, 1170, 1407 cm^{-1}) for the four F_{1u} infrared-active mode frequencies were found to be slightly lower (1–22 cm^{-1}), when compared to the mode frequencies in Table 11.1 for the solid state.

To evaluate the status of various theoretical calculations, we compare in Table 11.2 various theoretical values for the mode frequencies for the 14 first-order Raman (A_g and H_g) and infrared-active (F_{1u}) modes which are known to ± 1 cm^{-1}. Overall agreement between experiment (see Table 11.1) and some of the calculations in Table 11.2 is quite good [see calculations by Negri *et al.* [11.43] (*ab initio*), Jishi *et al.* [11.22] (adjustable force constants),

Table 11.2

Comparison of experimental and calculated Raman, and infrared mode frequencies for various theoretical models. Further comparison of calculated models to experimental determinations can be made by reference to Tables 11.1 and 11.5.

Mode	Expt.	Wu[a]	W[b]	S[c]	A[d]	N[e]	J[f]	Q[g]	F[h]
$\omega_1(A_g)$	496 ± 2	548	510	610	537	513	492	478	483
$\omega_2(A_g)$	1470 ± 2	1627	1830	1667	1680	1442	1468	1499	1470
$\omega_1(H_g)$	271 ± 2	272	274	263	249	258	269	258	268
$\omega_2(H_g)$	433 ± 3	428	413	447	413	440	439	439	438
$\omega_3(H_g)$	710 ± 2	552	526	771	681	691	708	727	692
$\omega_4(H_g)$	774 ± 2	780	828	924	845	801	788	767	782
$\omega_5(H_g)$	1099 ± 2	1160	1292	1261	1209	1154	1102	1093	1094
$\omega_6(H_g)$	1250 ± 2	1398	1575	1407	1453	1265	1217	1244	1226
$\omega_7(H_g)$	1426 ± 2	1688	1910	1596	1624	1465	1401	1443	1431
$\omega_8(H_g)$	1575 ± 3	1831	2085	1722	1726	1644	1575	1576	1568
$\omega_1(F_{1u})$	526.5	491	478	577	494	544	505	547	547
$\omega_2(F_{1u})$	575.8	551	618	719	643	637	589	570	578
$\omega_3(F_{1u})$	1182.9	1374	1462	1353	1358	1212	1208	1176	1208
$\omega_4(F_{1u})$	1429.2	1655	1868	1628	1641	1437	1450	1461	1445

[a]Calculated mode frequencies of Ref. [11.9].
[b]Calculated mode frequencies of Ref. [11.7].
[c]Calculated mode frequencies of Ref. [11.5].
[d]Calculated mode frequencies of Ref. [11.38].
[e]Calculated mode frequencies of Ref. [11.6].
[f]Calculated mode frequencies of Ref. [11.22].
[g]Calculated mode frequencies of Ref. [11.23].
[h]Calculated mode frequencies of Ref. [11.24].

Quong *et al.* [11.23] (first principles local density approximation, LDA), and Feldman *et al.* [11.24] (adjustable force constants)]. It should be noted that the phonon mode calculations in Table 11.2 represent only a subset of the various vibrational mode calculations that have thus far been reported.

11.3.2. Theoretical Crystal Field Perturbation of the Intramolecular Modes of C_{60}

The intermolecular interactions between C_{60} molecules in the solid state are weak, and therefore the crystal field resulting from the C_{60} crystal lattice may be treated as a perturbation on the molecular modes [11.44–46]. Since the observed splittings and frequency shifts of vibrational modes related to solid state effects are very small in fullerene solids, most of these observations do not require crystal field symmetry-lowering effects for their explanation. In fact, most of the experimentally observed splittings that have been attributed to crystal field symmetry-lowering effects in the literature are theoretically inconsistent with such an explanation. A detailed treatment of the symmetry of crystal field effects is given in §7.1.2.

For temperatures above the orientational ordering transition T_{01} (see §7.1.3), the C_{60} molecules are rotating rapidly, at a frequency comparable to the librational frequencies (10–40 cm^{-1}) observed below T_{01}. Since the rotational frequencies are low compared with the intramolecular vibrational frequencies, the C_{60} molecules in the solid above T_{01} are orientationally disordered. The effect of the crystal field is to make all the vibrational modes in the solid phase weakly Raman and infrared-active, with a large linewidth due to the molecular orientational disorder; this effect would be observable in terms of a background Raman scattering or infrared absorption. The reported Breit–Wigner–Fano Raman lineshapes (see Fig. 11.23) would represent a signature of this broad Raman background.

For temperatures below T_{01}, the crystal field would be expected to exhibit two effects, if all the C_{60} molecules were to align orientationally with respect to the T_h site symmetry of the crystal field. The first effect is associated with the splitting of the icosahedral I_h modes as the symmetry is lowered from I_h to cubic T_h symmetry. From §7.1.2 and Table 7.10, we have the result that in the cubic crystal field (where the irreducible representations have A, E, and T symmetries), the icosahedral modes exhibit the following splitting pattern:

$$\begin{array}{ccc} A & \rightarrow & A \\ F_1 & \rightarrow & T \\ F_2 & \rightarrow & T \\ G & \rightarrow & A+T \\ H & \rightarrow & E+T \end{array} \qquad (11.2)$$

in which parity is conserved in a cubic symmetry-lowering centrosymmetric crystal field perturbation. Group theory requires that only the G and H modes are split in the cubic crystal field. Since in T_h symmetry, A_g, E_g, and T_g are all Raman-active, cubic crystal field effects can give rise to new Raman-active modes. Since the infrared-active modes have T_u symmetry for the group T_h and F_{1u} symmetry in the I_h group, no crystal field–induced mode splitting is expected, although the F_{2u}, G_u, and H_u modes can be made infrared-active by crystal field effects in accordance with Eq. (11.2). The symmetry-lowering effect discussed above would be expected to give rise to mode splittings of a few wave numbers, depending on the magnitude of the intermolecular interactions.

In contrast, there is a second crystal field effect, associated with the increase in size of the crystalline unit cell by a factor of 4 in volume, giving rise to zone-folding effects which would be expected to result in former zone edge and midzone modes being folded into the zone center, and giving rise to Raman and infrared modes with much larger "splittings," because these mode splittings depend on the width of the branch of the phonon dispersion relations. The symmetries for the mode splittings under zone folding are determined by considering the direct product of the equivalence transformation with each of the modes in T_h symmetry, and the results are given in Table 7.10. Whereas experimental evidence for zone-folding effects is well documented for the intermolecular vibrational modes (see §11.4), there is no clear evidence for such effects in the intramolecular vibrational spectrum for solid C_{60}.

Having summarized the expected crystal field–induced effects, we now offer two reasons why crystal field symmetry-lowering effects are not important to a first approximation for the analysis of intramolecular vibrational spectra for C_{60}. The first argument relates to merohedral disorder and the second to Brillouin zone folding. Regarding merohedral disorder, we note that if a molecule is orientationally disordered, then both the misaligned molecule and all its nearest-neighbors lose the T_h crystal field symmetry. Thus a 10% merohedral disorder at low temperature (see §7.1.4) should be sufficient to essentially smear out all the predicted T_h crystal field splittings of the G and H symmetry modes.

The second argument relates to zone folding the phonon dispersion relations in reciprocal space giving rise to new Raman-active modes (see also §7.1.2). Even though the dispersion of the phonon branches is small for fullerene solids (~50–100 cm^{-1}), the zone folding would be expected to result in a multiplicity of lines with relatively large splittings in the Raman-active modes (~tens of cm^{-1}). Since large splittings of this magnitude are not seen experimentally, we conclude that the coupling of the dynamical matrix to the cubic crystal potential is weak.

11.3.3. Isotope Effects in the Vibrational and Rotational Spectra of C_{60} Molecules

As an illustration of the various isotope effects on the rotational and vibrational energy levels, we discuss here the most spectacular of the isotope effects, specifically the modification of the rotational levels of $^{12}C_{60}$ [11.47]. We then list a number of other isotope effects that are expected for C_{60}, some of which have not yet been reported experimentally. To treat these isotope effects, we consider the total wave function of the 60 carbon atoms (including their nuclei) Ψ, which can be expressed by the product wave function

$$\Psi = \Psi_{\rm el}\Psi_{\rm vib}\Psi_{\rm rot}\Psi_{\rm ns}, \tag{11.3}$$

where $\Psi_{\rm el}$, $\Psi_{\rm vib}$, $\Psi_{\rm rot}$ and $\Psi_{\rm ns}$ refer, respectively, to the electronic, vibrational, rotational, and nuclear spin factors for the molecule. The ground state electronic structure of a C_{60} molecule requires $\Psi_{\rm el}$ to have A_g symmetry.

In the case of $^{12}C_{60}$, the total wave function, Ψ, should be totally symmetric for any permutations of the ^{12}C nuclei within the molecule, because each ^{12}C nucleus is a boson with nuclear spin $\mathscr{I} = 0$. In contrast, the $^{13}C_{60}$ nuclei have totally antisymmetric states for an odd number of nuclear exchanges, since each ^{13}C nucleus is a fermion with nuclear spin $\mathscr{I} = 1/2$. For a C_{60} molecule which contains both ^{12}C and ^{13}C isotopes, the proper statistics should be applied to the permutations among the ^{12}C atoms and among the ^{13}C atoms. There are no restrictions regarding the statistics for permutations between ^{12}C and ^{13}C atoms, since each is considered to be a different particle, and permutation exchanges between unlike particles are not symmetry operations of a group. For illustrative purposes, we consider initially only $^{12}C_{60}$ and $^{13}C_{60}$ molecules, i.e., C_{60} molecules consisting only of ^{12}C isotopes or of ^{13}C isotopes.

Any symmetry operation of the icosahedral group I_h can be expressed by a permutation of 60 elements, and the group I_h is a subgroup of the symmetric group (or permutation group) $S(60)$. It is then necessary to consider symmetries of the wave function under the permutations which belong to group I_h.

The rotational energy levels of a C_{60} molecule in vacuum are given by

$$E_J(\mathrm{rot}) = \frac{\hbar^2}{2I}J(J+1) = 3.3 \times 10^{-2} J(J+1) \ \ [\mathrm{K}], \tag{11.4}$$

where I is the moment of inertia of a C_{60} molecule ($I = 1.0 \times 10^{-43}$ kg-m^2), J denotes the angular momentum quantum number of the molecule, and the rotational energies are given in degrees Kelvin. The $\Delta J = \pm 1$ transitions between the rotational states of the molecule $E_J(\mathrm{rot})$ are much lower in frequency than the lowest intramolecular vibration.

Evidence for P, Q, and R branches associated with the rotational–vibrational states of the C_{60} molecules comes from high temperature ($\sim 1060°$C) gas-phase infrared spectroscopy [11.41]. The rotational–vibrational spectra associated with the free C_{60} molecule are exceptional and represent a classic example of the effect of isotopes on the molecular vibrational–rotational levels of highly symmetric molecules [11.48].

Calculations show that isotope effects should be very important in the vibrational spectroscopy of the gas-phase of C_{60} at very low temperatures (< 1 K), where quantum effects are dominant [11.47]. Because of the large moment of inertia of C_{60}, classical behavior (which suppresses the isotope effect) becomes dominant at relatively low temperatures (> 1 K). However, differences in the symmetry of the ^{12}C and ^{13}C nuclei in a C_{60} molecule are expected to be relevant to the rotational motion of the molecule at low temperature, imposing severe restrictions on the allowed rotational states of $^{12}C_{60}$, which contains only ^{12}C isotopes [11.47, 49, 50]. The symmetry differences between ^{12}C and ^{13}C also serve to break down the icosahedral symmetry of C_{60} molecules which contain both ^{12}C and ^{13}C isotopes, and lead to the observation of weak symmetry-breaking features in the Raman and infrared spectra (see §11.5.3 and §11.5.4). As mentioned in §4.5, the natural abundance of the ^{13}C isotope is 1.1%, the remaining 98.9% being ^{12}C.

In the case of $^{12}C_{60}$, there is no nuclear spin for ^{12}C and thus the nuclear spin wave function $\Psi_{\rm ns}$ transforms as the irreducible representation A_g in icosahedral I_h symmetry. Using Eq. (11.3) and the fact that $\Psi_{\rm el}$, $\Psi_{\rm vib}$, $\Psi_{\rm ns}$ all transform as A_g at low temperature (< 1 K), the rotational states $\Psi_{\rm rot}$ for $^{12}C_{60}$ are also required to have A_g symmetry. Thus the rotational motion for $^{12}C_{60}$ is restricted to J values which contain the irreducible representation A_g of the point group I_h. From Table 4.3, where the decomposition of the angular momentum states J of the full rotation group into the irreducible representations of I_h is given, we obtain the J values which contain A_g symmetry, namely $J = 0, 6, 10, 12, 16, \ldots$. Thus the lowest Pauli-allowed rotationally excited state for $^{12}C_{60}$ must have $J = 6$. Using Eq. (11.4), we estimate the energy of the $J = 6$ level to be $\sim$1.6 K. It is interesting that all rotational states from $J = 1$ to $J = 5$ are not allowed in $^{12}C_{60}$ by the Pauli principle. Since the nuclear spin for ^{13}C is 1/2 and the total nuclear spin angular momentum for the $^{13}C_{60}$ molecule can vary between 0 and 30, $\Psi_{\rm ns}$, in general, contains all possible irreducible representations of I_h, and consequently there are no restrictions on $\Psi_{\rm rot}$ or on J for the free $^{13}C_{60}$ molecule. This symmetry requirement for $^{12}C_{60}$ leads to major differences between the low-lying rotational levels for $^{12}C_{60}$ relative to $^{13}C_{60}$ (and even relative to $^{12}C_{59}{}^{13}C$) for low J values.

Referring to Fig. 11.6 we see a schematic diagram for possible vibration–rotation spectra associated with an infrared-active F_{1u} mode for C_{60}. The top trace shows P, Q, and R branches for the $^{13}C_{60}$ molecule. The rotational structure shows mostly uniform splittings in the P ($\Delta J = +1$) and R ($\Delta J = -1$) infrared branches with spacings of $2B(1 - \xi) = 0.0064$, 0.0083, 0.0076, and 0.0062 cm^{-1}, respectively, for each of the four fundamental infrared-active modes $F_{1u}(1), \ldots, F_{1u}(4)$ (see Table 11.1). In contrast, the $^{12}C_{60}$ molecule shows a sparse spectrum, reflecting the stringent

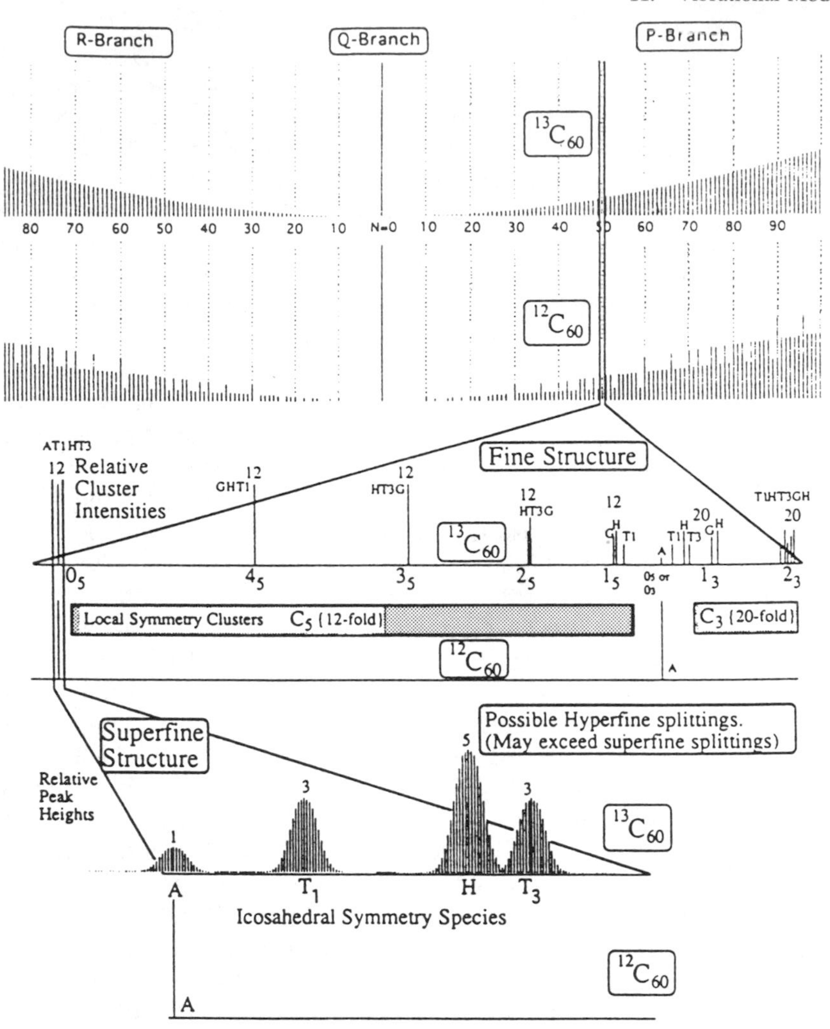

Fig. 11.6. Possible rotational–vibrational structure of an infrared-active F_{1u} fundamental mode of C_{60}. IR fundamental modes exist at 527, 576, 1183, and 1429 cm^{-1} (see Table 11.1), and for each F_{1u} mode, additional structure is expected, associated with icosahedral symmetry-lowering effects (fine structure) and the nuclear hyperfine interaction (superfine structure). Highlighted in this figure is the difference between the $^{12}C_{60}$ and $^{13}C_{60}$ spectra [11.47].

selection rule on allowed J values. These differences in the $^{12}C_{60}$ and $^{13}C_{60}$ rotational–vibrational spectra should be important in observations of these spectra at low temperatures.

To show the effect of the icosahedral symmetry on the rotational energy levels, the decomposition of a high angular momentum level (e.g., $J \sim 50$)

in full rotational symmetry into many states in the lower-symmetry icosahedral group is shown in Fig. 11.6 in the spectra labeled "fine structure." A summary of the decomposition of the irreducible representations of the full rotation group into the irreducible representations of the I_h group is contained in Table 4.3 up to $J = 34$. All of the icosahedral symmetries are permitted for rotational states in $^{13}C_{60}$, but only modes with A_g symmetry are allowed for $^{12}C_{60}$. If the icosahedral symmetry is explicitly considered in writing the wave functions for the angular momentum states J, then fine structure splittings result as shown in Fig. 11.6 [11.47]. These fine structure splittings vary as the sixth or higher power of the angular momentum J (where $\mathbf{J} = \mathbf{L} + \mathbf{S}$) and could range in frequency from a few kHz to several GHz (1 THz = 33.356 cm^{-1}). The accidental degeneracies that remain in the fine structure for $^{13}C_{60}$ are associated with the 12 C_5 local clusters and the 20 C_3 local clusters of the regular truncated icosahedron.

Since the high-angular-momentum lines for $^{12}C_{60}$ contain only states with A_g symmetry (e.g., the state $J = 50$ contains only two A_g states when considered as a reducible representation of the I_h group), it is the splitting of only these two A_g states that is seen in the fine structure of $^{12}C_{60}$.

In addition to the effect of the icosahedral symmetry on the rotational levels of the molecule which results in the fine structure, there is an interaction due to the nuclear spin which gives rise to superfine splittings, which are distinguished from the hyperfine interaction because the J values for the angular momentum in the superfine interaction are considered in terms of the irreducible representations of the icosahedral group I_h. Since the energy associated with the icosahedral distortion is much larger than that for the hyperfine interaction, arising from the magnetic field at the position of the nuclei associated with the rotational angular momentum J, the icosahedral distortion is considered first in perturbation theory. The superfine splittings vary over several orders of magnitude, from a few Hz to several MHz, and are particularly important for $^{13}C_{60}$.

In $^{13}C_{60}$, each ^{13}C atom has a nuclear spin $\mathscr{I} = 1/2$ and each molecule thus has 2^{60} nuclear spin states, with $\mathscr{I}_{\text{tot}}$ values ranging from $\mathscr{I}_{\text{tot}} = 0$ to $\mathscr{I}_{\text{tot}} = 30$. The decomposition of 2^{60} nuclear spin states into irreducible representations of all possible $\mathscr{I}_{\text{tot}}$ states has been examined by Harter and Reimer [11.47], noting that the totally antisymmetric states of $^{13}C_{60}$ belong to the A_g irreducible representation of I_h, because all operations of I_h are even permutations, and the totally antisymmetric wave function does not change sign under even permutations. Thus the statistical weight for each irreducible representation of I_h for all 2^{60} states of $^{13}C_{60}$ is well approximated by the dimension of the irreducible representation [11.50].

In the superfine structure for $^{13}C_{60}$ in Fig. 11.6 we see two effects. The first is the splitting of the 12-fold and 20-fold accidentally degenerate levels of the fine structure when the nuclear spin is included, since the nuclear spin can be up or down at each ^{13}C atomic site. For the example shown in Fig. 11.6, we see a 12-fold level split into the four irreducible representations of group I_h that are contained. For each of these four spectral lines, a second effect appears, explicitly associated with the hyperfine interaction between the orbital angular momentum and the total spin $\mathscr{I}_{tot}$ of the molecule. The large number of very closely spaced lines appearing with each fine structure line in the superfine spectrum is due to the large number ($0 \leq \mathscr{I}_{tot} \leq 30$) of values that $\mathscr{I}_{tot}$ may have for $^{13}C_{60}$, each $\mathscr{I}_{tot}$ value having its own weight, as discussed above.

In contrast to the complex superfine spectrum for $^{13}C_{60}$ in Fig. 11.6, there are no superfine splittings for $^{12}C_{60}$ because it has no nuclear spin [11.47, 50].

For ordinary C_{60} molecules, the distribution of the ^{12}C and ^{13}C isotopes is in proportion to their natural abundance (i.e., 1.108% of ^{13}C), so that approximately half the C_{60} molecules have one or more ^{13}C isotopes and approximately half have only ^{12}C isotopes (see §4.5). High isotopic purity is needed to study the vibrational spectrum of $^{12}C_{60}$, since a purity of 99.9%, 99.99%, or 99.999% in the ^{12}C atom content yields a purity of only 94.1%, 99.4%, or 99.9%, respectively, for the $^{12}C_{60}$ molecule (see §4.5). For the molecules containing both ^{12}C and ^{13}C isotopes, there are no symmetry restrictions for the allowed J values. Since approximately half of the molecules in ordinary (i.e., without isotopic enrichment) C_{60} samples are $^{12}C_{60}$ molecules, the isotope effects discussed above are expected to affect the line intensities of the rotational and rotational–vibrational modes significantly in ordinary C_{60} samples at very low temperatures.

Isotope effects also apply to other situations [11.50]. For both the $^{12}C_{60}$ and the $^{13}C_{60}$ molecules, the intensity ratios of both infrared and Raman transitions between the restricted excited rotational states are expected to be strongly affected by the isotope effect at very low temperatures. In addition to the rotational levels for the free molecules, the librational levels of $^{12}C_{60}$ and $^{13}C_{60}$ crystalline solids should experience symmetry selection rules, with restrictions also placed on the intensities for the excitation of specific librational states. Specifically, the rotation–vibration spectra in the solid state are expected to differ from the spectra observed for the free molecules, because of differences between the rotational and librational levels (see §11.4) [11.50].

Whereas the $^{12}C_{60}$ molecule exhibits the highest degree of symmetry and the most stringent selection rules, and the $^{13}C_{60}$ molecule shows the next highest stringency regarding selection rules, lower-symmetry molecules, such as $^{13}C_1{}^{12}C_{59}$ and $^{13}C_2{}^{12}C_{58}$, are still expected to show some isotope-dependent behavior [11.50]. If the vibrational and rotational levels and their occupations are modified by the isotope effect, it is expected that the corresponding infrared spectra, Raman spectra, and specific heat [11.50] would also be modified at very low temperature (below 5 K), where the required boson character of the wave function constrains the allowed states to an even number of phonons in $^{12}C_{60}$, since the vibrational wave function must have even parity.

Also of interest in relation to vibrational spectra is the predicted 8% increase in the transition temperature T_{01} for pure $^{12}C_{60}$ relative to that for the $^{13}C_{60}$ molecule, because of the higher-mass and therefore larger moment of inertia of $^{13}C_{60}$ relative to $^{12}C_{60}$. However, the difference between T_{01} values for $^{12}C_{59}{}^{13}C_1$ and $^{12}C_{60}$ is probably too small to be observed experimentally. The isotope effect might, however, be a source of line broadening for this phase transition. Closely related to this isotope effect is the softening of a given intramolecular vibrational mode by virtue of the effect of the isotopic mass on the $(\kappa/M)^{1/2}$ factor, which determines force constants in the dynamical matrix. Such an effect has been reported for the $A_g(2)$ Raman-active mode for $^{13}C_x{}^{12}C_{60-x}$ for various values of x [11.25], as discussed in §11.5.5.

11.4. Intermolecular Modes

Because of the molecular nature of solid C_{60} and the large difference in frequency between the inter- and intramolecular modes, one expects a decoupling of these two types of vibrational modes. Consistent with this view is the experimental observation of a large frequency gap between the lowest frequency intramolecular $[H_g(1)]$ mode at 273 cm^{-1} and the highest frequency intermolecular or "phonon" mode at $\sim$ 60 cm^{-1}. Thus, lattice dynamics models in which molecules are allowed to translate and rotate as rigid units about their equilibrium lattice positions would be expected to be reasonably successful for describing the intermolecular vibrations, as shown below.

Inelastic neutron scattering provides a powerful and almost unique experimental technique for probing the wave vector dependence (i.e., dispersion) of the vibrational modes in solids [11.12, 51–55]. These scattering experiments require large single crystals, and in the case of C_{60}, suitable samples have only become available recently (see §9.1.1). Information on the intermolecular mode vibrations has also been provided by infrared [11.56] and Raman scattering [11.57] experiments. In principle, Brillouin scattering experiments [11.58] should be possible and would provide information on the acoustic branch for the intermolecular modes. However, the laser power needed to do these experiments may be so large that it causes phototransformation of the C_{60}.

Most of the inelastic neutron scattering studies on librational excitations in C_{60} have been on polycrystalline samples. From this work a number of important findings have resulted. The major peak in the neutron scattering occurs at ~2.7 meV (21.8 cm^{-1}) and a smaller peak is observed at ~4 meV (32 cm^{-1}) [11.12, 13]: the more intense peak is identified with 9 of the intermolecular modes in Table 11.3 and the smaller peak with the remaining 3 intermolecular modes. The temperature dependence shows that the librational energy softens by ~35% in going from 20 K to 250 K and the peak broadens by a factor of 6 over this temperature range. Above T_{01}, no scattering intensity associated with librations is found, as expected. The large broadening effect of the librational line with temperature is attributed to a

Table 11.3

Zone center (Γ point) translational and librational mode frequencies for C_{60} below 261 K.[a] In most cases, symmetry assignments were not provided by the original references.

Symmetry	Experiment[b] (cm^{-1})	Theory (cm^{-1})		
		Li[c]	Yildirim[d]	Lipin[e]
T_u	27[f], 40[g], 41[h]	32	30	41.2
A_u	38[g]	30	31	39.5
E_u	38[g]	33	33	41.2
T_u	59[f], 54[g], 55[h], 50[i]	45	43	55.5
A_g	15[g]	13.0	7.7[k]	20.5
T_g	19[g], 19.2[j]	10.5	8.9[k]	19.3
E_g	22[g]	10.5	11.7[k]	19.6
T_g	22[g], 28.8[i], 21.6[j], 30.4[l]	14.1	12.5[k]	20.8
T_g	32[g], 36.8[i], 40.9[l]	20.0	17.9[k]	29.7

[a]The acoustic branch of the translational modes with T_u symmetry has $\omega = 0$ at $q = 0$ and is not listed. The A_u and E_u modes are neither Raman active nor infrared-active. The T_u mode is IR active and all gerade modes (i.e., A_g, E_g, and T_g) are Raman-active.

[b]The conversion between wave numbers and meV is 1 meV = 8.0668 cm^{-1}.

[c]Li *et al.* , Ref. [11.60].

[d]Yildirim and Harris, Ref. [11.11].

[e]Lipin and Mavrin, Ref. [11.61].

[f]IR spectra [11.62].

[g]Inelastic neutron scattering [11.13]. Estimated from dispersion curves.

[h]IR spectra [11.56].

[i]Inelastic neutron scattering [11.12]

[j]Inelastic neutron scattering [11.52].

[k]If the calculated librational modes in Ref. [11.63] are multiplied by 2, then good agreement is obtained with experiment.

[l]Raman spectra at 10 K [11.57].

softening of the orientational potential (see §7.1.3), since single-crystal inelastic neutron scattering [11.12] does not show a broadening of individual modes in this temperature range.

Axe *et al.* [11.51] have recently reviewed the structure and dynamics of solid C_{60}, emphasizing the contributions from neutron scattering studies to our understanding of the vibrational spectra and structural properties of fullerenes. One difficulty with interpreting many of the early Raman and neutron measurements relevant to the intermolecular vibrations is the paucity of polarization and mode identification information [11.12, 57]. Single-crystal data for the dispersion relations are now becoming available for low frequency intermolecular translational modes in C_{60} (see Fig. 11.7) from inelastic neutron scattering studies [11.12].

Since the C_{60} molecules rotate rapidly about their fcc equilibrium positions above the orientational ordering temperature $T_{01} \sim 261$ K [11.59], an orientational restoring force is ineffective above T_{01}, and the rotational intermolecular modes (i.e., librational modes or "librons") are lost from the phonon spectrum. Above T_{01}, solid C_{60} exhibits an fcc lattice which has only one molecule per primitive cell. Therefore, only three acoustic branches (all with T_u symmetry for the cubic group at the Γ-point) are expected, and these involve approximately rigid axial and radial displacements of the molecules. In Fig. 11.7(a), we display the phonon dispersion data (closed symbols) of Pintschovius and co-workers for solid C_{60} in the orientationally disordered (fcc) state [11.13], obtained by inelastic neutron scattering along the [100], [110], and [111] directions of a small 6 mm^3 single crystal at $T = 295$ K. The solid curves in the figure are the result of a two-parameter

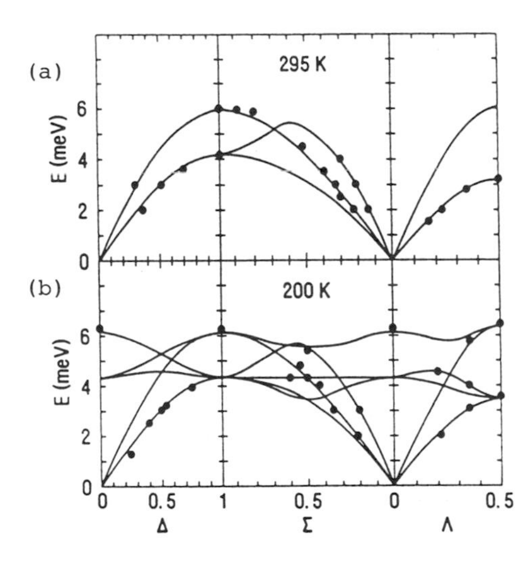

Fig. 11.7. Dispersion of the intermolecular translational modes of single-crystal C_{60} above (a) and below (b) the first-order phase transition temperature at 261 K. The solid curves are the result of a fit to the experimental points with a two-parameter force constant model. The results for 200 K are shown in an extended zone scheme corresponding to an fcc lattice [11.12].

Born–von Kárman model used to fit the data via the adjustment of first and second nearest-neighbor force constants. Rapid convergence for the force constant model was obtained, with the second neighbor force constant being only 10% of that of the nearest neighbor. As can be seen, this simple "rigid molecule" model describes very successfully the measured phonon dispersion in the orientationally disordered phase, with $\sim$50 cm^{-1} (or 6.2 meV) being the highest frequency observed. Referring to Fig. 11.7(a), we see that two Debye frequencies are needed to describe the acoustic modes in the high-T, orientationally disordered phase for C_{60}. Shown in Fig. 11.7(b) are the translational modes observed at 200 K by inelastic neutron scattering on the same sample as in Fig. 11.7(a) [11.13]. These data show clear evidence for additional phonon branches upon cooling below T_{01}, as explained below in more detail.

At temperatures well below the orientational ordering temperature $T_{01} \simeq$ 261 K, where there are four inequivalent molecules in the simple cubic cell of space group $Pa\bar{3}$ (see §7.1.3), both orientational and translational restoring forces are effective. In this regime, the three rotational and three translational degrees of freedom per molecule lead to $6 \times 4 = 24$ degrees of freedom per primitive unit cell. The symmetries of these 24 modes at the zone center can be obtained by taking the direct product of the molecular site symmetry in the T_h^6 space group (see Table 7.8), that is, $(A_g + T_u)$, with those of the rotations (T_g) and translations (T_u). This direct product yields $(A_g + T_u) \otimes (T_u + T_g) = (A_g + E_g + 3T_g) + (A_u + E_u + 3T_u)$ [11.45]. Thus 10 distinct zone center frequencies (including the modes at $\omega = 0$ which are the T_u-symmetry acoustic phonons) should be observable in this low-temperature structure. The gerade modes and ungerade modes are identified, respectively, with zone center (long-wavelength) librations and translational phonon modes. One of the T_u modes at the zone center corresponds to a pure translation, whereby $\omega = 0$ at $q = 0$, while the other two T_u optic phonon modes are IR active; all the gerade librational modes are Raman active. However, the observation of these particular modes by optical spectroscopy should be quite difficult, because of their low frequency (see Table 11.3).

We show in Fig. 11.8(a), the experimental phonon/libron dispersion relations for C_{60} in the low-temperature $Pa\bar{3}$ phase obtained from inelastic neutron scattering studies of a small single crystal (6 mm^3) along high-symmetry directions following the notation shown in Fig. 11.8(b) [11.13]. The data in Fig. 11.8(a) were taken along the $\Gamma - \Lambda - R$ direction in reciprocal space, where 16 branches are allowed, and along the zone edge $(R - T - M)$, where, because of mode degeneracies, only 6 branches are allowed [11.11] (see Table 11.4). Seven of the ten frequencies expected at

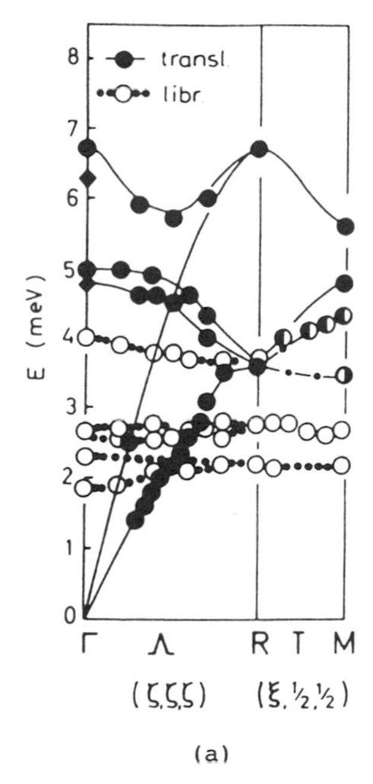

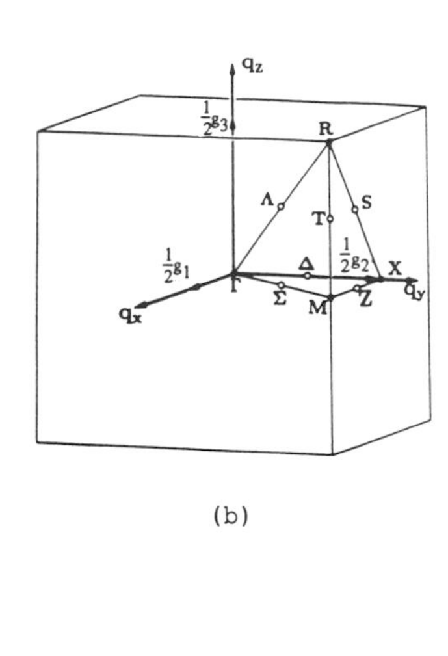

Fig. 11.8. (a) Experimentally determined low-frequency phonon (closed circles) and libron (open circles) dispersion relations, obtained from inelastic neutron scattering measurements in the simple cubic (sc) low-temperature phase [11.13]. The solid and dashed lines connecting the experimental points are guides to the eye. (b) Symmetry labels for high symmetry points and axes of the Brillouin zone for the low-temperature $Pa\overline{3}$ structure, consistent with labels in (a). Also shown are several half reciprocal lattice vectors, namely $g_1/2$, $g_2/2$, and $g_3/2$ [11.51].

the Γ-point on the basis of group theoretical considerations have thus far been reported. The modes in Fig. 11.8 that are assigned primarily to librational or translational phonons are indicated by the open and closed circles, respectively. As can be seen, the acoustic mode dispersion is broad enough to overlap all the optic branches and exhibits a maximum frequency of 6.7 meV (55 cm^{-1}). Furthermore, the librational modes (open circles) are the four lowest optic branches observed, with frequencies lower than would be expected from an analysis of specific heat measurements [11.64, 65]. These librational modes are seen to exhibit almost no dispersion. The upper two optic branches, which exhibit zone center frequencies of 5 and 6.7 meV, are in very good agreement with mode frequencies observed at 41 cm^{-1} (5.1 meV) and 55 cm^{-1} (6.8 meV), respectively, using IR spectroscopy on powder samples at $T = 1.5$ K [11.56]. Therefore, these upper two branches should be assigned to T_u symmetry. These features in the far-IR spectra [11.56] were first found at about 261 K, as degassed powder

Table 11.4

Point group symmetries and branches of the various dispersion curves along high-symmetry axes [see Fig. 11.8(b)] [11.51]

Points	$q(\pi/a)$	Point symmetry	Number of branches (degeneracy)
Γ	$(0,0,0)$	T_h	$2(1)+2(2)+6(3)$
X	$(0,1,0)$	D_{2h}	12(2)
M	$(1,1,0)$	D_{2h}	6(4)
R	$(1,1,1)$	T_h	6(4)
$\Delta(\Gamma-X)$	$(0,\alpha,0)$	$C_{2v}(y)$	24(1)
$\Sigma(\Gamma-M)$	$(\alpha,\alpha,0)$	$C_3(z)$	24(1)
$\Lambda(\Gamma-R)$	(α,α,α)	C_3	$8(1)+8(2)$
$S(X-R)$	$(\alpha,1,\alpha)$	$C_3(y)$	12(2)
$Z(X-M)$	$(\alpha,1,0)$	$C_{2v}(x)$	12(2)
$T(M-R)$	$(1,1,\alpha)$	$C_{2v}(z)$	6(4)

samples were cooled from room temperature. Another report for frequencies of the T_u modes by IR spectroscopy [11.66] yielded values of 27 and 59 cm^{-1}, which are not in good agreement with other IR measurements [11.56] or with the interpretation of the single-crystal neutron data [11.13].

As mentioned above, the zone center librations $(A_g + E_g + 3T_g)$ are all Raman active, and various Raman studies of the low-frequency modes have been reported [11.44, 57, 67]. Three low-frequency Raman lines were reported [11.44] at 56, 81, and 109 cm^{-1}. Whereas the line at 56 cm^{-1} might be a disorder-induced scattering signature of the otherwise IR-active T_u mode observed by neutron and IR techniques near 55 cm^{-1}, the remaining two peaks appear to exhibit much too high a frequency to be identified with any of the intermolecular phonon excitations. This view is also borne out by theoretical calculations [11.60], as discussed below.

Values for the zone center phonon frequencies, obtained from neutron scattering and IR spectroscopy, are collected together with selected theoretical values in Table 11.3. Symmetry assignments are made for all the zone center librational modes, assuming they appear on the experimental energy scale in the same order as calculated by Yildirim and Harris (see below) [11.10, 11, 61, 63]. It is interesting to note that if the values for the Γ-point librational (gerade) frequencies of Yildirim and Harris are simply scaled by a factor of 2, they are then found in good overall agreement with the neutron scattering data of Pintschovius *et al.* [11.13]. Furthermore, since the dispersion calculations find the Γ-point mode frequencies of the E_g and T_g modes quite close in value, it is reasonable at this stage to make a tentative assignment of the neutron peak at 22 cm^{-1} (2.7 meV) at the Γ-point to an unresolved pair of E_g and T_g modes.

For comparison to the dispersion relations for the intermolecular modes obtained by analysis of the inelastic neutron scattering data, we show, for example, the theoretical dispersion curves of Yildirim and Harris (Fig. 11.9) calculated in the limit of complete decoupling between the librations and the translational phonons. Note that the librational branches [upper panel (a)] and translational branches [lower panel (b)] are plotted on two different frequency scales. Their calculations are based on a phenomenological point charge model [11.68], in which a negative point charge is placed on each of the double bonds of the C_{60} molecule and positive point charges (for charge neutrality) are placed on the C atoms. This model reproduces qualitatively the experimentally derived [11.48] orientational potential. Several variations [11.60, 68–70] of this point charge model [11.68] have been used to calculate the phonon spectrum of C_{60} in which negative point charges are placed on the electron-rich double bonds, and the positive point charges, depending on the particular calculation, were positioned at various other locations on the molecule surface, i.e., either centered on

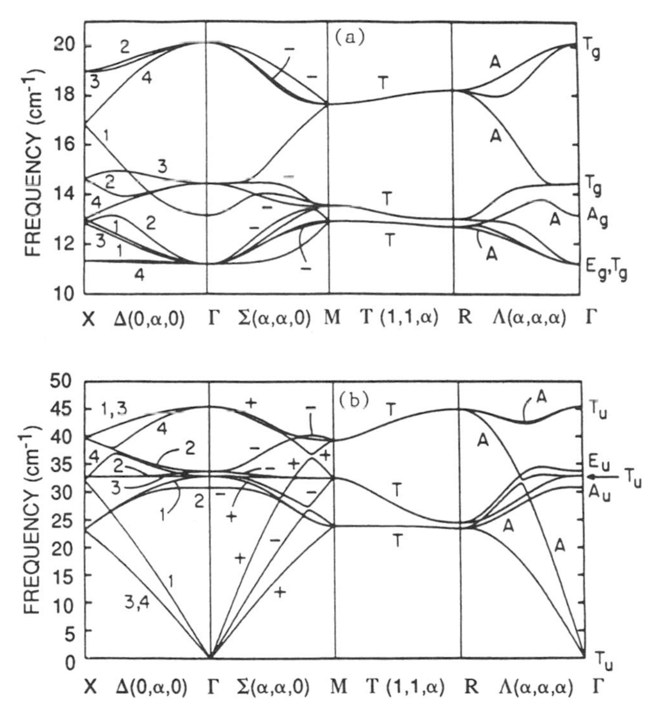

Fig. 11.9. Theoretical calculations of the low-frequency librations (a) and translational branches (b) of the intermolecular phonon dispersion relations [11.11]. See Fig. 11.8(b) for labels of symmetry points in the Brillouin zone.

the pentagonal face, centered on the short C–C bonds around the pentagonal faces, centered on the C atoms, or even positioned at the center of the C_{60} molecule. These point charges were introduced phenomenologically to harden the orientational potential, so that the resulting dispersion curves would reproduce qualitatively the features of the neutron data. However, the calculated librational frequencies are noticeably low, indicating that the potentials are still too soft. Models based on Lennard–Jones 6–12 potentials [11.71], similar to the point charge models, also underestimate the librational frequencies by about a factor of 2. Finally, it should be noted that the model of Yildirim and Harris [11.11] results in a small frequency difference between the A_u, E_u, and lowest T_u translational phonon branches [Fig. 11.9(b)], as do also other models. This suggests that the separate contributions from these branches may be difficult to obtain quantitatively by inelastic neutron studies until larger single crystals are available.

By comparing experimental data with theoretical calculations for the intermolecular modes (see Table 11.3), we can make several observations. Using level ordering arguments and Fig. 11.8, we tentatively assign the E_u mode to 38 cm^{-1}. The assignment of the symmetries for the librational modes is more difficult than for the odd-parity vibrational modes, because so many of the librational modes lie close to one another. Using the calculated intermolecular phonon model, some assignments of the experimental data to specific intermolecular modes can be made with some confidence (see Table 11.3). It is also of interest to note that fairly good agreement is obtained between the experimental dispersion relations (Fig. 11.8) and the corresponding calculations (Fig. 11.9) away from the zone center, as for example near the M and R points in the Brillouin zone, regarding the number of modes, their parity, and mode frequencies (after multiplying the calculated librational modes by a factor of 2, as mentioned above). It should also be mentioned that inelastic neutron scattering measurements on polycrystalline samples [11.51, 52, 55, 72] give a broad peak at ~2.3 meV (19 cm^{-1}), consistent with observations on single crystals (see Fig. 11.8).

Experimental specific heat studies suggest phonon bands at 21, 26, and 40 cm^{-1} [11.64], where the 21 cm^{-1} phonon band is identified with $A_g + E_g + 2T_g$ librations, the 26 cm^{-1} band with T_u phonons, and the 40 cm^{-1} band with $A_u + E_u + 2T_u + T_g$ modes [11.51]. These assignments appear to be consistent with other experiments (see Table 11.3).

A theoretical calculation of the specific heat [11.73], using the density of states of the 24 librational and intermolecular vibrational modes, reproduces the experimental specific heat very well, except for the excess heat which appears below T_{01} (see §7.1.1 and §14.8.1). An excess entropy is associated with the orientational disorder of the 30 double bonds (C=C) for a C_{60} molecule. The entropy for N molecules with random orientations is

$$\Delta S = k_{\mathrm{B}} \ln(30)^N = 28.3\ \mathrm{JK^{-1}mol^{-1}}, \tag{11.5}$$

which is very close to the experimental value of 30 $\mathrm{JK^{-1}mol^{-1}}$ [11.74]. The difference between these two values may be due to the contribution from the additional ratcheting states that are coupled to the nuclear spin states, because of the presence of the ^{13}C isotope in its natural abundance.

In addition to the detailed studies available for the intermolecular vibrations in C_{60}, a limited amount of information is also becoming available for C_{70} (see §11.7.2) and for doped C_{60} (see §11.6.4).

11.5. Experimental Results on C_{60} Solids and Films

In this section we gather experimental results for the intermolecular and intramolecular modes for solid C_{60} from Raman scattering, infrared spectroscopy, photoluminescence, and inelastic neutron scattering. The results are compared with theory. Raman and infrared spectroscopies provide the most quantitative methods for determining the vibrational mode frequencies and symmetries. In addition, these methods are sensitive for distinguishing C_{60} from higher-molecular-weight fullerenes with lower symmetry (e.g., C_{70} has D_{5h} symmetry as discussed in §4.4). Since most of the higher-molecular-weight fullerenes have lower symmetry as well as more degrees of freedom, they have many more infrared- and Raman-active modes.

11.5.1. Raman-Active Modes

In Fig. 11.10 we display the polarized Raman spectra of a vacuum-deposited C_{60} film (on glass) [11.19], and similar Raman spectra for C_{60} have been reported by many groups [11.3, 4, 8, 15, 75, 76]. The spectra in Fig. 11.10 were taken with 514 nm laser excitation at room temperature, and the spectra show both the $(\|, \|)$ and $(\|, \perp)$ scattering geometries, where the symbols indicate the optical E field direction for the incident and scattered light, respectively. Ten strong lines identified with first-order scattering were observed and identified with the 10 Raman-active modes ($2A_g + 8H_g$) expected for the isolated molecule. The frequencies of these lines are in good agreement with those listed in Table 11.1. Two of the lines in the spectra are identified with polarized (A_g) modes, and therefore they should disappear from the spectrum collected with crossed polarizers $(\|, \perp)$. As can be seen in the $(\|, \perp)$ spectrum, the purely radial A_g mode (496 $\mathrm{cm^{-1}}$) appears

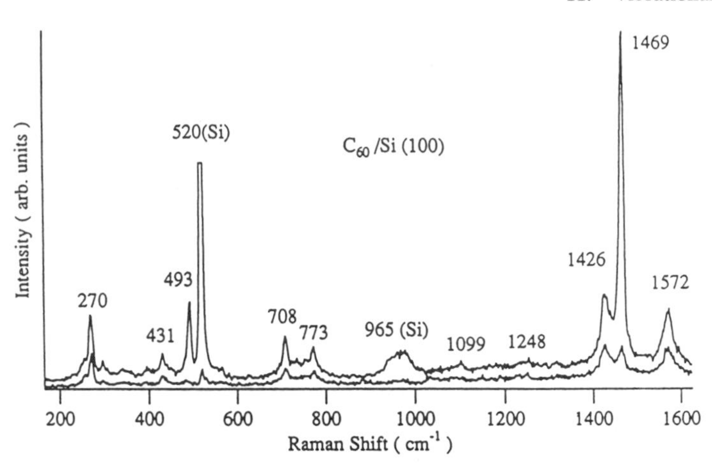

Fig. 11.10. Polarized Raman spectra for C_{60}. The upper trace is for the $(\|, \|)$ polarization and shows both A_g and H_g modes. The lower trace is for the $(\|, \perp)$ polarization and shows primarily H_g modes, the weak intensities for the A_g modes being attributed to small polarization leakage [11.19].

to become extinguished, yet the high-frequency, tangential A_g mode ("pentagonal pinch mode") at 1469 cm^{-1} exhibits considerable intensity. Complete extinction of the pentagonal pinch mode has since been reported [11.19]. Although the 1469 cm^{-1} line disappears in cross-polarizers, the nearby shoulder at 1458 cm^{-1} does not, confirming that 1469 cm^{-1} is the intrinsic A_g mode frequency. Early Raman studies by several groups [11.76–79] incorrectly identified the unpolarized 1458 cm^{-1} mode with the intrinsic pentagonal pinch mode, and later studies have shown that this mode should be identified with the photochemically induced polymeric state of C_{60} (see §11.8) [11.80].

11.5.2. Infrared-Active Modes

The first-order infrared spectrum for C_{60} contains only four strong lines, each identified with an intramolecular F_{1u} mode (see Table 11.5 and Fig. 11.11). The four strong F_{1u} lines at 526, 576, 1182, 1428 cm^{-1} in Fig. 11.11 [11.82] provide a convenient spectral identification for the molecule as C_{60}. This infrared spectrum, in fact, provided a historical identification of C_{60} in the first publication on solid C_{60} [11.82]. The infrared spectra also provide a sensitive characterization tool for measuring the compositional purity of C_{60} samples, especially with regard to their contamination with C_{70}. Because of its lower symmetry and larger number of carbon atoms, C_{70} shows a much more complex infrared spectrum [11.88] than C_{60} (see §11.7.1), with 31 distinct infrared-active mode frequencies

Table 11.5

Experimental normal mode vibrational frequencies of the C_{60} molecule from Raman, IR, photoluminescence (PL), neutron inelastic scattering (NIS), and high-resolution electron energy loss spectroscopy (HREELS) data.[a]

Even-parity		Odd-parity	
$\omega_i(\mathscr{R})$	Frequency (cm^{-1})	$\omega_i(\mathscr{R})$	Frequency (cm^{-1})
$\omega_1(A_g)$	492[b], 496[c], 497.5[d,e]	$\omega_1(A_u)$	1143.0[d,e]
$\omega_2(A_g)$	1467[f], 1470.0[c,d,e]		
		$\omega_1(F_{1u})$	525[f], 526.5[d,e,g,h], 527[i], 528[g], 532[b,f,j,k,l]
$\omega_1(F_{1g})$	502.0[d,e,h]	$\omega_2(F_{1u})$	575.8[d,e], 577[g,i], 581[l]
$\omega_2(F_{1g})$	960[k,l], 970[b], 975.5[d,e,h]	$\omega_3(F_{1u})$	1182.9[d,e,g,i], 1186[l]
$\omega_3(F_{1g})$	1355[k], 1357.5[d,e], 1363[b]	$\omega_4(F_{1u})$	1428[i], 1429.2[d,e,g]
$\omega_1(F_{2g})$	566[f], 566.6[d,e,h], 573[k], 581[l]	$\omega_1(F_{2u})$	347[l], 355[b,j], 355.5[d,e,h]
$\omega_2(F_{2g})$	865.0[d,e], 879[k]	$\omega_2(F_{2u})$	678[b], 680.0[d,e]
$\omega_3(F_{2g})$	914.0[d,e], 920[k]	$\omega_3(F_{2u})$	1026.0[d,e,h]
$\omega_4(F_{2g})$	1360.0[d,e]	$\omega_4(F_{2u})$	1201.0[d,e,h], 1202[k]
		$\omega_5(F_{2u})$	1576.5[d,e,h]
$\omega_1(G_g)$	483[f], 484[k], 486.0[d,e], 488[h]		
$\omega_2(G_g)$	621.0[d,e,k]	$\omega_1(G_u)$	399.5[d,e], 400[f], 403[b,k], 404[h]
$\omega_3(G_g)$	806.0[d,e]	$\omega_2(G_u)$	758[j,l], 760.0[d,e], 761[f], 765[h], 766[b]
$\omega_4(G_g)$	1065[k], 1073[l], 1075.5[d,e], 1077[f]	$\omega_3(G_u)$	924.0[d,e]
$\omega_5(G_g)$	1355[k], 1356.0[d,e], 1360	$\omega_4(G_u)$	960[l], 968[j], 970.0[d,e], 971[h]
$\omega_6(G_g)$	1520[h], 1524[f], 1524.5[d,e]	$\omega_5(G_u)$	1309[f], 1310.0[d,e]
		$\omega_6(G_u)$	1436[l], 1446.0[d,e], 1448[h], 1452[j]
$\omega_1(H_g)$	266[k], 271[h], 273.0[c,d,e], 274[b,j,l]		
$\omega_2(H_g)$	428[f], 432.5[d,e,h], 436[b,k,l], 436.5[c], 444[j]	$\omega_1(H_u)$	342.5[d,e], 344[h], 347[k,l]
$\omega_3(H_g)$	708[f], 710[c,k], 711.0[d,e]	$\omega_2(H_u)$	563.0[d,e,h], 565[b], 566[f], 581[l]
$\omega_4(H_g)$	771[f], 774[c,k], 775.0[d,e]	$\omega_3(H_u)$	686[j], 696.0[d,e]
$\omega_5(H_g)$	1097[f,j,k,l], 1099[c], 1101[d,e]	$\omega_4(H_u)$	801.0[d,e]
$\omega_6(H_g)$	1248[f], 1250[c], 1251.0[d,e], 1258[j], 1274[b,l]	$\omega_5(H_u)$	1117.0[d,e], 1121[k]
$\omega_7(H_g)$	1426.5[d,e], 1427[f], 1428[c], 1436[l]	$\omega_6(H_u)$	1385.0[d,e]
$\omega_8(H_g)$	1570[c], 1577.5[d,e], 1581[k]	$\omega_7(H_u)$	1557[l], 1559.0[d,e], 1565[j]

[a]Various models are given in Table 11.2. In this table only the IR and Raman experiments give information regarding symmetry assignments. The experimental peaks from other techniques have been assigned based on their proximity to the model predictions. Not all peaks reported in the various experimental references are listed in this table because some of the experimental features are identified with overtone and combination modes (see Table 11.6); [b]NIS [11.84]; [c]Raman (first order) [11.81]; [d]IR (first and higher order) [11.21]; [e]Raman (first and higher order) [11.20]; [f]PL [11.14]; [g]IR (first order) [11.82]; [h]NIS [11.83]; [i]IR absorption spectrum [11.81]; [j]HREELS [11.85]; [k]NIS [11.86]; [l]HREELS [11.87].

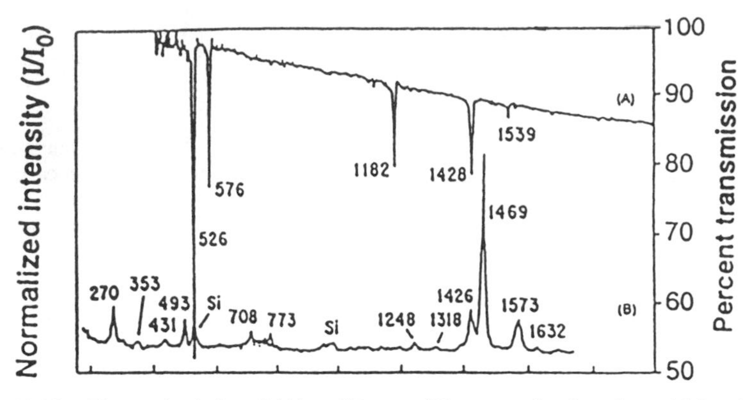

Fig. 11.11. First-order infrared (A) and Raman (B) spectra for C_{60} taken with low incident optical power levels (<50 mW/mm^2) [11.80].

[11.8, 88]. The IR spectrum of solid C_{60} remains almost unchanged relative to that of the isolated C_{60} molecule, with the most prominent addition being the weak feature at 1539 cm^{-1} [11.19, 31, 42]. However, more than 60 additional weak lines can be observed in the infrared spectrum in the 400 to 4000 cm^{-1} range. These weak features can be identified with higher-order processes, as discussed in §11.5.4.

The strong correspondence between the solution and/or gas-phase IR spectrum and the solid-state IR spectrum for C_{60} is indicative of the highly molecular nature of the crystalline phase of C_{60}. The infrared spectrum shown in Fig. 11.11 is for a thin film (~5000 Å) of C_{60} on a KBr substrate. We discuss below (§11.6.2) the effect of doping on the mode frequencies and intensities of the infrared-allowed modes in the C_{60} spectrum.

11.5.3. Higher-Order Raman Modes in C_{60}

Raman scattering measurements on a C_{60} film show a well-resolved second-order Raman spectrum (see Fig. 11.12) [11.20]. From group theoretical considerations (see §4.2), the expected number of second-order Raman lines is very large, consisting of a total of 151 modes with A_g symmetry and 661 modes with H_g symmetry [11.20]. The total number of second-order modes with A_g symmetry is found by taking the direct product $n_i\Gamma_i \otimes n_f\Gamma_f$ and counting the number of modes with A_g symmetry in that direct product. Here, the number of modes n_i and n_f with Γ_i and Γ_f symmetries, respectively, that yield modes with A_g symmetry are found in Table 4.6, and the direct products are given in Table 4.7. The same procedure is used to find the number of overtones and combination modes with H_g symmetry. The second-order Raman spectra include both overtones (i.e.,

Table 11.6

Overtone and combination mode frequencies (in cm^{-1}) in the range 250–1710 cm^{-1} observed in several experiments.

Assignment	Model[a]	IR, Raman	PL, NIS, HREELS
—	—	—	264[b], 265[c,d]
—	—	—	350[c,d]
—	—	—	436[e,f,g], 444[h]
—	—	—	484[f], 492[e]
$H_g(5) - H_g(2)$	668.5	—	665[d], 669[f]
$G_u(1) + H_g(1)$	672.5	673.5[i]	669[f], 673[b]
$2H_u(1)$	685.0	—	678[e], 686[e]
$G_u(1) + H_u(1)$	742.0	738[c]	739[c,d], 742[c,f]
$F_{2u}(1) + G_u(1)$	755.0	757[c]	755[c,d], 758[e,g,h], 761[d]
$2G_u(1)$	799.0	796[c]	796[d]
—	—	—	813[b,j]
$H_u(1) + G_g(1)$	828.5	822.9[i]	824[c], 827[d]
$F_{2g}(1) + H_g(1)$	839.5	—	839[b,f], 840[b,j]
$F_{2u}(2) + H_g(1)$	953.0	956.9[i]	950.8[c]
$F_{1u}(1) + H_g(2)$	959.0	956.9[i]	960[c,d,f]
$2G_g(1)$	972.0	972[c]	968[c,h], 971[j]
$F_{1u}(1) + F_{2u}(1)$	882.0	—	879[e,f], 879.1[e]
$2F_{1g}(1)$	1004.0	—	1000[f,i]
$H_g(1) + H_g(4)$	1048.0	—	1044[b,j]
$H_u(2) + F_{1g}(1)$	1065.0	—	1064[e], 1065[f], 1077[d]
$F_{1u}(1) + H_u(2)$	1089.5	—	1089[e,j], 1097[d,f,h]
$F_{2u}(2) + H_g(2)$	1112.5	1114.0[i]	1112[b]
$F_{1g}(1) + H_g(3)$	1213.0	—	1202[f], 1211[c,d]
$F_{1u}(1) + H_u(3)$	1222.5	—	1217[b,j]
$H_u(1) + F_{2g}(3)$	1256.5	1258.6[i]	1248[d], 1258[e,f,h]
$H_u(2) + H_g(3)$	1274.0	—	1270[d], 1274[e]
$G_u(2) + F_{2g}(2)$	1326.5	1329.9[i]	1327[b,j]
$H_u(2) + H_g(4)$	1338.0	1342.5[i]	1342[c,d], 1355[f]
$F_{1g}(1) + F_{2g}(3)$	1416.0	1417.0[c]	1419.5[b], 1420[f]
$F_{1u}(1) + G_u(3)$	1450.5	1450.5[c]	1448[i], 1452[e,h]
$F_{2u}(2) + G_g(3)$	1486.0	1483.2[i]	1484[e,f]
$G_u(1) + H_g(5)$	1500.5	1495.7[i]	1496[c,d]
$G_u(2) + G_g(3)$	1566.0	1570.9[i]	1563[b,j], 1565[g,h]
$H_u(4) + H_g(4)$	1576.0	1570.9[i]	1572[c,d], 1581[f]
$F_{2u}(1) + H_g(6)$	1606.5	1607.5[i]	1603[b,j]
$F_{2g}(4) + H_u(1)$	1702.5	—	1702[b,j]

[a]Mode frequencies are from a model obtained from a fit to the higher-order Raman and infrared spectra [11.20, 21]. [b]Neutron inelastic scattering (NIS), Neumann *et al.* [11.52]. [c]Raman, Dong *et al.* [11.20]. [d]PL, Nissen *et al.* [11.14]. [e]NIS, Cappelletti *et al.* [11.84]. [f]NIS, Prassides *et al.* [11.86]. [g]HREELS, Gensterblum *et al.* [11.87]. [h]HREELS, Lucas *et al.* [11.85]. [i]IR, Wang *et al.* [11.21]. [j]NIS, Coulombeau *et al.* [11.83].

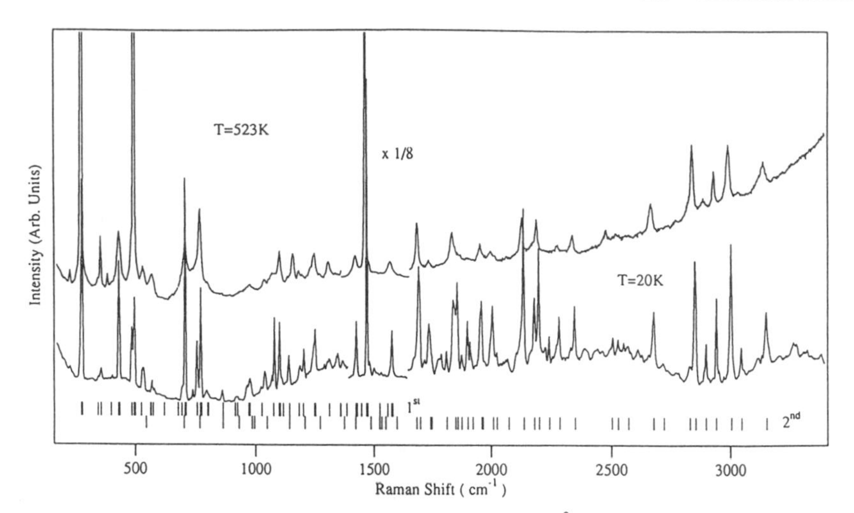

Fig. 11.12. Raman spectra for solid C_{60} films (~7000 Å thick) taken at temperatures $T = 523$ and 20 K, showing overtones, combination modes, and modes arising from isotopic symmetry-breaking effects. The data were taken using 488.0 nm Ar laser radiation. First-order and second-order mode frequencies are marked. Note that the intense structure near 1500 cm^{-1} has been rescaled [11.20].

integral multiples of the modes in Table 11.1) and combination modes (i.e., sums and differences of the modes in this table).

If we assume that the strongest second-order Raman features involve only modes which are also observed in first order, then the most intense second-order lines with A_g symmetry total 39, and the corresponding second-order lines with H_g symmetry total 88. To find this subset of overtones and combination modes, we calculate the number of modes with A_g and H_g symmetry that are contained in the direct product $(2A_g + 8H_g) \otimes (2A_g + 8H_g)$ and taking care not to double count. These special second-order modes have been marked on Fig. 11.12 and account for most of the highest-intensity features in the second-order spectrum. Somewhat weaker features in the second-order Raman spectrum are expected to arise from combination modes where only one of the components relates to the first-order Raman-active lines. In this case, modes with A_g or H_g symmetry arise from the direct product $(2A_g + 8H_g) \otimes n_f\Gamma_f$ where the n_f vibrational modes with Γ_f symmetry are silent in the first-order Raman spectrum. In this category, there are no modes with A_g symmetry in the second-order Raman spectrum, whereas 152 modes with H_g symmetry are expected.

Thus the study of the second-order Raman spectra of C_{60} provides a powerful technique for the determination of the frequencies of the 32 silent modes. Even weaker lines in the second-order Raman spectra could arise from either overtones or combination modes associated with terms in the direct product $n_i\Gamma_i \otimes n_f\Gamma_f$ with A_g and H_g symmetry, where Γ_i and Γ_f are both silent modes in the first-order Raman spectrum with the same parity.

Experimental studies of higher-order Raman lines show a large number of sharp features (see Fig. 11.12). Taking into account the strong, well-established features in the first-order Raman and infrared spectra, the higher-order features of the Raman spectra [11.20] are identified, as shown in Table 11.6 [11.21, 42]. In addition, some of the features in Fig. 11.12 are identified with symmetry-lowering effects associated with the random presence of the ^{13}C isotope (natural abundance 1.1%) in approximately 50% of the C_{60} molecules (see §4.5). From an analysis of the higher-order Raman and infrared spectra for C_{60}, values for the 32 silent mode frequencies and their symmetries have been obtained (see Table 11.1). Further refinements of these mode frequencies and assignments are needed as more experimental data become available. One check on the consistency of these mode frequencies is provided by a comparison of the spectroscopic results used to obtain the entries in Table 11.1 with the experimental determination of mode frequencies obtained by other techniques, such as photoluminescence [11.14, 89, 90], high-resolution electron energy loss spectroscopy [11.91], and inelastic neutron scattering [11.52] (see Table 11.6). The consistency of the mode assignments in Table 11.1 needs to be checked further by theoretical models, such as force constant models or first principles pseudopotential models, and by experimental studies of the temperature and pressure dependence of the spectral features in the higher-order Raman and infrared spectra.

11.5.4. Higher-Order Infrared Modes in C_{60}

In addition to the first-order infrared spectra discussed in §11.5.2, well-resolved higher-order infrared spectral features extending to $\sim$3200 cm^{-1} can be observed in thick films (see Fig. 11.13). Whereas isolated molecules generally exhibit narrow higher-order infrared features, most crystalline solids show broad first-order vibrational infrared features and only continuum features in the second-order spectra. Thus the observation of many well-resolved higher-order infrared lines in the high-resolution infrared spectrum of C_{60} films in Fig. 11.13 is unusual for studies of crystalline solids [11.21].

From group theoretical arguments, the expected number of second-order lines in the infrared molecular spectrum should be very large (a total of 380 combination modes). Since infrared-active modes have odd parity, no infrared overtone modes are expected in the second-order infrared spec-

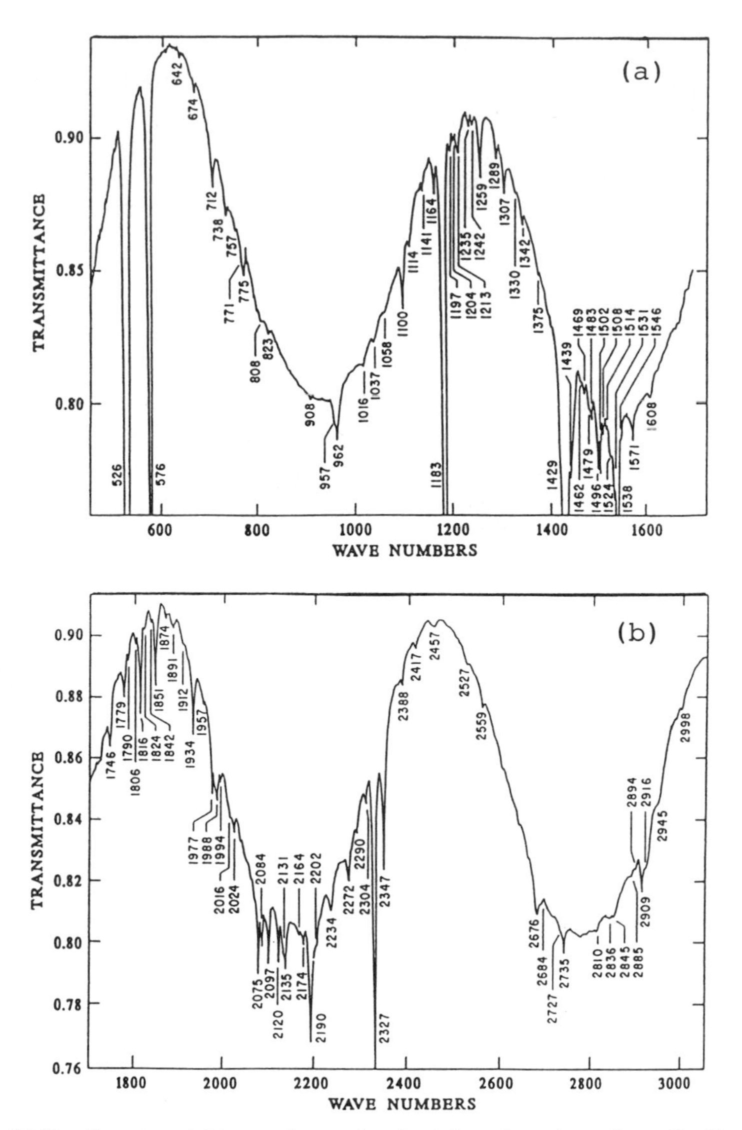

Fig. 11.13. Experimental trace of second-order infrared spectrum for a C_{60} film in the frequency range (a) of the first-order vibrations and (b) above the highest first-order vibration. The frequencies predicted by the second-order model described in the text are marked on the figure [11.21].

trum, in contrast to the situation for the higher-order Raman spectra, where overtones (harmonics) are commonly observed. The total number of combination modes in the infrared spectra can be found by counting the number of times the symmetry F_{1u} is contained in the direct product $n_i\Gamma_i \otimes n_f\Gamma_f$, where Γ_i denotes the symmetry of one of the irreducible representations of the I_h icosahedral group, and n_i is the number of vibrational modes with symmetry Γ_i, and likewise for the second irreducible representation Γ_f. The values for n_i and n_f are contained in Table 11.1 for the vibrational modes for C_{60} and the direct products $\Gamma_i \otimes \Gamma_f$ containing F_{1u} can be found from Table 4.7. For example, since the direct product $F_{1g} \otimes F_{1u} = A_u + F_{1u} + H_u$ contains F_{1u} once, one infrared-active combination mode arising from the F_{1u} and F_{1g} vibrations is predicted for the second-order infrared spectrum. Not all mode combinations are possible; for example, the direct product of H_g and F_{1u} does not result in infrared-active combination modes in the second-order spectrum, since $H_g \otimes F_{1u} = F_{2u} + G_u + H_u$ does not contain F_{1u}. If we assume that the strongest second-order vibrations must involve one of the four F_{1u} modes, then the most intense second-order infrared lines would be expected to arise by taking the direct product $4F_{1u} \otimes n_j\Gamma_j$. Table 4.7 shows that strong infrared combination modes occur only for Γ_j equal to either A_g, F_{1g}, or H_g, thereby giving rise to a total of 52 combination modes. These modes, which are expected to be more intense, account well for ~75% of the observed higher-order IR modes. Since the A_g and H_g mode frequencies are well known from Raman spectroscopy, it is only the silent modes with F_{1g} symmetry that are not as well established. The detailed analysis of the higher-order infrared spectra for thick C_{60} films (see Fig. 11.13) has, together with the corresponding higher-order Raman spectra (see Fig. 11.12), provided a complete determination of all the silent modes for C_{60}. A summary of the mode frequencies and symmetries thus obtained for C_{60} is given in Table 11.1. As further experiments and calculations are carried out, some refinements to the mode frequencies and their symmetry identifications are expected.

More recent work on single crystals [11.92–96] has yielded over 200 well-resolved features in the higher-order infrared spectrum for C_{60}. Use of synchrotron radiation sources provides a higher photon flux and therefore enhances the weak features of the higher-order infrared spectra [11.93, 95]. Temperature-dependent studies have been used to indicate that difference combination modes are not important for explaining the low-frequency features in the higher-order spectrum [11.93, 95]. Temperature-dependent infrared studies may also be useful for identifying specific features in the infrared spectra with (1) an isotope-induced symmetry-lowering effect, (2) the onset of crystal field effects below T_{01}, and (3) combination modes that give rise to infrared activity [11.92] (see §11.5.10).

11.5.5. Isotope Effects in the Raman Spectra

In addition to the symmetry-lowering phenomena associated with the presence of the ^{13}C isotope discussed in §4.5, §11.3.3, and §11.5.3, there is a direct isotopic effect on the mode frequencies arising from their dependence on $(\kappa/M)^{1/2}$, where κ and M are the force constant and atomic mass, respectively. Thus if the fullerene mass is increased by having one or two ^{13}C atoms on a fullerene molecule $^{13}C_x{}^{12}C_{60-x}$, the vibrational frequency is expected to be lowered by an amount proportional to the mass difference between $^{12}C_{60}$ and $^{13}C_x{}^{12}C_{60-x}$, which can be simply written as

$$\frac{\omega(x)^2}{\omega(0)^2} = 1 - \frac{x}{720}, \tag{11.6}$$

where the atomic mass of $^{12}C_{60}$ is 720 a.u. and x is the number of ^{13}C atoms on a C_{60} molecule [11.25].

By preparing high-purity C_{60} in a CS_2 solution, the natural Raman linewidth could be reduced, so that the Raman lines associated with $^{12}C_{60}$, $^{13}C_1{}^{12}C_{59}$, and $^{13}C_2{}^{12}C_{58}$ could be resolved with a separation of 1.00 ± 0.02 cm^{-1} between these Raman peaks, as shown in Fig. 11.14 for the spectrum taken at 30 K. Data taken at higher temperatures are shown in the inset [11.25]. By taking both the mass spectra and Raman spectra for a sample that was isotopically enriched with ^{13}C, but still containing many $^{12}C_{60}$ molecules, quantitative confirmation of this perturbative interpretation of the isotopic composition of the $A_g(2)$ Raman line could be verified [11.25]. A more detailed analysis of the mode shift in terms of first-order nondegenerate perturbation theory provides a sensitive probe of the mode mixing induced by the isotope effect. To fit the experimental data, it is predicted that the $G_g(6)$ mode should lie $\sim$20 cm^{-1} above the $A_g(2)$ mode at 1470 cm^{-1} [11.25]. This may be the best determination of the mode frequency at 1490.0 cm^{-1} for $G_g(6)$ presently available (see Table 11.1).

11.5.6. Silent Intramolecular Modes in C_{60}

In addition to Raman and infrared spectroscopy, several other experimental techniques such as inelastic neutron scattering (see §11.5.8), electron energy loss spectroscopy (see §11.5.9), and photoluminescence (see §11.5.7) provide information on the lattice modes of C_{60} (see Tables 11.5 and 11.6). These techniques traditionally provide information about the silent modes, not observed in the first-order Raman or infrared spectra, and also provide complementary information (although usually at lower resolution than Raman and infrared spectroscopy) about the optically active modes.

As discussed in §11.5.3 and §11.5.4, the higher-order Raman and infrared spectra for C_{60} provide an even more powerful means for determining the

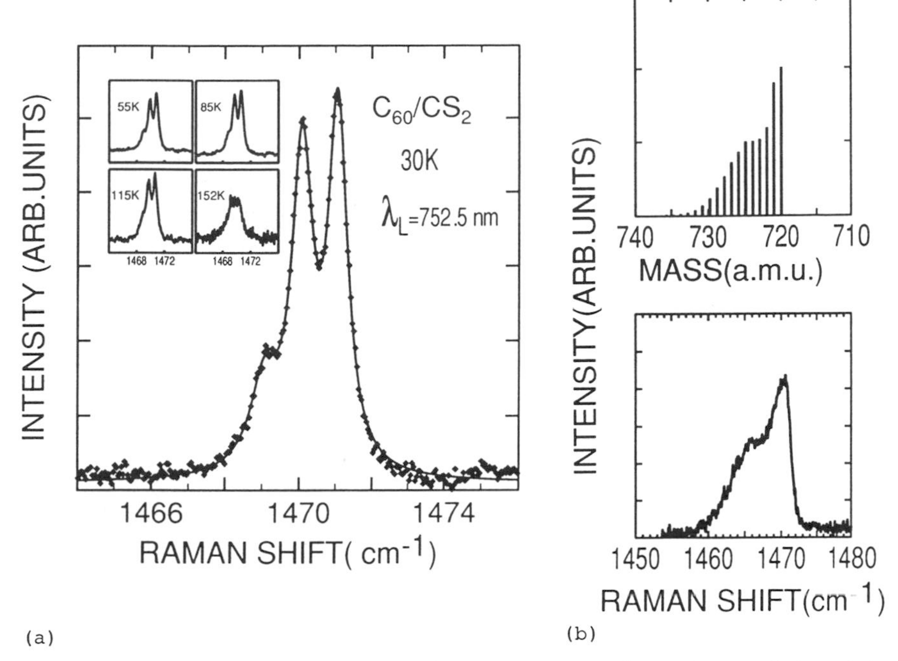

Fig. 11.14. (a)Unpolarized Raman spectrum near 1470 cm^{-1} of a frozen solution of C_{60} in CS_2 at 30 K. The solid line is a three-Lorentzian fit to the experimental data. The highest frequency peak is assigned to the totally symmetric pentagonal pinch $A_g(2)$ mode in $^{12}C_{60}$. The other two lines are assigned to the pentagonal pinch mode in molecules containing one and two ^{13}C atoms, respectively. The inset shows the evolution of these peaks as the frozen solution is heated. The three components for the Raman feature are no longer resolved as the frozen solution melts above 150 K. (b) Comparison between the mass spectra and the Raman spectra for a ^{13}C-enriched C_{60} sample [11.25].

mode frequencies for the vibrational modes that are not observed in the first-order Raman and infrared spectra. The two advantages of using the second-order Raman and infrared spectra to find mode frequencies for the silent modes are the symmetry information for the silent modes that is contained in these spectra and the sensitivity of the higher-order Raman and infrared spectra for determining all the silent modes to within $\sim$5 cm^{-1} (see §11.5.3 and §11.5.4). In general, it is not possible to observe sharp higher-order Raman and infrared spectra in solids because of dispersion arising from interatomic and/or intermolecular interactions in the crystalline phase. Because of the highly molecular nature of fullerenes in their crystalline phases, higher-order Raman and infrared spectra can be observed, in analogy to studies of molecules in solution or in the gas-phase.

Surface-enhanced Raman spectroscopy (SERS) could also provide information on silent modes. Such studies carried out for C_{60} on noble-metal surfaces show substantial shifts of the $A_g(2)$ pentagonal pinch mode, and these shifts are associated with charge transfer [11.97] (see §17.7). More recently, SERS experiments for C_{60} on Ag and In polycrystalline surfaces have also been carried out, showing large differences in the SERS spectra upon room temperature annealing of samples grown at low temperature. Different behaviors were observed for the Ag and In surfaces upon annealing [11.98].

11.5.7. Luminescence Studies of Vibrations in C_{60}

Singlet oxygen photoluminescence (PL) has been shown to provide a powerful probe of the intramolecular vibrational modes of C_{60}, independent of their symmetry, as shown in Fig. 11.15 [11.14]. The applicability of the photoluminescence technique to probe fundamental vibrational excitations of C_{60} is based on the weak vibronic coupling of O_2 molecules intercalated into the C_{60} lattice (see §13.3). Raman scattering studies had shown previously that this coupling is indeed weak, since no shifts in the Raman-active C_{60} mode frequencies were detected upon oxygen intercalation [11.75]. This

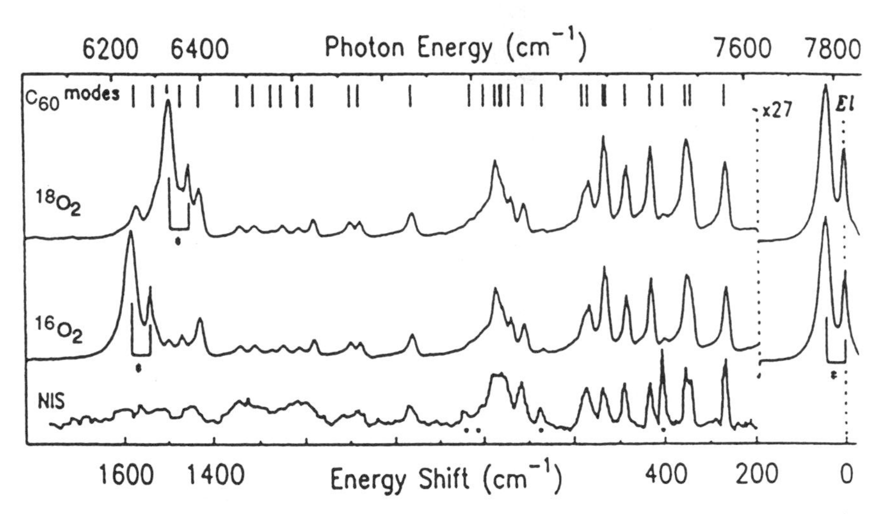

Fig. 11.15. Photoluminescence (PL) spectra of C_{60} films exposed to $^{18}O_2$, top, and $^{16}O_2$, middle, at $T = 4.2$ K, and excited by 720 nm photons [11.14]. The photon energy scale for the PL is shown on the top. On the bottom is a vibrational mode energy scale measured from the purely electronic principal transition, labeled *El*. The asterisks indicate the O_2 vibrational mode replica of the strong electronic doublet. The bottom curve (NIS) is the inelastic neutron scattering data of Ref. [11.83].

Raman study further implied the presence of oxygen as O_2, rather than as atomic oxygen bonded to the C_{60} molecules, which is directly confirmed by the singlet oxygen PL spectra (see Fig. 11.15). More important, the vibronic coupling between the O_2 and C_{60} allows the molecular vibrational frequencies of C_{60} to appear as side bands on the $T = 4.2$ K luminescence emission spectra, as dioxygen returns to the triplet ground state ($^3\Sigma_g^-$) from the first excited singlet state ($^1\Delta_g$). The upper two PL spectra in Fig. 11.15 are from $^{18}O_2$ and $^{16}O_2$, respectively [11.14], whereas the bottom vibrational spectrum is obtained from inelastic neutron scattering studies of microcrystalline C_{60} powders [11.83]. As can be seen from the figure, remarkably good agreement is obtained between the PL vibrational side bands and the more conventional inelastic neutron scattering probe. Furthermore, the higher-energy mode structure (> 700 cm^{-1}) is much sharper in the PL data than in the inelastic neutron scattering data, thereby yielding more reliable values for the mode frequencies. Thirty-two of the 46 vibrational modes for C_{60} are seen in the PL spectra, many of which are unobservable in the first-order Raman or IR spectra due to symmetry selection rules.

11.5.8. Neutron Scattering

Inelastic neutron scattering provides an important technique for the study of intramolecular vibrations, because the technique is sensitive to modes of all symmetries, providing mode frequencies for Raman-active, infrared-active, silent modes, overtones, and combination modes. In addition, neutron scattering studies provide direct information on the overall phonon density of states. Because of its lower energy resolution than infrared or Raman spectroscopy, inelastic neutron scattering is most useful for the study of silent modes and the phonon dispersion relations, rather than the infrared- or Raman-active mode frequencies. Since the C_{60} vibrations extend to relatively high frequencies, pulsed neutron spallation sources are used for $\hbar\omega_q > 100$ meV (800 cm^{-1}), while reactor sources are available for vibrational mode studies for $\hbar\omega_q < 100$ meV [11.54]. Also, the limited size of C_{60} single crystals makes the study of phonon dispersion over the entire Brillouin zone possible at the present time only for the low-energy intermolecular branches (see §11.4).

A number of workers have measured the intramolecular mode frequencies for C_{60} by inelastic neutron scattering [11.83, 84, 86, 99] and the results are summarized in Tables 11.5 and 11.6. In early neutron scattering work, the authors relied heavily upon experimental Raman and infrared spectroscopy as well as theoretical phonon mode calculations for their mode identifications. As seen in Table 11.5, good overall agreement is obtained between the mode frequencies studied by inelastic neutron scattering methods and by other methods. A plot of the inelastic neutron scattering intensity as a function of phonon energy ($\hbar\omega_q$) is shown in Fig. 11.16 for the energy range $\hbar\omega_q < 110$ meV and in Fig. 11.17 for $100 < \hbar\omega_q < 220$ meV [11.83].

Also shown in Fig. 11.16 are inelastic neutron scattering results for K_3C_{60} [11.86] and Rb_3C_{60} [11.100]. Although not shown in the figure, measurements on K_3C_{60} have been made up through 200 meV, indicating higher neutron scattering intensity for K_3C_{60} relative to C_{60} in the phonon energy range $130 < \hbar\omega_q < 200$ meV [11.86]. Since all mode symmetries can be observed by the inelastic neutron scattering technique, it is significant that the peaks in K_3C_{60} and Rb_3C_{60} are in general broader than for pristine C_{60}, but the mode broadening is not uniform, since certain spectral features appear to be broadened much more than others. In particular, the modes with H_g symmetry are broadened the most. Significant differences are reported in the 35 K spectrum for Rb_3C_{60} above the superconducting transition temperature T_c (29 K) in comparison to the 22 K spectrum below T_c [11.100]. Significant changes in the scattering near 135 and 180 meV are observed, as well as the disappearance of the features at 54 and 60 meV below T_c, suggesting strong broadening of these modes, presumably due to the electron–phonon interaction.

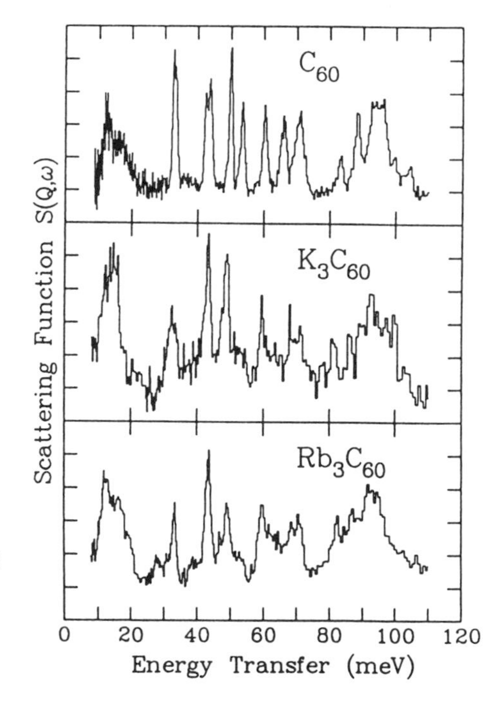

Fig. 11.16. Inelastic neutron scattering intensity *vs.* energy transfer (phonon energy $\hbar\omega_q$) for (a) C_{60}. For comparison the corresponding spectra are shown in (b) for K_3C_{60} (summed data at 5 K and 30 K) [11.86] and in (c) Rb_3C_{60} (summed data at 22 K and 35 K). The C_{60} peaks at 54 and 66 meV are essentially absent in K_3C_{60} and Rb_3C_{60} [11.100]. Regarding units, note that 1 meV = 8.0668 cm^{-1}.

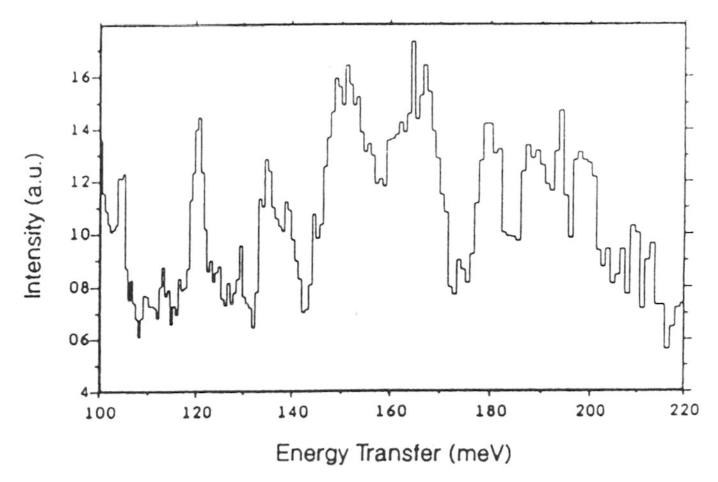

Fig. 11.17. Inelastic neutron scattering spectrum for C_{60} above 100 meV (807 cm^{-1}) measured with a pulsed neutron source [11.83].

11.5.9. HREELS Study of Vibrational Modes

Since the energy of the vibrational modes is generally very much smaller than the energies of electron beams used in electron energy loss spectroscopy (EELS), high-resolution detection of the energy of the electron beam is needed. Thus vibrational mode studies using EELS are normally carried out using the HREELS (high-resolution electron energy loss spectroscopy) version of this surface science technique (see §17.2).

If the sample surface is specular or nearly so, then LEED (low-energy electron diffraction) measurements can be carried out with the same electron beam to determine the crystal structure at the surface (see §17.2). In the EELS experiment, the inelastic scattering of electrons on the surface is used to measure the change in energy and momentum of the scattered electron beam relative to the incident beam. Two scattering mechanisms are possible for the scattered electron beam [11.91, 101]. In the impact scattering mode, the electron scattering is by the local atomic potentials of near-surface atoms, and the scattered electron beam is essentially isotropic, so that large momentum transfers can occur. In contrast, for the dipole scattering mechanism, the probe electron beam interacts with the electric field caused by a dipole in the near-surface region. Such a dipole field is associated with the excitation of infrared-active phonon modes with F_{1u} symmetry. The long-range nature of the Coulomb interaction implies that the probing electrons interact primarily with long-wavelength excitations having small momentum $\mathbf{q}$ and a small angular aperture $\psi = \hbar\omega_q/2E_p$, where E_p is the energy of the primary probing electron beam, which is typically 3.7 eV for studies of the vibrational modes. The infrared-active modes are therefore expected to dominate the HREELS spectrum for scattering angles close to specular reflection, while the other vibrational modes are expected to be more prominent outside this angular scattering range.

Experimental studies of the vibrational spectra using HREELS have been carried out primarily on C_{60}, using a variety of substrates, including Si (100) [11.87], GaSe (0001) [11.87], GeS (001) [11.87], Cu (111) [11.102], and Rh (111) [11.103]. Except for spectra taken to enhance the dipole scattering contribution, HREE'LS performed under the impact scattering mode provides an important method for observing silent modes, not seen in the IR or Raman spectra. The most intense of the 11 features that were reported in the HREELS spectra for C_{60} on Si (100) and GaSe (0001) is the peak near 66 meV (532 cm^{-1}), which is near the two 527 cm^{-1} and 577 cm^{-1} infrared-active modes that cannot be resolved individually by HREELS. The resolution is ~10 meV on an Si (100) surface and ~7.5 meV on a GaSe (0001) substrate. The observed HREELS features identified with vibrational excitations, in general, are resolution limited [11.91, 101].

By using a GaSe (0001) substrate, which is lattice matched to C_{60}, good specular surfaces can be achieved for the adsorbed C_{60} film, leading to a clear resolution between the dipole scattering and impact scattering features. For C_{60} on GaSe (0001), the $F_{1u}(3)$ and $F_{1u}(4)$ infrared-active modes can also be seen by HREELS, and the feature at 66 meV has the expected linewidth of $\hbar\omega_q/eE_0 \approx 10^{-2}$ radians (or 0.6°). An analysis of the relative intensities of the infrared-active lines in terms of Lorentz oscillators yield a ratio of the relative oscillator strengths of $F_{1u}(1)$:$F_{1u}(2)$:$F_{1u}(3)$:$F_{1u}(4)$ = 100:29:6:5 [11.91].

11.5.10. Vibrational Modes as a Probe of Phase Transitions

The structural changes associated with the phase transition at T_{01} (see §7.1.3) result in changes in the Raman and infrared spectra, as reported by several groups [11.42, 44, 57, 104, 105]. Working with single-crystal samples, a hardening of six of the Raman-active modes for the fcc → sc ($Fm\bar{3}m \rightarrow Pa\bar{3}$) phase transition at T_{01} was identified [11.44] , with magnitudes of the hardening ranging from 2 to 11 cm^{-1} as shown in Fig. 11.18. Since the Raman vibrational frequencies are large compared with the frequency of rotation, Raman spectroscopy takes a snapshot of the vibrating molecules showing the relative orientation of adjacent molecules. Above T_{01} the C_{60} molecules are randomly oriented with respect to each other, and below T_{01} there is some degree of orientational correlation between molecules. Certain Raman modes are sensitive to this local crystal site orientation effect. The mode frequency upshifts shown in Fig. 11.18 for the

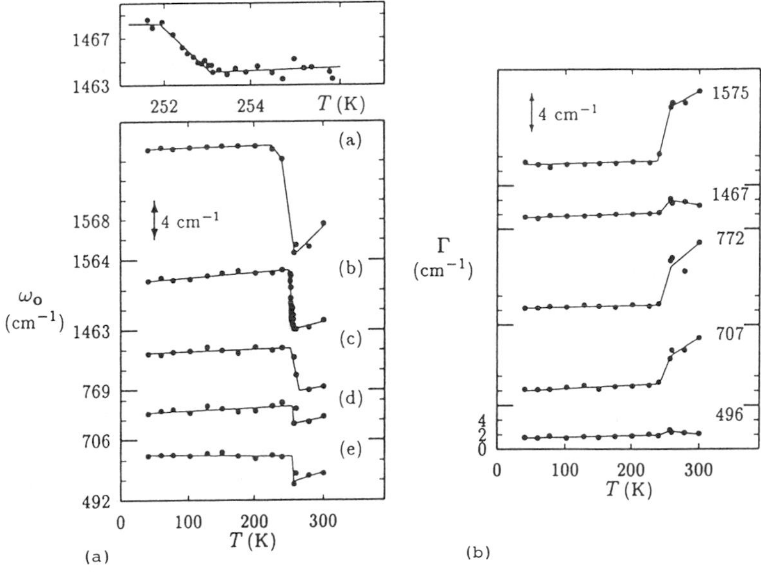

Fig. 11.18. On the left, the large figure shows the temperature dependence of the frequency of the Raman-active modes at (from top to bottom) (a) 1575 cm^{-1} [$H_g(8)$], (b) 1467 cm^{-1} [$A_g(2)$], (c) 772 cm^{-1} [$H_g(4)$], (d) 707 cm^{-1} [$H_g(3)$], and (e) 496 cm^{-1} [$A_g(1)$], respectively. The top panel on the left shows the temperature dependence of the 1467 cm^{-1} [$A_g(2)$] mode near the fcc $\rightarrow$ sc transition, indicating a discontinuous transition with a width for the transition of $\sim$1 K [11.44]. The figure on the right shows the temperature dependence of the linewidths of the observed Raman-active modes at (from top to bottom) 1575, 1467, 772, 707, and 496 cm^{-1}, respectively. For both figures on the left (a) and right (b), the zero point on the vertical scale is indicated for each curve by the thick tick marks. (The solid lines are a guide to the eye.) [11.44].

fcc $\rightarrow$ sc transition (see §14.15(b)) are consistent with the volume decrease that is observed [11.44]. A decrease in linewidth was reported for five of the Raman-active modes at the fcc–sc orientational phase transition (see Fig. 11.18(b)), with large decreases (factors of 3–4) in linewidth observed in the low-temperature phase for the $H_g(3)$, $H_g(4)$, and $H_g(8)$ modes [11.44]. These large changes in linewidth were identified with enhanced vibrational–rotational coupling in the fcc phase. Discontinuous changes in the intensity of the Raman lines at T_{01} were also reported [11.57].

Although the main effect of temperature variation on the Raman spectra occurs at T_{01}, significant changes in line intensities were observed upon lowering the temperature from 300 K down to low temperatures (10 K) [11.57]. Mode splittings of some of the Raman-active modes were observed as well as the appearance of combination modes (see §11.5.3) [11.57]. Both the mode frequencies and linewidths of specific Raman features show little temperature dependence in the sc phase for temperatures below T_{01} [11.44], although the increased intensities at low temperature bring out certain features in the Raman spectra that contribute to the background scattering near T_{01} [11.57]. To illustrate the type of mode splittings that are observed in the Raman spectrum, we show the effect of temperature on the $A_g(1)$ mode in Fig. 11.19. In this figure we see an asymmetric line at 300 K (above T_{01}), which shows well-resolved peaks at 10 K that appear to grow out of the scattering intensity in the low-frequency wing of the 300 K spectrum. Since crystal field splitting would not be expected above T_{01} for an A_g nondegenerate mode and the observed splittings below T_{01} are too large to be associated with an isotope effect, it is suggestive to identify the line asymmetry above T_{01} with local intermolecular orientational correlations, associated with fast ratcheting motion above T_{01} (see §7.1.3). At 10 K the local intermolecular orientational correlations become well defined and much stronger, giving rise to resolved Raman lines. The number of resolved Raman lines (three) in the 10 K spectrum is also suggestive of a local intermolecular correlation mechanism, since conventional crystal field splittings would give rise to only two mode frequencies because

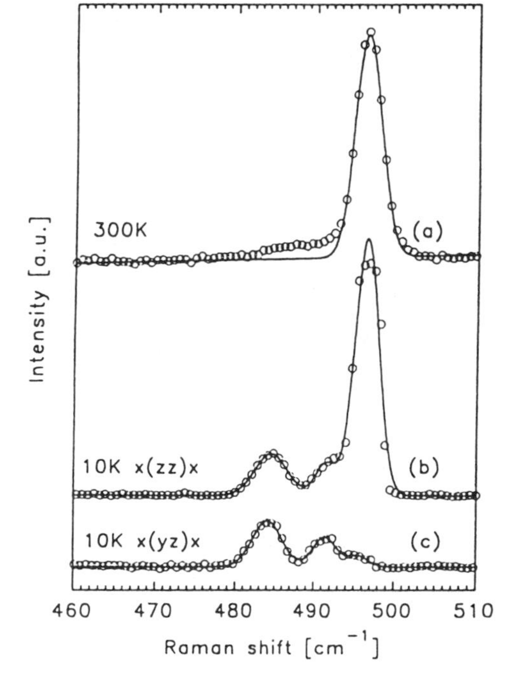

Fig. 11.19. Raman spectra of the $A_g(1)$ breathing mode of a C_{60} single crystal at (a) 300 K, (b) 10 K for the parallel scattering geometry, and (c) 10 K for the perpendicular scattering geometry. The solid lines are a Lorentzian fit to the data taken at 514 nm and 19 W/cm^2 [11.57].

$A_g \otimes \chi^{\text{sites}} = A_g + T_g$, taking into account the larger unit cell for the $Pa\bar{3}$ structure (see §7.1.3).

Temperature-dependent infrared studies on single-crystal C_{60} have revealed splittings of each of the $F_{1u}(1)$, $F_{1u}(3)$, and $F_{1u}(4)$ spectral features into a quartet of lines below the first-order phase transition at T_{01} (261 K), as shown in Fig. 11.20 [11.106]. The development of the quartet structure occurs gradually with decreasing temperature, and not sharply at T_{01}. The authors [11.106] attributed the splittings to cubic crystal field effects associated with orientational ordering of the molecules below T_{01} and with increasing orientational ordering as T is decreased further. Group theoretical analysis indicates that the appearance of four inequivalent sites in the $Pa\bar{3}$ structure below T_{01} results in a symmetry-lowering effect. If $\chi^{\text{sites}} = A_g + T_g$ describes the transformation properties of the four inequivalent sites, then the direct product $T_u \otimes \chi^{\text{sites}}$ is decomposed into the irreducible representations $A_u + E_u + 3T_u$ (in T_h symmetry). Thus, if crystal field effects were responsible for the observed quartet of lines, we would expect each F_{1u} mode to split into three infrared-active T_u modes, whose mode frequencies could be estimated by a zone-folding scheme applied to the phonon dispersion curves appropriate to the larger fcc Brillouin zone. The authors [11.106] attributed the remaining line of the quartet to either the activation of a silent mode due to crystalline disorder or the splitting of one of the T_u

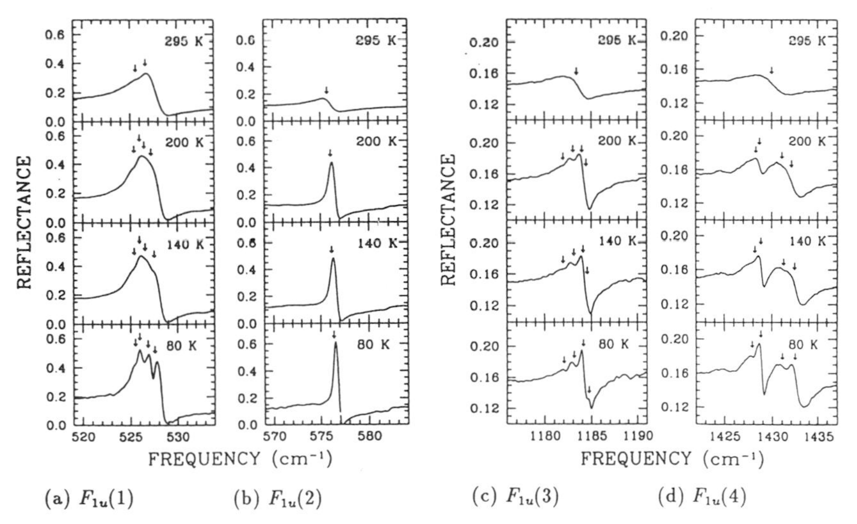

Fig. 11.20. The reflectance of a single-crystal C_{60} sample at 295, 200, 140, and 80 K in the vicinity of the four infrared-active modes. The fourfold splittings of the $F_{1u}(1)$, $F_{1u}(3)$ and $F_{1u}(4)$ infrared-active modes are indicated by the arrows, but no splitting is observed for the $F_{1u}(2)$ mode [11.106].

modes due to disorder. The absence of splitting in the $F_{1u}(2)$ feature and the narrowing of the $F_{1u}(2)$ line with decreasing temperature have not yet been explained [11.106].

If the crystal field effect were associated with zone folding of the phonon dispersion relations, it is unlikely that the dispersion over the entire fcc Brillouin zone would be as small as a few cm^{-1}, nor does the crystal field effect easily account for the fourth line in the quartet. Another possible explanation for the low-temperature structure is to associate the observed split IR modes with the effect of the local environment on the vibrational modes. In the high-temperature crystalline phase we would expect the molecules to see an average local environment. Below T_{01} the rotations of the molecules become restricted to a few axes [11.73] (see, for example, §7.1.5), and the molecules jump (or ratchet) between a few (e.g., five) orientational minima. We then identify each of the components of the low-temperature multiplet with a specific site orientation of the molecule relative to its neighbors. As the temperature is lowered, these specific sites become better defined and therefore the splittings are expected to become better resolved. The effect of the local molecular environment should be more pronounced for some atomic displacements (i.e., certain T_u modes), in agreement with experiment. Furthermore, the magnitude of the mode splittings arising from local environment effects is expected to be only a few cm^{-1}, also in agreement with observations. Crystal field effects associated with the local environment of C_{60} molecules also seem to be important in explaining temperature-dependent structure observed in the Raman spectrum (see Fig. 11.19).

Since the appearance of the quartet structure in the infrared spectrum occurs gradually as T is decreased, these experiments do not provide an accurate method for the determination of T_{01}. The results of Fig. 11.18, however, show that the Raman-active modes more sensitively determine T_{01}, since they are more sensitive to local environment changes near T_{01}.

The appearance of additional structure in the infrared-active spectra below T_{01} has also been verified [11.92] through temperature-dependent infrared spectroscopy studies on single-crystal C_{60}. This study [11.92] was focused on identifying which features in the infrared spectra are due to isotope effects, crystal field effects, and combination modes, by monitoring the change in lineshape for the many lines observed in the 200 or more resolved features in the high-resolution infrared spectra (see §11.5.4). For example, features associated with the isotope effect narrow at rather low temperatures (see §11.5.5 for the Raman analog). Crystal field effects are expected to be turned on below T_{01} as preferential orientational ordering sets in.

11.5.11. Vibrational States Associated with Excited Electronic States

Because of the metastability of the lowest-lying triplet electronic state above the Fermi level (see §13.2.3), it is possible to observe vibrational features [11.107] associated with the metastable triplet state [11.108, 109]. Experimental results are available for the $A_g(2)$ pentagonal pinch mode in the excited triplet electronic state. This mode is found to be downshifted relative to that for the vibration in the electronic ground state, because of the weaker binding and consequently reduced force constants for the excited state [11.107]. Above a laser excitation intensity of ~50 W/cm^2, a line downshifted from the 1468 cm^{-1} $A_g(2)$ mode appears in the Raman spectra taken at a temperature of 40 K and 514 nm laser excitation wavelength, as shown in Fig. 11.21 [11.107]. At these low temperatures, the probability for molecular alignment to induce polymerization is very low. Furthermore, the reversibility of the observations on lowering the laser intensity provides supporting evidence that the observed phenomena are not connected to polymerization. The identification of the downshifted feature with Raman scattering from the excited metastable triplet state is further supported by a quenching of this feature by the introduction of oxygen, which greatly decreases the lifetime of the excited triplet state (see §13.2.3). This excited state Raman line grows in intensity with increasing laser intensity (see Fig. 11.21) until about 300 W/cm^2; above this laser intensity, the process becomes irreversible [11.107].

The effect of 514 nm laser intensity on the Raman spectrum of C_{60} has also been studied at room temperature [11.109], showing a downshifted

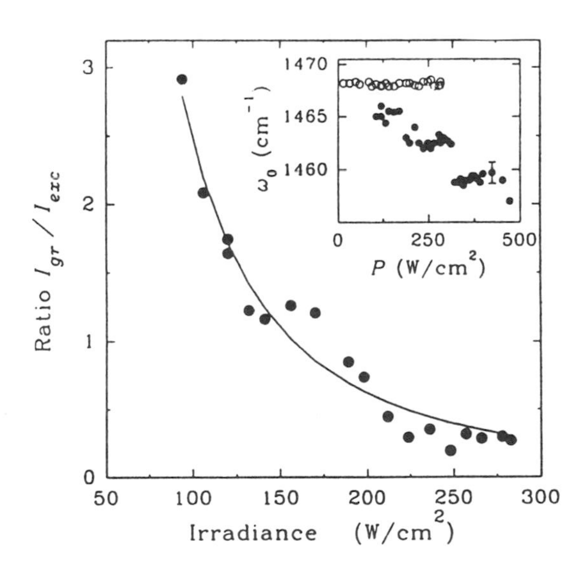

Fig. 11.21. Ratio of the peak intensities of the $A_g(2)$ mode (1468 cm^{-1}) in the electronic ground state to that in the triplet excited state as a function of the laser excitation intensity at $T = 40$ K (dots). The solid line shows the fit of an inverse square laser intensity dependence to the data. The inset shows the dependence of the frequency of the ground state 1468 cm^{-1} peak (open circles) and the corresponding triplet excited state mode (closed circles) on laser excitation intensity P [11.107].

Lorentzian line, which is reversible up to 15 mW of power, with a frequency downshift depending nonlinearly on the laser power. Since this effect is observed above T_{01}, where the molecules are rotating rapidly, and since the dependence on laser intensity is quite different from that shown in Fig. 11.21 [11.107], the physical origin of the phenomena at higher temperature may be different.

11.6. Vibrational Modes in Doped Fullerene Solids

Since the discovery of moderately high-temperature superconductivity in the alkali metal (M)–doped C_{60} solids M_3C_{60} (M = K, Rb), considerable activity has been expended to document the doping-induced changes in the vibrational modes of these materials and to investigate whether or not the superconducting pairing interaction is mediated by vibrational modes and, if so, by which modes. The contributions of the four distinct mode types (acoustic, librational, optic, molecular) to a vibration-mediated pairing mechanism for superconductivity have been considered both theoretically and experimentally. Insulating compositions of M_xC_{60}, such as M_6C_{60}, have also been studied for comparison to the behavior of C_{60} itself. Furthermore, the $x \sim 1$ solid solution system, which is found to be stable at elevated temperatures ($T > 200°C$) and has been reported to polymerize, has also been studied in Raman scattering experiments [11.19].

The availability of a variable range of alkali metal doping ($0 < x \leq 6$) presents the opportunity to consider whether or not the dopant may be treated as simply a perturbing influence on the C_{60}-derived modes (via charge transfer between the M_x and C_{60} sublattices). For example, many of the Raman- and IR-active molecular modes for the stoichiometries $x \sim 1$, and $x \sim 6$ remain reasonably sharp and are found experimentally to be related simply to the parent ($x = 0$) solid spectrum [11.19]. The mode frequencies of these M_xC_{60} compounds, although shifted from those in the parent C_{60} material, are, for fixed x, largely independent of the dopant species, dopant mass or the crystal structure [11.19]. This suggests that the principal effect of alkali metal doping is to produce C_{60} anions (C_{60}^{n-}, where $n = 1, 3, 4, 6$), which are weakly coupled to one another and also weakly coupled to the cation M^+ sublattice [11.110, 111].

The introduction of alkali metal dopants into the lattice also gives rise to low-frequency vibrational modes, whereby the alkali metal ions vibrate relative to the large fullerene molecules. Such modes should be accessible for investigation by Raman and infrared spectroscopy, as well as other techniques. The presence of alkali metal ions in the lattice also influences the characteristics of the intermolecular modes of C_{60}, as discussed in §11.6.4.

11.6.1. Doping Dependence of Raman-Active Modes

This section summarizes observed doping-induced shifts in frequency of the A_g and H_g modes and use of these frequency shifts for sample characterization regarding the doping concentration.

The addition of alkali metal dopants to form the superconducting M_3C_{60} (M = K, Rb) compounds and the alkali metal–saturated compounds M_6C_{60} (M = Na, K, Rb, Cs) perturbs the Raman spectra only slightly relative to the solution molecular spectra and to the spectra in the undoped solid C_{60}, as presented in Fig. 11.22, where the Raman spectra for solid C_{60} are shown in comparison to various M_3C_{60} and M_6C_{60} spectra [11.3, 15]. One can, in fact, identify each of the features in the M_xC_{60} spectra with those of pristine C_{60}, and very little change is found from one alkali metal dopant to another [11.82]. The small magnitude of the perturbation of the Raman spectrum by alkali metal doping and the insensitivity of the M_6C_{60} spectra to the specific alkali metal species indicate a very weak coupling between the C_{60} molecules and the M^+ ions.

The Raman spectra for the M_3C_{60} phase (M = K, Rb) in Fig. 11.22 are of particular interest. Here again, the spectra are quite similar to that of C_{60}, except for the greater sharpness of the A_g modes in the M_3C_{60} phases, the apparent absence in the M_3C_{60} spectra of several of the Raman lines derived from the H_g modes in C_{60}, and the broadening of other H_g lines. This is particularly true in the spectrum of Rb_3C_{60}, for which the same sample was shown resistively to exhibit a superconducting transition temperature of $T_c \sim 28$ K [11.75]. For the case of K_3C_{60} (see Fig. 11.23) [11.3, 114] and Rb_3C_{60} [11.88], the coupling between the phonons and a low-energy continuum strongly broadens the H_g-derived modes [11.3, 76, 115] and gives rise to Breit–Wigner–Fano modifications in the Raman lineshape [see Fig. 11.23 for this effect on the $H_g(1)$ Raman feature], which is Lorentzian for undoped C_{60} [11.114]. As discussed in §12.7.3, the electronic structure for K_3C_{60} and Rb_3C_{60} is described in terms of a band model, while that for the M_6C_{60} compounds is best described by levels for a molecular solid. Furthermore, the molecular ions in the M_3C_{60} compounds are expected to experience a Jahn–Teller distortion while the M_6C_{60} compounds do not (see §14.2.2 and §14.2.4). The symmetry lowering that results from the Jahn–Teller distortion may be a possible cause for the broadening (or splitting) of some of the degenerate vibrational modes in the M_3C_{60} spectra. The splitting of the $H_g(1)$ modes for M_6C_{60} in Fig. 11.23 is attributed to crystal field effects, which are expected to be more pronounced in M_6C_{60} compared to C_{60} itself because of the polarization effects of the alkali metal atoms in the M_6C_{60} lattice. The splitting of $H_g(1)$ into a doublet with cubic symmetries $E_g + T_g$ for both polarization geometries HH($\parallel, \parallel$) and HV($\parallel, \perp$) is consistent with §7.1.3.

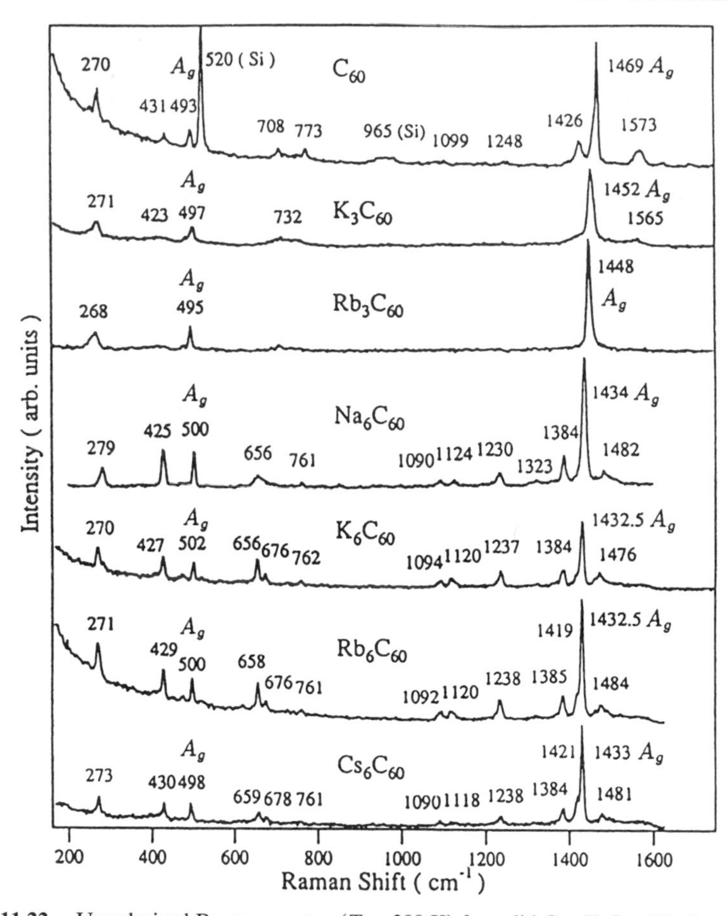

Fig. 11.22. Unpolarized Raman spectra ($T = 300$ K) for solid C_{60}, K_3C_{60}, Rb_3C_{60}, Na_6C_{60}, K_6C_{60}, Rb_6C_{60} and Cs_6C_{60} [11.3, 15]. The tangential and radial modes of A_g symmetry are identified, as are the features associated with the Si substrate. From these spectra it is concluded that the molecule–molecule interactions between the C_{60} molecules are weak, as are the interactions between the C_{60} molecules and the alkali metal ions.

As a result of alkali metal doping, electrons are transferred to the π electron orbitals on the surface of the C_{60} molecules, elongating the C–C bonds and downshifting the intramolecular tangential modes. A similar effect was noted in alkali metal–intercalated graphite, where electrons are transferred from the alkali metal M layers to the graphene layers [11.116]. The magnitude of the mode softening in alkali metal–doped C_{60} is comparable ($\sim$60% of that for alkali metal–doped graphite intercalation compounds) and can be explained semiquantitatively by a charge transfer model [11.117]. Referring to Fig. 11.24, the dependence of the Raman frequencies on alkali

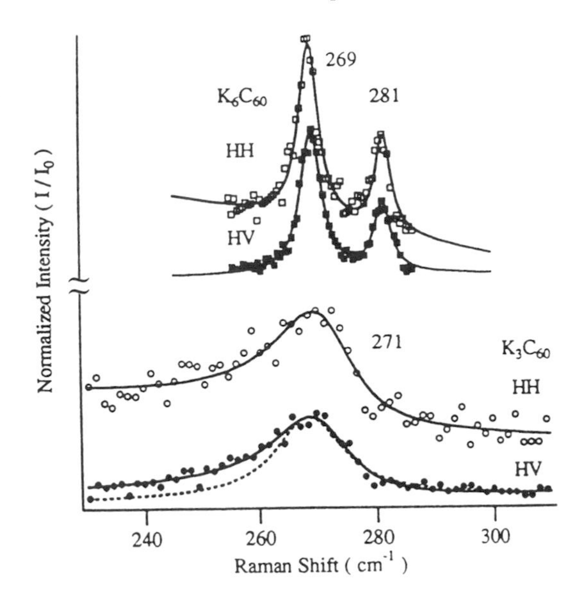

Fig. 11.23. Breit–Wigner–Fano lineshape observed for the Raman-active $H_g(1)$ mode for K_3C_{60} for both polarizations HH(‖, ‖) and HV(‖, ⊥). Also shown are two components for the $H_g(1)$ line in K_6C_{60} which are well described by a Lorentzian lineshape for both polarizations HH(‖, ‖) and HV(‖, ⊥) [11.3, 114].

metal M concentration x for three dominant modes in M_xC_{60} (M = K, Rb, Cs) is presented. The data in Fig. 11.24 come from several publications of the Kentucky, Vienna, and AT&T research groups [11.19, 112, 113]. Normal mode displacements for two of the three mode eigenvectors are also shown in the figure (see also Figs. 11.2 and 11.3).

The softening of the 1469 cm^{-1} tangential $A_g(2)$ mode by alkali metal doping is often used as a convenient method to characterize the stoichiometry x of K_xC_{60} samples. This mode in M_xC_{60} is downshifted by ~6 cm^{-1}/M atom, and since all end-point M_6C_{60} compounds (M = K, Rb, Cs) exhibit approximately the same value for this mode frequency (1432 cm^{-1}), it is reasonable to expect that a downshift of ~6 cm^{-1}/M atom for the $A_g(2)$ mode frequency is approximately applicable also for both K and Rb. It is found that the frequency downshifts of the $A_g(2)$ mode for x = 1, 3, 4, and 6 follow a linear dependence on x quite well [11.118]. The radial $A_g(1)$ mode, on the other hand, stiffens slightly with increasing x due to competing effects, associated with a mode *softening* arising from the charge transfer effect (similar to the situation for the tangential modes) and a larger mode *stiffening* effect due to electrostatic interactions between the charged C_{60} molecule and the surrounding charged alkali metal atoms, as their atomic separations change during a normal mode displacement [11.117].

Mode broadening occurs with increasing x, and this broadening is most pronounced near $x \sim 3$ in the metallic phases of M_3C_{60}. This line broadening has in part been attributed to enhanced electron–phonon coupling, and

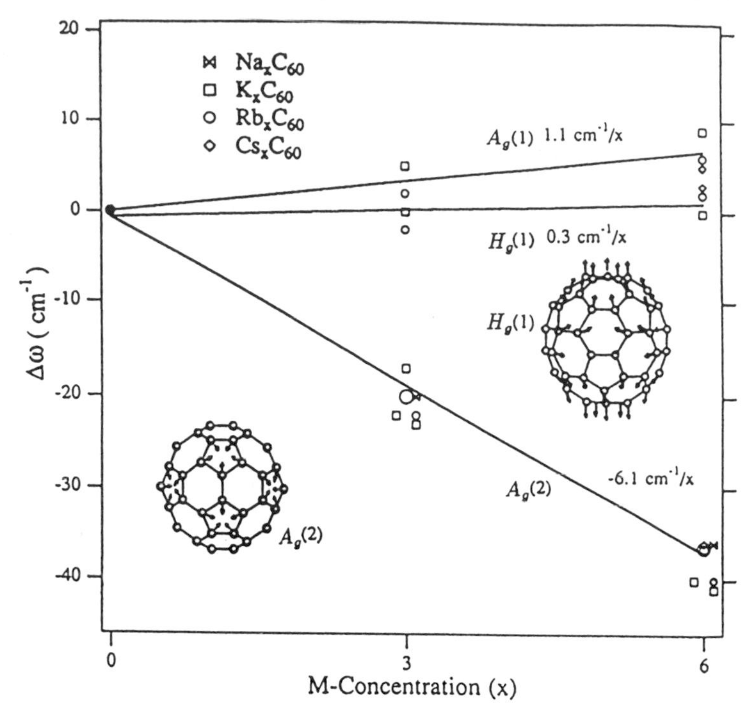

Fig. 11.24. Dependence of the frequencies of the $A_g(1)$, $A_g(2)$, and $H_g(1)$ modes on alkali metal concentration x in M_xC_{60}, where M = Na, K, Rb, Cs. The frequency shifts of these Raman-active modes are plotted relative to the frequencies $\omega[A_g(1)] = 493$ cm^{-1}, $\omega[A_g(2)] = 1469$ cm^{-1}, and $\omega[H_g(1)] = 270$ cm^{-1} for C_{60} at $T = 300$ K. A schematic of the displacements for the eigenvectors for the $A_g(2)$ pentagonal pinch and $H_g(1)$ modes are shown. All the C atom displacements for the $A_g(1)$ modes are radial and of equal magnitude. Data from Ref. [11.112] and Ref. [11.113] are artificially displaced to the left and right, respectively. The centered data are from Ref. [11.19].

a study of the electron–phonon interaction shows strong electron coupling for the $H_g(2)$, $H_g(7)$, and $H_g(8)$ Raman-active modes and enhanced coupling for $H_g(1)$. For other modes the broadening is so great that the lines cannot be resolved in the M_3C_{60} spectrum [11.118].

The Raman spectra for Rb_1C_{60} have been investigated by several groups [11.119, 120] and the spectra show some unusual features, not present in the corresponding spectra for the other doped M_xC_{60} compounds (see Fig. 11.25). These unusual features are related to the special structural phases of Rb_1C_{60} (see §8.5.2). At 125°C, where Rb_1C_{60} assumes a rock salt structure [11.119], the 10 Raman-active modes of C_{60} are clearly seen with little line broadening relative to the C_{60} spectrum (see Fig. 11.25). If the sample is slowly cooled to 25°C, it is believed that a polymerized phase is

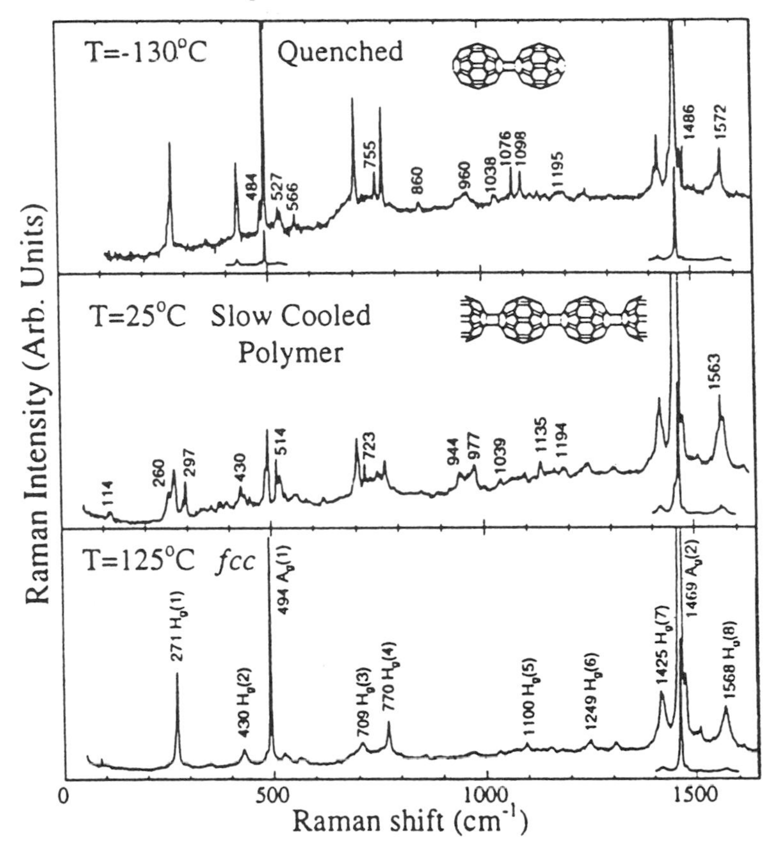

Fig. 11.25. Raman spectra of a Rb_1C_{60} sample in its fcc rock salt structure (bottom panel), its polymer state (middle), and its "quenched" state (top). The respective temperatures are indicated on the curves. The 10 Raman-active modes of pristine C_{60} and their identification are labeled in the bottom panel; the frequencies of modes activated by the reduction in symmetry are labeled for the quenched and polymer state spectra in the upper two panels. The inset sketches represent the dimer and polymer structures of the C_{60} molecules in the quenched and slow-cooled polymer states of Rb_1C_{60}, respectively [11.121].

formed (see §8.5.2); the lower symmetry of this polymerized phase causes many more modes to become Raman-active, although the observed spectrum suggests that inversion symmetry is conserved in the 25°C spectrum (see Fig. 11.25). If instead, the Rb_1C_{60} is rapidly quenched to -130°C, additional lines then appear in the Raman spectrum (see Fig. 11.25) and these additional lines are identified with the breakdown of inversion symmetry [11.121].

Resonant enhancement Raman effects have been observed using 6470 Å laser excitation. The low-frequency modes are enhanced near the main interband transitions: $h_u \rightarrow t_{1g}$ at 2.6 eV and $h_g \rightarrow t_{1g}$ at 3.5 eV (see §13.1.2). The low-frequency Raman lines show Fano lineshapes [11.19, 118] indicative of an interaction with a Raman-active continuum, most likely associated with doping-induced electronic transitions.

11.6.2. Doping Dependence of the F_{1u}-Derived Intramolecular Modes

As stated above, the four F_{1u} symmetry, IR-active, intramolecular modes for C_{60} are observed [11.81] at $\omega_1(F_{1u}) = 527$, $\omega_2(F_{1u}) = 576$, $\omega_3(F_{1u}) = 1183$ and $\omega_4(F_{1u}) = 1429$ cm^{-1} (see Table 11.1). When solid C_{60} is doped with alkali metals (K, Rb), several groups [11.77, 93, 119, 122] have found that these modes persist in the infrared spectrum, exhibiting doping-induced shifts for three of the modes (ω_1, ω_2, ω_4), whereas ω_3 is found to be largely insensitive to doping. Infrared spectra have been reported for $x = 0, 1, 3, 4, 6$ for Rb_xC_{60} [11.93, 121], and solid-state effects are found to be weak in these infrared spectra. Similar to results found for the Raman-active A_g and H_g symmetry intramolecular modes, the doping-induced changes in the IR activity or the frequencies of the F_{1u} modes do not appear to depend on the size or mass of the alkali metal dopant; i.e., the important doping effects are simply related to the number of electrons transferred to the C_{60} molecule from the alkali metal dopants.

The $F_{1u}(j)$ mode-integrated oscillator strength S_j is proportional to x^2 and is given by [11.123]

$$S_j = \int \sigma_1(\omega)d\omega \simeq (\pi/2)N\alpha\lambda_j x^2\omega_j^2 \tag{11.7}$$

where $\sigma_1(\omega)$ is the real part of the optical conductivity, N denotes the number of C_{60} molecules per unit volume, α is the contribution to the static electronic polarizability per added electron from the $t_{1u} \rightarrow t_{1g}$ electric dipole transitions, λ_j is a dimensionless coupling constant, and $\omega_j = \Omega_j(C_{60}, x = 0)$ is the unrenormalized mode frequency obtained by fitting the C_{60} spectrum. The phonon softening is linearly dependent on x and is given by

$$\Omega_j(x) \simeq \omega_j(1 - x\lambda_j/2) \tag{11.8}$$

where ω_j and Ω_j refer to the bare (pristine) and renormalized (doped C_{60}) mode frequencies, respectively. Values for Ω_j, λ_j, S_j for the four IR-active modes are given in Table 11.7 for Rb_3C_{60} and Rb_6C_{60} and for undoped C_{60} for comparison [11.124].

Optical transmission studies of saturation-doped ($x = 6$) K_xC_{60} films on salt substrates [11.127] show a large ($\sim 10^2$) doping-induced enhancement in the oscillator strength of two of the four F_{1u}-derived modes (ω_2, ω_4). Rice and Choi [11.123] pointed out the possibility that this enhancement

Table 11.7

Experimentally determined IR-active frequencies Ω_j for C_{60}, Rb_3C_{60}, and Rb_6C_{60}, their full width at half-maximum (FWHM) linewidths Γ_j (in cm^{-1}), the relative integrated oscillator intensities $S_{Rel}(\omega_j)$, and the electron–molecular vibration coupling constants λ_j (see text) for each of the four IR-active modes [11.124].

	C_{60}	Rb_3C_{60}	Rb_6C_{60}
Ω_1 (expt)	527	—	467
Γ_1	2.3	—	1.6
Ω_1 (model)	—	—	524
$S_{Rel}(\omega_1)^c$	—	—	—[d]
λ_1	—	—	0.0016 0.006[b]
Ω_2 (expt)	576	572	566
Γ_2	3.2	2.8	2.8
Ω_2 (model)	—	564	551
$S_{Rel}(\omega_2)^c$	—	—	4.2
λ_2	—	0.013 0.012[a] —	0.015 0.01[a] 0.039[b]
Ω_3 (expt)	1183	1184	1184
Γ_3	2.9	16.0	5.2
Ω_3 (model)	—	1182	1179
$S_{Rel}(\omega_3)^c$	—	—	1.08
λ_3	—	0.0017 —	0.0005 0.0005[b]
Ω_4 (expt)	1428	1365	1342
Γ_4	3.4	17.0	7.3
Ω_4 (model)	—	1412	1397
$S_{Rel}(\omega_4)^c$	—	—	4.5
λ_4	—	0.006 0.015[a]	0.008 0.016[a] 0.021[b]

[a] Experimental values reported for K_xC_{60} [11.125, 126].

[b] Values reported by Rice and Choi [11.123] and deduced from the IR transmission data of K_6C_{60} [11.127].

[c] $S_{Rel}(\omega_j)$ is defined as the relative integrated intensity $S_j(Rb_3C_{60})/S_j(Rb_6C_{60})$ for the jth IR-active mode of C_{60} and is equal to 4 in the charged phonon model of Rice and Choi [11.123].

[d] The $F_{1u}(1)$ mode in the metallic Rb_3C_{60} cannot be observed and therefore an experimental value for $S_{Rel}(\omega_1)$ cannot be obtained.

might be due to a coupling between the F_{1u} modes and electronic transitions in the M-doped C_{60} compounds (i.e., transitions between filled t_{1u}- and empty t_{1g}-derived orbitals). According to Rice and Choi, the electron–molecular vibration (EMV) coupling, in effect, allows for a transfer of oscillator strength from the electronic absorption band observed near ~1.2 eV, and identified in M_6C_{60} [11.125, 127–129] with t_{1u}–t_{1g} electronic absorption, to vibrational absorption bands identified with the otherwise weakly IR-active F_{1u} modes. Thus, the EMV coupling can be viewed as if it contributes "charge" to the mode, enhancing the dipole moment or IR activity. For moderate EMV coupling, these modes are referred to as "charged" phonons [11.123]. In the limit of very weak EMV or electron–phonon coupling, the phonon or vibration is not termed "charged," but the EMV mechanism may still induce a"dynamic" dipole moment responsible for IR-mode activity, even though no static dipole exists in the unit cell (e.g., graphite) or in the C_{60} molecule [11.130]. Furthermore, as Rice and Choi point out, this phenomenon has been appreciated by physical chemists since 1958 [11.131] and has been applied successfully to explain the remarkable infrared activity in one-dimensional systems such as the linear chain organic semiconductors (MEM)$(TCNQ)_2$ [11.132] and (TEA)$(TCNQ)_2$ [11.133].

In their paper, Rice and Choi report the results of model calculations for the doping dependence of the F_{1u} vibrational mode oscillator strength and frequency renormalization. Assuming icosahedral symmetry for the molecular ion, the EMV coupling between the t_{1u}–t_{1g} electronic transitions and the $F_{1u}(T_{1u})$ modes is symmetry allowed; i.e., the direct product $T_{1u} \otimes T_{1g} = A_u + T_{1u} + H_u$ contains $F_{1u}(T_{1u})$ symmetry. For simplification, Rice and Choi neglect weak electronic and vibrational coupling between C_{60}^{x-} molecules in the M_xC_{60} lattice and obtain, in the molecular limit and in the limit of weak EMV coupling, the results: (1) The $F_{1u}(j)$ mode oscillator strength $S_j(x)$ is proportional to x^2 [i.e., $S_j(x) \sim \lambda_j x^2$] where x denotes the charge transfer. (2) The frequency renormalization is linear in x or $\Delta\omega \sim \lambda_j x$, where λ_j is the EMV coupling constant for the mode j. In the molecular limit, x electrons occupy the lower-lying t_{1u} orbital, which is full (with six electrons) at the saturation doping limit ($x = 6$). Furthermore, it is noted that the result $S_j(x) \sim x^2$ ignores any intrinsic IR activity, that might be expected in the absence of an EMV effect. Such intrinsic IR activity might stem, for example, from a static interaction between the C_{60}^{x-} ion electronic orbitals and the electric field from the M^+ sublattice.

Since the original work on the saturated K- and Rb-doped compounds [11.127], several groups have carried out experiments to investigate the doping dependence of the F_{1u}-derived modes [$F_{1u}(i)$, $i = 1, 2, 3, 4$]. Kuzmany and co-workers [11.77, 125, 126, 134] obtained qualitative information from reflectance studies of ~1-μm-thick C_{60} films deposited on Si substrates;

the films were K doped *in situ* in a vacuum of 10^{-7} torr by repeated exposure to the vapors of the alkali metal. These sequential doping steps were continued until "spectroscopically clean" phases were obtained and equilibrated. Using this procedure, they reported the observation of spectra for the $x = 1, 3, 4, 6$ phases, the $x = 1$ cubic phase being stable only at elevated temperatures. Three of the four F_{1u} modes were found to exhibit a mode-softening effect that is linear in x, consistent with the theory of Rice, whereas the ω_3 mode exhibited little or no sensitivity to doping. No values for the oscillator strengths were obtained from the data, but the magnitude of the change ΔR in the reflectance associated with the $F_{1u}(2)$ and $F_{1u}(4)$ modes appears to indicate that a strong enhancement of the oscillator strength was observed for these modes, consistent with the charged-phonon model of Rice and Choi [11.123].

Since linear mode softening was also observed for some of the H_g (Raman-active) modes in M_xC_{60} [11.19], linear mode softening, by itself, should not be considered as strong evidence for "charged phonons" in M_xC_{60}. Furthermore, a static, doping-induced expansion of the intramolecular C–C bonds could also contribute to the mode softening, similar to that observed for the Raman-active, pentagonal pinch $A_g(2)$ mode, which in C_{60} is observed at 1469 cm^{-1}.

Two recent optical transmission studies were carried out [11.80, 93] on K- and Rb-doped films and these studies arrived at quantitative values for the oscillator strengths of the F_{1u} modes. Martin *et al.* [11.93] carried out transmission experiments similar to the original experiments of Fu *et al.* [11.127], but as a continuous function of doping. Similar to Kuzmany and co-workers [11.77, 134], Martin *et al.* [11.93] made an *in situ* observation of the four-point probe resistivity of their films. The spectra of Martin *et al.* [11.93] were taken with the substrate maintained at ~100°C, and as a result, their frequencies may exhibit a several wave number, temperature-induced, shift relative to data taken by others at room temperature. Since the doping dependence is much larger than the temperature dependence, the x dependence of the mode softening can still be obtained from their study. A series of spectra showing the evolution of the F_{1u} modes as the film is doped with Rb *in situ* are presented in Fig. 11.26. The data are exhibited in two panels, a lower-frequency panel containing ω_1 and ω_2 data and a higher-frequency panel containing ω_3 and ω_4 data. The spectrum labeled "min" corresponds to the dopant concentration where the four-point probe electrical resistivity was observed to be a minimum; presumably this trace corresponds to a value of $x \sim 3$. The frequency downshift and increased IR activity (oscillator strength) for the ω_2 and ω_4 modes are readily apparent in the figure, as is the lack of sensitivity of ω_3 to the Rb doping.

In the infrared studies by Rao *et al.* [11.80], the sample preparation was as follows. A three-layer C_{60}–M–C_{60} sandwich was vacuum-deposited onto

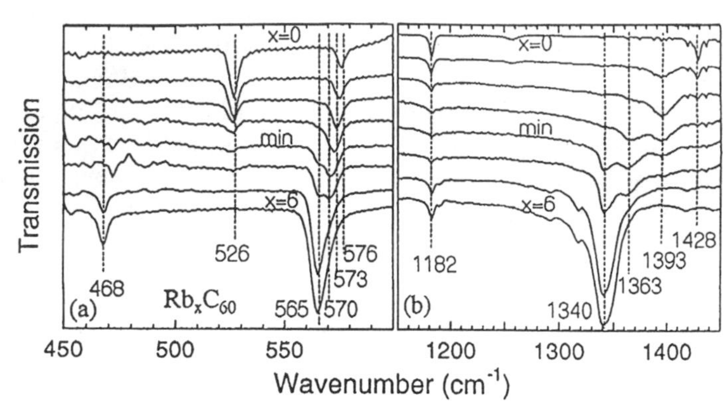

Fig. 11.26. Infrared transmission spectra showing the evolution of the F_{1u} modes as the C_{60} films are doped *in situ* with Rb. The left-hand panel shows the evolution of the lower-frequency ω_1 and ω_2 modes while the right panel shows the evolution of the higher-frequency ω_3 and ω_4 modes. The trace labeled "min" corresponds to the minimum-resistivity Rb concentration, presumably at $x = 3$ [11.93].

a salt substrate, which was removed from the deposition chamber under an Ar atmosphere and loaded immediately into a small, stainless steel IR cell, equipped with an ion pump. The samples were maintained thereafter at 10^{-5} torr during initial Raman experiments to determine the doping level and during subsequent detailed IR transmission studies. In this way, Rao *et al.* prepared $x = 3$ and $x = 6$ alkali metal–doped samples, as determined by the x dependence of the $A_g(2)$ pentagonal pinch mode frequency (see Fig. 11.24). Both Rao *et al.* [11.80] and Martin *et al.* [11.93] obtained values for the oscillator strength $S_j(x)$ by fitting a Lorentz oscillator dielectric function model to the optical transmission data (see Table 11.7). The ω_2 and ω_4 modes exhibit oscillator strength enhancements proportional to x^2, in reasonable agreement with the charged phonon model. The microscopic reasons why the oscillator strength for ω_1 does not appear to follow an x^2 dependence, or why ω_3 shows no x dependence in the M_xC_{60} compounds, have not yet been clarified. Plots of $S_j(x)$ *vs.* x do not exhibit any significant dependence on the particular alkali metal dopant (K or Rb), indicating that the C_{60} anions are only weakly coupled to the cations.

In Fig. 11.27 a summary of the currently available data [11.77, 80, 93] is displayed for the x dependence of the four F_{1u} mode frequencies. The data for both Rb- and K-doped films are plotted on the same graphs, since the mode frequencies are insensitive to the particular dopant. In Fig. 11.27, ω_1, ω_2, and ω_4 exhibit a linear frequency softening with increasing doping concentration x, whereas ω_3 appears insensitive to doping. The values of the slope [i.e., the wave number (cm^{-1}) change in mode frequency per added electron to the C_{60} molecule] are given for each plot in terms of

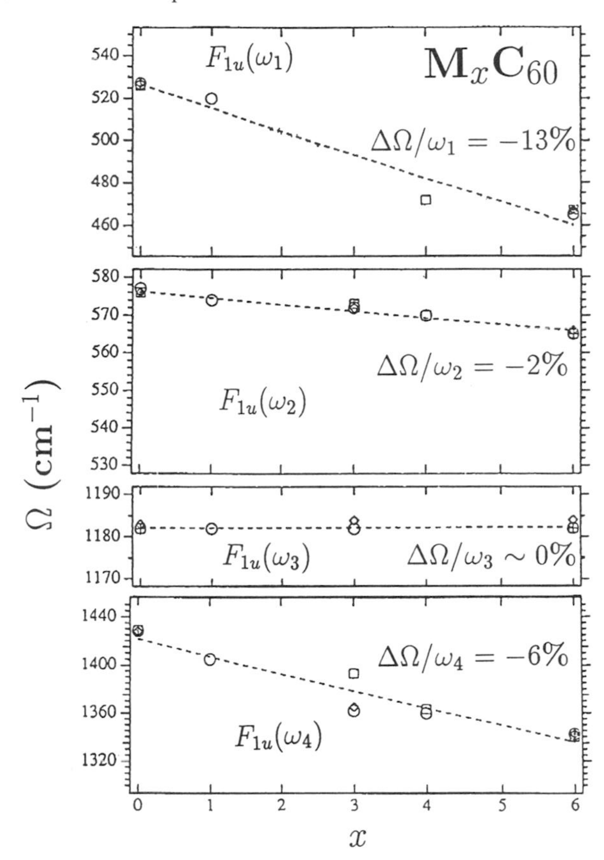

Fig. 11.27. Plots of the dependence of the four $F_{1u}(j)$ mode frequencies for K_xC_{60} and Rb_xC_{60} on dopant concentration x [11.93]. This plot includes data of Rao *et al.* (open diamonds) [11.80], Martin *et al.* (open squares) [11.93], Kuzmany *et al.* (open circles) [11.77], and Fu *et al.* (pluses) [11.127].

the percentage frequency change between $x = 0$ and $x = 6$. As the system cools below the phase transition, phase separation effects are observed in the infrared spectra [11.121, 135–137] for Na_1C_{60}, K_1C_{60}, and Rb_1C_{60}. Additional lines in the infrared spectra are identified with the breakdown of inversion symmetry in the dimer low-temperature quenched phase, as indicated in the inset of Fig. 11.25 [11.121]. Other experiments showing this phase transition include electron spin resonance (ESR) [11.138] and differential scanning calorimetry [11.119]. Crystallographic studies [11.139] have shown that the low-temperature phase of Rb_1C_{60} is orthorhombic with an unusually short separation of 9.1 Å between the centers of C_{60} molecules along the crystallographic *a* direction (see §8.5.2).

11.6.3. Other Doping-Dependent Intramolecular Effects

High energy electron energy loss spectroscopy (HREELS) has also been used to explore the doping dependence of the vibrational spectrum of M_xC_{60} [11.140]. EELS measurements done with a primary electron beam of 2.9 eV on Rb_xC_{60} samples with nominal compositions $x = 3, 4$, and 6 (although the near-surface composition may have been alkali metal deficient) show a strong attenuation in signal for the $x = 3$ stoichiometry for all the vibrational modes, and this attenuation was attributed to screening effects. Mode intensities for $x = 4$ and $x = 6$ are comparable to those for C_{60}, indicative of their semiconducting properties. Modes $F_{1u}(2)$ and $F_{1u}(4)$ could be followed as a function of alkali metal doping, while $F_{1u}(1)$ and $F_{1u}(3)$ could not be well resolved for $x = 6$, where the best EELS spectra for the doped samples were observed. The HREELS results are qualitatively similar to the infrared results summarized in §11.6.2, although significant quantitative differences between the IR and HREELS spectra have been reported [11.140].

11.6.4. Doping-Dependent Intermolecular Effects

A limited amount of information is now available through inelastic neutron scattering measurements on librations in the doped C_{60} compounds K_3C_{60} [11.141], Rb_3C_{60}, [11.72], $Rb_{2.6}K_{0.4}C_{60}$ [11.55], and $Na_2Rb_1C_{60}$ [11.142], using polycrystalline samples. Representative spectra for polycrystalline K_3C_{60} for temperatures from 12 to 675 K are shown in Fig. 11.28, where the dashed lines show the librational contribution at low energy. In contrast to the corresponding observation in C_{60}, only a small mode softening and broadening is observed in K_3C_{60} with increasing T over this wide temperature range, and these temperature-related effects are attributed to dispersion in the librational modes [11.55, 141]. The measurements further show little change in librational mode frequency or linewidth as T is lowered below the superconducting transition T_c, although a small sharpening of the librational peak below T_c may indicate some small librational contribution to the electron–phonon interaction [11.141]. The persistence of the librational mode to 675 K in Fig. 11.28 indicates that the C_{60} molecules in K_3C_{60} do not undergo rotational diffusion at high T, but rather continue to librate, consistent with the orientational alignment of the C_{60} molecules in K_3C_{60}. For a librational energy of 3.5 meV (or 28 cm^{-1}) at 300 K for C_{60} molecules in K_3C_{60} (see Fig. 11.28), the activation energy V_A for rotational jumps is estimated as $V_A \simeq 520$ meV for 44.5° jumps in the rotational potential, which is in reasonable agreement with estimates for V_A from other experiments. The observed temperature dependences of the librational modes for $Rb_{2.6}K_{0.4}C_{60}$ and K_3C_{60} are similar with regard to mode frequency and

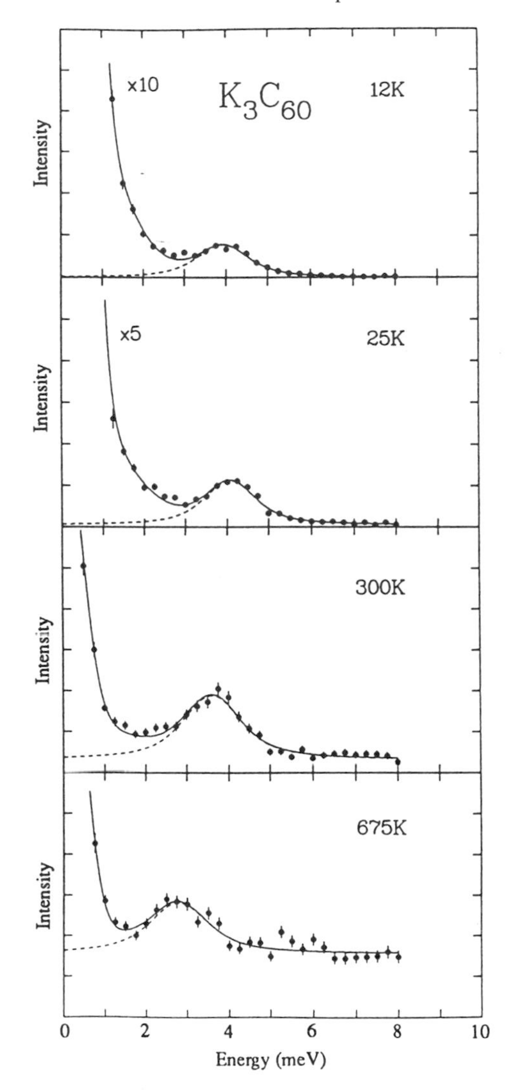

Fig. 11.28. Inelastic neutron scattering spectra for K_3C_{60} at $Q = 5.72$ Å^{-1}, at several temperatures. Solid circles are the experimental points and solid lines are best fits to the data. The dashed line is the small energy limit of the Lorentzian component which arises from librations. Note changes in the intensity scale [11.55].

linewidth [11.55]. In contrast, the librational energy for $Na_2Rb_1C_{60}$ shows a strong temperature dependence [11.142], with a strong decrease in frequency and a striking increase in linewidth with increasing T, which is attributed to the $Pa\bar{3}$ structure at low T for Na-containing compounds, in contrast to the fcc structure for K_3C_{60}.

A comparison between the librational mode frequency and its temperature dependence for various alkali metal-doped C_{60} compounds has been used to show the dependence of the rotational potential on alkali metal

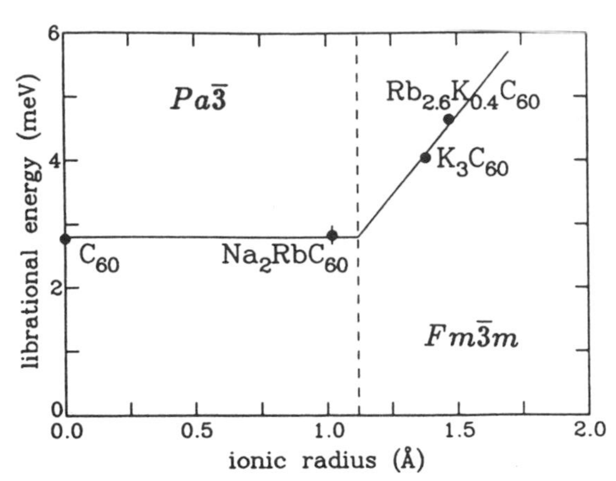

Fig. 11.29. Schematic diagram showing the dependence of the librational energy at low temperature on the ionic radius of the alkali metal ions occupying the tetrahedral interstices in M_3C_{60} fullerides: C_{60} [11.52], K_3C_{60} [11.141], $Rb_{2.6}K_{0.4}C_{60}$ [11.55], and $Na_2Rb_1C_{60}$ [11.142]. The dashed line at 1.12 Å corresponds to the size of the tetrahedral cavity in pristine C_{60} and separates simple cubic $Pa\bar{3}$ from fcc ($Fm\bar{3}m$) solids.

species (see §7.1.5). The experimental results are shown in Fig. 11.29 and are related to structural considerations which are determined by the orientational potential. When the size of the alkali metal ion in a tetrahedral site is equal to or less than the size of the site (1.12 Å), the behavior of the librations is similar to that of C_{60}, consistent with observations on $Na_2Rb_1C_{60}$ [11.142] and their $Pa\bar{3}$ crystal structure.

However, for K_3C_{60} and $Rb_{2.6}K_{0.4}C_{60}$, where the ionic size exceeds 1.12 Å, the crystal structure remains fcc. For these compounds the inelastic neutron scattering studies show larger librational energies which increase with the size of the tetrahedral ion, demonstrating that the repulsive part of the alkali metal–carbon (A–C) interaction is crucial for determining the orientational potential for K_3C_{60} and Rb_3C_{60}. In contrast, this repulsive term is not so important for the Na_xC_{60} compounds, and the orientational potential is most sensitive to the interaction between the C_{60} molecular anions as in the case of C_{60} itself, favoring a nesting of the double bonds between two hexagons on one C_{60} molecule opposite pentagons and hexagons on adjacent C_{60} molecules. When the alkali–carbon (A–C) repulsive interaction (due to the large size of the alkali metal ion) becomes important, the minimum in the orientational potential is achieved when the hexagonal faces of the C_{60} molecules face the tetrahedral sites, thereby maximizing the distance between the alkali metal and carbon sites for a given lattice constant, in agreement with molecular dynamics calculations [11.71].

11.7. Vibrational Spectra for C_{70} and Higher Fullerenes

Except for C_{70}, relatively little is known about the detailed vibrational spectra of the higher fullerenes. Because of their generally lower symmetry, larger number of degrees of freedom, and the many possible isomers, little systematic experimental or theoretical study has been undertaken on high-mass fullerene molecules or crystalline solids. The lack of adequate quantities of well-separated and characterized samples of higher fullerenes has curtailed experimental studies. For these reasons, we focus our attention on vibrations in C_{70} which have now been studied both theoretically and experimentally by Raman and infrared spectroscopy and by inelastic neutron scattering studies. Most of the studies on C_{70} thus far have been on the intramolecular vibrations, although a few studies on the intermolecular vibrations of C_{70} have more recently been carried out. The symmetry of the C_{70} molecule and of its molecular vibrations is discussed in §4.4.1 and symmetry tables useful for the discussion of molecular vibrations are Tables 4.12–4.24. Section §4.4.1 also provides symmetry information useful for describing the vibrations in higher-mass fullerenes.

11.7.1. Intramolecular Vibrational Spectra for C_{70}

The Raman and infrared spectra for C_{70} are much more complicated than for C_{60} because of the lower symmetry and the large number of Raman-active modes (53) and infrared-active modes (31), out of a total of 122 possible vibrational mode frequencies for C_{70}.

A number of *ab initio* calculations have been carried out using several different approaches [11.23, 24, 37, 143–146], reporting results for the 122 eigenfrequencies and normal modes. A phenomenological force constant model [11.88, 147] has also been published and the results for the predicted eigenfrequencies have been used in the interpretation of the experimental Raman and infrared spectra. In all models, the D_{5h} symmetry of C_{70} is explicitly treated.

Using the simplest possible approximation, the phenomenological treatment [11.147] assumes the same set of force constants for C_{70} as was used in the case of C_{60} [11.22], except that a perturbation is introduced to reduce the shear forces between the 10 additional belt atoms and the atoms on the adjacent rings. This perturbation is shown to lead to an improved set of calculated eigenfrequencies for C_{70}, while introducing only two additional parameters. This simple model can account for many of the observed Raman and infrared spectral features [11.147].

The D_{5h} symmetry for the C_{70} molecule implies that there are five inequivalent atomic sites on the C_{70} molecule (see § 3.2), which, in turn, allows for the existence of eight different bond lengths as well as 12 different angles between the bonds connecting nearest-neighbor atoms [11.143]. Thus, if only nearest-neighbor interactions are considered in constructing the dynamical matrix, the lower symmetry of the C_{70} molecule requires consideration of eight different bond-stretching and 12 different angle-bending force constants. In the simplified phenomenological treatment described above, only two distinct bond-stretching and two distinct angle-bending force constants are considered for bonds connecting nearest-neighbor atoms in C_{70} [11.147]. In this approximation, each bond is classified as either long or short, depending on whether its length is longer or shorter than the average bond length. Similarly, an angle is classified as belonging to either a pentagon or a hexagon. These simple approximations are sufficient to provide a qualitative description of the vibrational frequencies in C_{70}.

Well-resolved infrared spectra [11.8, 88] and Raman spectra [11.44, 81, 88, 148–150] have been observed experimentally for C_{70}, as shown in Fig. 11.30. For both the infrared and Raman spectra, many more lines are observed for C_{70} relative to C_{60} (see Fig. 11.11), consistent with the lower symmetry and larger number of degrees of freedom for C_{70}. Using polarization studies and a force constant model calculation [11.88, 147], a preliminary assignment of the mode symmetries has been made. A listing of the calculated C_{70} mode frequencies is given in Table 11.8 [11.147].

In this table, the mode frequencies are labeled by their irreducible representation of group D_{5h}, and modes associated predominantly with belt atom displacements are also labeled. It is found that mode frequencies below ~ 900 cm^{-1} tend to have predominantly radial displacements, while the higher-frequency modes have predominantly tangential displacements.

11.7.2. Intermolecular Vibrational Spectra for C_{70}

Inelastic neutron scattering measurements similar to those discussed in §11.3 for C_{60} have been carried out on C_{70} to study the librational spectra for solid C_{70} [11.72, 152] below the orientational phase transition at T_{01}. The study in C_{70} is more difficult to carry out because of the anisotropy of the C_{70} molecule, the lower crystal symmetry of the low-temperature phase, and the difficulty in obtaining single-crystal C_{70} samples of sufficient purity and size for detailed studies. The measurements thus far have been done on polycrystalline samples of reasonably good quality.

The inelastic neutron scattering measurements at 10 K show a broad peak centered at 1.82 meV (14.7 cm^{-1}) for the lowest librational energy with a full width at half-maximum (FWHM) linewidth of 1.8 meV [11.152]. This librational frequency is significantly lower than for the corresponding mode at 2.77 meV (22.4 cm^{-1}) for C_{60}, which has a much smaller FWHM linewidth of 0.38 meV (3.1 cm^{-1}) at 20 K. The softening of the librational mode between the lowest temperature (10–20 K) and 250 K (near the orientational phase transition) is much smaller in C_{70} ($\sim$17%) than in C_{60} ($\sim$35%). Since the authors could not explain the observed linewidth for

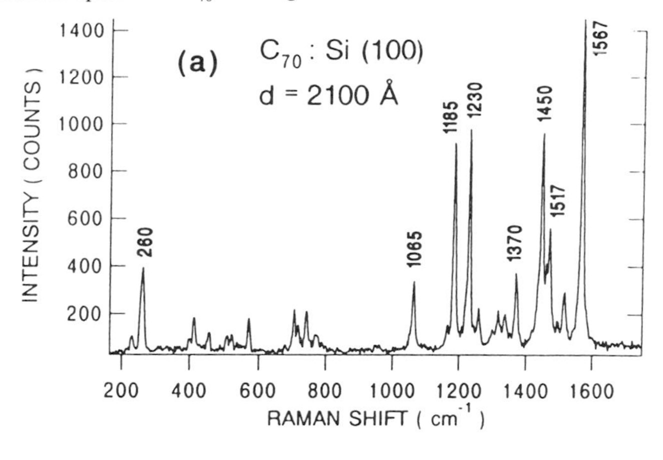

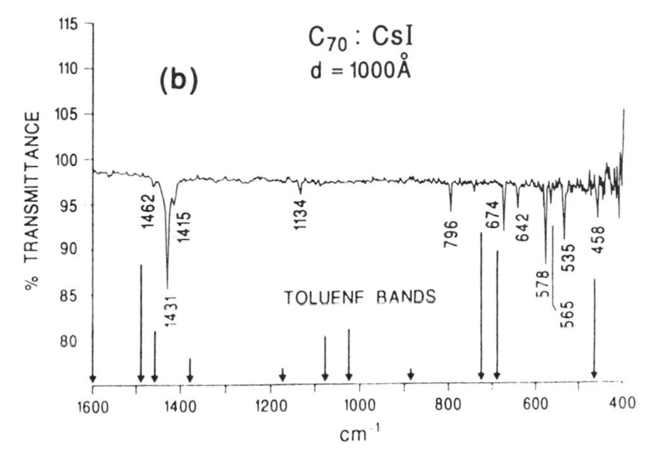

Fig. 11.30. (a) Room temperature first-order Raman spectrum for a C_{70} film on an Si (100) substrate [11.88]. The Raman peak at ≈520 cm^{-1} is associated with the Si substrate. (b) Room temperature first-order infrared spectrum of a C_{70} film on CsI. The arrows at the bottom of (b) denote infrared-active toluene modes, with the arrow length indicative of the mode intensity [11.8, 81, 151].

C_{70} in terms of the difference in the moments of inertia parallel and perpendicular to the fivefold axis of C_{70} (see Table 3.2), they suggested that the orientational potential for C_{70} was weaker and more anisotropic than for C_{60}.

A molecular dynamics calculation of the intermolecular vibrations for C_{70} [11.153] yields a low energy for the lowest peak in the density of states (∼1.2 meV) and a large linewidth (∼0.9 meV), consistent with the experiments, but the calculated values are at considerably lower energy than

Table 11.8

Calculated modes for C_{70} and M_6C_{70} by force constant model [11.147].

Frequencies of the modes, even under reflection, expressed in cm^{-1}							
A_1' modes		A_2' modes		E_1' modes		E_2' modes	
C_{70}	M_6C_{70}	C_{70}	M_6C_{70}	C_{70}	M_6C_{70}	C_{70}	M_6C_{70}
230[a]	269[a]	558	536	389[a]	328[a]	260	247
308	294	563[a]	540	397	381	348[a]	317[a]
475	437	618	563[a]	420	397	415[a]	371[a]
536	462	705[a]	619[a]	468	409	494	451
737	678	910	867	550[a]	503[a]	508[a]	456[a]
825	814	1123	1066	574	518	620	565
1160	1156	1226	1201	623	613	685	626
1253	1260	1396	1340	719	628	705	634
1290	1288[a]	1512	1437	744	684	809	769
1335[a]	1302			809	782	850	798
1483	1438			982	939	856	838
1585	1495			1026	1008	980	923
				1187[a]	1168[a]	1070	1058
				1213	1209	1182	1186
				1262	1259	1203	1193
				1316[a]	1282[a]	1305	1285
				1375	1325	1325[a]	1319
				1377	1356	1350	1326[a]
				1430	1387	1414	1340
				1503	1427	1415	1365
				1563	1473	1548	1469
						1571	1473

continues

the experimental values. As better samples become available for neutron scattering studies, it is expected that more detailed information on the intermolecular vibrations will become available.

11.7.3. Vibrational Modes in Doped C_{70} Solids

Studies of the vibrational modes of doped solid C_{70} have focused on the fully saturated insulating alkali metal compound M_xC_{70} ($x \sim 6$) and on the maximum-conductivity phase ($x \sim 4$). Only a small number of reports are presently available for the vibrational properties of doped M_xC_{70}.

Regarding the insulating M_6C_{70} phase (M = K, Rb, Cs), the Raman spectra shown in Fig. 11.31(b) exhibit many more spectral features than the corresponding spectrum for M_6C_{60}, which is shown in Fig. 11.31(a) [11.148]. A comparison between the Raman features for the M_xC_{70} and the pristine C_{70} film allows identification of the upshifts and downshifts of many of the

Table 11.8 *continued*

Frequencies of the modes, odd under reflection, expressed in cm^{-1}							
A_1'' modes		A_2'' modes		E_1'' modes		E_2'' modes	
C_{70}	M_6C_{70}	C_{70}	M_6C_{70}	C_{70}	M_6C_{70}	C_{70}	M_6C_{70}
363	350	374	349	285	271	358	340
597	543	525	444	460	419	458	428
742	670	668	664	555	510	466	438
784	744	741	678	592	551	563	515
979	934	1053	1051	632	590	656	603
1046	1003	1202	1190	716	655	703	668
1301	1280	1271	1269	796	735	768	707
1367	1338	1450	1385	839	825	803	760
1560[a]	1452[a]	1508[a]	1453[a]	874	847	859	799
		1588	1496	955	922	979	939
				1172	1167	1054	1043
				1202	1200	1179	1167
				1253	1241	1227	1225
				1345	1320	1264	1273
				1384	1349	1348	1322
				1422	1356	1371	1336
				1494[a]	1427[a]	1437	1366
				1540[a]	1445[a]	1502[a]	1415[a]
				1586	1485	1575	1482
						1591[a]	1490[a]

[a] Identified as a belt mode.

modes observed for the M_6C_{70} compounds. Furthermore, the Raman spectra for K_6C_{60}, Rb_6C_{70}, and Cs_6C_{70} show many common features [11.154], as do the corresponding M_6C_{60} Raman spectra [11.3, 15, 19], as previously noted. To aid with the mode identification, the calculated mode frequencies for M_6C_{70} (see Table 11.8) could be used [11.147], but the agreement is only fair between the observed and predicted mode frequencies. The observed Raman mode frequency shifts for the M_6C_{70} films relative to C_{70} show a strong correlation with the corresponding behavior of the M_6C_{60} films relative to C_{60} [11.154].

In addition, the Raman spectra of the maximum-conductivity K_4C_{70} phase were investigated to provide further evidence for enhanced electron–phonon coupling for a half-filled band (see §12.7.5). The experimental results for K_4C_{70} (Fig. 11.32) show only a small amount of line broadening [11.148] relative to the behavior in the half-filled t_{1u} band compound K_3C_{60}. In addition, a comparison between the Raman spectra of K_4C_{70} and C_{70} (see Fig. 11.32) allows identification of the experimental frequency shifts of the various modes upon alkali metal doping. Assuming that the electron–phonon coupling strength is comparable to the Coulomb repulsion, the

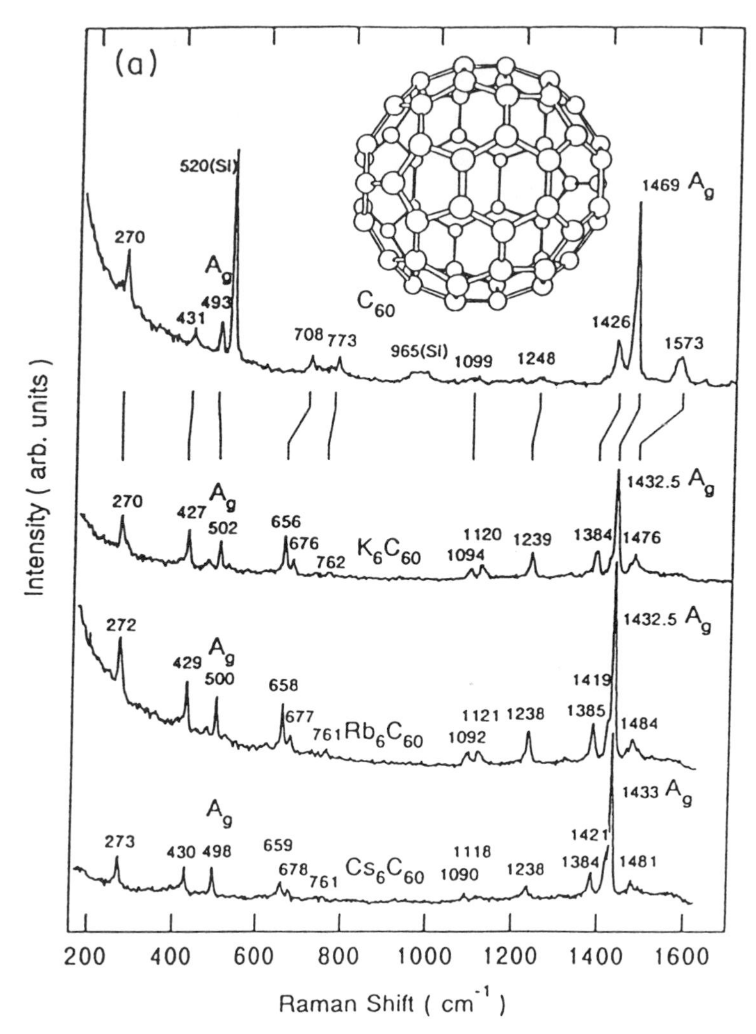

Fig. 11.31. Unpolarized room temperature Raman spectra (resolution ~6 cm^{-1}) for (a) C_{60} and M_6C_{60} and (b) C_{70} and M_6C_{70} (M = K, Rb, Cs) thin films on Si (100). The insets in (a) and (b) are models for the C_{60} and C_{70} molecules, respectively. The solid lines between the C_{60} and K_6C_{60} spectra in (a) [or the C_{70} and K_6C_{70} spectra in (b)] indicate schematically the charge transfer–induced frequency shifts due to the M dopant. As a guide to the eye, a portion of the K_6C_{70} spectrum ($\omega > 1100$ cm^{-1}) in (b) is replotted with an increased frequency of 50 cm^{-1} so that the correspondence between the tangential mode frequencies in pristine C_{70} and in K_6C_{70} can be noted [11.19, 154]. The symbols over the features in the trace for C_{70} (b) are identified in Ref. 11.154.

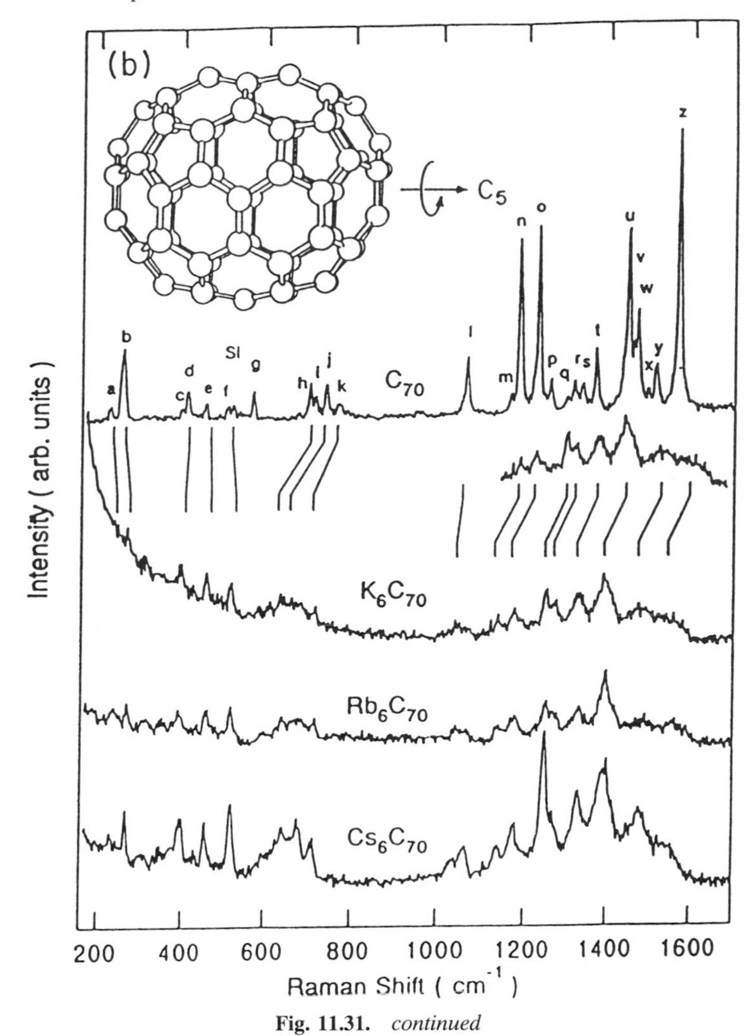

Fig. 11.31. *continued*

weaker electron–phonon coupling in K_4C_{70} relative to K_3C_{60} is believed to be responsible for the absence of superconductivity in K_4C_{70} down to 1.35 K [11.148].

11.8. Vibrational Spectra for Phototransformed Fullerenes

Because of the symmetry lowering of the phototransformed fullerene phases, Raman and infrared spectroscopies provide powerful tools for the characterization of the phototransformation process. Although Brillouin

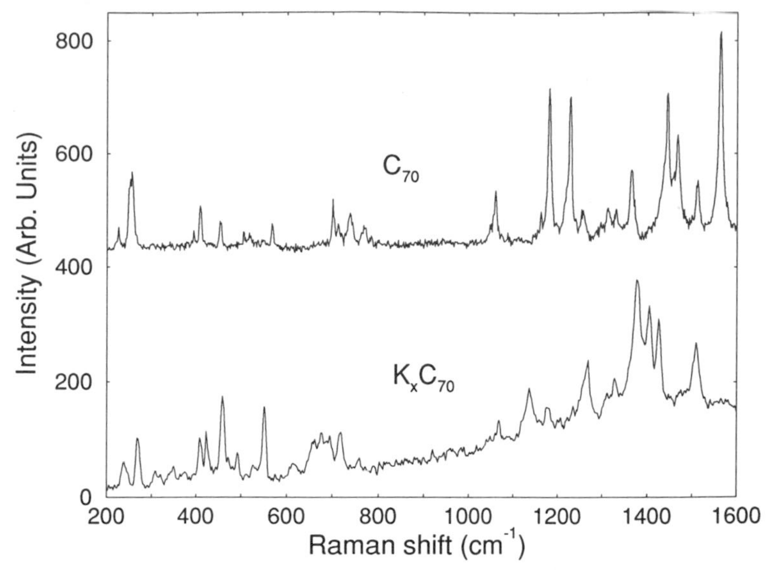

Fig. 11.32. Room temperature Raman spectra of pristine C_{70} and K_xC_{70} at the maximum-conductivity phase (nominally $x = 4$ in K_xC_{70}). The film thickness is ~2500 Å [11.148].

scattering can in principle be used for characterizing phototransformed C_{60}, the power levels required to carry out Brillouin scattering are so large that the Brillouin scattering probe itself introduces phototransformation of C_{60} [11.58]. Thus far, most of the vibrational spectra on phototransformed films have been taken on phototransformed C_{60}.

The overall effect of phototransformation on the vibrational spectra of solid C_{60} is shown in Fig. 11.33 [11.80], where the infrared transmission and Raman spectra of solid C_{60} films can be compared for the pristine fcc phase (upper panel (a)) and the phototransformed phase (lower panel (b)). Phototransformation occurs at a moderate optical flux (> 5 W/cm^2) in the visible and in the ultraviolet (UV) regions of the spectrum when light is incident on thin solid fullerite films [11.80]. One of the most important effects of phototransformation on the Raman spectrum of C_{60} is the quenching of the intensity of the $A_g(2)$ pentagonal pinch mode at 1469 cm^{-1} in pristine C_{60}, with a new mode appearing close by at 1458 cm^{-1} in the phototransformed material. The various effects observed in the Raman and infrared spectra in Fig. 11.33 have been interpreted [11.80] to be due to a photopolymerization of the lattice in which the C_{60} molecules become cross-linked by covalent bonds, rather than by the weaker van der Waals forces that normally bond C_{60} monomer molecules in the solid phase (see §7.5.1). The shift to lower frequencies of the pentagonal pinch mode has been

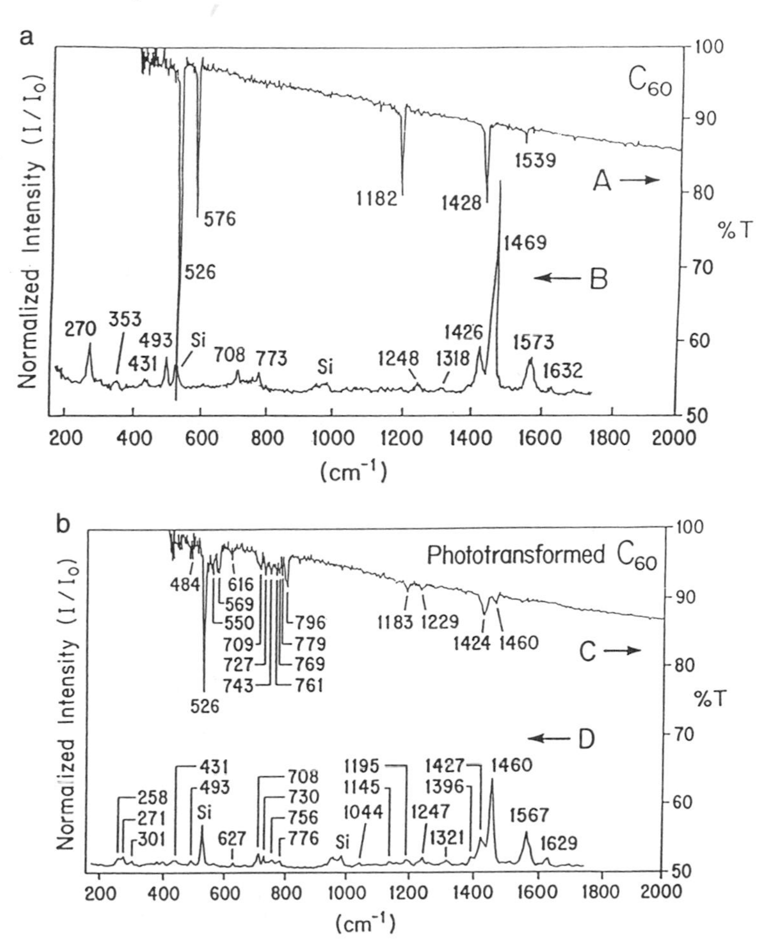

Fig. 11.33. Effect of photoinduced dimerization and polymerization on the infrared and Raman spectra of C_{60}. The upper panel (a) shows infrared (A) and Raman (B) spectra for C_{60} taken with low incident optical power levels (< 50 mW/mm^2). The lower panel (b) shows the corresponding infrared (C) and Raman (D) spectra taken after exposure of the C_{60} film to intense optical power levels for several h, until the Raman line at 1469 cm^{-1} disappeared. After the intense optical irradiation, many more infrared and Raman lines appear in these spectra, due to a lowering of the symmetry caused by the photoinduced covalent bonding between the fullerene molecules [11.80].

modeled using molecular dynamics [11.155] and is in good agreement with experiment.

In Fig. 11.34 we see the Raman spectra for polymerized C_{60} in the vicinity of the pentagonal pinch mode for various temperatures. At the lowest temperature shown (65°C), the trace exhibits the characteristic spec-

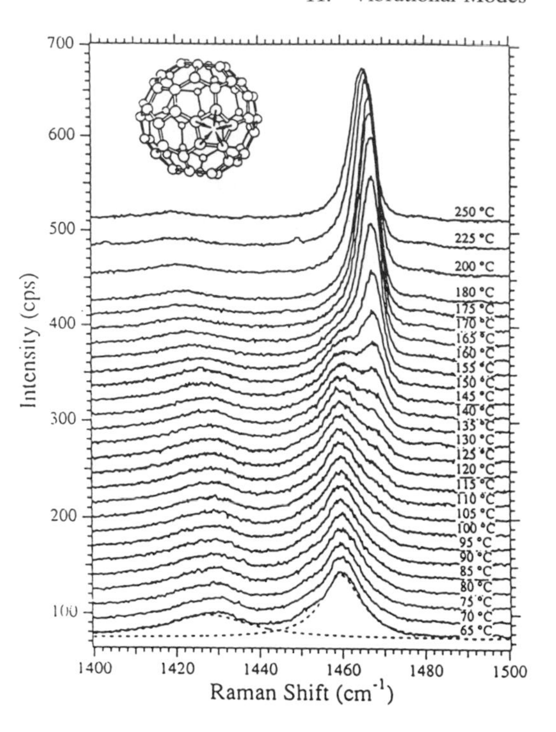

Fig. 11.34. Raman spectra in the vicinity of the pentagonal pinch mode for a photochemically polymerized C_{60} film ($d \approx 4500$ Å) on a Suprasil (fused silica) substrate for various temperatures. The dashed curves in the bottom spectrum represent a Lorentzian lineshape fit to each of the Raman lines in the spectrum at 65°C. As the temperature is increased, the features associated with the polymerized phase are quenched [11.156].

tral features of the phototransformed film, showing a broad Raman line at 1459 cm^{-1} associated with the $A_g(2)$ mode of C_{60}, perturbed by the formation of dimers, trimers, and higher oligomers [11.156]. As the temperature is increased, the narrow $A_g(2)$ line at 1469 cm^{-1} grows at the expense of the broad 1459 cm^{-1} line, indicating a transformation back to the monomeric solid by a thermal detachment of C_{60} monomers from the oligomers. The competition between the photoinduced attachment of C_{60} monomers to form oligomers and the thermal detachment process to form monomers is shown in Fig. 11.35 by plotting the temperature dependence of the integrated intensity of the 1459 cm^{-1} feature associated with the polymerized phase and the corresponding intensity of the 1469 cm^{-1} sharp Raman line associated with the C_{60} monomer [11.156]. The solid curves in this figure are a fit of the rate equations for oligomer formation by the phototransformation process which competes with the thermal dissociation process for monomer formation. Detailed fits of the model to the experimental data show that the rate constant for polymerization is dependent on the photon flux and wavelength, as indicated by the inset to Fig. 11.35 showing the dependence of the fractional monomer population (FMP) in the photopolymerized film on both temperature and incident light flux Φ_0 at 488 nm. An

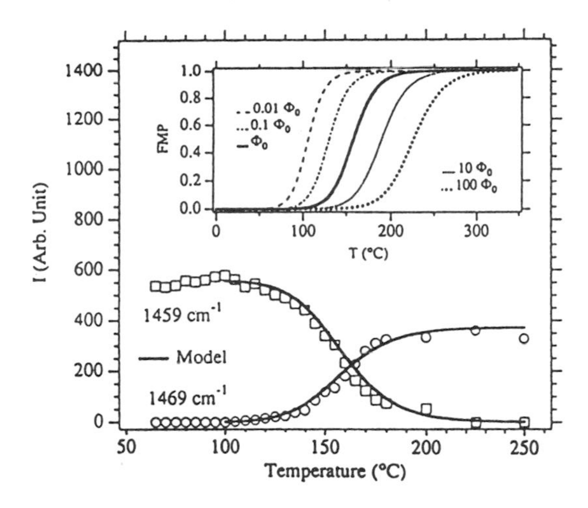

Fig. 11.35. Temperature dependence of the integrated intensity I obtained from analysis of the Raman spectra in Fig. 11.34 for the 1459 cm^{-1} (□) line (C_{60} polymer) and the 1469 cm^{-1} (○) line (C_{60} monomer) in comparison to rate equation model calculation (solid curves). Inset: the calculated fractional monomer population (FMP) in photopolymerized C_{60} as a function of temperature for several values of the incident light flux Φ_0 at 488 nm [11.156].

Arrhenius plot of the Raman intensity ratio (I_{1459}/I_{1469}) yields an activation energy of 1.25 eV for the thermal dissociation process [11.156].

Additional radial and tangential molecular modes are activated (Fig. 11.33) by the apparent breaking of the icosahedral symmetry, resulting from bonds that cross-link adjacent molecules (see Fig. 7.20). A new Raman-active mode is also observed at 118 cm^{-1} in phototransformed C_{60} [11.80], and this new mode is identified with a stretching of the cross-linking bonds *between* molecules [see Fig. 11.36] [11.156]. This frequency falls in the gap between the lattice and molecular modes of pristine C_{60} (see Fig. 11.1), and the value of this mode frequency is in good agreement with theoretical predictions for this mode at 104 cm^{-1} [11.157]. Figure 11.36 shows the temperature dependence of the intensity of the 118 cm^{-1} mode and the suppression of this mode as the thermal detachment of C_{60} monomers from oligomer clusters proceeds with increasing temperature. The decrease in intensity of the 118 cm^{-1} mode is closely correlated with the attenuation of the 1459 cm^{-1} Raman feature, discussed above and shown in Fig. 11.34 [11.156].

Because of the characteristic features of the Raman and IR spectra of phototransformed C_{60}, these spectroscopic tools are often used to characterize the phototransformed material, including the amount of phototransformation that has occurred when studies of other properties of phototransformed C_{60} are carried out.

Infrared spectroscopy has also been used to characterize C_{70} films regarding photoinduced oligomer formation [11.122]. Once again, the onset of additional spectral features and line broadening is used for the characterization of the polymerization process.

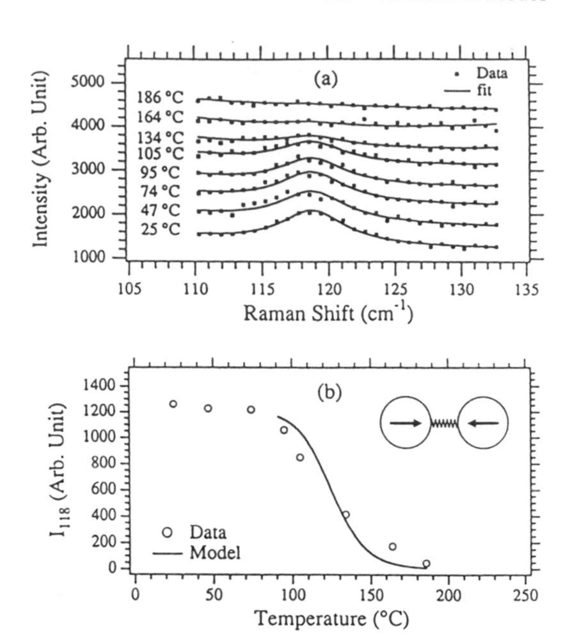

Fig. 11.36. Temperature dependence of the Raman-active 118 cm^{-1} intermolecular mode in a photopolymerized C_{60} film [$d \approx 4500$ Å, Si (100) substrate]. (a) Raman spectra taken in the 110–133 cm^{-1} range at the indicated temperatures. The solid lines are the results of a Lorentzian lineshape analysis in which the peak frequency and width were fixed at the values obtained at 25°C. (b) Integrated intensity I_{118} of the photoinduced intermolecular mode at 118 cm^{-1} *vs* temperature. The circles are obtained from a Lorentzian lineshape analysis and the solid curve is calculated from a rate equation model. The inset shows the intermolecular mode schematically [11.156].

11.9. Vibrational Spectra for C_{60} under Pressure

The pressure dependence of the vibrational spectra has been used in two ways to elucidate the phase diagram and stability of fullerenes. One approach has involved study of the effect of pressure on the ambient vibrational spectra of the fullerenes (see §11.5.1 and §11.5.2) and the second has involved use of spectroscopy to characterize some of the novel high-pressure phases of fullerenes (see §7.3).

A number of studies of the pressure dependence of the vibrational spectra have been carried out [11.78, 149], confirming the stability of the fullerene phase of carbon, providing information on the pressure dependence of the most prominent Raman-active modes, and identifying possible phase boundaries at higher pressures. Some of these studies were carried out with sufficient laser intensity to cause some phototransformation of the C_{60} films (see §11.8), resulting in photoinduced modification of the spectra in addition to pressure-induced effects.

By tracking the pressure dependence of the room temperature Raman spectra, the stability of the C_{60} phase up to 22 GPa has been confirmed [11.149], in comparison to graphite [11.158, 159]. C_{60} films show evidence for a phase transition in the 15–18 GPa range. Although C_{60} is a molecular solid with vibrational frequencies similar to those for the free molecule in solution, and the compressibility of the molecule itself is expected to be low, the pressure dependence of the vibrational modes is readily mea-

Table 11.9

Pressure dependence of selected Raman-active mode frequencies in C_{60} [11.149].

ω_i (cm^{-1})	$\partial\omega_i/\partial P$ (cm^{-1}/GPa)	Assignment
268	1.1	$H_g(1)$
421	2.4	$H_g(2)$
491	0.94	$A_g(1)$
707	−0.55	$H_g(3)$
772	−0.50	$H_g(4)$
1431	2.4	$H_g(7)$
1465	1.7	$A_g(2)$
1570	3.7	$H_g(8)$

sured, and some typical results are shown in Table 11.9. The high-frequency modes appear to harden with increasing pressure, in contrast to the low-frequency modes, some of which harden while others soften, showing less of a clear trend. Above 22 GPa, the characteristic Raman spectrum for C_{60} is no longer observed. On the basis of the vibrational spectra, an irreversible transition to an amorphous carbon phase occurs above this pressure [11.149]. The Grüneisen parameters (which describe how the frequency of a given Raman mode scales with the volume) are very small. These Grüneisen parameters are compatible with there being very little change in intramolecular bond length for a change in intermolecular spacing of crystalline C_{60} [11.78].

Similar to the spectral characteristics of the polymerized films (see §11.8), the spectral features for the high pressure–high temperature phases for C_{60} (see §7.3) also show more lines in the infrared and Raman spectra, as compared with crystalline C_{60} (see Figs. 11.37 and 11.38). Both pressure-induced forms of C_{60}, namely fcc(pC_{60}) and rh($p\,C_{60}$) (see §7.3), show structure in the infrared spectra in the 700–800 cm^{-1} range (see Fig. 11.37), similar to the behavior of polymerized films (see Fig. 11.33). The downshift of the $F_{1u}(4)$ mode from 1428 cm^{-1} in conventional fcc C_{60} to 1422 cm^{-1} in fcc(pC_{60}), to 1383 cm^{-1} for rh(pC_{60}), the line broadening, and the appearance of multiple spectral lines in the infrared spectra all suggest molecular distortions, changes in bonding, and changes in charge distribution in the pressure-induced phases of C_{60} relative to conventional crystalline C_{60} [11.160]. The Raman spectra for the pressure-induced phases (see Fig. 11.38) [taken at a low power level (2 W/cm^2) to avoid polymerization effects] likewise show broadened features, rich spectra containing many spectral features, and a downshift of the $A_g(2)$ mode from 1468 cm^{-1} in fcc C_{60}, to 1457 cm^{-1} in fcc(pC_{60}), and to 1447 cm^{-1} in rh(pC_{60}). These features in the Raman spectra further corroborate the indication that the C_{60}

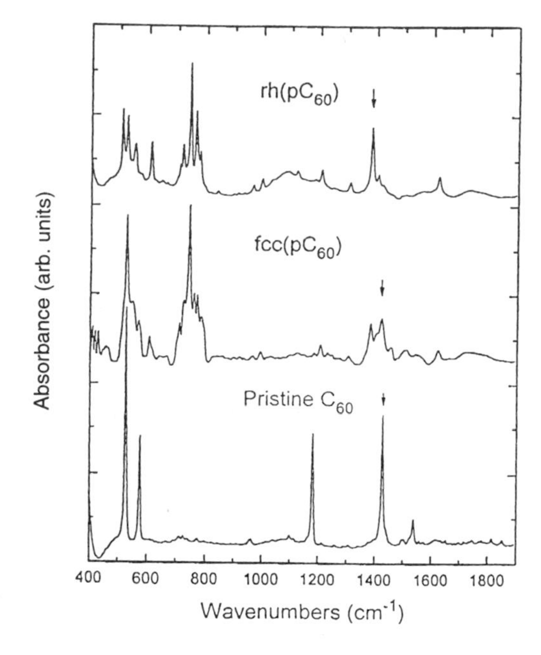

Fig. 11.37. Infrared absorption spectra of powders dispersed in KBr pellets for pristine C_{60}, and the pressure-induced phases fcc(p C_{60}), and rh(pC_{60}). The background signal has been subtracted [11.160].

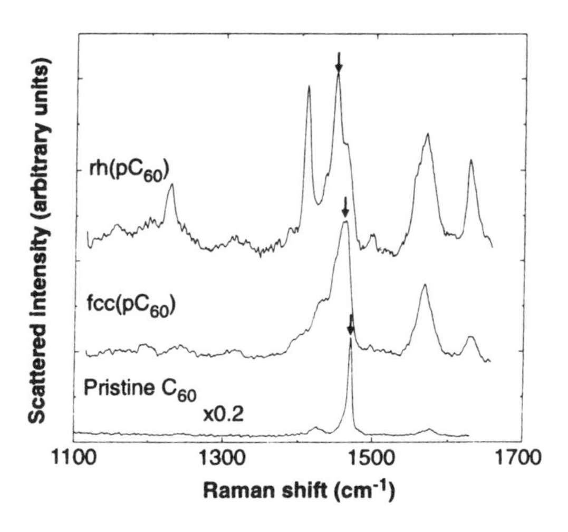

Fig. 11.38. Raman spectra of powder samples of pristine C_{60} and the pressure-induced phases fcc(p C_{60}) and rh(pC_{60}) [11.160].

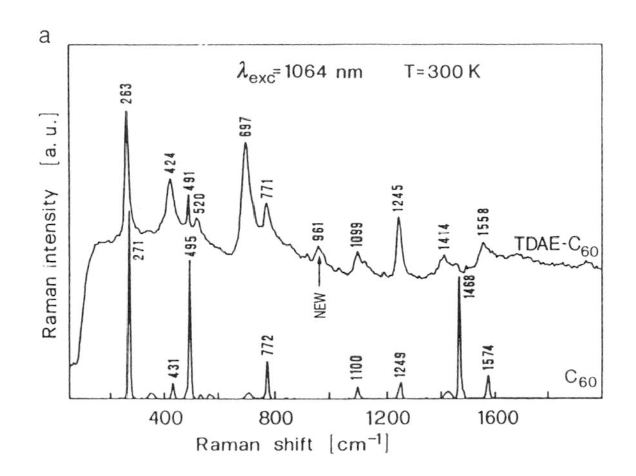

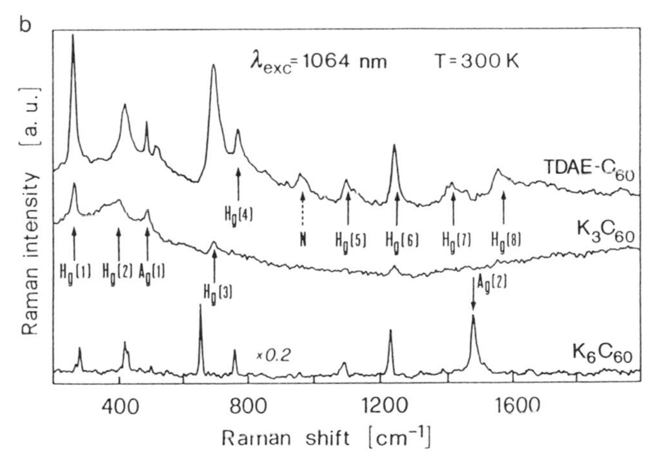

Fig. 11.39. (a) Room temperature Raman spectrum of TDAE–C_{60} at 1.16 eV laser excitation energy compared to that of undoped C_{60}. Note the small red shift of the spectral lines, the broadening of the H_g modes, and the new mode (labeled NEW) at 961 cm^{-1}. (b) Comparison of Raman spectra of TDAE–C_{60} with metallic K_3C_{60} and insulating K_6C_{60}. Qualitative similarity between the TDAE–C_{60} and the metallic K_3C_{60} spectra is observed and is attributed to charge transfer and electron–phonon coupling effects. Identifications are given in (a) for the most prominent features in the TDAE–C_{60} spectra with symmetry designations given in (b) [11.161].

molecules in the pressure-induced phases are distorted from icosahedral I_h symmetry and are cross-linked with one another [11.160].

11.10. Vibrational Spectra of Other Fullerene-Related Materials

In the section we give examples of the vibrational spectra of other fullerene-related materials, specifically TDAE–C_{60} and TDAE–C_{70} [11.161]. It is expected that the vibrational spectra for many other fullerene-based materials will be explored in the future, as the structure and properties for new fullerene derivatives are investigated.

Since TDAE–C_{60} is the organic ferromagnet with the highest transition temperature (see §18.5.2) ($T_c = 16.1$ K) [11.162, 163], many of the properties of TDAE–C_{60} have been studied, including its vibrational spectra and crystal structure (see §8.7.3). Shown in Fig. 11.39 are comparisons of the Raman spectrum of TDAE–C_{60} to those of C_{60} [Fig. 11.39(a)] and to those of K_3C_{60} and K_6C_{60} [Fig. 11.39(b)], showing significant line broadening and suppression of the $A_g(2)$ mode in TDAE–C_{60}, suggestive of significant electron–phonon coupling. A new line at 961 cm^{-1} is also seen in the TDAE–C_{60} spectrum [11.161]. The mode shifts of the H_g and A_g modes are quite similar to those for K_3C_{60} [11.164] but the broadening and damping of the Raman features are much less for TDAE–C_{60}. Studies of the Raman effect as a function of incident laser power show evidence for a phototransformed phase in this material [11.164]. Infrared measurements show negligible mode shifts or splittings for the IR modes, except possibly for the mode at 1428 cm^{-1} [11.164].

The Raman spectrum for TDAE–C_{70} shows much narrower Raman features than for TDAE–C_{60}, suggesting lower electron–phonon coupling for TDAE–C_{70}. The lower-frequency intramolecular modes for TDAE–C_{60} are significantly enhanced by cooling to low temperature (10 K), while air exposure of TDAE–C_{60} strongly damps all the A_g and H_g modes and gives rise to a new broad Raman feature at 1848 cm^{-1} [11.161, 165].

References

[11.1] A. F. Hebard. *Phys. Today*, **45**, 26 (1992). November issue.

[11.2] R. C. Haddon, A. F. Hebard, M. J. Rosseinsky, D. W. Murphy, S. J. Duclos, K. B. Lyons, B. Miller, J. M. Rosamilia, R. M. Fleming, A. R. Kortan, S. H. Glarum, A. V. Makhija, A. J. Muller, R. H. Eick, S. M. Zahurak, R. Tycko, G. Dabbagh, and F. A. Thiel. *Nature (London)*, **350**, 320 (1991).

[11.3] K. A. Wang, Y. Wang, P. Zhou, J. M. Holden, S. L. Ren, G. T. Hager, H. F. Ni, P. C. Eklund, G. Dresselhaus, and M. S. Dresselhaus. *Phys. Rev. B*, **45**, 1955 (1992).

[11.4] G. Meijer, D. S. Bethune, W. C. Tang, H. J. Rosen, R. D. Johnson, R. J. Wilson, D. D. Chambliss, W. G. Golden, H. Seki, M. S. de Vries, C. A. Brown, J. R. Salem, H. E. Hunziker, and H. R. Wendt. In R. S. Averback, J. Bernholc, and D. L. Nelson (eds.), *Clusters and Cluster-Assembled Materials, MRS Symposia Proceedings*, vol. 206, p. 619, Materials Research Society Press, Pittsburgh, PA (1991).

[11.5] R. E. Stanton and M. D. Newton. *J. Phys. Chem.*, **92**, 2141 (1988).

[11.6] F. Negri, G. Orlandi, and F. Zerbetto. *Chem. Phys. Lett.*, **144**, 31 (1988).

[11.7] D. E. Weeks and W. G. Harter. *J. Chem. Phys.*, **90**, 4744 (1989).

[11.8] D. S. Bethune, G. Meijer, W. C. Tang, and H. J. Rosen. *Chem. Phys. Lett.*, **174**, 219 (1990).

[11.9] Z. C. Wu, D. A. Jelski, and T. F. George. *Chem. Phys. Lett.*, **137**, 291 (1987).

[11.10] Y. Wang, D. Tománek, and G. F. Bertsch. *Phys. Rev. B*, **44**, 6562 (1991).

[11.11] T. Yildirim and A. B. Harris. *Phys. Rev. B*, **46**, 7878 (1992).

[11.12] L. Pintschovius, B. Renker, F. Gompf, R. Heid, S. L. Chaplot, M. Haluška, and H. Kuzmany. *Phys. Rev. Lett.*, **69**, 2662 (1992).

[11.13] L. Pintschovius, S. L. Chaplot, R. Heid, B. Renker, M. Haluška, and H. Kuzmany. New York, Berlin (1993). Springer-Verlag. Springer series in solid-state sciences, H. Kuzmany, J. Fink, M. Mehring, and S. Roth, (eds.), vol. 117, p. 162.

[11.14] M. K. Nissen, S. M. Wilson, and M. L. W. Thewalt. *Phys. Rev. Lett.*, **69**, 2423 (1992).

[11.15] P. Zhou, K.-A. Wang, Y. Wang, P. C. Eklund, M. S. Dresselhaus, G. Dresselhaus, and R. A. Jishi. *Phys. Rev. B*, **46**, 2595 (1992).

[11.16] G. Meijer and D. S. Bethune. *Chem. Phys. Lett.*, **175**, 1 (1990).

[11.17] M. S. Dresselhaus, G. Dresselhaus, and P. C. Eklund. *J. Mater. Res.*, **8**, 2054 (1993).

[11.18] E. Brendsdal, J. Brunvoll, B. N. Cyvin, and S. J. Cyvin. In I. Harguttai (ed.), *Quasicrystals, Network and Molecules of Five-fold Symmetry*, p. 277. VCH Publishers, New York (1990).

[11.19] P. C. Eklund, P. Zhou, K.-A. Wang, G. Dresselhaus, and M. S. Dresselhaus. *J. Phys. Chem. Solids*, **53**, 1391 (1992).

[11.20] Z. H. Dong, P. Zhou, J. M. Holden, P. C. Eklund, M. S. Dresselhaus, and G. Dresselhaus. *Phys. Rev. B*, **48**, 2862 (1993).

[11.21] K. A. Wang, A. M. Rao, P. C. Eklund, M. S. Dresselhaus, and G. Dresselhaus. *Phys. Rev. B*, **48**, 11375 (1993).

[11.22] R. A. Jishi, R. M. Mirie, and M. S. Dresselhaus. *Phys. Rev. B*, **45**, 13685 (1992).

[11.23] A. A. Quong, M. R. Pederson, and J. L. Feldman. *Solid State Commun.*, **87**, 535 (1993).

[11.24] J. L. Feldman, J. Q. Broughton, L. L. Boyer, D. E. Reich, and M. D. Kluge. *Phys. Rev. B*, **46**, 12731 (1992).

[11.25] S. Guha, J. Menendez, J. B. Page, G. B. Adams, G. S. Spencer, J. P. Lehman, P. Giannozzi, and S. Baroni. *Phys. Rev. Lett.*, **72**, 3359 (1994).

[11.26] M. Schlüter, M. Lannoo, M. Needels, G. A. Baraff, and D. Tománek. *J. Phys. Chem. Solids*, **53**, 1473 (1992). In this reference the normal mode models were prepared by M. Grabow.

[11.27] M. A. Schlüter, M. Lannoo, M. Needels, G. A. Baraff, and D. Tománek. *Materials Science and Engineering*, **B19**, 129 (1993).

[11.28] L. T. Chadderton. *J. Phys. Chem. Solids*, **54**, 1027 (1992).

[11.29] H. Lamb. *London, Math. Soc. Trans.*, **14**, 50 (1883).

[11.30] A. M. Vassallo (1992). Unpublished. Cited in Ref. [11.28].

[11.31] A. M. Vassallo, L. S. Pang, P. A. Cole-Clark, and M. A. Wilson. *J. Am. Chem. Soc.*, **113**, 7820 (1991).

[11.32] D. E. Weeks and W. G. Harter. *Chem. Phys. Lett.*, **144**, 366 (1988).

[11.33] E. Brendsdal, B. N. Cyvin, J. Brunvoll, and S. J. Cyvin. *Spectrosc. Lett.*, **21**, 313 (1988).
[11.34] S. J. Cyvin, E. Brendsdal, B. N. Cyvin, and J. Brunvoll. *Chem. Phys. Lett.*, **143**, 377 (1988).
[11.35] R. S. Ruoff and A. L. Ruoff. *Appl. Phys. Lett.*, **59**, 1553 (1991).
[11.36] M. D. Newton and R. E. Stanton. *J. Am. Chem. Soc.*, **108**, 2469 (1986).
[11.37] Z. Slanina, J. M. Rudzinski, M. Togasi, and E. Osawa. *J. Mol. Struct.*, **202**, 169 (1989).
[11.38] G. B. Adams, J. B. Page, O. F. Sankey, K. Sinha, J. Menendez, and D. R. Huffman. *Phys. Rev. B*, **44**, 4052 (1991).
[11.39] R. Jones, C. D. Latham, M. I. Heggie, V. J. B. Torres, S. Öberg, and S. K. Estreicher. *Phil. Mag. Lett.*, **65**, 291 (1992).
[11.40] J. Kohanoff, W. Andreoni, and M. Parrinello. *Phys. Rev. B*, **46**, 4371 (1992).
[11.41] C. I. Frum, R. Engleman Jr., H. G. Hedderich, P. F. Bernath, L. D. Lamb, and D. R. Huffman. *Chem. Phys. Lett.*, **176**, 504 (1991).
[11.42] B. Chase, N. Herron, and E. Holler. *J. Phys. Chem.*, **96**, 4262 (1992).
[11.43] F. Negri, G. Orlandi, and F. Zerbetto. *Chem. Phys. Lett.*, **190**, 174 (1992).
[11.44] P. H. M. van Loosdrecht, P. J. M. van Bentum, and G. Meijer. *Phys. Rev. Lett.*, **68**, 1176 (1992).
[11.45] G. Dresselhaus, M. S. Dresselhaus, and P. C. Eklund. *Phys. Rev. B*, **45**, 6923 (1992).
[11.46] S. P. Love, D. McBranch, M. I. Salkola, N. V. Coppa, J. M. Robinson, B. I. Swanson, and A. R. Bishop. *Chem. Phys. Lett.*, **225**, 170 (1994).
[11.47] W. G. Harter and T. C. Reimer. *Chem. Phys. Lett.*, **194**, 230 (1992).
[11.48] F. Negri, G. Orlandi, F. Zerbetto, G. Ruani, A. Zakhidov, C. Taliani, K. Kinuchi, and Y. Achiba. *Chem. Phys. Lett.*, **211**, 353 (1993).
[11.49] W. G. Harter. *Principles of Symmetry, Dynamics and Spectroscopy*. John Wiley & Sons, New York (1993).
[11.50] R. Saito, G. Dresselhaus, and M. S. Dresselhaus. *Phys. Rev. B*, **50**, 5680 (1994).
[11.51] J. D. Axe, S. C. Moss, and D. A. Neumann. In H. Ehrenreich and F. Spaepen (eds.), *Solid State Physics: Advances in Research and Applications*, vol. 48, pp. 149–224, New York (1994). Academic Press. Chapter 3.
[11.52] D. A. Neumann, J. R. D. Copley, W. A. Kamitakahara, J. J. Rush, R. L. Cappelletti, N. Coustel, J. E. Fischer, J. P. McCauley, Jr., A. B. Smith, III, K. M. Creegan, and D. M. Cox. *J. Chem. Phys.*, **96**, 8631 (1992).
[11.53] K. Prassides, C. Christides, M. J. Rosseinsky, J. Tomkinson, D. W. Murphy, and R. C. Haddon. *Europhys. Lett.*, **19**, 629 (1992).
[11.54] J. R. D. Copley, D. A. Neumann, R. L. Cappelletti, and W. A. Kamitakahara. *J. Phys. Chem. Solids*, **53**, 1353 (1992).
[11.55] D. A. Neumann, J. R. D. Copley, W. A. Kamitakahara, J. J. Rush, R. L. Paul, and R. M. Lindstrom. *J. Phys. Chem. Solids*, **54**, 1699 (1993).
[11.56] S. A. Fitzgerald and A. J. Sievers. *Phys. Rev. Lett.*, **70**, 3177 (1993).
[11.57] M. Matus and H. Kuzmany. *Appl. Phys. A*, **56**, 241 (1993).
[11.58] M. Manfredini, C. E. Bottani, and P. Milani. *Chem. Phys. Lett.*, **226**, 600 (1994).
[11.59] R. Tycko, G. Dabbagh, M. J. Rosseinsky, D. W. Murphy, A. P. Ramirez, and R. M. Fleming. *Phys. Rev. Lett.*, **68**, 1912 (1992).
[11.60] X. P. Li, J. P. Lu, and R. M. Martin. *Phys. Rev. B*, **46**, 4301 (1992).
[11.61] A. S. Lipin and B. N. Mavrin. *Phys. Stat. Sol. (b)*, **177**, 85 (1993).
[11.62] S. Huant, J. B. Robert, G. Chouteau, P. Bernier, C. Fabre, and A. Rassat. *Phys. Rev. Lett.*, **69**, 2666 (1992).
[11.63] T. Yildirim, O. Zhou, J. E. Fischer, N. Bykovetz, R. A. Strongin, M. A. Cichy, A. B. Smith III, C. L. Lin, and R. Jelinek. *Nature (London)*, **360**, 568 (1992).

[11.64] W. P. Beyermann, M. F. Hundley, J. D. Thompson, F. N. Diederich, and G. Grüner. *Phys. Rev. Lett.*, **68**, 2046 (1992).
[11.65] E. Grivei, M. Cassart, J.-P. Issi, L. Langer, B. Nysten, and J.-P. Michenaud. *Phys. Rev. B*, **48**, 8514 (1993).
[11.66] S. Huant, G. Chouteau, J. B. Robert, P. Bernier, C. Fabre, A. Rassat, and E. Bustarret. *C. R. Acad. Sci. Paris–Serie II*, **314**, 1309 (1992).
[11.67] M. Matus, J. Winter, and H. Kuzmany. vol. 117, Springer-Verlag, Berlin (1993). Springer Series in Solid-State Sciences, ed. H. Kuzmany, J. Fink, M. Mehring and S. Roth, p. 255.
[11.68] M. Sprik, A. Cheng, and M. L. Klein. *J. Phys. Chem.*, **96**, 2027 (1992).
[11.69] K. H. Michel. *J. Chem. Phys.*, **97**, 5155 (1992).
[11.70] J. P. Lu, X. P. Li, and R. M. Martin. *Phys. Rev. Lett.*, **68**, 1551 (1992).
[11.71] A. Cheng and M. L. Klein. *J. Phys. Chem.* **95**, 6750 (1992).
[11.72] B. Renker, F. Gompf, R. Heid, P. Adelmann, A. Heiming, W. Reichardt, G. Roth, H. Schober, and H. Rietschel. *Z. Phys. B–Cond. Mat.*, **90**, 325 (1993).
[11.73] R. Saito, G. Dresselhaus, and M. S. Dresselhaus. *Phys. Rev. B*, **49**, 2143 (1994).
[11.74] T. Matsuo, H. Suga, W. I. F. David, R. M. Ibberson, P. Bernier, A. Zahab, C. Fabre, A. Rassat, and A. Dworkin. *Solid State Commun.*, **83**, 711 (1992).
[11.75] P. Zhou, A. M. Rao, K. A. Wang, J. D. Robertson, C. Eloi, M. S. Meier, S. L. Ren, X. X. Bi, P. C. Eklund, and M. S. Dresselhaus. *Appl. Phys. Lett.*, **60**, 2871 (1992).
[11.76] S. J. Duclos, R. C. Haddon, S. H. Glarum, A. F. Hebard, and K. B. Lyons. *Science*, **254**, 1625 (1991).
[11.77] H. Kuzmany, M. Matus, T. Pichler, and J. Winter. In K. Prassides (ed.), *Phys. and Chemistry of Fullerenes*, Kluwer, Academic Press, Dordrecht. NATO ASI-C Series: Crete. vol. 443, p. 287 (1993).
[11.78] S. H. Tolbert, A. P. Alivisatos, H. E. Lorenzana, M. B. Kruger, and R. Jeanloz. *Chem. Phys. Lett.*, **188**, 163 (1992).
[11.79] S. J. Duclos, K. Brister, R. C. Haddon, A. R. Kortan, and F. A. Thiel. *Nature (London)*, **351**, 380 (1991).
[11.80] A. M. Rao, P. Zhou, K.-A. Wang, G. T. Hager, J. M. Holden, Y. Wang, W. T. Lee, X.-X. Bi, P. C. Eklund, D. S. Cornett, M. A. Duncan, and I. J. Amster. *Science*, **259**, 955 (1993).
[11.81] D. S. Bethune, G. Meijer, W. C. Tang, H. J. Rosen, W. G. Golden, H. Seki, C. A. Brown, and M. S. de Vries. *Chem. Phys. Lett.*, **179**, 181 (1991).
[11.82] W. Krätschmer, L. D. Lamb, K. Fostiropoulos, and D. R. Huffman. *Nature (London)*, **347**, 354 (1990).
[11.83] C. Coulombeau, H. Jobic, P. Bernier, C. Fabre, and A. Rassat. *J. Phys. Chem.*, **96**, 22 (1992).
[11.84] R. L. Cappelletti, J. R. D. Copley, W. Kamitakahara, F. Li, J. S. Lannin, and D. Ramage. *Phys. Rev. Lett.*, **66**, 3261 (1991).
[11.85] A. Lucas, G. Gensterblum, J. J. Pireaux, P. A. Thiry, R. Caudano, J. P. Vigneron, and P. Lambin. *Phys. Rev. B*, **45**, 13694 (1992).
[11.86] K. Prassides, T. J. S. Dennis, J. P. Hare, J. Tomkinson, H. W. Kroto, R. Taylor, and D. R. M. Walton. *Chem. Phys. Lett.*, **187**, 455 (1991).
[11.87] G. Gensterblum, L. M. Yu, J. J. Pireaux, P. A. Thiry, R. Caudano, J. M. Themlin, S. Bouzidi, F. Coletti, and J. M. Debever. *Appl. Phys. A*, **56**, 175 (1993).
[11.88] R. A. Jishi, M. S. Dresselhaus, G. Dresselhaus, K. A. Wang, P. Zhou, A. M. Rao, and P. C. Eklund. *Chem. Phys. Lett.*, **206**, 187 (1993).
[11.89] Y. Wang, J. M. Holden, A. M. Rao, P. C. Eklund, U. Venkateswaran, D. Eastwood, R. L. Lidberg, G. Dresselhaus, and M. S. Dresselhaus. *Phys. Rev. B*, **51**, 4547 (1995).

[11.90] W. Guss, J. Feldmann, E. O. Gobel, C. Taliani, H. Mohn, W. Muller, P. Haubler, and H. U. ter Meer. *Phys. Rev. Lett.*, **72**, 2644 (1994).
[11.91] G. Gensterblum, L.-M. Yu, J. J. Pireaux, P. A. Thiry, R. Caudano, P. Lambin, A. A. Lucas, W. Krätschmer, and J. E. Fischer. *J. Phys. Chem. Solids*, **53**, 1427 (1992).
[11.92] M. C. Martin, J. Fabian, J. Godard, P. Bernier, J. M. Lambert and L. Mihaly. *Phys. Rev. B*, **51**, 2844 (1995).
[11.93] M. C. Martin, D. Koller, and L. Mihaly. *Phys. Rev. B*, **50**, 6538 (1994).
[11.94] R. Winkler, T. Pichler, and H. Kuzmany. In H. Kuzmany, J. Fink, M. Mehring, and S. Roth (eds.), *Progress in Fullerene Research: International Winterschool on Electronic Properties of Novel Materials*, p. 349 (1994). Kirchberg Winter School, World Scientific Publishing Co., Ltd., Singapore.
[11.95] M. C. Martin, X. Du, and L. Mihaly. In H. Kuzmany, J. Fink, M. Mehring, and S. Roth (eds.), *Progress in Fullerene Research: International Winterschool on Electronic Properties of Novel Materials*, p. 353 (1994). Kirchberg Winter School, World Scientific Publishing Co., Ltd., Singapore.
[11.96] K. Kamarás, K. Matsumoto, M. Wojnowski, E. Schönherr, H. Klos, and B. Gotschy. In H. Kuzmany, J. Fink, M. Mehring, and S. Roth (eds.), *Progress in Fullerene Research: International Winterschool on Electronic Properties of Novel Materials*, p. 357 (1994). Kirchberg Winter School, World Scientific Publishing Co., Ltd., Singapore.
[11.97] S. J. Chase, W. S. Bacsa, M. G. Mitch, L. J. Pilione, and J. S. Lannin. *Phys. Rev. B*, **46**, 7873 (1992).
[11.98] A. Rosenberg and D. P. Dilella. *Chem. Phys. Lett.*, **223**, 76 (1994).
[11.99] C. Coulombeau, H. Jobic, P. Bernier, C. Fabre, D. Schültz, and A. Rassat. *C. Rend. Acad. Sci. Paris, Série II*, **313**, 1387 (1991).
[11.100] J. W. White, G. Lindsell, L. Pang, A. Palmisano, D. S. Sivia, and J. Tomkinson. *Chem. Phys. Lett.*, **191**, 92 (1992).
[11.101] A. A. Lucas. *J. Phys. Chem. Solids*, **53**, 1415 (1992).
[11.102] J. E. Rowe, R. A. Malic, and E. E. Chaban. *Bulletin Amer. Phys. Soc.*, **39**, 212 (1994). C23.2 at APS March meeting.
[11.103] A. Sellidj and B. E. Koel. *J. Phys. Chem.*, **97**, 10076 (1993).
[11.104] V. S. Babu and M. S. Seehra. *Chem. Phys. Lett.*, **196**, 569 (1992).
[11.105] K. Kamarás, L. Akselrod, S. Roth, A. Mittelbach, W. Hönle, and H. G. von Schnering. *Chem. Phys. Lett.*, **214**, 338 (1993).
[11.106] C. C. Homes, P. J. Horoyski, M. L. W. Thewalt, and B. P. Clayman. *Phys. Rev. B*, **49**, 7052 (1994).
[11.107] P. H. M. van Loosdrecht, P. J. M. van Bentum, and G. Meijer. *Chem. Phys. Lett.*, **205**, 191 (1993).
[11.108] H. J. Byrne, W. K. Maser, W. W. Rühle, A. Mittelbach, and S. Roth. *Appl. Phys. A*, **56**, 235 (1993).
[11.109] H. J. Byrne, L. Akselrod, C. Thomsen, A. Mittelbach, and S. Roth. *Appl. Phys. A*, **57**, 299 (1993).
[11.110] M. G. Mitch, S. J. Chase, and J. S. Lannin. *Phys. Rev. Lett.*, **68**, 883 (1992).
[11.111] M. G. Mitch and J. S. Lannin. *J. Phys. Chem. Solids*, **54**, 1801 (1993).
[11.112] T. Pichler, M. Matus, J. Kürti, and H. Kuzmany. *Phys. Rev. B*, **45**, 13841 (1992).
[11.113] S. H. Glarum, S. J. Duclos, and R. C. Haddon. *J. Am. Chem. Soc.*, **114**, 1996 (1992).
[11.114] P. Zhou, K. A. Wang, A. M. Rao, P. C. Eklund, G. Dresselhaus, and M. S. Dresselhaus. *Phys. Rev. B*, **45**, 10838 (1992).
[11.115] R. Danieli, V. N. Denisov, G. Ruani, R. Zambone, C. Taliani, A. A. Zakhindov, A. Ugawa, K. Imaeda, K. Yakushi, H. Inokuchi, K. Kikuchi, I. Ikemoto, S. Suzuki, and Y. Achiba. *Solid State Commun.*, **81**, 257 (1992).

[11.116] M. S. Dresselhaus and G. Dresselhaus. *Light Scattering in Solids III*, **51**, 3 (1982). edited by M. Cardona and G. Güntherodt, Springer-Verlag Berlin, Topics in Applied Physics.
[11.117] R. A. Jishi and M. S. Dresselhaus. *Phys. Rev. B*, **45**, 6914 (1992).
[11.118] M. G. Mitch and J. S. Lannin. *Phys. Rev. B*, **51**, 6784 (1995).
[11.119] Q. Zhu, O. Zhou, J. E. Fischer, A. R. McGhie, W. J. Romanow, R. M. Strongin, M. A. Cichy, and A. B. Smith, III. *Phys. Rev. B*, **47**, 13948 (1993).
[11.120] J. Winter and H. Kuzmany. In H. Kuzmany, J. Fink, M. Mehring, and S. Roth (eds.), *Progress in Fullerene Research: International Winterschool on Electronic Properties of Novel Materials*, p. 271 (1994). Kirchberg Winter School, World Scientific Publishing Co., Ltd., Singapore.
[11.121] M. C. Martin, D. Koller, A. Rosenberg, C. Kendziora, and L. Mihaly. *Phys. Rev. B*, **51**, 3210 (1995).
[11.122] A. M. Rao, M. Menon, K. A. Wang, P. C. Eklund, K. R. Subbaswamy, D. S. Cornett, M. A. Duncan, and I. J. Amster. *Chem. Phys. Lett.*, **224**, 106 (1994).
[11.123] M. J. Rice and H. Y. Choi. *Phys. Rev. B*, **45**, 10173 (1992).
[11.124] P. C. Eklund, A. M. Rao, Y. Wang, P. Zhou, K. A. Wang, J. M. Holden, M. S. Dresselhaus, and G. Dresselhaus. *Thin Solid Films*, **257**, 211 (1995).
[11.125] T. Pichler, J. Kürti, and H. Kuzmany. *Conds. Matter Commun.*, **1**, 21 (1993).
[11.126] T. Pichler, R. Winkler, and H. Kuzmany. *Phys. Rev. B*, **49**, 15879 (1994).
[11.127] K. J. Fu, W. L. Karney, O. L. Chapman, S. M. Huang, R. B. Kaner, F. Diederich, K. Holczer, and R. L. Whetten. *Phys. Rev. B*, **46**, 1937 (1992).
[11.128] X. K. Wang, R. P. H. Chang, A. Patashinski, and J. B. Ketterson. *J. Mater. Res.*, **9**, 1578 (1994).
[11.129] E. Sohmen, J. Fink, and W. Krätschmer. *Europhys. Lett.*, **17**, 51 (1992).
[11.130] S. Rice. Private communication (1993).
[11.131] E. E. Ferguson and F. A. Matsen. *J. Chem. Phys.*, **29**, 105 (1958).
[11.132] M. J. Rice, V. M. Yartsev, and C. S. Jacobsen. *Phys. Rev. B*, **21**, 3427 (1980).
[11.133] M. J. Rice, L. Pietronero, and P. Bruesch. *Solid State Commun.*, **21**, 757 (1977).
[11.134] T. Pichler, M. Matus, and H. Kuzmany. *Solid State Commun.*, **86**, 221 (1993).
[11.135] J. Winter and H. Kuzmany. *Solid State Commun.*, **84**, 935 (1993).
[11.136] D. Koller, M. C. Martin, X. Du, L. Mihaly, and P. W. Stephens. *Bulletin Amer. Phys. Soc.*, **39**, 213 (1994). Unpublished, paper C23.8 at APS.
[11.137] M. C. Martin, D. Koller, X. Du, P. W. Stephens, and L. Mihaly. *Phys. Rev. B*, **49**, 10818 (1994).
[11.138] A. Jànossy, O. Chauvet, S. Pekker, J. R. Cooper, and L. Forró. *Phys. Rev. Lett.*, **71**, 1091 (1993).
[11.139] O. Chauvet, G. Oszànyi, L. Forró, P. W. Stephens, M. Tegze, G. Faigel, and A. Jànossy. *Phys. Rev. Lett.*, **72**, 2721 (1994).
[11.140] M. G. Mitch, G. P. Lopinski, and J. S. Lannin. In P. Bernier, T. W. Ebbesen, D. S. Bethune, R. Metzger, L. Y. Chiang, and J. W. Mintmire (eds.), "Science and Technology of Fullerene Materials" MRS Symposium Proc. vol. 359, 493 (1995).
[11.141] C. Christides, D. A. Neumann, K. Prassides, J. R. D. Copley, J. J. Rush, M. J. Rosseinsky, D. W. Murphy, and R. C. Haddon. *Phys. Rev. B*, **46**, 12088 (1992).
[11.142] C. Christides, I. M. Thomas, T. J. S. Dennis, and K. Prassides. *Europhys. Lett.*, **22**, 611 (1993).
[11.143] K. Raghavachari and C. M. Rohlfing. *J. Phys. Chem.*, **95**, 5768 (1991).
[11.144] Y. Wang, J. M. Holden, A. M. Rao, W.-T. Lee, G. T. Hager, X. X. Bi, S. L. Ren, G. W. Lehman, G. T. Hager, and P. C. Eklund. *Phys. Rev. B*, **45**, 14396 (1992).
[11.145] F. Negri, G. Orlandi, and F. Zerbetto. *J. Amer. Chem. Soc.*, **113**, 6037 (1991).

[11.146] Y. Shinohara, R. Saito, T. Kimura, G. Dresselhaus, and M. S. Dresselhaus. *Chem. Phys. Lett.*, **227**, 365 (1994).
[11.147] R. A. Jishi, R. M. Mirie, M. S. Dresselhaus, G. Dresselhaus, and P. C. Eklund. *Phys. Rev. B*, **48**, 5634 (1993).
[11.148] Z. H. Wang, M. S. Dresselhaus, G. Dresselhaus, and P. C. Eklund. *Phys. Rev. B*, **48**, 16881 (1993).
[11.149] D. W. Snoke, Y. S. Raptis, and K. Syassen. *Phys. Rev. B*, **45**, 14419 (1992).
[11.150] P. H. M. van Loosdrecht, M. A. Verheijen, H. Meeks, P. J. M. van Bentum, and G. Meijer. *Phys. Rev. B*, **47**, 7610 (1993).
[11.151] S. L. Ren, K. A. Wang, P. Zhou, Y. Wang, A. H. Rao, M. S. Meier, J. Selegue, P. C. Eklund. *Appl. Phys. Lett.*, **61**, 124 (1992).
[11.152] C. Christides, T. J. S. Dennis, K. Prassides, R. L. Cappelletti, D. A. Neumann, and J. R. D. Copley. *Phys. Rev. B*, **49**, 2897 (1994).
[11.153] A. Cheng and M. L. Klein. *Phys. Rev. B*, **45**, 4958 (1992).
[11.154] K. A. Wang, P. Zhou, A. M. Rao, P. C. Eklund, M. S. Dresselhaus, and R. A. Jishi. *Phys. Rev. B*, **48**, 3501 (1993).
[11.155] G. B. Adams, J. B. Page, O. F. Sankey, and M. O'Keeffe. *Phys. Rev. B*, **50**, 17471 (1994).
[11.156] Y. Wang, J. M. Holden, Z. H. Dong, X. X. Bi, and P. C. Eklund. *Chem. Phys. Lett.*, **211**, 341 (1993).
[11.157] M. Menon, K. R. Subbaswamy, and M. Sawtarie. *Phys. Rev. B*, **49**, 13966 (1994).
[11.158] M. Hanfland, H. Beister, and K. Syassen. *Phys. Rev. B*, **39**, 12598 (1989).
[11.159] M. Hanfland, K. Syassen, and R. Sonnenschein. *Phys. Rev. B*, **40**, 1951 (1989).
[11.160] Y. Iwasa, T. Arima, R. M. Fleming, T. Siegrist, O. Zhou, R. C. Haddon, L. J. Rothberg, K. B. Lyons, H. L. Carter, Jr., A. F. Hebard, R. Tycko, G. Dabbagh, J. J. Krajewski, G. A. Thomas, and T. Yagi. *Science*, **264**, 1570 (1994).
[11.161] V. N. Denisov, A. A. Zakhidov, G. Ruani, R. Zamboni, C. Taliani, K. Tanaka, K. Yoshizaea, K. Okahara, T. Yamabe, and Y. Achiba. *Synth. Metals*, **56**, 3050 (1993).
[11.162] K. Awaga and Y. Maruyama. *Chem. Phys. Lett.*, **158**, 556 (1989).
[11.163] P.-M. Allemand, K. C. Khemani, A. Koch, F. Wudl, K. Holczer, S. Donovan, G. Grüner, and J. D. Thompson. *Science*, **253**, 301 (1991).
[11.164] D. Mihailovič, P. Venturini, A. Hassanien, J. Gasperič, K. Lutar, S. Miličev, and V. I. Srdanov. In H. Kuzmany, J. Fink, M. Mehring, and S. Roth (eds.), *Progress in Fullerene Research: International Winterschool on Electronic Properties of Novel Materials*, p. 275 (1994). Kirchberg Winter School, World Scientific Publishing Co., Ltd., Singapore.
[11.165] K. Okahara. Ph.D. thesis, Kyoto University (1994). Department of Chemistry: Studies on the Electronic and Magnetic Properties of C_{60} and Related Materials (in English).

CHAPTER 12

Electronic Structure

On the basis of the vibrational spectroscopic studies discussed in Chapter 11, it is concluded that fullerenes form highly molecular solids, which imply narrow electronic energy bands. In this chapter it will be shown that these bandwidths are only a factor of 2 to 3 wider than the highest vibrational energies (see §11.3.1). Thus, the electronic structures of the crystalline phases are expected to be closely related to the electronic levels of the isolated fullerene molecules [12.1].

For this reason we start the discussion of the electronic structure of fullerene solids by reviewing the electronic levels for free C_{60} molecules, the fullerene that has been most extensively investigated. We then discuss modification to the electronic structure of the free molecules through intermolecular interactions which occur in the solid state, as well as modifications to the electronic structure arising from doping. Because of the weak hybridization between the dopant and carbon levels, doping effects are first considered in terms of the electronic states of the $C_{60}^{n\pm}$ molecular ions, and their solid-state effects are then considered, including the case of the metallic alkali metal M_3C_{60} compounds. The molecular states are considered from both a one-electron and a many-electron point of view, and the relation between these approaches is discussed. Also, the electronic structure for the solid is considered from both the standpoint of a highly correlated molecular solid and as a one-electron band solid, and the relation between these approaches is also discussed.

Theoretical work on the electronic structure of fullerenes had relied heavily on comparisons of these calculations with photoemission experiments [12.2, 3], because, in principle, photoemission and inverse photoemission experiments are, respectively, sensitive to the HOMO (highest occupied molecular orbital) and LUMO (lowest unoccupied molecular orbital) electronic densities of states. Photoemission studies are reviewed in §17.1. Optical studies provide another powerful tool for studying the electronic structure (see §13.2). Since optical studies involve excitonic effects, the electronic structure of exciton states is reviewed in this chapter.

12.1. Electronic Levels for Free C_{60} Molecules

Each carbon atom in C_{60} has two single bonds along adjacent sides of a pentagon and one double bond between two adjoining hexagons (see §3.1). If these bonds were coplanar, they would be very similar to the sp^2 trigonal bonding in graphite. The curvature of the C_{60} surface causes the planar-derived trigonal orbitals to hybridize, thereby admixing some sp^3 character into the dominant sp^2 planar bonding. The shortening of the double bonds to 1.40 Å and lengthening of the single bonds to 1.46 Å in the hexagonal rings of the C_{60} molecule also strongly influence the electronic structure. The details of the structural arrangement are important for detailed calculations of the electronic levels.

In this section, we first review the general framework of the electronic structure imposed by symmetry considerations, which is followed by a discussion of the molecular ground state and the corresponding excited states, including the electronic structure of C_{60} as well as higher mass fullerene molecules.

12.1.1. Models for Molecular Orbitals

Several models for the electronic structure of fullerene molecules have been developed, ranging from one-electron Hückel calculations or tight-binding models [12.4] to first principles models [12.5]. While sophisticated models yield somewhat more quantitative agreement with optical, photoemission, and other experiments sensitive to the electronic structure, the simple Hückel models lead to the same level ordering near the Fermi level and for this reason are often used for physical discussions of the electronic structure.

In the simple models, each carbon atom in C_{60} is equivalent to every other carbon atom, each being located at an equivalent site at the vertices of the truncated icosahedron which describes the structure of the C_{60} molecule. By introducing the atomic potential for the isolated carbon atom at each of these sites, the one-electron energy levels in a tight-binding model can be calculated. Corresponding to each carbon atom, which is in column IV of the periodic table, are three distorted (sp^2) bonds which are predominantly confined to the shell of the C_{60} molecule, coupling each carbon atom to its three nearest neighbors and giving rise to three occupied bonding σ orbitals and one occupied bonding π orbital normal to the shell, thus accounting for the four valence electrons. The σ orbitals lie low in energy, roughly 3 to 6 eV below the Fermi level E_F (roughly estimated from the C–C cohesive energy [12.6]), and therefore do not contribute significantly to the electronic transport or optical properties, which are dominated by the π orbitals, lying near E_F.

The ordering of the π molecular energy levels and their icosahedral symmetry identifications based on a Hückel calculation [12.4] are shown in Fig. 12.1, where the filled levels are indicated as '+' and the unoccupied states by '−'. Figure 12.1 and the results of a local density calculation of the molecular energy levels for the C_{60} molecule shown in Fig. 12.2(a) are both widely used in the literature for discussion of the electronic structure of the C_{60} molecule. For example, the molecular orbital calculation in Fig. 12.2(a) implies that the bonding σ electrons lie more than 7 eV below the HOMO level of C_{60} and the antibonding σ electron states lie more than 6 eV above the Fermi level. According to the energy scale of the $\sigma(sp^2)$ bonding and antibonding states in fullerenes, the bandwidth of the molecular levels in Fig. 12.2 is large (~25 eV) and of similar magnitude to the corresponding states in graphite.

A number of calculations of the electronic states for the free C_{60} molecule have been carried out using a variety of calculational methods [12.5, 8–10], and general agreement is found with the level ordering shown in Figs. 12.1 and 12.2(a). Some models have even been successful in obtaining bond lengths of 1.45 Å for a_5 and 1.40 Å for a_6 [12.11], in good agreement with experimental values (see §3.1) [12.12, 13]. Another relevant energy is the ionization potential for the free molecule (7.6 eV), denoting the energy needed to remove an electron from neutral C_{60} to create the C_{60}^+ ion. This energy is to be compared with the much smaller electron affinity (2.65 eV [12.14]), the energy needed to add one electron, giving rise to the C_{60}^- ion. Whereas the electron affinity for C_{60} is similar to that for other acceptor molecules (e.g., TCNQ, for which the electron affinity is 2.82 eV), the large ionization potential for C_{60} makes it unlikely that C_{60} would behave as an electron donor.

Since C_{60} and other closed cage fullerene molecules have an approximately spherical shape, some authors have used a phenomenological approach to the electronic structure based on symmetry considerations [12.15, 16]. According to this spherical shell approach, the fullerene eigenstates can be simply described by spherical harmonics and classified by their angular momentum quantum numbers. The icosahedral (or lower) symmetry of the fullerene molecule is then imposed and treated as a perturbation, causing

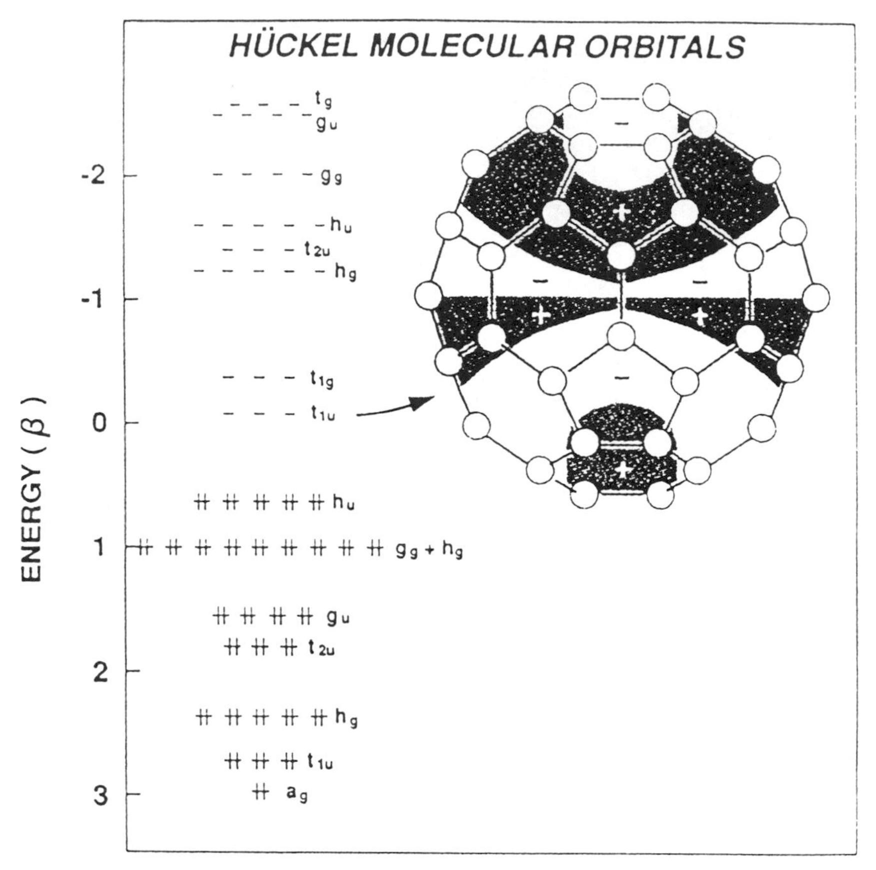

Fig. 12.1. Molecular orbital levels for C_{60} from a Hückel calculation. The filled states are labeled '+' and the empty states '−'. The shading on the icosahedron on the right-hand side of the figure pertains to whether the wave function is negative or positive, and the figure schematically shows three nodes and odd parity for the wave function for the t_{1u} (f_{1u}) triply degenerate conduction band level [12.4].

symmetry-imposed splittings of the molecular levels, which are then classified according to the symmetry of the icosahedral (or lower) symmetry group. In §12.1.2 we use this approach to give a physical picture of the level filling for the molecular states for fullerenes, and in §12.2 we review symmetry-based models that can be applied to account for experimental observations related to the electronic structure.

12.1.2. Level Filling for Free Electron Model

The simplest model that captures the essence of the electronic structure of fullerenes considers each C_{60} molecule to contain 180 σ electrons in

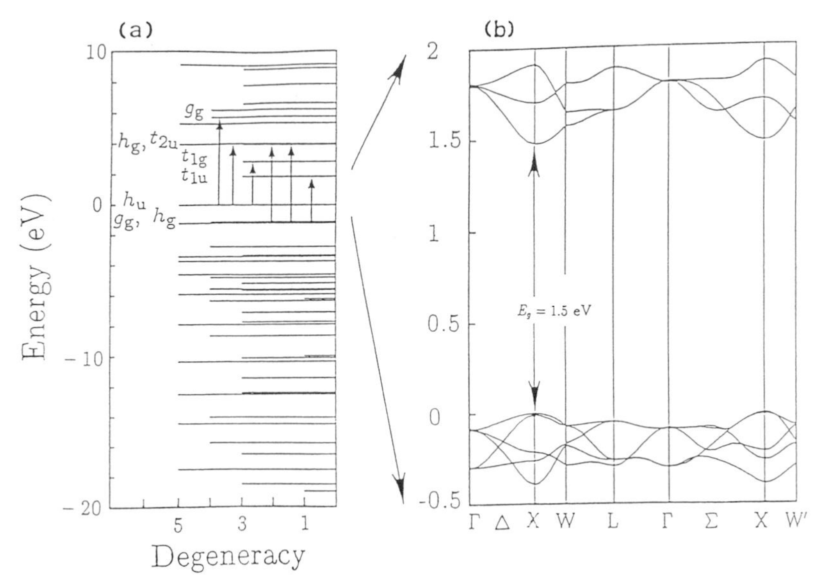

Fig. 12.2. Calculated electronic structure of an (a) isolated C_{60} molecule and (b) face-centered cubic (fcc) solid C_{60}, where the direct band gap at the X point is calculated to be 1.5 eV on the basis of a one-electron model [12.7].

strongly bonding sp^2 directed orbitals and 60 π electrons (i.e., one π electron per carbon atom) close to the Fermi level. The level filling and overview of the electronic structure can be obtained by filling the electronic states according to their orbital angular momentum quantum numbers, first assuming full rotational symmetry and then imposing level splittings in accordance with the icosahedral symmetry of C_{60}, treated as a perturbation.

These symmetry issues are discussed in §4.3 and the results for the level filling are given in Table 4.10, where the number of electrons filling each of the angular momentum states is listed together with the cumulative number of filled electronic states and with the splittings of each of the angular momentum states in an icosahedral field. Table 4.10 shows that 50 π electrons fully occupy the angular momentum states through $\ell = 4$, so that the 10 remaining π electrons of C_{60} are available to start filling the $\ell = 5$ state. In full spherical symmetry, the $\ell = 5$ state can accommodate 22 electrons, which would correspond to an accumulation of 72 electrons, assuming that all the $\ell = 5$ state levels are filled, before any $\ell = 6$ levels fill. However, the $\ell = 5$ state splits in icosahedral symmetry into the $H_u + F_{1u} + F_{2u}$ irreducible representations, as indicated in Table 4.10. The level of lowest energy is the fivefold H_u level, which is completely filled by the 10 available electrons in C_{60}, as indicated in Table 4.10 for 60 electrons. The resulting h_u^{10} ground state configuration, by Hund's rule, is nondegenerate, has angular momentum quantum numbers $L = 0$, $S = 0$, $J = 0$, and the many-electron h_u^{10} configuration has A_g symmetry within the point group I_h. We use the notation capital H_u to denote the symmetry of the one-electron level and the lowercase h_u^{10} to denote the many-electron configuration. Neglecting any thermal excitation, the two threefold F_{1u} and F_{2u} levels (some authors refer to these levels as T_{1u} and T_{2u}, and we use both notations in this volume) in icosahedral symmetry are empty. Whether C_{60} is considered in terms of a spherical approximation, or if its icosahedral symmetry is taken into account, the many-electron ground state is nondegenerate with a total angular momentum of $J = 0$. This special circumstance has been suggested as being in part responsible for the high stability of the C_{60} molecule [12.15].

Figure 12.1 shows that C_{60} has just enough electrons to fully fill the H_u molecular level, which forms the highest occupied molecular orbital for C_{60}, while the next higher F_{1u} level remains empty and becomes the lowest unoccupied molecular orbital. Molecular local density approximation calculations for C_{60} yield a value of 1.92 eV for the HOMO–LUMO gap in the free molecule [see Fig. 12.2(a)]. The results of Hückel calculations, such as shown in Fig. 12.1, have been especially useful for establishing the ordering of the molecular levels of C_{60} [12.8]. Higher-lying unoccupied states would include the F_{2u} level from the $\ell = 5$ state, as well as $A_g + F_{1g} + G_g + H_g$ levels from the $\ell = 6$ state, where the F_{1g} level has been calculated to lie lowest [12.8]. Hückel calculations (see Fig. 12.1) [12.7, 8, 17] show that the F_{1g} level of $\ell = 6$ lies lower than the F_{2u} level associated with $\ell = 5$, implying that the F_{1g} level would fill before the F_{2u} level in a doping experiment.

Furthermore, these ideas can be used to fill states for large fullerenes. Considerations for the filling of states for icosahedral molecules up to C_{980} on the basis of a spherical approximation are given in Table 4.11, showing that C_{80} has 8 electrons in the partially filled $\ell = 6$ shell and C_{140} has 12 electrons in the partially filled $\ell = 8$ shell [12.15]. These arguments can be extended to consider fullerenes with lower symmetry where the HOMO level splittings in Table 4.10 would have to be expressed in terms of the irreducible representations of the lower symmetry group.

Another example of state filling relates to the C_{70} fullerene with D_{5h} symmetry. Here we have 70 π electrons to place in the molecular orbitals. As discussed above, filling levels through $\ell = 4$ leaves 20 π electrons for the $\ell = 5$ shell, which is fully occupied with 22 electrons. In a spherical approximation, two possible electronic configurations which suggest themselves are $s^2p^6d^{10}f^{14}g^{18}h^{20}$, which has two holes in the $\ell = 5$ (or h) shell. If, on the other hand, we are guided by the level ordering in icosahedral C_{60}, the suggested ground state configuration would be $s^2p^6d^{10}f^{14}g^{18}h^{16}j^4$, which has six holes in the $\ell = 5$ state and four electrons in the $\ell = 6$ state. Group theoretical considerations cannot decide on which of the possible configurations represents the ground state. Considering C_{70} to consist of the two hemispheres which comprise C_{60} with a belt of five additional hexagons along the equator of the fullerene, then the level splittings arising from lowering of the symmetry of $I \rightarrow D_5$ are appropriate. In I_h symmetry the first 16 electrons for $\ell = 5$ fill the 10-fold H_u level and the 6-fold F_{1u} level, as shown in the Hückel level-filling picture shown in Fig. 12.1. The next four electrons would partly fill the F_{1g} level. In icosahedral I symmetry, this level, corresponding to the configuration f_{1g}^4, splits according to F_1 in I symmetry into $A_2 + E_1$ in D_5 symmetry, thus suggesting that the fourfold E_1 level forms the HOMO for C_{70}, while the twofold A_2 level forms the unoccupied LUMO. The simple argument given here is, however, not in agreement with molecular orbital calculations [12.7, 18, 19] for C_{70}, which show that the HOMO level is a nondegenerate level, as discussed further in §12.6.1. The LUMO level for C_{70} is also a nondegenerate level.

12.2. Symmetry-Based Models

As stated above, a simple approach that is useful for finding the energy levels for any fullerene C_{n_C} is based on symmetry considerations, exploiting the observation that the closed cage fullerene molecules are close to being spherical shells. The symmetry lowering imposed by the structure of the specific fullerene under consideration is then treated in perturbation theory and symmetry is used to find the form of the Hamiltonian. The resulting eigenvalues are expressed in terms of a set of expansion parameters which are evaluated either from experiments or from first principles calculations, or from a combination of experiment and theory. The approach is similar to that taken to calculate the crystal field splitting of the rare earth ions in a cubic crystal field [12.20].

The phenomenological one-electron symmetry–based Hamiltonian for a general fullerene is then written as [12.21–23]

$$\mathcal{H}_i = \mathcal{H}_0(i) + \mathcal{H}_\Gamma(i), \tag{12.1}$$

where $\mathcal{H}_0(i)$ is the spherically symmetrical part and is the dominant term, while $\mathcal{H}_\Gamma(i)$ is the symmetry-lowering perturbation for symmetry group Γ which describes the symmetry of the fullerene. In the case of icosahedral I_h fullerenes, $\mathcal{H}_\Gamma(i) \equiv \mathcal{H}_{I_h}(i)$ is the appropriate symmetry-lowering perturbation. As an example, we will consider the symmetry-lowering perturbation to be an icosahedral perturbation. For icosahedral molecules, $\mathcal{H}_{I_h}(i)$ can be written in the form

$$\mathcal{H}_{I_h}(i) = t_6 T^{(6)} + t_{10} T^{(10)} + t_{12} T^{(12)} + \cdots, \tag{12.2}$$

where $\mathcal{H}_{I_h}(i)$ is expressed in terms of expansion parameters t_j and the lowest-order spherical tensor component $T^{(6)}$ is written in terms of the icosahedral harmonics T_m^6 as

$$T^{(6)} = \left[\frac{\sqrt{11}}{5} T_0^6 + \frac{\sqrt{7}}{5} (T_5^6 - T_{-5}^6) \right], \tag{12.3}$$

using Table 4.3, where it is noted that only icosahedral harmonics T_m^6 with $m = 0, \pm 5$ couple in lowest order. Similar expressions can be written for the higher-order terms $T^{(10)}$, $T^{(12)}$, etc. [12.15]. It is of interest to note that the lowest-order icosahedral harmonic to enter Eq. (12.1) is for $\ell = 6$ (see Table 4.3), since this is the lowest-order icosahedral harmonic with A_g symmetry, and the second nonvanishing term in the expansion is for $\ell = 10$. The large values for ℓ which enter the expansion in Eq. (12.1) indicate the rapid convergence of the expansion and the validity of the spherical approximation as a good lowest-order approximation. The Hamiltonian [Eq. (12.1)] can thus be evaluated convergently using spherical harmonics as the basis functions. The diagonalization of Eq. (12.1) gives one-electron molecular orbital energy levels for the C_{60} molecule. Good agreement has been found for the lowest-energy states (up through $\ell = 4$) between the expansion of Eq. (12.1) using a three parameter fit [12.15] and first principles *ab initio* calculations [12.9, 24–26], as shown in Table 12.1. To obtain a satisfactory fit at higher energies (with regard to the energy of the HOMO level and to the level ordering of the f_{2u} and f_{1g} levels for $\ell = 5$ and $\ell = 6$, respectively) more terms in the expansion of Eq. (12.2) for icosahedral symmetry are needed.

The same basic Hamiltonian is applicable for all icosahedral fullerenes, while the Hamiltonians for the rugby ball–shaped fullerenes are fit using a perturbation in Eq. (12.1) $\mathcal{H}_\Gamma \equiv \mathcal{H}_{D_{5h}}$ with D_{5h} symmetry

$$\mathcal{H}_{D_{5h}}(i) = t_2 T^{(2)} + t_4 T^{(4)} + t_6^{(1)} T_1^{(6)} + t_6^{(2)} T_2^{(6)} + \cdots \tag{12.4}$$

corresponding to the A_1' entries in Table 4.16 (note that there are two distinct irreducible tensors of A_1' symmetry for $\ell = 6$). We note that for fullerenes with lower symmetry, perturbation terms such as $\mathcal{H}_{D_{5h}}(i)$ enter in lower order as a function of ℓ (i.e., $\ell = 2, 4$), indicating that more terms may be necessary to carry out the perturbation expansion for fullerenes with lower symmetry. However, we can consider the symmetry-lowering perturbation for C_{70} to arise from two perturbations: first an icosahedral perturbation $\mathcal{H}_{I_h}$ given by Eq. (12.2), which is the dominant term, and a smaller perturbation $\mathcal{H}_{D_{5h}}$ given by Eq. (12.4) due to the elongation along the fullerene fivefold axis. Then more rapid convergence of the symmetry-lowering Hamiltonian in Eq. (12.1) may be achieved.

Table 12.1

π energy states (in eV) of the C_{60} phenomenological Hamiltonian model.

ℓ	Symmetry	Model[a]	*Ab initio*[b]
0	a_g	−7.41	−7.41
1	f_{1u}	−6.87	−6.87
2	h_g	−5.87	−5.82
3	f_{2u}	−4.40	−4.52
	g_u	−4.13	−3.99
4	h_g	−2.21	−2.44
	g_g	−2.12	−2.37
5	h_u	−0.20	−1.27
	f_{1u}	0.88	0.62
	f_{2u}	1.82	2.71
6	f_{1g}	3.38	1.59
	h_g	3.43	2.78
	g_g	4.92	4.60
⋮	⋮	⋮	⋮

[a] Ref. [12.15]. To improve the model at higher ℓ values, higher-order terms in the phenomenological Hamiltonian must be included [Eq. (12.2)].

[b] Ref. [12.24].

12.3. Many-Electron States for C_{60} and Other Icosahedral Fullerenes

In the previous sections the molecular energy levels were considered from a one-electron point of view, whereby the energy levels of the electron are treated in terms of an average potential due to the other ions and to the other valence electrons in the molecule. This simplified treatment is appropriate to the case of filled-shell configurations or for the excitation of one electron (hole) from the filled-shell configurations. Because of the large number of valence electrons on a single fullerene molecule, it is often necessary to go beyond the one-electron Hamiltonian and to consider more explicitly the interaction between electrons through exchange and correlation effects. In enumerating the Pauli-allowed states for many-electron excitations, a many-electron approach is implicit.

The phenomenological Hamiltonian for the many-electron (N) problem can be written as

$$\mathcal{H} = \sum_{i=1}^{N} \mathcal{H}_i + \sum_{\substack{i,j=1 \\ i<j}}^{N}{}' \mathcal{H}_{ij}, \tag{12.5}$$

where $\mathscr{H}_i$ refers to the one-electron Hamiltonian, $\mathscr{H}_{ij}$ is the interaction Hamiltonian between electrons i and j, and the prime on the sum avoids double counting. Without loss of generality $\mathscr{H}_{ij}$ can be written as

$$\mathscr{H}_{ij} = J_{ij}\mathscr{S}_i \cdot \mathscr{S}_j, \tag{12.6}$$

where $\mathscr{S}_i$ is the spin angular momentum operator and J_{ij} is an exchange integral. For two-electron states based on C_{60}, this exchange interaction gives rise to a singlet–triplet splitting which is very important in discussing excited states (§12.4.2) and optical properties (see §13.1.2) [12.17].

In this section we discuss many-electron states from the point of view of forming the ground state for molecular fullerenes. The specification of the excited states then follows. We argue that it is necessary that a ground state be nondegenerate, because if it were degenerate, the molecule would then distort, causing a level splitting of this degenerate state and the lowering of the energy of the system. Such distortions are important for considering the electronic states of charged fullerenes, as discussed below.

In Table 4.10, which we have already discussed in terms of the filling of the angular momentum states, we can see that in full rotational symmetry the filling of each angular momentum state ℓ results in a many-body nondegenerate state with A_g symmetry, which is a good candidate for a ground state. Also listed in Table 4.10 are the symmetries of these filled levels in icosahedral symmetry. Thus the filling of the $\ell = 5$ state in full rotational symmetry corresponds to the filling of levels h_u^{10}, f_{1u}^6, and f_{2u}^6 in the lower-symmetry icosahedral group with 10, 6, and 6 electrons, respectively. From strictly symmetry considerations, filled icosahedral levels for $\ell = 5$ could occur for various numbers of electrons, such as 56, 60, 62, or 66. However, when the energetics of the problem are also considered (see Fig. 12.1 for the Hückel calculation), only the cases of 60 electrons and perhaps 66 electrons correspond to actual ground states for icosahedral fullerenes.

Another view of the many-electron ground state problem is provided by Table 4.11, which considers the filling of states for all icosahedral fullerenes up to C_{980}, with the full electronic configurations given for the spherical model approximation, as discussed in §12.2. Using Hund's rule, the quantum numbers for the ground state for each electronic configuration are found, and the symmetries of the resulting multiplet levels in icosahedral symmetry are given in the right-hand column. It is seen that only a few of the icosahedral fullerenes correspond to nondegenerate states, namely C_{60}, C_{180}, and C_{420}, while the others listed in Table 4.11 would appear to correspond to degenerate ground states. For example, the ground state J value for icosahedral C_{20} is 4, which yields a ninefold degenerate multiplet in full rotational symmetry which in icosahedral symmetry splits into a $G_g(4)$ and a $H_g(5)$ level. In the case of a degenerate ground state, a Jahn–Teller distortion [12.1, 5] to fullerenes with lower symmetry is expected, resulting in a further splitting of the ground state level and a consequent lowering of the energy of the system. The advantage of using a spherical approximation for discussing the ground states of higher fullerenes is the simplicity of the approach, which enables qualitative arguments to be presented for complex systems, such as the large fullerenes shown in Table 4.11.

Before discussing the excited states for the neutral C_{60} molecule explicitly, we next consider the more general issue of the ground states and excited states for the related free molecular ions $C_{60}^{n\pm}$.

12.4. Multiplet States for Free Ions $C_{60}^{n\pm}$

The electronic states for C_{60} anions and cations are important for a variety of experiments in the gas-phase, in solutions, and in the solid state. For example, experiments in the gas-phase relating to the addition of electrons by the interaction of C_{60} with electron beams lead to negatively charged C_{60} molecules. As another example, experiments on C_{60} in electrolytic solutions could, for example, involve electrochemical processes such as voltammetry experiments, where charge is transferred (see §10.3.2) or optical studies, whereby electrons and holes are simultaneously introduced as a result of photon excitation (see §13.1). Since solid C_{60} is a highly molecular solid, the electronic states of the molecule [described by the intramolecular Hamiltonian $\mathcal{H}_{\text{intra}}$ of Eq. (12.1)] form the basis for developing the electronic structure for both solid C_{60} and doped C_{60}, for which intermolecular interactions $\mathcal{H}_{\text{inter}}$ are also of importance. In the limit that $\mathcal{H}_{\text{intra}} \gg \mathcal{H}_{\text{inter}}$, the molecular approach is valid, while for cases where $\mathcal{H}_{\text{intra}} \approx \mathcal{H}_{\text{inter}}$, a band approach is necessary. Thus for alkali metal M_3C_{60} compounds, which exhibit superconductivity (see §15.1), band theory is expected to be applicable. Whether band theory with correlated electron–electron interactions or a molecular solid with weakly interacting molecules is the most convergent approach for specific situations is further discussed in §12.7.

In qualitative discussions of physical phenomena, such as the filling of states by doping, use is made of the electronic states of the molecular ions $C_{60}^{n\pm}$. For example, referring to Fig. 12.1, where the level ordering of C_{60} is presented, it is often argued that the maximum conductivity is expected when the LUMO t_{1u} (f_{1u}) level is half filled with electrons at the stoichiometry M_3C_{60} and that an insulator is again achieved when the t_{1u} level is completely filled by six electrons at the stoichiometry M_6C_{60}, assuming that one electron is transferred to each C_{60} molecule per alkali metal atom uptake. These simple arguments are, in fact, substantiated experimentally. An extension of this argument would imply that 12 electrons would fill both the t_{1u} and t_{1g} levels, following the level ordering of Fig. 12.1. The evidence regarding the filling of the t_{1g} levels in accordance with these simple ideas has been presented by experiments on alkaline earth doping of C_{60} with Ca, Ba, and Sr [12.27–30] as well as by theoretical studies [12.31]. These simple level-filling concepts have also been used in a more limited way to account for the alkali metal doping of C_{70}.

12.4.1. Ground States for Free Ions $C_{60}^{n\pm}$

More quantitative information about the electronic states for the $C_{60}^{n\pm}$ ions can be found first by considering the Pauli-allowed states associated with these ions, for both their ground states and excited states. The ground state configurations for the various $C_{60}^{n\pm}$ ions can be specified by Hund's rule starting from a spherical model, and the results are summarized in Table 12.2 for C_{60}^{n-} anions ($-6 \leq n \leq 0$) and in Table 12.3 for C_{60}^{n+} cations ($0 \leq n \leq 5$). On the left-hand side of these tables, the angular momentum values for S, L, and J from Hund's rule are given, assuming spherical symmetry for the $C_{60}^{n\pm}$ ions, $-6 \leq n \leq +5$. In going from the spherical ground state approximation, specified by J in spherical symmetry, to icosahedral I_h symmetry, level splittings occur, as indicated on the right-hand side of Tables 12.2 and 12.3. Here the electronic orbital configuration for each $C_{60}^{n\pm}$ ion is listed in I_h symmetry, together with levels of the ground state Pauli-allowed multiplet of this configuration in I_h symmetry, and the hyperfine structure for each level of the multiplet. Since the spin–orbit interaction in carbon is small (see Table 3.1), the $L \cdot S$ coupling scheme is valid for fullerene molecules. This has an important bearing on the optical transitions which are spin-conserving ($\Delta S = 0$), as described in §13.1.3.

Referring to Table 12.2, we see that neutral C_{60} has a nondegenerate A_g ground state, and so does C_{60}^{6-}. However, the other ions have degenerate ground states. For example, C_{60}^{3-} corresponding to the half-filled t_{1u} (f_{1u}) level (mentioned above) has a Hund's rule quartet ${}^4A_u$ ground state. We would therefore expect a Jahn–Teller symmetry-lowering distortion to occur for C_{60}^{3-} [12.32], as discussed further in §13.4.1. Likewise, Jahn–Teller symmetry-lowering distortions would be expected for the other degenerate $C_{60}^{n\pm}$ ground states listed in Tables 12.2 and 12.3 [12.5]. The effect of strong interactions between electrons and the intramolecular vibrations has been shown to lead to an electron pairing for the anions [12.33].

12.4.2. Excited States for Negative Molecular Ions C_{60}^{n-}

In general, the ground states discussed in §12.4.1 are the lowest-energy states of a ground state multiplet. In this section we discuss the excited

Table 12.2

Various Pauli-allowed states associated with the ground state configurations for the icosahedral C_{60}^{n-} anions.

n	$SU(2)^a$	S	L	J^b	I_h Sym.	I_h	I_h State^d	HOMO
6−	$\ldots h^{16}$	3	15	18^c	A_g F_{1g} $2F_{2g}$ $3G_g$ $3H_g$	$..f_{1u}^6$	${}^1A_g{}^c$	$A_g{}^c$
5−	$\ldots h^{15}$	$\frac{7}{2}$	14	$\frac{35}{2}{}^c$	Γ_6^- Γ_7^- $2\Gamma_8^-$ $4\Gamma_9^-$	$..f_{1u}^5$	${}^2F_{1u}{}^c$	$\Gamma_6^-\ \Gamma_8^{-c}$
4−	$\ldots h^{14}$	4	12	16^c	A_g $2F_{1g}$ F_{2g} $2G_g$ $3H_g$	$..f_{1u}^4$	${}^1A_g$ ${}^3F_{1g}^c$ ${}^1H_g$	A_g A_g, F_{1g}, H_g^c H_g
3−	$\ldots h^{13}$	$\frac{9}{2}$	9	$\frac{27}{2}{}^c$	Γ_7^- $2\Gamma_8^-$ $3\Gamma_9^-$	$..f_{1u}^3$	${}^4A_u^c$ ${}^2F_{1u}$ ${}^2H_u$	Γ_8^{-c} Γ_6^-, Γ_8^- Γ_8^-, Γ_9^-
2−	$\ldots h^{12}$	5	5	10^c	A_g F_{1g} F_{2g} G_g $2H_g$	$..f_{1u}^2$	${}^1A_g$ ${}^3F_{1g}^c$ ${}^1H_g$	A_g A_g^c, F_{1g}, H_g H_g
1−	$\ldots h^{11}$	$\frac{11}{2}$	0	$\frac{11}{2}{}^c$	Γ_6^- Γ_8^- Γ_9^-	$..f_{1u}^1$	${}^2F_{1u}{}^c$	$\Gamma_6^{-c}, \Gamma_8^-$
0	$\ldots h^{10}$	5	5	0^c	A_g	$..h_u^{10}$	${}^1A_g{}^c$	$A_g{}^c$

[a] Assumes spherical symmetry.
[b] Only state of maximum multiplicity $S = n/2$ is listed.
[c] Hund's rule ground state. The ordering in this table is according to the J values for the free molecule.
[d] All Pauli-allowed states are listed.

states associated with these multiplets for the negatively charged free ion states, while in §12.4.3 we discuss the corresponding states for the positively charged C_{60} ions, and in §12.5 we discuss excited states formed by excitation of electrons to higher-lying electronic configurations (e.g., by excitation across the HOMO–LUMO gap), such as would occur in the optical excitation of neutral C_{60}.

The simplest example of an excited state within the ground state configuration comes from the spin–orbit splitting of the ground state configuration

Table 12.3

Various Pauli-allowed states associated with the ground state configurations for the icosahedral C_{60}^{n+} cations.

n	$SU(2)^a$	S	L	J^b	I_h Sym.	Val.	I_h state^d	HOMO
0	$\ldots h^{10}$	5	5	0^c	A_g	$\ldots h_u^{10}$	${}^1A_g{}^c$	$A_g{}^c$
1+	$\ldots h^9$	$\frac{9}{2}$	9	$\frac{9}{2}{}^c$	Γ_8^- Γ_9^-	$\ldots h_u^9$	${}^2H_u{}^c$	$\Gamma_8^-, \Gamma_9^{-c}$
2+	$\ldots h^8$	4	12	8^c	F_{2g} G_g $2H_g$	$\ldots h_u^8$	${}^1A_g$ ${}^3F_{1g}$ ${}^1H_g$ ${}^3F_{2g}$ ${}^3G_g^c$ ${}^1G_g$ ${}^1H_g$	A_g A_g, F_{1g}, H_g H_g G_g, H_g^c F_{2g}, G_g, H_g^c G_g H_g
3+	$\ldots h^7$	$\frac{7}{2}$	14	$\frac{21}{2}{}^c$	Γ_6^- $2\Gamma_8^-$ $2\Gamma_9^-$	$\ldots h_u^7$	${}^2F_{1u}$ ${}^4F_{1u}$ $2[{}^2H_u]$ $({}^2F_{2u}, {}^2G_u)$ $({}^4F_{2u}, {}^4G_u)^c$ $({}^2G_u, {}^2H_u)$ $({}^2F_{1u}, {}^2F_{2u}, {}^2H_u)$	Γ_6^-, Γ_8^- $\Gamma_6^-, \Gamma_8^-, \Gamma_9^-$ $2[\Gamma_8^-, \Gamma_9^-]$ $\Gamma_9^-, (\Gamma_7^-, \Gamma_9^-)$ $(\Gamma_8^-, \Gamma_9^-, \Gamma_7^-, \Gamma_9^-), (\Gamma_8^-, \Gamma_9^-)^c$ $(\Gamma_7^-, \Gamma_9^-), (\Gamma_8^-, \Gamma_9^-)$ $(\Gamma_8^-, \Gamma_9^-), (\Gamma_6^-, \Gamma_8^-, \Gamma_9^-)$
4+	$\ldots h^6$	3	15	12^c	A_g F_{1g} F_{2g} $2G_g$ $2H_g$	$\ldots h_u^6$	$2[{}^1A_g]$ $2[{}^3F_{1g}]$ ${}^5H_g^c$	$2[A_g]$ $2[A_g, F_{1g}, H_g]$ $A_g, F_{1g}, H_g, (F_{2g} + G_g), (G_g + H_g)^c$
							⋮	⋮
5+	$\ldots h^5$	$\frac{5}{2}$	15	$\frac{25}{2}{}^c$	Γ_6^- Γ_7^- Γ_8^- $3\Gamma_9^-$	$\ldots h_u^5$	${}^2A_u$ ${}^6A_u$	Γ_6^- Γ_9^{-c}
							⋮	⋮

[a] Assumes spherical symmetry.

[b] Only state of maximum multiplicity is listed, i.e., $S = n/2$.

[c] Hund's rule ground state. The ordering in this table is according to the J values for the free molecule.

[d] All Pauli-allowed states for the valence configuration (Val.) are listed for $n = 0, 1+, 2+, 3+$. For $n = 4+, 5+$ only selected states are listed.

$h_u^{10} f_{1u}^1$, which arises from the addition of a single electron to C_{60} (see Table 12.2). Information about this unpaired electron can be obtained from EPR studies (see §16.2.2) The $SU(2)$ notation in Table 12.2 denotes the electronic configuration in spherical symmetry. Physically, the electron in the f_{1u}^1 configuration is delocalized on the shell of the C_{60} molecule and interacts with the other 60 π valence electrons of C_{60}^-. The orbital part of this electronic configuration transforms as f_{1u} and the spin ($S = 1/2$) as Γ_6^+ so that on taking the direct product, we obtain the symmetries for the j states

$$F_{1u} \otimes \Gamma_6^+ \rightarrow \Gamma_6^- + \Gamma_8^-, \tag{12.7}$$

where Γ_6^- forms the ground state ($j = 1/2$) and Γ_8^- forms the excited state ($j = 3/2$) of the multiplet, as listed in Table 12.2. The label HOMO for the last column in Table 12.2 denotes levels which include the spin–orbit interaction in icosahedral I_h symmetry; i.e., C_{60}^- is described by double group representations for configurations with an odd number of electrons.

A more interesting case is that found for the ground state multiplet for C_{60}^{2-} which has an $h_u^{10} f_{1u}^2$ electronic configuration. Here the orbital part of the multiplet is described by the states contained in the direct product

$$F_{1u} \otimes F_{1u} \rightarrow A_g + F_{1g} + H_g. \tag{12.8}$$

These orbital states must be combined with spin states through the spin–orbit interaction to form Pauli-allowed singlet ($S = 0$) and triplet ($S = 1$) states, for which the total wave functions are antisymmetric under interchange of any two electrons. Since the F_{1g} state is orbitally antisymmetric ($L = 1$), whereas the A_g and H_g states are symmetric ($L = 0, 2$, respectively), the only Pauli-allowed states for which the total wave functions are antisymmetric under interchange of electrons are the ${}^1A_g$, ${}^3F_{1g}$, and ${}^1H_g$ states indicated in Table 12.2 under C_{60}^{2-}. States in this multiplet are split in energy by relatively small interactions (perhaps ~0.2 eV). Extending the Hund's rule arguments to icosahedral symmetry, we argue that the ground state is the one where S is a maximum ($S = 1$), suggesting ${}^3F_{1g}$ as the ground state. Spin–orbit interaction splits the ${}^3F_{1g}$ level into a multiplet, via the direct product of spin and orbit ($F_{1g} \otimes F_{1g} = A_g + F_{1g} + H_g$), corresponding to $j = 0, 1, 2,$. where the minimum j value is identified with the A_g ground state for a less than half-filled shell. The singlet ${}^1H_g$ state is not split by the spin–orbit interaction since $S = 0$ for this state (see Table 12.2).

The three-electron configuration for C_{60}^{3-} corresponding to $h_u^{10} f_{1u}^3$ has symmetries for its ground state multiplet that are found by taking the direct products

$$F_{1u} \otimes F_{1u} \otimes F_{1u} \rightarrow A_u + 3F_{1u} + F_{2u} + G_u + 2H_u. \tag{12.9}$$

This configuration is important since it corresponds to M_3C_{60}, which is a relatively high-T_c superconductor for M = K or Rb (see §15.1). Then considering the spin wave functions for three electrons, we can have $S = 3/2$ and $S = 1/2$ spin states yielding the Pauli-allowed states for this configuration: ${}^4A_u$, ${}^2F_{1u}$, and ${}^2H_u$, which are listed in Table 12.2. Since the ${}^4A_u$ state has the maximum S value of 3/2, it is likely to be the ground state, via extended Hund's rule arguments. This level transforms as the fourfold irreducible representation Γ_8^- of the point group I_h. The ${}^2F_{1u}$ level corresponding to $L = 1$ transforms under the spin–orbit interaction as $F_{1u} \otimes \Gamma_6^+ = \Gamma_6^- + \Gamma_8^-$, resulting in a level splitting. Finally, the ${}^2H_u$ level under spin–orbit interaction transforms as $H_u \otimes \Gamma_6^+ = \Gamma_8^- + \Gamma_9^-$, also resulting in a spin–orbit splitting. In addition to the ground state multiplet associated with the configuration f_{1u}^3, many higher-lying multiplets are created by exciting one (or more) of the three electrons into a higher-lying state, as discussed below in §12.5.

As the f_{1u} shell becomes more than half full (see Table 12.2), as for example for C_{60}^{4-}, then the extended Hund's rule for the state j of the multiplet suggests that the maximum j value should be selected for the ground state configuration. In general, the ground state levels for a more than half-filled state can be considered as hole levels. These hole levels correspond to the same structure as the corresponding electron state of each multiplet under the electron–hole duality, except for the ordering of the levels in energy. Thus the multiplet levels for C_{60}^{4-} and C_{60}^{2-} are closely related, as are the multiplet levels for C_{60}^{5-} and C_{60}^{1-}. In the following section (§12.4.3) we consider hole multiplets for the HOMO states, in contrast to the hole multiplets considered in this section for the LUMO states.

12.4.3. Positive Molecular Ions C_{60}^{n+}

The electronic states for the positive ions C_{60}^{n+} can be treated similarly to the negative ions discussed in §12.4.2, and the Pauli-allowed states for C_{60}^{n+} are summarized in Table 12.3 for $n \leq 5$. Thus for positively charged C_{60}^{+}, the configuration in I_h symmetry is h_u^9, which corresponds to one hole in the HOMO level and thus has fivefold orbital degeneracy and a spin of 1/2. When spin–orbit interaction is taken into account, the hyperfine splitting in icosahedral symmetry corresponds to the level splitting

$$H_u \otimes \Gamma_6^+ \rightarrow \Gamma_8^- + \Gamma_9^-, \tag{12.10}$$

where Γ_9^- forms the ground state ($j = 5/2$), as listed in Table 12.3, and Γ_8^- is the split-off level, which being a hole level, is more negative in energy. As discussed earlier, the Jahn–Teller effect is expected to give rise to a lower-symmetry molecule.

For calculation of the electronic levels of C_{60}^{2+} [12.34], the h_u^8 configuration has orbital symmetries

$$H_u \otimes H_u \rightarrow A_g + F_{1g} + F_{2g} + 2G_g + 2H_g, \quad (12.11)$$

as listed in Table 12.3. When combined with the spin singlet ($S = 0$) and triplet ($S = 1$) states, we find that the Pauli-allowed states for h_u^8 have the following symmetries: ${}^1A_g$, ${}^3F_{1g}$, ${}^1H_g$, $({}^3F_{2g} + {}^3G_g)$, and $({}^1G_g + {}^1H_g)$ as shown in Table 12.3, corresponding to L values of $0, 1, 2, 3, 4$, respectively. The effect of spin–orbit interaction on these levels is given in the far right-hand column.

Since the ground state multiplets for the half-filled hole configuration h_u^5 and nearly half-filled h_u^6 configuration have many constituents, the levels for these multiplets are not written in detail in Table 12.3. For the configurations $h_u^1, \ldots, h_u^4$ where the h_u level is less than half filled with electrons, the same symmetries for the multiplets appear as for the less than half-filled h_u hole states using the electron–hole duality, except that the level orderings are expected to be different for n holes as compared with n electrons.

12.5. Excitonic States for C_{60}

Since the ground state multiplet h_u^{10} for the neutral C_{60} molecule consists of a single nondegenerate level, optically excited states are formed by promoting one of the electrons in the h_u shell to a higher configurational multiplet. For C_{60} the lowest excited multiplet would correspond to the $h_u^9 f_{1u}^1$ configuration and the next higher-lying excited multiplet to the $h_u^9 f_{1g}^1$ configuration, using the level ordering of the Hückel calculation [12.4] or of a more detailed calculation [12.7, 18]. The simple physical realization of these excited states is the electron–hole pair created by optical excitation through the absorption of a photon, as discussed further in §13.1.3, the hole corresponding to the h_u^9 term and the electron to either the f_{1u}^1 or f_{1g}^1 term.

To find the symmetries of the lowest-lying exciton multiplet, we take the direct product of the excited electron (F_{1u}) with that for the hole (H_u), to obtain $F_{1u} \otimes H_u = H_g + G_g + F_{2g} + F_{1g}$, as listed under the $h^9 f_1^1$ configuration in Table 12.4. Since both the electron and hole have a spin of 1/2, we can form singlet ($S = 0$, A_g symmetry) and triplet ($S = 1$, F_{1g} symmetry) states. Including the spin–orbit interaction, leads to Pauli-allowed singlet states ${}^1H_g + {}^1G_g + {}^1F_{2g} + {}^1F_{1g}$, and the corresponding triplet states are found by taking the direct product of F_{1g} (for $S = 1$) with each of the orbital terms $H_g + G_g + F_{2g} + F_{1g}$ yielding the triplet states given in Table 12.4 under $h_u^9 f_{1u}^1$. For example, the levels arising from the triplet ${}^3H_g$ state are found by taking the direct product $F_{1g} \otimes H_g = H_g + G_g + F_{2g} + F_{1g}$, as listed

Table 12.4

States for the neutral C_{60}^0 molecule allowed by the Pauli principle.

Config.	Singlet		Triplet	
h^{10}	^1A	A	–	–
h^9a^1	^1H	H	^3H	F_1+F_2+G+H
$h^9f_1^1$	^1H	H	^3H	F_1+F_2+G+H
	^1G	G	^3G	F_2+G+H
	$^1F_2$	F_2	$^3F_2$	$G+H$
	$^1F_1$	F_1	$^3F_1$	$A+F_1+H$
$h^9f_2^1$	^1H	H	^3H	F_1+F_2+G+H
	^1G	G	^3G	F_2+G+H
	$^1F_2$	F_2	$^3F_2$	$G+H$
	$^1F_1$	F_1	$^3F_1$	$A+F_1+H$
h^9g^1	^1H	H	^3H	F_1+F_2+G+H
	^1H	H	^3H	F_1+F_2+G+H
	^1G	G	^3G	F_2+G+H
	$^1F_2$	F_2	$^3F_2$	$G+H$
	$^1F_1$	F_1	$^3F_1$	$A+F_1+H$
h^9h^1	^1H	H	^3H	F_1+F_2+G+H
	^1H	H	^3H	F_1+F_2+G+H
	^1G	G	^3G	F_2+G+H
	^1G	G	^3G	F_2+G+H
	$^1F_2$	F_2	$^3F_2$	$G+H$
	$^1F_1$	F_1	$^3F_1$	$A+F_1+H$
	^1A	A	^3A	F_1

in Table 12.4. The corresponding set of levels are found for the excitonic configuration $h_u^9 f_{1g}^1$ except for a parity change for all orbital levels of the multiplet. Table 12.4 in addition lists Pauli-allowed states for all symmetry types that are possible starting with optical excitation from the C_{60} HOMO denoted by h_u^{10}. The Jahn–Teller distortion associated with the lowest-lying $^1F_{1g}$ state arising from the $h_u^9 f_{1u}^1$ multiplet has been calculated by several groups [12.35].

As another illustration of excited configurational states, we consider the excitonic states for the C_{60}^{3-} ion. In this case we can excite an electron from the f_{1u}^3 configuration to a higher-lying configuration $f_{1u}^2 f_{1g}^1$ or to a yet higher-lying configuration such as $f_{1u}^2 f_{2u}^1$. The corresponding doublet and quartet Pauli-allowed states ($S = 1/2$ and $S = 3/2$) for these and many other configurations are listed in Table 12.5, following the same notation

Table 12.5

States for the C_{60}^{3-} anion allowed by the Pauli principle.

Config.	Doublet		Quadruplet	
f_1^3	^2H	$\Gamma_8+\Gamma_9$	–	–
	$^2F_1$	$\Gamma_6+\Gamma_8$	–	–
	–	–	^4A	Γ_8
$f_1^2a^1$	^2H	$\Gamma_8+\Gamma_9$	–	–
	$^2F_1$	$\Gamma_6+\Gamma_8$	$^4F_1$	$\Gamma_6+\Gamma_8+\Gamma_9$
	^2A	Γ_6	–	–
$f_1^2f_1^1$	^2H	$\Gamma_8+\Gamma_9$	–	–
	^2H	$\Gamma_8+\Gamma_9$	^4H	$\Gamma_6+\Gamma_7+\Gamma_8+2\Gamma_9$
	^2G	$\Gamma_7+\Gamma_9$	–	–
	$^2F_2$	Γ_9	–	–
	$^2F_1$	$\Gamma_6+\Gamma_8$	$^4F_1$	$\Gamma_6+\Gamma_8+\Gamma_9$
	$^2F_1$	$\Gamma_6+\Gamma_8$	–	–
	$^2F_1$	$\Gamma_6+\Gamma_8$	–	–
	^2A	Γ_6	^4A	Γ_8
$f_1^2f_2^1$	^2H	$\Gamma_8+\Gamma_9$	–	–
	^2H	$\Gamma_8+\Gamma_9$	^4H	$\Gamma_6+\Gamma_7+\Gamma_8+2\Gamma_9$
	^2G	$\Gamma_7+\Gamma_9$	^4G	$\Gamma_8+2\Gamma_9$
	^2G	$\Gamma_7+\Gamma_9$	–	–
	$^2F_2$	Γ_9	–	–
	$^2F_2$	Γ_9	–	–
	$^2F_1$	$\Gamma_6+\Gamma_8$	–	–
$f_1^2g^1$	^2H	$\Gamma_8+\Gamma_9$	–	–
	^2H	$\Gamma_8+\Gamma_9$	^4H	$\Gamma_6+\Gamma_7+\Gamma_8+2\Gamma_9$
	^2H	$\Gamma_8+\Gamma_9$	–	–
	^2G	$\Gamma_7+\Gamma_9$	^4G	$\Gamma_8+2\Gamma_9$
	^2G	$\Gamma_7+\Gamma_9$	–	–
	^2G	$\Gamma_7+\Gamma_9$	–	–
	$^2F_2$	Γ_9	$^4F_2$	$\Gamma_7+\Gamma_8+\Gamma_9$
	$^2F_2$	Γ_9	–	–
	$^2F_1$	$\Gamma_6+\Gamma_8$	–	–
$f_1^2h^1$	^2H	$\Gamma_8+\Gamma_9$	–	–
	^2H	$\Gamma_8+\Gamma_9$	^4H	$\Gamma_6+\Gamma_7+\Gamma_8+2\Gamma_9$
	^2H	$\Gamma_8+\Gamma_9$	–	–
	^2H	$\Gamma_8+\Gamma_9$	–	–
	^2G	$\Gamma_7+\Gamma_9$	^4G	$\Gamma_8+2\Gamma_9$
	^2G	$\Gamma_7+\Gamma_9$	–	–
	^2G	$\Gamma_7+\Gamma_9$	–	–
	$^2F_2$	Γ_9	$^4F_2$	$\Gamma_7+\Gamma_8+\Gamma_9$
	$^2F_2$	Γ_9	–	–
	$^2F_1$	$\Gamma_6+\Gamma_8$	$^4F_1$	$\Gamma_6+\Gamma_8+\Gamma_9$
	$^2F_1$	$\Gamma_6+\Gamma_8$	–	–
	^2A	Γ_6	–	–

as for Table 12.4. Excitonic states can also be formed with a $h_u^9 f_{1u}^4$ configuration where the notation describes the electron–hole pair for the C_{60}^{3-} ion. Similar arguments can be applied in this case also to get the symmetries for multiplets corresponding to the $h_u^9 f_{1u}^4$ configuration. We note that the allowed f_{1u}^4 states for the $h_u^9 f_{1u}^4$ configuration are the same as the f_{1u}^2 holes in the $h_u^9 f_{1u}^2$ configuration (although the physical ordering of the states is expected to be quite different).

12.6. Molecular States for Higher-Mass Fullerenes

The electronic states for other fullerenes can be found using techniques similar to those described above for C_{60} except that in general the number of interacting electrons increases, the symmetry of the fullerene molecule decreases, and the possibility of large numbers of isomers is introduced.

12.6.1. Molecular States for C_{70}

The filling of the electronic levels for C_{70} on the basis of a spherical shell electron model is described in §12.1.2. However, molecular orbital calculations for a spherical shell model show that significant level mixing occurs between $\ell = 5$ and $\ell = 6$ orbital angular momentum states, and more detailed calculations are needed.

Calculations for the electronic structure for a C_{70} molecule using the local density approximation and density functional theory have been carried out and the calculated density of states [12.18, 36] has been compared with measured photoemission spectra [12.37–39]. These calculations indicate that C_{70} has a slightly higher binding energy per C atom relative to C_{60} (by ~20 meV) and that five distinct carbon atomic sites and eight distinct bond lengths must be considered for C_{70} (see §3.2). A bunching of levels is found for the C_{70} molecular states, just as for C_{60}, and in both cases the molecular levels show only small splittings of the spherical harmonic states, yielding a higher density of states for certain energy ranges. The energy width (~19 eV) from the deepest valence state to the highest-occupied valence state in C_{70} is similar to that in C_{60}, graphite, and diamond. A HOMO–LUMO gap of 1.65 eV is predicted for the free molecule on the basis of a local density approximation (LDA) calculation [12.18], somewhat smaller than the results for similar LDA calculations for the HOMO–LUMO gap for C_{60} (see Fig. 12.2).

Hückel calculations for C_{70} show that the states near the Fermi level have the symmetries and degeneracies shown in Fig. 12.3, where the level orderings for the neutral C_{70} molecule and for the negatively charged C_{70}^- and C_{70}^{2-} ions are shown. Here it is seen that in neutral C_{70} the highest-lying filled state is orbitally nondegenerate (with a_2'' symmetry), and similarly the lowest-lying unfilled state is also nondegenerate (with a_1'' symmetry). This electronic structure suggests that half-filled levels for electron-doped C_{70} correspond to the C_{70}^{1-} and C_{70}^{4-} ion species, in contrast to the half-filling of the LUMO level in C_{60}, which corresponds to C_{60}^{3-} anions.

The lower symmetry of C_{70} (D_{5h}) implies a more limited utility of group theoretical arguments. The energies of the calculated molecular orbitals for the neutral C_{70} molecule are shown in Fig. 12.3(a). However, the convergence of the perturbation scheme in §12.2 suggests that we consider the 70 electron configuration to have the following occupied states near the Fermi level $h_u^{10} f_{1u}^6 f_{1g}^4$, implying that the molecular states associated with two F_{1g} holes will have symmetries given by their direct product, yielding $F_{1g} \otimes F_{1g} = A_g + F_{1g} + H_g$ states. When spin is included, the Pauli-allowed states have ${}^1A_g + {}^3F_{1g} + {}^1H_g$ symmetries. The splitting of these states in D_{5h}

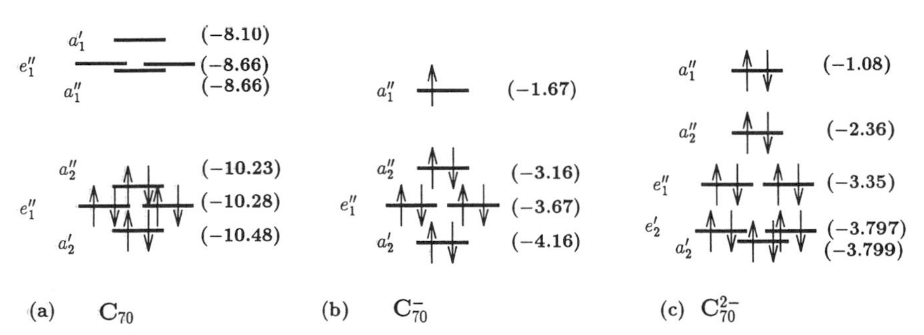

Fig. 12.3. Electronic levels (in units of eV) for (a) the neutral C_{70} molecule [12.22, 40], (b) the molecular ion C_{70}^- [12.19], and (c) the molecular ion C_{70}^{2-} [12.22, 41]. The various calculational schemes have largely focused on the bond lengths and the position of the LUMO levels (particularly for the neutral molecule). Calculations for the LUMO levels must be regarded as more uncertain than for the HOMO levels. The levels for the neutral molecule have been shifted to coincide with the anions.

symmetry suggests an ${}^1A_1'$ ground state with excited triplet states ${}^3A_2'$ and ${}^3E_2'$ and excited singlet states ${}^1A_1'$, ${}^1E_1'$, and ${}^1E_2'$, referring to the discussion of §4.4.1.

From the above discussion, we expect the ground state of C_{70} to have an $\ldots a_2'^2 e_1'^4 a_2''^2$ configuration with ${}^1A_1'$ symmetry (see Table 12.6). Also listed in this table are the level designations for the ground state multiplets for the various ions $C_{70}^{n\pm}$ ($-6 \leq n \leq +6$), first neglecting spin-orbit coupling (column 3) and then including spin-orbit coupling (column 4) under the heading Fine Structure. Excitonic excited states are thus expected to have an $a_2' a_1''^1$ configuration, which would have a ${}^1A_2'$ and a ${}^3A_2'$ excitonic excited state symmetry. The electronic structure of the C_{70}^- ion has been calculated

Table 12.6

Multiplet structures of the ground states of C_{70}^n (low spin state) [12.22].

n	Configuration	D_{5h} state[a]	Fine Structure
6−	$\ldots e_1''^4 a_2''^2 a_1''^2 e_1''^4$	${}^1A_1'(g)$	A_1'
5−	$\ldots e_1''^4 a_2''^2 a_1''^2 e_1''^3$	${}^2E_1''(g)$	Γ_7,Γ_8
4−	$\ldots e_1''^4 a_2''^2 a_1''^2 e_1''^2$	${}^1A_1'$ ${}^1E_2'$ ${}^3A_2'(g)$	A_1' E_2' A_1',E_1'
3−	$\ldots e_1''^4 a_2''^2 a_1''^2 e_1''^1$	${}^2E_1''(g)$	Γ_7,Γ_8
2−	$\ldots e_1''^4 a_2''^2 a_1''^2$	${}^1A_1'(g)$	A_1'
1−	$\ldots e_1''^4 a_2''^2 a_1''^1$	${}^2A_1''(g)$	Γ_7^-
0	$\ldots a_2'^2 e_1''^4 a_2''^2$	${}^1A_1'(g)$	A_1'
1+	$\ldots a_2'^2 e_1''^4 a_2''^1$	${}^2A_2''(g)$	Γ_7^-
2+	$\ldots a_2'^2 e_1''^4$	${}^1A_1'(g)$	A_1'
3+	$\ldots a_2'^2 e_1''^3$	${}^2E_1''(g)$	Γ_7,Γ_8
4+	$\ldots a_2'^2 e_1''^2$	${}^1A_1'$ ${}^1E_2'$ ${}^3A_2'(g)$	A_1' E_2' A_1',E_1'
5+	$\ldots a_2'^2 e_1''^1$	${}^2E_1''(g)$	Γ_7,Γ_8
6+	$\ldots a_2'^2$	${}^1A_1'(g)$	A_1'

[a]The Hund's rule ground state is denoted by (g).

[12.19], yielding the level ordering shown in Fig. 12.3(b), which closely relates to that calculated for C_{70} in Fig. 12.3(a). The breakdown of the simple symmetry arguments for C_{70} and C_{70}^{-} given in §4.4.1 indicates that detailed calculations are needed to specify the level ordering for the lower symmetry fullerenes.

12.6.2. *Metallofullerenes*

Calculations of the electronic structure for the endohedral metallofullerenes K@C_{60} and Cs@C_{60} [12.42–44] show charge transfer of one electron, from the endohedral metal atom to the C_{60} shell. More detailed calculations for a variety of metallofullerenes show that the ions tend to lie off-center and the negative charge on the fullerene cage is delocalized [12.45, 46] (see §5.4 and §8.2). The amount of off-center displacement has been shown to depend on the charge transfer and the size of the ion. The dependence on the ion size is estimated by comparing the equilibrium positions of Na and K in C_{60} and La and Y in C_{60} [12.46]. Electronic energy level calculations for Al@C_{60} show a single electron charge transfer from the Al to C_{60} and very little off-center displacement of Al^+ within the fullerene cage. The corresponding results for La@C_{60} (see Fig. 12.4) show a lifting of the degeneracy of the HOMO and LUMO levels of C_{60} and a significant lowering of the energy of the LUMO levels as the La shifts to an off-center position. A similar result was reported for Y@C_{60} [12.46, 47], although for Y@C_{60} the d levels of the Y are not close to the C_{60} LUMO levels, as is the case for La@C_{60}, shown in Fig. 12.4.

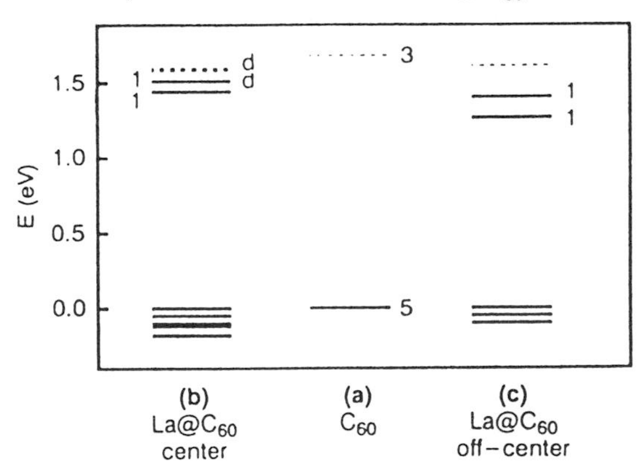

Fig. 12.4. Energy level diagram at the HOMO–LUMO gap region: for C_{60} (a), La@C_{60} with La at the center of the shell (b), and the La at an off-center position corresponding to the lowest energy state (c) [12.46]. In this diagram the solid lines denote filled states and the dashed lines denote empty states. The presence of the endohedral La splits the degenerate t_{1u} and h_u levels of C_{60}. The numbers 1, 3, 5 refer to orbital level degeneracy and the letter d denotes admixture of La $5d$ wavefunction to the indicated fullerene levels.

Experimental photoemission experiments on the $Ca@C_{60}$ cluster [12.48] support the calculated results (see Fig. 12.13), showing that two electrons are transferred to the carbon shell per alkaline earth dopant. In contrast, calculations of the electronic structure for $Cl@C_{60}$ and $O@C_{60}$ [12.43, 49] do not show such charge transfer.

Motivated by the extraction and isolation of higher-mass endohedrally doped fullerenes, namely $La@C_{82}$, $Sc_2@C_{82}$, $Sc_2@C_{84}$, $Sc_3@C_{84}$, etc., from the soot (see §5.4) and by experimental EPR spectra in $Sc_3@C_{82}$ and other metallofullerenes [12.50–54] (see §16.2.1), calculations of the electronic structure of metallofullerenes have been carried out, for specific fullerene isomers. The emphasis of the calculations has been directed to identifying the most stable isomers supporting endohedral doping [12.55, 57, 58, 130], identifying the amount of charge transfer, and the location of the dopant species within the fullerene shell [12.46]. These calculations have also been directed toward obtaining the best agreement with experimental EPR spectra [12.59], photoemission data [12.48, 60], and other physical measurements. Self-consistent calculations, for which the carbon shell atoms are allowed to relax in response to the endohedral dopant, show La^{3+} to be the stable charge state for forming $La@C_{82}$. In contrast, La^{2+} is stable for exohedral doping, although endohedral doping yields a lower overall energy state. The bonding between the internal La^{3+} ion and the carbon cage atoms leads to extensive differences between the properties of the doped and undoped fullerenes [12.60]. These calculations also show significant differences in electronic properties, depending on whether the La is on an on-center or off-center position within the metallofullerene. Since La^{3+} donates three electrons to the fullerene shell, two electrons can be paired, but one remains unpaired, forming a singly occupied molecular orbital (SOMO) for the isolated $La@C_{82}$ metallofullerene. It is reported [12.60] that in the solid phase relatively strong intermolecular interactions split the SOMO-derived band and the occupied portion of this band is centered 0.64 eV below E_F, so that $La@C_{82}$ is expected to be nonmetallic. The La $5d$ states do not strongly mix with the π-cage orbitals and the La $5d$ states remain localized and empty about 1.2 eV above the SOMO level [12.57]. However, the La $5p$ core states do hybridize with the π-cage orbitals. The intermolecular bonding for $La@C_{82}$ is found to be considerably stronger than for C_{82} itself, leading to a greatly reduced vapor pressure of $La@C_{82}$ [12.61]. Regarding the molecular distortions, the calculations (see Fig. 12.5) show that the largest displacements of the carbon distances on the C_{82} shell occur for carbon atoms close to the off-center La ion [12.60]. Because of the large amount of experimental activity in this area, the calculation of the electronic structure of various metallofullerene isomers is expected to remain an active field for some time.

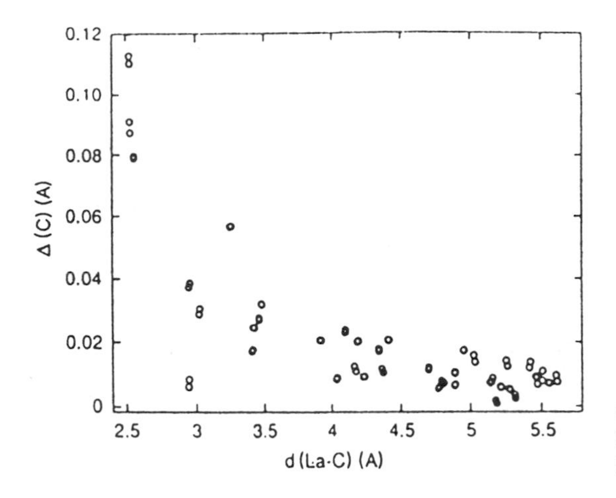

Fig. 12.5. Displacement of the carbon atoms from C_{82} to La@C_{82}, as a function of the La–carbon distance [12.60].

12.7. Electronic Structure of Fullerenes in the Solid State

In this section we first review unusual aspects of the electronic structure for fullerenes in the solid state and approaches that have been taken to treat the electronic structure of these unusual materials. In the previous sections it has been emphasized that C_{60} is a molecular solid, thereby implying that the electronic structure is primarily determined by that of the free molecule, although the intermolecular interaction is of course significant, as evidenced by the magenta color of C_{60} in solution and the yellow color of the solid film [12.62]. To date, most of the experimental and theoretical work on the electronic structure has been directed to C_{60} and doped C_{60}, and this historical emphasis is reflected in the review presented here. A summary of progress in understanding the electronic structure of higher fullerenes is also included.

12.7.1. Overview of the Electronic Structure in the Solid State

Numerous experiments regarding the lattice vibrations (§11.4), optical properties (§13.3), photoemission and inverse photoemission (§17.1), and Auger spectroscopy (§17.3) indicate the molecular nature of fullerite solids, based on the narrow bandwidths, the weak dispersion of these energy bands, and the strong intramolecular Coulomb interaction of holes and electrons. From an experimental standpoint, the highly molecular nature of fullerite solids makes it possible to carry out classes of experiments that normally are not possible in other crystalline materials, such as the study of harmonics and combination modes in the vibrational spectra (§11.5.3 and §11.5.4).

To treat the electronic structure of fullerite solids, two approaches have been taken. One approach involves one-electron band calculations, exploiting the periodicity of the crystal lattice of fullerenes in the solid state. In this approach the molecular levels of the fullerenes are treated much as the electronic levels of atoms in more conventional solids, and the energy bands are calculated on the basis of these molecular states. Thus the highest occupied (h_u) state of the C_{60} cluster forms the valence band and the lowest unoccupied (t_{1u}) state forms the conduction band. This approach has the advantage that sophisticated energy band calculational techniques are presently available and many computer codes have been developed, so that many fundamental issues which relate to structure–property relations can be addressed [12.63]. For these reasons band structure calculations have added in a major way toward increasing our understanding of experiments pertaining to the electronic structure and the density of electronic states. *Ab initio* calculations, using for example the local density approximation (LDA), allow calculation also of crystal structures and lattice constants, although most calculations that have thus far been reported assume experimentally determined crystal structures and lattice constants. Differences between the many *ab initio* calculations [12.25, 64–71] are likely due to small differences in the selection of lattice constants and structural geometry [12.72], and also the calculational techniques that are used [12.5]. Modifications to the calculated band structure due to Coulomb interactions, narrow band features, intramolecular effects, merohedral disorder (see §7.1.4), and other effects described by interaction energies comparable to the C_{60}–C_{60} cohesive energy of 1.6 eV are then considered.

A second approach considers the intramolecular interactions to be dominant and band effects to be less important. This more chemical approach has been most helpful in gaining physical understanding of experimental results, such as the relation between optical absorption and luminescence spectra, high-resolution Auger spectra, and photoemission–inverse photoemission spectra [12.73]. The chemical approach has also been helpful in elucidating the magnitudes of many-body effects, such as the Coulomb onsite repulsion U between two holes [12.73] (see §12.7.7). Both approaches have thus provided complementary insights into the electronic structure of fullerite solids, as reviewed below. At the present time, neither approach has by itself been able to explain all the experimental observations.

Both approaches to the calculation of the electronic structure emphasize the narrow band nature of this highly correlated electron system with an electron density of states close to that of the free molecule. Both approaches emphasize the value of photoemission and inverse photoemission experiments for providing determinations of the HOMO–LUMO gap, without complications from excitonic and vibronic effects associated with interband excitations. Arguments, based on the low density of dangling bonds, suggest that surface reconstruction is not as important for fullerite surfaces as for typical semiconductors. Although photoemission and inverse photoemission experiments are generally highly surface-sensitive, they are expected to yield information on bulk properties for the case of fullerenes, because of the low density of dangling bonds and surface states.

A totally different kind of electronic structure calculation is the phenomenological tight-binding approach that employs crystal symmetry and uses experimental measurements to determine the coefficients (band parameters) of the model. Such calculations have now been carried out for C_{60} [12.74, 75] and the expansion coefficients of these models have been evaluated by comparison to experiment [12.74–77]. Because of the relative simplicity of the tight-binding calculation, more complicated physical cases can be handled, such as the quadridirectional orientational ordering of C_{60} below T_{01} (see Fig. 7.5). Optimization of intermolecular bond lengths and geometrical representations can be handled relatively easily by such an approach [12.78–80]. Since tight binding calculations are so closely related to group theoretical approaches, they can be used in conjunction with *ab initio* calculations in desired ways, by parameterizing the *ab initio* calculations and extending them to include additional interactions, using parameters that are evaluated by comparison with experiment.

The discussion given here for the various approaches for calculating the electronic structure of crystalline C_{60} also applies to calculating the electronic structure for both lower-mass and higher-mass fullerenes (see §12.7.6), as well as to doped fullerenes (see §12.7.3), as discussed below.

12.7.2. Band Calculations for Solid C_{60}

The band approach to the electronic structure of C_{60} gained early favor because it provided important insights into important experimental progress in the field [12.25, 81]. Early band calculations found a band gap for C_{60} of ~1.5 eV, close to the onset of the experimental optical absorption and close to the low-energy electron energy loss peak at 1.55 eV [12.82–84]. Although later experiments showed the optical absorption edge to be more complicated and not simply connected to the HOMO–LUMO gap (see §13.1.2), the apparent early successes of the LDA approach stimulated many detailed band calculations on fullerene-based solids, which have since provided many valuable insights, even if answers to all the questions raised by the many experiments reported thus far have not yet been found.

A typical and widely referenced band calculation for C_{60} is the total energy electronic structure calculation using a norm-conserving pseudopotential, an LDA in density functional theory, and a Gaussian-orbital basis set

[12.25]. The results of this calculation for C_{60} are shown in Fig. 12.2(b), including the dispersion relations associated with the HOMO and LUMO levels of the C_{60} crystalline solid, where we note the calculated HOMO–LUMO band gap is 1.5 eV, and the bandwidth is ~0.4 eV for each of these bands. The calculation of the dispersion relations is sensitive to the crystal structure of solid C_{60} and the lattice constant. This small calculated bandwidth is in agreement with optical and transport experiments, although the calculated band gap by the LDA method is now known to be underestimated [12.5, 9, 25]. One success of the LDA approach to the C_{60} molecule has been the good agreement of the bond lengths along the pentagon edges a_5 and those shared by two hexagons a_6 from the LDA calculation ($a_5 = 1.45$ Å and $a_6 = 1.40$ Å) [12.25] and from nuclear magnetic resonance (NMR) measurements ($a_5 = 1.46$ Å and $a_6 = 1.40$ Å) [12.85, 86]. Since these bond lengths are unchanged as the molecule enters the solid phase, this result shows that the band calculations retain the molecular integrity of the fullerenes, as experiment requires.

The agreement of the calculated C_{60}–C_{60} cohesive binding energy (1.6 eV) with thermodynamic experiments (1.7 eV/C_{60}) [12.87] for the heat of sublimation/vaporization is another correct feature of band theory. Furthermore, LDA band theory gives a value of the C_{60}–C_{60} intermolecular distance that is only 4% smaller [12.25] and a pseudopotential approach yields a distance 1 to 2% smaller [12.64] than measured by x-rays [12.88]. Using the energy band dispersion relations calculated from a pseudopotential approach, an equation of state has been calculated [12.64] from which a bulk modulus of 16.5 to 18.5 GPa was obtained, in good agreement with early experimental results, but larger than experimental values reported later (see Table 7.1).

Electronic energy band calculations [12.25, 64, 69, 89] also offer predictions valuable for the interpretation of experiments, such as the energies of one-electron optical transitions and electron density contours (see Fig. 12.6). The charge density contours are important for understanding intermolecular charge transport during conduction (§14.1). Figure 12.2(b) further predicts C_{60} to be a direct gap semiconductor (gap at the X-point in the fcc Brillouin zone), in agreement with other LDA calculations, yielding values for the band gaps of 1.5 eV [12.7], 1.4 eV [12.90] and 1.2 eV [12.91]. These and other related calculations all locate the band extrema at the X-point in the Brillouin zone.

Band structure calculations for simple cubic (the low-temperature structure below T_{01}) and fcc C_{60} (the room temperature phase) have been carried out using the tight-binding calculation, showing the sensitivity of the band dispersion to the crystal structure [12.25, 75, 92]. These calculations show a narrowing of the h_u and t_{1u}-derived bands below T_{01}, and the density of

Fig. 12.6. Charge density map for a unidirectionally oriented array of C_{60} molecules normal to the (110) direction, showing weak coupling between neighboring C_{60} molecules [12.25].

states of the t_{1u}-derived band is found to split into two subbands, the lower subband with eight states and the upper level with four states for most regions of the Brillouin zone [12.75], as discussed further below.

Below 261 K each C_{60} molecule tries to align with respect to its 12 nearest-neighbors, so that its 12 electron-poor pentagons are located, as much as possible, opposite an electron-rich double bond on the nearest-neighbor C_{60} (see §7.1.3). Because of the difference in geometry between the 12 pentagons on a regular truncated icosahedron and the 12 nearest-neighbors of an fcc lattice, the alignment between the double bonds and pentagons on adjacent C_{60} molecules cannot be done perfectly, and some misalignment persists down to low temperatures, as discussed in §7.1.4. This issue has been addressed in band calculations, where the highest symmetry compatible with both the icosahedral symmetry of C_{60} and the fcc lattice ($Fm\bar{3}$ or T_h^3) has been considered [12.5, 9]. The effect of orientational disorder on the band structure has been considered by several groups [12.25, 93], showing that details in the band structure and density of states are sensitive to orientational disorder [12.94]. In particular, it has been shown that orientational disorder, finite resolution effects, and the multiband nature of the bands near E_F all lead to a reduction of the predicted dispersion in the angle-resolved photoemission [12.95] consistent with experimental findings [12.96, 97].

From the dispersion relations, effective mass estimates for the conduction and valence bands of C_{60} have been calculated ($m_e^l = 1.33m_0$, $m_e^t = 1.15m_0$ for the longitudinal and transverse effective electron masses at the X-point, and $m_h^l = 3.31m_0$, $m_h^t = 1.26m_0$ for the two valence band masses) [12.64]. Although one-electron band calculations are not able to explain the details of the optical spectra or the magnitude of the HOMO–LUMO gap (see

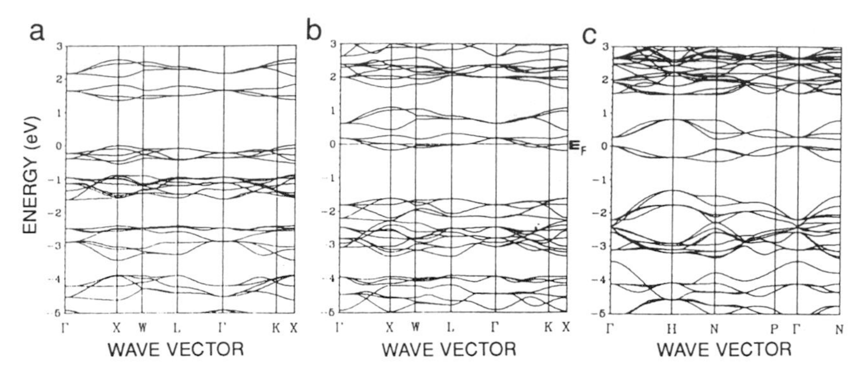

Fig. 12.7. Band structure of (a) fcc C_{60}, (b) fcc K_3C_{60}, and (c) body-centered cubic (bcc) K_6C_{60}. The zero energy is either at the top of the valence band or at the Fermi level [12.90].

§13.1.2), good agreement was obtained between the calculated density of states and photoemission and inverse photoemission measurements (§17.1) [12.7, 37, 38] with regard to the number of peaks and their relative placements in energy. The good agreement with the photoemission spectra may in part be due to the close relation between the one-electron bands and the molecular levels.

Charge density maps for C_{60} normal to the (100) direction (see Fig. 12.6), based on an LDA energy band calculation and the $Fm\overline{3}$ structure, show almost no charge density at the center of the fullerene (over a diameter of ~4 Å) and very little charge density between adjacent C_{60} molecules, even along the close-packed direction, consistent with the weak coupling between C_{60} molecules. Even lower charge density is found for crystalline C_{60} in the direction of the octahedral sites.

Band calculations have placed an octahedral site state at 6.5 eV above the valence band maximum and two levels at 7.0 and 7.2 eV for the tetrahedral site states [12.64]. These levels are expected to be significantly perturbed on filling of the states with electrons (see Fig. 12.7). The detailed bandwidths and density of states profiles are sensitive to the details of the structural symmetries and to the disorder [12.75]. The narrow bands and weak dispersion found in these calculations of the electronic structure are consistent with the molecular nature of C_{60}.

12.7.3. Band Calculations for Alkali Metal–Doped C_{60}

Local density approximation band calculations have also been extended to treat the electronic energy band structure of doped fullerene solids called fullerides. Most calculations for the doped fullerene compounds also refer to experimental determinations of the lattice constants and structural geometry, as was the case for pure C_{60}. For the doped fullerenes, distortions of the icosahedral shell and perturbations of bond lengths and structural geometry due to charge transfer effects may be important in some cases, such as the M_3C_{60} compounds (M = K, Rb) [12.72]. Such effects have been shown to be less important for K_6C_{60} and Rb_6C_{60} [12.67], where the t_{1u} band is completely filled with electrons.

Parallel calculations of the electronic structure for K_3C_{60} and K_6C_{60} [12.65, 90] are shown in Fig. 12.7(b) and (c), for the fcc and bcc crystal structures, respectively [12.98]. It is shown in this section that the one-electron approach to the electronic structure is capable of explaining a number of key experimental observations. Band calculations show that the cohesive energy between the fullerenes increases upon the addition of an alkali metal. Furthermore, the addition of the alkali metal atom into a tetrahedral site increases the cohesive energy of the fullerene solid more than when the alkali atom is in an octahedral site. Thus alkali metals tend to fill tetrahedral sites before octahedral sites, which in turn have a higher cohesive energy than an endohedral location for an alkali metal atom. These band calculations show that a large charge transfer (close to one electron per alkali metal atom) occurs upon alkali metal addition [12.25, 99]. Figure 12.8 shows that the charge transferred to the fullerenes serves to enhance the bonding between fullerenes. The difference between the charge density contours for C_{60} (Fig. 12.6) and for K_3C_{60} (Fig. 12.8) is striking in terms of the charge coupling between the fullerenes, suggesting that a band picture for K_3C_{60} has merit, and this is also supported by the observation of superconductivity in K_3C_{60}. The calculated electronic density of states for M_3C_{60} (M = K, Rb) (see Fig. 12.9) shows a double hump structure with roughly two thirds of the states in the lower-energy hump of Fig. 12.9. Because of the location of the Fermi level on the falling side of the lower-energy hump, it is expected that the various calculations

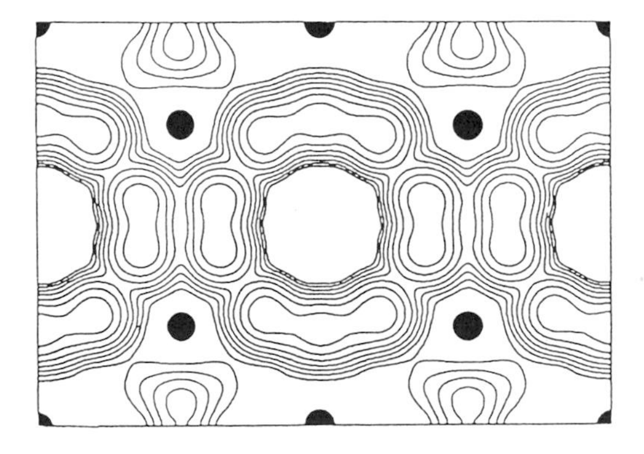

Fig. 12.8. Contour maps of the valence electron densities of K_3C_{60} on the (110) plane. Solid circles denote the tetrahedral and octahedral sites at which K atoms are located. The highest-density contour is 0.90 electrons/Å^3, and each contour represents half or twice the density of the adjacent contour [12.25].

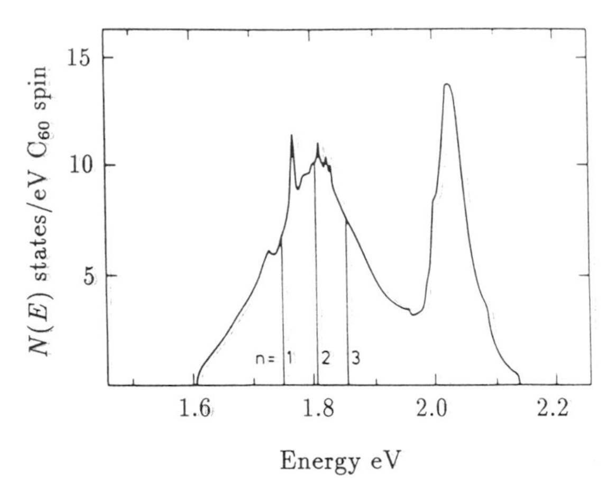

Fig. 12.9. Calculated density of states for unidirectionally ordered fcc C_{60}. The position of the Fermi level for M_nC_{60} with $n = 1, 2,$ and 3 is indicated, where M denotes an alkali metal such as K or Rb [12.75].

of $N(E_F)$ would give relatively high values for $N(E_F)$ for M_3C_{60} compounds (6.6–9.8 states/eV C_{60} spin) [12.66, 68, 75, 100–102], as well as a fairly large spread in values of $N(E_F)$. It should be mentioned that the oft-quoted $N(E_F) = 12.5$ states/eV C_{60} spin [12.25] is somewhat larger than that for other calculations [12.5, 72]. Despite the wide spread of values for $N(E_F)$ among the various calculations for K_3C_{60}, the ratio of $N(E_F)$ for Rb_3C_{60}, and K_3C_{60}, shows a much smaller spread (1.14–1.27) [12.72]. Another source of unreliability in the density of states calculations $N(E_F)$ stems from the common neglect of issues relating to merohedral disorder, insofar as the calculations are generally carried out for unidirectionally oriented fullerene molecules [12.72]. Tight binding calculations directed at exploring the effect of merohedral disorder on $N(E_F)$, however, find little change in the value of $N(E_F)$, although the density of states curve $N(E_F)$ is significantly broadened by the effect of merohedral disorder [12.93, 103, 104], in contrast to GW quasiparticle calculations which show a decrease in $N(E_F)$ with merohedral disorder, but little broadening of the t_{1u}-derived band [12.9]. Reasons why the tight-binding model works as well as it does for the M_3C_{60} compounds have been attributed to a very weak contribution to the LUMO wave functions and energy eigenvalues by the alkali metal s orbitals and to the lack of overlap of the LUMO t_{1u} band with higher-lying bands [12.72, 75]. It is further found that the density of states is insensitive to the dopant (whether K or Rb), except for differences in the lattice constant between K_3C_{60} and Rb_3C_{60} (see §14.5).

In the case of K_6C_{60} [see Fig. 12.7(c)], the Fermi level is located between the $t_{1u}(f_{1u})$ and the $t_{1g}(f_{1g})$-derived excited state levels, giving rise to a narrow gap semiconductor, in agreement with experiment. Although the connection to the molecular C_{60} levels is similar for $E(\mathbf{k})$ in the C_{60} and

M_6C_{60} crystalline phases, some differences arise because of the bcc crystal structure for the M_6C_{60} compounds (K_6C_{60}, Rb_6C_{60} and Cs_6C_{60}). Some interesting features about the M_6C_{60} bands are the calculated reduced band gap (0.8 eV) between the h_u and t_{1u}-derived bands and between the t_{1u} and t_{1g}-derived levels (0.3–0.5 eV), as the Fermi level for K_6C_{60} and other M_6C_{60} compounds lies between the t_{1u} and t_{1g} energy bands [12.65, 90]. Some calculations of $E(\mathbf{k})$ for M_6C_{60} give an indirect band gap [12.65], while others [12.90] suggest a direct gap at the P point of the bcc Brillouin zone. In both cases, the bands are broadened versions of the molecular levels, with little hybridization of the carbon-derived bands with the alkali metal levels. Some relaxation of the K ions in K_6C_{60} has been suggested, with lattice distortions that do not lower the lattice symmetry [12.67].

The most interesting case for the alkali metal-doped C_{60} occurs for M_3C_{60}, which corresponds to a half-filled band and a maximum in the conductivity σ versus composition x curve for M_xC_{60} (see §14.1.1). The band calculations further show that the Fermi surface for K_3C_{60} has both electron and hole orbits [12.81, 105] and that the superconducting transition temperature T_c depends approximately linearly on the density of states for alkali metal-doped C_{60} (see §15.2) [12.106]. To show the relative stability of various K_xC_{60} stoichiometries, the calculated total energy is plotted as a function of lattice constant in Fig. 12.10. The figure shows the relative stabilities of K_xC_{60}, with K_3C_{60} being the most stable (largest binding energy). As we will see below, the situation for the alkaline earth dopants is different because of the filling of the t_{1g} bands and the greater hybridization of the alkaline earth levels with the $E(\mathbf{k})$ for the fullerene bands near E_F.

Figure 12.11 shows the density of states for Na_2CsC_{60} as an example of an M_3C_{60} compound where the two small sodium atoms are in tetrahedral sites, while the larger Cs atom is in an octahedral site, as determined from structural measurements [12.107]. This figure is typical of LDA density of states calculations for M_3C_{60} compounds, showing narrow bands with some structure in the density of states $N(E)$. The magnitude of $N(E_F) =$ 22.4 states/eV-C_{60} (including both spin orientations) [12.25, 81, 106] is large compared to other calculations, $N(E_F) = 13.2$ states/eV/C_{60} [12.66], indicating a strong sensitivity of $N(E_F)$ to the details of the band calculation. The Fermi surface for K_3C_{60} is shown in Fig. 12.12 [12.66], where two inequivalent sheets are seen. One sheet is a closed spheroidal surface with protrusions in the $\langle 111 \rangle$ directions, while the second sheet is a multiply connected surface, reflecting contact of the Fermi surface with the Brillouin zone boundary and showing two (rather than four) necks along the $\langle 100 \rangle$ directions, indicative of the T_h^6 ($Pa\bar{3}$) lower-symmetry space group.

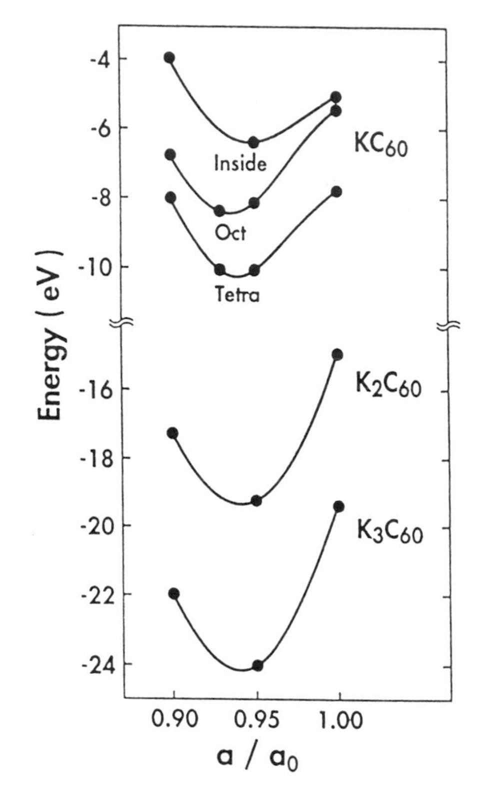

Fig. 12.10. Calculated total energy of K_3C_{60} as a function of lattice constant a, normalized to the experimental lattice constant a_0 of pristine C_{60}. The total energies are measured from the sum of the energies of the isolated C_{60} cluster and the isolated x potassium atoms. The total energy is the negative of the cohesive energy [12.25].

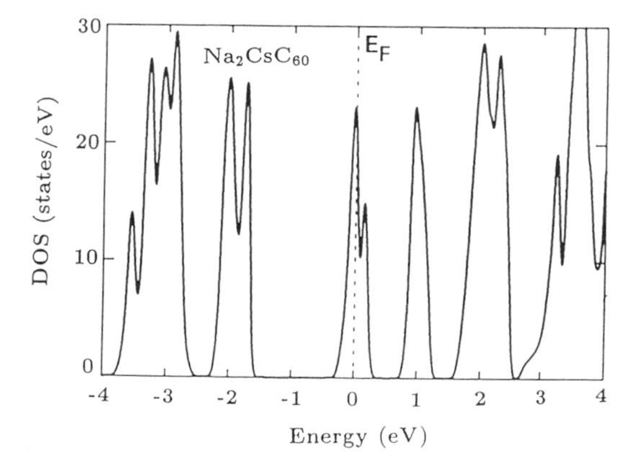

Fig. 12.11. Electronic density of states of the alkali metal compound Na_2CsC_{60} as a member of the M_3C_{60} family [12.25], and showing the location of the Fermi level.

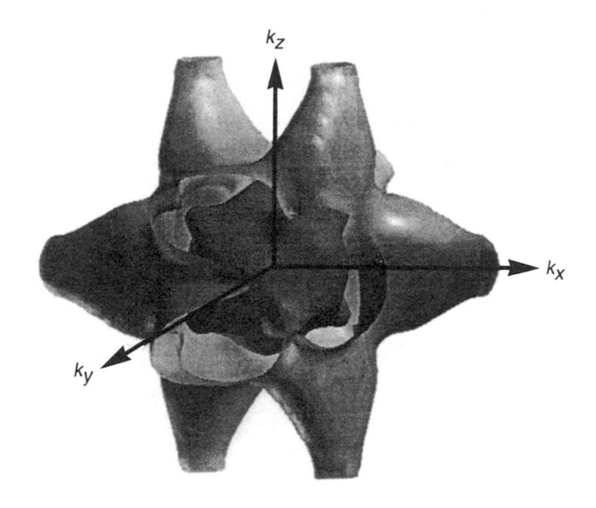

Fig. 12.12. Calculated Fermi surfaces of K_3C_{60} in the ordered $Pa\bar{3}$ crystal structure. The multiply connected outer surface has two arms at the Brillouin zone boundary along each of the crystal axes k_x, k_y, and k_z, reflecting twofold symmetry [12.5].

Calculations for the band structure of K_4C_{60} have been carried out [12.130], for several different structural models, but in all cases the calculation finds metallic behavior near E_F, inconsistent with experimental measurements showing K_4C_{60} to be semiconducting. Further theoretical work is needed to account for the experimental observations, as discussed in §12.7.7.

12.7.4. Electronic Structure of Alkaline Earth–Doped C_{60}

Band calculations using the local density approximation have also been carried out for the alkaline earth compounds, Ca_5C_{60}, Ba_6C_{60}, and Sr_6C_{60} [12.36, 44], as well as for Ca_3C_{60} [12.44, 63]. The results for the electronic structures for these alkaline earth compounds show three important features, as described below. First, at low doping concentrations each alkaline earth atom generally transfers two electrons to the C_{60} shell. Second, because of the strong hybridization of the 4s Ca levels with the C_{60} t_{1g} energy levels, as shown in Fig. 12.13, where the energy levels of a Ca@C_{60} cluster are presented, the electrons fill the Ca 4s-derived level with two electrons before the t_{1g} levels are occupied. With further doping, the 4s alkaline earth level rises above the t_{1g} levels, and only C_{60}-derived levels are filled. Third, the equilibrium position for one Ca in the octahedral site is off-center, thereby making room for additional calcium ions in an octahedral site. Moreover, different crystal structures are found for these compounds, with Ca_5C_{60} crystallizing in the fcc structure, while Ba_6C_{60} and Sr_6C_{60} both crystallize in the bcc structure (see §8.6). For all three compounds Ca_5C_{60},

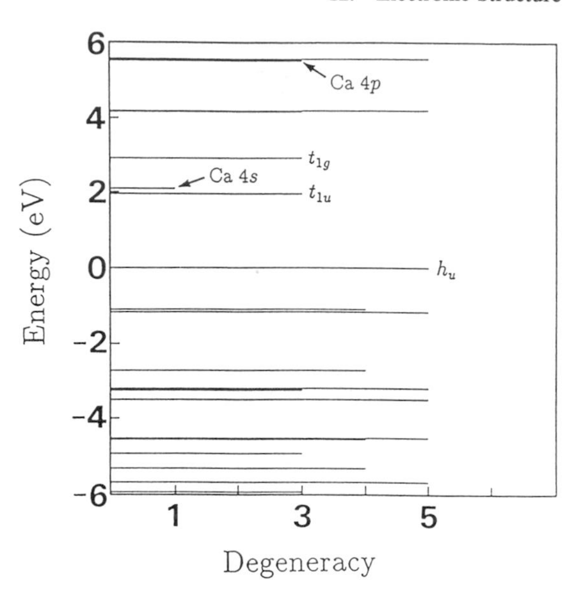

Fig. 12.13. Electronic energy levels of a Ca@C_{60} cluster, where the Ca atom is located at the center of the C_{60} cage. The energies are measured from the h_u state, the highest occupied state of the C_{60} cluster. The t_{1u} state, the lowest unoccupied state of the C_{60} cluster, is now occupied by two electrons. The a_g and t_{1u} states derived mainly from Ca 4s and 4p states, respectively, appear at higher energies than the t_{1u} state [12.25, 44].

Ba_6C_{60}, and Sr_6C_{60}, the t_{1g}-hybridized levels are partly occupied in the solid state, and these are the states responsible for the conductivity and superconductivity in these crystals. For the case of Ca_5C_{60}, the hybridization between the three Ca orbitals associated with the octahedral site and the adjacent C_{60} molecules is strong, and at the stoichiometry Ca_5C_{60}, the t_{1u} level is totally filled, while the (Ca^{2-} ion hybridized) t_{1g} level of C_{60} is half filled [12.44]. Experimental evidence for the partial filling of the t_{1g} level comes from both photoemission spectra [12.108, 109] and transport measurements [12.30].

Closely related to the electronic dispersion relations $E(\mathbf{k})$ for Ca_5C_{60}, discussed above, are the dispersion relations for Sr_6C_{60} and Ba_6C_{60} in the crystalline phase shown in Fig. 12.14(a) and (c), respectively, and both are compared in Fig. 12.14(b) with the corresponding $E(\mathbf{k})$ for a hypothetical bcc form of C_{60}, calculated in a similar way [12.36]. Of particular interest, both Ba_6C_{60} and Sr_6C_{60} are semimetals, with equal volumes enclosed by their electron and hole Fermi surfaces [12.44, 76]. The semimetallic behavior arises from the hybridization of the t_{1g} valence bands with conduction bands of the alkaline earth. Both Sr_6C_{60} and Ba_6C_{60} have strong hybridization between the molecular-derived C_{60} levels and the alkaline earth (Sr 4d or Ba 5d) states, which for the M_6C_{60} compounds lie lower than the Sr 5s or Ba 6s valence states for the free atoms. This hybridization induces considerable dispersion in the t_{1g} band near the Fermi level and also in higher bands, as seen in Fig. 12.14(a) and (c). Plots of the contour

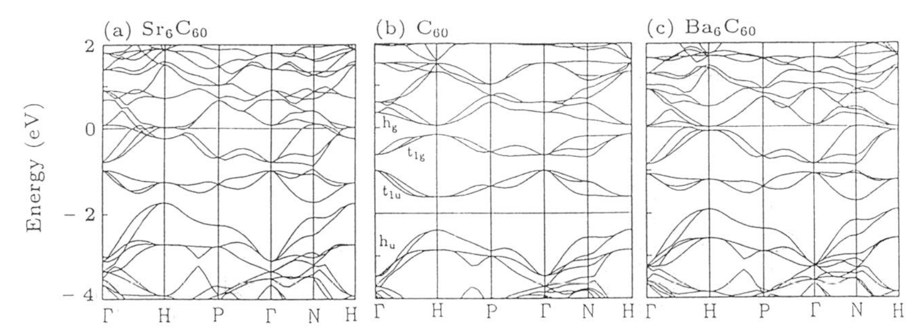

Fig. 12.14. Electronic band structure for (a) Sr_6C_{60}, (b) a hypothetical undoped bcc solid C_{60}, and (c) Ba_6C_{60} [12.25]. Energy is measured from the Fermi level at 0 eV in (a) and (c), which is denoted by the horizontal lines. In (b), the energy bands are shifted downward by 2 eV for the sake of easier comparison to (a) and (c) and each bunched band is labeled according to the corresponding state of the C_{60} cluster.

maps for Sr_6C_{60} and Ba_6C_{60} show considerable electron charge distribution around the metal sites, indicative of incomplete charge transfer to the C_{60} and consistent with the strong hybridization between the metal d levels and the t_{1g} level discussed above [12.36]. The semimetallic nature of Ba_6C_{60} can be seen by noting the hole pockets at the N point and the electron pockets at the H point [see Fig. 12.14(c)] [12.36, 76]. The density of states at the Fermi level $N(E_F)$ is calculated to be lower in Sr_6C_{60} (11.6 states/eV-C_{60}) and Ba_6C_{60} (4.3 states/eV-C_{60}), as compared with that for many of the M_3C_{60} alkali metal compounds, although their superconducting T_c values are 7 K for Ba_6C_{60} and 4 K for Sr_6C_{60} (see §15.1). In contrast to the behavior of the M_3C_{60} alkali metal compounds, there is no clear relation between T_c and $N(E_F)$ for the alkaline earth–derived superconducting compounds. Since E_F for Sr_6C_{60} and Ba_6C_{60} lies in a broad, strongly hybridized band, the Coulomb repulsion between the superconducting carriers is expected to be smaller than for the M_3C_{60} compounds [12.36], but μ^* for Sr_6C_{60} is expected to be larger than μ^* for Ba_6C_{60} based on the more uniform charge distribution calculated for Ba_6C_{60}, thereby perhaps accounting for the higher T_c in Ba_6C_{60}, despite its lower $N(E_F)$ value relative to Sr_6C_{60} [12.130].

One-electron band calculations have significantly influenced progress by experimentalists by providing insights and explanations for observed phenomena, by making predictions for new phenomena, and by suggesting systems of theoretical interest that experimentalists may be able to study, such as alkaline earth–alkali metal alloy compounds (e.g., Ca_2CsC_{60}) for study of hybridization effects and their potential as superconductors [12.25]. The calculated density of states for Ca_2CsC_{60} (with doubly charged Ca^{2+} in each of the tetrahedral sites and Cs in the octahedral site) gives a relatively high value for $N(E_F)$, suggesting Ca_2CsC_{60} as an attractive candidate for superconductivity, and since the t_{1u} band in this case could be filled with five out of six electrons, this would be expected to be a hole-type superconductor. Other suggested candidates are the $M_3M'_3C_{60}$ fcc structure where M is an alkaline earth and M′ is an alkali metal whereby ~9 electrons would be transferred to C_{60} to half fill the t_{1g} band, with the alkaline earth ions going into a distorted tetrahedral site surrounded by four pentagon-oriented C_{60} clusters and the alkali metal going into distorted tetrahedral sites surrounded by four hexagon-oriented C_{60} clusters. Also, it has been suggested that the study of intermetallic $Ba_3Sr_3C_{60}$ might prove interesting with regard to charge transfer issues [12.130].

12.7.5. Band Calculations for Other Doped Fullerenes

One promising band structure calculation is for an intercalation compound of graphite in which a ⟨111⟩ layer of C_{60} or K_4C_{60} is introduced between the graphite layers to stoichiometries $C_{60}C_{32}$ and $K_4C_{60}C_{32}$ [12.110]. In the case of $C_{60}C_{32}$, a small charge transfer is found from the graphite layer to the C_{60} intercalate because of the curvature of the C_{60} molecules which admixes some s state character into the π orbitals and thus lowers the energy of the C_{60}-derived π state. The resulting charge transfer stabilizes the C_{60} in a symmetric position with respect to the surrounding graphite layers. The coupling between the C_{60} and graphite layers in this case is weak, so that the energy bands of the isolated constituents are only weakly perturbed. Since all the atoms of this structure are carbon, this graphite intercalation compound is expected to be stable under ambient conditions.

Even more interesting is the case of $K_4C_{60}C_{32}$ [Fig. 12.15(a)], where potassium atoms go into interstitial positions between the C_{60} and the surrounding graphene sheets. The shortest K–K distance in this model calculation is 5.7 Å and the shortest C–K distance to C_{60} is 3.1 Å and to the graphene plane is 3.3 Å. In this case the calculations show that complete charge transfer takes place from the K to both the C_{60} and the graphene sheets. From Fig. 12.15(b), it is seen that the resulting intercalation compound is predicted to be metallic, with a high density of states at the Fermi level [$N(E_F)$ = 24.9 states/eV], which is similar in magnitude to that calculated for K_3C_{60} using the same technique. The coupling to the graphite layers gives rise to a graphite conduction band with a large bandwidth that serves to reduce the Coulomb repulsion energy μ^*. The large $N(E_F)$ and small μ^* are favorable for high-T_c superconductivity. In the graphene layers electrons are attractively coupled by vibrations associated with molecular-type C_{60} vibrations. A rough cohesive energy calculation indicates that $C_{60}C_{32}$ should be stabilized by a very weak attractive van

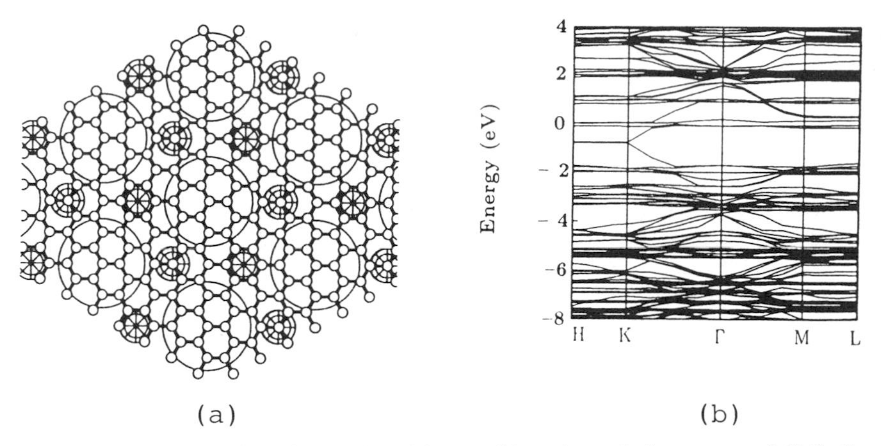

Fig. 12.15. (a) Top view of the K-doped C_{60}–graphite cointercalation compound, $K_4C_{60}C_{32}$. Large open spheres and small decorated spheres denote C_{60} clusters and K ions, respectively. (b) Electronic band structure of the K-doped C_{60}–graphite cointercalation compound, $K_4C_{60}C_{32}$, with a c-axis repeat distance $c = 12.530$ Å. The energy is measured from the Fermi level of the system, which is at the anticrossing point of the C_{60} t_{1u} conduction band and the graphite conduction band [12.110].

der Waals interaction, while $K_4C_{60}C_{32}$ is further stabilized by electrostatic forces [12.110]. If $K_4C_{60}C_{32}$ were found to be superconducting, a whole class of materials, obtained by varying either the dopant (K) or the number of graphene sheets (stage) would become of interest.

Band calculations for halogen doping of C_{60} have also been carried out, showing a cohesive energy of ~2–3 eV for solid C_{60}Br, and indicating that bromine doping should be possible [12.111]. Experimental observation of the formation of brominated C_{60} supports these conclusions (see §10.5). However, because of the high electronegativity of C_{60}, it is not likely that holes would be created in the H_u-derived C_{60} HOMO level, also in agreement with experiment [12.112]. The calculations for Br on a tetrahedral site or on an octahedral site both show that the Fermi level lies in the Br $4p$ state, which lies between the h_{1u} HOMO and t_{1u} LUMO levels, thereby indicating that no charge transfer occurs, and a large charge accumulation takes place near the Br site [12.113]. Similar results were obtained for C_{60}Cl [12.130].

A local density approximation calculation for a BC_{59} molecule has been carried out [12.113] showing the fivefold degenerate h_u band to bunch into four lower-lying bands and one higher-lying band which is half empty, thereby indicating hole doping with $E_F \sim 0.2$ eV below the top of the valence band (see Fig. 12.16). The possibility of formation of substitutionally doped BC_{59} has been demonstrated experimentally in mass spectra [12.114],

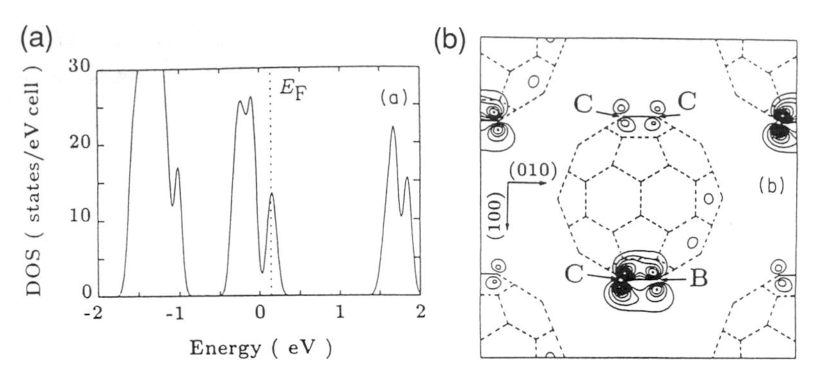

Fig. 12.16. (a) Electronic density of states and (b) distribution of states near the Fermi level calculated for solid BC_{59} [12.113].

but this species has been neither isolated nor purified, so that no measurements on the physical properties are yet available. The results of the band calculation suggest that BC_{59} might be an interesting candidate for a hole-doped fullerene system.

12.7.6. Band Calculations for Other Fullerenes

Regarding smaller fullerenes, the electronic structure for a solid constructed from a C_{28} molecule with T_d symmetry has been calculated [12.115, 116], and the electronic structure in this case was found to be semiconducting. Other calculations have also been reported for small fullerenes [12.25].

With measurements now becoming available for higher fullerenes, including NMR experiments on C_{84} [12.117], photoemission spectroscopy results on C_{76} and C_{84} [12.118–120], scanning tunneling microscopy on C_{76}, C_{80}, C_{82}, and C_{84} [12.121, 122], and photoabsorption spectroscopy on C_{76} and C_{84} [12.117, 123], there is great interest in determining the electronic band structure and density of states for such higher-mass fullerenes. There is particular interest in identifying the most probable isomers for each C_{n_C}; such information would also provide insights into the growth model for fullerenes.

Using a tight-binding and an LDA approach, the electronic structure and density of states for several of the C_{76}, C_{82} and C_{84} isomers have been calculated [12.124, 125]. For C_{76}, the electronic structure and density of states (DOS) of the D_2 isomer have been calculated using the tight binding model [12.126, 127] and by a local density approximation [12.128], yielding good agreement with the experimental photoemission spectrum [12.118], including consideration of both occupied and unoccupied bands. The DOS of T_d C_{76} has also been calculated [12.129]. For C_{84}, tight binding [12.125, 130] and LDA [12.131] electronic structure and density of states calculations have been reported on four D_2 candidates for the major C_{84} isomer, as well as on the minor D_{2d} C_{84} isomer. The density of states for three possible candidates for the major C_2 isomer of C_{82} has also been calculated [12.129]. Comparison has also been made between measured properties (such as photoemission and inverse photoemission spectra [12.118, 132, 133]) and model calculations, using both the LDA and tight-binding methods for finding distributions $N(E)$ in the density of states. Such studies have already provided sensitive identifications for the most likely isomers for C_{76}, C_{82}, and C_{84} [12.25]. As more experimental data on higher fullerenes become available we can expect more detailed calculations to be made on the various isomers of interest.

In the limit that the number of carbon atoms on a fullerene becomes very large, the electronic level picture for such a large molecule approaches that of a graphene sheet where each carbon atom is in an sp^2 σ bonding configuration. In this limit, the π electron states are separated into an occupied bonding π band and an empty antibonding π band with no band gap separating these two bands, and electron–electron correlation effects are not expected to be of importance.

12.7.7. *Many-Body Approach to Solid* C_{60}

Many-body treatments of the electronic structure of crystalline C_{60} have been strongly stimulated by measurements of the luminescence and absorption phenomena associated with the optical absorption edge in solution and in the solid phase (see §13.2 and §13.3). On this basis, considerable experimental evidence has been demonstrated for the presence of strong electron–electron intramolecular Coulomb interactions [12.73, 134–138]. In addition, experimental determinations of the magnitude of the HOMO–LUMO band gap through photoemission studies have significantly exceeded LDA calculations, and furthermore experimental angle-resolved photoemission data show very little dispersion.

These observations have stimulated theoretical work in two main directions [12.33, 139–147]. Along one approach, efforts have been made to correct the LDA underestimation of the energy gap in solid C_{60} by properly treating the electron excitations with an *ab initio* quasiparticle approach. Work along these lines has been done by Shirley and Louie [12.9, 89]. In their GW approach, the Dyson equation is solved within a Greens function (G) formalism using a dynamically screened Coulomb interaction W to obtain the quasiparticle energy spectrum. The unperturbed LDA wave function and eigenvalues used as a basis set are in good agreement with other LDA calculations [12.64, 99, 100, 148]. A large increase in the calculated energy gap (from 1.04 to 2.15 eV) is found using this GW quasiparticle approach yielding better agreement with photoemission data (see Fig. 12.17), as measured by the energy differences for the *onsets* of the photoemission and inverse photoemission spectra (see §17.1.1). Reasonably good agreement was also obtained between the quasiparticle GW calculations (3.0 eV) and the *peak-to-peak* energy in the photoemission and inverse photoemission spectra [12.9]. Finally, the GW quasiparticle technique was used to calculate the angle-resolved photoemission spectra along ΓM and ΓK for the h_u, t_{1u}, t_{2u}, t_{1g}, and t_{2g} bands, thereby yielding an explanation for the very weak dispersion found experimentally for these energy bands [12.96, 149]. This calculational technique has also been very useful for predicting the effect of merohedral disorder on the observed density of states (see Fig. 12.18) and on the angle-resolved photoemission spectra [12.9]. In Fig. 12.18 we see a comparison between the density of states calculated by the quasiparticle GW technique for an orientationally ordered $Fm\bar{3}$ structure (the highest symmetry compatible with icosahedra on a cubic fcc lattice), an ordered version of the $Pa\bar{3}$ structure (which is observed experimentally in C_{60} below T_{01}), and the merohedrally disordered $Pa\bar{3}$ structure. A comparison of these densities of states shows that merohedral disorder washes out structure in the t_{1u} density of states without

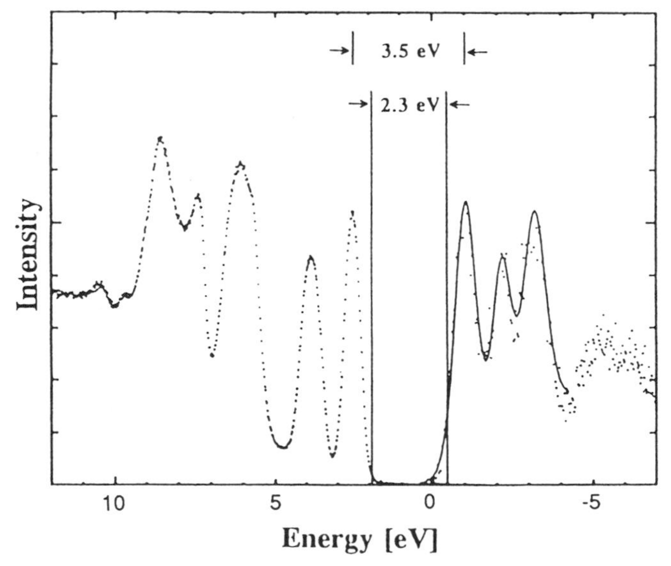

Fig. 12.17. Photoemission (left) and inverse-photoemission (right) spectra of solid C_{60}. The zero of energy is at the Fermi level of an Ag reference sample. The vertical lines indicate the onsets for the gap determination [12.73].

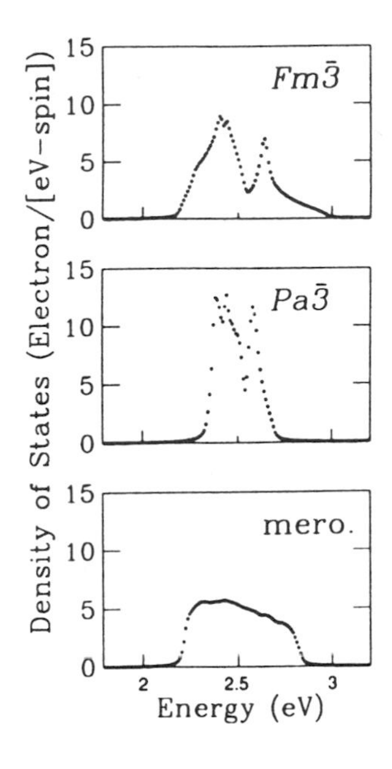

Fig. 12.18. The density of states of the t_{1u} LUMO band for the $Fm\bar{3}$, $Pa\bar{3}$, and isotropically merohedrally disordered (labeled mero) structures, computed from quasiparticle GW band structure calculations [12.9].

much change in the width of the density of states feature [12.9]. While the quasiparticle GW approach considers many-body corrections to the LDA approach, little work has thus far been published linking this approach to calculations starting from a many-body point of view, discussed below.

The second approach used to consider many-body effects starts from a many-body point of view. A number of many-body calculations have focused on calculations of quantities such as the Coulomb repulsion U of two holes on the same C_{60} molecule. Many different calculational techniques have been used by different groups, yielding values of U in the ~0.8–2.1 eV range [12.33, 73, 138, 140–147].

A widely quoted determination of U comes from analysis of the valence photoemission spectra with the Auger electron spectra, which was first carried out for C_{60} [12.73] and then extended to K_3C_{60} and K_6C_{60}, yielding similar Hubbard correlation energies $U = 1.4 \pm 0.2$ eV for C_{60} and K_3C_{60} and $U = 1.5 \pm 0.2$ eV for K_6C_{60} [12.150]. In the initial work on C_{60} [12.73], a shift of 0.6 ± 0.3 eV was measured between the Auger spectra and the self-convoluted valence photoemission spectrum for C_{60}. If the core hole lifetime broadening is much smaller than the half-width of the LUMO band, the screening of the core hole will occur before the Auger process takes place [12.150]. The authors thus conclude that the intermolecular screening does not occur during the XPS process, but it occurs before the Auger decay [12.150].

Many direct calculations of U have been reported. Both theoretical studies and the more empirical investigations which combine experimental measurements with many-body theory conclude that U is greater than the electronic bandwidth W, which is in the 0.2–0.5 eV range, indicating that fullerenes are strongly correlated systems and that level degeneracies and electron filling may significantly affect the electronic properties of fullerite and fulleride materials.

Realistic LDA and semiempirical calculations have been used to obtain theoretical estimations for the effective on-site Coulomb repulsion U. Antropov *et al.* [12.143] performed linear-muffin-tin-orbital (LMTO) calculations within the atomic-sphere approximation (ASA) and the local density approximation (LDA) to obtain an unscreened value of $U_0 = 2.7$ eV for the isolated C_{60} molecule. To obtain U for the fcc solid, the screening due to polarization of the surrounding molecules was then included by assigning a polarizability to each molecule. A corrected value for $U = 0.8$–1.3 eV was then obtained for fcc C_{60}, in good agreement with the semiempirical approach discussed above. Explicit investigation of screening effects was made [12.144] by calculating the dielectric function for fcc C_{60} within the LDA technique and a value of $U = 2.1$ eV was obtained, somewhat larger than the values quoted above. Semiempirical modified-neglect-of-differential-overlap (MNDO) quantum chemistry calculations have yielded $U \sim 3$ eV for the isolated C_{60} molecule [12.145], which was then corrected for the polarization in the solid state, using an empirical model for the polarization, to obtain $U = 1.3$ eV for K_3C_{60}, in very good agreement with the semiempirical approach [12.73] discussed above.

Referring to Fig. 12.17 for the photoemission (PES) and inverse photoemission (IPES) spectra for C_{60} [12.73], we see an energy difference of 3.5 eV between the IPES and PES peaks and an energy difference of 2.3 eV between the onsets, which are determined as the point of maximum change in slope in the onset region. This difference in energy (1.2 eV) is close to the calculated value of U. If we then use the value of 1.84 eV for the optical absorption edge (see §17.1.1), a value of 0.46 eV is obtained for the exciton binding energy.

Merohedral disorder or local misorientation of the C_{60} icosahedra has been shown [12.93] to give rise to large shifts in the energy bands ($\sim$1 eV), also indicating a need to consider many-body effects. The disorder in molecular orientations leads to random hopping matrices, which tend to smooth out the spikes in the density of states curves (as shown in Fig. 12.18), and these orientational effects have a significant bearing on measurement of various electronic and transport properties. Because of localization and correlation effects, the Fermi surface of doped C_{60} may not be well defined.

The short electron lifetime ($\hbar/\tau \sim E_F$) also introduces uncertainty into the quasiparticle concept used to describe the superconducting state.

One many-body approach to the fullerene electronic structure describes the optical absorption or emission process in terms of a two-band Hubbard Hamiltonian

$$\mathscr{H} = \sum_{k,\sigma} \epsilon_k c^{\dagger}_{k,\sigma} c_{k,\sigma} + \sum_{k} \eta_k d^{\dagger}_{k,\sigma} d_{k,\sigma} + U \sum_{l,\sigma} (n^{\dagger}_{l,\sigma} n_{l-\sigma} + m^{\dagger}_{l,\sigma} m_{l-\sigma}), \quad (12.12)$$

where $n_{k,\alpha} = c^{\dagger}_{k,\sigma} c_{k,\sigma}$ and $m_{k,\alpha} = d^{\dagger}_{k,\sigma} d_{k,\sigma}$ are, respectively, the number densities for the h_u-derived hole band and the t_{1u}-derived electron band, where ϵ_k and η_k are the corresponding energies for these levels and U is the on-site Coulomb interaction between two electrons on the same fullerene molecule. This Hubbard model in Eq. (12.12) has been applied to account for the intensities of the various features in the KVV Auger spectrum (see §17.3) and to the interpretation of the optical spectra in solid C_{60} (see §13.3) in terms of excitonic states in strongly correlated bands where $U \gg W$, and W is the level width [12.73]. The Coulomb repulsion U is related to the ionization potential E_I, the electron affinity E_A, and the HOMO–LUMO splitting in the free molecule Δ by

$$U = E_I - E_A - \Delta, \quad (12.13)$$

in which $\Delta = \overline{n_k} - \overline{\epsilon_k}$, where the bars denote average values. To obtain estimates for U in the solid state, it is necessary to introduce polarization effects, since $E_I = 7.6$ eV and $E_A = 2.65$ eV are usually measured in the gas-phase. For the free molecule, $E_I = 7.6$ eV [12.151–153] and $E_A = 2.65$ eV [12.154], and using an estimate of $\Delta = 1.9$ eV [12.7] from Fig. 12.2 for the free molecule, we obtain $U_{\text{mol}} = 3.1$ eV.

Polarization effects lower the ionization energy E_I by the polarization energy $E_P = ze^2\alpha/2R^4$, where z is the number of nearest-neighbors (in the fcc lattice $z = 12$), $\alpha \sim 85$ Å^3 is the polarizability [12.155–159], and $R = 10.02$ Å is the distance between C_{60} molecules, yielding $E_P = 0.69$ eV. Correspondingly, the polarization of the medium increases the electron affinity E_A by the same polarization energy E_P. Thus U in the solid is decreased by $2E_P = 1.4$ eV relative to $U_{\text{mol}} = 3.1$ eV for the free molecule, $U = U_{\text{mol}} - 2E_P$, yielding a value for U in the solid of 1.7 eV, in good agreement with the measured value of 1.6 eV [12.73].

Furthermore, the band gap E_{gap} in the solid is related to the HOMO–LUMO splitting in the molecule Δ through the relation [12.73]

$$E_{\text{gap}} \simeq \Delta + U - W \quad (12.14)$$

where it is assumed that the bandwidths for the HOMO- and LUMO-derived bands are approximately equal, $W_{h_u} \approx W_{t_{1u}}$. Average values from the literature for the pertinent energies for C_{60} are: $\Delta = 1.8$ eV [12.140], $U = 0.8$ to 1.6 eV [12.73, 141, 160–166], $E_A = 2.65$ eV [12.154], $E_I = 7.58$ eV [12.167] (where E_I and E_A are given for free C_{60} molecules), and $W = 0.2 - 0.5$ eV. Thus using $U = 1.2$ eV, $\Delta = 1.8$ eV, and $W = 0.4$ eV, we obtain an estimated value for $E_{\text{gap}} \sim 2.6$ eV, roughly consistent with the value of 2.3 eV obtained from the photoemission and inverse photoemission data shown in Fig. 12.17 [12.140]. For discussion of the application of Auger spectroscopy for the determination of U see §17.3.

REFERENCES

[12.1] G. Herzberg. *Molecular Spectra and Molecular Structure*, vol. 3, Van Nostrand, Princeton, NJ (1945).

[12.2] J. H. Weaver. *Acc. Chem. Res.*, **25**, 143 (1992).

[12.3] J. H. Weaver and D. M. Poirier. In H. Ehrenreich and F. Spaepen (eds.), *Solid State Physics*, vol. 48, p. 1, Academic Press, New York (1994). Chapter 1.

[12.4] R. C. Haddon. *Accounts of Chemical Research*, **25**, 127 (1992).

[12.5] W. E. Pickett. In H. Ehrenreich and F. Spaepen (eds.), *Solid State Physics*, vol. 48, p. 225, Academic Press, New York (1994).

[12.6] C. Kittel. *Introduction to Solid State Physics*. John Wiley and Sons, New York, NY, 6th ed. (1986).

[12.7] S. Saito and A. Oshiyama. *Phys. Rev. Lett.*, **66**, 2637 (1991).

[12.8] R. C. Haddon, L. E. Brus, and K. Raghavachari. *Chem. Phys. Lett.*, **125**, 459 (1986).

[12.9] S. G. Louie and E. L. Shirley. *J. Phys. Chem. Solids*, **54**, 1767 (1993).

[12.10] B. I. Dunlap, D. W. Brenner, J. W. Mintmire, R. C. Mowrey, and C. T. White. *J. Phys. Chem.*, **95**, 5763 (1991).

[12.11] Z. Zhang, C.-C. Chen, S. P. Kelty, H. Dai, and C. M. Lieber. *Nature (London)*, **353**, 333 (1991).

[12.12] Q. M. Zhang, J. Y. Yi, and J. Bernholc. *Phys. Rev. Lett.*, **66**, 2633 (1991).

[12.13] P. Damay and F. Leclercq. *Phys. Rev. B*, **49**, 7790 (1994).

[12.14] L.-S. Wang. *Chem. Phys. Lett.*, **182**, 5 (1992).

[12.15] R. Saito, G. Dresselhaus, and M. S. Dresselhaus. *Phys. Rev. B*, **46**, 9906 (1992).

[12.16] P. W. Fowler and J. Woolrich. *Chem. Phys. Lett.*, **127**, 78 (1986).

[12.17] R. Saito, G. Dresselhaus, and M. S. Dresselhaus. *Chem. Phys. Lett.*, **210**, 159 (1993).

[12.18] S. Saito and A. Oshiyama. *Phys. Rev. B*, **44**, 11532 (1991).

[12.19] K. Tanaka, M. Okada, K. Okahara, and T. Yamabe. *Chem. Phys. Lett.*, **202**, 394 (1993).

[12.20] K. R. Lea, M. J. M. Leask, and W. P. Wolf. *J. Phys. Chem. Solids*, **23**, 1381 (1962).

[12.21] W. G. Harter and T. C. Reimer. *J. Chem. Phys.*, **94**, 5426 (1991).

[12.22] R. Saito (1994). Private communication.

[12.23] M. R. Savina, L. L. Lohr, and A. M. Francis. *Chem. Phys. Lett.*, **205**, 200 (1993).

[12.24] K. Tanigaki, T. W. Ebbesen, S. Saito, J. Mizuki, J. S. Tsai, Y. Kubo, and S. Kuroshima. *Nature (London)*, **352**, 222 (1991).

[12.25] A. Oshiyama, S. Saito, N. Hamada, and Y. Miyamoto. *J. Phys. Chem. Solids*, **53**, 1457 (1992).

[12.26] P. W. Fowler. *Chem. Phys. Lett.*, **131**, 444 (1986).

[12.27] A. R. Kortan, N. Kopylov, S. H. Glarum, E. M. Gyorgy, A. P. Ramirez, R. M. Fleming, F. A. Thiel, and R. C. Haddon. *Nature (London)*, **355**, 529 (1992).

[12.28] A. R. Kortan, N. Kopylov, R. M. Fleming, O. Zhou, F. A. Thiel, R. C. Haddon, and K. M. Rabe. *Phys. Rev. B*, **47**, 13070 (1993).
[12.29] R. C. Haddon, G. P. Kochanski, A. F. Hebard, A. T. Fiory, R. C. Morris, and A. S. Perel. *Chem. Phys. Lett.*, **203**, 433 (1993).
[12.30] R. C. Haddon, G. P. Kochanski, A. F. Hebard, A. T. Fiory, and R. C. Morris. *Science*, **258**, 1636 (1992).
[12.31] M. S. Dresselhaus, G. Dresselhaus, and R. Saito. *Mater. Sci. Eng.*, **B19**, 122 (1993).
[12.32] Z. H. Dong, P. Zhou, J. M. Holden, P. C. Eklund, M. S. Dresselhaus, and G. Dresselhaus. *Phys. Rev. B*, **48**, 2862 (1993).
[12.33] L. Bergomi and T. Jolicœur. *C. R. Acad. Sci. Paris*, **318, Série II**, 283 (1994).
[12.34] J. Hrusak and H. Schwarz. *Chem. Phys. Lett.*, **205**, 187 (1993).
[12.35] S. Suzuki, D. Inomata, N. Sashide, and K. Nakao. *Phys. Rev. B*, **48**, 14615 (1994).
[12.36] S. Saito and A. Oshiyama. *Phys. Rev. Lett.*, **71**, 121 (1993).
[12.37] J. H. Weaver, J. L. Martins, T. Komeda, Y. Chen, T. R. Ohno, G. H. Kroll, N. Troullier, R. E. Haufler, and R. E. Smalley. *Phys. Rev. Lett.*, **66**, 1741 (1991).
[12.38] M. B. Jost, N. Troullier, D. M. Poirier, J. L. Martins, J. H. Weaver, L. P. F. Chibante, and R. E. Smalley. *Phys. Rev. B*, **44**, 1966 (1991).
[12.39] M. B. Jost, P. J. Benning, D. M. Poirier, J. H. Weaver, L. P. F. Chibante, and R. E. Smalley. *Chem. Phys. Lett.*, **184**, 423 (1991).
[12.40] K. Nakao, N. Kurita, and M. Fujita. *Phys. Rev. B.*, **49**, 11415 (1994).
[12.41] Y. Shinohara. Ph.D. thesis, University of Electro-Communications, Tokyo (1994). Department of Electronics Engineering: Master of Science Thesis (in Japanese).
[12.42] S. Saito. In *MRS Abstracts, Fall Meeting*, p. 208. Materials Research Society Press, Pittsburgh, PA (1990).
[12.43] S. Saito. In R. S. Averback, J. Bernholc, and D. L. Nelson (eds.), *Clusters and Cluster-Assembled Materials, MRS Symposia Proceedings, Boston*, vol. 206, p. 115. Materials Research Society Press, Pittsburgh, PA (1991).
[12.44] S. Saito and A. Oshiyama. *Solid State Commun.*, **83**, 107 (1992).
[12.45] S. C. Erwin. In W. Billups and M. Ciufolini (eds.), *Buckminsterfullerene*, p. 217, VCH Publishers, New York (1993).
[12.46] W. Andreoni and A. Curioni. In H. Kuzmany, J. Fink, M. Mehring, and S. Roth (eds.), *Proceedings of the Winter School on Fullerenes*, pp. 93–101 (1994). Kirchberg Winter School, World Scientific Publishing Co., Ltd., Singapore.
[12.47] A. Rosén and B. Wästberg. *Z. Phys. D*, **12**, 387 (1989).
[12.48] L. S. Wang, J. M. Alford, Y. Chai, M. Diener, J. Zhang, S. M. McClure, T. Guo, G. E. Scuseria, and R. E. Smalley. *Chem. Phys. Lett.*, **207**, 354 (1993).
[12.49] A. Rosén. *Z. Phys. D*, **25**, 175 (1993).
[12.50] H. Shinohara, H. Sato, Y. Saito, A. Izuoka, T. Sugawara, H. Ito, T. Sakurai, and T. Matsuo. *J. Phys. Chem.*, **96**, 3571 (1992).
[12.51] H. Shinohara, H. Yamaguchi, N. Hayashi, H. Sato, M. Inagaki, Y. Saito, S. Bandow, H. Kitagawa, T. Mitani, and H. Inokuchi. *Mater. Sci. Eng.*, **B19**, 25 (1993).
[12.52] C. S. Yannoni, M. Hoinkis, M. S. de Vries, D. S. Bethune, J. R. Salem, M. S. Crowder, and R. D. Johnson. *Science*, **256**, 1191 (1992).
[12.53] R. D. Johnson, D. S. Bethune, and C. S. Yannoni. *Acc. Chem. Res.*, **25**, 169 (1992).
[12.54] R. D. Johnson, M. S. de Vries, J. Salem, D. S. Bethune, and C. S. Yannoni. *Nature (London)*, **355**, 239 (1992).
[12.55] K. Laasonen, W. Andreoni, and M. Parrinello. *Science*, **258**, 1916 (1992).
[12.56] S. Saito and A. Oshiyama. *Unpublished* (1994).
[12.57] T. Suzuki, Y. Maruyama, T. Kato, K. Kikuchi, and Y. Achiba. *J. Am. Chem. Soc.*, **115**, 11006 (1993).

[12.58] S. Nagase and K. Kobayashi. *Chem. Phys. Lett.*, **214**, 57 (1993).
[12.59] K. Kikuchi, S. Suzuki, Y. Nakao, N. Nakahara, T. Wakabayashi, H. Shiromaru, I. Ikemoto, and Y. Achiba. *Chem. Phys. Lett.*, **216**, 67 (1993).
[12.60] D. M. Poirier, M. Knupfer, J. H. Weaver, W. Andreoni, K. Laasonen, M. Parrinello, D. S. Bethune, K. Kikuchi, and Y. Achiba. *Phys. Rev. B*, **49**, 17403 (1994).
[12.61] D. S. Bethune, R. D. Johnson, J. R. Salem, M. S. de Vries, and C. S. Yannoni. *Nature (London)*, **366**, 123 (1993).
[12.62] W. Krätschmer, L. D. Lamb, K. Fostiropoulos, and D. R. Huffman. *Nature (London)*, **347**, 354 (1990).
[12.63] S. Saito and A. Oshiyama. *J. Phys. Chem. Solids*, **54**, 1759 (1993).
[12.64] N. Troullier and J. L. Martins. *Phys. Rev. B*, **46**, 1754 (1992).
[12.65] S. C. Erwin and M. R. Pederson. *Phys. Rev. Lett.*, **67**, 1610 (1991).
[12.66] S. C. Erwin and W. E. Pickett. *Science*, **254**, 842 (1991).
[12.67] W. Andreoni, F. Gygi, and M. Parrinello. *Phys. Rev. Lett.*, **68**, 823 (1992).
[12.68] D. L. Novikov, V. A. Gubanov, and A. J. Freeman. *Physica C*, **191**, 399 (1992).
[12.69] W. Y. Ching, M.-Z. Huang, Y.-N. Xu, W. G. Harter, and F. T. Chan. *Phys. Rev. Lett.*, **67**, 2045 (1991).
[12.70] Y. N. Xu, M. Z. Huang, and W. Y. Ching. *Phys. Rev. B*, **44**, 13171 (1992).
[12.71] B. Szpunar. *Europhys. Lett.*, **18**, 445 (1992).
[12.72] M. P. Gelfand. *Superconductivity Review*, **1**, 103 (1994).
[12.73] R. W. Lof, M. A. van Veenendaal, B. Koopmans, H. T. Jonkman, and G. A. Sawatzky. *Phys. Rev. Lett.*, **68**, 3924 (1992).
[12.74] E. Manousakis. *Phys. Rev. B*, **44**, 10991 (1991).
[12.75] S. Satpathy, V. P. Antropov, O. K. Andersen, O. Jepson, O. Gunnarsson, and A. I. Liechtenstein. *Phys. Rev. B*, **46**, 1773 (1992).
[12.76] S. C. Erwin and M. R. Pederson (1994). Private communication to W. E. Pickett. [12.5].
[12.77] D. A. Papaconstantopoulos, W. E. Pickett, M. R. Pederson, and S. C. Erwin (1994). Unpublished.
[12.78] B. L. Zhang, C. H. Xu, C. Z. Wang, C. T. Chan, and K. M. Ho. *Phys. Rev. B*, **46**, 7333 (1992).
[12.79] M. Menon and K. R. Subbaswamy. *Chem. Phys. Lett.*, **201**, 321 (1993).
[12.80] M. Menon and K. R. Subbaswamy. *Phys. Rev. Lett.*, **67**, 3487 (1991).
[12.81] A. Oshiyama, S. Saito, Y. Miyamoto, and N. Hamada. *J. Phys. Chem. Solids*, **53**, 1689 (1992).
[12.82] G. Gensterblum, J. J. Pireaux, P. A. Thiry, R. Caudano, J. P. Vigneron, P. Lambin, A. A. Lucas, and W. Krätschmer. *Phys. Rev. Lett.*, **67**, 2171 (1991).
[12.83] A. Skumanuch. *Chem. Phys. Lett.*, **182**, 486 (1991).
[12.84] S. L. Ren, Y. Wang, A. M. Rao, E. McRae, G. T. Hager, K. A. Wang, W. T. Lee, H. F. Ni, J. Selegue, and P. C. Eklund. *Appl. Phys. Lett.*, **59**, 2678 (1991).
[12.85] R. D. Johnson, G. Meijer, and D. S. Bethune. *J. Am. Chem. Soc.*, **112**, 8983 (1990).
[12.86] R. D. Johnson, N. Yannoni, G. Meijer, and D. S. Bethune. In R. M. Fleming *et al.* (eds.), *Videotape of Late News Session on Buckyballs, MRS, Boston*. Materials Research Society Press, Pittsburgh, PA (1990).
[12.87] C. Pan, M. P. Sampson, Y. Chai, R. H. Hauge, and J. L. Margrave. *J. Phys. Chem.*, **95**, 2944 (1991).
[12.88] R. M. Fleming, A. P. Ramirez, M. J. Rosseinsky, D. W. Murphy, R. C. Haddon, S. M. Zahurak, and A. V. Makhija. *Nature (London)*, **352**, 787 (1991).
[12.89] E. L. Shirley and S. G. Louie. *Phys. Rev. Lett.*, **71**, 133 (1993).
[12.90] Y.-N. Xu, M.-Z. Huang, and W. Y. Ching. *Phys. Rev. B*, **44**, 13171 (1991).

[12.91] R. O. Jones and O. Gunnarsson. *Rev. Mod. Phys.*, **61**, 689 (1989).
[12.92] N. Hamada and S.-I. Sawada. In *Proc. Int. Workshop on Electronic Properties and Mechanisms in High T_c Superconductors, Tsukuba*, Amsterdam (1991). Elsevier, North-Holland.
[12.93] M. P. Gelfand and J. P. Lu. *Phys. Rev. Lett.*, **68**, 1050 (1992).
[12.94] B. L. Gu, Y. Maruyama, J. Z. Yu, K. Ohno, and Y. Kawazoe. *Phys. Rev. B*, **49**, 16202 (1994).
[12.95] S. G. Louie. In H. Kuzmany, J. Fink, M. Mehring, and S. Roth (eds.), *Progress in Fullerene Research: International Winterschool on Electronic Properties of Novel Materials*, p. 303 (1994). Kirchberg Winter School, World Scientific Publishing Co., Ltd., Singapore.
[12.96] J. M. Themlin, S. Bouzidi, F. Coletti, J. M. Debever, G. Gensterblum, L. M. Yu, J. J. Pireaux, and P. A. Thiry. *Phys. Rev. B*, **46**, 15602 (1992).
[12.97] G. Gensterblum, J. J. Pireaux, P. A. Thiry, R. Caudano, T. Buslaps, R. L. Johnson, G. L. Lay, V. Aristov, R. Günther, A. Taleb-Ibrahimi, G. Indlekofer, and Y. Petroff. *Phys. Rev. B*, **48**, 14756 (1993).
[12.98] Y. N. Xu, M. Z. Huang, and W. Y. Ching. *Phys. Rev. B*, **46**, 4241 (1992).
[12.99] J. L. Martins and N. Troullier. *Phys. Rev. B*, **46**, 1766 (1992).
[12.100] S. C. Erwin and W. E. Pickett. *Phys. Rev. B*, **46**, 14257 (1992).
[12.101] M. Z. Huang, Y. N. Xu, and W. Y. Ching. *Phys. Rev. B*, **46**, 6572 (1992).
[12.102] N. Troullier and J. L. Martins. *Phys. Rev. B*, **46**, 1766 (1992).
[12.103] M. P. Gelfand and J. P. Lu. *Phys. Rev. B*, **46**, 4367 (1992). Erratum: Phys. Rev. B **47**, 4149 (1992).
[12.104] M. P. Gelfand and J. P. Lu. *Appl. Phys. A*, **56**, 215 (1993).
[12.105] N. Hamada, S.-I. Sawada, Y. Miyamoto, and A. Oshiyama. *Jpn. J. Appl. Phys.*, **30**, L2036 (1991).
[12.106] A. Oshiyama and S. Saito. *Solid State Commun.*, **82**, 41 (1992).
[12.107] I. Hirosawa, K. Tanigaki, J. Mizuki, T. W. Ebbesen, Y. Shimakawa, Y. Kubo, and S. Kuroshima. *Solid State Commun.*, **82**, 979 (1992).
[12.108] G. K. Wertheim, D. N. E. Buchanan, and J. E. Rowe. *Science*, **258**, 1638 (1992).
[12.109] Y. Chen, D. M. Poirier, M. B. Jost, C. Gu, T. R. Ohno, J. L. Martins, J. H. Weaver, L. P. F. Chibante, and R. E. Smalley. *Phys. Rev. B*, **46**, 7961 (1992).
[12.110] S. Saito and A. Oshiyama. *Phys. Rev. B*, **49**, 17413 (1994).
[12.111] Y. Miyamoto, A. Oshiyama, and S. Saito. *Solid State Commun.*, **82**, 437 (1992).
[12.112] J. E. Fischer, P. A. Heiney, and A. B. Smith III. *Acc. Chem. Res.*, **25**, 112 (1992).
[12.113] Y. Miyamoto, N. Hamada, A. Oshiyama, and S. Saito. *Phys. Rev. B*, **46**, 1749 (1992).
[12.114] T. Guo, C.-M. Jin, and R. E. Smalley. *J. Phys. Chem.*, **95**, 4948 (1991).
[12.115] T. Guo, M. D. Diener, Y. Chai, J. M. Alford, R. E. Haufler, S. M. McClure, T. R. Ohno, J. H. Weaver, G. E. Scuseria, and R. E. Smalley. *Science*, **257**, 1661 (1992).
[12.116] D. M. Bylander and L. Kleinman. *Phys. Rev. B*, **47**, 10967 (1993).
[12.117] K. Kikuchi, N. Nakahara, T. Wakabayashi, M. Honda, H. Matsumiya, T. Moriwaki, S. Suzuki, H. Shiromaru, K. Saito, K. Yamauchi, I. Ikemoto, and Y. Achiba. *Chem. Phys. Lett.*, **188**, 177 (1992).
[12.118] S. Hino, K. Matsumoto, S. Hasegawa, H. Inokuchi, T. Morikawa, T. Takahashi, K. Seki, K. Kikuchi, S. Suzuki, I. Ikemoto, and Y. Achiba. *Chem. Phys. Lett.*, **197**, 38 (1992).
[12.119] M. S. Golden, M. Kunpfer, M. Merkel, and M. Sing. In A. Y. Vul and V. V. Lemanov (eds.), *Fullerenes and Atomic Clusters*, p. 12 (1993). Int. Workshop, St. Petersburg.
[12.120] D. M. Poirier, J. H. Weaver, K. Kikuchi, and Y. Achiba. *Zeitschrift für Physik D: Atoms, Molecules and Clusters*, **26**, 79 (1993).

[12.121] X.-D. Wang, T. Hashizume, Q. Xue, H. Shinohara, Y. Saito, Y. Nishina, and T. Sakurai. *Jpn. J. Appl. Phys.*, **B32**, L866 (1993).
[12.122] Y. Z. Li, J. C. Patrin, M. Chander, J. H. Weaver, K. Kikuchi, and Y. Achiba. *Phys. Rev. B*, **47**, 10867 (1993).
[12.123] F. Diederich, R. Ettl, Y. Rubin, R. L. Whetten, R. Beck, M. Alvarez, S. Anz, D. Sensharma, F. Wudl, K. C. Khemani, and A. Koch. *Science*, **252**, 548 (1991).
[12.124] S. Saito, S. I. Sawada, N. Hamada, and A. Oshiyama. *Mater. Sci. Eng.*, **B19**, 105 (1993).
[12.125] S. Saito, S. I. Sawada, N. Hamada, and A. Oshiyama. *Jpn. J. Appl. Phys.*, **32**, 1438 (1993).
[12.126] S. Saito, S. Sawada, and N. Hamada. *Phys. Rev. B*, **45**, 13845 (1992).
[12.127] S. Saito, S. Sawada, and N. Hamada. Unpublished (1993).
[12.128] H. P. Cheng and R. L. Whetten. *Chem. Phys. Lett.*, **197**, 44 (1992).
[12.129] S. Saito, S. Sawada, and N. Hamada. *Chem. Soc. Japan* (1993). In 5th General Symposium of the C_{60} Res. Soc.
[12.130] S. Saito, S. I. Sawada, N. Hamada, and A. Oshiyama. *Mater. Sci. Eng.*, **B19**, 105 (1993).
[12.131] X. Q. Wang, C. Z. Wang, B. L. Zhang, and K. M. Ho. *Phys. Rev. Lett.*, **69**, 69 (1992).
[12.132] S. Hino, K. Matsumoto, S. Hasegawa, K. Kamiya, H. Inokuchi, T. Morikawa, T. Takahashi, K. Seki, K. Kikuchi, S. Suzuki, I. Ikemoto, and Y. Achiba. *Chem. Phys. Lett.*, **190**, 169 (1992).
[12.133] S. Hino, K. Matsumoto, S. Hasegawa, K. Iwasaki, K. Yakushi, T. Morikawa, T. Takahashi, K. Seki, K. Kikuchi, S. Suzuki, I. Ikemoto, and Y. Achiba. *Phys. Rev. B*, **48**, 8418 (1993).
[12.134] J. H. Weaver. *J. Phys. Chem. Solids*, **53**, 1433 (1992).
[12.135] P. J. Benning, F. Stepniak, D. M. Poirier, J. L. Martins, J. H. Weaver, R. E. Haufler, L. P. F. Chibante, and R. E. Smalley. *Phys. Rev. B*, **47**, 13843 (1993).
[12.136] F. Stepniak, P. J. Benning, D. M. Poirier, and J. H. Weaver. *Phys. Rev. B*, **48**, 1899 (1993).
[12.137] C. Gu, F. Stepniak, D. M. Poirier, M. B. Jost, P. J. Benning, Y. Chen, T. R. Ohno, J. L. Martins, J. H. Weaver, J. Fure, and R. E. Smalley. *Phys. Rev. B*, **45**, 6348 (1992).
[12.138] M. D. Seta and F. Evangelisti. *Phys. Rev. B*, **51**, 1096 (1995).
[12.139] J. P. Lu. *Phys. Rev. B.*, **49**, 5687 (1994).
[12.140] R. W. Lof, M. A. van Veenendaal, B. Koopmans, A. Heessels, H. T. Jonkman, and G. A. Sawatzky. *Int. J. Modern Phys. B*, **6**, 3915 (1992).
[12.141] O. Gunnarsson, D. Rainer, and G. Zwicknagl. *Int. J. Mod. Phys. B*, **6** 3993 (1992).
[12.142] B. Friedman and J. Kim. *Phys. Rev. B*, **46**, 8638 (1992).
[12.143] V. P. Antropov, O. Gunnarsson, and O. Jepsen. *Phys. Rev. B*, **46**, 13647 (1992).
[12.144] D. P. Joubert. *Journal of Phys. Condensed Matter*, **5**, 8047 (1993).
[12.145] R. L. Martin and J. P. Ritchie. *Phys. Rev. B*, **48**, 4845 (1993).
[12.146] L. Bergomi, J. P. Blaizot, T. Jolicoeur, and E. Dagotto. *Phys. Rev. B*, **47**, 5539 (1993).
[12.147] R. T. Scalettar, A. Moreo, E. Dagotto, L. Bergomi, T. Jolicoeur, and H. Monien. *Phys. Rev. B*, **47**, 12316 (1993).
[12.148] S. E. Erwin and W. E. Pickett. *Science*, **254**, 842 (1992).
[12.149] J. Wu, Z. X. Shen, D. S. Dessau, R. Cao, D. S. Marshall, P. Pianetta, I. Lindau, X. Yang, J. Terry, D. M. King, B. O. Wells, D. Elloway, H. R. Wendt, C. A. Brown, H. Hunziker, and M. S. de Vries. *Physica C*, **197**, 251 (1992).
[12.150] P. A. Brühwiler, A. J. Maxwell, A. Nilsson, N. Mårtensson, and O. Gunnarsson. *Phys. Rev. B*, **48**, 18296 (1993).

[12.151] J. de Vries, H. Steger, B. Kamke, C. Menzel, B. Weisser, W. Kamke, and I. V. Hertel. *Chem. Phys. Lett.*, **188**, 159 (1992).
[12.152] J. A. Zimmermann, J. R. Eyler, S. B. H. Bach, and S. W. McElvany. *J. Chem. Phys.*, **94**, 3556 (1991).
[12.153] D. L. Lichtenberger, K. W. Nebesny, C. D. Ray, D. R. Huffman, and L. D. Lamb. *Chem. Phys. Lett.*, **176**, 203 (1991).
[12.154] L. S. Wang, J. Conceicao, C. Jin, and R. E. Smalley. *Chem. Phys. Lett.*, **182**, 5 (1991).
[12.155] R. L. Hettich, R. N. Compton, and R. H. Ritchie. *Phys. Rev. Lett.*, **67**, 1242 (1991).
[12.156] P. C. Eklund, A. M. Rao, Y. Wang, P. Zhou, K. A. Wang, J. M. Holden, M. S. Dresselhaus, and G. Dresselhaus. *Thin Solid Films*, **257**, 211 (1995).
[12.157] M. R. Pederson and A. A. Quong. *Phys. Rev. B*, **46**, 13584 (1992).
[12.158] D. Ostling, P. Apell, and A. Rosen. *Zeits. Physik D*, **25**, 1 (1992).
[12.159] N. Matsuzawa and D. A. Dixon. *J. Phys. Chem.*, **96**, 6872 (1992).
[12.160] P. A. Brühwiler, A. J. Maxwell, and N. Mårtensson. Unpublished. Cited by Lof *et al.* [12.73].
[12.161] B. Wästberg, S. Lunell, C. Enkvist, P. A. Brühwiler, A. J. Maxwell, and N. Mårtensson. *Phys. Rev. B*, **50**, 13031 (1994).
[12.162] A. J. Maxwell, P. A. Brühwiler, A. Nilsson, N. Mårtensson, and P. Rudolf. *Phys. Rev. B*, **49**, 10717 (1994).
[12.163] C. Enkvist, S. Lunell, B. Sjögren, S. Svensson, P. A. Brühwiler, A. Nilsson, A. J. Maxwell, and N. Mårtensson. *Phys. Rev. B*, **48**, 14629 (1993).
[12.164] P. A. Brühwiler, A. J. Maxwell, P. Rudolf, C. D. Gutleben, B. Wästberg, and N. Mårtensson. *Phys. Rev. Lett.*, **71**, 3721 (1993).
[12.165] P. A. Brühwiler, A. J. Maxwell, and N. Mårtensson. *Int. J. Mod. Phys. B*, **6**, 3923 (1992).
[12.166] P. A. Brühwiler, A. J. Maxwell, A. Nilsson, R. L. Whetten, and N. Mårtensson. *Chem. Phys. Lett.*, **193**, 311 (1992).
[12.167] E. E. B. Campbell and I. V. Hertel. *Carbon*, **30**, 1157 (1992).

CHAPTER 13

Optical Properties

In this chapter we discuss primarily the optical response of fullerenes over a photon energy range from the near infrared (IR) through the ultraviolet (UV) region of the spectrum (1–6 eV). This region of the spectrum is used to explore the nature of the electronic states of fullerenes, whether they are studied in the gas phase, in solution, or assembled in the solid state. Results of experiments to obtain the linear, as well as the nonlinear, optical response are considered and compared, when possible, with theoretical models and calculations. Most of the work to date has focused on C_{60}. A much smaller body of information is available on the optical properties of C_{70} and higher fullerenes. Also included in this chapter is a review of current knowledge of the optical properties of doped and phototransformed fullerenes.

13.1. Optical Response of Isolated C_{60} Molecules

This section provides a framework for discussion of the absorption and luminescence spectra for isolated C_{60} molecules, either in the gas phase or in solution. The symmetries of the ground state and excitonic states associated with photoabsorption are considered. It is shown that vibronic states which couple the excitonic states to vibrational states are needed for electric dipole transitions near the optical absorption edge, and the symmetries for vibronic states involved in absorption, luminescence, intersystem crossings, and phosphorescence are discussed. Optical absorption at higher photon energies to dipole-allowed states is also reviewed.

13.1.1. Introduction to Molecular Photophysics

In this subsection we discuss the photophysical properties of isolated C_{60} molecules. This molecular isolation can be realized in the gas phase and can also be achieved approximately for molecules in solution. Since C_{60} is soluble in a wide variety of solvents (see Table 5.1), numerous optical studies on solutions have been carried out to investigate the "photophysics" of this unusual molecule. By a "photophysical" process, it is usually meant that an electronic excitation is achieved through the absorption of a photon, without a resulting chemical change (e.g., fragmentation or bond rearrangement). Photophysical processes can either be unimolecular or bimolecular (i.e., involve one or multiple molecules), and they can involve the absorption or emission of one or several photons. In the case of C_{60}, and to some degree C_{70}, photochemical transformation to a dimer or oligomer has been reported [13.1–5]. However, these reports refer to fullerenes in the solid state, where the molecules are in closer contact. Thus, these photochemical changes are much less likely (but also possible) in solution.

Photoexcited electronic states of molecules, in general, include spin triplet ($S = 1$) and spin singlet ($S = 0$) states arising from the excited electron and the hole that is left behind. The notion of triplet states (e.g., spin-paired electronic states with a multiplicity of $2S + 1 = 3$) is quite familiar to chemists, and possibly less familiar to solid-state physicists and material scientists. The conventional notation denoting these states in photophysical processes is as follows [13.6]:

S_0		ground singlet state
S_1		first excited singlet state
S_p	$(p > 1)$	higher excited singlet states
T_1		first excited triplet state
T_q	$(q > 1)$	higher excited triplet states.

Because of the coupling of vibrational and electronic levels by the electron–molecular vibration (EMV) interaction, a ladder of "vibronic" states with approximate energy

$$E_n = E_i + (n + 1/2)\hbar\omega_j \tag{13.1}$$

is realized, where E_i and $\hbar\omega_j$ are, respectively, particular (renormalized) electronic and vibrational energies, and n is an integer $n = 0, 1, 2, \ldots$. In Fig. 13.1, we show schematically optical transitions between the ground state vibronic ladder and an excited state vibronic ladder for a generic molecule. The transition $1 \rightarrow 0$ is commonly called a "hot band" because elevated temperatures are needed to populate the $n = 1$ vibrational level in the ground electronic state (e.g., a 250 cm^{-1} molecular vibration corre-

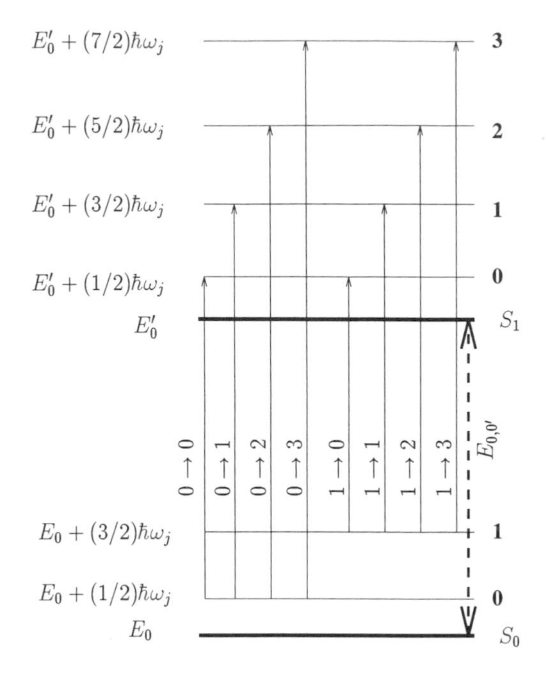

Fig. 13.1. Schematic level diagram for vibronic multiphonon absorption transitions from the singlet S_0 electronic ground state to the first excited singlet electronic state S_1. The symbols $n \to m$ denote the optical transitions between vibronic levels associated with the initial and final electronic states E_0 and E'_0, respectively.

sponds to a temperature of 360 K). Note that the vibrational frequency of a particular vibrational mode in the electronic ground state should differ slightly from the frequency of this mode in the excited electronic states, because of a softening of the harmonic potential due to the excited state electronic configuration. The energy $E_{0,0'}$ (see Fig. 13.1) denotes the energy difference between the excited and ground electronic zero vibrational states, while $E_{n,m'}$ denotes the energy difference between the vibrational states of E_n and E'_m, where the prime notation is used to denote the excited electronic state (as shown in Fig. 13.1). These vibronic ladders can be generated from either triplet or singlet electronic states, and, as we shall see in §13.1.3, these vibronic states may be required in the case of C_{60} to understand the low-lying weak, optical absorption, near the absorption edge.

In Fig. 13.2, we display a generic energy level diagram for an isolated molecule, showing only the $n = 0$ vibronic states (i.e., only the base of each vibronic ladder of Fig. 13.1). For clarity, the singlet and triplet states are formally separated in Fig. 13.2; singlet and triplet states are drawn to the left and right, respectively. The molecule depicted in Fig. 13.2 has a singlet ground state (S_0), as does C_{60}. Of particular interest to us here are the following transitions:

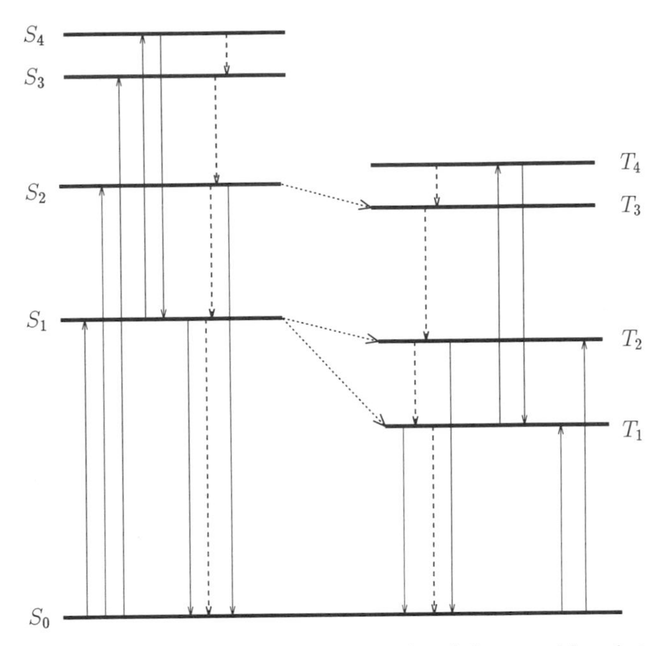

Fig. 13.2. Level diagram for optical absorption and emission transitions between singlet and triplet molecular states. Solid lines denote optical transitions, and dashed lines denote radiationless transitions. The dotted lines denote intersystem crossings.

1. Excited state absorption ($S_1 \rightarrow S_p$ and $T_1 \rightarrow T_p$; $p > 1$), which are spin-conserving transitions, and therefore can be quite strong (shown in Fig. 13.2 as solid lines).
2. Fluorescence ($S_1 \rightarrow S_0$).
3. Phosphorescence ($T_1 \rightarrow S_0$).
4. Internal conversion ($S_p \rightarrow S_{p-1}$ and $T_p \rightarrow T_{p-1}$), which are radiationless transitions (shown as dashed vertical lines in Fig. 13.2).
5. Intersystem crossing ($S_p \rightarrow T_q$ for $p > 0$, or $T_q \rightarrow S_p$ for $q > 0$), which requires the spin–orbit interaction to mix levels with differing spin [the selection rule is that there is no symmetry change for the electronic wave function in an intersystem crossing, i.e., $\Gamma_i(S) \rightarrow \Gamma_i(T)$].

Typically, internal conversion is a rapid, spin-conserving process or series of processes which returns the photoexcited molecule to either an S_1 or T_1 state. From these states, luminescence, either fluorescence (S_1) or phosphorescence (T_1), can occur. Because phosphorescence is not a spin-conserving transition (and the spin–orbit interaction is needed to couple the T_1 and S_0 levels), the radiative lifetime of the T_1 state is long. Duration times for phosphorescent emission from a collection of molecules are typically ~ 1 ms to 10 s, while typical duration times for fluorescent luminescence are considerably shorter, ~ 1 ns to 1 μs.

In thinking about the nature of the photoexcited states discussed above, it is important to realize that, in principle, all relevant interactions are included in determining the eigenfunctions and eigenenergies, such as electronic many-body effects. We discuss this point further in §13.1.2 in connection with the appropriate excited states in C_{60}.

13.1.2. Dipole-Allowed Transitions

In this section we consider the photoexcited states for molecular C_{60}. Near the threshold for optical absorption, the oscillator strengths for optical absorption are very weak because the optical transitions are forbidden by the electric dipole absorption process. In this regime, the absorption and luminescence processes are described in terms of vibronic states associated with excitonic ladders shown in Figs. 13.1 and 13.2. At higher photon energies, allowed electric dipole transitions can occur and a simple approach is used to describe the dipole-allowed optical transitions. To illustrate the two regimes for the optical transitions (forbidden transitions at low photon energies and allowed transitions at higher energies), it is instructive to consider the photon energy dependence of the electric dipole oscillator strength which denotes the square of the matrix element between initial and final states in an optical transition.

In Fig. 13.3 we compare the electric dipole oscillator strength for transitions from the ground state of a C_{60} molecule as calculated by Westin and Rosén [13.7, 8] and by Negri *et al.* [13.9]. The Westin and Rosén calculation uses one-electron wave functions (i.e., molecular orbitals) while the Negri calculation is a quantum chemical calculation using the full molecular many-electron wave function with configuration mixing. In these calculations, the area under an optical absorption band for an isolated molecule is approximately proportional to $\hbar\omega f$ where $\hbar\omega$ is the photon energy and f is the oscillator strength of the particular transition as shown in Fig. 13.3. The photon energy weight magnifies the absorption in the UV relative to the visible region of the spectrum. Optical bands identified with a particular excitation may be broadened by several interactions: molecule–solvent, electron–molecular vibration, or intermolecular interactions, the latter being more important in the solid state.

Both calculations in Fig. 13.3 consider only first-order optical processes; i.e., the excited state has one hole and one electron. The calculations of Negri *et al.* explicitly take into account the electron–hole interaction in the excited state, whereas the single-particle calculations of Westin and Rosén

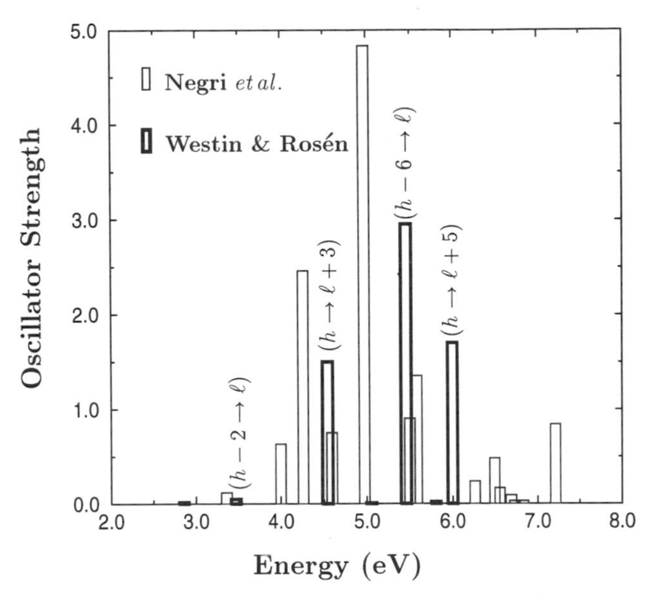

Fig. 13.3. Oscillator strength of allowed transitions according to Westin and Rosén [13.7, 8] and to Negri *et al.* [13.9]. In the diagram, the one-electron transitions are labeled by ℓ and h for the LUMO and HOMO levels, respectively, while $h-2$ refers to two levels below the HOMO level and $\ell+3$ refers to the one-electron band three levels above the LUMO level.

do not. In the many-particle formalism discussed in §12.3, the electronic ground state for a C_{60} molecule has A_g symmetry, and the electric dipole transitions are only allowed from this ground state to excited state configurations with T_{1u} symmetry. In the single-particle approach, an electron in a filled molecular orbital is excited by the photon to another, higher-lying empty orbital, whose parity must be different from that of the initial state. The many-particle calculations of Negri *et al.* show that the electronic eigenenergies are stable after reaching a configuration size of 14 (or 196 singly excited configurations). However, an accurate determination of the oscillator strengths required a larger configuration [13.9]. The dipole-allowed transitions for C_{60} in solution are further discussed in §13.2.1.

13.1.3. Optical Transitions between Low-Lying Herzberg–Teller Vibronic States

In a one-electron electronic structure model (see Fig. 12.1), a C_{60} molecule in the ground state has two spin states for each of the five degenerate h_u symmetry orbitals (HOMO) [13.10]. The lowest unoccupied molecular orbitals (or LUMO) involve the threefold degenerate t_{1u} one-electron states, and each orbital t_{1u} state can hold a spin-up and a spin-down electron. Since both the h_u and t_{1u} states have the same odd [or ungerade (u)] parity, an electric dipole transition between these one-electron states ($h_u \rightarrow t_{1u}$) involving the absorption of a single photon is forbidden by parity considerations alone. The first (lowest energy) allowed electric dipole optical absorption would then be $h_u \rightarrow t_{1g}$ (to the "LUMO + 1 state," as shown in Fig. 12.1). Of course, transitions from other filled ungerade (gerade) to empty gerade (ungerade) one-electron molecular orbitals are allowed, subject to satisfying the selection rules for electric dipole transitions (e.g., $h_u \rightarrow a_g$ is forbidden, even though the initial and final states are of opposite parity). The theoretical selection rules for a "dipole-allowed" transition from an initial state with symmetry Γ_i require that the decomposition of the direct product $\Gamma_i \otimes F_{1u}$ contains the symmetry of the excited state level Γ_j. The magnitude of the nonvanishing matrix element can often be inferred by invoking the "f-sum rule" [13.11]. This rule relates to the physical fact that wave functions for the allowed states which are close to the initial state in energy will have nearly the same number of nodes as the initial state and hence have the largest matrix elements and show the strongest optical transitions. This argument would say that the transitions between the angular momentum states $\ell = 4$ and $\ell = 5$ (or $g_g \rightarrow t_{1u}$ and $h_g \rightarrow t_{1u}$ in icosahedral symmetry) would have a larger optical oscillator strength than the $\ell = 5$ to $\ell = 6$ transitions ($h_u \rightarrow t_{1g}$ and $h_u \rightarrow t_{2g}$) [13.7, 8, 12], because the former transitions have fewer nodes in the wave functions and hence a greater overlap in phase between the wave functions for the initial and final states.

To assign the weak absorption band at the absorption edge to transitions between HOMO- and LUMO-derived states, a symmetry-breaking interaction must be introduced which alters the symmetry of either the initial or the final state, so that electric dipole transitions can occur. Similar to the behavior of other aromatic molecules [13.6], this symmetry-breaking mechanism can be supplied by the electron–vibration interaction, and this effect is sometimes referred to as "Herzberg–Teller" (H–T) coupling [13.6]. The coupled electron–vibration state is called a "vibronic" state. Group theory can be used to identify the symmetry of these vibronic states, and the optical selection rules follow from the symmetries of the vibronic states and that of the dipole operator. The first step in the process is to identify the symmetries of the many-electron states which result from a configurational mixing of the one-electron states, and the second step is the determination of the particular vibrational mode symmetries which can participate in H–T coupling to these many-electron molecular states.

The mathematical formulation of this problem follows from the arguments given in §11.3.3. The Herzberg–Teller states (vibronic states) are written as

$$\Psi_{H-T}(\Gamma_{tot}) = \Psi_{el}(\Gamma_{el})\Psi_{vib}(\Gamma_{vib}) \tag{13.2}$$

where the rotational and nuclear terms that appear in Eq. (11.3) are assumed to have A_g symmetry, i.e., no rotation or isotopic symmetry breaking is considered here, and therefore the corresponding terms in the total wave function do not appear in Eq. (13.2). The important term in the Hamiltonian that gives rise to the H–T coupling is the electronic–vibrational mode coupling, denoted by $\mathscr{H}_{el-vib}$, which has A_g symmetry because of the general invariance of the Hamiltonian under symmetry operations of the group. There is, however, an admixture of other electronic symmetries associated with the group theoretical direct product $\Gamma_{el} \otimes \Gamma_{vib}$ which is brought about by the $\mathscr{H}_{el-vib}$ term in the Hamiltonian that turns on the H–T terms in the optical absorption.

To discuss optical transitions we must investigate symmetry-allowed matrix elements of the type $\langle\psi(\Gamma_i)|\mathbf{p}\cdot\mathbf{A}|\psi(\Gamma_f)\rangle$ where Γ_i and Γ_j denote the symmetries of the initial and final states, respectively, and $\mathbf{p}\cdot\mathbf{A}$ is proportional to the electric dipole matrix element. To illustrate the calculation of matrix elements for vibronic transitions explicitly, we select the transition ${}^1A_g;0 \rightarrow {}^1F_{1g};\nu(A_u)$, which denotes an absorption from a singlet ground state with no vibrational excitations and A_g symmetry to an excited vibronic singlet state with F_{1g} symmetry for the electronic state and A_u symmetry for the vibration. This vibronic transition is allowed since $F_{1g} \otimes A_u = F_{1u}$ so that the final state couples to A_g via an electric dipole transition.

The matrix element for the ${}^1A_g;0 \rightarrow {}^1F_{1g};\nu(A_u)$ vibronic transition is written in terms of the electric dipole interaction p_α and the electron–vibration (electron–phonon for a crystalline solid) interaction $\mathscr{H}_{el-vib}$ as

$$\begin{aligned}&\langle\Psi_{el}({}^1A_g);0|p_\alpha|\Psi_{el}({}^1F_{1g});\nu(A_u)\rangle\\ &\quad= \langle\Psi_{el}({}^1A_g);0|p_\alpha|\Psi_{el}({}^1F_{1u});0\rangle\frac{\langle\Psi_{el}({}^1F_{1u});0|\mathscr{H}_{el-vib}|\Psi_{el}({}^1F_{1g});\nu(A_u)\rangle}{E({}^1F_{1g})+\hbar\omega(A_u)-E({}^1F_{1u})}.\end{aligned} \tag{13.3}$$

The same interaction that couples the initial and final states also gives rise to a shift in the energy for the transition given by

$$\begin{aligned}E[{}^1F_{1g};\nu(A_u)] &= E({}^1F_{1g})+\hbar\omega(A_u)\\ &\quad+\frac{|\langle\Psi_{el}({}^1F_{1u});0|\mathscr{H}_{el-vib}|\Psi_{el}({}^1F_{1g});\nu(A_u)\rangle|^2}{E({}^1F_{1g})+\hbar\omega(A_u)-E({}^1F_{1u})}+\cdots\end{aligned} \tag{13.4}$$

In some cases a “dipole-forbidden” transition, which occurs via an H–T coupling, can show a larger coupling than a higher-energy “dipole-allowed” transition, because the admixed state [see Eq. (13.3)] can contribute a larger matrix element.

We now present a more explicit discussion of the application of the Herzberg–Teller theory to C_{60}, reviewing the description of the ground state and the excited excitonic state, before discussing the vibronic states in more detail. As discussed in §12.1.1, the ground state electronic configuration for C_{60} is h_u^{10}, and because all the states in h_u^{10} are totally occupied, the h_u^{10} many-electron configuration has A_g symmetry and is a singlet ground state ^1A_g with quantum numbers $S = 0$ and $L = 0$. This ^1A_g state is also commonly labeled S_0 by chemists (see Figs. 13.2 and 13.4). The higher-lying excited singlet states of various symmetries are referred to by chemists as $S_1, S_2, \ldots$. The lowest group of excited states stem from the promotion of one of the h_u electrons to the higher-lying t_{1u} state, leaving the molecule in an excited state configuration $h_u^9 t_{1u}^1$. This orbital configuration then describes an exciton with one electron (t_{1u}) and one hole (h_u), each with spin 1/2. As described in §13.1.1, this excited exciton state can be either a singlet state S_i ($S = 0$) or a triplet state T_i ($S = 1$), where S is the spin angular momentum quantum number. In Fig. 13.4, we show a schematic energy level diagram of singlet and triplet molecular excitonic states of C_{60} built from the $h_u^9 t_{1u}^1$ configuration (bold horizontal lines) which lie above the A_g ground state. When spin–orbit coupling is considered, then the "spin" and "orbital" parts of the wave function determine the symmetry of the coupled wave functions through the direct product $\Gamma_{\text{spin}} \otimes \Gamma_{\text{orbit}}$, where $\Gamma_{\text{spin}} = A_g$ denotes the symmetry for the singlet state and $\Gamma_{\text{spin}} = F_{1g}$ for the triplet state (see §12.5). Figure 13.4 shows that the triplet states are generally lower in energy than the singlet states formed from the same molecular configuration. Because of the Pauli principle, two electrons in a triplet state will have different orbital wave functions, and because of wave function orthogonality considerations, these two electrons will tend to repel each other. Thus, the effect of the Coulomb repulsion in the triplet state is reduced relative to that of the singlet state where the orbital wave functions are the same. Excitonic states can also be formed by promoting an electron from the h_g^{10} HOMO−1 level (see Fig. 12.1) to the t_{1u} LUMO level, as indicated in the figure legend to Fig. 13.4.

Since the electric dipole operator does not contain a spin operator in the absence of spin–orbit coupling, the selection rule $\Delta S = 0$ is in force, so that optical transitions from the singlet ^1A_g ground state must therefore be to an excited singlet state (and not to an excited triplet state). The construction of excitonic triplet and singlet states is discussed in §12.5. In Fig. 13.4 the singlet and triplet states have been separated (by convention) to the left and right in the diagram, respectively. The dotted arrows represent the important radiationless intersystem crossings from the singlet manifold to

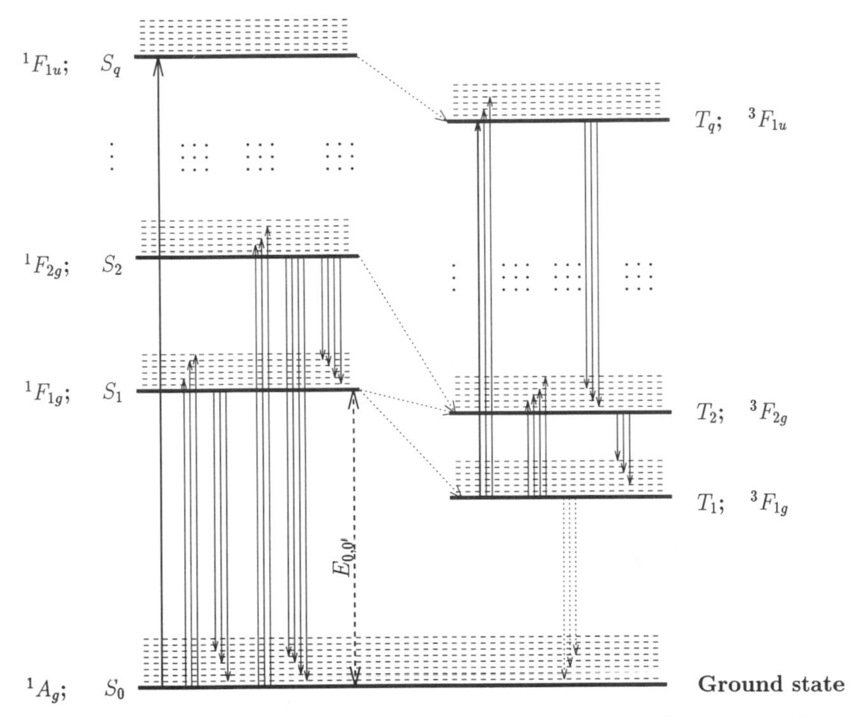

Fig. 13.4. Schematic diagram for two of the lowest excitonic levels (boldface lines) associated with the $h_u^9 t_{1u}^1$ excitonic configuration of C_{60}. Associated with each excitonic level are a number of vibronic levels some of which couple to the ground state by an absorption or emission (luminescence) process. Also shown schematically are intersystem crossings between the singlet and triplet vibronic levels, as well as transitions from a triplet exciton state to the electronic ground state in a phosphorescence transition. The S_0 ground state together with its associated vibronic levels are shown across both left and right columns for convenience. The $^1F_{1u}$; S_q and $^3F_{1u}$; T_q states in the figure are associated with the $h_g^9 h_u^{10} t_{1u}^1$ excitonic configuration of the HOMO–1 and LUMO exciton and represent states for allowed optical transitions (heavier full vertical lines). The transition to the $^1F_{1u}$ state is expected to occur in C_{60} at about 3 eV.

the triplet manifold [13.13]. The superscript on the symmetry label indicates whether the state is a singlet (e.g., $^1F_{2g}$ for antiparallel spins of the electron and hole) or a triplet (e.g., $^3F_{2g}$ for spins parallel to each other).

Intersystem crossings (see Fig. 13.4) involve a singlet–triplet transition. The singlet and triplet states are coupled by the spin-orbit interaction $\mathcal{H}_{s-o}$ which in carbon is very small (see Table 3.1). Since $\mathcal{H}_{s-o}$ has A_g symmetry, the spin flip transition couples states of the same symmetry. The triplet spin state ($S = 1$), however, transforms as F_{1g}, which means that the singlet state with orbital symmetry Γ_S will make an intersystem crossing to those triplet states which contain the symmetry Γ_S in the direct product $F_{1g} \otimes \Gamma_S$. Likewise, a phosphorescent transition from the lowest triplet state ${}^3F_{1g}(T_1)$ would couple weakly to a vibronic level in the ${}^1A_g$ ground state manifold if its symmetry is contained in the direct product $F_{1g} \otimes F_{1g} \otimes F_{1u}$ between the orbital (F_{1g}) and spin (F_{1g}) wave functions and the electric dipole matrix element (F_{1u}).

Using group theory, the symmetries of the excitonic states built from the configuration $h_u^9 t_{1u}^1$ can be worked out from the direct product describing the hole and the electron: $H_u \otimes F_{1u} = F_{1g} + F_{2g} + G_g + H_g$ (see §4.2, §12.1, and §12.5). These four symmetries apply to both singlet and triplet excitonic states that are formed from the $h_u^9 t_{1u}^1$ configuration in C_{60} [13.14]. The symmetries of the two lowest-lying singlet excitonic levels have been identified as F_{1g} and F_{2g} [13.9, 15], as shown in Fig. 13.4, and the two excitonic levels lie very close to each other in energy. The energy separation between these levels is not drawn to scale in the figure in order to show the vibronic states discussed below. Theoretical calculations of these excitonic energy levels and the associated oscillator strengths for the free molecules have been made by Negri *et al.* [13.9] using a quantum mechanical extension to π-electrons of the "consistent force field method" (QCFF/PI method). They find that the ${}^1F_{1g}$ and ${}^1F_{2g}$ excitonic states are nearly degenerate and are the lowest-lying singlet states, but to explain magnetic circular dichroism measurements in C_{60} [13.16], Negri *et al.* [13.9] have placed the ${}^1F_{1g}$ level slightly lower in energy than the ${}^1F_{2g}$ level.

Since all the states in the initial h_u^{10} state and final $h_u^9 t_{1u}$ excitonic configurations have gerade parity, electric dipole transitions between these singlet excitonic states and the A_g ground state are forbidden. This conclusion follows whether using the many-electron configuration notation or the one-electron notation given above. Thus configurational mixing, by itself, is insufficient to explain the weak optical absorption observed in C_{60} below 2.6 eV [13.17], denoting the threshold for allowed dipole transitions (see §13.2). However, with a sufficiently strong electron–vibration interaction, it is possible to admix enough ungerade parity into the excitonic state wave function to form a "vibronic" state to which optical transitions can occur, and whose energy is, to lowest order [Eq. (13.1)], the sum of an electronic contribution (E_i) and a molecular vibration energy ($\hbar\omega_j$). Thus, we associate with each excitonic level a manifold of vibronic levels, for which the excitonic level is the zero vibration state (see Fig. 13.4).

Shown in Fig. 13.4 are the vibronic manifolds for the A_g ground state (h_u^{10}) and for the various excitonic states. The vibronic manifold for the ground state is involved in the photoluminescent transitions from the vibrationless exciton states shown in Fig. 13.4 by the solid downward arrows [13.6]. The absorption transitions between the vibrationless ground state and selected vibronic excited states are indicated in Fig. 13.4 by solid upward arrows. The dashed and heavy horizontal solid lines, respectively, indicate the positions of vibronic states ($E_i + \hbar\omega_j$) and zero-vibration electronic states (E_i), where E_i denotes both the ground state (E_0) and the excitonic states. The phosphorescent transitions from the T_1 triplet state to the S_0 ground vibronic states are indicated by vertical downward dotted arrows. The S_0 ground vibronic state in Fig. 13.4 is shown as extending across the two columns, showing singlet and triplet excited states, respectively.

The symmetries of vibrations which couple the ground state S_0 to a vibronic excited state are given in Table 13.1. As an example of using this table, consider the symmetries of the vibronic states arising from an H_u vibrational mode coupling the ground state (A_g) to a ${}^1F_{2g}$ symmetry excitonic state. The symmetries of these vibronic states are found by taking the direct product $F_{2g} \otimes H_u = F_{1u} + F_{2u} + G_u + H_u$ and examining the irreducible representations contained in this direct product. If the direct product contains the representation for the electric dipole operator (F_{1u}), as it does in this case, then the electric dipole transition between the ${}^1A_g$ ground state and the vibronic states $|{}^1F_{2g}; \nu[H_u(j)]\rangle$ is allowed for each of the $j = 1, \ldots, 7$ vibrational modes with H_u symmetry listed in Table 11.1. Therefore, as listed in Table 13.1, vibrations with H_u symmetry activate a transition between the ${}^1F_{2g}$ excitonic state and the A_g ground state either in an absorption process or in a luminescence process. In Table 13.1 are tabu-

Table 13.1

Symmetries of vibrations which activate an electronic transition between the ground state with A_g symmetry and an excitonic state with symmetry Γ_i for icosahedral symmetry.[a] The symmetry of the vibronic states is found from the direct product $\Gamma_{ex} \otimes \Gamma_{vib}$ for the exciton Γ_{ex} and the normal mode vibration Γ_{vib}. Listed are those vibrational modes which contain F_{1u} symmetry in the direct product $\Gamma_{ex} \otimes \Gamma_{vib}$.

Γ_{ex}	Γ_{vib}	Γ_{ex}	Γ_{vib}
A_g	F_{1u}	A_u	F_{1g}
F_{1g}	A_u, F_{1u}, H_u	F_{1u}	A_g, F_{1g}, H_g
F_{2g}	F_{2u}, G_u, H_u	F_{2u}	F_{2g}, G_g, H_g
G_g	F_{2u}, G_u, H_u	G_u	F_{2g}, G_g, H_g
H_g	F_{1u}, F_{2u}, G_u, H_u	H_u	F_{1g}, F_{2g}, G_g, H_g

[a]Note that "u" vibrational modes couple with "g" excitonic states (and vice versa) in order to achieve an excited vibronic state with "u" parity. Vibronic states with "u" parity couple via an electric dipole transition to the "g" ground state (${}^1A_g$), subject to the selection rules given in this table.

lated all the vibrational modes which lead to a vibronic state containing the F_{1u} irreducible representation for excitonic states of every possible symmetry. It should be noted that many of the intramolecular mode frequencies in the fullerenes are quite high, and therefore they can make a significant contribution to the energy of the vibronic state. For C_{60}, the highest first-order normal mode frequency is 1577 cm^{-1}, which corresponds to ~0.2 eV (see Table 11.1).

13.2. Optical Studies of C_{60} in Solution

To obtain an adequate signal-to-noise ratio, the study of the electronic levels of the "free" molecule is most conveniently carried out in solution or by using matrix isolation techniques [13.16, 18]. A summary of results obtained from absorption, luminescence and time-resolved measurements on C_{60} molecules in solution is presented in this section. Features near the absorption edge and at higher photon energies are both considered. The weak absorption at the absorption edge has a number of special features associated with excitonic states which are activated into electronic transitions through a vibronic Herzberg–Teller (H–T) coupling mechanism (as discussed in §13.2.1). The absorption at higher photon energies is, however, dominated by dipole-allowed transitions which are described by more conventional treatments.

13.2.1. Absorption of C_{60} in Solution

In Fig. 13.5(c), we show experimental results from Leach *et al.* [13.17] for the optical density as a function of wavelength for C_{60} dissolved in *n*-hexane in the UV–visible region of the spectrum. Since C_{60} in other organic solvents yields almost the same spectrum, the structure in the figure is not identified with solvent-related absorption bands, but rather with the intrinsic spectrum for C_{60}. The optical density (OD) is defined as $\log_{10}(1/\mathscr{T})$, where $\mathscr{T}$ is the optical transmission coefficient. Neglecting corrections for reflection loss at the optical cell walls, the optical density is proportional to the optical absorption coefficient. As can be seen in Fig. 13.5(c), a region of weak π–π^* absorption extends from ~ 640 nm (1.9 eV) to 440 nm (2.8 eV), where π and π^* refer to *p*-function-derived occupied and empty levels (or bands), respectively. Since the threshold for this π–π^* absorption (640 nm) is close to the values for the HOMO–LUMO gap E_{H-L} (1.9 eV) calculated on the basis of a one-electron approach ([13.19]), a number of authors have attempted to describe the optical properties of C_{60} over a wide range of photon energies in terms of a one-electron treatment. However, as discussed above, the π–π^* transitions at the absorption edge should be

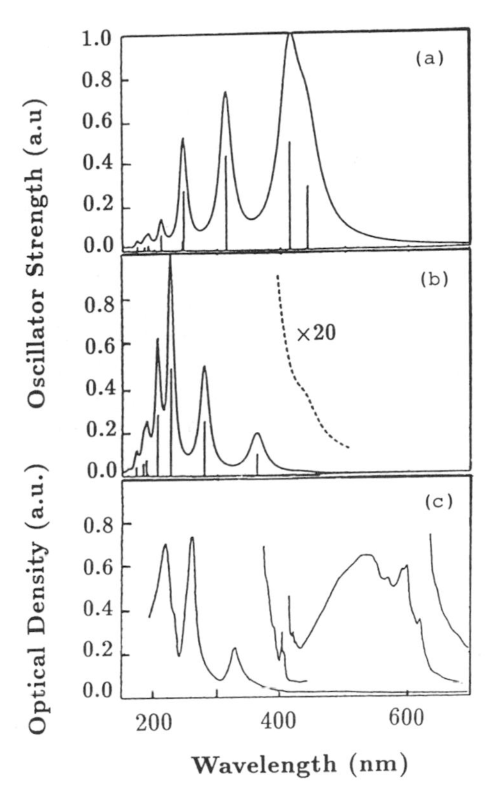

Fig. 13.5. (a) Calculated free response oscillator strength of C_{60} [13.7]. (b) Calculated screened response oscillator strength [13.7]. (c) Experimental absorption spectra for C_{60} in *n*-hexane solution [13.17]. The three experimental spectra to the right are magnified with respect to the first as discussed in Ref. [13.17].

forbidden by symmetry considerations, and a more detailed treatment including strong electron–electron and electron–phonon effects is therefore needed.

As discussed in §13.1.3, the weak absorption band between 440 and 640 nm can be explained in terms of vibronic transitions from the HOMO-derived, A_g symmetry ground state to excitonic states, as shown schematically by the upward arrows in Fig. 13.4. Negri *et al.* [13.9] have calculated the oscillator strengths for various electric dipole transitions involving important vibronic states using the "complete neglect of differential overlap for spectroscopy" or the CNDO/S method [13.20]. Accordingly, they have used their results to assign the structure in the weak optical absorption (OA) band spanning the 440–640 nm range to transitions between the ^1A_g (S_0) level and the vibronic manifold associated with the zero vibration $^1F_{1g}$

(S_1) state (Fig. 13.4). Consistent with Table 13.1, the H–T active vibrational modes were shown to be the $F_{1u}(4)$, $H_u(7)$, and $A_u(1)$ symmetry vibrational modes, where the number of mode frequencies of each symmetry type is indicated in parentheses (see Table 11.1). Of course, numerous combination and overtone modes (see §11.5.3 and §11.5.4) also have the F_{1u}, H_u, and A_u symmetries and they too give rise to higher-order H–T active excitations. The relation between the optical absorption spectra and the photoluminescence spectra is discussed in §13.2.2.

We next turn our attention to the much stronger optical absorption observed experimentally for C_{60} in *n*-hexane [Fig. 13.5(c)] at wavelengths in the UV for $\lambda \leq 440$ nm. As can be seen in Fig. 13.5(c), three, reasonably sharp, bands are observed, with absorption maxima at 3.7 eV (335 nm), 4.6 eV (270 nm), and 5.8 eV (215 nm) [13.17]. Electron impact studies on free C_{60} have been interpreted to show allowed dipole transitions at 3.78 and 4.84 eV and a collective π–π^* state at 6.1 eV [13.21]. The origin of these absorption bands can be understood using the approximation of one-electron orbitals (see Fig. 12.1). Theoretical calculations of the excited one-electron state energies have been carried out by several authors [13.7, 19, 22, 23] and these are summarized in Table 13.2. Electric dipole absorption is allowed between these states as long as the initial (occupied) and final (empty) states have different parity and are coupled by the electric dipole matrix element which transforms as T_{1u} (F_{1u}), using the notation of Table 13.1. Of course, depending on the nature of the wave functions for the ground and excited states, the calculated oscillator strengths for

Table 13.2

Transition frequencies (in eV), oscillator strengths, and proposed assignments of interband transitions in solid C_{60}. Results for the transition frequencies and oscillator strengths are given for both the free molecules and the screened (scr) response [13.2.4]. Here h and ℓ denote abbreviations for the HOMO and LUMO levels, respectively [13.7].

Transition				Unscreened		Screened	
HOMO		LUMO		ω (eV)	f_{nm}	ω^{scr}(eV)	f_{nm}^{scr}
h	h_u	$\ell+1$	t_{1g}	2.80	4.13	2.86	0.018
$h-2$	h_g	ℓ	t_{1u}	2.98	7.22	3.47	0.58
h	h_u	$\ell+3$	h_g	3.94	6.28	4.54	1.50
$h-1$	g_g	$\ell+2$	t_{2u}	5.00	3.90	5.05	0.01
$h-2$	h_g	$\ell+2$	t_{2u}	5.06	0.62	5.07	0.002
$h-6$	g_g	ℓ	t_{1u}	5.07	0.08	5.45	2.95
$h-4$	t_{1u}	$\ell+1$	t_{1g}	5.80	0.03	5.80	0.02
h	h_u	$\ell+5$	g_g	5.88	1.02	6.00	1.70

these dipole-allowed transitions may vary by several orders of magnitude [13.7, 19]. This may be seen in Fig. 13.3, where, for example, the strength of the weak $(h-2) \rightarrow \ell$ transition may be compared to that of the strong $(h-6) \rightarrow \ell$ transition.

The absorption coefficient for a dilute molecular system (i.e., gas phase or solution) may be calculated according to

$$\epsilon(\omega) = 1 - 4\pi N\chi \tag{13.5}$$

where $\epsilon(\omega)$ is the optical dielectric function, N is the molecular density, and χ is the molecular polarizability given by

$$\chi = \frac{e^2}{m} \sum_j \frac{f_j}{(\omega_j^2 - \omega^2) - i\Gamma_j \omega}. \tag{13.6}$$

In this expression, f_j is the quantum mechanical oscillator strength, ω_j is the transition frequency given by $\hbar\omega_j = (E_{fj} - E_{0j})$, and Γ_j is the broadening parameter of the jth transition between initial and final electronic states with energies E_{0j} and E_{fj}, respectively. The absorption coefficient $\alpha(\omega)$ can be calculated directly from Eq. (13.5) using the relation [13.25]

$$\alpha = 2(\omega/c)\mathrm{Im}[\epsilon^{1/2}]. \tag{13.7}$$

For a dense collection of molecules, Eq. (13.6) must be corrected using the Clausius–Mossotti relation (see §13.3.1) to account for local electric field contributions to the applied field [13.26].

Calculations [13.7, 22] of the optical response of an isolated C_{60} molecule have shown that the spectrum is strongly affected by electron correlations. These correlations were found to lead to strongly screened oscillator strengths and renormalized transition energies. In Fig. 13.5 we show the results of the calculations of Westin and Rosén for the optical response of an isolated C_{60} molecule for the case of: (a) free response (without screening) and (b) the screened response. The Westin–Rosén model employs a one-electron LCAO approach in the local density approximation, and the results were corrected for intramolecular screening using a random phase approximation (RPA) "sum over states" method. The transitions implied by curves (a) and (b) of Fig. 13.5 simulate the optical absorption coefficient $\alpha = 2(\omega/c)\mathrm{Im}[\epsilon^{1/2}]$ for C_{60} in the gas phase or in solution, and the transitions implied by these curves are summarized in Table 13.2. The curves in Fig. 13.5 can be calculated according to Eqs. (13.5) and (13.6), using values for the resonance frequencies (ω_j) and oscillator strengths (f_j) found in Table 13.2. In calculating these curves, Γ_j was set equal to 0.3 eV to simulate line-broadening interactions, such as intramolecular vibrations and solvent–C_{60} interactions. The narrow vertical lines in Figs. 13.5(a) and (b) represent the respective oscillator strengths for the cases of free (f_{nm}) and screened

(f_{nm}^{scr}) optical response [13.7]. In Fig. 13.5(c), the experimental optical absorption data [13.17] for C_{60} in solution (hexane) are shown for comparison. In this figure, the calculations and the data are plotted *vs.* wavelength rather than photon energy. As stated above, the observed absorption band between 700 and 430 nm [which has been magnified in Fig. 13.5(c)] is associated with dipole-forbidden transitions between the HOMO (h_u) and LUMO (t_{1u}) levels and therefore does not appear in the calculated spectra.

The dramatic effect of screening on the electromagnetic response of C_{60} is shown in Fig. 13.5(a) and (b) and in Table 13.2, where it can be seen that the oscillator strength of the lowest-energy transition is screened by a factor of ~200. Some renormalization of the various transition frequencies in Table 13.2 also occurs. In this table the specific optical resonances are also identified with one-electron transitions. Comparing the calculated results in Fig. 13.5(b) for the screened response of C_{60} to the experimental data in (c), we can see that the number and relative strengths of the experimental absorption bands are qualitatively reproduced in the calculation. However, the calculated absorption band positions are red-shifted relative to the data. The calculated results of Bertsch *et al.* [13.22] are similar to those of Westin and Rosén [13.7] shown in Fig. 13.5, in that they also find significant screening of the low-frequency dipole-allowed transitions. However, their calculated band positions are in better agreement with the experimental data. Unfortunately, they only make assignments for their lowest three resonances to specify optical transitions. These three assignments, however, are in agreement with those of Westin and Rosén [13.7].

13.2.2. Photoluminescence of C_{60} in Solution

The luminescence spectra discussed in this section relate to the recombination of excited excitons in the S_1 exciton state with a transition to the ground S_0 state. The conclusion by Negri *et al.* [13.9], that the experimental data favor the ${}^1F_{1g}$ state as the lowest excited singlet (S_1) state, is particularly important for the interpretation of luminescence data, since molecular systems most often emit radiation as they make a transition from the S_1 zero-vibration exciton state, even though the molecules may have been optically pumped to a much higher energy singlet state. This follows as a result of successive, rapid, nonradiative transitions down through the excited singlet state ladder to S_1 [13.6], as shown in Fig. 13.2.

In the literature, the term photoluminescence (PL) refers to either luminescence or phosphorescence [13.6]. By definition, the "luminescence" is associated with radiative decay from an excited singlet state (usually S_1) and the "phosphorescence" with radiative decay from an excited triplet state (usually the T_1 state, using the notation in Fig. 13.2). In the absence of significant spin–orbit coupling, the decay from a triplet state to the singlet ground state is slow but can be enhanced by a spin flip mechanism [possibly the magnetic dipole–dipole interaction of the exciton triplet state with a magnetic impurity such as a triplet ($S = 1$) oxygen molecule]. Since the spin flip cannot be provided by the electric dipole operator, the radiative lifetimes for phosphorescence are typically much longer (> 1 ms) when compared to luminescence (< 1 ms). In C_{60}, both luminescence [13.15, 27, 28] and phosphorescence [13.29] emission has been observed. By virtue of the nearly 100% efficient intersystem crossing [13.13, 30, 31] observed for a C_{60} molecule, electrons excited to the singlet states rapidly descend into the triplet states, unless they first leave the S_1 excitonic state in a luminescence process. The high sensitivity of photomultiplier detectors allows detection of the luminescence from the S_1 singlet excitonic state.

In Fig. 13.6(a), we display the luminescence spectra for C_{60} in a 6.6×10^{-5} molar methylcyclohexane (MCH) solution at 77 K using laser excitation wavelengths of $\lambda_{ex} = 460$ nm and $\lambda_{ex} = 330$ nm [13.15]. The corresponding optical absorption (OA) spectrum for the same solution, shown at the right in Fig. 13.6(a), is displayed in the wavenumber range between 15,000 cm^{-1} and 18,500 cm^{-1}, while the PL spectra are displayed between 11,700 cm^{-1} and 16,000 cm^{-1} [13.15]. To make an identification between corresponding features in the PL and OA spectra of Fig. 13.6(a), we label the features M_i and $M_{i'}$, respectively, in order of increasing frequency. First, it is found that the PL and OA features in this figure are only weakly sensitive to the solvent [13.15, 17, 32] (e.g., toluene or MCH), implying that the solution spectra probe the molecular states of C_{60} molecules, isolated from one another. Second, the energy of the vibronic features in the PL spectra depends only weakly on excitation energies, as shown by comparison of the two PL traces in Fig. 13.6(a) [13.15].

But most important, the assignments labeled with symbols from $M_{0'}$ to $M_{9'}$ in the observed OA spectrum are identified as mirror images of the vibronic transitions in the PL spectrum labeled with symbols from M_0 to M_9, implying that transitions M_i and $M_{i'}$ involve the same H–T active vibrational mode, consistent with the PL and OA processes described in Fig. 13.4. The vibrational frequencies ω_{vib} are determined from the optical data as the difference between the observed position of the vibronic feature and the zero-phonon transition energy, $E_{0,0'}$, indicated by the solid vertical lines in Fig. 13.6(a). For a given sample and set of operating conditions, the electronic origins for the PL and OA processes in solution are determined, and for the data shown in Fig. 13.6(a), values of $E^{PL}_{0,0'} = 15{,}200$ cm^{-1} and $E^{OA}_{0,0'} = 15{,}510$ cm^{-1} are obtained. The resulting values obtained for ω_{vib} (for features M_0 to M_6 in PL and $M_{0'}$ to $M_{6'}$ in OA) are generally in good agreement with first-order Raman [13.33] and infrared [13.34] mode

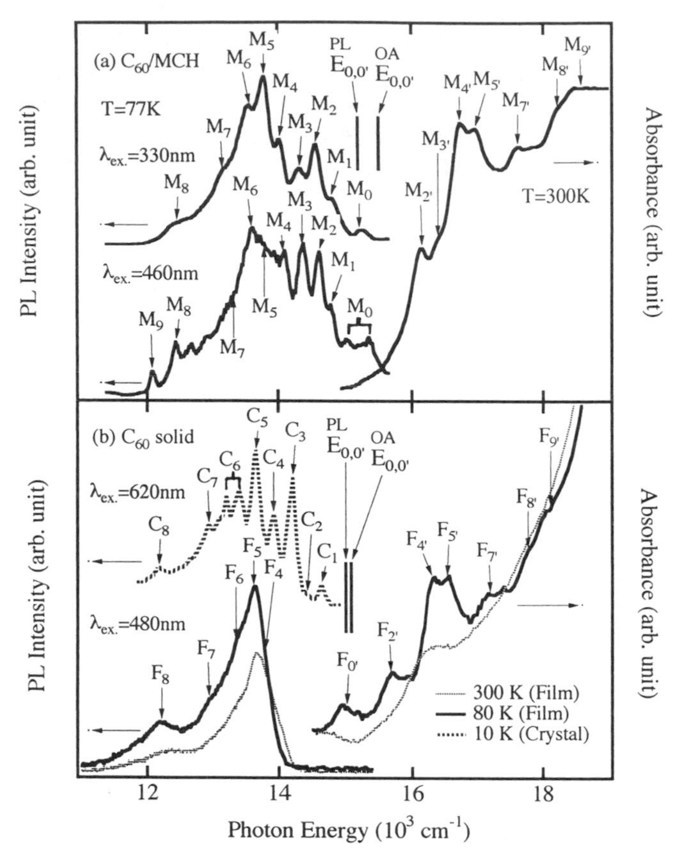

Fig. 13.6. (a) Photoluminescence (PL) and optical absorption (OA) spectra of C_{60} in 6.6×10^{-6} molar methylcyclohexane (MCH) solution (features labeled M_i for the PL spectra and $M_{i'}$ for the OA spectra). The PL spectra were taken with 330 nm (top) and 460 nm (bottom) excitation, respectively. (b) PL (features labeled F_j) and OA (features labeled $F_{j'}$) spectra at 80 K for oxygen-free pristine ~4500 Å thick C_{60} films on a Suprasil (fused silica) substrate. The PL excitation wavelength was 488 nm at ~ 0.06 W/cm^2 [13.15]. The upper curve is the 10 K PL spectrum of C_{60} single crystal (features labeled C_i) taken from Guss *et al.*[13.28].

frequencies of the C_{60} molecule [13.15]. These choices of the zero-phonon transition energy $E_{0,0'}$ and the assignments of the PL and OA features agree quite well with those of Negri *et al.* [13.35]. The difference between $E_{0,0'}^{\rm PL}$ and $E_{0,0'}^{\rm OA}$ may be due to impurities and defects which localize the excitons [13.35, 36]. The vibrational frequencies obtained for features labeled M_7 to M_9 in PL, and $M_{7'}$ to $M_{9'}$ in OA, are identified with H–T active overtone $(2\omega_i)$ and combination $(\omega_i + \omega_j)$ vibrational modes.

In general, vibrational modes involved in strong first-order vibronic transitions also participate in strong combination and overtone vibronic transitions. Since the $\nu_4(F_{1u})$ mode is found to be the strongest H–T active vibrational mode [13.35], we identify features labeled M_7 to M_9 and $M_{7'}$ to $M_{9'}$ with combination modes and overtones involving the $\nu_4(F_{1u})$ mode. Although the transition between the zero vibration levels (at $E_{0,0'}$) is forbidden by symmetry, a weak peak at this energy is seen in the PL spectra [Fig. 13.6(a)] [13.15] and also in the OA spectra shown in the same figure [13.17]. These $E_{0,0'}$ transitions are denoted by M_0 in Fig. 13.6(a).

13.2.3. Dynamics of Excited States in Isolated C_{60} Molecules

We now review efforts to determine the dynamics of the low-lying excited states (S_1 and T_1) in C_{60} and C_{70}. Our discussion will center around the schematic energy level diagram shown in Fig. 13.7, which summarizes pertinent parameters for these low-lying excited states, as detected in transient

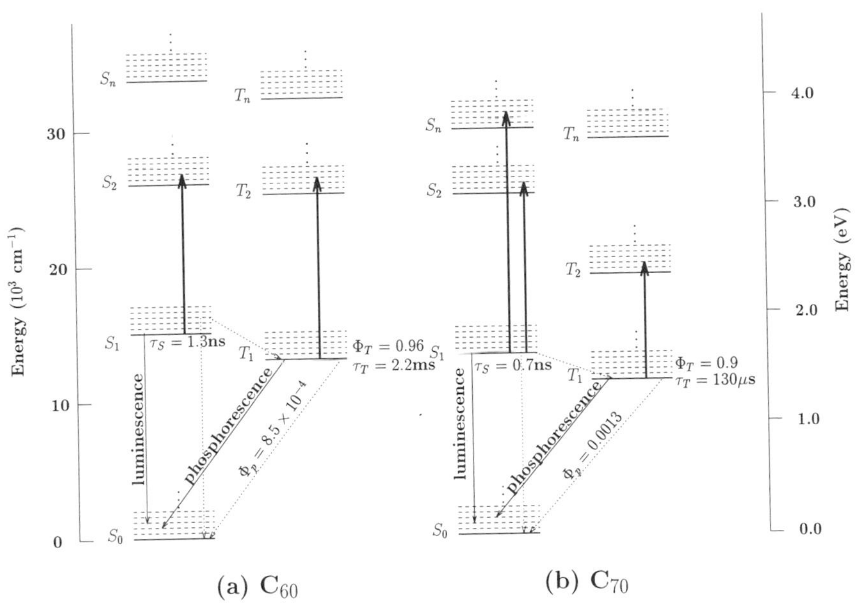

Fig. 13.7. Energy diagram for electronic states of (a) C_{60} and (b) C_{70} based on experimental photophysical and transient absorption spectra [13.30, 37]. Indicated in the figure are room temperature values for lifetimes τ_S of singlet states and τ_T of triplet states, quantum efficiencies Φ_T, and branching ratios for phosphorescence Φ_p. In the figure, S_1 and T_1 denote the manifold of excitonic and vibronic states associated with the lowest excited configuration $h_u^9 t_{1u}$, and the higher-lying states S_2, T_2 refer to higher energy electronic configurations. Less is known about the photodynamics of the higher excited states. (See Table 13.3.)

absorption and luminescence studies. Experimental values for the level positions and the lifetimes for radiative and nonradiative transitions are listed in Table 13.3. In pump–probe experiments [13.46], the C_{60} molecule is first excited by a short-duration optical pulse from the electronic ground state (S_0) to some high-lying level from which nonradiative transitions to the lowest excited singlet state S_1 occur rapidly. Upon populating S_1 by this "pump" pulse, a very rapid ($\sim$ 1.2 ns) radiationless intersystem crossing to T_1 occurs, with a quantum efficiency Φ_T close to unity, thereby depopulating S_1 [13.30, 31, 43] (see Table 13.3). By virtue of this pump pulse and the subsequent intersystem crossing, transient, time-dependent populations in S_1 and T_1 states are thereby created which allow the molecule to be further excited to higher-lying ($S_1 \rightarrow S_n$) and ($T_1 \rightarrow T_n$) states by a second, time-delayed, short duration optical pulse called the "probe" pulse. The higher cross sections for absorption from the excited singlet S_1 state ($\sigma_S = 1.57 \times 10^{-17}$ cm^2) and the excited triplet T_1 state ($\sigma_T = 9.22 \times 10^{-18}$ cm^2) relative to the ground state S_0 absorption cross section ($\sigma_0 = 2.87 \times 10^{-18}$ cm^2) have stimulated studies of absorption processes originating from optically excited states [13.37, 41]. Whereas transient absorption experiments provide powerful techniques for studying the energies of the excited states, emission studies are most useful for providing information on the lifetimes of the

Table 13.3

Photophysical and dynamical parameters for absorption and luminescence of C_{60} and C_{70} in solution.

	C_{60}	C_{70}
S_1 energy (eV)	2.00[a]	1.86–1.95[a,b,c]
T_1 energy (eV)	1.62–1.69[a,d]	1.54–1.60[b,d,e]
S_2 energy (eV)	3.4[e]	3.4[e]
T_2 energy (eV)	3.3[e]	2.9[e]
S_n energies (eV)	3.6, 4.4[f]	3.7[e]
T_n energies (eV)	4.1, 4.4,[f] 5.4[g]	4.4,[f]
Φ_T quantum efficiency	~1.0[a,h,i]	~0.9[a]
Φ_p branching ratio	8.5×10^{-4}[a]	1.3×10^{-3}[a]
$\sigma(S_0)$ (cm^2)	1.57×10^{-17}[f]	—
$\sigma(S_1)$ (cm^2)	9.22×10^{-18}[f]	—
$\sigma(T_1)$ (cm^2)	2.87×10^{-18}[f]	—
$\alpha_{mol}(S_1)$@920 nm (M^{-1}cm^{-1})	8000[e]	3800[j]
$\alpha_{mol}(T_1)$@747 nm (M^{-1}cm^{-1})	15000[e]	2000[e]
S_1 lifetime (ns)	1.2–1.3[a,e,f]	0.7[e]
T_1 lifetime (ms)	55[a]	51–53[k,l]

[a][13.30]; [b][13.38]; [c][13.39]; [d][13.40]; [e][13.41]; [f][13.37]; [g][13.42]; [h][13.31]; [i][13.43]; [j][13.44]; [k][13.13]; [l][13.45].

excited states. In experimental determinations of the decay times for the various excited states, it is important to take into account the very large decrease in the measured decay times of excited states caused by rapid relaxation via impurities, especially oxygen.

In Fig. 13.8 we show the transient absorption spectra for C_{60} and C_{70} in toluene solution obtained for different delay times between the pump and probe pulses [13.41]. In these experiments, the "pump" laser emitted 0.8 ps pulses at 587 nm (0.3 mJ/pulse at 20 Hz), and the "probe" pulse continuum was generated by focusing the same pump pulse into water. The arrows in Fig. 13.8(a) indicate whether an increase (↑) or decrease (↓) of excited state absorption occurs with increasing time delay between the pump and probe pulses. Two peaks are apparent in the spectra; one at 920 nm (1.35 eV) and the other at the 747 nm (1.66 eV). As a function of pump–probe delay time, the 747 nm peak is seen to grow at the expense of the 920 nm peak, indicating that the 920 nm peak should be identified with $S_1 \rightarrow S_n$ absorption, since the decrease in absorption is consistent with the depopulation of the S_1 state by the $S_1 \rightarrow T_1$ intersystem crossing. Accordingly, the 747 nm peak is identified with $T_1 \rightarrow T_n$ absorption. From fits to the data in Fig. 13.8(a) the lifetime of the S_1 state was found to be 1.3 $\pm$0.2 ns, as obtained from the time evolution of the 920 nm peak above [13.37, 41], in good agreement with other determinations of the lifetime of the S_1 state [13.47–50], including single photon counting fluorescence decay studies yielding a lifetime of 1.17 ns for the S_1 state [13.48]. The S_1 lifetime should be approximately equal to the intersystem crossing time (which has been reported as 1.2 ns [13.37]), since the quantum efficiency for this process is nearly unity, and almost complete $S_1 \rightarrow T_1$ transfer has been observed after a time of 3.1 ns [13.41]. The energy separation between the S_1 and S_2 levels [see Fig. 13.8(a)] has been reported by different groups as 1.35 eV [13.41] and as 1.40 eV [13.37], as summarized in Table 13.3. In this table we see that the peak molar absorption coefficient $\alpha_{\rm mol}$ for the triplet $T_1 \rightarrow T_2$ transition is found to be twice as large as for the $S_1 \rightarrow S_2$ transition. Several groups have also assigned the 747 nm (1.66 eV) transient absorption feature to $T_1 \rightarrow T_n$ transitions [13.37, 42, 48, 51] and transitions to higher-lying triplet and singlet states have also been reported (see Table 13.3). We return to a discussion of the transient spectroscopy of C_{60} in the solid state in §13.7.1.

Whereas there is general agreement that the lifetime of the singlet S_1 state is 1.2–1.3 ns, and that the lifetime of the triplet T_1 state is much longer (by several orders of magnitude) than that for S_1, a wide range of values has been quoted for the triplet lifetime. Analysis of Raman intensities for oxygen-free C_{60} thin films has indicated values of 55 $\pm$ 5 ms for the triplet lifetime, falling to 30 $\pm$ 5 ms for oxygenated C_{60} films [13.45], in

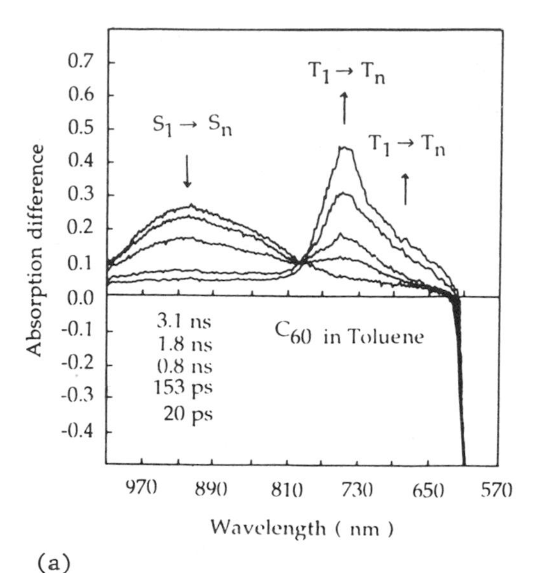

Fig. 13.8. Transient absorption spectra of (a) C_{60} in toluene obtained for delay times of 20, 153 ps, 0.8, 1.8, and 3.1 ns between the pump and probe pulses, and of (b) C_{70} in toluene for delay times of 20, 153 ps and 0.8, 1.8 ns. The pump wavelength is 587 nm. The arrow $\uparrow$ and (or $\downarrow$) indicates an increase (or decrease) of the transient absorption intensity with an increase in delay time [13.41].

good agreement with luminescent lifetimes of ~50 ms reported for C_{60} in a frozen matrix solution [13.52]. Very long triplet lifetimes have also been reported for C_{70} based on phosphorescent lifetime and time-resolved low-temperature (9 K) electron paramagnetic resonance (EPR) studies, yielding values of 53 ± 0.1 ms and 51 ± 2 ms, respectively [13.13]. It is known that the presence of dioxygen rapidly relaxes the excited triplet state and that C_{60}–C_{60} collisions in liquid solutions also relax the excited triplet state through triplet–triplet interactions. At present, the long triplet lifetimes for C_{70} appear to be well established especially by the phosphorescent lifetime studies, although more work may yet be needed to clarify the intrinsic magnitude of the triplet lifetime in C_{60}, which has not yet been observed in phosphorescent lifetime studies.

Transient absorption spectra for C_{70} in toluene [Fig. 13.8(b)] show absorption peaks at 960 nm (1.29 eV) associated with triplet absorption and at 675 nm (1.84 eV) and 840 nm (1.48 eV) associated with singlet absorption. From the time response of the growth of the 960 nm peak and the decay of the peaks at 675 and 840 nm, a lifetime of 0.7±0.05 ns for the S_1 singlet state was determined for C_{70}, about a factor of 2 shorter than for C_{60} measured in a similar way [13.41]. The molar absorption coefficient α_{mol} for C_{70} in both the singlet S_1 and triplet T_1 excited states is much lower than for the corresponding states in C_{60} (see Table 13.3), indicating lower excited state absorption for C_{70} relative to C_{60}.

Because of the high efficiency of the intersystem crossing (see above), the long lifetime of the triplet state, and the low probability of the $T_1 \rightarrow S_0$ transition, it is possible to build up a population in the T_1 triplet state. By populating this triplet T_1 state, it is possible to enhance the absorption coefficient of C_{60} by making electric dipole–allowed transitions from the odd-parity T_1 state to the appropriate higher-lying even-parity states satisfying the selection rule for such transitions. The practical application based on this excited state enhanced absorption is called "optical limiting" [13.53–56], and this application exploits the increase in the absorption coefficient with increasing intensity of the incident light (see §20.1.1). The physical process supporting the optical limiting phenomenon is the enhanced absorption from the T_1 excited state to higher-lying triplet states as the photon intensity increases, and consequently the occupation of the T_1 state increases.

Optical power limiting processes in C_{60} (and also for C_{70}) solutions have been measured in a double-pass geometry designed to produce high beam attenuation [13.57]. In the optical limiting process, an electron from a singlet ground state (S_0) is excited to a singlet (S_1) vibronic transition followed by an intersystem crossing to populate the metastable triplet states (T_1) (see Fig. 13.4). Once the T_1 triplet state is populated, its long lifetime allows optical excitations to excited triplet states (T_n for $n > 1$). Since the cross section for allowed optical transitions from the T_1 state to higher-lying T_n states is high (an order of magnitude greater than the absorption from the ground state), strong absorption occurs in the visible and near-IR spectral regions. Experimentally it is found that during one single nanosecond pulse, each C_{60} molecule can absorb 200–300 eV of energy, or up to

100 photons per molecule. This magnitude of optical limiting requires that the intersystem crossing transfers the initial excitation to the triplet state within a few picoseconds. Because of the differences in the ground state and excited state absorption coefficients between C_{60} and C_{70} mentioned above, C_{60} is expected to be more suitable for optical limiting applications, in agreement with observations. Excited state absorption in fullerene films is further discussed in §13.7.

Values for some of the photophysical parameters in Table 13.3 show many similarities between C_{60} and C_{70}, but also significant differences are found. The addition of ten atoms to the belt of C_{60} to form the elongated, lower symmetry C_{70} molecule alters the photophysical properties with regard to the energies of the excited states, the relative intensities for excited state transitions, and their lifetimes. For both C_{60} and C_{70} the quantum efficiency Φ_T for the intersystem crossing $S_1 \rightarrow T_1$ is very high: $\Phi_T \sim 1.0$ for C_{60} and $\Phi_T \sim 0.9$ for C_{70}. As for the case of C_{60}, the triplet state lifetime for C_{70} in solution or in a frozen matrix is highly sensitive to O_2 impurities, and for this reason wide variations for the triplet state lifetime appear in the literature for both C_{60} and C_{70} [13.30, 58].

It is interesting to note that the exchange splitting between the S_1 and T_1 levels derived from optical measurements is small (~0.33 eV) because of the large diameter of the molecule and the small electron–electron repulsion energy [13.58]. The optical determination of the exchange splitting (S_1–T_1) in solution spectra is in good agreement with more accurate determination of S_1–T_1 ~0.29 eV by electron energy loss spectroscopy in films (see §17.2.2).

13.2.4. Optically Detected Magnetic Resonance (ODMR) for Molecules

Optically detected magnetic resonance (ODMR) studies [13.59] are especially sensitive to the dynamics of spin-dependent recombination processes, thereby identifying photoexcitation processes associated with spin 0, 1/2, and 1 states and providing information about excited triplet states [13.60–64]. The $S = 0$ states are identified with singlet states, the $S = 1$ with triplet states, and the $S = 1/2$ with charged polarons [13.65]. With the ODMR technique, pump laser excitation (e.g., at 488 nm, as supplied by an argon ion laser) induces a transition in the fullerene molecule from the ground state to an excited singlet state. Because of the highly efficient intersystem crossing (see Fig. 13.2), the C_{60} molecule reaches a metastable triplet T_1 state (at ~1.6 eV) relative to the singlet ground state S_0 (see Table 13.3). Microwave power at a fixed frequency (e.g., 9.35 GHz) is simultaneously introduced. When a magnetic field is then applied, the triplet ($S = 1$) state splits into three levels corresponding to $m_S = 1, 0, -1$, with a splitting that is controlled by the magnetic field. There is, in addition, a small zero-field splitting due to the Jahn–Teller induced distortion of the molecule in the excited state, thereby giving rise to an angular dependence of the electronic wave function on the fullerene shells (see §16.2.3). Since all the singlet ($S = 0$) exciton states are nondegenerate, the magnetic field does not introduce any magnetic field-dependent splitting of the singlet spin states. As the magnetic field is swept, a resonance condition is met when the separation between the adjacent m_S levels is equal to the microwave frequency. Under the resonance condition, microwave power is resonantly absorbed, resulting in a change in population of the various m_S levels and in a magnetic field–induced admixture of singlet and triplet state wave functions (the spin–orbit interaction $\mathscr{H}_{s-o}$ could also cause such an admixture, but $\mathscr{H}_{s-o}$ is too small in carbon-based systems to be effective). This admixture of singlet wave function to the triplet state increases the emission probability of the triplet excited state to the singlet ground state. Correspondingly, at resonance the probability of absorption from S_0 to the T_1 state by a probe optical beam is also increased. The ODMR technique is thus a magnetic resonance (MR) technique insofar as the magnetic field is swept to achieve resonant absorption of microwave power. However, optical detection (OD) techniques are used, including absorption (ADMR) and photoluminescence (PLDMR) of both the fluorescent or phosphorescent varieties. The pump and probe optical beams in the ODMR experiments need not be at the same frequency. The microwave absorption is usually modulated by a low frequency perturbation to permit phase-sensitive detection of small changes in transmission, absorption, or photoluminescence.

A number of ODMR studies of fullerenes in solution and in their crystalline phases have been carried out [13.61–63, 66–70]. Since ODMR spectra are usually taken at low temperatures, frozen solution samples are typically used for molecular spectroscopy, while films and single crystals are used for probing the solid state. Referring to Fig. 13.9 [13.63], we see typical ODMR spectra as a function of magnetic field for C_{60} and C_{70} in a frozen toluene/polystyrene solution at 10 K using 488 nm laser excitation and 9.35 GHz microwave excitation. For the case of C_{60}, only a fluorescent emission signal is seen, whereas for C_{70} both fluorescent and phosphorescent (see §13.1.1) emission signals are observed. From the absence of a phosphorescent emission signal for C_{60}, it was concluded that C_{70} has a longer lifetime in the triplet state [13.63]. The details of the ODMR spectra for C_{60} and C_{70} are very different (Fig. 13.9), implying important differences for their triplet excited states. For the C_{60} frozen solution spectrum, the separation between the two strong peaks at $\sim$120 G is attributed to a triplet exciton delocalized over the whole molecule, while the weaker and broader features separated by $\sim$230 G in the C_{60} spectrum are attributed

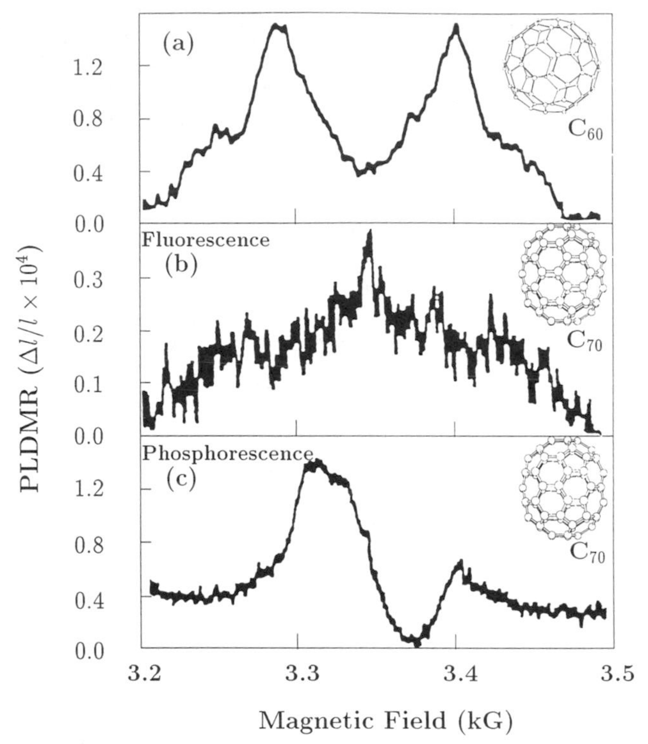

Fig. 13.9. Experimental ODMR spectra of C_{60} and C_{70} in frozen toluene/polystyrene solution at 10 K observed through photoluminescence (PLDMR) using 488 nm laser excitation and 9.35 MHz microwave excitation. The excited triplet state is observed for: (a) triplet C_{60} fluorescence, (b) triplet C_{70} fluorescence, and (c) triplet C_{70} phosphorescence [13.63].

to a triplet exciton state localized over a pentagon or hexagon face. A third type of triplet exciton with peaks separated by only 37 G is attributed to a triplet exciton delocalized over several coupled C_{60} molecules (as discussed below) and is most clearly seen in the solid state [13.63].

In contrast, the ODMR spectra for C_{70} show a strong signal and an asymmetric lineshape. The triplet exciton delocalized over the whole molecule is identified with the pair of structures separated by ~200 G, and a strong asymmetric phosphorescent structure is observed for the triplet C_{70} exciton [13.63]. The linewidth of the ODMR lines for the delocalized triplet exciton in C_{70} is narrower than in triplet C_{60} because of the larger size of the C_{70} molecule. The ODMR spectra are strongly temperature dependent, reflecting the temperature dependence of the excited state lifetimes. The temperature dependence of the ODMR features is used to identify the physical origins of the ODMR spectral features. For example, the delocalized triplet features are more rapidly quenched for both C_{60} and C_{70} by increasing the temperature than is the localized exciton [13.63]. Variation of the modulation frequency for the magnetic resonance features is used to provide information on lifetimes associated with various ODMR structures.

The ODMR spectra associated with films show some features that are present in the frozen solution ODMR spectra, and other features which are different. The ODMR spectra associated with spin 1 are closely related to the solution spectra and show two features at 1.1 and 1.8 eV in the absorption ADMR spectra, and they are associated with long-lived neutral triplet excitons excited to higher-lying triplet exciton states, in good agreement with induced photoabsorption measurements (see §13.7.1) [13.64].

In the solid state, an additional sharp feature, not present in the solution ODMR spectra, is observed for both C_{60} and C_{70} and is attributed to a long-lived polaron with spin 1/2 and $g = 2.0017$ for C_{60} and $g = 2.0029$ for C_{70} [13.63]. Excited isolated molecules decay rapidly to an excitonic state, thereby accounting for the absence of the polaron feature in solution ODMR spectra. In films, the electrons and holes can separate in a charge transfer process between adjacent fullerenes, thereby stabilizing polarons. Evidence for polarons is also found in EPR spectra (see §16.2.3) and in photoconductivity spectra (see §14.7). This sharp feature identified with polaron formation is observed both in the photoluminescence (PLDMR) spectrum [13.63] and in the absorption ADMR spectra [13.61, 64]. The spin 1/2 ADMR spectrum contains three features (at 0.8, 2.0, and 2.4 eV), in good agreement with the induced photoabsorption spectrum (see §13.2.3) that was measured independently [13.64] and with the optical spectra of doped C_{60} films (see §13.4) [13.71, 72].

Photopolymerization of the C_{60} film greatly reduces the intensity of the features in the ODMR spectrum associated with the localized triplet exciton state and strongly enhances the features associated with the triplet exciton (with 37 G separation) that is delocalized over more than one fullerene [13.63].

13.3. Optical Properties of Solid C_{60}

Since the absorption and luminescence spectra of solid C_{60} resemble to a considerable degree the corresponding spectra observed for C_{60} in solution, it is believed that molecular states should be a good starting point for the description of the electronic bands in solid C_{60}.

Thus the discussion of the optical properties of solid C_{60}, as in solution, concerns both forbidden transitions near the absorption edge and dipole-allowed transitions at higher photon energies. In solution, the forbidden electronic transitions are described by weak phonon-assisted excitonic transitions, while conventional, one-electron molecular calculations are more successful in accounting for the allowed transitions at higher photon energies. In §13.3.1 we present an overview of the optical properties of solid C_{60}, which is followed in §13.3.2 by a discussion of both the forbidden transitions near the absorption edge and the allowed transitions at higher energies. In §13.3.3 the low-frequency dielectric response of C_{60} is briefly summarized. Finally, in §13.3.4 the optical properties of phototransformed films are summarized.

13.3.1. Overview

Many of the observed optical properties of solids are expressed in terms of the complex optical dielectric function $\epsilon(\omega) = \epsilon_1(\omega) + i\epsilon_2(\omega)$, so that $\epsilon(\omega)$ provides a convenient framework for discussing the optical properties over a broad frequency range. It is also convenient to express $\epsilon(\omega)$ in terms of the complex refractive index $N(\omega) = [n(\omega) + ik(\omega)]$:

$$\epsilon(\omega) = \epsilon_1(\omega) + i\epsilon_2(\omega) = [N(\omega)]^2 = [n(\omega) + ik(\omega)]^2 \tag{13.8}$$

where n and k are the frequency-dependent optical constants and are, respectively, the refractive index and extinction coefficient, while ϵ_1 and ϵ_2 are the real and imaginary parts of the dielectric function. The optical absorption coefficient α is then defined by

$$\alpha = 2(\omega/c)k(\omega), \tag{13.9}$$

and the intensity of a plane electromagnetic wave is attenuated exponentially according to $e^{-\alpha z}$, where z is distance in the medium along the line of wave propagation (Beer's law).

In Fig. 13.10 is shown the frequency dependence of the dispersive part [$\epsilon_1(\omega)$; upper panel] and the absorptive part [$\epsilon_2(\omega)$; lower panel] of $\epsilon(\omega)$ for solid C_{60} films at room temperature, as obtained from a variety of optical experiments, as discussed below [13.1]. The data are plotted on a semilog scale and cover the frequency range from the IR to the ultraviolet (~0.05 to 5.5 eV). Above 7 eV $\epsilon_1(\omega)$ obtained from electron energy loss spectroscopy (EELS) studies is included. We now discuss $\epsilon(\omega)$ for various frequency ranges. The behavior at very low frequencies (below 0.05 eV) has special properties that are singled out and discussed in §13.3.3.

Using the experimental results for $\epsilon(\omega)$, the molecular polarizability $\alpha_M(\omega)$ can be calculated from the Clausius–Mossotti relation

$$\alpha_M(\omega) = \frac{3}{4\pi\rho_N}\frac{\epsilon - 1}{\epsilon + 2} \tag{13.10}$$

where $\rho_N = 4/a^3$ is the number density of C_{60} molecules in the face-centered cubic (fcc) unit cell with a lattice constant $a = 14.17$ Å. The real

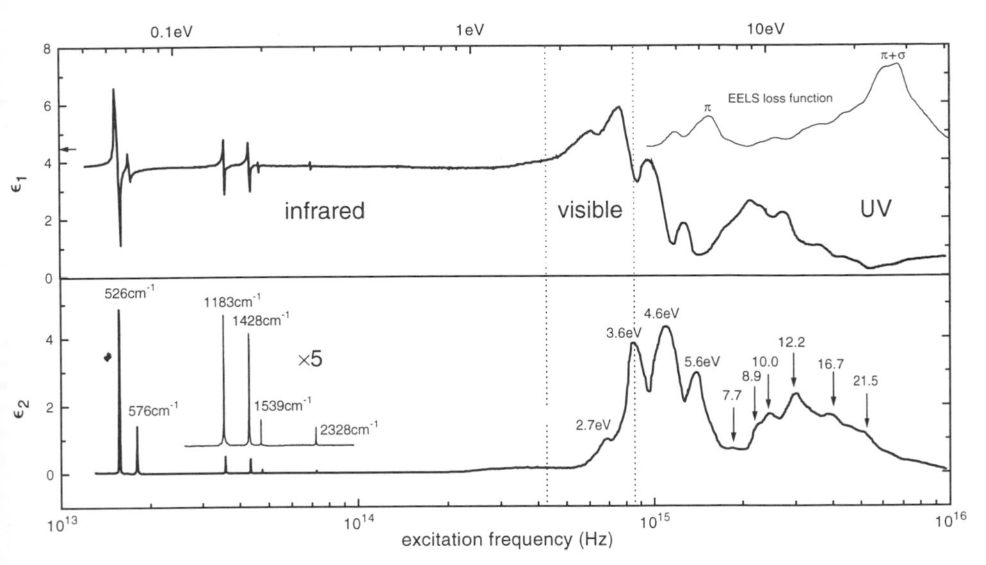

Fig. 13.10. Summary of real $\epsilon_1(\omega)$ and imaginary $\epsilon_2(\omega)$ parts of the dielectric function for C_{60} vacuum-sublimed solid films at room temperature over a wide frequency range. The data between 0.05 and 0.5 eV (mid- to near-infrared) were collected using the Fourier transform infrared (FTIR) transmission technique [13.1]. The visible–UV range was investigated by variable angle spectroscopic ellipsometry (VASE) [13.73] and near-normal-incidence reflection and transmission experiments [13.74]. UV data above ~7 eV were obtained using electron energy loss spectroscopy (EELS) [13.75] by Kramers–Kronig analysis of the EELS loss function (inset). The arrow at the left axis points to $\epsilon_1 = 4.4$, the low-frequency value of the dielectric constant determined by capacitance measurements [13.76].

part of the resulting molecular polarizability $\alpha_M(\omega)$ up to 5 eV is shown in Fig. 13.11 on a semilog plot [13.24, 77]. The application of Eq. (13.10) assumes minimal intermolecular interaction, which is strictly valid for an ideal molecular solid. The contributions to the low-frequency $\alpha_M(\omega)$ from vibrational and electronic processes are estimated to be ~2 and ~83 Å^3, respectively, as shown in Fig. 13.11, leading to a value of $\alpha_M \sim 85$ Å^3 [13.24]. This value for α_M compares well with some other determinations of the low-frequency polarizability [13.78–80] and not as well with others [13.22, 81, 82].

Between 0.05 and 0.5 eV (1 eV = 8066 cm^{-1}), the experimental data of Fig. 13.10 (solid curves) were collected by measuring the optical transmission through an ~6000 Å thick film of C_{60} on a KBr substrate. To obtain quantitative values for $\epsilon_1(\omega)$ and $\epsilon_2(\omega)$ in the infrared region of the spectrum, the contributions to the reflectance and transmission from multiple internal reflections within the film were taken into account.

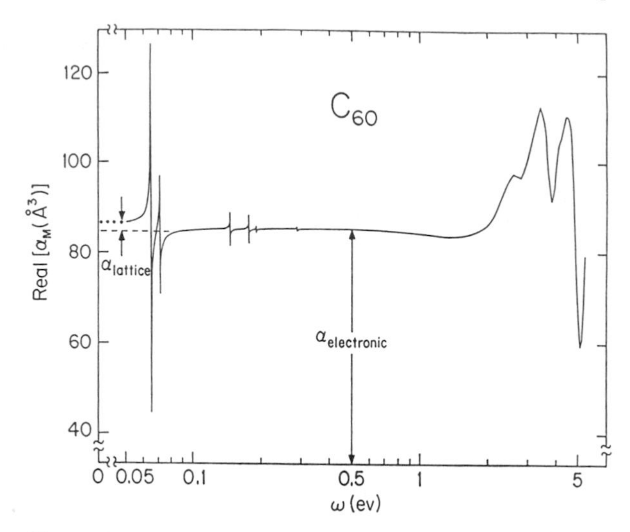

Fig. 13.11. The real part of the complex molecular polarizability for solid C_{60} obtained from the data of Fig. 13.10 using the Clausius–Mossotti relation [Eq. (13.10)] [13.24].

The features in the infrared spectrum (see Fig. 13.10) below ~0.3 eV are due to molecular vibrations, including the four strong first-order F_{1u} intramolecular modes at 526, 576, 1183, and 1428 cm^{-1}, together with two strong combination ($\omega_1 + \omega_2$) modes at 1539 cm^{-1} and 2328 cm^{-1} (see §11.5). Transmission experiments carried out on thicker (~2–3 μm) films reveal numerous (>100) additional weaker features attributed to other intramolecular combination modes [13.15], as discussed in §11.5.3 and §11.5.4.

Values for $\epsilon_1(\omega)$ and $\epsilon_2(\omega)$ in the IR ($\hbar\omega < 0.5$ eV) in Fig. 13.10 were calculated by a fit of the experimental reflectivity and transmission data according to the Lorentz model, in which the complex dielectric function $\epsilon(\omega)$ is given by the sum of contributions from a background core (real) constant ϵ_∞ and a phonon term $\epsilon_{\text{phonon}}(\omega)$ approximated as a sum of Lorentz oscillators whose parameters were adjusted to best fit the data:

$$\epsilon(\omega) = \epsilon_1(\omega) + i\epsilon_2(\omega) = \epsilon_\infty + \epsilon_{\text{phonon}}(\omega) \tag{13.11}$$

where

$$\epsilon_{\text{phonon}}(\omega) = \sum_{j=1}^{n} \frac{f_j \omega_j^2}{\omega_j^2 - \omega^2 - i\Gamma_j \omega}. \tag{13.12}$$

For the jth phonon contribution, f_j, ω_j, and Γ_j are the oscillator strength, frequency, and damping, respectively, for each IR mode. Values of the first-order IR mode Lorentz oscillator parameters are listed in Table 11.7.

As discussed in Chapter 11, there are also two first-order IR-active intermolecular modes with T_{1u} symmetry, which contribute at low frequencies. They correspond approximately to an out-of-phase displacement of "rigid" C_{60} molecules and are observed in the far IR at 40.9 cm^{-1} and 54.7 cm^{-1} [13.83].

13.3.2. *Optical Absorption in C_{60} Films*

Contributions to the dielectric function of solid C_{60} above ~1.7 eV come from electronic transitions from filled states, i.e., states derived from the HOMO$-j$ orbitals, to empty states derived from the LUMO$+j'$ orbitals, where $j, j' = 0, 1, 2, \ldots$. If C_{60} were an ideal molecular crystal, then the optical properties of the solid might be calculated directly from the Clausius–Mossotti relation [Eq. (13.10)]. In this limit, the solid state, or intermolecular interaction, does not significantly affect the electronic molecular wave functions, but rather the molecules interact with each other via an electric dipole–dipole interaction which introduces a local electric field correction. Higher-order, quadrupole or octapole terms can be added to describe the intermolecular interaction, if necessary. However, the interaction between next-neighbor molecules in the real crystalline solid will broaden the molecular orbitals into electronic energy bands, and these effects have been calculated by several groups [13.84, 85] (see §12.1.2). Although using different computational techniques, these calculations arrive at qualitatively the same result, namely, that the intermolecular interaction broadens the molecular orbitals into bands with widths on the order of 0.5 to 1.0 eV. As calculated by Gelfand and Lu [13.86, 87], the molecular orientational disorder in the high temperature solid state ($T > T_{01} = 261$ K) does not seriously affect the bandwidth of the electronic energy levels, but rather rounds off the sharp features in the electronic density of states.

In Fig. 13.12 we display experimental data showing the threshold for optical absorption at 1.7 eV associated with electronic transitions in C_{60} films [13.15]. Shown in Fig. 13.12 are $\epsilon_2(\omega)$ data for two thin solid C_{60} films of distinctly different thicknesses. The dielectric function data were obtained by fitting simultaneously the transmission and reflectance for samples of C_{60} films on quartz substrates to values for $\epsilon_1(\omega)$ and $\epsilon_2(\omega)$, and the contributions to the observed spectra from the quartz (Suprasil) substrate were taken into account in the analysis.

More detailed information is obtained by comparing the optical absorption (OA) and photoluminescence (PL) spectra for pristine C_{60} films and in further comparing each of these film spectra with their corresponding solution spectra in the vicinity of the absorption edge (see Fig. 13.6). Structures observed in both the OA and PL thin-film spectra near the absorption

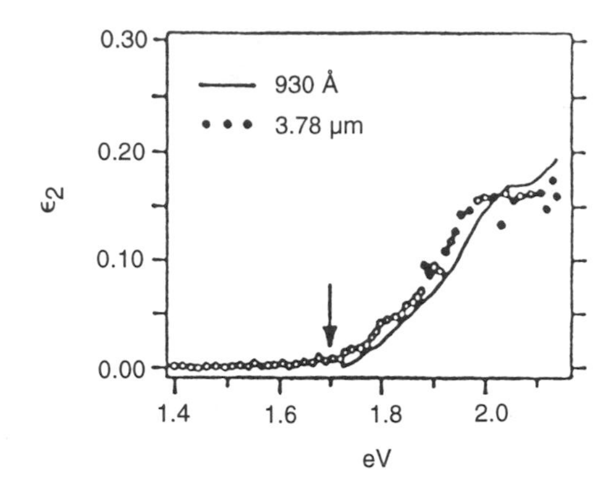

Fig. 13.12. $\epsilon_2(\omega)$ for two thin (930 Å and 3.78 μm thick) solid C_{60} films vacuum deposited on Suprasil (quartz) substrates ($T = 300$ K). The arrow marks the onset for absorption at 1.70 eV [13.24].

edge have been identified primarily with Herzberg–Teller vibronic coupling of the respective singlet states to vibrational modes. This basic interpretation of the optical spectrum near the absorption edge (~1.7 eV) in the solid state is based on the similarity between the OA spectra in the film (as for the spectra at 80 K) and in solution (see Fig. 13.6). This similarity suggests the same origin for the transitions in both C_{60} solid films and C_{60} in solution and provides strong evidence for the molecular character of the solid. Therefore the features ($F_{i'}$) in the OA spectrum of the film (see Fig. 13.6) are classified using the index i' to relate the $F_{i'}$ labels to the $M_{i'}$ labels for the OA spectrum of the C_{60}/MCH (methylcyclohexane) solution. Although the features in the room temperature OA spectrum of the film are very broad relative to those in the $T = 300$ K C_{60}/MCH solution, the 80 K OA spectrum in the film near the optical threshold is remarkably similar to the $T = 300$ K OA spectrum for C_{60} in solution (C_{60}/MCH). The entire 80 K OA spectrum of the C_{60} solid film is, however, red-shifted by ~430 cm^{-1} with respect to that of C_{60} in solution, corresponding to a shift in the zero-phonon transition energy $E_{0,0'}^{OA}$. This shift in $E_{0,0'}^{OA}$ is not due to changes in the vibrational mode frequencies themselves, which are observed to exhibit little change between the solid phase and that of C_{60} in solution [13.88]. Furthermore, the red shift of the 80 K OA film spectrum relative to the 300 K solution spectrum is four times larger than the ~100 cm^{-1} blue shift of this spectrum due to lowering the temperature.

It is also instructive to correlate the features in the film PL spectrum (labeled by symbols F_4 to F_8 [Fig. 13.6(b)]) with the corresponding features observed in the solution PL spectrum [Fig. 13.6(a)]. Luminescence spectra appear to be more sensitive to sample quality and to show wider sample-to-sample variation than absorption spectra [13.89]. The 80 K film PL spectrum downshifts by ~ 200 cm^{-1} with respect to the 77 K solution

PL spectrum, yielding $E^{PL}_{0,0'} = 15,200$ cm^{-1} (1.88 eV) for C_{60} in solution and $E^{PL}_{0,0'} = 15,000$ cm^{-1} (1.86 eV) for the C_{60} solid film. It is found that the PL-deduced values for the vibrational frequencies ω_{vib} in the film are close to those deduced from the PL spectrum of C_{60} in solution for PL transitions from the $E^{PL}_{0,0'}$ vibrationless initial state to the various final states in the spectrum.

Despite the similarities between the film and solution PL and OA spectra, there are three noteworthy differences: (1) the film OA and PL spectra are red-shifted in energy with respect to the solution OA and PL spectra and this red shift is attributed to solid state interactions; (2) the solution (C_{60}/MCH) features labeled M_0 to M_3 are not observed in the PL spectrum of the film, but the corresponding C_1, C_2 and C_3 features are indeed found in the single-crystal PL spectra [13.28] (the features C_i for the single-crystal PL data in Fig. 13.6 are labeled similarly to those in solution); (3) the PL features in the solid film [Fig. 13.6(b)] are broadened considerably with respect to those in the solution spectrum [Fig. 13.6(a)]. This broadening is not simply due to intermolecular C_{60}–C_{60} interactions in the solid state, as can be seen by comparing the 10 K single-crystal PL data [13.28] to that of the 80 K film and the 77 K solution PL data shown in Fig. 13.6(b). If the broadening were due to C_{60}–C_{60} interactions, the single-crystal PL spectrum would not be so remarkably similar to that of C_{60} in solution. Since the PL spectrum of the single crystal not only contains most of the features seen in the solution PL spectrum, but also exhibits similar relative intensities of the various features, a clear connection is seen between the molecular (H–T vibronic) origin of the PL structure in the solution PL spectrum and that in the crystal and film PL spectra. Furthermore, the single-crystal PL spectrum shows that the intrinsic intermolecular interactions in fcc C_{60} are not responsible for the broadening of the vibronic features in the film PL spectrum. The observed broadening is therefore identified with grain boundary defects in the film. Consistent with this view is the observation that the PL spectrum of a polycrystalline powder[13.90, 91] also shows similar broad vibronic features.

Since the PL spectra of both C_{60} single crystals and films downshift by about the same amount relative to the solution spectrum, it is concluded that it is the lattice potential that introduces the spectral downshift in the solid-state spectra. In aromatic, molecular crystals, a spectral shift relative to solution data has been connected with two mechanisms [13.6]: the *solvent shift* and the *exciton shift*. The *solvent shift* is due to the polarizability of the environment, which is quite different for C_{60} in solution and for C_{60} in the pristine solid. The *exciton shift* is due to the electron–hole interaction spread over translationally equivalent molecules. Within a given set of vibronic transitions, one usually finds that the solvent shift is constant.

Since the crystal grain size in the solid films of Fig. 13.6(b) is $\sim$ 20 nm [13.15], many C_{60} molecules are within one or next-neighbor distances of a grain boundary (the C_{60} diameter $\sim$1 nm). Grain boundary defects will lower the local symmetry of C_{60} and lift the high degeneracy of the intramolecular modes. Therefore, for molecules located near grain boundaries, many more vibrational modes might participate in vibronic transitions, thus accounting for the broadening of the PL film spectrum. Another factor which might contribute to the broadening of the film PL spectrum is the presence of X-traps [13.9, 28, 92] associated with grain boundary defects. An X-trap yields a PL spectrum with the same vibronic features as the intrinsic (undisturbed) crystal, but the features are red-shifted by the trap depth ΔE [13.28]. A range of trap depths ΔE ($25 < \Delta E < 200$ cm^{-1}) could also explain the observed broadening of the PL spectra of the films.

As we move away from the absorption edge to higher photon energies, we see a large increase in absorption coefficient. In this higher energy regime the electronic energy band dispersion is expected to broaden the optical absorption bands in the solid significantly over those observed in solution or in the gas phase. In Fig. 13.13 we compare the UV–visible optical absorption of C_{60} in solution (decalin) to that in a polycrystalline C_{60} film on a Suprasil substrate [13.73] in a frequency range higher than the data shown near the absorption edge in Fig. 13.12. As can be seen in Fig. 13.13, the optical spectra of C_{60} in solution and in the solid state are quite similar, and the solid-state broadening is not dramatic. Four strong optical absorption bands, identified with dipole-allowed transitions, exhibit peaks at $\sim$3.0, 3.6, 4.7, and 5.7 eV for both the solid film and for C_{60} in solution [13.73]. Note that the three highest-energy optical bands exhibit an increase in their FWHM linewidths (full widths at half-maximum) of $\sim$0.3 to 0.5 eV and the peak at 3.0 eV is also broadened. Furthermore, the intensities of the lowest two bands at 3.0 and 3.6 eV, measured relative to the higher two bands at 4.7 and 5.7 eV, exhibit a significant increase in intensity in the solid phase. EELS spectra (taken at 60 meV resolution) in the region 1 to 8 eV on C_{60} in the gas phase and in the solid phase are quite similar to the UV–visible optical spectrum of C_{60} in solution, exhibiting similar width peaks in both the gas and solid phases at 2.24, 2.9, 3.8, 4.9, and 5.8 eV [13.93]. Consistent with the similarity of the UV–visible spectra for C_{60} in solution and in the solid state, the EELS data also indicate that most of the intrinsic width in the optical absorption bands stems from intramolecular processes, most likely due to electron–molecular vibration scattering, as suggested by theoretical calculations in both C_{60} and C_{70}, where intramolecular bond disorder is introduced to simulate the effect of molecular vibrations on the dipole absorption [13.94, 95]. By carrying out *in situ* optical absorption measurements in thin films of C_{60} as a function of alkali metal doping, it has been possi-

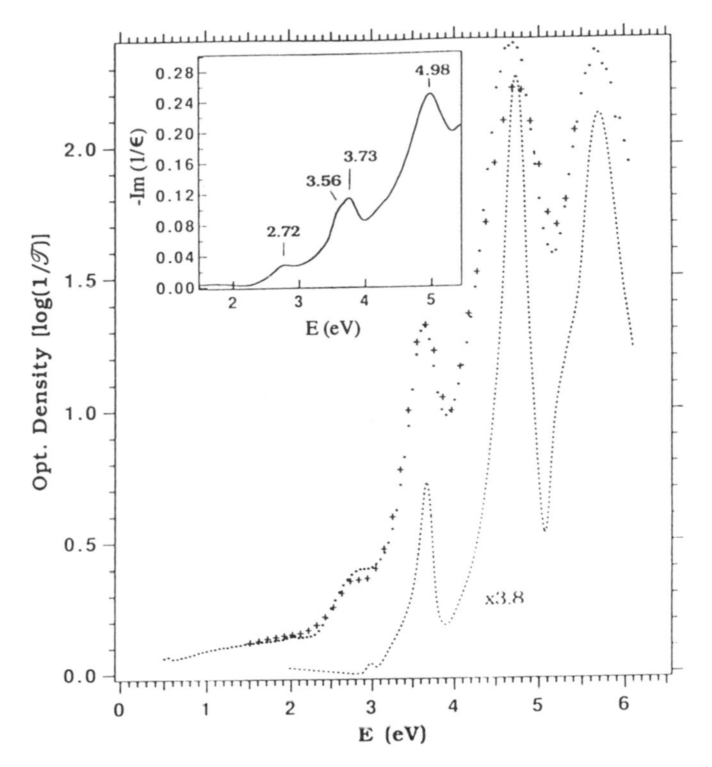

Fig. 13.13. Optical density of solid C_{60} on a Suprasil substrate derived from ellipsometry measurements (+) together with the frequency dependent optical density measured directly for C_{60} on Suprasil by normal incidence transmission spectroscopy (·). For comparison, the solution spectrum for C_{60} dissolved in decalin (dotted curve) is shown below the spectra for the films [13.73]. The inset is a plot of $-\text{Im}(1/\epsilon)$ *vs.* E comparing the peaks in the optical data with the peaks in the high-resolution electron energy loss spectrum (HREELS).

ble to identify the origin of the interband transitions in C_{60} [13.96–99]. In these experiments the stoichiometry was monitored by simultaneous electrical conductivity measurements [13.100] and the interband transitions were identified by noting the transitions that were induced by doping and those that were quenched by doping [13.96–99]. The resulting identifications are given in Table 13.4.

Two theoretical approaches have been applied to calculate the contribution to the optical dielectric function for solid C_{60}. The first approach is based on the electronic band structure of C_{60} calculated self-consistently using an orthogonalized linear combination of atomic orbitals in the local density approximation (LDA) [13.104]. The second approach considers the solid to be composed of weakly coupled molecules with an intermolecular

Table 13.4

Energies (eV) of the electronic transitions in thin films of pristine C_{60} [13.99].

Identification	Transition energy (eV)		
$t_{1u} \rightarrow t_{1g}$	I[a]	—	1.1[b]
$h_u \rightarrow t_{1u}$	~2.0[a]	1.7[c]	1.8[d]
$t_{1u} \rightarrow t_{2u}$	I[a]	—	—
$h_u \rightarrow t_{1g}$	2.8[a]	3.0[c]	2.7[b,d,e]
$h_g + g_g \rightarrow t_{1u}$	3.6[a]	3.6[c]	3.6[b,d,e]
$t_{1u} \rightarrow h_g$	I[a]	—	2.7[b,d,e]
$h_u \rightarrow h_g$	4.7[a]	4.7[c]	4.5[b,d,e]
$h_g + g_g \rightarrow t_{2u}$	5.5[a]	5.7[c]	5.5[b,d,e]

[a]Ref. [13.97, 99]. "I" denotes transitions induced by alkali doping. [b]Ref. [13.101]. [c]Ref. [13.73]. [d]Ref. [13.102]. [e]Ref. [13.103].

hopping integral $t_w \sim 0.1t$, where t is a C–C hopping integral on a single molecule ($t = 1.8$ eV). Dipole-allowed transitions for a four-molecule cluster were examined to determine to what extent the charge in the photoexcited state is localized on the same molecule (Frenkel exciton) or delocalized over neighboring molecules [charge transfer (CT) exciton] in the photon absorption process [13.95]. An example of the first (or band) approach can be found in the work of Xu and co-workers [13.105], who determined the contribution to the dielectric function by calculating the matrix elements for vertical (k-conserving) transitions throughout the Brillouin zone, in addition to calculating the joint density of states.

In the band approach, the band structure for solid C_{60} was computed in the local density approximation using an orthogonalized, linear combination of atomic orbitals (OLCAO), and the results are in good agreement with other LDA calculations [13.12, 80, 106–109]. In Fig. 13.14 [13.104] we show the resulting electronic density of states (DOS), where the labels V_j and C_j in the figure refer to the various valence and conduction bands. Note that the electronic energy bandwidths are small ~0.5–0.6 eV and that the bands are well separated by gaps. A direct gap of 1.34 eV was calculated between the bottom of the conduction band and the top of the valence band at the X-point in the Brillouin zone. However, direct transitions are forbidden at X between the V_1 and C_1 bands of Fig. 13.14. These calculations [13.104] show that the t_{1u}-derived band splits into three bands in the vicinity of the X-point and that transitions are allowed between the top of the valence band and the second highest conduction band near X, giving rise to a calculated optical threshold at 1.46 eV. It is common for LDA-based

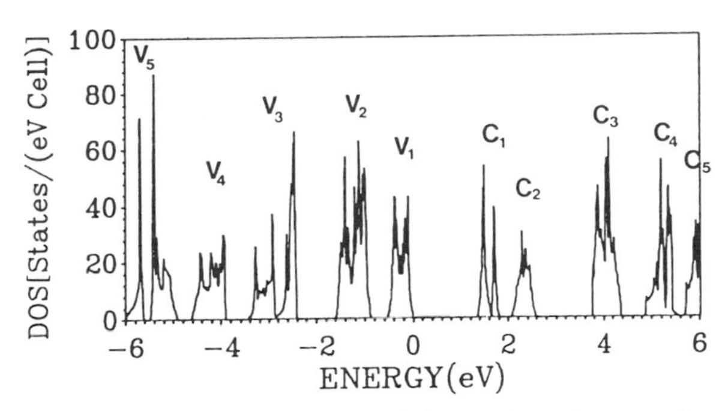

Fig. 13.14. Calculated density of states (DOS) for fcc C_{60} using an orthogonalized linear combination of atomic orbitals within the LDA approximation [13.104]. The valence (V_i) and conduction (C_i) bonds are labeled relative to the Fermi level which is located between V_1 and C_1.

band structure calculations to yield absorption thresholds at lower energies than experimental observations [13.87]. In Fig. 13.15 [13.104] we show the results of these calculations for the imaginary part $\epsilon_2(\omega)$ (a) and the real part $\epsilon_1(\omega)$ (b) of the dielectric function using both the matrix elements and DOS derived from their band structure calculations. Labeled A through E in Fig. 13.15(a) are five distinct optical absorption bands in the visible–UV range which were assigned by Xu and co-workers [13.106] to particular groups of interband transitions following the notation of Fig. 13.14: (A) $V_1 \rightarrow C_1$; (B) $V_1 \rightarrow C_2$ and $V_2 \rightarrow C_1$; (C) $V_1 \rightarrow C_3$ and $V_2 \rightarrow C_2$; (D) $V_2 \rightarrow C_3$ and $V_1 \rightarrow C_4$; (E) $V_1 \rightarrow C_5$ and $V_2 \rightarrow C_4$. In the inset to Fig. 13.15(a), a comparison is made between their calculated optical conductivity $\sigma_1(\omega) = (\omega/4\pi)\epsilon_2(\omega)$ (solid line, arbitrary units) and the experimental absorption spectrum (dashed line, arbitrary units) for a thin solid C_{60} film vacuum-deposited on quartz. The relation between the assignments of Xu *et al.* [13.106] in Fig. 13.14 and Srdanov *et al.* [13.99] in Table 13.4 is: $V_1 = h_u$, $V_2 = h_g + g_g$, $C_1 = t_{1u}$, $C_2 = t_{1g}$, $C_3 = t_{2u}$, $C_4 = h_g$, and the C_5 level is not identified with a one-electron band (see Fig. 12.1). However, the assignments of the optical absorption bonds are still tentative.

It should be noted that the observed oscillator strength for the $V_1 \rightarrow C_1$ transition is much lower than that identified with feature A in Fig. 13.15, where V_1 and C_1 are derived from the HOMO (h_u) and LUMO (t_{1u}) molecular orbitals, respectively. The results of Fig. 13.15 [13.104] show that the intermolecular (or solid-state) interaction also admixes gerade character into the conduction and valence band states, thereby activating transitions between HOMO- and LUMO-derived states in the fcc solid. We recall that in the free molecule, a vibronic interaction was needed to activate transitions between the h_u and t_{1u} orbitals because of optical selection rules. It

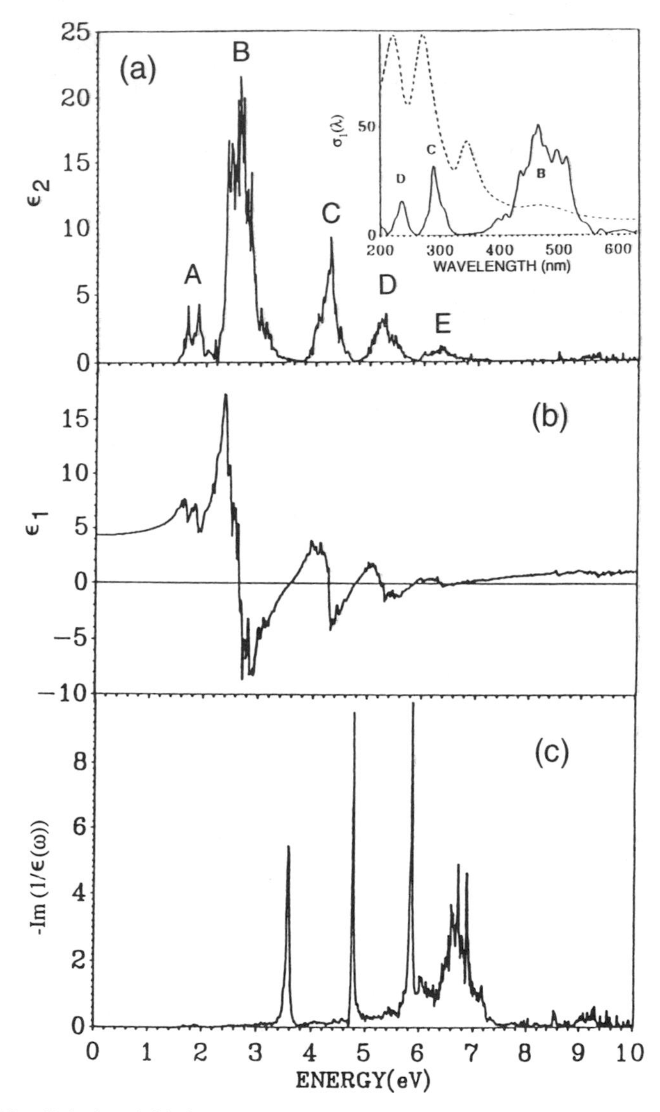

Fig. 13.15. Calculated (a) imaginary part of the dielectric function $\epsilon_2(\omega)$, (b) real part of the dielectric function $\epsilon_1(\omega)$, and (c) energy-loss function $\mathrm{Im}[-1/\epsilon(\omega)]$, all plotted as a function of photon energy. The inset to (a) shows the calculated optical conductivity $\sigma_1(\lambda)$ curve (solid line) compared with experimental visible–UV absorption spectra from Krätschmer and Huffman [13.110] (dashed line) in arbitrary units [13.104], and plotted as a function of wavelength.

seems reasonable, therefore, to expect that both vibronic and intermolecular interactions contribute to the dipole oscillator strength between narrow h_u- and t_{1u}-derived bands in solid C_{60}.

Turning to the higher-energy features B, C, D, E in Fig. 13.15 for the *calculated* $\epsilon_2(\omega)$ curve, the optical absorption bands are strongest at low energy and diminish in strength with increasing photon energy. Comparison between theory and experiment is shown in the inset to Fig. 13.15(a) [13.104], where it is seen that the calculations overestimate the relative oscillator strength of feature B and underestimate the strength of the higher-energy features C and D. Interestingly, this situation was also encountered in theoretical calculations of the oscillator strengths in the free molecule (see §13.1.2). As shown previously in Fig. 13.5, the calculated free-molecule *unscreened* response also overestimates significantly the optical response in the visible and underestimates the UV response. However, as pointed out by both Westin and Rosén [13.7] and Bertsch *et al.* [13.22], the inclusion of electronic correlations (or screening) in the calculations for the oscillator strengths for the isolated molecule leads to a dramatic shift of oscillator strength from the visible to the UV, and thus reasonably good agreement with experiment is achieved for the free molecules (see Fig. 13.5). It is reasonable to expect that this theoretical observation for the free molecule applies to C_{60} in the solid phase as well, suggesting that a proper treatment of electronic correlations in the solid will also shift oscillator strength to the UV region and thereby improve considerably the agreement between theory and experiment for C_{60} films.

Louie and co-workers have also calculated the energy band structure in relation to the implied optical properties of solid C_{60} using an *ab initio*, quasiparticle approach [13.87, 111, 112]. Their results for the excited band states also include many-electron corrections to the excitation energy, which they find in solid C_{60} leads to an increase in the direct gap from 1.04 eV to 2.15 eV. One important advance of this calculation is to show how to correct for the underestimate in the X-point band gap in C_{60} which is generally found in LDA calculations [13.111]. The band gap value of 2.15 eV is in better agreement with the gap of 2.3 to 2.6 eV obtained from combined photoemission and inverse photoemission data [13.113–115]. Furthermore, Louie and co-workers find that although the empty t_{1u} and t_{1g} bands exhibit 0.8–1.0 eV dispersion, the angular dependence of the inverse photoemission associated with these bands is surprisingly weak, in agreement with experiment [13.114, 116].

It is widely believed that the photoemission gap should be larger than an experimental optical gap based on optical absorption edge measurements [13.111]. This is a direct consequence of the attractive interaction between excited electron–hole pairs, which creates excitons and leads to excitation energies less than the band gap between the valence and conduction bands because of the binding energy of the exciton ($\sim$0.5 eV in C_{60}). While excitons are excited optically, they are not probed in photoemission and inverse photoemission experiments, which instead probe the density of states for the addition of a quasi-hole or quasi-electron to the solid, or equivalently probe the density of states of the valence band and the conduction band. By inclusion of the Coulomb interaction between the quasi-electrons and the quasi-holes in the conduction and valence bands, Louie has calculated a series of narrow exciton bands in the gap between the valence (H_u) and conduction bands (T_{1u}) [13.112] and has assigned the appropriate symmetries to these excitons using the relation $H_u \otimes T_{1u} = T_{1g} + T_{2g} + G_g + H_g$, as discussed in §12.5. The excited state wave functions are found to be largely localized on a single molecule; i.e., they are Frenkel excitons. For singlet excitons they find the lowest exciton (T_{2g} symmetry) at 1.6 eV, close to the threshold observed for optical absorption in solid C_{60} (see Fig. 13.12) and leading to an exciton binding energy of $(2.15 - 1.60) = 0.55$ eV. The onset of a series of triplet excitons is reported starting 0.26 eV below the T_{2g} exciton, in good agreement with the experimental value of 0.28 eV for the splitting of the lowest excited singlet and triplet states from EELS measurements [13.103]. However, since the ground state of solid C_{60} in the many-electron notation has A_g symmetry, electric dipole transitions from the ground state A_g to the excitonic states T_{1g}, T_{2g}, G_g, or H_g are forbidden by parity (see §13.1.3), so that a Herzberg–Teller vibronic interaction mechanism must be invoked to admix ungerade symmetry into the excitonic state, as described above for the isolated C_{60} molecule. However, it should be recalled that the energy of the final excitonic (vibronic) state includes both the electronic ($E_{0,0}$) and the vibrational mode energy ($\hbar\omega$), which can range from 30 meV to 190 meV. These vibrational energies represent a significant fraction of the singlet exciton binding energy (0.55 eV) and the singlet–triplet exciton splitting (0.28 eV). The exciton binding energy is determined experimentally from the difference between the HOMO–LUMO gap found by photoemission/inverse photoemission experiments and the electronic energy $E_{0,0}$ determined optically.

Harigaya and Abe [13.94] have also analyzed the excitonic structure for a linear four-molecule cluster to simulate solid-state effects. Their goal was to separate the contributions to the optical absorption for solid C_{60} from excitons localized on single molecules from somewhat less localized excitations, where charge is transferred to nearest-neighbor molecules. To model their cluster they required an intermolecular hopping integral $t_w \sim 0.18$ eV. In Fig. 13.16(a) their calculated density of states for occupied and unoccupied states is presented, while Fig. 13.16(b) shows the excitation density *vs.* excitation energy. The shaded vertical bars in Fig. 13.16(b) correspond

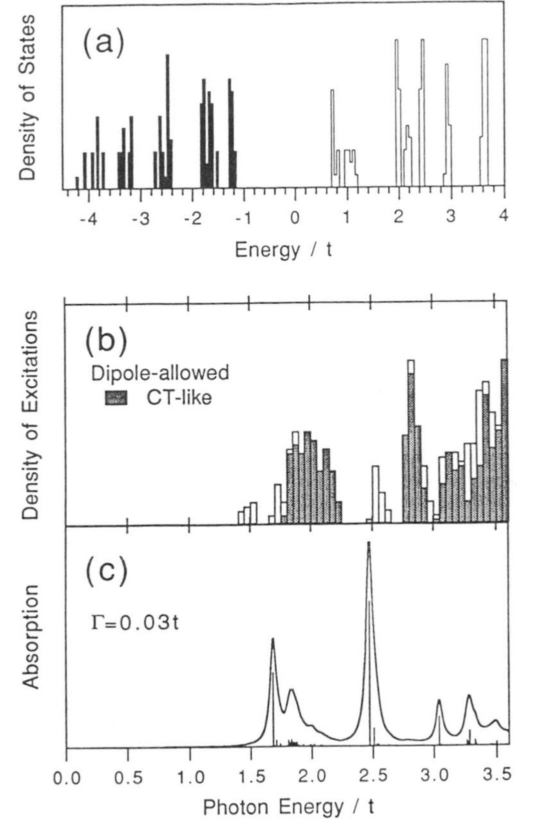

Fig. 13.16. (a) Calculated density of electronic states for a cluster of four C_{60} molecules by the Hartree–Fock approximation. The occupied (unoccupied) states are shown by the black (white) bars. (b) Density of excitations as a function of the excitation energy, showing both localized Frenkel excitons (indicated by open bars) and delocalized charge transfer–like excitons (shown as shaded bars). (c) The absorption spectrum where the Lorentzian broadening is taken as $\Gamma = 0.03t$ (t is a scaling factor taken as 1.8 eV) [13.95].

to excitons which exhibit at least a 50% probability of being delocalized over more than one molecule, while unshaded vertical bars correspond to Frenkel excitons, i.e., those that are localized primarily on one molecular site. The delocalized excitons are referred to as charge transfer (CT) excitons because the absorption of the photon to create this excited state must necessarily involve the transfer of an electron localized in the ground state on one molecule to an excited state delocalized over neighboring molecules. Similar to the calculations of Louie *et al.* [13.87, 111], Harigaya and Abe find that near the absorption edge, the largest oscillator strength in the spectrum lies primarily with localized (Frenkel) excitons. They therefore conclude that most of the observed photoexcited states at the absorption threshold should be identified with Frenkel-like excitons but that charge transfer excitons become important a few hundred meV above the absorption threshold.

13.3.3. *Low-Frequency Dielectric Properties*

Both intrinsic and extrinsic effects contribute to the low-frequency ($< 10^{11}$ Hz) dielectric properties of fullerene materials, below the frequency of any intermolecular vibrations. In this low-frequency regime, the effects of oxygen and other interstitial adsorbates are particularly important.

If the interstitial sites of solid C_{60} are partially occupied by impurities that can transfer charge to nearby C_{60} molecules, then the resulting dipole moments can be significant because of the large size of fullerene molecules. Such effects have been detected by ac impedance measurements of polarization effects in M–C_{60}–M (M = metal) trilayer structures [13.117].

Referring to Fig. 13.10, for every loss process represented by a peak in the $\epsilon_2(\omega)$ spectrum, there is a rise (sometimes accompanied by a small oscillation) in $\epsilon_1(\omega)$, as prescribed by the Kramers–Kronig relations [13.26]. As we move below 10^{13} Hz in frequency, $\epsilon_1(\omega)$ approaches its dc value. The small difference between $\epsilon_1(10^{13}\,\text{Hz}) \approx 3.9$ [13.77] and $\epsilon_1(10^5\,\text{Hz}) \approx 4.4$ [13.76] is perhaps due to the losses associated with the rapid rotation of the C_{60} molecules (at $\sim 10^9$ Hz) at room temperature ($T > T_{01}$) (see §7.1.3 and §16.1.4). From the Kramers–Kronig relations, the dc dielectric constant $\epsilon_1(0)$ can be related to $\epsilon_2(\omega)$ over a broad frequency range

$$\epsilon_1(0) = \frac{2}{\pi}\int_0^\infty \frac{\epsilon_2(\omega)}{\omega}\,d\omega = \frac{2}{\pi}\int_0^\infty \epsilon_2(\omega)\,d(\ln\omega), \tag{13.13}$$

where the low-frequency $\epsilon_2(\omega)$ contributes most importantly to $\epsilon_1(0)$.

Measurements of the dielectric properties of solids below $\sim$1 MHz are usually made by placing a sample between closely spaced parallel conducting plates and monitoring the ac equivalent capacitance $C(\omega)$ and the dissipation factor $D(\omega)$ of the resulting capacitor. The capacitance is proportional to $\epsilon_1(\omega)$, the real part of the relative dielectric function, according to the relation $C(\omega) = \epsilon_1(\omega)\epsilon_0 A/d$, where A is the cross-sectional area of the capacitor, d is the separation between the plates, and ϵ_0 is the absolute permittivity of free space (8.85×10^{-12} F/m). Measurement of the dissipation factor (also known as the loss tangent) $D(\omega) = \epsilon_2(\omega)/\epsilon_1(\omega)$ can then be used to extract the imaginary part of the dielectric function $\epsilon_2(\omega)$.

To avoid problems arising from changes in the plate spacing d of the C_{60} capacitor due to the thermal expansion of the C_{60} lattice, or to the structural first-order phase transition at $T_{01} \approx 261$ K, or during the intercalation of oxygen or other species into the interstitial spaces of C_{60} solid, a microdielectrometry technique [13.118–120] has been used to measure the dielectric properties of C_{60} at low frequencies (from 10^{-2} to 10^5 Hz) [13.121]. Instead of using a parallel plate geometry, both electrodes in the microdielectric measurement are placed on the same surface of an integrated circuit (see Fig. 13.17), and the medium to be studied (C_{60} film) is

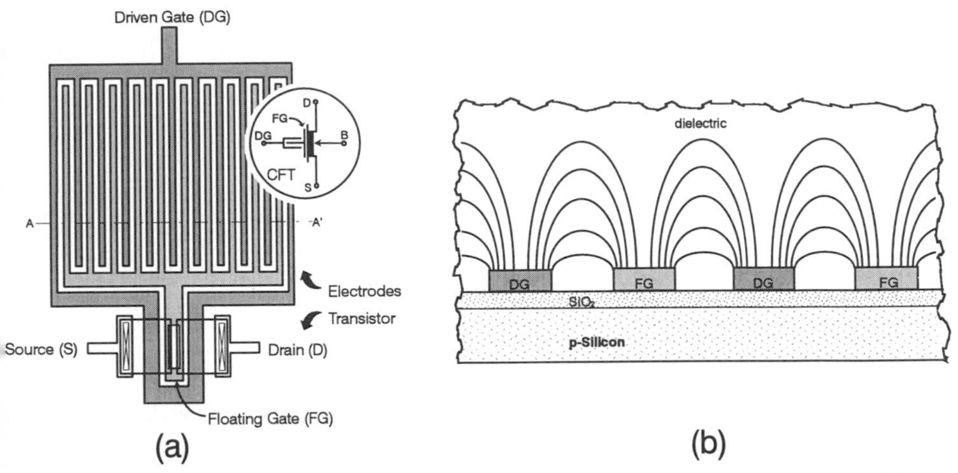

Fig. 13.17. (a) Schematic view of the active portion of a microdielectrometer sensor chip. CFT refers to "floating-gate charge-flow transistor," whose gate electrode (or floating gate [FG]) is one of the two interdigitated electrodes. Since the sensing electrode is electrically floating, and because of the proximity of the CFT amplifier to that electrode, a good signal-to-noise ratio can be achieved at very low frequencies (down to $\sim 10^{-2}$ Hz). (b) Schematic cross section through the electrode region [denoted by AA′ in (a)] showing the electric field pattern between the interdigitated driven gate (DG) and floating gate (FG) electrodes [13.119].

placed over the electrodes by thermal sublimation in vacuum [13.76]. The comb electrodes in Fig. 13.17, in contrast to parallel plates, provide a fixed calibration for both ϵ_1 and ϵ_2. Such an open-face configuration also makes it possible to monitor the frequency response and dielectric properties of the C_{60} film dynamically as a function of doping with selected gases and other species [13.121].

A small amount of charge transfer between the oxygen and C_{60} molecules is responsible for creating electrical dipoles, which can be coupled by the applied ac (low-frequency) electric field. This charge transfer imposes preferential orientations of the molecules, limiting their degrees of freedom, and giving rise to dielectric loss peaks and enhanced polarization. Because of the large size of the C_{60} molecule, only a small amount of charge transfer is required to create a significant dipole moment, resulting in a big increase in polarization $\epsilon_1(\omega)$ [13.121]. Broad loss peaks due to dipolar reorientations associated with interstitial hopping of oxygen impurity atoms can be thermally activated over a 0.5 eV energy barrier and a frequency prefactor of 2×10^{12} Hz. An interstitial diffusion constant for oxygen in solid C_{60} of $\sim 4 \times 10^{-11}$ cm^2 s^{-1} at 300 K has been inferred [13.121]. With increasing oxygenation, the interstitial sites become nearly fully occupied, interstitial hopping is inhibited, and the loss peaks together with the enhanced polarization disappear.

13.3.4. Optical Transitions in Phototransformed C_{60}

C_{60} solid films can be readily transformed into a polymeric phase under irradiation with visible or ultraviolet light [13.1, 122]. In this phototransformed phase, it has been proposed that C_{60} molecular shells are coupled together by covalent bonds to form a polymeric solid or "polyfullerene" [13.1]. Some experimental results on phototransformed C_{60} suggest the formation of four-membered rings located between adjacent C_{60} molecules through "2 + 2" cycloaddition, as discussed in §7.5 [13.2, 123]. Recent theoretical calculations on the C_{60} dimer find that a four-membered ring is the lowest-energy configuration [13.124, 125].

In this subsection we summarize results for the OA and PL spectra of phototransformed C_{60} solid films. The features in the PL spectra are observed to broaden and downshift by $\sim$330 cm^{-1}, whereas the OA features upshift and broaden considerably. These results are consistent with the reduced symmetry imposed by the introduction of covalent bonds between C_{60} molecules (see §7.5.1). Photopolymerization by the exciting source may also be the cause for the previously reported [13.126] problems in the spot-to-spot reproducibility in the photoluminescence spectra of C_{60}.

In Fig. 13.18(a) we compare the room temperature absorbance [$\mathscr{A} = -\log_{10}(\mathscr{T})$ where $\mathscr{T}$ is the transmission] of a C_{60} film (d $\sim$500 Å) on a Suprasil substrate in the pristine (solid curve) and phototransformed (dashed curve) phases. In §13.3.2 and in Table 13.4 the four prominent absorption peaks in Fig. 13.18(a) at $\sim$2.7, 3.6, 4.7, and 5.6 eV for the pristine phase are identified with predominantly "dipole-allowed" electronic transitions (see §13.1.2) [13.77].

After phototransformation of the sample, the optical bands associated with these optical transitions are noticeably broadened and reduced in peak intensity [Fig. 13.18(a)]. This broadening is attributed to a random photochemical cross-linking of C_{60} molecules, giving rise to inhomogeneous broadening. Cross-linking the molecular shells completely removes the degeneracies of the electronic energy levels in the C_{60} monomers. Also, the phototransformation generates a distribution of C_{60} oligomeric units. Both of these changes in the system should contribute to line broadening. It was found by fitting Lorentz oscillator functions to the absorption spectra that the width of the peak at $\sim$3.6 eV in the "polyfullerene" phase is $\sim$0.3 eV wider than that in the pristine phase. This $\sim$0.3 eV broadening is consistent with the results of photoelectron spectroscopy [13.127], where the electronic energy levels (HOMO and HOMO–1 states) in the photoemission spectra

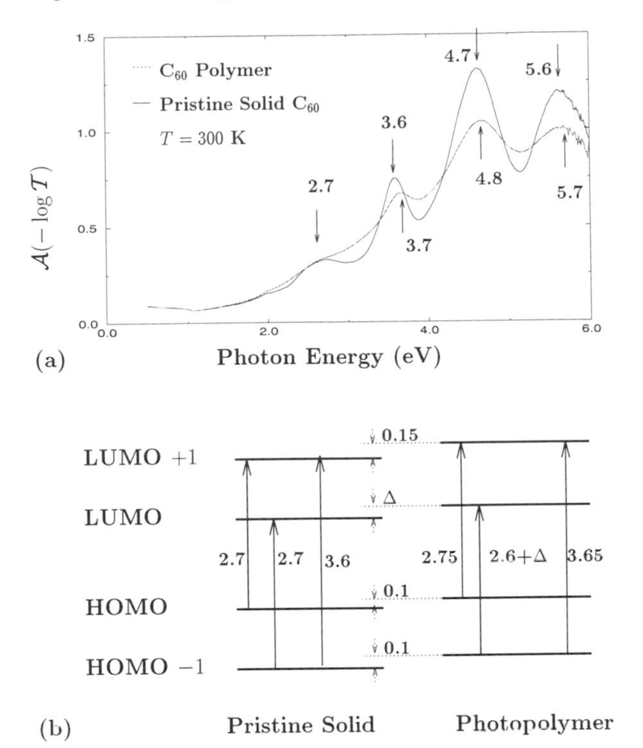

Fig. 13.18. (a) Optical absorbance $\mathscr{A}$ for pristine and phototransformed C_{60} (called C_{60} polymer and Photopolymer in the figure) solid films (thickness ~ 500 Å) on a Suprasil (fused silica) substrate in the photon energy range 0.5–6.0 eV. The same C_{60} film was used for both spectra. (b) Schematic diagram of the electronic states and optical transitions of pristine and phototransformed solid C_{60} near the Fermi level using simple molecular orbital notation. The numbers next to the arrows denote the transition energies (in eV) at the center of the optical absorption bands [13.77].

of phototransformed C_{60} were found to be ~0.1–0.15 eV wider than those in pristine C_{60}. It is expected that the LUMO+1 states are broadened by about the same amount as the HOMO states.

In addition to the photoinduced broadening of the absorption bands in polyfullerene, the peaks at ~3.6, 4.7, and 5.6 eV in pristine solid C_{60} are noticeably blue-shifted in the polymer. It is difficult to judge whether the ~2.7 eV feature is shifted because the absorption band in the polymer is too broad to locate its center. To visualize the effect of phototransformation on the lowest energy states of crystalline C_{60}, we refer to the schematic diagram in Fig. 13.18(b) that was used to interpret the absorption spectra in Fig. 13.18(a). The variances between the level energies in Fig. 13.18(b) and Table 13.4 are indicative of present uncertainties in the assignments of the optical transitions in C_{60}. The solid horizontal lines on the left in Fig. 13.18(b) represent the peak positions of the calculated Gaussian density of state (DOS) for the contributing band of electronic states [13.104], and allowed optical transitions and transition energies (in eV) are indicated in Fig. 13.18(b). To explain the observed OA blue shift upon phototransformation, and to be consistent with the photoemission data on the photopolymer [13.127], the electronic levels of the phototransformed C_{60} are upshifted with respect to the electronic levels of pristine C_{60}. In the figure, Δ is an energy shift of the LUMO level in the phototransformed phase which can be estimated from the data in Fig. 13.18(a). Since the optical band in the phototransformed C_{60} solid that is assigned to electronic transitions between the HOMO–1 and LUMO+1 states (at ~3.7 eV) is blue-shifted by ~0.05 eV, the LUMO+1 states are upshifted by ~0.15 eV to be consistent with the photoemission results [13.127]. If the LUMO state is not also upshifted by ~0.15 eV (i.e., $\Delta \neq \sim 0.15$ eV), the energies of the HOMO $\rightarrow$ LUMO+1 and HOMO–1 $\rightarrow$ LUMO transitions will be different [see Fig. 13.18(b)], and the peak at ~2.7 eV in the pristine C_{60} spectrum would split into two poorly resolved peaks in the phototransformed C_{60} spectrum. Since there is no evidence for an unresolved doublet at ~2.7 eV, it is assumed that $\Delta \sim 0.1$–0.2 eV.

In Fig. 13.19, we display the 80 K PL spectra near the absorption edge for the same C_{60} film on a Suprasil fused silica substrate before (solid curve)

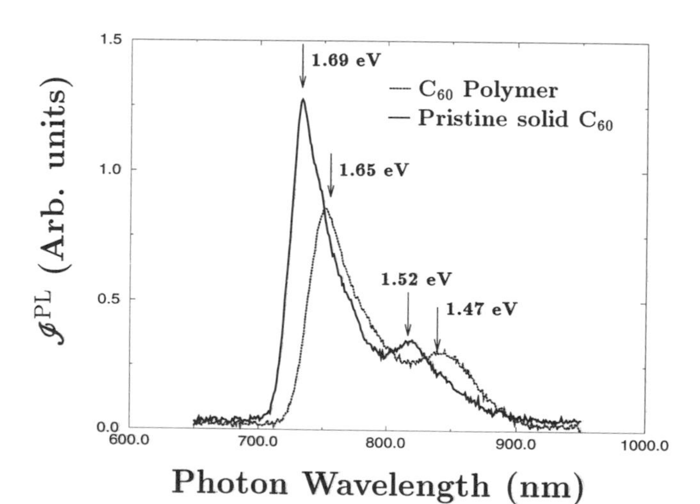

Fig. 13.19. Photoluminescence intensity $\mathcal{I}^{PL}$ at 80 K for a pristine (solid line) and a phototransformed (dotted line) C_{60} film (~4500 Å thick) on a Suprasil fused silica substrate. The same C_{60} film was used for both spectra [13.128].

and after (dashed curve) the phototransformation [13.128]. A ~330 cm^{-1} spectral red shift and peak broadening of the vibronic transitions in the whole spectrum is observed upon phototransformation. Because of the similarity of the phototransformed spectrum to that of monomeric C_{60}, it is concluded that the features in the photopolymerized spectrum are still closely related to those observed in Fig. 13.6 for the C_{60} film prior to phototransformation. The red shift of the PL features for the polymer relative to the peak energies for the pristine film is consistent with the luminescence red shifts observed for oligomerization (i.e., monomer → dimer; dimer → trimer, etc.), such as is observed in the helicene series and in polyacenes [13.6]. Furthermore, it can be seen that the two predominant peaks at ~1.47 eV (841 nm or 11,900 cm^{-1}) and 1.65 eV (752 nm or 13,300 cm^{-1}) and the shoulders in the spectrum of the phototransformed sample are slightly broadened compared to those at 1.53 eV (818 nm or 12,200 cm^{-1}) and 1.69 eV (734 nm or 13,600 cm^{-1}) in the pristine spectrum. This broadening effect is consistent with a lifting of the degeneracies in the vibrational modes, as seen in the Raman and IR vibrational spectroscopy of the photopolymer (§11.7) [13.1], and is also consistent with the inhomogeneous broadening discussed above in connection with the OA spectrum of the phototransformed C_{60}.

13.4. Optical Properties of Doped C_{60}

In this section we discuss the optical properties of the M_xC_{60} alkali metal–doped phases for C_{60}. Of these, the M_3C_{60} phase has been studied most extensively, and measurements have been made in both the normal and superconducting phases. Since the optical spectra of these alkali metal compounds do not depend on the alkali metal species M in most cases, it is believed that the optical properties of the alkali metal–doped materials are closely related to the optical properties of the relevant C_{60}^{n-} anions themselves, which can be prepared electrolytically and studied as prepared in solution. For this reason we start this section with a summary of the experimental results on the optical spectra of C_{60}^{n-} anions in solution in §13.4.1, which is followed by a review of the optical properties of the M_6C_{60} compounds in §13.4.2 because of the relatively close relation of the M_6C_{60} compounds to crystalline C_{60}. The presentation continues with a review of the optical properties of the M_3C_{60} compounds in the normal and superconducting phases (§13.4.3 and §13.4.4), and the section concludes with a brief summary of the optical properties of the M_1C_{60} compounds (§13.4.5), which have some unique properties.

13.4.1. Optical Absorption of C_{60} Anions in Solution

C_{60} has been reduced electrochemically in solution into various oxidation states: C_{60}^{n-}, $n = 1, 2, 3, 4$ [13.18, 129–132] (see §10.3.2). Although neutral C_{60} in solution does not exhibit optical absorption in the near IR, distinct absorption bands have been identified in the near IR with specific C_{60}^{n-} ions in solution [13.133]. As we shall see, these new absorption bands are a direct consequence of the electron transfer to the t_{1u} (LUMO) level of C_{60} to form the anions. The Jahn–Teller distortion mechanism is invoked to split the orbitally degenerate C_{60}-derived t_{1u} and t_{1g} levels, and specific transitions between these Jahn–Teller split levels are proposed to explain the observed spectra. Since the $t_{1u} - t_{1g}$ splitting in C_{60} is ~1.1 eV [13.30, 37, 41, 42, 58], new near-IR lines would be expected near ~1000 nm in the absorption spectra for C_{60}^{n-} ions. The spectra observed for C_{60} and the molecular anions C_{60}^{n-} $(n = 1, 2, 3, 4)$ are shown in Figs. 13.20 and 13.21 [13.134]. The near-IR spectrum of the C_{60}^{-} anion has been reported by several authors [13.134–138].

For the spectra shown in Figs. 13.20 and 13.21, the electrochemical reduction of C_{60} was carried out within the optical spectrometer [13.139]. The C_{60} (~0.15 nM) was dissolved in benzonitrile with 0.1 M Bu_4NPF_6 used as a supporting electrolyte. Au and Pt were used as the working and counter electrodes, respectively, and the cell potentials were measured relative to a standard electrode [Ag/Ag^+ (DMSO)] [13.134]. Potentials of −1.1, −1.6, −2.2, and −2.9 V were found to generate the $n = 1, 2, 3$, and 4

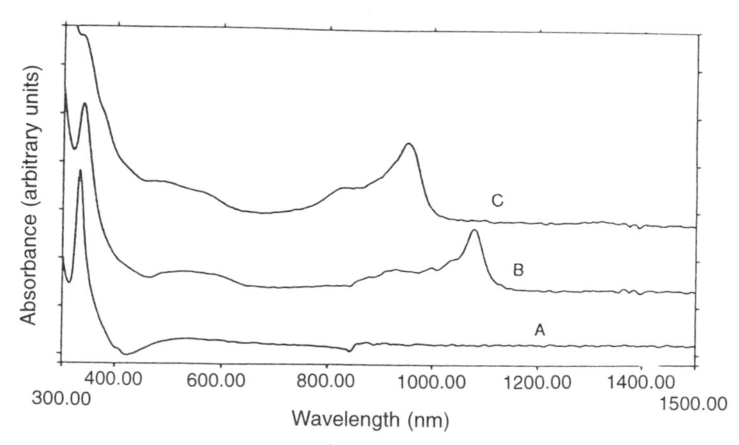

Fig. 13.20. UV–visible near-IR absorption spectra for C_{60} (trace A), C_{60}^{1-} (trace B), and C_{60}^{2-} (trace C). A blank sample of the supporting electrolyte solution was used for background subtraction. The feature at 840 nm in all three spectra corresponds to an instrument detector change [13.134].

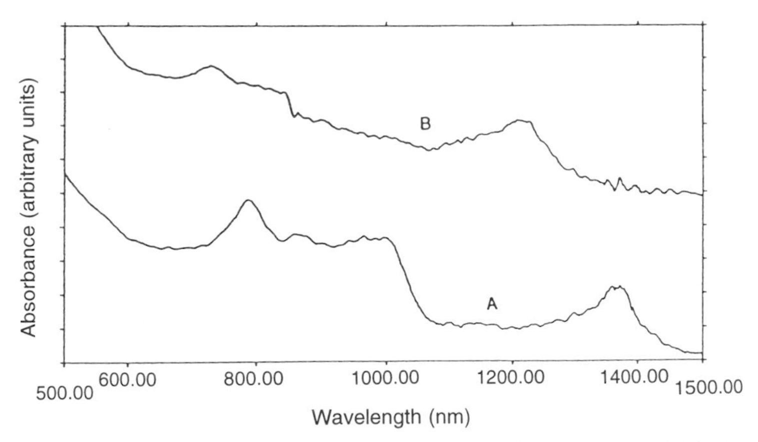

Fig. 13.21. UV–visible near-IR absorption spectra for C_{60}^{3-} (trace A) and C_{60}^{4-} (trace B). A blank sample of the supporting electrolyte solution was used for background subtraction. The feature at 840 nm in both spectra corresponds to an instrument detector change. The absorbance scale in this figure is 0.7 relative to that in Fig. 13.20 [13.134].

molecular anions, respectively, in good agreement with the literature (see §10.3.2).

To provide a quantitative optical basis for the determination of the concentrations of these C_{60}^{n-} species, Lawson *et al.* [13.134] determined the molar extinction coefficients at wavelengths corresponding to the positions of the absorption band maxima (λ_{max}) in the near-IR and in the red regions of the spectra shown in Figs. 13.20 and 13.21. Their results for λ_{max} (corresponding to the peaks in the absorption) and the molar extinction ratios for C_{60}^{n-} ($n = 1, 2, 3, 4$) are summarized in Table 13.5 [13.134].

Lawson *et al.* [13.133, 134] have proposed the schematic molecular orbital and optical transition diagrams shown in Fig. 13.22(A), (B) and (C) for C_{60}, C_{60}^{1-} and C_{60}^{2-} and in Fig. 13.23(A) and (B) for C_{60}^{3-} and C_{60}^{4-}, respectively, to qualitatively explain their observed spectra. In these figures, the solid vertical arrows indicate allowed dipole transitions and the dashed arrows identify transitions which are not dipole allowed, but are activated by vibronic coupling, as described in §13.1.3 for the $h_u \to t_{1u}$ transitions in neutral C_{60}. Above each level diagram is found a simulated spectrum, although the oscillator strengths and the widths of the various bands in the simulated spectra are not derived from a calculation. The optical response of isolated neutral C_{60} molecules in solution has been previously described in detail in §13.2.1 and §13.2.2.

The level diagram for the monoanion C_{60}^{1-} in Fig. 13.22(A) shows the threefold degenerate t_{1u} LUMO of C_{60} split into a_{2u} and e_{1u} states by a Jahn–Teller distortion, thus removing the degeneracy of the ground state.

Table 13.5

Spectral data for various C_{60} species in benzonitrile solution [13.134].

Species	λ_{max} (nm)	Extinction coeff. $(\text{mol cm})^{-1}$
C_{60}	330	48,000
C_{60}^{1-}	1078	12,000
C_{60}^{2-}	952	16,000
	830	7,000
C_{60}^{3-}	1367	6,000
	956	9,000
	878	7,000
	788	14,000
C_{60}^{4-}	1209	6,000
	728	4,000

The additional electron in the C_{60}^{1-} anion occupies the lower a_{2u} state. The t_{1g} orbital must also split under this distortion, $t_{1g} \rightarrow e_{1g} + a_{2g}$. Three transitions (one of these being weak) are identified with the optical absorption bands in trace B of Fig. 13.20. The transitions $a_{2u} \rightarrow e_{1g}$ in Fig. 13.22(B) are allowed and are identified with the peak at 1078 nm in trace B of Fig. 13.20. The weaker transitions $a_{2u} \rightarrow a_{2g}$ are forbidden and are identified with a vibronic manifold between 800 and 1000 nm. These groups of transitions are labeled and correspond to the peaks identified in the simulated spectrum above the level diagram.

In column C of Fig. 13.22, the level diagram proposed to explain spectrum C in Fig. 13.20 for C_{60}^{2-} is displayed. This anion has been reported by some authors to have a triplet ground state ($S = 1$) from EPR studies (see §16.2.2) [13.140]. Thus, the doubly degenerate e_{1u} state is placed below the singly degenerate a_{2u} level. Two dipole-allowed transitions are expected: $e_{1u} \rightarrow e_{1g}$ (b), which is identified with the absorption band at 952 nm; and $e_{1u} \rightarrow a_{2g}$ (a), which is identified with the absorption band at 830 nm.

The spectra for the C_{60}^{3-} and C_{60}^{4-} anions are more complex, and the assignments are more speculative. Turning to the trianion first, the proposed level scheme is shown in column A of Fig. 13.23. Based on observed EPR spectra, C_{60}^{3-} has been reported to resemble a spin 1/2 system (see §16.2.2) [13.141]. It is assumed that a Jahn–Teller distortion occurs for the C_{60}^{3-} anion to remove the ground state degeneracy and lower the system energy. A C_{2v}

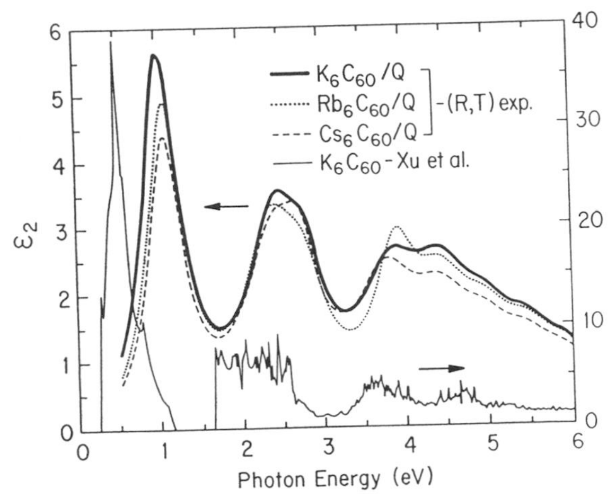

Fig. 13.26. Comparison of the room temperature experimental $\epsilon_2(\omega)$ data (left scale) for M_6C_{60} (M = K, Rb, Cs) films on quartz substrates [13.77] with the calculations of Xu *et al.* [13.105] (thin solid noisy line at the bottom of the figure) (right scale).

was obtained as follows using standard expressions for the optical reflectivity and transmission $\mathscr{R}$ and $\mathscr{T}$, involving the dielectric function of both the Suprasil substrate and the sample. The M_6C_{60} dielectric function $\epsilon(\omega)$ was expressed in terms of contributions from a sum of Lorentz oscillators,

$$\epsilon(\omega) = \epsilon_1(\omega) + i\epsilon_2(\omega) = \epsilon_{\text{core}} + \sum_{j=1}^{n} \frac{\omega_{Pj}^2}{\omega_{0j}^2 - \omega^2 - i\Gamma_j\omega} \tag{13.14}$$

where ω_{0j}, ω_{Pj}, and Γ_j, respectively, denote the frequency, strength, and damping of the jth oscillator, and ϵ_{core} is the core dielectric constant used to approximate interband absorption well beyond the energy range of the experimental data. Values for these Lorentz oscillator parameters were adjusted to obtain the best fit to the $\mathscr{R}$ and $\mathscr{T}$ spectra simultaneously. From this analysis the experimental curves for $\epsilon_2(\omega)$ shown in Fig. 13.26 were obtained. The parameter values found from these fits are summarized in Table 13.6. The left-hand vertical scale for $\epsilon_2(\omega)$ in Fig. 13.26 refers to the three closely spaced experimental curves (heavy solid curve for K_6C_{60}, dotted curve for Rb_6C_{60}, and dashed curve for Cs_6C_{60}) [13.77].

As can be seen in Fig. 13.26, the optical $\epsilon_2(\omega)$ data for the M_6C_{60} fullerides are found to be insensitive to the particular dopant (M = K, Rb, Cs), similar to observations regarding the intramolecular vibrational mode frequencies in M_6C_{60} compounds described in §11.6. Thus, it is concluded that no significant hybridization of the alkali metal and C_{60} states occurs in

Table 13.6

Interband optical parameters of Eq. (13.14) (in eV) for M_6C_{60} (M = K, Rb, Cs) where $V_1 = t_{1u}$, $V_2 = h_u$, $C_1 = t_{1g}$, and $C_2 = t_{2u}$ or h_g and where $\epsilon_{core} = 3.0$ [13.77].

		$V_1 \rightarrow C_1$ [a]	$V_1 \rightarrow C_2$ [b] $V_2 \rightarrow C_1$	$V_2 \rightarrow C_2$	Oscillator ensemble			
Rb_6C_{60}	ω_0	1.10[c]	2.40, 2.80[d]	3.90	4.47	5.0	5.5	6.0
	ω_P	1.62	2.2, 1.5	2.1	2.5	1.8	1.8	1.8
	Γ	0.52	0.8, 0.6	0.6	0.92	0.8	0.8	0.8
Cs_6C_{60}	ω_0	1.10	2.42, 2.75	3.80	4.47	5.0	5.5	6.0
	ω_P	1.5	2.2, 1.4	2.4	2.25	1.7	1.7	1.7
	Γ	0.5	0.85, 0.5	0.9	0.92	0.8	0.8	0.8
K_6C_{60}	ω_0	1.05	2.45, 2.80	3.85	4.47	5.0	5.5	6.0
	ω_P	1.75	2.3, 1.3	2.4	2.5	1.8	1.8	1.8
	Γ	0.55	0.8, 0.5	0.9	0.92	0.8	0.8	0.8

[a]Theoretical calculations give 1.0 eV [13.105] and 0.8 eV [13.145].
[b]Theoretical calculations give (1.9, 2.7) eV [13.105] and (2.4, 2.4) eV [13.145].
[c]EELS measurements yield 1.3 eV [13.101].
[d]EELS measurements yield 2.75 eV [13.101].

M_6C_{60} and that these compounds are therefore strongly ionic. This conclusion is in agreement with that reached theoretically by Erwin and Pederson [13.145], who calculated only a 4% admixture of K and C_{60} states in the electronic band structure of K_6C_{60}. The insensitivity of the optical data in Fig. 13.26 to the particular dopant also rules out the assignment of any of the spectral features to charge transfer excitations between C_{60}^{6-}- and M-derived states or to transitions between filled and empty alkali metal states. All the structure in the $\epsilon_2(\omega)$ data is therefore identified with dipole transitions between states associated with the C_{60} hexanion C_{60}^{6-}.

Although it has not yet been well established that M_xC_{60} solids can be described as ordinary band solids, a comparison between energy band theory for K_6C_{60} [13.105, 145] and the experimental data can be made. Such a comparison is useful for identifying the interband transitions responsible for the peaks in the $\epsilon_2(\omega)$ spectrum. The thin solid curve in Fig. 13.26 (right-hand scale) represents theoretical LDA calculations for the interband contribution to $\epsilon_2(\omega)$ by Xu *et al.* [13.105]. The theoretical $\epsilon_2(\omega)$ is based on an electronic band structure calculation similar to that reported by Erwin and Petersen and includes the effects of the $\mathbf{p} \cdot \mathbf{A}$ matrix elements for the k points throughout the Brillouin zone. In agreement with theoretical calculations [13.105], the data in the figure show that the M_6C_{60} compounds exhibit a series of distinguishable peaks in $\epsilon_2(\omega)$ on the order of ~1 eV in width. It is, however, seen that the positions of the experimentally determined absorption bands are consistently upshifted by ~0.5 eV relative to the theoretical peak positions. Nevertheless, the shape of the theoretical $\epsilon_2(\omega)$ spectral features reflects well that obtained by experiment, although the magnitude of the calculated $\epsilon_2(\omega)$ (right-hand scale) is somewhat higher than the experimental $\epsilon_2(\omega)$ (left-hand scale) in Fig. 13.26.

To assign peaks within the simple band picture in the low energy $\epsilon_2(\omega)$ data to particular band-to-band transitions, $\epsilon_2(\omega)$ peak frequencies are compared to the energy difference between the centers of a particular valence band (V_i) and conduction band (C_i). For M_6C_{60} the major contributions to $\epsilon_2(\omega)$ come from the following bands: $V_1 = t_{1u}$, $V_2 = h_u$, $C_1 = t_{1g}$ and $C_2 = t_{2u}$, h_g [13.77], in accordance with band structure calculations [13.105, 145]. Consistent with this notation, the four lowest energy features in the $\epsilon_2(\omega)$ curve for M_6C_{60} are identified with the transitions $V_1 \rightarrow C_1$ (1.1 eV), $V_1 \rightarrow C_2$ (2.4 eV), $V_2 \rightarrow C_1$ (2.8 eV), and $V_2 \rightarrow C_2$ (3.85 eV), essentially independent of M, as summarized in Table 13.6. The t_{2u}- and h_g-derived bands in M_6C_{60} are too close together in energy to give rise to separate peaks in $\epsilon_2(\omega)$ and therefore are considered to be quasidegenerate. Of the two possible groups of transitions in $V_1 \rightarrow C_2$, higher oscillator strength is expected for the $t_{1u} \rightarrow h_g$ transitions, unless solid-state interactions and/or vibronic coupling are particularly effective in activating the $t_{1u} \rightarrow t_{2u}$ transitions. Furthermore, both the $V_1 \rightarrow C_2$ and $V_2 \rightarrow C_1$, transitions lie too close in energy and are not separately resolved in the $\epsilon_2(\omega)$ spectrum of Fig. 13.26. Finally, the Lorentz oscillators introduced to describe the optical reflection and transmission of the M_6C_{60} compounds above 4.5 eV (see Table 13.6 and Fig. 13.26) are not associated with any sharp spectral features in the optical spectra, but rather were introduced to simulate the broad absorption between 3.5 eV and 6.0 eV in the $\mathscr{R}$ and $\mathscr{T}$ spectra. No attempts were made to compare the positions of this broad absorption with theory [13.77].

Using *in situ* K doping of C_{60} films on transparent substrates, optical absorption of K_xC_{60} films has been recorded as a function of exposure to K vapor [13.96–99]. Simultaneous optical and electrical resistivity measurements were made in the vacuum chamber in which the C_{60} films were doped, and the evolution of the optical transmission and electrical resistivity from K exposure were recorded. On the basis of the observed optical spectra it was concluded [13.98] that for $0 < x < 3$, the K_xC_{60} system exhibits mixed phase behavior. Specifically, it was shown that in the region 300–800 nm, the optical density of the K-doped film could be fit first as a linear combination of C_{60} and K_3C_{60} spectra until the resistivity reached a minimum, signaling the formation of the pure metallic K_3C_{60} phase. Further exposure to K beyond the K_3C_{60} stoichiometry resulted in a spectrum well fit by a linear combination of the spectra for K_3C_{60} and K_6C_{60}. The optical transmission spectra for the semiconducting K_6C_{60} phase were analyzed in some detail and the peaks in the optical density were identified with specific interband transitions by noting the structures which were inhanced in intensity upon doping with K (i.e., transitions from the t_{1u} band) and those that were suppressed or quenched (such as $h_u \rightarrow t_{1u}$ and the $h_g + g_g \rightarrow t_{1u}$ transitions). The results for interband transitions of three *in situ* optical studies of K_6C_{60} as a function of alkali metal doping are summarized in Table 13.7 [13.96–99]. Also included in Table 13.7 for comparison are the interband transitions given in Table 13.6 for K_6C_{60} [13.77], as obtained from the $\epsilon(\omega)$ analysis discussed above. While overall agreement between the experimental results from the four groups of investigators is excellent, the identification of the interband transitions in some cases remains tentative. Good agreement is also achieved between the optical data summarized in Table 13.7 and EELS data for K_6C_{60} [13.147].

In the above discussion the optical spectra were considered in terms of a simple band picture. An alternative discussion of the optical spectra for M_6C_{60} can be made in terms of excitons, which emphasize the importance of the electron–hole interaction in the optically excited state. Harigaya and Abe [13.148] have considered molecular excitons in the M_6C_{60} compounds using the same model they used for solid C_{60} (see §13.3.2). As for the case of pristine C_{60}, bond disorder is used to simulate lattice fluctuation effects in the M_6C_{60} compounds. Their model calculation was carried out in the Hartree–Fock approximation using a single excitation configuration interaction, and the excitons were localized on a single anion as Frenkel-

Table 13.7

Energies (eV) of the electronic transitions in thin films of pristine K_6C_{60} as identified from optical transmission studies.

Transitions	Transition energies (eV)			
$t_{1u} \rightarrow t_{1g}$	1.1[a]	1.3[b]	1.2[c]	1.05[d]
$t_{1u} \rightarrow t_{2u}$	—	2.0[b]	1.9[c]	2.45[d]
$h_u \rightarrow t_{1g}$	—	2.5[b]	2.2[c]	2.8[d]
$t_{1u} \rightarrow h_g$	2.9[a]	2.8[b]	3.2[c]	2.8[d]
$t_{1u} \rightarrow ?$	4.1[a]	—	3.8[c]	3.85[d]
$h_u \rightarrow h_g + t_{2u}$	4.6[a]	4.6[b]	—	4.5[d]
$h_g + g_g \rightarrow t_{2u}$	5.5[a]	5.5[b]	—	5.5[d]

[a]Ref. [13.96]. [b]Ref. [13.97]. [c]Ref. [13.98]. [d]Ref. [13.77, 146].

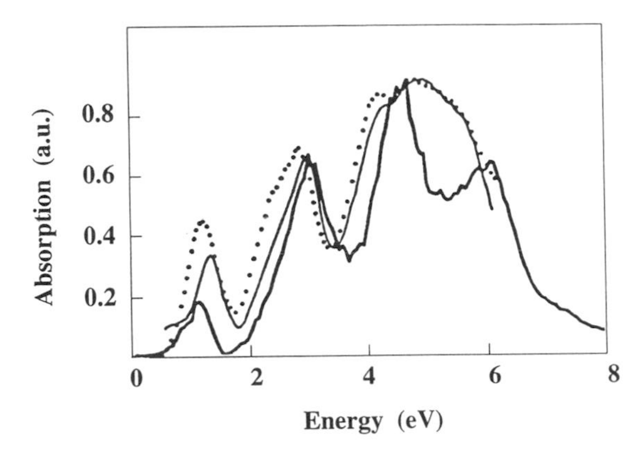

Fig. 13.27. Calculated optical absorption spectra for C_{60}^{6-}, shown in arbitrary units, (thick line) [13.148], and compared with experimental optical data for K_6C_{60} are taken from Pichler *et al.* [13.97] (thin line), and with experimental EELS data for Rb_6C_{60} from Sohmen and Fink [13.101] (dots).

type excitons. Electron–electron interactions were treated using the Ohno potential form

$$W(r) = \{(1/U)^2 + (r/[r_0 V])^2\}^{-1/2} \qquad (13.15)$$

which is parameterized by an on-site (C-atom) Coulomb strength $U = 2t$ and a long-range component $V = t$, where $t = 2.0$ eV and r_0 is an average bond length. Values for t and r_0 were optimized for the C_{60}^{6-} ion, to fit the optical spectrum. Their calculated results (thick line) are shown in Fig. 13.27 for a bond disorder broadening of $0.2t$ (0.4 eV). Also shown in Fig. 13.27 is the thin-film K_6C_{60} optical absorption spectrum of Pichler *et al.* (thin line) [13.97] and the small momentum transfer EELS data by Sohmen and Fink [13.101] for Rb_6C_{60} (dotted line) derived from the electron energy loss function. The EELS data yield a value of $\epsilon_1(0) = 7.1$ for the low-frequency dielectric constant, which compares very well to the value 7.2 obtained optically [13.74]. As can be seen in Fig. 13.27, the calculated optical spectrum is in good agreement with the EELS-derived result [13.101] and the experimental optical spectrum of Pichler *et al.* [13.97] for K_6C_{60} in Fig. 13.27 and also the optical spectra for the M_6C_{60} films (M = K, Rb, Cs) in Fig. 13.26. The theoretical results are seen to be in good agreement with experiment up to ~5.0 eV, and the deviations at higher energies have been ascribed [13.148] to the omission of σ-electron excitations in the theoretical model.

13.4.3. *Normal State Optical Properties of M_3C_{60}*

Several optical studies of alkali metal-doped M_3C_{60} (M = K, Rb) have been reported showing these materials to be highly absorbing in the infrared, in contrast to pristine C_{60}, and consistent with the metallic nature of the transport properties of M_3C_{60} (see §14.1.1). Reflectivity ($\mathscr{R}$) measurements have been carried out on opaque samples, e.g., pressed powders [13.149–151], doped single crystals [13.152], and thick, vacuum-deposited films [13.153]. Transmission ($\mathscr{T}$) studies [13.153–155] have also been reported on thinner, vacuum-deposited films. Kramers–Kronig (KK) analyses have been carried out on reflectivity data for M_3C_{60} films to obtain quantitative information on the electronic contribution to the optical conductivity $\sigma(\omega)$. Below ~0.5 eV, the free electron contribution to the optical properties is dominant, and above ~0.5 eV, interband transitions from the partially occupied t_{1u} level to the higher-lying t_{1g} level take place. These contributions to the dielectric function are also sensitively studied by electron energy loss spectroscopy (EELS) (see §17.2.2). In addition, the optical conductivity data show interesting structure in the far IR ($\omega <$ 400 cm^{-1}). Transmission studies have been reported in the mid-IR region (400–4000 cm^{-1}) and have emphasized the doping-induced changes in the intramolecular IR-active (F_{1u}) phonons (see §11.6.2). These F_{1u} phonons were also observed in single-crystal reflectance studies [13.152], but were not observed in pressed powder samples (3 mm diameter) [13.149, 150]. Reflectivity data on M_3C_{60} extending to the far IR region of the spectrum (~ 1 meV = 8 cm^{-1}) are available only for pressed powder samples [13.149, 150], though single-crystal optical studies have also been carried out.

In Fig. 13.28 we show the low-temperature, normal state reflectivity ($\mathscr{R}$) spectra for pressed-powder samples of K_3C_{60} (a) and Rb_3C_{60} (b) [13.151]. The $\mathscr{R}(\omega)$ data are plotted on a log frequency scale, covering four decades of frequency (10–10^5 cm^{-1}). These samples were reported to have been excluded rigorously from oxygen and, furthermore, were found to exhibit superconducting transition temperatures T_c= 19 K (K) and 29 K (Rb) and diamagnetic shielding fractions of ~40–50%. Upon cooling these samples through T_c, large increases were observed in the reflectance below ~100 cm^{-1}, indicative of the onset of superconductivity (see §13.4.4). The frequency-dependent reflectivity data $\mathscr{R}(\omega)$ in Fig. 13.28 for the normal state of K_3C_{60} and Rb_3C_{60} show a characteristic spectrum of a metal in the "dirty limit," i.e., high $\mathscr{R}(\omega)$ at low frequency, and a broad Drude edge marking the position of the decline in $\mathscr{R}(\omega)$ to lower values (~20–30%), the decline in $\mathscr{R}(\omega)$ being associated with contributions from electronic interband transitions. The broad width of the Drude edge for both K_3C_{60} and Rb_3C_{60} indicates a short mean free path for the normal state electrons, consistent with transport measurements (see §14.1.2). The Drude edge in the reflectance data of K_3C_{60} reported for doped single crystals [13.152] is also broad and in good agreement with the data shown in Fig. 13.28. (For a comparison between the single crystal and pressed-powder K_3C_{60} reflectivity spectra, see Fig. 5 in Ref.[13.151].)

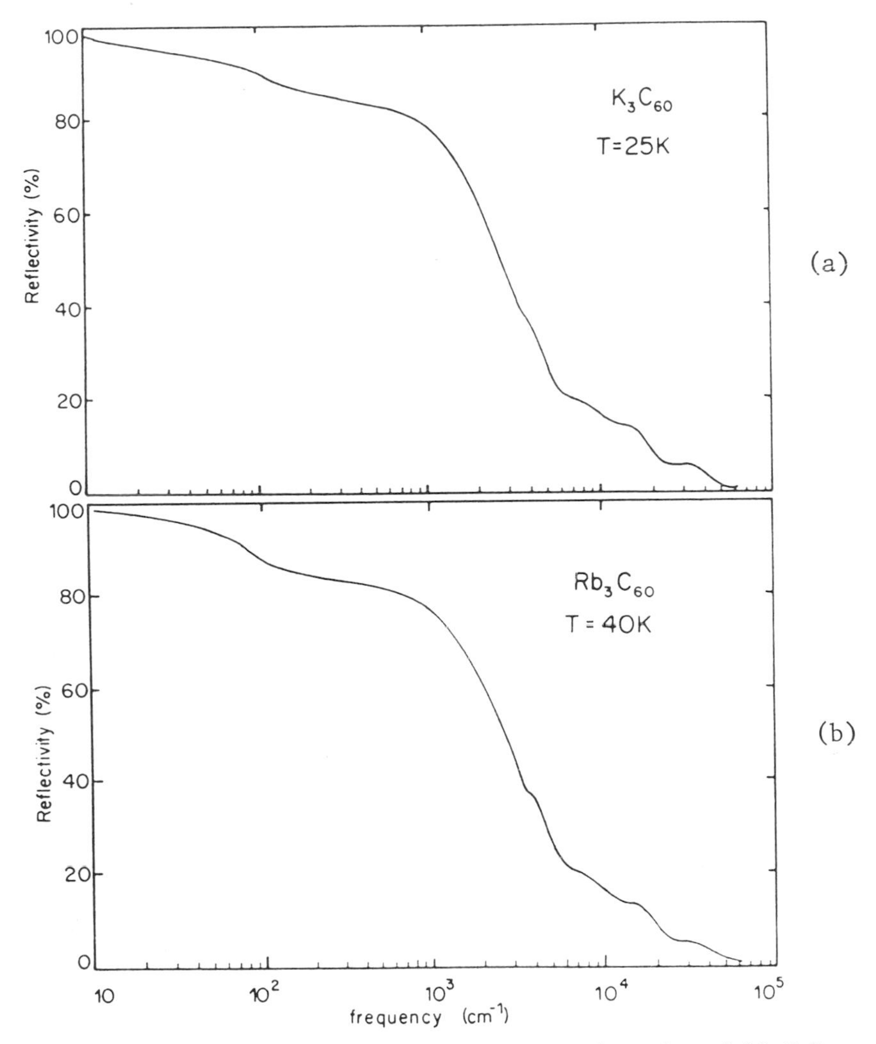

Fig. 13.28. Normal state reflectivity spectra on pressed powders of (a) K_3C_{60} and (b) Rb_3C_{60} at 25 and 40 K, respectively [13.151].

A Kramers–Kronig (KK) integral transform of the $\mathscr{R}(\omega)$ data in Fig. 13.28 was performed to determine the optical conductivity $\sigma(\omega) = \sigma_1(\omega) + i\sigma_2(\omega)$ for K_3C_{60} (a) and Rb_3C_{60} (b) [13.151]. The results thus obtained for the real part, $\sigma_1(\omega)$, which is associated with optical absorption, are plotted in Fig. 13.29 on a log frequency scale. The solid and dash-dotted curves represent, respectively, the experimental $\sigma_1(\omega)$

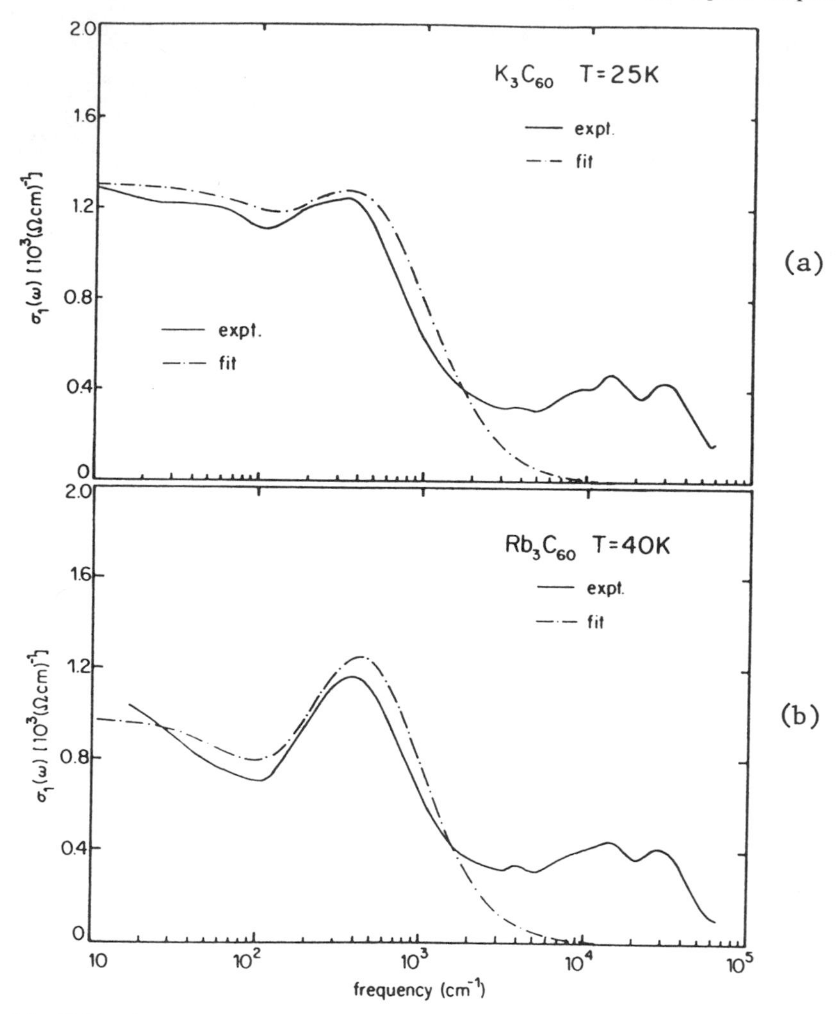

Fig. 13.29. Normal state optical conductivity for (a) K_3C_{60} and (b) Rb_3C_{60} at 25 and 40 K, respectively, as evaluated from a Kramers–Kronig analysis of the $\mathscr{R}(\omega)$ reflectivity data in Fig. 13.28. The phenomenological fits to Eq. (13.16) are the dashed-dotted curves, and values for the fitting parameters are summarized in Table 13.8 [13.151].

obtained from the KK analysis and that calculated on the basis of a model discussed below. At $\omega = 0$, the $\sigma_1(0)$ data match the dc transport value, consistent with the low-frequency data extension in Fig. 13.29 [13.151]. Above 400 cm^{-1}, the $\sigma_1(\omega)$ data in Fig. 13.29 for Rb_3C_{60} and K_3C_{60} are similar. Below 400 cm^{-1}, the Rb_3C_{60} data exhibit a more noticeable dip at $\sim$100 cm^{-1}, or equivalently, the peak at $\sim$400 cm^{-1} is more pronounced in the Rb_3C_{60} data. This peak is reminiscent of the so-called "mid-IR" band exhibiting maxima in the range 2000–4000 cm^{-1} in high T_c cuprate materials [13.156]. IR-active, F_{1u}-derived phonons would appear as narrow peaks riding on the electronic background. They have been observed in films and crystals in the range $\omega < 1700$ cm^{-1} [13.153–155], but these peaks are apparently too weak to be detected in the powder samples of Fig. 13.29 [13.151]. Finally, the experimentally observed structure above $\sim$3000 cm^{-1} is quite similar for both compounds and therefore must be identified with interband transitions between C_{60}-derived states, rather than with "charge transfer" excitations involving initial or final states (but not both) associated with the alkali metal ions.

The model calculation results for $\sigma_1(\omega)$ in Fig. 13.29 are based on the dielectric function given by

$$\epsilon(\nu) = \epsilon_\infty + \nu_p^2 \left[\frac{f_G}{\nu_G^2 - \nu^2 - i\nu\gamma_G} - \frac{1 - f_G}{\nu(\nu + i\gamma_D)} \right] \tag{13.16}$$

where ϵ_∞ is the core dielectric constant associated primarily with higher energy interband transitions, and ν_p is the electronic plasma frequency. In the notation of Eq. (13.16) the optical conductivity is, in general, related to the dielectric function by

$$\sigma(\nu) = \frac{-i\nu[\epsilon(\nu) - 1]}{4\pi}. \tag{13.17}$$

The first term in the brackets of Eq. (13.16) is a harmonic oscillator function which describes the experimentally observed "mid-IR" absorption band, and the second term is the Drude expression for the dielectric response of conduction electrons. As written, the oscillator strength f_G (or spectral weight) of the mid-IR band is borrowed from the Drude term. The electrons described by Eq. (13.16) behave approximately as free electrons for $\nu < \gamma_D$ and as bound electrons for $\nu > \gamma_D$. An alternative, and more conventional approach is to decouple the two terms of Eq. (13.16) in brackets, i.e., to set $f_G = 0$ in the second term [13.151].

As can be seen from Fig. 13.29, the model dielectric function given by Eq. (13.16) leads to a reasonably good description of the experimental $\sigma_1(\omega)$ data below $\sim$1000 cm^{-1}, and the values for the parameters in Eq. (13.16) which correspond to the dash-dotted curves in Fig. 13.29 are summarized in Table 13.8. In this table the plasma frequency ν_p is found from $\nu_p^2 = (ne^2)/\pi m_b$, where m_b is the band mass and n is the free electron concentration. Using Eq. (13.16) and choosing a value for $n = 4.1 \times 10^{21}$ cm^{-3}, consistent with three electrons in the conduction band per C_{60} molecule, a band mass of $m_b \sim 4m_0$ is obtained, where m_0 is the free electron mass. This value is in reasonable agreement with that obtained by

Table 13.8

Parameters for the phenomenological fit to the optical data for K_3C_{60} and Rb_3C_{60} using Eq. (13.16): the total plasma frequency ν_p and the damping γ_D for the Drude term, the resonance frequency ν_G and the damping γ_G for the harmonic oscillator, the high-frequency contribution to the dielectric function ϵ_∞, and f_G the weight factor between the Drude and harmonic oscillator terms [13.151].

	ν_p (cm^{-1})	γ_D (cm^{-1})	ν_G (cm^{-1})	γ_G (cm^{-1})	ϵ_∞	f_G
K_3C_{60}	9491	147	444	1210	5	0.878
Rb_3C_{60}	9220	100	444	1089	6	0.931

spin susceptibility measurements ($m_b = 6.5m_0$) [13.157] and band structure calculations [13.158], giving $h\nu_p = 1.3$ eV $= 1.0 \times 10^4$ cm^{-1}, or $m_b \sim 3m_0$, while the EELS plasmon studies yield $h\nu_p \sim 1.1$ eV and $m_b \sim 4m_0$. However, as seen in Table 13.8, a value $f_G \sim 0.9$ is obtained, indicating that almost all the oscillator strength normally assigned to the Drude term has shifted to the mid-IR band. As a consequence, values for $m_b \sim 4$ are obtained, rather than much higher values for m_b. If instead, the free carriers and mid-IR band are decoupled [i.e., $f_G = 0$ in the second term in the brackets of Eq. (13.16)], then a band mass of $m_b \sim 30\ m_0$ (K_3C_{60}) and $m_b \sim 60\ m_0$ (Rb_3C_{60}) would be obtained. These band masses are clearly much larger than those obtained by present band structure calculations and larger than those found in other experimental work. If this interpretation of the optical data is correct, then strong electron correlations and/or a strong electron–phonon interaction might be implicated.

Thus, it would appear that either a theoretical basis for the shared oscillator strength in Eq. (13.16) must be established, or experimental tests of the electronic band structure and the associated Fermi surface (e.g., through Shubnikov–de Haas or de Haas–van Alphen data) are needed to provide further evidence for lower effective mass values ($m_b \sim 4$). If the oscillator strength of the mid-IR band is indeed borrowed from the free electrons, a more detailed study of the temperature dependence of f_G might be in order. It may be of interest to note that the mid-IR bands observed in high-T_c cuprate materials have been recently identified with the photoassisted hopping of polarons [13.156], where it is pointed out that the existence of polarons does not necessarily require an activated electrical resistivity.

13.4.4. Superconducting State Optical Properties

In Fig. 13.30 we display reflectivity $\mathscr{R}(\omega)$ data in the far-infrared for K_3C_{60} (a) and Rb_3C_{60} (b) for several temperatures above and below T_c [13.151]. A Kramers–Kronig analysis of these data is given in Fig. 13.31 for one

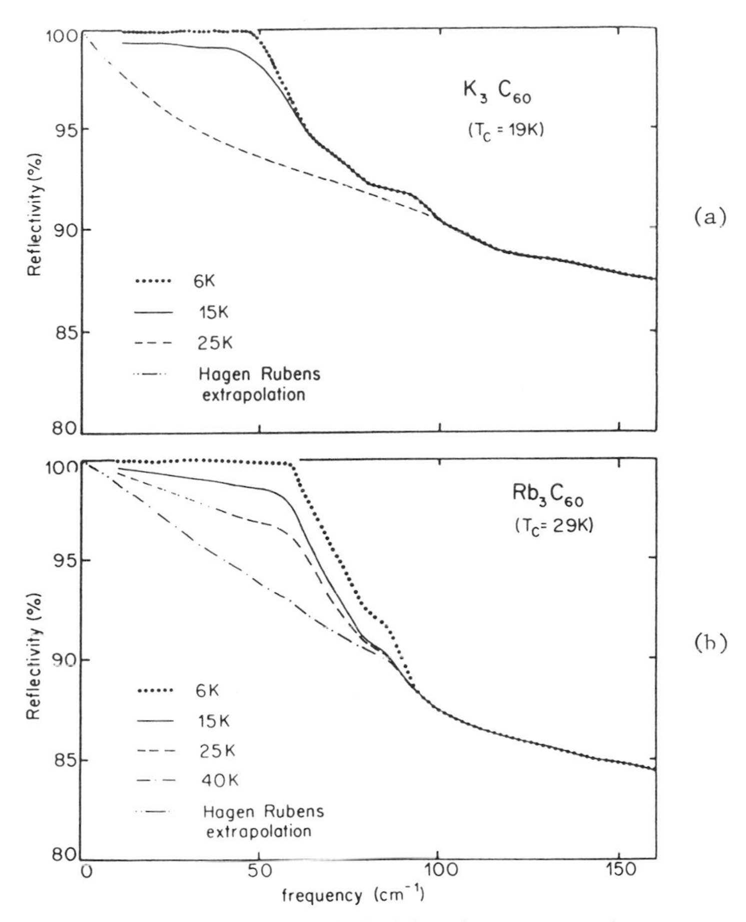

Fig. 13.30. Optical reflectivity spectra in the infrared range at several temperatures (below and above T_c) for (a) K_3C_{60} and (b) Rb_3C_{60}. Also shown in the figure is the Hagen–Rubens extrapolation for the Kramers–Kronig analysis of the $\mathscr{R}(\omega)$ reflectivity data at low frequencies in (a) for the 25 K data, and in (b) for the 25 K and 40 K data [13.151].

temperature above T_c and another below T_c (6 K) for both K_3C_{60} (a) and Rb_3C_{60} (b). In the superconducting ground state, $\sigma_1(\omega) = 0$ for $\hbar\omega < 2\Delta$, where 2Δ is the superconducting gap. Thus the $\sigma_1(\omega)$ data in Fig. 13.31 indicate that $2\Delta = 48$ cm^{-1} (K_3C_{60}) and 60 cm^{-1}(Rb_3C_{60}). Normalizing these values to T_c yields $2\Delta/k_BT_c = 3.6$ (K_3C_{60}) and 2.98 (Rb_3C_{60}) [13.151], in good agreement with the weak-coupling Bardeen–Cooper–Schrieffer (BCS) value $2\Delta/kT_c = 3.52$.

To satisfy the conductivity sum rule, the oscillator strength associated with the area A in Fig. 13.31 between the $\sigma_1(\omega)$ curves in the normal

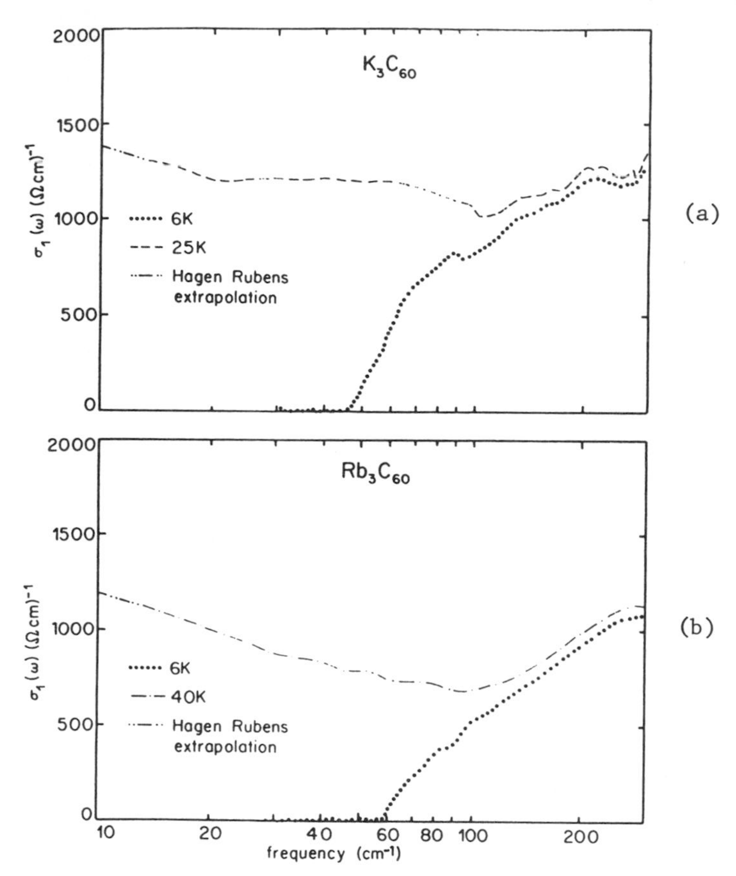

Fig. 13.31. The optical conductivity $\sigma_1(\omega)$ presented on a logarithmic frequency scale above and below T_c for (a) K_3C_{60} and (b) Rb_3C_{60} evaluated from the $\mathcal{R}(\omega)$ reflectivity data of Figs. 13.28 and 13.30 [13.151]. The normal state $\sigma_1(\omega)$ data in the low frequency limit is determined from the Hagen Rubens extrapolation.

and superconducting states is shifted to a δ function at $\omega = 0$ for the superconducting phase [13.159]. Here, the area A is given by

$$A = \int_0^{\infty} [\sigma_{1,n}(\nu) - \sigma_{1,s}(\nu)]d\nu \tag{13.18}$$

where the subscripts n and s refer, respectively, to the normal and superconducting states. Furthermore, the integral A in Eq. (13.18) is related to the London penetration depth λ_L [13.159] by the relation $\lambda_L^2 = (c^2/8)A$, where c is the speed of light, yielding the value $\lambda_L = 8000 \pm 500$ Å for both K_3C_{60} and Rb_3C_{60}. This value of λ_L is somewhat higher than values for λ_L obtained from other experimental probes (see §15.3).

More detailed fits of the far-infrared conductivity in the superconducting and normal states provide strong evidence for an electron–phonon coupling mechanism involving intramolecular phonon modes (see §15.4) [13.151].

13.4.5. *Optical Properties of M_1C_{60} Compounds*

Since the structural properties of M_1C_{60} compounds are distinctly different from those of the other M_xC_{60} alkali metal–doped fullerides, the optical properties are also expected to exhibit unique features. Depending on the heat treatment conditions of the sample, several M_1C_{60} phases can be reached as a function of measurement temperature. Using IR transmission as a measure of the physical state of the M_1C_{60} system, a schematic view of various phases of Rb_1C_{60} is shown in Fig. 13.32 [13.160]. This model is consistent with thc measurements of the optical properties shown in Fig. 13.33. Upon slow cooling, the optical transmission through a C_{60} film in the fcc (rock salt) phase that is stable at high temperatures (225°C) is essentially independent of temperature, with a spectrum similar to that shown in Fig. 13.33 (bottom curve) for the *slow-cooled* sample. Upon further slow cooling, the optical transmission for Rb_1C_{60} drops at the fcc–orthorhombic phase transition [13.161], indicative of a lower electrical resistivity phase, which is found to be orthorhombic in structure [13.162], where the C_{60} anions are closely linked along a proposed polymer chain direction (see §8.5.2). As the temperature continues to decrease, the optical transmission

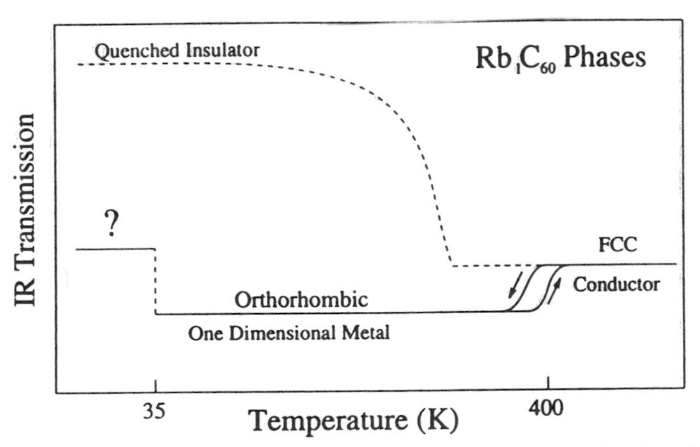

Fig. 13.32. Schematic illustration of the various phases of Rb_1C_{60} as a function of temperature, as probed by the IR transmission technique. The upper (dashed) and lower (solid) curves correspond to the rapidly quenched and slow-cooled Rb_1C_{60} samples, respectively [13.160].

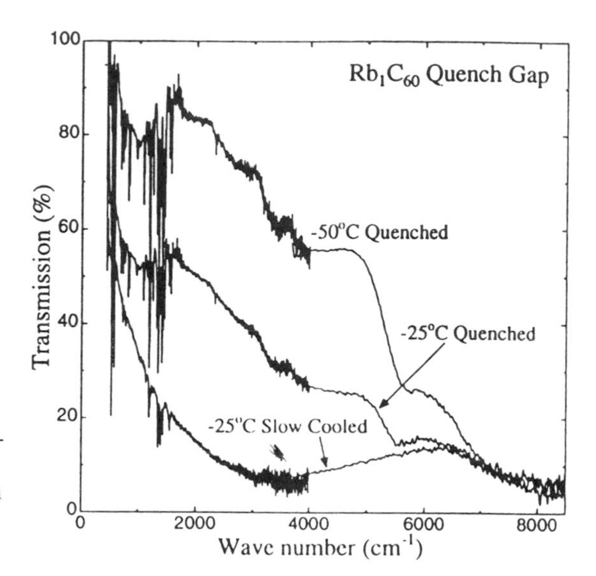

Fig. 13.33. Three representative IR transmission spectra recorded on a thin Rb_1C_{60} film in the slow-cooled and rapidly quenched phases [13.160].

in the orthorhombic phase remains independent of temperature down to 34 K, where the transmission again increases, indicative of another phase transition, as is also suggested by EPR measurements [13.162]. A fit to the IR transmission spectrum for the slow-cooled sample in Fig. 13.33 suggests a dc conductivity of ~10 mΩ-cm for that phase. When the rapidly quenched state is prepared by rapid cooling from 225°C to room temperature, the optical transmission spectrum is similar to that shown for the −25°C slow-cooled spectrum in Fig. 13.33. However, rapid quenching of the Rb_1C_{60} film from 225°C to lower temperatures (~–25°C) [13.161] produces a phase with higher optical transmission at low frequencies, corresponding to a more insulating phase associated with a lower carrier density (see Fig. 13.32). In this phase two adjacent C_{60}^- anions may become linked in a shortened intermolecular distance, which has been identified with dimer formation in structural studies [13.162]. However, since the dimers are formed in random directions, it is not believed that the rapidly quenched dimer phase is a precursor for the chain-like structures of the slow-cooled more conductive phase [13.160]. More detailed analysis of the transmission spectra for the K_1C_{60} samples that were rapidly quenched to −50°C indicate a semiconducting phase with an energy gap of ~5000 cm^{-1} (0.63 eV) in the electronic system [13.160, 161]. The cutoff in transmission seen above 8000 cm^{-1} for all the transmission spectra in Fig. 13.33 is attributed to the $t_{1u} \rightarrow t_{1g}$ interband transition, also seen in other alkali metal–doped fullerides at about 1 eV [13.143, 144].

13.5. Optical Properties of C_{60}–Polymer Composites

The first report of very high photoconductivity from a fullerene–polymer composite was made by Ying Wang at the duPont Research Laboratories [13.163]. This discovery may lead someday to applications for fullerenes in xerography, photovoltaic cells, and photorefractive devices as discussed in §20.1.2, §20.1.3, §20.2.3, and §20.2.4. Wang's work was followed by a series of papers from researchers at University of California (UC) Santa Barbara [13.164–170] who explored the underlying physical mechanism behind the enhanced photoconductivity, which led them to investigate the enhanced photoconductivity phenomenon in a variety of C_{60}-doped polymers. Concurrently, optical absorption and luminescence studies on C_{60}-doped polyhexylthiophene (PHT) were reported by Morita and co-workers [13.171], who showed that PHT can be doped with acceptor molecular dopants to produce a conducting polymer [13.172].

Wang's discovery was made in polyvinylcarbazole (PVK) films doped with a few ($\sim$2.7) weight % of fullerenes C_{60} and C_{70} in the ratio 85:15 [13.163]. The PVK and fullerenes were both dissolved in toluene, and the solution was then spin-coated onto an aluminum substrate to form a thin solid film after subsequent solvent evaporation. The resulting films, 1–28 μm thick, were found to be air-stable and optically clear. Photoconductivity in these films was studied by a photoinduced discharge method in which the surface of the film is corona-charged, either positively or negatively. This surface charge induces a voltage difference between the front surface of the film and the substrate. When the film is exposed to light, photogenerated electron–hole pairs are produced in the film. If the exciton (electron–hole) binding energy is sufficiently small, mobile electrons or holes, depending on the sign of the surface charge, drift to the surface of the film, thereby neutralizing the surface charge and reducing the voltage across the film. The rate at which the film voltage V decays to zero is a measure of the charge generation efficiency ϕ of the film

$$\phi = -[\epsilon/4\pi eLI](dV/dt) \tag{13.19}$$

where ϵ and L are, respectively, the dielectric constant and thickness of the film, e is the electronic charge, I is the incident photon flux, and the initial discharge rate of the surface potential dV/dt is calculated at the moment when the light is first incident on the film. A discharge process of this type is the basis for xerography [13.163].

In Fig. 13.34, we display the photoinduced discharge curves for a pure PVK film (upper curve) and a PVK film doped with a few weight percent of fullerenes (lower curve) [13.163]. Both samples were exposed to broadband light from a tungsten lamp beginning at time $t = 0$. From the data in

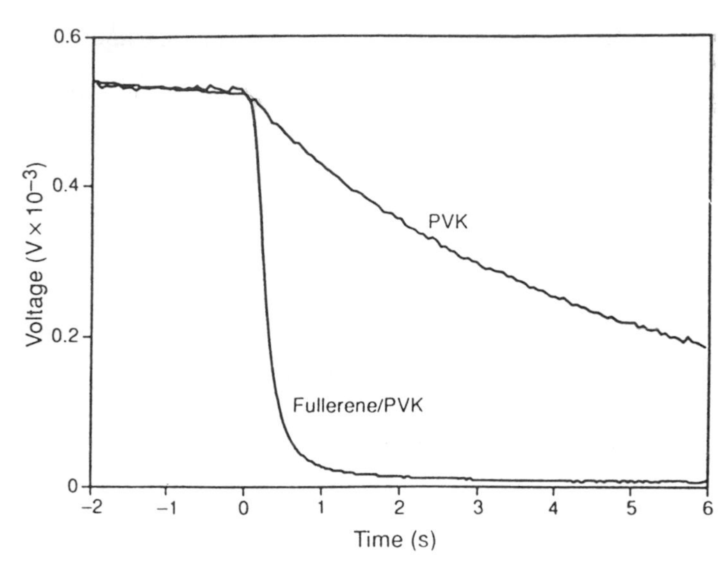

Fig. 13.34. A qualitative comparison of the photoinduced discharge curves for pure PVK and fullerene-doped PVK under the same experimental conditions. A tungsten lamp (50 mW/cm^2) is used as the light source [13.163].

Fig. 13.34 at times $t < 0$, the finite slope in dV/dt is due to the dark conductivity of the respective films, and it can be seen that the fullerene-doped PVK film has a low dark conductivity, comparable to that of the pristine PVK. However, under illumination ($t > 0$) the two films behave very differently. As shown in the figure, the discharge rate of the surface potential for the fullerene-doped PVK is much larger than that for the pristine PVK; i.e., the charge generation efficiency ϕ for PVK has been greatly enhanced by the addition of a few (e.g., 2.7%) weight percent of fullerenes.

To elucidate the mechanism for enhanced photoconductivity, Wang studied the wavelength (λ) dependence of the charge generation efficiency ϕ and observed that the long-wavelength threshold for ϕ was very near to the threshold for optical absorption in molecular C_{60}. This result suggests that the enhanced photoconductivity in the polymer composite involves the photoexcitation of C_{60}, and through this process an electron is exchanged between the fullerene and the nearby carbazole unit. Thus mobile carriers are generated in the PVK polymer chains by the incident light and these carriers can drift to the film surface, neutralizing the applied surface charge. In the UV, it was noted [13.163] that the charge generation efficiency ϕ was much larger for positive, than for negative, surface charge neutralization.

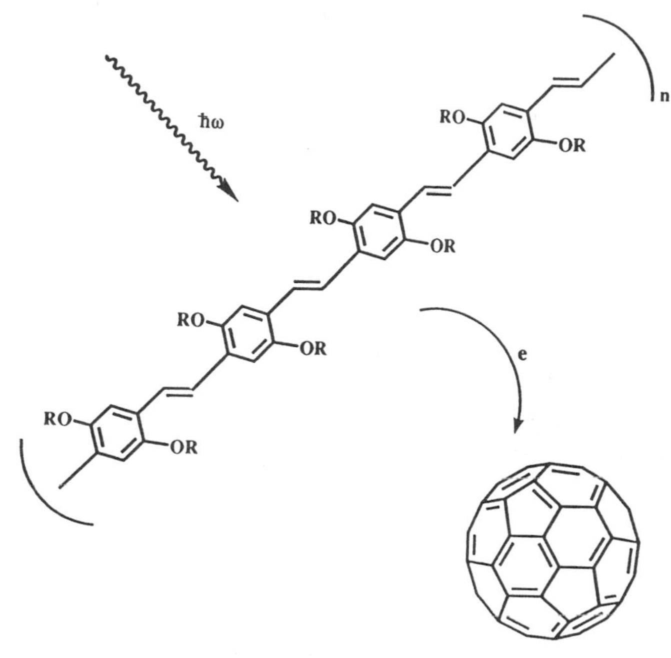

Fig. 13.35. Schematic illustration of the photoinduced electron transfer from semiconducting polymers to form C_{60} anions [13.168].

In a series of experiments on a variety of C_{60}–polymer composites, the UC Santa Barbara researchers obtained results indicating that a photoinduced electron transfer from a semiconducting polymer to a nearby C_{60} molecule occurs, forming metastable C_{60} anions (consistent with the high electron affinity of C_{60}) and mobile holes in the polymer [13.164, 166, 168, 169, 173]. A schematic illustration of this process is presented in Fig. 13.35, where it is shown that the photoexcitation occurs on the polymer (rather than on the C_{60} molecule), followed by electron transfer to the C_{60}. For example, using 2.4 eV excitation, the intense luminescence from the photoexcited polymer [e.g., poly(2-methoxy, 5-12′-ethyl-hexyloxyl)-*p*-phenylene vinylene, or MEH-PPV] was quenched by nearly three orders of magnitude (see Fig. 13.36) when C_{60} was introduced into MEH-PPV at 1:1 by weight [13.164]. This observation indicates a strong *excited state* interaction between MEH-PPV and C_{60}. It was also shown that at low light intensity the absorption spectrum of a 1:1 C_{60}:MEH-PPV sample appears to be a simple superposition of spectra from C_{60} and from MEH-PPV, indicating that in the ground state, the C_{60}–polymer composite is a simple, weakly interacting mixture [13.164]. A schematic band diagram useful for understanding

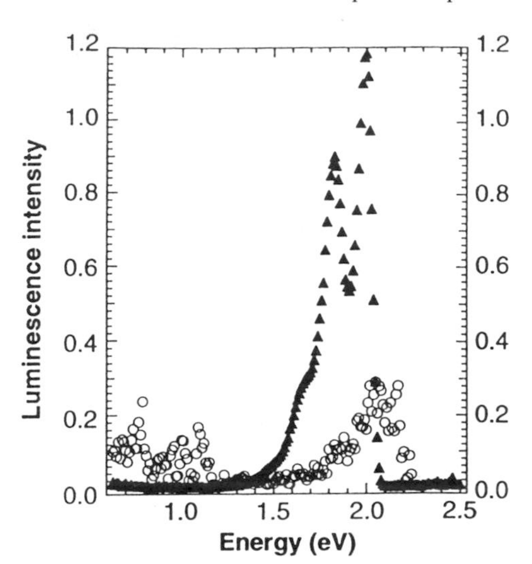

Fig. 13.36. Luminescence of the polymer MEH-PPV (solid triangles) (left axis, $\times 10^{-2}$) and of the C_{60}-(MEH-PPV) composite (open circles) (right axis, $\times 10^{-4}$). The unit for luminescence in this figure is the output voltage of the detector [13.164].

the mechanism for the photoinduced electron transfer reaction between a polymer semiconductor and C_{60} is shown in Fig. 13.37. As proposed [13.164, 168], the incident photon is absorbed across the semiconducting gap of the polymer. This excitation drives a rapid structural relaxation of the polymer ($< 10^{-13}$ s) due to electron–phonon coupling, thereby creating self-trapped polarons whose energy levels are in the semiconducting gap. The upper polaron level strongly couples to the LUMO of C_{60}, and electron transfer on a time scale of $\sim 10^{-12}$ s produces a metastable charge separation. The hole (positive polaron) then is free to drift away from the C_{60} anion that was produced, provided that the interaction between the hole and C_{60} anion is sufficiently screened. Thus this process represents a metastable photodoping process.

13.6. Optical Properties of Higher-Mass Fullerenes

Whereas most optical studies on fullerenes have dealt with C_{60}, some studies have been carried out on C_{70}, and a very few studies have been reported for higher-mass fullerenes (C_{n_C}; $n_C = 76, 78, 82, 84, 90, 96$) [13.174–177]. Many similarities are found between the observed spectra for all fullerenes because the spectra relate to the fundamental molecular electronic structure of fullerenes, which are all closed cage molecules of similar basic design. For each of these fullerene species, the solution and solid-state spectra are very similar, and all involve excitons near the absorption edge. However, with increasing n_C, the HOMO–LUMO gap decreases, as expected since in the limit $n_C \rightarrow \infty$, the 2D zero-gap graphene semiconductor is reached.

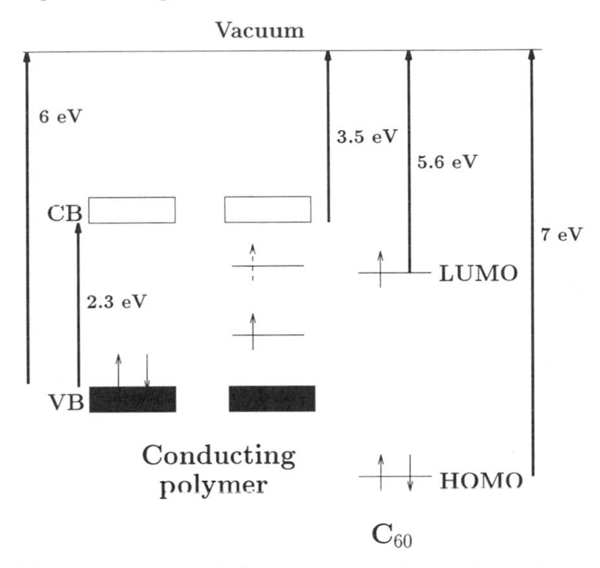

Fig. 13.37. Schematic energy level diagram for the photo-induced electron transfer from a semiconducting polymer on to C_{60} [13.168]. The left column shows the carrier occupation prior to photoexcitation and the right column shows the formation of polaron states in the band gap by photoexcitation. Transfer of the electron from the upper polaron state produces a C_{60} anion and a mobile hole in the polymer.

The color differences between the C_{60} and C_{70} solutions are consistent with a smaller gap for C_{70} than for C_{60}. The deeper, reddish-orange color observed for similarly concentrated C_{70} solutions stems from both a red-shifted HOMO–LUMO gap for C_{70} relative to C_{60} and stronger optical absorption from electronic transitions across this gap, associated with the lower symmetry of C_{70}. Molar extinction coefficients for C_{60} and C_{70} have also been reported at several wavelengths [13.178]. The availability of reliable molar extinction data is essential for *quantitative* high-performance liquid chromatography (HPLC) analysis of C_{60}/C_{70} mixtures.

13.6.1. Optical Properties of C_{70}

To date, most of the optical studies on pristine and doped C_{70} solids have been carried out in transmission on thin solid films deposited on various substrates such as quartz, Si, or KBr. Typically C_{70} films have been prepared by sublimation of microcrystalline C_{70} powder onto a substrate without careful attention given to the choice of substrate, nor to lattice matching issues.

The room temperature ($T = 300$ K) dielectric function of nanocrystalline solid C_{70} films (i.e., grain size on the order of 10–50 nm) was first reported

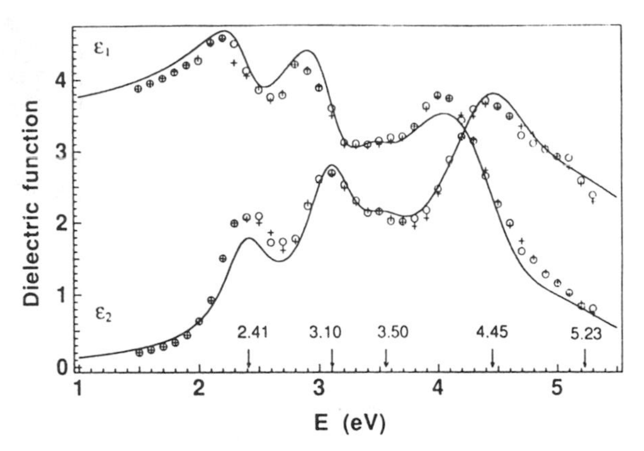

Fig. 13.38. Ellipsometrically derived values of $\epsilon_1(\omega)$ and $\epsilon_2(\omega)$ for C_{70} on Si (100) (+++) and C_{70} on Suprasil (∘ ∘ ∘) as a function of photon energy. The solid curves are calculated according to Eq. (13.14) for the parameters listed in Table 13.9 [13.179].

by Ren *et al.* [13.179] using a combination of variable angle spectroscopic ellipsometry (VASE) and near-normal incidence reflection/transmission ($\mathscr{R}/\mathscr{T}$) optical techniques. The data are shown in Fig. 13.38 for the real $\epsilon_1(\omega)$ and imaginary $\epsilon_2(\omega)$ parts of the dielectric function in the range 1.0–5.8 eV for films vacuum-deposited on either room temperature Si (100) (+++) or on Suprasil quartz (∘ ∘ ∘) substrates. The arrows on the horizontal axis indicate the position of Lorentz oscillators used to fit the dielectric function [see Eq. (13.14)]. Values for the oscillator parameters are given in Table 13.9 and $\epsilon_{\text{core}} = 1.67$.

Table 13.9

Lorentz oscillator parameters for $\epsilon(\omega)$ for C_{70}[a] [13.179].

j	ω_{Pj} (eV)	ω_{0j} (eV)	Γ_j (eV)
1	1.18	2.41	0.47
2	1.72	3.10	0.50
3	1.10	3.50	0.55
4	3.40	4.45	0.99
5	3.10	5.23	1.70
6	3.60	5.86	2.30

[a] The parameters ω_{Pj}, ω_{0j}, and Γ_j (eV) denote the Lorentz oscillator strength, frequency, and linewidth [see Eq. (13.14)].

From these data a static refractive index $n(0) = 1.94$ [13.179] or a static dielectric constant $\epsilon_1(0) = n^2(0) = 3.76$ was determined, somewhat lower than the corresponding values for solid C_{60}.

It can be seen in Fig. 13.39 that the optical density $\log_{10}(1/\mathscr{T})$ of C_{70}:Suprasil (d = 280 Å) predicted from VASE measurements (+++) compares favorably to that obtained directly via normal incidence transmission ($\mathscr{T}$) data (solid curve). The dashed curve in Fig. 13.38 shows the optical density for molecular C_{70} in decalin (a solution spectrum), which was obtained from transmission measurements [13.180]. Overall, a striking similarity is found between the absorption spectrum in the C_{70} solid film and C_{70} in a decalin solution [13.180], consistent with the view that solid C_{70} is a molecular solid. Presumably, the intermolecular interaction accounts for the broadening in the film spectrum of the narrower features observed in solution. The inset to Fig. 13.39 shows the absorbance spectrum $\mathscr{A}(\omega) = 1 - [\mathscr{R}(\omega) + \mathscr{T}(\omega)]$ (solid curve) for solid C_{70}:Suprasil ($d = 280$ Å) near the threshold for optical absorption E_0 in C_{70}. The rise of $\mathscr{A}(\omega)$ above the baseline indicates that $E_0 = 1.25 \pm 0.05$ eV for solid C_{70}. The calculated HOMO–LUMO gap between molecular orbitals for an isolated C_{70} molecule is higher than the observed absorption edge threshold: 1.76 eV [13.181], 1.68 eV [13.182], and 1.65 eV [13.183]. Intermolecular interactions in the solid state, and the on-ball, electron–hole (exciton) interaction would both be expected to lower the optical absorption threshold

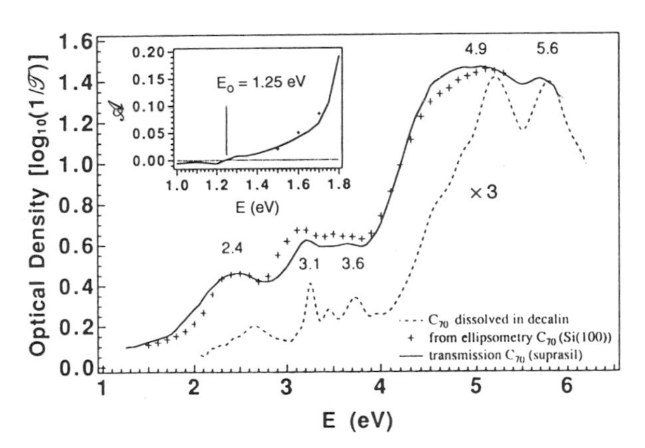

Fig. 13.39. Optical density of C_{70} dissolved in decalin (dashed curve) [13.180], the optical density of a C_{70} film (280 Å thick) deduced from ellipsometry (+++) and that measured directly (solid line) for the same film by normal incidence transmission spectroscopy [13.179]. The inset shows the absorbance [$\mathscr{A}(\omega)$] on a magnified energy scale near the absorption edge for a thin solid C_{70} film on quartz. The onset for absorption E_0 is seen to be near 1.25 ± 0.05 eV for solid C_{70} [13.179].

below the calculated HOMO–LUMO gap. Localization of the electron and hole on a single C_{70} molecule leads to a relatively strong exciton binding energy, just as for C_{60} films (see §13.3.2). Molecular orbital calculations for C_{70} [13.12] reveal a large number of closely spaced orbitals both above and below the calculated HOMO–LUMO gap [13.12]. The large number of orbitals for higher-mass fullerenes with lower symmetry makes it difficult to assign particular groups of transitions to the structure that is observed in the solution spectra of C_{70}.

13.7. Dynamic and Nonlinear Optical Properties of Fullerenes

In this section we discuss dynamic optical properties of C_{60} and related materials in the solid state (§13.7.1). The corresponding discussion for fullerenes in solution is presented in §13.2.3. The nonlinear optical properties of fullerenes in both solution and crystalline phases are summarized in §13.7.2.

13.7.1. Dynamical Properties

A variety of behavior has been reported for the dynamics of the optical response of C_{60} in the solid state. The two most important techniques that have been used for these studies are pump–probe transmission/reflection and photoinduced optical absorption. These two techniques have also been important for studies of the dynamic properties of C_{60} in solution (see §13.2.3). In pump–probe experiments, a system is excited by a short light pulse (the pump pulse) and is then interrogated at a later time (the delay time) by the probe pulse. The probe pulse is normally much weaker in intensity and can be at the same or at a different frequency from the pump pulse. In photoinduced absorption spectroscopy, the absorption spectrum of the photoexcited system is probed. Photoinduced optical absorption is of particular interest in fullerenes because the absorption from the ground state is forbidden, while certain excited states are both long-lived and participate in subsequent dipole-allowed transitions. Reasons for the differences in dynamic behavior reported by the various groups are presumably related to the oxygen (and other impurity) content of the fullerene samples and the magnitude of the light intensity of the excitation pulses as discussed below.

Room temperature pump–probe experiments on C_{60} films near 2 eV have been carried out by several groups [13.184–186]. Whereas pump–probe experiments normally show an exponential decay of the photoexcited state which is expressed in terms of a single decay time for subpicosecond pulses, the pump–probe results for C_{60} films are unusual and show a nonexponential decay of the pump excitation. Analysis of the decay profile for the excited states of C_{60} shows contributions ranging from fast decay times ($\tau \sim 1$ ps) to decay times longer by several orders of magnitude, indicative of a number of relaxation processes for the excited state, including carrier trapping, carrier tunneling, carrier hopping, singlet–triplet decay, polaron formation, and triplet–triplet annihilation [13.184] (see §13.2.3). Decay times faster than 10^{-9} s have been identified with the decay of the singlet exciton state, while slow components ($\tau > 10^{-6}$ s) have been identified with decay of triplet states, consistent with results for C_{60} in solution (see §13.2.3). Studies of decay times as a function of laser fluence at constant wavelength (such as 597 nm) show a strong decrease in decay times with increasing laser fluence [13.46, 187], which may be connected with a phototransformation of the sample. Such a phototransformation lowers the symmetry, thereby enhancing the cross section for dipole-allowed radiative transitions. It is of interest to note that the pump pulse causes a decrease in optical transmission for both C_{60} and alkali metal–doped C_{60}, consistent with the dominance of laser-induced excited state absorption processes in pump–probe experiments for these materials [13.46].

In contrast to observations in solid C_{60}, the decay in solid C_{70} is significantly faster at the same laser excitation frequency ω' and intensity, where ω' for both materials is measured relative to their respective absorption thresholds. This effect was already reported for the dynamical properties of fullerenes in solution and is related both to the lower symmetry of C_{70} which provides more dipole-allowed decay channels, and to higher impurity and defect densities in C_{70} samples. Whereas the decay of C_{70} is only weakly dependent on laser fluence for wavelengths well below the absorption edge, the excited state decay in C_{70} is strongly dependent on laser intensity above the absorption edge [13.187].

However, low-temperature (5–10 K) time-resolved photoluminescence and pump–probe measurements on C_{60} in the solid state show a simple exponential decay [13.50] with a relaxation time of 1.2 ns for the excited state, corresponding to the lifetime of the singlet S_1 state. The low-temperature luminescence varies as the cube of the laser excitation intensity [13.50, 186] and exhibits an Arrhenius behavior, yielding an activation energy of 8–13 meV. This activation energy is similar in magnitude to the energy difference in the orientational potential between adjacent C_{60} molecules aligned so that an electron-rich double bond is opposite a hexagon rather than opposite a pentagon (see §7.1.3). The nonexponential decay mentioned above is observed at higher temperatures and at higher laser excitation intensities, where the emission spectrum is strongly shifted to lower photon energies. C_{60} films show different time-resolved luminescence behavior than single crystals [13.50].

Time-resolved absorption studies have also been carried out for cases where the pump and probe are at different photon energies. In these studies the fullerenes are promoted to an excited state and optical transitions to higher-lying states are made before the fullerenes relax to their ground states. Relaxation time measurements of the induced photoabsorption also reveal a fast component of ~2 ps and a slow component > 2 ns [13.188]. The relaxation time for the fast component was found to decrease and the intensity of the induced photoabsorption to increase with increasing laser excitation intensity as shown in Fig. 13.40. Here the induced photoabsorption is found to saturate at a laser excitation energy corresponding to the excitation of one electron–hole pair per C_{60} molecule [13.89]. Little temperature dependence was found, indicating that the induced photoabsorption is associated with the free molecule. It was suggested that the induced photoabsorption arises from a molecular distortion of the excited state (see §16.2.3) which lowers the symmetry and therefore increases the transition probability for optical absorption [13.89].

We now discuss photoinduced absorption studies originating from optically pumped excited states. Figure 13.41 shows a broad photoinduced absorption band from 0.8 to 2.2 eV with a maximum at ~1.8 eV in a C_{60} film [13.64]. Since the induced photoabsorption showed no initial photobleaching and no spectral diffusion, it was concluded that the induced photoabsorption response is dominated by photoinduced absorption originating from previously pumped excited states. Since the photoluminescence and induced photoabsorption exhibit the same dependence on the density of excitations, it appears that the induced photoabsorption occurs on a time scale of 10^{-12} s and is associated with Frenkel singlet excitons. The onset of

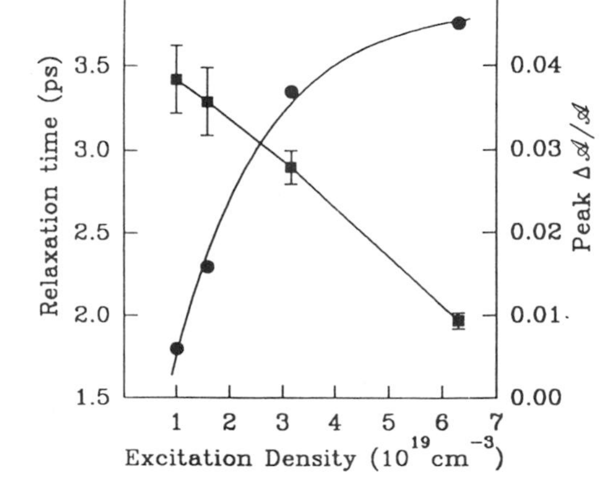

Fig. 13.40. Relaxation time (solid squares) and peak induced absorption ($\Delta\mathscr{A}/\mathscr{A}$) (solid circles) plotted as a function of the density of excited electron–hole pairs in C_{60} films [13.89].

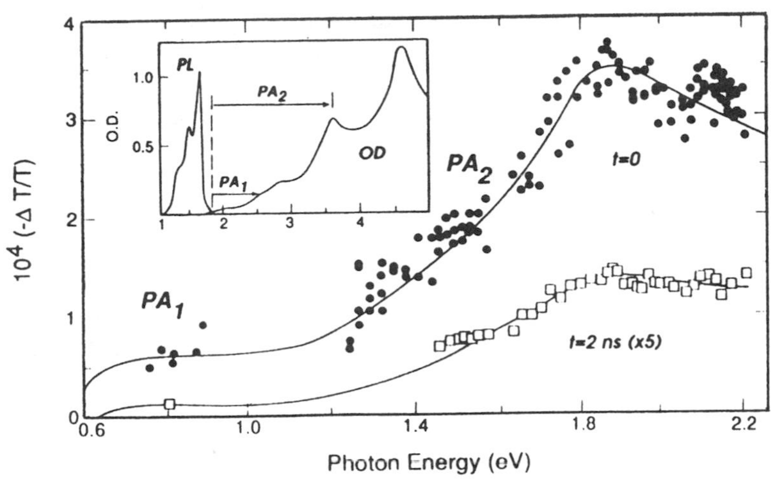

Fig. 13.41. Picosecond photoinduced absorption spectra of C_{60} film at $t = 0$ (5 ps resolution) and $t = 2$ ns (note change of scale by a factor of 5); the lines through the data points are to guide the eye. The inset shows the optical density (OD) spectrum and the photoluminescence (PL) emission band of the film. Horizontal arrows in the inset indicate energy separations between initial (PA_1) and final (PA_2) states in the photoinduced absorption spectrum [13.64].

the induced photoabsorption band at 0.8 eV (see Fig. 13.41) was attributed to a transition from the singlet S_1 exciton state to the LUMO+1 derived band, which is consistent with a $t_{1g} - t_{1u}$ separation of ~1 eV that has been observed in many experiments (see §13.3.2). The broad maximum in the induced photoabsorption band at 1.8 eV corresponds to the next allowed optical transition to the LUMO+2 derived band, and this transition is much stronger than the transition to the LUMO+1 level [see Fig. 13.18(a)].

By identifying the long-lived excitations in the μs range with triplet excitons and noting their strong temperature dependence, the two induced photoabsorption features with peaks at 1.1 and 1.8 eV are identified with excitations from the lowest triplet state to higher-lying neutral triplet states. Regarding the short-lived excitations, the two features at 0.8 and 2.0 eV are identified with charge polaron excitations because of their strong dependence on rf modulation frequency in ODMR experiments using both absorption [13.64] and emission [13.63] optical processes. Independent evidence for polaron formation is found in photoinduced EPR studies (see §16.2.3).

A few pump–probe studies of the electron dynamics of alkali metal–doped C_{60} have also been carried out [13.46, 189]. The earliest experiments were reported for an uncharacterized Rb_xC_{60} film and yielded a subpicosecond relaxation time (0.6 ps) which is independent of temperature ($10 < T < 300$ K) and of laser pump intensity ($<$ 40 mJ/cm^2, for 80 fs pulses at 620 nm or 2.0 eV) [13.46]. Subsequent pump–probe studies on well-characterized K_3C_{60} and Rb_3C_{60} films revealed the same relaxation time τ for both materials, but τ was found to be an order of magnitude faster in the alkali metal–doped compounds ($\sim$0.2 ps) relative to C_{60} for 50 fs pulses at 625 nm (1.98 eV), using very low power levels ($\sim$20 μJ/cm^2) [13.189]. These fast relaxation times are consistent with the suppression of exciton behavior through metallic screening in the M_3C_{60} compounds. By populating the excited states which are dipole-coupled to lower lying states, a more rapid return of the excited system to lower energy configurations can be realized.

13.7.2. Nonlinear Absorption Effects

Because of the high delocalization of electrons on the shell of fullerenes and the relatively high isolation between molecules, fullerene-based materials are expected to exhibit large nonlinear optical effects such as third-order nonlinear optical susceptibility, comparable to those for π-conjugated organic polymers [13.190]. Since the nonlinear polarizability for polymers depends on the fourth power of the length of the polymer chain, C_{70} would be expected to exhibit an even larger nonlinear response than C_{60} [13.191], in agreement with experiment. These nonlinear effects have been studied by degenerate four-wave mixing (DFWM) [13.191–195], third harmonic generation [13.196–199], and electric field–induced second harmonic generation [13.200], both in solution and in high-quality films. Since third harmonic generation and electric field–induced second harmonic generation are usually carried out above the absorption threshold, these techniques tend to be more surface sensitive because of the shorter optical skin depth. Therefore, more emphasis has been given in the literature to nonlinear studies using the DFWM technique.

In the DFWM experiment (see Fig. 20.4 for the geometry of the four light waves in the DFWM experiment), a polarization $\mathbf{P}_i$ is induced in the sample by the interaction between three light waves and the electrons in the medium through the relation

$$\mathbf{P}_i = \overleftrightarrow{\overleftrightarrow{\overleftrightarrow{\chi}}}\, \mathbf{E}_j \mathbf{E}_k \mathbf{E}_l^* \tag{13.20}$$

where two light waves with electric field amplitudes $\mathbf{E}_j$ and $\mathbf{E}_k$ are incident at the same region on the sample and have equal and opposite wave vectors (see Fig. 20.4). The waves $\mathbf{E}_j$ and $\mathbf{E}_k$ from, for example, a He–Ne laser set up an interference grating which interacts with a third light beam $\mathbf{E}_l^*$ coming from the back side of a transparent sample as shown in Fig. 20.4. A fourth beam (also shown in Fig. 20.4), resulting from this interaction with wave vector equal and opposite to that of $\mathbf{E}_l^*$, is detected, and the polarization vector $\mathbf{P}_i$ in Eq. (13.20) corresponds to the fourth signal light beam. The tensor, which relates $\mathbf{P}_i$ and the three interacting light waves $\mathbf{E}_j$, $\mathbf{E}_k$, and $\mathbf{E}_l^*$, is the third-order nonlinear susceptibility $\chi_{ijkl}^{(3)}$ which is a fourth-rank tensor and the subscripts *ijkl* refer to the components of the vectors $\mathbf{P}_i$, $\mathbf{E}_j$, $\mathbf{E}_k$, and $\mathbf{E}_l^*$, each having independent x, y, z components. It can be shown that DFWM experiments measure the *magnitude* of the complex third-order nonlinear susceptibility $\chi_{ijkl}^{(3)}$, so that additional measurements are needed to determine the real and imaginary parts of $\chi_{ijkl}^{(3)}$, as discussed below.

In most of the DFWM experiments that have been done thus far on fullerenes (both in solution and in films), the polarizations of all four beams have been the same, so that most of the measurements relate to the tensor component $\chi_{xxxx}^{(3)}$. A few measurements of $\chi_{xyyx}^{(3)}$ have also been made, when the pump and probe beams were cross polarized [13.195]. The cubic dependence of the DFWM signal on incident laser intensity verifies the third-order nonlinearity probed in the DFWM experiment as shown in Fig. 13.42. However, a wide range of magnitudes have been reported in the literature for $\chi_{xxxx}^{(3)}$, perhaps due to the sensitivity of this nonlinear coefficient to the presence of oxygen and the magnitude of the light intensity.

A frequently used wavelength for DFWM experiments in C_{60} and C_{70} has been 1.064 μm, where the optical absorption is negligibly small, and where resonant enhancement associated with a two-photon process may occur [13.191–194, 201]. A wide range of pulse widths (from the ns to fs range) have been used for nonlinear optics measurements, and the results obtained may be sensitive to the pulse widths because the contributions from the excited molecules become increasingly important as the pulse width increases.

Values of $\chi_{xxxx}^{(3)}$ are sensitive to laser excitation intensity and to pulse width. One estimate for the value of $\chi_{xxxx}^{(3)}$ for C_{60} at 1.064 μm is ~60 times greater than for benzene for which $\chi_{xxxx}^{(3)} = 8.3 \times 10^{-14}$ esu when each is normalized to the number of molecules [13.191, 195]. When normalized to the number of carbon atoms per molecule, the magnitude for $\chi_{xxxx}^{(3)}$ per carbon atom is within an order of magnitude equal to that for benzene. Evidence for a fifth-order nonlinear response has also been reported at higher laser excitation intensities for both C_{60} [13.187] and C_{70} (see Fig. 13.42) [13.195]. The two-photon absorption, which is the dominant absorption process at 1.064 μm, is believed to be responsible for forming a two-photon excited-

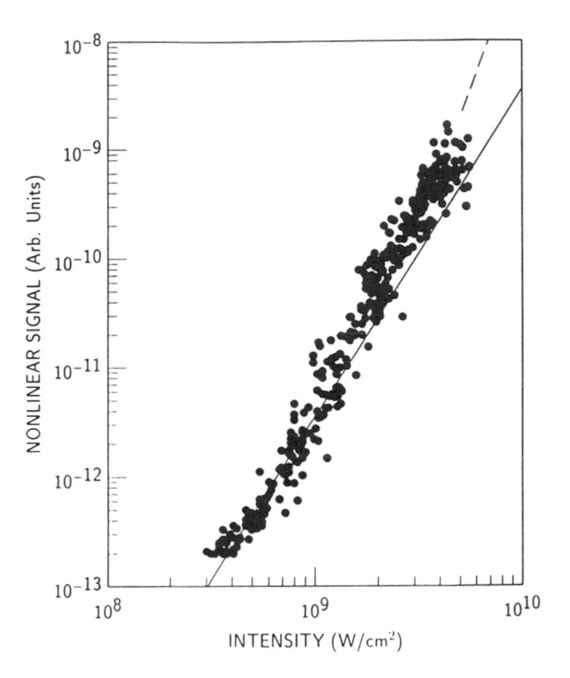

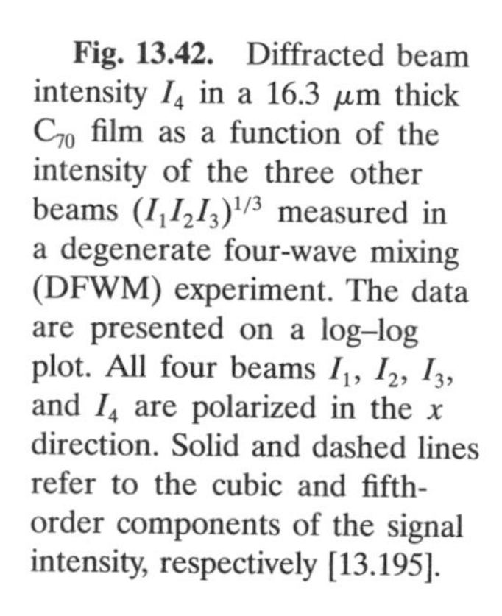
Fig. 13.42. Diffracted beam intensity I_4 in a 16.3 μm thick C_{70} film as a function of the intensity of the three other beams $(I_1 I_2 I_3)^{1/3}$ measured in a degenerate four-wave mixing (DFWM) experiment. The data are presented on a log–log plot. All four beams I_1, I_2, I_3, and I_4 are polarized in the x direction. Solid and dashed lines refer to the cubic and fifth-order components of the signal intensity, respectively [13.195].

state grating (see Fig. 20.4), thereby giving rise to a fifth-order contribution to the nonlinear response [13.195].

Increasing the photon energy into a regime of greater linear optical absorption leads to an enhanced nonlinear coefficient, as, for example, from $\chi^{(3)}_{xxxx} \sim 7 \times 10^{-12}$ esu at 1064 nm (1.16 ev) to $\sim 82 \times 10^{-12}$ esu at 675 nm (1.83 eV) to $\chi^{(3)}_{xxxx} \sim 380 \times 10^{-12}$ esu at 597 nm (2.08 eV) [13.187]. Although the reported magnitudes of $\chi^{(3)}_{xxxx}$ differ considerably from one group to another, there is general agreement regarding the dependence of $\chi^{(3)}_{xxxx}$ on laser excitation frequency above the absorption edge [13.185, 202]. The variation in the magnitudes for $\chi^{(3)}_{xxxx}$ in the literature is large for both C_{60} and C_{70}, with the variation of $\chi^{(3)}_{xxxx}$ for similar nominal samples so large that it is difficult to distinguish between values of $\chi^{(3)}_{xxxx}$ for fullerenes in solution or in films. Laser-induced phonon generation in DFWM experiments on C_{60} and C_{70} dissolved in toluene has also been reported for laser excitation at 532 nm using 20 ps pulses [13.202].

The linear and nonlinear optical parameters of C_{60} and C_{70} films at 1.064 μm have been compared. As shown in Table 13.10, an increase in linear absorption coefficient α is found with increasing fullerene mass, and this can be explained by the more stringent selection rules for dipole forbidden transitions at the absorption edge for the highest-

Table 13.10

Selected values of nonlinear optical coefficients for C_{60} and C_{70}[a] [13.195, 203].

Coefficient	C_{60}	C_{70}
α (cm^{-1})	6.0	20.6
β (cm/GW)	30–40	75
$\chi^{(3)}_{xxxx}$ (esu)	7×10^{-12}	1.2×10^{-12}
Im$\chi^{(3)}_{xxxx}$ (esu)	8.6×10^{-12}	1.6×10^{-12}
$\chi^{(3)}_{xyyx}$ (esu)	1×10^{-12}	1×10^{-12}

[a]Here α and β are, respectively, the linear and nonlinear absorption coefficients and $\chi^{(3)}$ is the tensor which relates to the third-order nonlinear susceptibility. All data pertain to the laser wavelength of 1.064 μm.

symmetry C_{60} molecules. An increase is also found in the two-photon absorption coefficient β and in the nonlinear susceptibility $\chi^{(3)}_{xxxx}$ for C_{70} relative to C_{60} as shown in Table 13.10 [13.195, 203], although similar values for $\chi^{(3)}_{xyyx}$ have been reported based on DFWM experiments [13.195].

From measurements of the two-photon absorption coefficient β and use of the relation [13.204]

$$\mathrm{Im}(\chi^{(3)}_{xxxx}) = \frac{9 \times 10^8 n^2 c^2 \varepsilon_0}{4\pi\omega}\beta \qquad (13.21)$$

where ε_0 is the permittivity of free space, ω is the frequency the incident of light, n is the index of refraction, and c is the speed of light, estimates for Im$(\chi^{(3)}_{xxxx})$ can be made, yielding the values given in Table 13.10 [13.195, 203]. The results show that the imaginary part of $\chi^{(3)}_{xxxx}$ exceeds the corresponding real part. In view of the uncertainties about values for $\chi^{(3)}_{xxxx}$, this conclusion requires further confirmation.

The origin of the optical nonlinearity is probably connected with the high efficiency (~100%) in transferring electrons from the excited singlet state manifold S_1 to the T_1 triplet excited states, and also with the larger optical matrix elements for triplet–triplet transitions (which are important for high light power levels) relative to the singlet–singlet transitions, which dominate the spectrum at low light power levels and very short times. The dependence of $\chi^{(3)}_{xxxx}$ on the laser excitation intensity shows that the nonlinear optical coefficients depend strongly on the populations of the singlet and triplet excited states [13.187].

Measurements of the third-order nonlinear optical response have also been made using third harmonic generation techniques [13.196, 198, 205], yielding magnitudes of $\chi^{(3)}$ [13.185, 191, 195] that also vary considerably from group to group, and in comparison with $\chi^{(3)}$ values obtained from DFWM experiments. Some studies of the third-order nonlinear harmonic generation also report a small second-order harmonic generation signal, which should be forbidden in centro-symmetric systems such as C_{60}. Large laser excitation intensities are typically used to obtain large enough non-linear signals, and these high laser intensities may lead to phototransformation effects, which lower the symmetry and perhaps turn on the second-harmonic generation signal. By carrying out measurements of the optical density as a function of incident laser intensity for 300 fs pulses at 612 nm on C_{60} films, a value of 2×10^{11} W/cm^2 was identified as a damage threshold for nonlinear optics studies on C_{60} films, although no characterization of the damage was carried out [13.203]. When a pulse repetition rate of 1 kHz was used for the pulses, the damage threshold power was reduced by one order of magnitude [13.203, 206]. It is generally believed that nonlinear optics experiments are more efficiently and reproducibly carried out using short pulses.

References

[13.1] A. M. Rao, P. Zhou, K.-A. Wang, G. T. Hager, J. M. Holden, Y. Wang, W. T. Lee, X.-X. Bi, P. C. Eklund, D. S. Cornett, M. A. Duncan, and I. J. Amster. *Science*, **259**, 955 (1993).

[13.2] P. Zhou, Z. H. Dong, A. M. Rao, and P. C. Eklund. *Chem. Phys. Lett.*, **211**, 337 (1993).

[13.3] A. M. Rao, M. Menon, K. A. Wang, P. C. Eklund, K. R. Subbaswamy, D. S. Cornett, M. A. Duncan, and I. J. Amster. *Chem. Phys. Lett.*, **224**, 106 (1994).

[13.4] C. Yeretzian, K. Hansen, F. N. Diederich, and R. L. Whetten. *Nature (London)*, **359**, 44 (1992).

[13.5] P. C. Eklund, A. M. Rao, P. Zhou, Y. Wang, and J. M. Holden. *Thin Solid Films*, **257**, 185 (1995).

[13.6] J. B. Birks. *Photophysics of Aromatic Molecules*. John Wiley & Sons, London (1970). A general review of the molecular spectroscopy of aromatic molecules.

[13.7] E. Westin and A. Rosén. *Zeitschrift für Physik D: Atoms, Molecules and Clusters*, **26**, S276 (1993).

[13.8] E. Westin and A. Rosén. *Zeitschrift für Physik D: Atoms, Molecules and Clusters*, **26**, S273 (1993).

[13.9] F. Negri, G. Orlandi, and F. Zerbetto. *Chem. Phys. Lett.*, **196**, 303 (1992).

[13.10] R. C. Haddon. *Acc. Chem. Res.*, **25**, 127 (1992).

[13.11] E. U. Condon and G. H. Shortley. *Theory of Atomic Spectra*. Cambridge University Press, Cambridge (1935).

[13.12] S. Saito and A. Oshiyama. *Phys. Rev. Lett.*, **66**, 2637 (1991).

[13.13] M. R. Wasielewski, M. P. O'Neil, K. R. Lykke, M. J. Pellin, and D. M. Gruen. *J. Am. Chem. Soc.*, **113**, 2772 (1991).

[13.14] K. Yabana and G. Bertsch. *Phys. Rev. B*, **46**, 14263 (1992).

[13.15] Y. Wang, J. M. Holden, A. M. Rao, P. C. Eklund, U. Venkateswaran, D. Eastwood, R. L. Lidberg, G. Dresselhaus, and M. S. Dresselhaus. *Phys. Rev. B*, **51**, 4547 (1995).

[13.16] Z. Gasyna, P. N. Schatz, J. P. Hare, T. J. Dennis, H. W. Kroto, R. Taylor, and D. R. M. Walton. *Chem. Phys. Lett.*, **183**, 283 (1991).

[13.17] S. Leach, M. Vervloet, A. Desprès, E. Bréheret, J. P. Hare, T. J. Dennis, H. W. Kroto, R. Taylor, and D. R. M. Walton. *Chem. Phys.*, **160**, 451 (1992).

[13.18] R. E. Haufler, J. J. Conceicao, L. P. F. Chibante, Y. Chai, N. E. Byrne, S. Flanagan, M. M. Haley, S. C. O'Brien, C. Pan, Z. Xiao, W. E. Billups, M. A. Ciufolini, R. H. Hauge, J. L. Margrave, L. J. Wilson, R. F. Curl, and R. E. Smalley. *J. Phys. Chem.*, **94**, 8634 (1990).

[13.19] A. Oshiyama, S. Saito, N. Hamada, and Y. Miyamoto. *J. Phys. Chem. Solids*, **53**, 1457 (1992).

[13.20] J. J. P. Stewart. Fujitsu Limited, Tokyo, Japan (1993). Semiempirical quantum chemistry library.

[13.21] R. Abouaf, J. Pommier, and S. Cvejanovic. *Chem. Phys. Lett.*, **213**, 503 (1993).

[13.22] G. F. Bertsch, A. Bulgac, D. Tománek, and Y. Wang. *Phys. Rev. Lett.*, **67**, 2690 (1991).

[13.23] W. E. Pickett. In H. Ehrenreich and F. Spaepen (eds.), *Solid State Physics*, vol. 48, p. 225, Academic Press, New York (1994).

[13.24] P. C. Eklund, A. M. Rao, Y. Wang, P. Zhou, K. A. Wang, J. M. Holden, M. S. Dresselhaus, and G. Dresselhaus. *Thin Solid Films*, **257**, 211 (1995).

[13.25] F. Wooten. *Optical Properties of Solids*. Academic Press, New York (1972).

[13.26] C. Kittel. *Introduction to Solid State Physics*. John Wiley & Sons, New York, 6th edn. (1986).

[13.27] M. K. Nissen, S. M. Wilson, and M. L. W. Thewalt. *Phys. Rev. Lett.*, **69**, 2423 (1992).

[13.28] W. Guss, J. Feldmann, E. O. Gobel, C. Taliani, H. Mohn, W. Muller, P. Haubler, and H. U. ter Meer. *Phys. Rev. Lett.*, **72**, 2644 (1994).

[13.29] L.-S. Wang. *Chem. Phys. Lett.*, **182**, 5 (1992).

[13.30] J. W. Arbogast, A. P. Darmanyan, C. S. Foote, Y. Rubin, F. N. Diederich, M. M. Alvarez, S. J. Anz, and R. L. Whetten. *J. Phys. Chem.*, **95**, 11 (1991).

[13.31] L. Biczok, H. Linschitz, and R. I. Walter. *Chem. Phys. Lett.*, **195**, 339 (1992).

[13.32] Y. Wang. *J. Phys. Chem.*, **96**, 764 (1992).

[13.33] Z. H. Dong, P. Zhou, J. M. Holden, P. C. Eklund, M. S. Dresselhaus, and G. Dresselhaus. *Phys. Rev. B*, **48**, 2862 (1993).

[13.34] K. A. Wang, A. M. Rao, P. C. Eklund, M. S. Dresselhaus, and G. Dresselhaus. *Phys. Rev. B*, **48**, 11375 (1993).

[13.35] F. Negri, G. Orlandi, and F. Zerbetto. *J. Chem. Phys.*, **97**, 6496 (1992).

[13.36] A. Andreoni, M. Bondani, and G. Consolati. *Phys. Rev. Lett.*, **72**, 844 (1994).

[13.37] T. W. Ebbesen, K. Tanigaki, and S. Kuroshima. *Chem. Phys. Lett.*, **181**, 501 (1991).

[13.38] S. P. Sibley, S. M. Argentine, and A. H. Francis. *Chem. Phys. Lett.*, **188**, 187 (1992).

[13.39] H. Ajie, M. M. Alvarez, S. J. Anz, R. D. Beck, F. Diederich, K. Fostiropoulos, D. R. Huffman, W. Krätschmer, Y. Rubin, K. E. Schriver, D. Sensharma, and R. L. Whetten. *J. Phys. Chem.*, **94**, 8630 (1990).

[13.40] R. E. Haufler, L.-S. Wang, L. P. F. Chibante, C.-M. Jin, J. J. Conceicao, Y. Chai, and R. E. Smalley. *Chem. Phys. Lett.*, **179**, 449 (1991).

[13.41] M. Lee, O. K. Song, J. C. Seo, D. Kim, Y. D. Su, S. M. Jin, and S. K. Kim. *Chem. Phys. Lett.*, **196**, 325 (1992).

[13.42] Y. Kajii, T. Nakagawa, S. Suzuki, Y. Achiba, K. Obi, and K. Shibuya. *Chem. Phys. Lett.*, **181**, 100 (1991).

[13.43] R. R. Hung and J. J. Grabowski. *J. Phys. Chem.*, **95**, 6073 (1991).

[13.44] K. Tanigaki, T. W. Ebbesen, and S. Kuroshima. *Chem. Phys. Lett.*, **185**, 189 (1991).

[13.45] L. Akselrod, H. J. Byrne, C. Thomsen, A. Mittelbach, and S. Roth. *Chem. Phys. Lett.*, **212**, 384 (1993).

[13.46] S. D. Brorson, M. K. Kelly, U. Wenschuh, R. Buhleier, and J. Kuhl. *Phys. Rev. B*, **46**, 7329 (1992).
[13.47] I. E. Kardash, V. S. Letokhov, Yu. E. Lozovik, Yu. A. Matveets, A. G. Stepanov, and V. M. Farztdinov. *JETP Lett.*, **58**, 138 (1993).
[13.48] D. H. Kim, M. Y. Lee, Y. D. Suh, and S. K. Kim. *J. Am. Chem. Soc.*, **114**, 4429 (1992).
[13.49] D. K. Palit. *Chem. Phys. Lett.*, **195**, 1 (1992).
[13.50] H. J. Byrne, W. Maser, W. W. Rühle, A. Mittelbach, W. Hönle, H. G. von Schmering, B. Movaghar, and S. Roth. *Chem. Phys. Lett.*, **204**, 461 (1993).
[13.51] R. J. Sension, C. M. Phillips, A. Z. Szarka, W. J. Romanow, A. R. McGhie, J. P. McCauley, Jr., A. B. Smith, III, and R. M. Hochstrasser. *J. Phys. Chem.*, **95**, 6075 (1991).
[13.52] T. W. Ebbesen, Y. Mochizuki, K. Tanigaki, and H. Hiura. *Europhys. Lett.*, **25**, 503 (1994).
[13.53] B. L. Justus, Z. H. Kafafi, and A. L. Huston. *Optics Lett.*, **18**, 1603 (1993).
[13.54] M. P. Joshi, S. R. Mishra, H. S. Rawat, S. C. Mehendale, and K. C. Rustagi. *Appl. Phys. Lett.*, **62**, 1763 (1993).
[13.55] A. Kost, L. Tutt, and M. B. Klein. *Optics Letters*, **18**, 334 (1993).
[13.56] F. Henari, J. Callaghan, H. Stiel, W. Blau, and D. J. Cardin. *Chem. Phys. Lett.*, **199**, 144 (1992).
[13.57] J. E. Wray, K. C. Liu, C. H. Chen, W. R. Garrett, M. G. Payne, R. Goedert, and D. Templeton. *Appl. Phys. Lett.*, **64**, 2785 (1994).
[13.58] J. W. Arbogast, and C. S. Foote. *J. Amer. Chem. Soc.*, **113**, 8886 (1991).
[13.59] R. H. Clarke. In R. H. Clarke (ed.), *Triplet State ODMR Spectroscopy: Techniques and Applications to Biophysical Systems*, John Wiley & Sons, New York (1982). Chapter 2.
[13.60] M. Matsushita, A. M. Frens, E. J. J. Groenen, O. G. Polucktov, J. Schmidt, G. Meijer, and M. A. Verheijen. *Chem. Phys. Lett.*, **214**, 349 (1993).
[13.61] P. A. Lane, L. S. Swanson, Q.-X. Ni, J. Shinar, J. P. Engel, T. J. Barton, and L. Jones. *Phys. Rev. Lett.*, **68**, 887 (1992).
[13.62] P. A. Lane, L. S. Swanson, J. Shinar, J. P. Engel, T. J. Barton, J. Wheelock, and L. Jones. *Synthetic Met.*, **55-57**, 3086 (1993).
[13.63] P. A. Lane and J. Shinar. In Z. Kafafi (ed.), *Fullerenes and Photonics: Proceedings of the International Society for Optical Engineering (SPIE)*, vol. 2284, pp. 21–32, Bellingham, WA (1994). SPIE Optical Engineering Press. San Diego, CA, July 24–29, 1994.
[13.64] D. Dick, X. Wei, S. Jeglinski, R. E. Benner, Z. V. Vardeny, D. Moses, V. I. Srdanov, and F. Wudl. *Phys. Rev. Lett.*, **73**, 2760 (1994).
[13.65] M. Matus, H. Kuzmany, and E. Sohmen. *Phys. Rev. Lett.*, **68**, 2822 (1992).
[13.66] X. Wei, S. Jeglinski, O. Paredes, Z. V. Vardeny, D. Moses, V. I. Srdanov, G. D. Stucky, K. C. Khemani, and F. Wudl. *Solid State Commun.*, **85**, 455 (1993).
[13.67] X. Wei, Z. V. Vardeny, D. Moses, V. I. Srdanov, and F. Wudl. *Synthetic Metals*, **54**, 273 (1993).
[13.68] A. Angerhofer, J. U. von Schutz, D. Widman, W. H. Muller, H. U. ter Meer, and H. Sixl. *Chem. Phys. Lett.*, **217**, 403 (1994).
[13.69] E. J. J. Groenen *et al. Mol. Crys. Liq. Crys.*, **in press** (1995).
[13.70] X. Wei and Z. V. Vardeny. *Mol. Crys. Liq. Crys.*, **256**, 307 (1994).
[13.71] B. Friedman. *Phys. Rev. B*, **48**, 2743 (1993).
[13.72] K. Harigaya. *Phys. Rev. B*, **45**, 13676 (1992).
[13.73] S. L. Ren, Y. Wang, A. M. Rao, E. McRae, G. T. Hager, K. A. Wang, W. T. Lee, H. F. Ni, J. Selegue, and P. C. Eklund. *Appl. Phys. Lett.*, **59**, 2678 (1991).

[13.74] P. C. Eklund. *Bull. Amer. Phys. Soc.*, **37**, 191 (1992); abstract C24.1.
[13.75] E. Sohmen, J. Fink, and W. Krätchmer. *Z. Phys. B*, **86**, 87 (1992).
[13.76] A. F. Hebard, R. C. Haddon, R. M. Fleming, and A. R. Kortan. *Appl. Phys. Lett*, **59**, 2109 (1991).
[13.77] Y. Wang, J. M. Holden, A. M. Rao, W.-T. Lee, X. X. Bi, S. L. Ren, G. W. Lehman, G. T. Hager, and P. C. Eklund. *Phys. Rev. B*, **45**, 14396 (1992).
[13.78] M. R. Pederson, K. A. Jackson, and L. L. Boyer. *Phys. Rev. B*, **45**, 6919 (1992).
[13.79] D. Ostling, P. Apell, and A. Rosen. *Zeits. Physik D*, **26**, 5282 (1993).
[13.80] N. Matsuzawa and D. A. Dixon. *J. Phys. Chem.*, **96**, 6872 (1992).
[13.81] P. W. Fowler, P. Lasseretti, and R. Zanasi. *Chem. Phys. Lett.*, **165**, 79 (1990).
[13.82] Z. Shuai and J. L. Bredas. *Synth. Metals*, **56**, 2973 (1993).
[13.83] S. A. Fitzgerald and A. J. Sievers. *Phys. Rev. Lett.*, **70**, 3177 (1993).
[13.84] S. Saito, S. Sawada, and N. Hamada. *Phys. Rev. B*, **45**, 13845 (1992).
[13.85] C. H. Xu and G. E. Scuseria. *Phys. Rev. Lett.*, **72**, 669 (1994).
[13.86] M. P. Gelfand and J. P. Lu. *Phys. Rev. Lett.*, **68**, 1050 (1992).
[13.87] S. G. Louie and E. L. Shirley. *J. Phys. Chem. Solids*, **54**, 1767 (1993).
[13.88] B. Chase, N. Herron, and E. Holler. *J. Phys. Chem.*, **96**, 4262 (1992).
[13.89] T. N. Thomas, J. F. Ryan, R. A. Taylor, D. Mihailovič, and R. Zamboni. *Int. J. Mod. Phys. B*, **6**, 3931 (1992).
[13.90] J. Feldmann, R. Fischer, E. Gobel, and S. Schmitt-Rink. *Phys. Stat. Sol. (b)*, **173**, 339 (1992).
[13.91] J. Sauvajol, Z. Hricha, N. Coustel, A. Zahab, and R. Aznar. *J. Phys.: Condens. Matter*, **5**, 2045 (1993).
[13.92] W. Andreoni, F. Gygi, and M. Parrinello. *Phys. Rev. Lett.*, **68**, 823 (1992).
[13.93] C. Bulliard, M. Allan, and S. Leach. *Chem. Phys. Lett.*, **209**, 434 (1993).
[13.94] K. Harigaya and S. Abe. *Phys. Rev. B*, **49**, 16746 (1994).
[13.95] K. Harigaya and S. Abe. *Mol. Cryst. Liq. Crist.*, **256**, 825 (1994).
[13.96] V. I. Srdanov, A. P. Sabb, D. Margolese, E. Poolman, K. C. Khemani, A. Koch, F. Wudl, B. Kirtman, and G. D. Stucky. *Chem. Phys. Lett.*, **192**, 243 (1992).
[13.97] T. Pichler, M. Matus, J. Kürti, and H. Kuzmany. *Solid State Commun.*, **81**, 859 (1992).
[13.98] W. L. Wilson, A. F. Hebard, L. R. Narasimhan, and R. C. Haddon. *Phys. Rev. B*, **48**, 2738 (1993).
[13.99] V. I. Srdanov, C. H. Lee, and N. S. Sariciftci. *Thin Solid Films*, **257**, 233 (1995).
[13.100] R. C. Haddon, A. F. Hebard, M. J. Rosseinsky, D. W. Murphy, S. J. Duclos, K. B. Lyons, B. Miller, J. M. Rosamilia, R. M. Fleming, A. R. Kortan, S. H. Glarum, A. V. Makhija, A. J. Muller, R. H. Eick, S. M. Zahurak, R. Tycko, G. Dabbagh, and F. A. Thiel. *Nature (London)*, **350**, 320 (1991).
[13.101] E. Sohmen and J. Fink. *Phys. Rev. B*, **47**, 14532 (1993).
[13.102] S. Modesti, S. Cerasari, and P. Rudolf. *Phys. Rev. Lett.*, **71**, 2469 (1993).
[13.103] G. Gensterblum, J. J. Pireaux, P. A. Thiry, R. Caudano, J. P. Vigneron, P. Lambin, A. A. Lucas, and W. Krätschmer. *Phys. Rev. Lett.*, **67**, 2171 (1991).
[13.104] W. Y. Ching, M.-Z. Huang, Y.-N. Xu, W. G. Harter, and F. T. Chan. *Phys. Rev. Lett.*, **67**, 2045 (1991).
[13.105] Y.-N. Xu, M.-Z. Huang, and W. Y. Ching. *Phys. Rev. B*, **44**, 13171 (1991).
[13.106] Y. N. Xu, M. Z. Huang, and W. Y. Ching. *Phys. Rev. B*, **46**, 4241 (1992).
[13.107] N. Troullier and J. L. Martins. *Phys. Rev. B*, **46**, 1766 (1992).
[13.108] M. R. Pederson and A. A. Quong. *Phys. Rev. B*, **46**, 13584 (1992).
[13.109] B. I. Dunlap, D. W. Brenner, J. W. Mintmire, R. C. Mowrey, and C. T White. *J. Phys. Chem.*, **95**, 5763 (1991).

[13.110] W. Krätschmer and D. R. Huffman. *Carbon*, **30**, 1143 (1992).
[13.111] E. L. Shirley and S. G. Louie. *Phys. Rev. Lett.*, **71**, 133 (1993).
[13.112] S. G. Louie. In H. Kuzmany, J. Fink, M. Mehring, and S. Roth (eds.), *Progress in Fullerene Research: International Winterschool on Electronic Properties of Novel Materials*, p. 303 (1994). Kirchberg Winter School, World Scientific Publishing, Singapore.
[13.113] J. Wu, Z. X. Shen, D. S. Dessau, R. Cao, D. S. Marshall, P. Pianetta, I. Lindau, X. Yang, J. Terry, D. M. King, B. O. Wells, D. Elloway, H. R. Wendt, C. A. Brown, H. Hunziker, and M. S. de Vries. *Physica C*, **197**, 251 (1992).
[13.114] P. J. Benning, J. L. Martins, J. H. Weaver, L. P. F. Chibante, and R. E. Smalley. *Science*, **252**, 1417 (1991).
[13.115] J. H. Weaver and D. M. Poirier. In H. Ehrenreich and F. Spaepen (eds.), *Solid State Physics*, vol. 48, p. 1, Academic Press, New York (1994).
[13.116] J. M. Themlin, S. Bouzidi, F. Coletti, J. M. Debever, G. Gensterblum, L. M. Yu, J. J. Pireaux, and P. A. Thiry. *Phys. Rev. B*, **46**, 15602 (1992).
[13.117] B. Pevzner, A. F. Hebard, and M. S. Dresselhaus. Unpublished.
[13.118] S. D. Senturia and S. L. Garverick. Method and apparatus for microdielectrometry. U.S. Patent No. 4,423,371, Dec. 27, 1983.
[13.119] S. D. Senturia, J. N. F. Sheppard, H. L. Lee, and D. R. Day. *J. Adhesion*, **15**, 69 (1982).
[13.120] S. D. Senturia and N. F. Sheppard. *Dielectric Analysis of Thermoset Cure*, vol. 80 of *Advances in Polymer Science*, pp. 1–47. Springer-Verlag, Berlin (1986).
[13.121] B. Pevzner, A. F. Hebard, R. C. Haddon, S. D. Senturia, and M. S. Dresselhaus. In P. Bernier, T. W. Ebbesen, D. S. Bethune, R. M. Metzger, L. Y. Chiang, and J. W. Mintmire (eds.), *Science and Technology of Fullerene Materials: MRS Symposia Proceedings, Boston, Fall 1994*, vol. 359, p. 423, Materials Research Society Press, Pittsburgh, PA (1995).
[13.122] P. Zhou, A. M. Rao, K. A. Wang, J. D. Robertson, C. Eloi, M. S. Meier, S. L. Ren, X. X. Bi, and P. C. Eklund. *Appl. Phys. Lett.*, **60**, 2871 (1992).
[13.123] Y. Wang, J. M. Holden, X. X. Bi, and P. C. Eklund. *Chem. Phys. Lett.*, **217**, 413 (1994).
[13.124] M. Menon, K. R. Subbaswamy, and M. Sawtarie. *Phys. Rev. B*, **49**, 13966 (1994).
[13.125] D. L. Strout, R. Murry, C. Xu, W. Eckhoff, G. Odom, and G. E. Scuseria. *Chem. Phys. Lett.*, **214**, 576 (1993).
[13.126] C. Reber, L. Yee, J. McKiernan, J. I. Zink, R. S. Williams, W. M. Tong, D. A. A. Ohlberg, R. L. Whetten, and F. Diederich. *J. Phys. Chem.*, **95**, 2127 (1991).
[13.127] A. Ito, T. Morihawa, and T. Takahashi. *Chem. Phys. Lett.*, **211**, 333 (1993).
[13.128] U. D. Venkateswaran, M. G. Schall, Y. Wang, P. Zhou, and P. C. Eklund. *Solid State Commun.*, **96**, 951 (1995).
[13.129] D. M. Cox, S. Behal, M. Disko, S. M. Gorun, M. Greaney, C. S. Hsu, E. B. Kollin, J. M. Millar, J. Robbins, W. Robbins, R. D. Sherwood, and P. Tindall. *J. Am. Chem. Soc.*, **113**, 2940 (1991).
[13.130] P.-M. Allemand, A. Koch, F. Wudl, Y. Rubin, F. Diederich, M. M. Alvarez, S. J. Anz, and R. L. Whetten. *J. Am. Chem. Soc.*, **113**, 1050 (1991).
[13.131] D. DuBois, K. Kadish, S. Flanagan, R. E. Haufler, L. P. F. Chibante, and L. J. Wilson. *J. Am. Chem. Soc.*, **113**, 4364 (1991).
[13.132] D. DuBois, K. Kadish, S. Flanagan, and L. J. Wilson. *J. Am. Chem. Soc.*, **113**, 7773 (1991).
[13.133] D. R. Lawson, D. L. Feldheim, C. A. Foss, P. K. Dorhout, C. M. Elliott, C. R. Martin, and B. Parkinson. *J. Phys. Chem.*, **96**(18), 7175 (1992).

[13.134] D. R. Lawson, D. L. Feldheim, C. A. Foss, P. K. Dorhout, C. M. Elliott, C. R. Martin, and B. Parkinson. *J. Electrochem. Soc.*, **139**, L68 (1992).
[13.135] T. Kato, T. Kodama, T. Shida, T. Nakagawa, Y. Matsui, S. Suzuki, H. Shiromaru, K. Yamauchi, and Y. Achiba. *Chem. Phys. Lett.*, **180**, 446 (1991).
[13.136] T. Kato, T. Kodama, M. Oyama, S. Okasaki, T. Shida, T. Nakagawa, Y. Matsui, S. Suzuki, H. Shiromaru, K. Yamauchi, and Y. Achiba. *Chem. Phys. Lett.*, **186**, 35 (1991).
[13.137] Z. Gasyna, L. Andrews, and P. N. Schatz. *J. Phys. Chem.*, **183**, 1525 (1992).
[13.138] J. W. Arbogast, C. S. Foote, and M. Kao. *J. Amer. Chem. Soc.*, **114**, 2277 (1992).
[13.139] A. J. Bard and L. Faulker. *Electrochemical Methods*. John Wiley & Sons, New York, (1980).
[13.140] M. A. Greaney and S. M. Gorun. *J. Phys. Chem.*, **95**, 7142 (1991).
[13.141] D. DuBois, M. T. Jones, and K. Kadish. In L. Y. Chiang, A. F. Garito, and D. J. Sandman (eds.), *Electrical, Optical and Magnetic Properties of Organic Solid State Materials, MRS Symposia Proceedings, Boston*, vol. 247, p. 247, Materials Research Society Press, Pittsburgh, PA (1992).
[13.142] V. de Coulon, J. L. Martins, and F. Reuse. *Phys. Rev. B*, **45**, 13671 (1992).
[13.143] I. Y, S. Watanabe, T. Kaneyasu, T. Yasuda, T. Koda, M. Nagate, and N. Mizutani. *J. Phys. Chem. Solids*, **54**, 1795 (1993).
[13.144] K. J. Fu, W. L. Karney, O. L. Chapman, S. M. Huang, R. B. Kaner, F. Diederich, K. Holczer, and R. L. Whetten. *Phys. Rev. B*, **46**, 1937 (1992).
[13.145] S. C. Erwin and M. R. Pederson. *Phys. Rev. Lett.*, **67**, 1610 (1991).
[13.146] Y. Wang. "Optical Properties of Fullerene-based Materials." Ph.D. Thesis. University of Kentucky (1993).
[13.147] E. Sohmen, J. Fink, and W. Krätschmer. *Europhys. Lett.*, **17**, 51 (1992).
[13.148] K. Harigaya and S. Abe. *Synth. Metals*, **70**, 1415 (1995).
[13.149] L. Degiorgi, G. Grüner, P. Wachter, S. M. Huang, J. Wiley, R. L. Whetten, R. B. Kaner, K. Holczer, and F. Diederich. *Phys. Rev. B*, **46**, 11250 (1992).
[13.150] L. Degiorgi, P. Wachter, G. Grüner, S. M. Huang, J. Wiley, and R. B. Kaner. *Phys. Rev. Lett.*, **69**, 2987 (1992).
[13.151] L. Degiorgi, E. J. Nicol, O. Klein, G. Grüner, P. Wachter, S. M. Huang, J. Wiley, and R. B. Kaner. *Phys. Rev. B*, **49**, 7012 (1994).
[13.152] Y. Iwasa, K. Tanaka, T. Yasuda, T. Koda, and S. Koda. *Phys. Rev. Lett.*, **69**, 2284 (1992).
[13.153] T. Pichler, M. Matus, and H. Kuzmany. *Solid State Commun.*, **86**, 221 (1993).
[13.154] M. C. Martin, D. Koller, and L. Mihaly. *Phys. Rev. B*, **47**, 14607 (1993).
[13.155] A. M. Rao, K. A. Wang, G. W. Lehman, and P. C. Eklund. Unpublished.
[13.156] X. X. Bi and P. C. Eklund. *Phys. Rev. Lett.*, **70**, 2625 (1993).
[13.157] W. H. Wong, M. E. Hanson, W. G. Clark, K. Holczer, G. Gruner, J. D. Thompson, R. L. Wetton, S. M. Huang, R. B. Kaner, and F. Diederich. *Europhys.*, **18**, 79 (1992).
[13.158] S. E. Erwin and W. E. Pickett. *Science*, **254**, 842 (1992).
[13.159] M. Tinkham. *Introduction to Superconductivity*, McGraw–Hill, New York (1980).
[13.160] L. Mihaly, D. Koller, and M. C. Martin. In H. Kuzmany, J. Fink, M. Mehring, and S. Roth (eds.), *Progress in Fullerene Research: International Winterschool on Electronic Properties of Novel Materials*, p. 265, World Scientific, Singapore (1994).
[13.161] M. C. Martin, D. Koller, X. Du, P. W. Stephens, and L. Mihaly. *Phys. Rev. B*, **49**, 10818 (1994).
[13.162] O. Chauvet, G. Oszànyi, L. Forró, P. W. Stephens, M. Tegze, G. Faigel, and A. Jànossy. *Phys. Rev. Lett.*, **72**, 2721 (1994).
[13.163] Y. Wang. *Nature (London)*, **356**, 585 (1992).

[13.164] N. S. Sariciftci, L. Smilowitz, A. J. Heeger, and F. Wudl. *Science*, **258**, 1474 (1992).
[13.165] N. S. Sariciftci, D. Braun, C. Zhang, V. I. Srdanov, A. J. Heeger, C. Stucky, and F. Wudl. *Appl. Phys. Lett.*, **62**, 585 (1993).
[13.166] K. Lee, R. A. J. Janssen, N. S. Sariciftci, and A. J. Heeger. *Phys. Rev. B*, **49**, 5781 (1994).
[13.167] C. H. Lee, G. Yu, D. Moses, K. Pakbaz, C. Zhang, N. S. Sariciftci, A. J. Heeger, and F. Wudl. *Phys. Rev. B*, **48**, 15425 (1993).
[13.168] N. S. Sariciftci and A. J. Heeger. *Int. J. Mod. Phys. B*, **8**, 237 (1994).
[13.169] N. S. Sariciftci, B. Kraabel, C. H. Lee, K. Pakbaz, and A. J. Heeger. *Phys. Rev. B*, **50**, 12044 (1994).
[13.170] N. S. Sariciftci, L. Smilowitz, C. Zhang, V. I. Srdanov, A. J. Heeger, and F. Wudl. *Proc. SPIE - Int. Soc. Opt. Eng.*, **1852**, 297 (1993).
[13.171] S. Morita, A. A. Zakhidov, and K. Yoshino. *Solid State Commun.*, **82**, 249 (1992).
[13.172] A. J. Heeger, S. Kivelson, J. R. Schrieffer, and W. P. Su. *Rev. Mod. Phys.*, **60**, 781 (1988).
[13.173] B. Kraabel, C. H. Lee, D. McBranch, D. Moses, N. S. Sariciftci, and A. J. Heeger. *Chem. Phys. Lett.*, **213**, 389 (1993).
[13.174] F. Diederich, R. Ettl, Y. Rubin, R. L. Whetten, R. Beck, M. Alvarez, S. Anz, D. Sensharma, F. Wudl, K. C. Khemani, and A. Koch. *Science*, **252**, 548 (1991).
[13.175] F. Diederich and R. L. Whetten. *Acc. Chem. Res.*, **25**, 119 (1992).
[13.176] K. Kikuchi, N. Nakahara, T. Wakabayashi, S. Suzuki, H. Shiramaru, Y. Miyake, K. Saito, I. Ikemoto, M. Kainosho, and Y. Achiba. *Nature (London)*, **357**, 142 (1992).
[13.177] K. Kikuchi, N. Nakahara, T. Wakabayashi, M. Honda, H. Matsumiya, T. Moriwaki, S. Suzuki, H. Shiromaru, K. Saito, K. Yamauchi, I. Ikemoto, and Y. Achiba. *Chem. Phys. Lett.*, **188**, 177 (1992).
[13.178] P.-M. Allemand, K. C. Khemani, A. Koch, F. Wudl, K. Holczer, S. Donovan, G. Grüner, and J. D. Thompson. *Science*, **253**, 301 (1991).
[13.179] S. L. Ren, K. A. Wang, P. Zhou, Y. Wang, A. M. Rao, M. S. Meier, J. Selegue, and P. C. Eklund. *Appl. Phys. Lett.*, **61**, 124 (1992).
[13.180] J. B. Howard, J. T. McKinnon, Y. Makarovsky, A. L. Lafleur, and M. E. Johnson. *Nature (London)*, **352**, 139 (1991).
[13.181] W. Andreoni, F. Gygi, and M. Parrinello. *Chem. Phys. Lett*, **190**, 159 (1992).
[13.182] K. Harigaya. *Chem. Phys. Lett.*, **189**, 79 (1992).
[13.183] S. Saito and A. Oshiyama. *Phys. Rev. B*, **44**, 11532 (1991).
[13.184] R. A. Cheville and N. J. Halas. *Phys. Rev. B*, **45**, 4548 (1992).
[13.185] M. J. Rosker, H. O. Marcy, T. Y. Chang, J. T. Khoury, K. Hansen, and R. L. Whetten. *Chem. Phys. Lett.*, **195**, 427 (1992).
[13.186] T. Juhasz, H. Hu, C. Suarex, W. E. Bron, E. Maikwn, and P. Taborek. *Phys. Rev. B*, **48**, 429 (1993).
[13.187] S. R. Flom, R. G. S. Pong, F. J. Bartoli, and Z. H. Kafafi. *Phys. Rev. B*, **46**, 15598 (1992).
[13.188] S. L. Dexelheimer, W. A. Vareka, D. Mittleman, and A. Zettl. *Chem. Phys. Lett.*, **235**, 552 (1995).
[13.189] S. B. Fleischer, E. P. Ippen, G. Dresselhaus, M. S. Dresselhaus, A. M. Rao, P. Zhou, and P. C. Eklund. *Appl. Phys. Lett.*, **62**, 3241 (1993).
[13.190] P. Prasad. *Nonlinear Optical Properties of Organic Materials*. Plenum, New York (1991).
[13.191] N. Tang, J. P. Partanen, R. W. Hellwarth, and R. J. Knize. *Phys. Rev. B*, **48**, 8404 (1993).

[13.192] Z. H. Kafafi, F. J. Bartoli, J. R. Lindle, and R. G. S. Pong. *Phys. Rev. Lett.*, **68**, 2705 (1992).

[13.193] W. J. Blau, H. J. Byrne, D. J. Cardin, T. J. Dennis, J. P. Hare, H. W. Kroto, R. Taylor, and D. R. M. Walton. *Phys. Rev. Lett.*, **67**, 1423 (1991).

[13.194] Q. H. Gong, Y. X. Sun, Z. J. Xia, Y. H. Zou, Z. N. Gu, X. H. Zhou, and D. Qiang. *J. Appl. Phys.*, **71**, 3025 (1992).

[13.195] J. R. Lindle, R. G. S. Pong, F. J. Bartoli, and Z. H. Kafafi. *Phys. Rev. B*, **48**, 9447 (1993).

[13.196] H. Hoshi, N. Nakamura, Y. Maruyama, T. Nakagawa, S. Suzuki, H. Shiromaru, and Y. Achiba. *Jpn. J. Appl. Phys.*, **30**, L1397 (1991).

[13.197] J. S. Meth, H. Vanherzeele, and Y. Wang. *Chem. Phys. Lett.*, **197**, 26 (1992).

[13.198] F. Kajzar, C. Taliani, R. Zamboni, S. Rosini, and R. Danieli. *Synth. Metals*, **54**, 21 (1993).

[13.199] D. Neher, G. I. Stegeman, and F. A. Tinker. *Opt. Lett.*, **17**, 1491 (1992).

[13.200] Y. Wang and L. T. Cheng. *J. Phys. Chem.*, **96**, 1530 (1992).

[13.201] S. C. Yang, Q. H. Gong, Z. J. Xia, Y. H. Zou, Y. Q. Wu, and D. Qiang. *Applied Phys. Part B. Photophysics and Laser Chemistry*, **55**, 51 (1992).

[13.202] H. M. Liu, B. Taheri, and W. Y. Jia. *Phys. Rev. B*, **49**, 10166 (1994).

[13.203] I. V. Bezel, S. V. Chekalin, Y. A. Matveets, A. G. Stepanov and A. F. Yatsev and V. S. Letokhev, *Chem. Phys. Lett.*, **218**, 475 (1994). Erratum: ibid, vol. 221, p. 332, 1994.

[13.204] E. W. V. Stryland, H. Vanherzeele, M. A. Woodall, M. J. Soileau, A. L. Smirl, S. Guha, and T. F. Bogess. *Opt. Eng.*, **24**, 613 (1985).

[13.205] X. K. Wang, T. G. Zhang, W. P. Lin, Z. L. Sheng, G. K. Wong, M. M. Kappes, R. P. H. Chang, and J. B. Ketterson. *Appl. Phys. Lett.*, **60**, 810 (1992).

[13.206] S. V. Chekalin, V. M. Farztdinov, E. Akesson, and V. Sundstrom. *JETP Lett.*, **58**, 295 (1993).

CHAPTER 14

Transport and Thermal Properties

In this chapter the transport and thermal properties of fullerenes in the normal state are discussed. The superconducting properties are reviewed in Chapter 15. Since pristine solid C_{60} and C_{70} have no free carriers for transport, either doping, photoexcitation or some other mechanism must be used to generate carriers for electrical conduction.

The transport properties of doped fullerene solids are unusual for several reasons. First, the movement of an excess electron on the surface of a fullerene anion suffers relatively little scattering compared to that for movement of electrons from one fullerene anion to an adjacent neutral fullerene molecule. Increasing the distance between adjacent ions by selective donor doping gives rise to two competing processes regarding carrier transport: one process decreases the wave function overlap integral between adjacent fullerene ions and thereby sensitively decreases electronic transport, while the other increases the density of states at the Fermi level, thereby enhancing electronic transport. Because of the many types of vibrational modes in the lattice, the identification of the modes that dominate the scattering processes is of particular interest.

Second, metallic conduction seems to be restricted to a small range of stoichiometries close to a half-filled LUMO-derived band, as discussed in §14.1.1. Simple one-electron arguments suggest that any partial occupation of the LUMO-derived conduction band would give rise to metallic conduction, thus indicating that many-body electron–electron correlation effects are important for describing the electronic states in alkali metal– and alkaline earth–doped fullerenes. These arguments are consistent with the large relative value of the Hubbard U ($\sim$1–2 eV), which is large compared to the electronic bandwidth W of the LUMO-derived bands (0.2–0.5 eV) in the undoped fullerite crystal (see §12.7.7). A third special feature of the doped-fullerene bands is the fact that there appears to be no observed metal–insulator transition for a half-filled band, as occurs for many highly correlated systems [14.1]. Yet another special feature of transport in the doped fullerenes is the role of merohedral disorder as a scattering mechanism in the fullerene host material, thereby reducing the low-temperature electrical conductivity. Disorder of the dopant also contributes to additional carrier scattering in doped fullerenes. These novel features of the transport properties of fullerenes are further discussed throughout this chapter.

The transport properties of fullerene solids that have been studied include electrical conductivity (dc up through microwave frequencies), Hall effect, magnetoresistance, photoconductivity, and thermoelectric power. Also included in this chapter is a review of the thermal properties, including specific heat, thermal conductivity, the temperature coefficient of lattice expansion, and differential scanning calorimetry.

14.1. Electrical Conductivity

In previous discussions of the electronic structure, the semiconducting behavior of crystalline C_{60} and C_{70} was emphasized (see §12.7). The room temperature resistivity of the undoped fullerene solids is high with $\rho \sim 10^{14}$ Ω-cm reported for undoped mixtures of C_{60} and C_{70} films exposed to air [14.2] and $\rho \sim 10^{8}$ Ω-cm for oxygen-free C_{60} films [14.3]. Almost no carriers are available for transport in C_{60} unless they are thermally or optically excited, or more importantly if carriers are introduced by the doping of donor (or perhaps acceptor) species. Until now, the most intensively studied donor dopants are the alkali metals, which are efficient in providing electron charge transfer to the fullerenes and in creating carriers near the Fermi level. In addition, a few studies have been reported on the electrical conductivity in the alkaline earth–doped C_{60} compounds (§14.1.3) as well as in the alkali metal–doped C_{70} compounds (§14.1.4). Because of the degenerate ground states of the fullerene anions and cations (see §12.4.2 and §12.4.3), Jahn–Teller distortions of the molecule may occur, thereby contributing to the overlap of the wave functions between adjacent fullerenes and to the enhancement of their transfer integral.

14.1.1. Dependence on Stoichiometry

Because of the high-resistivity of undoped C_{60}, doping with alkali metals decreases the electrical resistivity ρ of C_{60} by many orders of magnitude.

As x in M_xC_{60} increases, the resistivity $\rho(x)$ decreases and eventually approaches a minimum at $x = 3.0\pm0.05$ [14.4, 5], corresponding to a half-filled f_{1u}-derived (t_{1u}-derived) conduction band. Then, upon further increase in x from 3 to 6, ρ again increases, as is shown in Fig. 14.1 [14.5]. It should be noted that stable crystallographic K_xC_{60} phases occur only for $x = 0, 1, 3, 4$, and 6 (see §8.5). The compounds corresponding to filled molecular levels (C_{60} and M_6C_{60}) are the most stable and exhibit maxima in the resistivity of M_xC_{60} (M = K, Rb) as a function of x, consistent with a filled band.

The measurements shown in Fig. 14.1 are also significant with regard to whether or not a band gap is formed at $x = 3$, corresponding to a half-filled band [14.6]. The $\rho(x)$ measurements in Fig. 14.1 yield a minimum resistivity very close to $x = 3$, where the stoichiometry was independently determined by the Rutherford backscattering (RBS) technique. The experimental results for $\rho(x)$ do not give evidence for a band gap at this stoichiometry, although many-body theoretical arguments suggest that a strongly electron-correlated material ($U > W$) should undergo a metal–insulator transition at a half-filled band (see §12.7.7) [14.1, 6, 7]. More recent experiments than in Fig. 14.1 using a thin C_{60} film with a much larger grain size of $\sim$1 μm as the starting material before the doping [14.8] show a shoulder in $\rho(x)$

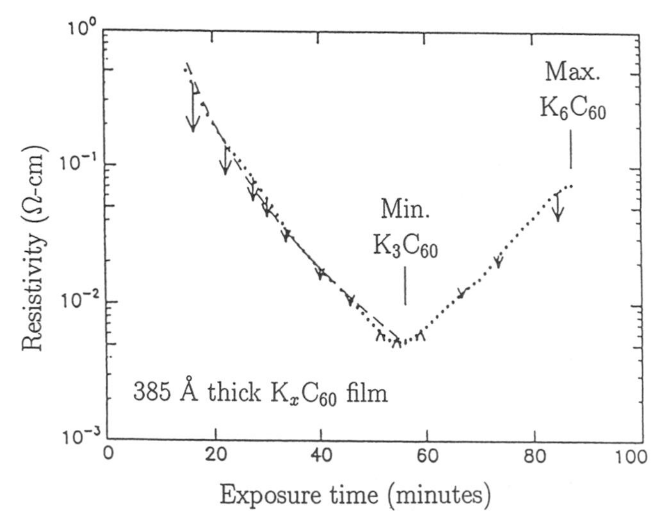

Fig. 14.1. Dependence of the resistivity of a K_xC_{60} film on exposure time to a K molecular beam in ultrahigh vacuum (UHV) at ambient temperature near 74°C. For this particular sample only the end point stoichiometry was determined explicitly; the stoichiometry at the minimum was determined from other similarly prepared samples. The arrows indicate changes in the resistivity of a similar sample, as it was heated from 60°C to 134°C. The dashed curve is a model fit to the data points [14.5].

at $x = 4$ and a pronounced maximum at $x = 6$. Overdoping beyond $x = 6$ leads to a metallic overcoat on the sample but no further uptake of K or Rb into the crystal lattice.

Somewhat similar behavior is shown in Fig. 14.2 for several alkali metal dopants where the resistivity is plotted against alkali metal concentration x in M_xC_{60} for M = Na, K, Rb, and Cs as determined by photoemission spectroscopy [14.3]. Furthermore, even at the minimum resistivity in M_xC_{60}, the value of ρ found in Fig. 14.2 for K_3C_{60} (2.5×10^{-3} Ω-cm) is high, typical of a high-resistivity metal [14.5]. Conduction is believed to occur by charge transfer from one C_{60} molecular ion to another along (110) directions through the weak overlap of the wave functions on adjacent C_{60} anions, perhaps enhanced by a Jahn–Teller distortion of the molecule to lower the degeneracy of the ground state (see §14.2.4).

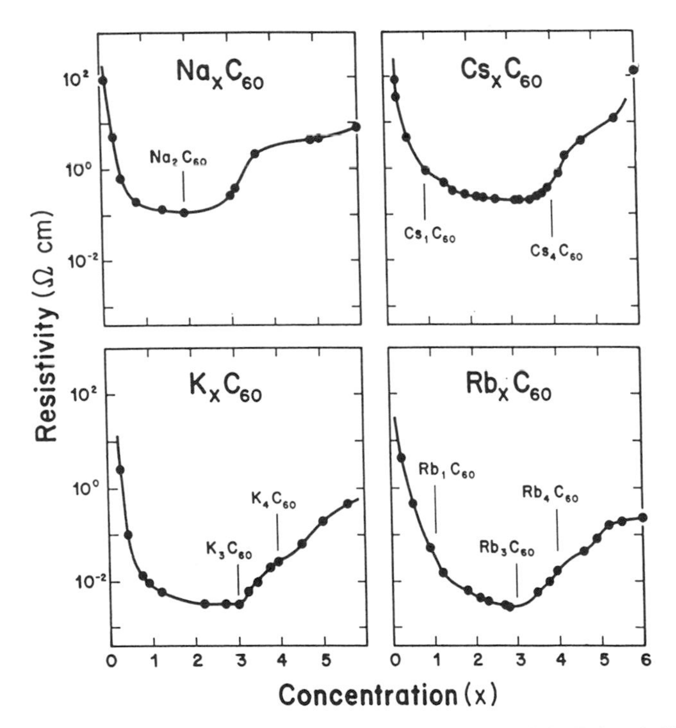

Fig. 14.2. $\rho(x)$ for thick films of C_{60} doped with Na, K, Rb, and Cs. Points indicate where exposure to the alkali-metal source was stopped and x-ray and ultraviolet photoemission spectra were acquired to determine the concentration x. The labels indicate the known fulleride phases at 300 K. The minima in $\rho(x)$ occur for stoichiometries corresponding to Na_2C_{60}, K_3C_{60}, Rb_3C_{60}, and $Cs_{3.5}C_{60}$. Structure in $\rho(x)$ can be associated with the development of different stable phases [14.3].

Consistent with band structure studies [14.9], nuclear magnetic resonance (NMR) Knight shift studies of ^{39}K [14.10] indicate that the charge transfer from the alkali metal dopants of the C_{60} is complete, so that the alkali metal ions do not participate in electrical conduction. From the charge transfer and the stoichiometry of the compound, the carrier density can be inferred. Then assuming three electrons per C_{60}^{3-} anion in K_3C_{60}, the carrier density is expected to be $\sim 4 \times 10^{21}/\text{cm}^3$, which is quite low for conducting systems. Very weak structure is observed (Fig. 14.2) in the $\rho(x)$ data for K_xC_{60} at $x = 4$ [14.3], suggesting that K_4C_{60} might form as a minority phase in this experiment. Whereas the dependence of the room temperature resistivity on alkali metal concentration is similar for potassium and rubidium (including the magnitudes of ρ_{min} in Fig. 14.2, which are 3.1×10^{-3} Ω-cm for K_3C_{60} and 2.8×10^{-3} Ω-cm for Rb_3C_{60}), the behavior of $\rho(x)$ is qualitatively different in three respects for sodium and cesium doping: the magnitude of ρ_{min}, the value of x where ρ_{min} occurs, and the shape of the ρ *vs.* x curves. The stable phases for the M_xC_{60} compounds are indicated on the figure. At all other x values, measurements are made on multiphase samples [14.3].

Of the various stoichiometries and alkali metal species for the bulk compounds indicated on Fig. 14.2, only the compounds close to the K_3C_{60} and Rb_3C_{60} stoichiometry (see Table 14.1) exhibit metallic temperature coefficients of the resistivity (see §14.1.2). It is significant that metallic conductivity occurs over a very narrow stoichiometry range near $x = 3$ in K_xC_{60}

Table 14.1

Physical constants for M_3C_{60} (M = K, Rb).

Quantity	K_3C_{60}	Rb_3C_{60}	Reference
Space group	$Fm3\underline{3}$	$Fm3\underline{3}$	[14.10]
C_{60}—C_{60} distance (Å)	10.06	10.20	[14.10]
M-C_{60} closest distance (Å)	3.27	3.33	[14.10]
Volume per C_{60} (cm^3)	7.22×10^{-22}	7.50×10^{-22}	[14.10]
fcc lattice constant a (Å)	14.253	14.436	[14.11]
$(-d \ln a/dP)$ (GPa^{-1})	1.20×10^{-2}	1.52×10^{-2}	[14.12]
Bulk modulus (GPa)	28	22	[14.13]
Thermal expansion coefficient (Å/K)	2×10^{-5}	—	[14.10]
Cohesive energy (eV)	24.2	—	[14.13]
Heat of formation (eV)	4.9	—	[14.13]
Density of states[a] [states/(eV/C_{60})]	25	35	[14.14]
Carrier density[a] (10^{21}/cm^3)	4.155	4.200	[14.10]
Electron effective mass (m_e)	1.3	—	[14.13]
Hole effective mass (m_e)	1.5, 3.4	—	[14.13]

[a] Assumes 3 electrons per C_{60}.

and Rb_xC_{60} and that K_xC_{60} and Rb_xC_{60} compounds are not metallic for all x values corresponding to a partially filled LUMO-derived band. Since the splittings of the levels in the LUMO-derived band are small compared with the bandwidth, the conduction band is considered to be degenerate, and the stoichiometry x in an alkali metal–doped compound M_xC_{60} corresponds to the filling of the band. Photoemission and optical studies show that the Hubbard U is large and does not change significantly upon variation of the dopant concentration x, indicating that doped fullerenes are strongly correlated systems (see §12.7.7). From the dopant dependence of ρ we conclude that one-electron theory is insufficient and that many-body electron–electron correlation effects must be important. Further theoretical work is needed to gain an understanding of why the transport behavior is so sensitive to level degeneracies and electron-filling effects and why the half-filled band does not give rise to a metal–insulator transition [14.6].

Another important aspect of the metallic phase is the high value for the ρ_{min} that is found experimentally at $x = 3$. For a given dopant, the doping (or intercalation) conditions and the nature of the fullerene host material both strongly influence the value of the minimum resistivity ρ_{min} (or the maximum conductivity σ_{max}) that is achieved, as well as the magnitude of the superconducting transition temperature T_c and the width of the superconducting transition ΔT_c for the case of the alkali metal M_3C_{60} (M = K, Rb) compounds. Slower intercalation under well-controlled conditions generally increases T_c, while decreasing ΔT_c and ρ_{min}. The highest T_c and the lowest ΔT_c and ρ_{min} are achieved with single-crystal C_{60} host materials [see Fig. 14.3(a)], and ρ_{min} values of $\sim 0.5 \times 10^{-3}\Omega$-cm have been reported at low T [14.17, 18]. Thin-film samples have also been widely studied and tend to have somewhat lower T_c values and larger ΔT_c widths and higher ρ_{min} values [14.16, 19, 20]. Powder samples have also been studied regarding their transport properties, and they show relatively high ρ_{min} values.

One cogent explanation for the high value of ρ_{min} arises from merohedral disorder which is present because icosahedral symmetry has no fourfold axis, thus giving rise to the random occurrence of the two "standard" orientations for the C_{60} molecules shown in Fig. 7.6. The conduction bands for a doped C_{60} crystal arise from the overlap of the LUMO-derived levels of the molecule. The intermolecular hopping matrix element and the electronic states both depend strongly on relative intermolecular orientations. Merohedral disorder significantly reduces the intermolecular hopping matrix element and therefore increases the resistivity. Calculations show that the merohedral disorder itself leads to a large residual resistivity of $\rho_0 \sim 300\ \mu\Omega$-cm in the absence of impurities and defects, indicating that merohedral disorder is the dominant scattering mechanism for doped C_{60} at low temperature [14.21–23]. Because of the interaction between the pos-

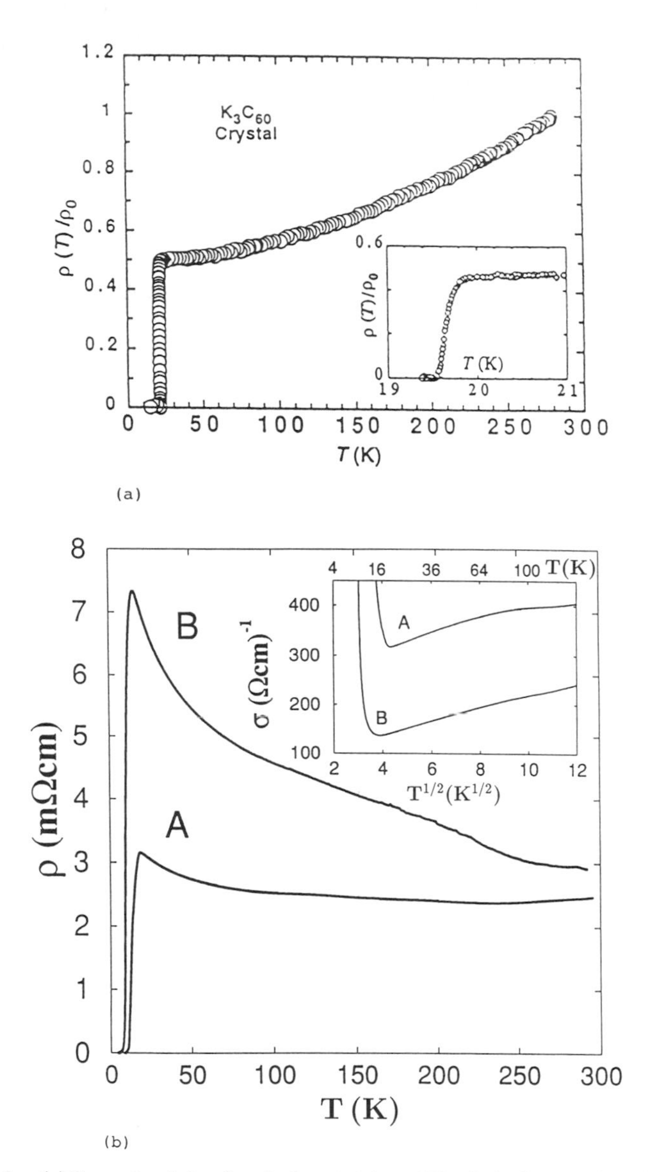

Fig. 14.3. (a)Normalized dc electrical resistivity $\rho(T)$ of single-crystal K_3C_{60}. The inset shows the $\rho(T)$ behavior near the superconducting transition temperature $T_c = 19.8$ K [14.15]. (b) Temperature dependence of ρ for two K_3C_{60} films. Sample A has less disorder than sample B. The inset plots the same data as $\sigma(T)$ *vs.* $T^{1/2}$ over a smaller range of T (top scale) [14.16]. Samples showing $(\partial\rho/\partial T) > 0$ tend to have sharp normal–superconducting transitions.

itively charged alkali metal dopants and the negatively charged C_{60} molecular ions, the energy of the LUMO-derived levels is sensitive to the alkali ion positions. Thus disorder in the position of the alkali ions introduces spatial fluctuations in the LUMO energies. For many M_xC_{60} samples, disorder in the position of the alkali metal M ions commonly occurs because the alkali metal ions are small in size compared to the octahedral sites (see §8.5) and the electrostatic interactions between the alkali metal ions and C_{60} anions cause ionic displacements from their ideal interstitial positions. Additional ionic displacements can also arise from the presence of neutral ternary species such as ammonia [14.24, 25]. The random occupation of octahedral and tetrahedral sites for systems which ideally have site ordering (e.g., Na_2KC_{60}, Na_2C_{60}, RbC_{60}; see §8.5) is also a source of disorder. Vacancies in the dopant site occupation are another source of disorder. In the case of strong disorder associated with dopant site occupation, carrier localization may occur [see Fig. 14.3(b)]. Such effects have been reported in both alkali metal–doped C_{60} and C_{70} [14.16, 26, 27]. Calculations show that both the orientational disorder of the C_{60} molecules and the positional disorder of the alkali metal ions contribute to the large carrier scattering in doped fullerenes [14.21–23]. In transport calculations, the hopping matrix is found by evaluating the overlap of LUMO wave functions on adjacent molecules.

The ρ_{min} values for Na_xC_{60} and Cs_xC_{60} in Fig. 14.2 are 0.11 Ω-cm and 0.20 Ω-cm, respectively, much higher than ρ_{min} for K_xC_{60} and Rb_xC_{60}. The different behavior for Na_3C_{60} relative to K_3C_{60} and Rb_3C_{60} has been related to the small size of the Na^+ ion, thereby allowing some fluctuations in the site positions of the Na^+ ions and perhaps occupation by some Na_2^+ molecular ions which provide less charge transfer. For the case of the Cs^+ ion, anomalous behavior arises from the large size of the Cs^+ ion, which is responsible for the stabilization of a body-centered cubic (bcc) structure at the Cs_3C_{60} stoichiometry, the face-centered cubic (fcc) structure being stable only under externally applied pressure [14.28, 29]. All compounds and phases in Fig. 14.2 show a local maximum in resistivity at $x = 6$, corresponding to the filling of the t_{1u} band, if one electron per alkali metal dopant is transferred to the C_{60} anions.

Since the alkali metal–doped fullerenes are chemically unstable in air and are reactive with other chemical species, great care needs to be exercised in sample handling and in the execution of transport measurements. In general, the electrodes are attached prior to alkali metal doping, and the doping procedure is carried out under the constraints of maintaining the integrity of the electrodes. To minimize the effects of phase separation on the transport measurements of the maximum conductivity phase, great care must be exercised to prepare the M_3C_{60} stoichiometry accurately in the transport samples and to ensure that the samples are well annealed.

Because of the small magnitude of the mean free path ℓ (~2–31 Å) in the M_3C_{60} compounds and because ℓ tends to be comparable to the lattice constant a (14 Å) and to the superconducting coherence length ξ_0 (~25 Å), the effects of crystal defects and grain boundaries are very important, so that film and single-crystal samples tend to exhibit somewhat different detailed transport behavior, especially regarding the magnitude of ρ_{min} and ℓ and the temperature dependence of $\rho(T)$, as discussed in §14.1.2. A theoretical estimate has been given for the mean free path ℓ of 18 Å in M_3C_{60}, considering the effect of strong orientational fluctuations [14.30].

14.1.2. Temperature Dependence

Studies of the temperature dependence of the resistivity of polycrystalline alkali metal–doped M_xC_{60} samples in the normal state show that conduction is by a thermally activated hopping process except for a small range of x near 3 where the conduction is metallic [14.4, 15, 31], as discussed in §14.1.1. As indicated by the arrows in Fig. 14.1, the magnitude of the activation energy E_a increases as x deviates further and further from the resistivity minimum at $x = 3$. For example, E_a has a magnitude of 0.12 eV for $x = 1$ in K_xC_{60} [14.5]. In contrast, the temperature dependence of $\rho(T)$ reported for Na_xC_{60} showed a metallic dependence in the higher temperature range (200–500 K) [14.32], which does not seem consistent with photoemission spectra [14.33, 34] and susceptibility measurements [14.35]. Further work is needed to clarify the behavior of $\rho(T)$ for the Na_xC_{60} compounds.

In the metallic regime, results for $\rho(T)$ for a superconducting single-crystal K_3C_{60} sample [see Fig. 14.3(a)] show a T dependence of $\rho(T)$ with a positive slope [14.15, 31, 36, 37]. A linear temperature dependence for $\rho(T)$ is found over a wide temperature range, when the $\rho(T)$ measurements are made on a sample at constant volume [14.38]. Closer to T_c, some authors have also reported a T^2 dependence for $\rho(T)$ [14.8, 31], and closer examination shows that the T^2 dependence arises from measurements at constant pressure [14.38]. At T_c the mean free path ℓ is comparable to the C_{60}–C_{60} intermolecular distance, and at room temperature ℓ is smaller than the C_{60}–C_{60} intermolecular distance for most film samples, implying relatively easy carrier transport on a singular molecular shell but strong scattering upon transfer to an adjacent shell.

The linear T dependence for the alkali metal–doped fullerenes is much weaker than that for the high T_c-cuprates [14.39]. There are presently significant differences in the functional form of $\rho(T)$ and the magnitudes of ρ at various temperatures from one research group to another for the same nominal single-crystal M_3C_{60} stoichiometry, depending on measurement conditions and crystalline quality [14.15,31,36,37]. Larger differences in the functional dependence of $\rho(T)$ are found when comparisons are made between reported single-crystal data and data on polycrystalline films and powder samples [14.5,16,19,36,40,41]. Disordered films often show a negative slope for $\rho(T)$ above T_c [14.16], as discussed further below in connection with localization effects. Measurements of $\rho(T)$ on K_3C_{60} and Rb_3C_{60} thin films with a grain size of $\sim$1 μm and a sharp superconducting transition show a linear T dependence in the temperature range 280–500 K [14.8], rather similar to the trends observed for $\rho(T)$ for single-crystal samples.

Microwave determinations of $\rho(T)$ at 60 GHz on pressed powder samples have also been reported [14.42], showing a low value of $\rho(T_c^+) \sim 0.5 \times 10^{-3}$ Ω-cm, a functional form for $\rho(T)$ that emphasizes the T^2 dependence and a much larger difference between $\rho(T_c^+)$ and $\rho(300\text{ K})$ than shown in Fig. 14.3. Here $T_c^+ \equiv T_c + \epsilon$ is used to denote temperatures just above the superconducting transition temperature for the superconducting–normal state transition.

Attempts to fit the observed $\rho(T)$ of single-crystal samples in the normal state with theoretical models [14.18,43] indicate that electron–phonon scattering is the dominant scattering mechanism near room temperature and that the lower-frequency intramolecular vibrations play an important role in the carrier scattering. Analysis of infrared measurements [14.44] implies a value of $\sim 0.4 \times 10^{-3}$ Ω-cm for $\rho(T_c^+)$, in good agreement with the microwave result and recent single-crystal dc $\rho(T)$ experiments [14.17,18].

Film samples with small grain size ($\ll$1 μm) tend to exhibit a negative temperature coefficient of $\rho(T)$ above T_c [see Fig. 14.3(b)] while single-crystal samples exhibit a positive $\partial\rho(T)/\partial T$ above T_c [see Fig. 14.3(a)]. The observed temperature dependence of the resistivity of films of M_xC_{60} (for $0 < x \leq 6$) has been interpreted by some authors in terms of a granular conductor with grain sizes in the 60–80 Å range [14.45] with hopping conduction between grains. This granularity strongly affects the superconducting properties of the M_3C_{60} films, as well as their properties in the normal state [14.45]. For more crystalline samples, merohedral disorder is the dominant scattering mechanism at low-temperature, as discussed above.

The normal state of K_3C_{60} is highly resistive, and if the Drude model is used to describe the transport properties, the mean free path is given by

$$\ell = \pi^2 \frac{\hbar/e^2}{\rho k_F^2}, \tag{14.1}$$

yielding an unreasonably short mean free path ℓ of 1–3 Å at room temperature for film samples. A value as small as $\ell = 0.63$ Å has been quoted for Rb_3C_{60} at 525 K based on thin-film samples with a grain size of $\sim$1 μm

[14.8]. Such a small ℓ value is less than the nearest-neighbor carbon–carbon distance and much less than the distance between adjacent C_{60} molecules. From the lattice constant of 14.24 Å and assuming three electrons per C_{60}^{3-} anion, a carrier density of $4.1\times10^{21}/\mathrm{cm}^3$ is obtained, implying $k_F = (\pi^2 n)^{1/3} = 3.4\times10^7\ \mathrm{cm}^{-1}$ on the basis of a nearly free electron model. More detailed calculations yield even lower estimates for k_F (e.g., $1.8\times10^7\ \mathrm{cm}^{-1}$ [14.9]). The short mean free path thus yields $\ell k_F < 1$, which implies a breakdown in the one-electron band picture, and further implies that the carriers are strongly localized. In such a case we should then write the electrical conductivity as

$$\sigma = \sigma_{\mathrm{Drude}} + \Delta\sigma \tag{14.2}$$

where $\Delta\sigma$ is given by localization theory [14.46]. Some authors have identified the apparent short mean free path with a strong reflection of electron waves at the crystallite boundaries (the intermolecular distance), so that the electrons move freely within a crystallite (e.g., a single molecule) but the transmission probability of an electron to a neighboring crystallite (or molecule) is small. As mentioned above, the physical mechanism for the intermolecular scattering is likely the merohedral disorder discussed above. The short mean free path suggests that the normal state of K_3C_{60} might be highly correlated (for instance, due to strong electron–electron interactions and localization effects), so that use of the Boltzmann theory is invalid. The merohedral disorder mechanism is also consistent with an enhanced role for electron–electron interactions in a highly correlated electronic system. In a disordered system, electron–electron interactions play a more important role due to localization effects. An independent estimate of ℓ can then be obtained by comparing experimental $\rho(T)$ results with predictions based on localization theory. Through such an analysis, an electron mean free path of 10–20 Å is estimated for the conduction electrons in K_3C_{60} [14.16], which is of similar magnitude to that obtained for high-quality single-crystal samples [14.17], consistent with a merohedral disorder mechanism. Near T_c, superconducting fluctuations are observed in K_3C_{60} in both single crystals and thin films, and the results for K_3C_{60} film samples are in good agreement with the fluctuation-induced tunneling model of Sheng [14.47].

14.1.3. Alkaline Earth–Doped C_{60}

The dependence of the resistivity on dopant concentration appears to be more complex in the alkaline earth compounds than for the alkali metal dopants for several reasons: the divalence of the alkaline earth ions, their small physical size, and the variety of crystal structures which they exhibit in the solid state. Thus far, special attention has been given to the dependence of the transport properties on stoichiometry, and explicit measurements are available for Ca, Sr, and Ba doping. For example, plots similar to those in Fig. 14.2 show that Sr addition to C_{60} gives rise to resistivity minima at $x = 2$ and 5, and maxima at $x = 3$ and 6.5, as shown in Fig. 14.4. Similar transport results were obtained with the other alkaline earths [14.48, 49], and results for the stoichiometries for the minimum and maximum resistivities are given in Table 14.2, together with the corresponding activation energies for Sr_xC_{60}, Ca_xC_{60}, and Ba_xC_{60} at the listed resistivity extrema. Referring to Fig. 14.4, two minima in the resistivity are observed in Sr_xC_{60} with increasing x, as well as two maxima. The first minimum in $\rho(x)$ occurs at $x_{\min} \simeq 2$ for both Sr_xC_{60} and Ca_xC_{60}, corresponding to the occupation of all the tetrahedral sites with alkaline earth dopants; both the resistivities and activation energies are relatively high at $\rho_{\min 1}$. Upon further addition of alkaline earth dopant, the t_{1u} bands become filled at $x = 3$, corresponding to two electrons transferred to the C_{60} anion per alkaline earth ion. Further addition of Sr and Ca results in occupation of the t_{1g} bands (see Fig. 12.2). At the stoichiometries $x \simeq 5$, a second minimum in resistivity $\rho_{\min 2}$ is observed, corresponding to much lower values of both resistivity and activation energy (see Table 14.2) and implying an approximately half filled t_{1g} band [14.50, 51]. At higher doping levels, it is believed that somewhat less than two electrons per alka-

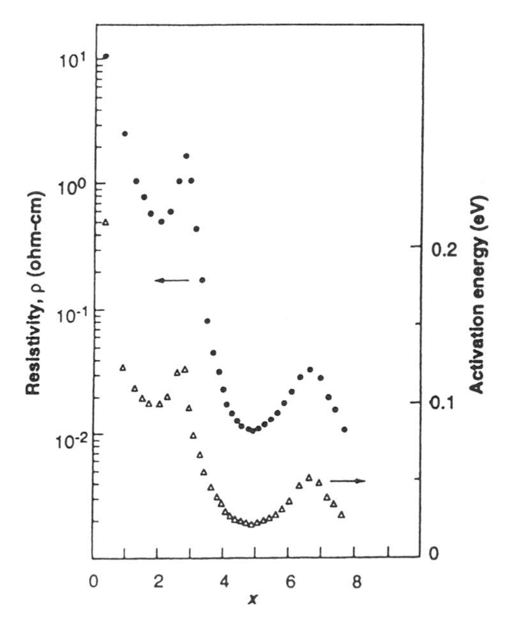

Fig. 14.4. The resistivity and the activation energy of the conductivity of an annealed Sr_xC_{60} film 190 Å thick at 60°C in ultrahigh vacuum (UHV) as a function of x. The stoichiometry was determined by use of a quartz crystal microbalance and *ex situ* RBS (Rutherford backscattering) measurements [14.48].

Table 14.2

Summary of transport data for alkaline earth–doped fullerenes M_xC_{60} [14.48, 49].

Alkaline earth	Stoichiometry	Resistivity (Ω-cm)	Activation Energy (eV)
Sr	$x_{min1} = 2.0$	0.5	0.10
	$x_{max1} = 2.9$	2.5	0.12
	$x_{min2} = 4.9$	1.1×10^{-2}	0.021
	$x_{max2} = 6.6$	3.2×10^{-2}	0.05
Ca	$x_{min1} = 2.0$	1.0	0.14
	$x_{max1} = 2.3$	1.1	0.15
	$x_{min2} = 5.0$	5.8×10^{-3}	0.018
	$x_{max2} = 7.1$	2.2×10^{-2}	0.05
Ba	$x_{min1} = 1.8$	0.25	0.07 ($x \sim 1.2$)
	$x_{max1} = 3.0$	$\gg$0.5	0.10 ($x \sim 2.5$)
	$x_{min2} = 5.0$	2.9×10^{-3}	0.0

line earth dopant atom are transferred to the C_{60} molecule, due to hybridization of the t_{1g}-derived band with the s band of the alkaline earth (see §12.7.4).

The behavior of Ba_xC_{60} (see Fig. 14.5) has many similar features to Sr_xC_{60} and Ca_xC_{60}, but also some differences, attributed to the larger size of the Ba^{2+} ion relative to the size of the tetrahedral vacancy, while the opposite is true for Ca^{2+} and Sr^{2+} (see Table 8.1). Thus, a stable phase for Ba_xC_{60} is believed to occur at $x \sim 1$, where Ba^{2+} occupies the octahedral site [14.49] in the fcc structure. At $x = 3$ where the t_{1u} band is filled, Ba_3C_{60} exhibits the A15 crystal structure, while the superconducting compound Ba_6C_{60} exhibits the bcc crystal structure [14.52] (see §8.6.2). The resistivity at the composition Ba_6C_{60} is the lowest thus far reported for an alkaline earth–doped fullerene solid (see Table 14.2).

Studies of the Ba_xC_{60} system demonstrate the importance of high temperature annealing for achieving greater stoichiometric homogeneity and better crystallinity [14.49]. It is also of interest to see that for an annealing temperature of 220°C, metallic conduction ($E_a = 0$) is achieved over a wide range of x values near $x = 5$. The samples used for superconductivity studies (see §15.1) were at a stoichiometry Ba_6C_{60} and were annealed well above 220°C [14.52].

It has also been reported that partial occupation of the t_{1g} band could be achieved by observing photoemission spectra from C_{60} on potassium surfaces [14.53].

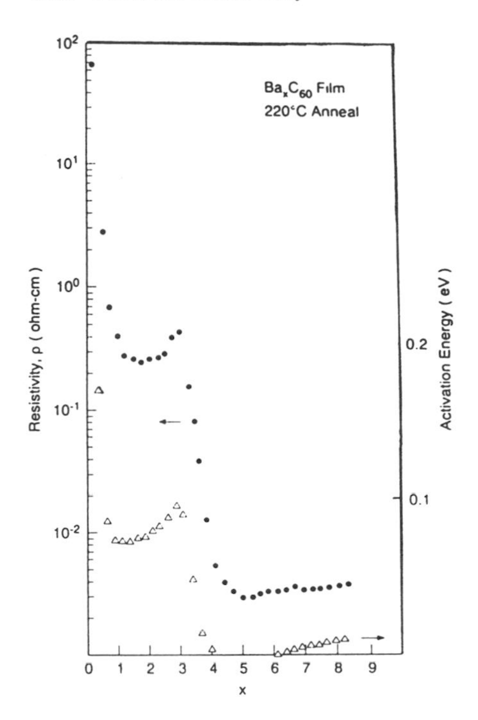

Fig. 14.5. The resistivity (filled circles) and the activation energy (open triangles) of the conductivity of a 212 Å thick Ba_xC_{60} film at 60°C in UHV as a function of x after annealing at 220°C. The stoichiometry was determined by use of a quartz crystal microbalance and *ex situ* RBS measurements [14.48].

14.1.4. *Alkali Metal–Doped C_{70}*

Electrical resistivity measurements have also been reported on M_xC_{70} [14.16, 41]. Pristine C_{70} is insulating, just as is undoped C_{60}, and this has been established both electrically and optically. Optical measurements have yielded an edge in the optical absorption at ~1.4 eV [14.54]. Unlike the behavior found for $\rho(x)$ in M_xC_{60} compounds, where there is only one minimum, and this minimum is at $x = 3$ (see Fig. 14.2), two minima in $\rho(x)$ are found for M_xC_{70} [14.26, 27], in good agreement with the electron paramagnetic resonance (EPR) studies on K_xC_{70} [14.55], where the spin concentration was found to go through two maxima as a function of doping. The susceptibility at the second maximum in the doping process showed Pauli-like behavior and the stoichiometry at this doping level was found to be K_4C_{70}. These results are also consistent with band structure calculations [14.55, 56], which show that the D_{5h} symmetry of the C_{70} molecule results in unfilled A_1'' and E_1'' levels above the Fermi level (see Fig. 12.3), which are orbitally singly and doubly degenerate, respectively [14.56]. One thus expects from one-electron band theory that K_1C_{70} and K_4C_{70} should have conductivity minima, due to the half-filling of their respective conduction bands. If the A_1'' states are indeed lower in energy, as indicated by band calculations (see Fig. 12.3), then K_4C_{70} is supposed to be more conducting, owing to its higher density of states at the Fermi level [14.56]. Values observed for ρ_{min} at 300 K for the two resistivity minima are $\rho_{min} \sim 0.5$ Ω-cm and 1.7 mΩ-cm at the first (K_1C_{70}) and second (K_4C_{70}) resistivity minimum, respectively, consistent with the trends predicted by the theoretical calculations.

At the second, much deeper, resistivity minimum, the temperature dependence of $\rho(T)$ was investigated [14.26] and a nonmetallic T dependence was found for disordered film K_xC_{70} samples. The observed functional form of $\rho(T)$ for K_4C_{70} [14.26] was well fit by the fluctuation-induced-tunneling (FIT) model [14.47] over the temperature range $4 < T < 300$ K. These results were interpreted in terms of a microstructure characteristic of a heterogeneous system in which small insulating barriers separate metallic regions, and within the metallic regions, the electron states are weakly localized. No superconducting transition was observed for $T > 1.35$ K. No electrical resistivity measurements are yet available on alkali metal–doped C_{70} single crystals.

14.1.5. Transport in Metal–C_{60} Multilayer Structures

Metal–C_{60} multilayers provide another class of fullerene-based materials (see §8.7.4) with interesting transport properties [14.57]. Several experiments, including x-ray scattering, *in situ* resistance, and Raman scattering measurements, have now been performed on metal–C_{60} multilayers, showing charge transfer between the metal layers and the C_{60} layers and a general decrease in resistance with increasing numbers of superlattice unit cells.

To illustrate the transport behavior of metal–C_{60} multilayers, consider the Al/C_{60} system shown in the inset of Fig. 14.6. The metal species for all multilayer systems studied thus far was selected to have stronger bonding to itself than to C_{60}, so that the metal does not intercalate into the C_{60} lattice. The inset to Fig. 14.6 shows C_{60} deposited on a substrate, in this case a polished (100) single crystal of yttrium-stabilized zirconia (YSZ), onto which contacts are deposited for four-terminal Van der Pauw transport measurements. The resistance measurements in Fig. 14.6 show that the presence of a C_{60} layer between the Al and the substrate significantly changes the resistance of a given thickness of Al deposited on the substrate. Figure 14.7 shows the decrease in resistance measured after depositing increasing numbers (N) of Al/C_{60} superlattice unit cells. When Al is deposited on C_{60}, it is likely that some Al diffuses into the C_{60} layers (marked in the inset to Fig. 14.6 as Al_xC_{60}). The mixed Al_xC_{60} layer is conducting because of the Al clusters and the charge transfer to the C_{60}, arising through the strong Al–C_{60} interaction. At the C_{60}–Al interface a doped monolayer (DML) of

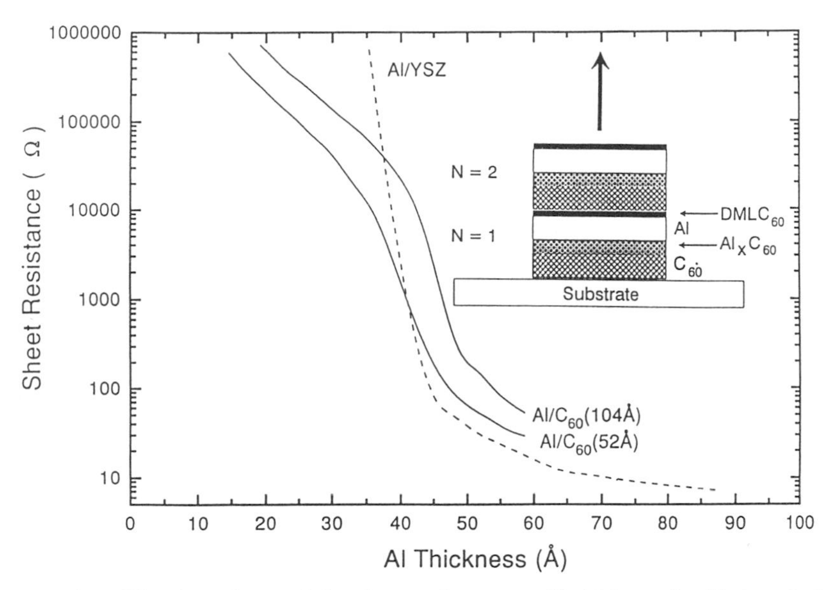

Fig. 14.6. The dependence of the sheet resistance on Al thickness for Al deposited directly on yttrium-stabilized zirconia (YSZ, dashed line) and on an underlying layer of C_{60} (solid lines). The inset shows schematically the Al/C_{60} superlattice, including the doped C_{60} monolayer (DML) and the Al_xC_{60} phases at the Al–C_{60} interfaces [14.57].

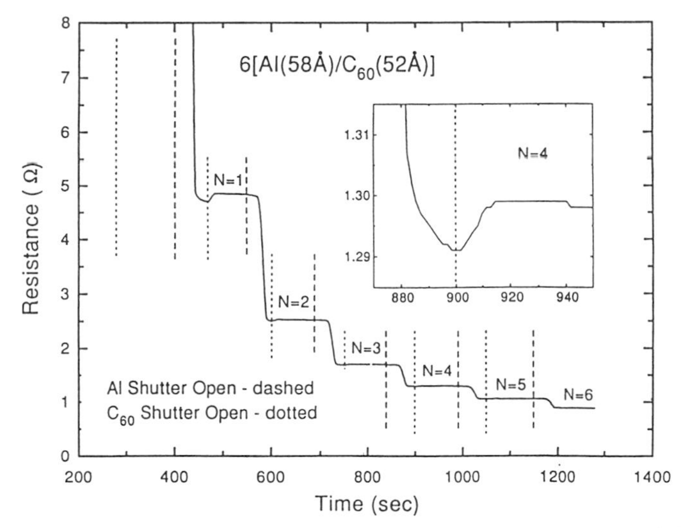

Fig. 14.7. The dependence of resistivity on deposition time for a six-layer Al/C_{60} superlattice sample, consisting of an alternating sequence of 52 Å of C_{60} followed by 58 Å of Al. The vertical dashed (dotted) lines delineate the Al (C_{60}) shutter openings. The expanded data of the inset show the increase in resistance at the beginning of the $N = 4$ plateau when C_{60} is deposited after the C_{60} shutter is opened [14.57].

C_{60} is formed with up to six electrons transferred from the Al to the C_{60} and this doped monolayer also contributes to the conduction process. Referring to Fig. 14.7, it is seen that the resistance decreases only when the Al is deposited but remains constant while the C_{60} is deposited. The resistance values at each plateau show a dependence $(1/R_N) = N/R_0$ where N is the number of superlattice periods and the change in resistance δR_N at each C_{60}/Al interface is proportional to $1/N^2$. These observations suggest that each Al layer contributes equally to the conductance, and the effect of the addition of each C_{60} layer is similar. In the limit of large N, the resistivity of the superlattice approaches that of aluminum.

Another group [14.58] used Sn, Ba, Ga, In, and Ag as metals for forming metal/C_{60} multilayer films on sapphire substrates and these authors used four-terminal resistance measurements similar to those described above [14.57]. Typical superlattice thicknesses were C_{60} (5 nm) and metal (1.5 nm). Whereas Ag and In showed little effect on sample resistance, the normalized surface resistance of the multilayers with Sn, Ba, and Ga showed a decrease in resistance with increasing numbers of superlattice layers, qualitatively similar to the behavior shown in Fig. 14.7, but with the magnitudes of the decrease in resistance strongly dependent on the metal species [14.58]. Further studies on metal bilayers [14.59] showed that the resistance drop occurred mainly within the deposition of the first C_{60} monolayer and the decrease in resistance was greater when the metal species had a higher intrinsic resistance. The metal species for very thin layers normally deposits in disconnected islands, and the presence of the C_{60} provides a conduction path between the islands through charge transfer to the fullerenes. Measurements on the Cu–C_{60} bilayer system showed a metallic temperature dependence of resistance and that the strongest metal–C_{60} interaction was with the interface doped C_{60} monolayer [14.59].

14.1.6. Lewis Acid–Doped C_{60}

Although it is generally reported that acceptor dopants do not transfer charge to fullerenes, there is one report of good conductivity of C_{60} reacted with the Lewis acids $SbCl_5$, AsF_5, and SbF_5 [14.60]. Measurements of the temperature dependence of the resistance (see Fig. 14.8) over a small temperature range (200–260 K) were fit to an Arrhenius plot yielding values of the thermal band gap of 1.1 eV for $SbCl_5$, 0.31 eV for SbF_5, and 0.16 eV for AsF_5. Because of the tendency for these Lewis acids to disproportionate [14.61, 62], the species in the compound may well be different from the initial dopant. By analogy with the behavior of these Lewis acids as acceptor intercalate species for graphite intercalation compounds [14.61], the implication is that conduction in these Lewis acid–doped fullerides is

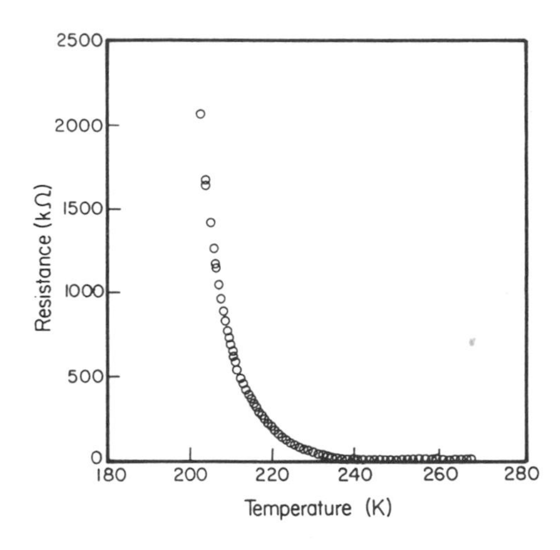

Fig. 14.8. The resistance of C_{60} reacted with $SbCl_5$ as a function of temperature showing good conduction near room temperature for this Lewis acid–doped C_{60} material [14.60].

by holes [14.60], although direct verification of hole conduction has not yet been reported.

14.2. Electron–Phonon Interaction

Since the electron–phonon interaction is presently believed to play an important role in electron pairing for superconducting doped fullerenes [14.63], the electron–phonon interaction for doped fullerenes has been intensively discussed. This discussion has focused on the unique properties of doped fullerenes regarding the electron–phonon interaction, estimates for the magnitude of this interaction, and the relative importance of the various phonons in their coupling to the electrons. In this section we also raise the possibility that transport properties are more strongly influenced by intermolecular charge scattering than by intramolecular vibrations, whereas the superconducting pairing is believed to be predominantly associated with intramolecular processes.

14.2.1. Special Properties of Fullerenes

The electron–phonon coupling in superconducting compounds such as K_3C_{60} has some unusual properties. The first unusual property arises from the low average electron density, the unusually low Thomas–Fermi screening length of these electrons, and the narrow widths of the energy bands. Although the nearest-neighbor C–C intramolecular and intermolecular distances are small (with average values of 1.4 Å and 3.0 Å, respectively), the lattice constants are large (14.24 Å for K_3C_{60} and 14.38 Å for Rb_3C_{60}), leading to a low average density of conduction electrons $n = 4.2 \times 10^{21}/\text{cm}^3$ for K_3C_{60} and $n = 4.0 \times 10^{21}/\text{cm}^3$ for Rb_3C_{60} (see Table 14.1).

On the basis of a simple free electron model, the Fermi energy of an electron gas of this density would be $E_F^0 = (\hbar^2/2m)(3\pi^2 n)^{2/3} = 0.9$ eV, with a density of states at the Fermi energy of $N^0(E_F) = 0.5$ states/eV-C_{60}, a Fermi velocity $v_F^0 = 6 \times 10^7$ cm/s, and a Fermi wave vector of $k_F = 3.4 \times 10^7$ cm^{-1}. The electron density of the crystalline K_3C_{60} is, however, not homogeneously distributed, but is largely confined to the icosahedral surfaces of the fullerene molecules. This distinction is of great importance in discussing both the normal state and superconducting properties, and this inhomogeneous charge distribution has not always been fully appreciated in the literature. Thus local density approximation (LDA) band structure calculations yield rather different values from those quoted above for a uniform free electron gas of the same density. For example, LDA calculations yield values of $E_F \sim 0.25$ eV, $N(E_F) = 13$ states/eV-C_{60} and $\langle v_F^2 \rangle^{1/2} = 1.8 \times 10^7$ cm/s [14.9], consistent with the narrow widths of the energy bands implied by the molecular nature of this highly correlated solid. Despite the low average electron density, the high density of states at the Fermi level suggests that screening effects are important. An estimate of the static Thomas–Fermi screening length ℓ_{sc} is given by

$$\ell_{sc} = [4\pi e^2 N(E_F)]^{-1/2} \tag{14.3}$$

yielding $\ell_{sc} \sim 0.5$ Å for K_3C_{60} as compared to that ($\ell_{sc}^0 \sim 2.5$ Å) for a homogeneous electron gas of the same electron density, which has an average electron–electron separation of ~ 7.5 Å.

A second unusual aspect of doped fullerenes that is important to the electron–phonon interaction relates to the unusual properties of the vibrational spectra, which include intramolecular vibrations spanning a large frequency range and intermolecular vibrations and librations at very low energies (see §11.4). Furthermore, the highest vibrational frequencies ($\sim$0.2 eV) are comparable to the occupied bandwidth ($\sim$0.25 eV), so that phonons not only scatter carriers on the Fermi surface but also can scatter carriers far away from E_F. Thus, carrier scattering in M_3C_{60} compounds may, in fact, involve most of the electron states of the t_{1u}-derived bands and states throughout the Brillouin zone. The similarity of the energy scales for the electrons and intramolecular phonons may violate Migdal's theorem, as has been discussed by various authors [14.64–68]. The Migdal criterion, that the phonon energy satisfy $\hbar\omega \ll \hbar \mathbf{v}_F \cdot \mathbf{q}$, is not met for some high-energy phonons with $\mathbf{q}$ near the Brillouin zone boundary [14.69]. This criterion is, however, satisfied for most phonon branches and for most $\mathbf{q}$ values. Detailed calculations show that the average electron velocity is small only near the band edge and zone boundaries and that orientational (merohedral) disorder (see §7.1.4 and §8.5) does not strongly affect the mean carrier velocity [14.70].

Thus, although the Migdal approximation is not strictly satisfied, many of the electrons and phonons in doped fullerenes do satisfy the criterion, and it is thought that the expressions usually used for the electron–phonon interaction are likely to provide a useful approximation also for the superconducting fullerene-derived compounds [14.9]. It is, however, believed that because the Migdal criterion is not strictly met, caution is needed in interpreting certain experimental data, such as the pressure dependence of T_c and the isotope effect [14.71].

Finally, we briefly mention questions concerning the validity of Bardeen–Cooper–Schrieffer (BCS) theory for C_{60} in terms of Migdal's theory. Migdal's theory states that vertex correction of the electron–phonon interaction may be small (~1/100) in ordinary BCS superconductivity. The factor that specifies the magnitude of the vertex correction is $\hbar\omega_D/E_F$ or $(m/M)^{1/2}$ where m and M are masses of electrons and ions, respectively [14.72]. The vertex correction is due to the fact that the sound velocity is much smaller than the Fermi velocity. If we neglect this vertex correction, then the integral equation for the self-energy (or free energy) can be solved approximately using Green's functions without any vertex correction, thus giving the result of the original BCS theory. In the case of C_{60}, the ratio of $(m/M)^{1/2}$ may still be small, even though $\hbar\omega_D/E_F$ is on the order of unity. Based on this fact, Pietronero and Strässler [14.73] have calculated a vertex correction of the electron–phonon interaction, and they show that a nonadiabatic effect produces a strong enhancement of T_c and implies the breakdown of Migdal's theorem. However, it is still unknown whether vertex corrections from other vertices could contribute to the electron–phonon interaction. This possibility should be examined in the future. Furthermore, in the nonadiabatic case, the renormalization of the electron–phonon interaction and the validity of the Fermi-liquid picture in the nonadiabatic regime based on a realistic electronic structure are important subjects for further investigation.

14.2.2. Magnitudes of the Coupling Constant

Taking into account some of the unusual properties of the doped fullerenes, a number of attempts have been made to estimate the magnitude of the electron–phonon coupling coefficient λ_{ep} and to identify the phonons that are most important in the electron pairing that gives rise to superconductivity. The magnitude of λ_{ep} enters directly into the BCS equation for T_c and depends strongly on the magnitude of the electron–phonon interaction discussed in this section and the electronic density of states at the Fermi level discussed in §14.5.

In the various estimates of the average electron–phonon coupling coefficient λ_{ep}, it is customary to sum over the contributions λ_{q_ν} from various phonon branches and wave vectors q_ν

$$\lambda_{ep} = \frac{1}{3n_b}\sum_{\nu=1}^{3n_b}\frac{1}{N}\sum_{q=1}^{N}\lambda_{q_\nu} \tag{14.4}$$

where n_b denotes the number of phonon branches and N is the number of wave vectors in the Brillouin zone [14.74].

In the BCS theory of superconductivity, the dimensionless coupling coefficient, λ_{ep}, is given by the simplest vertex of the electron–phonon coupling,

$$\lambda_{ep} = \frac{2}{N(E_F)}\sum_{p,\mathbf{q}}\frac{1}{2k_p(\mathbf{q})}\sum_{n,n',\mathbf{k},\mathbf{k}'}|I_{n\mathbf{k},n'\mathbf{k}'}(p,\mathbf{q})|^2\delta(\epsilon_{n\mathbf{k}})\delta(\epsilon_{n'\mathbf{k}'}) \tag{14.5}$$

where $\epsilon_{n\mathbf{k}}$ denotes the energy of a Bloch electron in band n and wave vector $\mathbf{k}$, and $I_{n\mathbf{k},n'\mathbf{k}'}(p,\mathbf{q})$ is the matrix element of the electron–phonon interaction between electronic states $n\mathbf{k}$ and $n'\mathbf{k}'$ mediated by the pth phonon with wave vector $\mathbf{q} = \mathbf{k} - \mathbf{k}'$ [14.75, 76]. The delta functions in Eq. (14.5) restrict the sum so that both the wave vectors for the initial and final electronic states $\mathbf{k}$ and $\mathbf{k}'$ are on the Fermi surface $\epsilon = 0$. Also in Eq. (14.5), $N(E_F)$ is the density of states at the Fermi level per spin orientation, $K_p(\mathbf{q}) = M\omega_p^2(\mathbf{q})$ is the force constant of the pth phonon, $\omega_p(\mathbf{q})$ is the frequency of a phonon on the pth branch with wave vector $\mathbf{q}$, and M is the reduced mass. Schlüter *et al.* [14.76] proceeded to evaluate ϵ_{nk} using the tight-binding representation for the Bloch states and to show that the derivative of an atomic hopping matrix element of an atom at a specific site is proportional to the matrix element $I_{n\mathbf{k},n'\mathbf{k}'}(p,\mathbf{q})$ in Eq. (14.5). These authors further show that all contributions to the electron–phonon coupling that contribute to pairing are small except for the phonons that modulate the strong intramolecular π–π overlap of the t_{1u} states. The calculations also show that the intramolecular vibrations are dominant, especially those with high frequencies and H_g symmetry, modes that contribute to Jahn–Teller distortions and to symmetry lowering for the M_3C_{60} compounds, as discussed further below.

In the limit that only the intramolecular transfer integral contributes to the electron–phonon coupling coefficient λ_{ep}, Schlüter *et al.* [14.76] have shown that

$$\lambda_{ep} = N(E_F)V = N(E_F)\sum_p V_p = N(E_F)\sum_{p,\mu}\frac{\mathrm{Trace}(I^2)_{p,\mu}}{9M\omega_p^2} \tag{14.6}$$

where V_p is the average electron–phonon coupling energy in the (3×3) matrix of the subspace associated with the threefold degenerate t_{1u} LUMO bands, and the trace appearing in Eq. (14.6) is taken over the same (3×3) space. The sum on p in Eq. (14.6) is over the normal modes of an isolated C_{60}^{3-} ion and the sum on μ is over the degeneracy of each mode [14.77]. The matrix elements in Eq. (14.6) were calculated by noting the connection of the electron–phonon coupling problem to the Jahn–Teller distortion of a molecule with a degenerate ground state (t_{1u}) in an icosahedral field. It has been argued that the intramolecular Jahn–Teller modes are the ones which directly modulate nearest-neighbor wave function overlaps, so that the screening effect [14.78], which strongly reduces other contributions to the electron–phonon coupling, has a relatively small effect on the intramolecular Jahn–Teller (Hg) modes.

Results for the contribution to $V = \sum_p V_p$ from each of the H_g modes are given in Table 14.3. The first column givcs rcsults obtained from an LDA calculation [14.75] for which the weighted average V and $\bar{\omega}_{\text{ph}}$ for the eight H_g modes are taken to be 52.0 meV and 950 cm^{-1}, respectively [14.76]. Here $\bar{\omega}_{\text{ph}}$ is an average phonon frequency used in McMillan's formula (see §15.2). A comparison of these values with results obtained

Table 14.3

Contribution of H_g and A_g intramolecular modes to the electron–phonon coupling potential.

	ω_p	V_p				
Mode (p)	(cm^{-1})	(meV)^a	(meV)^b	(meV)^c	(meV)^d	(meV)^e
$H_g(1)$	273.0	8.0	3.0	0.7	3.0	3.0
$H_g(2)$	432.0	7.0	2.4	0.2	1.0	6.0
$H_g(3)$	711.0	4.0	6.0	1.8	1.0	3.0
$H_g(4)$	775.0	7.0	4.8	4.5	0.0	3.0
$H_g(5)$	1101.0	1.0	0.0	7.9	6.0	3.0
$H_g(6)$	1251.0	3.0	0.6	5.4	0.0	8.0
$H_g(7)$	1426.5	13.0	7.0	7.8	34.0	20.0
$H_g(8)$	1577.5	9.0	8.4	4.6	11.0	22.0
$A_g(1)$	497.0	—	0.0	1.0	—	—
$A_g(2)$	1470.0	—	5.0	1.6	—	—

^{a}LDA method of Ref. [14.75, 77].
^{b}Bond charge method of Ref. [14.75, 79].
^{c}Force constant method of Ref. [14.80].
^{d}Deformation potential coupling contributions from various phonons in Ref. [14.81].
^{e}Linear muffin tin orbital (LMTO) method of Ref. [14.82].

by several other models is also included in the table. The results for several calculations cover a range of V from 32.2 to 82.2 meV and $\bar{\omega}_{\rm ph}$ from 786 to 1320 cm^{-1} [14.77], where the logarithmically averaged phonon frequency, $\bar{\omega}_{\rm ph}$ is defined by

$$\ln \bar{\omega}_{\rm ph} = \frac{1}{\lambda_{ep}} \sum_p \lambda_p \ln \omega_p. \tag{14.7}$$

Here λ_p is the contribution of the pth vibration to the electron–phonon coupling coefficient λ_{ep} (see Table 14.3) and is given by

$$\lambda_p = N(E_F)V_p. \tag{14.8}$$

We will return to a discussion of the parameters V, $\bar{\omega}_{\rm ph}$, and λ_{ep} in Chapter 15 (§15.7) in connection with the superconducting properties of doped fullerenes.

14.2.3. Symmetry Considerations

From a symmetry standpoint, the coupling of two electrons in the LUMO t_{1u} level is through vibrations with symmetries arising from the direct product of the corresponding irreducible representations $T_{1u} \otimes T_{1u} = A_g + T_{1g} + H_g$, using the irreducible representations of the icosahedral I_h group, where T and F are used interchangeably to label the threefold irreducible representations. Since T_{1g} is asymmetric under interchange of two electrons, the electrons do not in lowest order couple through vibrations with T_{1g} symmetry, and for this reason only the H_g and A_g modes for C_{60} are listed in Table 14.3. Coupling to A_g modes does not change the symmetry of the molecule and therefore cannot give rise to Jahn–Teller distortions, but the eight H_g modes can deform the fullerene dynamically. It is for this reason that the discussion of the phonon contribution to the electron–phonon interaction has focused on the H_g modes. Furthermore, two electrons in the t_{1g} level (as occurs for the alkaline earth compounds) couple predominantly through the same vibrations as do two t_{1u} electrons, since the direct product for the two t_{1g} electrons yields the same irreducible representations as for two t_{1u} electrons: $T_{1g} \otimes T_{1g} = A_g + T_{1g} + H_g$. At the present time there is no strong consensus about whether it is the high- or low-frequency phonons that are dominant in determining λ_{ep}, or whether both low- and high-frequency modes might perhaps contribute in a significant way to λ_{ep}. Almost all the attention given to the electron–phonon interaction in doped fullerene solids has been directed toward gaining a better understanding of the superconductivity pairing mechanism. Relatively less attention has been given to the electron–phonon interaction as it affects normal state transport in these systems.

14.2.4. Jahn–Teller Effects

Regarding the significance of the Jahn–Teller (JT) effect for the electron–phonon interaction, both static and dynamic JT effects must be considered. In the static JT effect, a structural distortion lowers the symmetry of the system and lifts the degeneracy of the ground state. For a partially filled band, such a distortion leads to a lowering of the energy of the system as the lower energy states of the multiplet are occupied and the higher-lying states remain empty. An example of the static JT effect is the bond alternation or Kekulé structure of the hexagonal rings in neutral C_{60} and its ions, which has been discussed [14.83–85] in terms of the Su–Schrieffer–Heeger (SSH) model, originally developed for polyacetylene $(CH)_x$ [14.86]. It should, however, be noted that the distortions of the Kekulé structure due to bond alternation in neutral C_{60} do not lower the icosahedral symmetry of the molecule. In the case of the C_{60} ions, a change of bond alternation does, however, change the degeneracy of the electronic levels. The degeneracy is lowered when additional electrons are introduced to partially fill the t_{1u} levels.

The dynamic JT effect [14.87] can occur when there is more than one possible distortion that could lead to a lowering of the symmetry (and consequently also a lowering of the energy) of the system. If the potential minima of the adiabatic potential are degenerate for some symmetry-lowered states of a molecule, the electrons will jump from one potential minimum to another, utilizing their vibrational energy, and if this hopping occurs on the same time scale as atomic or molecular vibrations, then no static distortion will be observed by most experimental probes for the dynamic JT effect. Those vibrational modes which induce the dynamic JT effect contribute strongly to the electron–phonon coupling.

One of the early discussions of the dynamic Jahn–Teller effect for doped fullerenes was given by Johnson *et al.* [14.78], who proposed the dynamic JT effect as a pairing mechanism for superconductivity. The Jahn–Teller effect (whether static or dynamic) plays an important role in determining the electronic states for the charged C_{60} molecules and has been widely discussed as an important mechanism for electron–phonon coupling in the doped fullerenes [14.75, 77, 81, 88]. For the case of the addition of three electrons to each fullerene by alkali metal doping, model calculations [14.89] show that the degenerate t_{1u}-derived LUMO ground state is split into three non-degenerate levels. This splitting can be understood in terms of a static Jahn–Teller effect for degenerate t_{1u} levels. No geometrical optimization of the structure of the alkali-doped C_{60} solid has yet been carried out. As for a single C_{60} molecule, a "Modified Intermediate Neglect of Differential Overlap, version 3.0, Molecular Orbital " (MINDO/3 MO) [14.90] calculation with the unrestricted Hartree–Fock (UHF) approach shows that the singly charged anion C_{60}^{-1} is deformed from I_h to D_{3d} symmetry by the Jahn–Teller effect [14.91]. It is noted here that the splitting of the t_{1u} levels comes not only from the lattice deformation, but also from the symmetry-lowering effect of the static Coulomb interaction caused by the additional electrons on the fullerene anions due to charge transfer and by the alkali metal ion dopants. In fact, "Modified Neglect of Diatomic Overlap, Configuration Interaction" (MNDO-CI) calculations for C_{60}^{n-} $(0 \leq n \leq 5)$, have been carried out, and the results show that a symmetry-lowering effect on the electronic structure occurs for the various ground states that were degenerate in I_h symmetry [14.92]. Anisotropy in the EPR lineshapes provides evidence for Jahn–Teller distortions for negatively charged fullerene anions (see §16.2.2).

14.3. Hall Coefficient

For simple one-carrier type conductors, the Hall coefficient provides a direct measurement of the carrier density through the relation $R_H = 1/ne$ (in MKS units), where e is the charge on the electron. Since pristine C_{60} and C_{70} have a very low carrier density, no Hall effect has been measured on these materials. Thus Hall effect measurements have been limited to doped fullerenes near their minimum resistivity stoichiometries (see §14.1).

Temperature-dependent Hall effect measurements have been carried out in the temperature range 30 to 260 K on a K_3C_{60} thin film [14.45]. Assuming three electrons per C_{60}, the expected Hall coefficient R_H based on a one carrier model $(-1.5 \times 10^{-9} m^3/C)$ would be about a factor of 4 greater in magnitude than the experimentally observed value $R_H = -0.35 \times 10^{-9}$ m^3/C at low T [14.45]. Experimentally R_H changes sign from negative below 220 K to positive above 220 K (see Fig. 14.9) [14.45]. The small value of the observed Hall coefficient suggests multiple carrier pockets including both electrons and holes, consistent with band structure calculations (see §12.7.3). A reduction in the magnitude of the Hall coefficient could also arise from merohedral disorder effects (see §8.5) [14.23]. Since three electrons occupy states in the three f_{1u} (t_{1u}) bands, it is reasonable to expect both electron and hole pockets in the Brillouin zone from elementary considerations. Multiple carrier types are also consistent with the Fermi surface calculations by Erwin and Pederson [14.93] and by Oshiyama *et al.* [14.13], who also found both electron and hole orbits on the Fermi surface.

Measurements of $R_H(T)$ on K_4C_{70} show similar behavior to that found for K_3C_{60} with small magnitudes of R_H $(-2 \times 10^{-9} m^3/C$ at low $T)$ and a crossover from negative to positive R_H at ~80 K [14.27]. Incidentally, the Hall coefficient R_H of K-intercalated graphite intercalation compounds (GICs) (see §2.13) also changes sign in the same temperature range [14.94].

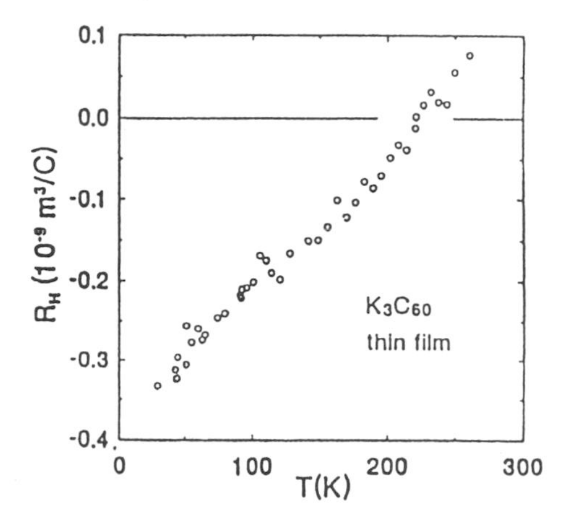

Fig. 14.9. Temperature dependence of the Hall coefficient $R_H(T)$ for K_3C_{60} [14.45].

For many materials, the Hall coefficient and the Seebeck coefficient provide complementary information on the sign of the dominant carrier and the carrier density. Since the M_xC_{60} compounds pertain to carrier occupation of several bands, the connection between the Hall coefficient and the Seebeck coefficient (see §14.12) is less direct than for simple one-band metals. Nevertheless, it is of interest to observe the similarity of behaviors of $R_H(T)$ for K_3C_{60} and K_4C_{70}, on one hand, and the temperature-dependent thermopower $S(T)$ behavior for the same compounds (see §14.12).

14.4. MAGNETORESISTANCE

Most of the emphasis thus far given to study the magnetic field dependence of the resistivity has gone into study of the shift of the superconducting transition temperature with magnetic field [14.17, 18, 45], and this topic is discussed in §15.3. Few experiments have, however, been done to study the temperature and magnetic field dependence of the magnetoresistance in the normal state. The magnetoresistance is typically defined as $(\Delta\rho/\rho_0) \equiv [\rho(H)-\rho(0)]/\rho(0)$. The most extensive normal state magnetoresistance studies presently available are for K_3C_{60} [14.16] and K_4C_{70} [14.26] film samples, where the emphasis was to use the T and H dependence of $\Delta\rho/\rho_0$ to gain understanding of the nature of the disorder for K_3C_{60} and K_4C_{70} thin films.

As a function of magnetic field, the magnetoresistance $\Delta\rho/\rho_0$ observed for thin-film K_3C_{60} samples (see Fig. 14.10) for temperatures up to 30 K and fields up to 15 tesla was divided into two terms

$$\frac{\Delta\rho}{\rho_0} = \frac{(\Delta\rho)_C}{\rho_0} + \frac{(\Delta\rho)_{L,I}}{\rho_0} \tag{14.9}$$

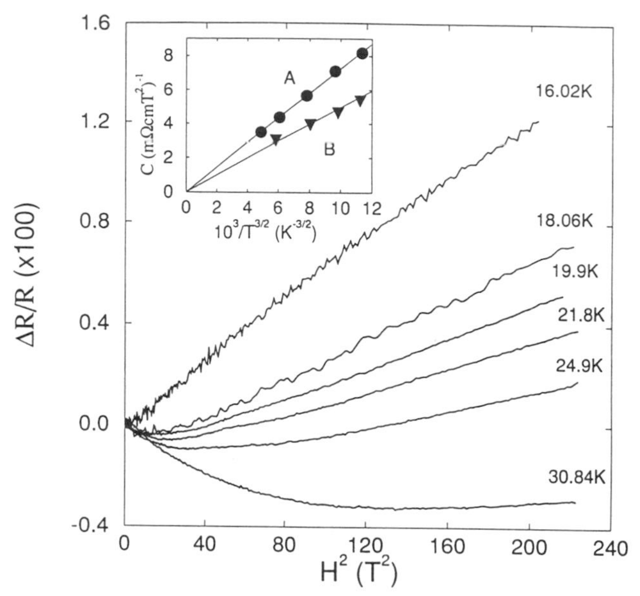

Fig. 14.10. Transverse magnetoresistance of a thin-film K_3C_{60} sample plotted as $\Delta R/R$ in % *vs.* H^2 for various temperatures $T > T_{max}$. The quadratic coefficient of the magnetoconductance C for two K_3C_{60} samples (A and B) is plotted *vs.* $T^{-3/2}$ in the inset [14.16].

where the first term $(\Delta\rho)_C/\rho_0$ (denoting classical magnetoresistance) is positive and quadratic at high H, while the second term $(\Delta\rho)_{L,I}/\rho_0$ due to 3D weak localization and strong electron–electron interaction is more complicated and is discussed below. The positive quadratic field dependence of $(\Delta\rho)_C/\rho_0$ implies that the corresponding magnetoconductivity is negative and quadratic and of the form $\Delta\sigma_I = -CH^2$, where the quadratic coefficient C varies as $T^{-3/2}$ as shown in the inset to Fig. 14.10 [14.16]. This positive quadratic term in $(\Delta\rho)_C/\rho_0$ may be due to conventional magnetoresistance mechanisms. The second term in Eq. (14.9) $(\Delta\rho)_{L,I}/\rho_0$ shows a predominantly negative magnetoresistance which is quadratic in H at low H and has an $H^{1/2}$ dependence at high magnetic field. A $T^{1/2}$ temperature dependence is found for the magnetoconductivity corresponding to the $(\Delta\rho)_{L,I}/\rho_0$ term, and this T dependence has been identified with 3D weak localization effects and carrier–carrier interaction effects. These effects are found to persist to high temperatures, which is attributed to the condition $\tau_i \ll \tau$, where τ_i and τ are, respectively, the inelastic and total scattering times. The 3D localization effects arise from disorder, which is associated in part with merohedral disorder and in part with missing or excess K^+ at the tetrahedral or octahedral sites, where the K^+ ions are normally found in the ideal K_3C_{60} compound. The effect of superconducting fluctuations on the magnetoresistance has been observed in both single crystals and film samples of K_3C_{60} near the superconducting transition T_c. This observation in doped fullerenes is the first time that 3D superconducting fluctuations have been observed in a 3D weakly localized system [14.16, 43].

The magnetoresistance of thin K_4C_{70} films [14.26] up to 30 K and in magnetic fields up to 15 tesla is qualitatively similar to that for thin K_3C_{60} films [14.16] and disordered carbon fibers [14.62], and the mechanism for the corresponding magnetoresistance in K_4C_{70} is identified with 3D weak localization and electron–electron interaction phenomena [14.26]. The electrical resistivity at zero magnetic field for the samples used in the magnetoresistance measurements was well fit by a fluctuation-induced tunneling model for metallic grains separated by poorly conducting material [14.26] (see §14.1.2).

14.5. Electronic Density of States

Experimentally, the reported density of states at the Fermi level $N(E_F)$ for the M_3C_{60} compounds varies from 1–2 states/eV per C_{60} per spin [14.95] to over 20 states/eV-C_{60}-spin [14.96]. The theoretical situation also shows a wide range of values, 6.6–12.5 states/eV-C_{60}-spin [14.97] for $N(E_F)$ (see §12.7.3), but not as wide a range as for the experiments. An elementary estimate of the electronic density of states at the Fermi level $N(E_F)$ can be made by assuming a constant $N(E)$ throughout the t_{1u} band which has 3 electrons per spin state and a bandwidth W, yielding $N(E_F) = 3/W$ per spin state per C_{60} or 6 states/eV-C_{60}-spin using a value of $W = 0.5$ eV. Detailed calculations by Erwin and Pickett [14.98] yield a value of $N(E_F) = 6.6$ states/eV-C_{60}-spin at a lattice constant of 14.24 Å, corresponding to that of K_3C_{60} (see Table 8.3), in good agreement with the rough physical estimate for $N(E_F)$ given above. This approximate relation is also in rough agreement with the estimate $N(E_F) = 3.5$ states/eV-C_{60}-spin that includes the effect of merohedral disorder [14.97]. Other LDA calculations yield values for the density of states of 7.2 and 8.1 states/eV-C_{60}-spin for K_3C_{60} and Rb_3C_{60}, respectively [14.99], and a value of 12.5 states/eV-C_{60}-spin for K_3C_{60} [14.13]. [It should be noted that various authors give different normalizations in reporting $N(E_F)$ values. These differences must be kept in mind when comparing results between different authors, using various experimental and theoretical techniques.] Theoretical analyses of experimental results have also been helpful in narrowing the range for estimates of $N(E_F)$ for K_3C_{60} and Rb_3C_{60} [14.97].

Photoemission studies are typically used in solid state physics to obtain measurements of the density of states for occupied levels, and inverse photoemission data are used to obtain the corresponding information for the unoccupied states. It is commonly assumed that the final state for an electron in the inverse photoemission is at the LUMO level and the initial electronic state in the photoemission process is at the HOMO level. These assumptions may not be applicable for highly correlated systems such as fullerenes [14.100]. Using the photoemission technique and its conventional interpretation, values of 1–2 states/eV-C_{60}-spin have been obtained [14.95], as mentioned above. It is, however, generally believed that the photoemission estimates for $N(E_F)$ for doped fullerenes are low because the spectral weight in the photoemission spectra is smeared out due to strong electron–phonon and electron–plasmon coupling, leading to a reduction in the photoemission intensity at E_F [14.101]. The high surface sensitivity of the photoemission technique, the small electron penetration depth, the large lattice constant, and the chemical instability of the K_3C_{60} and Rb_3C_{60} compounds further contribute to the difficulty in obtaining reliable bulk values for the density of states using the photoemission technique.

Pauli susceptibility (χ_P) measurements provide another fairly direct measurement of $N(E_F)$ for simple metals, since χ_P^0 neglecting many-body effects is simply $\chi_P^0 = 2\mu_B^2 N(E_F)$. The results for χ_P obtained for K_3C_{60} and Rb_3C_{60} by susceptibility measurements are, respectively, 14 and 19 states/eV-C_{60}-spin [14.102]. Many-body enhancement effects calculated by LDA methods suggest correction factors to the magnetic susceptibility determination of $N(E_F)$ so that $(\chi_P/\chi_P^0) = [1 - IN(E_F)]^{-1}$ where $I^{-1} =$ 26 states/eV-C_{60}-spin [14.103]. Including many-body effects thus leads to $N(E_F)$ values of 9.3 and 11.9 states/eV-C_{60}-spin for K_3C_{60} and Rb_3C_{60}, respectively, using the many-body correction factor (χ_P/χ_P^0) of 1.5 and 1.6 for K_3C_{60} and Rb_3C_{60}, respectively. ESR measurements of the bulk susceptibility [14.104] of 15.5 states/eV-C_{60}-spin have been reported, and if the many-body correction factor given above is used, a value of 10.3 states/eV-C_{60}-spin is obtained. When these many-body interpretations are considered, the susceptibility measurements overall are in reasonable agreement with calculations of the density of states.

Since thermopower measurements (see §14.12) are sensitive to the bandwidth W, estimates of $N(E_F)$ for K_3C_{60} and Rb_3C_{60} can be obtained using the relation $N(E_F) \sim 3.5/W$ states/eV-C_{60}-spin given above [14.97]. From analysis of the thermopower measurements of §14.12, estimates for W of 650 meV and 400 meV, respectively, were obtained for K_3C_{60} and Rb_3C_{60}, yielding the corresponding rough $N(E_F)$ estimate of 5.4 and 8.8 states/eV-C_{60}-spin for these compounds.

Measurements of the low-temperature specific heat ($T < 25$ K) provide a sensitive tool for measuring the electronic density of states in conventional metals. Specific heat measurements made on K_3C_{60} yield a value of

14 states/eV-C_{60}-spin, neglecting strong coupling enhancement effects and effective mass enhancement effects [14.102] (see §14.8). Since the magnitude of these many-body effects is not well established, the magnitude of the enhancement factor responsible for the large value of $N(E_F)$ is not presently known.

From the Fermi contact term in the nuclear magnetic resonance interaction, a determination of $N(E_F)$ can be made from NMR measurements of the temperature dependence of the nuclear spin relaxation rate (see §16.1.7). Using this approach, values of $N(E_F) = 17$ and 22 states/eV-C_{60}-spin were reported for K_3C_{60} and Rb_3C_{60}, respectively [14.105]). Recent calculations [14.99] have, however, shown that the NMR determinations of the density of states are likely to be high, because the experimental determinations have attributed all the spin relaxation to the Fermi contact term, while *ab initio* calculations have shown that the spin dipolar relaxation mechanism dominates over the orbital and Fermi contact mechanisms [14.99]. With this reinterpretation of the NMR measurements, values for $N(E_F)$ of 8.7 and 10.9 states/eV-C_{60}-spin are obtained for K_3C_{60} and Rb_3C_{60}, respectively.

By bringing theoretical effort to bear on the interpretation of experimental measurements, the range of values for $N(E_F)$ for K_3C_{60} and Rb_3C_{60} has been narrowed and brought into rough agreement with theoretical calculations. Based on presently available results, an estimate of $N(E_F)$ in the range 6–9 states/eV-C_{60}-spin is reached for K_3C_{60} [14.97], and $N(E_F)$ for Rb_3C_{60} is estimated to be a factor of 1.2–1.6 higher. Although the absolute value of $N(E_F)$ for the M_3C_{60} compounds is not firmly established, the percentage increase in $N(E_F)$ with lattice constant for the fcc alkali metal–doped fullerenes is quite well established (see Fig. 14.11).

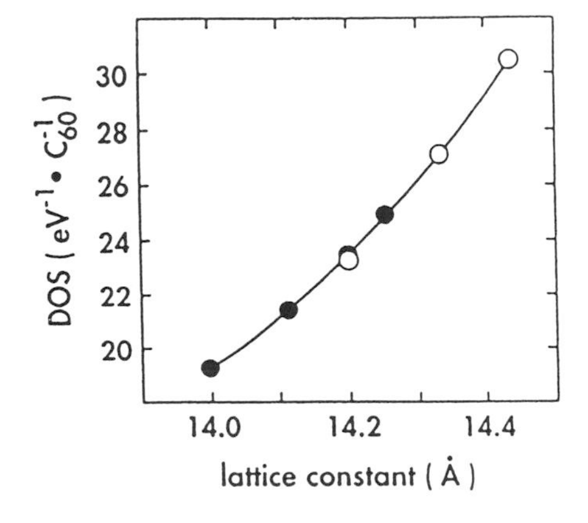

Fig. 14.11. Calculated Fermi level density of states for K_3C_{60} (solid circles) and for Rb_3C_{60} (open circles) as a function of lattice constant. The solid line is a parabolic fit to the calculated values [14.106]. The percentage increase in the density of states with lattice constant is more accurate than the absolute values given in this figure.

14.6. Pressure Effects

Transport measurements have been used primarily to probe pressure-dependent behavior in the C_{60} crystalline phases that are stable at atmospheric pressure and to study pressure-induced shifts of the phase boundary of the structural phase transition at T_{01} (see §7.1.3).

The high electrical resistivity of C_{60} is reduced by the application of high pressure [14.107–109], as expected because of the increased interaction between the C_{60} molecules as the intermolecular distance decreases with increasing pressure. The rapid decrease in electrical resistance with pressure is shown in Fig. 14.12 for C_{60}, C_{70}, and $C_{60}I_4$ pressed powder samples. Since the resistance of C_{60} was too high to measure accurately at ambient pressure, measurements were made on the less resistive $C_{60}I_4$ samples, which show a more than five orders of magnitude decrease in resistance with pressures up to 10 GPa [14.109]. The temperature dependence of the resistance for C_{60}, C_{70}, and $C_{60}I_4$ could all be fit by the 3D variable-range hopping Mott formula

$$R = R_0 \exp[2(T_0/T)^{1/4}] \tag{14.10}$$

with the disorder attributed to impurities such as oxygen, lattice defects, stacking faults, and orientational disorder of the molecules with respect to the crystalline axes. The reported pressure dependence of the parameter $T_0 = [\xi^3/k_B N(E_F)]$ could provide valuable information about transport in these samples if independent measurements could be carried out to mea-

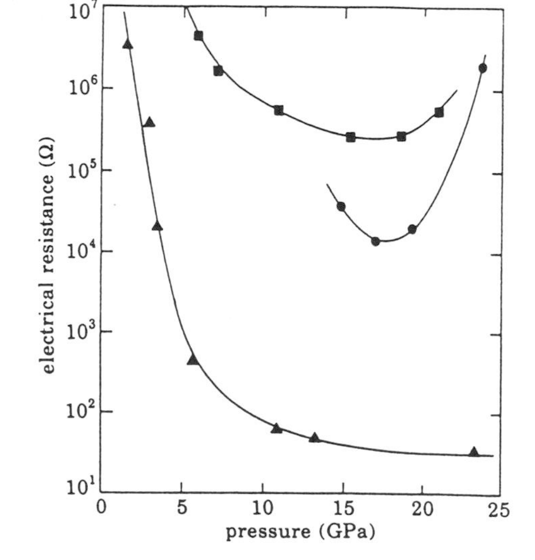

Fig. 14.12. Variation of the resistance (plotted on a log scale) at room temperature of fullerenes with pressure: C_{60} (solid squares), C_{70} (solid circles), and $C_{60}I_4$ (solid triangles). Note the very steep variation with pressure at low pressures for $C_{60}I_4$, almost reaching metallic values [14.109].

sure the localization length ξ or the density of states at the Fermi level $N(E_F)$. The results shown in Fig. 14.12 further demonstrate an increase in resistance for C_{60} and C_{70} at pressures in the 15–20 GPa range, which has been correlated with an irreversible phase transition to a more insulating state [14.109].

Electrical resistance measurements have also been carried out over a much narrower range of temperature and for pressures up to 13 kbar (1 GPa = 9.9 kbar) to study the pressure dependence of the transition temperature T_{01} associated with the loss of full rotational freedom, yielding a value of 9.9 ± 1 K/kbar [14.110] in good agreement with other measurements of this quantity (see Table 7.1). Two features were identified in the $R(T)$ curve near T_{01} and the temperature separation between these peaks was found to increase rapidly with increasing pressure. One proposed explanation for this effect is a small difference in the temperature of the rotational alignment of the C_{60} molecules at the cube edges and cube face centers [14.111]. The presence of the two phase transitions is not believed to be due to residual solvent in the sample [14.110]. By fitting their $R(T)$ data to an Arrhenius relation as a function of temperature, $R(T) = R(0)\exp(-E_a/2k_BT)$, values of E_a as a function of pressure were obtained. From these data an estimate of $E_a \simeq 2.0$ eV at 1 atm was reported [14.110], which is consistent with optical measurements of the band gap in C_{60} (see §13.3).

As the intermolecular C–C distance decreases upon application of pressure and becomes comparable to the intramolecular C–C distance, an electronic transition might be expected to occur, as discussed in §7.3. However, transport studies in the high pressure phases of C_{60} [14.28] have not yet been reported.

The dependence of the superconducting parameters on pressure is also interesting and is reviewed in §15.6, where the pressure dependence of T_c for K_3C_{60} and Rb_3C_{60} is discussed in some detail.

14.7. Photoconductivity

Many measurements of the photoconductivity of undoped C_{60} films have been published [14.2, 112–116], and a few measurements have also been reported on undoped C_{70} films [14.115, 117] and potassium-doped C_{70} films [14.115]. In addition, photoconductivity studies have been carried out on polymers that have been doped with small amounts of fullerenes [14.118–125]. For example, it has been reported that photoconducting films of polyvinylcarbazole (PVK) doped with fullerenes (a mixture of C_{60} and C_{70}) show high xerographic performance (see §13.5 and §20.1.2), comparable with that for some of the best photoconductors available commercially [14.113, 118]. Thus, practical applications may emerge from the special properties of the photoconductivity of fullerene-doped polymers.

Before reviewing the photoconductive properties of fullerenes in detail, it is important to mention the wide disagreement in the published results on this topic. There are four dominant reasons for these discrepancies. First, the magnitude of the photoconductivity signal is reduced by several orders of magnitude by the presence of oxygen, which acts as a trapping center for carriers; it is therefore important to work with oxygen-free fullerene films for measurements of intrinsic photoconductivity behavior [14.126–128]. Second, the persistent photoconductivity effect described below [14.129–131] implies that photocarrier relaxation times in fullerenes can extend to days. These extremely long relaxation times must be taken into account when using light pulses for photoexcitation and detection. Third, excessive incident light intensities can lead to phototransformation (see §13.3.4) of the fullerene film [14.132], thereby modifying the material during the measurement process. Finally, photoconductivity measurements show sensitivity to the degree of crystallinity and to the concentration of defects. For these four reasons, this review of the photoconducting properties of fullerenes focuses on selected references which have had more success toward making intrinsic measurements.

In preparing a typical photoconductivity sample, thin (~2000 Å) fullerene films are deposited on glass substrates with partially precoated silver electrodes, and on top of the fullerene films, semitransparent aluminum electrodes (~200 Å thick) are deposited. Light illumination is through these top electrodes and into the fullerene film, allowing both photoconductivity and dark conductivity measurements to be carried out, when a voltage (e.g., 1 V) is applied to the top electrodes, while the bottom electrodes are grounded [14.115]. To measure intrinsic photoconducting properties, attention must be given to address the four critical issues described above.

Oxygen-free C_{60} has a very high quantum efficiency (~55%), which is defined as the number of photogenerated carriers per absorbed photon [14.114, 116]. This high quantum efficiency, which is attributed to the long lifetimes of excited carriers in C_{60}, can be reduced by several orders of magnitude by oxygen uptake. Electrons are easily trapped at the interstitial octahedral sites where the oxygen molecules collect [14.133], giving rise to shallow traps for electron–hole recombination [14.116]. In this review, we emphasize reported results on nominally oxygen-free samples.

In general, both photoexcited electrons and holes should contribute to the photoconductivity. However, for most of the common photoconductors the lifetime for one of the carrier types tends to be much greater, so that the photoconductivity is often dominated by a single carrier. Experiments such as the Dember effect indicate that electrons are the dominant photocarriers in C_{60} films [14.128, 134]. The Dember effect arises from the different mobilities of electrons and holes produced by light, so that a net excess of positive charge is observed at the illuminated electrode relative to the dark electrode which has a net negative charge; this charge separation gives rise to the Dember voltage which has been measured in C_{60} films [14.128].

The carrier generation can be explained in terms of the following relations:

$$\Delta n = \frac{g}{2S_n n_0} \tag{14.11}$$

for a monomolecular (or unimolecular) process, and

$$\Delta n = \left(\frac{g}{S_n}\right)^{1/2} \tag{14.12}$$

for a bimolecular process [14.135], where Δn is the density of photoexcited excess carriers, n_0 the density of thermal carriers, S_n the capture coefficient by recombination, and g the carrier generation rate. Because the quantum efficiency η is defined by

$$\eta = \Delta n / n_p, \tag{14.13}$$

where n_p is the absorbed photon density which is proportional to g, then η becomes constant in a monomolecular process and proportional to $n_p^{-1/2}$ in a bimolecular process. Thus we see that photocarrier generation enhances photoconduction, while recombination processes inhibit photoconductivity.

The photoconductivity shows a wide range of rise times and decay times ranging from hundreds of picoseconds to days, with different mechanisms involved as the time scale is varied. As the different time scales are probed with various experimental techniques, different photocarrier excitation mechanisms come into play.

Figure 14.13 summarizes the transient photoconductivity response on the picosecond scale in pristine and oxygenated C_{60} films measured at a photon energy of $\hbar\omega = 2.0$ eV at room temperature [14.136]. Of particular significance is the very large magnitude of the transient photoconductivity, which is reported to be as much as nine orders of magnitude greater than the steady-state photoconductivity [14.137]. The transient photoconductivity of a pristine C_{60} film (open squares □ in Fig. 14.13) consists of short- and longer-lived components with relaxation times of 693 ps and 7.2 ns, respectively. From the temperature-dependent photoconductivity studies, it was shown [14.136] that these two components are dominated by two distinct transport mechanisms. The carrier relaxation times, as well as the magnitudes of the photo and dark conductivities, decrease dramatically with oxygen exposure, indicating that oxygen in the C_{60} film creates efficient deep traps for the photocarriers, which diminish the probability of

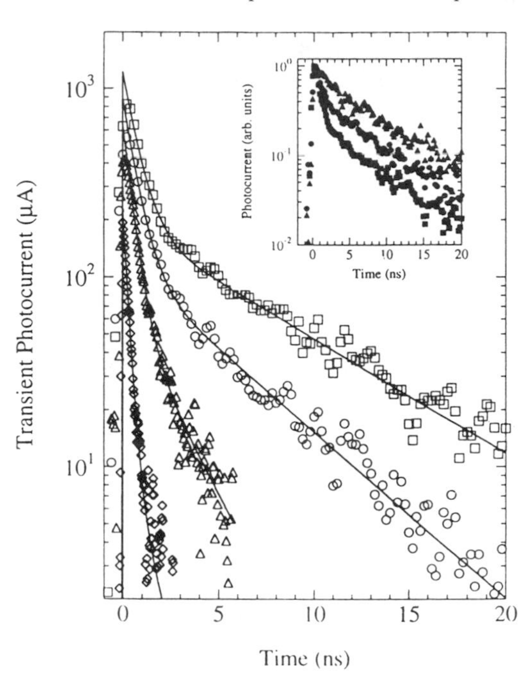

Fig. 14.13. The time-resolved transient photocurrent (T =300 K, $\hbar\omega$ = 2.0 eV) in a pristine C_{60} film and in the film at various levels of oxygen content. The transient photocurrent data for the oxygen-free sample are represented by the □ points, and each sample was characterized by I_d its dark current: □ I_d = 3.6 nA; ○ I_d = 5.67 pA; △ I_d = 0.24 pA; ◇ I_d < 0.01 pA, where 1 pA = 10^{-12}A. The solid lines are the best fit of the experimental data to a double-exponential function mentioned in the text. The inset shows the normalized transient photocurrent of pristine C_{60} at different laser intensities: 2.7 × 10^{14} (▲), 1.1 × 10^{15} (•), and 5.4 × 10^{15} photons/cm² (■) [14.136].

carrier release into the extended band states in the photoconductivity process [14.136]. The curve denoted by open diamonds in Fig. 14.13 was taken after the C_{60} film was exposed to air for over a week, showing a decrease in dark conductivity by more than five orders of magnitude from the pristine value, and the time constants associated with the short- and long-lived components of the transient photocurrent were 238 ps and 1.0 ns, respectively.

In contrast to the short relaxation times obtained in transient photoconductivity experiments, a persistent photoconductivity (PPC) effect has been observed in C_{60} films corresponding to photogenerated carrier decay times as long as ~10^6 s [14.130]. The PPC effect is observed as a metastable increase in dark conductivity by up to an order of magnitude upon exposure of the sample to light (see Fig. 14.14). The PPC effect in C_{60} persists to 260°C (which is the highest recorded temperature for this effect for any material system), and it is excited by photons with energies above the weak absorption edge [14.130].

Photoconductivity can be a very sensitive probe of changes in the density of gap states in a semiconductor. To ascertain these changes, the photoconductivity $\Delta\sigma$ can be studied as function of temperature and light intensity by identifying the parameter γ in the relation

$$\Delta\sigma \propto I_0^{\gamma}, \tag{14.14}$$

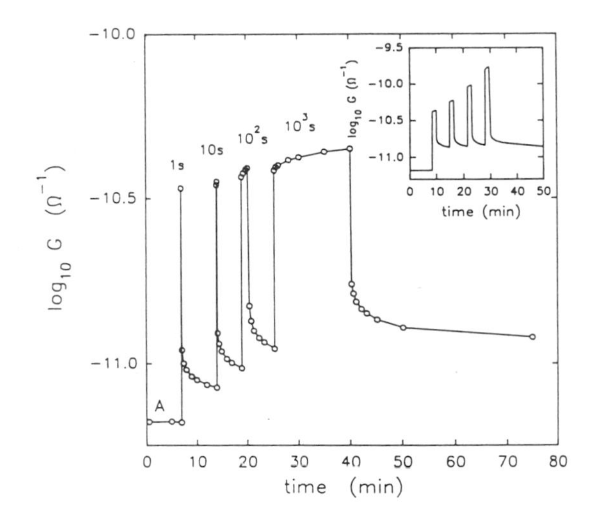

Fig. 14.14. Conductance G (on a log scale) as a function of time for a C_{60} thin film before, during, and after exposure to 2 mW/cm^2 heat-filtered white light for increasing exposure times from 1 s to 1000 s. The inset shows the effect on the conductance of illuminating the film for the same duration (1 min) with increasing light intensities of 2, 6, 20, and 60 mW/cm^2 [14.130].

where I_0 is the incident light intensity. This relation between the incident light intensity I_0 and the photoconductivity $\Delta\sigma$ determines the dominant recombination process. If $\gamma = 1$, the recombination of the carriers is called monomolecular and is predominantly with dark carriers, while $\gamma = 1/2$ is associated with bimolecular recombination of photogenerated electrons with photoholes. Since the presence of oxygen decreases the dark conductivity of C_{60} films by four orders of magnitude and decreases the photocurrent by a factor of ~400 [14.128], the relative importance of dark carriers and photocarriers in the recombination process will sensitively depend on oxygen content. A number of authors [14.115, 138] have reported that carrier recombination in C_{60} is dominated by the monomolecular recombination process ($\gamma \approx 1$) in the low-light-intensity regime and by the bimolecular recombination process ($\gamma \approx 1/2$) in the high-light-intensity regime. A temperature-dependence study of the γ coefficient in Eq. (14.14) [14.139] reveals that γ monotonically increases with decreasing temperature from $\gamma \sim 0.2$ at 500 K to $\gamma \sim 0.75$ at ~260 K, goes through a pronounced maximum at $T_{01} = 260$ K, followed by a minimum at ~200 K, and then increases further as T is decreased below ~200 K. It was also found in detailed studies of the transient photoconductivity of C_{60} films [14.136] that γ tends to decrease as the photon energy of the incident light $\hbar\omega$ increases (e.g., $\gamma = 0.95$ at 2.0 eV, whereas at 3.1 eV, $\gamma = 0.76$). On this basis we conclude that the carrier recombination process is complicated and depends strongly on oxygen content, light intensity, temperature, and photon energy. It is for these reasons that differing reports appear in the literature regarding the dominant recombination process for photoexcited carriers in C_{60}

[14.115, 128]. At low light intensity levels, low temperature, and low photon energies where the photocarrier density is low, monomolecular recombination is important.

Although the magnitude of the photoconductivity signal is highly sensitive to oxygen exposure, the normalized spectral dependence of the photoconductivity for C_{60} and C_{70} shown in Fig. 14.15 is almost independent of oxygen exposure [14.116]. The dominant feature in Fig. 14.15 is the photoconductivity threshold at ~1.8 eV for C_{60}, consistent with the measurements of the optical absorption edge [14.113, 126–128, 140]. Studies of the spectral dependence show that the photocarriers in C_{60} are generated directly by optical absorption. Also seen in this figure is a much weaker photoconductivity band below 1.5 eV, associated with carriers bound to non-oxygen-related defects, but this low-energy photoconductivity band can be suppressed by heat treatment and is therefore believed to be extrinsic and related to gap states [14.126]. Large variations in behavior are reported by various groups regarding the photoconductivity for photon energies below the absorption edge. A shift of the onset for photoconductivity from 1.81 eV for the fcc phase to 1.85 eV for the simple cubic $Pa\bar{3}$ phase has also been reported [14.127].

The temperature dependence of the photoconductivity $\Delta\sigma$ and dark conductivity σ for C_{60} fits a variable-range hopping model with an exponential $T^{-1/4}$ law [14.113, 127] indicative of strong carrier localization. Some authors have reported structure in the $\Delta\sigma(T)$ curves near T_{01} associated with the structural phase transition of C_{60} at this temperature [14.136], but others either do not observe this feature or question this identification [14.126]. Changes in the magnitude of the photoconductivity connected with

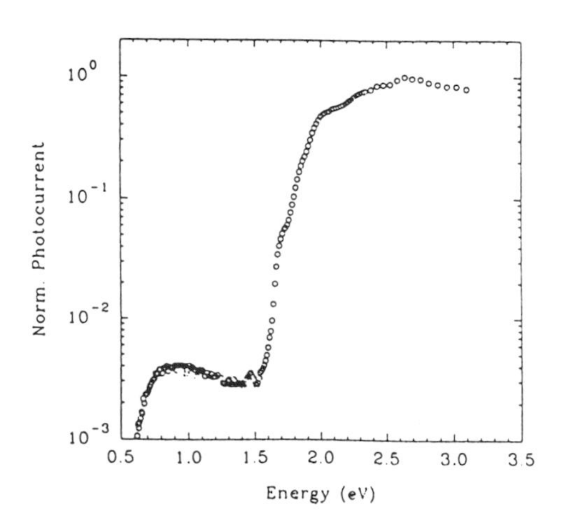

Fig. 14.15. Spectral dependence of the normalized photoconductivity for nominally oxygen-free C_{60} [14.116].

the lower-temperature structural phase transition have also been reported near 155 K [14.131].

The model that emerges for the photoconductivity of C_{60} is that carrier generation is via a direct absorption process. In the presence of oxygen, rapid carrier electron–hole recombination occurs at shallow oxygen trap levels, lowering the free carrier lifetime and yielding a low quantum efficiency. When oxygen is rigidly excluded from the sample, the quantum efficiency is high and carrier transport between C_{60} molecules occurs via a hopping mechanism, consistent with the dominant mechanism for dark carrier transport. The explanation of why the spectral dependence of the photoconductivity follows that of the absorption coefficient presents a challenge, since no energy seems to be required to overcome the exciton binding energy which lowers the energy of the absorption edge relative to the HOMO–LUMO gap in an optical absorption or luminescence process. One explanation for the absence of the exciton binding energy from the photoconductivity process also relates to optically detected magnetic resonance (ODMR) and electron paramagnetic resonance (EPR) measurements, which give evidence for the rapid formation of a self-trapped polaron by the photoexcited hole (see §13.2.4). This self-trapped polaron is strongly localized on a single fullerene molecule and in the polaron state the hole is screened from its photo-generated electron by a charged cloud from the π electrons on the C_{60} molecule. Thus the photogenerated electron can participate in electrical conduction using the same transport mechanism as for the dark carriers.

Much less is known about the photoconductivity of C_{70} relative to C_{60}, though a few measurements have been reported [14.115, 117]. The lower interest regarding photoconductivity studies in C_{70} is attributed to the similar behavior between C_{60} and C_{70} films, as has been confirmed in detailed studies of the persistent photoconductivity effect [14.117]. Since the photoconductivity for C_{70} also follows the spectral dependence of the optical absorption, the range of photosensitivity for C_{70} is extended to the technologically important longer-wavelength regime [14.115].

The temperature dependence of the photoconductivity for a potassium-doped C_{70} film with a room temperature dark conductivity of 8.4 S/cm has also been reported [14.115]. The measurements were carried out under a constant current of 1 A between two silver electrodes with a gap of 1 mm between the electrodes, on which a C_{70} film was deposited. Because of the high optical absorption of the K_xC_{70} film, a semitransparent top electrode is used for the K_xC_{70} photoconductivity measurements. Doping was stopped at three different conductivity values of 40, 8.4, and 1.5 S/cm (at 300 K) in order to measure the temperature dependence at each doping level. Similar to the results for C_{70}, an activation behavior of the dark conductivity at high temperatures and an almost constant photoconductivity at low-temperatures are also observed in K_xC_{70}. However, the magnitude of the photoconductivity for K_xC_{70} is much smaller than that of the dark conductivity (approximately two orders of magnitude smaller at room temperature), even under an illumination which is 200 times greater than that used for the $\Delta\sigma$ measurement of pristine C_{70}. The activation energy E_a of the dark conductivity at high temperature decreases as the room temperature conductivity increases, with values of $E_a = 77$ meV, 55 meV, and 10 meV, respectively, having been reported for samples with conductivity values of 1.5, 8.4, and 40 S/cm [14.115].

14.8. Specific Heat

Measurement of the temperature dependence of the heat capacity at constant pressure $C_p(T)$ provides the most direct method for the study of the temperature evolution of the degrees of freedom, phase transitions, the enthalpy change and the entropy change associated with phase transitions if they are first order, the identification of higher-order phase transitions, and other thermal consequences of structural rearrangements. Because of the different structures of the various fullerene molecules and the different crystal structures possible in the solid state, the various fullerene-derived materials experience different phase transitions and hence distinct $C_p(T)$ curves. Thus far, most of the specific heat measurements have been carried out on undoped C_{60} and C_{70}. A small amount of specific heat data is also available for K_3C_{60}. These specific heat results for C_{60}, C_{70}, and the doped compound K_3C_{60} are reviewed below.

14.8.1. Temperature Dependence

In our discussion of the temperature dependence of the heat capacity $C_p(T)$ for C_{60} we emphasize the existence of three basic temperature regimes: (1) low-temperatures, well below the temperature where the internal degrees of freedom of the fullerene molecule are excited. Here $C_p(T)$ is dominated by the excitation of translational degrees of freedom associated with a fullerene molecule considered as a giant atom. (2) Intermediate temperatures, where the rotational, librational, and intermolecular vibrational degrees of freedom of these giant atoms are activated and where phase transitions are turned on. (3) Very high temperatures where the internal degrees of the fullerene molecules are activated, if the fullerenes do not first transform to other phases. The various regimes of the intermolecular and intramolecular vibrational excitations are discussed in §11.1. It is of interest that the temperature range (up to the Debye temperature of $\sim$70 K) over which the translational modes contribute overlaps with the range over which the librational and intermolecular vibrations also contribute. The process of activation of the internal modes is initiated at rather high temperature ($\sim$350 K) and continues to contribute to the specific heat up to very high temperatures ($\sim$2700 K), assuming fullerenes remain stable at these high T values. Above this temperature, we are left with the degrees of freedom of the individual carbon atoms, which yield the same $C_p(T)$ per C atom for all allotropic forms of carbon. Except for the translational modes, the intramolecular and intermolecular excitations can be treated in terms of Einstein modes [14.141].

If the specific heat anomalies associated with phase transitions are removed from the experimental $C_p(T)$ data for C_{60}, then the temperature dependence of the residual specific heat contribution for C_{60} is in good agreement with that of graphite [14.142, 143] for $T \geq T_{01}$ (261 K) (see §7.1.3). This similarity with graphite has been noted by various authors [14.144, 145]. The corresponding residual specific heat for C_{70} is higher than that for graphite, consistent with the larger number of degrees of freedom. It is expected that in the high-temperature limit where $k_B T$ is well above the vibrational frequencies of the fullerenes and of graphite (>2000 K), the specific heat per carbon atom has the same $3R$ value per carbon atom for C_{60}, C_{70}, and graphite. Since the lattice modes for C_{60} cover approximately the same frequency range as for graphite, the contribution of the Einstein modes to the specific heat should be approximately equal, consistent with a $C_p(T)$ similar to that observed for graphite. From the experimental and theoretical values for the intermolecular modes, we would expect contributions from the intermolecular modes in C_{60} to be dominant below $\sim$85 K.

The dominant feature in the experimental temperature dependence of the specific heat for C_{60} is the large specific heat anomaly that is observed at $T_{01} = 261$ K (see Fig. 14.16) [14.144–149]. The most accurate measurement of this specific heat anomaly was done at a 10 K/min heating rate, yielding a transition temperature $T_c = 257.6$ K, with an associated enthalpy change for a highly crystalline sample of $\Delta H = 7.54$ kJ/mol and an entropy change of $\Delta S = 30.0$ J/K mol [14.147]. It has been suggested that the best available C_{60} crystalline samples give a transition temperature $T_{01} = 261$ K [14.149]. This specific heat anomaly is associated with a first-order transition in the solid phase, corresponding to the structural transition from the fcc space group $Fm\bar{3}m$ for the higher-temperature phase where all the C_{60} molecules are rotating nearly freely, to the low-temperature $Pa\bar{3}$ structure where each of four C_{60} molecules within the cubic unit cell is rotating about a different $\langle 111 \rangle$ axis (see §7.1.3). The model calculation described in §7.1.5 has been applied to yield the temperature dependence of the heat capacity shown

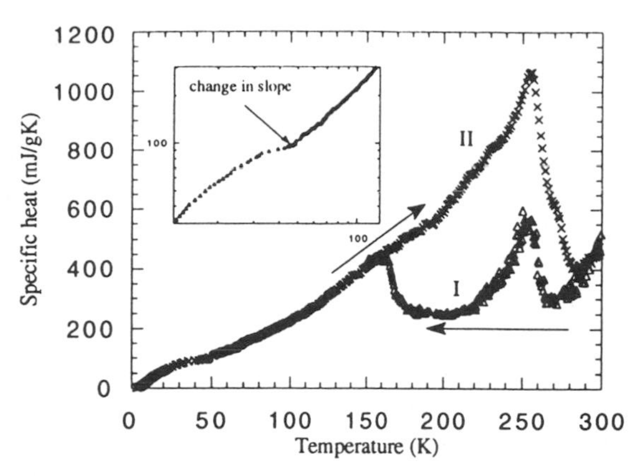

Fig. 14.16. Comparison between the temperature dependence of the specific heat for C_{60} on cooling (Δ) and heating ($\times$). The inset shows the low-temperature specific heat on a magnified scale and the change in slope near 50 K is indicated explicitly [14.146].

in Fig. 14.17 [14.150]. The model predicts a first-order phase transition at T_{01} with agreement obtained between theory and experiment regarding the latent heat and entropy change at T_{01} (see §7.1.5).

Upon further cooling of crystalline C_{60}, a second anomaly in $C_p(T)$ is observed, with a peak in the specific heat at 165 K [14.146]. This peak in the specific heat is identified with a Schottky anomaly associated with the

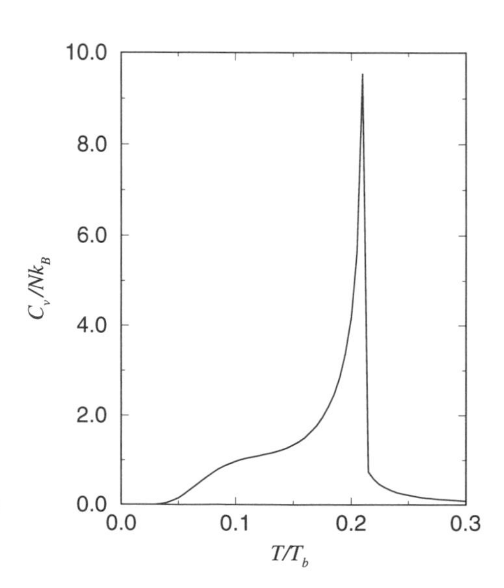

Fig. 14.17. Calculated temperature dependence of the heat capacity at constant volume C_v/Nk_B in dimensionless units with a sharp peak in the heat capacity occurring at T_{01} [14.150].

alignment of the electron-rich double bonds of one C_{60} molecule opposite the centers of either the electron-poor pentagons or the electron-rich hexagons. As discussed in some detail in §7.1.3, these two states differ in energy by only 11.4 meV.

The hysteresis in the $C_p(T)$ data in Fig. 14.16 suggests that on heating from below 165 K, the C_{60} molecules in the crystal phase remain in their low-temperature frozen state until a temperature of ~261 K is reached, where a first-order transition to the fcc structural phase occurs [14.146]. On cooling, the C_{60} molecules gradually go from a rotating motion to a ratcheting motion, spending increasingly more time in one of the two stable states discussed above and in §7.1.3. The large magnitude of the anomaly in $C_p(T)$ at 261 K on heating, in comparison to the two smaller anomalies at 261 K and 165 K on cooling, supports this interpretation [14.146]. Below 165 K where the C_{60} molecules freeze, the $C_p(T)$ behavior is the same upon heating and cooling. Associated with the specific heat anomaly at $T_{02} = 165$ K is an enthalpy change of 22.2 kJ/mol, a change in the heat capacity of 7 J/K mol, and a relaxation time of 4×10^{-11} s associated with the two potential minima.

At low-temperatures, a change in the slope of the $C_p(T)$ curve is observed at 50 K [14.145–148], with an excess specific heat that corresponds to an activation energy of 40 meV and is associated with the degrees of freedom for the intermolecular vibrations. A more detailed analysis of the low-temperature Einstein contribution yields $\Theta_E = 35$ K (or 24 cm^{-1}) [14.141], in good agreement with the neutron scattering value of 2.8 meV (21.6 cm^{-1}) [14.151]. In the low-temperature regime, the C_{60} molecules should be considered as giant atoms of 720 atomic mass units which acquire rotational and translational macroscopic degrees of freedom (see §11.4). At very low T (below 8 K), a T^3 dependence of $C_p(T)$ is observed, yielding a Debye temperature Θ_D of 70 K [14.146] associated with the low-frequency intermolecular acoustic modes of this lattice of "giant atoms." More detailed studies at very low-temperature [14.141, 148] in C_{60}(85%)/C_{70}(15%) compacts showed the low-T heat capacity to follow a

$$C_p(T) = C_1 T + C_3 T^3 \tag{14.15}$$

behavior. From the T^3 term a Debye temperature $\Theta_D = 80$ K was obtained (Debye velocity of sound 2.4×10^5 cm/s), while the linear T contribution was attributed to low-temperature disorder, presumably associated with the merohedral disorder and having an excitation energy of 400 GHz (13.2 cm^{-1}) [14.141]. The magnitude of the linear term was found to be similar to that in amorphous SiO_2 and amorphous As_2S_3 [14.141].

14.8.2. *Specific Heat for C_{70}*

We have already commented in §14.8.1 on the high-temperature behavior of $C_p(T)$ for C_{60}. From that discussion, it is expected that the intermediate temperature range would be of particular interest regarding thermal properties. For C_{70}, two specific heat anomalies are observed at 337 K and 280 K, corresponding to enthalpy changes of 2.2 J/g and 3.2 J/g, respectively [14.144]. Structural studies [14.152–154] show a phase transition from an isotropic fcc structure with nearly freely rotating C_{70} molecules to an anisotropic structure at 337 K where the fivefold axes of the C_{70} molecule become aligned (see §7.2). At ~276 K, structural studies indicate that rotations about the fivefold axes cease, as the molecules align in a monoclinic structure corresponding to a distorted hexagonal close-packed (hcp) structure with $c/a = 1.82$.

An anomaly in the specific heat for C_{70} is observed at 50 K and is likely due to the onset of vibrational degrees of freedom of the C_{70} intermolecular normal mode. The low-temperature measurements (< 8 K) show a T^3 law from which a Debye temperature of 70 K is estimated. The high value for the low-temperature $C_p(T)$ in C_{70} is attributed [14.144] to a high density of librational modes for the whole C_{70} molecule [14.155, 156]. In addition, the 30 intramolecular modes associated with the 10 additional belt atoms tend to have relatively low vibrational frequencies. These additional intramolecular modes thus tend to contribute to the specific heat more effectively at lower temperatures, thus raising the $C_p(T)$ of C_{70} per atom above that for C_{60}, in agreement with experiment [14.146].

14.8.3. *Specific Heat for K_3C_{60}*

Very few studies of the heat capacity for doped fulleride solids have thus far been reported. Using a conventional adiabatic heat pulse method and helium exchange gas for thermal contact [14.102], the low-temperature specific heat of K_3C_{60} was measured. The resulting data shown in Fig. 14.18 were analyzed in terms of four contributions: (1) a contribution linear in T (and labeled TL) arising from structural disorder, (2) a cubic term identified with a Debye contribution ($\Theta_D = 70$ K determined by scaling the bulk moduli data for copper and M_3C_{60}) [14.102], (3) an Einstein term attributed to the librational mode at 34 K [14.157], and (4) a small electronic term associated with the specific heat jump near the superconducting transition temperature T_c. The results show that doping does not strongly affect the Debye and Einstein contributions to the specific heat. Most of the emphasis of this work [14.102] was given to estimating the electronic density of states and structure in the heat capacity associated with the superconducting phase transition.

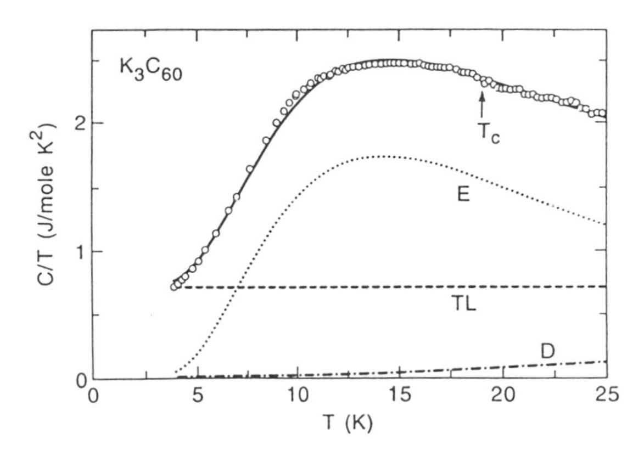

Fig. 14.18. The temperature dependence of the specific heat divided by temperature, $C(T)/T$, for K_3C_{60}. The data were least squares fitted to an Einstein term (dotted line), a linear term (dashed line), and a Debye term (dash-dotted line), the sum of which is shown by the solid line through the data points [14.102].

The linear term in the specific heat determines $N(E_F)$ for a simple metal, but in the case of doped fullerenes there are also many-body mass enhancement effects associated with both the electron–phonon interaction and spin fluctuations, so that the specific heat measurements and analysis carried out thus far do not yield a definitive determination of $N(E_F)$ [14.102]. Also, merohedral disorder is expected to broaden the peak structure in the density of states, thereby lowering the value of $N(E_F)$ expected for a perfectly ordered crystal. After subtracting the Einstein and Debye phonon contributions to the heat capacity, the residual heat capacity $\delta C(T)$ divided by T was obtained and the results for K_3C_{60} are plotted *vs.* T in Fig. 14.19 at zero magnetic field and at a field of 6 T. The results show a specific heat jump at T_c [14.102], of magnitude $(\Delta C/T_c) = 68 \pm 13$ mJ/mol K^2, which was interpreted in terms of the expression

$$\frac{\Delta C}{T_c} = \frac{\pi^2}{3} k_B^2 N(E_F) \Big[1.43 + g(\lambda_{ep}) \Big] \Big[1 + \lambda_{ep} + \lambda_s \Big], \tag{14.16}$$

where $g(\lambda_{ep})$ is a strong coupling correction term, λ_{ep} is the electron–phonon coupling coefficient, and λ_s is a spin fluctuation contribution to the quasiparticle mass enhancement [14.102]. If the terms $g(\lambda_{ep})$, λ_{ep} and λ_s in Eq. (14.16) are neglected, then $N(E_F)$ would become 28 states/eV-C_{60}. Since the values of $g(\lambda_{ep})$, λ_{ep}, and λ_s are not accurately known, it is difficult at present to obtain a quantitative value for $N(E_F)$ directly from measurements of $\Delta C/T$ for K_3C_{60}. A comparison between the $H = 0$ and $H = 6$ T data in Fig. 14.19 yields $dH_{c2}/dT = -3.5 \pm 1.0$ T/K, in agreement with magnetization measurements [14.158–160], as discussed in §15.3.

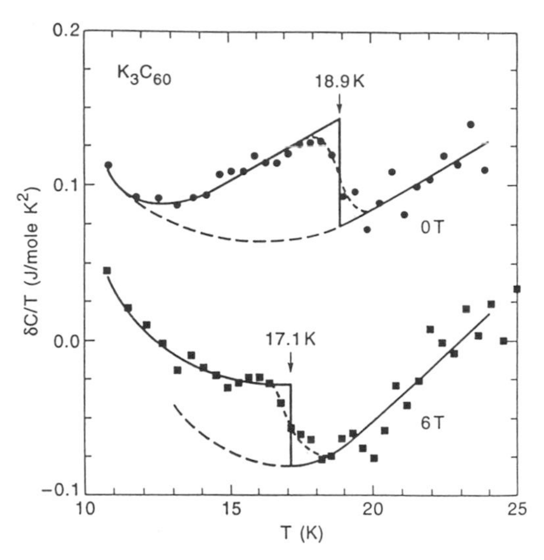

Fig. 14.19. The specific heat contribution, $\delta C(T)/T$, divided by temperature after subtraction of the phonon contribution shown in Fig. 14.18. The solid lines are to guide the eye and show a jump $\Delta C(T)/T_c$ due to the superconducting transition. Also shown are data taken in an applied magnetic field of 6 T. The two data sets are offset for clarity [14.102].

14.9. Scanning Calorimetry Studies

Differential scanning calorimetry (DSC) has been widely applied to study phase transitions in C_{60}, the main focus of this work being directed to establishing the order of the transition, the transition temperature, and the enthalpy changes associated with the transition [14.145, 161–163]. The DSC results of Fig. 14.20 show a latent heat, indicating that the phase transition for C_{60} near 261 K is first order. For the particular sample that was studied in Fig. 14.20, the measured transition temperature was $T_{01} = -15.34°C$ (257.66 K) and an enthalpy change of 9.1 J/g was measured [14.163, 164]. Also shown in Fig. 14.20 is the effect of residual solvent in the sample, giving rise to a small downshift in the transition temperature T_{01} and a sizable decrease in the enthalpy change at T_{01}. These results show the importance of sample quality for thermodynamic measurements on fullerenes. Calorimetric studies have also been used to establish the pressure dependence of the transition temperature T_{01}, which is observed to increase at a rate of 10.4–11.7 K/kbar (see §7.3) [14.111, 165]. With this value of dT_{01}/dp, the Clausius–Clapeyron equation has been used [14.165] to obtain a value for the isobaric volume expansion of the C_{60} lattice at T_{01} (see §14.10). Scanning calorimetric spectra of samples are often taken to test whether undoped C_{60} material is present in a given sample, by looking for the characteristic DSC structure for C_{60} in the vicinity of T_{01}, since the alkali metal–doped M_xC_{60} does not show structure in this temperature region.

DSC studies of C_{70} have also been carried out showing two first-order phase transitions (see §7.2), a heat of transition of 3.5 J/g at the lower

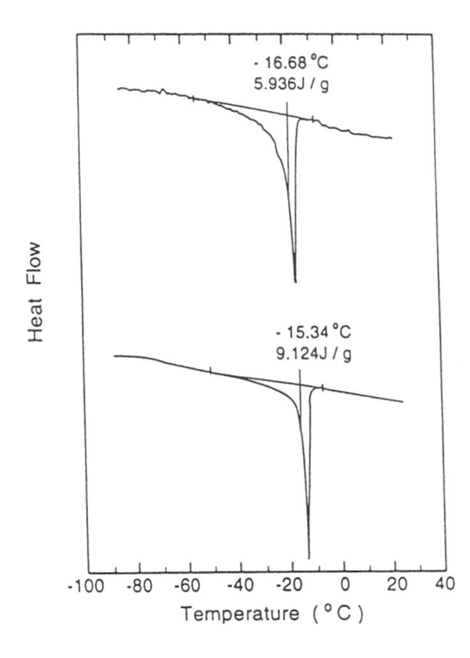

Fig. 14.20. Differential scanning calorimetry of chromatographed C_{60} samples, heated at 5°C/min under N_2 flow. Top: scan performed immediately after room temperature evaporation of solvent and showing a broad precursor 10–20°C below the main peak. Bottom: scan performed after 200°C annealing for 18–24 h, showing a decrease in the magnitude of the precursor, an increase in transition temperature T_{01}, and an increase in area under the peak (measured between the fiducial marks) [14.163].

transition temperature T_{02} (276 K) and of 2.7 J/g at the higher temperature T_{01} (337 K) [14.166].

14.10. Temperature Coefficient of Thermal Expansion

The temperature dependence of the lattice constant a_0 for C_{60} (see Fig. 14.21) shows a large discontinuity at the structural phase transition temperature $T_{01} = 261$ K [14.153, 167, 168] (see §7.1.3 and §7.4). The change in a_0 at T_{01} is found to be 0.044±0.004 Å on heating [14.153], indicative of a first-order phase transition and consistent with other experiments, especially the heat capacity (see §14.8). From the Clausius–Clapeyron equation, measurement of the latent heat at the T_{01} phase transition yields a value for the fractional change in the volume of the unit cell of $\Delta V/V = 7.5 \times 10^{-3}$ [14.165], which is in reasonable agreement with the direct structural measurement of $9.3 \pm 0.8 \times 10^{-3}$ [14.153].

The average isobaric volume coefficient of thermal expansion α both below and above the 261 K transition is $6.2 \pm 0.2 \times 10^{-5}$/K [14.153, 167, 168], which is relatively large compared to values for typical ionic crystals. Surprisingly, measurements of α taken to high temperature (1180 K) show a somewhat lower value of 4.7×10^{-5}/K than for the lower temperature range [14.169]. The thermal expansion of K_3C_{60} has also been measured

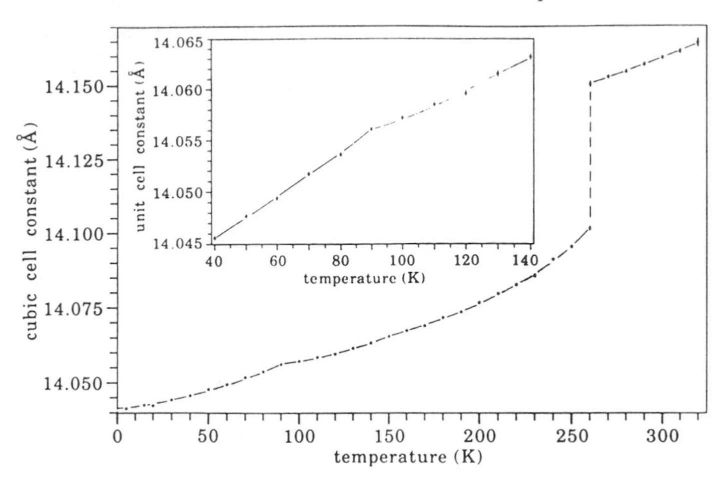

Fig. 14.21. Temperature variation of the cubic lattice constant a_0 for solid C_{60}. The inset shows the low-temperature data including the anomaly at $\sim$ 90 K on an expanded scale [14.167, 168].

and found to be about 80% of that for C_{60} prior to doping. The compressibilities for K_3C_{60} and Rb_3C_{60} are similar but slightly smaller than for C_{60} itself, corresponding to about 70% and 80% of the C_{60} compressibility, respectively [14.153, 170].

14.11. Thermal Conductivity

Because of the low carrier concentration in C_{60} and related compounds, the dominant contribution to the thermal conductivity κ is due to lattice vibrations. No experimental information is yet available on the electronic contribution κ_e to the thermal conductivity of C_{60}. However, as shown theoretically by Gelfand and Lu [14.22], if there were an electronic contribution to $\kappa(T)$, one would expect the Wiedemann–Franz law $\kappa_e/\sigma T = L$ to hold, where L is the Lorenz number. The lattice contribution given by the classical relation $\kappa_L = \frac{1}{3}C_v v\Lambda$, where v is the velocity of sound and Λ is the phonon mean free path, suggests that the contribution κ_L should be small as well.

Results on the temperature dependence of the thermal conductivity $\kappa(T)$ have been reported [14.171] for a single-crystal sample of C_{60} over the temperature range from 30 K to 300 K. These results show an anomaly in $\kappa(T)$ near 261 K associated with the first-order structural phase transition at T_{01} (see §7.1.3). The $\kappa(T)$ results further show the magnitude of the maximum thermal conductivity $\kappa_{\max}$ to be less than that for graphite (in-plane) [14.172, 173] and diamond [14.174] by more than three orders of magnitude. The low value of the thermal conductivity κ is attributed to both a high density of defect states associated with the merohedral disorder (see §14.2.1) and a low Debye temperature ($\Theta_D = 70$ K) in C_{60} compared to graphite ($\Theta_D = 2500$ K) [14.175].

To model the observed temperature dependence of $\kappa(T)$, two nearly degenerate molecular orientations are considered (differing by ~12 meV in energy), and separated by an energy barrier of ~290 meV (see §7.1.3). One interpretation of this result, consistent with specific heat measurements (see §14.8.1) and structural measurements (see §7.1.3), suggests that the energy difference of 11.4 meV may relate to the energy difference between the orientation of the double bonds opposite pentagonal faces relative to hexagonal faces, and the larger energy of ~290 meV may correspond to the energy barrier that must be overcome in making the transition from the higher-lying hexagonal face orientation to the lower-energy pentagonal face orientation opposite the double bond in the adjacent C_{60} molecule (see §7.1.3 and §14.8.1).

A thermal conductivity study of a C_{60}(85%)/C_{70}(15%) compacted sample going down to ~0.04 K shows an approximately T^2 dependence below ~1 K and an almost temperature-independent $\kappa(T)$ above ~10 K. A comparison of the measured $\kappa(T)$ (points) with an Einstein model with $\Theta_E = 35$ K and no adjustable parameters (dashed curve) is shown in Fig. 14.22(a), where the Einstein contribution to the thermal conductivity is

$$\kappa_{\mathrm{Ein}} = 2\frac{k_{\mathrm{B}}^2}{\hbar}\frac{n_v^{1/3}}{\pi}\Theta_E\frac{x^2 e^x}{(e^x - 1)^2} \tag{14.17}$$

in which n_v is the number density of C_{60} molecules and $x = \Theta_E/T$. In Fig. 14.22(b) are shown results for single-crystal samples of diamond, C_{60}, and graphite for heat flow in-plane and along the c-axis. For $T < 10$ K, solids such as SiO_2 and As_2S_3 [14.141] also show a T^2 dependence for $\kappa(T)$ below 1 K [see Fig. 14.22(a)]. Sundqvist has modeled $1/\kappa(T)$ *vs.* T in terms of phonon–phonon and phonon–defect scattering and finds that the $\kappa(T)$ data for C_{60} can be explained by standard theories [14.176].

14.12. Thermopower

The Seebeck coefficient or thermoelectric power S is defined as the electric field E produced by a temperature gradient ∇T under the condition of no current flow, $S(E) \equiv E/\nabla T$, where S represents the amount of heat per electric charge carried by the carriers as they move from a hot to a cold junction. At the hot junction there are relatively more energetic carriers than at the cold junction, thereby giving rise to a redistribution of carriers. This charge redistribution causes a small pileup of negative charge at the cold junction and of positive charge at the hot junction, thereby causing a voltage (or an electric field) to develop, which impedes further diffusion of

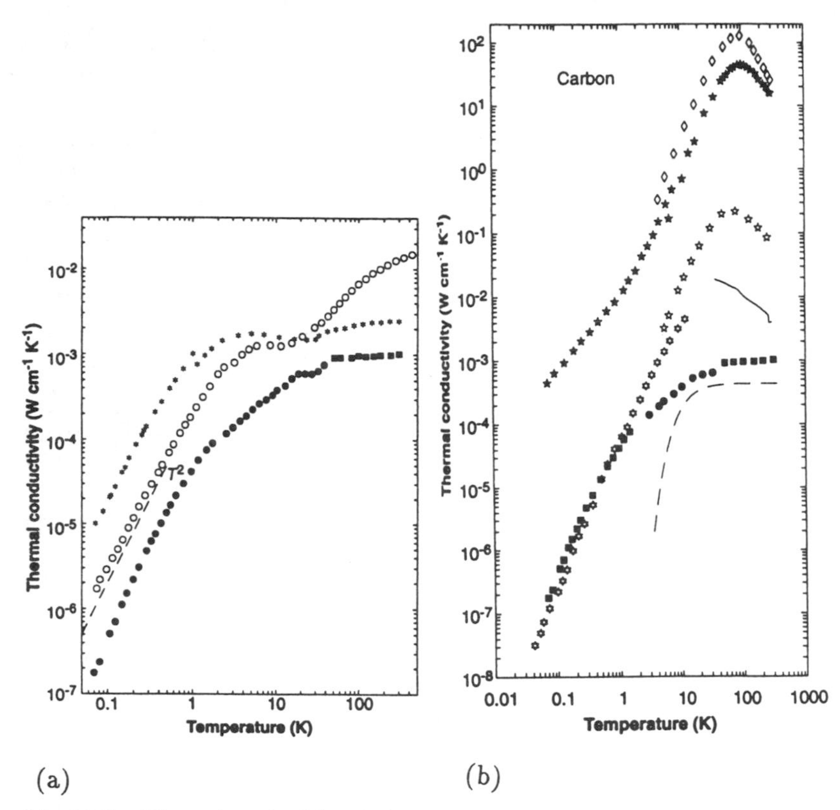

Fig. 14.22. Thermal conductivity of (a) polycrystalline C_{60}/C_{70} compacts (filled circles and squares) compared with that of amorphous SiO_2 (open circles) and amorphous As_2S_3 (asterisks) and a T^2 dependence (dashed line). (b) Carbon in its three phases: diamond (open diamonds), single-crystal C_{60} (solid line), and single-crystal graphite, heat flow in the *ab*-plane (filled stars) and parallel to *c*-axis (open stars). The dashed line corresponds to the thermal conductivity based on the Einstein model (Eq. 14.17). [14.141].

carriers. In addition to this diffusion process, which gives rise to a diffusion term S_d in the thermopower, there is also a phonon drag contribution S_p arising from the flow of phonons from the hot to the cold junction, dragging carriers with them as they are transported. Thus we normally write the following expression for the temperature dependence of the thermopower:

$$S(T) = S_d(T) + S_p(T). \tag{14.18}$$

The thermopower for undoped C_{60} and C_{70} is negligible because of the absence of carriers. Thus physical thermopower measurements of fullerenes have focused on M_3C_{60} (M = K, Rb) and K_xC_{70} ($x \sim 4$).

The temperature dependence of the thermopower $S(T)$ of single-crystal K_3C_{60} and Rb_3C_{60} (prepared from CS_2 solution) has been measured from

300 K down to low T [14.37]. The results (see Fig. 14.23) show that $S(T)$ is negative (consistent with conduction by electrons) and nearly linear in T for both K_3C_{60} (above ~150 K) and Rb_3C_{60} (above ~100 K) [14.32, 37, 177]. The results have been used to obtain an estimate for the Fermi energy and bandwidth using the arguments given below.

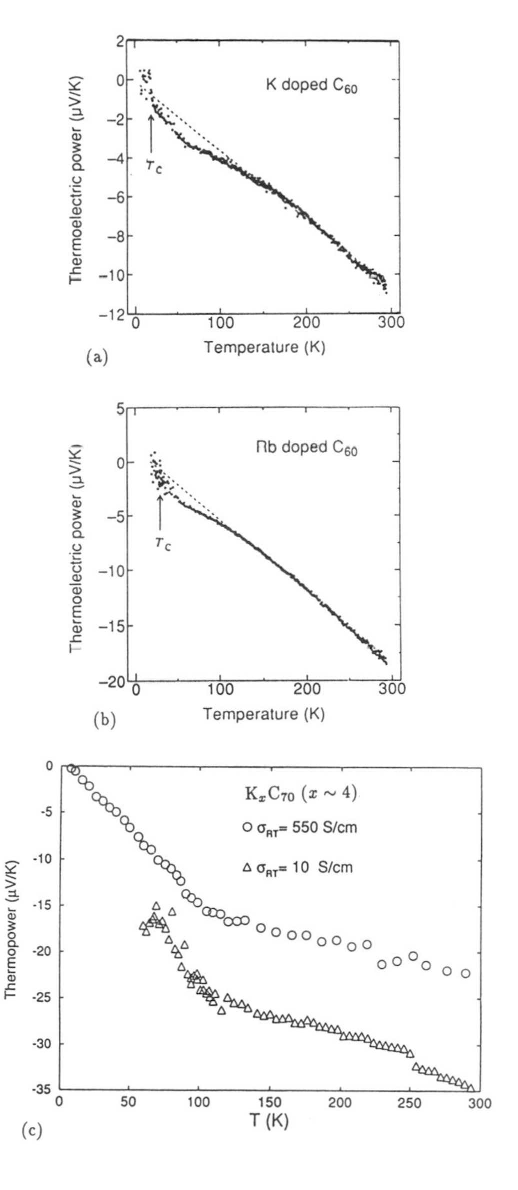

Fig. 14.23. Temperature dependence of the thermoelectric power of (a) K-doped C_{60} and (b) Rb-doped C_{60}. The dashed lines in (a) and (b) are an extrapolation of the linear $S(T)$ behavior to show the low temperature hump associated with phonon drag more clearly. In both (a) and (b) the superconducting T_c is indicated. (c) The temperature dependence of the thermopower of K_xC_{70} at the stoichiometry of the maximum-conductivity phase ($x \sim 4$). The circles and triangles are for the data from samples with room temperature electrical conductivities $\sigma_{RT} = 575$ S/cm and 10 S/cm, respectively [14.27].

For the M_3C_{60} compounds, carriers from the three t_{1u} LUMO-derived bands make a contribution $S_i(T)$ to the total thermopower $S(T)$ in accordance with

$$S(T) = \sum_{i=1}^{3} \sigma_i S_i(T) \Big/ \sum_{i=1}^{3} \sigma_i \tag{14.19}$$

in which $S_i(T)$ is the contribution from carriers in each band i and σ_i is the corresponding conductivity. Assuming an equal contribution from each band to $S(T)$, we can write for the diffusion thermopower

$$S_i(T) = \frac{\pi^2 k_B^2 T}{eE_F} \left[\frac{\partial \ln \sigma(E)}{\partial \ln E} \right]_{E=E_F}. \tag{14.20}$$

Assuming a parabolic dispersion relation for the three bands, and $\tau_i(E) \propto E^{p_i}$, then the diffusion thermopower can be written as

$$S_d(T) = -\frac{\pi^2 k_B^2 T}{3e} \sum_{i=1}^{3} \frac{1}{E_i} \left(\frac{3}{2} + p_i \right) r_i \tag{14.21}$$

where E_i refers to the ith band at the Fermi level and $r_i = \sigma_i / \sigma$ [14.177]. For impurity scattering $r_i \simeq 1/3$, $p_i = -1/2$ and $1/E_F = (1/3) \sum 1/E_i$. Using a 3D free electron model for the carriers, Eq. (14.21) becomes

$$S_d(T) = \frac{\pi^2 k_B^2 T}{6eE_F} \tag{14.22}$$

which is linear in T. Furthermore, the phonon drag term $S_p(T)$ is proportional to T^4 since

$$S_p = \sum_q C(q) R(q) \propto T^4 \tag{14.23}$$

where the specific heat term $C(q) \propto T^3$ (since the phonon density of states at low T has a T^3 dependence). The scattering term $R(q)$ in Eq. (14.23) can be expressed in terms of the scattering times $t(q)$ and $t_c(q)$ by

$$R(q) = \frac{t(q)}{t_c(q)} < 1 \tag{14.24}$$

so that at low-temperature the scattering rate $1/t(q)$ is dominated by crystallite boundary scattering and is independent of T, since the transition probabilities are given by

$$1/t(q) = \sum_i 1/t_i(q) \tag{14.25}$$

while

$$1/t_c(q) = aq \tag{14.26}$$

where aq is the scattering rate of the qth phonon due to the electron–phonon interaction, so that $R(q) \simeq aq/b \propto T$. At higher temperatures $S_p(T) \sim 1/T^n$ where $n > 1$, since at higher temperatures $1/t(q)$ is controlled by phonon–phonon interaction processes. Thus the S_p contribution gives rise to a hump in the $S(T)$ curve. Below this hump, the temperature dependence of the diffusion thermopower is found from Eq. (14.22), which in turn provides an estimate for E_F, yielding E_F values of 0.32 ± 0.05 eV and 0.20 ± 0.02 eV, respectively, for K_3C_{60} and Rb_3C_{60} [14.177].

The $S(T)$ results of Fig. 14.23 further show room temperature values of -11μV/K and -18μV/K for K_3C_{60} and Rb_3C_{60}, respectively, and the linear part of $S(T)$ extrapolates to anomalies at ~19 K and ~28 K, respectively, for K_3C_{60} and Rb_3C_{60}, in good agreement with the superconducting T_c values for these compounds, as indicated on Fig. 14.23(a) and (b). From the E_F values given above, the ratio of the electronic density of states for these compounds is obtained

$$\frac{N^{K_3C_{60}}(E_F)}{N^{Rb_3C_{60}}(E_F)} = \frac{E_F^{K_3C_{60}}}{E_F^{Rb_3C_{60}}} \sim 1.5\text{–}1.8, \tag{14.27}$$

which is in good agreement with the observed ratio of $[T_c(Rb_3C_{60})/T_c(K_3C_{60})] = 1.55$ and provides additional confirmation of thc cmpirical relation shown in Fig. 15.3 between T_c and $N(E_F)$ [14.37]. If it is assumed that E_F is approximately at midband, a bandwidth of ~0.65 eV and 0.4 eV is estimated for K_3C_{60} and Rb_3C_{60}, respectively, from the thermopower determination of E_F [14.177].

Gelfand and Lu [14.21, 97] have considered the effect of orientational disorder and have found the relation

$$\frac{dS}{dT} \approx -\frac{\pi^2 k_B^2}{6eW} \tag{14.28}$$

to apply to the doped fullerenes where W is the bandwidth. The same values of 5.4 and 8.8 states/eV-C_{60}-spin were reported for K_3C_{60} and Rb_3C_{60} when merohedral disorder was considered.

Below ~100 K a hump in $S(T)$ is observed for the M_3C_{60} compounds. This behavior has been attributed to a phonon drag contribution S_p to $S(T)$ [14.177]. The magnitude of S_p from Fig. 14.23(a) is ~1–1.5 μV/K for K_3C_{60}, indicative of a relatively weak electron–phonon interaction. Since the intramolecular vibrations occur at temperatures (~400 K) far above the hump in $S_p(T)$, the phonon drag effect has been identified with electron coupling to intermolecular vibrational modes. This explanation for the hump in $S(T)$ at low T does not explain the change in sign observed in the Hall coefficient at ~220 K [14.45]. If the samples contain conducting particles separated by low activation barriers, then the carrier transport likely involves hopping between conducting particles. Physically, the conducting particles may be C_{60} molecules and the barrier may be the intermolecular distances over which carriers must hop. If charge transport involves hopping, a fluctuation-induced-tunneling mechanism, rather than a free electron model, should probably be used.

The temperature dependence of the thermopower has also been measured for K_xC_{70} at the maximum conductivity phase (which nominally occurs at the composition K_4C_{70}), and the results are shown in Fig. 14.23(c) [14.27]. In this case, $S(T)$ shows an approximately linear T dependence at high T and another linear T dependence with a larger slope at low T. In comparison with S(300 K) for K_3C_{60} given above, the room temperature value of S(300 K) for K_4C_{70} is $\sim$ 20–35μV/K, and the slope at low T gives a value for $E_F \sim 0.2$ eV and a bandwidth of ~0.4 eV, using similar assumptions as for K_3C_{60}. It is of interest to note that the temperature at which the slope of $S(T)$ shows a discontinuity is in good agreement with the temperature where the Hall coefficient changes sign, in contrast to the observations for the M_3C_{60} compounds. The data for $S(T)$ in K_4C_{70} show no clear evidence for a phonon drag anomaly, from which we conclude that the electron–phonon coupling in K_4C_{70} is weaker than for the M_3C_{60} compounds, in agreement with Raman scattering studies on K_4C_{70} and with the absence of a superconducting transition down to 1.35 K [14.178].

For K_4C_{70} the thermopower (S) is negative and metallic, indicative of charge carriers that are electrons. An anomaly in $S(T)$ for K_4C_{70} is found at approximately 100 K [14.178]. For the same sample as was used for the thermopower measurements, the temperature dependence of the magnetoconductance was found to fit weak localization (WL) and electron–electron (e–e) interaction theories [14.179–181].

14.13. Internal Friction

Internal friction experiments are sensitive to vibrational damping, and measurements of the internal friction can be carried out by preparing a fullerene film on a quartz piezoelectric oscillator [14.182]. The temperature dependence of the internal friction shows temperature-independent behavior at low-temperature and a peak in the magnitude of the internal friction at higher temperatures. This peak has been related to the magnitude of one of the elastic constants. The magnitude of the low-temperature internal friction decreases with increasing crystallite size. With films having a crystallite size of ~1500 Å, as observed by scanning tunneling microscopy (STM), the magnitude of the internal friction becomes too small to support a glassy phase at low-temperature, consistent with other studies (see §7.1.3).

References

[14.1] N. F. Mott. *Metal Insulator Transitions*. Taylor & Francis, New York (1990).

[14.2] J. Mort, R. Ziolo, M. Machonkin, D. R. Huffman, and M. I. Ferguson. *Chem. Phys. Lett.*, **186**, 284 (1991).

[14.3] F. Stepniak, P. J. Benning, D. M. Poirier, and J. H. Weaver. *Phys. Rev. B*, **48**, 1899 (1993).

[14.4] A. F. Hebard. In P. Jena, S. N. Khanna, and B. K. Rao (eds.), *Physics and Chemistry of Finite Systems: From Clusters to Crystals*, vol. 374, pp. 1213–1220, Kluwer Academic Publishers, Dordrecht (1991). NATO ASI Series C: Mathematical and Physical Sciences.

[14.5] G. P. Kochanski, A. F. Hebard, R. C. Haddon, and A. T. Fiory. *Science*, **255**, 184 (1992).

[14.6] R. W. Lof, M. A. van Veenendaal, B. Koopmans, H. T. Jonkman, and G. A. Sawatzky. *Phys. Rev. Lett.*, **68**, 3924 (1992).

[14.7] C. Gu, F. Stepniak, D. M. Poirier, M. B. Jost, P. J. Benning, Y. Chen, T. R. Ohno, J. L. Martins, J. H. Weaver, J. Fure, and R. E. Smalley. *Phys. Rev. B*, **45**, 6348 (1992).

[14.8] A. F. Hebard, T. T. M. Palstra, R. C. Haddon, and R. M. Fleming. *Phys. Rev. B*, **48**, 9945 (1993).

[14.9] W. E. Pickett. In H. Ehrenreich and F. Spaepen (eds.), *Solid State Physics*, vol. 48, p. 225, Academic Press, New York (1994).

[14.10] K. Holczer and R. L. Whetten. *Carbon*, **30**, 1261 (1992).

[14.11] A. R. Kortan, N. Kopylov, S. H. Glarum, E. M. Gyorgy, A. P. Ramirez, R. M. Fleming, F. A. Thiel, and R. C. Haddon. *Nature (London)*, **355**, 529 (1992).

[14.12] P. Zhou, A. M. Rao, K. A. Wang, J. D. Robertson, C. Eloi, M. S. Meier, S. L. Ren, X. X. Bi, P. C. Eklund, and M. S. Dresselhaus. *Appl. Phys. Lett.*, **60**, 2871 (1992).

[14.13] A. Oshiyama, S. Saito, N. Hamada, and Y. Miyamoto. *J. Phys. Chem. Solids*, **53**, 1457 (1992).

[14.14] R. Tycko, G. Dabbagh, M. J. Rosseinsky, D. W. Murphy, R. M. Fleming, A. P. Ramirez, and J. C. Tully. *Science*, **253**, 1419 (1991).

[14.15] X.-D. Xiang, J. G. Hou, G. Briceño, W. A. Vareka, R. Mostovoy, A. Zettl, V. H. Crespi, and M. L. Cohen. *Science*, **256**, 1190 (1992).

[14.16] Z. H. Wang, A. W. P. Fung, G. Dresselhaus, M. S. Dresselhaus, K. A. Wang, P. Zhou, and P. C. Eklund. *Phys. Rev. B*, **47**, 15354 (1993).

[14.17] J. G. Hou, V. H. Crespi, X. D. Xiang, W. A. Vareka, G. Briceño, A. Zettl, and M. L. Cohen. *Solid State Commun.*, **86**, 643 (1993).

[14.18] M. L. Cohen. *Mater. Sci. Eng.*, **B19**, 111 (1993).

[14.19] H. Ogata, T. Inabe, H. Hoshi, Y. Maruyama, Y. Achiba, S. Suzuki, K. Kikuchi, and I. Ikemoto. *Jpn. J. Appl. Phys.*, **31**, L166 (1992).

[14.20] H. Ogata. *Electronic Structures of Exotic Molecular Solids*. Ph.D. thesis, The Graduate University for Advanced Studies, Okazaki, Japan (1992). Department of Functional Molecular Science, School of Mathematical and Physical Science.

[14.21] M. P. Gelfand and J. P. Lu. *Phys. Rev. Lett.*, **68**, 1050 (1992).

[14.22] M. P. Gelfand and J. P. Lu. *Phys. Rev. B*, **46**, 4367 (1992). Erratum: *Phys. Rev. B*, **47**, 4149 (1992).

[14.23] M. P. Gelfand and J. P. Lu. *Appl. Phys. A*, **56**, 215 (1993).

[14.24] O. Zhou, R. M. Fleming, D. W. Murphy, M. J. Rosseinsky, A. P. Ramirez, R. B. van Dover, and R. C. Haddon. *Nature (London)*, **362**, 433 (1993).

[14.25] M. J. Rosseinsky, D. W. Murphy, R. M. Fleming, and O. Zhou. *Nature (London)*, **364**, 425 (1993).

[14.26] Z. H. Wang, K. Ichimura, M. S. Dresselhaus, G. Dresselhaus, W. T. Lee, K. A. Wang, and P. C. Eklund. *Phys. Rev. B*, **48**, 10657 (1993).
[14.27] Z. H. Wang, M. S. Dresselhaus, G. Dresselhaus, K. A. Wang, and P. C. Eklund. *Phys. Rev. B*, **49**, 15890 (1994).
[14.28] Y. Iwasa, T. Arima, R. M. Fleming, T. Siegrist, O. Zhou, R. C. Haddon, L. J. Rothberg, K. B. Lyons, H. L. Carter, Jr., A. F. Hebard, R. Tycko, G. Dabbagh, J. J. Krajewski, G. A. Thomas, and T. Yagi. *Science*, **264**, 1570 (1994).
[14.29] T. T. M. Palstra, O. Zhou, Y. Iwasa, P. Sulewski, R. Fleming, and B. Zegarski. *Solid State Commun.*, **93**, 327 (1995).
[14.30] M. S. Deshpande, S. C. Erwin, S. Hong, and E. J. Mele. *Phys. Rev. Lett.*, **71**, 2619 (1993).
[14.31] X.-D. Xiang, J. G. Hou, V. H. Crespi, A. Zettl, and M. L. Cohen. *Nature (London)*, **361**, 54 (1993).
[14.32] Y. Maruyama, T. Inabe, M. Ogata, Y. Achiba, K. Kikuchi, S. Suzuki, and I. Ikemoto. *Mater. Sci. Eng.*, **B19**, 162 (1993).
[14.33] T. Takahashi, T. Morikawa, S. Hasegawa, K. Kamiya, H. Fujimoto, S. Hino, K. Seki, H. Katayama-Yoshida, H. Inokuchi, K. Kikuchi, S. Suzuki, K. Ikemoto, and Y. Achiba. *Chem. Soc. Jpn.* (1992). 2nd General Symposium of the C_{60} Res. Soc.
[14.34] T. Takahashi, T. Morikawa, S. Hasegawa, K. Kamiya, H. Fujimoto, S. Hino, K. Seki, H. Katayama-Yoshida, H. Inokuchi, K. Kikuchi, S. Suzuki, K. Ikemoto, and Y. Achiba. *Physica C*, **190**, 205 (1992).
[14.35] Y. Iwasa. Unpublished (1993). US–Japan Seminar on Fullerenes. Waikiki, Hawaii, July 20, 1993.
[14.36] Y. Maruyama, T. Inabe, H. Ogata, Y. Achiba, S. Suzuki, K. Kikuchi, and I. Ikemoto. *Chem. Lett.*, **10**, 1849 (1991). The Chemical Society of Japan.
[14.37] T. Inabe, H. Ogata, Y. Maruyama, Y. Achiba, S. Suzuki, K. Kikuchi, and I. Ikemoto. *Phys. Rev. Lett.*, **69**, 3797 (1992).
[14.38] W. A. Vareka and A. Zettl. *Phys. Rev. Lett.*, **72**, 4121 (1994).
[14.39] Y. Iye. *Physical Properties of High Temperature Superconductors III*, edited by D. M. Ginsberg, World Scientific Series on High Temperature Superconductors, World Scientific Publishing, Singapore (1992). Vol. III, p. 285 (1992). Chapter 4.
[14.40] A. F. Hebard, M. J. Rosseinsky, R. C. Haddon, D. W. Murphy, S. H. Glarum, T. T. M. Palstra, A. P. Ramirez, and A. R. Kortan. *Nature (London)*, **350**, 600 (1991).
[14.41] R. C. Haddon, A. F. Hebard, M. J. Rosseinsky, D. W. Murphy, S. J. Duclos, K. B. Lyons, B. Miller, J. M. Rosamilia, R. M. Fleming, A. R. Kortan, S. H. Glarum, A. V. Makhija, A. J. Muller, R. H. Eick, S. M. Zahurak, R. Tycko, G. Dabbagh, and F. A. Thiel. *Nature (London)*, **350**, 320 (1991).
[14.42] N. Klein, U. Poppe, N. Tellmann, H. Schulz, W. Evers, U. Dähne, and K. Urban. *IEEE Trans. Applied Superconductor*, **AS-2**, (1992).
[14.43] V. H. Crespi, J. G. Hou, X. D. Xiang, M. L. Cohen, and A. Zettl. *Phys. Rev. B*, **46**, 12064 (1992).
[14.44] L. D. Rotter, Z. Schlesinger, J. P. McCauley, Jr., N. Coustel, J. E. Fischer, and A. B. Smith III. *Nature (London)*, **355**, 532 (1992).
[14.45] T. T. M. Palstra, R. C. Haddon, A. F. Hebard, and J. Zaanen. *Phys. Rev. Lett.*, **68**, 1054 (1992).
[14.46] B. L. Altshuler, A. G. Aronov, and P. A. Lee. *Phys. Rev. Lett.*, **44**, 1288 (1980).
[14.47] P. Sheng. *Phys. Rev. B*, **21**, 2180 (1980).
[14.48] R. C. Haddon, G. P. Kochanski, A. F. Hebard, A. T. Fiory, and R. C. Morris. *Science*, **258**, 1636 (1992).

[14.49] R. C. Haddon, G. P. Kochanski, A. F. Hebard, A. T. Fiory, R. C. Morris, and A. S. Perel. *Chem. Phys. Lett.*, **203**, 433 (1993).

[14.50] G. K. Wertheim, D. N. E. Buchanan, and J. E. Rowe. *Science*, **258**, 1638 (1992).

[14.51] J. M. Gildemeister and G. K. Wertheim. *Chem. Phys. Lett.*, **230**, 181 (1994).

[14.52] A. R. Kortan, N. Kopylov, R. M. Fleming, O. Zhou, F. A. Thiel, R. C. Haddon, and K. M. Rabe. *Phys. Rev. B*, **47**, 13070 (1993).

[14.53] L. Q. Jiang and B. E. Koel. *Phys. Rev. Lett.*, **72**, 140 (1994).

[14.54] S. L. Ren, K. A. Wang, P. Zhou, Y. Wang, A. M. Rao, M. S. Meier, J. Selegue and P. C. Eklund. *Appl. Phys. Lett.*, **61**, 124 (1992).

[14.55] K. Imaeda, K. Yakushi, H. Inokuchi, K. Kikuchi, I. Ikemoto, S. Suzuki, and Y. Achiba. *Solid State Commun*, **84**, 1019 (1992).

[14.56] S. Saito and A. Oshiyama. *Phys. Rev. Lett.*, **66**, 2637 (1991).

[14.57] A. F. Hebard, C. B. Eom, Y. Iwasa, K. B. Lyons, G. A. Thomas, D. H. Rapkine, R. M. Fleming, R. C. Haddon, J. M. Phillips, J. H. Marshall, and R. H. Eick. *Phys. Rev. B*, **50**, 17740 (1994).

[14.58] W. Zhao, K. Luo, J. Cheng, C. Li, D. Yin, Z. Gu, X. Zhou, and Z. Jin. *Solid State Commun.*, **83**, 853 (1992).

[14.59] X. D. Zhang, W. B. Zhao, K. Wu, Z. Y. Ye, J. L. Zhang, C. Y. Li, D. L. Yin, Z. N. Gu, X. H. Zhou, and Z. X. Jin. *Chem. Phys. Lett.*, **228**, 100 (1994).

[14.60] W. R. Datars, P. K. Ummat, T. Olech, and R. K. Nkum. *Solid. State Commun.*, **86**, 579 (1993).

[14.61] M. S. Dresselhaus and G. Dresselhaus. *Adv. Phys.*, **30**, 139 (1981).

[14.62] M. S. Dresselhaus, G. Dresselhaus, K. Sugihara, I. L. Spain, and H. A. Goldberg. *Graphite Fibers and Filaments*, vol. 5 of *Springer Series in Materials Science*. Springer-Verlag, Berlin (1988).

[14.63] M. S. Dresselhaus, G. Dresselhaus, and R. Saito. In D. M. Ginsberg (ed.), *Physical Properties of High Temperature Superconductors IV*, World Scientific Series on High Temperature Superconductors, Singapore (1994). World Scientific Publishing Co. Vol. IV, Chapter 7, p. 471.

[14.64] Y. Takada. *J. Phys. Chem. Solids*, **54**, 1779 (1993).

[14.65] G. M. Eliashberg. *Zh. Eksp. Teor. Fiz.*, **38**, 966 (1960). Sov. Phys.-JETP **11**, 696 (1960).

[14.66] G. M. Eliashberg. *Zh. Eksp. Teor. Fiz.*, **39**, 1437 (1960). Sov. Phys.-JETP **12**, 1000 (1961).

[14.67] D. J. Scalapino, J. R. Schrieffer, and J. W. Wilkins. *Phys. Rev.*, **148**, 263 (1966).

[14.68] A. B. Migdal. *Zh. Eksp. Teor. Fiz.*, **34**, 1438 (1958). Sov. Phys.-JETP **7**, 996 (1958).

[14.69] W. E. Pickett. *Phys. Rev. B*, **26**, 1186 (1982).

[14.70] S. C. Erwin. In W. Billups and M. Ciufolini (eds.), *Buckminsterfullerene*, p. 217, VCH Publishers, New York (1993).

[14.71] H. Zheng and K. H. Bennemann. *Phys. Rev. B*, **46**, 11993 (1992).

[14.72] J. R. Schrieffer. *Theory of Superconductivity*. Benjamin, New York (1983).

[14.73] L. Pietronero and S. Strässler. In C. Taliani, G. Ruani, and R. Zamboni (eds.), *Proc. of the First Italian Workshop on Fullerenes: Status and Perspectives*, vol. 2, p. 225, World Scientific, Singapore (1992).

[14.74] P. B. Allen. *Phys. Rev. B*, **6**, 2577 (1972).

[14.75] M. A. Schlüter, M. Lannoo, M. Needels, G. A. Baraff, and D. Tománek. *Phys. Rev. Lett.*, **68**, 526 (1992).

[14.76] M. A. Schlüter, M. Lannoo, M. Needels, G. A. Baraff, and D. Tománek. *Mater. Sci. Eng.*, **B19**, 129 (1993).

[14.77] M. Schlüter, M. Lannoo, M. Needels, G. A. Baraff, and D. Tománek. *J. Phys. Chem. Solids*, **53**, 1473 (1992).
[14.78] K. H. Johnson, M. E. McHenry, and D. P. Clougherty. *Physica C*, **183**, 319 (1991).
[14.79] G. Onida and G. Benedek. *Europhys. Lett.*, **18**, 403 (1992).
[14.80] R. A. Jishi and M. S. Dresselhaus. In C. L. Renschler, J. Pouch, and D. M. Cox (eds.), *Novel Forms of Carbon: MRS Symposia Proceedings, San Francisco*, vol. 270, p. 161, Materials Research Society Press, Pittsburgh, PA (1992).
[14.81] C. M. Varma, J. Zaanen, and K. Raghavachari. *Science*, **254**, 989 (1991).
[14.82] V. P. Antropov, O. Gunnarsson, and A. I. Liechtenstein. *Phys. Rev. B*, **48**, 7651 (1993).
[14.83] B. Friedman. *Phys. Rev. B*, **45**, 1454 (1992).
[14.84] K. Harigaya. *Phys. Rev. B*, **45**, 13676 (1992).
[14.85] K. Harigaya. *Synth. Met.*, **56**, 3202 (1993).
[14.86] W. P. Su, J. R. Schrieffer, and A. J. Heeger. *Phys. Rev. Lett.*, **42**, 1698 (1979). see also Phys. Rev. *B*, **22**, 2099 (1980).
[14.87] H. Kamimura. *Theoretical aspects of band structures and electronic properties of pseudo-one-dimensional solids*. Kluwer Academic Publishers, Hingham, MA, USA (1985).
[14.88] M. Lannoo, G. A. Baraff, M. Schlüter, and D. Tománek. *Phys. Rev. B*, **44**, 12106 (1991).
[14.89] H. Scherrer and G. Stollhoff. *Phys. Rev. B*, **47**, 16570 (1993).
[14.90] R. C. Bingham, M. J. S. Dewar, and D. H. Lo. *J. Am. Chem. Soc.*, **97**, 1285 (1975).
[14.91] K. Tanaka, M. Okada, K. Okahara, and T. Yamabe. *Chem. Phys. Lett.*, **191**, 469 (1992).
[14.92] R. Saito, G. Dresselhaus, and M. S. Dresselhaus. *Chem. Phys. Lett.*, **210**, 159 (1993).
[14.93] S. C. Erwin and M. R. Pederson. *Phys. Rev. Lett.*, **67**, 1610 (1991).
[14.94] H. Suematsu, K. Higuchi, and S. Tanuma. *J. Phys. Soc. Jpn.*, **48**, 1541 (1980).
[14.95] C. T. Chen, L. H. Tjeng, P. Rudolf, G. Meigs, J. F. Rowe, J. Chen, J. P. McCauley, A. B. Smith, A. R. McGhie, W. J. Romanow, and C. Plummer. *Nature (London)*, **352**, 603 (1991).
[14.96] R. Tycko, G. Dabbagh, M. J. Rosseinsky, D. W. Murphy, R. M. Fleming, A. P. Ramirez, and J. C. Tully. *Science*, **253**, 884 (1991).
[14.97] M. P. Gelfand. *Superconductivity Review*, **1**, 103 (1994).
[14.98] S. C. Erwin and W. E. Pickett. *Science*, **254**, 842 (1991).
[14.99] V. P. Antropov, I. I. Mazin, O. K. Andersen, A. I. Liechtenstein, and O. Jepson. *Phys. Rev. B*, **47**, 12373 (1993).
[14.100] S. Rice. Private communication (1993).
[14.101] M. Knupfer, M. Merkel, M. S. Golden, J. Fink, O. Gunnarsson, and V. P. Antropov. *Phys. Rev. B*, **47**, 13944 (1993).
[14.102] A. P. Ramirez, M. J. Rosseinsky, D. W. Murphy, and R. C. Haddon. *Phys. Rev. Lett.*, **69**, 1687 (1992).
[14.103] S. Satpathy, V. P. Antropov, O. K. Andersen, O. Jepson, O. Gunnarsson, and A. I. Liechtenstein. *Phys. Rev. B*, **46**, 1773 (1992).
[14.104] W. H. Wong, M. Hanson, W. G. Clark, G. Grüner, J. D. Thompson, R. L. Whetten, S. M. Huang, R. B. Kaner, F. Diederich, P. Pettit, J. J. Andre, and K. Holczer. *Europhys. Lett.*, **18**, 79 (1992).
[14.105] R. Tycko, G. Dabbagh, M. J. Rosseinsky, D. W. Murphy, A. P. Ramirez, and R. M. Fleming. *Phys. Rev. Lett.*, **68**, 1912 (1992).
[14.106] A. Oshiyama and S. Saito. *Solid State Commun.*, **82**, 41 (1992).

[14.107] M. N. Regueiro, P. Monceau, A. Rassat, P. Bernier, and A. Zahab. *Nature (London)*, **354**, 289 (1991).

[14.108] M. N. Regueiro, P. Monceau, and J.-L. Hodeau. *Nature (London)*, **355**, 237 (1992).

[14.109] M. N. Regueiro, O. Bethoux, J. M. Mignot, P. Monceau, P. Bernier, C. Fabre, and A. Rassat. *Europhys. Lett.*, **21**, 49 (1993).

[14.110] S. K. Ramasesha and A. K. Singh. *Solid State Commun.*, **91**, 25 (1994).

[14.111] G. A. Samara, J. E. Schirber, B. Morosin, L. V. Hansen, D. Loy, and A. P. Sylwester. *Phys. Rev. Lett.*, **67**, 3136 (1991).

[14.112] J. Mort, M. Machonkin, R. Ziolo, D. R. Huffman, and M. I. Ferguson. *Appl. Phys. Lett.*, **60**, 1735 (1992).

[14.113] M. Kaiser, J. Reichenbach, H. J. Byrne, J. Anders, W. Maser, S. Roth, A. Zahab, and P. Bernier. *Solid State Commun.*, **81**, 261 (1992).

[14.114] H. Yonehara and C. Pac. *Appl. Phys. Lett.*, **61**, 575 (1992).

[14.115] M. Hosoya, K. Ichimura, Z. H. Wang, G. Dresselhaus, M. S. Dresselhaus, and P. C. Eklund. *Phys. Rev. B*, **49**, 4981 (1994).

[14.116] N. Minami and M. Sato. *Synth. Met.*, **56**, 3092 (1993).

[14.117] A. Hamed, H. Rasmussen, and P. H. Hor. *Appl. Phys. Lett.*, **64**, 526 (1994).

[14.118] Y. Wang. *Nature (London)*, **356**, 585 (1992).

[14.119] N. S. Sariciftci, L. Smilowitz, A. J. Heeger, and F. Wudl. *Science*, **258**, 1474 (1992).

[14.120] N. S. Sariciftci, D. Braun, C. Zhang, V. I. Srdanov, A. J. Heeger, C. Stucky, and F. Wudl. *Appl. Phys. Lett.*, **62**, 585 (1993).

[14.121] K. Lee, R. A. J. Janssen, N. S. Sariciftci, and A. J. Heeger. *Phys. Rev. B*, **49**, 5781 (1994).

[14.122] C. H. Lee, G. Yu, D. Moses, K. Pakbaz, C. Zhang, N. S. Sariciftci, A. J. Heeger, and F. Wudl. *Phys. Rev. B*, **48**, 15425 (1993).

[14.123] N. S. Sariciftci and A. J. Heeger. *Int. J. Mod. Phys. B*, **8**, 237 (1994).

[14.124] N. S. Sariciftci, L. Smilowitz, C. Zhang, V. I. Srdanov, A. J. Heeger, and F. Wudl. *Proc. SPIE - Int. Soc. Opt. Eng.*, **1852**, 297 (1993).

[14.125] B. Kraabel, D. W. McBranch, K. H. Lee, and N. S. Sariciftci. In Z. H. Kafafi (ed.), *Proceedings of the International Society for Optical Engineering (SPIE). "Fullerenes and Photonics I; Proc. v. 2284*, pp. 194–207, Bellingham, WA (1994). SPIE Optical Engineering Press. San Diego, CA, July 24-29, 1994.

[14.126] M. Kaiser, W. K. Maser, H. J. Byrne, A. Mittelbach, and S. Roth. *Solid State Commun.*, **87**, 281 (1993).

[14.127] R. Könenkamp, J. Erxmeyer, and A. Weidinger. *Appl. Phys. Lett.*, **65**, 758 (1994).

[14.128] S. Kazaoui, R. Ross, and N. Minami. *Solid State Commun.*, **90**, 623 (1994).

[14.129] A. Hamed, Y. Y. Sun, Y. K. Tao, R. L. Meng, and P. H. Hor. *Phys. Rev. B*, **47**, 10873 (1993).

[14.130] A. Hamed, H. Rasmussen, and P. H. Hor. *Phys. Rev. B*, **48**, 14760 (1993).

[14.131] A. Hamed, D. Y. Xing, C. C. Chow, and P. H. Hor. *Phys. Rev. B*, **50**, 11219 (1994).

[14.132] A. M. Rao, P. Zhou, K.-A. Wang, G. T. Hager, J. M. Holden, Y. Wang, W. T. Lee, X.-X. Bi, P. C. Eklund, D. S. Cornett, M. A. Duncan, and I. J. Amster. *Science*, **259**, 955 (1993).

[14.133] R. A. Assink, J. E. Schirber, D. A. Loy, B. Morosin, and G. A. Carlson. *J. Mater. Res.*, **7**, 2136 (1992).

[14.134] D. Sarkar and N. J. Halas. *Solid State Commun.*, **90**, 261 (1994).

[14.135] M. H. Brodsky. *Amorphous Semiconductors*. Springer-Verlag, Berlin (1985). 2nd edition, vol 36 in Topics in Applied Phys., edited by M. H. Brodsky.

[14.136] C. H. Lee, G. Yu, B. Kraabel, D. Moses, and V. I. Srdanov. *Phys. Rev. B*, **49**, 10572 (1994).

[14.137] C. H. Lee, G. Yu, D. Moses, V. I. Srdanov, X. Wei, and Z. V. Vardeny. *Phys. Rev. B*, **48**(11), 8506 (1993).
[14.138] H. Yonehara and C. Pac. *Appl. Phys. Lett.*, **61**, 575 (1992).
[14.139] A. Hamed, R. Escalante, and P. H. Hor. *Phys. Rev. B*, **50**, 8050 (1994).
[14.140] A. Skumanuch. *Chem. Phys. Lett.*, **182**, 486 (1991).
[14.141] J. R. Olson, K. A. Topp, and R. O. Pohl. *Science*, **259**, 1145 (1993).
[14.142] W. DeSorbo and W. W. Tyler. *J. Chem. Phys.*, **21**, 1660 (1953).
[14.143] P. H. Keesom and N. Pearlman. *Phys. Rev.*, **99**, 1119 (1955).
[14.144] E. Grivei, B. Nysten, M. Cassart, A. Demain, and J. P. Issi. *Solid State Commun.*, **85**, 73 (1993).
[14.145] T. Atake, T. Tanaka, R. Kawaji, K. Kikuchi, K. Saito, S. Suzuki, I. Ikemoto, and Y. Achiba. *Physica C*, **185-189**, 427 (1991).
[14.146] E. Grivei, M. Cassart, J. P. Issi, L. Langer, B. Nysten, J. P. Michenaud, C. Fabre, and A. Rassat. *Phys. Rev. B*, **48**, 8514 (1993).
[14.147] T. Matsuo, H. Suga, W. I. F. David, R. M. Ibberson, P. Bernier, A. Zahab, C. Fabre, A. Rassat, and A. Dworkin. *Solid State Commun.*, **83**, 711 (1992).
[14.148] W. P. Beyermann, M. F. Hundley, J. D. Thompson, F. N. Diederich, and G. Grüner. *Phys. Rev. Lett.*, **68**, 2046 (1992).
[14.149] J. E. Fischer and P. A. Heiney. *J. Phys. Chem. Solids*, **54**, 1725 (1993).
[14.150] R. Saito, G. Dresselhaus, and M. S. Dresselhaus. *Phys. Rev. B*, **49**, 2143 (1994).
[14.151] D. A. Neumann, J. R. D. Copley, W. A. Kamitakahara, J. J. Rush, R. L. Cappelletti, N. Coustel, J. E. Fischer, J. P. McCauley, Jr., A. B. Smith, III, K. M. Creegan, and D. M. Cox. *J. Chem. Phys.*, **96**, 8631 (1992).
[14.152] M. A. Verheijen, H. Meekes, G. Meijer, P. Bennema, J. L. de Boer, S. van Smaalen, G. V. Tendeloo, S. Amelinckx, S. Muto, and J. van Landuyt. *Chem. Phys.*, **166**, 287 (1992).
[14.153] P. A. Heiney, G. B. M. Vaughan, J. E. Fischer, N. Coustel, D. E. Cox, J. R. D. Copley, D. A. Neumann, W. A. Kamitakahara, K. M. Creegan, D. M. Cox, J. P. McCauley, Jr., and A. B. Smith, III. *Phys. Rev. B*, **45**, 4544 (1992).
[14.154] B. J. Nelissen, P. H. M. van Loosdrecht, M. A. Verheijen, A. van der Avoird, and G. Meijer. *Chem. Phys. Lett.*, **207**, 343 (1993).
[14.155] C. Coulombeau, H. Jobic, P. Bernier, C. Fabre, and A. Rassat. *J. Phys. Chem.*, **96**, 22 (1992).
[14.156] S. Huant, G. Chouteau, J. B. Robert, P. Bernier, C. Fabre, A. Rassat, and E. Bustarret. *C. R. Acad. Sci. Paris - Serie II*, **314**, 1309 (1992).
[14.157] D. A. Neumann, J. R. D. Copley, W. A. Kamitakahara, J. J. Rush, R. L. Paul, and R. M. Lindstrom. *J. Phys. Chem. Solids*, **54**, 1699 (1993).
[14.158] K. Holczer, O. Klein, G. Grüner, J. D. Thompson, F. Diederich, and R. L. Whetten. *Phys. Rev. Lett.*, **67**, 271 (1991).
[14.159] C. Politis, V. Buntar, W. Krauss, and A. Gurevich. *Europhys. Lett.*, **17**, 175 (1992).
[14.160] C. Politis, A. I. Sokolov, and V. Buntar. *Mod. Phys. Lett. B*, **6**, 351 (1992).
[14.161] J. S. Tse, D. D. Klug, D. A. Wilkinson, and Y. P. Hanada. *Chem. Phys. Lett.*, **183**, 387 (1991).
[14.162] A. Dworkin, H. Szwarc, S. Leach, J. P. Hare, T. J. Dennis, H. W. Kroto, R. Taylor, and D. R. M. Walton. *C.R. Acad. Sci., Paris Serie II*, **312**, 979 (1991).
[14.163] A. R. McGhie, J. E. Fischer, P. A. Heiney, P. W. Stephens, R. L. Cappelletti, D. A., Neumann, W. H. Mueller, H. Mohn, and H.-U. ter Meer. *Phys. Rev. B.*, **49**, 12614 (1994).
[14.164] P. A. Heiney. *J. Phys. Chem. Solids*, **53**, 1333 (1992).

[14.165] G. Kriza, J. C. Ameline, J. Jerome, A. Dworkin, H. Szwarc, C. Fabre, D. Schutz, A. Rassat, and P. Bernier. *J. Phys. I, France*, **1**, 1361 (1991).
[14.166] J. E. Fischer. *Mater. Sci. Eng.*, **B19**, 90 (1993).
[14.167] W. I. F. David, R. M. Ibberson, T. J. S. Dennis, J. P. Hare, and K. Prassides. *Europhys. Lett.*, **18**, 219 (1992).
[14.168] W. I. F. David, R. M. Ibberson, T. J. S. Dennis, J. P. Hare, and K. Prassides. *Europhys. Lett.*, **18**, 735 (1992).
[14.169] G. B. M. Vaughan, P. A. Heiney, D. E. Cox, J. E. Fischer, A. R. McGhie, A. L. Smith, R. M. Strongin, M. A. Cichy, and A. B. Smith III. *Chem. Phys.*, **178**, 599 (1993).
[14.170] F. Gugenberger, R. Heid, C. Meingast, P. Adelmann, M. Braun, H. Wühl, M. Haluska, and H. Kuzmany. *Phys. Rev. Lett.*, **60**, 3774 (1992).
[14.171] R. C. Yu, N. Tea, M. B. Salamon, D. Lorens, and R. Malhotra. *Phys. Rev. Lett.*, **68**, 2050 (1992).
[14.172] T. Nihira and T. Iwata. *J. Appl. Phys.*, **14**, 1099 (1975).
[14.173] D. T. Morelli and C. Uher. *Phys. Rev. B*, **31**, 6721 (1985).
[14.174] R. Berman, P. R. W. Hudson, and M. Martinez. *J. Phys. C: Solid State*, **8**, L430 (1975).
[14.175] J. Krumhansl and H. Brooks. *J. Chem. Phys.*, **21**, 1663 (1953).
[14.176] B. Sundqvist. *Phys. Rev. B*, **48**, 14712 (1993).
[14.177] K. Sugihara, T. Inabe, Y. Maruyama, and Y. Achiba. *J. Phys. Soc. Japan*, **62**, 2757 (1993).
[14.178] Z. H. Wang, M. S. Dresselhaus, G. Dresselhaus, and P. C. Eklund. *Phys. Rev. B*, **48**, 16881 (1993).
[14.179] A. Kawabata. *Solid State Commun.*, **34**, 431 (1980).
[14.180] P. A. Lee and T. V. Ramakrishnan. *Phys. Rev. B*, **26**, 4009 (1982).
[14.181] B. L. Altshuler and A. G. Aronov. In A. L. Efros and M. Pollack (eds.), *Electron-electron interactions in disordered systems*, North Holland, Amsterdam (1985).
[14.182] B. E. White, J. E. Freund, K. A. Topp, and R. O. Pohl. In P. Bernier, T. W. Ebbesen, D. S. Bethune, R. M. Metzger, L. Y. Chiang, and J. W. Mintmire (eds.), *MRS Symposium Proc. "Science and Technology of Fullerene Materials"* vol. 359, p. 411 (1995).

CHAPTER 15

Superconductivity

The first observation of superconductivity in doped C_{60} attracted a great deal of attention because of the relatively high transition temperature T_c that was observed, the first observation being in the alkali metal–doped K_3C_{60} with a T_c of 18 K [15.1]. Ever since, superconductivity of doped fullerenes has remained a very active research field, as new superconducting fullerene materials were discovered and attempts were made to understand the unusual aspects of superconductivity in these fascinating materials and the pairing mechanism for the electrons. At the present time, superconductivity has been reported only for C_{60}-based solids, although transport properties down to $\sim$1 K have been measured on alkali metal–doped C_{70} [15.2] and perhaps higher-mass fullerenes as well.

In this chapter the experimental observations of superconductivity in various fullerene-related compounds are reviewed, followed by a discussion of critical temperature determinations and how T_c relates to the electronic density of states at the Fermi level. Magnetic field phenomena are then summarized, followed by a discussion of the temperature dependence of the energy gap, the isotope effect, and pressure-dependent effects. The leading contenders for the pairing mechanism (the electron–phonon and electron–electron interactions) are then discussed, although it is by now generally agreed that the electron–phonon interaction is the dominant pairing mechanism, giving rise to a mostly conventional Bardeen–Cooper–Schrieffer (BCS) fullerene superconductor [15.3].

15.1. Experimental Observations of Superconductivity

The first observation of superconductivity in a carbon-based material goes back to 1965, when superconductivity was observed in the first-stage alkali metal graphite intercalation compound (GIC) C_8K (see §2.14) [15.4]. Except for the novelty of observing superconductivity in a compound having no superconducting constituents, this observation did not attract a great deal of attention, since the superconducting transition temperature T_c in C_8K was very low (~140 mK) [15.5]. Later, higher T_c values were observed in GICs using superconducting intercalants (e.g., second-stage $KHgC_8$, for which $T_c = 1.9$ K [15.6, 7]) and by subjecting the alkali metal GICs to pressure (e.g., first-stage C_2Na, for which $T_c \sim 5$ K) [15.8].

As stated above, the early observation of superconductivity in a doped fullerene solid attracted interest because of its relatively high T_c (18 K in K_3C_{60}) [15.1]. This work was soon followed by observations of superconductivity at even higher temperatures: in Rb_3C_{60} ($T_c = 30$ K) [15.9, 10], in $RbCs_2C_{60}$ ($T_c = 33$ K) [15.11, 12], and in Cs_3C_{60} under pressure ($T_c = 40$ K) [15.13]. It is interesting to note that if superconductivity had been found in these alkali metal–doped C_{60} compounds prior to the discovery of superconductivity in the cuprates [15.14], the alkali metal–doped fullerenes would have been the highest-T_c superconductors studied up to that time. At present, the highest-T_c organic superconductor is an alkali metal–doped fullerene, namely Cs_3C_{60} under pressure (~12 kbar) with $T_c \sim 40$ K [15.13], breaking a record previously held by the tetrathiafulvalene derivative $(BEDT\text{-}TTF)_2CuN(CN)_2Cl$ with $T_c = 12.8$ K under 0.3 kbar pressure [15.15]. Table 8.3 lists the transition temperatures for various fullerene-based superconductors.

For the alkali metal–doped fullerenes, a metallic state is achieved when the t_{1u}-derived level is approximately half-filled, corresponding to the stoichiometry M_3C_{60} for a single metal (M) species or $M_xM'_{3-x}C_{60}$ for a binary alloy dopant, as discussed in §14.1. Increases in T_c relative to that of K_3C_{60} have been achieved by synthesizing compounds of the type $M_xM'_{3-x}C_{60}$, but with larger intercalate atoms, resulting in unit cells of larger size and with larger lattice constants as seen in Table 8.3 and in Fig. 15.1. As the lattice constant increases, the intermolecular C_{60}–C_{60} coupling decreases, narrowing the width of the t_{1u}-derived LUMO band and thereby increasing the corresponding density of states. Simple arguments based on BCS theory of superconductivity yield the simplest BCS estimate for T_c

$$k_B T_c = 1.13\overline{\omega}_{\rm ph} \exp\left(-\frac{1}{N(E_F)V}\right), \tag{15.1}$$

where $\overline{\omega}_{\rm ph}$ is an average phonon frequency for mediating the electron pairing, $N(E_F)$ is the density of electron states at the Fermi level, and V is the superconducting pairing interaction, where the product $N(E_F)V$ is equal to λ_{ep}, the electron–phonon coupling constant. The Debye temperature as measured from a specific heat experiment for the case of fullerenes refers

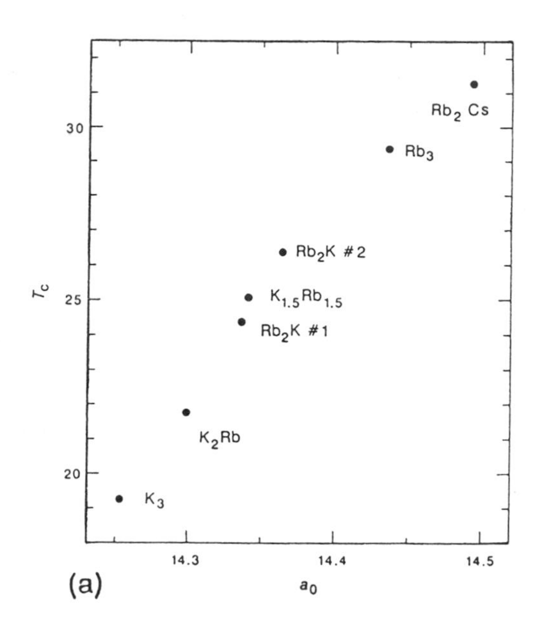

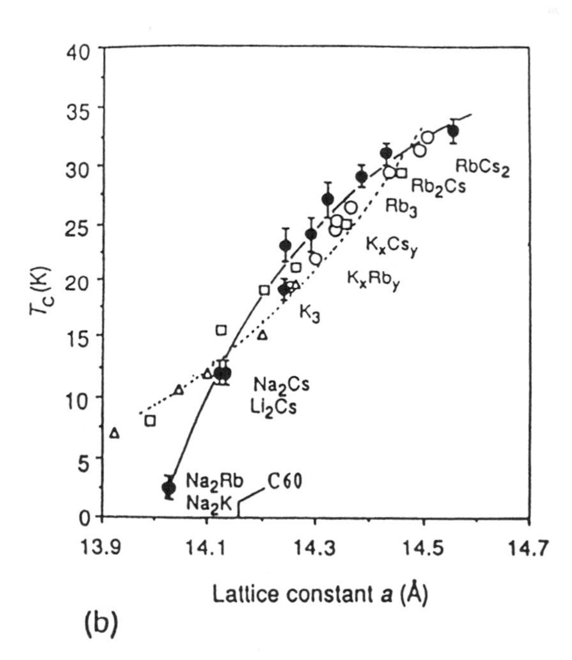

Fig. 15.1. (a) Early reports on the dependence of T_c for various $M_xM'_{3-x}C_{60}$ compounds on the fcc lattice constant a_0 [15.16]. (b) More complete summary of the dependence of T_c on a_0 [15.12], including points provided by pressure-dependent studies of T_c [15.17].

to the intermolecular degrees of freedom, while the pairing is generally believed to be dominated by high-frequency intramolecular vibrations, although the detailed pairing mechanism still remains an open question at this time (see §15.7). Some typical values for the parameters in Eq. (15.1) are $\overline{\omega}_{\mathrm{ph}} \sim 1200\ \mathrm{cm}^{-1}$, $V \sim 50$ meV, and $N(E_F) \sim 9$ states/eV/C_{60}/spin which are consistent with $T_c \sim 18$ K [15.3].

The BCS formula for T_c [Eq. (15.1)] suggests that higher T_c values might be achievable by increasing $N(E_F)$. A high density of electron states in the doped fullerenes occurs primarily because of the highly molecular nature of this solid, resulting in a narrow width (0.5–0.6 eV) for the electronic LUMO bands in K_3C_{60} and Rb_3C_{60}. Two main approaches have been used experimentally to increase T_c. The first approach has been directed toward extensive studies of the $M_xM'_{3-x}C_{60}$ alkali metal system to maximize the lattice constant while maintaining the face-centered cubic (fcc) phase, and the second has been directed toward the introduction of neutral spacer molecules such as NH_3, along with the dopant charge transfer species, to increase the lattice constant. The first step taken toward increasing T_c was the experimental verification that T_c could be increased by increasing the lattice constant. This was done in two ways, the first being a study of the relation between T_c and the lattice constant a_0 within the fcc structure, and the second involved study of the effect of pressure on T_c. The ternary $M_xM'_{3-x}C_{60}$ approach [15.11, 16, 18–21] provided many data points for evaluating the empirical relation between T_c and a_0 (see Fig. 15.1). Early work along these lines validated the basic concept [15.16] and led to the discovery of the highest present T_c value for a binary alloy compound, $RbCs_2C_{60}$ [15.12, 18]. Studies of the pressure dependence of T_c for K_3C_{60} and Rb_3C_{60} (see §15.5) provided further confirmation of the basic concept and provided points (denoted by open symbols) for the plots in Fig. 15.1 [15.17, 22, 23]. The crystal structure of the highest-T_c doped fullerene Cs_3C_{60} under pressure ($T_c = 40$ K at 12 kbar) is not fcc, and is believed to be a mixed A15 and body-centered tetragonal (bct) phase [15.13], and for this reason cannot be plotted on Fig. 15.1, which is only for the fcc structure.

Much effort has gone into discussing deviations from the simple empirical linear relation between T_c and a_0 for small a_0 values and strategies to enhance T_c. Regarding the low a_0 range, it was found that the introduction of the small alkali metals Li and Na for one of the M or M′ metal constituents lowered a_0 below that for undoped C_{60}, indicating the importance of electrostatic attractive interactions in the doping process. Some alkali metal $M_xM'_{3-x}C_{60}$ compounds with these small a_0 values exhibit superconductivity with low T_c values and nevertheless tend to fall on the solid curve of Fig. 15.1(b). Other compounds show anomalously low T_c values relative to their lattice constants, as also indicated in the figure.

A systematic study of the relation between T_c and the lattice constant for $Cs_xRb_{3-x}C_{60}$ was carried out [15.12] in an effort to maximize T_c, since Cs_2RbC_{60} is the fcc alloy superconductor with the highest T_c for the M_3C_{60} series. Direct measurement of T_c *vs.* alloy composition for this family of compounds presented in Fig. 15.2 [15.12] showed that the maximum T_c occurs near the stoichiometry Cs_2RbC_{60}. This work eventually led to the discovery of the Cs_3C_{60} phase under pressure with a higher T_c value but having a different crystal structure [15.13]. It should be noted that as the lattice constant increases, the overlap of the electron wave function from one C_{60} anion to that on adjacent anions decreases, and this effect tends to lower T_c, in contrast to the effect discussed above. Thus there is a limit to the maximum T_c that can be achieved merely by increasing the lattice constant.

A vivid demonstration of the effect of the increase of lattice constant a_0 on T_c is found in the uptake of NH_3 by Na_2CsC_{60} to form the ternary intercalation compound $(NH_3)_4Na_2CsC_{60}$, thereby increasing the lattice constant from 14.132 Å to 14.473 Å while increasing the T_c from 10.5 to 29.6 K [15.24]. We thus see that the T_c value of 29.6 K achieved for $(NH_3)_4Na_2CsC_{60}$ fits the empirical T_c *vs.* a_0 relation of Fig. 15.1 and suggests that the insertion of spacer molecules to further increase a_0 within the fcc structure [15.12] may lead to further enhancement of T_c. However, the results of Fig. 15.2 for the $Cs_xRb_{3-x}C_{60}$ system suggest that further enhancement of T_c by simply increasing a_0 may not be rewarding. Structural studies for the $(NH_3)_4Na_2CsC_{60}$ compound show that all of the (NH_3) groups and half of the Na ions go into octahedral sites in the form of large

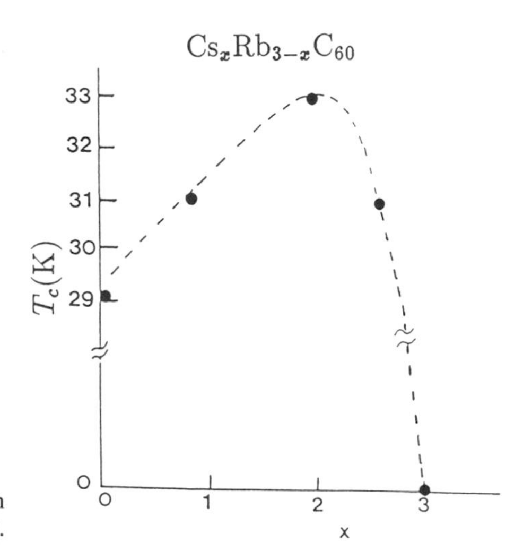

Fig. 15.2. Dependence of T_c on Cs concentration for $Cs_xRb_{3-x}C_{60}$. The discontinuity in the dashed curve at $x = 2.7$ indicates a change in crystal structure from fcc for $x \leq 2.7$ to bcc close to $x = 3$ [15.12].

$(NH_3)_4Na$ cations [15.24]. The remaining bare Na and Cs ions are found in tetrahedral sites [15.24]. This is somewhat unusual, since in the pristine Na_2CsC_{60} the Cs ion occupies an octahedral site while the Na ions preferentially occupy tetrahedral sites. The $(NH_3)_4Na_2CsC_{60}$ compound is analogous to ternary graphite intercalation compounds of the type M-C_6H_6 GICs, M-THF GICs, or M-NH_3 GICs, where M denotes the alkali metal. Such compounds provided a powerful vehicle for tailoring the physical properties of GICs [15.25, 26]. Ternary fullerene compounds of this kind may likewise open up new possibilities for modifying the properties of fullerene host materials.

Since the addition of $(NH_3)_4$ to Na_2CsC_{60} was effective in increasing T_c, ammonia was added to K_3C_{60} in an attempt to enhance T_c further. A stoichiometry of $(NH_3)K_3C_{60}$ was achieved upon NH_3 addition, without additional charge transfer, but this compound was not found to be superconducting [15.27] even though the unit cell volume was almost the same as that for Rb_2CsC_{60} with a $T_c = 31.3$ K (see Table 8.3). The introduction of (NH_3) produced a 6% volume expansion of the unit cell, accompanied by a distortion of the crystal structure away from cubic symmetry. The absence of superconductivity in $NH_3K_3C_{60}$ was attributed to either departures from a cubic crystal structure or charge localization on the C_{60}^{3-} anions due to correlation effects [15.27].

Since the T_c of a superconductor such as $K_xRb_{3-x}C_{60}$ varies continuously with x [15.18], it is believed that the superconductivity in alkali metal alloy systems $M_{3-x}M'_xC_{60}$ is not sensitive to superconducting fluctuations [15.20]. This is in contrast to organic superconductors such as (BEDT-TTF)$_2$X, where a mixture of two different ions suppresses superconductivity [15.28, 29].

It is generally believed that a relation between T_c and the electronic density of states at the Fermi level $N(E_F)$ is more fundamental than between T_c and the lattice constant a_0, and efforts have been made to present the literature data in this form (see Fig. 15.3) [15.22]. On the other hand, there is considerable uncertainty presently regarding the magnitude of the experimental electronic density of states for specific compounds (see §14.5), while the lattice constants can be more reliably measured. It is for this reason that plots of T_c *vs.* lattice constant a_0 are more commonly used at the present time by workers in the field.

Superconductivity has also been observed in alkaline earth–doped superconductors with Ca, Ba, and Sr. Although the T_c value for these compounds is much lower than for alkali metal dopants, the superconductivity in these compounds is interesting in its own right and merits attention. Since the alkaline earth ions tend to have smaller ionic radii than the alkali metal ions (see Table 8.1), they would be expected to result in less lattice expansion

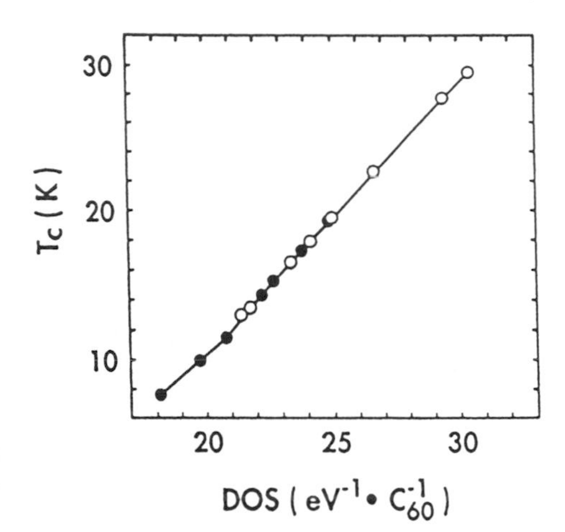

Fig. 15.3. Dependence of T_c on the density of states at the Fermi level for K_3C_{60} (closed circles) and Rb_3C_{60} (open circles) [15.22] using pressure-dependent measurements of T_c on these compounds [15.17, 23].

and therefore lower T_c values generally, and this is indeed found experimentally [15.30–32]. Also because of their smaller size, multiple alkaline earth metal ions can enter the larger octahedral sites for the case of the fcc structure. In fact, the stoichiometries that exhibit superconductivity from magnetic susceptibility and microwave measurements are Ca_5C_{60} [15.30], Ba_6C_{60} [15.31], and Sr_6C_{60} [15.33], and neither Ba_6C_{60} nor Sr_6C_{60} exhibits the fcc structure. Shown in Fig. 15.4 are measurements of the magnetization *vs.* temperature for Ca_5C_{60} in a field-cooled (FC) and zero-field-cooled (ZFC) situation, with the intersection between the two curves yielding the T_c value.

Two aspects of the superconducting alkaline earth compounds bear special attention. The first relates to the difference in crystal structures relative to the alkali metal compounds and the greater variety of crystal structures

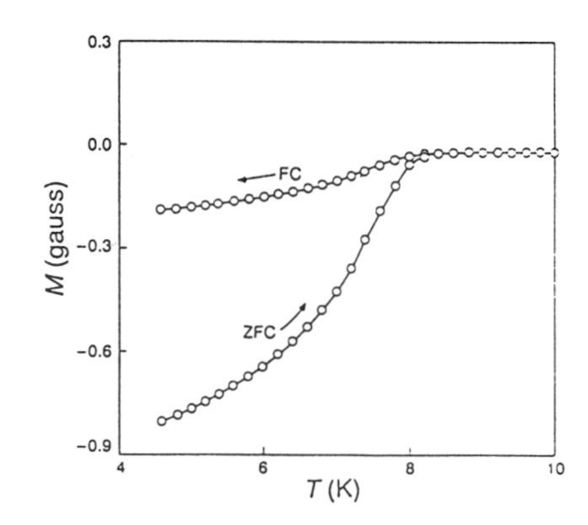

Fig. 15.4. Magnetization *vs.* temperature for field-cooled (FC) and zero-field-cooled (ZFC) cycles for Ca_5C_{60} [15.30].

that support superconductivity in the alkaline earth–doped C_{60}. The second relates to the divalence of the alkaline earth ions, where up to two electrons per dopant atom are donated to the C_{60}-derived states. The t_{1u}-derived band is filled at three alkaline earth dopants per C_{60} unit, and upon further metal ion addition, the t_{1g}-derived bands start to fill, reaching a metallic phase when they are about half-filled at a stoichiometry between five and six alkaline earth ions per C_{60} unit (see §14.1.3), since the metal-derived bands significantly hybridize with the C_{60} bands above the t_{1u}-derived band.

Figure 15.5 shows that the T_c for the fcc superconducting alkaline earth Ca_5C_{60} follows the same relation between T_c and the C_{60}–C_{60} separation that is found for the alkali metal M_3C_{60} compounds, whereas the body-centered cubic (bcc) alkaline earth compounds Sr_6C_{60} and Ba_6C_{60} also show correlation between T_c and the C_{60}–C_{60} distance, but show a different functional relation relative to the alkali metal compounds.

Regarding the issue of crystal structures, superconductivity in the alkali metal $M_xM'_{3-x}C_{60}$ compounds corresponds to the fcc $Fm\bar{3}m$ space group, while superconductivity in the alkaline earth compounds corresponds to the following structures: Ca_5C_{60} has a simple cubic structure where each of the four C_{60} anions of the fcc structure is distinct, while Ba_6C_{60} and Sr_6C_{60} both have the bcc $Im\bar{3}$ crystal structure (which is also the structure exhibited by the semiconductors K_6C_{60} and Rb_6C_{60}) (see §7.1.1 and §8.5). Since the superconductivity for all of these compounds is similar, we conclude that the superconductivity is predominantly influenced by the C_{60}-derived states and to a lesser degree by the dopant species. With regard to band filling, the observation of superconductivity in Ca_5C_{60}, Ba_6C_{60}, and Sr_6C_{60} indicates a transfer of electron charge carriers to C_{60}, beyond the six electrons needed to fully occupy the t_{1u} bands. These observations indicate that superconductivity is possible for both a half-filled t_{1u} band and a partially (perhaps half) filled t_{1g} band. Symmetry arguments (see §14.2.2) [15.30] suggest that if the

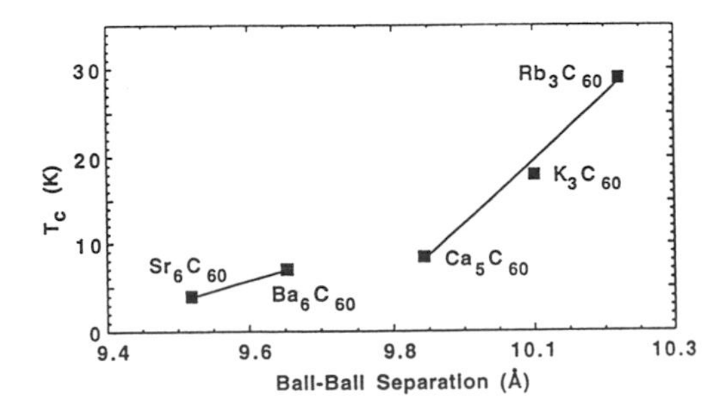

Fig. 15.5. Plot of the superconducting transition temperature T_c dependence on the intermolecular C_{60}–C_{60} separation of fcc superconductors (including Ca_5C_{60}) is extended to include the bcc superconductors Ba_6C_{60} and Sr_6C_{60}. The bcc-based superconductors also exhibit a positive slope in the T_c *vs.* C_{60}–C_{60} plot, but with a different magnitude for the two slopes [15.33].

electron–phonon interaction is the dominant electron-pairing mechanism, then the electron–phonon matrix elements would be expected to be similar for electrons in the t_{1u} or t_{1g} bands (see §14.2.2).

It is also interesting to note that alkali metal–doped C_{70} [15.34–38] and alkali metal–doped higher-mass fullerenes C_{76} [15.38, 39], C_{78} [15.38], C_{82} [15.40], C_{84} [15.38, 41, 42], and C_{90} [15.38] do not show superconductivity. Further, the addition of C_{70} to C_{60} results in a rapid decrease in T_c for the $M_x(C_{60})_{1-y}(C_{70})_y$ compounds [15.38, 43]. In fact, doped C_{70} never becomes fully metallic, showing activated conduction even for the concentration K_4C_{70} where the magnitude of the conductivity is a maximum (see §14.1). Furthermore, it is not expected on the basis of curvature arguments that doped carbon nanotubes would be attractive as a high-T_c superconducting material (see §19.11). It has been argued that the lower curvature of the higher-mass fullerenes suppresses the electron–phonon interaction and thus suppresses the occurrence of superconductivity. It would thus appear that doping might enhance the electron–phonon coupling for the lower-mass fullerenes more than for C_{60}. But at the same time, the smaller lattice constant would be expected to lead to a lower density of states at the Fermi level. Thus it is not clear whether higher T_c values could be achieved by the doping of smaller-mass fullerenes.

15.2. Critical Temperature

The onset of superconductivity is typically measured in one of three ways: (1) by the loss of resistivity in temperature-dependent $\rho(T)$ curves (see Fig. 14.3), (2) the temperature where the field-cooled and the zero-field-cooled magnetic susceptibilities merge (see Fig. 15.4), and (3) the onset of the decrease in microwave losses in temperature-dependent surface resistance plots $R_s(T)$. In the initial report of superconductivity in K_3C_{60}, evidence for superconductivity by all three methods was reported [15.1]. From measurements using one or more of the three techniques mentioned above, T_c is found for specific doped fullerene compounds, and the results for a variety of doped fullerene compounds are given in Table 8.3. A plot of the field-cooled and zero-field-cooled magnetic susceptibility *vs.* T is probably the most convenient method for the measurement of T_c (see Fig. 15.4 [15.30]), since the slope of the magnetization curve $M(H)$ also gives the shielding fraction or superconducting fraction x_{sh} through the relation

$$x_{sh} = -4\pi \left(\frac{\partial M}{\partial H} \right)_{H \to 0}, \tag{15.2}$$

and an ideal type I superconductor is a perfect diamagnet (i.e., $x_{sh} = 1$). Since the shielding fraction is dependent on the flux exclusion, x_{sh} is

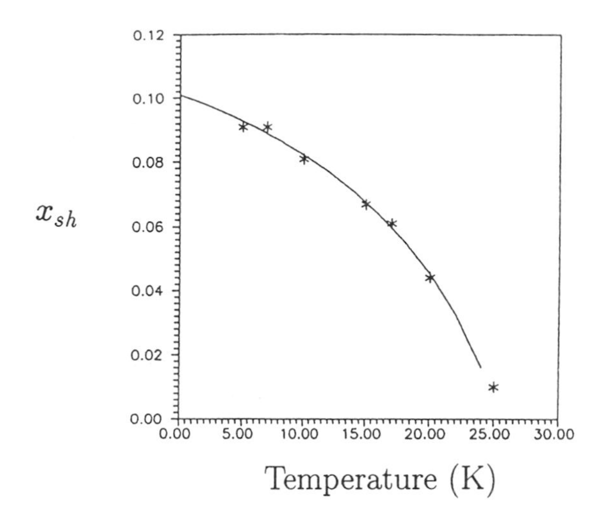

Fig. 15.6. Temperature dependence of the shielding fraction x_{sh} for an Rb_3C_{60} powder sample [15.44].

temperature dependent, as shown in Fig. 15.6 [15.44]. The value for x_{sh} that is quoted when the sample characterization specifications are given is $x_{sh}(T \to 0)$. Many of the superconducting fullerene samples that were prepared initially had low values of x_{sh} (< 0.5), although more recent samples typically have higher values, such as $x_{sh} > 0.6$, reported for K_3C_{60}, $CsRb_2C_{60}$, and Cs_2RbC_{60} [15.45]. Much effort continues to be given to improving synthesis techniques for preparing M_3C_{60} compounds with high values of x_{sh}, high T_c values, and small transition widths ΔT_c [15.46–48].

Referring to the BCS formula [Eq. (15.1)], if the dominant phonon frequency responsible for the electron–phonon coupling is known, then λ_{ep} can be found from Eq. (14.5), assuming that the fullerene superconductors are in the weak coupling limit. However, if the fullerene superconductors are strong coupling superconductors, we should instead use the McMillan solution [15.49] of the Eliashberg equations

$$T_c = \frac{\hbar\overline{\omega}_{\rm ph}}{1.45k_{\rm B}} \exp\left\{\frac{-1.04(1+\lambda_{ep})}{\lambda_{ep} - \mu^*(1+0.62\lambda_{ep})}\right\} \tag{15.3}$$

where $\overline{\omega}_{\rm ph}$ denotes the dominant phonon frequency for the electron–phonon coupling, and where the effective electron–electron repulsion is denoted by μ^* and is related to the short-range Coulomb repulsion μ by

$$\mu^* = \frac{\mu}{1 + \mu \ln(E_F/\overline{\omega}_{\rm ph})} \tag{15.4}$$

for $\mu < 1$. In Eq. (15.4) the Coulomb interaction may be reduced because the screening of one electron by another is faster than the effective vibrational frequency of the electron–phonon coupling. Most superconductors have values of μ^* between 0.1 and 0.3, and values of μ^* in this range have also been suggested for fullerene-based superconductors [15.50].

To gain further understanding of the McMillan formula with regard to doped fullerenes, numerical evaluations of λ_{ep} and μ^* have been reported by several groups based on phonon mode and electronic structure calculations, since it is difficult to calculate T_c from first principles [15.50, 51]. Referring to Eq. (15.1), one problem that is encountered in estimating $\lambda_{ep} = VN(E_F)$ for crystalline C_{60} is the large range of experimental values for $N(E_F)$ (see §14.5) and V in the literature. Referring to Eq. (15.3), there is also a large range of values in the literature for $\overline{\omega}_{\text{ph}}$ (see §14.2.2). Thus the theoretical understanding behind the relatively high T_c values in the doped fullerenes is still in a formative state. We summarize below the range of values in the literature for V and $\overline{\omega}_{\text{ph}}$.

In §15.7, results are given for a number of models for calculating V, the pairing interaction energy in Eq. (15.1) arising from the electron–phonon interaction. Assuming that the electron pairing interaction is via intramolecular phonons, the summary of values reported in the literature (between 32.2 and 82.3 meV [15.52]) indicates the need for further systematic work, especially because of the high sensitivity of T_c to V through the exponential relation in Eq. (15.1). A value of $V \sim 50$ meV provides a rough estimate for this interaction energy [15.3].

In §15.7, some discussion is given of the various viewpoints on the specific phonon modes that are most important in coupling electrons and phonons in the pairing mechanism. As noted in §15.7, $\overline{\omega}_{\text{ph}}$ is determined by Eq. (14.7) from the contribution to λ_{ep} from each vibrational mode. A better determination of $\overline{\omega}_{\text{ph}}$ awaits a definitive answer about the modes that contribute most importantly to λ_{ep}, and this subject is also considered further in §15.7.

15.3. Magnetic Field Effects

The most important parameters describing type II superconductors in a magnetic field are the upper critical field H_{c2} and the lower critical field H_{c1}. The lower critical field denotes the magnetic field at which flux penetration into the superconductor is initiated and magnetic vortices start to form.

The lower critical field $H_{c1}(0)$ in the limit $T \to 0$ is also related to the Ginzburg–Landau coherence length ξ_0 and the London penetration depth λ_L through the relation

$$H_{c1}(0) = \frac{\phi_0}{4\pi\lambda_L^2} \ln(\lambda_L/\xi_0). \tag{15.5}$$

Table 15.1

Experimental values for the macroscopic parameters of the superconducting phases of K_3C_{60} and Rb_3C_{60}.

Parameter	K_3C_{60}	Rb_3C_{60}
fcc a_0 (Å)	14.253[a]	14.436[a]
T_c (K)	19.7[b]	30.0[b]
$2\Delta(0)/kT_c$	5.2[c], 4.0[e], 3.6[g], 3.6[h]	5.3[d], 3.1[e], 3.6[f], 3.0[g], 2.98[h]
$(dT_c/dP)_{p=0}$ (K/GPa)	−7.8[i]	−9.7[i]
$H_{c1}(0)$ (mT)	13[j]	26[j], 19[k]
$H_{c2}(0)$ (T)	26[j], 30[l], 29[m], 17.5[b]	34[j], 55[l], 76[b]
$H_c(0)$ (T)	0.38[i]	0.44[i]
J_c (10^6 A/cm²)	0.12[j]	1.5[j]
ξ_0 (nm)	2.6[i], 3.1[l], 3.4[n], 4.5[b]	2.0[i], 2.0[b], 3.0[n]
λ_L (nm)	240[j], 480[o], 600[p], 800[q]	168[j], 370[f], 460[p], 800[q], 210[k]
$\kappa = \lambda_L/\xi_0$	92[j]	84[j], 90[k]
dH_{c2}/dT (T/K)	−1.34[b], −3.5[r]	−3.8[b]
ξ_{00} (nm)	9.5[b], 12.0[t], 15.0[s]	4.0–5.5[b]
ℓ (nm)	3.1[b], 1.0[t]	0.9[b]

[a]Ref. [15.30]; [b]Ref. [15.50]; [c]STM measurements in Ref. [15.53]; [d]STM measurements in Ref. [15.54]; [e]NMR measurements in Ref. [15.55, 56]; [f]μSR measurements in Ref. [15.57]; [g]Far-IR measurements in Ref. [15.58]; [h]Far-IR measurements in Ref. [15.59]; [i]Ref. [15.17]; [j]Ref. [15.60]; [k]Ref. [15.44]; [l]Ref. [15.61]; [m]Ref. [15.62]; [n]Ref. [15.63]; [o]Ref. [15.64]; [p]Ref. [15.55]; [q]Ref. [15.65, 66]; [r]Ref. [15.67]; [s]Ref. [15.68]; [t]Ref. [15.69].

where ϕ_0 is the magnetic flux quantum. The value of $H_{c1}(0)$ is typically very low for fullerene superconductors (see Table 15.1).

The upper critical field denotes the magnetic field above which full magnetic flux penetration takes place and a transition from the superconducting to the normal state occurs. The determination of $H_{c2}(0)$ in the limit $T \to 0$ is also of importance, since $H_{c2}(0)$ is related to the coherence length ξ_0 of the superconducting wave function through the relation

$$H_{c2}(0) = \frac{\phi_0}{2\pi\xi_0^2}. \tag{15.6}$$

The temperature dependence of the resistivity of a superconductor in a magnetic field provides one method for studying the magnetic field penetration in the superconductor. Such data are shown in Fig. 15.7 for a single crystal of K_3C_{60} (where the resistivity is normalized to its room temperature value) for various values of magnetic field (up to 7.3 T) [15.70]. The results of Fig. 15.7 show that the application of a magnetic field decreases T_c and increases the transition width ΔT_c, as is also observed for high-T_c cuprate superconductors. One can obtain the T dependence of $H_{c2}(T)$ from ρ/ρ_0

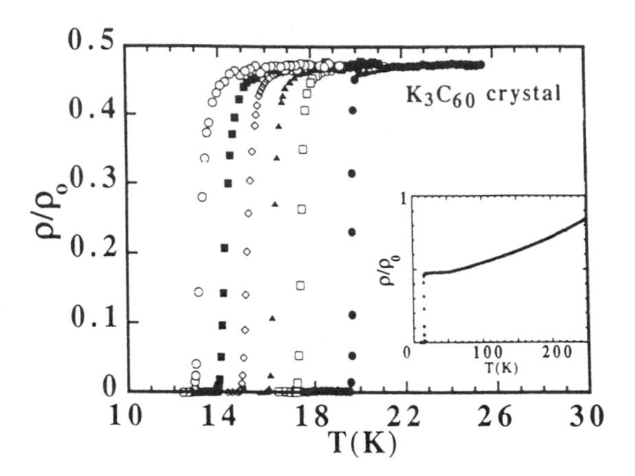

Fig. 15.7. Normalized resistivity of single-crystal K_3C_{60} near T_c for varying values of the applied magnetic field: ●, 0.0 T; □, 1.6 T; ▲, 3.1 T; ◇, 4.6 T; ■, 6.1 T; and ○, 7.3 T. ρ_0 is the room temperature value and the inset shows the zero-field temperature dependence of the normalized resistivity [15.50, 70].

vs. T data such as those shown in Fig. 15.7. For a given H, the temperature for which $(\rho/\rho_0) \to 0$ is used to obtain $H_{c2}(T)$. Results of this data analysis for K_3C_{60} are shown in Fig. 15.8.

Assuming the validity of BCS theory, one can obtain the zero-temperature value of the critical field $H_{c2}(0)$ from the slope of the $H_{c2}(T)$ data near T_c by means of the Werthamer, Helfand, and Hohenberg (WHH) formula [15.71]:

$$H_{c2}(0) = 0.69\left[\frac{\partial H_{c2}}{\partial T}\right]_{T_c} T_c. \tag{15.7}$$

According to the WHH formula values of $H_{c2}(0)$ at $T = 0$ can be estimated from H_{c2} measurements predominately near T_c. The results presented in Fig. 15.8 show that $(\partial H_{c2}/\partial T)$ is constant near T_c, and H_{c2} has a linear T dependence near T_c in accordance with Ginzburg–Landau theory. Justifi-

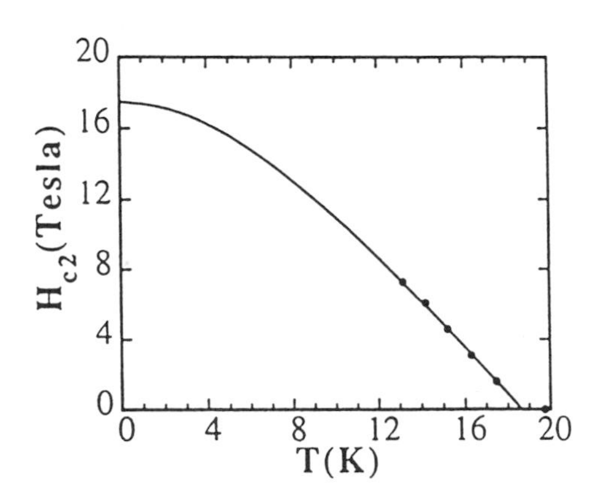

Fig. 15.8. Temperature dependence of the upper critical field H_{c2} for K_3C_{60}. The dots are experimental points taken at relatively low magnetic fields and the solid curve is a fit of these data to the WHH model (see text) [15.70].

cation for use of the WHH formula for fullerene superconductors comes from measurements on powder samples of K_3C_{60} and Rb_3C_{60} extending to very high magnetic fields and including Pauli-limiting spin effects [15.61]. An independent determination of the dH_{c2}/dT has been made from analysis of the measured jump in the specific heat at T_c, yielding a value of $dH_{c2}/dT = -3.5 \pm 1.0$ T/K (see §14.8.3) [15.67]. The high-field determinations of $H_{c2}(0)$ yielding 30 T and 55 T for K_3C_{60} and Rb_3C_{60}, respectively, represent the most detailed experimental determinations of $H_{c2}(0)$ that are presently available, although the high-field measurements were not made on high-quality single crystals. Although prior work suggested departures from the WHH formula on samples taken to high fields [15.62] due to sample granularity and Fermi surface anisotropy, it seems to be generally believed on the basis of more detailed high-field measurements [15.61] that the WHH model provides a good first approximation for describing the functional form of $H_{c2}(T)$ for fullerene superconductors. Generally, $H_{c2}(0)$ for single-crystal Rb_3C_{60} is reported to be higher than that for an Rb_3C_{60} film [15.17, 23, 61, 72], although the opposite trend was reported for K_3C_{60} (see Table 15.1). Further systematic studies are needed to clarify the dependence of the superconducting parameters on grain size and microstructure.

Typical values for $H_{c2}(0)$ are very high, as given in Table 15.1 for powder, film, and single-crystal K_3C_{60} and Rb_3C_{60} samples. Also given in Table 15.1 is the range of values for ξ_0, which are typically found from $H_{c2}(0)$ [see Eq. (15.6)] and therefore reflect the corresponding range of values for $H_{c2}(0)$. Also shown in Table 15.1 are values for the London penetration depth λ_L which is measured through magnetic field penetration studies. Microwave experiments for K_3C_{60} and Rb_3C_{60} have yielded values near 2000 Å [15.60, 73], while muon spin resonance studies [15.57, 64, 74] have yielded values of λ_L about a factor of 2 greater (~4000 Å), and far-infrared optical experiments have reported even larger values of λ_L (~8000 Å) [15.65]. The reason for this discrepancy in the value for λ_L is not presently understood, and this remains one of the areas where further systematic work is needed.

Measurements for $H_{c2}(0)$ show on the basis of Eq. (15.5) that ξ_0 is only slightly larger than a lattice constant for the fcc unit cell of the fullerene superconductors. In contrast, the measured λ_L values are very much larger than both ξ_0 and ℓ (see Table 15.1), especially near T_c. Magnetization studies on powder samples of Rb_3C_{60} show that the temperature dependence of the London penetration depth is well fit by the empirical relation [15.44]

$$\lambda_L(T) = \lambda_{GL}(0)\left[1 - \frac{T}{T_c}\right]^{-1/2}, \tag{15.8}$$

where $\lambda_{GL}(0)$ is the Ginzburg–Landau penetration depth at $T = 0$. The values of λ_L in Table 15.1 correspond to extrapolations of the measured $\lambda_L(T)$ to $T \to 0$, i.e., to $\lambda_{GL}(0)$, if Eq. (15.8) is used to fit the data.

Measurements of the lower critical field H_{c1}, where magnetic flux penetration initially occurs, show that $H_{c1}(0)$ for superconducting fullerene solids has a very small value [15.17]. Microwave measurements indicate that $H_{c1}(T)$ follows the simple empirical formula [15.17, 73]

$$[H_{c1}(T)/H_{c1}(0)] = 1 - (T/T_c)^2. \tag{15.9}$$

Values for $H_{c1}(0)$ are also given in Table 15.1, and through Eq. (15.5) the values for $H_{c1}(0)$ provide consistency checks for determinations of λ_L and ξ_0.

More detailed studies of the superconducting state for the M_3C_{60} type II superconductors reveal a phase diagram akin to that of high-T_c cuprates, consisting of a Meissner phase (with no vortices), a vortex glass phase with pinned vortices, and a vortex fluid phase [15.75]. The inset to Fig. 15.9 shows the temperature dependence of the magnetic susceptibility $\chi(T)$ for a powder sample of K_3C_{60} ($T_c = 19.0$ K) under zero-field-cooling (ZFC) and field-cooling (FC) conditions for an applied field of 10 G. Comparing

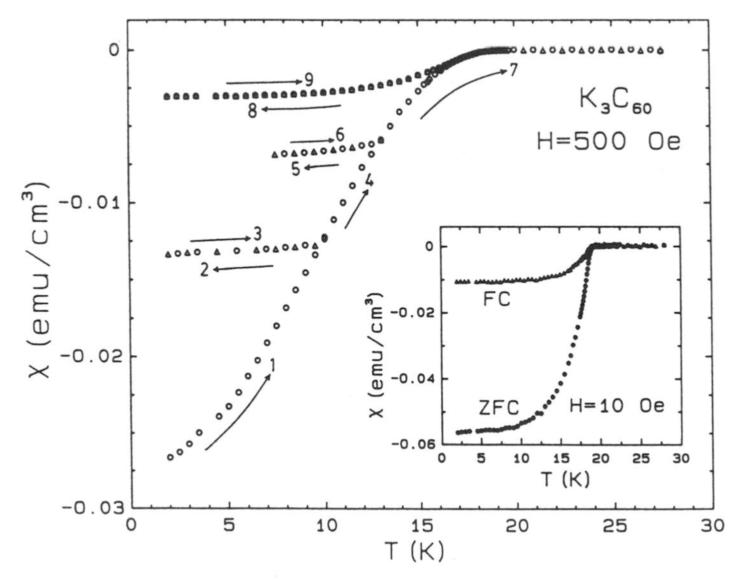

Fig. 15.9. Temperature dependence of the magnetic susceptibility of K_3C_{60} measured at $H = 500$ Oe. Open circles for heating (numbered arrows: 1, 3, 4, 6, 7, and 9); open triangles for cooling (numbered arrows: 2, 5, and 8). The inset shows field-cooled (FC) and zero-field-cooled (ZFC) susceptibility data for K_3C_{60} measured at $H = 10$ Oe [15.75].

the measured low-temperature ZFC and FC values of χ to the ideal χ value of $-1/4\pi$ yields values of 85% for the shielding fraction (ZFC) and 19% for the Meissner fraction (FC), consistent with an irreversible ZFC diamagnetism that is more than a factor of 4 greater than the reversible FC diamagnetism [15.75]. Referring to the main plot in Fig. 15.9 for the same K_3C_{60} sample but now in a field of 500 Oe, we see irreversible behavior for the $1 \rightarrow 2$, $4 \rightarrow 5$, $7 \rightarrow 8$ segments and reversible behavior for the $2 \rightarrow 3$, $5 \rightarrow 6$, $8 \rightarrow 9$ segments. The intersection points between the reversible and irreversible behaviors converge to a value T^* for a given H [i.e., each $\chi(T)$ diagram such as Fig. 15.9 yields a single (T^*, H) pair], which determines a point on the phase boundary between the vortex glass and vortex fluid phases shown in Fig. 15.10; the entire phase boundary in Fig. 15.10 is constructed by measuring T^* for a range of H values [15.75]. The results in Fig. 15.10 show similarities in the superconducting phase diagram for K_3C_{60} and Rb_3C_{60}, except for differences in scale arising from the different T_c values for these compounds. The phase boundaries for both K_3C_{60} and Rb_3C_{60} are fit by the relation

$$H = H_0[1 - T^*(H)/T^*(0)]^{\gamma} \tag{15.10}$$

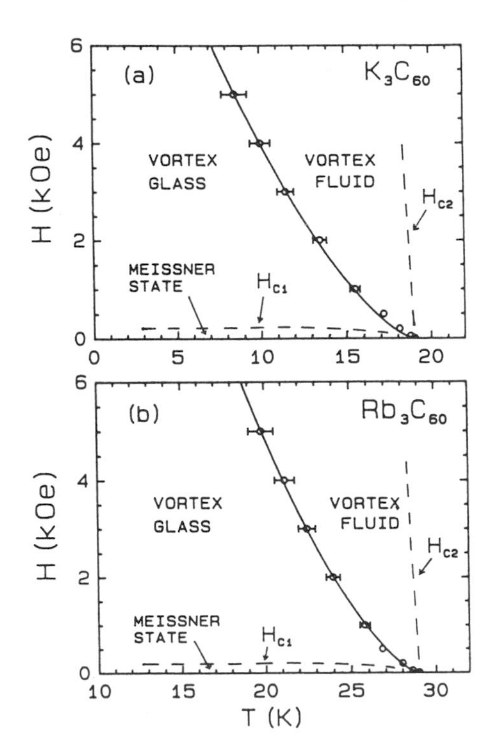

Fig. 15.10. Magnetic phase diagrams determined for (a) K_3C_{60} and (b) Rb_3C_{60}. Solid lines are fits to the de Almeida–Thouless relation [Eq. (15.10) with $\gamma = 3/2$], which represents the demarcation between vortex glass and vortex fluid phases. Dashed lines indicate schematically the critical fields $H_{c1}(T)$ and $H_{c2}(T)$ [15.75].

Table 15.2

Parameters for the phase boundary between the vortex fluid and vortex glass phases[a] [15.75].

Materials	γ	T^* (K)	H_0 (kOe)
K_3C_{60}	1.41 ± 0.10	19.2 ± 0.1	12.2 ± 0.07
Rb_3C_{60}	1.64 ± 0.14	29.1 ± 0.1	27.2 ± 1.7

[a] Values for T^* and H_0 are calculated using Eq. (15.10).

which defines a second-order phase transition. Values of the parameters γ, $T^*(0)$ and H_0 of Eq. (15.10) are given in Table 15.2 for K_3C_{60} and Rb_3C_{60}. The values of γ found experimentally are consistent with the de Almeida–Thouless relation [15.76] given by Eq. (15.10) with $\gamma = 3/2$. The vortex pinning mechanism for K_3C_{60} and Rb_3C_{60} is associated with a large density of grain boundaries in these powder samples. While T^* coincides with T_c for $H = 0$, the value of H_0 differs from $H_{c2}(0)$, since H_0 relates to the boundary between the vortex glass and the vortex fluid, while $H_{c2}(0)$ relates to the boundary between the vortex fluid and normal phases.

The granular nature of many of the M_3C_{60} films that have been studied gives rise to much vortex pinning, which often results in an overestimate of H_{c1}. The thermodynamic critical field $H_c(0)$ listed in Table 15.1 is found from the relation

$$H_c^2(0) = H_{c1}(0)H_{c2}(0)/\ln\kappa \tag{15.11}$$

where

$$\kappa = \frac{\lambda_L}{\xi_0}. \tag{15.12}$$

The large values of κ (see Table 15.1) indicate that M_3C_{60} compounds are strongly type II superconductors [15.17].

Reports on values for the mean free path ℓ vary from $\sim$2 to 30 Å and ℓ is often reported to be smaller than the superconducting coherence length ξ_0 for single crystals, films, and powder samples. The small value of ℓ is mainly attributed to the merohedral disorder, which is present in all doped fullerene samples, including the best available single crystals (see §7.1.4). In addition, ℓ is smaller for films and powders than for single crystals, since ℓ is highly sensitive to carrier scattering from crystal boundaries. For some single-crystal samples, ℓ is comparable to, or greater than, the diameter of a C_{60} molecule. But since $\ell \ll \xi_0$ in most cases, or at best $\ell \sim \xi_0$, fullerene superconductors are considered to be in the dirty limit defined by $\ell \leq \xi_0$.

This conclusion, that the M_3C_{60} compounds are dirty superconductors, is supported by the vortex glass behavior [15.75].

The coherence length ξ_0, which enters the formulae for $H_{c1}(0)$ and $H_{c2}(0)$, is the Ginzburg–Landau coherence length, and ξ_0 is sensitive to materials properties. The Pippard coherence length ξ_{00} in the clean limit is an intrinsic parameter, not sensitive to materials properties, and is related to ξ_0 by

$$\xi_0 \simeq (\xi_{00}\ell)^{1/2}, \tag{15.13}$$

and the BCS theory provides the following estimate for ξ_{00}:

$$\xi_{00} = \frac{1}{\pi}\frac{\hbar v_F}{\Delta} \tag{15.14}$$

yielding ξ_{00} in the 120–130 Å range for the superconducting fullerenes [15.51, 69], consistent with estimates using Eq. (15.13) and mean free path values ℓ in the 5–10 Å range (see Table 15.1). Available data thus suggest that a fullerene superconductor may be described as a strongly type II superconductor, which is in the dirty limit and has weak to moderate electron–phonon coupling.

15.4. Temperature Dependence of the Superconducting Energy Gap

Several different experimental techniques have been applied to measure the superconducting energy gap for M_3C_{60} superconductors, including scanning tunneling microscopy, nuclear magnetic resonance, optical reflectivity, specific heat, and muon spin resonance measurements. The various results that have been obtained are reviewed in this section.

It would appear that scanning tunneling microscopy should provide the most direct method for measuring the superconducting energy gap. Using this technique, the temperature dependence of the energy gap has been investigated for both K_3C_{60} and Rb_3C_{60} [15.53, 54]. Figure 15.11 shows dI/dV, the derivative of the tip current I with respect to bias voltage V, plotted *vs.* bias voltage for an Rb_3C_{60} sample with a 60% Meissner fraction at 4.2 K. The sample in this study was characterized by zero-field-cooled and field-cooled magnetization measurements at very low magnetic field (~5 Oe) [15.53]. The experimental data for (dI/dV) are compared in Fig. 15.11 to the expression for (dI/dV)

$$dI/dV = \mathrm{Re}\{(eV - i\Gamma)/[(eV - i\Gamma)^2 - \Delta^2]^{1/2}\}, \tag{15.15}$$

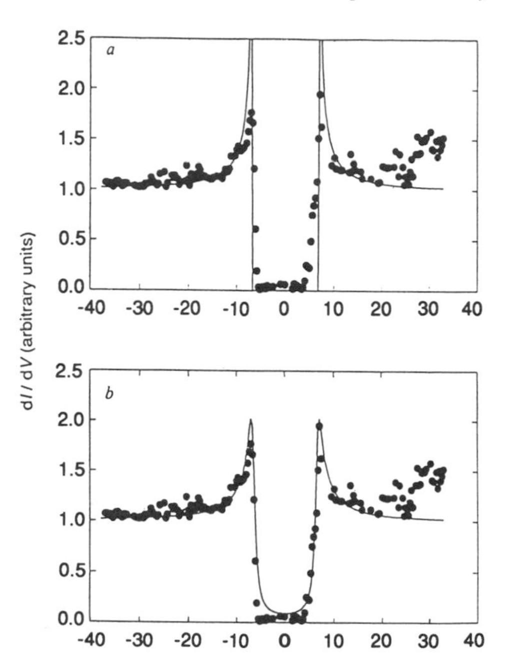

Fig. 15.11. Plot of dI/dV *vs.* bias voltage V (meV) for an Rb_3C_{60} sample at a temperature of 4.2 K [15.53]. The experimental data for the conductance (solid circles) were calculated numerically from I–V measurements. The data in (a) are fit with the expression $dI/dV = eV/[(eV)^2 - \Delta^2]^{1/2}$ (solid curve) with $\Delta = 6.8$ meV and assuming no broadening. The data in (b) are fit with the expression given in Eq. (15.15) which includes a phenomenological broadening parameter Γ ($\Delta = 6.6$ meV and $\Gamma = 0.6$ meV).

which is plotted in Fig. 15.11 for the superconducting state normalized to that in the normal state [15.54]

$$\left[\frac{(dI/dV)_S}{(dI/dV)_N}\right] = \mathrm{Re}\left[\frac{(E - i\Gamma)}{[(E - i\Gamma)^2 - \Delta^2]^{1/2}}\right], \tag{15.16}$$

where $E = eV$ is the energy of the tunneling electrons relative to E_F, 2Δ is the superconducting energy gap, and Γ is a phenomenological broadening parameter. A good fit between the experimental points and the model is obtained. From data such as in Fig. 15.11, the temperature dependence of the superconducting energy gap $\Delta(T)$ was determined and a good fit for $\Delta(T)$ to BCS theory was obtained (see Fig. 15.12) [15.53] for $\Delta(0) =$ 6.6 meV and $\Gamma = 0.6$ meV, from which $(2\Delta(0)/k_BT_c)$ was found to be 5.3 for K_3C_{60} and 5.2 for Rb_3C_{60}, well above the value 3.52 for the case of a BCS superconductor. These results were interpreted to indicate that the M_3C_{60} compounds are strong coupling superconductors [15.54]. The discrepancy between the experimental points and the theoretical curves in Fig. 15.11 has been cited as an indication of systematic uncertainties in the data sets, and a reexamination of the determination of 2Δ by tunneling spectroscopy seems necessary. Also, other experiments, discussed below, yield values of $2\Delta(0)/k_BT_c$ close to the BCS value.

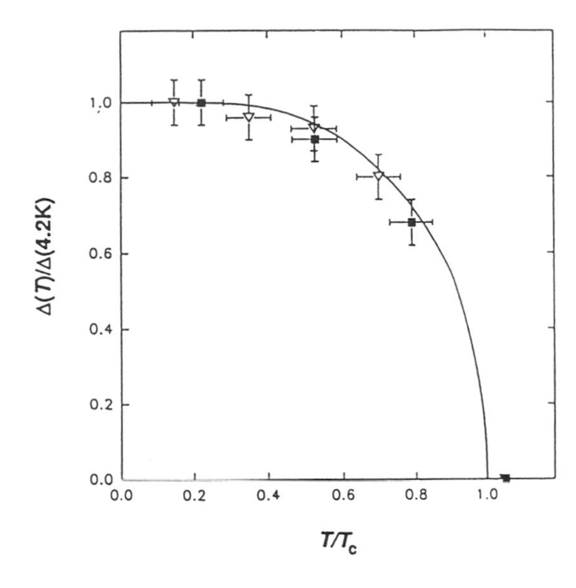

Fig. 15.12. The temperature dependence of the normalized superconducting energy gap for K_3C_{60} (dark squares) and Rb_3C_{60} (open triangles) plotted as a function of reduced temperature T/T_c. The solid line corresponds to the temperature dependence of $\Delta(T)/\Delta(0)$ calculated by BCS theory [15.54].

Two independent determinations of the superconducting energy gap were made using the NMR technique. In the first method, the temperature dependence of the spin–lattice relaxation rate $(1/T_1)$ in the superconducting state was measured [15.55, 56], and the results were interpreted using the relation [15.77]

$$\frac{1}{T_1} \sim \exp\left(-\frac{\Delta}{T}\right). \tag{15.17}$$

The interpretation of these experiments is further discussed in §16.1.8. Early work yielded values for $2\Delta(0)/k_B T_c$ of 4.0 and 3.1 for K_3C_{60} and Rb_3C_{60}, respectively, but a later determination yielded $(2\Delta/k_B T_c) = 3.5$ for both compounds [15.55, 78], in excellent agreement with conventional BCS theory.

Closely related to the NMR spin relaxation experiments are the muon spin relaxation studies (see §16.3) on Rb_3C_{60} powder samples with T_c = 29.4 K [15.57] into which muonium was introduced endohedrally into the C_{60} anions, denoted by Mu@C_{60} [15.79]. The conduction electrons associated with the C_{60} anions screen the positive charge on the muon, so that no local dipole moment is present on the muon. The spin relaxation time T_1 of the muon due to the conduction electrons in the normal and superconducting states was measured as a function of temperature for various values of magnetic field, the field being necessary for observation of the muon spin resonance phenomenon. By making measurements as a function of magnetic field, the effect of the magnetic field on the superconducting energy gap could be taken into account by an extrapolation procedure to the small-field limit. The coupling between the muon spin and the conduction electrons arises because of the small but nonzero value of the π-electron wave functions for the conduction electrons at the center of the C_{60} anion. The muon spin relaxation results show a good fit of the temperature dependence of $(1/T_1)$ to the theory of Hebel and Slichter for spin relaxation in the superconducting state (see Fig. 15.13), yielding a superconducting energy gap value of $\Delta/k_B = 53 \pm 4$ K corresponding to $2\Delta/k_B T_c = 3.6 \pm 0.3$, which is in good agreement with the very closely related NMR spin relaxation experiments [15.55] mentioned above and discussed more fully in §16.1.8. A broadening of the density of states associated with the presence of field inhomogeneities due to trapped magnetic flux was reported in the muon spin resonance experiment [15.79]. Broadening effects in the density of states near T_c were also reported in the scanning tunneling microscopy (STM) studies of the density of states [15.54]. The coherence peak predicted in the Hebel–Slichter theory [15.78] is clearly observed in the muon spin res-

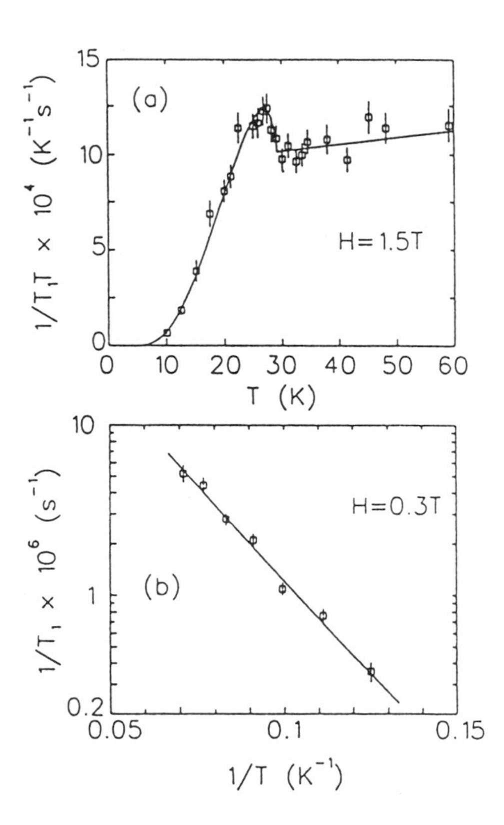

Fig. 15.13. The temperature dependence of the muonium spin resonance relaxation time T_1 in Rb_3C_{60}. In (a) $(T_1 T)^{-1}$ *vs.* T is plotted in a magnetic field of 1.5 T. The solid curve is a fit to the theory of Hebel and Slichter with a broadened density of states, while (b) shows an Arrhenius plot of T_1^{-1} in a magnetic field of 0.3 T [15.57].

onance experiment as shown in Fig. 15.13 and arises from the divergence of the superconducting density of states at the energy gap edges and from the reduced density of states of excited quasiparticles at low-temperature. The observation of the Hebel–Slichter peak in $(1/T_1)$ in the muon spin relaxation studies is consistent with a weak coupling superconductor.

The successful observation of a Hebel–Slichter peak in the muon spin resonance (μSR) experiment was important historically [15.57], because in early NMR work the Hebel–Slichter peak was not observed in the NMR measurements of the temperature dependence of the spin–lattice relaxation time T_1 [15.55]. The successful observation of this peak using the μSR technique suggested that small magnetic fields should be used to look for the Hebel–Slichter peak with the NMR technique, and this strategy eventually led to its successful observation in NMR spin–lattice relaxation studies [15.80]. The μSR technique has also been applied to study $\lambda_L(T)$, and from these studies values of $\lambda_L(0)$ have been found for K_3C_{60} and Rb_3C_{60}, as given in Table 15.1. It was also concluded that superconducting fullerenes have an isotropic energy gap and could follow an s-wave BCS model [15.64].

Analysis of the Knight shift of the NMR resonance for three different nuclei in the same sample as the temperature is lowered below T_c provides a determination of the temperature dependence of the spin susceptibility which yields a sensitive measure of $2\Delta/k_BT_c$ [15.80, 81] (see §16.1.8). Explicit application of this approach provided a second NMR technique for sensitive determination of $2\Delta/k_BT_c$. Measurements of the frequency shift in the NMR spectra for the ^{13}C, ^{87}Rb, and ^{133}Cs nuclei in the Rb_2CsC_{60} compound below T_c [15.80, 81] (see §16.1.8) yield a good fit for $2\Delta/k_BT_c = 3.52$ for both solids (see Fig. 16.10) and further show that $2\Delta/k_BT_c = 4$ is outside the error bars for the Rb_2CsC_{60} measurements (see §16.1.8).

Analysis of infrared reflectivity spectra above and below T_c also provides information on the superconducting energy gap. From a Kramers–Kronig fit to the frequency-dependent reflectivity measurements, the optical conductivity $\sigma_1(\omega)$ in the far-infrared region of the spectrum is deduced (see Fig. 13.31). From the frequency at which $\sigma_1(\omega) \to 0$, the superconducting energy gap is determined. Such measurements have also been carried out [15.58] for both K_3C_{60} and Rb_3C_{60} (see §13.4.4) and show good agreement with predictions of the BCS model with values of $(2\Delta/k_BT_c) = 3.6$ and 3.0 for K_3C_{60} and Rb_3C_{60}, respectively [15.58, 59]. The absolute reflectivity measurements [15.59] yielded $(2\Delta/k_BT_c) = 3.6$ and 2.98 for K_3C_{60} and Rb_3C_{60}, respectively, in good agreement with the previous infrared measurements [15.58].

The ratio of the frequency-dependent optical conductivities in the superconducting and normal states $(\sigma_{1,s}/\sigma_{1,n})$ was also fit to predictions from the BCS theory [15.66] using the response function of the Mattis and Bardeen theory for the high-frequency conductivity in the superconducting state [15.82], which depends parametrically only on 2Δ. A reasonable fit to the Mattis–Bardeen theory was obtained for the superconducting energy gap values, using parameters determined directly from the Kramers–Kronig analysis [15.66]. However, using the more complete Eliashberg theory [15.83], which incorporates the excitation spectrum $\alpha_e^2 F(\omega)$ that is responsible for the electron pairing, a substantial improvement was obtained in fitting the experimental optical conductivity data [15.66]. These fits employed a Lorentzian function for $\alpha_e^2 F(\omega)$ peaked at $\omega_0 \sim 1200$ cm^{-1} with a full width at half-maximum (FWHM) intensity of $\sim$320 cm^{-1}, which simulates the high-frequency intramolecular mode spectrum of C_{60}. The fits remained good for various positions of ω_0 for this excitation band, as long as $\hbar\omega_0 \gg k_B T_c$. This good agreement was interpreted as providing strong evidence for an effective pairing interaction, mediated by these high-frequency intramolecular vibrational modes [15.66]. Furthermore, the possibility of important contributions to the pairing from low-frequency intermolecular or librational modes [$\omega < 100$ cm^{-1} (see §11.4)] was considered, and it was concluded that, within the Eliashberg theory, these low-frequency modes led to strong coupling, in sharp contrast to the main experimental result for $\sigma_1(\omega)$ indicating that $2\Delta \sim 3.5 k_B T_c$ [15.66].

In summary, NMR and muon spin relaxation measurements, NMR Knight shift measurements, and detailed far-infrared reflectivity determinations of the superconducting energy gap are consistent with the predictions of BCS theory for a weak coupling superconductor. The STM measurements of the superconducting energy gap agree in a number of ways with the other energy gap determinations, except for the values of $2\Delta/k_B T_c$, where the STM data suggest that the superconducting fullerenes are strong coupling superconductors. A two-peak model for the Eliashberg spectral function $\alpha_e^2 F(\omega)$ utilizing both the high-frequency intramolecular vibrations and the low-frequency intermolecular vibrations [15.84] can be made to account for a value of $(2\Delta/k_B T_c) \sim 5$. Nevertheless, the present consensus supports the lower value of $(2\Delta/k_B T_c) = 3.5$, consistent with a weak coupling BCS superconductor.

15.5. Isotope Effect

There have been several measurements of the isotope effect in superconducting M_3C_{60} compounds. The main objective of these measurements was to provide insight into whether or not the pairing mechanism responsible for superconductivity in the doped fullerenes involves electron–electron or electron–phonon coupling. In the experiments reported to date, the isotopic enrichment has mainly involved ^{13}C substituted for ^{12}C [15.69, 85, 86]. In measurements of the isotope effect a fit is made to the relation $T_c \propto M^{-\alpha}$, where M is the isotopic mass and α is the isotope shift exponent. In analyzing experiments on the isotope effect, the observed α values are compared with the BCS prediction of $\alpha = 0.5$. The observation of an isotope shift of T_c for the carbon isotopes but not for the alkali metal isotopes indicates that carbon atom vibrations play an important role in the superconductivity of the alkali metal–doped fullerenes.

Magnetization experiments using a SQUID magnetometer on Rb_3C_{60} where up to 75% of the carbon was the ^{13}C isotope gave $\alpha = 0.37 \pm 0.05$ [15.69]. The magnetization results of Chen and Lieber [15.85] on K_3C_{60} prepared from 99% ^{13}C powder are shown in Fig. 15.14 in comparison with the corresponding data for $K_3{}^{12}C_{60}$, from which a value of $\alpha = 0.30 \pm 0.06$ is obtained. From these two experiments one could conclude that the measured values imply a somewhat smaller α value than expected for a BCS model superconductor. A comparison with α determinations for other superconductors (see Table 15.3) shows α values close to 0.5 for classical BCS superconductors, but significant departures from $\alpha = 0.5$ are found for transition metal superconductors and high-T_c materials. It is of interest to note that a large range of values for α have been reported for high-T_c cuprate materials (see Table 15.3). Shown in Table 15.3 are results for α for various stoichiometries of the $La_{2-x}Sr_xCuO_4$ and $La_{2-x}Ca_xCuO_4$ systems. The results for the $La_{2-x}Sr_xCuO_4$ system show relatively high values for α (0.4–0.6) for $x < 0.15$, but much lower values of α for $x \geq 0.15$ [15.88]. Some variations of α in samples with similar nominal stoichiometries were reported [15.88, 89], and these variations are reflected by the range of α

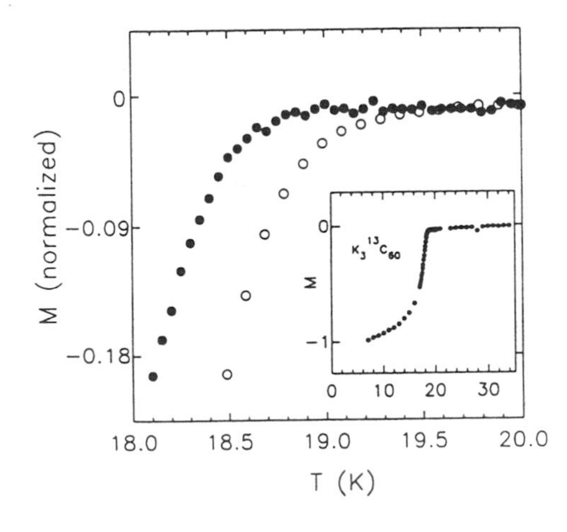

Fig. 15.14. High-resolution temperature-dependent magnetization measurements on $K_3{}^{13}C_{60}$ (filled circles) and $K_3{}^{12}C_{60}$ (open circles) samples, highlighting the depression in T_c for the $K_3{}^{13}C_{60}$ isotopically substituted material. The inset shows a full magnetization curve for a $K_3{}^{13}C_{60}$ sample [15.87].

Table 15.3

Isotope effect for various superconductors.

Materials	T_c^a (K)	α	Isotopic abundance	Reference
Sn	3.7	0.47		[15.95]
Pb	7.2	0.49		[15.95]
Hg	4.1	0.50		[15.95]
Ru	0.51	0.0		[15.95]
Os	0.66	0.15		[15.95]
$La_{1.887}Sr_{0.113}CuO_4$	29.6	0.60–0.64	85–90% ^{18}O, 15–10% ^{16}O	[15.88]
$La_{1.85}Sr_{0.15}CuO_4$	37.8	0.08–0.37	85–90% ^{18}O, 15–10% ^{16}O	[15.88]
$La_{1.925}Ca_{0.075}CuO_4$	9.2	0.81	92%^{18}O, 8% ^{16}O	[15.89]
$La_{1.9}Ca_{0.1}CuO_4$	19.5	0.50–0.55	17–93% ^{18}O, 83–7% ^{16}O	[15.89]
K_3C_{60}	19.3	0.30 ± 0.06	1% ^{12}C, 99% ^{13}C	[15.85]
K_3C_{60}	19.3	0.37 ± 0.05	25% ^{12}C, 75% ^{13}C	[15.69]
K_3C_{60}	19.3	1.4 ± 0.5	67% ^{12}C, 33% ^{13}C	[15.86, 93]

[a]The listed T_c value corresponds to 100% abundance of dominant isotope.

values quoted in Table 15.3. These data are significant in showing the wide deviations from $\alpha = 0.5$ that are observed in high-T_c cuprate materials. These measurements, taken for a wide range of ^{18}O and ^{16}O isotopic abundances, show that α is not sensitive to isotopic oxygen substitutions [15.89]. Similar results have been obtained using the $La_{1.85}Sr_{0.15}Cu_{1-x}Ni_xO_4$ system as x was varied from $x = 0$ ($\alpha = 0.12$) to $x = 0.3$ ($\alpha = 0.45$), with a monotonic increase in α observed with increasing x [15.90]. Other pertinent results on the dependence of α on stoichiometry of high-T_c cuprate materials have been reported for the $YBa_2Cu_3O_{7-\delta}$ system giving α values in the range 0.55–0.10 as Pr was substituted for Y [15.91] and for the Nd-Ce-CuO system $\alpha \leq 0.05$ where the ^{18}O:^{16}O ratio was varied [15.92].

Measurements of the isotope effect on doped C_{60} samples with only a 33% enrichment of ^{13}C gave a much larger value of $\alpha = 1.4 \pm 0.5$ [15.86, 93]. The large range in the values of the exponent α reported by the various groups seems to be associated with the difficulty in determining T_c with sufficient accuracy when the normal–superconducting transition is not sharp. The importance of using samples with high isotopic enrichment can be seen from the following argument. Since $T_c = CM^{-\alpha}$, we can differentiate this expression and write

$$\frac{\Delta T_c}{T_c} = -\alpha \left(\frac{\Delta M}{M} \right) f \tag{15.18}$$

where M is the isotopic mass and f is the fractional isotopic substitution $0 \leq f \leq 1$, assuming that ΔM is related to the difference between the average elemental mass and that of a particular isotope. Equation (15.18) thus shows that ΔT_c is measured more accurately when the isotope effect is measured on light mass species and when the fractional abundance f of the isotope under investigation is large.

To provide another perspective on the experimental determination of the isotope effect, two distinct substitutions of ~50% ^{13}C were prepared in the Rb_3C_{60} compound [15.94]. In one sample $Rb_3(^{13}C_{1-x}{}^{12}C_x)_{60}$ called the mass-averaged sample, each C_{60} molecule had ~50% of ^{12}C and ^{13}C, while in the other sample $Rb_3[(^{13}C_{60})_{1-x}(^{12}C_{60})_x)]$ called the mass-differentiated sample, approximately half of the C_{60} molecules were prepared from the ^{12}C isotopes and the other half from ^{13}C isotopes. The samples thus prepared were characterized by infrared spectroscopy to confirm that the C_{60} molecules were mass averaged in the first sample and mass differentiated in the second sample. Measurements of the isotope effect in the mass-averaged sample confirmed the $\alpha = 0.3$ previously reported for K_3C_{60} [15.85], while the mass-differentiated sample yielded a value of $\alpha = 0.7$ [15.94]. No explanation has yet been given for the different values of α obtained for these two kinds of samples.

On the basis of the superconducting energy gap equation [Eq. (15.9)], all of the reported values of the exponent α suggest that phonons are involved in the pairing mechanism for superconductivity and that the electron–phonon coupling constant is relatively large [15.69, 86]. Future work is needed to clarify the experimental picture of the isotope effect in the M_3C_{60} compounds, and if the large value of $\alpha = 1.4$ were to be confirmed experimentally, detailed theoretical work is needed to explain the significance of the large α value.

15.6. Pressure-Dependent Effects

Of interest also is the dependence of the superconducting parameters on pressure. Closely related to the high compressibility of C_{60} [15.96] and M_3C_{60} (M = K, Rb) [15.17] is the large (approximately linear) decrease in T_c with pressure observed in K_3C_{60} [15.17] and Rb_3C_{60} [15.17] for measurements up to a pressure of 1.9 GPa (see Table 15.1). Figure 15.15 is a plot of the temperature-dependent susceptibility for several values of pressure for a pressed powder sample of K_3C_{60} [15.23]. This figure clearly shows a negative pressure dependence of T_c, which is more clearly seen in the pressure-dependent plot of T_c in Fig. 15.16. The pressure coefficients of T_c for K_3C_{60} and Rb_3C_{60} are similar to those measured for the high-T_c superconductors in the $La_{2-x}(Ba,Sr)_xCuO_4$ system where $(\partial T_c/\partial p) \sim +0.64$ K/kbar [15.97–99], much greater than those for the A15 superconductors (+0.024 K/kbar), but also less than those for BEDT-TTF salt organic superconductors for which $(\partial T_c/\partial p) \sim -3$ K/kbar [15.29].

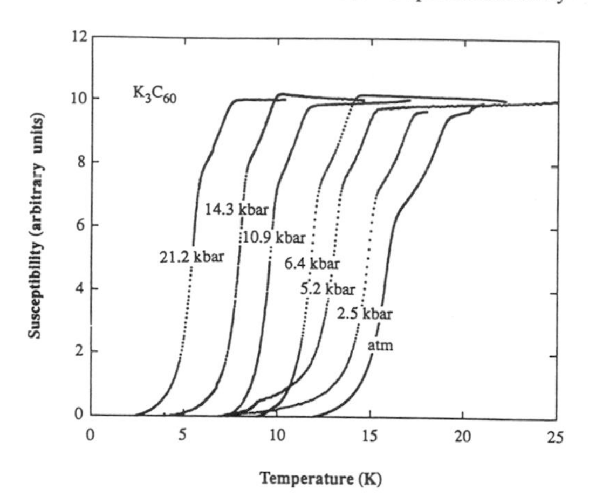

Fig. 15.15. Plot of the temperature dependence of the susceptibility of a pressed K_3C_{60} sample at the various pressures indicated [15.23].

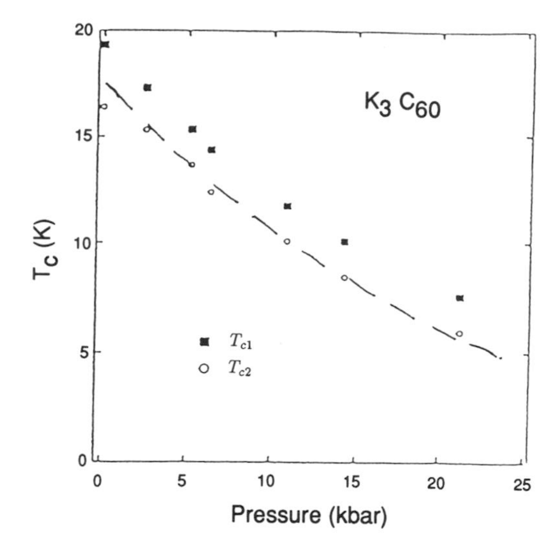

Fig. 15.16. Pressure dependence of T_{c1} (transition onset) and of T_{c2} [the morphology-dependent kink in the $\chi(T)$ curves] (see Fig. 15.15) for a pressed powder sample of K_3C_{60} [15.23].

When the pressure dependence of T_c for K_3C_{60} and Rb_3C_{60} is expressed as $(1/T_c)(\partial T_c/\partial p)_{p=0}$ in the limit of $p = 0$, a common value of -0.35 GPa^{-1} is obtained for both K_3C_{60} and Rb_3C_{60} [15.60, 100]. In this interpretation, the smaller size of the K^+ ion relative to Rb^+ (0.186 Å) is attributed to an effective relative "chemical pressure" of 1.06 GPa, which is found by considering the compressibilities of the two compounds (see Table 15.1). By displacing the pressure scale of the $T_c(p)$ data for K_3C_{60} by 1.06 GPa, the pressure dependence of T_c for K_3C_{60} and Rb_3C_{60} could be made to coincide, and the results for both compounds could be fit to the same functional form

$$T_c(p) = T_c(0) \exp(-\gamma p) \tag{15.19}$$

with the same value of $\gamma = 0.44 \pm 0.03$ GPa^{-1} [15.17]. A plot of T_c *vs.* pressure obtained from curves (such as in Fig. 15.15) is presented in Fig. 15.16, showing a nearly linear decrease of T_c with pressure. The result given by Eq. (15.19) can be interpreted in terms of a T_c that depends only on the lattice constant for the M_3C_{60} compound and not on the identity of the alkali metal dopant (see §15.1). The large effect of pressure on T_c is due to the sensitivity of the intermolecular coupling to the overlap between nearest-neighbor carbon atoms on adjacent molecules, which are separated by 3.18 Å in the absence of pressure. When this intermolecular carbon–carbon distance becomes comparable to the nearest-neighbor intramolecular C–C distance (~1.5 Å), a transition to another structural phase takes place [15.17].

15.7. Mechanism for Superconductivity

The observation of a ^{13}C isotope effect on T_c indicates that C-related vibrational modes are involved in the pairing mechanism. Several other experimental observations suggest that the role of the alkali metal dopant is simply to transfer electrons to C_{60}-derived states (t_{1u} band) and to expand the lattice. This is consistent with the absence of an isotope effect for Rb in Rb_3C_{60} and with the strong correlation of T_c with lattice constant rather than, for example, with the mass of the alkali metal ion. If metal ion displacements were important in the pairing mechanism, then Rb_3C_{60} would be expected to have a lower T_c than K_3C_{60}, contrary to observations.

Which C_{60} vibrational modes are most important to T_c is not completely resolved. The dependence of T_c not on the specific alkali metal species but rather on the lattice constant a_0 of the crystal implies that the coupling mechanism for superconductivity is likely through lattice vibrations and is more closely related to intramolecular than to intermolecular processes. Raman scattering studies (see §11.6.1) of the H_g-derived, intramolecular modes in M_3C_{60} show large increases in the Raman linewidths of many of these modes relative to their counterparts in the insulating parent material C_{60} or in the doped and insulating phase M_6C_{60}. This linewidth broadening strongly suggests that an important contribution to the electron–phonon interaction comes from the intramolecular modes.

Theoretical support for the broadening of certain intramolecular vibrational modes as a result of alkali metal doping has been provided by considering the effect of the Jahn–Teller mechanism on the line broadening through both an adiabatic and a nonadiabatic electron–intramolecular vibration coupling process [15.101]. Broadening was found in modes with H_g symmetry, as well as in some modes with H_u and T_{1u} symmetries, and the broadening results from a reduction in symmetry due to a Jahn–Teller distortion. An especially large effect on the highest H_g mode due to doping has been predicted [15.101]. The broadening effect appears to be in agreement with phonon spectra on alkali metal–doped fullerenes as observed in neutron scattering studies (see §11.5.8).

While keeping these fundamental characteristics in mind, theorists have been trying to identify the dominant electron pairing mechanism for superconducting fullerenes. A number of approaches have been attempted and are summarized below. In the first, only the electron–phonon interaction is considered and electron–electron interactions are neglected. This main approach has been followed by most workers, with different possible side branches pursued, as outlined below. The second approach considers a negative electron–electron interaction whereby superconductivity is explained by an attractive Hubbard model. Returning to the various branches of the first approach, one branch considers a BCS version of the electron–phonon interaction without vertex corrections, while another branch includes vertex corrections for the enhancement of T_c.

Within the first approach, much effort has gone into the identification of the phonon modes which are most important in the electron–phonon interaction. Much of the discussion has focused on the use of the Eliashberg equations to yield the frequency-dependent $\alpha_e^2 F(\omega)$ spectral functions from which the electron–phonon coupling parameter λ_{ep} can be determined by the relation

$$\lambda_{ep} = 2\int_0^\infty d\omega \frac{\alpha_e^2 F(\omega)}{\omega} \tag{15.20}$$

and the average phonon frequency $\overline{\omega}_{\text{ph}}$ is then determined by

$$\overline{\omega}_{\text{ph}} = \exp\left\{\frac{2}{\lambda_{ep}}\int_0^\infty \frac{d\omega}{\omega}\alpha_e^2 F(\omega)\ln\omega\right\}. \tag{15.21}$$

It is readily seen that a given spectral weight $\alpha_e^2 F(\omega)$ at low frequency contributes more to λ_{ep} than the corresponding weight at high ω because of the ω factor in the denominator of Eqs. (15.20) and (15.21).

Since T_c depends predominantly on the lattice constant, which in turn depends on the density of states, as discussed in §15.1, T_c is most sensitive to molecular properties. Thus, it is generally believed that the intramolecular phonons are dominant in the electron–phonon interaction [15.50]. This conclusion is based on detailed studies of both the temperature dependence of the transport properties [15.50] (see §14.1) and the electron–phonon coupling constant directly [15.51, 52]. Although most authors favor dominance by the high-frequency intramolecular vibrations, some authors have emphasized the low-frequency intramolecular vibrations [15.102] for special reasons, such as an effort to explain possible strong coupling in K_3C_{60} and Rb_3C_{60} as implied by the STM tunneling experiments to determine the superconducting energy gap [15.54]. Thus, it may be said that no firm conclusion about the dominant modes has yet been reached [15.84, 103, 104]. If both high- and low-frequency phonons were to contribute strongly to the electron–phonon interaction, then two peaks in the Eliashberg spectral function $\alpha_e^2F(\omega)$ would occur in Eqs. (15.20) and (15.21), such that

$$\overline{\omega}_{\rm ph} = \omega_1^{\lambda_1/\lambda_{ep}} \omega_2^{\lambda_2/\lambda_{ep}}, \tag{15.22}$$

where λ_{ep} is obtained from Eq. (15.20) by integration of $2\alpha_e^2F(\omega)/\omega$ over all phonon frequencies and where λ_1 and λ_2 are related to the contribution from each peak in $\alpha_e^2F(\omega)$ [15.84]. According to this two-peak approach, phonons with energies $\hbar\omega_{ph}$ greater than $\pi k_B T_c$ are effective at pair breaking, while the low-energy phonons are not.

In narrowband systems, such as in M_3C_{60}, we expect a Coulomb repulsion to be present between electrons that are added to the C_{60} molecule in the degenerate f_{1u} (or t_{1u}) levels or energy bands. In the limit of a strong Coulomb repulsive interaction, the t_{1u} energy band is split into upper and lower Hubbard bands, and thus the solid becomes a magnetic insulator [15.105, 106] or acts as a heavy fermion system. If M_3C_{60} is a Mott insulator, then the superconducting phase must have off-stoichiometric compositions $M_{3-\delta}C_{60}$, where a value of $\delta = 0.001$ could account for the normal state conductivity. It has been argued that the success of LDA band calculations for K_3C_{60} and Rb_3C_{60} in predicting a linear dependence of T_c on pressure [15.107] and a variety of experimental efforts to look for anomalous behavior very close to the $x = 3$ stoichiometry indicates that K_3C_{60} and Rb_3C_{60} are metals and not Mott insulators.

In the case of solid C_{60}, the competition between the electron–electron interaction and the transfer energy from molecule to molecule must be considered in treating the superconductivity of π electrons in the f_{1u} (t_{1u}) energy band. Below we also review efforts to treat the pairing of two electrons on the basis of the electron–electron interaction.

White and his co-workers discussed the attraction between two electrons localized on a single C_{60} anion by defining the pair-binding energy, $E_{\rm pair}^i$ [15.108],

$$E_{\rm pair}^i = 2\Phi_i - \Phi_{i-1} - \Phi_{i+1} \qquad (i = 1, 3, 5), \tag{15.23}$$

where Φ_i is the total energy of a molecule when the molecule has i additional electrons. If $E_{\rm pair}^i$ is positive, it is energetically favorable for two adjacent molecules to have $(i+1)$ electrons on one molecule and $(i-1)$ electrons on the other. This transfer of electrons has been proposed as a possible mechanism for superconductivity in fullerenes [15.109, 110].

This situation has been examined by a numerical calculation using the extended Hubbard Hamiltonian,

$$\mathcal{H} = -\sum_{\langle i,j\rangle,\sigma} t_{ij}(C^{\dagger}_{i\sigma}C_{j\sigma} + \text{H.c.}) + \frac{U}{2}\sum_{i,\sigma} n_{i\sigma}n_{i-\sigma} + V\sum_{\langle i,j\rangle,\sigma,\sigma'} n_{i\sigma}n_{j\sigma'} \quad (15.24)$$

where $C^{\dagger}_{i\sigma}$ creates an electron with spin σ on the ith site, and $n_{i\sigma} = C^{\dagger}_{i\sigma}C_{i\sigma}$ is the density of electrons with spin σ on site i, and t_{ij} denotes the transfer integral for the transfer of charge. U and V represent on-site and off-site electron–electron interactions, the sums in Eq. (15.24) are taken over orbital and spin states, and H.c. denotes Hermitian conjugate. The exact diagonalization of $\mathcal{H}$ given by Eq. (15.24) has been done for small clusters with N electrons on a one-dimensional ring [15.111], torus ($N = 16$), cube ($N = 8$), and truncated tetrahedron ($N = 12$) [15.108], which are all smaller than the case of $N = 60$ for C_{60}. The calculated results show that E^{i}_{pair} of Eq. (15.23) can be positive or negative, depending on the relative ratios of the parameters t, U, and V. If an electron–electron pairing mechanism were to be important, it would need to explain not only superconductivity arising from the partial filling of the t_{1u}-derived band but also, for the case of the alkaline earth compounds, the partial filling of the t_{1g}-derived band, including the strong hybridization of the t_{1g} electronic states with the alkaline earth metal s and d states.

Recently, the total energy of the C_{60}^{n-} ion has been calculated using semi-empirical self-consistent field (SCF) and configurational interaction (CI) programs from the library MOPAC [15.112]. The calculated results show that the total energy as a function of n has a positive curvature, which implies that E^{i}_{pair} is always negative. In this sense the effective attractive interaction between two electrons in a molecule might not be relevant as a mechanism for superconductivity.

When the electron–phonon interaction screens the Coulomb repulsion between two electrons, the effective Coulomb interaction, U_{eff}, can be negative. In this case we can apply an attractive Hubbard model to such a system. Starting from an attractive Hubbard model, it has been shown [15.113] that the superconducting state is more stable than the charge density wave (CDW) state in the fcc structure of K_3C_{60}, similar to the case in $BaBiO_3$. The effective intramolecular interaction between π electrons, U_{eff}, can be negative for the case in which the optic phonons involving the K^+ ions couple strongly to the π electron carriers on C_{60} molecules of the K_3C_{60} crystal. The absolute value of U_{eff} caused by two optic modes that are related to octahedral–tetrahedral and octahedral–octahedral sites (estimated as $U_{\text{eff}} \sim -0.65$ and -0.27 eV, respectively) may be larger than that of the repulsive interaction caused by the screened intramolecular Coulomb repulsion U_{e-e}(1.2–1.7 eV) or the negative values caused by a bond dimerization effect (~ -0.026 eV) or the Jahn–Teller interaction ($\sim$ –25 meV) [15.114]. Within mean field theory, the superconducting state is more stable than the CDW state for any negative U_{eff}. This fact can be understood both by the calculation of T_c in the weak coupling limit of U_{eff} and by the perturbation expansion in t of the total energy for the two states in the strong coupling limit of $|U_{\text{eff}}/t| \gg 1$.

A number of proposed mechanisms for superconductivity in doped fullerenes involve the Hubbard model. Therefore criticisms of using a Hubbard model should be mentioned. The validity of the Hubbard model requires that the intra-site screening be much smaller than the intersite screening and that the bare intrasite Coulomb repulsion should be much larger than the bare intersite repulsion. Since the nearest-neighbor C–C distance between adjacent C_{60} molecules is $\sim$3 Å and of the same order as the diameter of C_{60} ($\sim$7 Å), the bare intrasite and intersite repulsions should be of comparable magnitude. The large number of on-site electrons (240) for C_{60} suggests that the intrasite screening is important. Thus use of the Hubbard model for describing the electronic structure of fullerenes has been questioned. It has also been argued that the energy needed to transport an electron should be the difference $U_{\text{intra}} - U_{\text{inter}}$ rather than the intracluster repulsion U_{intra}, and it is expected that $U_{\text{intra}} - U_{\text{inter}} \ll U_{\text{intra}}$ [15.107, 115].

Another proposed theory of superconductivity for doped C_{60} is based on a parity doublet [15.116] formed from an $h_u^{10} t_{1u}^2 t_{1g}^1$ configuration in the doped C_{60} material. In this case, the $t_{1u}^1 t_{1g}^1$ electron pair forms the parity doublet that triggers the superconducting transition. Also, the possibility of plasmons mediating an attractive interaction between electrons has been suggested as a pairing mechanism for low-carrier-density systems [15.117].

Although many attempts have been made to explain the superconducting state in doped fullerenes as special cases of a Hubbard system with U ranging from negative to positive values, none of the theories are yet conclusive, partly because of their inability to account for many of the published experimental results. Likewise, many first principles calculations elucidate certain fundamental issues but do not make systematic predictions which can be tested. Most experiments suggest that the transfer energy is comparable to the phonon energy and to the electron–electron interaction. Quantitative applications of the theory using realistic parameters are needed. Furthermore, justification is also needed of the methodology that was used to determine reliable values of the parameters that are employed in theoretical calculations. Particular insights into the dominant superconducting mechanism are expected through gaining a better understanding of the isotope effect and pressure-dependent phenomena.

While it is generally believed that superconductivity in doped fullerenes arises from some kind of pairing mechanism between electrons, the detailed nature of the pairing mechanism is not well established. Although the electron–phonon mechanism has been most widely discussed, it appears that electron–electron pairing mechanisms have not yet been ruled out. For those who believe that the electron–phonon interaction is the dominant pairing mechanism, heated discussion currently focuses on which phonons play a dominant role in the coupling. To produce pairing, many authors have invoked a dynamic Jahn–Teller mechanism induced by pertinent intramolecular modes, and further clarification is needed of the role of the Jahn–Teller effect in fullerene superconductivity. Experimental evidence in support of Jahn–Teller distortions for C_{60}^{3-} anions comes from both EPR studies (§16.2.2) and optical studies (§13.4.1).

A host of other fundamental theoretical issues remain to be clarified. Since the charge distribution in the superconducting fullerenes is mainly on the surface of the icosahedral C_{60}^{3-} anions, in contrast to the more distributed charge distribution found in conventional solids, modifications are probably necessary to the traditional interpretation of a variety of measurements, such as the plasma frequency, specific heat, magnetic susceptibility, and temperature- and magnetic–field dependent transport measurements, to mention but a few. While concern about the validity of the Migdal theorem has been raised, since the vibrational frequencies are comparable in magnitude to the Fermi energy and to the bandwidth of the LUMO levels, no clear picture has yet emerged on which aspects of the Migdal theory need modification. While many workers in the field agree that many-body effects are important, no consensus has been reached on how to handle strong correlations between electrons in a system where the diameter of the C_{60} molecules and the nearest-neighbor distances are of comparable magnitude and the electronic energy bandwidths are very narrow.

REFERENCES

[15.1] R. C. Haddon, A. F. Hebard, M. J. Rosseinsky, D. W. Murphy, S. J. Duclos, K. B. Lyons, B. Miller, J. M. Rosamilia, R. M. Fleming, A. R. Kortan, S. H. Glarum, A. V. Makhija, A. J. Muller, R. H. Eick, S. M. Zahurak, R. Tycko, G. Dabbagh, and F. A. Thiel. *Nature (London)*, **350**, 320 (1991).

[15.2] Z. H. Wang, K. Ichimura, M. S. Dresselhaus, G. Dresselhaus, W. T. Lee, K. A. Wang, and P. C. Eklund. *Phys. Rev. B*, **48**, 10657 (1993).

[15.3] M. P. Gelfand. *Superconductivity Review*, **1**, 103 (1994).

[15.4] N. B. Hannay, T. H. Geballe, B. T. Matthias, K. Andres, P. Schmidt, and D. MacNair. *Phys. Rev. Lett.*, **14**, 225 (1965).

[15.5] Y. Koike, K. Higuchi, and S. Tanuma. *Solid State Commun.*, **27**, 623 (1978).

[15.6] A. Chaiken, M. S. Dresselhaus, T. P. Orlando, G. Dresselhaus, P. Tedrow, D. A. Neumann, and W. A. Kamitakahara. *Phys. Rev. B*, **41**, 71 (1990).

[15.7] S. Tanuma. In H. Zabel and S. A. Solin (eds.), *Graphite Intercalation Compounds II: Transport and Electronic Properties*, Springer Series in Materials Science, pp. 163–193, Springer-Verlag, Berlin (1992). Vol. 18.

[15.8] I. T. Belash, A. D. Bronnikov, O. V. Zharikov, and A. V. Pal'nichenko. *Synth. Metals*, **36**, 283 (1990).

[15.9] K. Holczer, O. Klein, G. Grüner, S.-M. Huang, R. B. Kaner, K.-J. Fu, R. L. Whetten, and F. Diederich. *Science*, **252**, 1154 (1991).

[15.10] A. F. Hebard. In P. Jena, S. N. Khanna, and B. K. Rao (eds.), *Physics and Chemistry of Finite Systems: From Clusters to Crystals*, vol. 374, pp. 1213–1220, Dordrecht (1991). Kluwer Academic Publishers. NATO ASI Series C: Mathematical and Physical Sciences.

[15.11] K. Tanigaki, T. W. Ebbesen, S. Saito, J. Mizuki, J. S. Tsai, Y. Kubo, and S. Kuroshima. *Nature (London)*, **352**, 222 (1991).

[15.12] K. Tanigaki, S. Kuroshima, J. Fujita, and T. W. Ebbesen. *Appl. Phys. Lett.*, **63**, 2351 (1993).

[15.13] T. T. M. Palstra, O. Zhou, Y. Iwasa, P. Sulewski, R. Fleming, and B. Zegarski. *Solid State Commun.*, **93**, 327 (1995).

[15.14] J. G. Bednorz and K. A. Müller. *Z. Physik. B*, **64**, 189 (1986).

[15.15] J. M. Williams, A. J. Shultz, U. Geiser, K. D. Carlson, A. M. Kini, H. H. Wang, W. K. Kwok, M. H. Whangbo, and J. E. Schirber. *Science*, **252**, 1501 (1991).

[15.16] R. M. Fleming, A. P. Ramirez, M. J. Rosseinsky, D. W. Murphy, R. C. Haddon, S. M. Zahurak, and A. V. Makhija. *Nature (London)*, **352**, 787 (1991).

[15.17] G. Sparn, J. D. Thompson, R. L. Whetten, S.-M. Huang, R. B. Kaner, F. Diederich, G. Grüner, and K. Holczer. *Phys. Rev. Lett.*, **68**, 1228 (1992).

[15.18] C.-C. Chen, S. P. Kelty, and C. M. Lieber. *Science*, **253**, 886 (1991).

[15.19] D. W. Murphy, M. J. Rosseinsky, R. M. Fleming, R. Tycko, A. P. Ramirez, R. C. Haddon, T. Siegrist, G. Dabbagh, J. C. Tully, and R. E. Walstedt. *J. Phys. Chem. Solids*, **53**, 1321 (1992).

[15.20] K. Tanigaki, I. Hirosawa, T. W. Ebbesen, J. Mizuki, Y. Shimakawa, Y. Kubo, J. S. Tsai, and S. Kuroshima. *Nature (London)*, **356**, 419 (1992).

[15.21] O. Zhou, G. Vaughan, B. M. Gavin, Q. Zhu, J. E. Fischer, P. A. Heiney, N. Coustel, J. P. McCauley, Jr., and A. B. Smith III. *Science*, **255**, 833 (1992).

[15.22] A. Oshiyama, S. Saito, N. Hamada, and Y. Miyamoto. *J. Phys. Chem. Solids*, **53**, 1457 (1992).

[15.23] G. Sparn, J. D. Thompson, S.-M. Huang, R. B. Kaner, F. Diederich, R. L. Whetten, G. Grüner, and K. Holczer. *Science*, **252**, 1829 (1991).

[15.24] O. Zhou, R. M. Fleming, D. W. Murphy, M. J. Rosseinsky, A. P. Ramirez, R. B. van Dover, and R. C. Haddon. *Nature (London)*, **362**, 433 (1993).

[15.25] M. S. Dresselhaus and G. Dresselhaus. *Adv. Phys.*, **30**, 139 (1981).

[15.26] S. A. Solin and H. Zabel. *Adv. Phys.*, **37**, 87 (1988).

[15.27] M. J. Rosseinsky, D. W. Murphy, R. M. Fleming, and O. Zhou. *Nature (London)*, **364**, 425 (1993).

[15.28] J. M. Williams, H. H. Wang, M. A. Bena, T. J. Emge, L. M. Sowa, P. T. Copps, F. Behrooze, L. N. Hall, K. D. Carlson, and G. W. Crabtree. *Inorg. Chem.*, **23**, 3839 (1984).

[15.29] M. Tokumoto, H. Anzai, K. Murata, K. Kajimura, and T. Ishiguro. *Synth. Metals*, **27**, 251 (1988).

[15.30] A. R. Kortan, N. Kopylov, S. H. Glarum, E. M. Gyorgy, A. P. Ramirez, R. M. Fleming, F. A. Thiel, and R. C. Haddon. *Nature (London)*, **355**, 529 (1992).

[15.31] A. R. Kortan, N. Kopylov, S. H. Glarum, E. M. Gyorgy, A. P. Ramirez, R. M. Fleming, O. Zhou, F. A. Thiel, P. L. Trevor, and R. C. Haddon. *Nature (London)*, **360**, 566 (1992).

[15.32] A. R. Kortan, N. Kopylov, R. M. Fleming, O. Zhou, F. A. Thiel, R. C. Haddon, and K. M. Rabe. *Phys. Rev. B*, **47**, 13070 (1993).

[15.33] A. R. Kortan, N. Kopylov, E. Özdas, A. P. Ramirez, R. M. Fleming, and R. C. Haddon. *Chem. Phys. Lett.*, **223**, 501 (1994).

[15.34] Z. H. Wang, M. S. Dresselhaus, G. Dresselhaus, and P. C. Eklund. *Phys. Rev. B*, **48**, 16881 (1993).

[15.35] T. Takahashi. In T. Oguchi, K. Kadowaki, and T. Sasaki (eds.), *Electronic Properties and Mechanisms of High T_c Superconductors*, p. 91, Amsterdam (1992). Elsevier, North-Holland.

[15.36] T. Takahashi, T. Morikawa, S. Hasegawa, K. Kamiya, H. Fujimoto, S. Hino, K. Seki, H. Katayama-Yoshida, H. Inokuchi, K. Kikuchi, S. Suzuki, K. Ikemoto, and Y. Achiba. *Physica C*, **190**, 205 (1992).

[15.37] E. Sohmen and J. Fink. *Phys. Rev. B*, **47**, 14532 (1993).

[15.38] A. A. Zakhidov, K. Yakushi, K. Imaeda, H. Inokuchi, K. Kikuchi, S. Suzuki, I. Ikemoto, and Y. Achiba. *Mol. Cryst. Liq. Cryst.*, **218**, 299 (1992).

[15.39] S. Hino, K. Matsumoto, S. Hasegawa, H. Inokuchi, T. Morikawa, T. Takahashi, K. Seki, K. Kikuchi, S. Suzuki, I. Ikemoto, and Y. Achiba. *Chem. Phys. Lett.*, **197**, 38 (1992).

[15.40] S. Hino, K. Matsumoto, S. Hasegawa, K. Iwasaki, K. Yakushi, T. Morikawa, T. Takahashi, K. Seki, K. Kikuchi, S. Suzuki, I. Ikemoto, and Y. Achiba. *Phys. Rev. B*, **48**, 8418 (1993).

[15.41] S. Hino, K. Matsumoto, S. Hasegawa, K. Kamiya, H. Inokuchi, T. Morikawa, T. Takahashi, K. Seki, K. Kikuchi, S. Suzuki, I. Ikemoto, and Y. Achiba. *Chem. Phys. Lett.*, **190**, 169 (1992).

[15.42] D. M. Poirier, J. H. Weaver, K. Kikuchi, and Y. Achiba. *Zeitschrift für Physik D: Atoms, Molecules and Clusters*, **26**, 79 (1993).

[15.43] A. A. Zakhidov, K. Imaeda, K. Yakushi, H. Inokuchi, Z. Iqbal, R. H. Baughman, B. L. Ramakrishna, and Y. Achiba. *Physica.*, **185C**, 411 (1991).

[15.44] A. I. Sololov, Y. A. Kufaev, and E. B. Sonin. *Physica C*, **212**, 19 (1993).

[15.45] K. Mizoguchi. *J. Phys. Chem. Solids*, **54**, 1693 (1993).

[15.46] H. H. Wang, J. A. Schlueter, A. C. Cooper, J. L. Smart, M. E. Whitten, U. Geiser, K. D. Carlson, J. M. Williams, U. Welp, J. D. Dudek, and M. A. Caleca. *J. Phys. Chem. Solids*, **54**, 1655 (1993).

[15.47] M. Tokumoto, Y. Tanaka, N. Kinoshita, T. Kinoshita, S. Ishibashi, and H. Ihara. *J. Phys. Chem. Solids*, **54**, 1667 (1993.).

[15.48] K. Tanigaki, I. Hirosawa, T. W. Ebbesen, J. I. Mizuki, and J. S. Tsai. *J. Phys. Chem. Solids*, **54**, 1645 (1993).

[15.49] W. L. McMillan. *Phys. Rev.*, **167**, 331 (1968).

[15.50] M. L. Cohen. *Mater. Sci. Eng.*, **B19**, 111 (1993).

[15.51] M. A. Schlüter, M. Lannoo, M. Needels, G. A. Baraff, and D. Tománek. *Mater. Sci. Eng.*, **B19**, 129 (1993).

[15.52] M. Schlüter, M. Lannoo, M. Needels, G. A. Baraff, and D. Tománek. *J. Phys. Chem. Solids*, **53**, 1473 (1992). In this reference the normal mode models were prepared by M. Grabow.

[15.53] Z. Zhang, C.-C. Chen, S. P. Kelty, H. Dai, and C. M. Lieber. *Nature (London)*, **353**, 333 (1991).

[15.54] Z. Zhang, C. C. Chen, and C. M. Lieber. *Science*, **254**, 1619 (1991).

[15.55] R. Tycko, G. Dabbagh, M. J. Rosseinsky, D. W. Murphy, A. P. Ramirez, and R. M. Fleming. *Phys. Rev. Lett.*, **68**, 1912 (1992).

[15.56] K. Holczer, O. Klein, H. Alloul, Y. Yoshinari, and F. Hippert. *Europhys. Lett.*, **23**, 63 (1993).

[15.57] R. F. Kiefl, W. A. MacFarlane, K. H. Chow, S. Dunsiger, T. L. Duty, T. M. S. Johnston, J. W. Schneider, J. Sonier, L. Brard, R. M. Strongin, J. E. Fischer, and A. B. Smith III. *Phys. Rev. Lett.*, **70**, 3987 (1993).

[15.58] L. D. Rotter, Z. Schlesinger, J. P. McCauley, Jr., N. Coustel, J. E. Fischer, and A. B. Smith III. *Nature (London)*, **355**, 532 (1992).

[15.59] L. Degiorgi, G. Grüner, P. Wachter, S. M. Huang, J. Wiley, R. L. Whetten, R. B. Kaner, K. Holczer, and F. Diederich. *Phys. Rev. B*, **46**, 11250 (1992).

[15.60] K. Holczer and R. L. Whetten. *Carbon*, **30**, 1261 (1992).

[15.61] S. Foner, E. J. McNiff, Jr., D. Heiman, S. M. Huang, and R. B. Kaner. *Phys. Rev. B*, **46**, 14936 (1992).

[15.62] G. S. Boebinger, T. T. M. Palstra, A. Passner, M. J. Rosseinsky, D. W. Murphy, and I. I. Mazin. *Phys. Rev. B*, **46**, 5876 (1992).

[15.63] R. D. Johnson, D. S. Bethune, and C. S. Yannoni. *Acc. Chem. Res.*, **25**, 169 (1992).

[15.64] Y. J. Uemura, A. Keren, L. P. Le, G. M. Luke, B. J. Sternlieb, W. D. Wu, J. H. Brewer, R. L. Whetten, S. M. Huand, S. Lin, R. B. Kaner, F. Diederich, S. Donovan, and G. Grüner. *Nature (London)*, **352**, 605 (1991).

[15.65] L. Degiorgi, P. Wachter, G. Grüner, S. M. Huang, J. Wiley, and R. B. Kaner. *Phys. Rev. Lett.*, **69**, 2987 (1992).

[15.66] L. Degiorgi, E. J. Nicol, O. Klein, G. Grüner, P. Wachter, S. M. Huang, J. Wiley, and R. B. Kaner. *Phys. Rev. B*, **49**, 7012 (1994).

[15.67] A. P. Ramirez, M. J. Rosseinsky, D. W. Murphy, and R. C. Haddon. *Phys. Rev. Lett.*, **69**, 1687 (1992).

[15.68] T. T. M. Palstra, R. C. Haddon, A. F. Hebard, and J. Zaanen. *Phys. Rev. Lett.*, **68**, 1054 (1992).

[15.69] A. P. Ramirez, A. R. Kortan, M. J. Rosseinsky, S. J. Duclos, A. M. Mujsce, R. C. Haddon, D. W. Murphy, A. V. Makhija, S. M. Zahurak, and K. B. Lyons. *Phys. Rev. Lett.*, **68**, 1058 (1992).

[15.70] J. G. Hou, V. H. Crespi, X. D. Xiang, W. A. Vareka, G. Briceño, A. Zettl, and M. L. Cohen. *Solid State Commun.*, **86**, 643 (1993).

[15.71] N. R. Werthamer, E. Helfand, and P. C. Hohenberg. *Phys. Rev.*, **147**, 295 (1966).

[15.72] K. Holczer, O. Klein, G. Grüner, J. D. Thompson, F. Diederich, and R. L. Whetten. *Phys. Rev. Lett.*, **67**, 271 (1991).

[15.73] M. Dressel, L. Degiorgi, O. Klein, and G. Grüner. *J. Phys. Chem. Solids*, **54**, 1411 (1993).

[15.74] Y. J. Uemura, L. P. Le, and G. M. Luke. *Synthetic Metals*, **56**, 2845 (1993).

[15.75] C. L. Lin, T. Mihalisin, N. Bykovetz, Q. Zhu, and J. E. Fischer. *Phys. Rev. B*, **49**, 4285 (1994).

[15.76] J. R. L. de Almeida and D. Thouless. *J. Phys. A*, **11**, 983 (1978).

[15.77] C. P. Slichter. *Principles of Magnetic Resonance*. Springer-Verlag, Berlin, 3rd edition (1990).

[15.78] L. C. Hebel and C. P. Slichter. *Phys. Rev.*, **113**, 1504 (1959).

[15.79] R. F. Kiefl, T. L. Duty, J. W. Schneider, A. MacFarlane, K. Chow, J. W. Elzey, P. Mendels, G. D. Morris, J. H. Brawer, E. J. Ansaldo, C. Niedermayer, D. R. Nokes, C. E. Stronach, B. Hitti, and J. E. Fischer. *Phys. Rev. Lett.*, **69**, 2005 (1992).

[15.80] V. A. Stenger, C. H. Pennington, D. R. Buffinger, and R. P. Ziebarth. *Phys. Rev. Lett.*, **74**, 1649 (1995).

[15.81] V. A. Stenger, C. Recchia, J. Vance, C. H. Pennington, D. R. Buffinger, and R. P. Ziebarth. *Phys. Rev. B*, **48**, 9942 (1993).

[15.82] D. C. Mattis and J. Bardeen. *Phys. Rev.*, **111**, 412 (1958).

[15.83] G. M. Eliashberg. *Zh. Eksp. Teor. Fiz.*, **39**, 1437 (1960). Sov. Phys.-JETP **12**, 1000 (1961).

[15.84] I. I. Mazin, O. V. Dolgov, A. Golubov, and S. V. Shulga. *Phys. Rev. B*, **47**, 538 (1993).

[15.85] C.-C. Chen and C. M. Lieber. *J. Am. Chem. Soc.*, **114**, 3141 (1992).

[15.86] T. W. Ebbesen, J. S. Tsai, K. Tanigaki, J. Tabuchi, Y. Shimakawa, Y. Kubo, I. Hirosawa, and J. Mizuki. *Nature (London)*, **355**, 620 (1992).

[15.87] C. M. Lieber and C. C. Chen. In H. Ehrenreich and F. Spaepen (eds.), *Solid State Physics*, **48**, p. 109, New York (1994). Academic Press.

[15.88] M. K. Crawford, M. N. Kunchur, W. E. Farneth, E. M. McCarron III, and S. J. Poon. *Phys. Rev. B*, **41**, 282 (1990).

[15.89] M. K. Crawford, W. E. Farneth, R. Miao, R. L. Harlow, C. C. Torardi, and E. M. McCarron. *Physica C*, **185–189**, 1345 (1991).

[15.90] N. Babyshkina, A. Inyushkin, V. Ozhogin, A. Taldenkov, K. Kobrin, T. Vorobe'va, L. Molchanova, L. Damyanets, T. Uvarova, and A. Kuzakov. *Physica C*, **185–189**, 901 (1991).

[15.91] J. P. Franck, S. Gygax, G. Soerensen, E. Altshuler, A. Hnatiw, J. Jung, M. A. K. Mohamed, M. K. Yu, G. I. Sproule, J. Chrzanowski, and J. C. Irwin. *Physica C*, **185–189**, 1379 (1991).

[15.92] B. Batlogg, S. W. Cheong, G. A. Thomas, S. L. Cooper, L. W. Rupp, Jr., D. H. Rapkine, and A. S. Cooper. *Physica C*, **185–189**, 1385 (1991).

[15.93] A. A. Zakhidov, K. Imaeda, D. M. Petty, K. Yakushi, H. Inokuchi, K. Kikuchi, I. Ikemoto, S. Suzuki, and Y. Achiba. *Phys. Lett. A*, **164**, 355 (1992).

[15.94] C. C. Chen and C. M. Lieber. *Science*, **259**, 655 (1993).

[15.95] C. Kittel. *Introduction to Solid State Physics*. John Wiley & Sons, New York, 6th ed. (1985).

[15.96] J. E. Fischer, P. A. Heiney, A. R. McGhie, W. J. Romanow, A. M. Denenstein, J. P. McCauley, Jr., and A. B. Smith III. *Science*, **252**, 1288 (1991).

[15.97] A. Driessen, R. Griessen, N. Koemon, E. Salomons, R. Brouwer, D. G. de Groot, K. Heeck, H. Hemmes, and J. Rector. *Phys. Rev. B*, **36**, 5602 (1987).

[15.98] I. D. Parker and R. H. Friend. *J. Phys. C: Solid State Phys.*, **21**, L345 (1988).

[15.99] C. W. Chu, L. Gao, F. Chen, Z. J. Huang, R. L. Meng, and Y. Y. Xue. *Nature*, **365**, 323 (1993).

[15.100] K. Holczer. *Int. J. Mod. Phys. B*, **6**, 3967 (1992).

[15.101] Y. Asai. *Phys. Rev. B*, **49**, 4289 (1994).

[15.102] K. H. Johnson, M. E. McHenry, and D. P. Clougherty. *Physica C*, **183**, 319 (1991).

[15.103] G. H. Chen and W. A. Goddard III. *Proc. Nat. Acad. Sci.* **90**, 1350 (1993).

[15.104] O. V. Dolgov and I. I. Mazin. *Solid State Commun.*, **81**, 935 (1992).

[15.105] R. W. Lof, M. A. van Veenendaal, B. Koopmans, H. T. Jonkman, and G. A. Sawatzky. *Phys. Rev. Lett.*, **68**, 3924 (1992).

[15.106] R. W. Lof, M. A. van Veenendaal, B. Koopmans, A. Heessels, H. T. Jonkman, and G. A. Sawatzky. *Int. J. Mod. Phys. B*, **6**, 3915 (1992).

[15.107] S. Saito, S. I. Sawada, N. Hamada, and A. Oshiyama. *Mater. Sci. Eng.*, **B19**, 105 (1993).
[15.108] S. R. White, S. Chakravarty, M. P. Gelfand, and S. A. Kivelson. *Phys. Rev. B*, **45**, 5062 (1992).
[15.109] S. Chakravarty, M. P. Gelfand, and S. A. Kivelson. *Science*, **254**, 970 (1991).
[15.110] S. Chakravarty, L. Chayes, and S. A. Kivelson. *Lett. Math. Phys.*, **23**, 265 (1992).
[15.111] R. M. Fye, M. J. Martins, and R. T. Scalettar. *Phys. Rev. B*, **42**, 6809 (1990).
[15.112] R. Saito, G. Dresselhaus, and M. S. Dresselhaus. *Chem. Phys. Lett.*, **210**, 159 (1993).
[15.113] F. C. Zhang, M. Ogata, and T. M. Rice. *Phys. Rev. Lett.*, **67**, 3452 (1991).
[15.114] V. de Coulon, J. L. Martins, and F. Reuse. *Phys. Rev. B*, **45**, 13671 (1992).
[15.115] M. B. J. Meinders, , L. H. Tjeng, and G. A. Sawatzky. *Phys. Rev. Lett.*, **73**, 2937 (1994).
[15.116] R. Friedberg, T. D. Lee, and H. C. Ren. *Phys. Rev. B*, **46**, 14150 (1992).
[15.117] Y. Takada. *Physica C*, **185–189**, 419 (1991).

CHAPTER 16

Magnetic Resonance Studies

Various magnetic resonance techniques have been used to characterize the fullerene-based materials. These techniques include nuclear magnetic resonance (NMR), electron paramagnetic resonance (EPR), and muon spin resonance (μSR). The major results obtained by magnetic resonance studies of fullerenes are reviewed in this chapter. In the early experimental studies of these materials, NMR was exploited to authenticate the C_{60} and C_{70} structures. More recently, NMR techniques have been applied to illuminate many important fullerene properties relating to bond lengths, molecular dynamics, dopant locations, stoichiometric characterization, determinations of the superconducting energy gap, and others. The EPR technique has provided a major tool for studying spin states and endohedrally doped fullerenes, molecular anions and paramagnetic doped fullerene systems. Finally, μSR has provided a special tool useful for illuminating special aspects of the superconductivity of doped fullerenes.

16.1. Nuclear Magnetic Resonance

Nuclear magnetic resonance has provided a powerful tool for the study of C_{60}, C_{70} and doped fullerenes [16.1]. Most of the nuclear resonance studies have been made on the ^{13}C nucleus, which has a natural abundance of 1.1%, a nuclear spin of 1/2, and no electric quadrupole moment. The probability for m isotopic substitutions to occur on an n_C atom fullerene C_{n_C} is given in Eqs. (4.8) and (4.9) of §4.5 and explicit values for these probabilities for the natural abundance of ^{13}C in C_{60} and C_{70} are given in Table 4.25. Since the more abundant ^{12}C nucleus (98.9%) has no nuclear spin, ^{12}C is not a suitable nucleus for NMR studies. NMR studies on doped fullerenes have also been made on nuclei associated with various dopant species, such as ^{39}K, ^{133}Cs, and ^{87}Rb.

In this section we review the major ways in which NMR has been used to gain new and in some cases unique knowledge about fullerenes and fullerene-related materials. Some of the main findings regarding the structure and properties of C_{60} have been achieved using NMR techniques.

16.1.1. Structural Information about the Molecular Species

NMR was one of the earliest spectroscopies used to show conclusively that the icosahedral structure for the C_{60} molecule was correct. The 60 carbon atoms in C_{60} are now known to be located at the vertices of a regular truncated icosahedron where every site is equivalent to every other site (see Fig. 3.1). Such a structure (with a single ^{13}C nuclear substitution) is consistent with a single sharp line in the NMR spectrum [16.3–13]. For characterization purposes, the ^{13}C NMR line for C_{60} in benzene solution is at 142.7 ppm (parts per million frequency shift relative to tetramethylsilane) [see Fig. 16.1(a)] [16.4, 7]. In contrast, C_{70} in benzene shows five lines at 150.7, 148.1, 147.4, 145.4 and 130.9 ppm in a 1:2:1:2:1 intensity ratio (see Fig. 16.1(b)), indicative of the five different site symmetries on the C_{70} rugby ball structure shown in Fig. 3.5 and discussed in §3.2 [16.4]. NMR has also been useful for identifying the C_{76} molecule with a chiral structure (see §3.2), which explains the complicated NMR spectrum with 19 lines [16.14]. NMR techniques are also being applied to establish details about the shapes and chiralities of various higher-mass fullerenes C_{n_C} and the variety of isomers associated with each C_{n_C} molecule.

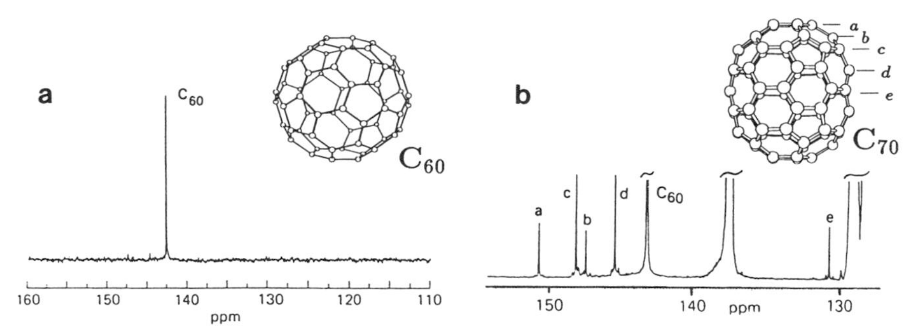

Fig. 16.1. (a) Room temperature NMR spectrum for C_{60} which has one unique carbon site. (b) NMR spectrum for C_{70} in which the five lines labeled a, b, c, d, e are identified with the five unique carbon sites in the C_{70} molecule (see Fig. 3.5) [16.2].

16.1.2. Bond Lengths

By determining the ^{13}C–^{13}C magnetic dipolar coupling in the NMR experiment, which depends on the inverse cube of the nearest-neighbor C–C distance a_{C-C} [16.15], the C–C bond distances in C_{60} have been measured [16.16]. Such measurements were made at 77 K on an isotopically enriched (6% ^{13}C) sample using a Carr–Purcell–Meiboom–Gill sequence to yield a doublet in the wings of the central line of the Fourier transform of the ^{13}C NMR signal [16.16]. The strong central line in this spectrum is due to ^{13}C nuclei with no nearest-neighbor ^{13}C nuclei, while the doublet lines in the wings are due to ^{13}C–^{13}C neighbors. The doublet structure shows that there are two types of C–C bonds in a C_{60} molecule, and the frequency of the two doublet peaks gives the C–C distance through the $(a_{C-C})^{-3}$ dependence of the magnetic dipolar coupling noted above, yielding a_{C-C} distances of 1.45 ± 0.075 Å and 1.40 ± 0.015 Å for neutral C_{60} [16.16]. Such experiments have been important in establishing the small difference in the bond lengths of the single and double bonds in the neutral C_{60} molecule (see Fig. 3.1). The differences in the bond lengths decrease for the doped C_{60} anions (see Fig. 8.2 in §8.1). From these bond lengths the diameter of the C_{60} molecule has been calculated to be 7.1 ± 0.07 Å [16.16].

16.1.3. Structural Information about Dopant Site Locations

NMR provides a powerful tool for site identification of the dopant species. For this use of the NMR technique, the NMR resonance is typically probed with respect to a convenient dopant nucleus. Because of the distinct local environment of the NMR resonant nucleus for each distinct crystalline site, measurements of the chemical shift (Fig. 16.2) of the ^{39}K NMR line in K_3C_{60} show that there are two different ^{39}K NMR lines corresponding to the two distinct chemical shifts (see Fig. 16.2) associated with each of the

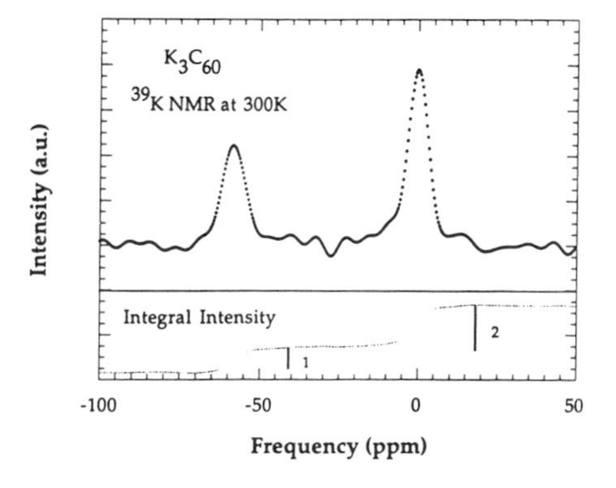

Fig. 16.2. The room temperature ^{39}K NMR spectrum of the K_3C_{60} superconductor referenced to KF in water solution, demonstrating the distinct chemical shifts for occupation of the tetrahedral and octahedral lattice sites in the superconducting K_3C_{60} compound. The integrated intensity plot indicates a 2:1 site occupation for the tetrahedral and octahedral sites [16.17].

two distinct tetrahedral and octahedral site locations of crystalline K_3C_{60} (see §8.5.1). Measurements of the integrated NMR line intensities of the two lines provide information about the probability of occupation of the distinct sites in the crystalline structure. Specifically, experimental measurements of the integrated intensities of the two NMR features in Fig. 16.2 show a 2:1 occupancy probability for the tetrahedral to octahedral sites in K_3C_{60} [16.17].

As another example of site specification, NMR has been used to show that specific species favor one of the distinct crystalline sites relative to other sites. For example, ^{133}Cs NMR resonance studies have been carried out on $CsRb_2C_{60}$ and Cs_2RbC_{60} [16.18] to show that since the Cs^+ ion is larger than the Rb^+ ion (see §8.5), the Cs^+ ion favors the larger octahedral site relative to the smaller tetrahedral site in certain cases. For $CsRb_2C_{60}$, a single broad ^{133}Cs NMR line is observed extending from –150 to –400 ppm frequency shift, indicative of a Cs^+ ion that is preferentially in an octahedral site, while for Cs_2RbC_{60} two ^{133}Cs NMR lines are observed, consistent with occupation of both tetrahedral and octahedral sites by the Cs^+ ion. Because of the large size of the Cs^+ ion, it expands and distorts the lattice when occupying a tetrahedral site. The distortions implied by NMR spectra for these compounds are consistent with those implied by their corresponding x-ray spectra [16.19]. Also related to this work are NMR spectra taken on Rb_3C_{60} and Rb_2CsC_{60} by examining the ^{87}Rb nucleus [16.20]. In this case three lines are observed: one for an Rb ion in an octahedral site, another for occupation of a tetrahedral site, and a third for an Rb ion in a distorted tetrahedral site [16.21].

NMR techniques have also been used to obtain information on the distinct site locations for the carbon atoms in doped C_{60}. In this case the NMR measurements are done on the ^{13}C nucleus and ^{13}C magic angle rotation spectra for Rb_3C_{60} show three distinct lines with intensities in ratios close to the expected values of 12:24:24, for the three distinct carbon sites on the C_{60}^{3-} ion in the $Fm\bar{3}m$ (fcc) structure of the doped Rb_3C_{60} (see Fig. 7.3 in §7.1.2) [16.22].

16.1.4. Molecular Dynamics in C_{60} and C_{70}

The NMR technique has been used to study the molecular dynamics of C_{60}, C_{70}, and doped fullerenes, by studying the linewidth and lineshapes of the NMR lines themselves and by studying the temperature dependence of the spin–lattice relaxation time T_1. At high temperatures, the molecules tend to rotate almost freely, giving rise to very narrow NMR linewidths (see Fig. 16.3) through the motional narrowing effect [16.23]. As T decreases, the molecular rotations become hindered (see §7.1) and the NMR lines

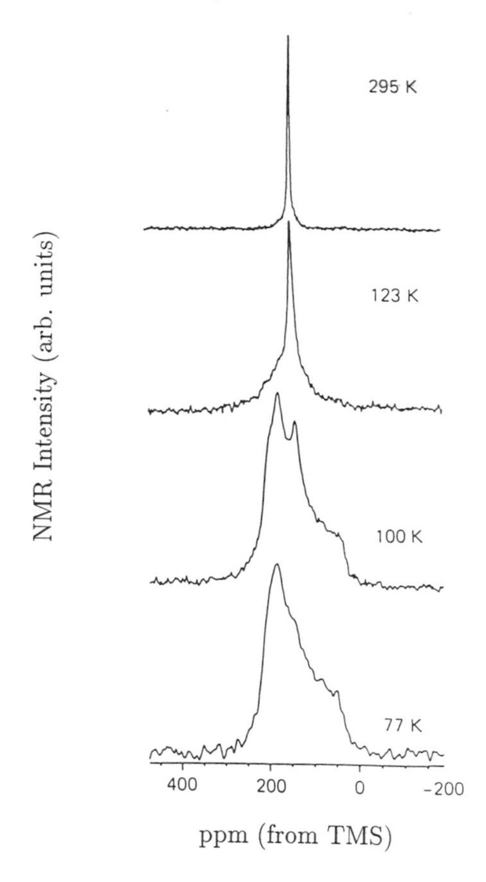

Fig. 16.3. Temperature variation of the ^{13}C NMR spectra of crystalline C_{60} taken at 15 MHz and the NMR line positions are measured relative to TMS (tetramethylsilane) [16.5].

broaden and develop asymmetric lineshapes as the nuclei probe specific sites, thereby providing information about the motions of the molecules containing the NMR nuclei.

In the gas phase the molecules rotate with a characteristic reorientation time of 3.1 ps and the corresponding value in solution (1,1,2,2-tetrachloroethane) at 283 K is 15.5 ps [16.2]. In the solid phase at 340 K, the reorientation time is 5 ps, similar to the values in the gas phase and in solution, but below T_{01}, the reorientation times in the solid phase become much slower ($\tau \sim 2$ ns at 250 K) because of the ratcheting motion of the molecules in the $Pa\bar{3}$ phase [16.2, 24–26].

The molecular dynamics of undoped C_{60} and C_{70} have been studied by several research groups by measuring the temperature-dependent broadening of the ^{13}C NMR lineshape and the temperature dependence of the spin–lattice relaxation time T_1 [16.5, 18, 24]. Because of the differences of the shapes of the C_{60} and C_{70} molecules, major differences in the molecular dynamics occur in their respective crystalline solids. A typical sequence of NMR spectral lines for C_{60} is shown in Fig. 16.3 at various temperatures, where a sharp NMR line is seen down to about 125 K, below which broadening due to the chemical shift anisotropy is observed. The corresponding temperature dependence of the NMR lineshapes for solid C_{70} is shown in Fig. 16.4 [16.18]. The full chemical shift anisotropy spectrum for both C_{60} and C_{70} is seen at 77 K and below. Observation of the full chemical shift is attributed to a slowing down of the molecular reorientation rate below that of the widths of the NMR spectral line [16.18]. In this limit the ^{13}C nuclei can experience a great variety of local sites during the slow molecular motions of the molecules to which they are attached. At intermediate temperatures (e.g., T between 100 K and 123 K in Fig. 16.3), the chemical shift anisotropy is associated with the ratcheting motion of the C_{60} molecule around the various threefold axes (see §7.1.3). However, for the case of a C_{70} single crystal (see Fig. 16.4), the intermediate temperature range occurs between 180 and 280 K, where the chemical shift anisotropy is associated with the molecular motion around the long fivefold axis of C_{70} (see §7.2) [16.6, 18].

Measurement of the temperature dependence of the spin–lattice relaxation time T_1 provides a sensitive probe of the dynamic motion in the lattice. For example, measurement of the temperature dependence of T_1 has been used to monitor the first-order structural phase transition occurring in solid C_{60} near 261 K (see §7.1.3 and §14.8) [16.24, 27]. The sharp discontinuity in T_1 associated with this phase transition is clearly seen in Fig. 16.5, where T_1 for a single-crystal C_{60} sample is plotted *vs.* $1/T$ on a semilog

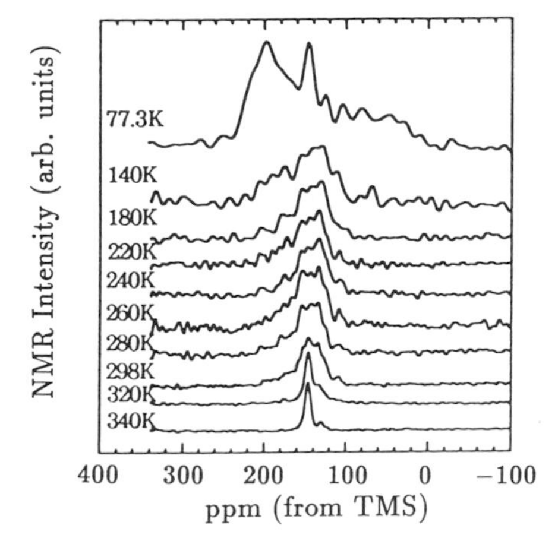

Fig. 16.4. The ^{13}C NMR spectra in a C_{70} crystal at various temperatures from 77 to 340 K and a frequency of 66.5085 MHz. The purity of the sample is more than 99% C_{70} for all temperatures except T = 77.3 K, where the sample includes a 6% C_{60} impurity. The NMR line positions are measured relative to that for TMS (tetramethylsilane) [16.18].

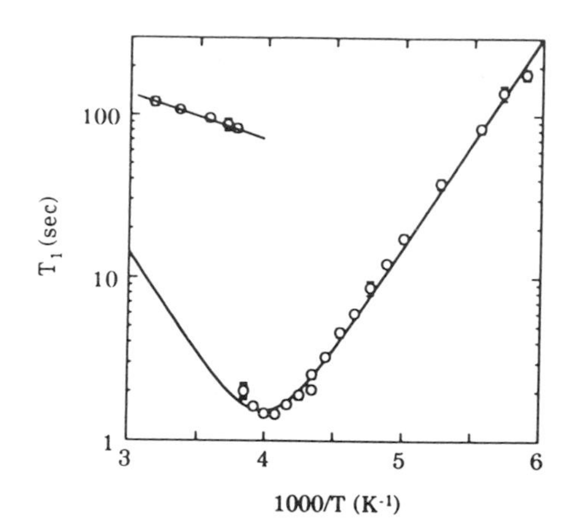

Fig. 16.5. Temperature dependence of the spin–lattice relaxation time T_1 for a crystalline C_{60} sample [16.18].

plot. The behavior of the $T_1(T)$ data near the structural phase transition temperature $T_{01} \sim 261$ K (see §7.1.3) provides a quantitative measure of T_{01} for highly crystalline samples, as well as sensitive characterization information about the crystalline microstructure of C_{60} samples [16.5, 18, 24, 28].

The spin–lattice relaxation time T_1 is described by the relation

$$\frac{1}{T_1} = \gamma_n^2 \langle h_m^2 \rangle \frac{2\tau}{1+\omega_0^2\tau^2} \tag{16.1}$$

where ω_0 is the nuclear Larmor frequency, γ_n is the nuclear gyromagnetic ratio, $\langle h_m^2 \rangle$ is the time average of the squared local field seen by the nucleus, and τ is the correlation time of the precessing nucleus which follows a thermal activation model

$$\tau = \tau_0 \exp(E_a/k_B T) \tag{16.2}$$

where E_a denotes an activation energy for this process. Because of the functional form of T_1 given in Eq. (16.1), it is easy to determine the value of τ at the temperature where T_1 is a minimum. For example, measurement of T_1 at a frequency of 75 MHz yields a minimum in T_1 at 233 K where $\tau \sim 1/\omega_0 \sim 10^{-9}$ s [16.5]. By varying the resonant frequency ω_0 (through variation of the externally applied field), the correlation time for the molecular motion can be determined as a function of temperature. The quantity $\langle h_m^2 \rangle$ is particularly sensitive to the axis of rotation of the molecule during the T_1 measurement, since h^2 is given by the relation

$$h^2 = \frac{3}{20} H_0^2 \Delta^2 (1+\eta^{2/3}) \tag{16.3}$$

where H_0 is the resonant magnetic field, Δ depends of the principal values σ_{ii} of the chemical shift tensor through $\Delta = 2[\sigma_{33} - (\sigma_{11} + \sigma_{22})/2]/3$, and $\eta = (\sigma_{11} - \sigma_{22})/\Delta$. In fact, the NMR lineshape shown in Fig. 16.4 for C_{70} is strongly dependent on the effects of the anisotropy of the chemical shift, which is presented schematically in Fig. 16.6 for the cases of a C_{70} molecule in isotropic rotation (high temperature), uniaxial rotation (intermediate temperature), and static orientation (low-temperature).

The changes in the local magnetic field due to the structural phase transition near T_{01} (see §7.1.3) are sensitively measured by the temperature dependence of T_1 as shown in Fig. 16.5. However, the molecular motion is too fast to provide information about this phase transition from the NMR linewidth itself. Detailed measurements of the temperature dependence of the decay of a saturation pulse below and above T_{01} have been made for C_{60} [16.18], yielding $E_a \sim 690$ K (59.5 meV) for temperatures $T > T_{01}$ and 2980 K (257 meV) for $T < T_{01}$, where T_{01} is the temperature of the phase transition near 261 K. The observed dependence of $1/T_1$ on the external magnetic field H goes as $1/T_1 \sim \tau H^2$ for $\tau < \omega_0^{-1}$ and $T_1 \sim \tau$

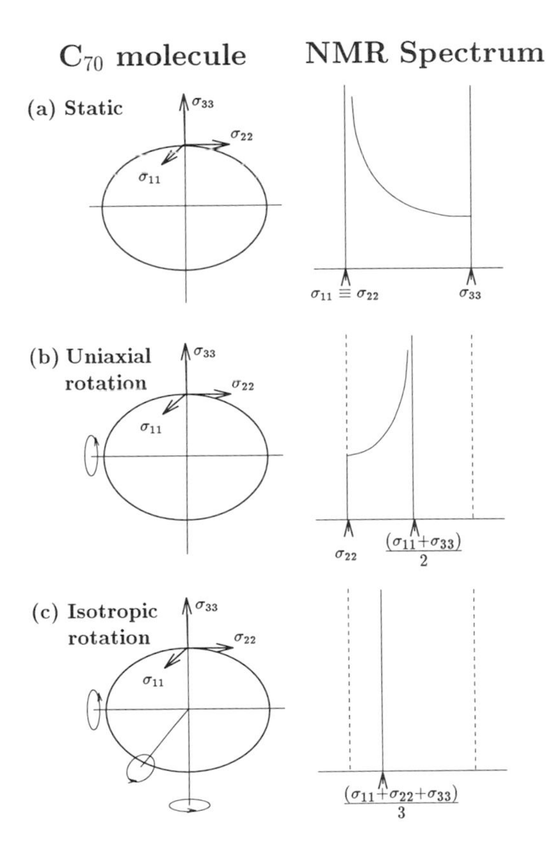

Fig. 16.6. Schematic illustration of the anisotropy of the chemical shift tensor σ_{ij}, leading to a qualitative understanding of the NMR lineshape change as a function of temperature in C_{70}. (a) When the molecule is static, one expects a fully broadened spectrum from the chemical shift anisotropy. (b) When the C_{70} molecule rotates about σ_{22} uniaxially, then σ_{11} and σ_{33} are expected to be averaged by rotation, while σ_{22} is almost unaffected by the uniaxial rotation. (c) When the molecule rotates isotropically, all the principal values of the chemical shift tensor should be averaged out to $(\sigma_{11} + \sigma_{22} + \sigma_{33})/3$ [16.18].

for $\tau > \omega_0^{-1}$ which is characteristic of the chemical shift anisotropy mechanism for the line broadening. In contrast, the behavior characteristic of the nuclear dipole-coupling mechanism for line broadening is not observed [16.28], namely $T_1 \sim \tau H^2$ at low-temperature for $\tau > \omega_0^{-1}$ and $T_1 \sim 1/\tau$ at high temperature for $\tau < \omega_0^{-1}$.

16.1.5. Stoichiometric Characterization of M_xC_{60} Phases

Through measurement of the chemical shift of ^{13}C, NMR spectral studies have been used to get detailed information about the stable phases of doped fullerides and about phase separation information more generally. Based on these findings, NMR spectra have been used to provide stoichiometric characterization for specific doped C_{60} samples. The NMR spectra for alkali metal–doped C_{60} show that K_3C_{60} and K_6C_{60} form stable phases, while $K_{1.5}C_{60}$ and K_2C_{60} do not [16.29], in good agreement with the phase diagram for K_xC_{60} shown in Fig. 8.5.

Specifically, the room temperature NMR spectra for C_{60} and K_3C_{60} show sharp single resonance lines at 143 ppm and 186 ppm, where the NMR shifts are expressed in parts per million frequency shifts for the ^{13}C nucleus with respect to the reference compound tetramethylsilane (TMS). The NMR spectra for samples with nominal stoichiometries $K_{1.5}C_{60}$ [Fig. 16.7(b)] and $K_{2.0}C_{60}$ [Fig. 16.7(c)] show the same two lines as for C_{60} and K_3C_{60}, which is interpreted as indicating that at the stoichiometries $K_{1.5}C_{60}$ and $K_{2.0}C_{60}$, phase separation occurs into the C_{60} and K_3C_{60} stable phases, and a determination of the stoichiometry is made by measuring the relative intensities of the NMR lines at 143 ppm for C_{60} [Fig. 16.7(a)] and at 186 ppm for K_3C_{60} [Fig. 16.7(d)]. More generally, the integrated intensities of the two peaks, as shown in Fig. 16.7, provide a characterization of the sample stoichiometry for K_xC_{60} samples with x in the range $0 \leq x \leq 3$ [16.6]. Figure 16.7(e), however, shows a broad NMR line for K_6C_{60} indicative of the chemical shift anisotropy of the ^{13}C NMR line due to the slow molecular reorientation of the C_{60}^{6-} ions in the body-centered cubic (bcc) K_6C_{60} structure. Thus the NMR spectra at room temperature cannot be easily used to provide stoichiometric information for K_xC_{60} compounds with $x > 3$. Likewise, structure in the K_3C_{60} line [see Fig. 16.7(d)] has been interpreted to indicate that the C_{60}^{3-} anions in K_3C_{60} are not freely rotating at room temperature but rather ratchet between symmetry-equivalent orientations, in accordance with structural studies (see §8.5). More detailed studies of the NMR lineshapes for K_3C_{60} and Rb_3C_{60} as a function of temperature [16.27, 30, 31] show motional narrowing of the ^{13}C NMR lines of Rb_3C_{60} and K_3C_{60} above 350 K and 296 K, respectively [16.1]. The NMR linewidths for these samples increase

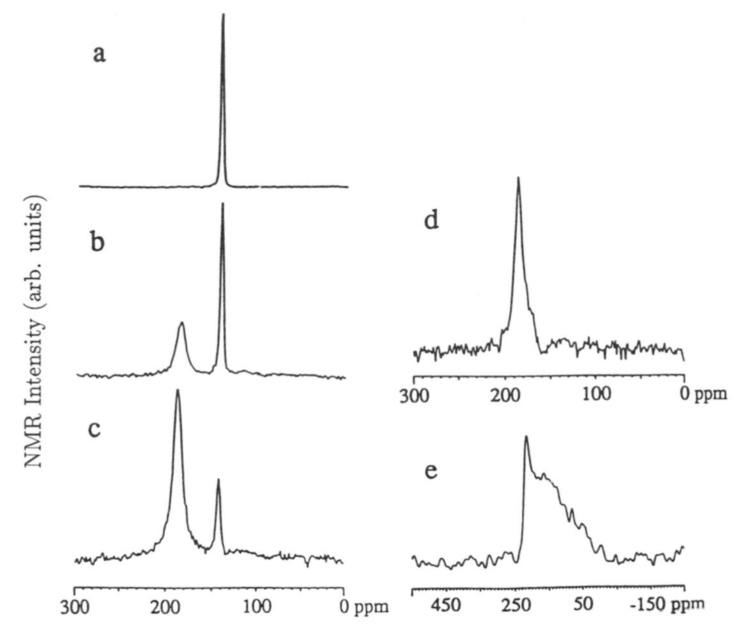

Fig. 16.7. Room temperature ^{13}C NMR spectra of C_{60} powder (a), and K_xC_{60} samples with nominal values of $x = 1.5$ (b), $x = 2.0$ (c), $x = 3.1$ (d), and $x = 6.0$ (e) [16.30].

gradually with decreasing T until just above the superconducting transition temperature T_c, where the linewidths are 80 ppm and 110 ppm for K_3C_{60} and Rb_3C_{60}, respectively. Additional broadening occurs below T_c due to the inhomogeneous local magnetic fields associated with magnetic vortices [16.1].

NMR measurements confirm that Rb_4C_{60} is not a metal because of the observed temperature dependence of T_1T, although the magnitude of T_1 is quite small, suggesting rapid spin fluctuations. On the other hand, T_1 for K_6C_{60} is much larger, and for this semiconductor, relaxation is believed to be due to structural defects. The magic angle spinning spectrum of K_6C_{60} is compatible with three inequivalent carbon sites (see Fig. 7.3) in the bcc structure [16.1].

The NMR spectrum has also been used to identify the M_1C_{60} rock salt phase (M = K, Rb) that is stable at high temperature [16.1, 32, 33]. ^{87}Rb and ^{133}Cs NMR spectra in Rb_1C_{60} and Cs_1C_{60} show a single sharp line in the high temperature phase, indicative of a nonconductive state although all the nearly free electron charge is transferred from the alkali metal to the C_{60}^- anions [16.33]. Analysis of the relaxation rate data for T_1 as a function of temperature for Rb_1C_{60} and Cs_1C_{60} indicates that for the M_1C_{60}

compounds, the dipolar contribution to the hyperfine interaction dominates over the contact term [16.33], as is also the case for K_3C_{60} and Rb_3C_{60}. Both Rb_1C_{60} and Cs_1C_{60} also have stable, noncubic low-temperature phases.

However, as the temperature of K_1C_{60} is reduced slowly below the stability point (425 K) for the high-temperature K_1C_{60} phase (see §8.5.2), phase separation into K_3C_{60} and C_{60} phases is clearly observed in the NMR spectrum [16.32]. However, when the sample was quickly quenched from the high-temperature C_{60} phase to a temperature below 355 K, a low-temperature K_1C_{60} phase was identified by NMR. Interpretation of the magic angle spinning NMR spectrum and a long spin–lattice relaxation time T_1 supports a polymerization of K_1C_{60} in the low-T phase, consistent with the interpretation of x-ray diffraction patterns. Many similarities are found between the NMR spectra of K_1C_{60}, Rb_1C_{60}, and Cs_1C_{60} [16.32, 33].

16.1.6. Knight Shift and the Metallic Phase of M_3C_{60}

The experimentally observed downshift of the ^{13}C NMR resonance by 43 ppm for K_3C_{60} relative to that for undoped C_{60} has been attributed to a Knight shift [16.23] identified with the hyperfine coupling between ^{13}C nuclei through the conduction electrons [16.6]. Since the observation of a Knight shift indicates the presence of conduction electrons, NMR measurements confirm the metallic behavior for K_3C_{60} and Rb_3C_{60}. Moreover, measurements of the temperature dependence of the spin–lattice relaxation time T_1 show a good fit to the Korringa relation $T_1T = \kappa$ [16.23], with $\kappa = 140$ and 69 K-s for K_3C_{60} and Rb_3C_{60}, respectively [16.6], for $T < 273$ K [16.6]. On the other hand, Knight shift experiments on the ^{39}K NMR line revealed no Knight shift for K_3C_{60}, consistent with a model of complete charge transfer from the potassium to the C_{60}^{3-} ion [16.34]. ^{13}C NMR results similar to those found for K_3C_{60} are also obtained for $CsRb_2C_{60}$ and Cs_2RbC_{60}, again providing evidence that these materials are metallic [16.18].

Study of the Knight shift in a superconductor also provides information about the superconducting state. As T falls below T_c, the density of excited quasiparticles decreases and approaches zero as $T \to 0$. Measurements of the functional form of the temperature dependence of the Knight shift $K(T)$ thus have been used to provide information on the magnitude of the superconducting gap $2\Delta/k_BT_c$ [16.23]. Such Knight shift experiments have been carried out for superconducting Rb_3C_{60} and Rb_2CsC_{60} [16.20] (see §16.1.8).

In contrast to the alkali metal intercalation of M_3C_{60}, which shows a Knight shift in the ^{13}C NMR line, the acceptor intercalation of iodine into C_{60} to yield a stoichiometry $I_{2.28}C_{60}$ shows no chemical shift [16.34]. Furthermore, no metallic relaxation, as would appear in the Korringa relation [16.35], is reported for $I_{2.28}C_{60}$, thereby supporting the absence of charge transfer in the iodine-doped fulleride (see §8.7.1).

16.1.7. *Density of States Determination by NMR*

Early studies of the spin–lattice relaxation in K_3C_{60} and Rb_3C_{60} made use of the Korringa relation $T_1T = \kappa$ where κ is related to the spin–lattice relaxation time T_1 by the hyperfine contact interaction [16.36]

$$\frac{1}{\kappa} = \frac{64}{9}\pi^3 k_B \hbar^3 \gamma_e^3 \gamma_n^3 |\psi(0)|^4 [N(E_F)]^2 \tag{16.4}$$

in which γ_e and γ_n are the gyromagnetic ratios of the electron and the nucleus. In this work, measurement of T_1T was used to show that these fullerides were indeed metals, and the results for T_1T were used to determine the electronic density of states $N(E_F)$, making appropriate assumptions for the electronic wave function at the position of the nucleus $|\psi(0)|$.

A simple analysis of these data yielded a large value for the density of states of 34 states/molecule/eV for K_3C_{60} [16.6, 27, 37] and a value of about 44 states/molecule/eV for Rb_3C_{60}, the higher density of states being consistent with the higher observed T_c in Rb_3C_{60} [16.37]. Subsequent work has, however, shown that the dipolar term in the hyperfine interaction is much larger than the contact term [Eq. (16.4)] [16.38], thereby indicating that the previously published values for $N(E_F)$ from analysis of NMR data were much too large. Other interactions complicating the interpretation of the NMR spin lattice relaxation rate data include effects of electron–electron interaction and merohedral disorder [16.1, 18, 37].

Although the use of measurements of the T dependence of T_1 may not yield reliable values of $N(E_F)$ (see §14.5), the trends in the values of $N(E_F)$ that have been reported for K_3C_{60}, Rb_3C_{60}, $CsRb_2C_{60}$, and Cs_2RbC_{60} [16.18] based on NMR studies are consistent with an increase in $N(E_F)$ as the lattice constant increases (see §15.1).

16.1.8. *Determination of Superconducting Gap by NMR*

Since the spin–lattice relaxation rate $(1/T_1)$ in a superconductor exhibits a temperature dependence

$$T_1^{-1} = W_A \exp\left(-\frac{\Delta}{T}\right), \tag{16.5}$$

where W_A denotes the spin–lattice relaxation rate at high T, the temperature dependence of $(1/T_1)$ can be used to determine the superconducting gap 2Δ.

Referring to Fig. 16.8, we see that the product $(T_1T)^{-1}$ is nearly constant above T_c [16.27, 37] for both K_3C_{60} and Rb_3C_{60}. However, below T_c

the magnitude of the product $(T_1T)^{-1}$ falls rapidly [16.27] with decreasing temperature, signifying the onset of superconducting behavior. The corresponding temperature dependence of T_1 is shown in the Arrhenius plots of Fig. 16.8(b) and (d) for K_3C_{60} and Rb_3C_{60}, respectively. From these plots, the values for $(2\Delta/k_BT_c) = 3.5$, consistent with Bardeen–Cooper–Schrieffer (BCS) theory, are obtained. Also contributing to T_1 are spin diffusion and vortex motion in the superconducting phase.

It should be noted that Fig. 16.8(a) and (c) do not exhibit any Hebel–Slichter peak in the $(T_1T)^{-1}$ *vs.* T plot just below T_c, as would be expected for an ideal BCS superconductor [16.39, 40]. Recent reports have, however, shown direct evidence for the Hebel–Slichter peak in NMR measurements at lower magnetic fields and the suppression of this peak with increasing magnetic field [16.41].

Another feature of importance to the NMR spectra in the superconducting phase is the sharp decrease in the resonant frequency of the ^{13}C line below T_c as shown in Fig. 16.9. The decrease in resonant frequency ω_0 may be in part due to a sharp decrease in the Knight shift below T_c as the

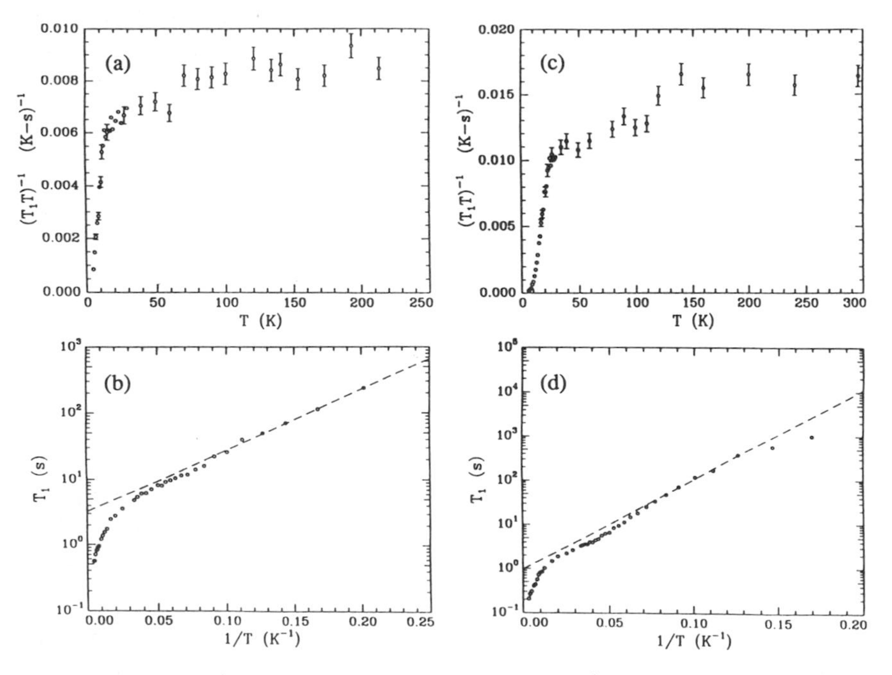

Fig. 16.8. Shown in (a) and (c) are the plots of $(T_1T)^{-1}$ *vs.* T for K_3C_{60} and Rb_3C_{60} respectively [16.27], where T_1 is the spin-lattice relaxation time for these samples. Arrhenius plot for the ^{13}C spin–lattice relaxation rate T_1 in (b) K_3C_{60} and (d) Rb_3C_{60}, demonstrating the temperature dependence, $(1/T_1) \approx \exp(-\Delta/T)$, where Δ is the superconducting energy gap.

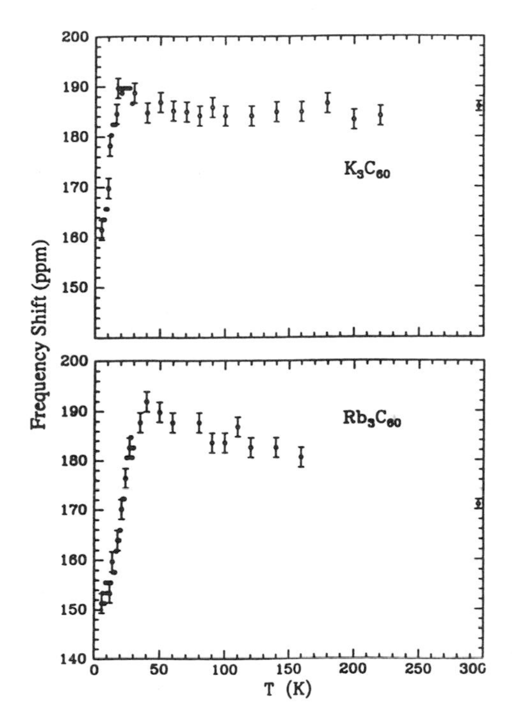

Fig. 16.9. Temperature dependence of the ^{13}C NMR frequency in K_3C_{60} and Rb_3C_{60}, reported as the frequency shift relative to a tetramethylsilane reference signal [16.27].

density of excited conduction electrons decreases. The decrease in resonant frequency may also be due to the formation of an internal magnetic field associated with the diamagnetism of the sample below T_c [16.27].

A more detailed investigation of the superconducting gap was made in the Rb_3C_{60} and Rb_2CsC_{60} systems by measuring the Knight shift for different NMR nuclei in the same sample [16.20, 41]. In this work the frequency shifts for the ^{13}C, ^{87}Rb, and ^{133}Cs nuclei were all measured for Rb_2CsC_{60} as T was decreased below T_c. The results for the Knight shift were analyzed in terms of three contributions for each NMR nucleus i

$$K_i(T) = K_i^{\text{spin}}(T) + K_i^{\text{orb}} + \Delta H(T)/H \tag{16.6}$$

where $K_i^{\text{spin}}(T)$ is identified with the Knight shift which is proportional to the electronic Pauli spin susceptibility $\chi^{\text{spin}}(T)$ associated with the conduction electrons (see §16.1.6), K_i^{orb} is attributed to the temperature-independent orbital or chemical shift, and $\Delta H(T)/H$ is the fractional change in local magnetic field in the sample which results from Meissner screening currents and can be used to determine the superconducting penetration depth λ_L [16.20]. By taking the difference in $K_i(T)$ for two different NMR nuclear species, the temperature-dependent $K_i^{\text{spin}}(T)$ can

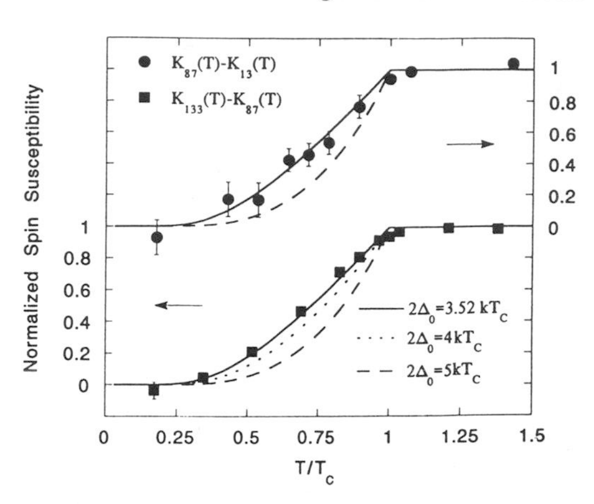

Fig. 16.10. NMR Knight shift differences ($K_{87} - K_{13}$) for Rb_3C_{60} (filled circles) and ($K_{133} - K_{87}$) for Rb_2CsC_{60} (filled squares) *vs.* T/T_c. Also shown are BCS spin susceptibility predictions for $2\Delta/k_BT_c = 3.52$ (solid line), 4 (dotted line), and 5 (dashed line) [16.20].

be determined [16.41], since the terms in K_i^{orb} and $\Delta H/H$ in Eq. (16.6) cancel upon subtraction. Such measurements of the frequency shifts K_i for two nuclei have been carried out for the ^{13}C and ^{87}Rb nuclei in Rb_3C_{60} and for the ^{87}Rb and ^{133}Cs nuclei in Rb_2CsC_{60} [16.20, 41]. The resulting analysis of the temperature dependence of the difference between the K_i for ^{13}C and ^{87}Rb in Rb_3C_{60} and for ^{87}Rb and ^{133}Cs in Rb_2CsC_{60} is shown in Fig. 16.10. The results for $(K_{87} - K_{133})$ were found to be quite insensitive to the magnetic field. The resulting plots of the normalized spin susceptibility *vs.* T/T_c (see Fig. 16.10) show conventional Knight shift behavior and provide a good fit to the weak coupling BCS theory for $2\Delta = 3.52k_BT_c$. The sensitivity of the analysis to the determination of the superconducting gap is seen in Fig. 16.10, where the fit for $2\Delta/k_BT_c = 4$ is clearly outside the experimental error for the Rb_2CsC_{60} data, while the fit to $2\Delta/k_BT_c = 5$ is far outside the experimental error. Other methods for determining $2\Delta/k_BT_c$ for superconducting fullerenes are discussed in §15.4.

From the broadening of the NMR lineshape as T goes through T_c from above, estimates can be made of the London magnetic field penetration depth λ_L. Such an analysis for K_3C_{60} and Rb_3C_{60}, respectively, leads to estimates of $\lambda_L = 6000$ Å and 4600 Å for K_3C_{60} and Rb_3C_{60}, respectively [16.27]. These values for λ_L are in good agreement with those found in the μSR experiments [16.42, 43].

16.1.9. NMR Studies of Magnetic Fullerenes

The magnetic fullerene compound TDAE–C_{60} (see §8.7.3 for structure and §18.5.2 for magnetic properties) has been investigated using NMR techniques, studying both the ^{13}C NMR line associated with C_{60} and the proton line associated with TDAE. In fact, two ^{13}C NMR lines (A and B) have been identified (one with C_{60} and one with TDAE) and two proton lines (A and B) associated with two hydrogen sites in TDAE when attached to C_{60} [16.44, 45].

For both ^{13}C and ^{1}H, the position of the NMR B lines is almost independent of temperature, while the A lines in both cases show a strong paramagnetic shift upon approaching T_c from above, with the magnitude of the ^{13}C paramagnetic shift one order of magnitude smaller than that for the ^{1}H shift [16.44–47]. The T_1 value for the ^{13}C A line is $\sim$25 ms between 25 K and 300 K while T_1 for the proton A line is $\sim$1 ms. The B lines for ^{13}C and ^{1}H both show-temperature dependences, and these dependences are quite different from one another [16.45]. The dynamics of TDAE–C_{60} has been studied via the chemical shift and linewidth ΔH_{pp} of the ^{13}C NMR line [16.44]. The results for $\Delta H_{pp}(T)$ show motional narrowing above 180 K and an increased linewidth below, indicative of the freezing of the rotation of the molecules, thereby yielding a value of $T_{01} \sim 180$ K [16.44], in good agreement with $T_{01} \sim 170$ K obtained by other experiments (see §18.5.2). All four NMR lines show inhomogeneous broadening effects at low-temperature. The same Curie temperature behavior is observed for all NMR linewidths upon lowering the temperature. The recovery of the magnetization after exposure to an inverting pulse shows a simple exponential time dependence at high field and $T > T_{01}$. However, the recovery time follows a stretched exponential for $T < T_{01}$, while at low fields (0.3 T), a stretched exponential recovery is observed for all temperatures, consistent with a spin glass model.

16.2. Electron Paramagnetic Resonance

Electron paramagnetic resonance (EPR) and electron spin resonance (ESR) techniques have also yielded valuable information about the structure and electronic properties of fullerenes. In this section, we review the information these techniques have provided for our understanding of fullerenes and related compounds. The fullerene systems that have been studied by EPR include endohedrally doped fullerenes (§16.2.1), C_{60}-related anions (§16.2.2), and doped fullerenes (§16.2.4).

Electron paramagnetic resonance sensitively measures the hyperfine interaction between the nuclear moment of a nucleus and the magnetic field at the nucleus arising from nearby electrons [16.48]. The hyperfine interaction is described by a Hamiltonian

$$\mathcal{H}_{\text{hpf}} = -\mu_I \cdot H_I \sim \gamma\hbar\mu_B|\psi(0)|^2\mathbf{I}\cdot\mathbf{S} = A\mathbf{I}\cdot\mathbf{S}, \tag{16.7}$$

in which μ_I is the magnetic moment of the nucleus, H_I is the magnetic field at the position of the nucleus, γ is the gyromagnetic ratio, μ_B is the

Bohr magneton, $\psi(0)$ is the electron wave function at the position of the nucleus, and **I**, **S** are the nuclear and electron spin vectors, respectively. In the case of fullerenes and doped fullerenes the electron in the hyperfine interaction is associated with the fullerene molecules and the interacting nuclei may bc associated with the dopant. The parameters of interest are the number of hyperfine lines $(2I+1)$, their separation in energy (A), their linewidth (ΔH), and the g-value for the electron that is interacting with the nucleus. For the case of electron spin resonance, the magnetic field–induced transitions are between the spin-up and spin-down states for $S = 1/2$.

16.2.1. EPR Studies of Endohedrally Doped Fullerenes

Of special interest have been the EPR studies of endohedrally doped fullerenes where the spin on the endohedral nucleus is coupled to an electron on the surrounding fullerene shell by the hyperfine interaction [Eq. (16.7)]. EPR spectra have been identified for several endohedrally doped fullerenes (see §5.5 and §8.4.1), and the EPR technique has even been utilized as a real-time characterization method for the extraction and purification of endohedral fullerenes [16.49, 50]. The most widely studied metallofullerene is La@C_{82} [16.51–54], for which EPR spectra are shown in Fig. 16.11 [16.51], both in the solid state [Fig. 16.11(a)] and in solution [Fig. 16.11(b)]. From the solution spectrum, eight hyperfine lines are easily identified, consistent with a spin of 7/2 on the endohedral ^{139}La nucleus. The coupling between the electron spin and the nuclear spin in endohedral

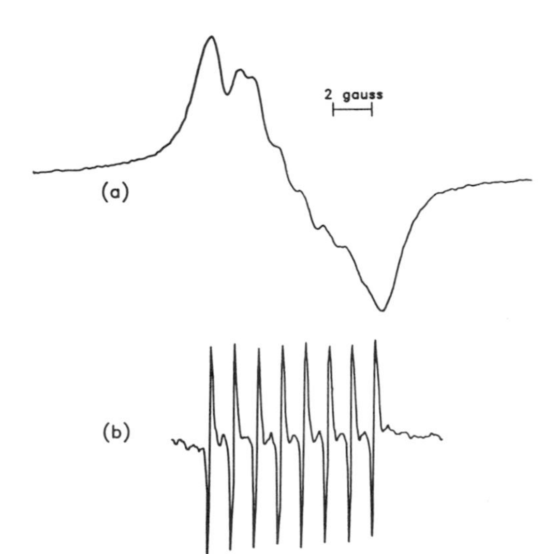

Fig. 16.11. EPR hyperfine spectra of endohedrally doped La@C_{82} mixed with C_{60} and C_{70} (a) in the solid state (powder) and (b) in 1,1,2,2-tetrachloroethane solution. The spectra are taken at 9.112 GHz and at ambient temperature [16.51].

fullerenes is weak compared with typical hyperfine spectra [16.48]. In the solid, crystal field effects at room temperature are not averaged out by the molecular motion of the fullerene, so a broad hyperfine line appears [see Fig. 16.11(a)]. This is in contrast with the solution spectrum, where the molecular motion gives rise to motional narrowing [16.48], and sharp spectral features for each of the $(2I + 1)$ components are observed. The large peaks in Fig. 16.11(b) are due to $\mathscr{H}_{hpf}$ associated with the coupling of the ^{139}La nuclear spin to an electron on the surrounding C_{60} shell, while the small peaks between the large peaks are identified with interactions between La guest ions and ^{13}C nuclei on the C_{60} shell. The model emerging from these measurements is that the La^{3+} ion (with a nuclear spin of 7/2) is stable within the fullerene shell, and the electrons are transferred to the LUMO level of the molecular ion [16.51]. More detailed studies have shown EPR octet spectra for two different isomers of $La@C_{82}$ [16.54] (see Table 16.1). Other octet EPR spectra have also been identified for La in other fullerene shells such as $La@C_{76}$ and $La@C_{84}$.

In addition, EPR spectra have been identified for Sc-endohedral metallofullerenes where the nuclear spin of ^{45}Sc (100% natural abundance) is also 7/2. The EPR spectrum for $Sc@C_{82}$ shows eight narrow lines, while that for $Sc_3@C_{82}$ shows 22 hyperfine lines (see Fig. 16.12). These 22 lines are identified with three interacting Sc nuclei and correspond to a maximum total nuclear spin of $3I = 21/2$ [16.55–57]. Values for the hyperfine parameters (the hyperfine coupling coefficient A, the electron g-value, and the linewidth ΔH of each hyperfine line) for several La and Sc metallofullerenes are given in Table 16.1. The magnitudes of the coupling coef-

Table 16.1

EPR hyperfine parameters for several metallofullerenes [16.54].

Species	No. lines	A (G)	g-value	ΔH (G)
$La@C_{82}{}^{a}$	8	1.20	2.0008	0.049
$La@C_{82}{}^{b}$	8	0.83	2.0002	0.052
$La@C_{76}$	8	0.44	2.0043	0.061
$Sc@C_{82}$	8	3.83	1.9998	0.45
$Sc_3@C_{82}$	22	6.25	1.9986	0.78
$Y@C_{82}{}^{c}$	2	0.48	2.0003	0.10
$Y@C_{82}{}^{d}$	2	0.36	1.9998	0.11

[a,b]Two distinct octet EPR hyperfine spectra are observed and are identified with two isomers of C_{82}. The metallofullerene (a) is more stable than (b).

[c]Higher-intensity doublet of the $Y@C_{82}$ spectrum [16.53].

[d]Lower-intensity doublet of the $Y@C_{82}$ spectrum [16.53].

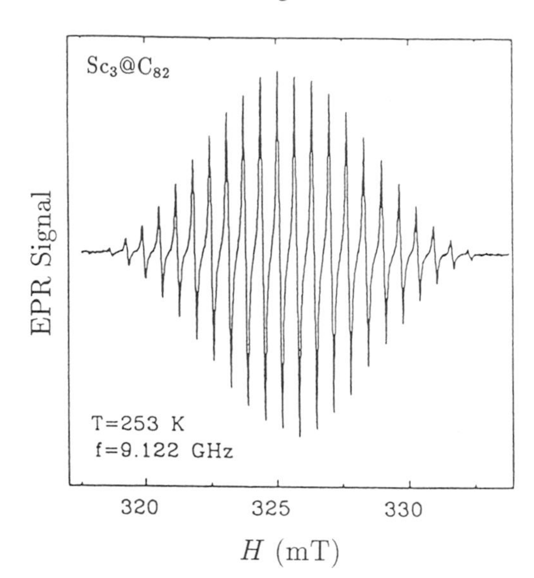

Fig. 16.12. EPR spectrum at $T = 253$ K of a highly purified endohedrally doped $Sc_3@C_{82}$ sample dissolved in toluene. The 22 lines observed in this spectrum are due to the hyperfine coupling to three equivalent Sc ($I = 7/2$) ions in the C_{82} cage [16.55].

ficients A tend to be small for all the metallofullerenes in the table. The linewidth for the EPR line for $Sc_3@C_{82}$ is exceptionally large and is temperature dependent because of the internal motion of the three equivalent Sc ions within the cage. Furthermore, not all metallofullerenes yield EPR spectra, $Sc_2@C_{82}$ and $Sc_2@C_{84}$ being examples of diamagnetic EPR-silent metallofullerenes, where the two nuclear spins are paired to give a total $I = 0$. When multiple endohedral metal ions are located within a single fullerene cage, they are believed to occupy off-central positions [16.54] (see §8.2). The off-central endohedral ions are therefore closer to the carbon atoms in the fullerene cage and thus the endohedral nuclei would be expected to interact more strongly with an electron on the fullerene cage. The EPR studies of the endohedral fullerenes indicate that three electrons are transferred to the fullerene cage, whether one or multiple metallic species are inside the fullerene cage [16.54]. The sharp lines in Fig. 16.11(b) suggest a weak coupling between the La nucleus and the s-electrons on the C_{82} fullerene. The EPR results support the endohedral location of the La^{3+} ion inside the C_{82} shell, but not necessarily at the center of the C_{82} shell.

EPR spectra for other families of metallofullerenes have been reported, such as for $Y@C_{82}$ [16.53, 58, 59] and $Er@C_{82}$ [16.56]. For $Y@C_{82}$, the ESR spectra [16.53] are more complex, showing a pair of doublets as dominant features of the spectra, indicative of $I = 1/2$ for the ^{89}Y nucleus, which has a 100% natural abundance. Although several tentative explanations have been offered for the pair of doublets, a detailed explanation of the origin of the pair of doublet lines with a 3:1 intensity ratio remains an open issue [16.53]. Included in Table 16.1 are the EPR hyperfine parameters for the higher- and lower-intensity doublets of the Y@C_{82} EPR spectrum.

Preliminary EPR results for Er@C_{82} have been reported, with $g \simeq 2.008$, $\Delta H_{pp} = 1.75$ G for the solution EPR spectrum and $g = 2.005$, $\Delta H_{pp} = 1.5$ G for the EPR spectrum in solid extracts for the ^{167}Er ($I = 7/2$) species, which has a natural abundance of ~20%. A broad EPR line associated with crystal field effects has also been observed for the Er@C_{82} metallofullerene with an effective g-factor of 8.5; the broadening of this line was attributed to interactions between the Er $4f$ electrons and the fullerene cage. In the ground state, a free Er^{3+} ion has a $^4I_{15/2}$ ground state and a spin of 3/2. The spin–spin interactions of the Er $4f$ electrons with an electron on the fullerene cage leads to a broad EPR line (1.5 G at 300 K, 9.3 GHz) and to the lack of observable hyperfine splitting in the EPR spectrum [16.56].

16.2.2. EPR Studies of Neutral C_{60} and Doped C_{60}

Electron paramagnetic resonance (EPR) signals have been observed for neutral C_{60}, for C_{60} anions, and for doped C_{60}. For example, EPR spectra for the C_{60}^-, C_{60}^{3-} and C_{60}^{5-} radical ions have been reported with g-factors of $g^{1-} = 2.00062$, $g^{3-} = 2.0009$, and $g^{5-} = 2.0012$, respectively [16.60–62], and many papers have been published on the EPR spectra of C_{60}^{1-}, C_{60}^{2-}, and C_{60}^{3-} anions, as discussed below.

Neutral C_{60} molecules have no unpaired spins and therefore do not give rise to an EPR signal [16.63]. However, EPR signals can be observed from a nominally neutral C_{60} sample because of the presence of C_{60} anions or cation impurities, such as an ionized C_{60}^+ or a molecule with a bound electron C_{60}^- [16.60, 63–65], or from a C_{60} molecule in an optically excited metastable triplet state (see §13.1.2 and §16.2.3). The large spatial separation between the C_{60} molecules in a crystal (or in solution) implies that the coupling between fullerenes is weak and that the spin excitations in the EPR experiment arise from an electron spin localized on a single C_{60} molecule. EPR signals appearing with g-factors greater than the free electron value of $g_0 = 2.0023$ are associated with bound holes and are most commonly observed in the EPR spectrum of nominally pure C_{60} (see Fig. 16.13), although EPR signals with $g < g_0$ are also observed and are attributed to bound electrons. For both electrons and holes, the EPR lines for undoped C_{60} are narrow (see Fig. 16.13), which implies that the hole (or the electron) is delocalized on all the carbon atoms of the C_{60} cage, so that the contributions to the EPR line are dynamically averaged over all the site locations on the C_{60} ion. These EPR lines show a very weak temperature

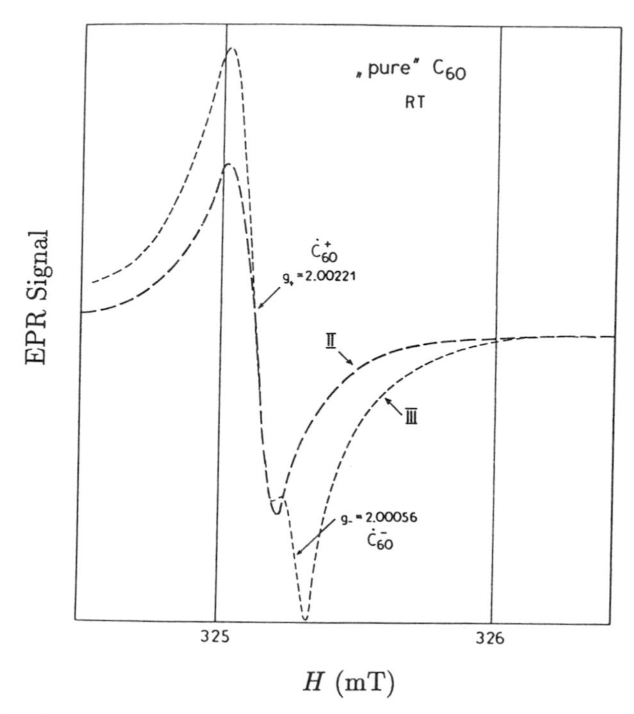

Fig. 16.13. One-component (denoted II) and two-component (denoted III) EPR room temperature spectra of pristine C_{60} fullerene samples. (II and III denote different samples of C_{60}.) The spectra indicate frequent occurrence of hole states C_{60}^{+} with $g^{+} = 2.00221$ and rare occurrence of electron states C_{60}^{-} with $g^{-} = 2.00056$ in nominally pure samples of C_{60} [16.66].

dependence, and the g-factor for an electron on a C_{60} surface is expected to be similar to $g_{\perp}$ for graphite ($g_{\parallel} = g_a = g_b = 2.0026$, $g_{\perp} = g_c = 2.050$) [16.67]. Both electron and hole EPR signals are bleached (made to disappear) when the sample is exposed to ultraviolet (UV) light and reappear when the UV irradiation is stopped [16.66].

Most of the EPR spectra on C_{60} anions have been taken on electrochemically prepared and characterized samples [16.68]. Most of the effort has gone into studies of the temperature dependence of the EPR spectra for the C_{60}^{-}, C_{60}^{2-}, and C_{60}^{3-} anions. An example of EPR spectra for the C_{60}^{-} anion, which has been most widely studied [16.69–73], is shown in Fig. 16.14. The effect of trapping an electron to form C_{60}^{-} produces a small distortion to the molecule, which lowers its energy by ~0.76 eV, making C_{60}^{-} a stable species in a crystalline C_{60} sample [16.74]. The low-temperature spectra show two components to the EPR line with different temperature dependences of their relaxation rates [16.68]. Analysis of the EPR line in various

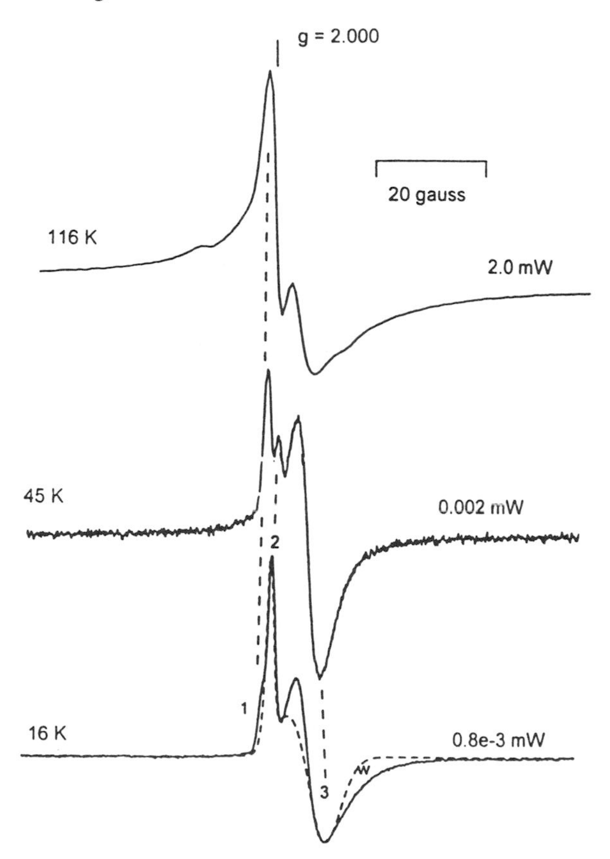

Fig. 16.14. EPR spectra of C_{60} in frozen toluene-acetonitrile solution for several values of the temperature and microwave power at 9.3898 GHz (116 K) or 9.2339 GHz (16 and 45 K). Spectra taken at different frequencies were shifted horizontally to align corresponding g values (see vertical dashed lines). The microwave power is shown to the right of each spectrum. The simulated spectrum (- - -) was calculated for anisotropic values of the g-factor $g_\perp = 1.9937$ and $g_\parallel = 1.9987$ and for the anisotropic linewidths $\sigma_\perp = 6.2$ G and $\sigma_\parallel = 1.0$ G [16.68].

solvents in the temperature range 77–210 K yields a g-factor of 1.998 and spin 1/2 [16.75]. Anisotropy is observed in the EPR spectra of C_{60}^- where the anisotropic g-factors are $g_\perp = 1.9937$ and $g_\parallel = 1.9987$ and the anisotropic linewidths are $\sigma_\perp = 6.2$ G and $\sigma_\parallel = 1.0$ G [16.68]. Anisotropies in the EPR lineshape have been attributed to a Jahn–Teller distortion of the molecule, arising from the lifting of the degeneracy of the ground state [16.76, 77]. Referring to Table 12.2, a ground state with Γ_6^- symmetry is expected under Jahn–Teller distortion and this is supported by optical spectra for C_{60}^- in solution [16.78] (see §13.4.1).

The EPR spectra reported for electrochemically prepared C_{60}^{2-} [16.75, 79–83] are rather different from the corresponding spectra for C_{60}^{-} and have been interpreted in terms of a triplet $S = 1$ ground state (consistent with Hund's rule as shown in Table 12.2), a g-value of 2.00, and a separation of 12.6 Å between the two unpaired electron spins [16.75]. The optical spectra for C_{60}^{2-} in solution [16.78] are also consistent with a triplet spin ground state (see §13.4.1). Other authors have, however, concluded that, because of the dynamic Jahn–Teller effect, the C_{60}^{2-} anion should be diamagnetic, and that the reported EPR spectra should be identified with impurities [16.68, 70].

The spectrum of C_{60}^{3-} is shown in Fig. 16.15 and differs significantly from that shown in Fig. 16.14 for C_{60}^{-} [16.75, 80–83]. The EPR spectra for electrochemically prepared C_{60}^{3-} show a sharp line with $g = 2.000$ (attributed to other species), superimposed on a broad anisotropic signal (identified with C_{60}^{3-}) $g_{\perp} = 1.997$ and $g_{\parallel} = 2.008$ and a spin of 1/2 [16.68]. The broad component of the spectra in Fig. 16.15 is believed to be intrinsic to C_{60}^{3-}, while the sharp component, which is less temperature dependent, is believed to be due to other species. An increase in EPR linewidth with increasing temperature due to spin–lattice relaxation has been reported by several groups [16.68, 75, 80, 81]. The anisotropic g-factor in the C_{60}^{3-} spectrum is consistent with molecular distortions arising from a dynamic Jahn–Teller effect, which is also supported by zero-field splitting effects observed for C_{60}^{3-} [16.68]. Hund's rule arguments (Table 12.2) imply a fourfold degenerate spin ground state, but this does not seem consistent with either the EPR spectra or optical spectra (see §13.4.1) which have been interpreted in terms of a Jahn–Teller symmetry-lowering distortion and a single unpaired electron.

The EPR spectra observed for the alkali-doped fullerides show rather different properties (see Fig. 16.16) relative to the undoped case. First, the EPR lines in K_3C_{60} and Rb_3C_{60} are much broader than for the undoped case (see Fig. 16.16), and the line broadening is attributed to the localization of the unpaired electrons by the $[M_x^+C_{60}^{3-}]$ complex. The characteristics of the EPR lines in doped samples show a high sensitivity to the alkali metal concentration x and a strong temperature dependence [16.66]. For example, at room temperature the g-factor is 2.0017 for a K_3C_{60} sample, while at 4 K, $g = 2.0008$. The proximity of the g-value to 2 has been interpreted to imply that the orbital state is nondegenerate [16.70], which is consistent with optical spectra for C_{60}^{3-} anions in solution. The inverse intensity increases linearly with T, indicating that the EPR transition is associated with the ground state [16.70]. Also the dominant EPR linewidth decreases by an order of magnitude over a similar temperature range (see Fig. 16.17). Room temperature EPR spectra for K_3C_{60} show a single line with a g-value of 2.0002 (as compared with the free electron value of 2.0023). Since the

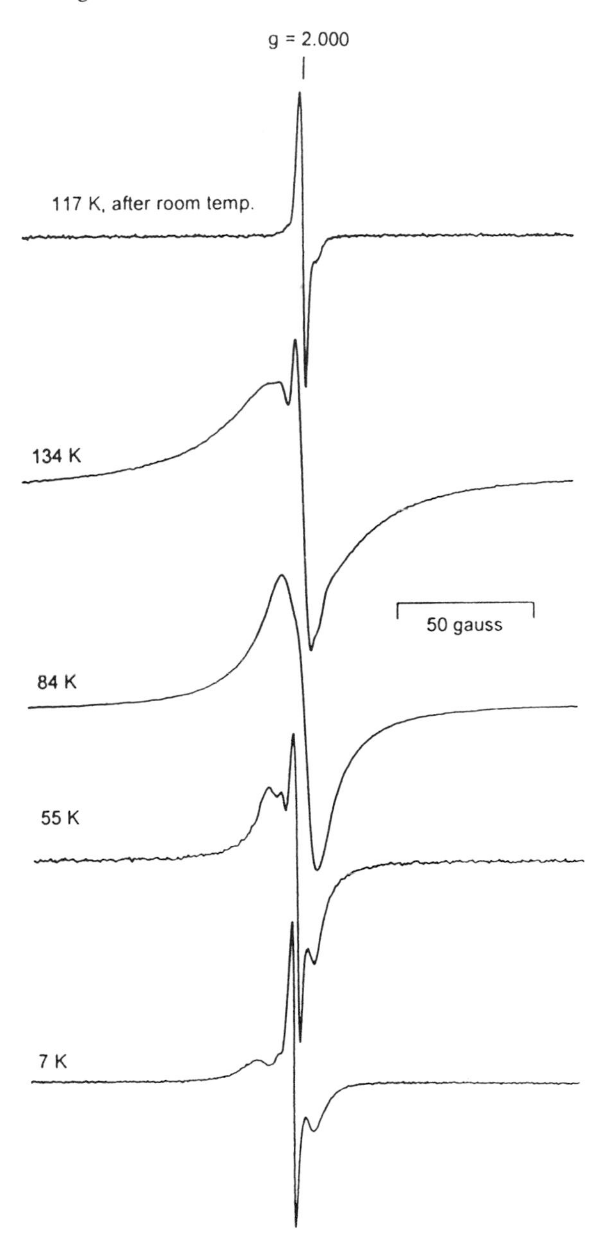

Fig. 16.15. Temperature dependence of the EPR spectra for C_{60}^{3-} in frozen toluene–acetonitrile solution at various temperatures and at a frequency of 9.3895 GHz. Also shown is an EPR spectrum for a sample frozen at 117 K and left to stand at room temperature for about 15 min and then refrozen (top trace). Spectra taken at different frequencies were shifted horizontally to align corresponding g-values [16.68].

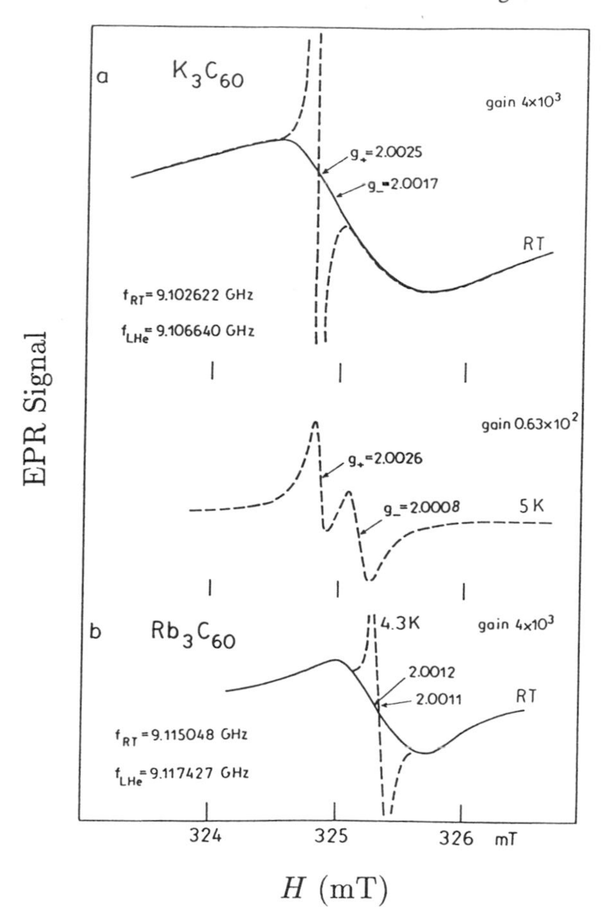

Fig. 16.16. EPR spectra of (a) K_3C_{60} at room temperature (RT) and at 5 K and (b) Rb_3C_{60} at RT and 4.3 K [16.66].

g-value of 2.0002 appears to be characteristic of all C_{60} radical anions, we can conclude that the spins on K_3C_{60} are located on the C_{60}^{3-} ions [16.17, 71, 84]. The EPR lines have a relatively large intensity and their linewidths do not vary with magnetic field, from which it has been concluded that the EPR linewidth is related to the intrinsic lifetime of the conduction electrons [16.33].

16.2.3. Time-Resolved EPR Study of Triplet State of Fullerenes

Time-resolved EPR techniques have been applied to study the excited triplet state of C_{60} [16.85–94] and of C_{70} [16.89, 90, 93]. The experiments are typically done using a laser pulse to excite the triplet state, and

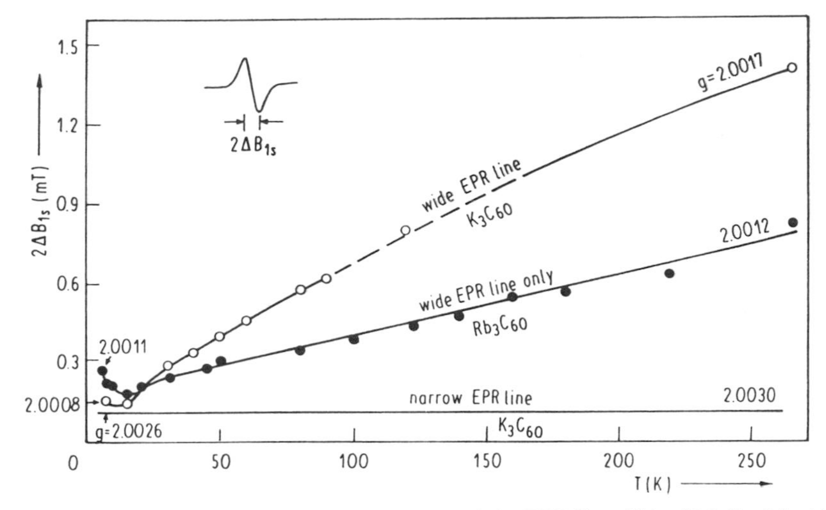

Fig. 16.17. Observed temperature dependence of the EPR linewidth of M_xC_{60} fullerides. For Rb_3C_{60} only the wide component is found; for K_3C_{60} EPR spectra show both a wide component and a narrow component of $2\Delta B_{1s} \approx 0.1$ mT whose width does not depend on temperature [16.66].

time-resolved EPR techniques are used to measure the EPR signal while it decays. Pulse-echo experiments are especially useful for lifetime measurements, and many of the studies have been made as a function of temperature.

Observation of zero-field splittings in the excited triplet state show that the fullerene molecules become distorted in the excited state (due to the Jahn–Teller effect), with static distortions occurring at low T (e.g., 3–10 K) and dynamic distortions at higher T (e.g., 77 K) [16.90]. If there were no distortions of the fullerene molecules, there would be no time-resolved EPR signal. These zero-field splittings affect the frequency for transitions with $\Delta m_S = \pm 1$ between the three levels $m_S = -1, 0, 1$ of the lowest triplet excited state, giving rise to an angular dependence

$$\omega(\theta, \phi) = \omega_0 + \frac{1}{2} D(3\cos^2\theta - 1) + \frac{3}{2} E \sin^2\theta(\cos^2\phi - \sin^2\phi) \qquad (16.8)$$

where ω_0 is the Larmor frequency, (θ, ϕ) are the Euler angles, and (D, E) are fine structure parameters, where D is associated with an axial elongation of the molecule and E with nonaxial distortions. A negative value for D, corresponding to a prolate spheroidal shape for distorted C_{60} in the excited triplet state, is verified by comparisons of the zero-field splittings for C_{60} with those for C_{70}, which is known to be a prolate spheroid from x-ray studies (see §3.2) [16.90].

Most of the time-resolved EPR experiments have been done in solution, or in frozen solution going down to very low-temperature. In solution, the time-resolved EPR spectrum shows a g-value of 2.00135 and a very narrow line (only 0.14 G linewidth) [16.89], due to motional narrowing which averages out the contributions of D and E in Eq. (16.8). The identification of the motional narrowing mechanism is supported by the much larger linewidth (6.6 G) of C_{70} in solution, due to the prolate spheroidal shape of the C_{70} molecule [16.89]. In studies of the pulsed EPR signal of C_{60} in solid benzonitrile as a function of temperature, it is found that the intersystem crossing produces about equal populations for the $|\pm 1\rangle$ triplet states, but an underpopulation of the $|0\rangle$ triplet state (for $D < 0$) [16.87]. The microwave resonance serves to change these relative populations, and these population changes are then probed by the EPR measurement. The triplet decay process is strongly temperature dependent and includes contributions from both spin–lattice relaxation (T_1), which is relatively more important at higher temperatures, and triplet decay, which is relatively more effective at low-temperatures. The observation of significant spin–lattice relaxation ($T_1 \sim 5$ μs) at low-temperatures (8 K) gives evidence for molecular motion down to very low-temperatures [16.87, 89]. The experimental time-resolved EPR spectra also indicate the presence of two different triplet states with different D and E values, one corresponding to a delocalized triplet state over the whole fullerene molecule where $D = 117 \times 10^{-4}$ cm^{-1} and $E = 6.9 \times 10^{-4}$ cm^{-1}, and a second more localized triplet state with smaller D and E values ($D = 47 \times 10^{-4}$ cm^{-1} and $E \simeq 0$) [16.87]; evidence for more than one triplet state is also provided by the optically detected magnetic resonance (ODMR) technique (see §13.2.4).

16.2.4. EPR Studies of Strong Paramagnets

Electron paramagnetic resonance experiments provide four important quantities— (1) absorption intensity, (2) g-factors, (3) lineshape, and (4) linewidth—which are of use for the study of paramagnetic systems and systems undergoing a transition to a magnetically ordered state. From this standpoint, EPR experiments have been carried out on TDAE–C_{n_C} when $n_C = 60, 70, 84, 90, 96$ [16.45, 95–99].

Shown in Fig. 16.18 are the EPR spectra for TDAE–C_{60} as a function of temperature. The EPR spectra show a Lorentzian lineshape at room temperature and a narrowing of the linewidth as the temperature is lowered, and this narrowing is attributed to the increased exchange interaction between spins [16.95], particularly as the molecules become orientationally frozen below $T_{01} \sim 170$ K [16.99]. As TDAE–C_{60} undergoes a magnetic phase transition, a broadening of the EPR line occurs, and is attributed to

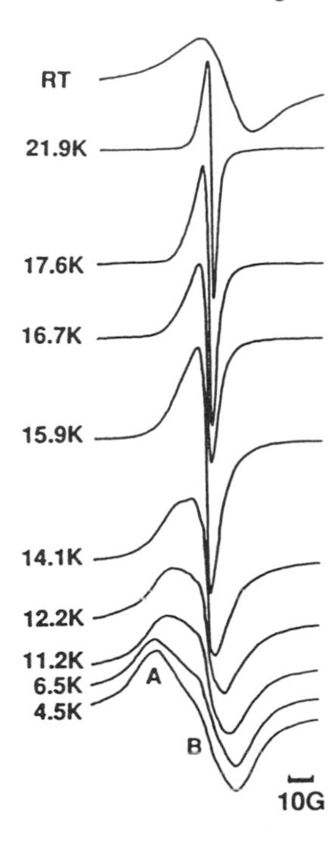

Fig. 16.18. Measured EPR spectra for TDAE–C_{60} in the low-temperature region. "A" signifies the broad peak appearing below 16.7 K, whereas "B" denotes the narrow peak associated with Curie–Weiss-type behavior [16.95].

enhanced spin–spin correlations. This broad component increases in intensity as T decreases further below T_c. The intensity of the EPR line implies one unpaired electron per C_{60} unit [16.45, 95], and the same result is obtained for TDAE–C_{70} [16.96]. The relatively broad EPR line is attributed to the localization of the spins on the C_{60}^- and other $C_{n_C}^-$ anions. The detailed dependence of the linewidth on applied magnetic field is consistent with a field dependence of T_c [16.45], as discussed further in §18.5.2. The g-value of TDAE–C_{60} is between 2.0003 and 2.0008 at room temperature [16.95, 100], close to the values reported for electrochemically prepared C_{60}^- [16.78]. The high g-factor (2.0017–2.0025) below T_c for TDAE–C_{60} is attributed to the broad EPR line which appears below T_c. In contrast, the g-factors for the TDAE–C_{n_C} compounds for $n_C = 70, 84, 90, 96$ (see Fig. 16.19) show little temperature dependence and are close to the free electron g-factor of 2.0023. The EPR linewidths for the higher fullerene derivatives also show only a weak temperature dependence, especially in comparison to that for TDAE–C_{60}, as can be seen in Fig. 16.20.

The transition at T_c is also connected with a large ($\sim 10^3$) increase in the intensity of the ESR line for TDAE–C_{60} [16.45], indicative of a large increase in local correlations and in the magnetic susceptibility. Pulsed

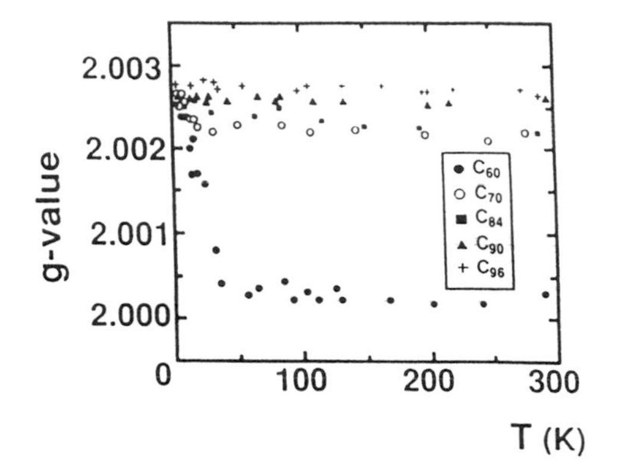

Fig. 16.19. Experimental temperature dependence of the g-values of TDAE–fullerides C_{60}, C_{70}, C_{84}, C_{90}, C_{96} [16.95, 96].

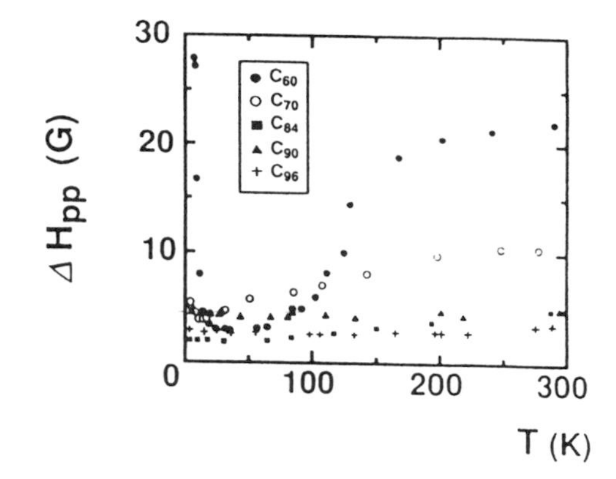

Fig. 16.20. Temperature dependence of the peak-to-peak linewidths (ΔH_{pp}) of TDAE-fullerides C_{60}, C_{70}, C_{84}, C_{90}, C_{96} [16.95, 96].

ESR measurements yield a peak in the electron spin-echo amplitude at $T_c = 16.5$ K and yield a phase memory time $T_M \sim 3$ ns which increases slowly between T_c and 9 K and then increases sharply to ~8 ns at about 9 K. Analysis of the temperature and magnetic field dependence of the magnetization lead to a spin glass model at low-temperature with spin clusters containing $10^2 - 10^3$ spins [16.45]. The isothermal decay of the magnetization $M(t)$ was measured in terms of the zero field ESR signal after a small external field ($H_0 \sim 20$ G) was switched off. A time-dependent signal $M(t)$ was only observed below T_c where $M(t)$ was found to be of the form

$$M(t) = M_0 + M^* \exp[-(t/\tau)^\alpha] \tag{16.9}$$

in which the temperature-dependent remanent magnetization M_0 was found to decrease with increasing T (M_0 was unobservably small above 12 K) and α varied from $\alpha = 0.14$ for $4 < T < 7$ K to $\alpha = 0.07$ above 8 K, while $\tau = 75$ s at 4 K and τ is thermally activated. This behavior is consistent with a spin glass phase. From the behavior of $M(t)$ and T_M, it was proposed that upon cooling, TDAE–C_{60} enters a ferromagnetic phase and below ~9 K a spin glass phase occurs [16.45]. At present the detailed nature of the low-temperature magnetic phase of TDAE–C_{60} remains an open issue.

Integration of the ESR signal yields the spin susceptibility χ. Measurements of $\chi(T)$ and of the linewidth ΔH_{pp} of the ESR line as a function of temperature have been instrumental in establishing the structural ordering temperature $T_{01} \approx 170$ K at which the TDAE–C_{60} molecular rotational motion freezes and in demonstrating a correlation between the spin ordering on adjacent TDAE–C_{60} molecules with the orientational ordering of these molecules [16.99], as is further discussed in §18.5.2.

16.3. Muon Spin Resonance (μSR)

In a μSR experiment, a spin-polarized muon beam is thermalized within a nonmagnetic solid by spin-conserving collisions. The thermalized spin-polarized muon usually becomes localized at a site in the material before it finally decays into two γ rays. The angular distribution of the 2-γ coincidence gives a direct measure of the muon spin direction and the momentum of the muon at the time of the decay. Thus the μSR probes the crystalline environment of the muon prior to the decay process. When an external magnetic field is applied, the precession of the muon spin can be directly measured.

There are several reasons why muon spin resonance has been very useful for probing low energy interactions in solids. Since the muon is found to be stable at an endohedral site in the fullerene, the introduction of a muon probe imposes very little perturbation to the crystal lattice and to the electronic structure of C_{60}-based materials. For these reasons, μSR provides a gentle, noninvasive probe for magnetic resonance, unlike EPR, where chemical doping is needed to introduce the spin probe through a paramagnetic impurity species. Since the muon magnetic moment interacts with magnetic fields, μSR provides a very sensitive probe of the magnetic field at the muon position, which can then provide information on the spatial dependence of the local magnetic fields. Since the muon spin (1/2) interacts with the electron spin, the muon spin relaxation can be used to provide information on the electron carrier density, including its temperature and spatial dependence.

μSR has proved to be a complementary tool to EPR and other characterization techniques for C_{60} and related materials. Of particular importance thus far have been μSR studies of the superconductivity of fullerenes, mainly concerning superconductivity in the K_3C_{60} and Rb_3C_{60} compounds. Two areas in particular that have been investigated are the temperature dependences of the London penetration depth $\lambda_L(T)$ and of the muon spin relaxation in these compounds. In the case of studies on $\lambda_L(T)$ for the superconducting phase, strong support was found for an s-wave singlet pairing mechanism for superconductivity in the M_3C_{60} alkali metal compounds [16.42, 101]. Of interest also has been the large London penetration depth (λ_L) as $T \to 0$ for K_3C_{60} relative to Rb_3C_{60}, consistent with measurements of NMR line broadening in the same compounds [16.27].

The muon spin relaxation studies of the superconducting K_3C_{60} and Rb_3C_{60} compounds have had an important impact on the field (see §15.4). Initially, the NMR spin relaxation studies yielded no Hebel–Slichter peak, in contrast to the behavior observed for conventional BCS superconductors (see §16.1.8). The coherence peak in the temperature dependence of the spin–lattice relaxation time T_1, predicted by Hebel–Slichter theory [16.39], arises from the divergence of the superconducting density of states at the Fermi level and the reduced density of states of excited quasiparticles at low temperature. The observation of the Hebel–Slichter peak in the μSR experiment (see Fig. 15.13) was achieved by using a low magnetic field for observing the μSR resonance. When a similar approach was used in the NMR experiment, the Hebel–Slichter peak was indeed found (see §16.1.8), thereby resolving one of the early problems in understanding the superconductivity mechanism in this class of carbon-based superconductors.

Muon spin relaxation has also been studied as a function of temperature for TDAE–C_{60} (see §8.7.3) which has a magnetic ordering temperature of 16.1 K (see §18.5.2). The time evolution of the muon polarization has been interpreted in terms of muons predominantly in a diamagnetic environment with two distinct sites identified for the muons. Motional narrowing is observed above 180 K, indicative of a freezing of the rotational motion, yielding $T_{01} \sim 180$ K, much lower than that for C_{60} itself, consistent with the large TDAE group that is attached to the fullerene. The strong magnetic depolarization observed below ~10 K is indicative of short range order and a spin glass model, with a nonvanishing order parameter below 8 – 10 K [16.102, 103].

References

[16.1] R. Tycko. *J. Phys. Chem. Solids*, **54**, 1713 (1993).

[16.2] R. D. Johnson, D. S. Bethune, and C. S. Yannoni. *Acc. Chem. Res.*, **25**, 169 (1992).

[16.3] R. D. Johnson, G. Meijer, J. R. Salem, and D. S. Bethune. *J. Am. Chem. Soc.*, **113**, 3619 (1991).

[16.4] R. Taylor, J. P. Hare, A. K. Abdul-Sada, and H. W. Kroto. *J. Chem. Soc. Chem. Commun.*, **20**, 1423 (1990).

[16.5] C. S. Yannoni, R. D. Johnson, G. Meijer, D. S. Bethune, and J. R. Salem. *J. Phys. Chem.*, **95**, 9 (1991).

[16.6] R. Tycko, R. C. Haddon, G. Dabbagh, S. H. Glarum, D. C. Douglass, and A. M. Mujsce. *J. Phys. Chem.*, **95**, 518 (1991).

[16.7] R. D. Johnson, G. Meijer, and D. S. Bethune. *J. Am. Chem. Soc.*, **112**, 8983 (1990).
[16.8] K. Balasubramanian, *J. Phys. Chem.*, **97**, 6990 (1993).
[16.9] C. W. S. Conover, Y. A. Yand, and L. A. Bloomfield. *Phys. Rev. B*, **38**, 3517 (1988).
[16.10] E. C. Honea, M. L. Homer, P. Labastie, and R. L. Whetten. *Phys. Rev. Lett.*, **63**, 394 (1989).
[16.11] G. Rajagopal, R. N. Barnett, A. Nitzan, U. Landman, E. C. Honea, P. Labastie, M. L. Homer, and R. L. Whetten. *Phys. Rev. Lett.*, **64**, 2933 (1990).
[16.12] P. J. Ziemann and A. W. Castleman. *J. Chem. Phys.*, **94**, 718 (1991).
[16.13] T. P. Martin and T. Bergmann. *J. Chem. Phys.*, **90**, 6664 (1989).
[16.14] R. Ettl, I. Chao, F. Diederich, and R. L. Whetten. *Nature (London)*, **353**, 443 (1991).
[16.15] G. E. Pake. *J. Chem. Phys.*, **16**, 327 (1948).
[16.16] C. S. Yannoni, P. P. Bernier, D. S. Bethune, G. Meijer, and J. R. Salem. *J. Am. Chem. Soc.*, **113**, 3190 (1991).
[16.17] K. Holczer and R. L. Whetten. *Carbon*, **30**, 1261 (1992).
[16.18] K. Mizoguchi, Y. Maniwa, and K. Kume. *Mater. Sci. Eng.*, **B19**, 146 (1993).
[16.19] I. Hirosawa, K. Tanigaki, J. Mizuki, T. W. Ebbesen, Y. Shimakawa, Y. Kubo, and S. Kuroshima. *Solid State Commun.*, **82**, 979 (1992).
[16.20] V. A. Stenger, C. Recchia, J. Vance, C. H. Pennington, D. R. Buffinger, and R. P. Ziebarth. *Phys. Rev. B*, **48**, 9942 (1993).
[16.21] R. E. Walstedt, D. W. Murphy, and M. Rosseinsky. *Nature (London)*, **362**, 611 (1993).
[16.22] C. S. Yannoni, H. R. Wendt, M. S. de Vries, R. L. Siemens, J. R. Salem, J. Lyerla, R. D. Johnson, M. Hoinkis, M. S. Crowder, C. A. Brown, D. S. Bethune, L. Taylor, D. Nguyen, P. Jedrzejewski, and H. C. Dorn. *Synth. Metals*, **59**, 279 (1993).
[16.23] C. P. Slichter. *Principles of Magnetic Resonance*. Springer-Verlag, Berlin, 3rd edition (1990).
[16.24] R. Tycko, G. Dabbagh, R. M. Fleming, R. C. Haddon, A. V. Makhija, and S. M. Zahurak. *Phys. Rev. Lett.*, **67**, 1886 (1991).
[16.25] R. D. Johnson, C. S. Yannoni, H. C. Dorn, J. R. Salem, and D. S. Bethune. *Science*, **255**, 1235 (1992).
[16.26] R. F. Kiefl, J. W. Schneider, A. MacFarlane, K. Chow, T. L. Duty, S. R. Kreitzman, T. L. Estle, B. Hitti, R. L. Lichti, E. J. Ansaldo, C. Schwab, P. W. Percival, G. Wei, S. Wlodek, K. Kojima, W. J. Romanow, J. P. McCauley, Jr., N. Coustel, J. E. Fischer, and A. B. Smith, III. *Phys. Rev. Lett.*, **68**, 1347 (1992). ibid **69**, 2708 (1992).
[16.27] R. Tycko, G. Dabbagh, M. J. Rosseinsky, D. W. Murphy, A. P. Ramirez, and R. M. Fleming. *Phys. Rev. Lett.*, **68**, 1912 (1992).
[16.28] Y. Maniwa, K. Mizoguchi, K. Kume, K. Kikuchi, I. Ikemoto, S. Suzuki, and Y. Achiba. *Solid State Commun.*, **80**, 609 (1991).
[16.29] R. Tycko, G. Dabbagh, M. J. Rosseinsky, D. W. Murphy, R. M. Fleming, A. P. Ramirez, and J. C. Tully. *Science*, **253**, 1419 (1991).
[16.30] R. Tycko, G. Dabbagh, M. J. Rosseinsky, D. W. Murphy, R. M. Fleming, A. P. Ramirez, and J. C. Tully. *Science*, **253**, 884 (1991).
[16.31] S. E. Barrett and R. Tycko. *Phys. Rev. Lett.*, **69**, 3754 (1992).
[16.32] T. Kälber, G. Zimmer, and M. Mehring. *Phys. Rev. B*, **51**, 16471 (1995).
[16.33] R. Tycko, G. Dabbagh, D. W. Murphy, Q. Zhu, and J. E. Fischer. *Phys. Rev. B*, **48**, 9097 (1993).
[16.34] Y. Maniwa, T. Shibata, K. Mizoguchi, K. Kume, K. Kikuchi, I. Ikemoto, S. Suzuki, and Y. Achiba. *J. Phys. Soc. Jpn.*, **61**, 2212 (1992).
[16.35] A. Abragam. *Principles of Nuclear Magnetism*. Oxford University Press, London (1961).

[16.36] H. J. Zeiger and G. W. Pratt. *Magnetic Interactions in Solids*. Clarendon Press, Oxford (1973).

[16.37] K. Holczer, O. Klein, H. Alloul, Y. Yoshinari, and F. Hippert. *Europhys. Lett.*, **23**, 63 (1993).

[16.38] V. P. Antropov, I. I. Mazin, O. K. Andersen, A. I. Liechtenstein, and O. Jepson. *Phys. Rev. B*, **47**, 12373 (1993).

[16.39] L. C. Hebel and C. P. Slichter. *Phys. Rev.*, **113**, 1504 (1959).

[16.40] S. Sasaki, A. Matsudo, and C. W. Chu. *J. Phys. Soc. Japan*, **63**, 1670 (1994).

[16.41] V. A. Stenger, C. H. Pennington, D. R. Buffinger, and R. P. Ziebarth. *Phys. Rev. Lett.*, **74**, 1649 (1995).

[16.42] Y. J. Uemura, A. Keren, L. P. Le, G. M. Luke, B. J. Sternlieb, W. D. Wu, J. H. Brewer, R. L. Whetten, S. M. Huand, S. Lin, R. B. Kaner, F. Diederich, S. Donovan, and G. Grüner. *Nature (London)*, **352**, 605 (1991).

[16.43] R. F. Kiefl, W. A. MacFarlane, K. H. Chow, S. Dunsiger, T. L. Duty, T. M. S. Johnston, J. W. Schneider, J. Sonier, L. Brard, R. M. Strongin, J. E. Fischer, and A. B. Smith III. *Phys. Rev. Lett.*, **70**, 3987 (1993).

[16.44] L. Cristofolini, M. Ricco, R. De Renzi, G. P. Ruani, S. Rossini, and C. Taliani. In H. Kuzmany, J. Fink, M. Mehring, and S. Roth (eds.), *Proceedings of the Winter School on Fullerenes*, pp. 279–282 (1994). Kirchberg Winter School, World Scientific Publishing, Singapore.

[16.45] R. Blinc, P. Cevc, D. Arčon, J. Dolinšek, D. Mihailovič, and P. Venturini. In H. Kuzmany, J. Fink, M. Mehring, and S. Roth (eds.), *Proceedings of the Winter School on Fullerenes*, pp. 283–288 (1994). Kirchberg Winter School, World Scientific Publishing, Singapore.

[16.46] R. Blinc, J. Dolinšek, D. Arčon, D. Mihailovič, and P. Venturini. *Solid State Commun.*, **89**, 487 (1994).

[16.47] P. Venturini, D. Mihailovič, R. Blinc, P. Ceve, J. Dolinšek, D. Abramic, B. Zalar, H. Oshio, P. M. Allemand, A. Hirsch, and F. Wudl. *J. Mod. Phys. B*, **6**, 3947 (1992).

[16.48] C. Kittel. *Introduction to Solid State Physics*. John Wiley & Sons, New York, 6th ed. (1985).

[16.49] S. Stevenson, H. C. Dorn, P. Burbank, K. Harich, Z. Sun, C. H. Kiang, J. R. Salem, M. S. DeVries, P. H. M. van Loosdrecht, R. D. Johnson, C. S. Yannoni, and D. S. Bethune. *Annal. Chem.*, **66**, 2680 (1994).

[16.50] S. Stevenson, H. C. Dorn, P. Burbank, K. Harich, J. Haynes, Jr., C. H. Kiang, J. R. Salem, M. S. DeVries, P. H. M. van Loosdrecht, R. D. Johnson, C. S. Yannoni, and D. S. Bethune. *Annal. Chem.*, **66**, 2675 (1994).

[16.51] R. D. Johnson, M. S. de Vries, J. Salem, D. S. Bethune, and C. S. Yannoni. *Nature (London)*, **355**, 239 (1992).

[16.52] S. Suzuki, S. Kawata, H. Shiromaru, K. Yamauchi, K. Kikuchi, T. Kato, and Y. Achiba. *J. Phys. Chem.*, **96**, 7159 (1992).

[16.53] M. Hoinkis, C. S. Yannoni, D. S. Bethune, J. R. Salem, R. D. Johnson, M. S. Crowder, and M. S. de Vries. *Phys. Chem. Lett.*, **198**, 461 (1992).

[16.54] H. Shinohara, H. Yamaguchi, N. Hayashi, H. Sato, M. Inagaki, Y. Saito, S. Bandow, H. Kitagawa, T. Mitani, and H. Inokuchi. *Mater. Sci. Eng.*, **B19**, 25 (1993).

[16.55] C. S. Yannoni, M. Hoinkis, M. S. de Vries, D. S. Bethune, J. R. Salem, M. S. Crowder, and R. D. Johnson. *Science*, **256**, 1191 (1992).

[16.56] P. H. M. van Loosdrecht, R. D. Johnson, R. Beyers, J. R. Salem, M. S. de Vries, D. S. Bethune, P. Burbank, J. Haynes, T. Glass, S. Stevenson, H. C. Dorn, M. Boonman, P. J. M. van Bentum, and G. Meijer. In K. M. Kadish and R. S. Ruoff (eds.), *Recent Advances in the Chemistry and Physics of Fullerenes and Related Materials:*

Electrochemical Society Symposia Proceedings, San Francisco, May 1994, vol. 94–24, p. 1320, Electrochemical Society, Pennington, N.J. (1994).

[16.57] H. Shinohara, H. Sato, M. Ohkohchi, Y. Ando, T. Kodama, T. Shida, T. Kato, and Y. Saito. *Nature (London)*, **357**, 52 (1992).

[16.58] J. H. Weaver, Y. Chai, G. H. Kroll, C. Jin, T. R. Ohno, R. E. Haufler, T. Guo, J. M. Alford, J. Conceicao, L. P. Chibante, A. Jain, G. Palmer, and R. E. Smalley. *Chem. Phys. Lett.*, **190**, 460 (1992).

[16.59] H. Shinohara, H. Sato, Y. Saito, A. Izuoka, T. Sugawara, H. Ito, T. Sakurai, and T. Matsuo. *J. Phys. Chem.*, **96**, 3571 (1992).

[16.60] S. C. Kukolich and D. R. Huffman. *Chem. Phys. Lett.*, **182**, 263 (1991).

[16.61] A. A. Zakhidov, A. Ugawa, and K. Imaeda. *Solid State Commun.*, **79**, 939 (1991).

[16.62] A. A. Zakhidov, K. Imaeda, K. Yakushi, H. Inokuchi, Z. Iqbal, R. H. Baughman, B. L. Ramakrishna, and Y. Achiba. *Physica.*, **185C**, 411 (1991).

[16.63] D. M. Cox, S. Behal, M. Disko, S. M. Gorun, M. Greaney, C. S. Hsu, E. B. Kollin, J. M. Millar, J. Robbins, W. Robbins, R. D. Sherwood, and P. Tindall. *J. Am. Chem. Soc.*, **113**, 2940 (1991).

[16.64] A. R. West. *Solid State Chemistry and its Applications*. John Wiley & Sons, New York (1984).

[16.65] P. Byszewski, J. Stankowski, Z. Trybula, W. Kempiński, and T. Żuk. *J. Mol. Structures*, **269**, 175 (1992).

[16.66] J. Stankowski, P. Byszewski, W. Kempiński, Z. Trybula, and T. Żuk. *Phys. Stat. Sol. (b)*, **178**, 221 (1993).

[16.67] G. Wagoner. *Phys. Rev.*, **118**, 647 (1960).

[16.68] M. M. Khaled, R. T. Carlin, P. C. Trulove, G. R. Eaton, and S. S. Eaton. *J. Am. Chem. Soc.*, **116**, 3465 (1994).

[16.69] A. J. Schell-Sorokin, F. Mehran, G. R. Eaton, S. S. Eaton, A. Viehbeck, T. R. O'Toole, and C. A. Brown. *Chem. Phys. Lett.*, **195**, 225 (1992).

[16.70] F. Mehran, A. J. Schell-Sorokin, and C. A. Brown. *Phys. Rev. B*, **46**, 8579 (1992).

[16.71] P.-M. Allemand, G. Srdanov, A. Koch, K. Khemani, F. Wudl, Y. Rubin, F. Diederich, M. M. Alvarez, S. J. Anz, and R. L. Whetten. *J. Am. Chem. Soc.*, **113**, 2780 (1991).

[16.72] P. N. Keizer, J. R. Morton, K. F. Preston, and A. K. Sugden. *J. Phys. Chem.*, **95**, 7117 (1991).

[16.73] T. Kato, T. Kodama, M. Oyama, S. Okasaki, T. Shida, T. Nakagawa, Y. Matsui, S. Suzuki, H. Shiromaru, K. Yamauchi, and Y. Achiba. *Chem. Phys. Lett.*, **186**, 35 (1991).

[16.74] K. Tanaka, M. Okada, K. Okahara, and T. Yamabe. *Chem. Phys. Lett.*, **191**, 469 (1992).

[16.75] D. DuBois, M. T. Jones, and K. Kadish. *J. Am. Chem. Soc.*, **114**, 6446 (1991).

[16.76] J. Stinchcombe, A. Penicaud, P. B. P. D. W. Boyd, and C. A. Reed. *J. Am. Chem. Soc.*, **115**, 5212 (1993).

[16.77] T. Kato, T. Kodama, and T. Shida. *Chem. Phys. Lett.*, **205**, 405 (1993).

[16.78] D. R. Lawson, D. L. Feldheim, C. A. Foss, P. K. Dorhout, C. M. Elliott, C. R. Martin, and B. Parkinson. *J. Electrochem. Soc.*, **139**, L68 (1992).

[16.79] P. Boulas, R. Subramanian, W. Kutner, M. T. Jones, and K. M. Kadish. *J. Electrochem. Soc.*, **140**, L130 (1993).

[16.80] M. Baumgarten, A. Gugel, and L. Gherghel. *Adv. Mater.*, **5**, 458 (1993).

[16.81] C. Bossard, S. Rigaut, D. Astruc, M. H. Delville and G. Felix. *J. Chem. Soc., London Chem. Commun.*, pp. 333–334 (1993).

[16.82] D. DuBois, K. Kadish, S. Flanagan, R. E. Haufler, L. P. F. Chibante, and L. J. Wilson. *J. Am. Chem. Soc.*, **113**, 4364 (1991).

[16.83] R. D. Rataiczak, W. Koh, R. Subramanian, M. T. Jones, and K. M. Kadish. *Synth. Metals*, **56**, 3137 (1993).
[16.84] W. H. Wong, M. Hanson, W. G. Clark, G. Grüner, J. D. Thompson, R. L. Whetten, S. M. Huang, R. B. Kaner, F. Diederich, P. Pettit, J. J. Andre, and K. Holczer. *Europhys. Lett.*, **18**, 79 (1992).
[16.85] M. A. Greaney and S. M. Gorun. *J. Phys. Chem.*, **95**, 7142 (1991).
[16.86] P. J. Krusic, E. Wasserman, P. N. Keizer, J. R. Morton, and K. F. Preston. *Science*, **254**, 1183 (1991).
[16.87] M. Bennati, A. Grupp, M. Mehring, K. P. Dinse, and J. Fink. *Chem. Phys. Lett.*, **200**, 440 (1992).
[16.88] G. H. Goudsmit and H. Paul. *Chem. Phys. Lett.*, **208**, 73 (1993).
[16.89] G. L. Closs, P. Gautam, D. Zhang, P. J. Krusic, S. A. Hill, and E. Wasserman. *J. Phys. Chem.*, **96**, 5228 (1992).
[16.90] M. Terazima, N. Hirota, H. Shinohara, and Y. Saito. *Chem. Phys. Lett.*, **195**, 333 (1992).
[16.91] E. J. J. Groenen, O. G. Poluektov, M. Matsushita, J. Schmidt, J. H. van der Waals, and G. Meijer. *Chem. Phys. Lett.*, **197**, 314 (1992).
[16.92] M. Matsushita, A. M. Frens, E. J. J. Groenen, O. G. Poluektov, J. Schmidt, G. Meijer, and M. A. Verheijen. *Chem. Phys. Lett.*, **214**, 314 (1993).
[16.93] M. R. Wasielewski, M. P. O'Neil, K. R. Lykke, M. J. Pellin, and D. M. Gruen. *J. Am. Chem. Soc.*, **113**, 2772 (1991).
[16.94] A. Regev, D. Gamliel, V. Meiklyar, S. Michaeli, and H. Levanon. *J. Phys. Chem.*, **97**, 3671 (1993).
[16.95] K. Tanaka, A. A. Zakhidov, K. Yoshizawa, K. Okahara, T. Yamabe, K. Yakushi, K. Kikuchi, S. Suzuki, I. Ikemoto, and Y. Achiba. *Phys. Rev. B*, **47**, 7554 (1993).
[16.96] K. Tanaka, A. A. Zakhidov, K. Yoshizawa, K. Okahara, T. Yamabe, K. Kikuchi, S. Suzuki, I. Ikemoto, and Y. Achiba. *Solid State Commun.*, **85**, 69 (1993).
[16.97] K. Yoshizawa, T. Sato, K. Tanaka, T. Yamabe, and K. Okahara. *Chem. Phys. Lett.*, **213**, 498 (1993).
[16.98] D. Mihailovič, P. Venturini, A. Hassanien, J. Gasperič, K. Lutar, S. Miličev, and V. I. Srdanov. In H. Kuzmany, J. Fink, M. Mehring, and S. Roth (eds.), *Progress in Fullerene Research: International Winterschool on Electronic Properties of Novel Materials*, pp. 275–278 (1994). Kirchberg Winter School, World Scientific Publishing Co., Singapore.
[16.99] D. Mihailovič, D. Arčon, P. Venturini, R. Blinc, A. Omerzu, and P. Cevc. *Science*, **268**, 400 (1995).
[16.100] K. Okahara. Ph.D. thesis, Kyoto University (1994). Department of Chemistry: Studies on the Electronic and Magnetic Properties of C_{60} and Related Materials (in English).
[16.101] Y. J. Uemura, L. P. Le, and G. M. Luke. *Synth. Metals*, **56**, 2845 (1993).
[16.102] Y. J. Uemura, T. Yamazaki, D. R. Harshman, M. Senba, and E. J. Ansaldo. *Phys. Rev. B*, **31**, 546 (1985).
[16.103] H. Pinkvos, A. Kalk, and C. Schwink. *Phys. Rev. B*, **41**, 590 (1990).

CHAPTER 17

Surface Science Studies Related to Fullerenes

Surface science studies of fullerenes have been used in many unique ways to advance our knowledge of fullerenes and doped fullerenes and their interaction with substrates. The interpretation of surface science studies frequently depends on measurements using complementary surface science techniques. Photoemission and inverse photoemission studies (§17.1) have provided unique information on the density of states and, in conjunction with optical absorption and luminescence spectra (see §13.2 and §13.3), have clarified fundamental aspects of the electronic structure of fullerenes in the solid state (see §12.7). Electron energy loss spectra (EELS) provide complementary information to optical studies regarding interband transitions, the frequency and wave-vector dependence of the dielectric function $\epsilon(\omega, q)$, as well as information on fundamental excitations such as plasmons and phonons (§17.2). Auger spectroscopy is very important for identifying the species that is desorbed from a surface and for monitoring the quantity that is desorbed (see §17.3). Auger spectroscopy studies have also yielded unique information about the on-site hole–hole Coulomb interaction (§17.3), while scanning tunneling microscopy and spectroscopy studies have also yielded valuable information about the crystalline structure, the surface reconstruction and the electronic structure of fullerenes (§17.4). Temperature-programmed desorption has proved especially useful for the identification of sublimation energies and the relative bonding of fullerenes to each other and to the substrate species (§17.5). Of importance to many potential applications of fullerenes (§20.2.6, §20.2.7, and §20.3.2) is the nature of the interface between fullerenes and various semiconductors and metallic substrates, and this topic is discussed in §17.9.

17.1. Photoemission and Inverse Photoemission

Photoemission and inverse photoemission spectroscopies (PES and IPES) have been extensively applied [17.1–4] to study the density of states within a few eV of the Fermi level for both undoped and doped fullerenes, and many calculations of the density of states have been compared to these measurements [17.4–6]. This work has been extensively reviewed [17.2, 4, 7].

In the standard photoemission experiment, an incident photon of energy $\hbar\omega$ excites an electron from an occupied state to the vacuum level and the electron kinetic energy E_e is measured to determine $\hbar\omega - E_e$, which is plotted with reference to the Fermi level. Comparison of the photoemission intensity *vs.* $(\hbar\omega - E_e)$ taken at a variety of incident photon energies is used to relate the PES spectra to a density of occupied states (see upper inset to Fig. 17.1(a)). Correspondingly, for the inverse photoemission experiment, an incident electron enters a bound excited state, emitting a photon, whose energy is measured, so that the spectrum provided by IPES measurements yields the density of unoccupied states. In the PES and IPES spectra of fullerenes, there are few complications due to excitonic and vibronic interactions, which dominate studies of the optical absorption and luminescence spectra near the fundamental absorption edge (see §13.1.3).

Ultraviolet (UV) photoemission and inverse photoemission experiments are highly sensitive to the first two or three nanometers of material near the surface and because of surface reconstruction effects, photoemission measurements of the density of states of semiconductors do not usually provide an accurate determination of the density of states in bulk compounds. The low density of electron charge found on the surface of fullerene solids indicates a low density of surface states. Thus, surface reconstruction effects may not be important for fullerite surfaces, thereby making PES and IPES especially useful tools for studying fullerenes in the solid state. Nevertheless, the large unit cell dimensions of fullerite crystals imply a sensitivity of the PES and IPES techniques to only a few unit cells at the surface of the crystalline solid, thereby emphasizing the surface properties relative to bulk properties.

With regard to doped fullerenes, surface states, screening effects, and surface reconstruction effects may be more important, so that PES and IPES measurements may not provide as accurate a measurement of the density of states for M_xC_{60} compounds as for the case of undoped fullerenes. Furthermore upon addition of an alkali metal such as K, spatially separated phases such as K_3C_{60} and K_6C_{60} are formed, and these phases may be inhomogeneously distributed throughout the sample. In particular, the surface layer may have different stability regimes for these phases relative to the bulk. Such inhomogeneities may present significant difficulties with the interpretation of PES and IPES spectra in the near-surface layers (20–30 Å thick) of doped fullerenes.

17.1.1. UV Spectra for C_{60}

Typical photoemission spectra ($E < E_F$, where E_F is the Fermi energy) and inverse photoemission [17.3, 8] spectra ($E > E_F$) are shown in the lowest trace in Fig. 17.1(a) for C_{60}, where the intensity maxima in the spectra correspond to peaks in the density of states, and the energy difference between the thresholds for the photoemission and inverse photoemission spectra yield important information on the HOMO–LUMO gap in solid C_{60} [17.9, 10]. Also shown in Fig. 17.1(a) are PES and IPES spectra for K_xC_{60} or potassium-doped C_{60}, and these spectra are further discussed in §17.1.3.

For the undoped fullerenes, the systems under investigation for the PES and IPES spectra contain C_{n_C} molecules with $(n_C - 1)$ π-electrons (for PES spectra) and with $(n_C + 1)$ π electrons (for IPES spectra), where n_C is the number of carbon atoms in the fullerene, as shown schematically in the inset to Fig. 17.1(a) and in Fig. 17.1(b). In contrast, the states excited in an optical absorption or EELS experiment correspond to molecules with n_C π electrons for which an excited electron and hole are located on the same or neighboring fullerenes and are bound in a Frenkel-type exciton state, if the electron and hole are on the same molecule, or in a charge transfer state, if the electron and hole are on different molecules [see Fig. 17.1(b)]. If the binding energy of the exciton is supplied, so that the electron and hole are separated to infinity, then the $n_C + 1$ and $n_C - 1$ states in the IPES and PES experiments, respectively, can be realized. Using a value of 3.7 eV for the separation between the peaks for the PES and IPES spectra [17.9] and a value of 1.4 eV for the Hubbard U [17.4], we obtain a value of 2.3 eV for the intrinsic band gap Δ. With this value of the band gap Δ and a value of 1.84 eV for the singlet absorption edge, a value of 0.46 eV is obtained for the exciton binding energy as shown in Fig. 17.2. The bulk value of U for C_{60} may be lower than that measured near the surface because of surface screening deficiencies [17.12]. The Hubbard U is the energy needed to transfer an electron from one neutral fullerene C_{n_C} to create two distinct molecules with n_C+1 and n_C-1 π electrons, respectively [see Fig. 17.1(b)]. The onset energy (2.3–2.6 eV) which is equal to $(U + \Delta - W)$ is consistent with a bandwidth of $\sim$1 eV for the LUMO-derived band, and an energy width W of ~ 0.5 eV [17.9].

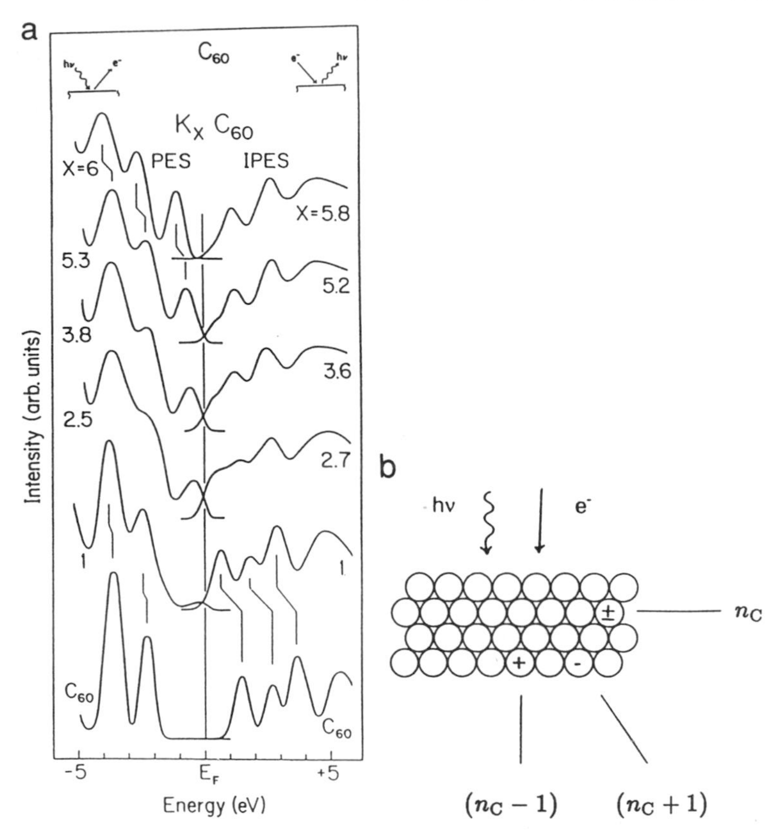

Fig. 17.1. (a) Photoemission (PES) and inverse photoemission (IPES) spectra for C_{60} and K-doped C_{60} films [17.2]. The spectra show the effects of K incorporation into C_{60} films. Adding more K causes an increase of the emission intensity below E_F in the PES spectra and a decrease of intensity above E_F in the IPES spectra. The top spectrum shows that E_F shifts into the gap when the LUMO t_{1u} band is filled. (b) An array of molecules exposed to a photon and an electron beam. The figure shows schematically a molecule with n_C electrons (open circles), a localized Frenkel exciton (±), and an electron (−) and a hole (+) on different molecules containing $n_C + 1$ and $n_C - 1$ electrons, respectively [17.4].

The peaks in the photoemission spectra tend to be wide compared with other determinations of the bandwidths (~0.5 eV). The reason for the larger PES widths is not presently understood, although it is believed that molecular rotations at room temperature are in part responsible for the large PES and IPES linewidth for C_{60}. Band calculations of the density of states, whether based on a one-electron picture [17.5, 13, 14] or a many-body picture [17.9, 10], provide a good fit to the observed photoemission and inverse photoemission spectra for C_{60}, at least at the level of detail

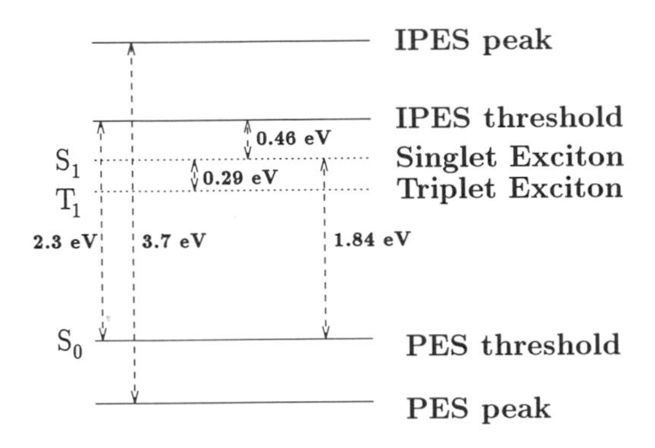

Fig. 17.2. Schematic energy levels showing the energy separation between the IPES and PES peaks which yields the HOMO–LUMO gap (Δ) plus the Hubbard U (1.4 eV). Using a value of 1.84 eV for the singlet optical absorption edge yields an exciton binding energy of 0.46 eV, where the zero of energy is for convenience placed at the S_0 level. The triplet level (T_1) at 1.55 eV [17.11] indicates an exchange energy of 0.29 eV.

that has been used in these comparisons. For example, since the density of states obtained from PES and IPES experiments is nearly an order of magnitude less than that obtained in typical calculations (see §12.7), the comparisons between theory and experiment have focused on the energy of the spectral features and on the relative intensity of these features [17.4].

To study the dispersion of the various bands, angle-resolved photoemission experiments have been carried out. The first angle-resolved photoemission study of C_{60} [17.15] indicated little band dispersion, suggesting that the PES linewidth might arise from vibronic side bands. Subsequent more detailed angle-resolved studies have indicated some dispersion in the HOMO-derived level ($\sim$0.4 eV) [17.16], and good agreement between experiment and theoretical calculations [17.17] has been obtained for this amount of dispersion in the C_{60} valence bands. These experiments and calculations show that the measured dispersion is sufficient to account for the PES linewidth, without need to identify vibronic side bands as a broadening mechanism [17.11, 18–20], although vibronic side bands may indeed contribute.

Although most of the PES and IPES measurements at UV photon energies have been done on C_{60}, a significant body of work has already been carried out on C_{70} [17.21, 22] and on higher fullerenes C_{76}, C_{82}, and C_{84} [17.16, 23–25]. The results largely show a decreasing band gap for fullerenes as n_C increases, consistent with other experimental probes of the electronic structure and with theoretical expectations [17.5, 6]. The photoemission spectra observed for C_{60}, C_{70}, and C_{84} (see Fig. 17.3) show some remarkable similarities, with a set of bands between 1.5–4.0 eV below the Fermi level E_F, another group located about 6 eV below E_F, and finally a third set of bands in the 7–9 eV range below E_F [17.26]. The broadening in the density of states is also seen in Fig. 17.3 with increasing n_C in C_{n_C} associ-

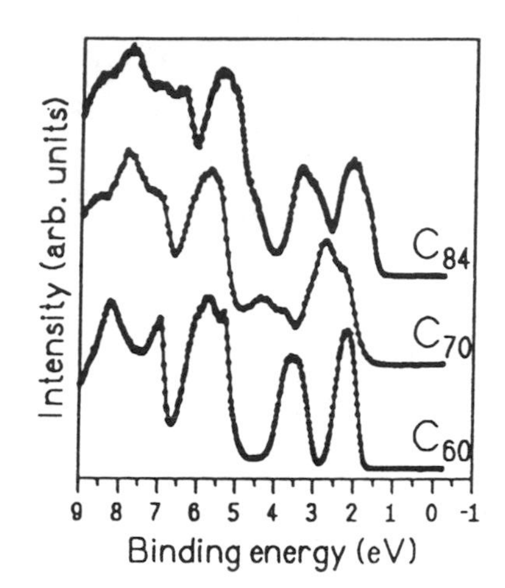

Fig. 17.3. Photoemission spectra of C_{60}, C_{70}, and C_{84} fullerene thin films deposited on gold. Photoemission measurements are made at a resolution of 25 meV and at 300 K [17.26].

ated with the lower crystal symmetry of the higher fullerenes. The higher fullerenes (e.g., C_{84}) also show line broadening associated with the presence of isomers [17.27].

17.1.2. XPS Spectra for C_{60}

X-ray photoelectron spectroscopy (XPS) refers to the PES process when the excitation photon is in the x-ray range and the electron excitation is from a core level. XPS spectra have been obtained for a variety of fullerenes. For example, Fig. 17.4 shows the XPS spectrum for C_{60}. In this spectrum, we can see well-defined peaks which show an intense, narrow main line (peak #1 in Fig. 17.4) identified with the emission of a photoexcited electron from the carbon 1*s* state, which has a binding energy of 285.0 eV and a very small linewidth of 0.65 eV at half-maximum intensity [17.28]. The sharpest side band feature in the downshifted XPS spectrum (labeled 2) is identified with an on-site molecular excitation across the HOMO–LUMO gap at 1.9 eV [17.28]. Features 3, 4 and 5 are the photoemission counterparts of dipole excitations seen in optical absorption, while features 6 and 10 represent on-molecule plasmon collective oscillations of the π and σ charge distributions, respectively. Plasma excitations are also prominently featured in core level electron energy loss spectra (EELS) discussed in §17.2.3.

XPS spectra for C_{60} have been obtained by a number of workers [17.29, 30] and have been compared with theory, as well as with XPS spectra for various aromatic compounds.

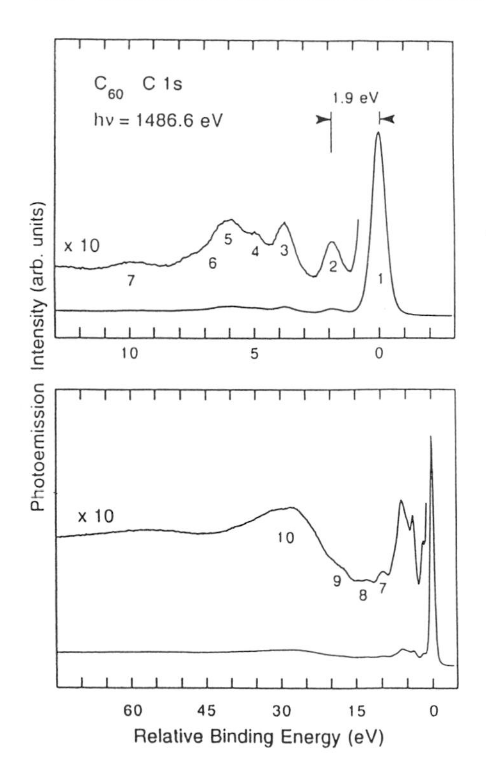

Fig. 17.4. X-ray photoelectron spectra for the carbon 1*s*-derived satellite structures for C_{60} [17.28]. The sharp feature 1 is the central XPS line, while the downshifted features 2–10 refer to C_{60} excitations (see text). The XPS data are shown on two energy scales to emphasize features near E_F (upper spectrum) and features farther away in energy (lower spectrum) [17.28].

17.1.3. PES and IPES Spectra for Doped Fullerenes

For the case of doped fullerenes, surface states and dangling bonds associated with nonstoichiometry are more important than for pristine C_{60} and the photoemission experiments may not reflect bulk properties as accurately as for their undoped counterparts. Nevertheless, photoemission and inverse photoemission data for doped fullerenes do provide convincing evidence for charge transfer and for band filling as x in K_xC_{60} increases. Referring to Fig. 17.1(a), we can see that the density of states peak associated with the t_{1u}-derived band just above the Fermi level E_F in the C_{60} trace moves closer to E_F as K is added to C_{60} [17.2]. Upon further addition of potassium, the t_{1u}-related peak crosses the Fermi level and eventually for $5.8 \leq x \leq 6.0$, the t_{1u}-derived peak falls below E_F, indicating complete filling of the t_{1u}-derived level; the data for this composition further show a small band gap to the next higher lying t_{1g}-derived level (see Fig. 12.2). PES and IPES studies on doped fullerenes have been carried out by many groups, and more recent work has tended to be done under higher resolution and with more control in sample preparation [17.24, 31–35]. Photoemission and inverse photoemission [17.8] studies have further confirmed strong similarities between the electronic structures of K_xC_{60} and both Rb_xC_{60} and Cs_xC_{60} [17.3, 36–38]. Weak features at 0.3 and 0.7 eV below E_F in the photoemission spectra of K_3C_{60} [17.39] have been identified with electrons coupling to vibrational modes (~0.3 eV) and to plasma excitations (~0.7 eV) [17.35, 40]. Differences have also been clearly demonstrated between the density of states for K_xC_{60} and those for both Na_xC_{60} and Ca_xC_{60}, for which multiple metal ions can be accommodated in the octahedral sites, because of the small size of the Na and Ca ions (see §8.5).

Calculations to explain the photoemission and inverse photoemission spectra [17.3, 13, 28, 41–44] have been based on band models for the electronic states, and good agreement between theory and experiment has been reported by many authors [17.4]. It is not believed that the agreement between the experimental and calculated densities of states favors any particular approach to the calculation of the electronic levels, whether by band calculation approaches or by many-body approaches. Because of the molecular nature of solid fullerene films, the electronic states up to ~100 eV above the Fermi level which form the final states in the photoemission process relate to the free molecule to some degree. Photoemission experiments thus show that details in the density of states depend to some degree on the photon energy used for the photoemission probe. For this reason, the PES measurements of the occupied density of states are not expected to correspond in detail to generic calculations.

UV photoemission spectra taken on K_3C_{60} [17.33, 36, 45] show that the spectral peaks are not sensitive to the electron mean free path. Furthermore, the Auger spectra and the photoelectron spectroscopy self-convolution for K_6C_{60} and K_3C_{60} have common features and the analysis of these spectra yields the same values for the Hubbard $U = 1.4 \pm 0.2$ eV for both K_3C_{60} and K_6C_{60} [17.45]. A similar experiment for undoped C_{60} yielded $U = 1.6$ eV [17.9] (see §17.3).

Regarding other doped fullerene systems, photoemission studies have been carried out both on M_xC_{60} and M_xC_{70} systems for various metals M. For example, PES studies on K_xC_{70} [17.22, 46] have shown regions of stability for K_1C_{70}, K_4C_{70}, and K_6C_{70} and have demonstrated band filling at $x = 6$, but no metallic phases for the composition range $0 \leq x \leq 6$. Photoemission studies have also been carried out on Rb_xC_{70} [17.47]. Regarding alkaline earth–doped fullerides, photoemission spectra for Ca_5C_{60} [17.48] and Ba_6C_{60} [17.33, 49] have been reported showing partial filling of the t_{1g} LUMO+1 band for these systems.

17.1.4. XPS Studies of Adsorbed Fullerenes on Substrates

XPS spectra of the carbon 1*s* core level for C_{60} on noble metal substrates (Ag, Cu, Au) show a downshift relative to bulk C_{60} in the 1*s* binding energy, as illustrated in Fig. 17.5(a). This shift rapidly disappears with increasing C_{60} coverage, showing that about two monolayers (ML) are sufficient to yield bulk values [see Fig. 17.5(b)] [17.50]. This result is consistent with temperature-programmed desorption studies (see §17.5) and scanning tunneling microscopy (STM) studies (see §17.4), showing a different behavior for the first monolayer relative to subsequent C_{60} layers.

A more detailed study of C_{60} on Cu (100) surfaces [17.51] shows a downshift in the carbon 1*s* XPS line between multilayer and monolayer C_{60} coverage as shown in Fig. 17.5(c). This downshift (0.56 eV) is comparable to that for K_3C_{60} films (0.66 eV) and is attributed to (1) charge transfer from the Cu substrate to the adsorbed C_{60} molecule, and (2) hybridization between the Cu (4*s*) and C (2*p*) states [17.51]. Improved calibration of the surface coverage of deposited C_{60} is achieved using single crystal substrates and flashing off all but the first monolayer [17.52], which adheres more strongly to the surface (see §17.4.1).

Photoemission has also been used to study the interaction between C_{60} and transition metals such as tungsten (100) [17.53] and Rh (111) [17.54], showing that the binding of C_{60} to these transition metal surfaces is stronger than the C_{60}–C_{60} bond, just as for the noble metals discussed above. Measurement of the binding energy of the HOMO level shows charge transfer from the W (100) surface to C_{60}, while for the Rh (111) surface it is claimed in one report [17.54] that electrons from C_{60} flow into the transition metal surface.

17.2. Electron Energy Loss Spectroscopy

Electron energy loss spectroscopy (EELS) is a complementary tool to optical probes for studying excitations in solids. The emphasis of EELS studies is in surface science because inelastic electron scattering experiments typically are carried out in an energy range where the electron penetration depth is only two or three nanometers. Because the interaction between the incoming electron and the fullerenes is much stronger than that of the photon, selection rules are less important in the EELS experiment. In this way the EELS technique is complementary to optical probes. Although most of the publications on fullerenes deal with the inelastic scattering of electrons, a few structural studies using elastic electron scattering have also been reported, including both electron diffraction and low-energy electron diffraction (LEED) to examine bulk and surface structural features

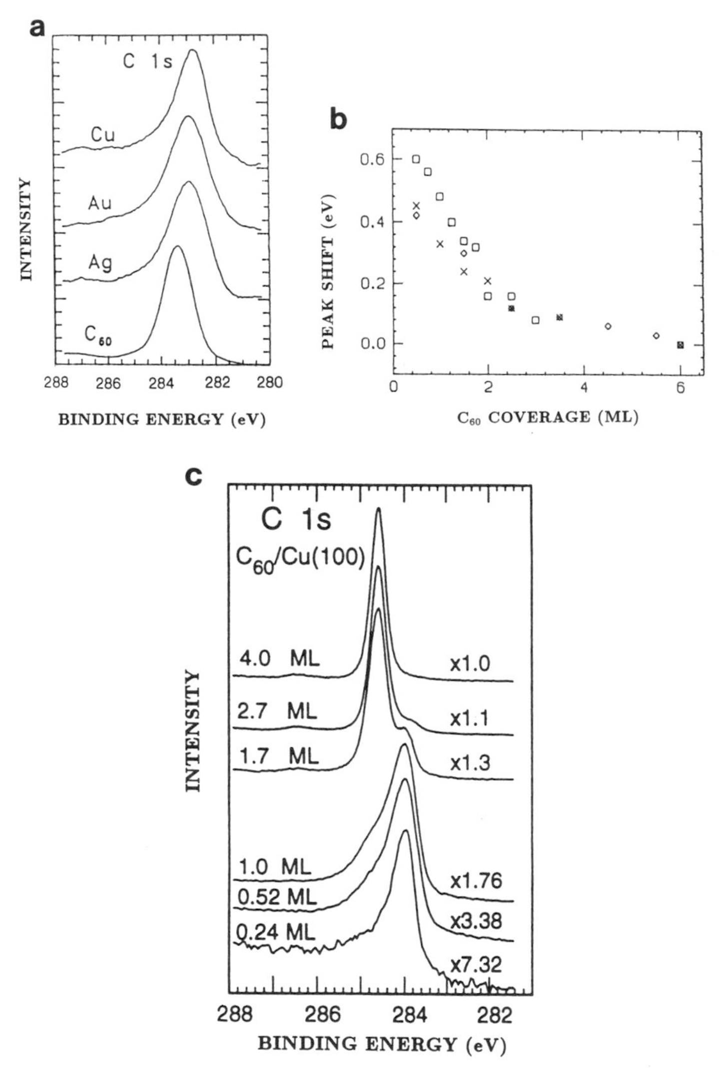

Fig. 17.5. (a) XPS spectra showing the XPS carbon 1s peak (taken with Mg K_α radiation) of 0.5-ML C_{60} on Cu, Au, and Ag substrates compared to bulk C_{60}. (b) Shift in the C 1s peak with increasing C_{60} coverage, compared to bulk C_{60}. The points indicate an Ag substrate (◇), Cu substrate (□), Au substrate (×). On all three substrates, bulk C_{60} and six or more ML of C_{60} yield the same 1s peak energy, as shown in the figure [17.50]. (c) X-ray photoemission spectra of the C (1s) core level for C_{60} films at coverages of 0.24, 0.52, 1.0, 1.7, 2.7, and 4 ML of C_{60} deposited on Cu (100). The curves for the various coverages are displaced vertically for clarity and the scale factors indicated for each curve compensate for the relative coverage change and attenuation [17.51].

(§17.2.1). Electron diffraction studies are directed toward elucidating either the structure within the first few atomic layers from the surface or the structure of very small samples, utilizing the high interaction cross section of electron probes with matter. Regarding inelastic electron energy loss spectroscopy (EELS), most of the effort has been directed toward study of interband transitions within a few electron volts of the Fermi level or transitions from core levels to levels near the Fermi level. EELS also provides a powerful tool for the investigations of plasmons and their dispersion relations. In such measurements, the electron energy loss function $\mathrm{Im}[-1/\epsilon(\omega)]$ is determined because of the sensitivity of $\mathrm{Im}[-1/\epsilon(\omega)]$ to plasmons, since the plasma frequency is defined by the vanishing of $\epsilon_1(\omega)$. The function $\mathrm{Im}[-1/\epsilon(\omega)]$ is determined either at fixed electron momentum (wave vector) transfer as a function of electron energy ($\hbar\omega$) or at fixed energy as a function of momentum transfer. EELS thus provides information on the dielectric function over a range of q values, whereas optical probes are confined to q values very close to zero. Using high-resolution techniques, vibrational spectra can also be explored with the EELS technique in a complementary way to infrared and Raman scattering, allowing many modes that are silent using infrared or Raman spectroscopy to become observable, but the vibrational modes observed by the EELS technique have much lower energy resolution. Progress with the EELS techniques for the investigation of fullerenes and related materials is reviewed in the following subsections.

17.2.1. Elastic Electron Scattering and Low-Energy Electron Diffraction Studies

Because of the strong interaction between electron probes and matter, electron diffraction is a method of choice for the structural analysis of materials only available in very small quantities, such as higher-mass fullerenes, or for the structural analysis of only a few atomic layers on a surface. Electron diffraction data are now available for the structure of crystalline C_{60}, C_{70}, C_{76}, and C_{84} (see Fig. 17.6) [17.23]. The results show that each of these molecules crystallizes in a face-centered cubic (fcc) structure with a lattice constant $a_{\mathrm{fcc}} \sim \sqrt{n_{\mathrm{C}}}$, where n_{C} is the number of carbon atoms in the fullerene. The relation between a_{fcc} and n_{C} is consistent with a shell model for fullerenes and a diffraction lineshape determined by a simple spherical form factor $\sin(qr)/(qr)$ for these fullerenes, where r is the average radius of the fullerene and q is a distance in reciprocal space. The distance d_f between fullerenes in the fcc lattice is about the same for all of these fullerenes ($d_f \sim 2.9$ Å) and is related to the lattice constant a_{fcc} of the fcc

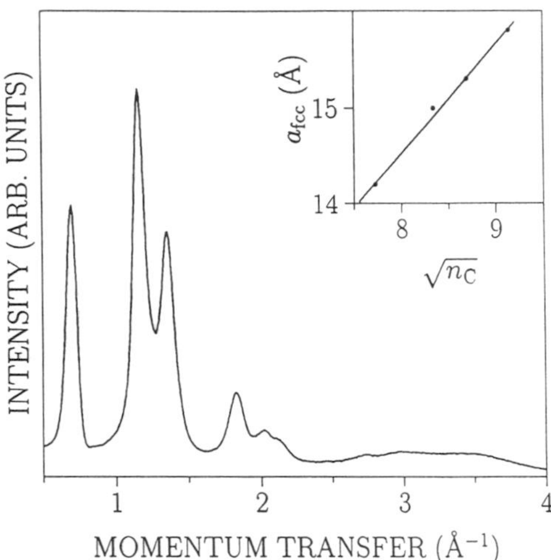

Fig. 17.6. Electron diffraction profile of a thin film of C_{76} recorded in the EELS spectrometer with a momentum transfer resolution of 0.06 Å^{-1}. The inset is a plot of the fcc lattice constant a_{fcc} as determined from such measurements for various fullerenes with n_C carbon atoms, all crystallizing in the fcc structure [17.23].

lattice and to the fullerene radius r by

$$d_f = \frac{a_{fcc}}{\sqrt{2}} - 2r. \tag{17.1}$$

Good agreement is obtained between the EELS diffraction data and selected area diffraction studies done by high-resolution transmission electron microscopy (TEM) techniques [17.23] yielding values of $a_0 = 14.2$, 15.0, 15.3, and 15.8 Å for C_{60}, C_{70}, C_{76}, and C_{84}, respectively, to an accuracy of ±0.1 Å [17.26]. The deviation of the data point for C_{70} in the inset to Fig. 17.6 from the straight line may be associated with the nonspherical shape of the C_{70} molecule.

Low-energy electron diffraction has also been used to probe the surface structure of crystalline fullerene surfaces adsorbed to various substrates. LEED studies of C_{60} on unheated Cu surfaces showed poor LEED patterns for less than two or three monolayers of C_{60} on Cu (100), Cu (110), and Cu (111) surfaces. However, for Cu substrates heated to ~300°C, epitaxial, single-domain growth occurred for thicker C_{60} growth on Cu (111). Good epitaxy was also achieved for the heated Cu (100) and Cu (110) surfaces, but with a strongly preferred (111) orientation for C_{60} growth, since the Cu (111) surface has the lowest surface energy [17.51].

17.2.2. Electronic Transitions near the Fermi Level

Low-energy electronic transition studies by the EELS technique for primary electrons of 10 eV or less have provided complementary information to optical probes. EELS results have been especially important for understanding the electronic structure near the HOMO–LUMO gap. Because of differences in the interaction Hamiltonian for electron probes, spin flipping transitions can occur in the electronic excitation from the ground state (h_u^{10}) to the lowest excited state ($h_u^9 t_{1u}^1$), allowing direct identification of the lowest triplet exciton level (denoted as T_1) at an excitation energy of ~1.55 eV [17.11] (see §13.1). The EELS spectrum in Fig. 17.7 for a C_{60} multilayer, in fact, shows two sharp lines (one at 1.55 eV and one at 1.84 eV). The onset of the more intense line at 1.84 eV is identified with the absorption edge to the singlet exciton manifold of the $h_u^9 t_{1u}^1$ configuration, which peaks at 2.2 eV, the intensity at higher energies coming from larger momentum transfer, with the peak corresponding to a momentum transfer of

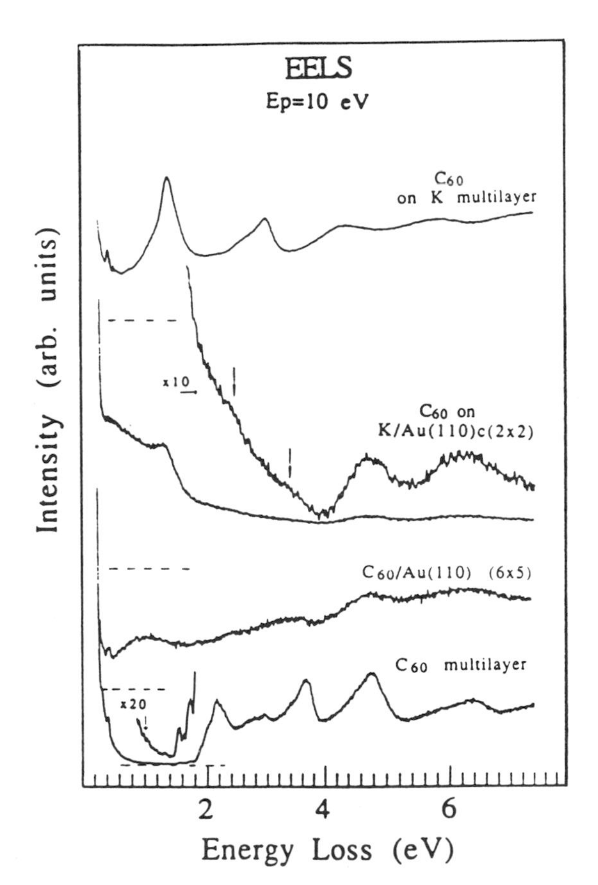

Fig. 17.7. EELS spectra of C_{60} multilayers and monolayers on different substrates using incident electrons of 10 eV. The dashed line under each spectrum marks the zero level for the electron energy loss intensity. The broad feature below 1 eV in the third spectrum from the top is identified with the t_{1u} plasmon. The second trace from the top is related to the third trace by a scale factor of ten [17.55].

~0.9 Å^{-1} [17.21]. The energy separation between the triplet and singlet excitons (0.29 eV) thus gives the exchange energy associated with this exciton. Other band-to-band transitions in the EELS spectra are reported at 2.7, 3.6, 4.5, and 5.5 eV [17.11, 21, 55], and the energies of these peaks are consistent with optical spectra (see §13.3.1). The EELS spectra for C_{70} are rather similar to those for C_{60} with structure in the optical conductivity derived from the C_{70} EELS spectra reported at 2.5, 3.2, 4.4, and 5.4 eV [17.21]. Since the EELS and optical spectra both result in an excited electron and a hole that is left behind, both types of excitations give rise to similar excitonic states, except for differences in the selection rules cited above [17.28].

Because of the stronger interaction of electron probes with matter, EELS can also be used to investigate the electronic structure of monolayers on surfaces, while the more weakly interacting photon probes are more useful for studying bulk properties. Referring to Fig. 17.7, we see the effect of charge transfer from the Au (110) surface to an adsorbed monolayer of C_{60} through the large plasma background below 1 eV and the broad features in the EELS spectra at higher energies.

Deposition of a monolayer of K on the Au (110) surface prior to the C_{60} deposition greatly enhances the magnitude of the plasma absorption and shows the presence of a strong peak at 1.3 eV, which is within 0.1 eV of the EELS peaks in K_3C_{60} [17.40] and in Rb_3C_{60} [17.21], suggesting comparable charge transfer to that for the M_3C_{60} alkali metal–doped fullerenes (namely, three electrons transferred to each adsorbed C_{60} molecule). When multiple K multilayers are deposited on the Au (110) surface prior to the monolayer C_{60} deposition, the EELS pattern in Fig. 17.7 changes drastically relative to the K monolayer spectrum, showing a much lower background absorption level and sharp EELS lines at 1.35, 2.6, 3.0, 4.3, and 5.8 eV, with the 1.35 eV peak in good agreement with EELS spectra for K_6C_{60}, Rb_6C_{60} and Cs_6C_{60} [17.55], suggesting a complete filling of the t_{1u} band of C_{60} by charge transfer of up to six electrons from the multiple K layers on the Au (110) substrate [17.55].

EELS spectra, similar to the C_{60} multilayer spectrum in Fig. 17.7, are observed for thick C_{60} films on various substrates, such as Si (100) [17.56] and Rh (111) [17.54]. Some differences in the energies of the lower-lying electronic peaks have been reported for the case of monolayer C_{60} coverage on Rh (111) as compared with Si (100) [17.54], insofar as the blue shift of high-resolution electron energy loss spectroscopy (HREELS) vibrational features for one ML C_{60} coverage on Rh (111) has been interpreted in terms of C_{60} cation formation [17.54], indicative of electron transfer from C_{60} to the Rh (111) surface.

17.2.3. Core Level Studies

EELS spectra taken with high-energy electrons can probe excitations from the 1*s* carbon core level to available states near the Fermi level. Such core spectra (shown in Fig. 17.8 for C_{60}, C_{70}, C_{76}, and C_{84}) indicate four distinct spectral features in the figure below 289 eV. These features are associated with transitions from the carbon 1*s* core level to π^* bands just above E_F, and the structure above 290 eV is identified with the corresponding transitions to σ^* bands [17.23]. A broadening of the core level transitions is seen for both the C_{76} and C_{84} spectra in Fig. 17.8 and is attributed to differences in the structure and electronic states for the various isomers of C_{76} and C_{84} and to their lower symmetries. Although the spectra in Fig. 17.8 are strongly related to one another, shifts are seen in the frequency of the onset of the core level excitations from ~284.1 eV for C_{60} to ~283.7 eV for C_{84}, as well as variations in the details of the spectra from one fullerene

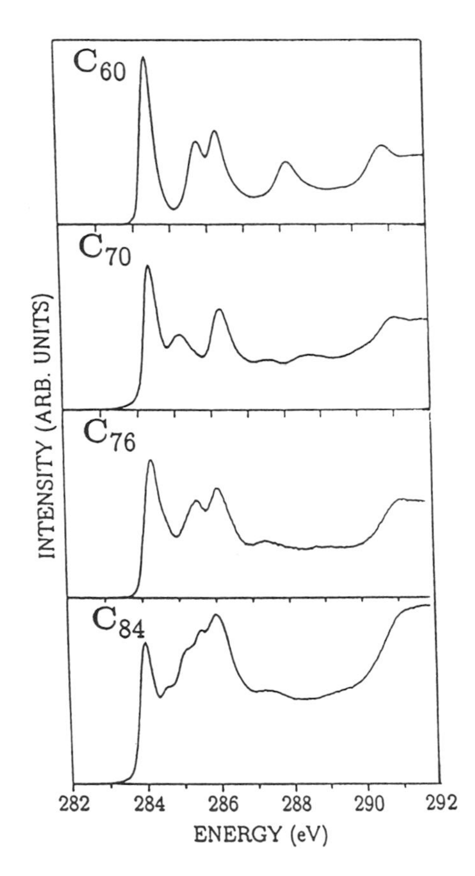

Fig. 17.8. Carbon 1*s* core level excitation spectra of various fullerenes as observed by the EELS technique on transmission [17.23].

to another [17.23, 25]. This shift is identified with the lower curvature of the larger fullerenes as they become closer to the core level spectra for graphite. Comparisons have been made between core level spectra in the solid state to molecular spectra [17.58, 59] and between experiments and theory [17.60].

17.2.4. Plasmon Studies

Because of the differences in selection rules, the EELS technique is able to detect electron energy loss due to longitudinal excitations, such as plasmon excitations, which are not generally accessible by optical techniques. Typically, in an EELS experiment (see Fig. 17.9), the frequency dependence of the electron energy loss function $\mathrm{Im}[-1/\epsilon(\omega)]$ is measured at a fixed electron momentum transfer (see §17.2). From a Kramers–Kronig analysis of the loss function in conjunction with optical data or by using a sum rule, $\epsilon_1(\omega)$, $\epsilon_2(\omega)$, and the complex conductivity $\sigma(\omega)$ are obtained [17.16] for

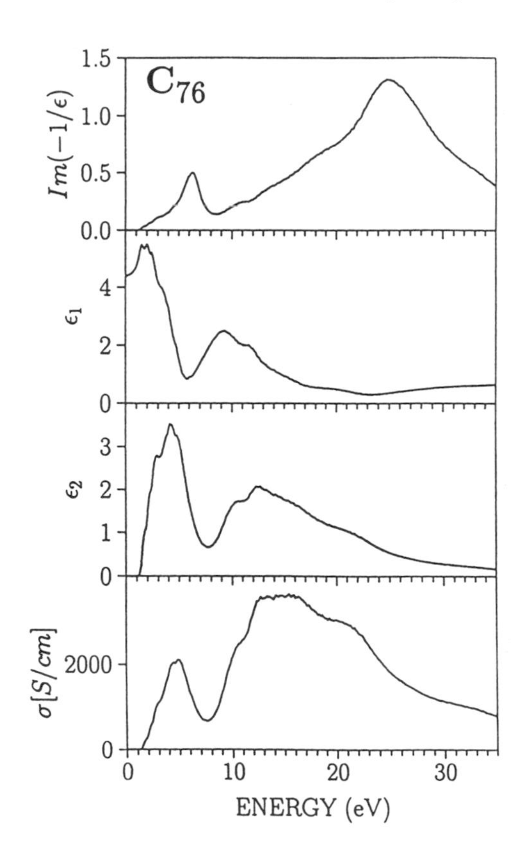

Fig. 17.9. Analysis of EELS measurements for C_{76} to yield the frequency-dependent: (a) $\mathrm{Im}[-1/\epsilon(\omega)]$, (b) $\epsilon_1(\omega)$, (c) $\epsilon_2(\omega)$, and (d) the optical conductivity $\sigma(\omega)$. A momentum transfer of 0.1 Å^{-1} was used to take these data [17.23].

that fixed momentum transfer value. Use of a sum rule leads to $\epsilon_1(0) = 3.6$ for C_{60} and 4.4 for C_{76} [17.23]. Reasonable agreement is obtained between the value of $\epsilon_1(0)$ from EELS and values for $\epsilon_1(0)$ for C_{60} using optical (3.9) and low-frequency (4.4) measurement techniques (see §13.3.3).

Referring to Fig. 17.9(a) we see two dominant peaks in the frequency dependence of the loss function $\mathrm{Im}[-1/\epsilon(\omega)]$ for C_{76}, the lower peak occurring at ~6 eV for the π plasmon and an even more prominent peak occurring at ~25 eV for the $(\pi+\sigma)$ plasmon. These peaks in $\mathrm{Im}[-1/\epsilon(\omega)]$ are typical of the spectra for the π and $(\pi+\sigma)$ plasmons in fullerenes and are identified with the π plasmon and $(\pi+\sigma)$ plasmon frequencies ω_π and $\omega_{\pi+\sigma}$. Also shown in this figure are $\epsilon_1(\omega)$ and $\epsilon_2(\omega)$ for C_{76}, which were used to determine $\mathrm{Im}[-1/\epsilon(\omega)]$, and the optical conductivity $\sigma(\omega)$, which is derived from $\epsilon_2(\omega)$. Several groups have reported values for the frequencies $\omega_{\sigma+\pi}$ for C_{60} [17.11,16,23,56,61–66] and a few values have also been reported for ω_π and $\omega_{\pi+\sigma}$ for the higher-mass fullerenes C_{70}, C_{76}, and C_{84} [17.16,23,25,61,62,67,68]. The results summarized in Table 17.1 show that ω_π and $\omega_{\pi+\sigma}$ have only a weak dependence on n_C and are also close to the corresponding values reported for graphite (see Table 17.1). Although the variations of these plasma frequencies with n_C are bracketed by the range of values reported by the various groups for C_{60} alone, trends in the dependence of ω_π and $\omega_{\pi+\sigma}$ on n_C can be inferred by comparing results using the same experimental technique and method of analysis. This approach (see Fig. 17.10) suggests that ω_π may shift to lower energy as the fullerene mass increases, whereas $\omega_{\pi+\sigma}$ for the various fullerenes shifts to higher energies. The spectra in Fig. 17.10 further show well-resolved peaks in the π plasmon of C_{60} which are attributed to the multiple narrow energy bands in C_{60} associated with π–π^* transitions. These transitions are not so well resolved in the EELS spectrum for the higher-mass fullerenes. Figure 17.10 also shows that the onset of the EELS loss function decreases from 1.8 eV for C_{60} to 1.2 eV for C_{84}, consistent with a smaller

Table 17.1

Plasmon parameters for various fullerenes and for graphite.

	Plasmon (eV)			Binding energy (eV)
	ω_π	$\omega_{\sigma+\pi}$	ϵ_0	C1s
C_{60}	6.3	26.4	3.6	285
C_{70}	6.2	25.2	3.8	285
C_{76}	6.1	24.8	4.4	284.8
C_{84}	6.0	24.6	5–6	284.6
Graphite	6.5	23.0	—	285.2

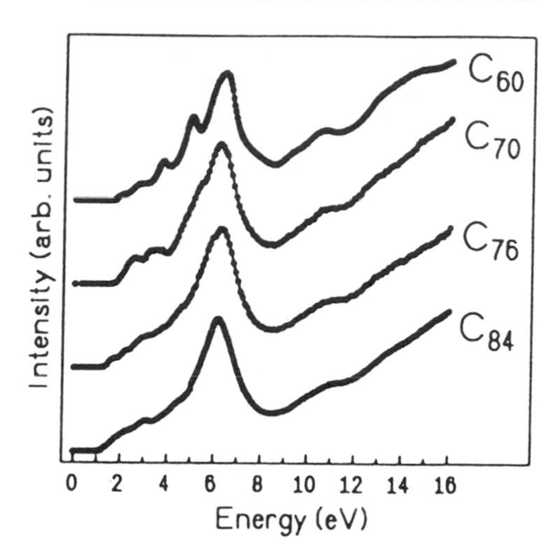

Fig. 17.10. Electron energy loss function of C_{60}, C_{70}, C_{76}, and C_{84} fullerene thin films recorded using EELS in transmission with an electron momentum transfer of $q = 0.15$ Å^{-1} [17.26].

band gap for the higher-mass fullerenes. In all cases the onset energies are much lower than the HOMO–LUMO gap derived from PES/IPES experiments (see §17.1.1), consistent with the excitonic nature of the absorption edge in EELS. From the difference between the EELS and PES/IPES gap values, the exciton binding energy is estimated to be between ~0.5 and 0.8 eV for C_{60}, C_{70}, and C_{84} [17.26]. Also given in Table 17.1 are the values for the static dielectric constant ϵ_0 and the 1*s* carbon core level energy for various fullerenes in comparison to graphite.

17.2.5. HREELS and Vibrational Spectra

High-resolution electron energy loss spectroscopy (HREELS) has been used to study the vibrational spectra of fullerenes (see §11.5.9), because this technique is also sensitive to the so-called silent modes, which are forbidden in the first-order Raman and infrared spectra (see §11.5.3 and §11.5.4). Since the energy of the vibrational modes is much smaller than the energies of electron beams used in electron energy loss spectroscopy (EELS), the high-resolution version of EELS is used to study the vibrational spectra.

The high-resolution EELS technique has been used to study the vibrational spectra of C_{60} and doped C_{60} compounds, as well as for monolayers of C_{60} on surfaces. Measurements of the red shifts or blue shifts of these modes for monolayer C_{60} coverage on substrates indicate whether charge is transferred to the C_{60} or from the C_{60} to the surface. This topic is further discussed in §11.5.9.

17.2.6. EELS Spectra for Alkali Metal–Doped Fullerenes

Electron energy loss spectra have been reported up to 40 eV for alkali metal–doped Rb_3C_{60} and Rb_6C_{60} as well as for Rb_3C_{70} and Rb_6C_{70}, where the nominal stoichiometries for doped films were inferred from x-ray characterization [17.21]. The energy loss spectra for Rb_xC_{60} (see Fig. 17.11) show a marked dependence on alkali metal concentration, although the spectra for Rb_xC_{60} and Rb_xC_{70} for similar x values show many similarities. As shown in Fig. 17.11, analysis of M_xC_{60} data to yield $\epsilon_1(\omega, q)$ and $\epsilon_2(\omega, q)$ indicates a large negative $\epsilon_1(0)$ for Rb_3C_{60}, as expected for a metallic system, while the energy loss spectra for C_{60} and Rb_6C_{60} indicate semiconducting behavior. In contrast, the EELS spectra for the nominal composition Rb_3C_{70} do not indicate conducting behavior, and this is attributed to the much lower maximum in the electrical conductivity (by about two orders of magnitude) of alkali metal–doped C_{70} relative to that of Rb_3C_{60}. Also of interest is the very similar behavior of the energy loss spectra for different alkali metals M_xC_{60} for similar x values, indicative of the small hybridization between the fullerene and alkali metal wave functions, as is also observed in many other experimental studies. Of interest are the relatively high values for the low-frequency dielectric function for Rb_6C_{60} [$\epsilon_1(0) = 7.1$] and for Rb_6C_{70} [$\epsilon_1(0) = 8.0$] [17.21]. For all Rb_xC_{60} and Rb_xC_{70} thin films that were studied ($x = 0, 3, 6$), $\pi \rightarrow \pi^*$ structure was observed below 8 eV and $\sigma \rightarrow \sigma^*$ or $(\pi + \sigma) \rightarrow (\pi + \sigma)^*$ structure above 8 eV [17.21]. Using the data in Fig. 17.11, the frequency dependence of the optical conductivity can be calculated, and the results for Rb_3C_{60} show a peak at 0.5 eV and a second smaller peak at 1.1 eV. The lower-energy peak is identified with

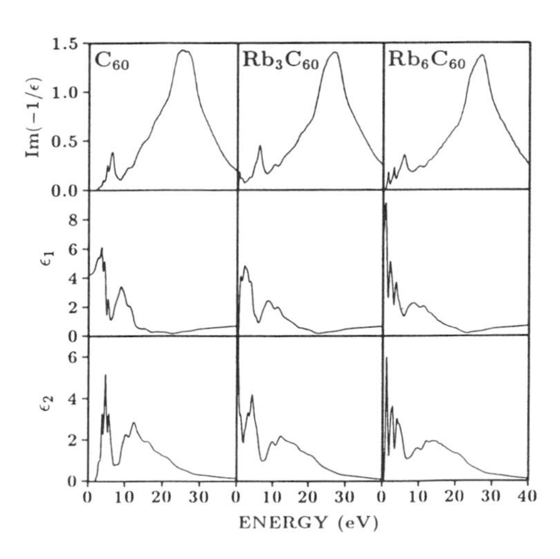

Fig. 17.11. Electron energy loss function, $\text{Im}(-1/\epsilon)$, the real part of the dielectric function $\epsilon_1(\omega, q)$, and the imaginary part of the dielectric function $\epsilon_2(\omega, q)$ for Rb_xC_{60} ($x = 0, 3,$ and 6) plotted vs. electron energy $\hbar\omega$ up to 40 eV and for a momentum transfer of $q = 0.15$ Å^{-1} [17.21].

a heavy mass ($m^*/m_0 \sim 4$) charge carrier plasmon and a Fermi energy of ~0.25 eV, consistent with a narrow width for the t_{1u}-derived band for Rb_3C_{60}. The peak at ~1.1 eV is identified with $t_{1u} \rightarrow t_{1g}$ interband transitions in Rb_3C_{60}, consistent with calculated and experimentally observed optical properties for K_3C_{60} (see §13.4.3) [17.69]. Similar transitions near 1 eV are also observed for Rb_3C_{70}, Rb_6C_{60}, and Rb_6C_{70}. Regarding the π plasmon and the $(\pi + \sigma)$ plasmon, many similarities are found between these features among the various Rb_xC_{60} and Rb_xC_{70} films [17.21]. Whereas some differences appear in the detailed spectral features for the π electron structures in the Rb_xC_{60} and Rb_xC_{70} systems, the σ electron structures appear to be unchanged upon doping with Rb.

17.3. Auger Electron Spectroscopy

Auger electron spectroscopy (AES) is a surface science technique which yields a unique spectrum for each chemical species and therefore is useful for identifying the chemical species on the surface. The intensity of the Auger lines for a given chemical element further gives the concentration of that species on the surface. Scanning Auger spectroscopy allows a map to be made of the sample surface regarding its stoichiometry. In AES experiments, a high-energy incident photon excites an electron from a core level to an ionization state, inducing a transition from a higher-lying electron state to fill the core hole level and the simultaneous emission of an electron whose energy is measured. The energy of the emitted electron is determined by the energy levels of the atom, and since the atomic energy levels for each atom are unique, the AES spectrum gives a unique identification of each atom. Each Auger line for a given species is labeled by three symbols, such as KLM, indicating the atomic shell of the core level (K), of the electron filling the core level (L), and of the emitted Auger electron that is detected (M). We discuss in this section the use of the AES technique to provide information on the surface composition, as well as on many-body interactions used to determine the repulsive Hubbard U interaction energy for two holes on the same C_{60} molecule.

An example of the analysis of the surface composition of a C_{60} film on an Si substrate is shown in Fig. 17.12. By monitoring the intensity of the Auger electron peak for carbon (272 eV) and for the substrate [e.g., Si (100) or Si (111) with an Si peak at 90 eV] (see Fig. 17.12), characterization information can be obtained on the surface composition of samples used in other surface science studies, such as the ratio of the carbon/silicon AES line intensities as a function of surface temperature [17.70, 71]. At low-temperature, below the desorption temperature for C_{60}, the C/Si ratio is high (~1.2). Once desorption of C_{60} occurs (at ~600 K), the carbon sur-

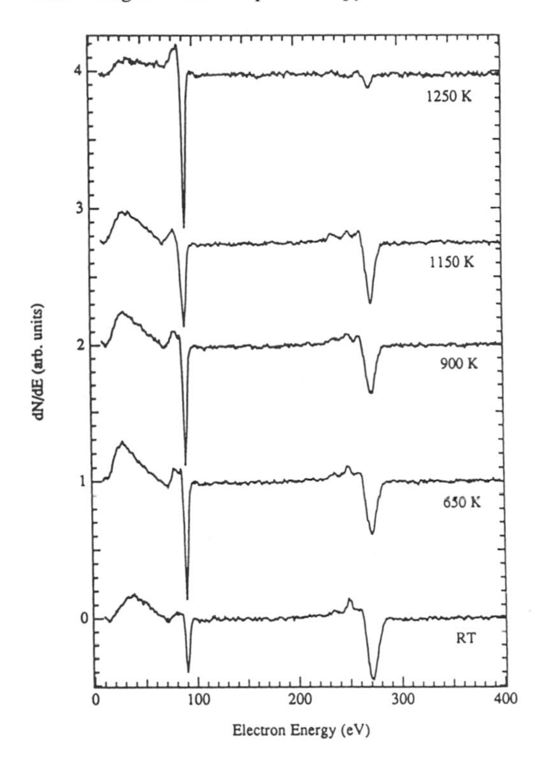

Fig. 17.12. Series of Auger electron spectra (AES) at 2 keV primary beam energy for as-deposited C_{60} on an Si (100) (1 × 1) surface at room temperature (RT) and the C_{60}/Si (100) surface annealed to 650, 900, 1150, and 1250 K. The silicon AES feature appears at 90 eV and the carbon feature at 272 eV. The surface is heated at 3 K/s to the annealing temperature and is then rapidly cooled to less than 350 K, at which temperature each AES spectrum is taken. The various spectra are diplaced for clarity [17.70].

face depletion shows a C/Si ratio of ~0.4, associated with the residual C_{60} molecules that are tightly bound to the clean Si (100) surface [17.70, 71]. The AES C/Si ratio for an Si (111) surface is a factor of 2 higher than for an Si (100) surface, indicative of the larger number of binding sites for C_{60} on Si (111). In the temperature range 900–1150 K, the C_{60} molecules were reported to rupture, yielding carbon atoms and clusters on the surface which start reacting with the Si surface above 1150 K to form SiC [17.72]. The weakly bound carbon atoms desorb more readily than the C_{60} itself, thereby enhancing the C/Si ratio that is measured by desorption. The increase in the C/Si ratio in the AES detector occurs at a slightly lower temperature for an Si (111) substrate, indicative of the higher activity and larger C_{60} surface concentration of the Si (111) surface relative to Si (100) [17.71]. Above 1150 K, only residual C impurities remain on the Si (100) and Si (111) surfaces, as SiC forms and diffuses into the bulk yielding a low C/Si surface ratio that is nearly temperature independent [17.70, 71]. These results can be explained by the fact that the bonding of C_{60} to a clean (100) Si surface is much stronger (~56 kcal/mol) than for C_{60} to itself (32 kcal/mol), so that the heating of C_{60} on an Si (100) surface yields a strongly adhering C_{60} monolayer above about 600 K. Although Si is not a metal, strong bonding between C and Si occurs, in accordance with the strong C–Si bond in SiC and the high density of dangling bonds on the Si surface. At high temperatures above 1200 K, only a few carbon atoms remain on the Si (100) surface [17.70, 72].

A detailed observation of cage opening of C_{60} at 1020 K has been observed on an Si (111) (7 × 7) surface using a variety of surface science techniques, including scanning tunneling microscopy (STM) in an ultrahigh vacuum (UHV) chamber, temperature-programmed desorption (TPD), and Auger electron spectroscopy (AES) [17.71]. At 1020 K, the carbon clusters appear under STM examination with an open cup-like structure and at yet higher temperatures, almost all of the carbon desorbs from the surface. The STM studies (see §17.4 and §17.9) suggest that C_{60} cages are mobile on Si surfaces at high T and tend to agglomerate, as also occurs for the fragments after the cages are opened. The thermal opening of the C_{60} cages may provide an explanation for the enhanced diamond nucleation on an Si surface [17.73, 74] (see §20.3.1).

Auger spectroscopy is commonly used to monitor whether the growth pattern of C_{60} on a metal surface is layer-by-layer or islandic. In this application of AES, Auger compositional maps are made during the adsorption process. For example, by monitoring the Rh (302 eV) Auger line, the layer-by-layer adsorption of C_{60} on a Rh (111) surface was established [17.54].

Another closely related example is the use of the AES technique to monitor the interaction of C_{60} with metal surfaces such as aluminum, where it is found that the C_{60} adheres much more strongly to an Al surface than to another C_{60} molecule [17.75]. Because of this strong adhesion, a complete C_{60} monolayer tends to form before the second C_{60} layer is initiated in the growth of a C_{60} film on Al, and correspondingly, multiple C_{60} layers are easily desorbed from a metal surface such as Al to leave a continuous adsorbed C_{60} monolayer for $T = 570$ K. This strongly bound C_{60} layer passivates the Al surface, inhibiting Al_2O_3 formation [17.75] (see §20.2.6). Similar behavior is found for C_{70} on an Al surface, with the C_{70} multilayers leaving the surface at 620 K. Both C_{60} and C_{70} fullerene monolayers are stable on a clean Al surface until ~650 K (as shown in Fig. 17.13 for C_{60} on Al) and above this temperature the fullerenes begin to diffuse rapidly into the bulk metal [17.75]. Further study of desorption from clean, oriented single crystal faces of Al would be of great interest.

Auger electron spectroscopy has also been applied to study many-body effects in fullerenes, based on an analysis of the Auger lineshape [17.76, 77]. From a many-body standpoint, the magnitude of the on-site Coulomb repulsion energy U between two holes on the same C_{60} molecule can be determined from a high-resolution KVV Auger spectrum, such as that shown in

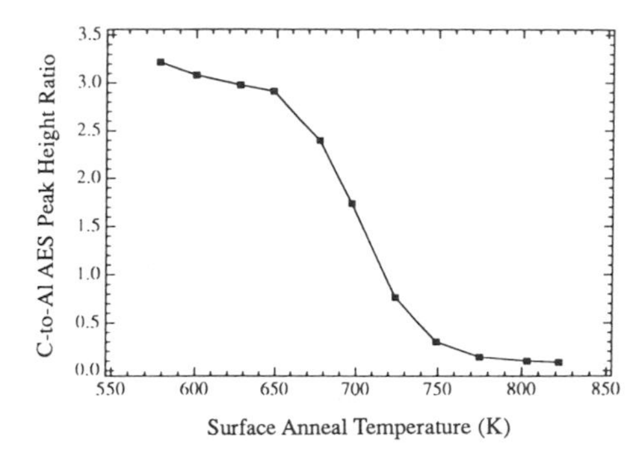

Fig. 17.13. Plot of the carbon (272 eV) to aluminum (68 eV) AES peak height ratio as a function of surface anneal temperature from 570 to 825 K for a single C_{60} monolayer on a clean Al surface [17.75].

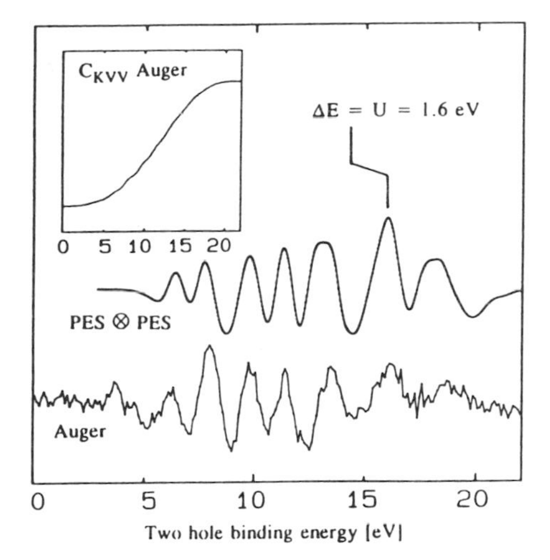

Fig. 17.14. Comparison of the carbon C_{KVV} Auger spectrum of solid C_{60} with the self-convolution of the photoelectron spectrum after subtraction of a smoothing curve from the original Auger spectrum shown in the inset. The central curve is obtained from the self-convolution of the photoelectron spectrum after also subtracting a smooth curve and shifting the whole spectrum to higher energies by 1.6 eV. This energy shift is identified with the Hubbard U interaction energy [17.9, 10].

Fig. 17.14. In this Auger process, the valence (V) electron emitted from the surface into the electron collector and the valence electron (V) dropping down to the carbon K hole level are both excited by the incident photon [17.9]. Analysis of the KVV Auger lineshape indicates a similar electron correlation energy for C_{60}, benzene, and graphite [17.76]. By studying the Auger and photoemission spectra on the same sample, analysis of the KVV Auger spectrum in Fig. 17.14 shows that the spectrum must be shifted by an energy $U = 1.6 \pm 1$ eV to match the self-convolution of the photoelectron spectra (labeled PES⊗PES in Fig. 17.14), thereby giving the density of valence states for the two valence electrons (V) of the Auger KVV process. These two valence electrons are excited when an electron in the K shell is ionized by the incident x-ray. This interpretation of the KVV Auger spectrum thus shows that the Coulomb repulsion energy U is much larger than the bandwidth of either the valence or conduction bands [17.9]. It should be noted that the nearest neighbor correlation energy V has also been determined ($V = 0.3$ eV) by the AES technique [17.78, 79].

17.4. Scanning Tunneling Microscopy

Scanning tunneling microscopy has provided an excellent technique for studying the epitaxial growth of C_{60} films on various substrates and for identifying the surface crystal structures, crystalline domain sizes, and the common defect structures that occur in epitaxial fullerene films [17.2]. It is found that STM is an especially powerful technique for studying the interaction between fullerene molecules and surfaces and between fullerene molecules adsorbed on various substrates, as is discussed in §17.4.1 and more extensively in §17.9. Also of importance is the use of STM to identify the crystal structure of nanometer dimension crystallites prepared from higher-mass fullerenes (§17.9.5). The STM tip has also been used to manipulate individual C_{60} molecules on the surface. The imaging of individual carbon atoms on the C_{60} molecules using STM has been difficult to achieve because of the rotational degrees of freedom of fullerenes near 300 K, although images showing internal structure have been obtained (see §17.9). It has also been found that the addition of C_{60} to an STM tip increases the resolution of the STM instrument [17.80], as is further discussed in §20.5.2.

An STM tip has also been used for electron beam irradiation of a C_{60} surface to stimulate the polymerization of C_{60} [17.81]. The same STM tip operating at a lower voltage was used to probe the polymerized material (see §7.5.2) and to write fine lines on a fullerene surface. In addition, STM has been used to carry out tunneling experiments on K_3C_{60} and Rb_3C_{60} to measure the temperature dependence of the superconducting energy gap [17.82, 83], and these results are summarized in §15.4. The STM technique has also been used extensively to study the structure of carbon nanotubes (see §19.6.1) [17.84, 85].

17.4.1. STM Studies of the Fullerene–Surface Interaction

Scanning tunneling microscopy provides an excellent tool for studying the interaction between fullerenes and surfaces and has been used extensively for such studies (see §17.9). The STM results show that C_{60} interacts strongly with metal surfaces, such as Al [17.75], Cu (111) [17.86], Ag (111) [17.87], Au (111) [17.88], Au (001) and Au (100) [17.89], and weakly with oxide surfaces [17.72, 90–93]. The interaction with these metal surfaces is much stronger than the C_{60}–C_{60} interaction, but the opposite applies to these oxide surfaces. Although a range of values are given in the literature for the binding energies of C_{60} to specific surfaces, the values given in

Table 17.2

Desorption energy and desorption temperature for C_{60} from various substrates.

	Desorption	Desorption energy			
Substrate	T_D (K)	(kcal/mol)	(eV/C_{60})	Technique[a]	Reference
C_{60}	500–600	32	1.4	TPD	[17.94]
SiO_2	460	23	1.0	MB, TPD	[17.70, 95]
α-Al_2O_3	435–450	25	1.1	TPD	[17.94]
Al[b]	~750	>44	>1.9	AES	[17.75]
Au (111)	~773	~48	~2.0	AES, STM	[17.96]
Ag (111)	~823	~51	~2.2	AES, STM	[17.96]
Steel[c]	~650	>36	>1.5	AES	[17.97]
Si (100)[c]	~900	>56	>2.4	AES	[17.70]
Si (111)[c]	~850	>53	>2.3	AES, STM	[17.71]
HOPG	550–620	38–40	1.6–1.7	TPD, MB	[17.97]

[a]Notation: temperature-programmed desorption (TPD), molecular beam spectroscopy (MB), Auger electron spectroscopy (AES), and scanning tunneling microscopy (STM).

[b]Dissolution of C_{60} into surface of the aluminum foil occurred. Therefore no desorption could be observed.

[c]Reaction of C_{60} with the surface occurred to form carbide. Therefore no desorption was observed.

Table 17.2 serve to distinguish between surfaces which bind more strongly to C_{60} than C_{60} to itself, and surfaces which bind less strongly to C_{60}.

For semiconductor surfaces, the interaction can be very strong, as for the case of C_{60} on Si (111) [17.90]. The strong C_{60}–Si (111) interaction is attributed to the high concentration of dangling bonds at the surface. The interaction of C_{60} with a semiconductor surface can also be much weaker than that for C_{60} on Si and comparable to that of C_{60} with itself, as, for example, for C_{60} with the GaAs (110) surface [17.91, 92].

The growth pattern of the fullerene films is strongly affected by the strength of the C_{60}–substrate interaction, the substrate corrugation, and the lattice mismatch [17.88]. For example, for a GaAs (110) substrate which exhibits a weak van der Waals interaction with C_{60} as mentioned above, the C_{60} molecules are mobile and the film growth occurs initially through formation of large two-dimensional islands, with a second layer forming before the first layer growth is complete. The growth of C_{60} on GaAs (110) is highly temperature dependent, with the room temperature growth determined by the structure of the GaAs surface, yielding commensurate monolayer growth and high-index facets for multilayer C_{60} films. At higher temperatures (e.g., 200°C), the C_{60}–C_{60} interaction dominates and incommensurate (111) oriented C_{60}films are formed [17.91, 92]. For the case of the Si (111)

(7×7) surface, the bonding of C_{60} to the substrate is very strong, so that the C_{60} has low surface mobility, bonds easily to the Si, and, under a variety of experimental conditions, shows no preferred bonding sites [17.71, 90, 98]. Although, no long-range well-ordered C_{60} growth takes place on Si (111) at room temperature [17.90], more crystalline C_{60} films can be grown on heated Si substrates [17.99].

17.4.2. Atomic Force Microscope Studies

In addition to STM, atomic force microscopy (AFM) has been applied, but to a lesser degree, to the study of fullerene molecules, films, and the interaction of fullerenes with substrates.

An AFM tip has been used to image single fullerene molecules [17.100]. These AFM imaging measurements were made under ethanol to provide the very low contact forces required to prevent destruction or deformation of the soft C_{60} or C_{70} crystal surface. Under ethanol, three benefits for AFM observations can be realized: the tip–fullerene interaction is very weak (dominated by van der Waals forces), the AFM cantilever noise is low, and the fullerene surface is cleaned of soluble contaminants. The AFM tip under ethanol can distinguish the rugby ball shape of C_{70} relative to the soccer ball shape of C_{60}. The maximum diameters for the C_{60} (9.4 ± 1.3 Å) and the C_{70} (11.2 ± 1.3 Å) molecules can be measured, based on the calibration of the AFM instrument by imaging the structure of mica [17.100]. The AFM tip was used successfully to determine the local crystal structure of a C_{60} film, including the rapid alternation of fcc and hexagonal close-packed (hcp) structures within a few nm and the presence of numerous lattice defects and faults [17.100], as is also observed with the STM [17.2]. Elliptically shaped fullerene molecules in an fcc structure were not observed with the AFM, nor were individual carbon atoms within the fullerene molecules imaged. Three reasons were offered for the absence of internal structure in the AFM image: (1) the absence of topographic details that can be sensed by an AFM probe, (2) a reduction in AFM resolution due to the softness of fullerenes, and (3) the averaging of the internal structure due to the rapid rotation of fullerenes [17.100].

We now give two examples of the interaction of fullerenes with surfaces. In one study [17.101], the spiral growth of C_{60} on a cleaved highly oriented pyrolytic graphite (HOPG) surface (see §2.3) at 100°C and at a growth rate of 0.1 Å/min was investigated by AFM. Triangular pyramid-shaped islands were found to form from multilayered terraces, and these structures were found to nucleate at substrate step edges.

As a second example, the atomic force microscope was used to investigate the morphological and frictional characteristics of C_{60} adsorbed on Si (111) and mica substrates [17.102]. Microcrystallites of C_{60} were observed on both Si (111) (~450 Å size) and mica (~600 Å size) with mostly random molecular arrangements. Force measurements taken while scanning the tip over the surface indicate an increase in friction by about a factor of 5 for the adsorbed fullerene-coated surface relative to bare Si (111) or mica [17.102], consistent with similar findings using macroscopic measurement techniques [17.103].

Atomic-scale friction measurements, made by sliding an Si_3N_4 AFM tip against a fullerene surface, yield a coefficient of friction of ~0.07, slightly less than the coefficient of friction (~0.08) for the same Si_3N_4 AFM tip on a diamond surface [17.104]. Nanoindentation studies of such an AFM tip on a clean substrate and after the tip had made contact with a C_{60} surface give clear evidence for the transfer of fullerenes by the AFM tip from a fullerene surface to a substrate. Fullerene transfer thus allows material manipulation of fullerenes on a molecular scale using an AFM tip [17.104].

17.5. Temperature-Programmed Desorption

The temperature-programmed desorption technique is especially sensitive for determining the desorption rate of C_{60} from various surfaces as a function of temperature (see Fig. 17.15), yielding information on the binding energy of fullerenes to one another (for multilayer fullerene coverage) and to various surfaces (for submonolayer coverage). The heat of desorption for C_{60} multilayers is measured by making an Arrhenius plot of the loss of material from the surface *vs.* reciprocal temperature. Temperature-programmed desorption (TPD) studies thus provide a sensitive technique for studying the bonding of fullerenes to clean surfaces. In Table 17.2 both the desorption energy/C_{60} and desorption temperature are listed for C_{60} on various substrates, which are compared to the desorption of C_{60} from a C_{60} surface, also termed sublimation ($T_D \sim 500$–600 K).

The desorption process for C_{60} likely involves long-range surface diffusion as well as thermal excitation of the C_{60} molecule to overcome the surface barrier upon emission. As a C_{60} molecule diffuses over a weakly interacting 2D surface, the C_{60} molecule, because of its large size, is hardly perturbed by the corrugation potential of the surface and the C_{60} continues to diffuse over the surface until colliding with another C_{60} molecule [17.95]. In temperature-programmed desorption measurements, C_{60} molecules are introduced on a clean surface at room temperature. The surface is then heated and the number of C_{60} anions desorbed from the surface as a function of temperature is monitored by a quadrupole mass spectrometer, which is set to accept a predetermined mass-to-charge ratio. A peak in the TPD spectrum as a function of temperature indicates the binding energy of the fullerene molecule to the surface.

Temperature-programmed desorption measurements on multilayer C_{60} surfaces yield a value of 32.4 kcal/mol [17.70, 71, 94, 105, 106], independent of the substrate. This value for the heat of desorption is in the range of published experimental values for the heat of sublimation (21.5–40.1 kcal/mol) [17.71, 105–108] and for the cohesive energy (36.9 kcal/mol) [17.13]. The temperature at which C_{60} desorption from C_{60} multilayers is complete is independent of the underlying substrate, for all substrates which have a strong bonding to a C_{60} monolayer [17.71]. Values of the temperature at which C_{60} molecules desorb from various substrates are given in Table 17.2, where it is seen that higher desorption temperatures T_D correlate well with higher activation energies for desorption.

The heat of desorption for submonolayer C_{60} coverage is highly surface dependent, thereby providing detailed information about the bonding of C_{60} to various substrates. For some substrates such as SiO_2 and sapphire (α-Al_2O_3), the binding of C_{60} to the substrate is weaker than to another C_{60}, and for such surfaces, the C_{60} molecules at low coverage tend to form islands that upon heating are completely desorbed from the substrate, leaving a clean substrate surface behind [17.94]. For example, the desorption energy for C_{60} from a sapphire $(1\bar{1}02)$ surface is 25 ± 6 kcal/mol, where one monolayer of C_{60} corresponds to 1.21×10^{14} C_{60} molecules/cm^2, and this desorption energy is 7 kcal/mol less than the C_{60} binding to itself. For the case of a clean Si (100) surface, C_{60} molecules remain on the surface until high temperatures (900–1150 K) are reached. At these elevated temperatures, the C_{60} molecules burst, thereby depositing a large number of carbon atoms or clusters on the surface, which then react with the hot Si substrate to form SiC (see §17.3). As Table 17.2 shows, the binding energy of the C_{60} to the Si (100) surface is $>$ 56 kcal/mol, well above the C_{60}–C_{60} activation energy for desorption (32.4 kcal/mol) [17.70].

Closely related to the sublimation energy is the temperature dependence of the equilibrium vapor pressure of C_{60} given by

$$P_v(T) = (2\pi m k_B T)^{1/2} \frac{\nu_0}{\alpha} \exp\left(- \frac{\Delta H_{\text{sub}}}{RT} \right) \tag{17.2}$$

where α is the sticking coefficient for the molecule on the substrate, ν_0 is the preexponential factor, m is the molecular mass, and k_B and R are, respectively, the Boltzmann constant and the gas constant. Assuming $\alpha = 1$, estimates for P_v are 10^{-9} torr at 490 K, 10^{-6} torr at 610 K, 10^{-3} torr at 800 K, and 1 torr at 1200 K [17.105], using $\nu_0 = 4.3 \times 10^{12}$ monolayer/s for the desorption rate of C_{60} from a C_{60} surface [17.94].

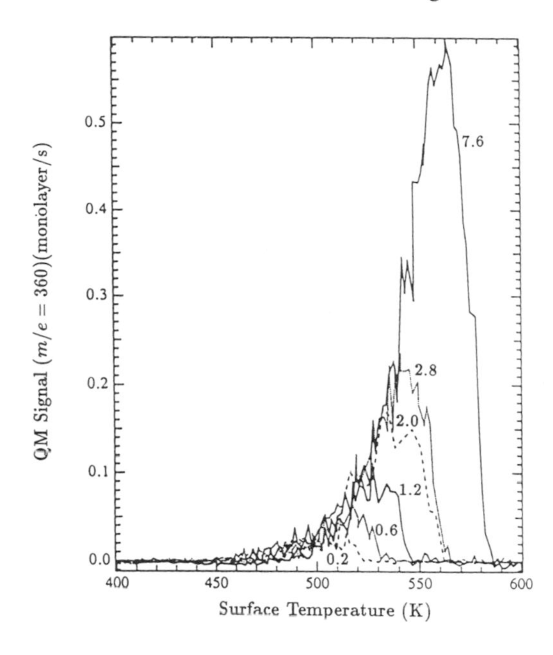

Fig. 17.15. Temperature-programmed desorption (TPD) of C_{60} multilayers from an Si (100) (2×1) surface as the C_{60} coverage is increased. The amount of C_{60} that is desorbing increases from 0.2 to 7.6 monolayers (ML) as shown in the figure. Doubly ionized C_{60}^{2-} is monitored with a quadrupole mass spectrometer (QMS). An Arrhenius analysis of the rising portion of the zeroth-order TPD curves yields an activation energy of 31 ± 1 kcal/mol and a preexponential factor of $(2 \pm 2)\times 10^{12}$ monolayer/s [17.70].

17.6 Work Function

The work function ϕ denotes the energy necessary to take an electron from its highest-lying bound state into the vacuum level and is usually measured by the photoemission technique. The work function for C_{60} has been determined using high resolution EELS measurements [17.54, 55, 109] and from measurements of the secondary cut-off in the photoemission spectra [17.110, 111], by measuring the change in work function of a metal surface when a monolayer, submonolayer or multilayer of C_{60} is deposited on the surface. The results obtained from a variety of metal substrates [Au (110), Rh (111), Ta (110), GeS (001)] yield values of $\phi = 4.8 \pm 0.1$ eV for the work function for a C_{60} multilayer sample, and $\phi = 4.9 \pm 0.1$ eV for a monolayer or submonolayer C_{60} sample. It has also been shown that the presence of alkali metal atoms on a clean metal surface lowers the work function for C_{60} subsequently deposited on this surface [17.111–114]. The lowering of the work function of a metal by the presence of adsorbed C_{60} indicates charge transfer between C_{60} and the metal surface [17.54].

17.7. Surface-Enhanced Raman Scattering

Surface-enhanced Raman scattering (SERS) provides yet another informative probe of the interaction between fullerenes and substrates. Although most of the SERS studies have been carried out on noble metal surfaces (gold [17.115, 116], silver [17.116, 117], copper [17.116]), one study has also been done on Al_2O_3 [17.50]. In some of the SERS experiments, sufficiently high light intensities were used [17.116, 117] so that the resulting C_{60} spectra are suggestive of phototransformation effects (see §7.5.1); nevertheless, interesting qualitative conclusions are provided by these SERS experiments. In this early SERS work, polycrystalline (nonoriented) substrates were used for the SERS experiments.

For a monolayer of C_{60} on noble metal substrates, downshifts in the Raman-active mode frequencies were generally observed, with some modes downshifting more than others. Also, additional mode frequencies were identified in the SERS spectra. Using the strong Raman $A_g(2)$ pentagonal pinch mode as a calibration standard, downshifts between 15–18 cm^{-1} were observed on Au surfaces, between 26–28 cm^{-1} on Ag surfaces, and 23 cm^{-1} on Cu surfaces, indicative of the increasing interaction between C_{60} and the Au, Cu, and Ag surfaces in that order, and consistent with the sequence of work functions for the substrates Ag (4.3 eV), Cu (4.7 eV), and Au (5.1 eV) [17.116]. Essentially no shift in the $A_g(2)$ SERS Raman line was observed for C_{60} on Al_2O_3 [17.116], consistent with the very weak interaction between C_{60} and oxide surfaces (see §17.4.1).

The nature of the fullerene surface interaction is attributed to an electron charge transfer mechanism from the metal surface to the fullerenes, with frequency downshifts for the $A_g(2)$ mode of approximately the same magnitude as those observed for M_3C_{60} (M = K, Rb). This charge transfer effect is consistent with photoemission experiments [17.118] and calculations [17.119], and the relative magnitudes of the C_{60}–metal interactions found by SERS measurements are consistent with STM studies (see §17.4 and §17.9). On the other hand, the spectral shifts of the various mode frequencies for M_3C_{60} relative to C_{60} in comparison to the SERS spectra of C_{60} on noble metal surfaces do not follow a close analogy, indicating that effects other than simply charge transfer are present in the interaction of C_{60} with noble metal surfaces. In addition, a moderate intensity vibrational mode has been identified at 340 cm^{-1} for C_{60} on Au substrates [17.115] and 341 cm^{-1} on Ag [17.116, 117], and this mode frequency has been identified with a C_{60}–metal vibrational mode. The appearance of additional modes in the SERS spectra was attributed to a symmetry-lowering effect induced by the surface [17.115–117]. Line-broadening effects were also observed in the SERS spectra [17.115].

SERS experiments for C_{70} on an Ag surface also show a softening of the pentagonal pinch mode, but by a lesser amount (17 cm^{-1}) than for C_{60} (26 cm^{-1}) [17.116], indicative of a weaker interaction between C_{70} and the Ag substrate and of a lower electronegativity in C_{70} relative to C_{60} [17.117]. Corroborative UV photoemission studies indicate that a metallic surface is obtained from a monolayer of C_{60} on noble metal substrates. X-ray photoemission spectroscopy studies of the carbon 1s core level for C_{60} on noble metal surfaces, however, yield trends somewhat inconsistent with SERS measurements, also suggesting that other interactions, in addition to charge transfer between C_{60} and noble metal interfaces, may be important [17.116].

17.8. Photoionization

Photoionization studies in the gas phase and in the solid state show similar spectral features, providing direct measures of the ionization potential and of the density of states [17.120, 121]. The similarity of the photoionization spectra for the gas phase and the solid state (see Fig. 17.16) confirms the molecular nature of C_{60}, as does also the small magnitude of the broadening in the solid-state photoionization spectrum.

17.9. Fullerene Interface Interactions with Substrates

The bonding of fullerenes to some substrates is strong while to others it is weak. Furthermore, metals tend to have a lower work function than semiconductors. Because of the band gap between the occupied valence band

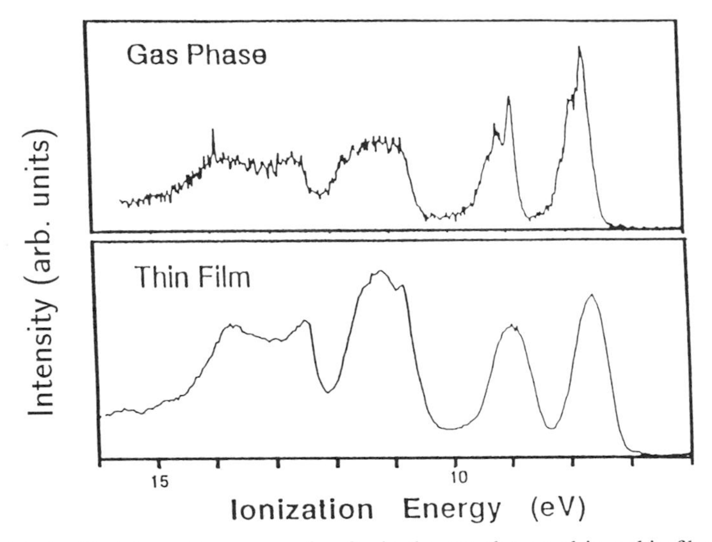

Fig. 17.16. Photoionization spectra for C_{60} in the gas phase and in a thin film. The gas phase results show fine structure associated with the coupling of electronic and vibronic channels. The broadening of electronic levels in the solid is not very different from that for isolated molecules [17.120, 121].

and the empty conduction band in semiconductors, more energy is usually required to remove an electron from its highest-lying occupied state to the vacuum level, as compared to a metal. Thus it is expected that charge is more easily transferred between a fullerene and the substrate in the case of a metal. Consequently, fullerenes should bind more strongly to metal surfaces than to semiconductor or insulating surfaces. This simple argument is in general applicable to real C_{60} interfaces with substrates, except in cases where the semiconductor has a high density of surface bonds, such as the Si (111) and Si (100) surfaces. The binding of C_{60} to oxides such as SiO_2 and sapphire (α-Al_2O_3) is weak, while the binding to metals such as Al, Au (111), Ag (111), and Cu (111) is strong. The most informative probes for these studies have been scanning tunneling microscopy and temperature-programmed desorption, which are, respectively, reviewed in §17.4 and §17.5.

Substrate temperature plays an important role in film growth on semiconductor surfaces. Low-temperatures tend to favor the growth of commensurate epilayers of small grain size in cases where the bonding is strong and approximate lattice matching can occur. Such growth, however, is often accompanied by a great deal of strain that is relieved by the formation of a large number of defects in the epilayer. Because of enhanced surface diffusion, higher temperatures favor growth that exhibits much larger structural perfection on incommensurate epilayers. An example of incommensurate growth is the formation at higher temperatures (470 K) of nearly incommensurate defect-free C_{60} (111) regions of ~1000 Å in size on GaAs(110) [17.91]. An example of commensurate overlayer formation is the $c(2 \times 4)$ overlayer of C_{60} on GaAs (110) [17.92]. Since the sublimation of C_{60} molecules from a surface becomes appreciable in the temperature range 300–500°C, or lower from some surfaces, film growth is normally carried out at temperatures below which sublimation is pronounced.

A wide variety of substrates have been used for C_{60} deposition and many of these have been investigated by STM techniques (see §17.4), including Si (111), GaAs (110), GeS (001), sapphire, Au (111), Au (100), Cu (111), CaF_2 (111), mica, H-terminated Si (111), cleaved MoS_2, NaCl, and other alkali halides [17.4]. Suitable substrates for C_{60} require that the cohesion energy between substrate atoms is large so that intercalation does not occur and that the integrity of the metal surface is preserved upon fullerene adsorption. Furthermore, studies of the fullerene–substrate interface interaction have been carried out for a variety of fullerenes including C_{60}, C_{70}, C_{76}, C_{78}, C_{82}, and C_{84} [17.4, 25], showing that the reduction of symmetry tends to result in a preferred alignment of the fullerenes during film growth, as imposed by steric considerations.

Temperature-programmed desorption studies (see §17.5) provide direct information on the binding energy of C_{60} to a substrate, as well as the

C_{60}–C_{60} binding in the solid phase. It is of particular interest to distinguish between those substrates to which C_{60} binds more strongly than C_{60} binds to itself and substrates to which C_{60} is more weakly bound. Scanning tunneling microscopy (see §17.4) is especially informative regarding the growth characteristics of C_{60} on various surfaces, the ordering of the C_{60} molecules on these surfaces, and surface reconstruction effects, should they be present.

In this section we review the interaction of C_{60} with various types of substrates.

17.9.1. C_{60} on Noble Metal Surfaces

For noble metal surfaces, Cu, Ag, and Au, the magnitude of the C_{60}–substrate interaction is between that for the weakly bound GaAs surface and the strongly bound Si surface. Most of the studies of C_{60} on the fcc noble metal Cu, Ag, and Au surfaces have been on (111) faces because they are close packed. For these metal surfaces, the impinging C_{60} molecules are mobile, but once the C_{60} binds to the metal surface, the C_{60} molecule is not mobile up to ~500°C. For the noble metal surfaces, the film growth is strongly dependent on the adsorption site and the growth conditions [17.87].

Evidence for the strong bonding of C_{60} to noble metal surfaces is provided by electronic, structural, and thermal studies. From an electronic standpoint, the strong bonding of C_{60} to noble metal surfaces is supported by the large shift of the Fermi level observed in the inverse photoemission spectra [17.86], in photoemission studies [17.51], and in theoretical calculations of the electronic structure [17.86], since C_{60} is chemisorbed rather than physisorbed to the metal surface. Shown in Fig. 17.17 is the resulting upshift in the Fermi level from the gap between band #180 (on this diagram the HOMO level is at band $3 \times 60 = 180$) and band #181 (the LUMO level). Here E_F is seen to lie within band #181 (the LUMO band), giving theoretical justification for electron charge transfer to C_{60} from the metal surface, thereby forming C_{60} anions. Thermal evidence for the strong binding of C_{60} to noble metal surfaces is provided by the observation that C_{60} desorbs from a C_{60} surface at ~300°C, but much higher temperatures (of ~500°C) are required to remove a C_{60} monolayer from an Ag (111) or Au (111) surface [17.87, 88]. This effect allows the preparation of single-monolayer C_{60} films on Ag (111) and Au (111) surfaces by annealing a multilayer film at about 300°C. However, a high density of steps on the metal substrate often leads to a somewhat nonideal monolayer surface. Structural evidence for the strong binding between C_{60} and a noble metal substrate is provided by the observation of growth initiation at fcc sites at the edges of monatomic surface steps on narrow terraces for both Ag (111) and Au

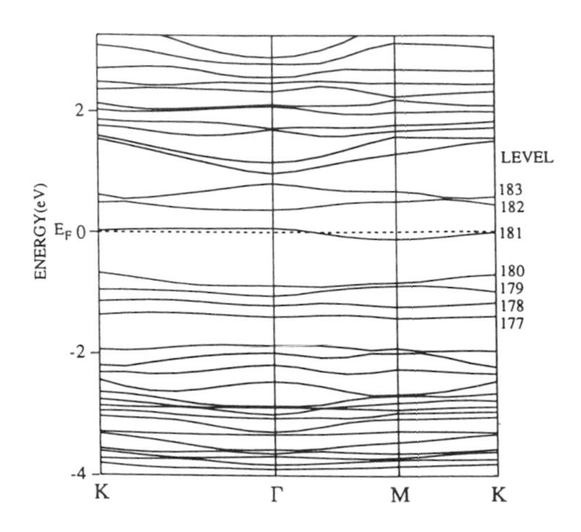

Fig. 17.17. Two-dimensional energy band structure of the C_{60} triangular lattice on a Cu (111) substrate, which allows for charge transfer from the substrate to the C_{60} adsorbate. The triangular lattice with no coupling between the C_{60} and the substrate would place the Fermi energy between the #180 and #181 levels. It is seen that charge transfer moves E_F into band #181 (the LUMO-derived band) [17.86].

(111). On these surfaces the C_{60} molecules are mobile and easily traverse steps, so that the C_{60} molecules can travel large distances prior to binding strongly to preferred sites [17.87, 88].

A close-packed monolayer growth is initiated at the step edges, but with a different structure for each of the noble metals. For Cu (111) with the smallest lattice constant, a commensurate (4 × 4) superlattice is formed, because the nearest-neighbor Cu–Cu distance on the substrate is close (lattice mismatch <1%) to one fourth of the C_{60}–C_{60} distance in its fcc lattice (10.2 Å) [17.86] (see Table 17.3). This superlattice is to be compared with that formed on the Ag (111) substrate, which has a nearest-neighbor distance of 2.88 Å in its fcc lattice. C_{60} on Ag (111) forms an ordered $(2\sqrt{3} \times 2\sqrt{3})$ commensurate superlattice [17.87], where we note that $2\sqrt{3} \times 2.88$ Å ~ 10.0 Å is approximately the same as the nearest-neighbor C_{60}–C_{60} distance of 10.18 Å (see Table 17.3). Four types of domains are observed

Table 17.3

Structures of C_{60} adsorbed monolayers on various noble metal substrates[a] [17.86].

Substrate	a (Å)	Superlattice
Cu (111)	2.55	(4×4)
Ag (111)	2.89	$(2\sqrt{3} \times 2\sqrt{3})$R30°
Au (111)	2.88	$(2\sqrt{3} \times 2\sqrt{3})$R30°
Au (111)	2.88	(38×38)

[a] Additional commensurate structures have been observed, depending on surface conditions.

for the C_{60} overlayer on Cu (111), two of which are displaced from each other by one third of the Cu–Cu distance in the $\langle 0\bar{1}1 \rangle$ direction, consistent with a hexagon of C_{60} located over the hollow site of the Cu (111) substrate, as shown in Fig. 17.18, where the locations of the three surrounding pentagons are indicated [17.86].

Interestingly, the Au (111) surface with $a_0 = 2.89$ Å shows a minority phase $(2\sqrt{3} \times 2\sqrt{3})$R30° commensurate structure for the epitaxial C_{60} monolayer, but C_{60} on Au (111) shows a majority phase with a large unit cell relative to the substrate (38×38), although having poor lattice matching to the substrate. The unit cell for this majority phase contains (11 × 11) C_{60} molecules. A second epitaxial phase $[(7/2) \times (\sqrt{3}/2)]$R13.9° for C_{60} adsorption on Ag (111) has also been reported [17.122]; this phase can coexist with the $(2\sqrt{3} \times \sqrt{3})$R30° structure, depending on the substrate temperature and the film growth conditions. The C_{60} molecule is more likely to adsorb to specific sites on Ag (111) than on Au (111), indicating a somewhat stronger binding of C_{60} to the Ag (111) surface. This stronger binding is also confirmed by measurement of the reorientation times of C_{60} molecules on Ag (111) and Au (111) surfaces [17.88], but thermal measurements indicate that the binding of C_{60} to Ag (111) is only slightly greater than to Au (111). The structures listed in Table 17.3 for C_{60} adsorbed on various substrates are clearly only a partial list, since other structures are expected to occur under different growth conditions.

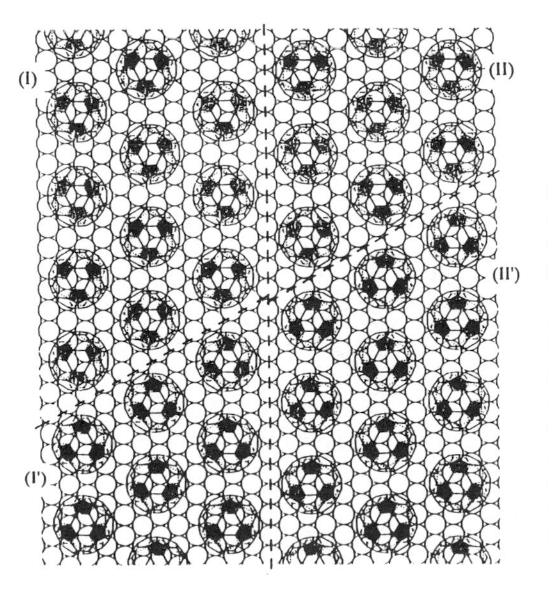

Fig. 17.18. Proposed adsorption model of the C_{60} overlayer on Cu (111), showing a (4 × 4) superlattice and threefold symmetry. The intermolecular C_{60}–C_{60} distance is 10.2 Å, equal to four times the Cu-Cu nearest-neighbor distance (see Table 17.3). Three pentagonal rings located in the upper portion of the C_{60} molecules are shaded for better viewing of the rotational freedom of each C_{60} molecule. Four distinct domains are shown in (*I*), (*I'*), (*II*), and (*II'*), where each C_{60} is adsorbed over a different hollow site in the (4 × 4) superlattice [17.86].

In the Au (111) surface, steps tend to run parallel to the $[1\bar{1}0]$ direction and perpendicular to the reconstruction lines. The orientation of the steps determines the orientation of the C_{60} overlayer for the Au (111) substrate and to a lesser degree for Ag (111). In the growth process on noble metal surfaces, the first layer of the C_{60} growth tends to be completed before C_{60} molecules are added to the second layer [17.88]. The molecules in the second layer tend to occupy the threefold hollow sites of the first layer. The second layer molecules show little affinity for steps, in contrast with the first layer. The second layer C_{60} molecules spin rapidly (10^9/s), while the first layer molecules are tightly bound. By the time the third layer forms, there is little explicit memory of the substrate, as discussed below. Well-ordered C_{60} overlayers of thickness >2 monolayers can be grown on all three Cu surfaces, but only the Cu (111) substrate yields single-domain epitaxial growth. For ordered C_{60} growth, elevated substrate temperatures ($\sim$300°C) are necessary [17.51]. The surface energy for C_{60} on Cu (111) is much lower than that for Cu (100) and Cu (110).

When a gold overlayer was deposited on a single monolayer of C_{60} on Au (111), clustering on top of the C_{60} molecules was observed by STM [17.123]. However, when an Au overlayer was deposited on two layers of C_{60}, then the Au diffused under the second monolayer to form 2D clusters, resulting in a granular structure for the second C_{60} layer.

Interesting charge transfer studies were also carried out using the EELS technique (see §17.2) for C_{60} on Au (110) surfaces [17.55]. By monitoring the EELS peaks in the 1–5 eV range, in the carbon 1*s* core excitation spectra, and using high-resolution EELS to measure the vibrational spectra, a charge transfer of $1.3 \pm 0.7e$ from the Au (110) surface to the C_{60} was obtained. By depositing a monolayer of K on the Au (110) surface before introducing the C_{60}, the charge transfer was increased to $2.3 \pm 0.7e$ (almost comparable to that observed in K_3C_{60}), and when several K layers were deposited on Au (110) before introducing the C_{60}, a value of 5.3 ± 0.7 electrons of charge transfer per C_{60} molecule was reported (almost comparable to that observed for K_6C_{60}) [17.55]. A C_{60} monolayer on Au (001) is commensurate with the gold substrate, but because of the lattice mismatch, a uniaxial stress along the $\langle 110 \rangle$ direction develops. This stress causes the charge density around the C_{60} molecule to deform, and a deformed charge density is observed in scanning tunneling spectroscopy (STS) data [17.89]. These experiments point toward a way to control charge transfer to adsorbed fullerenes by control of the substrates (with different work functions) prior to the introduction of the fullerenes. These observations also point to the importance of using STM to characterize substrates before and after the deposition of the fullerenes to determine charge transfer.

One unique feature of the field ion scanning tunneling microscopy study of C_{60} on Cu (111) is the observation of a unique threefold symmetry pattern [17.86], which agrees well with the calculated charge density around the molecule (see §7.1.2). After all the step sites are occupied, two-dimensional C_{60} islands start to grow with a close-packed arrangement from the step edges toward the upper terraces, eventually forming a monolayer, before the second layer starts to grow; this growth pattern seems to be characteristic for C_{60} on (111) noble metal surfaces. STM determination of the height of the C_{60} monolayer on Cu (111) ranges from 4.2 to 5.0 Å at a bias voltage between –3 V and +3 V. The monolayer step height is significantly smaller than the height of the second layer (8.2 Å), reflecting the difference in the local density of states between the Cu (111) surface and the C_{60} molecule. Higher-magnification images reveal a threefold symmetry (see Figs. 17.19 and 7.3) for a single C_{60} molecule on a Cu (111) surface, indicative of an excess of surface states near the surrounding three pentagonal rings and a deficiency of states [indicated by the doughnut shape pattern of Fig. 17.19(a)] over the central hexagon of this cluster. The patterns of Fig. 17.19 suggest that a hexagonal ring of the C_{60} molecule lies face down on the Cu (111) surface, flanked by three pentagons forming a triangular pattern about this hexagon. If a single carbon atom or the carbon double bond of a C_{60} molecule were to make contact with the surface [as occurs for an Si (100)(2 × 1) surface], twofold rather than threefold symmetry would be seen [17.119].

17.9.2. C_{60} on Transition Metal Surfaces

Thus far, only a few studies have been done on the interaction between C_{60} and transition metal surfaces. Two surfaces which have been studied are W (100) [17.53] and Rh (111) [17.54]. The available data on transition metal

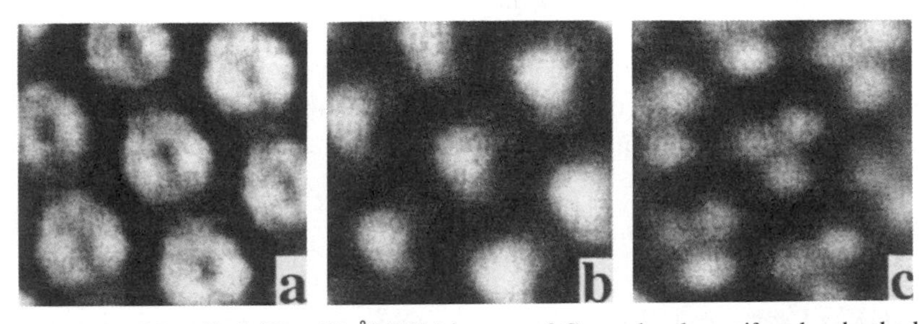

Fig. 17.19. Magnified (22 × 22 Å) STM images of C_{60} molecules uniformly adsorbed on the Cu (111) surface at bias voltages of (a) –2.0 V, (b) –0.10 V, and (c) 2.0 V. STM images reveal a strong dependence of the images on the bias voltage, but all three images exhibit the threefold symmetry of the adsorbed layer [17.86].

surfaces are not yet sufficient for forming general conclusions, although relatively strong binding seems to occur between C_{60} and transition metal substrates, as is the case for the noble metal surfaces, and no tendency for exohedral intercalation of transition metals into crystalline phase C_{60} interstices is observed.

Photoemission studies of C_{60} on W (100) surfaces provide evidence for electron charge donation from the metal surface to C_{60} to form C_{60} anions, as for the case of C_{60} monolayers on noble metal surfaces. Annealing a multilayer C_{60} film at a temperature as low as 100°C appears to result in sublimation of the C_{60} overlayers, leading to monolayer C_{60} coverage on the transition metal surface, consistent with the stronger bonding of C_{60} to W (100) as compared with C_{60}–C_{60} bonding. Consistent with the behavior of C_{60} on noble metal surfaces, it is concluded that the Fermi energy is pinned to the fullerene LUMO level for C_{60} on the W (100) surface [17.53].

Studies of the interaction between C_{60} and the Rh (111) surface have been carried out using several complementary measurement techniques, including Auger electron spectroscopy, high-resolution electron energy loss spectroscopy, work function measurements, and ultraviolet photoelectron spectroscopy [17.54]. For Rh (111), the C_{60}–Rh interaction is stronger than the C_{60}–C_{60} binding, as determined from temperature-programmed desorption studies, although no peaks associated with the C_{60}–Rh interaction could be identified in the ultraviolet photoelectron spectroscopy (UPS) spectra [17.54]. As shown by Auger electron spectroscopy studies, the growth mechanism favors layer-by-layer growth rather than island growth, at least up through 3 ML. However, some spectral features in the optical spectra show upshifts in energy toward E_F. These energy upshifts have been interpreted to imply that electrons are transferred from the C_{60} to the Rh substrate, in contrast to the sign of the charge transfer for W and the noble metals. This sign of the charge transfer is further supported by the blue shift of the vibrational modes of C_{60}, as observed by HREELS [17.54], and decreases in the work function of the Rh (111) surface by 0.35 eV, upon C_{60} adsorption, although no fundamental explanation has been given for this sign of charge transfer. From the HREELS study, a positive induced dipole moment of 2.7×10^{-3} Cm was deduced.

17.9.3. C_{60} on Si Substrates

Because of the high density of dangling bonds on Si (111) 7×7 [17.124] and Si (100) 2×1 surfaces [17.125], a considerable amount of charge transfer can be expected for C_{60} on Si (111) and Si (100) surfaces, and such effects have indeed been observed [17.71, 119, 126]. STM results for a C_{60} monolayer on an Si (100) 2×2 surface heated to 330°C show strong bonding between C_{60}

and the Si substrate and show that the preferred adsorption site for the C_{60} is the trough site surrounded by four Si dimers or eight Si atoms, as shown in Fig. 17.20, without any preference for step edges or defect sites. (Other studies of C_{60} on Si (111) 7×7 [17.71, 90, 98] show no preferred absorption sites.) The STM scans show that two different structures form for C_{60} on Si (100) 2×1, namely centered $c(4 \times 3)$ and $c(4 \times 4)$ ordered overlayer structures, as shown in Fig. 17.20. For the $c(4 \times 4)$ overlayer, the nearest-neighbor C_{60}–C_{60} distance is 10.90 Å (in-plane density of $8.42 \times 10^{13}/cm^2$), where the C_{60}–C_{60} distance is somewhat larger than the 10 Å separation in the fcc bulk phase [17.127]. In contrast, the $c(4 \times 3)$ structure (see Fig. 17.20) has a distance of 11.5 Å between A–A molecules and 9.6 Å between A–B′ molecules, leading to a planar molecular density of $1.13 \times 10^{14}/cm^2$. We note in Fig. 17.20 that the A site has four dangling bonds nearby, while the B′ site has eight dangling bonds. In comparison, the C_{60} density on an fcc (111) surface of crystalline C_{60} is $1.15 \times 10^{14}/cm^2$. The strong bonding of C_{60} to the Si (100) 2×1 surface leads to strained overlayer structures. These strains are relieved by islandic growth on multiple C_{60} layers in the Stranski–Krastanov growth regime with C_{60}–C_{60} distances of 10.4 Å. In this islandic C_{60} growth (see Fig. 17.21), the dominant structures are fcc (111) surfaces, which appear clearly after three ML of C_{60} have grown. The first and second ML of the island image, as obtained by STM, are somewhat

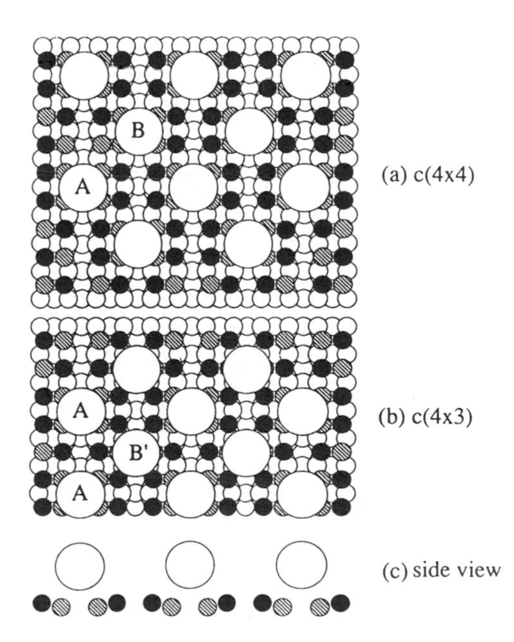

Fig. 17.20. Schematic diagram of the adsorption position of the first layer of C_{60} on the Si (100) 2×1 surface: (a) this arrangement represents the $c(4 \times 4)$ configuration, while (b) represents the $c(4 \times 3)$ configuration, and (c) is the side view of (a) and (b) at position A. Small open circles are the bulk Si atoms, hatched and solid circles are buckled Si surface dimers, with the solid circles representing the upper and the hatched circles representing the lower Si atoms in the dimers. The large open circles are C_{60} molecules [17.126].

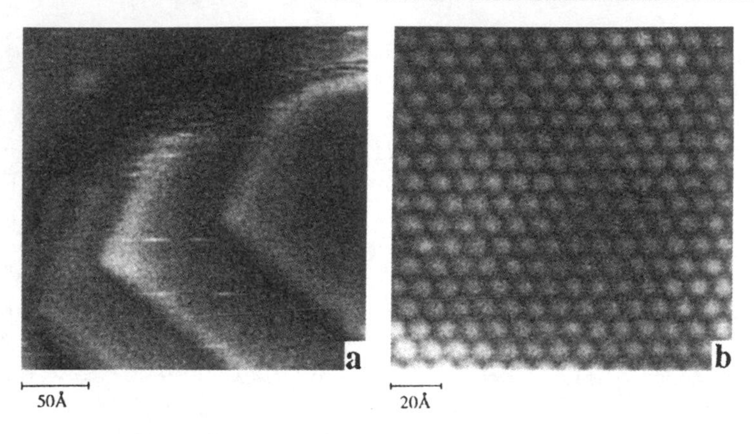

Fig. 17.21. (a) STM image of a five-layer C_{60} island showing the close-packed fcc (111) surface. (b) An enlarged STM image of the top layer of the surface shown in (a), reveals uniform brightness of individual molecules (tip voltage and current being $V_s = -3.5$ V, $I_t =$ 20 pA) [17.126].

disordered. Typical defect structures are vacancies and screw dislocations [17.126].

The local electron density of states as measured by STM shows internal structures consisting of four stripes running normal to the dimer rows [17.126], in agreement with calculations of the density of states at the HOMO level [17.128] for the $c(4 \times 3)$ structure. These calculations further show that the Fermi level is shifted toward the LUMO band because of charge transfer from the Si surface. However, overall the strong coupling between C_{60} and the Si (100) 2×1 substrate does not result in major changes in the overall electronic structure, but predominantly affects the Fermi level, similar to observations for K doping of C_{60}. The literature also cites various unsuccessful attempts to grow ordered films of C_{60} on Si (100) 2×1 [17.129] and on Si (111) 7×7 surfaces [17.90, 98, 119].

Single-crystal domains of C_{60} on clean Si (111) 7×7 surfaces have in fact been achieved [17.99] by careful growth of C_{60} on heated (~200°C) substrates. We show in Fig. 17.22 schematic diagrams of the two crystalline orientations of C_{60} on Si (111) 7×7, each making an angle of $\pm 11^\circ$ between the C_{60} $[\bar{1}10]$ axis and the Si $[\bar{2}11]$ axis. These two phases become stabilized after a sufficient number of C_{60} layers have been deposited on a clean Si (111) substrate [17.99]. In the ordered phases that are shown in Fig. 17.22, the C_{60} molecules favor locations at either the corner hole site or bridge sites of the Si (111) 7×7 structure as shown in Fig. 17.22(c). Also shown in the figure are the related preferred C_{60} sites on the same Si (111) 7×7 surface for submonolayer coverage.

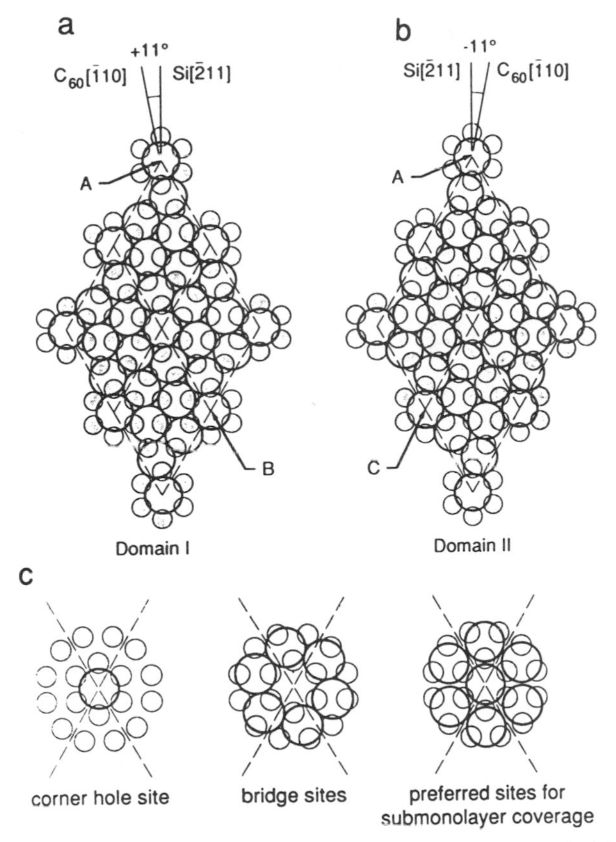

Fig. 17.22. Top: (a) and (b) show the proposed double domain interfacial structure for C_{60} crystals grown on Si (111) 7×7. Bottom: (c) shows the two basic configurations for the proposed interfacial structure beneath C_{60} crystals and also shows the preferred bonding sites for C_{60} on Si (111) 7×7 at submonolayer coverage [17.99].

When potassium is evaporated on a C_{60} film grown on an Si (100) 2×1 surface, no change is found in the STM image. However, the STS spectra for the K-doped C_{60} film show an appreciable density of states near the Fermi level, consistent with the XPS measurements of alkali metal–doped C_{60} [17.91, 92].

Auger electron spectroscopy studies (§17.3) show that the number of binding sites for C_{60} is about twice as high on an Si (111) 7×7 surface, as compared to an Si (100) 2×1 surface. The C(272 eV)/Si(90 eV) Auger peak height ratio increases in the 620–1070 K temperature range, indicative of the opening of the C_{60} cage; this cage opening occurs at a lower temperature for Si (111) 7×7 than for Si (100) 2×1, indicative of the greater activity of the Si (111) 7 × 7 surface [17.71]. The opening of the C_{60} cage has been successfully imaged by STM for C_{60} on Si (111) 7 × 7 [17.71].

17.9.4. C_{60} on GaAs

For a GaAs (110) substrate, STM studies show that C_{60} adsorption occurs initially at surface steps and because of the large number of inequivalent structures formed at 300 K, a complex surface structure is observed. Multilayer growth shows close-packed layer sequencing. At 450 K a uniform first layer forms, followed by thermodynamically favored growth in the (111) direction [17.91]. Smooth, well-ordered surfaces were preserved upon potassium doping of C_{60} to K_3C_{60} [17.91]. The surfaces for the K_3C_{60}-doped sample appeared to be conducting based on *I–V* measurements, and the surface ordering was found to be consistent with potassium in tetrahedral and octahedral sites. The absence of potassium-related STM features also provides evidence for complete charge transfer in K_3C_{60}. Additional potassium incorporation eventually led to a nonmetallic overlayer, with a great deal of disorder, consistent with a crystalline transformation from an fcc to a bcc structure and with a high density of grain boundaries and defects [17.91].

17.9.5. Higher-Mass Fullerenes on Various Substrates

Soon after the discovery of C_{60} in the solid phase, STM experiments on fullerenes on gold surfaces were undertaken, showing images with dimensions comparable to those of fullerenes [17.130]. Although no distinction could be made between the images of C_{60} and C_{70} in this early work, an image was reported with a diameter of ~1.5 nm (possibly associated with a spherical giant fullerene C_{n_C} corresponding to $n_C \sim 180$–240 carbon atoms). Subsequently, STM studies on deposits characterized by mass spectra, which had a high content of higher-mass fullerenes, showed large spherical objects in the 10–20 Å range, suggestive of C_{n_C} with $60 < n_C < 330$ carbon atoms [17.131].

Experimental STM results show that good-quality C_{70} thin films can be grown on an Si (100) 2 × 1 surface at a substrate temperature of 120°C. The behavior of C_{70} on the Si (100) 2 × 1 surface is similar to that of C_{60}, showing strong bonding of the first monolayer and weaker bonding of subsequent C_{70} layers. The first strongly bonded layer is stable up to 1000°C, but reacts at higher temperatures to form SiC (100) 3 × 2 islands, which are desorbed for $T > 1300$°C, leaving a rough substrate surface behind [17.132]. Band calculations show that the band dispersion curves for C_{70} on Si (100) 2 × 1 are similar to those of bulk crystalline C_{70}, at least much more so than for the corresponding case of C_{60} on Si (100) 2×1 [17.128]. One early report of C_{70} on GaAs (110) shows that C_{70} favors growth initiation near a substrate step. No intramolecular features could be resolved, indicating that the molecules are rotating near room temperature [17.133]. The characteristics of the growth of C_{70} and other higher-mass fullerenes on GaAs (110) are summarized in Table 17.4.

Several studies of C_{84} on semiconductor substrates have also been reported [17.126, 134]. C_{84} can easily be distinguished from C_{60} by STM because of its 10% larger size. The binding of C_{84} to an Si (100) 2×1 substrate is at the trough sites just as for C_{60} (see Fig. 17.20), with a C_{84}–C_{84} distance of ~14 Å at low coverage and with less local ordering than for C_{60} because of the larger misfit between the size of C_{84} and an Si (100) 2×1 unit cell [17.126]. At room temperature, no ordered structure for C_{84} on Si (100) 2×1 has yet been reported. However, in the 100–150°C range, an ordered monolayer can be observed with a C_{84}–C_{84} nearest-neighbor distance of 12.1 Å and a lattice constant of 17.1 Å [17.126]. Unlike the case of C_{60}, individual C_{84} molecules appear with different brightness on an Si (100) 2×1 surface, with height differences of ~1.2 Å between the lowest and highest imaged C_{84} molecules, indicative of the presence of more than one C_{84} isomer. In fact, there are 24 different isomers of C_{84} satisfying the isolated pentagon rule (see §3.2) [17.134], with each isomer expected to bind differently to Si (100). A preliminary identification has been made of the observed internal structure of the C_{84} isomer #22 (using the notation of Ref. [17.134]) with D_2 symmetry, which was identified as the dominant isomer present in the STM images [17.126].

By analyzing the geometrical relation between two C_{84} domains on an Si (100) surface, it is concluded that the island growth of C_{84} on Si (100) 2×1 is

Table 17.4

Characteristics of (4×2) island structure of C_{n_C} fullerenes on GaAs (110) substrates [17.133].

n_C	Monolayer height (Å)	C_{n_C}–C_{n_C} (Å)	Comments and Intramolecular features
60	8.9	10.2	used as a benchmark
70	9.0	10.2	none
76	9.2	11.3	incommensurate
78	9.9	11.0	none
82	10.2	11.7	random order, 2 Å height differences
84	11.0	12.1[a]	ridge-like

[a]The C_{84}–C_{84} distance is for a thin film of C_{84} grown on Si (100) 2×1 [17.126].

an fcc structure [17.126]. The STM image of C_{84} on Si (100) shows internal stripe structures similar to those found for C_{60}. It is also concluded that at room temperature the C_{84} molecules are not rotating. STM images show that the orientation of the individual C_{84} molecules is somewhat random, and the cohesive energy among the C_{84} molecules is different, depending on the molecular orientation [17.126]. The desorption temperature T_D of C_{n_C} molecules from an Si (100) surface increases with increasing fullerene mass, so that the T_{01} phase transition temperature for internal rotations for C_{84} is expected to increase relative to that for C_{60}, likely to values above room temperature [17.126].

Temperature-programmed desorption studies indicate that a temperature of $\sim$400°C is needed to desorb C_{84} from multilayers of C_{84}, while a temperature of $\sim$1000°C is needed to remove the last C_{84} monolayer from Si (100). Above 1000°C, SiC bonds are formed, and the carbon cannot easily be removed. Multiple layer islands form after the first monolayer of C_{84} on Si (100) is complete [17.126].

Fullerene overlayer structures on GaAs (110) are basically different from similar structures on Si (100) because, in the case of the GaAs (110) substrate, the interaction between fullerenes is stronger than the interaction between the fullerene and the GaAs (110) substrate. As a consequence, C_{60} at room temperature rotates freely on a GaAs (110) surface, so that intramolecular features are time averaged over the fullerene motion, and no resolved internal structure for C_{70}, C_{76}, and C_{78} is observed [17.135, 136]. On the other hand, C_{82} and C_{84} have a sufficiently high mass density (moment of inertia) that their rotational motion is highly suppressed and some intramolecular features can be observed by STM. For C_{84} on GaAs (110) substrates, a suitable substrate temperature for adsorption is 400 K [17.133]. For all fullerenes, thus far studied on GaAs (110), namely C_{60}, C_{70}, C_{76}, C_{80}, C_{82}, and C_{84}, island growth is observed, consistent with a stronger binding between the C_{n_C}–C_{n_C} fullerenes as compared to the fullerene–substrate binding. With submonolayer amounts of fullerenes deposited on GaAs (110) at 460 K, the molecules diffuse on the surface and nucleate into patches, establishing monolayer-high two-dimensional islands with grain boundaries at substrate steps [17.133], and an approximately close-packed structure is formed to minimize the free energy. While the steps on the substrate nucleate island formation for C_{60} on GaAs (110), the steps are not favored for island formation for larger fullerene molecules. This can in part be understood in terms of the step height on the substrate becoming small in comparison with the C_{n_C} diameter as n_C increases.

The various fullerenes C_{60}, C_{70}, C_{76}, C_{80}, C_{82}, and C_{84} on GaAs (110) form similar 4×2 "A-type" adsorption structures (see Fig. 17.21) [17.135, 136], but the details of the structure depend on the fullerene dimensions, as shown in

Table 17.4, where, for each fullerene C_{n_C}, the measured height of the monolayer is given, along with the nearest-neighbor C_{n_C}–C_{n_C} distance. Comments on the intramolecular features are also given in Table 17.4 [17.137].

It is expected that the greater availability of higher-mass fullerenes in purified form will stimulate further detailed surface science studies of the structure of monolayer films on various substrates.

17.9.6. Metallofullerenes Adsorbed on Surfaces

Study of the interface interaction between metallofullerenes (see §5.5 and §8.2) and the surface to which they are adsorbed is still at an early stage. One study has already been reported using scanning tunneling microscopy (STM) to probe the adsorbate–surface interaction, and the systems studied were $Sc@C_{74}$ and $Sc_2@C_{74}$ on an Si (100) 2×1 surface [17.138]. Similar to the behavior of C_{60} and other fullerenes on Si surfaces, once the metallofullerenes hit the Si (100) 2×1 surface, they immediately bond to the surface, because the adsorbate–surface bonding is very strong, and thus almost no surface migration of the metallofullerenes occurs. Also, there is no evidence for the nucleation of metallofullerene clusters on the substrate or their segregation at step edges of the substrate. The STM images for the metallofullerenes show approximately spherical structures, indicating that the metal species is contained within the fullerene cage. Since good STM images can be found at a low tip voltage ($\sim$ 0.7 eV), it is concluded that the metallofullerenes $Sc@C_{74}$ and $Sc_2@C_{74}$ on an Si (100) 2×1 surface are either metallic or have a small band gap [17.138]. Using C_{60} as a reference with regard to both the density of states and the fullerene diameter (7.1 Å), diameters of 8.8 Å and 9.5 Å were obtained for the metallofullerenes $Sc@C_{74}$ and $Sc_2@C_{74}$, respectively [17.138]. These larger diameters should be compared with that for the fullerene C_{84}, with a calculated diameter of 9.4 Å [17.139] and an experimental value of 8.6 Å [17.126]. The measured diameter for $Sc_2@C_{74}$ is large, suggesting that charge transfer occurs from the endohedral metal atoms to the surrounding carbon shell and perhaps also from the substrate to the fullerene shell.

References

[17.1] J. E. Rowe, P. Rudolf, L. H. Tjeng, R. A. Malic, G. Meigs, C. T. C. J. Chen, and E. W. Plummer. In C. Taliani, G. Ruani, and R. Zamboni (eds.), *Proc. of the First Italian Workshop on Fullerenes: Status and Perspectives*, vol. 2, p. 133, World Scientific, Singapore (1992).

[17.2] J. H. Weaver. *Acc. Chem. Res.*, **25**, 143 (1992).

[17.3] T. Takahashi, S. Suzuki, T. Morikawa, H. Katayama-Yoshida, S. Hasegawa, H. Inokuchi, K. Seki, K. Kikuchi, S. Suzuki, K. Ikemoto, and Y. Achiba. *Phys. Rev. Lett.*, **68**, 1232 (1992).

[17.4] J. H. Weaver and D. M. Poirier. In H. Ehrenreich and F. Spaepen (eds.), *Solid State Physics*, vol. 48, p. 1, New York (1994). Academic Press. Chapter 1.

[17.5] W. E. Pickett. In H. Ehrenreich and F. Spaepen (eds.), *Solid State Physics*, vol. 48, p. 225, Academic Press, New York (1994).

[17.6] R. Saito (1994). Private communication.

[17.7] J. H. Weaver. *J. Phys. Chem. Solids*, **53**, 1433 (1992).

[17.8] M. B. Jost, N. Troullier, D. M. Poirier, J. L. Martins, J. H. Weaver, L. P. F. Chibante, and R. E. Smalley. *Phys. Rev. B*, **44**, 1966 (1991).

[17.9] R. W. Lof, M. A. van Veenendaal, B. Koopmans, H. T. Jonkman, and G. A. Sawatzky. *Phys. Rev. Lett.*, **68**, 3924 (1992).

[17.10] R. W. Lof, M. A. van Veenendaal, B. Koopmans, A. Heessels, H. T. Jonkman, and G. A. Sawatzky. *Int. J. Modern Phys. B*, **6**, 3915 (1992).

[17.11] G. Gensterblum, J. J. Pireaux, P. A. Thiry, R. Caudano, J. P. Vigneron, P. Lambin, A. A. Lucas, and W. Krätschmer. *Phys. Rev. Lett.*, **67**, 2171 (1991).

[17.12] P. A. Brühwiler. Private communication.

[17.13] S. Saito and A. Oshiyama. *Phys. Rev. Lett.*, **66**, 2637 (1991).

[17.14] A. Oshiyama, S. Saito, N. Hamada, and Y. Miyamoto. *J. Phys. Chem. Solids*, **53**, 1457 (1992).

[17.15] J. Wu, Z. X. Shen, D. S. Dessau, R. Cao, D. S. Marshall, P. Pianetta, I. Lindau, X. Yang, J. Terry, D. M. King, B. O. Wells, D. Elloway, H. R. Wendt, C. A. Brown, H. Hunziker, and M. S. de Vries. *Physica C*, **197**, 251 (1992).

[17.16] E. Sohmen, J. Fink, and W. Krätschmer. *Z. Phys. B*, **86**, 87 (1992).

[17.17] S. G. Louie and E. L. Shirley. *J. Phys. Chem. Solids*, **54**, 1767 (1993).

[17.18] G. Gensterblum, J. J. Pireaux, P. A. Thiry, R. Caudano, T. Buslaps, R. L. Johnson, G. L. Lay, V. Aristov, R. Günther, A. Taleb-Ibrahimi, G. Indlekofer, and Y. Petroff. *Phys. Rev. B*, **48**, 14756 (1993).

[17.19] P. J. Benning, C. G. Olson, D. W. Lynch, and J. H. Weaver. *Phys. Rev. B*, **50**, 11239 (1994).

[17.20] G. Gensterblum, L.-M. Yu, J. J. Pireaux, P. A. Thiry, R. Caudano, P. Lambin, A. A. Lucas, W. Krätschmer, and J. E. Fischer. *J. Phys. Chem. Solids*, **53**, 1427 (1992).

[17.21] E. Sohmen and J. Fink. *Phys. Rev. B*, **47**, 14532 (1993).

[17.22] P. J. Benning, D. M. Poirier, T. R. Ohno, Y. Chen, M. B. Jost, F. Stepniak, G. H. Kroll, and J. H. Weaver. *Phys. Rev. B*, **45**, 6899 (1992).

[17.23] J. F. Armbruster, H. A. Romberg, P. Schweiss, P. Adelmann, M. Knupfer, J. Fink, R. H. Michel, J. Rockenberger, F. Hennrich, H. Schreiber, and M. M. Kappes. *Z. Phys. B*, **95**, 469 (1994).

[17.24] S. Hino, K. Matsumoto, S. Hasegawa, K. Iwasaki, K. Yakushi, T. Morikawa, T. Takahashi, K. Seki, K. Kikuchi, S. Suzuki, I. Ikemoto, and Y. Achiba. *Phys. Rev. B*, **48**, 8418 (1993).

[17.25] D. M. Poirier, J. H. Weaver, K. Kikuchi, and Y. Achiba. *Zeitschrift für Physik D: Atoms, Molecules and Clusters*, **26**, 79 (1993).

[17.26] M. S. Golden, M. Knupfer, J. Fink, J. F. Armbruster, T. R. Cummins, H. A. Romberg, M. Roth, R. Michel, J. Rockenberger, F. Hennrich, H. Schreiber, and M. M. Kappes. In H. Kuzmany, J. Fink, M. Mehring, and S. Roth (eds.), *Progress in Fullerene Research: International Winterschool on Electronic Properties of Novel Materials*, p. 309 (1994). Kirchberg Winter School, World Scientific, Singapore.

[17.27] S. Hino, K. Matsumoto, S. Hasegawa, K. Kamiya, H. Inokuchi, T. Morikawa, T. Takahashi, K. Seki, K. Kikuchi, S. Suzuki, I. Ikemoto, and Y. Achiba. *Chem. Phys. Lett.*, **190**, 169 (1992).

[17.28] J. H. Weaver, J. L. Martins, T. Komeda, Y. Chen, T. R. Ohno, G. H. Kroll, N. Troullier, R. E. Haufler, and R. E. Smalley. *Phys. Rev. Lett.*, **66**, 1741 (1991).

[17.29] C. Enkvist, S. Lunell, B. Sjögren, S. Svensson, P. A. Brühwiler, A. Nilsson, A. J. Maxwell, and N. Mårtensson. *Phys. Rev. B*, **48**, 14629 (1993).

[17.30] S. Lunell, C. Enkvist, M. Agback, S. Svensson, and P. A. Brühwiler, *Int. J. Quant. Chem.*, **52**, 135 (1994).

[17.31] P. J. Benning, F. Stepniak, D. M. Poirier, J. L. Martins, J. H. Weaver, R. E. Haufler, L. P. F. Chibante, and R. E. Smalley. *Phys. Rev. B*, **47**, 13843 (1993).

[17.32] J. H. Weaver, P. J. Benning, F. Stepniak, and D. M. Poirier. *J. Phys. Chem. Solids*, **53**, 1707 (1992).

[17.33] M. Merkel, M. Knupfer, M. S. Golden, J. Fink, R. Seemann, and R. L. Johnson. *Phys. Rev. B*, **47**, 11470 (1993).

[17.34] J. Fink, E. Sohmen, M. Merkel, A. Masaki, H. Romberg, A. M. Alexander, M. Knupfer, M. S. Golden, P. Adelmann, and B. Renker. In C. Taliani, G. Ruani, and R. Zamboni (eds.), *Proc. of the First Italian Workshop on Fullerenes: Status and Perspectives*, vol. 2, p. 161, World Scientific, Singapore (1992).

[17.35] M. Knupfer, M. Merkel, M. S. Golden, J. Fink, O. Gunnarsson, and V. P. Antropov. *Phys. Rev. B*, **47**, 13944 (1993).

[17.36] C. T. Chen, L. H. Tjeng, P. Rudolf, G. Meigs, J. F. Rowe, J. Chen, J. P. McCauley, A. B. Smith, A. R. McGhie, W. J. Romanow, and C. Plummer. *Nature (London)*, **352**, 603 (1991).

[17.37] M. J. Rosseinsky, D. W. Murphy, R. M. Fleming, R. Tycko, A. P. Ramirez, T. Siegrist, G. Dabbagh, and S. E. Barrett. *Nature (London)*, **356**, 416 (1992).

[17.38] C. Gu, F. Stepniak, D. M. Poirier, M. B. Jost, P. J. Benning, Y. Chen, T. R. Ohno, J. L. Martins, J. H. Weaver, J. Fure, and R. E. Smalley. *Phys. Rev. B*, **45**, 6348 (1992).

[17.39] P. J. Benning, F. Stepniak, and J. H. Weaver. *Phys. Rev. B*, **48**, 9086 (1993).

[17.40] E. Sohmen, J. Fink, and W. Krätschmer. *Europhys. Lett.*, **17**, 51 (1992).

[17.41] R. C. Haddon, L. E. Brus, and K. Raghavachari. *Chem. Phys. Lett.*, **125**, 459 (1986).

[17.42] J. L. Martins, N. Troullier, and J. H. Weaver. *Chem. Phys. Lett.*, **180**, 457 (1991).

[17.43] Q. M. Zhang, J. Y. Yi, and J. Bernholc. *Phys. Rev. Lett.*, **66**, 2633 (1991).

[17.44] B. P. Feuston, W. Andreoni, M. Parrinello, and E. Clementi. *Phys. Rev. B*, **44**, 4056 (1991).

[17.45] P. A. Brühwiler, A. J. Maxwell, A. Nilsson, N. Mårtensson, and O. Gunnarsson. *Phys. Rev. B*, **48**, 18296 (1993).

[17.46] M. Knupfer, D. M. Poirier, and J. H. Weaver. *Phys. Rev. B*, **49**, 8464 (1994).

[17.47] T. Takahashi, T. Morikawa, S. Hasegawa, K. Kamiya, H. Fujimoto, S. Hino, K. Seki, H. Katayama-Yoshida, H. Inokuchi, K. Kikuchi, S. Suzuki, K. Ikemoto, and Y. Achiba. *Physica C*, **190**, 205 (1992).

[17.48] Y. Chen, D. M. Poirier, M. B. Jost, C. Gu, T. R. Ohno, J. L. Martins, J. H. Weaver, L. P. F. Chibante, and R. E. Smalley. *Phys. Rev. B*, **46**, 7961 (1992).

[17.49] M. Knupfer, F. Stepniak, and J. H. Weaver. *Phys. Rev. B*, **49**, 7620 (1994).

[17.50] B. Chase, N. Herron, and E. Holler. *J. Phys. Chem.*, **96**, 4262 (1992).

[17.51] J. E. Rowe, P. Rudolf, L. H. Tjeng, R. A. Malic, G. Meigs, C. T. Chen, J. Chen, and E. W. Plummer. *Int. J. Mod. Phys.*, **B6**, 3909 (1992).

[17.52] A. J. Maxwell, P. A. Brühwiler, A. Nilsson, N. Mårtensson, and P. Rudolf. *Phys. Rev. B*, **49**, 10717 (1994).

[17.53] G. K. Wertheim and D. N. E. Buchanan. *Solid State Commun.*, **88**, 97 (1993).

[17.54] A. Sellidj and B. E. Koel. *J. Phys. Chem.*, **97**, 10076 (1993).

[17.55] S. Modesti, S. Cerasari, and P. Rudolf. *Phys. Rev. Lett.*, **71**, 2469 (1993).

[17.56] A. Lucas, G. Gensterblum, J. J. Pireaux, P. A. Thiry, R. Caudano, J. P. Vigneron, and P. Lambin. *Phys. Rev. B*, **45**, 13694 (1992).
[17.57] M. B. J. Meinders. L. H. Tjeng and G. A. Sawatzky, *Phys. Rev. Lett.*, **73**, 2937 (1994). Comment on Ref. 17.79.
[17.58] S. Krummacher, M. Biermann, M. Neeb, A. Liebsch, and W. Eberhardt. *Phys. Rev. B*, **48**, 8424 (1993).
[17.59] M. Biermann, M. Neeb, F. P. Johnson, and S. Krummacher. In K. M. Kadish and R. S. Ruoff (eds.), *Recent Advances in the Chemistry and Physics of Fullerenes and Related Materials: Electrochemical Society Symposia Proceedings, San Francisco, May 1994*, vol. 94–24, p. 952, Electrochemical Society, Pennington, N.J. (1994).
[17.60] B. Wästberg, S. Lunell, C. Enkvist, P. A. Brühwiler, A. J. Maxwell, and N. Mårtensson. *Phys. Rev. B*, **50**, 13031 (1994).
[17.61] V. I. Rubtsov and Y. M. Shul'ga. *JETP*, **76**, 1026 (1993).
[17.62] V. I. Rubtsov, Y. M. Shul'ga, A. P. Moravsky, and A. S. Lobach. *Synthetic Metals*, **56**, 2961 (1993).
[17.63] Y. M. Shul'ga, V. I. Rubtsov, A. S. Lobach, N. G. Spitsyna, and E. B. Yagubskii. *Phys. Solid State*, **36**, 987 (1994).
[17.64] P. L. Hansen, P. J. Fallon, and W. Kraätschmer. *Chem. Phys. Lett.*, **181**, 367 (1991).
[17.65] A. A. Lucas. *J. Phys. Chem. Solids*, **53**, 1415 (1992).
[17.66] G. Gensterblum, J. J. Pireaux, P. A. Thiry, R. Caudano, P. Lambin, and A. A. Lucas. *Journal of Electron Spectroscopy and Related Phenomena, V64-5*, **64/65**, 835 (1993).
[17.67] R. Kuzuo, M. Terauchi, M. Tanaka, Y. Saito and H. Shinohara. *Phys. Rev. B*, **49**, 5054 (1994).
[17.68] J. F. Armbruster, M. Roth, H. A. Romberg, M. Sing, M. Schmitt, M. S. Golden, P. Schweiss, P. Adelmann, J. Fink, R. H. Michel, J. Rockenberger, F. Hennrich, and M. M. Kappes. *Phys. Rev. B*, **51**, 4933 (1995).
[17.69] Y.-N. Xu, M.-Z. Huang, and W. Y. Ching. *Phys. Rev. B*, **44**, 13171 (1991).
[17.70] A. V. Hamza and M. Balooch. *Chem. Phys. Lett.*, **201**, 404 (1993).
[17.71] M. Balooch and A. V. Hamza. *Appl. Phys. Lett.*, **63**, 150 (1993).
[17.72] A. V. Hamza, M. Balooch, and M. Moalem. *Surface Science*, **317**, L1129 (1994).
[17.73] R. Meilunas, R. P. H. Chang, S. Z. Lu, and M. M. Kappes. *Appl. Phys. Lett.*, **59**, 3461 (1991).
[17.74] R. Meilunas and R. P. H. Chang. *J. Mater. Res.*, **9**, 61 (1994).
[17.75] A. V. Hamza, J. Dykes, W. D. Mosley, L. Dinh, and M. Balooch. *Surface Science*, **318**, 368 (1994).
[17.76] D. E. Ramaker, N. H. Turner, and J. Milliken. *J. Phys. Chem.*, **96**, 7627 (1992).
[17.77] W. M. Tong, D. A. A. Ohlberg, H. K. You, R. S. Williams, S. J. Anz, M. M. Alvares, R. L. Whetten, Y. Rubin, and F. Diederich. *J. Phys. Chem.*, **95**, 4709 (1991).
[17.78] P. A. Brühwiler, A. J. Maxwell, P. Rudolf, C. D. Gutleben, B. Wästberg, and N. Mårtensson. *Phys. Rev. Lett.*, **71**, 3721 (1993).
[17.79] P. A. Brühwiler, A. J. Maxwell, P. Rudolf, and N. Mårtensson, *Phys. Rev. Lett.*, **73**, 2938 (1994). Reply to Ref. 17.57.
[17.80] J. Resh, D. Sarkar, J. Kulik, J. Brueck, A. Ignatiev, and N. J. Halas. *Surface Science*, **316**, L1061 (1994).
[17.81] Y. B. Zhao, D. M. Poirier, R. J. Pechman, and J. H. Weaver. *Appl. Phys. Lett.*, **64**, 577 (1994).
[17.82] Z. Zhang, C.-C. Chen, S. P. Kelty, H. Dai, and C. M. Lieber. *Nature (London)*, **353**, 333 (1991).
[17.83] Z. Zhang, C. C. Chen, and C. M. Lieber. *Science*, **254**, 1619 (1991).
[17.84] C. H. Olk and J. P. Heremans. *J. Mater. Res.*, **9**, 259 (1994).

[17.85] M. Ge and K. Sattler. *Science*, **260**, 515 (1993).

[17.86] T. Hashizume, K. Motai, X. D. Wang, H. Shinohara, Y. Saito, Y. Maruyama, K. Ohno, Y. Kawazoe, Y. Nishina, H. W. Pickering, Y. Kuk, and T. Sakura. *Phys. Rev. Lett.*, **71**, 2959 (1993).

[17.87] E. I. Altman and R. J. Colton. *Surface Science*, **295**, 13 (1993).

[17.88] E. I. Altman and R. J. Colton. *Surface Science*, **279**, 49 (1992).

[17.89] Y. Kuk, D. K. Kim, Y. D. Suh, K. H. Park, H. P. Noh, S. J. Oh, and K. S. Kim. *Phys. Rev. Lett.*, **70**, 1948 (1993).

[17.90] Y. Z. Li, M. Chander, J. C. Patrin, J. H. Weaver, L. P. F. Chibante, and R. E. Smalley. *Phys. Rev. B*, **45**, 13837 (1992).

[17.91] Y. Z. Li, M. Chander, J. C. Patrin, J. H. Weaver, L. P. F. Chibante, and R. E. Smalley. *Science*, **253**, 429 (1991).

[17.92] Y. Z. Li, J. C. Patrin, M. Chander, J. H. Weaver, L. P. F. Chibante, and R. E. Smalley. *Science*, **252**, 547 (1991).

[17.93] T. R. Ohno, Y. Chen, S. E. Harvey, G. H. Kroll, J. H. Weaver, R. H. Haufler, and R. E. Smalley. *Phys. Rev. B*, **44**, 13747 (1991).

[17.94] A. V. Hamza and M. Balooch. *Chem. Phys. Lett.*, **198**, 603 (1992).

[17.95] M. Moalem, M. Balooch, A. V. Hamza, W. J. Siekhaus, and D. R. Olander. *J. Chem. Phys.*, **99**, 4855 (1993).

[17.96] E. I. Altmann and R. J. Colton. *Surface Science*, **295**, 13 (1993).

[17.97] M. Balooch and A. V. Hamza. Unpublished. Private communication.

[17.98] X. D. Wang, T. Hashizume, H. Sinohara, Y. Saito, Y. Nishina, and T. Sakurai. *Jpn. J. Appl. Phys.*, **31**, L880 (1992).

[17.99] H. Xu, D. M. Chen, and W. N. Creager. *Phys. Rev. Lett.*, **70**, 1850 (1993).

[17.100] P. Dietz, P. Hansma, K. Fostiropoulos, and W. Krätschmer. *Appl. Phys. Part A. Solids and Surfaces*, **56**, 207 (1993).

[17.101] J. Fujita, S. Kuroshima, T. Satoh, J. S. Tsai, and T. W. Ebbesen. *Appl. Phys. Lett.*, **63**, 1008 (1993).

[17.102] T. Thundat, R. J. Warmack, D. Ding, and R. N. Compton. *Appl. Phys. Lett.*, **63**, 891 (1993).

[17.103] P. J. Blau and C. E. Haberlin. *Thin Solid Films*, **219**, 129 (1992).

[17.104] J. Ruan and B. Bhushan. *J. Mater. Res.*, **8**, 3019 (1993).

[17.105] A. Tokmakoff, D. R. Haynes, and S. M. George. *Chem. Phys. Lett.*, **186**, 450 (1991).

[17.106] J. Abrefah, D. R. Olander, M. Balooch, and W. J. Siekhaus. *Appl. Phys. Lett.*, **60**, 1313 (1992).

[17.107] R. E. Haufler, J. J. Conceicao, L. P. F. Chibante, Y. Chai, N. E. Byrne, S. Flanagan, M. M. Haley, S. C. O'Brien, C. Pan, Z. Xiao, W. E. Billups, M. A. Ciufolini, R. H. Hauge, J. L. Margrave, L. J. Wilson, R. F. Curl, and R. E. Smalley. *J. Phys. Chem.*, **94**, 8634 (1990).

[17.108] C. Pan, M. P. Sampson, Y. Chai, R. H. Hauge, and J. L. Margrave. *J. Phys. Chem.*, **95**, 2944 (1991).

[17.109] G. Gensterblum, K. Hevesi, B. Y. Han, L. M. Yu, J. J. Pireaux, P. A. Thiry, D. Bernaerts, S. Amelinckx, G. V. Tendeloo, G. Bendele, T. Buslaps, R. L. Johnson, M. Foss, R. Feidenhans'l, and G. LeLay. *Phys. Rev. B*, **50**, 11981 (1994).

[17.110] M. W. Ruckman, B. Xia, and S. L. Qui. *Phys. Rev. B*, **48**, 15457 (1993).

[17.111] P. Rudolf and G. Gensterblum. *Phys. Rev. B*, **50**, 12215 (1994).

[17.112] L. Q. Jiang and B. E. Koel. *Phys. Rev. Lett.*, **72**, 140 (1994).

[17.113] L. Q. Jiang and B. E. Koel. *Chem. Phys. Lett.*, **223**, 69 (1994).

[17.114] P. Rudolf (1995). Private communication.

[17.115] R. L. Garrell, T. M. Herns, C. A. Szafranski, F. Diederich, R. Ettl, and R. L. Whetten. *J. Am. Chem. Soc.*, **113**, 6302 (1991).
[17.116] S. J. Chase, W. S. Bacsa, M. G. Mitch, L. J. Pilione, and J. S. Lannin. *Phys. Rev. B*, **46**, 7873 (1992).
[17.117] K. L. Akers, L. M. Cousins, and M. Moskovits. *Chem. Phys. Lett.*, **190**, 614 (1992).
[17.118] Y. Zhang, G. Edens, and M. J. Weaver. *J. Am. Chem. Soc.*, **113**, 9395 (1991).
[17.119] T. Hashizume, X. D. Wang, Y. Nishina, H. Shinohara, Y. Saito, Y. Kuk, and T. Sakurai. *Jpn. J. Appl. Phys.*, **31**, L880 (1992).
[17.120] D. L. Lichtenberger, K. W. Nebesny, C. D. Ray, D. R. Huffman, and L. D. Lamb. *Chem. Phys. Lett.*, **176**, 203 (1991).
[17.121] D. L. Lichtenberger, M. E. Jatcko, K. W. Nebesny, C. D. Ray, D. R. Huffman, and L. D. Lamb. In R. S. Averback, J. Bernholc, and D. L. Nelson (eds.), *Clusters and Cluster-Assembled Materials, MRS Symposia Proceedings, Boston*, vol. 206, p. 673. Materials Research Society Press, Pittsburgh, PA (1991).
[17.122] A. Fartash. *Appl. Phys. Lett.*, **64**, 1877 (1994).
[17.123] Z. Y. Rong and L. Rokhinson. *Phys. Rev. B*, **49**, 7749 (1994).
[17.124] K. Takayanagi, Y. Tanishiro, M. Takahashi, and S. Takahashi. *J. Vac. Sci. Technol. A*, **3**, 1502 (1985).
[17.125] R. J. Hamers, R. M. Tromp, and J. E. Demuth. *Phys. Rev. Lett.*, **56**, 1972 (1986).
[17.126] X.-D. Wang, T. Hashizume, H. Shinohara, Y. Saito, Y. Nishina, and T. Sakurai. *Phys. Rev. B*, **47**, 15923 (1993).
[17.127] X.-D. Wang, T. Hashizume, Q. Xue, H. Shinohara, Y. Saito, Y. Nishina, and T. Sakurai. *J. Appl. Phys.*, **32**, L147 (1993).
[17.128] Y. Kawazoe, H. Mamiyama, Y. Maruyama, and K. Ohno. *Jpn. J. Appl. Phys.*, **32**, 1433 (1993).
[17.129] G. Gensterblum, L. M. Yu, J. J. Pireaux, P. A. Thiry, R. Caudano, J. M. Themlin, S. Bouzidi, F. Coletti, and J. M. Debever. *Appl. Phys. A*, **56**, 175 (1993).
[17.130] L. Wragg, J. E. Chamberlain, H. W. White, W. Krätschmer, and D. R. Hoffman. *Nature (London)*, **348**, 623 (1990).
[17.131] L. D. Lamb, D. R. Huffman, R. K. Workman, S. Howells, T. Chen, D. Sarid, and D. F. Ziolo. *Science*, **255**, 1413 (1992).
[17.132] X. D. Wang, Q. Xue, T. Hashizume, H. Shinohara, Y. Nishina, and T. Sakurai. *Phys. Rev. B*, **49**, 7754 (1994).
[17.133] Y. Z. Li, J. C. Patrin, M. Chander, J. H. Weaver, K. Kikuchi, and Y. Achiba. *Phys. Rev. B*, **47**, 10867 (1993).
[17.134] D. E. Manolopoulos and P. W. Fowler. *J. Phys. Chem.*, **96**, 7603 (1992).
[17.135] Y. Z. Li, M. Chander, J. C. Patrin, J. H. Weaver, L. P. F. Chibante, and R. E. Smalley. *Science*, **252**, 547 (1992).
[17.136] Y. Z. Li, M. Chander, J. C. Patrin, J. H. Weaver, L. P. F. Chibante, and R. E. Smalley. *Science*, **253**, 429 (1992).
[17.137] K. Kikuchi, N. Nakahara, T. Wakabayashi, S. Suzuki, H. Shiramaru, Y. Miyake, K. Saito, I. Ikemoto, M. Kainosho, and Y. Achiba. *Nature (London)*, **357**, 142 (1992).
[17.138] X.-D. Wang, T. Hashizume, Q. Xue, H. Shinohara, Y. Saito, Y. Nishina, and T. Sakurai. *Jpn. J. Appl. Phys.*, **B32**, L866 (1993).
[17.139] S. Saito, S. I. Sawada, N. Hamada, and A. Oshiyama. *Jpn. J. Appl. Phys.*, **32**, 1438 (1993).

CHAPTER 18

Magnetic Properties

The magnetic properties of fullerenes are interesting and surprising. Materials with closed-shell configurations are diamagnetic. Thus C_{60} itself is expected to be diamagnetic, since by Hund's rule the ground state for the C_{60} molecule is a nondegenerate $J = 0$ state [18.1]. The diamagnetic behavior of C_{60} is, however, unique, distinctly different from that of graphite, and interesting in its own right, because of an unusual cancellation of ring currents in the molecule. This is discussed in §18.1.

There are several methods for introducing paramagnetic behavior into fullerenes, including methods for producing magnetically ordered phases. For example, the introduction of a magnetic dopant ion with an unfilled *d* or *f* shell (either endohedrally of exohedrally) is expected to lead to Curie paramagnetism. Evidence for such effects is presented in §18.5. The introduction of conduction electrons delocalized on a fullerene shell would be expected to lead to Pauli paramagnetism (see §18.4). Furthermore, the addition (or subtraction) of electrons to (or from) fullerene molecules could lead to unfilled valence shells and thus give rise to magnetic states (see §18.3). In fact, paramagnetic behavior due to unfilled *p*-bands has also been reported for this exceptional system (see §18.5), arising because of the electron localization on the molecular sites. At low-temperatures, phase transitions to magnetically ordered phases could occur (see §18.5). Many of these effects have already been observed in fullerenes and fullerene-derived compounds. In this chapter we review present knowledge about the various types of magnetic behavior and magnetic phases that have been observed in fullerene-based systems.

18.1. DIAMAGNETIC BEHAVIOR

The large number of aromatic rings in fullerenes suggests that they might be highly diamagnetic, setting up shielding screening currents upon application of a magnetic field [18.2, 3]. More detailed analysis shows that the hexagonal rings contribute a diamagnetic term to the total susceptibility χ, while the pentagonal rings contribute a paramagnetic term of almost equal magnitude, thereby leading to an unusually small diamagnetic susceptibility for C_{60} in comparison with other ring compounds [18.4–6] and with graphite [18.7] itself, which has only hexagonal rings. This approach to the analysis of the magnetic susceptibility is in good agreement with χ measurements on both C_{60} and C_{70}, as discussed below [18.8–10].

The points in Fig. 18.1 show the measured temperature dependence of the magnetic susceptibility for a C_{60} powder sample [18.8]. The results show a temperature-independent contribution to the susceptibility at high T of $\chi_g = -0.35 \times 10^{-6}$ emu/g, which is identified with the intrinsic χ for C_{60}, where the subscript g refers to the susceptibility per gram of sample. In addition, a temperature-dependent paramagnetic contribution is observed at low-temperature (see Fig. 18.1) which is identified with 1.5×10^{-4} unpaired electron spins per C atom, in agreement with typical electron spin resonance (ESR) measurements on C_{60} (see §16.2.1) [18.8]. Measurements of $\chi(T)$ on a C_{60} single crystal, carefully prepared and handled to avoid oxygen contamination, show no low-temperature Curie term [18.11], from which it is concluded that the low-temperature Curie contribution is likely due to oxygen contamination. Detailed measurements of $\chi(T)$ near the phase transition temperature T_{01} (see §7.1.3) show a 1.2% change in $\chi(T)$ at T_{01}, with $\chi(T_{01}^-)$ more negative than $\chi(T_{01}^+)$ for the susceptibility at either side of the phase transition [18.11].

The results of $\chi(T)$ for C_{70} are similar to those for C_{60} in functional form, except that the temperature-independent magnitude of χ for C_{70} is about twice as large, $\chi_g = -0.59 \times 10^{-6}$ emu/g [18.8], consistent with the larger number (5 additional) of hexagons in C_{70} as compared with C_{60}. These values for χ are compared in Table 18.1 with the corresponding diamagnetic susceptibility for graphite, diamond, and carbon nanotubes (see § 19.6).

The remarkably small value of the temperature-independent diamagnetic susceptibility for C_{60} has been explained by calculating the ring current in pentagonal and hexagonal rings in the C_{60} structure [18.6], following the London theory [18.17] where the current $\mathbf{J}_{ij}$ from site $\mathbf{R}_i$ to the nearest-neighbor site $\mathbf{R}_j$ is expressed in terms of eigenvectors C_i^n of the Hamiltonian matrix as

$$\mathbf{J}_{ij} = \left[\sum_n (C_i^n)^*(C_j^n)\right] \exp\left\{\frac{ie}{2\hbar c}[\mathbf{A}(\mathbf{R}_i) - \mathbf{A}(\mathbf{R}_j)] \cdot (\mathbf{R}_i + \mathbf{R}_j)\right\} \mathbf{v}_{ij} \quad (18.1)$$

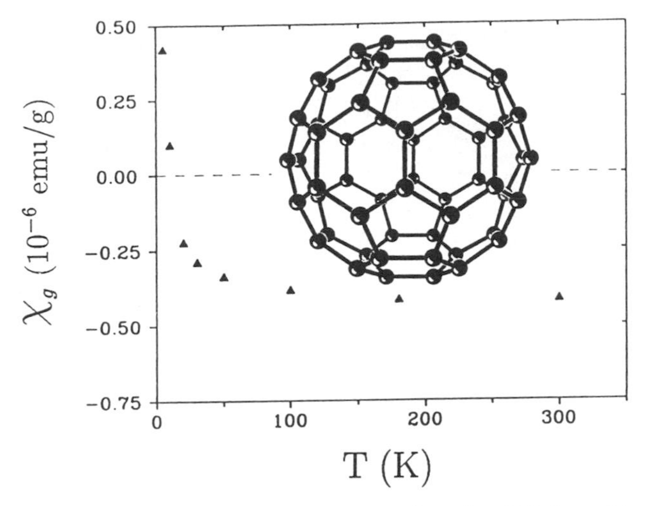

Fig. 18.1. Measured magnetic susceptibility of C_{60} (in units of 10^{-6} emu/g) as a function of temperature. After subtracting the contribution from the capsule, a high-temperature value $\chi_g = -0.35 \times 10^{-6}$ emu/g is found for C_{60} [18.8].

where $\mathbf{A}(\mathbf{r})$ is the vector potential, and $\mathbf{v}_{ij}$ is a vector along the bond connecting lattice vectors $\mathbf{R}_i$ and $\mathbf{R}_j$ with $\mathbf{v}_{ij}$ having a magnitude related to the bond strength. The total current in the ij bond is found by summing $\mathbf{J}_{ij}+\mathbf{J}_{ji}$ to yield a real value for this observable.

The results of the detailed calculation for C_{60} are shown in Fig. 18.2, where it is seen that the electron ring current in a hexagonal ring is dia-

Table 18.1

Diamagnetic susceptibility for fullerenes and related materials.

Material		χ (10^{-6} emu/g)	Reference
C_{60}		−0.35	[18.8, 9]
C_{70}		−0.59	[18.8, 9]
Graphite	$H \parallel c$-axis	−21.1	
	$H \perp c$-axis	− 0.4	[18.12]
	average[a]	− 7.3	
Diamond		−0.49	[18.13]
Benzene		−0.61	[18.14]
Carbon nanotubes		−10.2	[18.15, 16]

[a]The average susceptibility $(\chi_\parallel + 2\chi_\perp)/3$, where $\chi_\parallel$ and $\chi_\perp$, respectively, refer to the magnetic field $\parallel$ and $\perp$ to the c-axis of graphite.

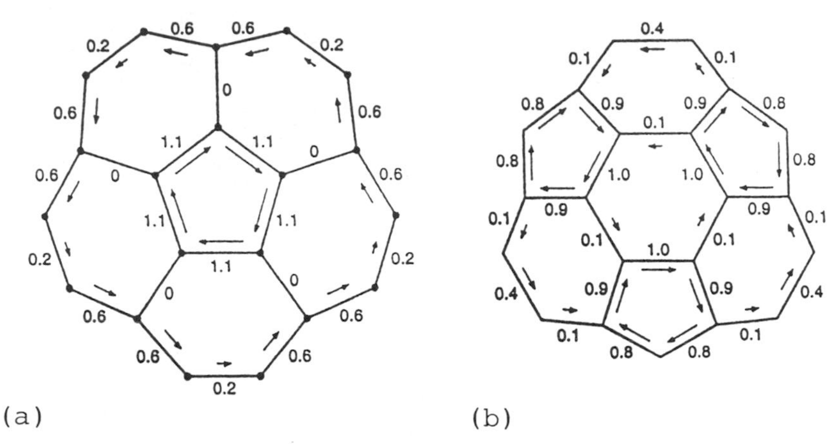

Fig. 18.2. Electron ring currents in C_{60} for a magnetic field oriented perpendicular to a plane containing (a) a pentagon and (b) a hexagon, respectively. The electron ring currents in the pentagons are paramagnetic, while the currents in the hexagons are diamagnetic. The ring current strength is given with respect to that in benzene [18.6].

magnetic and of about the same magnitude as that for benzene, while the adjacent pentagons make a paramagnetic contribution of similar magnitude. Thus for C_{60}, the diamagnetic and paramagnetic contributions nearly cancel, yielding a very small diamagnetic value for χ. Since C_{70} has five additional diamagnetic hexagons, χ for C_{70} is thereby somewhat more diamagnetic than for C_{60}, but still the diamagnetic behavior is weak and roughly comparable to that of diamond. As the number of hexagons in a fullerene is further increased, as for example to form a carbon nanotube, the magnitude of χ becomes large and highly diamagnetic (see Table 18.1), approaching about half the value of graphite for $\mathbf{H} \parallel c$-axis, when the magnetic field is directed normal to the tubule axis (see § 19.6). The value of χ for C_{60} is, however, roughly comparable to that for graphite with $\mathbf{H} \perp c$-axis [18.12], as is also seen in Table 18.1.

18.2. Magnetic Endohedral and Exohedral Dopants

The introduction of endohedral or exohedral dopants with unfilled d or f shells into fullerene host materials (see §5.4) is expected to result in paramagnetic materials. For example, in Table 18.2 we present a list of rare earth ions that might be doped endohedrally into C_{60}. For each rare earth ion, the electronic configuration and basic level designation for the S, L, and J quantum numbers are given on the basis of Hund's rule. Also given in this table are the corresponding magnetic moments μ_m calculated

Table 18.2

Level configurations and magnetic moments expected for various rare earth endohedral species in C_{60}.

Ion	Configuration	Basic level[a]	Moment[b]
Ce^{3+}	$4f^1 5s^2 p^6$	$^2F_{5/2}$	2.54
Pr^{3+}	$4f^2 5s^2 p^6$	$^3H_4$	3.58
Nd^{3+}	$4f^3 5s^2 p^6$	$^4I_{9/2}$	3.62
Pm^{3+}	$4f^4 5s^2 p^6$	$^5I_4$	2.68
Sm^{3+}	$4f^5 5s^2 p^6$	$^6H_{5/2}$	0.84
Eu^{3+}	$4f^6 5s^2 p^6$	$^7F_0$	0
Gd^{3+}	$4f^7 5s^2 p^6$	$^8S_{7/2}$	7.94
Tb^{3+}	$4f^8 5s^2 p^6$	$^7F_6$	9.72
Dy^{3+}	$4f^9 5s^2 p^6$	$^6H_{15/2}$	10.63
Ho^{3+}	$4f^{10} 5s^2 p^6$	$^5I_8$	10.60
Er^{3+}	$4f^{11} 5s^2 p^6$	$^4I_{15/2}$	9.59
Tm^{3+}	$4f^{12} 5s^2 p^6$	$^3H_6$	7.57
Yb^{3+}	$4f^{13} 5s^2 p^6$	$^2F_{7/2}$	4.54

[a]For the basic level configuration, the S value is found through equating the left superscript to $2S+1$, the letter gives the L value, and the lower right subscript gives the J value according to Hund's rule.

[b]The magnetic moment per ion is calculated according to Eq. (18.2).

according to

$$\mu_m = \mu_B g[J(J+1)]^{1/2} \tag{18.2}$$

where μ_B is the Bohr magneton and g is the Landé g-factor given by

$$g = 1 + \frac{J(J+1) + S(S+1) - L(L+1)}{2J(J+1)}. \tag{18.3}$$

Although progress has been made in preparing endohedrally doped fullerene molecules, the yield has not yet been sufficient to study their magnetic properties in the solid state. From Table 18.2, it is seen that there are a variety of rare earth ions which can introduce magnetism into fullerenes by endohedral doping.

In considering the solid state, a magnetic phase could also be achieved by the exohedral doping of magnetic ions into tetrahedral or octahedral sites (see §8.5), in analogy with the magnetic properties observed in intercalation compounds based on graphite or transition metal dichalcogenide host materials (e.g., C_6Eu).

18.3. Magnetic Properties of Fullerene Ions

Interestingly, the three ions that have been most commonly used as endohedral dopants into fullerenes (La^{3+}, Y^{3+}, and Sc^{3+}) all have $J = 0$ (non-degenerate) ground states by Hund's rule (see §12.4.1) and therefore have no magnetic moments. Assuming no charge transfer to the C_{60} shell, no Jahn–Teller distortion would be expected for a single endohedral atom placed at the center of a C_{60} molecule. However, the insertion of La^{3+}, Y^{3+} or Sc^{3+} gives rise to charge transfer (up to three electrons for each +3 ion). Since the ground state for a charged C_{60} ion differs from that of a neutral C_{60} molecule, a magnetic moment on the shell can result from charge transfer. In Table 18.3 are listed the ground state configurations for $C_{60}^{n\pm}$ molecular ions in icosahedral symmetry, the corresponding ground state

Table 18.3

Pauli-allowed states associated with the ground state configurations for icosahedral $C_{60}^{n\pm}$ ions.

n	Config.[a]	I_h state[b]	J value[c]	Icosahedral symmetry of configurations
6−	$h_u^{10} f_{1u}^6$	^1A_g	0	$A_g{}^d$
5−	$h_u^{10} f_{1u}^5$	$^2F_{1u}$	$\frac{3}{2}$	$\Gamma_6^-, \Gamma_8^{-d}$
4−	$h_u^{10} f_{1u}^4$	$^3F_{1g}$	2	$A_g, F_{1g}, H_g{}^d$
3−	$h_u^{10} f_{1u}^3$	^4A_u	$\frac{3}{2}$	Γ_8^{-d}
2−	$h_u^{10} f_{1u}^2$	$^3F_{1g}$	0	$A_g{}^d, F_{1g}, H_g$
1−	$h_u^{10} f_{1u}^1$	$^2F_{1u}$	$\frac{1}{2}$	$\Gamma_6^{-d}, \Gamma_8^-$
0	h_u^{10}	^1A_g	0	$A_g{}^d$
1+	h_u^9	^2H_u	$\frac{5}{2}$	$\Gamma_8^-, \Gamma_9^{-d}$
2+	h_u^8	$(^3F_{2g}, {}^3G_g)$	4	$H_g, (F_{2g}, G_g), (G_g, H_g)^d$
3+	h_u^7	$(^4F_{2u}, {}^4G_u)$	$\frac{9}{2}$	$\Gamma_8^-, \Gamma_9^-, (\Gamma_7^-, \Gamma_9^-), (\Gamma_8^-, \Gamma_9^-)^d$
4+	h_u^6	$(^3F_{2g}, {}^3G_g)$	4	$H_g,$
5+	h_u^6	$(^3F_{2g}, {}^3G_g)$	4	$H_g,$
6+	h_u^4	$(^3F_{2g}, {}^3G_g)$	4	$H_g,$
7+	h_u^3	$(^4F_{2u}, {}^4G_u)$	$\frac{9}{2}$	$\Gamma_8^-, \Gamma_9^-, (\Gamma_7^-, \Gamma_9^-), (\Gamma_8^-, \Gamma_9^-)^d$
8+	h_u^2	$(^3F_{2g}, {}^3G_g)$	4	$H_g,$
9+	h_u^1	^2H_u	$\frac{5}{2}$	$\Gamma_8^-, \Gamma_9^{-d}$
10+	h_u^0	^1A_g	0	$A_g{}^d$

[a] The configurations are given for icosahedral symmetry. The emptying of the HOMO (h_u) level and the filling of the LUMO (f_{1u}) level are indicated.

[b] Pauli-allowed states are labeled by their spin degeneracy and irreducible representations of group I_h (see §4.1).

[c] J values for Hund's rule ground state.

[d] In specifying the hyperfine structure, the symmetry of the Hund's rule ground state is explicitly identified by the symbol d.

designations, the J values, and the hyperfine structures for $C_{60}^{n\pm}$ molecules ($-6 < n < 10$) that is expected on the basis of Hund's rule. We see that $J = 3/2$ for a C_{60}^{3-} molecular ion, which has three additional electrons on the C_{60} shell. Because of the Jahn–Teller effect, the resulting degenerate ground state would be expected to give rise to a lattice distortion, leading to a nondegenerate ground state, and a carbon cage of lower symmetry surrounding the endohedral dopant. The various entries in Table 18.3 correspond to the removal of electrons from the HOMO level ($1 \leq n \leq 10$) and to the addition of electrons to the LUMO level ($-6 \leq n \leq -1$).

In the above discussion the fullerene ion was considered as a large charged shell with icosahedral symmetry and localized electrons. If instead, for some doping stoichiometries, the transferred charge is delocalized and forms a free electron gas, then Pauli paramagnetism could result, as discussed in §18.4 [18.18–20].

18.4. Pauli Paramagnetism in Doped Fullerenes

Whereas undoped C_{60} is characterized by a weak temperature-independent diamagnetic susceptibility (see §18.1), a greater variety of magnetic behavior can occur in doped C_{60}. Thus far, only a few systems of this kind have been studied with regard to their magnetic properties. One system that has been studied is the Cs_xC_{60} system, where the magnetic susceptibility can be described by a sum of contributions, including a temperature-independent term χ_0 discussed in the present section, a Curie–Weiss contribution $\chi_{CW}(T, H)$ from localized moments (see §18.5), and a ferromagnetic contribution $\chi_f(T, H)$ described in §18.5. For the doped CsC_{60} solid, the C_{60} molecules couple to each other through the dopant species and show no ring currents.

In this section we focus on the observation of Pauli paramagnetism associated with conduction electrons. Pauli paramagnetism is observed in materials with conduction electrons which give rise to positive, temperature-independent contributions to the susceptibility, and the corresponding ESR signals have intensities that are temperature-independent and linewidths that have a linear T dependence.

For the Cs_xC_{60} system, the Pauli paramagnetism contribution to χ is strongly dependent on the Cs concentration x. Shown in Fig. 18.3(b) is a plot of the temperature-independent susceptibility term χ_0 *vs.* Cs concentration x in Cs_xC_{60} [18.18] over a range of x for which χ_0 is paramagnetic, with a maximum value of $\chi_0 \sim 5 \times 10^{-7}$ emu/g occurring for $x \simeq 1$. Also shown in the figure (dashed line) is the core diamagnetism, which is the only contribution to χ_0 for undoped C_{60} and is almost independent of Cs concentration. On the basis of susceptibility, ESR, and temperature-dependent

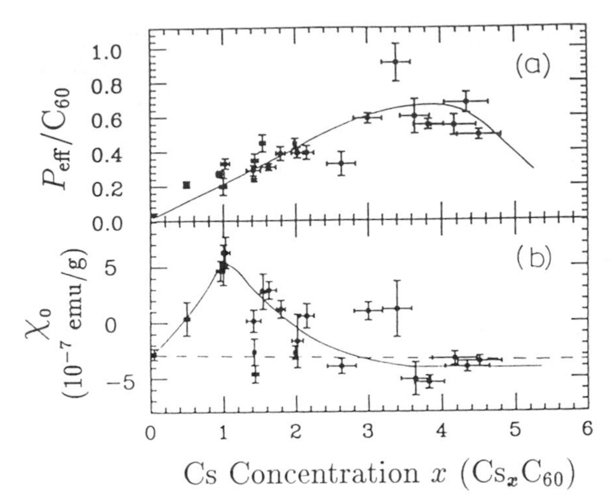

Fig. 18.3. (a) Cs concentration dependence of the effective number of Bohr magnetons per C_{60} molecule obtained from the Curie–Weiss part of the susceptibility for Cs_xC_{60}. The full line is a guide for the eye. (b) Cs concentration dependence of the temperature-independent part of the magnetic susceptibility χ_0 of Cs_xC_{60}. The broken line indicates the core diamagnetic susceptibility due to filled shells up through the h_u^{10} shell [18.18].

electrical conductivity measurements [18.18], it is concluded that the compound CsC_{60} corresponding to the maximum χ_0 in Fig. 18.3(b) (see §8.5.2) is metallic. From the magnitude of χ_0, a density of states of 32±5 states/eV per C_{60} molecule at E_F has been reported. Strong electron correlation is known to reduce the effective carrier concentration and give rise to a linear T term in the heat capacity and an enhancement of the Pauli susceptibility [18.21], so that the determination of $N(E_F)$ for CsC_{60} from measurement of χ_0 may not be reliable.

In addition to this study on Cs_xC_{60}, evidence for Pauli paramagnetism has come from studies on the K_xC_{70} system near $x = 4$. Transport studies on alkali metal–doped C_{70} [18.22–24] first show that the resistivity for K_xC_{70} initially decreases with doping and reaches a minimum ($\rho_{min} \sim 0.5$ Ω-cm), then increases with further doping and goes through a second minimum with lower resistivity ($\rho_{min} = 1.7$ mΩ-cm) before increasing once again (see §14.1.4). Electron paramagnetic resonance (EPR) studies on K_xC_{70} show that the spin concentration goes through two maxima as a function of doping [18.25]. The susceptibility at the second maximum in the doping process showed Pauli-like behavior and the stoichiometry at this doping level was found to be K_4C_{70}, consistent with the band structure calculations [18.25, 26], which show that C_{70} with D_{5h} symmetry has two lowest unfilled levels: A_1'' and E_1'', which are singly and doubly degenerate, re-

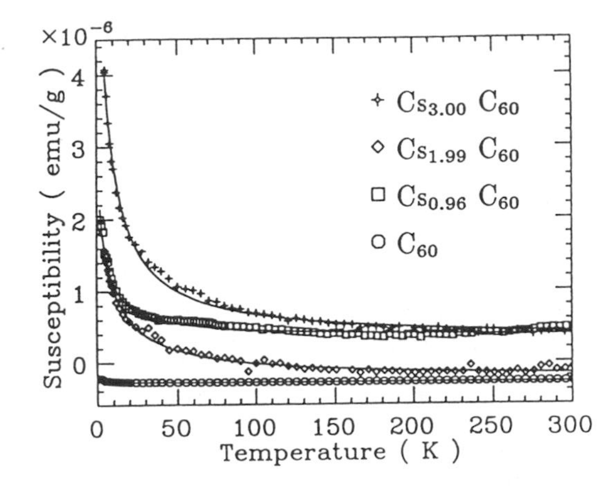

Fig. 18.4. Temperature dependence of the magnetic susceptibility of Cs_xC_{60}. Full lines are the best fit to the experimental points [18.18].

spectively [18.26] (see §12.7.6). It is expected that K_1C_{70} and K_4C_{70} would have maxima in spin concentration due to the half-filling of their LUMO and LUMO+1 bands, respectively. The K_4C_{70} phase is supposed to be more conducting than the K_1C_{70} phase owing to its higher density of states [18.26].

In addition to the Pauli paramagnetism due to conduction electrons, a Curie–Weiss paramagnetic contribution due to localized moments is observed (see Fig. 18.4) and is discussed in §18.5.

18.5. *p*-LEVEL MAGNETISM

In the case of fullerenes, there is also the possibility of observing unusual magnetic behavior, associated with unfilled carbon *p* levels, because of the localization of electrons and spins on fullerene molecules. Normally Curie paramagnetism is associated only with unfilled *d* and *f* levels. Two examples of this unusual phenomenon of *p*-level magnetism have been reported and are described below. In the first case, the magnetic effect is associated with an insulator (Cs_4C_{60}), while in the second example, the magnetic species is an organic conductor (TDAE–C_{60}).

18.5.1. Observations in Alkali Metal–Doped C_{60}

Measurements of the magnetic properties of Cs-doped C_{60} [18.18, 19] show unexpected behavior in the susceptibility, which implies that for some Cs concentrations, the molecular ions have magnetic dipole moments. As discussed in §18.4, the magnetic susceptibility for negatively charged fullerene ions C_{60}^{n-} is described by a constant term χ_0 and a Curie–Weiss term χ_{CW}

[18.18, 19]. The temperature dependence of χ_{CW} is shown in Fig. 18.4 for various values of x in Cs_xC_{60}. From the measurements in Fig. 18.4, the Curie–Weiss temperature can be obtained yielding $\theta_{CW} = -5$ K, independent of the Cs concentration for $1 \leq x \leq 4$. The χ_0 term, however, is strongly dependent on x, as discussed in §18.4, having a negative value for $x = 0$, a maximum positive value at $x = 1$, then decreasing with increasing x and becoming negative for $x \geq 3$ [see Fig. 18.3(b)]. The maximum χ_0 at $x = 1$ is identified with a state of metallic conduction, with occupation of the Cs ions only on octahedral sites, and the positive χ_0 contribution is attributed to Pauli paramagnetism of itinerant electrons.

At Cs concentration $x > 3$ the Cs_xC_{60} system is no longer metallic and the electrons become localized. Using a localized picture for the electrons on the fullerene anions, the magnetic moment P_{eff} per C_{60} anion is determined from the Curie–Weiss contribution to the susceptibility [see Fig. 18.3(a)], where we see that the maximum magnetic moment in Cs_xC_{60} is observed experimentally at $x = 4$. This experimental result is consistent with the Hund's rule prediction for the various molecular ion ground states in Table 18.3, showing that a maximum J value ($J = 2$) for the C_{60}^{n-} ions ($0 \leq n \leq 6$) occurs at C_{60}^{4-} [18.1]. For acceptor-doped C_{60}, the maximum J value is predicted to occur for the +3 ion (see Table 18.3) [18.1], although this effect has not yet been confirmed experimentally. Although the maximum magnetic moment in Fig. 18.3(a) occurs for the C_{60}^{4-} ion, the measured effective Bohr magneton is much smaller than the theoretically calculated effective Bohr magneton given by $g[J(J+1)]^{1/2}\mu_B \simeq 2\sqrt{6}\mu_B$. The large discrepancy between the observed and expected Bohr magneton values may arise from a thermal averaging of J (since the spin–orbit interaction for fullerenes is so small) and from a partly itinerant character of the electron wave functions between the molecules in the fcc lattice. On the other hand, the magnetic moment near room temperature is almost independent of T, from which it has been concluded that $T_c \sim 600$ K, an extraordinarily high T_c for a molecular crystal with no unfilled d or f levels [18.18, 19].

18.5.2. Observations in TDAE–C_{60}

Another example of a magnetic system with only s- and p-orbitals is obtained by attaching the organic donor TDAE (tetrakis-dimethylamino-ethylene) complex to C_{60} [18.27–29]. The crystal structure for TDAE–C_{60} is described in §8.7.3. Since this compound exhibits the highest magnetic ordering temperature (ferromagnetic transition temperature $T_c = 16.1$ K) of any organic magnet, this material has been extensively studied for its magnetic properties [18.27, 30–36], ESR spectra (see §16.2.4) [18.31–33, 36–41], nuclear magnetic resonance (NMR) spectra (see §16.1.9) [18.39, 42, 43], optical properties [18.36, 41, 44], crystal structure (see §8.7.3) [18.45], and vibrational spectra [18.41] (see §11.10).

It is expected that the reduced intermolecular C_{60}–C_{60} distance in TDAE–C_{60} (see Fig. 8.16) results in greater overlap of the intermolecular wave functions. This increased wave function overlap and electron charge transfer from the TDAE to C_{60} both lead to a more conducting state. This is borne out by reports of conductivity values of $\sim 10^{-2}$S/cm in TDAE–C_{60} [18.27, 45, 46]. Microwave conductivity measurements [18.47] have confirmed a magnetic transition at 16 K, but it is found that the temperature dependence of the conductivity is not metallic, but instead shows activated behavior down to $T \sim 80$ K with an activation energy of ~60 meV [18.40, 47]. The implied electron localization and hopping transport mechanism is consistent with the absence of a Drude absorption in the optical conductivity [18.41]. The anisotropic crystal structure of TDAE–C_{60} is expected to lead to anisotropic magnetic properties, but since single crystals of this compound have not yet been prepared, the anisotropic magnetic properties have not yet been studied.

According to ESR and susceptibility studies (see §16.2.2), a charge transfer of one electron from the TDAE to C_{60} takes place, giving rise to an unpaired spin on each C_{60} anion in the lattice (see Fig. 8.16). This charge transfer is consistent with optical studies showing the t_{1u} level as the lowest unoccupied state [18.36, 41]. Other experiments providing confirmation for charge transfer include a shift in the Raman pentagonal pinch mode frequency, and this frequency shift is almost as large as that for KC_{60} and RbC_{60} [18.36] (although there are significant differences in their respective Raman spectra [18.41]). A similar softening of the low-frequency infrared-active modes is observed, and as with the Raman spectra, there are significant differences in the detailed spectra [18.41]. A shift in the ^{13}C NMR line from 142.7 ppm in C_{60} to 188 ppm has also been reported for TDAE–C_{60} [18.36], close to the values of 177 ppm in CsC_{60} and 173 ppm in RbC_{60} [18.48].

This unpaired spin gives rise to an effective magnetic moment of $\mu_{\rm eff} = 1.72\mu_B/C_{60}^-$ as determined by high-temperature ($T > 50$ K) magnetic susceptibility $\chi(T)$ measurements (see Table 18.4). The broad and intense ESR signal that is observed for TDAE–C_{60} in the magnetically ordered state (see §16.2.4) suggests that the spins are localized on each C_{60} molecule and are ferromagnetically correlated. Coordinated temperature-dependent susceptibility [$\chi'(T)$ and $\chi''(T)$] and ESR studies have shown a correlation between spin ordering in the low-temperature magnetic phase and orientational ordering of the C_{60} molecules in TDAE–C_{60} [18.36]. The orientational ordering temperature $T_{01} = 170$ K was established by temperature-

Table 18.4

Parameters relevant to the magnetism for TDAE–C_{60}.

Property	Value	Reference
T_c	16.1 K	[18.30]
M (5 K)	0.1 μ_B/C_{60}	[18.27]
χ_0 Pauli suspectibility	3.6×10^{-7} emu/g	[18.49]
Remanent magnetization (5 K)	0.3226×10^{-3} emu/g	[18.50]
Coercive field (5 K)	1.6 G	[18.50]
Spin concentration	4.9×10^{20}/mol-TDAE–C_{60}	[18.49]
Spin concentration	8.3×10^{22} cm^{-3}	[18.49]
dT_c/dH	0.17 K/kOe	[18.46]
dT_c/dp	−8.8 K/kbar	[18.46]
Spin	1/2	[18.46]
g-factor (paramagnetic phase)	2.0003–2.0008	[18.33, 46]
μ_{eff}	1.72 μ_B/C_{60}	[18.46]
θ_{CW}	0, −22.5 K, −58 K	[18.33, 36, 46]
C_{CW}	3.1×10^{-4} emu K/g	[18.49]
Lattice constants[a]	$a = 15.874$Å, $b = 12.986$Å, $c = 9.981$Å, $\beta = 93.31°$	[18.45]
NMR ^{13}C frequency shift	188 ppm	[18.48]
Rotational freezing (T_{01})	170 K	[18.36]

[a]C-centered monoclinic structure (see Fig. 8.16). Values given in this table are for room temperature and values given in §8.7.3 pertain to 11 K [18.45].

dependent ESR [18.36], NMR [18.42], and μSR [18.42] linewidth studies. The lower value of T_{01} for TDAE–C_{60} relative to C_{60} (for which T_{01} = 261 K) can be understood in terms of the hindered rotation of the C_{60} molecules to which the bulky TDAE groups are attached (see §8.7.3). The correlation between the magnetic properties with the orientational molecular ordering was established by showing large differences in the measured susceptibility $\chi'(T)$ and $\chi''(T)$ below 20 K for TDAE–C_{60} samples slowly cooled through T_{01} and samples that had been quenched from 300 K to 30 K [18.36]. Increased orientational ordering by slow cooling through T_{01} resulted in a sharper onset of the magnetization, a higher value of T_c, and a lower internal dissipation χ'' than for the quenched sample. Further confirmation for the correlation between orientational molecular ordering and spin ordering was provided by ESR experiments on samples cooled to 4 K in an external magnetic field of 9 T, where T_c was found to increase from 16 K to 28 K [18.36]. When a TDAE–C_{60} sample, which had been previously cooled to 4 K in a 9 T magnetic field, was subsequently heated above T_{01}, the sample lost its memory of prior low-temperature molecular ordering.

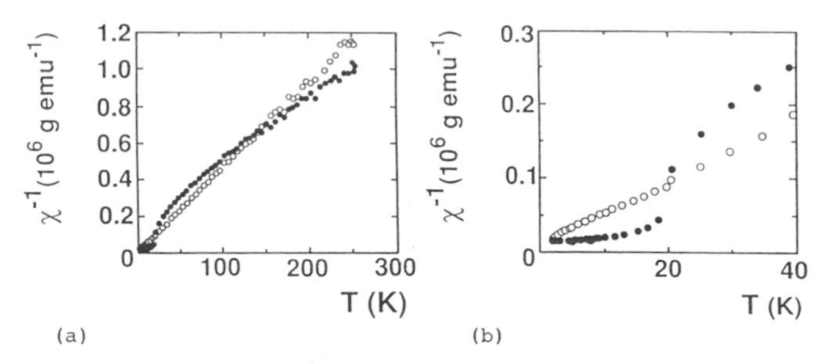

Fig. 18.5. Plot of $1/\chi$ *vs.* T for TDAE–C_{60} (closed circles) and TDAE–C_{70} (open circles) from (a) 0–300 K of relevance to the Curie–Weiss fitting of the data, and (b) 0–40 K for showing the phase transition in TDAE–C_{60} and its absence in TDAE–C_{70} [18.49].

From plots of $\chi(T)$ *vs.* $1/T$ in the paramagnetic regime, a Curie–Weiss temperature of $\theta_{\mathrm{CW}} = -22.5$ K was initially deduced [18.46], but because of deviations of the experimental points from the Curie–Weiss law (see Fig. 18.5), values in the range $0 \geq \theta_{\mathrm{CW}} \geq -58$ K have been reported [18.33, 36, 49, 51], though all reports give a negative or zero value for θ_{CW}. Corresponding to the fit of the susceptibility data to $\theta_{\mathrm{CW}} = -58$ K is the value of 3.1×10^{-4} emu K/g for the Curie–Weiss constant C_{CW} [18.49]

$$C_{\mathrm{CW}} = N_{\mathrm{CW}} \frac{g^2 \mu_B^2 S(S+1)}{3k_{\mathrm{B}}} \tag{18.4}$$

where N_{CW} is the density of unpaired spins on the C_{60} anions. A negative value for θ_{CW} is anomalous, since dc magnetic susceptibility and magnetization measurements show that TDAE–C_{60} undergoes ferromagnetic ordering below a temperature of $T_c = 16$ K under ambient pressure (see Fig. 18.6). Furthermore, it has been shown that by cooling a TDAE–C_{60} sample through T_{01} in a magnetic field (9 T), a good fit to a Curie law is obtained from 4.3 K to 300 K, whereas zero-field cooled samples show com-

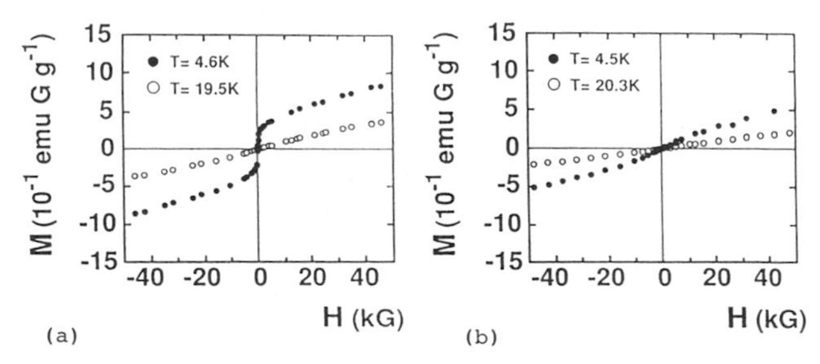

Fig. 18.6. Plot of magnetization *vs.* magnetic field H for (a) TDAE–C_{60} and (b) TDAE–C_{70} at ~4.5 K (closed circles) and ~20 K (open circles) showing evidence for a magnetic phase transition in TDAE–C_{60} and no phase transition in TDAE–C_{70} down to 4.5 K [18.49].

plicated history-dependent $\chi(T)$ behavior below $T_{01} \sim 170$ K [18.36]. The external field is believed to order the C_{60} molecules preferentially into one of the potential wells of the double-well potential associated with the merohedral disorder, leading to paramagnetic behavior in magnetic field-cooled samples corresponding to a single paramagnetic center [18.36].

Although early magnetization studies showed the TDAE–C_{60} material to have no magnetic hysteresis on cooling and heating through T_c [18.27], thereby indicating a vanishingly small remanent field, later studies [18.51] showed both hysteresis behavior and a small remanent magnetization of $\sim 3 \times 10^{-4}$ emuK/g (see Table 18.4). Furthermore, the T_c value of TDAE–C_{60} can be increased by cooling the sample in an external magnetic field (e.g., T_c is increased to 24.3 K at 0.5 T and to 28 K at 9 T [18.36, 46]), while the application of pressure reduces T_c remarkably [18.52], so that by an applied pressure of 1.6 kbar, T_c has been reduced below 2 K. Under pressure, the magnetic moment decreases even more rapidly than the transition temperature. On the basis of this rapid pressure dependence of T_c, it has been assumed that the observed magnetic state is associated with weak itinerant ferromagnetism [18.46], which is supported by the magnitudes of the observed conductivity ($\sim 10^{-2}$ S/cm) and saturation magnetization (0.1 μ_B/C_{60}). Since many papers were published before the hysteresis and remanent magnetization were found [18.50], early magnetic studies of TDAE–C_{60} identified the magnetic phase with superparamagnetism [18.33]. The correlation between orientational order of the C_{60} molecules and the spin ordering in TDAE–C_{60} at low-temperatures offers insight into the large pressure dependence observed for TDAE–C_{60}. Since the application of pressure strongly affects the molecular orientation, it shifts the energy of the minima of the orientational double-well potential, thereby changing the overlap of the π-electron orbitals on neighboring C_{60} molecules [18.36], which would be expected to strongly affect the magnetic properties of TDAE–C_{60}.

The magnetic contribution to the specific heat of TDAE–C_{60} shows a broad maximum at about 12 K [18.34], consistent with magnetic ordering of a finite number of spins, as would, for example, be characteristic of a spin glass [18.36]. A number of other experiments also support a spin glass model for TDAE–C_{60} at low-temperature. Three special properties define the characteristic behavior of a spin glass: (i) frozen-in magnetic moments below a certain freezing temperature T_f accompanied by a peak in the susceptibility $\chi(T)$, (ii) lack of periodic long-range magnetic order, and (iii) remanence and nonexponential time dependence of the magnetic relaxation below T_f [18.53]. Peaks in the ac susceptibility $\chi'(T)$ are indeed observed [18.40, 41] with characteristic behavior that depends on the ac frequency ($75 < f < 1000$ Hz). The large increase in intensity (by a factor of $\sim 10^3$) of the low field ESR line below T_c (rather than a divergent susceptibility) also supports spin-glass behavior [18.43]. The lack of long-range order follows from the absence of a large internal field in zero field ESR measurements below T_c [18.41], from the strong magnetic depolarization effects observed in the μSR spectra below 10 K [18.42], and from the stretched exponential relaxation time dependence of one of the proton NMR lines [18.42, 43]. Evidence for magnetic relaxation comes from the strong frequency dependence of χ' and χ'' [18.42], and the stretched exponential decay of the ESR signal after the field is switched off [18.43]. This is also consistent with the pulsed ESR spin echo measurements on TDAE–C_{60} which show strong evidence for inhomogeneous line broadening effects, also suggestive of spin-glass behavior [18.43].

As discussed in §16.2.2, the anisotropy of the g-factor of TDAE–C_{60} has not yet been measured, although an average g-value of 2.0003–2.0008 has been detemined by the electron spin resonance (ESR) technique at high temperature in the paramagnetic phase. A sharp increase in g-value is observed at low T for TDAE–C_{60}, while the g-factors for the TDAE–C_{n_C} ($n_C = 70, 84, 90, 94$) are all approximately independent of temperature and close to the free electron g-value of 2.0023.

The magnetic susceptibility of the related compounds TDAE–C_{70}, TDAE–C_{84}, TDAE–C_{90}, and TDAE–C_{94} has been investigated from room temperature down to 5 K [18.54], using ESR techniques, but all of these compounds show strictly paramagnetic behavior with no indication of a phase transition to a magnetically ordered state to the lowest measured temperature (5 K).

References

[18.1] R. Saito, G. Dresselhaus, and M. S. Dresselhaus. *Phys. Rev. B*, **46**, 9906 (1992).

[18.2] P. W. Fowler, P. Lasseretti, and R. Zanasi. *Chem. Phys. Lett.*, **165**, 79 (1990).

[18.3] R. C. Haddon and V. Elser. *Chem. Phys. Lett.*, **169**, 362 (1990).

[18.4] V. Elser and R. C. Haddon. *Nature (London)*, **325**, 792 (1987).

[18.5] V. Elser and R. C. Haddon. *Phys. Rev. A*, **36**, 4579 (1990).

[18.6] A. Pasquarello, M. Schlüter, and R. C. Haddon. *Science*, **257** (1992).

[18.7] B. T. Kelly. *Physics of Graphite*. Applied Science (London) (1981).

[18.8] R. S. Ruoff, D. Beach, J. Cuomo, T. McGuire, R. L. Whetten, and F. Diederich. *J. Phys. Chem.*, **95**, 3457 (1991).

[18.9] R. C. Haddon, L. F. Schneemeyer, J. V. Waszczak, S. H. Glarum, R. Tycko, G. Dabbagh, A. R. Kortan, A. J. Muller, A. M. Mujsce, M. J. Rosseinsky, S. M. Zahurak, A. V. Makhija, F. A. Thiel, K. Raghavachari, E. Cockayne, and V. Elser. *Nature (London)*, **350**, 46 (1991).

[18.10] M. Prato, T. Suzuki, F. Wudl, V. Lucchini, and M. Maggini. *J. Am. Chem. Soc.*, **115**, 7876 (1993).

[18.11] W. Luo, H. Wang, R. Ruoff, and J. Cioslowski. *Phys. Rev. Lett.*, **73**, 186 (1994).

[18.12] N. Ganguli and K. S. Krishnan. *Proc. Roy. Soc. (London)*, **A177**, 168 (1941).

[18.13] Y. G. Dorfman. *Diamagnetism and the Chemical Bond*. Elsevier, New York (1966).

[18.14] R. C. Weast. *CRC Handbook of Chemistry and Physics*. CRC Press, West Palm Beach, Florida (1978). 59th edition.

[18.15] X. K. Wang, R. P. H. Chang, A. Patashinski, and J. B. Ketterson. *J. Mater. Res.*, **9**, 1578 (1994).

[18.16] J. Heremans, C. H. Olk, and D. T. Morelli. *Phys. Rev. B*, **49**, 15122 (1994).

[18.17] F. London. *J. Phys. Rad.*, **8**, 397 (1937).

[18.18] H. Suematsu, Y. Murakami, T. Arai, K. Kikuchi, Y. Achiba, and I. Ikemoto. *Mater. Sci. Eng.*, **B19**, 141 (1993).

[18.19] Y. Murakami, T. Arai, H. Suematsu, K. Kikuchi, N. Nakahara, Y. Achiba, and I. Ikemoto. *Fullerene Sci. Tech.*, **1**, 351 (1993).

[18.20] M. Y, T. Shibata, K. Okuyama, T. Arai, H. Suematsu, and Y. Yoshida. *J. Phys. Chem. Solids*, **54**, 1861 (1993).

[18.21] N. F. Mott. *Metal Insulator Transitions*. Taylor & Francis, New York (1990).

[18.22] R. C. Haddon, A. F. Hebard, M. J. Rosseinsky, D. W. Murphy, S. J. Duclos, K. B. Lyons, B. Miller, J. M. Rosamilia, R. M. Fleming, A. R. Kortan, S. H. Glarum, A. V. Makhija, A. J. Muller, R. H. Eick, S. M. Zahurak, R. Tycko, G. Dabbagh, and F. A. Thiel. *Nature (London)*, **350**, 320 (1991).

[18.23] Z. H. Wang, K. Ichimura, M. S. Dresselhaus, G. Dresselhaus, W. T. Lee, K. A. Wang, and P. C. Eklund. *Phys. Rev. B*, **48**, 10657 (1993).

[18.24] Z. H. Wang, M. S. Dresselhaus, G. Dresselhaus, and P. C. Eklund. *Phys. Rev. B*, **48**, 16881 (1993).

[18.25] K. Imaeda, K. Yakushi, H. Inokuchi, K. Kikuchi, I. Ikemoto, S. Suzuki, and Y. Achiba. *Solid State Commun*, **84**, 1019 (1992).

[18.26] S. Saito and A. Oshiyama. *Phys. Rev. Lett.*, **66**, 2637 (1991).

[18.27] P.-M. Allemand, K. C. Khemani, A. Koch, F. Wudl, K. Holczer, S. Donovan, G. Grüner, and J. D. Thompson. *Science*, **253**, 301 (1991).

[18.28] P. W. Stephens. *Nature (London)*, **356**, 383 (1992).

[18.29] H. Klos, I. Rystan, W. Schütz, B. Gotschy, A. Skiebe, and A. Hirsch. *Chem. Phys. Lett.*, **224**, 333 (1994).

[18.30] K. Awaga and Y. Maruyama. *Chem. Phys. Lett.*, **158**, 556 (1989).

[18.31] K. Tanaka, A. A. Zakhidov, K. Yoshizawa, K. Okahara, T. Yamabe, K. Yakushi, K. Kikuchi, S. Suzuki, I. Ikemoto, and Y. Achiba. *J. Mod. Phys. B*, **6**, 3953 (1992).

[18.32] K. Tanaka, A. A. Zakhidov, K. Yoshizawa, K. Okahara, T. Yamabe, K. Yakushi, K. Kikuchi, S. Suzuki, I. Ikemoto, and Y. Achiba. *Phys. Lett.*, **A164**, 221 (1992).

[18.33] K. Tanaka, A. A. Zakhidov, K. Yoshizawa, K. Okahara, T. Yamabe, K. Yakushi, K. Kikuchi, S. Suzuki, I. Ikemoto, and Y. Achiba. *Phys. Rev. B*, **47**, 7554 (1993).

[18.34] K. Tanaka, T. Tanaka, T. Atake, K. Yoshizawa, K. Okahara, T. Sato, and T. Yamabe. *Chem. Phys. Lett.*, **230**, 271 (1994).

[18.35] A. Lappas, K. Prassides, K. Vavekis, D. Arcon, R. Blinc, P. Cevic, A. Amato, R. Feyerherm, F. N. Gygax, and A. Schenck. *Science*, **267**, 1799 (1995).

[18.36] D. Mihailovič, D. Arčon, P. Venturini, R. Blinc, A. Omerzu, and P. Cevc. *Science*, **268**, 400 (1995).

[18.37] K. Tanaka, A. A. Zakhidov, K. Yoshizawa, K. Okahara, T. Yamabe, K. Kikuchi, S. Suzuki, I. Ikemoto, and Y. Achiba. *Solid State Commun.*, **85**, 69 (1993).

[18.38] R. Seshadri, A. Rastogi, S. V. Bhat, S. Ramasesha, and C. N. R. Rao. *Solid State Commun.*, **85**, 971 (1993).

[18.39] P. Venturini, D. Mihailovič, R. Blinc, P. Ceve, J. Dolinšek, D. Abramic, B. Zalar, H. Oshio, P. M. Allemand, A. Hirsch, and F. Wudl. *J. Mod. Phys. B*, **6**, 3947 (1992).

[18.40] H. Klos, W. Brütting, A. Schilder, W. Schütz, B. Gotschy, G. Völkl, B. Pilawa, and A. Hirsch. In H. Kuzmany, J. Fink, M. Mehring, and S. Roth (eds.), *Proceedings of the Winter School on Fullerenes*, pp. 297–300 (1994). Kirchberg Winter School, World Scientific Publishing Co., Singapore.

[18.41] D. Mihailovič, P. Venturini, A. Hassanien, J. Gasperič, K. Lutar, S. Miličev, and V. I. Srdanov. In H. Kuzmany, J. Fink, M. Mehring, and S. Roth (eds.), *Proceedings of the Winter School of Fullerenes*, pp. 275–278 (1994). Kirchberg Winter School, World Scientific, Singapore.

[18.42] L. Cristofolini, M. Ricco, R. De Renzi, G. P. Ruani, S. Rossini, and C. Taliani. In H. Kuzmany, J. Fink, M. Mehring, and S. Roth (eds.), *Proceedings of the Winter School on Fullerenes*, pp. 279–282 (1994). Kirchberg Winter School, World Scientific, Singapore.

[18.43] R. Blinc, P. Cevc, D. Arčon, J. Dolinšek, D. Mihailovič, and P. Venturini. In H. Kuzmany, J. Fink, M. Mehring, and S. Roth (eds.), *Proceedings of the Winter School on Fullerenes*, pp. 283–288 (1994). Kirchberg Winter School, World Scientific, Singapore.

[18.44] F. Bommeli, L. Digiorgi, P. Wachter, and D. Mihailovic. *Phys. Rev. B*, **51**, 1366 (1995).

[18.45] P. W. Stephens, D. E. Cox, J. W. Lauher, L. Mihaly, J. B. Wiley, P. Allemand, A. Hirsch, K. Holczer, Q. Li, J. D. Thompson, and F. Wudl. *Nature (London)*, **355**, 331 (1992).

[18.46] F. Wudl and J. D. Thompson. *J. Phys. Chem. Solids*, **53**, 1449 (1992).

[18.47] A. Schilder, H. Klos, I. Rystaau, W. Schütz, and B. Gotschy. *Phys. Rev. Lett.*, **73**, 1299 (1994).

[18.48] R. Tycko, G. Dabbagh, D. W. Murphy, Q. Zhu, and J. E. Fischer. *Phys. Rev. B*, **48**, 9097 (1993).

[18.49] K. Okahara. Ph.D. thesis, Kyoto University (1994). Department of Chemistry: Studies on the Electronic and Magnetic Properties of C_{60} and Related Materials (in English).

[18.50] A. Suzuki, T. Suzuki, R. J. Whitehead, and Y. Maruyama. *Chem. Phys. Lett.*, **223**, 517 (1994).

[18.51] T. Suzuki, Y. Maruyama, T. Akasaka, W. Ando, K. Kobayashi, and S. Nagase. *J. Am. Chem. Soc.*, **116**, 1359 (1994).

[18.52] G. Sparn, J. D. Thompson, R. L. Whetten, S.-M. Huang, R. B. Kaner, F. Diederich, G. Grüner, and K. Holczer. *Phys. Rev. Lett.*, **68**, 1228 (1992).

[18.53] K. Binder and A. P. Young. *Rev. Mod. Phys.*, **58**, 801 (1986).

[18.54] K. Tanaka, M. Okada, K. Okahara, and T. Yamabe. *Chem. Phys. Lett.*, **202**, 394 (1993).

CHAPTER 19

C_{60}-*Related Tubules and Spherules*

In addition to ball-like fullerenes, it is possible to synthesize tubular fullerenes and nested concentric fullerenes (see §19.2 and §19.10). The field of carbon tubule research was greatly stimulated by the initial report of the existence of carbon tubules or nanotubes [19.1] and the subsequent report of conditions for the synthesis of large quantities of nanotubes [19.2, 3]. Various experiments carried out thus far [transmission electron spectroscopy (TEM), scanning tunneling microscopy (STM), resistivity, Raman scattering, and susceptibility] are consistent with identifying the carbon nanotubes with cylindrical graphene sheets of sp^2-bonded carbon atoms. In this chapter we review the present state of knowledge of carbon nanotubes (both monolayer and multilayer) and nested concentric fullerenes.

Formally, carbon nanotubes and fullerenes have a number of common features and also many differences. In reviewing the theoretical literature, the focus is on single-wall tubules, cylindrical in shape, either infinite in length or with caps at each end, such that the two caps can be joined to form a fullerene. Formally, the cylindrical portions of the tubules consist of a single graphene sheet, rolled to form the cylinder. The various types of cylindrical shells that can be formed are reviewed in §19.1, which is followed in §19.2 and §19.2.2 by a review of experimental observations in multiwall and single-wall tubules, respectively. Following a discussion of the synthesis (§19.2.5), possible growth mechanisms (§19.3) are considered. Next, the symmetry properties of carbon tubules are summarized (§19.4), followed by the remarkable electronic structure as predicted (§19.5) and observed (§19.6). The phonon dispersion relations are reviewed in §19.7, including the infrared and Raman spectroscopy in §19.7.3, and elastic properties in §19.8. The opening and filling of carbon nanotubes are discussed in §19.9 and nested carbon spherules in §19.10. The possibility of superconductivity in C_{60}-related tubules is considered in §19.11.

19.1. RELATION BETWEEN TUBULES AND FULLERENES

In this section we consider first two simple examples of single-wall carbon nanotubes based on the C_{60} fullerene. The concept of a single-wall nanotube is then generalized to specify the idealized structure of single-wall nanotubes in general.

In analogy to a C_{60} molecule, we can specify a single-wall C_{60}-derived tubule by bisecting a C_{60} molecule at the equator and joining the two resulting hemispheres with a cylindrical tube one monolayer thick and with the same diameter as C_{60}. If the C_{60} molecule is bisected normal to a fivefold axis, the "armchair" tubule shown in Fig. 19.1(a) is formed, and if the C_{60} molecule is bisected normal to a threefold axis, the "zigzag" tubule in Fig. 19.1(b) is formed [19.4]. Armchair and zigzag carbon nanotubules of larger diameter, and having correspondingly larger caps, are defined below.

In addition to the armchair and zigzag tubules, a large number of chiral carbon nanotubes can be formed with a screw axis along the axis of the tubule and with a variety of "hemispherical"-like caps. These general carbon nanotubules can be specified mathematically in terms of the tubule diameter d_t and chiral angle θ, which are shown in Fig. 19.2(a), where the chiral vector $\mathbf{C}_h$

$$\mathbf{C}_h = n\mathbf{a}_1 + m\mathbf{a}_2 \tag{19.1}$$

is shown, as well as the basic translation vector $\mathbf{T}$ for the tubule, which is discussed below. In Fig. 19.2(a), the vector $\mathbf{C}_h$ connects two crystallographically equivalent sites O and A on a two-dimensional (2D) graphene sheet where a carbon atom is located at each vertex of the honeycomb structure [19.4]. The construction in Fig. 19.2(a) shows the chiral angle θ of the nanotube with respect to the zigzag direction ($\theta = 0$) and the unit vectors $\mathbf{a}_1$ and $\mathbf{a}_2$ of the hexagonal honeycomb lattice. The armchair tubule [Fig. 19.1(a)] corresponds to $\theta = 30°$ on this construction. An ensemble of possible chiral vectors can be specified by Eq. (19.1) in terms of pairs of integers (n, m) and this ensemble is shown in Fig. 19.2(b) [19.6]. Each pair of integers (n, m) defines a different way of rolling the graphene sheet to form a carbon nanotube. We now show how the construction in Fig. 19.2(a) specifies the geometry of the carbon nanotube.

The cylinder connecting the two hemispherical caps of Fig. 19.1 is formed by superimposing the two ends OA of the vector $\mathbf{C}_h$. The cylinder joint is made by joining the line AB' to the parallel line OB in Fig. 19.2(a),

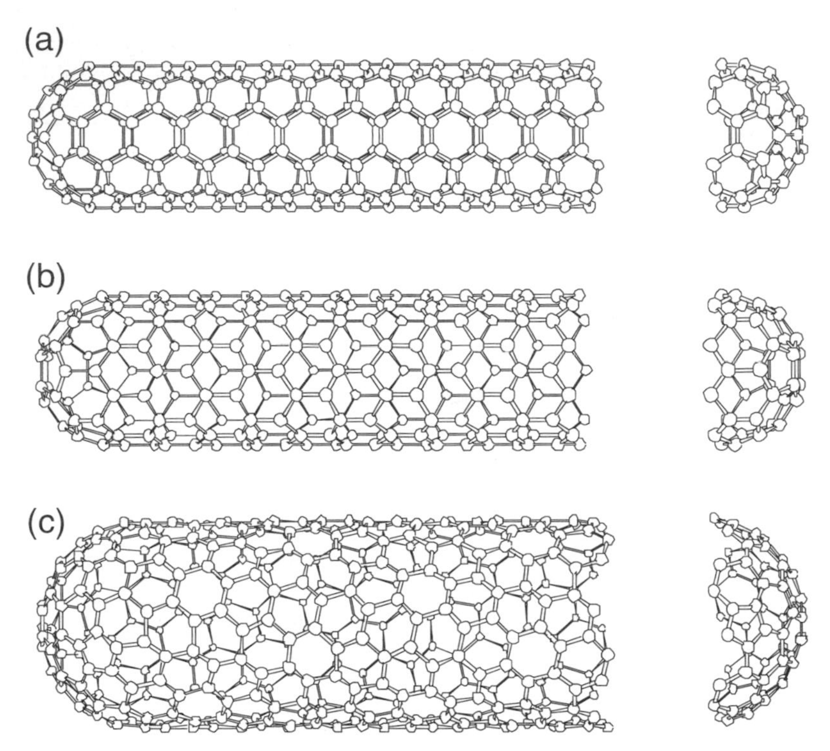

Fig. 19.1. By rolling a graphene sheet (a single layer from a 3D graphite crystal) into a cylinder and capping each end of the cylinder with half of a fullerene molecule, a "fullerene-derived tubule," one atomic layer in thickness, is formed. Shown here is a schematic theoretical model for a single-wall carbon tubule with the tubule axis normal to: (a) the $\theta = 30°$ direction (an "armchair" tubule), (b) the $\theta = 0°$ direction (a "zigzag" tubule), and (c) a general direction $0 < \theta < 30°$ (see Fig. 19.2) (a "chiral" tubule). The actual tubules shown in the figure correspond to (n, m) values of: (a) (5, 5), (b) (9, 0), and (c) (10, 5) [19.5].

where lines OB and AB' are perpendicular to the vector $\mathbf{C}_h$ at each end [19.4]. The chiral tubule thus generated has no distortion of bond angles other than distortions caused by the cylindrical curvature of the tubule. Differences in chiral angle θ and in the tubule diameter d_t give rise to differences in the properties of the various carbon nonotubes. In the (n, m) notation for specifying the chiral vector $\mathbf{C}_h$ in Eq. (19.1), the vectors $(n, 0)$ denote zigzag tubules and the vectors (n, n) denote armchair tubules, and the larger the value of n, the larger the tubule diameter. Both the $(n, 0)$ and (n, n) nanotubes have especially high symmetry, as discussed in §19.4.3, and exhibit a mirror symmetry plane normal to the tubule axis. All other vectors (n, m) correspond to chiral tubules [19.6]. Since both right- and left-handed

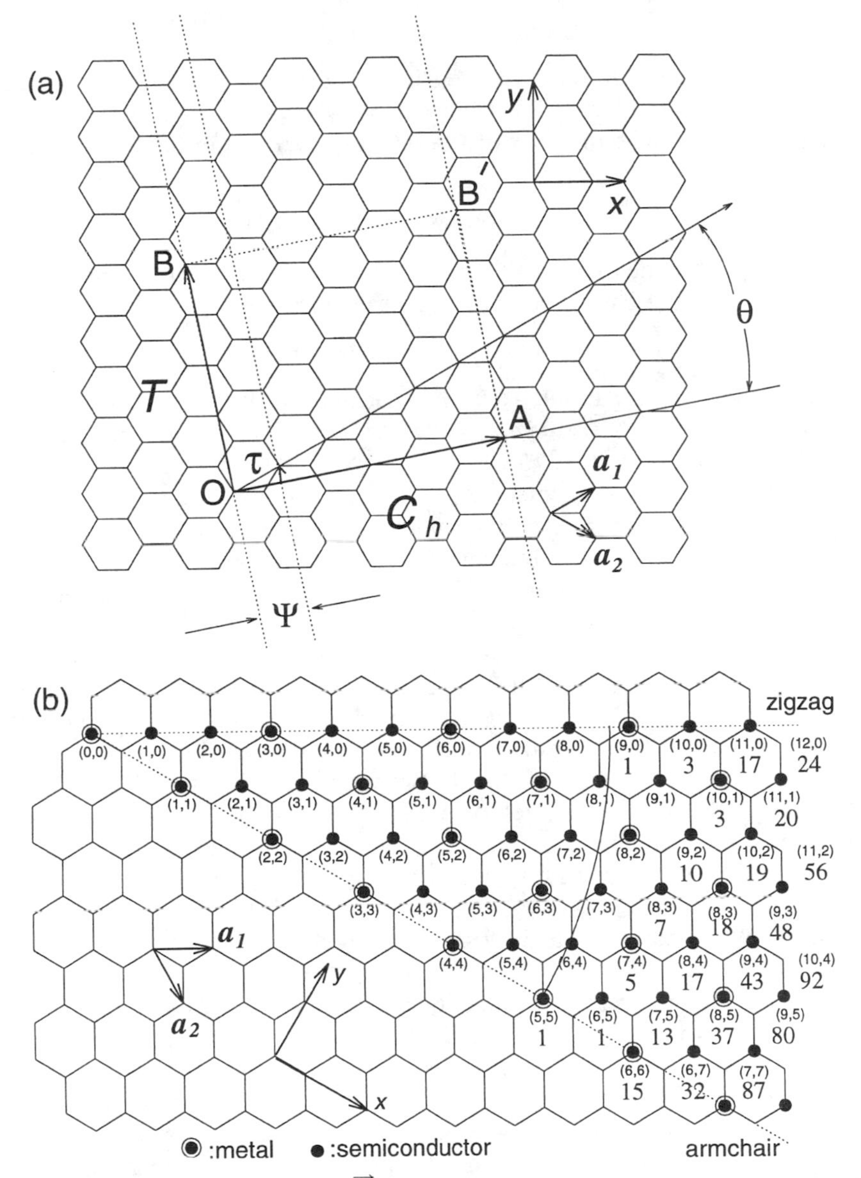

Fig. 19.2. (a) The chiral vector $\overrightarrow{OA}$ or $\mathbf{C}_h = n\mathbf{a}_1 + m\mathbf{a}_2$ is defined on the honeycomb lattice of carbon atoms by unit vectors $\mathbf{a}_1$ and $\mathbf{a}_2$ and the chiral angle θ with respect to the zigzag axis. Along the zigzag axis, $\theta = 0°$. Also shown are the lattice vector $\overrightarrow{OB} = \mathbf{T}$ of the 1D tubule unit cell and the rotation angle ψ and the translation τ which constitute the basic symmetry operation $R = (\psi|\tau)$ for the carbon nanotube. The diagram is constructed for $(n, m) = (4, 2)$. (b) Possible vectors specified by the pairs of integers (n, m) for general carbon tubules, including zigzag, armchair, and chiral tubules. Below each pair of integers (n, m) is listed the number of distinct caps that can be joined continuously to the carbon tubule denoted by (n, m) [19.4], as discussed in §19.2.3. The encircled dots denote metallic tubules while the small dots are for semiconducting tubules.

chirality is possible for chiral tubules, it is expected that chiral tubules are optically active to either right or left circularly polarized light propagating along the tubule axis. In terms of the integers (n, m), the tubule diameter d_t is given by

$$d_t = C_h/\pi = \sqrt{3}a_{C-C}(m^2 + mn + n^2)^{1/2}/\pi \tag{19.2}$$

where a_{C-C} is the nearest-neighbor C–C distance (1.421 Å in graphite), C_h is the length of the chiral vector $\mathbf{C}_h$, and the chiral angle θ is given by

$$\theta = \tan^{-1}[\sqrt{3}m/(m+2n)]. \tag{19.3}$$

For example, a zigzag tubule ($\theta = 0°$) specified by $(9,0)$ has a theoretical tubule diameter of $d_t = 9\sqrt{3}a_{C-C}/\pi = 7.15$ Å, while an armchair tubule specified by (5,5) has $d_t = 15a_{C-C}/\pi = 6.88$ Å, both derived from hemispherical caps for the C_{60} molecule and assuming an average $a_{C-C} = 1.44$ Å appropriate for C_{60}. If the graphite value of $a_{C-C} = 1.421$ Å is used, slightly smaller values for d_t are obtained. Substitution of $(n, m) = (5,5)$ into Eq. (19.3) yields $\theta = 30°$ while substitution of $(n, m) = (9,0)$ and $(0,9)$ yields $\theta = 0°$ and $60°$, respectively. The tubules $(0,9)$ and $(9,0)$ are equivalent, because of the sixfold symmetry of the graphene layer. Because of the point group symmetry of the honeycomb lattice, several different integers (n, m) will give rise to equivalent nanotubes. To define each nanotube once and once only, we restrict ourselves to consideration of nanotubes arising from the 30° wedge of the 2D Bravais lattice shown in Fig. 19.2(b). Because of the small diameter of a carbon nanotube (~10 Å) and the large length-to-diameter ratio (> 10^3), carbon nanotubes provide an important system for studying one-dimensional physics, both theoretically and experimentally.

Many of the experimentally observed carbon tubules are multilayered, consisting of capped concentric cylinders separated by ~ 3.5 Å. In a formal sense, each of the constituent cylinders can be specified by the chiral vector $\mathbf{C}_h$ in terms of the indices (n, m) of Eq. (19.1), or equivalently by the tubule diameter d_t and chiral angle θ. Because of the different numbers of carbon atoms around the various concentric tubules, it is not possible to achieve the ABAB... interlayer stacking of graphite in carbon nanotubes. Thus, an interlayer spacing closer to that of turbostratic graphite (3.44 Å) is expected, subject to the quantized nature of the (n, m) integers, which determine $\mathbf{C}_h$. We illustrate this quantum constraint by considering the nesting of the $(9,0)$ zigzag tubule (which has $d_t = 7.15$ Å) within an adjacent zigzag tubule of larger diameter. From the turbostratic constraint, we know that the adjacent nested tubule must have a diameter in excess of [7.15+2(3.44)] Å= 14.03 Å. From Eq. (19.2), the zigzag tubule $(18,0)$ has a diameter 14.29 Å which meets the turbostratic constraint, while the $(17,0)$ tubule has a diameter of only 13.50 Å, which is too small to accommodate a concentric (9, 0) tubule. In fact, Eq. (19.2) shows that the (18, 0) tubule has the smallest diameter subject to the turbostratic constraint ($d_t > 13.93$ Å), even when all possible (n, m) indices are considered, thus leading to a minimum interlayer separation distance of 3.57 Å. In the next two sections, we review experimental observations on multiwall (§19.2) and single-wall (§19.4.1) carbon nanotubes. Further discussion of Fig. 19.2(a) is given in §19.4.1 and §19.4.2 in terms of the symmetry of carbon nanotubes.

19.2. Experimental Observation of Carbon Nanotubes

Most of the experimental observations have been on multiwall carbon nanotubes and bundles of nanotubes, which are discussed initially. We then proceed to review the more recent work on single-wall tubules, which relate more closely to the theoretical calculations. Since much attention has been given to the structure of tubule caps, this topic is also reviewed. Finally, methods for synthesis of carbon nanotubes are discussed.

19.2.1. Observation of Multiwall Carbon Nanotubes

The earliest observations of carbon tubules with very small (nanometer) diameters [19.1, 7, 8] were based on high-resolution transmission electron microscopy (TEM) measurements on material produced in a carbon arc. This work provided strong evidence for μm-long tubules, with cross sections showing several coaxial tubes and a hollow core. In Fig. 19.3, the first published observations of carbon nanotubes are shown [19.1]. Here we see only multilayer carbon nanotubes, but one tubule has only two coaxial carbon cylinders [Fig. 19.3(b)], and another has an inner diameter of only 23 Å [Fig. 19.3(c)] [19.1]. Typically, the outer diameter of carbon nanotubes prepared by a carbon arc process ranges between 20 and 200 Å and the inner diameter ranges between 10 and 30 Å [19.9]. Typical lengths of the arc-grown tubules are $\sim$1 μm, giving rise to an aspect ratio (length-to-diameter ratio) of 10^2 to 10^3. Because of their small diameter, involving only a small number of carbon atoms, and because of their large aspect ratio, carbon nanotubes are classified as 1D carbon systems. Most of the theoretical work on carbon nanotubes emphasizes their 1D properties. In the multiwall carbon nanotubes, the measured (by high-resolution TEM) interlayer distance is 3.4 Å [19.1], in good agreement with the value of 3.39 Å for the average equilibrium interlayer separation, obtained from self-consistent electronic structure calculations [19.10, 11].

Scanning tunneling microscopy (STM) [19.3, 12–15] and atomic force microscopy (AFM) [19.15, 16] have also provided powerful local probes of

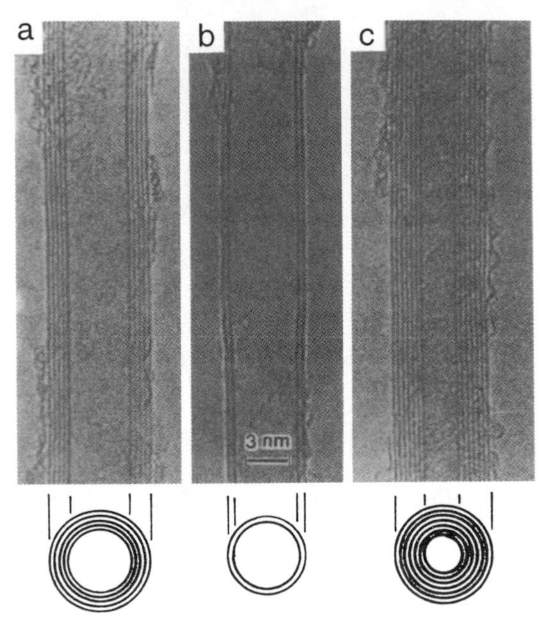

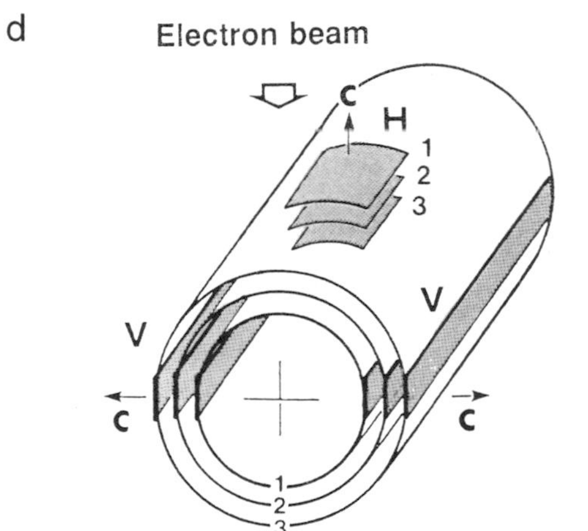

Fig. 19.3. The observation of N coaxial carbon tubules with various inner diameters d_i and outer diameters d_o reported by Iijima using TEM: (a) $N = 5$, $d_o = 67$ Å, (b) $N = 2$, $d_o = 55$ Å, and (c) $N = 7$, $d_i = 23$ Å, $d_o = 65$ Å. The sketch (d) indicates how the interference patterns for the parallel planes labeled H are used to determine the chiral angle θ, which in turn is found from the orientation of the tubule axis relative to the nearest zigzag axis defined in Fig. 19.2(a). The interference patterns that are labeled V determine the interplanar distances [19.1].

the topographical structure. STM techniques have also been used to study the electronic structure through measurements of the electronic density of states and AFM has been used to study the elastic properties of carbon nanotubes.

Carbon nanotubes grown by vapor growth methods have also been reported. In Fig. 19.4(a) we see a thin carbon tubule (b) nucleated on a con-

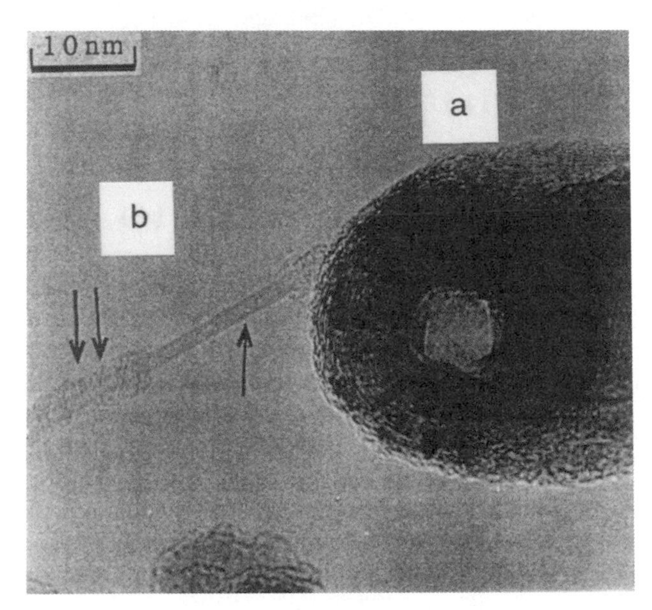

Fig. 19.4. Two kinds of vapor-grown carbon fibers (VGCF) observed in the as-grown sample: (a) a thick hollow fiber formed by a catalytic metal particle, (b) an exposed nanotube (↑) and a pyrolytically-coated segment of the nanotube (↓↓) [19.17].

ventional vapor-grown carbon fiber (a) during the growth process [19.7, 8]. The region denoted by the double arrows shows a pyrolytic carbon deposit on the as-grown nanotube. In contrast, the carbon tubules of Fig. 19.3(a), (b), and (c) were nucleated on the surface of the negative carbon electrode in a carbon arc discharge apparatus [19.1]. It is believed that the properties of carbon nanotubes grown by the arc discharge method are similar to those grown from the vapor phase [19.17] (see §19.2.5).

Although very small diameter carbon filaments, such as shown in Fig. 19.5, were observed many years earlier [19.18, 19], no detailed systematic studies of such very thin filaments were reported in the 1970s and 1980s. A direct stimulus to a more systematic study of carbon filaments of very small diameters came from the discovery of fullerenes by Kroto, Smalley, and co-workers [19.22] and subsequent developments resulting in the synthesis of gram quantities of fullerenes by Krätschmer, Huffman, and co-workers [19.23]. These developments heralded the entry of many scientists into the field, together with many ideas for new carbon materials, and bringing new importance to carbon systems of nanometer dimensions. Iijima's important contribution [19.1] included an appreciation for the importance of his nanotube work in relation to ongoing fullerene

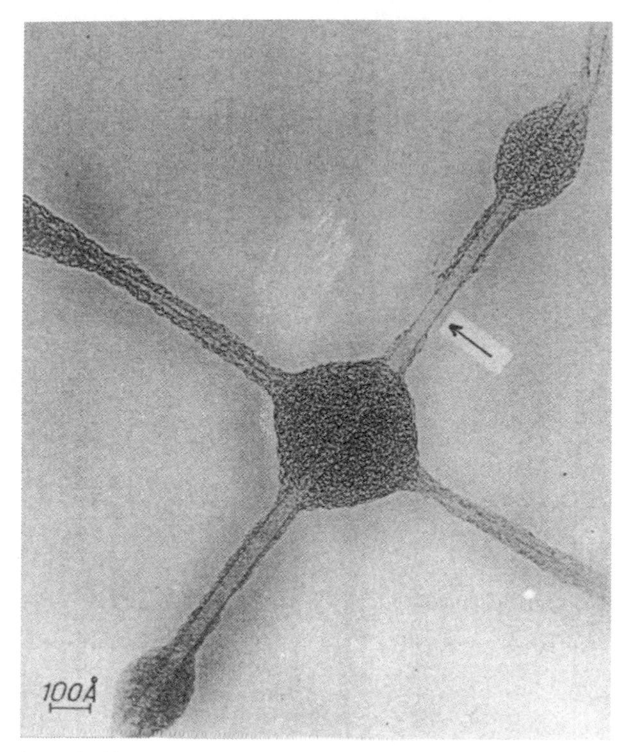

Fig. 19.5. An early high-resolution TEM micrograph showing carbon nanotubes with diameters less than 10 nm [19.18–21].

studies. Quite independently, Russian workers also reported the discovery of carbon tubules and bundles, but generally having much smaller aspect ratios, and hence they called their tubules "barrelenes" [19.24, 25]. These barrelenes have similarities to cylindrical fullerenes reported by Wang and Buseck [19.26] with length-to-diameter ratios or 10 or less.

Soon after the report of carbon nanotube synthesis by the arc discharge method [19.1], a second report was published giving conditions for the synthesis of copious amounts of fullerene tubules [19.27, 28] using the carbon arc discharge method. The availability of large quantities of carbon nanotubules greatly stimulated experimental activity in this field [19.27]. One interesting and useful characteristic of the growth of the carbon nanotubules is the tendency for large numbers of nanotubules to grow parallel to each other, forming a bundle of nanotubules [19.27], perhaps 50 nm in diameter, and reminiscent of the winding of carbon commercial fibers in continuous tows.

It should also be mentioned that the fullerene tubules of Fig. 19.3 differ in a fundamental way from the scroll-like graphite whiskers, synthesized by Bacon many years ago in a dc carbon arc discharge [19.29], under current and voltage conditions similar to those used for the growth of carbon nanotubules [19.27], but operating at much higher gas pressures [19.29]. The main difference is that graphite whiskers have been reported to form a scroll-like tube, while the nanotubes are coaxially stacked cylinders (see Fig. 19.3), as has been verified by high-resolution STM observations [19.12], which always show the same number of layers on the left and right hand sides of a carbon nanotube image, as seen in Figs. 19.3(a), (b), and (c). As mentioned above, highly elongated fullerenes with shapes between the traditional icosahedral fullerenes and the carbon nanotubules have been observed by high-resolution transmission electron microscopy [19.26]. It has also been shown that the carbon nanotubes are unstable under high-intensity electron beam irradiation and can thus be transformed into concentric spherical shells (sometimes called onions) [19.30], as discussed in §19.10.

19.2.2. Observation of Single-Wall Carbon Nanotubes

We now describe a few of the interesting characteristics of single-wall carbon nanotubes. The single-wall nanotubes, just like the multiwall nanotubes, tend to form bundles of nearly parallel tubules (Fig. 19.6), although single, isolated tubules are also found [19.31]. Soot commonly deposits on the surface of these tubules and also on the inside of tubules [19.31]. Deposits of amorphous carbon on vapor-grown carbon fibers are well known (see Figs. 19.4 and 19.5), and such deposits are used to thicken vapor-grown fibers, after appropriate heat treatment procedures (see §2.5). Furthermore, single-wall tubules are remarkably flexible, and bend into curved arcs with radii of curvature as small as 20 nm, as shown in Fig. 19.7. This flexibility suggests excellent mechanical properties, consistent with the high tensile strength and bulk modulus of commercial and research-grade vapor-grown carbon fibers (see §2.5). The longest reported nanotube length is 700 nm for a 0.9 nm diameter tubule, yielding a large length-to-diameter ratio (aspect ratio) of $\sim$800, although not as large as the aspect ratio ($\sim 10^4$) found in the larger ($>$100 nm diameter) vapor-grown carbon fibers [19.32]. The single-wall nanotubes, just like the multiwall nanotubes and the conventional vapor-grown carbon fibers, have hollow cores along the axis of the tubule.

The diameter distribution of single-wall carbon nanotubes is of great interest for both theoretical and experimental reasons. Theoretical studies have shown that the physical properties of carbon nanotubes are strongly

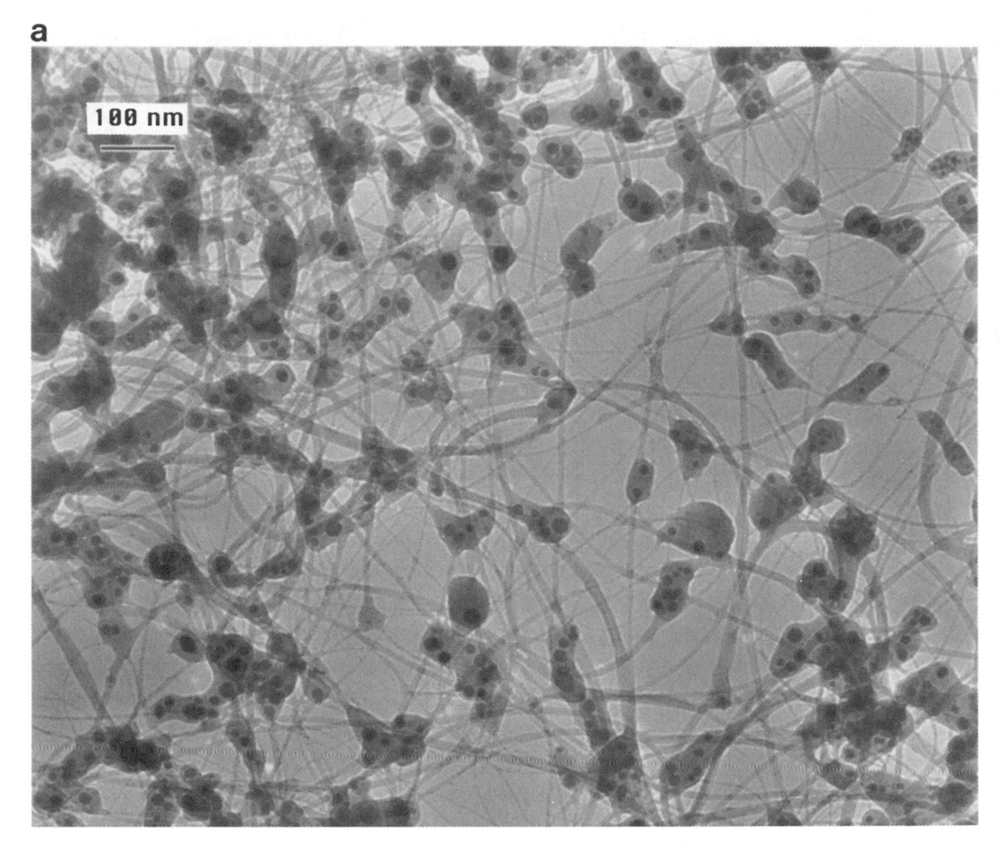

Fig. 19.6. (a) Transmission electron micrographs of the Co-catalyzed soot for a concentration of 4% Co. The threads in the figure are apparently individual nanotubules or bundles of nanotubes such as shown in (b). Most of the as-prepared nanotubes are covered with carbon soot [19.31].

dependent on tubule diameter. These predictions remain to be well tested experimentally. Because of the difficulty of making physical measurements on individual single-wall nanotubes, a number of exploratory studies have been made on bundles of tubules. Early results on the diameter distribution of Fe-catalyzed single-wall nanotubes [Fig. 19.8(a)] show a diameter range between 7 Å and 16 Å, with the largest peak in the distribution at 10.5Å, and with a smaller peak at 8.5Å [19.33]. The smallest reported diameter for the single-wall carbon nanotubes is 7 Å [19.33], the same as the smallest diameter expected theoretically (7.0 Å for a tubule based on C_{60}) [19.4], and this is further discussed in §19.2.3. Qualitatively similar results, but differing in detail were obtained for the Co-catalyzed

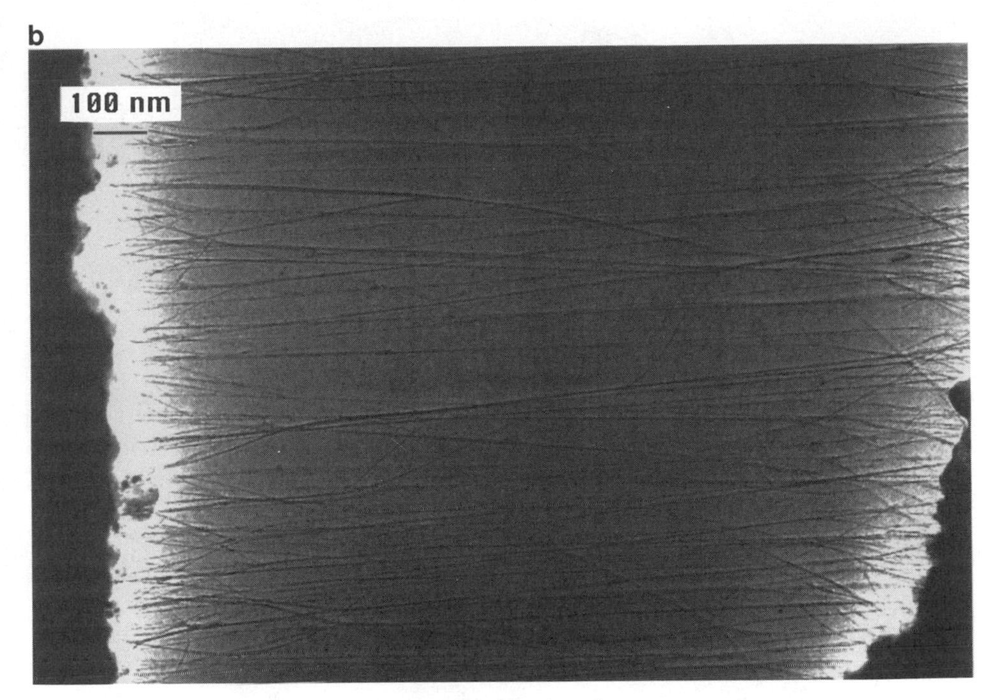

Fig. 19.6. (Continued).

nanotubes, with a peak in the distribution at 13Å, as shown in Fig. 19.8(b) [19.31]. These experimental results indicate that the diameter distribution of the single-wall nanotube involves predominantly small diameter tubules. The single-wall nanotubes have a much narrower diameter distribution which is dependent on the catalyst and preparation conditions, so that explicit measurements of the diameter distribution will be needed for samples used for properties measurements on tubule bundles.

The second characteristic parameter of great importance for properties measurements on single-wall carbon nanotubes is the chiral angle θ. Two experimental problems contribute to the difficulty in determining the chiral angle θ. Because of the small number of carbon atoms available for carrying out a diffraction experiment on a single-wall nanotube, electron diffraction is the favored technique. Furthermore, the low atomic number for carbon ($Z = 6$) results in a very low electron scattering cross section. Nevertheless, it has been demonstrated [19.33] that electron diffraction measurements with a suitable TEM instrument and using micro-diffraction techniques can provide detailed structural information on an individual single-wall nanotube, including measurement of the tubule diameter d_t and the chiral an-

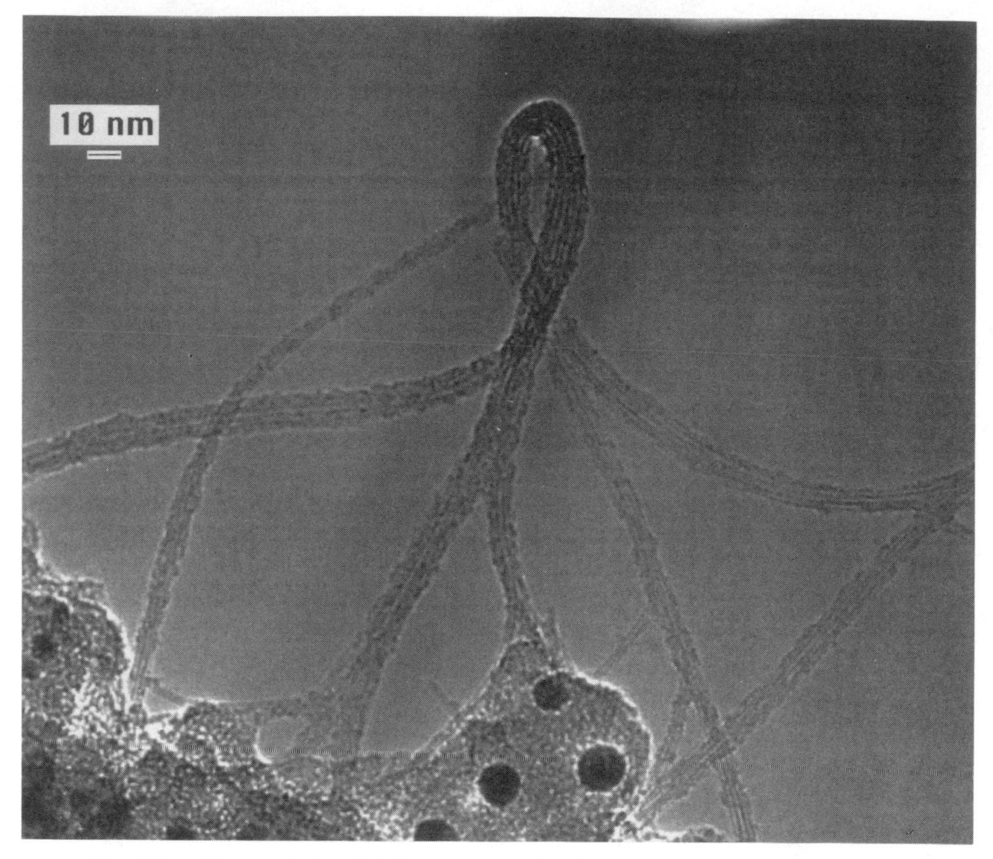

Fig. 19.7. High-resolution TEM micrographs showing nanotubes with a 20 nm radius of curvature [19.31].

gle θ. Based on their observations, Iijima and co-workers claim that most single-wall nanotubes show chirality [19.33] as was previously also claimed for the multiwall nanotubes. Atomic resolution STM techniques also provide sensitive techniques for the measurement of the chiral angle θ [19.15].

While the ability to measure the diameter d_t and chiral angle θ of individual single-wall tubules has been demonstrated, it remains a major challenge to determine d_t and θ for specific tubules used for an actual property measurement, such as electrical conductivity, magnetoresistance or Raman scattering. It should be emphasized that such properties measurements on individual carbon nanotubes are already very difficult, and adding to these difficulties are further challenges of characterizing each tubule regarding its d_t and θ values.

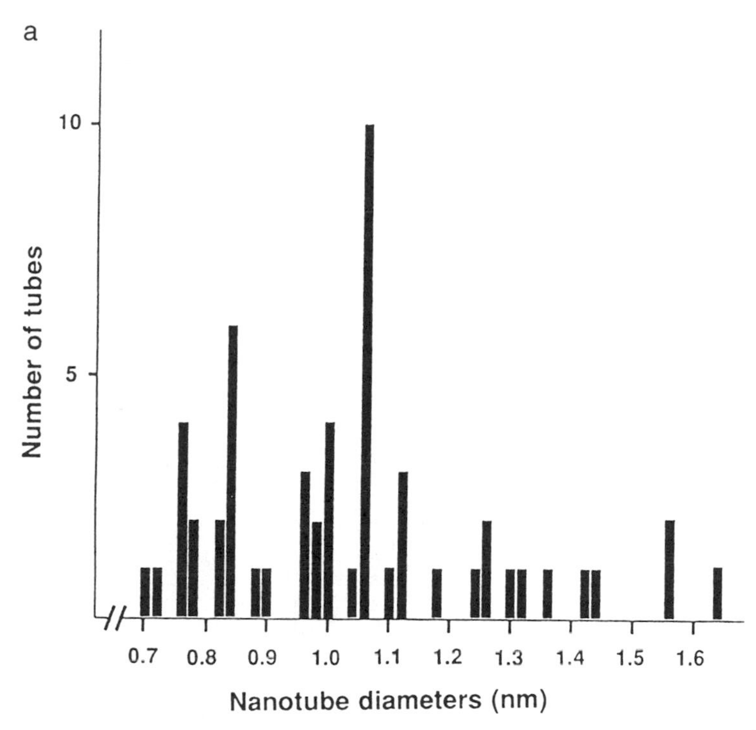

Fig. 19.8. Histograms of the single-wall nanotube diameter distribution for (a) Fe-catalyzed nanotubes [19.33] and (b) Co (4%)-catalyzed nanotubes [19.31].

19.2.3. Tubule Caps and Chirality

Most carbon nanotubes are capped by carbon shells that fit continuously on the ends of the long cylinder. This subsection reviews some of the properties of these caps. The number of possible caps containing only hexagons and pentagons that can be continuously attached to each carbon nanotube defined by (n, m) is given in Fig. 19.2(b), and is found conveniently by using the concept of projection mapping discussed in §3.3. In the enumeration given in Fig. 19.2(b), permutations of a given cap are not counted and only caps satisfying the isolated pentagon rule are included. From this figure we can conclude that the armchair tubule $(5, 5)$ which joins smoothly to a C_{60} hemisphere is the smallest diameter tubule that can be capped. The $(9, 0)$ zigzag tubule, which also joins smoothly to a C_{60} hemisphere, is the next smallest diameter tubule. Moving further out in diameter, we come to tubule $(6, 5)$ which is the smallest diameter chiral tubule for which there exists a unique cap containing only pentagons and hexagons, satisfying the isolated pentagon rule. In contrast, the chiral tubule $(7, 5)$ can be joined to

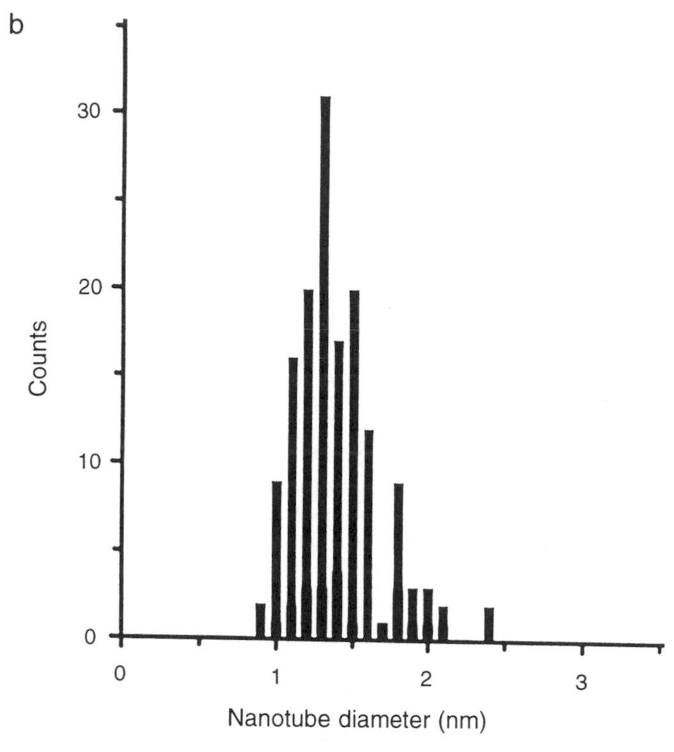

Fig. 19.8. (Continued).

13 different caps, two of which are shown in Fig. 19.9. As another example, the projection mapping of an icosahedral cap for a (10, 5) tubule is shown in Fig. 3.9 and 3.10(c). A schematic diagram of the (10, 5) tubule is shown in Fig. 19.10, where the chirality of this tubule can be clearly seen.

By adding many rows of hexagons parallel to *AB* and *CD* in Fig. 3.9, a properly capped chiral graphene tubule is obtained [19.34], as shown in Fig. 19.10. The projection method illustrated in Fig. 3.10(c) can be extended to generate all possible chiral tubules specified in Fig. 19.2(b). As noted above, many of the chiral tubules can have a multiplicity of caps, each cap joining smoothly to the same vector $\mathbf{C}_h = n\mathbf{a}_1 + m\mathbf{a}_2$ and hence specifying the same chiral tubule [19.6, 35]. We note that for nanotubes with smaller diameters than that of C_{60}, there are no caps containing only pentagons and hexagons that can be fit continuously to such a small carbon nanotube (n, m). For this reason it is expected that the observation of very small diameter (< 7 Å) carbon nanotubes is unlikely. For example, the (4, 2) chiral vector shown in Fig. 19.2(a) does not have a proper cap and therefore

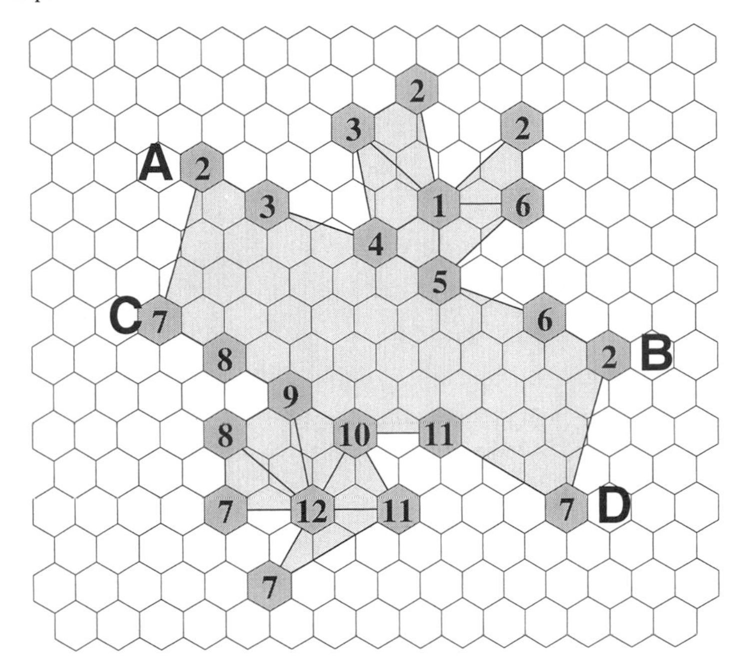

Fig. 19.9. A projection mapping of two possible caps that join continuously to a (7,5) chiral tubule and satisfy the isolated pentagon rule. (see §3.3).

is not expected to correspond to a physical carbon nanotube. One reason why tubules of larger diameter (> 10 Å) are predominantly observed may be related to the larger number of ways that caps can be formed for the larger-diameter tubules; this probability should increase very rapidly with increasing tubule diameter, just as the number of isomers obeying the isolated pentagon rule increases rapidly as the number of carbon atoms in the fullerene increases [19.36].

Most of the tubules that have been discussed in the literature have closed caps. Open-ended tubules have also been reported using high-resolution TEM [19.37] and STM [19.15] techniques. The open ends of tubules appear as "highlighted" edges in the STM pattern due to the dangling bonds at the open ends. A number of recurring cap shapes have been reported in the literature for the closed-end tubules [19.33, 38, 39]. Some typical examples are shown in Fig. 19.11. Following the discussion of Euler's theorem in §3.1, six pentagons (each corresponding to a disclination of $-\pi/3$) are needed to form a single cap crowning a carbon tubule. Shown in Fig. 19.11 are a symmetric cap (a), an asymmetric cap (b), and a flat cap (c) to a carbon nanotube. The caps for tubules of larger diameter tend to be more flat than

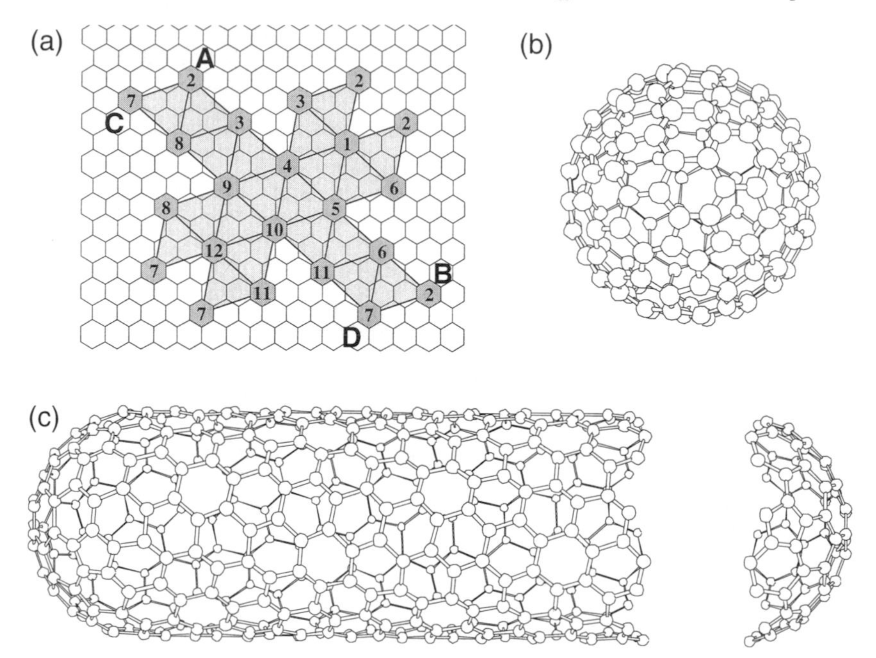

Fig. 19.10. (a) A projection mapping of the C_{140} fullerene. (b) A model of C_{140} with icosahedral I symmetry. (c) A schematic diagram of the $(10, 5)$ single-wall tubule capped by hemispheres of the icosahedral C_{140} fullerene at both ends [19.28].

for those with smaller diameters, which are more round. For example, the cap of a tubule of diameter $d_t = 42$ Å corresponds to the "hemisphere" of a large fullerene (approximately C_{6000}). Some of the common caps, such as in Fig. 19.11(a), have the shape of a cone, which is discussed further in §19.2.4.

The bill-like cap shown in Fig. 19.12 is explained in terms of the placement of a pentagon at "B" (introducing a $+\pi/3$ disclination) and a heptagon at C (with a disclination of $-\pi/3$). The narrow tube termination above points "B" and "C" is a typical polyhedral cap with six pentagons. We can understand the relation of the heptagon and pentagon faces by noting that a cap normally is formed by six pentagons each contributing a disclination of $\pi/3$, yielding a total disclination of 2π, which is the solid angle of a hemisphere. Each heptagon contributes a disclination of $-\pi/3$, so that an additional pentagon is needed to offset each heptagon in forming a closed surface.

A rare although important tubule termination is the semitoroidal type shown in Fig. 19.13, which has also been described theoretically [19.40]. Figure 19.13 shows a few of the inner cylinders with normal polyhedral

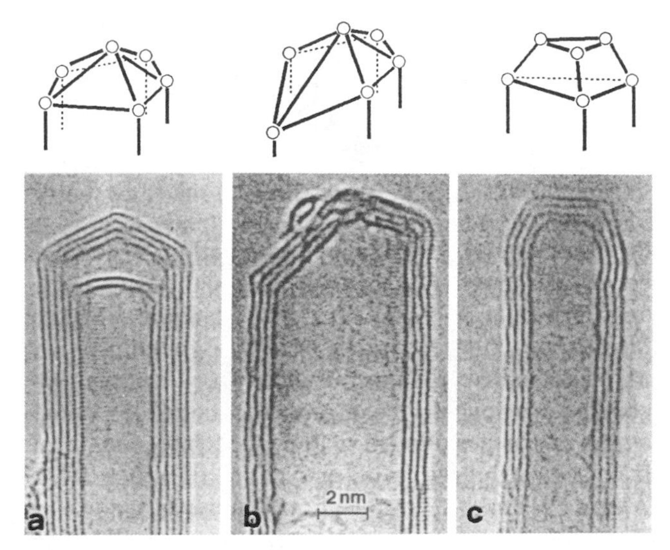

Fig. 19.11. Transmission electron microscopy pictures of carbon nanotubes with three common cap terminations: (a) a symmetric polyhedral cap, (b) an asymmetric polyhedral cap, (c) a symmetrical flat cap [19.37].

terminations and six cylinders with semitoroidal terminations, creating a lip on the tubule cap, which is clarified by the schematic diagram shown in Fig. 19.13(b).

Iijima reports [19.1, 9] that the majority of the carbon tubules, which he has observed by electron diffraction in a transmission electron microscope, have screw axes and chirality [i.e., θ is neither 0 nor $\pi/6$ in Eq. (19.3)]. It was further reported that in multiwall tubules there appears to be no particular correlation between the chirality of adjacent cylindrical planes [19.1, 9]. Iijima has also reported an experimental method for measurement of the chiral angle θ of the carbon nanotubules based on electron diffraction techniques [19.1, 37]. Dravid *et al.* [19.38], using a similar technique to Iijima, have also found that most of their nanotubes have a chiral structure. However, the TEM studies by Amelinckx and colleagues [19.41, 42] produced the opposite result, that nonchiral tubules ($\theta = 0$ or $\pi/6$) are favored in multiwall nanotubes.

Atomic resolution STM measurements (see §17.4) done under high-resolution conditions have been used to measure the chiral angle of the cylindrical shell on the surface of carbon nanotubes [19.12, 43]. Both chiral and nonchiral tubules have been observed. For the case of a single-wall tube, a zigzag nanotube with an outer diameter of 10 Å has been

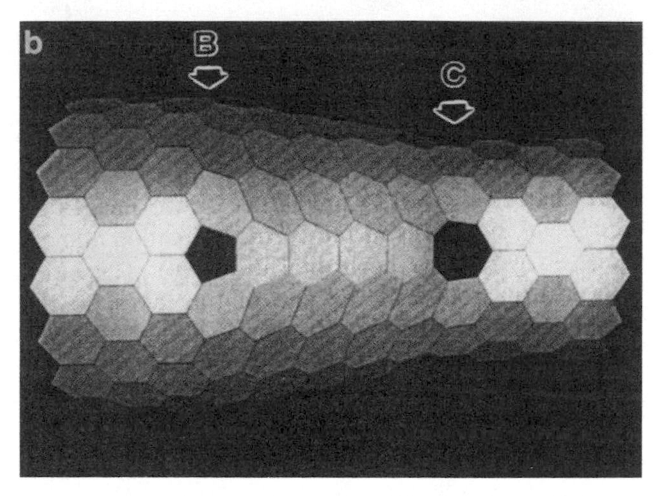

Fig. 19.12. (a) Transmission electron micrograph showing a bill-like termination of a nanotube consisting of three parts, a tube, a cone, and a smaller tube. Transitions in the shape of the tube are caused by a pentagon at "B" and a heptagon at "C." (b) Schematic illustration of the transition region showing the pentagon and heptagon responsible for the curvatures at "B" and "C" [19.37].

imaged directly [19.12]. Using a moiré pattern technique which gives the interference pattern between the stacking of adjacent layers, it was shown that the tubule cylinder and the adjacent inner cylinder were both helical with chiral angles of $\theta = 5°$ and $\theta = -4°$, respectively, yielding a relative chiral angle of 9° as shown in Fig. 19.14(c) [19.12]. Atomic resolution STM observations are very important in demonstrating networks of perfect honeycomb structures (see Fig. 19.14) and showing that a multilayer nanotube may consist of cylinders with different chiral angles [19.4, 12].

Several theoretical arguments favoring chiral nanotubes have been given [19.4, 6, 46, 47], based on the many more combinations of caps that

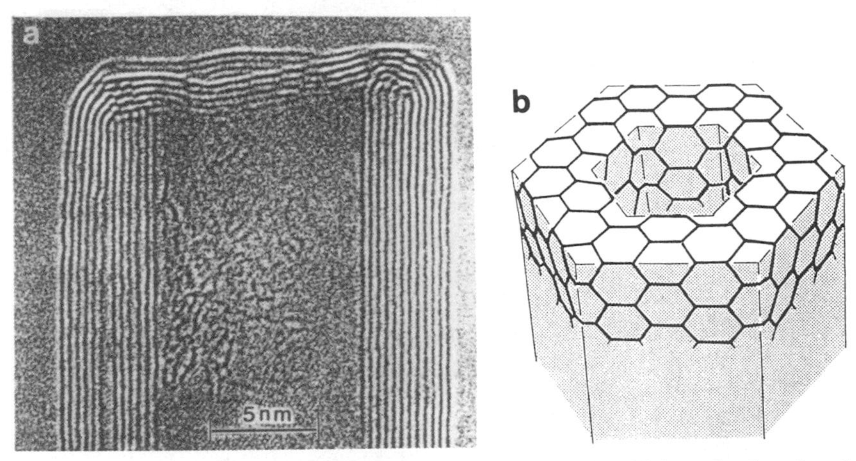

Fig. 19.13. (a) Transmission electron micrograph of the semitoroidal termination of a tube which consists of six graphene shells. (b) Schematic illustration of a semitoroidal termination of a tubule which is caused by six pentagon–heptagon pairs in a hexagonal network [19.37].

can be formed for chiral rather than nonchiral nanotubes, especially at larger diameters d_t, and the more favorable growth opportunities that are possible for chiral nanotubes (see §19.3). Although most workers have discussed carbon nanotubes with circular cross sections normal to the tubule axis, some reports show evidence for faceted polygonal cross sections, including cross sections with fivefold symmetry [19.48]. Polygonal cross sections have been known for vapor-grown carbon fibers, upon heat treatment to ~3000°C [19.17, 49]. For the case of carbon fibers with high structural order, the faceting introduces interlayer long-range order which results in a lower free energy, despite the strong curvatures at the polygon corners. It would seem that polygonal cross sections are more common in large-diameter tubules, where most of the carbon atoms would be on planar surfaces of a faceted nanotube. Nanotubes with polygonal cross sections could be of interest as possible hosts for intercalated guest species between planar regions on adjacent tubules.

Because of the special atomic arrangement of the carbon atoms in a C_{60}-based tubule, substitutional impurities are inhibited by the small size of the carbon atoms. Furthermore, the screw axis dislocation, the most common defect found in bulk graphite, should be inhibited by the monolayer structure of the C_{60} tubule. Thus, the special geometry of C_{60} and of the C_{60}-related carbon nanotubes should make these structures strong and stiff along the tubule axis and relatively incompressible to hydrostatic stress, when compared to other materials [19.50] (see §19.8).

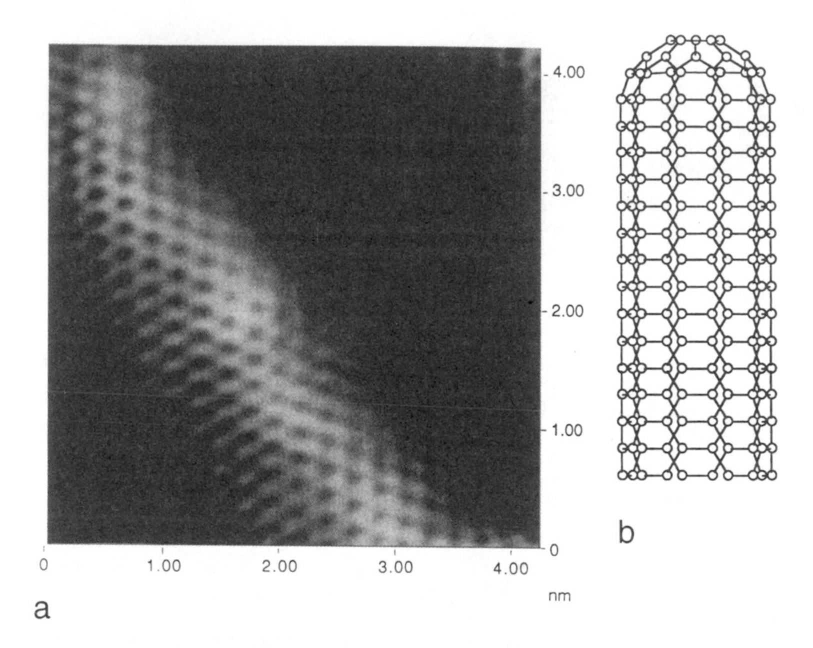

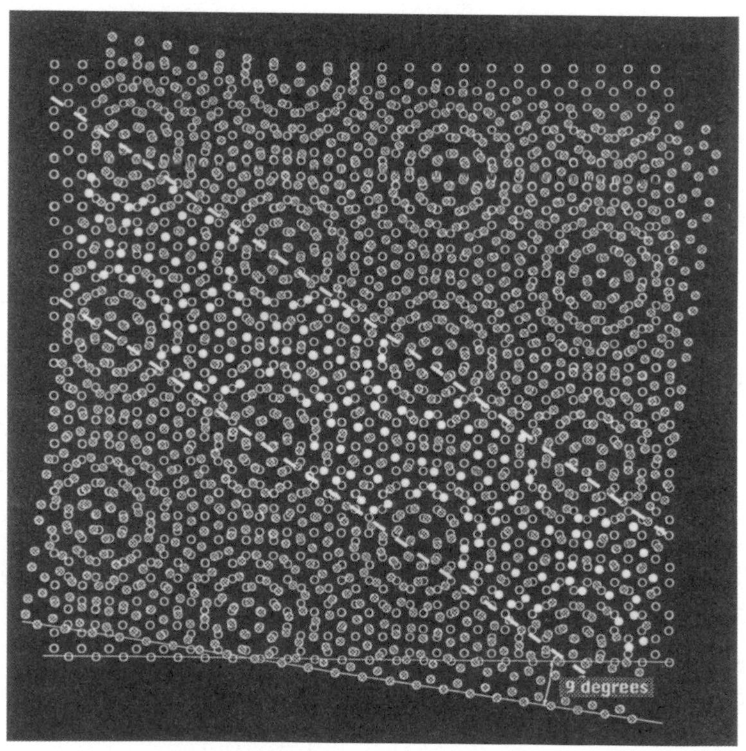

c

19.2.4. Carbon Nanocones

Carbon nanocones are found on the caps of nanotubes (see §19.2.3) and also as free-standing structures generated in a carbon arc. These cones have been studied predominantly by high-resolution TEM [19.39, 51] and by high-resolution STM [19.52]. The observed cones normally have opening angles of $\sim 19^\circ$ and can be quite long ($\sim$240 Å) [19.15]. Cones are formed from hexagons of the honeycomb lattice by adding fewer pentagons than the six needed by Euler's theorem to form a cylinder, which is the basic constituent of a nanotube. The addition or removal of a pentagon thus corresponds to a change in solid angle of $4\pi/12$ or $\pi/3$. Thus the addition of 12 pentagons corresponds to a solid angle of 4π and the formation of a closed cage. The solid angle Ω subtended by the cone is given by

$$\Omega = 2\pi(1 - \cos\theta_p) = 2\pi - n_p\pi/3 \tag{19.4}$$

where n_p is the number of pentagons in the cone, $2\theta_p$ is the cone angle, and $\pi - 2\theta_p$ is the opening angle of the cone. The cone angle $2\theta_p$ is then given by

$$2\theta_p = 2\cos^{-1}\left(\frac{n_p}{6}\right) \tag{19.5}$$

where θ_p is the angle of the cone with respect to the z axis, which is taken as the symmetry axis of the cone. Values for the cone angles $2\theta_p$ and for the opening angles $\pi - 2\theta_p$ in degrees are given in Table 19.1, and each of the entries for $\pi - 2\theta_p$ ($n_p = 1, 2, 3, 4, 5$) is shown in Fig. 19.15. The limiting cases for the cones in Table 19.1 are $\pi - 2\theta_p = 180^\circ$ corresponding to a flat plane and $\pi - 2\theta_p = 0^\circ$ corresponding to $n_p = 6$ and the formation of a cylindrical carbon nanotube rather than a cone. An example of a cone structure, as observed under high-resolution TEM, is shown in Fig. 19.16(a), and the schematic representation of this cone in the context of Eq. (19.4) is shown in Fig. 19.16(b).

Fig. 19.14. (a) Atomic resolution STM image of a carbon nanotube 35 Å in diameter. In addition to the atomic honeycomb structure, a zigzag superpattern along the tube axis can be seen. (b) A ball-and-stick structural model of a C_{60}-based carbon tubule. (c) Structural model of a giant superpattern produced by two adjacent misoriented graphene sheets. The carbon atoms in the first layer are shaded and the second layer atoms are open. Between the two dashed lines are highlighted those first layer white atoms that do not overlap with second layer atoms. Because of their higher local density of states at the Fermi level, these atoms appear particularly bright in STM images [19.44, 45]. This moiré pattern results in a zigzag superpattern along the tube axis within the two white dashed lines as indicated [19.12].

Table 19.1

Cone angles in degrees for nanocones with various numbers of pentagons.

n_p	$2\theta_p$	$\pi - 2\theta_p$
0	180.00	0.00
1	160.81	19.19
2	141.05	39.95
3	120.00	60.00
4	96.37	83.63
5	67.11	112.89
6	0	180.00

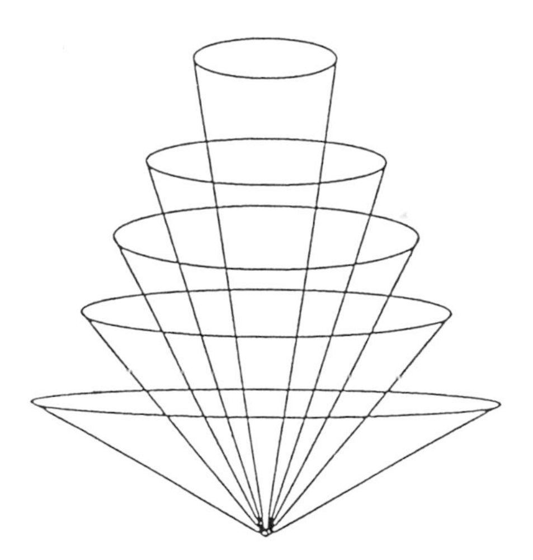

Fig. 19.15. Allowed opening angles ($\pi - 2\theta_p$) for carbon nanocones [19.15].

19.2.5. Nanotube Synthesis

There are currently three main methods used to synthesize carbon nanotubes: carbon arc synthesis, chemical vapor deposition, and ion bombardment, although some nanotube synthesis has been reported using flames [19.54, 55].

Most of the work thus far reported has used the carbon arc method for the preparation of carbon nanotubes [19.2, 3], which is similar to the carbon arc method for the synthesis of fullerenes (see §5.1). Typical synthesis conditions for nanotubes employ a dc current of 50–100 A and a voltage of 20–25 V operating in an inert (e.g., He) atmosphere (see Fig. 5.1). The magnitude of the current required should scale with the electrode diameter (i.e. larger currents are needed to vaporize larger electrodes); the same rule

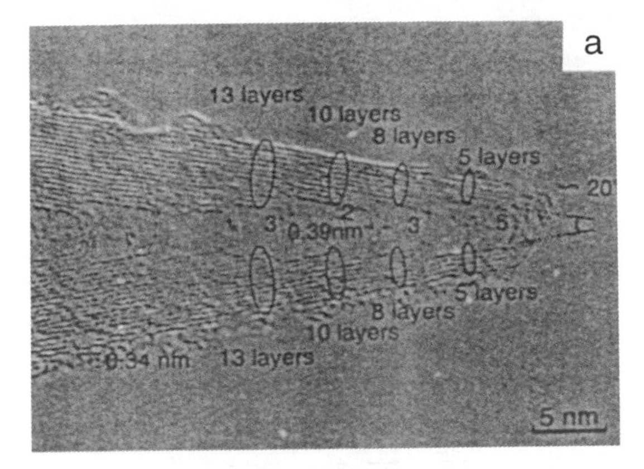

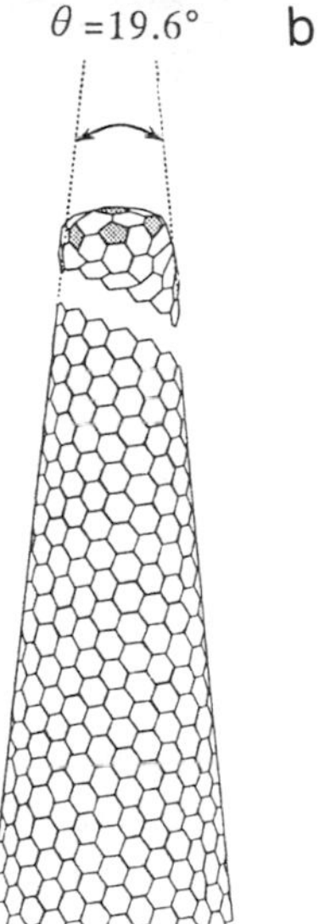

Fig. 19.16. (a) The cone-like tip of a carbon nanocone prepared by the vapor-growth method [19.17]. (b) A structural model of a conical nanotube based on a rolled-up hexagonal network [19.53].

applies to the arc synthesis of fullerenes. Some workers report optimum results for an He pressure of ~500 torr while others use lower pressures [19.56]. Once the arc between the cathode and anode is struck, the rods (typically about 6–7 mm diameter for the translatable anode and 9–20 mm for the fixed cathode) are kept about 1 mm apart. The deposit forms at the rate of ~1 mm per minute on the larger negative electrode (cathode), while the smaller positive electrode (anode) is consumed. The nanotubes form only where the current flows. The inner region of the electrode, where the most copious tubule harvest is made, has an estimated temperature of 2500–3000°C. In this inner region of the electrode, columnar growth tex-

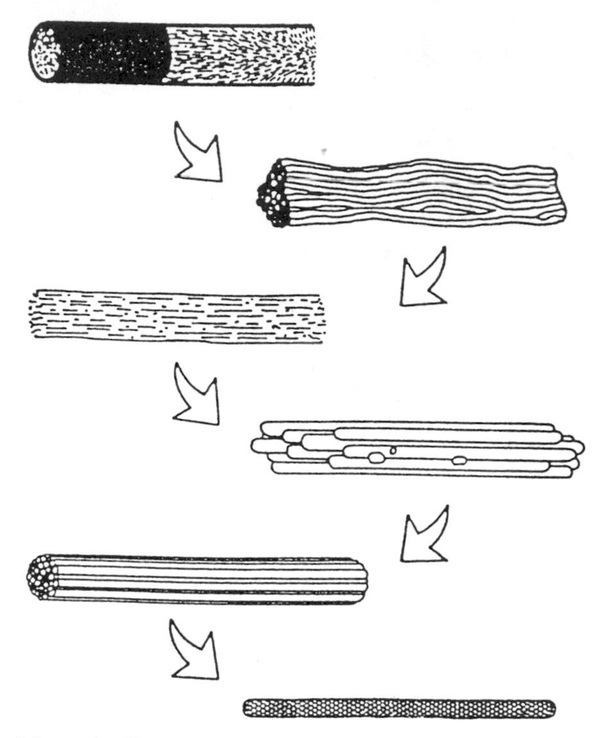

Fig. 19.17. Schematic diagram of the fractal-like organization of the carbon nanotube bundles, from the largest bundles down to individual carbon nanotubes [19.3].

ture of nanotube bundles containing smaller bundles of nanotubes (see Fig. 19.17) has been reported by many workers [19.27, 37, 38]. The smallest bundle is the microbundle, which consists of 10 to 100 aligned nanotubes of nearly the same length [19.3, 9]. Along with the tubule bundles is a growth of well-ordered carbon particles and other disordered carbonaceous material. The arc deposit typically consists of a hard gray outer shell (composed of fused nanotubes and nanoparticles) and a soft fibrous black core which contains about two thirds nanotubes and one third nanoparticles. Adequate cooling of the growth chamber is necessary to maximize the nanotube yield and their ordering. The growth of carbon tubules appears to be unfavorable under the conditions that are optimized to synthesize fullerene molecules.

The nanotubes can be separated from each other by sonication in solvents such as ethanol. A TEM picture of the core material of the carbon arc deposit is shown in Fig. 19.18(a), where both nanoparticles and nanotubes (20–200 Å outer diameter, length $\sim$1 μm) can be seen. The separation of the nanoparticles from the nanotubes can be accomplished by burning

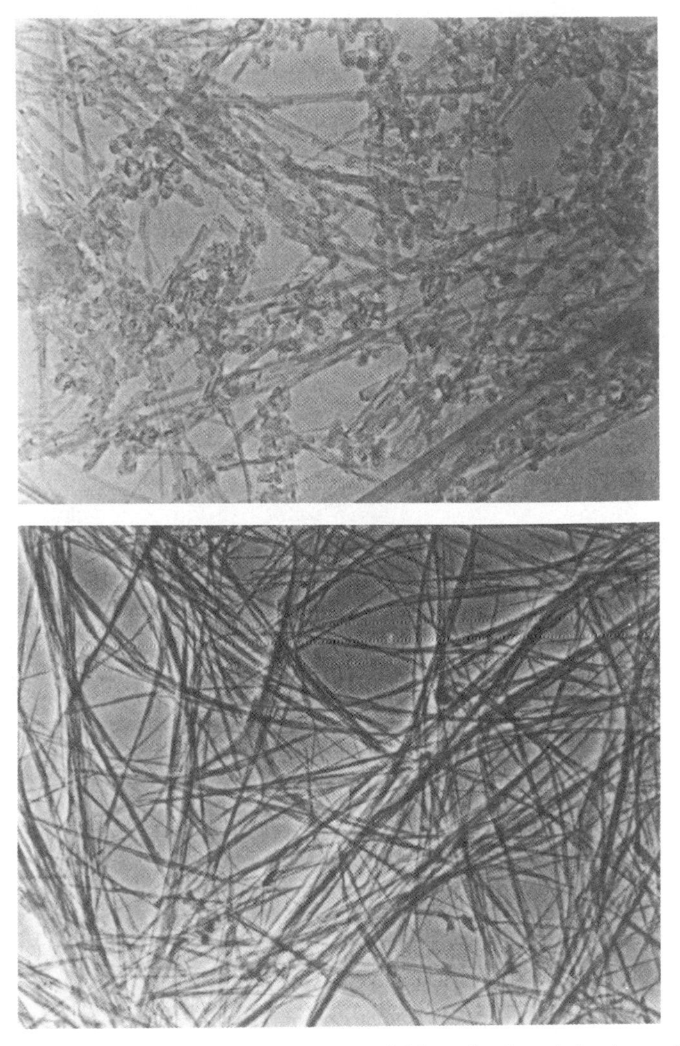

Fig. 19.18. TEM pictures of standard core material from the deposit (top) containing both nanotubes and nanoparticles and (bottom) purified nanotubes [19.9].

away the nanoparticles in oxygen while still leaving behind some of the nanotubes, which also vaporize by oxygen treatment but at a slower rate than the nanoparticles [19.9]. The oxygen-burning technique which is valuable for the elimination of nanoparticles also tends to eliminate most of the nanotubes, so that only about 1% of the initial deposit remains after the oxygen treatment.

The synthesis of single-shell nanotubes utilizes a variant of the carbon arc method. In this case, a hole is made in the carbon anode (6 mm diameter), which is filled with a composite mixture of transition metal material (Co, Fe, or Ni) [19.31, 33, 57–59] and graphite powder, while the 6-mm-diameter cathode rod is pure carbon. The transition metal serves as a catalyst and yields single shell tubules of small diameter with a narrow size distribution (see §19.2.2). Best results for the Co-catalyzed tubules [see Fig. 19.6(a)] were obtained when the Co–graphite mixture had a 4% Co stoichiometry. Typical operation conditions for the arc are a dc current of 95–115 A, a voltage of 20–25 V, with 300–500 torr He gas, and a flow rate of 5–15 ml/s [19.31]. For the Fe-catalyzed single-wall nanotubes, two vertical electrodes were used, and Fe filings were inserted into a cup-shaped indentation in the lower-lying cathode electrode (20 mm diameter). The arc was operated between the cathode and a 10 mm diameter anode at a dc current of 200 A at 20 V, and a gas mixture of 10 torr methane and 40 torr argon was used [19.33]. In this system the Fe and carbon were vaporized simultaneously. The single-wall tubules are found in web-like rubbery sooty deposits on the walls of the evaporation chamber and away from the electrodes. In the arc process, the iron or cobalt catalyst also forms nanometer-size carbide particles, around which graphene layers form, as well as metal clusters encapsulated within graphene layers [19.31, 33]. Single-wall carbon tubules are seen to grow out from the carbon nanoparticles.

The second synthesis method for carbon nanotubes is from the vapor phase and utilizes the same apparatus as is used for the preparation of vapor-grown carbon fibers (see §2.5), with the furnace temperature also held at 1100°C, but with a much lower benzene gas pressure [19.17, 21, 32, 60]. Carbon nanotubes can grow at the same time as conventional vapor-grown carbon fibers, as is seen in Fig. 19.4. Vapor-grown carbon nanotubes also grow in bundles. These bundles have been studied by high-resolution TEM in both their as-grown form and after heat treatment in argon at 2500–3000°C. The as-grown nanotubes generally show poor crystallinity. The crystallinity, however, is much improved after heat treatment to 2500–3000°C in argon, as seen in high-resolution TEM studies [19.17]. On the basis of the very large difference in the diameter of the hollow core between typical vapor-grown carbon fibers and carbon nanotubes and the appearance of internal bamboo-shaped structures [see Fig. 19.12(a)], it is suggested that the growth mechanism for the nanotubes may be different from that of the vapor-grown carbon fibers. Referring to the bamboo structure of Fig. 19.19 for carbon nanotubes, Endo [19.17] argues that the capping off of an inner layer terminates its growth, so that the exposed cap layer provides growth along the length and the epitaxial layers follow this growth while at the same

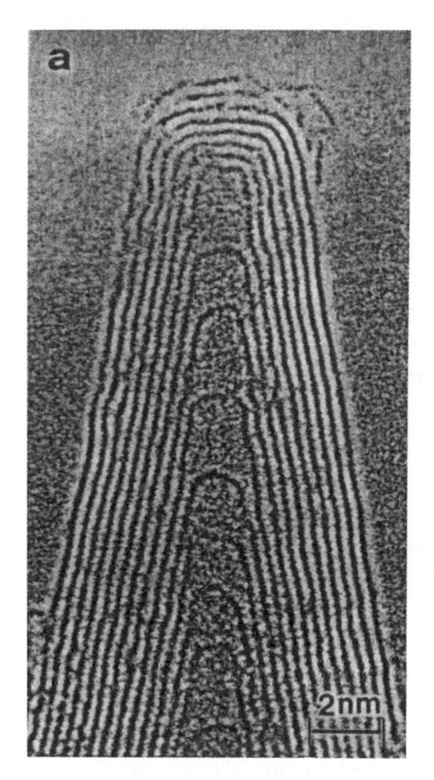

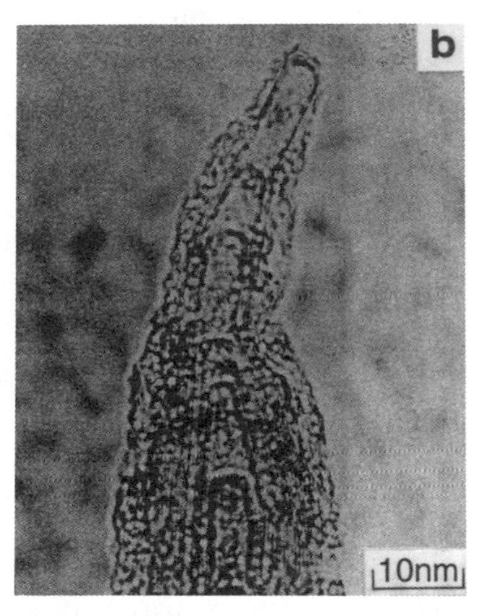

Fig. 19.19. (a) Transmission electron micrograph of a cone containing only single conical shells. The nearly periodic structures of the conical shells appear inside the cone tips [see part (a)], which are attributed to overshooting growth on the basis of the open tube growth model [19.37]. (b) Commonly observed nanotube structure for the cap region of vapor-grown carbon nanotubes heat treated at 2800°C in Ar. Here a number of bamboo-like structures are observed in the core region near the cap [19.17].

time adding to the tubule diameter. Bending of the growth axis of the nanotubes has been reported by a number of workers [19.17, 37] and is related to the introduction of a heptagon–pentagon defect pair at the bend location.

A third method of nanotube synthesis relates to the use of carbon ion bombardment to make carbon whiskers [19.61, 62]. In this method, carbon is vaporized in vacuum using either an electron beam [19.24] or resistive heating, and the deposit is collected on a cold surface. The deposit contains carbon nanotubes, along with other structures. Of the three techniques, less is known about the optimization of the ion bombardment technique as well as the characteristics of the material that is prepared. With all three methods, sample characterization and separation of the nanotubes from other structures remain challenges for workers in the field.

A totally different approach to the physical realization of single-wall carbon nanotubes has been through oxidation of capped multiwall nanotubes in CO_2 at ~850°C [19.63] and in air in the 700–800°C range [19.64], much higher than the temperature range for oxidation of C_{60} (500–600°C). Carbon nanoparticles and graphite both oxidize at higher temperatures than nanotubes, presumably due to the curvature of the tubes and the consequent presence of lattice strain. In fact, amorphous carbon and poorly ordered carbons are removed at even lower temperatures than any of the ordered structures, because of their weaker carbon–carbon bonding. Thus oxidation at elevated temperatures can be used to clean up a carbon tubule sample, removing amorphous carbon and carbon onions (see §19.10) [19.9, 65]. Both of these studies show that the cap is more reactive and is etched away first. Much slower layer-by-layer removal of the cylindrical layers follows, until eventually a region of single-wall tubules is seen under TEM imaging [19.63, 64]. An increase in the surface area by ~50% was observed in the oxidized tubules, which may be due to a reduction in average tubule diameter, and an increased accessibility of the inner tubule surface to probing gas molecules, after the cap has been removed and the tube is opened [19.63]. The oxidation reaction is thermally activated with an energy barrier of 225 kJ/mol in air [19.64]. When the carbon nanotubes open up, carbonaceous material may be sucked up the tube.

Tube opening can also occur at lower temperatures (~400°C), using, for example, a lead metal catalyst in air [19.66]. Lead metal in the absence of air is not sucked up the tubule, from which it is concluded that the material that is sucked up is likely an oxide of lead [19.9].

The stability of a carbon nanotube relative to a graphene sheet or a carbon onion (see §19.10) has been widely discussed [19.67–71]. For small clusters of carbon atoms, closed surfaces or closed rings are preferred to reduce dangling bonds, thus favoring fullerenes and onions. The growth geometry is also believed to be involved in the stabilization of carbon tubules, including both kinetic and space-filling considerations. In this context, the diameter of the smallest tubule has been calculated by balancing the energy gained by stitching together the dangling bonds along the generator element of the tubule cylinder with the energy lost by the strain in bending the graphene sheet to form the cylinder [19.68, 69, 72]. The diameter which results from these estimates is 6.78 Å, which is very close to the diameter of C_{60}.

19.2.6. *Alignment of Nanotubes*

For a variety of experiments and applications it is desirable to align the carbon nanotubes parallel to each other. Two approaches to nanotube alignment have been reported [19.73, 74]. In the first method, the nanotubes are dispersed in a polymer resin matrix to form a composite, which is then sliced with a knife edge, causing the nanotubes to align preferentially along the direction of the cut [19.73]. In the second method nanotube films are prepared by dispersing the deposit on the cathode of the carbon arc in ethanol to prepare a suspension which is then passed through a 0.2 μm pore ceramic filter. The deposit left in the filter is transferred to a Delrin or Teflon surface by pressing the tube-coated side of the filter on the surface. After lifting the filter, the tubules remain attached to the surface. The surface is then lightly rubbed with a thin Teflon sheet or aluminum foil to produce tubules on the surface aligned in the direction of the rubbing [19.74]. Ellipsometry and resistivity measurements show large anisotropies, with maximum conduction in the tubule direction ($\alpha_{\parallel}$), and lower but different values of conductivity normal to $\alpha_{\parallel}$ but within the plane ($\alpha_{\perp}$) and normal to the plane surface (β).

High-resolution STM studies of tubule bundles show that all the outer shells of the tubes in nanotube bundles are broken, suggesting that the nanotubes are strongly coupled through the outer shell of the bundles [19.15]. The inner tubes of the bundles, however, were not disturbed, indicating stronger intratube interaction in comparison with the intertube interaction. After a tubule of a certain diameter is reached it may be energetically favorable to grow adjacent tubules, leading to the generation of tubule bundles [19.15].

19.3. Growth Mechanism

The growth mechanism for cylindrical fullerene tubules is especially interesting and has been hotly debated. One school of thought [19.35, 75] assumes that the tubules are always capped (see §19.2.3) and that the growth mechanism involves a C_2 absorption process that is assisted by the pentagonal defects on the caps. The second school [19.37, 51, 76] assumes that the tubules are open during the growth process and that carbon atoms are added at the open ends of the tubules. Since the experimental conditions for forming carbon nanotubes vary significantly according to growth method, more than one mechanism may be operative in producing carbon nanotubule growth.

The first school of thought focuses on tubule growth at relatively low temperatures ($\sim$1100°C) and assumes that growth is nucleated at active sites of a vapor-grown carbon fiber of about 1000 Å diameter. Although the parent vapor-grown carbon fiber is itself nucleated by a catalytic transition metal particle [19.32], the growth of the carbon nanotube is thought to be associated with the absorption of a C_2 dimer near a pentagon in the cap of the tubule. Referring to the basic model for C_2 absorption in Fig. 6.3(b), we see that sequential addition of C_2 dimers results in the addition of a row of hexagons to the carbon tubule. To apply the C_2 absorption mechanism described in §6.1, it is usually necessary to use the Stone–Wales mechanism to bring the pentagons into their canonical positions, as necessary for the execution of each C_2 absorption, in accordance with Figs. 6.2(b) and 19.20. For example, in Fig. 19.20(b) the pentagonal defect labeled 2 is the active pentagon for C_2 absorption, but after one C_2 dimer absorption has occurred, the active pentagon becomes site 3, as shown in Fig. 19.20(c). Thus, Fig. 19.20 shows a sequence of five C_2 additions, which result in the addition of one row of hexagons to a carbon nanotube based on a C_{60} cap. [19.17].

For the growth of carbon nanotubes by the arc discharge method, it has been proposed that the tubules grow at their open ends [19.51, 67]. If the tubule has chirality [see Fig. 19.21(a)], it is easily seen that the absorption of a single C_2 dimer at the active dangling bond edge site will add one hexagon to the open end. Thus the sequential addition of C_2 dimers will result in continuous growth of the chiral tubule. If carbon atoms should be added out of sequence, then addition of a C_2 dimer would result in the addition of a pentagon, which could lead to capping of the tubule, while the addition of a C_3 trimer out of sequence as shown in Fig. 19.21(a) merely adds a hexagon. In the case of an armchair edge, here again a single C_2 dimer will add a hexagon, as shown in Fig. 19.21(b). Multiple additions of C_2 dimers lead to multiple additions of hexagons to the armchair edge as shown in Fig. 19.21(b). Finally, for the case of a zigzag edge, initiation of growth requires one trimer C_3 [see Fig. 19.21(c)], which then provides the necessary edge site to complete one row of growth for the tubule through the addition of C_2 dimers, except for the last hexagon in the row, which requires only a C_1 monomer. If, however, a C_2 dimer is initially bonded at a zigzag edge, it will form a pentagon. Because of the curvature that is introduced by the pentagon, the open end of the tubule will likely form a cap, and growth of the tubule by the open end process will be terminated.

A schematic diagram for the open tube growth method is shown in Fig. 19.22 [19.37]. While the tubes grow along the length, they also grow in diameter by an epitaxial growth process, as shown in Fig. 19.22. The large aspect ratio of the tubules implies that growth along the tube axis is more likely than growth along the tubule diameter. Referring to Fig. 19.19, Iijima argues that the inner tubes are capped first with the capping providing a

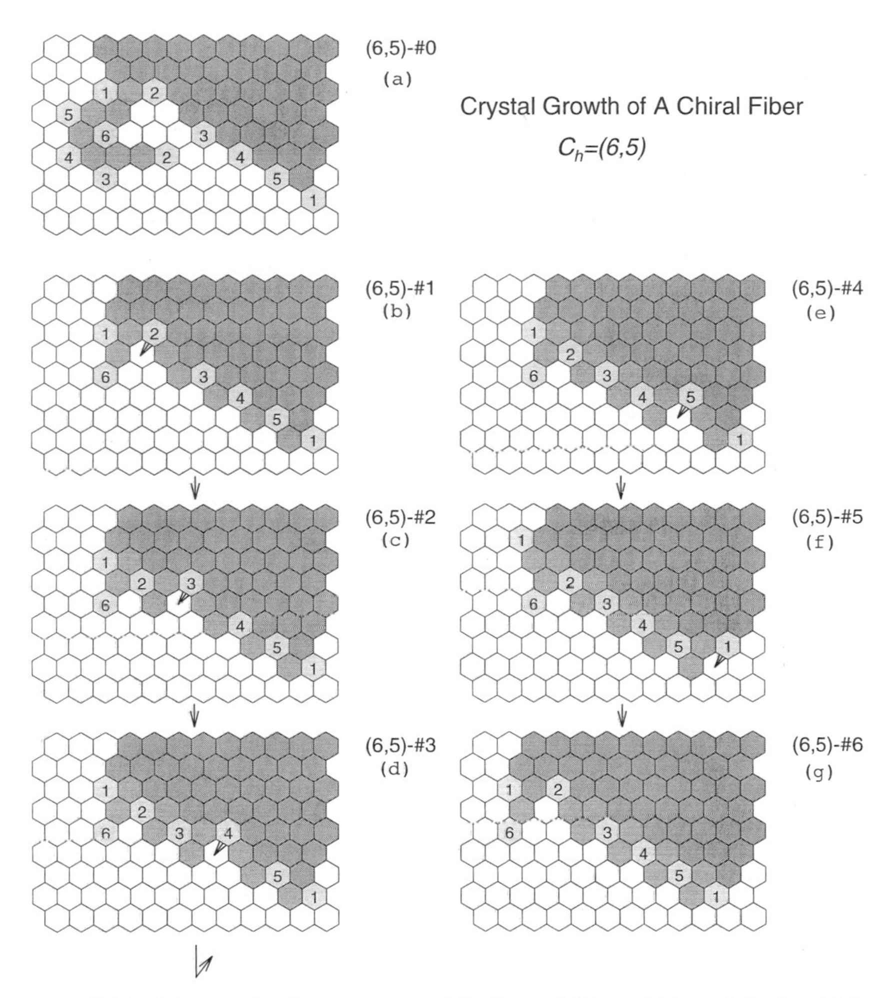

Fig. 19.20. Diagram showing a sequence of C_2 dimer additions which result in the addition of one row of hexagons to a carbon tubule. (a) A projection mapping of the cap atoms and the atoms on the cylinder of a $(6, 5)$ carbon tubule, where each pentagon defect is denoted by a light gray shaded hexagon containing a number (see §3.3). (b) The same projection mapping as in (a), except that the cap atoms do not appear explicitly. (c) The projection mapping of (b) after a C_2 dimer has been absorbed, resulting in the movement of the pentagonal defect 3. (d) The same as (c) except that 4 is the active pentagon which absorbs the C_2 dimer. (e) The same as (d) except that 5 is the active pentagon. (f) The same as (e) except that 1 is the active pentagon. (g) The final state implied by (f) which is geometrically identical to (b) except that one row of hexagons (10 carbon atoms) has been added to the tubule.

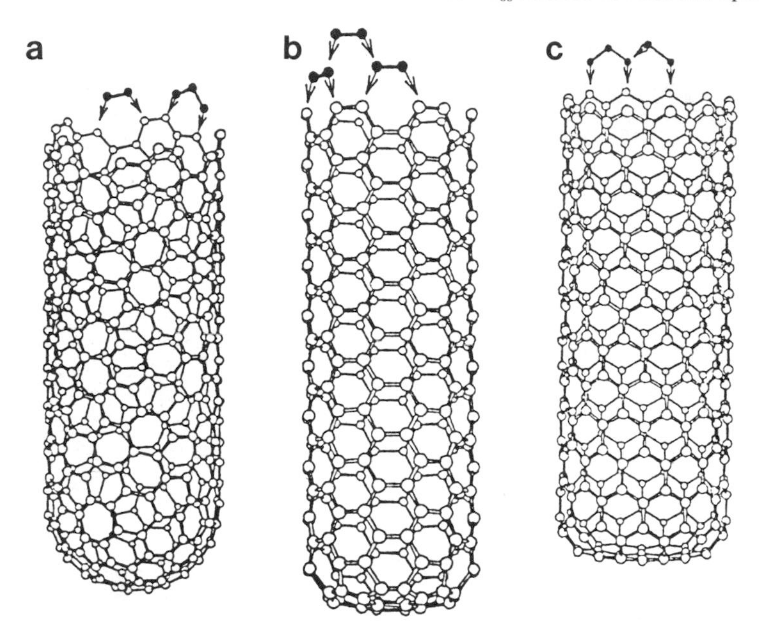

Fig. 19.21. Proposed growth mechanism of carbon tubules at an open end by the absorption of C_2 (dimers) and C_3 (trimers). (a) Absorption of C_2 dimers at the most active edge site of a chiral carbon tubule resulting in the addition of one hexagon. Also shown is an out-of-sequence absorption of a C_3 trimer. (b) Absorption of C_2 dimers at the open end of an armchair carbon tubule. (c) Absorption of a C_3 trimer at the open end of a zigzag carbon tubule and subsequent C_2 dimer absorption.

method for relieving strain in the cap region. The introduction of pentagons leads to positive curvature, while capping with heptagons lead to changes in tubule size (see Fig. 19.12) and orientation. One explanation for the semi-toroidal tubule termination (Fig. 19.13) is provided by the introduction of six heptagons at the periphery of an open tube. The subsequent addition of hexagons keeps the tubule diameter constant so that the toroidal surface can form. Thus the introduction of heptagon–pentagon pairs can produce a variety of tubule shapes. However, regarding this explanation of the semi-toroidal termination, it is hard to understand why suddenly six heptagons are introduced along the circumference at once, soon to be replaced by six pentagons which likely are strongly correlated with the prior introductions of the heptagons and the strain field associated with each heptagon.

At present the growth model for carbon nanotubes remains incomplete with regard to the role of temperature and helium gas. Since the vapor phase growth occurs at only 1100°C, any dangling bonds that might participate in the open tubule growth mechanism would be unstable, so that the

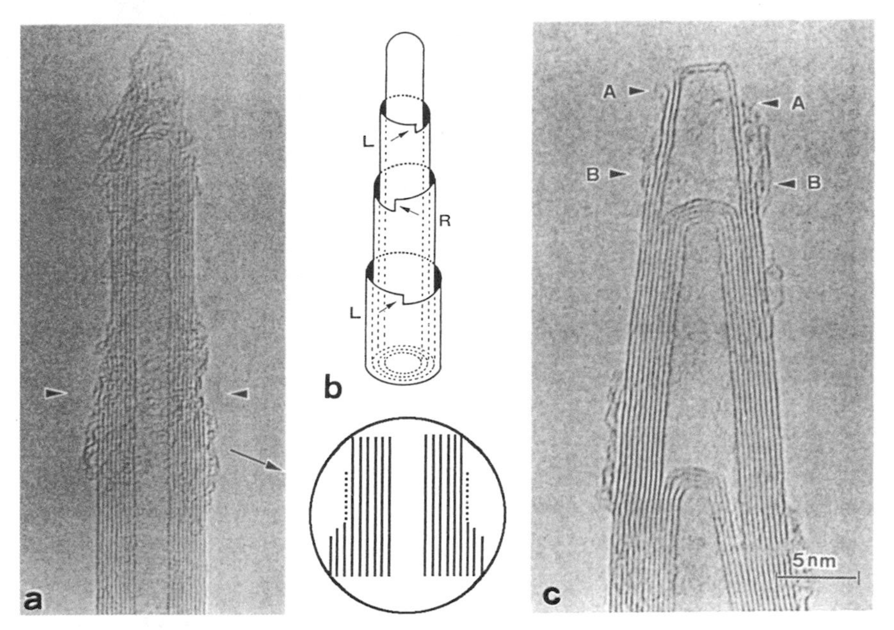

Fig. 19.22. Schematic diagram for open tube growth of nanotubes from a carbon supply. The figure shows the addition of carbon atoms to the open ends, the capping of the longest open end, and the initiation of new tubules. (a) Some shells are terminated at the positions indicated by arrowheads as illustrated in the circle. (b) Terminated shells carry left-handed or right-handed kink sites, owing to the helical tube structure. (c) Similar termination of the shells near the top of the tip, forming steps, one atom in height, as indicated by the arrowheads [19.37].

closed tube approach would be favored. In this lower temperature regime, the growth of the tubule core and the thickening process occur separately (see §2.5). In contrast, for the arc discharge synthesis method, the temperature where tubule growth occurs has been estimated to be about 3400°C [19.77], so that, in this case, the carbon is close to the melting point. At these high temperatures, tubule growth and the graphitization of the thickening deposits occur simultaneously, so that all the coaxial tubules grow at once at these elevated temperatures [19.35], and the open tubule growth may be favored.

An unexplained issue for the growth mechanism is the role of the helium gas. In most experiments helium gas and some other gases as well are used at about 100 torr for cooling the carbon system, since the tubule structure is not stable but is only a quasi-stable structure. Although the growth appears to be sensitive to the gas pressure, it is not clear how the He gas cooling of the carbon system causes growth of the quasi-stable tubule or

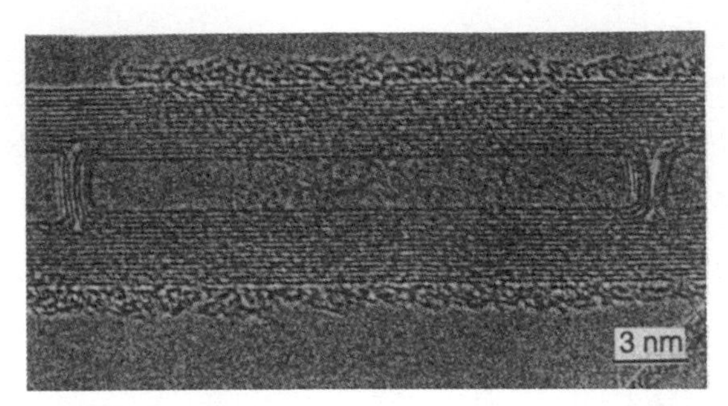

Fig. 19.23. Small carbon nanotube trapped inside a larger tubule [19.38].

fullerene phases. It is expected that future studies will provide a more detailed explanation of the growth mechanism, especially regarding the role of temperature, buffer gases, and electric fields.

One interesting growth feature, reported by several groups [19.37, 38, 78], is the containment of a small-diameter carbon tubule inside a larger-diameter tubule, as in Fig. 19.23, where it is seen that the inner tubule has no access to a carbon source. Such a feature seems to require growth by an open tube mechanism, although the nucleation of the cap from the terminated cap may be necessary for the growth of this feature.

Smalley [19.67] has suggested that the open ends of the tubules are stabilized by the electric fields that can be generated near the graphite surface in the arc discharge. Because of the high temperature of the particles in the arc discharge (up to 3400°C [19.77]), many of the species in the gas phase are expected to be charged, thereby screening the electrodes. Thus the potential energy drop associated with the electrodes is expected to occur over a distance of $\sim 1\mu$m or less, thereby causing very high electric fields. It is these high electric fields which may stabilize the open-ended tubes [19.67, 79], which ordinarily would represent a very high energy state because of the large number of dangling bonds.

It has been suggested that the catalyzed growth mechanism for the single-wall tubule is different from that for the multiwall tubule, due to the role of the catalyst. It is believed that nanosize iron particles act as catalysts for the formation of single-shell tubules, which are shown in Fig. 19.6, as they bridge carbide particles. In this growth process, the cementite particles eventually become coated with graphitic material, but the filaments themselves do not seem to contain any transition metal species after the growth is completed [19.79].

It is known from laser pyrolysis studies that transition metal carbide nanoscale particles such as cementite (FeC_3) catalyze the growth of graphene layers on their periphery [19.80], by a growth process similar to the formation of vapor-grown carbon fibers (see §2.5), whose growth is also catalyzed by transition metal nanoparticles [19.32]. It has also been shown [19.79, 81] using an arc discharge between two graphite electrodes that nanocrystalline LaC_2 particles surrounded by graphitic layers are formed when one of the two electrodes is packed with La-oxide (La_2O_3) powders in the center of the rod. The carbons formed on the surface of LaC_2 particles were found to be well graphitized, serving as a protection to the encapsulated pyrophoric LaC_2 particles. As another example, carbon encapsulated Co-carbide nanoparticles have also been formed using the arc discharge method [19.31].

19.4. Symmetry Properties of Carbon Nanotubes

Although the symmetry of the 2D graphene layer is greatly lowered in the 1D nanotube, the single-wall nanotubes have interesting symmetry properties that lead to nontrivial physical effects, namely a necessary degeneracy at the Fermi level for certain geometries.

19.4.1. Specification of Lattice Vectors in Real Space

To study the properties of carbon nanotubes as 1D systems, it is necessary to define the lattice vector **T** along the tubule axis and normal to the chiral vector $\mathbf{C}_h$ defined by Eq. (19.1) and Fig. 19.2(a). The vector **T** thus defines the unit cell of the 1D carbon nanotube. The length T of the translation vector **T** corresponds to the first lattice point of the 2D graphene sheet through which the vector **T** passes. From Fig. 19.2(a) and these definitions, we see that the translation vector **T** of a general chiral tubule as a function of n and m, can be written as [19.82]:

$$\mathbf{T} = [(2m+n)\mathbf{a}_1 - (2n+m)\mathbf{a}_2]/d_R \tag{19.6}$$

with a length

$$T = \sqrt{3}C_h/d_R \tag{19.7}$$

where the length C_h is given by Eq. (19.2), d is the highest common divisor of (n, m), and

$$d_R = \begin{cases} d & \text{if } n-m \text{ is not a multiple of } 3d \\ 3d & \text{if } n-m \text{ is a multiple of } 3d. \end{cases} \tag{19.8}$$

Thus for the $(5,5)$ armchair tubule $d_R = 3d = 15$, while for the $(9,0)$ zigzag tubule $d_R = d = 9$. The relation between the translation vector **T** and the symmetry operations on carbon tubules is discussed below and in §19.4.2. As a simple example, $T = \sqrt{3}a_{\text{C-C}}$ for a $(5,5)$ armchair nanotube and

$T = 3a_{\text{C-C}}$ for a $(9,0)$ zigzag nanotube, where $a_{\text{C-C}}$ is the nearest-neighbor carbon–carbon distance. We note that the length T is greatly reduced when (n, m) have a common divisor and when $(n - m)$ is a multiple of 3.

Having specified the length T of the smallest translation vector for the 1D carbon nanotube, it is useful to determine the number of hexagons, N, per unit cell of a chiral tubule specified by integers (n, m). From the size of the unit cell of the 1D carbon nanotube defined by the orthogonal vectors $\mathbf{T}$ and $\mathbf{C}_h$, the number N is given by

$$N = \frac{2(m^2 + n^2 + nm)}{d_R} \tag{19.9}$$

where d_R is given by Eq. (19.8) and we note that each hexagon contains two carbon atoms. As an example, application of Eq. (19.9) to the $(5,5)$ and $(9,0)$ tubules yields values of 10 and 18, respectively, for N. We will see below that these unit cells of the 1D tubule contain, respectively, five and nine unit cells of the 2D graphene lattice, each 2D unit cell containing two hexagons of the honeycomb lattice. This multiplicity is used in the application of zone-folding techniques to obtain the electronic and phonon dispersion relations in §19.5 and §19.7, respectively.

Referring to Fig. 19.2(a), we see that the basic space group symmetry operation of a chiral tubule consists of a rotation by an angle ψ combined with a translation τ, and this space group symmetry operation is denoted by $R = (\psi|\tau)$ and corresponds to the vector $\mathbf{R} = p\mathbf{a}_1 + q\mathbf{a}_2$ shown in Fig. 19.24. The physical significance of the vector $\mathbf{R}$ is that the projection of $\mathbf{R}$ on the chiral vector $\mathbf{C}_h$ gives the angle ψ scaled by $C_h/2\pi$, while the projection of $\mathbf{R}$ on $\mathbf{T}$ gives the translation vector τ of the basic symmetry operation of the 1D space group. The integer pair (p, q) which determines $\mathbf{R}$ is found using the relation

$$mp - nq = d \tag{19.10}$$

subject to the conditions $q < m/d$ and $p < n/d$. Taking the indicated scalar product $\mathbf{R} \cdot \mathbf{T}$ in Fig. 19.24 we obtain the expressions for the length of τ

$$\tau = Td/N \tag{19.11}$$

where d is the highest common divisor of (n, m), T is the magnitude of the lattice vector $\mathbf{T}$, and N is the number of hexagons per 1D unit cell, given by Eq. (19.9). For the armchair and zigzag tubules, Eq. (19.11) yields $\tau = T/2$ and $\tau = \sqrt{3}T/2$, respectively.

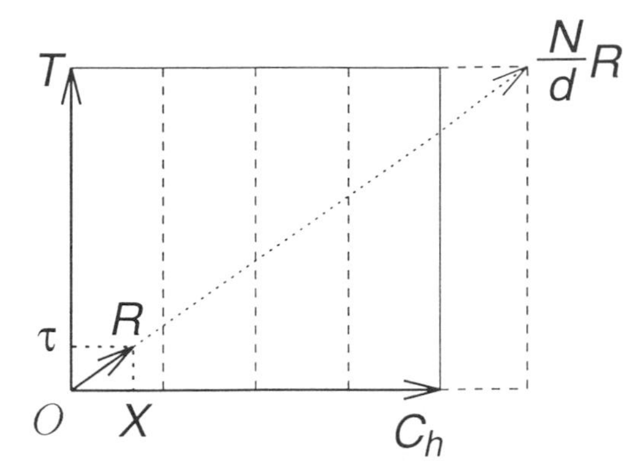

Fig. 19.24. Relation between the fundamental symmetry vector $\mathbf{R} = p\mathbf{a}_1 + q\mathbf{a}_2$ of the 1D unit cell and the two vectors that specify the carbon nanotube (n, m): the chiral vector $\mathbf{C}_h$ and translation vector $\mathbf{T}$. The projections of $\mathbf{R}$ on the $\mathbf{C}_h$ and $\mathbf{T}$ axes, respectively, yield ψ and τ, where X in the figure is ψ scaled by $(\mathbf{C}_h/2\pi)$. After (N/d) translations, $\mathbf{R}$ reaches a lattice point B'' (see text). The vertical dashed lines divide the vector $\mathbf{C}_h$ into d sectors [19.5].

Regarding the angle of rotation, the scalar product $\mathbf{R} \cdot \mathbf{C}_h$ yields

$$\psi = 2\pi\left[\frac{\Omega}{Nd} + \frac{\lambda}{d}\right] \tag{19.12}$$

where $\lambda = 0, 1, \ldots, d-1$ and

$$\Omega = [p(m+2n) + q(n+2m)](d/d_R) \tag{19.13}$$

in which the integers (p, q) determine the vector $\mathbf{R}$ (see Fig. 19.24). Thus (p, q) denotes the coordinates reached when the symmetry operation $(\psi|\tau)$ acts on an atom at (0,0), i.e., $(\psi|\tau)(0, 0) = (p, q)$.

If $(\psi|\tau)$ is a symmetry operation for the tubule, then $(\psi|\tau)^2$, $(\psi|\tau)^3, \ldots$ $(\psi|\tau)^{N/d}$ are all distinct symmetry operations where $(\psi|\tau)^{N/d} = E$ is the identity operation, bringing the lattice point O to an equivalent lattice point B'', where

$$\overrightarrow{OB}'' = (N/d)\mathbf{R} = (1/d)[\Omega\mathbf{C}_h + d\,\mathbf{T}]. \tag{19.14}$$

Referring to Fig. 19.24, we see that after N/d symmetry operations $(\psi|\tau)$, the vector $(N/d)\mathbf{R}$ along the zigzag direction reaches a lattice point, which we denote by B'' (not shown in figure). Correspondingly, after N/d symmetry operations $(\psi|\tau)^{N/d}$, the translation $(N/d)\tau$ yields one translation of the lattice vector $\mathbf{T}$ of the tubule and (Ω/d^2) revolutions of 2π around the tubule axis. Although Ω/d is an integer, Ω/d^2 need not be. In Table 19.2 we list the characteristic parameters of carbon tubules specified by (n, m), including d, the highest common divisor of n and m, and the related quantity d_R given by Eq. (19.8). Also listed in Table 19.2 are the tubule diameter d_t in units of Å, the translation repeat distance T of the 1D lattice in units of the lattice constant $a_0 = \sqrt{3}a_{\text{C-C}}$ for a 2D graphene sheet, the number N

Table 19.2

Values for characterization parameters[a] for selected carbon nanotubes labeled by (n, m) [19.5, 82].

(n,m)	d	d_R	d_t (Å)	L/a_0	T/a_0	N	$\psi/2\pi$	τ/a_0	Ω/d
(4,2)	2	2	4.14	5	$\sqrt{21}$	28	5/28	$\sqrt{21}/14$	5
(5,5)	5	15	6.78	$\sqrt{75}$	1	10	1/10	1/2	1
(9,0)	9	9	7.05	9	$\sqrt{3}$	18	−1/18	$\sqrt{3}/2$	−1
(6,5)	1	1	7.47	$\sqrt{91}$	$\sqrt{273}$	182	149/182	$\sqrt{3/364}$	149
(7,4)	1	3	7.55	$\sqrt{93}$	$\sqrt{31}$	62	17/62	$1/\sqrt{124}$	17
(8,3)	1	1	7.72	$\sqrt{97}$	$\sqrt{291}$	194	71/194	$\sqrt{3/388}$	71
(10,0)	10	10	7.83	10	$\sqrt{3}$	20	−1/20	$\sqrt{3}/2$	−1
(6,6)	6	18	8.14	$\sqrt{108}$	1	12	1/12	1/2	1
(10,5)	5	5	10.36	$\sqrt{175}$	$\sqrt{21}$	70	1/14	$\sqrt{3/28}$	5
(20,5)	5	15	17.95	$\sqrt{525}$	$\sqrt{7}$	70	3/70	$1/(\sqrt{28})$	3
(30,15)	15	15	31.09	$\sqrt{1575}$	$\sqrt{21}$	210	1/42	$\sqrt{3/28}$	5
⋮	⋮	⋮	⋮	⋮	⋮	⋮	⋮	⋮	⋮
(n,n)	n	$3n$	$\sqrt{3}na/\pi$	$\sqrt{3}n$	1	$2n$	$1/2n$	1/2	1
$(n,0)$	n	n	na/π	n	$\sqrt{3}$	$2n$	$-1/2n$	$\sqrt{3}/2$	−1

[a] Ω is given by Eq. (19.13); T is given by Eq. (19.7) with $a_0 = \sqrt{3}a_{C-C}$; L is the length of the chiral vector πd_t given by Eq. (19.2); N is given by Eq. (19.9); ψ is given by Eq. (19.12); τ is given by Eq. (19.11); Ω/d is the number of complete 2π revolutions about the tubule axis to reach a lattice point.

of hexagons per unit cell of the 1D tubule, the rotation angle ψ expressed in units of $C_h/2\pi$, and the translation τ of the symmetry operation $R = (\psi|\tau)$ expressed in units of a_0. The length of the chiral vector L ($L \equiv C_h = \pi d_t$) is listed in Table 19.2 in units of a_0. For cases where n and m have a common divisor d, the angle of rotation ψ is defined modulo $2\pi/d$ instead of 2π, so that the group $C'_{N/\Omega}$ has N/d elements. To illustrate use of Table 19.2 we refer to the (4, 2) tubule illustrated in Fig. 19.2(a), which has $N = 28$, $d = d_R = 2$, and $T = 14\tau = \sqrt{21}a_0$ and also $C_h = \sqrt{28}a_0$. Thus for the (4, 2) tubule, $(N/d)\psi = (5/2)(2\pi)$ corresponds to 5π rotations around the tubule axis, and (N/d) translations **R** reach a lattice point.

As a second example, we consider the case of tubule (7, 4) for which the phonon spectrum has been calculated (see §19.7) [19.83]. For this tubule, there are no common divisors, so $d = 1$, but since $n - m = 3$, we obtain $d_R = 3$. Solution of Eq. (19.10) for (p, q) yields $p = 2$ and $q = 1$, which yields the rotation angle (in units of 2π) $N\psi = \Omega/d = 17(2\pi)$, where $N = 62$. Thus, after 62 operations $(\psi|\tau)^N$, the origin O is transformed into a new lattice point a distance T from O along the T axis, after having completed 17 rotations of 2π around the tubule axis. For the armchair tubule (5, 5), the highest common divisor is 5, and since $n - m = 0$, we have

$d_R = 3 \times 5 = 15$, yielding $N = 10$ and $N\psi = 2\pi$. Regarding translations, use of Eq. (19.11) yields $\tau = a_0/2 = T/2$. Finally, we give the example of the smallest zigzag tubule (0, 9), for which $d = d_R = 9$ and $N = 18$. Using $p = 1$ and $q = 0$ and $N = 18$, we obtain $N\psi = 2\pi$ and $\tau = \sqrt{3}a_0/2 = T/2$.

19.4.2. Symmetry for Symmorphic Carbon Tubules

In this section we consider the symmetry properties of the highly symmetric armchair and zigzag carbon tubules, which can be described by symmorphic groups, and we then summarize the symmetry operations for the general chiral tubule.

For symmorphic groups, the translations and rotations are decoupled from each other, and we can treat the rotations simply by point group operations. Since symmorphic groups generally have higher symmetry than the nonsymmorphic groups, it might be thought that group theory plays a greater role in specifying the dispersion relations for electrons and phonons for symmorphic space groups and in discussing their selection rules. It is shown in §19.4.3 that group theoretical considerations are also very important for the nonsymmorphic groups, leading to important and simple classifications of their dispersion relations.

In discussing the symmetry of the carbon nanotubes, it is assumed that the tubule length is much larger than its diameter, so that the tubule caps, which were discussed in §19.2.3, can be neglected when discussing the electronic and lattice properties of the nanotubes. Hence, the structure of the infinitely long armchair tubule ($n = m$) or zigzag tubule ($m = 0$) is described by the symmetry groups D_{nh} or D_{nd} for even or odd n, respectively, since inversion is an element of D_{nd} only for odd n and is an element of D_{nh} only for even n [19.84]. Character tables for groups D_{5h} and D_{5d} are given in §4.4 under Tables 4.12 and 4.13. We note that group D_{5d} has inversion symmetry and is a subgroup of group I_h, so that compatibility relations can be specified between the lower symmetry group D_{5d} and I_h, as given in Table 4.15. In contrast, D_{5h} is not a subgroup of I_h, although D_5 is a subgroup of I. Character tables for D_{6h} and D_{6d} are readily available in standard group theory texts [19.85]. For larger tubules, appropriate character tables can be constructed from the generalized character tables for the D_n group given in Table 19.3 (for odd $n = 2j + 1$) and in Table 19.4 (for even $n = 2j$), and the basis functions are listed in Table 19.5. These tables are adapted from the familiar character table for the semi-infinite group $D_{\infty h}$ [19.85], and $\mathscr{R}$ denotes the irreducible representations.

The character table for group D_{nd} for odd integers $n = 2j + 1$ is constructed from Table 19.3 for group D_n [or group ($n2$) in the international notation] by taking the direct product $D_n \otimes i$ where i is the two element

Table 19.3

Character table for point group $D_{(2j+1)}$.

$\mathscr{R}$	E	$2C^1_{\phi_j}$ [a]	$2C^2_{\phi_j}$	...	$2C^j_{\phi_j}$	$(2j+1)C'_2$
A_1	1	1	1	...	1	1
A_2	1	1	1	...	1	−1
E_1	2	$2\cos\phi_j$	$2\cos 2\phi_j$	...	$2\cos j\phi_j$	0
E_2	2	$2\cos 2\phi_j$	$2\cos 4\phi_j$	...	$2\cos 2j\phi_j$	0
⋮	⋮	⋮	⋮	⋮	⋮	⋮
E_j	2	$2\cos j\phi_j$	$2\cos 2j\phi_j$	...	$2\cos j^2\phi_j$	0

[a] Where $\phi_j = 2\pi/(2j+1)$.

Table 19.4

Character Table for Group $D_{(2j)}$

$\mathscr{R}$	E	C_2	$2C^1_{\phi_j}$ [a]	$2C^2_{\phi_j}$	...	$2C^{j-1}_{\phi_j}$	$(2j)C'_2$	$(2j)C''_2$
A_1	1	1	1	1	...	1	1	1
A_2	1	1	1	1	...	1	−1	−1
B_1	1	−1	1	1	...	1	1	−1
B_2	1	−1	1	1	...	1	−1	1
E_1	2	−2	$2\cos\phi_j$	$2\cos 2\phi_j$	...	$2\cos(j-1)\phi_j$	0	0
E_2	2	2	$2\cos 2\phi_j$	$2\cos 4\phi_j$	...	$2\cos 2(j-1)\phi_j$	0	0
⋮	⋮	⋮	⋮	⋮	⋮	⋮	⋮	⋮
E_{j-1}	2	$(-1)^{j-1}2$	$2\cos(j-1)\phi_j$	$2\cos 2(j-1)\phi_j$	...	$2\cos(j-1)^2\phi_j$	0	0

[a] Where $\phi_j = 2\pi/(2j)$.

Table 19.5

Basis Functions for Groups $D_{(2j)}$ and $D_{(2j+1)}$

Basis functions		$D_{(2j)}$	$D_{(2j+1)}$	$C_{N/\Omega}$
(x^2+y^2, z^2)		A_1	A_1	A
	z	A_2	A_1	A
	R_z	A_2	A_2	A
(xz, yz)	(x, y), (R_x, R_y)	E_1	E_1	E_1
(x^2-y^2, xy)		E_2	E_2	E_2
		⋮	⋮	⋮

inversion group containing E and i. In addition to the identity class E, the classes of D_n in Table 19.3 constitute the nth roots of unity where $\pm\phi_j$ rotations belong to the same class, and in addition there is a class $(2j+1)C_2'$ of n twofold axes at right angles to the main symmetry axis C_{ϕ_j}. Thus, Table 19.3 yields the character tables for D_{nd}, or D_{5d}, D_{7d}, ... for symmorphic tubules with odd numbers of unit cells around the circumference [(5,5), (7,7), ... armchair tubules, and (9,0), (11,0), ... zigzag tubules]. Likewise, the character table for D_{nh}, or D_{6h}, D_{8h}, ... for even n is found from Table 19.4 by taking the direct product $D_n \otimes i = D_{nh}$. Table 19.4 shows two additional classes for group D_{2j} because rotation by π about the main symmetry axis is in a class by itself. Also the $2j$ twofold axes nC_2' form a class and are in a plane normal to the main symmetry axis C_{ϕ_j}, while the nC_2'' dihedral axes, which are bisectors of the nC_2' axes, also form a class for group D_n when n is an even integer. Correspondingly, there are two additional one-dimensional representations (B_1 and B_2) in D_{2j}, since the number of irreducible representations equals the number of classes. Table 19.5 lists the irreducible representations and basis functions for the various 1D tubule groups and is helpful for indicating the symmetries that are infrared active (transform as the vector x, y, z) and Raman active (transform as the symmetric quadratic forms).

19.4.3. Symmetry for Nonsymmorphic Carbon Tubules

The symmetry groups for carbon nanotubes can be either symmorphic (as for the special case of the armchair and zigzag tubules) or nonsymmorphic for the general case of chiral nanotubes. For chiral nanotubes the chiral angle in Eq. (19.3) is in the range $0 < \theta < 30^\circ$ and the space group operations $(\psi|\tau)$ given by Eqs. (19.11) and (19.12) involve both rotations and translations, as discussed in §19.4.1. Figure 19.24 shows the symmetry vector **R** which determines the space group operation $(\psi|\tau)$ for any carbon nanotubule specified by (n, m). From the symmetry operation $R = (\psi|\tau)$ for tubule (n, m), the symmetry group of the chiral tubule can be determined. If the tubule is considered as an infinite molecule, then the set of all operations R^j, for any integer j, also constitutes symmetry operators of the group of the tubule. Thus from a symmetry standpoint, a carbon tubule is a one-dimensional crystal with a translation vector **T** along the cylinder axis and a small number of carbon hexagons associated with the circumferential direction [19.82, 86, 87].

The symmetry groups for the chiral tubules are Abelian groups. All Abelian groups have a phase factor ϵ, such that all h symmetry elements of the group commute, and are obtained from any symmetry element by multiplication of ϵ by itself an appropriate number of times, such that $\epsilon^h = E$, the identity element. To specify the phase factors for the Abelian group, we introduce the quantity Ω of Eq. (19.13), where Ω/d is interpreted as the number of 2π rotations which occur after N/d rotations of ψ. The phase factor ϵ for the Abelian group for a carbon nanotube then becomes $\epsilon = \exp(2\pi i\Omega/N)$ for the case where (n,m) have no common divisors (i.e., $d = 1$). If $\Omega = 1$ and $d = 1$, then the symmetry vector **R** in Fig. 19.24 reaches a lattice point after a 2π rotation. As seen in Table 19.1, many of the actual tubules with $d = 1$ have large values for Ω; for example, for the (6,5) tubule, $\Omega = 149$, while for the (7,4) tubule $\Omega = 17$, so that many 2π rotations around the tubule axis are needed to reach a lattice point of the 1D lattice.

The character table for the Abelian group of a carbon nanotube is given in Table 19.6 for the case where (n,m) have no common divisors and is labeled by $C_{N/\Omega}$. The number of group elements is N, all symmetry elements commute with each other, and each symmetry operation is in a class by itself. The irreducible representation A in Table 19.6 corresponds to a 2π rotation, while representation B corresponds to a rotation by π. Except for the irreducible representations A and B, all other irreducible representations are doubly degenerate. The E_i representations correspond to two levels which stick together by time reversal symmetry, for which the corresponding eigenvectors are related to one another by complex conjugation. Since N can be quite large, there can be a large number of symmetry operations in the group $C_{N/\Omega}$, all of which can be represented in terms of a phase factor $\epsilon = \exp(i\psi) = \exp(2\pi i\Omega/N)$ and the irreducible represen-

Table 19.6

The character table for the group $C_{N/\Omega}$ for chiral nanotubes, where N and Ω have no common divisor, corresponding to (n,m) having no common divisor.[a]

$C_{N/\Omega}$	E	C^1	C^2	...	C^ℓ	...	C^{N-1}
A	1	1	1	...	1	...	1
B	1	−1	1	...	$(-1)^\ell$	...	−1
E_1	$\begin{Bmatrix}1\\1\end{Bmatrix}$	ϵ ϵ^*	ϵ^2 ϵ^{*2}	...	ϵ^ℓ $\epsilon^{*\ell}$	...	$\begin{Bmatrix}\epsilon^{N-1}\\\epsilon^{*(N-1)}\end{Bmatrix}$
E_2	$\begin{Bmatrix}1\\1\end{Bmatrix}$	ϵ^2 ϵ^{*2}	ϵ^4 ϵ^{*4}	...	$\epsilon^{2\ell}$ $\epsilon^{*2\ell}$	...	$\begin{Bmatrix}\epsilon^{2(N-1)}\\\epsilon^{*2(N-1)}\end{Bmatrix}$
⋮	⋮	⋮	⋮	⋮	⋮	⋮	⋮
$E_{\frac{N}{2}-1}$	$\begin{Bmatrix}1\\1\end{Bmatrix}$	$\epsilon^{\frac{N}{2}-1}$ $\epsilon^{*\frac{N}{2}-1}$	$\epsilon^{2(\frac{N}{2}-1)}$ $\epsilon^{*2(\frac{N}{2}-1)}$	...	$\epsilon^{\ell(\frac{N}{2}-1)}$ $\epsilon^{*\ell(\frac{N}{2}-1)}$	...	$\begin{Bmatrix}\epsilon^{(N-1)(\frac{N}{2}-1)}\\\epsilon^{*(N-1)(\frac{N}{2}-1)}\end{Bmatrix}$

[a]The complex number ϵ is $e^{2\pi i\Omega/N}$, so that $\epsilon^* = \exp(-2\pi i\Omega/N)$.

tations are related to the Nth roots of unity. The number of classes and irreducible representations of group $C_{N/\Omega}$ is thus equal to N, counting each of the partners of the E_i representations as distinct. Various basis functions for group $C_{N/\Omega}$ that are useful for determining the infrared and Raman activity of the chiral tubules are listed in Table 19.5. We note that the irreducible representations A and E_1 are infrared active and A, E_1, and E_2 are Raman active.

For tubules where (n, m) have a common divisor d, then after N translations τ, a length Td is produced, and Ω unit cells of area $(C_h T)$ are generated. For $d \neq 1$ the character table for the Abelian group in Table 19.6 contains $1/d$ as many elements, so that the order of the Abelian group becomes N/d. This case is discussed further below.

The space group for a chiral nanotube specified by (n, m) is given by the direct product of two Abelian groups [19.82, 87]

$$\mathscr{C} = C_{d'} \otimes C_{Nd/\Omega} \tag{19.15}$$

where d' is the highest common divisor of Ω/d and d, while d is the highest common divisor of (n, m). The symmetry elements in group $C_{d'}$ which is of order d' include

$$C_{d'} = \{E, C_{d'}, C_{d'}^2, ..., C_{d'}^{d'-1}\}, \tag{19.16}$$

and, correspondingly, the symmetry elements in group $C_{Nd/\Omega}$ include

$$C_{Nd/\Omega} = \{E, C_{Nd/\Omega}, C_{Nd/\Omega}^2, ..., C_{Nd/\Omega}^{(N/d')-1}\}, \tag{19.17}$$

and group $C_{Nd/\Omega}$ is of order N/d'. The irreducible representations of the groups $C_{d'}$ and $C_{Nd/\Omega}$ in Eq. (19.15) are given in Table 19.6 and are appropriate roots of unity. For the two-dimensional E_n irreducible representations, the characters of the symmetry operations $C_{d'}$ and $C_{Nd/\Omega}$ are given by

$$\chi_{E_n}(C_{d'}) = \begin{cases} e^{i2\pi n/d'} \\ e^{-i2\pi n/d'} \end{cases} \tag{19.18}$$

where $1 \leq n \neq d'$ and

$$\chi_{E_n}(C_{Nd/\Omega}) = \begin{cases} e^{i2\pi n\Omega/Nd} \\ e^{-i2\pi n\Omega/Nd} \end{cases}. \tag{19.19}$$

For the chiral nanotubes, a vector transforms according to the basis functions for the A and E_1 irreducible representations, whereas quadratic terms in the coordinates form basis functions for the A, E_1, and E_2 irreducible representations, as shown in the Table 19.5.

Referring to Fig. 19.2(a) for the (4, 2) tubule, we have $\psi = 2\pi(5/28)$ and $\Omega/d = 5$, so that N rotations produce a total rotation of $2\pi(5)$. Since

$d = 2$, N translations produce a distance $2T$, so that $N\mathbf{R}$ is the diagonal of a rectangle that contains $\Omega = 10$ times the area of the 1D nanotube unit cell. The first lattice point is, however, reached at $(N/d)\mathbf{R} = [(5/2)C_h, T]$ of the 1D tubule lattice. The number of elements in the Abelian group $C_{Nd/\Omega}$ is N/d, which is 14 in this case. The value of d' which is the highest common divisor of d and Ω/d is $d' = 1$, so that $C_{d'}$ is the identity group containing only one symmetry element. This example shows that for (n, m) pairs having a common divisor d, a lattice point is reached after $(N/d)\mathbf{R}$ translations, so that Ω/d^2 is not necessarily a multiple of 2π.

We now give some examples to show that if either $d > 1$ or $n - m = 3r$ (so that $d_R = 3d$) the 1D tubule unit cell is reduced in size, and the number of $\mathbf{R}$ vectors to reach a lattice point is reduced. For example, tubule (7, 5), for which caps are shown in Fig. 19.9, has a diameter of 8.18 Å, $N = 218$ symmetry operations, and 91 translations $\mathbf{R}$ are needed to reach a lattice point. Since the integers (7, 5) have no common divisor, and since $n - m \neq 3r$, the unit cell is large. If we now consider a tubule with $(n, m) = (8, 5)$ with a diameter of 8.89 Å, nearly 10% larger than the diameter for the (7, 5) tubule, we note that since $n - m$ is a multiple of 3, then $N = 86$ for the (8, 5) tubule, where N is much smaller than for the (7, 5) tubule. Likewise, for the (8, 5) tubule, the length $T = 16.1$ Å is much shorter than $T = 44.5$ Å for the (7, 5) tubule, and a lattice point is reached after 53 translations. An even smaller unit cell is obtained for the tubule corresponding to $(n, m) = (10, 5)$, despite the larger tubule diameter of 10.36 Å. For the (10, 5) tubule, N is only 70, and a lattice point is reached after only five translations $\mathbf{R}$. Although the number of symmetry operations of the nonsymmorphic groups tends to increase with tubule diameter, those tubules for which either $n - m = 3r$, where r is an integer, or (n, m) contains a common divisor d, the size of the unit cell and therefore the number of symmetry operations is reduced by factors of 3 and d, respectively.

The symmorphic groups corresponding to the armchair and zigzag tubules have relatively small unit cells with $T = a_0$ and $T = \sqrt{3}a_0$, respectively, where $a_0 = 2.46$ Å$= \sqrt{3}a_{C-C}$. The basic rotation angle ψ for both the (n, n) armchair tubule and the $(n, 0)$ zigzag tubule is $\psi = 2\pi/n$.

19.4.4. *Reciprocal Lattice Vectors*

To express the dispersion relations for electrons and phonons in carbon nanotubes, it is necessary to specify the basis vectors in reciprocal space $\mathbf{K}_i$ which relate to those in real space $\mathbf{R}_j$ by

$$\mathbf{R}_j \cdot \mathbf{K}_i = 2\pi\delta_{ij}. \tag{19.20}$$

The lattice vectors and the unit cells in real space are discussed in §19.4.1. In Cartesian coordinates we can write the two real space lattice vectors as

$$\mathbf{C}_h = (a_0/2d_R)\Big(\sqrt{3}(n+m), (n-m)\Big),$$

$$\mathbf{T} = (3a_0/2d_R)\Big(-(n-m)/\sqrt{3}, (n+m)\Big) \tag{19.21}$$

where d_R is defined by Eq. (19.8) and the length a_0 is the lattice constant of the 2D graphene unit cell, $a_0 = a_{C-C}\sqrt{3}$. From Eq. (19.21), it is easy to see that the lengths of $\mathbf{C}_h$ and $\mathbf{T}$ are in agreement with Eqs. (19.2) and (19.7). Then using Eqs. (19.20) and (19.21), we can write the corresponding reciprocal lattice vectors $\mathbf{K}_1$ and $\mathbf{K}_2$ in Cartesian coordinates as

$$\mathbf{K}_1 = (2\pi/Na_0)\Big(\sqrt{3}(n+m), (n-m)\Big),$$

$$\mathbf{K}_2 = (2\pi/Na_0)\Big(-(n-m)/\sqrt{3}, (n+m)\Big). \tag{19.22}$$

For the case of the armchair and zigzag tubules, it is convenient to choose a real space rectangular unit cell in accordance with Fig. 19.25. The area of each real space unit cell for both the armchair and zigzag tubules contains two hexagons or four carbon atoms. It should be noted that the real space unit cell defined by the vectors $\mathbf{T}$ and $\mathbf{C}_h$ is n times larger than the real space unit cells shown in Fig. 19.25. Also shown in this figure are the real space unit cells for a 2D graphene layer. It should be noted that the real space unit cell for the zigzag tubules follows from the definition given for the unit cells for chiral tubules [Fig. 19.2(a)], but the unit cell for the armchair tubule is specially selected for convenience.

Having specified the real space unit cells in Fig. 19.25, the corresponding unit cells in reciprocal space (or Brillouin zones) are determined by Eqs. (19.20) and (19.22) and are shown in Fig. 19.25 in comparison to the reciprocal space unit cell for a 2D graphene sheet. We note that for both the armchair and zigzag tubules, the 1D reciprocal space unit cells shown in Fig. 19.25 are half as large as those for the 2D graphene sheet. On the graphene sheet, the real space lattice vectors $\mathbf{C}_h$ and $\mathbf{T}$ form a rectangular unit cell for the zigzag or armchair tubules which has an area N times larger than the area of the corresponding primitive cell in the graphene sheet, where N is given by Eq. (19.9) [19.84], and this large unit cell spans the circumference of the cylinder of the tubule.

The one-dimensional Brillouin zone of the chiral tubule is a segment along the vector $\mathbf{K}_2$ of Eq. (19.22). The extended Brillouin zone for the tubule is a collection of N wave vector segments of length $|\mathbf{K}_2|$, each separated from the next segment by the vector $\mathbf{K}_1$. Thus, by zone folding the N

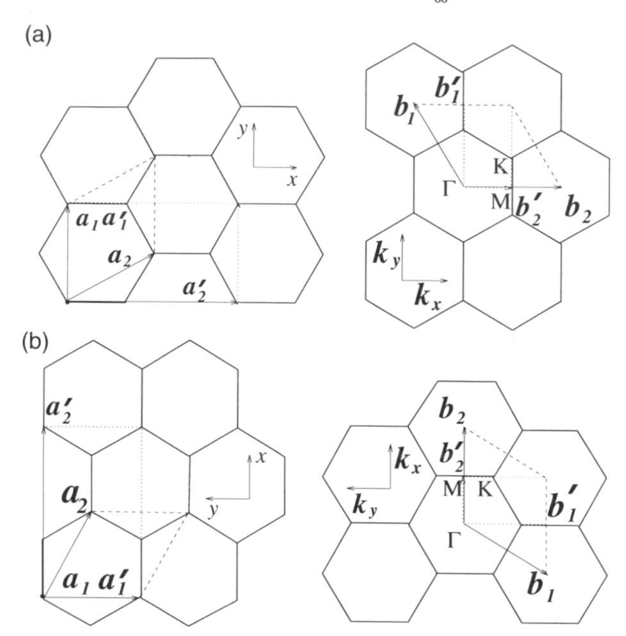

Fig. 19.25. Real space unit cell and Brillouin zone for (a) armchair and (b) zigzag tubules (dotted lines). Those for a 2D graphene sheet (dashed lines) are shown for comparison [19.6].

wave vector segments of the 2D dispersion relations of the graphene layer back to the first Brillouin zone of the 1D tubule, the 1D dispersion relations for the N electron energy bands (or phonon branches) of the tubule are obtained. For the special cases of the armchair and zigzag tubules, the real space unit cells in Fig. 19.25 correspond to four carbon atoms and two hexagons, while the reciprocal space unit cells also contain two wave vectors. Therefore each dispersion relation for the (n, n) armchair tubules and the $(n, 0)$ zigzag tubules is zone folded $n = N/2$ times in the extended Brillouin zone to match the 1D Brillouin zone of the tubule. We present the results of this zone folding in discussing the electronic structure in §19.5 and the phonon dispersion relations in §19.7.

19.5. Electronic Structure: Theoretical Predictions

Fullerene-based carbon tubules are interesting as examples of a one-dimensional periodic structure along the tubule axis. In this section we summarize the remarkable electronic properties predicted for single-wall carbon nanotubes. Specifically, it is predicted that small-diameter nanotubes will exhibit either metallic or semiconducting electrical conduction, depending on their diameter d_t and chiral angle θ, and independent of the presence of dopants or defects. It is further shown that this phenomenon is not quenched by the Peierls distortion or by intertubule interactions. Because of the great difficulty in making measurements on individual single-wall tubules, few detailed experiments have thus far been reported, although this is a very active current research area. In this section, we review theoretical predictions for the electronic structure in zero magnetic field of single-wall symmorphic and nonsymmorphic carbon nanotubes. The effect of interlayer interaction for multiwall nanotubes is then considered along with intertubule interactions. Finally, theoretical predictions for the electronic structure of carbon nanotubes in a magnetic field are discussed. The status of experimental work on the electronic structure of carbon nanotubes is reviewed in §19.6, and the effect of magnetic fields is considered in §19.5.4.

19.5.1. Single-Wall Symmorphic Tubules

Most of the calculations of the electronic structure of carbon nanotubes have been carried out for single-wall tubules. Confinement in the radial direction is provided by the monolayer thickness of the tubule. In the circumferential direction, periodic boundary conditions are applied to the enlarged unit cell that is formed in real space defined by $\mathbf{T}$ and $\mathbf{C}_h$, and as a consequence, zone folding of the dispersion relations occurs in reciprocal space. Such zone-folding calculations lead to 1D dispersion relations for electrons (described in this section) and phonons (see §19.7) in carbon tubules.

A number of methods have been used to calculate the 1D electronic energy bands for fullerene-based single-wall tubules [19.25, 47, 86, 88–93]. However, all of these methods relate to the 2D graphene honeycomb sheet used to form the tubule. The unit cells in real and reciprocal space used by most of these authors to calculate the energy bands for armchair and zigzag tubules are shown explicitly in Fig. 19.25. Furthermore, the armchair tubule can be described by a symmorphic space group (as is done in this section), but can also be described following the discussion for chiral tubules in §19.4.3 for a chiral angle of $\theta = 30°$.

We illustrate below the essence of the various calculations of the 1D electronic band structure, using the simplest possible approach, which is a tight binding or Hückel calculation that neglects curvature of the tubules. In making numerical evaluations of the energy levels and band gaps, it is assumed that the nearest-neighbor interaction energy γ_0 is the same as for crystalline graphite. Since the unit cells for the (n, m) armchair tubule and the $(n, 0)$ zigzag tubules in real space are a factor of n smaller than the area enclosed by $\mathbf{C}_h$ and $\mathbf{T}$ in Fig. 19.2(a), the size of the Brillouin zone is a factor of n larger than the reciprocal space unit cell defined by the reciprocal lattice vectors given by Eq. (19.21). Thus, zone folding of the large unit cell in reciprocal space introduces discrete values of the wave vector in the direction perpendicular to the tubule axis [19.46]. Using this procedure, the 2D energy dispersion relations for a single graphene layer are folded into the 1D Brillouin zone of the tubule. To illustrate this zone-folding technique, we start with the simplest form of the 2D dispersion relation for a single graphene sheet as expressed by the tight binding approximation [19.94]

$$E_{g_{2D}}(k_x, k_y) = \pm\gamma_0 \left\{1 + 4\cos\left(\frac{\sqrt{3}k_x a_0}{2}\right)\cos\left(\frac{k_y a_0}{2}\right) + 4\cos^2\left(\frac{k_y a_0}{2}\right)\right\}^{1/2} \qquad (19.23)$$

where $a_0 = 1.42 \times \sqrt{3}$ Å is the lattice constant for a 2D graphene sheet and γ_0 is the nearest-neighbor C–C overlap integral [19.95]. A set of 1D energy dispersion relations is obtained from Eq. (19.23) by considering the small number of allowed wave vectors in the circumferential direction. The simplest cases to consider are the nanotubes having the highest symmetry. Referring to Fig. 19.25, we see the unit cells and Brillouin zones for the highly symmetric nanotubes, namely for (a) an armchair tubule and (b) a zigzag tubule. The appropriate periodic boundary conditions used to obtain energy eigenvalues for the (N_x, N_x) armchair tubule define the small number of allowed wave vectors $k_{x,q}$ in the circumferential direction

$$N_x\sqrt{3}a_0 k_{x,q} = q2\pi, \qquad q = 1, \ldots, N_x. \qquad (19.24)$$

Substitution of the discrete allowed values for $k_{x,q}$ given by Eq. (19.24) into Eq. (19.23) yields the energy dispersion relations $E_q^a(k)$ for the armchair tubule [19.46]

$$E_q^a(k) = \pm\gamma_0 \left\{1 \pm 4\cos\left(\frac{q\pi}{N_x}\right)\cos\left(\frac{ka_0}{2}\right) + 4\cos^2\left(\frac{ka_0}{2}\right)\right\}^{1/2}, \quad (-\pi < ka_0 < \pi), \quad (q = 1, \ldots, N_x) \qquad (19.25)$$

in which the superscript a refers to armchair, k is a one-dimensional vector along the tubule axis, and N_x refers to the armchair index, i.e., $(n, m) \equiv (N_x, N_x)$. The resulting calculated 1D dispersion relations $E_q^a(k)$ for the (5, 5) armchair nanotube ($N_x = 5$) are shown in Fig. 19.26(a), where we see six dispersion relations for the conduction bands and an equal number for the valence bands. In each case, two bands are nondegenerate (thin lines) and four are doubly degenerate (heavy lines), leading to 10 levels in each

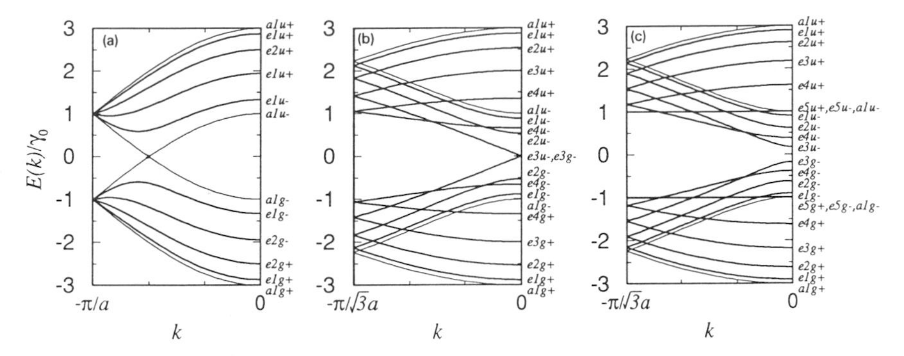

Fig. 19.26. One-dimensional energy dispersion relations for (a) armchair $(5,5)$ tubules, (b) zigzag $(9,0)$ tubules, and (c) zigzag $(10,0)$ tubules labeled by the irreducible representations of the point group $D_{(2n+1)d}$ at $k = 0$. The a-bands are nondegenerate and the e-bands are doubly degenerate at a general k-point in the 1D Brillouin zone [19.96].

case, consistent with the 10 hexagons around the circumference of the $(5,5)$ tubule. For all armchair tubules, the energy bands show a large degeneracy at the zone boundary, where $ka_0 = \pi$, so that Eq. (19.23) becomes

$$E_{g_{2D}}(k_x, \pi/a_0) = \pm\gamma_0 \tag{19.26}$$

for the 2D graphene sheet, independent of zone folding and independent of N_x. Although there are four carbon atoms in the unit cell for the real space lattice in Fig. 19.25(a), the two carbon atoms on the same sublattice of a graphene sheet are symmetrically equivalent, which causes a degeneracy of the energy bands at the boundary of the Brillouin zone. The valence and conduction bands in Fig. 19.26(a) cross at a k point that is two thirds of the distance from $k = 0$ to the zone boundary at $k = \pi/a_0$. The crossing takes place at the Fermi level and the energy bands are symmetric for $\pm k$ values.

Because of the degeneracy point between the valence and conduction bands at the band crossing, the $(5,5)$ tubule will exhibit metallic conduction at finite temperatures, because only infinitesimal excitations are needed to excite carriers into the conduction band. The $(5,5)$ armchair tubule is thus a zero-gap semiconductor, just like a 2D graphene sheet.

Similar calculations, as given by Eqs. (19.23), (19.24), and (19.25), show that all (n,n) armchair tubules yield dispersion relations similar to Eq. (19.25) with $2n$ conduction and $2n$ valence bands, and of these $2n$ bands, two are nondegenerate and $(n-1)$ are doubly degenerate. All (n,n) armchair tubules have a band degeneracy between the highest valence band and the lowest conduction band at $k = \pm 2\pi/(3a_0)$, where the bands cross the Fermi level. Thus, all armchair tubules are expected to exhibit metallic conduction, similar to the behavior of 2D graphene sheets [19.6, 46, 47, 89, 92, 97, 98].

The energy bands for the $(N_y, 0)$ zigzag tubule $E_q^z(k)$ can be obtained likewise from Eq. (19.23) by writing the periodic boundary condition on k_y as:

$$N_y a_0 k_{y,q} = q2\pi, \quad m = 1, \ldots, N_y, \tag{19.27}$$

to yield the 1D dispersion relations for the $2N_y$ states for the $(N_y, 0)$ zigzag tubule (denoted by the superscript z)

$$E_q^z(k) = \pm\gamma_0 \left\{1 \pm 4\cos\left(\frac{\sqrt{3}ka_0}{2}\right)\cos\left(\frac{q\pi}{N_y}\right) + 4\cos^2\left(\frac{q\pi}{N_y}\right)\right\}^{1/2},$$
$$\left(-\frac{\pi}{\sqrt{3}} < ka_0 < \frac{\pi}{\sqrt{3}}\right), \quad (q = 1, \ldots, N_y). \tag{19.28}$$

Referring to Fig. 19.26(b) we show for illustrative purposes the normalized energy dispersion relations for the zigzag (9, 0) tubule, where $N_y = 9$, corresponding to a tubule with a diameter equal to that of the C_{60} molecule. In the case of the (9, 0) zigzag tubule, the number of valence and conduction bands is 10, with 2 nondegenerate levels and 8 double degenerate levels yielding a total of 18 or $2N_y$ states, as expected from the number of hexagons for a circumferential ring in the 2D honeycomb lattice, which follows from Fig. 19.25(b). Most important is the band degeneracy that occurs at $k = 0$ between the doubly degenerate valence and conduction bands (symmetries e_{3u}^- and e_{3g}^-), giving rise to a fourfold degeneracy point at $k = 0$.

If we view the one-dimensional energy bands in the extended zone scheme, the dispersion for both the (5, 5) and (9, 0) tubules shown in Fig. 19.26(a) and (b) is just like "sliced" 2D energy dispersion relations, with slices taken along the directions $k_{x,q} = (q/N_x)(2\pi/\sqrt{3}a)$ and $k_{y,q} = (q/N_y)(2\pi/a)$ for the armchair and the zigzag tubules, respectively. For both types of tubules in Fig. 19.26(a) and (b), we have two 1D energy bands which cross at the Fermi energy ($E = 0$), giving rise to metallic conduction. In both cases, the band crossing at $E = 0$ occurs because the corresponding 2D energy bands cross at the K point (corners of a hexagon) of the 2D Brillouin zone (see Fig. 19.25), where the 2D graphene energy bands for the conduction and valence bands are degenerate. Although the density of states at the K point is always zero in the 2D case, the density of states in the 1D case can be finite, as shown below. Therefore, if there is no Peierls instability for this one-dimensional system (or if the Peierls gap is small compared to kT), carbon nanotubes can be metallic [19.46, 89].

If we carry out calculations similar to those described by Eqs. (19.27) and (19.28) for the (10, 0) zigzag tubule as shown in Fig. 19.26(c), we obtain 1D dispersion relations with some features similar to Fig. 19.26(b), but also some major differences are observed. In the (10, 0) case, the valence bands and the conduction bands each contain two nondegenerate levels and nine doubly degenerate levels to give a total of 20 states for each. Also, one degenerate level in the conduction band (e_{5u}^{+}, a_{1u}^{-}) and another in the valence band (e_{5g}^{+},a_{1g}^{-}) show little dispersion. However, the noteworthy difference between the (9, 0) and (10, 0) zigzag tubules is the appearance of an energy gap between the valence and conduction bands at $k = 0$ for the case of the (10, 0) tubule, in contrast to the degeneracy point that is observed at $k = 0$ for the (9, 0) tubule. Thus, the (10, 0) zigzag tubule is expected to exhibit semiconducting transport behavior, while the (9, 0) tubule is predicted to show metallic behavior.

The physical reason for this difference in behavior is that for the (10, 0) tubule, there are no allowed wave vectors $k_{y,q}$ from Eq. (19.27) which go through the K point in the 2D Brillouin zone of the graphene sheet. Referring to Fig. 19.25 it is seen that for all (n, n) armchair tubules, there is an allowed k vector going through the 2D zone corner K point, since the K point is always on the 1D zone boundary. However, for the $(n, 0)$ zigzag tubule, the allowed $k_{y,q}$ wave vectors only pass through the K point when n is divisible by 3, which is obeyed for the (9, 0) tubule, but not for the (10, 0) tubule. Thus, only one third of the $(n, 0)$ zigzag tubules would be expected to show metallic conduction. The other two thirds of the zigzag tubules are expected to be semiconducting. This is a remarkable symmetry-imposed result.

It is surprising that the calculated electronic structure can be either metallic or semiconducting depending on the choice of $(n, 0)$, although there is no difference in the local chemical bonding between the carbon atoms in the tubules, and no doping impurities are present [19.6]. These surprising results are of quantum origin and can be understood on the basis of the electronic structure of a 2D graphene sheet which is a zero-gap semiconductor [19.99] with bonding and antibonding π bands that are degenerate at the K point (zone corner) of the hexagonal 2D Brillouin zone of the graphene sheet. The periodic boundary conditions for the 1D tubules permit only a few wave vectors to exist in the circumferential direction. If one of these passes through the K-point in the Brillouin zone (see Fig. 19.25), then metallic conduction results; otherwise the tubule is semiconducting and has a band gap.

The density of states per unit cell using a model based on zone folding a 2D graphene sheet is shown in Fig. 19.27 for metallic (9, 0) and semiconducting (10, 0) tubules [19.6]. Here we see a finite density of states at

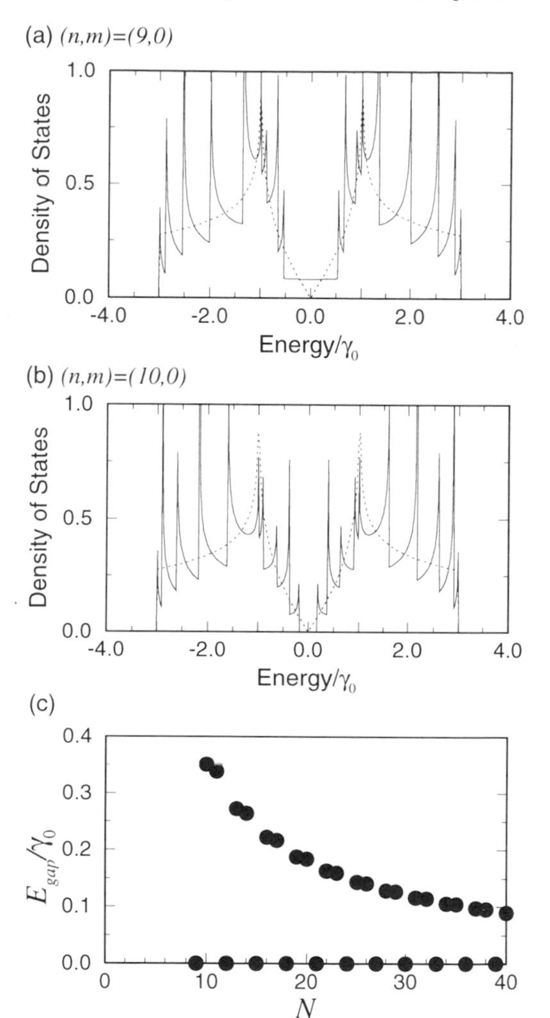

Fig. 19.27. Electronic 1D density of states per unit cell for two (n, m) zigzag tubules based on zone folding of a 2D graphene sheet: (a) the $(9,0)$ tubule which has metallic behavior, (b) the $(10,0)$ tubule which has semiconducting behavior. Also shown in the figure is the density of states for the 2D graphene sheet (dashed curves) [19.98]. (c) Plot of the energy gap for $(n,0)$ zigzag nanotubes plotted in units of γ_0 as a function of n, where γ_0 is the energy of the nearest-neighbor overlap integral for graphite [19.100].

the Fermi level for the $(9,0)$ tubules and a vanishing density of states for the $(10,0)$ tubules, consistent with the metallic nature of the $(9,0)$ tubules and the semiconducting nature of the $(10,0)$ zigzag tubules. Referring to Fig. 19.27, a singularity in the density of states appears as each energy band contributes to the density of states as a function of energy. An energy dependence of $(E - E_0)^{-1/2}$ is associated with each 1D singularity in the density of states shown in Fig. 19.27. Also shown in the figure (dashed curves) is the density of states for the 2D valence and conduction bands before zone folding. It is interesting to note that the $k = 0$ energy gap for the zigzag $(n, 0)$ tubules decreases with increasing n, as shown in Fig. 19.27(c), where it is noted that the tubules for which $n = 3q'$ (q' being an integer) are metallic and have no energy gap.

19.5.2. Single-Wall Nonsymmorphic Chiral Tubules

For the case of a chiral tubule, the generalization of the boundary conditions given in Eqs. (19.24) and (19.27) becomes [19.6]:

$$\sqrt{3}N_x k_x a_0 + N_y k_y a_0 = \mathbf{C}_h \cdot \mathbf{k} = 2\pi q, \tag{19.29}$$

in which $N_y a_0$ corresponds to the translation τ, and $\sqrt{3}N_x a_0$ corresponds to the rotation ψ in the space group symmetry operation $R = (\psi|\tau)$. The chiral vector $\mathbf{C}_h$ in Eq. (19.29) is defined by Eq. (19.1), and q is an integer for specifying inequivalent energy bands. Solutions for the 1D energy band structure for chiral tubules yield [19.6]

$$E_q(k) = \pm\gamma_0 \left\{ 1 + 4\cos\left(\frac{q\pi}{N_x} - \frac{N_y ka}{N_x 2}\right)\cos(ka_0/2) + 4\cos^2(ka_0/2) \right\}^{1/2} \tag{19.30}$$

where $-\pi \leq ka_0 \leq \pi$, the 1D wave vector $k = k_y$, while k_x in Eq. (19.23) is determined by Eq. (19.29).

The calculated results for the 1D electronic structure show that for small diameter graphene tubules, about one third of the tubules are metallic and two thirds are semiconducting, depending on the tubule diameter d_t and chiral angle θ. Metallic conduction in a general fullerene tubule is achieved when

$$2n + m = 3q \tag{19.31}$$

where n and m are pairs of integers (n, m) specifying the tubule diameter and chiral angle through Eqs. (19.2) and (19.3) and q is an integer. Tubules satisfying Eq. (19.31) are indicated in Fig. 19.2(b) as dots surrounded by large circles and these are the metallic tubules. The smaller dark circles in this figure correspond to semiconducting tubules. Figure 19.2(b) shows that all armchair tubules are metallic, but only one third of the possible zigzag and chiral tubules are metallic, as discussed above [19.6]).

A more detailed analysis of the $E(k)$ relations for the chiral nanotubes shows that they can be classified into three general categories [19.87, 101]. First, if Eq. (19.7.1) is not satisfied, or $n - m \neq 3q$, then the tubule is semiconducting with an energy band gap. This is the situation for approximately two thirds of the tubules [see Fig. 19.2(b)] and will be discussed further below. For those tubules where $n - m = 3q$ and where metallic conduction is thus expected, two further cases follow. If, in addition, $n-m=3rd$ is satisfied where d is the largest common divisor of n and m and r is an integer, then band degeneracies between the valence and conduction bands at the Fermi level develop at $k=\pm(2\pi/3T)$ where T is the length of the lattice vector of the 1D unit cell. The tight binding energy band structure for this case is illustrated in Fig. 19.28 for the (7,4) tubule for which $d=1$ and $r=1$ [19.87]. Armchair tubules denoted by (n,n) also fall into the category exemplified in Fig. 19.28. The third category satisfies both $n-m=3q$, which is the condition for metallic tubules, and $n-m\neq 3rd$. In this third case, the band degeneracy at the Fermi level occurs at $k=0$ and involves a fourfold band degeneracy. This case is illustrated in Fig. 19.29 for a (8,5) tubule [19.87]. From Eq. (19.31), metallic zigzag tubules $(0,3q)$ are in this category.

For all metallic tubules, independent of their diameter and chirality, it follows that the density of states per unit length along the tubule axis is a constant given by

$$N(E_F)=\frac{8}{\sqrt{3}\pi a_0\gamma_0} \tag{19.32}$$

where a_0 is the lattice constant of the graphene layer and γ_0 is the nearest-neighbor C–C tight binding overlap energy [19.87]. While the value of γ_0 for 3D graphite is 3.13 eV, a value of 2.5 eV is obtained for this overlap energy in the 2D case when the asymmetry in the bonding and antibonding

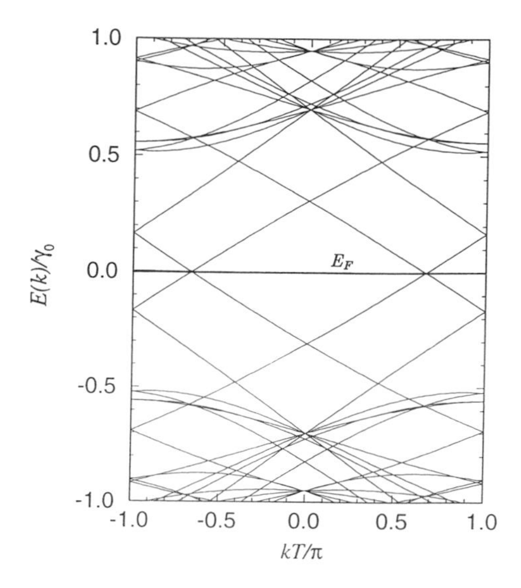

Fig. 19.28. Plot of the tight binding energy bands $E(k)$ for the 1D tubule for values of the energy between $-\gamma_0$ and γ_0, for the case of the metallic carbon nanotube specified by integers $(n,m)=(7,4)$. The general behavior of the two band degeneracies at $k=\pm(2/3)(\pi/T)$ (where T is the length of the 1D unit cell of the tubule and k is the 1D wavevector) is typical of the case where $n-m=3rd$ [19.101].

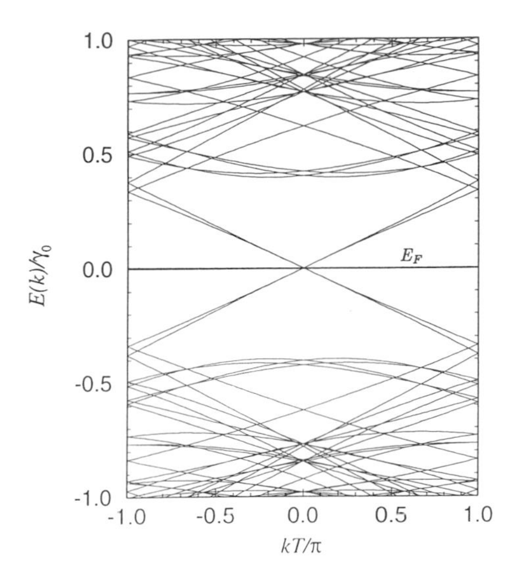

Fig. 19.29. Plot of the energy bands $E(k)$ for the 1D tubule for values of the energy, in dimensionless units $E(k)/\gamma_0$, for the case of the metallic carbon nanotube specified by integers $(n, m) = (9, 6)$ for which the largest common divisor is $d = 3$, and the Fermi level is at $E_F = 0$. The general behavior of the four energy bands intersecting at $k = 0$ is typical of the case where $n - m \neq 3rd$, and r is an integer [19.101].

states is averaged out, and this 2D value has been found to yield good agreement with first principles calculations [19.102].

Another important result pertains to the semiconducting tubules, showing that their energy gap depends upon the reciprocal tubule diameter d_t,

$$E_g = \frac{\gamma_0 a_{C-C}}{d_t}, \tag{19.33}$$

independent of the chiral angle of the semiconducting tubule, where a_{C-C} is the nearest-neighbor C–C distance on a graphene sheet. A plot of E_g *vs.* $100/d_t$ is shown in Fig. 19.30 for the graphite overlap integral taken as $\gamma_0 = 3.13$ eV. The results in Fig. 19.30 are important for testing the 1D model for the electronic structure of carbon nanotubes, because this result allows measurements to be made on many individual semiconducting tubules, which are characterized only with regard to tubule diameter without regard for their chiral angles (see §19.6). Using a value of $\gamma_0 = 2.5$ eV as given by the local density functional calculation [19.102], Eq. (19.33) suggests that the band gap exceeds thermal energy at room temperature for tubule diameters $d_t \leq 140$ Å. Furthermore, since about one third of the cylinders of a multiwall nanotube are conducting, certain electronic properties of nanotubes are dominated by contributions from these conducting constituents.

Metallic 1D energy bands are generally unstable under a Peierls distortion. However, the Peierls energy gap obtained for the metallic tubules is found to be greatly suppressed by increasing the tubule diameter, so that the Peierls gap quickly approaches the zero-energy gap of 2D graphite [19.46,

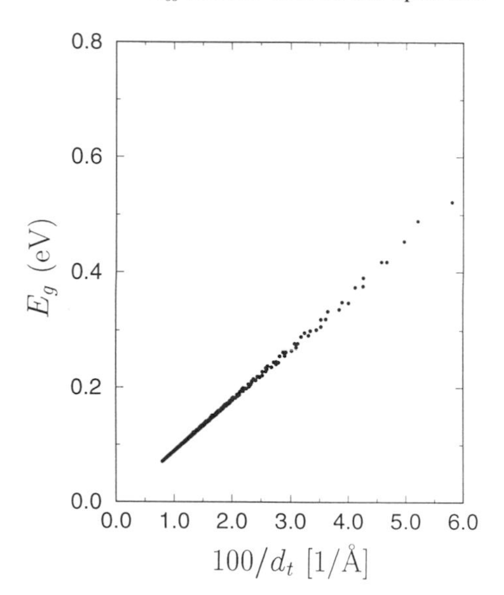

Fig. 19.30. The energy gap E_g calculated by a tight binding model for a general chiral carbon nanotubule as a function of 100 Å/d_t, where d_t is the tube diameter in Å. It is noted that the relationship becomes linear for large values of d_t [19.101].

89]. Thus if we consider finite temperatures or fluctuation effects, it is believed that such a small Peierls gap can be neglected. As the tubule diameter increases, more wave vectors become allowed for the circumferential direction, the tubules become more two-dimensional, and the semiconducting band gap decreases, as illustrated in Fig. 19.30.

The effect of curvature of the carbon nanotubes has been considered within the tight binding approximation [19.102]. This complicates the calculation considerably, by introducing four tight binding parameters, with values given by $V_{pp\pi} = -2.77$ eV, $V_{ss\sigma} = -4.76$ eV, $V_{sp\sigma} = 4.33$ eV, and $V_{pp\sigma} = 4.37$ eV [19.103, 104], instead of the single tight binding parameter $V_{pp\pi} = -2.5$ eV, which applies in the case where curvature is neglected. It should be mentioned that for both the armchair and zigzag tubules the band crossing at E_F is between energy bands of different symmetry (tubule curvature has no effect on symmetry) so that no interaction or band splitting would be expected at E_F. The effect of tubule curvature is further discussed in §19.8.

Some first principles calculations have been carried out for carbon nanotubes [19.92, 102, 105–107], yielding results in substantial agreement with the simple tight binding results which are described in this section. A comparison between the first principles local density functional determination of the band gap for the semiconducting tubules (open squares) in comparison with the tight binding calculation for ($V_{pp\pi} = -2.5$ eV) is shown in Fig. 19.31 as the solid curve [19.102]. In the limit that we average the

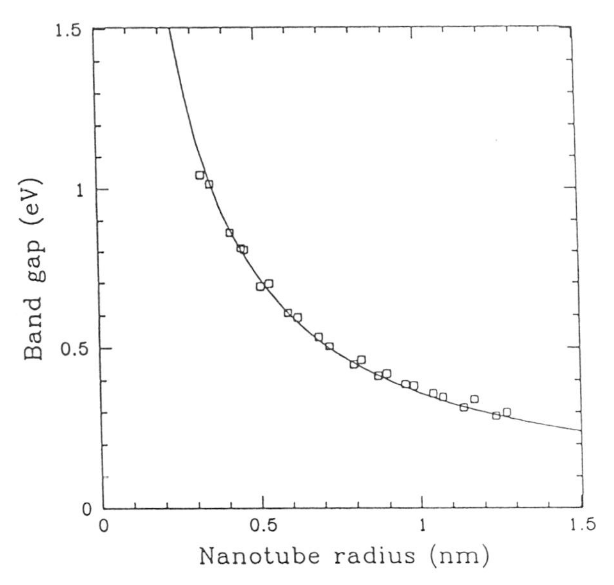

Fig. 19.31. Band gap as a function of nanotube radius using a first principles local density functional method. The solid line shows estimates using a graphene sheet model with $V_{pp\pi} = -\gamma_0 = 2.5$ eV [19.102].

asymmetry of the bonding and antibonding energies in graphite, the average overlap integral $\gamma_0 \rightarrow -2.5$ eV [19.102].

Another local density approximation (LDA) calculation shows that the large curvature of small single-wall carbon nanotubes leads to significant hybridization of σ^* and π^* orbitals, leading to the introduction of conduction band states into the energy gap. The effect of this hybridization is large for tubules of diameter less than that of C_{60} but is not so great for tubules in the range ($d_t > 0.7$ nm) that have been observed experimentally [19.105, 108]. The simple tight binding result for single-wall nanotubes that one third of the nanotubes are metallic and two thirds are semiconducting has a symmetry basis which is discussed in §19.4. There are several physical processes that tend to modify these simple considerations. A first principles LDA calculation [19.108] has determined that for low-diameter tubules the curvature of the graphene sheet results in band shifts which move their band edges into the semiconducting energy gap, hence suggesting that all small-diameter tubules should be metallic. The contrary conclusion could be reached if a Peierls distortion of the 1D conductor produced an energy gap at the Fermi level. Several authors [19.46, 89] have considered the Peierls distortion and have argued that it is suppressed in the carbon nanotube because of the large elastic energies that are associated with an in-plane lattice distortion. A splitting of the K-point degeneracy associated with a symmetry lowering (Jahn–Teller splitting) has been considered by some authors as a mechanism for removal of the metallic degeneracy. At the present time the experimental evidence is not clear about the nature of the conductivity of single-wall carbon nanotubes.

LDA-based calculations have also been carried out for BN nanotubes, and the results suggest that BN nanotubes should be stable and should have a band gap of ~5.5 eV independent of diameter [19.108, 109]. These calculations stimulated experimental work leading to the successful synthesis of pure multiwall BN nanotubes [19.110], with inner diameters of 1–3 nm and lengths up to 200 nm. The BN nanotubes are produced in a carbon-free plasma discharge between a BN-packed tungsten rod and a cooled copper electrode. Electron energy loss spectroscopy on individual nanotubes confirmed the BN stoichiometry [19.110, 111].

19.5.3. Multiwall Nanotubes and Arrays

Many of the experimental observations on carbon nanotubes thus far have been made on multiwall tubules [19.10, 11, 98, 112, 113]. This has inspired a number of theoretical calculations to extend the theoretical results initially obtained for single-wall nanotubes to observations in multilayer tubules. These calculations for multiwall tubules have been informative for the interpretation of experiments and influential for suggesting new research directions. Regarding the electronic structure, multiwall calculations have been done predominantly for double-wall tubules, although some calculations have been done for a four-wall tubule [19.10, 11, 112] and also for nanotube arrays [19.10, 11]

The first calculation for a double-wall carbon nanotube [19.98] was done with a tight binding calculation. The results showed that two coaxial zigzag nanotubes that would each be metallic as single-wall nanotubes yield a metallic double-wall nanotube when a weak interlayer coupling between the concentric nanotubes is introduced. Similarly, two coaxial semiconducting tubules remain semiconducting when a weak interlayer coupling is introduced [19.98]. More interesting is the case of coaxial metal–semiconductor and semiconductor–metal nanotubes, which also retain their metallic and semiconducting identities when a weak interlayer interaction is introduced. On the basis of this result, we conclude that it might be possible to prepare metal–insulator device structures in the coaxial geometry, as has been suggested in the literature [19.35, 114].

A second calculation was done with density functional theory in the local density approximation to establish the optimum interlayer distance between an inner (5, 5) armchair tubule and an outer armchair (10, 10) tubule. The result of this calculation yielded a 3.39 Å interlayer separation [19.10, 11], with an energy stabilization of 48 meV/carbon atom. The fact that the interlayer separation is somewhat less than the 3.44 Å separation expected for turbostratic graphite may be explained by the interlayer correlation between the carbon atom sites both along the tubule axis direction and circumferentially. A similar calculation for double-layered hyperfullerenes has been carried out, yielding an interlayer spacing of 3.524 Å between the two shells of $C_{60}@C_{240}$, with an energy stabilization of 14 meV/atom (see §19.10) [19.78]. In addition, for two coaxial armchair tubules, estimates for the translational and rotational energy barriers of 0.23 meV/atom and 0.52 meV/atom, respectively, were obtained, suggesting significant translational and rotational interlayer mobility for ideal tubules at room temperature. Of course, constraints associated with the cap structure and with defects on the tubules would be expected to restrict these motions. Detailed band calculations for various interplanar geometries for the two coaxial armchair tubules confirm the tight binding results mentioned above [19.10, 11].

Inspired by experimental observations on bundles of carbon nanotubes, calculations of the electronic structure have also been carried out on arrays of (6, 6) armchair nanotubes to determine the crystalline structure of the arrays, the relative orientation of adjacent nanotubes, and the spacing between them. Figure 19.32 shows one tetragonal and two hexagonal arrays that were considered, with space group symmetries $P4_2/mmc$ (D_{4h}^9), $P6/mmm$ (D_{6h}^1), and $P6/mcc$ (D_{6h}^2), respectively [19.10, 11, 106]. The calculation shows that the hexagonal $P6/mcc$ (D_{6h}^2) space group is the most stable, yielding an alignment between tubules that relates closely to the ABAB stacking of graphite, with an intertubule separation of 3.14 Å at closest approach, showing that the curvature of the tubules lowers the minimum interplanar distance (as is also found for fullerenes where the corresponding distance is 2.9 Å). The calculations further show that the hexagonal intertubule arrangement results in a gain in cohesive energy of 2.2 meV/carbon atom, while the atomic matching between adjacent tubules results in a gain in cohesive energy of 3.4 meV/carbon atom. The importance of the intertubule interaction can be seen in the reduction in the intertubule closest approach distance to 3.14 Å for the $P6/mcc$ (D_{6h}^2) structure, from 3.36 Å and 3.35 Å, respectively, for the tetragonal $P4_2/mmc$ (D_{4h}^9) and $P6/mmm$ (D_{6h}^1) space groups. A plot of the electronic dispersion relations for the most stable $P6/mmm$ (D_{6h}^1) structure is given in Fig. 19.33 [19.10, 11, 106], showing the metallic nature of this tubule array. It is expected that further calculations will consider the interactions between nested nanotubes having different symmetries, which on physical grounds should have a weaker interaction, because of a lack of correlation between near neighbors. Fur-

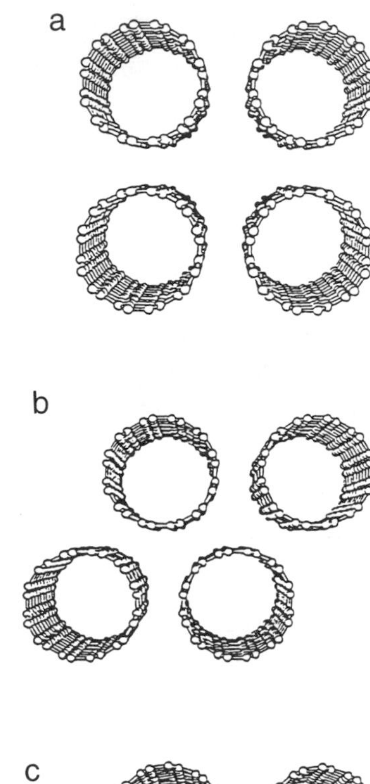

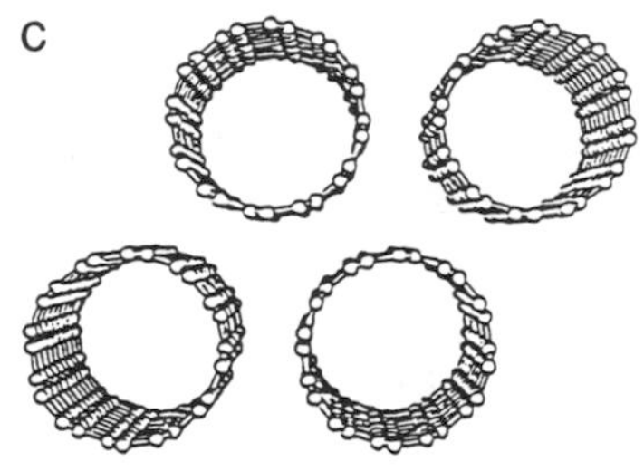

Fig. 19.32. Schematic array of carbon nanotubes with a common tubule axial direction in the (a) tetragonal [$P4_2/mmc$ (D_{4h}^9)], (b) hexagonal I [$P6/mmm$ (D_{6h}^1)], and (c) hexagonal II [$P6/mcc$ (D_{6h}^2)] space group arrangements. The reference nanotube is generated using a planar ring of 12 carbon atoms arranged in six pairs with the D_{6h} symmetry [19.10, 11, 106].

ther work remains to be done addressing the role of the strain energy in determining relative alignments of multilayer nanotubes and arrays, and in examining the possibility of a Peierls distortion in removing the coaxial nesting of carbon nanotubes.

It has been shown [19.115] that the double–single bond pattern of the graphite lattice is possible only for several choices of chiral vectors. The corresponding energy gap depends on the chiral vectors and the diameters of the tubules.

From these results one could imagine designing a minimum-size conductive wire consisting of a double-layered concentric carbon nanotube with

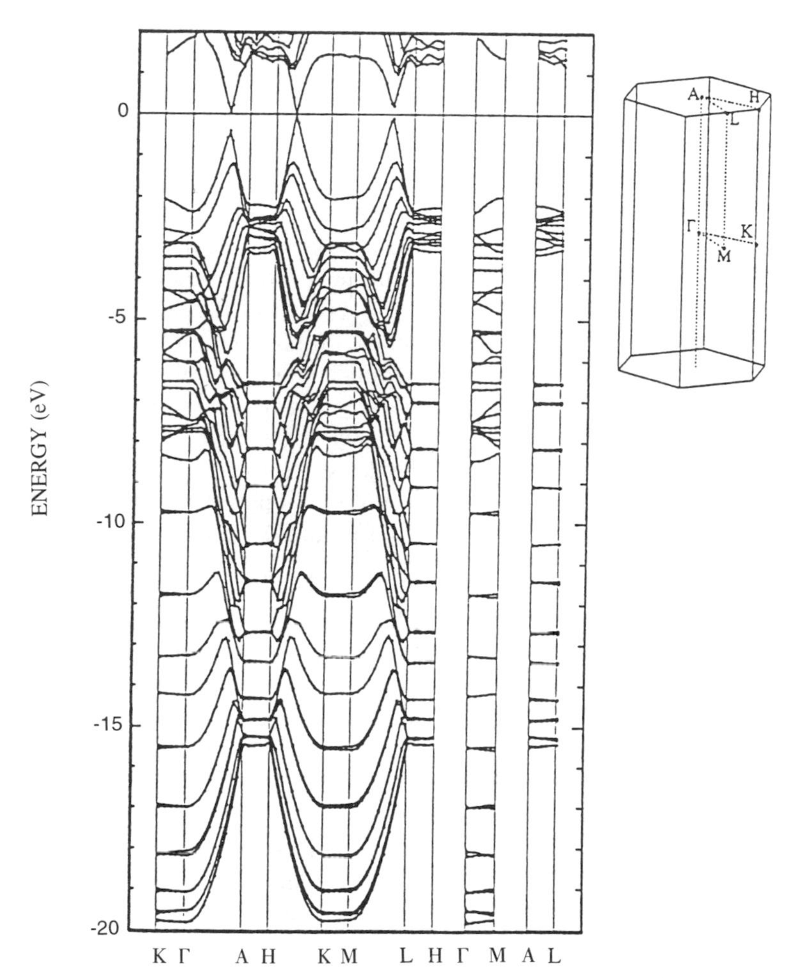

Fig. 19.33. Self-consistent $E(k)$ band structure (48 valence and 5 conduction bands) for the hexagonal II [$P6/mcc$ (D_{6h}^2)] arrangement of single-wall carbon nanotubes, calculated along different high-symmetry directions in the Brillouin zone. The Fermi level at $E_F = 0$ is positioned at the degeneracy point appearing between K and H, indicating metallic behavior for this tubule array [19.11].

a metallic inner tubule covered by a semiconducting (or insulating) outer tubule. This concept could be extended to the design of tubular metal–semiconductor devices without introducing any doping impurities [19.6]. From a conceptual standpoint, there are many possibilities for arranging arrays of metallic and semiconducting carbon nonotubes.

Modification of the conduction properties of semiconducting carbon nanotubes by B (p-type) and N (n-type) substitutional doping has also been discussed [19.116]. Modifications to the electronic structure by filling the capillaries of the tubes (see §19.9) have also been proposed [19.9]. Exohedral doping of the space between nanotubes in a tubule bundle could provide yet another mechanism for doping the tubules.

No superconductivity has yet been found in carbon nanotubes or nanotube arrays (see §19.11). Despite the prediction that 1D electronic systems cannot support superconductivity [19.117, 118], it is not clear that such theories are applicable to carbon nanotubes, which are tubular with a hollow core and have only a few unit cells around the circumference. It may perhaps be possible to align an ensemble of similar tubules to form a 3D solid, consisting of 1D constituents. In such a case we would expect the possibility of superconductivity to be highly sensitive to the density of states at the Fermi level. The appropriate doping of tubules may provide another possible approach to superconductivity for carbon nanotube systems.

19.5.4. 1D Electronic Structure in a Magnetic Field

The electronic structure of single-wall carbon nanotubes in a magnetic field has attracted considerable attention because of the intrinsic interest of the phenomena and the ability to make magnetic susceptibility and magnetoresistance measurements on bundles of carbon nanotubes. There are two cases to be considered, each having a different physical meaning, since the specific results for the magnetization and the magnetic susceptibility depend strongly on whether the magnetic field is oriented parallel or perpendicular to the tubule axis [19.119, 120]. Calculations for the magnetization and susceptibility have been carried out for these two field orientations using two different approaches: $\mathbf{k} \cdot \mathbf{p}$ perturbation theory and a tight binding formulation [19.119, 120]. In the case of the magnetic field parallel to the tubule axis, no magnetic flux penetrates the cylindrical graphene plane of the tubule, so that only the Aharonov–Bohm (AB) effect occurs.

Because of the periodic boundary conditions around a tubule, it is expected that the electronic states in the presence of a magnetic field parallel to the tubule axis will show a periodicity in the wave function

$$\Psi(r + \pi d_t) = \Psi(r) \exp(2\pi i \phi) \tag{19.34}$$

where $\phi = \Phi/\Phi_0$ defines a phase factor, Φ is the magnetic flux passing through the cross section of the carbon nanotube, and Φ_0 is the magnetic flux quantum $\Phi_0 = ch/e$ [19.119]. Thus, as the flux Φ is increased, Eq. (19.34) predicts a periodicity in magnetic properties associated with the phase factor $\exp(2\pi i \phi)$. This periodic behavior as a function of magnetic flux is expected for all types of tubules (metallic and semiconducting), independent of tubule diameter d_t and chiral angle θ, as shown in Fig. 19.34. At small magnetic fields, Fig. 19.34 shows that the magnetic moment is oriented along the direction of magnetic flux for metallic tubules (solid curve) and is opposite to the flux direction for the case of the semiconducting tubules. In the Aharonov–Bohm effect, the magnetic field changes the periodic boundary conditions which describe how to cut the 2D energy bands of the graphene layer and generate the 1D bands for the tubules [19.119]. When the magnetic field increases, the cutting line for the 2D bands is shifted in a direction perpendicular to the tubule axis. As a result, oscillations in the metallic energy bands periodically open an energy gap as a function of increasing magnetic field. Likewise, the energy band gap for the semiconducting tubules oscillates, thus closing the gap periodically [19.119].

For the magnetic field parallel to the tubule axis and $\phi = 1/6$, calculation of the magnetic moment and magnetic susceptibility for the metallic tubules as a function of tubule circumference (Fig. 19.35) shows almost no dependence on circumference, for tubule diameters d_t greater than about

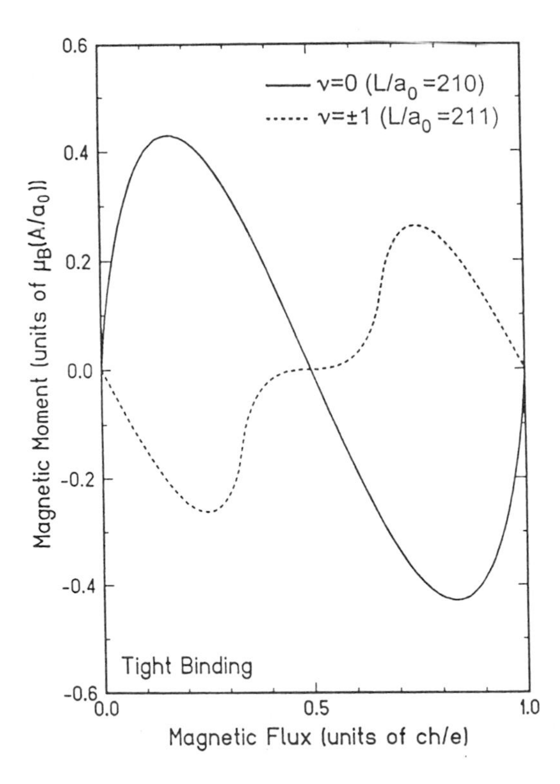

Fig. 19.34. The magnetic moment for **H** ∥ tubule axis calculated in a tight binding model *vs.* magnetic flux for a metallic (solid line, $\nu = 0$) and semiconducting (dotted line, $\nu = \pm 1$) carbon nanotube of diameter 16.5 nm. Here L is the circumference of the nanotube and $a_0 = 0.246$ nm [19.119].

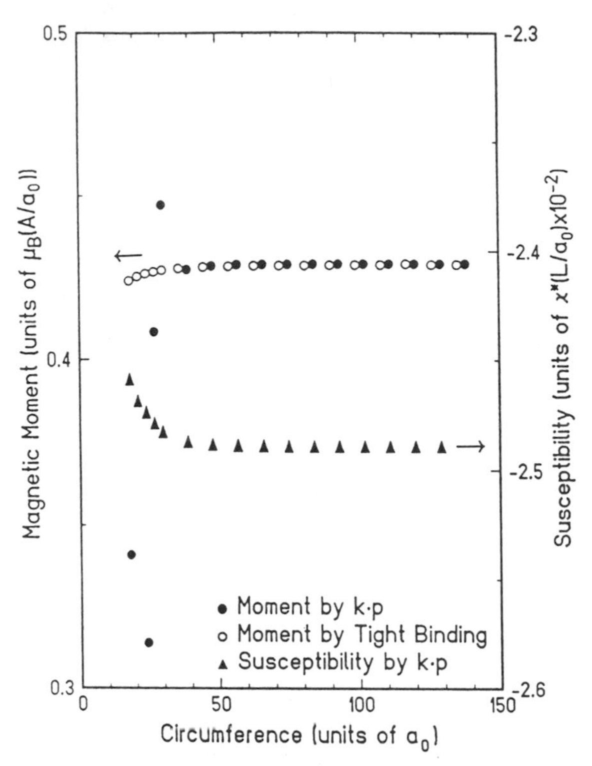

Fig. 19.35. Dependence on tubule circumference of the magnetic moment (closed circles) and magnetic susceptibility (triangles) of carbon nanotubes calculated according to the $\mathbf{k} \cdot \mathbf{p}$ approximation for $\mathbf{H} \parallel$ tubule axis and $\phi = 1/6$ and $a_0 = 0.246$ nm. The magnetic moment calculated according to the tight binding model is given by open circles. The magnetic moment is independent of the circumference L, except for very small diameter tubes [19.119].

30 Å. Also for the semiconducting tubules, the magnetic moment is almost independent of the tubule diameter. For these larger tubule diameters, it is shown that both $\mathbf{k} \cdot \mathbf{p}$ and tight binding calculations yield the same results for the magnetic moment, except for very small diameter tubules, where the effective-mass approximation used in the $\mathbf{k} \cdot \mathbf{p}$ calculation is no longer valid [19.119]. Referring to Fig. 19.35, the susceptibility is found to be diamagnetic (i.e., negative), with some decrease in the magnitude of χ for small-diameter tubules.

Plots of the calculated differential magnetic susceptibility obtained by differentiation of the magnetization by the magnetic field are shown in Fig. 19.36 for the field along the tubule axis and $(\pi d_t/a_0) = 210$, corresponding to a tubule diameter of 16.5 nm, a_0 being the lattice constant of a graphene layer (0.246 nm). The most striking features of Fig. 19.36 are

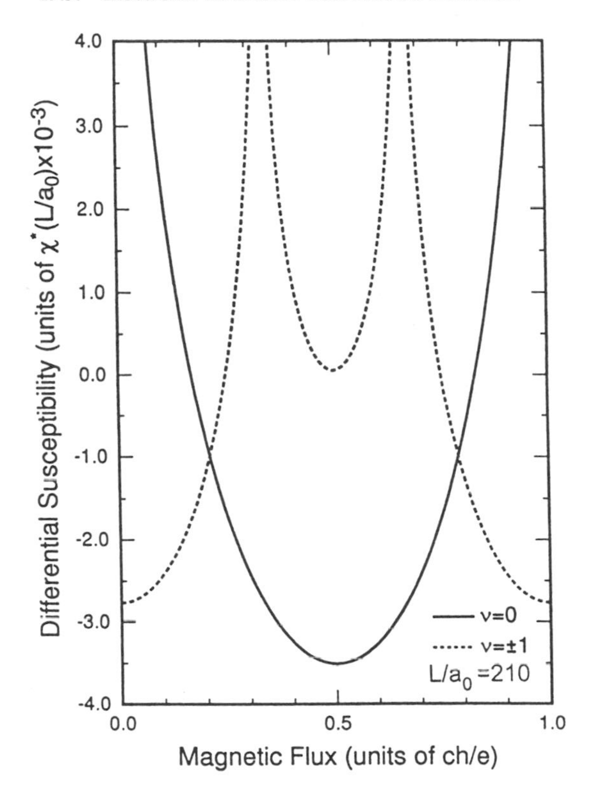

Fig. 19.36. Differential susceptibility *vs.* magnetic flux parallel to the tubule axis for metallic ($\nu = 0$, solid line) and semiconducting ($\nu = \pm 1$, dotted line) carbon nanotubes. The differential susceptibility is proportional to the tubule diameter, which is fixed at 16.5 nm for these plots [19.119].

the singularities in the differential susceptibility at $\phi = 0$ and $\phi = 1$ for metallic tubules and at $\phi = \pm 1/3$ for semiconducting tubules, and again showing a periodicity in the susceptibility as a function of the number of magnetic flux quanta [19.119].

For the magnetic field **H** normal to the tubule axis, Landau quantization occurs in the tubule plane. The magnetization and susceptibility are calculated by differentiating the total energy of the occupied electronic energy bands. When the Landau radius $r_L = \ell = (\hbar c/eH)^{1/2}$ or equivalently the magnetic length ℓ is small compared with the tubule radius, then, independent of whether the tubule is metallic or semiconducting, the calculated results for the magnetization and susceptibility approach those for graphite, which is highly diamagnetic. The magnitude of the susceptibility for **H** normal to the tubule axis, shown in Fig. 19.37 as the magnetic flux changes by one flux quantum, is very much larger (by about three orders of magnitude) than for **H** parallel to the tubule axis, shown in Fig. 19.36, but the dependence of χ on magnetic flux is a much smaller fraction of χ. The susceptibility of the carbon nanotube for **H** perpendicular to the tubule axis is

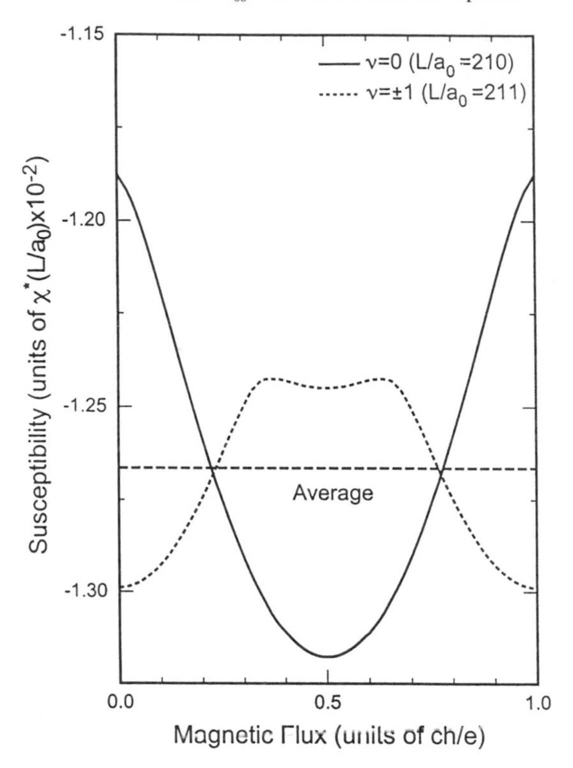

Fig. 19.37. Calculated susceptibility of metallic (solid line, $\nu = 0$) and semiconducting (dotted line, $\nu = \pm 1$) carbon nanotubes *vs.* magnetic flux for the magnetic field perpendicular to the tubule axis. The susceptibility is negative and is only weakly dependent on the flux. The dot-dashed line represents the average susceptibility as the magnetic flux ϕ changes by one quantum flux unit. Here the tubule diameter is $d_t = L/\pi$ and $a_0 = 0.246$ nm is the lattice constant of a 2D graphene sheet [19.119].

thus similar to the situation for a graphene sheet for which the susceptibility per unit area is given by

$$\chi = \frac{\gamma^2}{3\pi E_F}\left(\frac{e}{c\hbar}\right)^2, \tag{19.35}$$

in which $\gamma = \sqrt{3}\gamma_0 a_0/2$ and γ_0 is the nearest-neighbor tight binding overlap integral. For a graphene sheet, the magnetic susceptibility diverges ($\chi \to \infty$) as $T \to 0$. Because of the large graphite interlayer spacing in comparison with the 2D lattice constant a_0, the diamagnetic response of bulk graphite is dominated by that of the graphene layers. Since the interlayer coupling in bulk graphite is, however, nonvanishing, the susceptibility for 3D graphite is large, but not divergent, and is highly anisotropic, with small values of χ for the magnetic field oriented in the plane and large values for the magnetic field parallel to the c-axis.

Since the diamagnetic susceptibility for $\mathbf{H} \perp$ tubule axis is almost three orders of magnitude larger than that for $\mathbf{H} \parallel$ tubule axis, the magnetic response of a carbon nanotubule is dominated by the field component perpendicular to its tube axis, except in the case that this axis is accurately aligned in the magnetic field direction. This difference in magnitude relates to the difference in magnetic flux $L^2H/4\pi$ for a field **H** parallel to the tubule axis and $La_0H/\sqrt{2}$ for a perpendicular field. The factor $1/\sqrt{2}$ arises from an angular average of the field component perpendicular to the circumference of the tubule. The susceptibility for tubules averaged over the magnetic flux as it changes by a flux quantum (see Fig. 19.37) is obtained by replacing E_F in Eq. (19.35) by $(0.2\sqrt{3}\gamma_0 a_0)(2\pi/L)$.

A more recent calculation has made use of the magnetic levels in a full tight binding formulation of the electronic structure [19.120]. Landau quantization of the π energy bands of a carbon nanotube is calculated within the tight binding approximation. The electronic energy bands in a magnetic field do not show explicit Landau levels, but they do have energy dispersion for all values of magnetic field. The energy bandwidth shows new oscillations (see Fig. 19.38) with a period that is scaled by a cross section of the unit cell of the tubule which is specified by the symmetry of the nanotube. The magnetic response of the electronic structure for such one-dimensional materials with a two-dimensional surface is especially interesting and should be relevant to recent magnetoresistance [19.121] and susceptibility [19.122–124] experiments.

In a two-dimensional cosine band with lattice constant a_0, there is fractal behavior in the energy band spectra in a magnetic field H, depending on whether Ha_0^2/Φ_0 is a rational or irrational number, where $\Phi_0 = hc/e$ is a flux quantum [19.125]. All energy dispersion relations are a periodic function of integer values of Ha_0^2/Φ_0. However, the corresponding field is too large ($\sim 10^5$ T) to observe this fractal behavior explicitly. In relatively weak fields ($\sim 10^2$ T), we observe only Landau levels, except for Landau subbands near $E = 0$ in the case of a 2D cosine band, since the wave functions near $E = 0$ are for extended orbits [19.126].

Ajiki and Ando [19.119] have shown an Aharonov–Bohm effect in carbon nanotubes for $\mathbf{H} \parallel$ tubule axis and Landau level quantization for $\mathbf{H} \perp$ to the tubule axis. However, there is a limitation for using $\mathbf{k}\cdot\mathbf{p}$ perturbation theory around the degenerate point K at the corner of the hexagonal Brillouin zone in the case of a carbon nanotube, since the k values taken in the direction of the tubule circumference can be far from the K point, leading to incorrect values of the energy, especially for small carbon nanotubes and for energies far from that at the K point [19.120].

The magnetic energy band structure for carbon nanotubes within the tight binding approximation is shown in Fig. 19.38, in which it is assumed that the atomic wave function is localized at a carbon site and the magnetic field varies sufficiently slowly over a length scale equal to the lattice constant. This condition is valid for any large magnetic field that is currently available, even if we adopt the smallest-diameter carbon nanotube observed

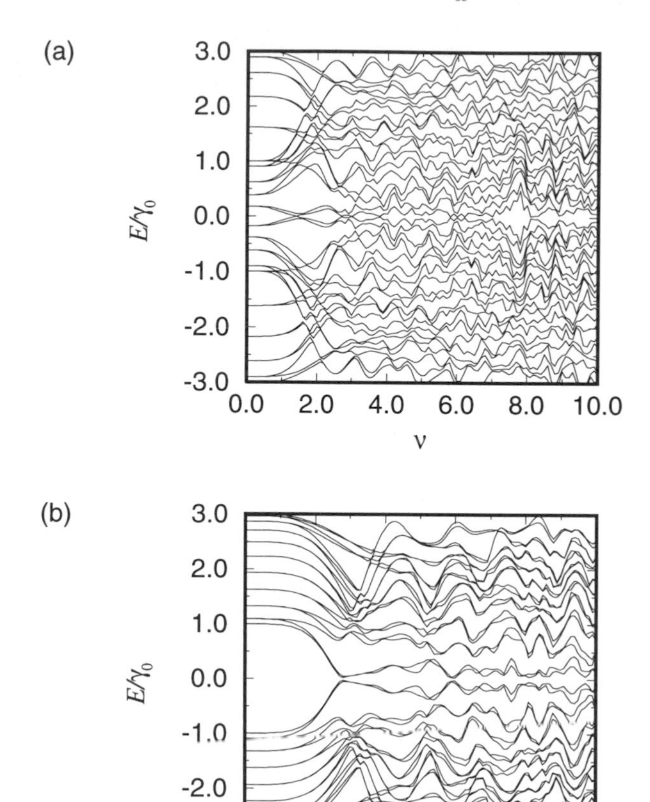

Fig. 19.38. The energy at $k_y = 0$ as a function of the inverse of a dimensionless magnetic length $\nu = L/2\pi\ell$ for (a) a zigzag tubule (10, 0) and (b) an armchair tubule (10, 10). Here $\ell = (\hbar c/eH)^{1/2}$ and $L = \pi d_t$ [19.120].

experimentally, whose diameter is five times as large as $a_{\text{C-C}} = 1.421$ Å, the nearest-neighbor C–C distance [19.4].

The calculations described above indicate that susceptibility measurements on an array of carbon nanotubes will be dominated by the field component perpendicular to the axis of the tubule, unless the axes for all the tubules are accurately aligned with respect to each other, and the applied magnetic field is accurately aligned along the tubule axis direction. For the case $\mathbf{H} \perp$ tubule axis, Fig. 19.37 shows that the magnitude of the differential susceptibility varies only weakly ($< 10\%$) as a function of the magnetic flux, for both the semiconducting and metallic tubules, although the dependence on magnetic flux differs in phase between the metallic and semiconducting tubules, as seen in Fig. 19.37.

The Pauli susceptibility has also been measured by electron spin resonance (ESR) for a purified tube, using the controlled burning technique (see §19.2.5). Since the nanotubes burn less readily than nanoparticles, mainly carbon nanotubes remain as a result of this procedure [19.127].

19.6. Electronic Structure: Experimental Results

Experimental measurements to test the remarkable theoretical predictions of the electronic structure of carbon nanotubes are difficult to carry out because of the strong dependence of the predicted properties on tubule diameter and chirality. The experimental difficulties arise from the great experimental challenges in making electronic or optical measurements on individual single-wall nanotubes, and further challenges arise in making such demanding measurements on individual single-wall nanotubes that have been characterized with regard to diameter and chiral angle (d_t and θ). Despite these difficulties, pioneering work has already been reported on experimental observations relevant to the electronic structure of individual nanotubes or on bundles of nanotubes, as summarized in this section.

19.6.1. Scanning Tunneling Spectroscopy Studies

The most promising present technique for carrying out sensitive measurements of the electronic properties of individual tubules is scanning tunneling spectroscopy (STS) because of the ability of the tunneling tip to probe the electronic density of states of either a single-wall nanotube [19.128], or the outermost cylinder of a multiwall tubule, or more generally a bundle of tubules. With this technique, it is further possible to carry out both STS and scanning tunneling microscopy (STM) measurements at the same location on the same tubule and therefore to measure the tubule diameter concurrently with the STS spectrum. It has also been demonstrated that the chiral angle θ of a carbon tubule can be determined using the STM technique [19.12] or high-resolution TEM [19.1, 14, 33, 38, 42].

Several groups have thus far attempted STS studies of individual tubules. The first report of I–V measurements by Zhang and Lieber [19.14] suggested a gap in the density of states below ~200 meV and semiconducting behavior in the smallest of their nanotubes (6 nm diameter).

Although still preliminary, the study which provides the most detailed test of the theory for the electronic properties of the 1D carbon nanotubes, thus far, is the combined STM/STS study by Olk and Heremans [19.129].

In this STM/STS study, more than nine individual multilayer tubules with diameters ranging from 1.7 to 9.5 nm were examined. STM measurements were taken to verify the exponential relation between the tunneling current and the tip-to-tubule distance in order to confirm that the tunneling measurements pertain to the tubule and not to contamination on the tubule surface. Barrier heights were measured to establish the valid tunneling range for the tunneling tip. Topographic STM measurements were made to obtain the maximum height of the tubule relative to the gold substrate, and these results were used for determination of the diameter of an individual tubule [19.129]. Then switching to the STS mode of operation, current–voltage (I–V) plots were made on the same region of the same tubule as was characterized for its diameter by the STM measurement. The I–V plots for three typical tubules are shown in Fig. 19.39. The results provide evidence for one metallic tubule with $d_t = 8.7$ nm [trace (1)] showing ohmic behavior, and two semiconducting tubules [trace (2) for a tubule with $d_t = 4.0$ nm and trace (3) for a tubule with $d_t = 1.7$ nm] showing plateaus at zero current and passing through $V = 0$. The dI/dV plot in the upper inset provides a crude attempt to measure the density of states, the peaks in the dI/dV plot being attributed to $(E_0 - E)^{-1/2}$ dependent singularities in the 1D density of states. Referring to Fig. 19.27, we see that, as the energy increases, the appearance of each new 1D energy band is accompanied by a singularity in the density of states.

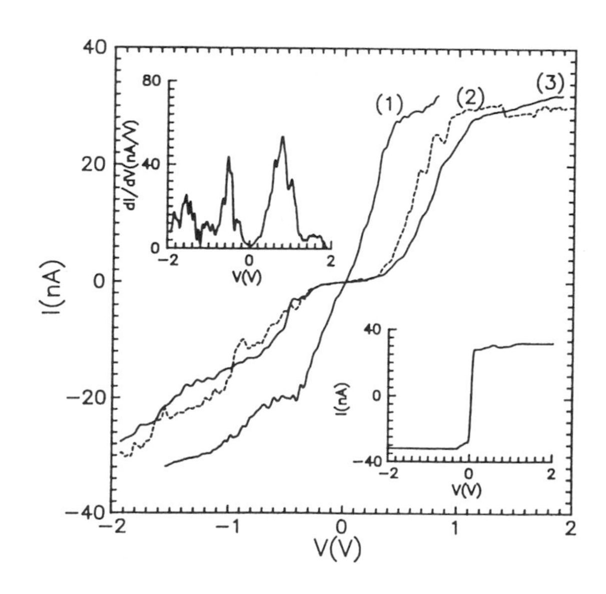

Fig. 19.39. Current–voltage I *vs.* V traces taken with scanning tunneling spectroscopy (STS) on individual nanotubes of various diameters: (1) $d_t = 87$ Å, (2) $d_t = 40$ Å, and (3) $d_t = 17$ Å. The top inset shows the conductance *vs.* voltage plot for data taken on the 17 Å nanotube. The bottom inset shows an I–V trace taken on a gold surface under the same conditions [19.129].

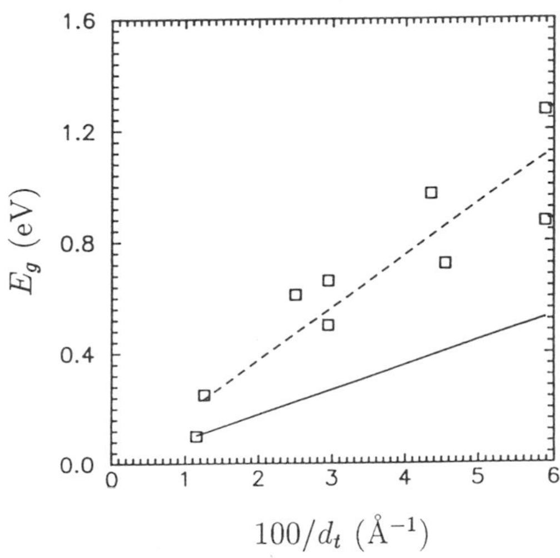

Fig. 19.40. Experimental points for the energy gap *vs.* $100/d_t$, the inverse nanotube diameter where d_t is in Å, for nine semiconducting carbon nanotubes. The dashed line is a fit to these points; the full line corresponds to a calculation [19.86, 98] for semiconducting zigzag nanotubes [19.129].

Finally, results for all the semiconducting tubules measured in this study by Olk and Heremans [19.129] are shown in Fig. 19.40, where the energy gaps determined for the individual tubules are plotted *vs.* $100/d_t$, the reciprocal tubule diameter [19.129]. Although the band gap *vs.* $100/d_t$ data in Fig. 19.40 are consistent with the predicted functional form shown in Fig. 19.30, the experimentally measured values of the band gaps are about a factor of 2 greater than that estimated theoretically on the basis of a tight binding calculation [19.86, 98]. Further experimental and theoretical work is needed to reach a detailed understanding of these phenomena.

19.6.2. Transport Measurements

The most detailed transport measurements that have been reported thus far have been done on a single bundle of carbon nanotubes to which two gold contacts were attached by lithographic techniques [19.121, 124]. In this way, the temperature dependence of the resistance $R(T)$ was measured from 300 K down to 200 mK, and the results of R *vs.* T are shown in Fig. 19.41 on a log T scale. The temperature dependence of the resistance for the tubule bundle was well fit from 2 to 300 K by a simple two-band semimetal model [19.130–132], yielding the electron and hole concentrations n and p as a function of temperature T

$$n = C_n k_B T \ln[1 + \exp(E_F/k_B T)],$$
$$p = C_p k_B T \ln\{1 + \exp[(\Delta - E_F)/k_B T]\}, \qquad (19.36)$$

where C_n and C_p are fitting parameters and Δ is the band overlap. The results of this fit are shown as a solid curve in Fig. 19.41, yielding a band

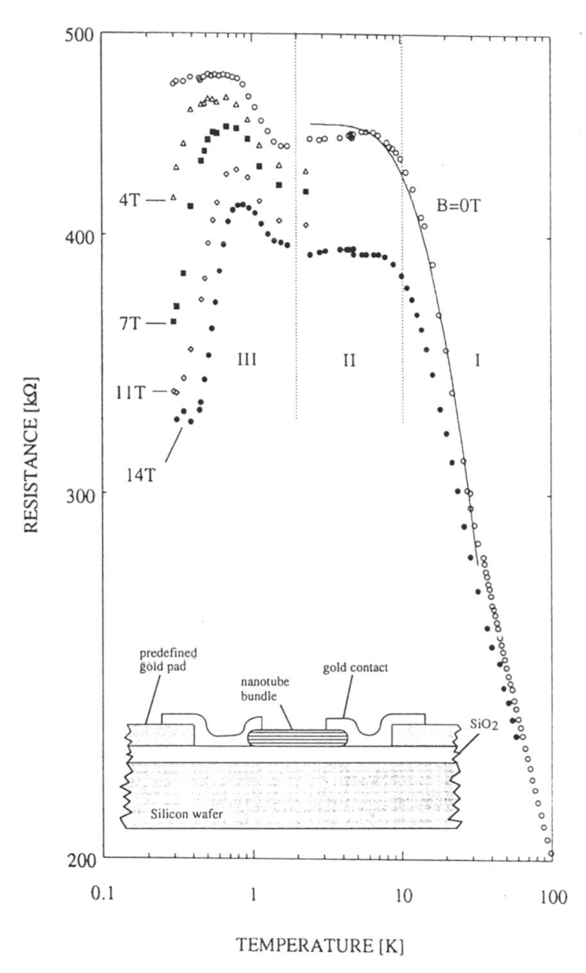

Fig. 19.41. Electrical resistance R as a function of temperature at the indicated magnetic fields for a single bundle of carbon nanotubes covering a range of tubule diameters and chiral angles. The two vertical lines separate the three temperature regimes: (I) showing that $R(T, H)$ for a tubule bundle is consistent with a two-band model for a semimetal; (II) showing saturation effects in $R(T, H)$; and (III) showing anomalous low-T behavior in $R(T, H)$. The continuous curve is a fit using the two-band model for graphite with a band overlap energy of $\Delta = 3.7$ meV and a Fermi level in the middle of the band overlap. The inset shows a schematic representation of the geometry of the sample and its electrical contacts (not to scale). The carbon nanotube bundle is electrically connected at both ends to two predefined gold pads [19.121].

overlap of $\Delta = 3.7$ meV, in contrast to $R(T)$ for 3D graphite, which corresponds to a band overlap of $\Delta = 40$ meV [19.130]. A smaller value of Δ for the tubules is expected, since the turbostratic stacking of the adjacent layers within a multiwall tubule greatly reduces the interlayer C–C interaction relative to graphite. Since this interlayer interaction is responsible for the band overlap in graphite, the carrier density for the nanotubes in the limit of low-temperature ($T \to 0$) is expected to be smaller by an order of magnitude than to that of graphite. Since the resistance measurement emphasizes the contribution of the metallic tubules in an ensemble of metallic and semiconducting tubules, this experiment provides very strong evidence for the presence of metallic tubules. The presence of metallic tubules is confirmed by numerical estimates of the magnitude of the room temperature resistance, assuming that one third of the layers of a multiwall tubule are metallic [19.121]. The corresponding resistivity value of 10^{-2} to 10^{-3} Ω-cm was calculated from the data in Fig. 19.41, in rough agreement with the best conductivity of ~100 S-cm reported for bulk arc deposited carbon material (containing about two thirds nanotubes and one third nanoparticles) [19.2], and with independent estimates of the resistivity of the black core material from the arc deposit of 10^{-2} Ω-cm [19.2]. Preliminary resistivity measurements using four current-voltage probes on a single nanotube 20 nm in diameter show a decrease in room temperature resistivity by an order of magnitude (to 10^{-4} Ω-cm) relative to a microbundle of carbon nanotubes 50 nm in diameter [19.133]. Although resistivity measurements are sensitive to the transport properties of the conducting nanotubes, such measurements provide little information about the semiconducting tubules, which hardly contribute to the resistivity measurements.

19.6.3. Magnetoresistance Studies

In addition to the temperature dependence of the zero-field resistance, Fig. 19.41 shows results for the temperature dependence of the resistivity in a magnetic field. According to the theory of Ajiki and Ando [19.119] described in §19.5.4, the application of a magnetic field normal to the tubule axis is expected to introduce a Landau level at the degeneracy point between the valence and conduction bands, thus increasing the density of states at the Fermi level. This increase in the density of states is expected to reduce the resistance of the carbon nanotubes, consistent with the experimental observations, which quantify the magnitude of the negative magnetoresistance. At low-temperatures below 2 K, an unexpected and unexplained behavior of large magnitude is observed both in the zero-field and high-field temperature-dependent resistance. A summary of these results for the magnetoresistance $\Delta R/R = [R(H) - R(0)]/R(0)$ at low tempera-

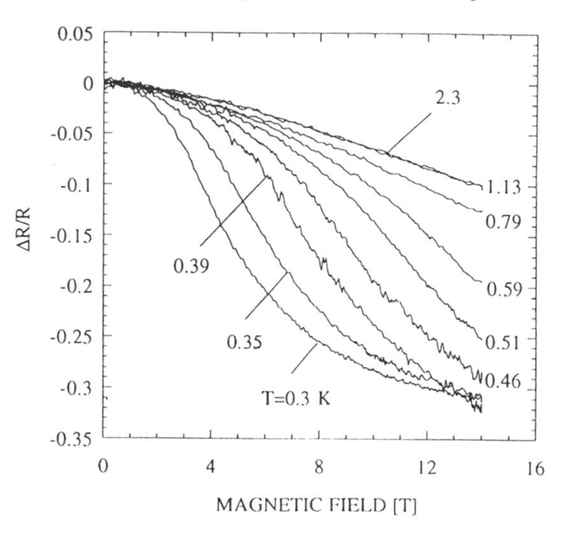

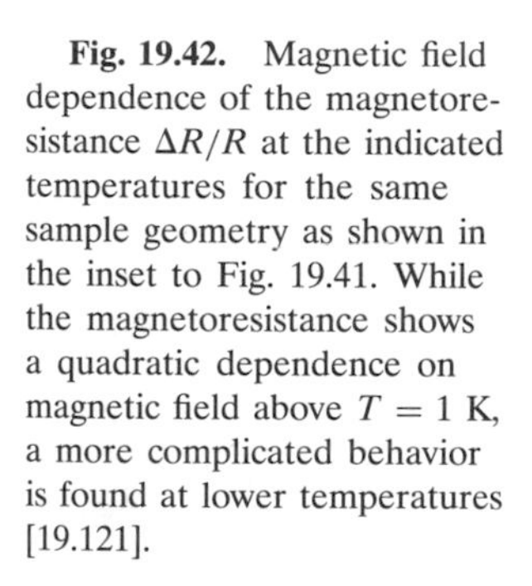

Fig. 19.42. Magnetic field dependence of the magnetoresistance $\Delta R/R$ at the indicated temperatures for the same sample geometry as shown in the inset to Fig. 19.41. While the magnetoresistance shows a quadratic dependence on magnetic field above $T = 1$ K, a more complicated behavior is found at lower temperatures [19.121].

ture is shown in Fig. 19.42, and may be related to the universal conductance fluctuations discussed below.

In contrast to the two-terminal transport experiments described above for a single bundle of nanotubes (< 1 μm diameter), four-terminal transport measurements, including $\rho(T)$, transverse magnetoresistance, and Hall effect experiments, have been reported on samples much larger in size (60 μm diameter and 350 μm in length between the two potential contacts) [19.134]. The magnetoresistance results of Fig. 19.43 are explained in a preliminary way in terms of a semimetal model. Regarding the transverse magnetoresistance and Hall effect measurements (where the current is along the axis of the nanotube bundles and the magnetic field is normal to the current), complicated behavior is observed as a function of temperature and magnetic field. At high temperature, the magnetoresistance can be described classically, yielding magnitudes for the electron and hole contributions to the conductivity and for the relative mobilities of holes and electrons ($\mu_p/\mu_n \sim 0.7$ and $\rho_{300\text{ K}} = 6.5 \times 10^{-3}$ Ω-cm). As T decreases, μ_p increases and the contributions to the transport properties of holes increase relative to that for electrons, so that the low-temperature transport is dominated by holes, as is common in disordered carbon-based systems [19.60].

Negative magnetoresistance is found at low T (<77 K) and low H (see Fig. 19.43), while positive magnetoresistance contributions dominate the measurements at high temperatures and also at high fields. The authors had some success in fitting the experimental results for the temperature-

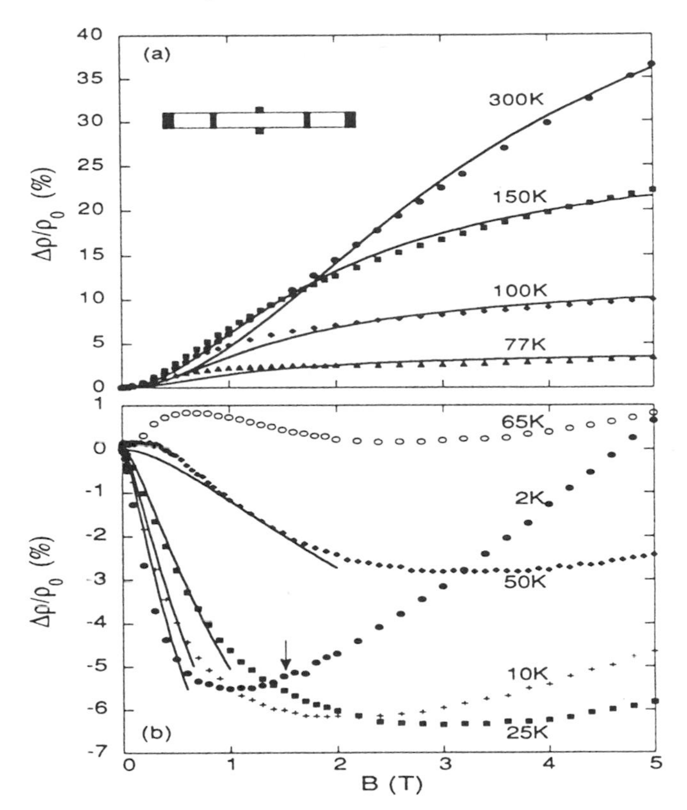

Fig. 19.43. Plots of magnetoresistance *vs.* magnetic field. The inset shows the arrangements of the four probes for resistivity and Hall effect measurements. The solid lines are calculated from 2D weak localization theory. The arrow shows the field at which the magnetic energy $g\mu_B H$ and thermal energy $k_B T$ are approximately equal at 2 K. The high-temperature data are presented in (a) and the low-temperature data in (b) [19.134].

dependent resistivity and the magnetoresistance at low T and H to 2D weak localization theory [19.135], showing a log T dependence for the weak localization contribution to the resistivity and a negative magnetoresistance with its characteristic T and H dependences [19.134]. A low-temperature sheet resistance of 5.5 kΩ/□ at 5 K is obtained from this 2D weak localization analysis, and this value of the sheet resistance is compatible with the diameter of the multiple tubule bundle sample.

Also of interest is that attempts to fit the low-temperature, low-field transport measurements on the tubule bundles give poor agreement with 1D weak localization theories [19.134]. Above 65 K, the magnetoresistance was found to be positive, and the conductivity was found to increase linearly with T, which was attributed mainly to an increase in carrier concentration with increasing T [19.134]. This is a reasonable conclusion in view of the low carrier concentration ($10^{19}/\text{cm}^3$) and the resultant low Fermi level (< 10 meV) of carbon nanotubes. The magnetoresistance behavior for this sample containing multiple tubule bundles shows notable differences relative to the magnetoresistance for pyrocarbons and individual disordered carbon fibers [19.60, 136] regarding their T- and H-dependent behaviors. Although at an early stage, these measurements show promise for providing new insights into the electronic structure and transport properties of carbon nanotubes.

Of particular interest to magnetoresistance studies is the recent report of the observation of conductance fluctuations in a single 20-nm-diameter nanotube [19.133] through measurement, using a four probe method, of sample-specific aperiodic structures in the magnetic field dependence of the resistance below 1 K. Referring to Fig. 19.44, we see aperiodic structures in the magnetoresistance at various magnetic field values. These aperiodic structures are independent of temperature in the low-temperature limit. The amplitudes of the two indicated fluctuations are constant in magnitude ($\Delta G = 0.2e^2/h$) at low-temperature (below 0.4 K) but decrease in magnitude above 0.4 K according to a $T^{-\alpha}$ ($\alpha \simeq 1/2$) law [19.133], consistent with the theory for universal conductance fluctuations [19.135, 138, 139]. At $T = 0$, a particular defect configuration in the sample gives rise to a specific pattern in the magnetoresistance (see Fig. 19.44), which remains

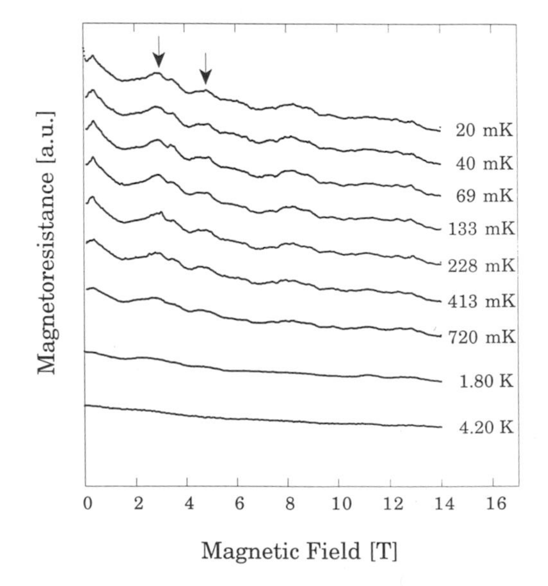

Fig. 19.44. Magnetic field dependence of the magnetoresistance of a 20-nm carbon nanotube at the indicated temperatures. The arrows indicate two unique temperature-independent peak positions [19.137].

essentially unchanged in amplitude until the temperature is increased to T_{CF}

$$T_{CF} = \frac{h\pi}{2k_B DL^2} \tag{19.37}$$

where D is the electron diffusion constant $D = 5 \times 10^{-3}$ m^2/s [19.134] and L is the tubule length. As T increases above T_{CF} the system averages over contributions from many uncorrelated patterns, thereby attenuating the signal due to conductance fluctuations.

19.6.4. Magnetic Susceptibility Studies

Closely related to the magnetoresistance study on a single tubule bundle are magnetization and magnetic susceptibility measurements carried out on samples containing multiple tubule bundles [19.122, 123]. In these studies, measurements were made on samples containing a distribution of tubule diameters with some variation in the orientation of the individual tubules, although attempts were made [19.123] to orient the magnetic field parallel or perpendicular to the axis of the tubule bundle. Because of the three orders of magnitude difference in the calculated susceptibility for **H** parallel or perpendicular to the tubule axis of a single-wall tubule [19.119], small misalignments of the field are expected to have an enormous effect on the measured χ for magnetic field directions close to the tubule axis. It may well be that the measurements shown in Fig. 19.45(f) labeled **H** $\parallel$ tubule axis [19.123] are in fact dominated by contributions from **H** $\perp$ tubule axis, even for small misalignment angles. On the other hand, with increasing tubule diameter, the susceptibility of the tubules should become more like that of graphite, which exhibits a large anisotropy in χ with $\chi_c = 22.0 \times 10^{-6}$ emu/g and $\chi_{ab} = 0.5 \times 10^{-6}$ emu/g [19.140].

These measurements nevertheless show that χ for carbon tubules is large and diamagnetic, being roughly half of the value for graphite with **H** $\parallel$ c-axis [19.122, 123], as shown in Fig. 19.45, where the room temperature χ for the tubules is compared with that for other carbons, including graphite (both **H** $\parallel$ c-axis and **H** $\perp$ c-axis), C_{60} powder, polycrystalline graphite anode material, and the gray-shell anode material [19.123]. The weak field dependence of χ in the 0.1 to 0.5 tesla field range (Fig. 19.45) is clarified by the plots of the low-temperature (5 K) magnetization and susceptibility shown in Fig. 19.46 over a wider field range [19.122].

Three magnetic field regimes for χ are identified in Fig. 19.46: the low-field regime, where χ rapidly becomes more diamagnetic with increasing field; the intermediate-field regime, where the diamagnetic susceptibility effect is largest and where χ is almost independent of field; and finally

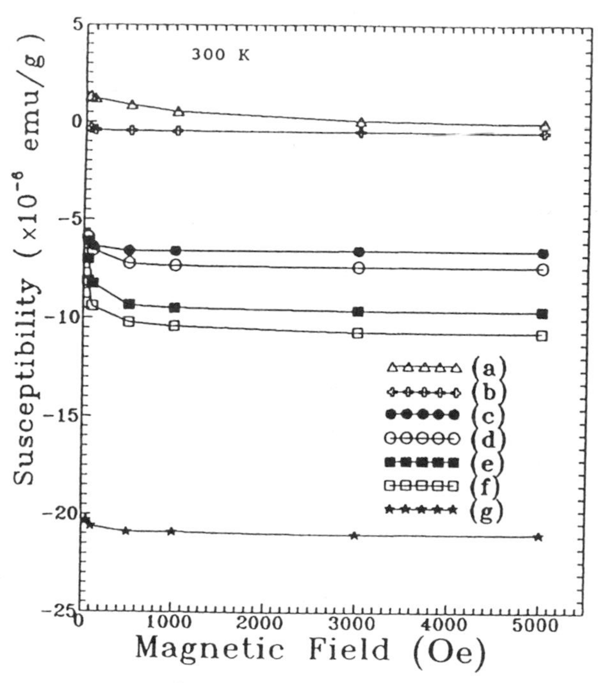

Fig. 19.45. Magnetic field dependence of the susceptibility measured up to 0.5 tesla at T = 300 K for (a) graphite with **H** perpendicular to the c-axis, (b) C_{60} powder, (c) polycrystalline graphite anode material, (d) gray-shell anode material, (e) a bundle of carbon nanotubes with **H** approximately $\perp$ to the tubule axis, (f) a bundle of carbon nanotubes with **H** approximately $\parallel$ to the tubule axis, and (g) graphite with **H** parallel to the c-axis [19.123].

the high-field regime, where χ becomes less diamagnetic with increasing magnetic field [19.122]. The behavior of the low-field regime may be due to the fact that the samples used for these measurements contain both metallic and semiconducting components. At the lowest magnetic fields, only the metallic constituents of the sample contribute to χ. But as the field is increased, the contribution from the semiconducting constituents increases until region II is reached, where the whole sample contributes to χ. If we identify the onset of region III with a regime where the magnetic radius $r_L = \ell = (\hbar c/eH)^{1/2}$ becomes comparable to the tubule radius, then the carriers are not much affected by the curvature of the tubule and the carriers then start to behave like the carriers in graphite, thereby explaining the decrease in the diamagnetism with increasing field, which is about the same percentage decrease as is observed for χ in graphite (**H** $\parallel$ c-axis). A detailed analysis of the results in Fig. 19.46 is not yet available.

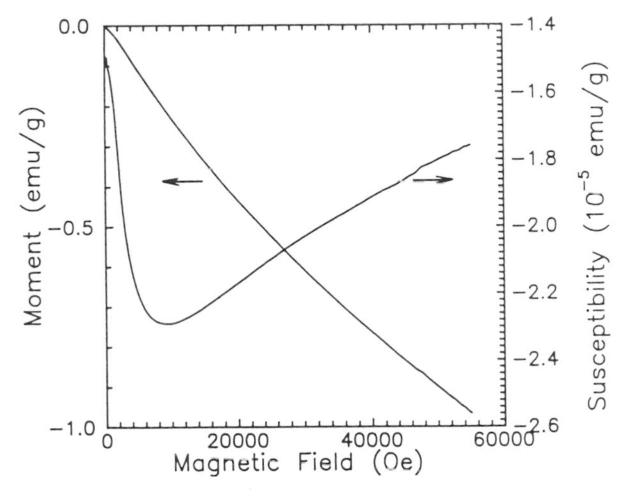

Fig. 19.46. Magnetic field dependence of the magnetic moment (left scale) and magnetic susceptibility (right scale) at high magnetic fields and at low temperature (5 K) [19.122].

To relate the measurements in Figs. 19.45 and 19.46 which pertain to different temperatures, we show for selected temperatures (see Fig. 19.47) the magnetic field dependence of the magnetic moment [19.122], which upon differentiation with respect to magnetic field yields χ. From Fig. 19.46, one would expect the room temperature χ to show anomalous behavior up to ~0.25 tesla. The behavior of χ from 0.5 to 6 tesla (not shown in detail)

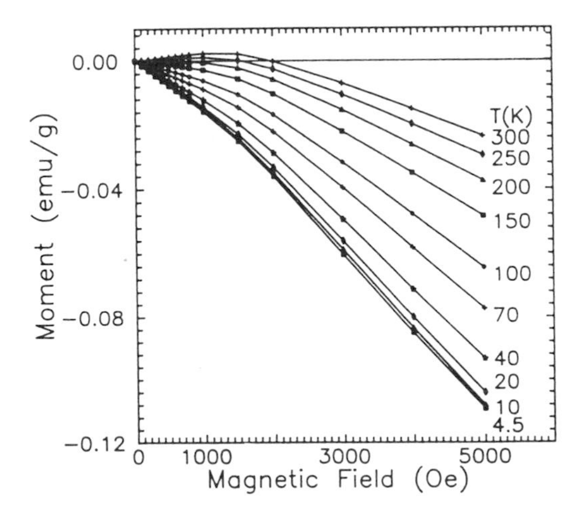

Fig. 19.47. Magnetic field dependence of the magnetic moment of carbon nanotubes at the indicated temperatures [19.122].

appears to be weakly dependent on H (see Fig. 19.48), but of course the magnitude of χ varies strongly with temperature [19.122].

The temperature dependence of χ for tubules is, however, remarkably different from that of graphite as shown in Fig. 19.48(i) where a large T dependence of χ for carbon nanotubes is seen, in contrast to the much weaker temperature dependence for graphite. In fact, χ for the tubules approaches the graphite value in the low-temperature limit [19.122, 123]. Whereas the susceptibility is nearly constant as a function of magnetic field for $0.5 < H < 1.5$ tesla, it is strongly temperature dependent, as shown in Fig. 19.48(ii), where the susceptibility χ is plotted *vs.* temperature for several values of magnetic field. Anomalous behavior is observed at low fields, as seen in Fig. 19.45 [19.123] and in more detail in Fig. 19.46 [19.122], suggesting a strong field dependence of χ at low H. No explanation of the anomalous low-temperature and low field dependences of χ for carbon nanotubes has yet been published, and these phenomena await further study.

Referring to Fig. 19.46, it is seen that the magnetic field of 2 T used to acquire the data in Fig. 19.48(i) corresponds to the intermediate-field regime of Fig. 19.46. In this regime the Landau radius $r_L = (c\hbar/eH)^{1/2}$ is becoming comparable to the largest tubule radius in the sample. Consequently, the data in Fig. 19.48(ii) show three different temperature-dependent behaviors [19.122], consistent with the three regimes of Fig. 19.46. Some possible explanation for the large temperature dependence of χ at low T may arise from the following two effects. Since the band gaps and band overlaps for some of the tubules in the sample are small, the derivatives of the Fermi function become temperature dependent at low temperatures. In addition, carrier scattering effects may contribute to the observed χ. In contrast, at high fields (e.g., 5 tesla), Fig. 19.48(ii) shows only a weak temperature dependence for χ [19.122]. At these high fields, r_L has become comparable to the tubule size, and the electron orbits become spatially localized, thereby not being sensitive to the tubule curvature. Thus in the high-field regime, graphitic behavior is expected. The results in Fig. 19.48(ii) at 5 tesla are indeed similar to observations in graphite [19.141].

Using ESR techniques, the g-values for aligned carbon nanotubes have been measured [19.142], showing strong evidence for a g-factor given by

$$g(\theta) = \left[(g_{\parallel} \cos\theta)^2 + (g_{\perp} \sin\theta)^2\right]^{1/2} \tag{19.38}$$

where $g_{\parallel} = 2.0137$ and $g_{\perp} = 2.0103$. These values for the g-factor yield an average value of $g_{av} = 2.012$ that is close to the average value for graphite, which is $g_{av} = (g_c + 2g_{ab})/3 = 2.018$, where $g_c = 2.050$ and $g_{ab} = 2.0026$. Although carbon nanotubes are significantly less anisotropic than graphite, the average g-factor for carbon nanotubes is quite similar to

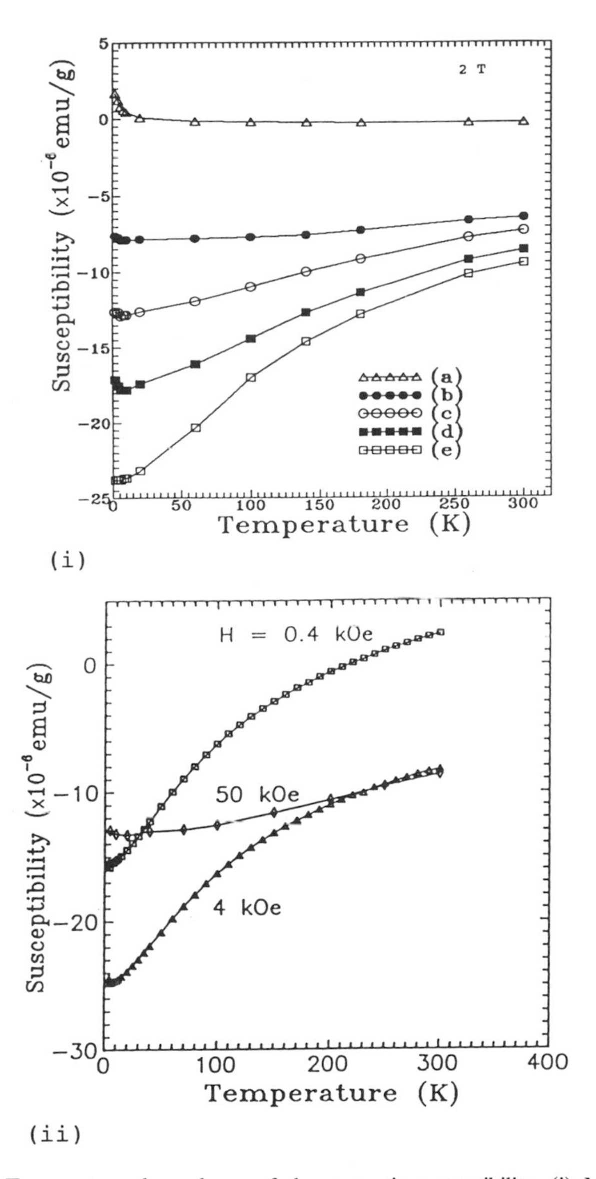

Fig. 19.48. Temperature dependence of the magnetic susceptibility. (i) Measured in a magnetic field of 2 T for various samples: (a) C_{60} powder; (b) polycrystalline graphite anode; (c) gray-shell anode material; (d) a bundle of carbon nanotubes: $\mathbf{H} \approx$ perpendicular to the axis of the bundle, (e) a bundle of carbon nanotubes: $\mathbf{H} \approx$ parallel to the axis of the bundle [19.123]. (ii) Measured at several magnetic fields for a bundle of carbon tubules [19.122].

that for graphite. Both $g_{\parallel}$ and $g_{\perp}$ exhibit a linear increase in magnitude as T is decreased from 300 K to 20 K, with low-temperature (20 K) values of $g_{\parallel} \sim 2.020$ and $g_{\perp} \sim 2.016$ [19.142]. From measurements of the spin–lattice relaxation time a value of 10^{-3} s was estimated for the room temperature resistive scattering time, yielding an estimate of 10^{-3} Ω-cm for the resistivity of carbon nanotubes, consistent with transport measurements (see §19.6.2).

19.6.5. Electron Energy Loss Spectroscopy Studies

Electron energy loss spectroscopy (EELS) measurements on samples removed from the black ring material from a carbon arc discharge showed differences regarding the corresponding spectrum of graphite. These early measurements further show polarization effects depending on whether the incident electron beam is parallel or perpendicular to the axis of the tubule bundle contained in the deposit (see Fig. 19.49). It is not surprising that the EELS spectra for the nanotube soot deposit are different from that of graphite, considering the variety of lower-dimensional materials present in the nanotube soot deposit, and these materials have lower carrier densities and conductivities than graphite. The EELS measurements on the tubule bundle in Fig. 19.49 are sensitive to differences between the spectrum for the incident electron beam along the tubule axis of the bundle and perpendicular to this axis [Fig. 19.49(c)]. These differences in the spectra show a shift of the EELS peaks for the nanotube soot to lower energies than in graphite, consistent with the lower carrier density of the as-prepared nanotubule deposit [19.38]. Based on these preliminary results, quantitative EELS measurements on tubules with well-characterized d_t and θ would be very interesting.

A step was taken toward achieving this goal when electron energy loss spectra were taken of individual carbon tubules, characterized for their outer diameter and number of cylindrical layers [19.143]. The results below 50 eV show a shift of the dominant electron energy loss peak due to collective σ bond excitations from 27 eV in graphite to lower energies in carbon nanotubes. As the tubule diameter decreases, the π electron peak is quenched (see Fig. 19.49), consistent with a lower carrier density for decreasing tubule diameter d_t, and consequently a smaller amount of screening. The greater average carrier localization, arising from the greater tubule curvature for the smaller-diameter tubules, may also contribute to the downshift of the electron energy loss peak [19.143]. Tubules of equal diameter, but having a greater number of layers, tend to have better structural order and therefore appear to have a more graphitic spectra. Physically, as the tubule diameter decreases, the curvature of the tubule surface increases, thereby reducing the interlayer correlation and changing the π

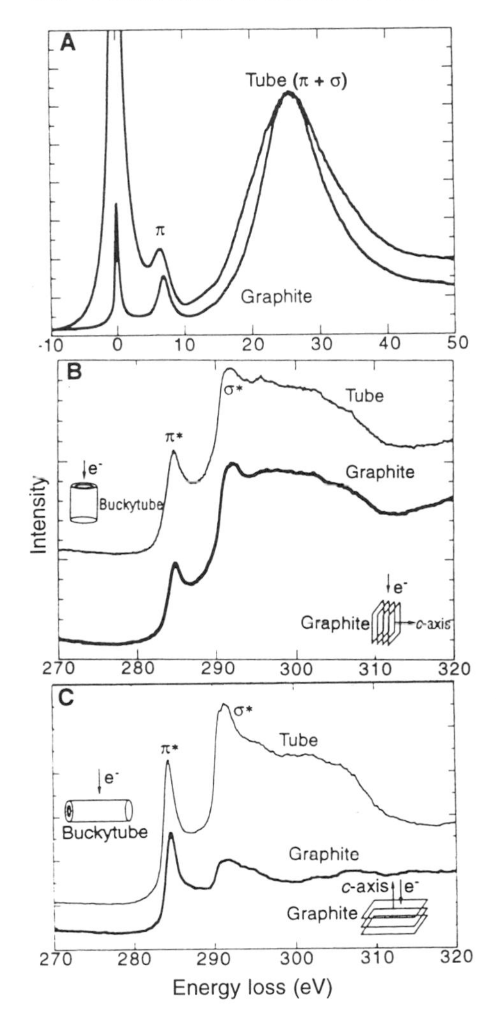

Fig. 19.49. (a) Low energy electron energy loss spectroscopy (EELS) traces at low electron energies for carbon nanotubes compared to those for graphite. (b) Core electron EELS spectra of graphite and carbon nanotubes in the case I geometry (the electron beam normal to the graphite c-axis). The spectra for graphite and the carbon nanotubes are virtually identical. (c) Core electron EELS spectra of graphite and carbon nanotubes in the case II geometry (the electron beam parallel to the c-axis of graphite and parallel to the tube axis). The additional σ^* contribution to the nanotube spectrum is attributed to the curved nature of the nanotube graphene sheets [19.38].

electron contribution to the interlayer coupling between concentric layers [19.143].

19.7. Phonon Modes in Carbon Nanotubes

In this section we summarize present knowledge of the phonon dispersion relations for carbon nanotubes, starting with a review of theoretical calculations of the dispersion relations (§19.7.1), followed by predictions for Raman and infrared activity in carbon nanotubes (§19.7.2). A summary is then given of the experimental observations of Raman spectra in carbon nanotubes (§19.7.3). No experiments have yet been reported on the infrared spectra of carbon nanotubes.

19.7.1. Phonon Dispersion Relations

The phonon modes for carbon nanotubes have been calculated based on zone folding of the phonon dispersion relations for a two-dimensional (2D) graphene layer [19.84], in analogy to the calculations previously described for the electronic energy band structure (see §19.5). Phonon dispersion relations have been obtained for nanotubes described by symmorphic and nonsymmorphic symmetry groups, and the results are discussed below.

The phonon dispersion relations for a 2D graphene sheet are well established (see Fig. 19.50) and both first principles [19.144] and phenomenological [19.145, 146] calculations are available. To obtain the general form of the dispersion relations, it is convenient to use a force constant model [19.84] reflecting the detailed symmetry of the carbon nanotubes. For such a force constant model calculation, the latest values for the force constants for graphite are used, thereby neglecting the effect of tubule curvature on the force constants. Values for the radial ($\phi_r^{(n)}$) and tangential [$\phi_{t_i}^{(n)}$ (in-plane) and $\phi_{t_0}^{(n)}$ (out-of-plane)] bending force constants for graphite going out to fourth-neighbor (n) distances for displacements in the plane and normal to the plane are given in Table 19.7 [19.84].

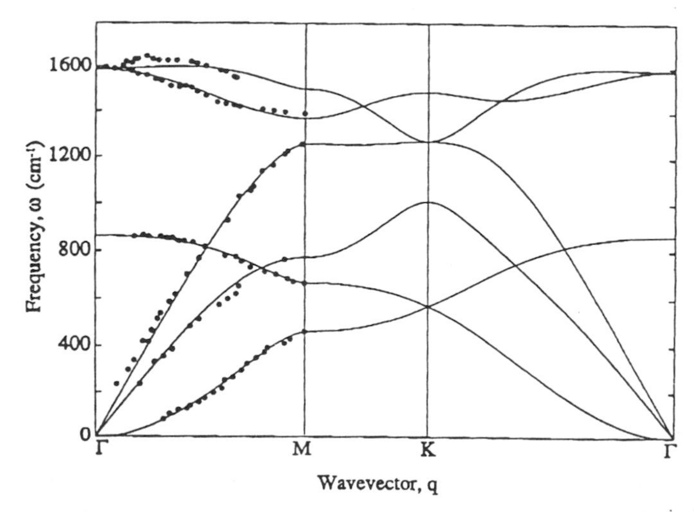

Fig. 19.50. The phonon dispersion relations for a 2D graphene sheet plotted along high-symmetry directions, using the force constants in Table 19.7. The points from neutron scattering studies in 3D graphite [19.147] were used in the modeling to obtain the phonon dispersion relations throughout the Brillouin zone [19.84].

Table 19.7

Values of the force constant parameters for graphite out to fourth-neighbor interactions in 10^4 dyn-cm [19.84].

Radial	Tangential	
$\phi_r^{(1)} = 36.50$	$\phi_{ti}^{(1)} = 24.50$	$\phi_{to}^{(1)} = 9.82$
$\phi_r^{(2)} = 8.80$	$\phi_{ti}^{(2)} = -3.23$	$\phi_{to}^{(2)} = -0.40$
$\phi_r^{(3)} = 3.00$	$\phi_{ti}^{(3)} = -5.25$	$\phi_{to}^{(3)} = 0.15$
$\phi_r^{(4)} = -1.92$	$\phi_{ti}^{(4)} = 2.29$	$\phi_{to}^{(4)} = -0.58$

As discussed in §19.4.1, the unit cell of each carbon tubule is defined by the integers (n, m), which specify the tubule diameter and chirality [see Eqs. (19.2) and (19.3)] and also specify the 1D unit cell for the tubule in terms of the chiral vector $\mathbf{C}_h$ and the lattice vector $\mathbf{T}$ along the cylindrical axis of the tubule [see Fig. 19.2(a)]. The normal mode frequencies of the carbon tubules at the Γ point ($k = 0$) in the Brillouin zone can be determined by using the method of zone folding, i.e., by determining the frequencies at specific points in the hexagonal Brillouin zone of the 2D graphene sheet which are equivalent to the Γ point of the 1D Brillouin zone of the tubule. Using this approach, the evaluation of the phonon mode frequencies in the tubule requires only the diagonalization of the dynamical matrix of a graphene sheet. This approach, however, neglects the effect of tubule curvature on the phonon dispersion relations.

The phonon dispersion relations in the tubule can thus be obtained from those of the 2D graphene sheet by using the relationship

$$\omega_{1D}(k) = \omega_{2D}(k\,\hat{K}_2 + \mu\,\hat{K}_1)\,,$$

$$\mu = 0, 1, 2, ..., N-1, \tag{19.39}$$

where the subscripts 1D and 2D refer, respectively, to the one-dimensional tubule and the two-dimensional graphene sheet, k is a wave vector in the direction $\hat{K}_2$ along the reciprocal space 1D periodic lattice of the tubule, μ is a nonnegative integer used to label the wave vectors or the states along the $\hat{K}_1$ reciprocal space direction normal to the tubule axis, and N, given by Eq. (19.9), denotes the number of hexagons in the 1D unit cell of the tubule. The result given by Eq. (19.39) includes both symmorphic and nonsymmorphic space groups.

In this section, we first present explicit dispersion relations for armchair and zigzag carbon nanotubes corresponding to the symmorphic space groups. We then extend the discussion to chiral nanotubes corresponding to nonsymmorphic space groups. In this discussion, we assume the length of the tubules is so much larger than their diameters that the tubules can be described in the 1D limit where the nanotubes have infinite length and the contributions from the carbon atoms in the caps can be neglected. In this discussion, we first enumerate the symmetry types of the various vibrational modes of the tubules and in the next subsection (§19.7.2) we comment on their optical activity.

The appropriate symmetry groups for the symmorphic armchair tubules ($n = m$) are D_{nh} or D_{nd}, respectively, depending on whether n is even or odd [19.4]. For armchair tubules with D_{nh} symmetry (n is even), the vibrational modes are decomposed according to the following irreducible representations (assuming that $n/2$ is even):

$$\begin{aligned}\Gamma_n^{\text{vib}} = {} & 4A_{1g} + 2A_{1u} + 4A_{2g} + 2A_{2u} + 2B_{1g} \\ & + 4B_{1u} + 2B_{2g} + 4B_{2u} + 4E_{1g} + 8E_{1u} + 8E_{2g} \\ & + 4E_{2u} + \cdots + 4E_{(n/2-1)g} + 8E_{(n/2-1)u}.\end{aligned} \tag{19.40}$$

If $n/2$ is odd [such as for $(n, m) = (6, 6)$], the 4 and 8 are interchanged in the last two terms in Eq. (19.40). The vibrational mode which describes rotation about the cylindrical axis has A_{2g} symmetry; the translation mode along the cylinder axis has A_{2u} symmetry, while the corresponding translations along the directions perpendicular to this axis have E_{1u} symmetry. For example, the (6, 6) armchair tubule for which $N = 12$ has 12 hexagons in the 1D unit cell and is described by 72 phonon branches. From Eq. (19.40) we see that the (6, 6) tubule has 48 distinct mode frequencies at the Brillouin zone center $k = 0$, while the (8, 8) armchair tubule has $N = 16$ and 96 phonon branches with 60 distinct mode frequencies at $k = 0$. The modes that transform according to the A_{1g}, E_{1g}, or E_{2g} irreducible representations are Raman active, while those that transform as A_{2u} or E_{1u} are infrared active. Hence there are 16 Raman-active mode frequencies ($4A_{1g} + 4E_{1g} + 8E_{2g}$) and 8 distinct infrared-active nonzero frequencies ($A_{1u} + 7E_{1u}$) for tubules with D_{nh} symmetry (n even). As an example, of the 48 distinct mode frequencies for the (6, 6) armchair tubule, 8 are IR active, 16 are Raman active, 2 have zero frequency, and 22 are silent.

Although the number of vibrational modes increases as the diameter of the carbon nanotube increases, the number of Raman-active and infrared-active modes remains constant. This is illustrated by comparing the normal modes of the (8, 8) and the (6, 6) armchair tubules. For the 60 distinct frequencies of the (8, 8) armchair tubule, only 8 are IR-active modes, 16 are Raman-active modes, 34 are silent modes, and 2 have zero frequency, yielding the same number of Raman-active and infrared-active modes as for the (6, 6) tubule. The $4E_{3g} + 8E_{3u}$ modes, present in the (8, 8) tubule but absent in the (6, 6) tubule, are all silent modes. The constant number of Raman- and infrared-active phonon tubule modes for a given symmetry group, or for a given chirality, is common to each category of carbon nanotube discussed below.

For armchair tubules with D_{nd} symmetry (n = odd integer), the vibrational modes of the carbon nanotube contain the following symmetries:

$$\begin{aligned}\Gamma_n^{\text{vib}} &= 3A_{1g} + 3A_{1u} + 3A_{2g} + 3A_{2u} \\ &+ 6E_{1g} + 6E_{1u} + 6E_{2g} + 6E_{2u} \\ &+ \cdots + 6E_{[(n-1)/2]g} + 6E_{[(n-1)/2]u}. \end{aligned} \tag{19.41}$$

As mentioned above, the zone center $k = 0$ vibrational frequencies are zero for one A_{2u} mode and for one E_{1u} mode, corresponding to the acoustic branches at wave vector $k = 0$. We therefore have 15 Raman-active modes and 7 infrared-active modes with distinct nonzero frequencies for armchair tubules with D_{nd} symmetry. As an example of a tubule with D_{nd} symmetry, the (5, 5) armchair tubule has $N = 10$, and 60 degrees of freedom per 1D unit cell, 12 nondegenerate phonon branches, 24 doubly degenerate phonon branches, with 7 nonvanishing IR-active mode frequencies, 15 that are Raman active, 2 that are of zero frequency and 12 silent mode frequencies, thereby accounting for the 36 distinct phonon branches.

To illustrate the phonon dispersion relations for carbon nanotubes, we show in Fig. 19.51 explicit results for the 36 phonon dispersion relations for

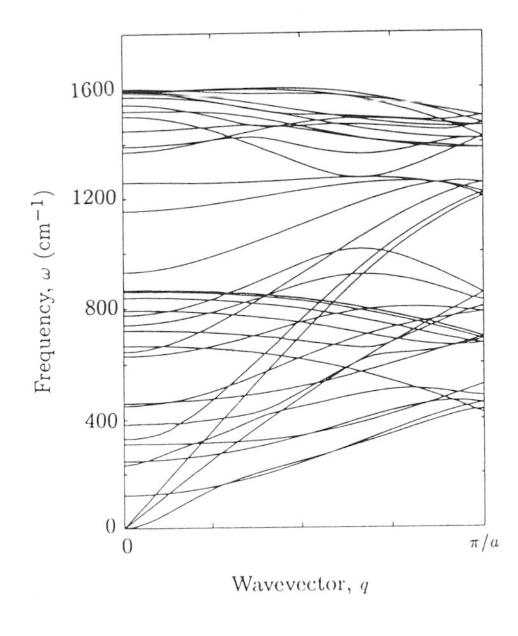

Fig. 19.51. Phonon dispersion relations for a (5, 5) carbon nanotubule (corresponding to the armchair tubule that fits on to a C_{60} hemispherical cap) [19.84].

a (5, 5) carbon nanotube (corresponding to the armchair tubule that fits on to a C_{60} hemispherical cap) [19.84]. This figure shows that many phonon branches result from the zone-folding procedure described by Eq. (19.41). We show that the phonon dispersion relations for a carbon nanotube depend on the (n, m) indices for the nanotube, or equivalently on the tubule diameter and chirality.

The structure of zigzag $(n, 0)$ tubules can also be described by groups with D_{nh} or D_{nd} symmetry for even or odd n, respectively. All zigzag tubules have, among others, $3A_{1g}$, $3A_{2u}$, $6E_{1g}$, $6E_{1u}$, and $6E_{2g}$ modes, irrespective of whether they have D_{nh} or D_{nd} symmetries. This gives 15 Raman-active modes and 7 infrared-active modes with distinct nonzero frequencies for zigzag tubules. As an example, consider the (9, 0) zigzag tubule which has $N = 18$ and 108 degrees of freedom and 60 distinct mode frequencies from Eq. (19.41). Of these distinct mode frequencies, 15 are Raman active, 7 are infrared active, two have zero frequency, and 36 are silent. For zigzag tubules (n odd, D_{nd} symmetry)

$$\begin{aligned}\Gamma_n^{\text{vib}} &= 3A_{1g} + 3A_{1u} + 3A_{2g} + 3A_{2u} \\ &+ 6E_{1g} + 6E_{1u} + 6E_{2g} + 6E_{2u} \\ &+ \cdots + 6E_{[(n-1)/2]g} + 6E_{[(n-1)/2]u}.\end{aligned} \tag{19.42}$$

For zigzag tubules (n even, D_{nh} symmetry)

$$\begin{aligned}\Gamma_n^{\text{vib}} &= 3A_{1g} + 3A_{1u} + 3A_{2g} + 3A_{2u} \\ &+ 3B_{1g} + 3B_{1u} + 3B_{2g} + 3B_{2u} \\ &+ 6E_{1g} + 6E_{1u} + 6E_{2g} + 6E_{2u} \\ &+ \cdots + 6E_{[(n-2)/2]g} + 6E_{[(n-2)/2]u}.\end{aligned} \tag{19.43}$$

We conclude this section with a summary of the phonon dispersion relations for chiral nanotubes. For the case where n and m are nonzero and have no common divisor, the point group of symmetry operations for chiral carbon nanotubes is given by the nonsymmorphic Abelian groups $C_{N/\Omega}$ discussed in §19.4.3. Here Ω is an integer [see Eq. (19.13)] which has no common divisor with N, and Ω denotes the number of 2π cycles traversed by the vector $\mathbf{R}$ before reaching a lattice point (see Fig. 19.24). Here N is the number of hexagons per 1D unit cell in the tubule and d is the highest common divisor of (n, m) [19.84]. Every tubule thus has $6N$ vibrational degrees of freedom, which are conveniently expressed in terms of their symmetry types [19.84]

$$\Gamma_N^{\text{vib}} = 6A + 6B + 6E_1 + 6E_2 + \cdots + 6E_{N/2-1}. \tag{19.44}$$

Of these modes, the Raman-active modes are those that transform as A, E_1, and E_2 and the infrared-active modes are those that transform as A or E_1. This gives 15 nonzero Raman-active mode frequencies and 9 nonzero infrared-active mode frequencies, after subtracting the modes associated with acoustic translations $(A + E_1)$ and with rotation of the cylinder (A). Zone-folding techniques can also be used to yield the phonon dispersion relations $\omega(k)$ for specific (n, m) chiral nanotubes, using Eq. (19.39). As an example of a chiral tubule, consider the $(n, m) = (7, 4)$ tubule. In this case $N = 62$, so that the 1D unit cell has 372 degrees of freedom. The phonon modes include branches with $6A + 6B + 6E_1 + 6E_2 + \cdots + 6E_{30}$ symmetries, 192 branches in all. Of these, 15 are Raman active at $k = 0$, while 9 are infrared active, 3 corresponding to zero-frequency modes at $k = 0$, and 162 are silent.

A major difference between the symmorphic armchair and zigzag tubules, on the one hand, and the nonsymmorphic chiral tubules, on the other hand, is that for symmorphic tubules, the M point of the 2D graphene sheet Brillouin zone is folded into the Γ point, while for the nonsymmorphic tubules the M point does not map into the Γ point. This difference in behavior with regard to zone folding causes a larger spread in the values of the Raman and infrared frequencies in armchair and zigzag tubules as compared with general chiral tubules. This feature is further discussed in §19.7.2.

Phonon dispersion relations, similar to those shown in Fig. 19.51 for the $(n, m) = (5, 5)$ tubule, have also been calculated for various zigzag and chiral tubules [19.83]. In general, the dispersion relations will have many more branches than are shown in Fig. 19.51 because of the larger size of the 1D unit cell for a general (n, m) nanotube. For many experiments involving phonon dispersion relations for carbon nanotubes, the measurements are made on a multitude of nanotubes, each having its own (n, m) indices. If the measurements involve spectroscopy, then the observed spectra will emphasize those mode frequencies which are common to many nanotubes. In §19.7.2, we will consider the Raman- and infrared-active frequencies for the various types of carbon nanotubes as a function of tubule diameter, to assist with the interpretation of spectroscopic measurements on multiple carbon nanotubes, arising both from multiwall nanotubes and from nanotubes with different tubule axes.

19.7.2. Calculated Raman- and Infrared-Active Modes

Of particular use for the interpretation of spectroscopic data is the dependence of the phonon frequencies on tubule diameter. Explicit experimental results are presented in §19.7.3.

The general method used to obtain the calculated dependence of the tubule mode frequencies on tubule diameter and chirality considers all tubules of a given chirality and then considers the effect of the tubule diameters on the mode frequencies. Results are now available for armchair, zigzag, and some representative chiral nanotubes [19.83, 84, 87].

In Figs. 19.52 and 19.53, we see the plots for the dependence of the Raman-active and infrared-active tubule modes, respectively, as a function of tubule diameter for the zigzag $(n, 0)$ tubules up to $n = 40$. In making these plots we make use of the fact that the number and symmetries of the optically active modes are independent of tubule diameter d_t, thereby allowing us to follow the evolution of the various optically active modes as a function of d_t. In these plots we see that many of the modes are strongly dependent on d_t, so that these modes do not contribute constructively to a single sharp Raman or infrared spectral line for an ensemble of tubules of varying diameters. Some features in Fig. 19.52 that are candidates for observed Raman lines include features at 840, 1350, and 1590 cm^{-1}. In addition, the Raman cross sections decrease strongly as the diameter increases,

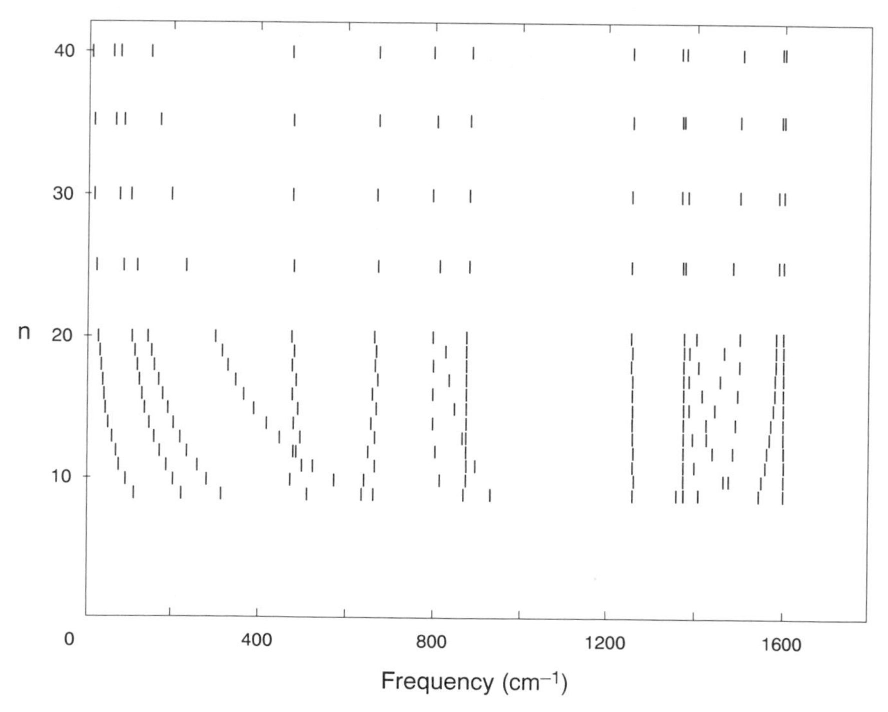

Fig. 19.52. Dependence of Raman-active zigzag $(n, 0)$ nanotube mode frequencies on tubule diameter where $d_t = na_0/\pi$ and $a_0 = 0.246$ nm [19.83, 87].

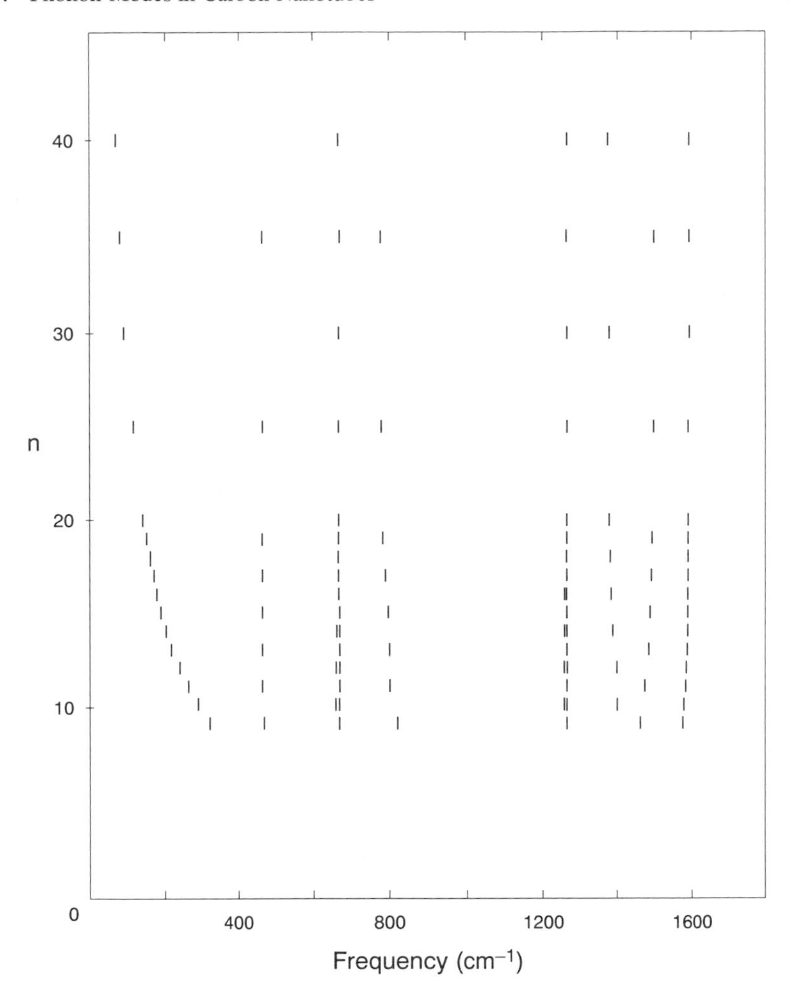

Fig. 19.53. Dependence of infrared-active zigzag $(n, 0)$ nanotube mode frequencies on tubule diameter where $d_t = na_0/\pi$ and $a_0 = 0.246$ nm [19.83, 87].

except for the 1590 cm^{-1} line, which is the strongest spectral feature in a 2D graphene sheet. Likewise, some candidates for observed spectral features in the infrared spectra (see Fig. 19.53) include 470, 670, 1250, and 1590 cm^{-1}. The corresponding results for the dependence of the infrared- and Raman-active mode frequencies of armchair tubules show a wide spread of mode frequencies, as illustrated in Figs. 19.52 and 19.53 for zigzag nanotubes, but with significant differences in the detailed behavior [19.83, 87].

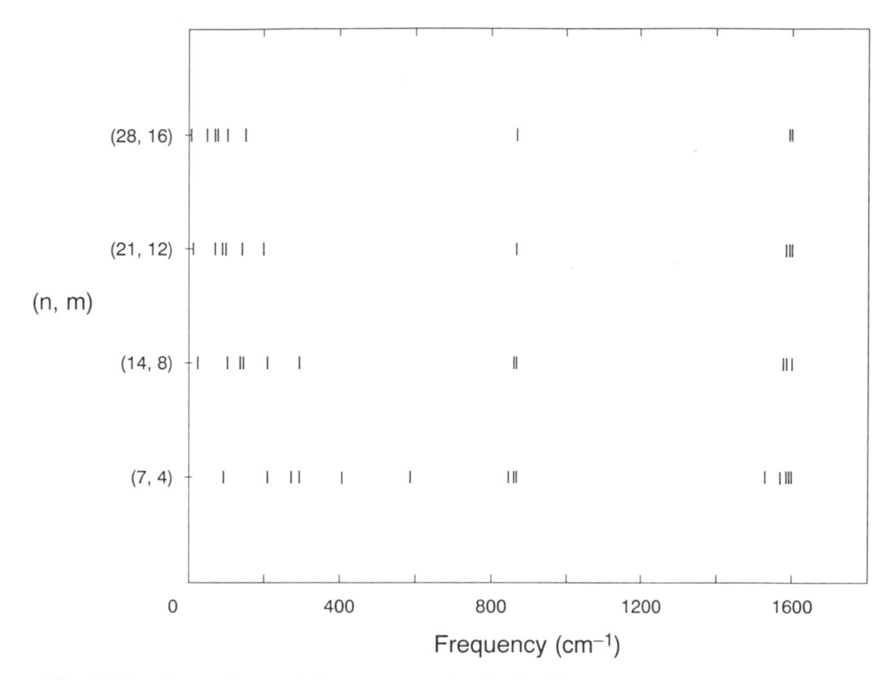

Fig. 19.54. Dependence of the Raman-active (7, 4) chiral nanotube mode frequencies on tubule diameter d_t [19.83, 87], which is found from the (n, m) index pairs by Eq. (19.2). The value of d_t for the $(n, m) = (7, 4)$ nanotube is 7.56 Å, and for $(n, m) = (28, 16)$ is 4(7.56) = 30.24 Å.

Finally, we show in Figs. 19.54 and 19.55 the dependence on tubule diameter d_t of the Raman-active and infrared-active modes for a (7, 4) chiral nanotube. This nanotube belongs to the category $n - m = 3$, which is a conducting nanotube, with $d = 1$, $d_R = 3$, $d_t = 7.56$ Å, and $\theta = \tan^{-1}[2\sqrt{3}/9]$ (see Table 19.2). Of interest is the difference of the spectral frequencies for nanotubes of different (n, m) values and particularly the absence of modes in the 1350 cm^{-1} region in Fig. 19.54.

Although the number of Raman-active and infrared-active modes is basically the same for the symmorphic and nonsymmorphic carbon nanotubes, Figs. 19.54 and 19.55 show more bunching of the optically active mode frequencies in comparison to the behavior for the symmorphic nanotubes shown in Figs. 19.52 and 19.53. This bunching effect arises from the very large real space unit cell of the 1D lattice for the chiral nanotubes, which results in a very small unit cell in reciprocal space. For example, the (7, 4) nanotube has $N = 62$, with 372 phonon branches. The zone folding of the graphene reciprocal space unit cell into the very small reciprocal space unit cell of the nanotube gives rise to many zone foldings so that the frequen-

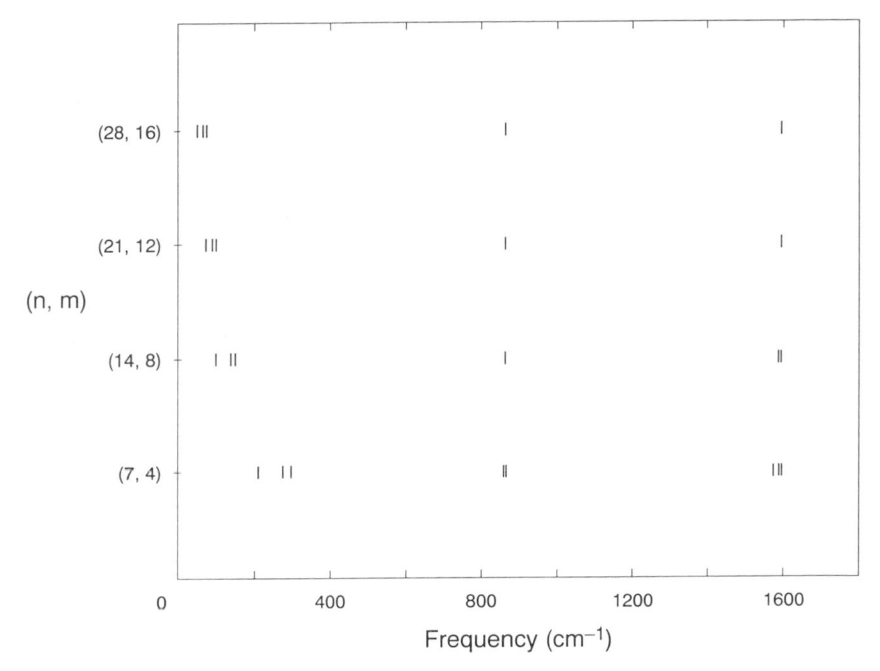

Fig. 19.55. Dependence of the infrared-active (7, 4) chiral nanotube mode frequencies on tubule diameter d_t [19.83, 87], which is found from the (n, m) index pairs by Eq. (19.2). The value of d_t for the $(n, m) = (7, 4)$ nanotube is 7.56 Å.

cies for the nanotube modes rapidly converge to their asymptotic limits for large d_t.

The most significant aspects of these predicted results are the detailed differences in the spectra between tubules of similar diameters but different symmetry, thereby providing, in principle, an independent means (other than electron diffraction characterization by TEM) for yielding information on the chirality of the nanotubes. It is seen in these spectra that certain mode frequencies are strongly dependent on tubule diameter, and others are either weakly dependent or almost independent of d_t. Thus, the Raman and infrared spectra will be dominated by sharp line features corresponding to Raman- or infrared-active modes with frequencies almost independent of d_t, while the weakly d_t-dependent mode frequencies will give rise to broad spectral features of low intensity [19.148].

As d_t becomes large, the mode frequencies for armchair and zigzag tubules approach the frequencies of the 2D graphene modes at the Γ and M points in the 2D Brillouin zone. For chiral tubules, only the Γ point modes are involved. The effects of the tubule geometry on the values of the Raman and infrared frequencies should be observable only for small-diameter (< 40 Å) tubules. It is convenient to refer to the effects related to the small tubule diameters as quantum effects, since they arise from the small number of carbon atoms around the tubule diameter. The Raman and infrared line intensities, furthermore, are expected to decrease, as the diameter increases, for all the modes except for the Raman-active mode at 1590 cm^{-1}, which is the only one which will have nonvanishing intensity in the limit of infinite diameter for single-wall nanotubes. For multiwall nanotubes, an infrared mode near 1590 cm^{-1} should retain significant intensity as well as an infrared mode near 860 cm^{-1} for out-of-plane vibrations as $n \to \infty$. For tubules with small diameters (~20 Å), a broad feature near 1350 cm^{-1} along with a weak peak near 860 cm^{-1} should be observable in the Raman spectra [19.84].

19.7.3. Experiments on Vibrational Spectra of Carbon Nanotubes

A number of experimental studies have been made of the Raman spectra of carbon nanotubes [19.149–151]. No papers have yet been published on the infrared spectra. In most of these Raman studies, the diameters of the nanotubes were sufficiently large that the observed spectra are similar to those from highly disordered graphite, yielding little information of relevance to the expected quantum effects in carbon nanotubes discussed in §19.7.2. Of the various published papers, the work by Holden *et al.* [19.149] provides the most information about quantum effects and for this reason is discussed first.

In the work of Holden *et al.* [19.149], Raman scattering measurements were carried out on carbonaceous material containing nanoscale soot, carbon-coated nanoscale Co particles, and nanotubes generated from a dc arc discharge between carbon electrodes in 300 torr of He [19.149], the conditions used to synthesize single-wall carbon nanotubes [19.31] (see §19.2.2). Raman microprobe measurements were used to obtain the distribution of the various carbon microstructures in the cathode of the carbon arc [19.152].

Of particular interest are the unique features of the Raman spectra obtained from the Co-catalyzed nanotube-containing soot, as shown in Fig. 19.56(e), in comparison with Raman spectra of other carbon materials shown on an expanded scale in Fig. 19.57. The distinctive features of spectrum in Fig. 19.56(e) include two sharp first-order lines at 1566 and 1592 cm^{-1}. Also prominent in the first-order Raman spectrum are a broad band centered at 1341 cm^{-1} and two second-order features at 2681 $cm^{-1} \cong$ 2(1341 cm^{-1}) and 3180 $cm^{-1} \cong$ 2(1592 cm^{-1}). The weak Raman peak near 1460 cm^{-1} [Fig. 19.56(d) and (e)] is identified with fullerene impurities in

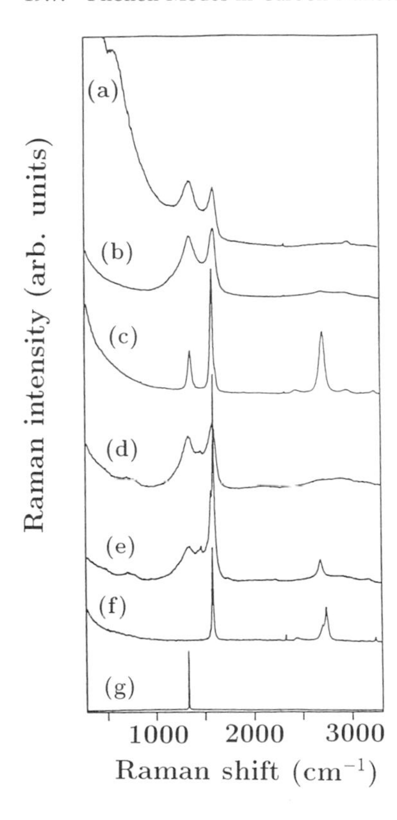

Fig. 19.56. Experimental Raman spectra (T = 300 K) of: (a) glassy carbon (phenol-formaldehyde precursor); (b) nanosoot, as synthesized by laser pyrolysis; (c) nanosoot from (b) after heat treatment at 2820°C; (d) soot obtained by the dc carbon arc method; (e) same as (d) but with Co added to anode; (f) highly oriented pyrolytic graphite (HOPG); and (g) single-crystal diamond (type IIa) [19.149].

the nanotube samples. In comparison, the Raman spectrum [Fig. 19.56(d)] for the dc arc-derived carbons prepared in the same way as that studied in Fig. 19.56(e), except for the absence of the cobalt catalyst, shows no evidence for the sharp 1566 and 1592 cm^{-1} lines, or the sharp second-order features at 2681 cm^{-1} and 3180 cm^{-1}, although other broad features in both spectra are strikingly similar in shape and frequency [19.149].

To show that the sharp features in Fig. 19.56(e) are not associated with ordinary amorphous soot, glassy carbon or graphitic carbons, etc., Raman spectra are included in Fig. 19.56 for a variety of sp^2 and sp^3 solid forms of carbon. These include the Raman spectrum for diamond [Fig. 19.56(g)], shown with its single sharp line at 1335 cm^{-1}, and for highly oriented pyrolytic graphite (HOPG) [Fig. 19.56(f)], showing a sharp first-order line at 1582 cm^{-1} and several features in the second-order Raman spectrum, including one at 3248 cm^{-1}, which is close to 2(1582 cm^{-1}) = 3164 cm^{-1}, but significantly upshifted, due to 3D dispersion of the uppermost phonon branch in graphite. The most prominent feature in the graphite second-

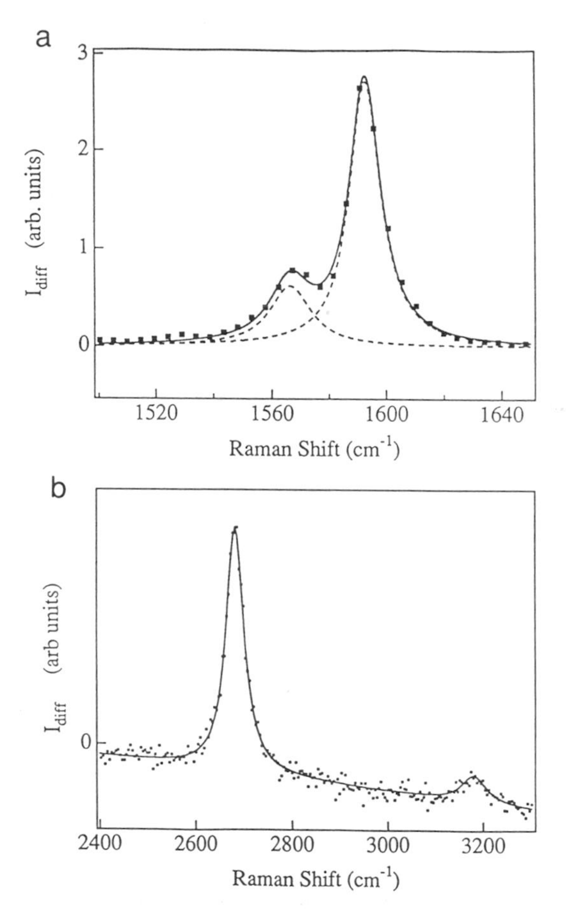

Fig. 19.57. Lorentzian lineshape analysis for (a) the first-order Raman doublet and (b) the second-order lines for the difference spectrum between the room temperature (300 K) Raman spectra of dc arc-derived carbons using 488-nm laser excitation with and without Co catalyst in the anode [19.149].

order spectrum is a peak close to 2(1360 cm^{-1}) = 2730 cm^{-1} with a shoulder at 2698 cm^{-1}, where the lineshape reflects the density of two-phonon states in 3D graphite. Another important benchmark spectrum shown in Fig. 19.56 is that for a typical glassy carbon specimen in Fig. 19.56(a) [19.149] showing a significantly broadened feature at 1600 cm^{-1} and a broad disorder-induced band with a maximum near 1359 cm^{-1}, both arising from a high density of phonon states in their respective spectral ranges. The strongest second-order feature for this glassy carbon sample is located at 2973 cm^{-1}, somewhat upshifted from a combination mode expected in graphite at (1359 + 1600) cm^{-1} = 2959 cm^{-1}, where the Raman intensity near 1600 cm^{-1} is associated with a midzone maximum of the uppermost optical phonon branch which has a Γ-point frequency of 1582 cm^{-1} [19.149].

The Raman spectrum [Fig. 19.56(b)] of the "as-synthesized" carbon nanosoot, prepared by laser pyrolysis of a mixture of benzene, ethylene, and iron carbonyl [19.153], is very similar to that of glassy carbon [Fig. 19.56(a)] and has peaks in the first-order Raman spectrum at 1359 and 1600 cm^{-1} and a broad second-order feature near 2950 cm^{-1}. In addition, the laser pyrolysis–derived carbon nanosoot has weak features in the second-order spectrum at 2711 and 3200 cm^{-1}, similar to HOPG, but appearing much closer to twice the frequency of the first-order lines, namely 2(1359 cm^{-1}) = 2718 cm^{-1} and 2(1600 cm^{-1}) = 3200 cm^{-1}, indicative of a much weaker 3D phonon dispersion in the carbon nanosoot than found in HOPG. Figure 19.56(c) shows the Raman spectrum of the laser pyrolysis–derived carbon black, heat treated to 2820°C, indicating enhanced crystallinity, consistent with a decrease in the intensity of the disorder-induced band at 1360 cm^{-1} [19.149].

The sharp features at 1566 and 1592 cm^{-1} (see Fig. 19.57) can be explained by the frequency-independent Raman-active modes near 1590 cm^{-1} in Figs. 19.52 and 19.54 which appear as a doublet for the pertinent range of tubule diameters. Figure 19.56 further suggests a broad feature near 1370 cm^{-1} and another weaker and broader feature in the 740–840 cm^{-1} range. Because of the reduced dimensionality of the carbon nanotubes, the second-order spectral features are expected to be located at harmonics of the first-order features, consistent with the observations in Fig. 19.57.

Raman spectra have also been reported by Hiura *et al.* [19.150] for the material in the core of their carbon arc cathode deposit, which contained predominantly carbon-coated bundles of nested nanotubes and likely had a different distribution of diameters from the material used for the spectrum in Fig. 19.56(e). Hiura *et al.* identified two Raman lines in their spectrum with these nanotubes: at 1574 (FWHM = 23 cm^{-1}) and at 2687 cm^{-1}. The linewidth of their first-order peak at 1574 cm^{-1} is more than twice as broad as either of the first-order lines in Fig. 19.56(c) and lies between the two sharp first-order lines at 1566 and 1592 cm^{-1} for Co-catalyzed carbons, consistent with a distribution of tube diameters in the tubule bundles. In addition, Hiura reported a wide peak at ~1346 cm^{-1}, close to the disorder-induced line at 1348 cm^{-1} in glassy carbon. While the second-order feature of Hiura *et al.* at 2687 cm^{-1} is slightly broader than, and upshifted by 6 cm^{-1} from, the second-order feature in Figs. 19.56(e) and 19.57, the second-order tube-related features in both spectra [19.149, 150, 154] are significantly downshifted from the corresponding features for other sp^2 carbons, consistent with the very weak dispersion expected for the phonon branches of carbon nanotubes. A weak feature at 2456 cm^{-1} was also reported by Hiura *et al.* [19.150] in their second-order spectrum for the carbon nanotubes, which they also identified in the second-order spectra for HOPG, glassy carbon, and the outer shell material in the carbon arc deposit, corresponding to the large density of phonon states near 860 and 1590 cm^{-1}.

Raman spectra were also reported by Chandrabhas *et al.* [19.151] for the central core deposit from a dc carbon arc discharge, which was characterized by x-ray and TEM measurements, showing tubules with diameters ranging from 15 to 50 nm and inner diameters down to 2 nm. For this tubule diameter distribution, the observed first-order Raman spectra would be expected to be close to that of graphite, and the observations showed first-order features at 49 cm^{-1}, a small feature at 58 cm^{-1}, 470 cm^{-1}, 700 cm^{-1}, 1353 cm^{-1}, and 1583 cm^{-1}, and second-order features at 2455 cm^{-1}, 2709 cm^{-1}, and 3250 cm^{-1}, which were interpreted in terms of disordered graphite, consistent with the x-ray and TEM data, which also indicated the presence of structural defects. The dominant features in the spectra of Chandrabhas *et al.* at 1583 cm^{-1} and 2709 cm^{-1} and a weaker feature at 1353 cm^{-1} are basically in agreement with the observations of Hiura *et al.* [19.150], although all peaks were upshifted relative to the frequencies observed by Hiura *et al.* These differences in behavior may be attributed to different distributions in the diameter and chirality of the nanotubes measured by the two groups [19.150, 151]. Most of the remaining features in the Raman spectrum reported by Chandrabhas *et al.* [19.151] can be identified with graphitic modes or local maxima in the density of states. The features reported by Chandrabhas *et al.* [19.151] are rather different from those reported by Holden *et al.* [19.149], presumably because of differences in the diameter distribution of the nanotube samples. Further progress awaits the preparation, characterization, and spectral measurements on nanotubes with diameters less than ~ 4 nm.

19.8. Elastic Properties

The elastic properties of fullerene tubules have been discussed both theoretically and experimentally. Direct observations, mostly using high-resolution TEM, have shown that small diameter carbon nanotubes are remarkably flexible. As shown in Fig. 19.58, even relatively large diameter ($\sim$10 nm) carbon nanotubes grown from the vapor phase can bend, twist, and kink without fracturing [19.17, 155]. The basic mechanical properties of the nanotubes are very different from those of conventional PAN-based and vapor-grown carbon fibers, which are much more fragile and are easily broken when bent or twisted. It is of interest to note that when bent or twisted, the nanotubes appear to flatten in cross section, especially single wall nanotubes with diameters greater than 2.5 nm [19.155, 156].

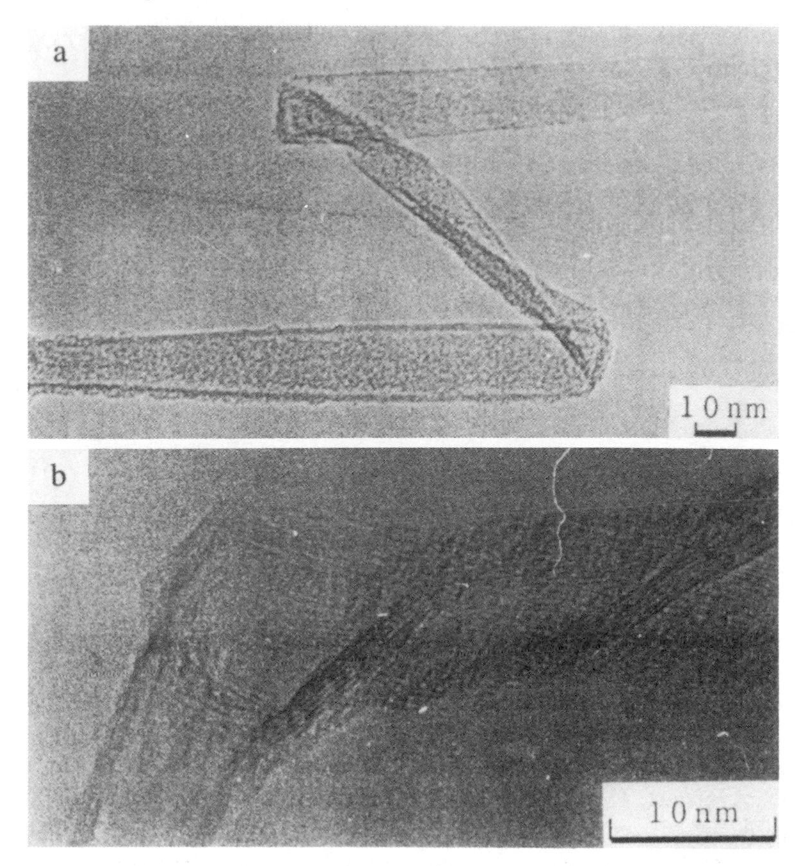

Fig. 19.58. High-resolution TEM images of bent and twisted carbon nanotubes. The length scales for these images are indicated [19.17].

Theoretical studies have focused on several issues, including the effect of curvature, tubule diameter, and chirality on the electronic structure, and the calculation of the bending modes for tubules. Regarding the dependence of the strain energy on the tubule diameter d_t, Mintmire and White [19.102] have shown from continuum elasticity theory that the strain energy per carbon atom σ/N is given by

$$\frac{\sigma}{N} = \frac{Ea^3\mathscr{A}_C}{6d_t^2} \tag{19.45}$$

where E is the elastic modulus, a is a length on the order of the interlayer graphite spacing (~3.35 Å), and $\mathscr{A}_C$ is the area associated with a single carbon atom on the tubule cylinder [19.69]. A plot of the strain energy per carbon atom using this simple model and a more detailed local density

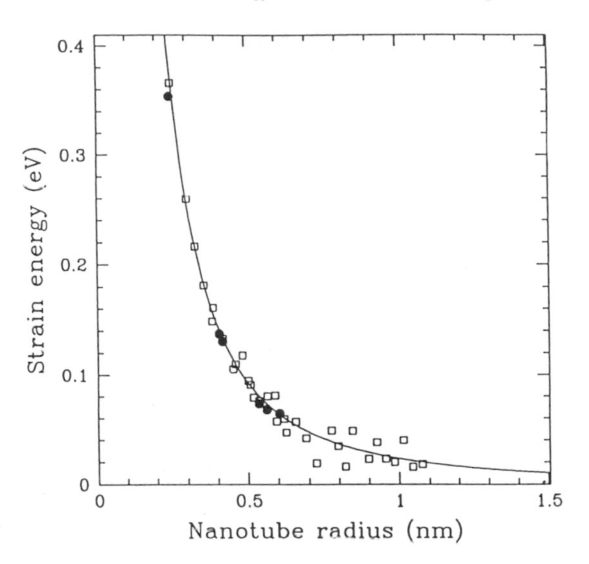

Fig. 19.59. Strain energy per carbon atom as a function of nanotube radius calculated for unoptimized nanotube structures (open squares) and optimized nanotube structures (solid circles). The solid line depicts the inverse square dependence of the strain energy on nanotube radius [see Eq. (19.45)] drawn through the point of smallest radius [19.102].

functional (LDF) calculation is shown in Fig. 19.59. This figure shows good agreement between the simple elastic model and the more detailed LDF calculation and further shows that for $d_t < 2$ nm the effect of strain energy exceeds that of the room temperature thermal energy. Thus it is only at small tubule diameters that the strain energy associated with tubule curvature is important. These calculations show that the strain energy per carbon atom for fullerenes of similar diameter is much larger than for the comparable nanotube. This larger strain energy for the fullerenes reflects their two-dimensional curvature in comparison with tubules which have only one-dimensional curvature. Calculations for the energetics of the stretching and compression of tubules show good agreement with an elastic continuum model based on the elastic constant of C_{11} for graphite [19.69, 157, 158]. It is also found that the tubules get softer with decreasing radius, and the tubule stiffness is found to depend on chiral angle, with the zigzag ($\theta = 0$) tubules having lower stiffness than the armchair ($\theta = 30°$) tubules [19.69].

The elastic continuum model also allows an estimate to be made for the stiffness of nanotubes E based on that for a graphene sheet E_0

$$E = E_0(A_0 - A_1)/A_0 = E_0(r_0^2 - r_1^2)/r_0^2, \tag{19.46}$$

where A_0 and A_1 are, respectively, the cross-sectional areas of the carbon nanotube and of the hole (r_0 and r_1 are the corresponding radii). Estimates for the single-wall nanotube of 1 μm diameter and an estimate of E_0 from C_{11} for graphite (1060 GPa) yields $E \simeq 800$ GPa for such a tubule. The large value for E accounts for the straightness of small diameter nanotubes in TEM micrographs [19.156]. As a consequence of the elastic limit model, the thermal conductivity of carbon nanotubes along the tubule axis is expected to be as high as the best observed for carbon fibers, while the thermal conductivity normal to the tubule axis for a tubule (or bundle of parallel tubules) is expected to be very low. Very low expansion coefficients are also expected tangential to the tubule surface. The absence of dangling bonds and defects on carbon nanotubes implies oxidation resistance, which is of interest for practical applications, including separation of the nanotubes from the soot.

Lattice mode calculations have been carried out on long fullerene tubules based on 30-carbon-atom hemispherical caps from C_{60} and consisting of 100, 200, and 400 carbon atoms for the tubule cylindrical sections [19.50]. These calculations, based on a Keating potential, show that the lowest bending mode frequency decreases as the length of the fullerene tube increases, saturating at a cluster size of about 200 carbon atoms. The magnitude of the beam rigidity of these tubules is found to exceed that of presently available materials [19.50], so that graphene tubules are expected to offer outstanding mechanical properties, consistent with observations shown in Fig. 19.58. Kosakovskaya *et al.* [19.24] have also reported extreme hardness for tows of carbon tubules, exceeding that of the toughest alloys which they used as substrates [19.24].

The effect of van der Waals forces between two adjacent tubules touching each other has been considered, and it has been reported that these intertubule forces distort the cylindrical shape of the adjacent nanotubes [19.159]. Experimental observation of the cross section of carbon nanotubes becoming oval under mechanical pressure have been reported [19.9], which indicates that although nanotubes are expected to be strong, they appear to be less rigid under certain stresses and more flexible than conventional carbon fibers, as is shown in Fig. 19.58. Flexibility in single-wall nanotubes has also been reported [19.160]. In the limit of large diameter, the coaxial carbon nanotubes can be related to vapor-grown carbon fibers [19.32] which possess a similar organizational pattern to the coaxial tubules, have excellent mechanical properties, and do not exhibit the elliptical distortions discussed above [19.60].

Recently, Ajayan and co-workers [19.73] have shown that by embedding carbon nanotubes in a polymer resin, they can line up the nanotubes with diameters ≤ 10 nm by cutting slices of the resin matrix less that 0.3 μm wide. Not only does the cutting process line up the nanotubes parallel to each other, but it also tends to separate the nanotubes from one another. The absence of broken tubules or tubules with distorted cross sections suggests excellent mechanical properties for the carbon tubules.

With regard to the formation of nanotube bundles, which are often observed, Harigaya and Fujita studied the dimerization of tubules [19.161]. The total energy of the tubule dimer depends on the dimerization pattern of the tubules, which is sensitive to their chirality.

19.9. Filled Nanotubes

The possibility of filling the hollow cores of carbon nanotubes with selected metals opens up exciting possibilities with regard to the physics of low-dimensional transport, magnetism, and superconductivity. To exploit the possibility of filling and emptying carbon nanotubes, as would be desirable to modify their 1D properties, the nanocapillarity of carbon nanotubes has been investigated. Simple experiments of putting a water droplet (surface tension ~73 mN/m) on the surface of a carbon nanotube show that the water droplet is readily sucked up the capillary, leaving a dry surface [19.9], consistent with the strong nanocapillary action predicted by Pederson and Broughton [19.162]. In contrast, mercury and lead droplets, which have high surface tensions (>400 mN/m), are not sucked up into the carbon nanotubes [19.9], although Pb in the presence of air is sucked up the capillary [19.64]. In the latter case, the authors claim that the first step in the filling process is the opening of the carbon nanotubes, which can be done by heating the tubules in air (or oxygen at 750°C) [19.63, 64]. It is believed that the opening of carbon nanotubes is a complex process that is aided by the high curvature and high stress at the cap region of the tubule, making these carbon atoms more highly reactive to the ambient oxygen. This enhanced reactivity at high stress regions of the tubule has also been used for tubule thinning [19.63] at elevated temperatures in the presence of oxygen.

To fill the carbon nanotubes it is believed that both capillarity and surface wetting of the nanotubes are needed [19.9]. In the case of the Pb filling experiment [19.163], it is believed that the presence of oxygen and the formation of an oxide or a carbide enhances the wetting of the carbon nanotube by the guest material [19.9]. In Fig. 19.60 we show a TEM image of nested carbon nanotubes containing a short length and a long length of an oxidized Pb "quantum wire" [19.66]. The nanoscale dimensions of the nanotube filling can be appreciated by noting that only about 20 gas molecules would be found in a 10-Å-diameter hollow tube 1μm long under conditions of standard temperature and pressure [19.9].

Nanotube filling studies have also been done on carbon tubules that were previously oxidized to remove their caps and to reduce the number of layers on the tubules, eventually reaching single-wall tubules [19.64]. It was found that tube filling is catalyzed by the presence of oxygen (e.g., Bi could be introduced into carbon tubules in air at ~850°C) [19.64]. It was also

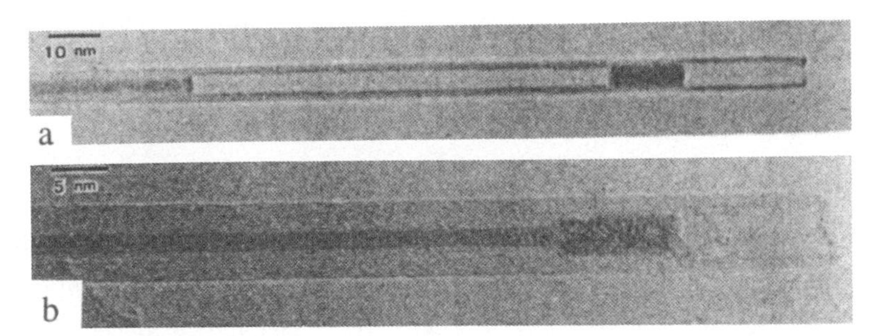

Fig. 19.60. TEM image of nested carbon nanotubes containing a Pb filled core [19.66].

found that when filling was attempted on opened tubules subsequent to the oxidation process, the filling of tubes in vacuum was more difficult. By applying high-intensity electron irradiation to hollow multilayer carbon nanotubes partially coated with Pb metal particles, beam-induced capillary action was observed, whereby columns of metal were sucked up the capillary tubules, partially filling the tubes with columns of high-density metal or metal-derived material [19.2].

Compounds of Gd, Yt, and Mn have also been encapsulated in nanotubes, simply by introducing these metals into the arc electrodes during tubule synthesis [19.164–166]. Using metal packed anodes, 15 metals and/or their carbides were introduced separately into the nanotube core, including Ti, Cr, Fe, Co, Ni, Cu, Zn, Mo, Pd, Sn, Ta, W, Gd, Dy, Yb [19.167]. Two types of nanotube filling are observed (see Fig. 19.60). In the first type, the tube filling in the core is a long wire which extends up to the closed tip of the tubule, and the core has a fixed diameter (Cr, Fe, Co, Ni, Pd, Gd, Dy, Yb). In the second type, much shorter sections of the tube are filled, and variable diameter swelling is often observed (Ti Cu, Zn, Mo, Pd, Sn, Ta, W) [19.167]. Wet chemical techniques have also been used to introduce oxides of Ni, Co, Fe, and U [19.168].

For the most part, electron diffraction shows that the filling is not the pure metal, but a carbide. In some cases, the carbide phase could be identified (Fe_3C, Ni_3C, TiC). Only for Co and Cu was a pure fcc metal filling reported. In one case, a Cr-carbide compound was observed to reside in the tubule core as a long single crystal. It was concluded that the higher the number of holes in the incomplete electronic shell of the metal, the greater the length and quantity of "nanowires" that it produces [19.167]. Other carbides reported to fill the hollow core of carbon nanotubes are LaC_2 [19.81] and yttrium carbide [19.164].

Electron energy loss spectra have been used to identify the foreign elements present in the filled nanotubes and to verify that the guest species is in the form of a carbide, not an oxide or a pure metal. Other useful characterization techniques for the filling material include TEM, x-ray diffraction powder patterns, and energy-dispersive x-ray analysis.

19.10. Onion-Like Graphitic Particles

Also known in the realm of common carbon-based materials are hollow carbon particles ranging from 100 Å to 1 μm in outer diameter and consisting of concentric spherical graphitic shells. These particles are classified under the heading of carbon blacks [19.169–171] (see §2.7).

Ugarte has recently reported [19.30, 172, 173] the formation of hollow concentric carbon spheres which are formed upon intense electron beam irradiation of carbon nanoparticles with faceted shapes (see Fig. 19.61). Of particular interest in the recent work is the report of an innermost sphere with an inner diameter of 7.1 Å, corresponding to the diameter of the C_{60} molecule. It is found that if enough energy is provided, the formation of concentric spherical shell structures is favored over the coaxial tubule structures for small numbers of carbon atoms [19.30, 172]. Using these techniques, spherical shells with diameters up to 100 Å have been synthesized (see Fig. 19.61), similar to the dimensions reported for spherical shells of small-sized carbon blacks. Although containing a large amount of strain energy, the spherical shells contain no dangling bonds and are stable under further electron bombardment, even when containing only a few (e.g., 2–4) spherical shells [19.174, 175]. Ugarte has speculated that the concentric spherical shells (or "onions" as they are often called) constitute another form of carbon, consisting of a fullerene at the center and epitaxial, concentric layers of carbon shells about the central core [19.30, 172]. Onion-like multilayer carbon shells can also be generated by shock wave treatment of carbon soot [19.176], from carbon deposits exposed to a plasma torch [19.177], by laser melting of carbon within a high-pressure (50–300 kbar) cell [19.178], and by annealing nanodiamonds at temperatures in the range 1100 to 1500°C [19.179].

It has been observed by many workers that carbon nanotubes, spherules, and other fullerene-like particles are formed in a carbon arc. The shape of the carbon structure that is formed depends strongly on the growth conditions. The lowest-energy state for large numbers of carbon atoms is that of 3D planar graphite. For structures involving few carbon atoms, the 3D planar graphite structure is not the most stable, because of the large concentration of dangling bonds. In this regime the small closed cage molecules gain stability because of the absence of dangling bonds, despite the high penalty of the curved surfaces and the concomitant strain energy. The stabilization of the concentric stacking of double-layered $C_{60}@C_{240}$ has been

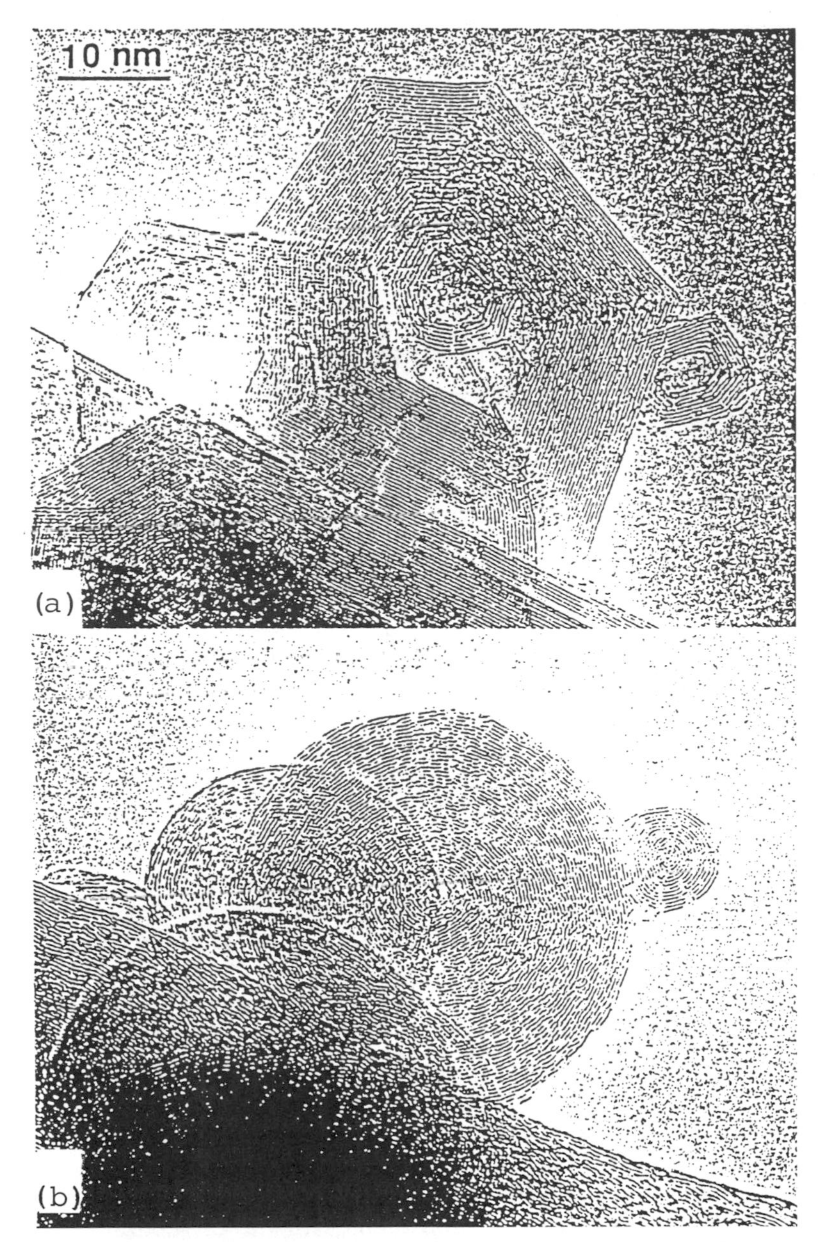

Fig. 19.61. High-resolution electron micrographs of graphitic particles: (a) as obtained from an electric arc deposit, the particles display a well-defined faceted structure and a large inner hollow space, and (b) the same particles after being subjected to intense electron irradiation. The particles now show a spherical shape and a much smaller central empty space [19.173].

calculated [19.78] showing an optimized interlayer separation of 3.52 Å and a stabilization energy of 14 meV per carbon atom. These calculations further show a low barrier (~0.16 meV per carbon atom) for rotations of the two shells with respect to each other, so that relative rotation between the shells should be possible at room temperature [19.10]. Also, a core–shell displacement of 0.4 Å results in an energy increase of 1.5 meV/atom, so that internal oscillations over this displacement should be possible [19.78].

The growth of carbon layers is believed to begin at the surface and progress toward the center [19.169, 172].The onion-like carbon particles are stabilized by the energy gain from the weak van der Waals interaction between adjacent carbon shells [19.175, 180, 181]. *Ab initio* calculations show that for C_{240}, a spherical fullerene is slightly more stable (–7.07 eV/carbon atom) than a polyhedral fullerene (–7.00 eV/carbon atom) [19.182] and that concentric spherical shells of C_{240}@C_{540} are more stable than their concentric icosahedral counterparts [19.173]. The size (number of carbon atoms) at which the total energy of a closed surface particle becomes equal to a planar graphite sheet remains unanswered both theoretically and experimentally. The synthesis and purification of macroscopic quantities of quasispherical onion-like particles with a small size distribution remain a major challenge to studying the properties of carbon onions.

Closely related to studies on the filling of carbon nanotubes (see §19.9) are reports of the encapsulation of single-crystal and amorphous metals, as well as carbide and other phases, all wrapped inside of carbon nanoparticles [19.79, 81, 164, 183–187] . One illustration of this encapsulation is the formation of crystalline α-LaC_2 [19.81] in some of the hollow volume of these carbon nanoparticles. In bulk form α-LaC_2 is metallic. The encapsulated nested material was prepared by introducing La into one of the carbon electrodes used in forming the arc. The encapsulation of the α-LaC_2 within concentric spheres adds to the chemical stability of the encapsulated material.

Carbon-coated YC_2 particles (30–70 nm diameter) [19.183] have also been prepared from a composite rod arc discharge (80% carbon and 20% Y_2O_3) in He gas at 50–100 torr, using 50 A and a potential drop of 25 V between the electrodes. The YC_2 particles showed faceted graphitic faces with an average turbostratic interlayer distance of 0.344 nm, containing crystalline YC_2 with lattice spacings corresponding to the (002) plane. These particles characteristically contain a void volume or cavity in addition to the single-crystal YC_2 material, consistent with the crystallization of the graphitic shell while the Y–C alloy was still molten, until the YC_2 composition was reached, followed finally by crystallization of the YC_2 [19.183]. It was further shown that the graphitic material chemically protects the YC_2 core over long times, although the protective shell could be penetrated by the intervention of an electron beam [19.183]. Wrapping a guest material inside a carbon shell could lead to applications utilizing the core material through its chemical stabilization or by strengthening the shell material for possible use, for example, as a lubricant [19.183].

A similar synthesis process using 20% Fe_2O_3 (weight percent) in the composite arc material led to two kinds of nanoparticles, one which was a faceted graphitic shell containing crystalline α-Fe and a void region, while the other was approximately round in shape and contained crystalline cementite Fe_3C without a void region [19.184]. The particle size for these nanoparticles is in the 20–200 nm range.

19.11. Possible Superconductivity in C_{60}-Related Tubules

Superconductivity in a carbon tubule would be of great interest as an example of a 1D superconductor. However, thus far no reports of superconductivity in carbon tubules have been published. Some thought has, however, been given to the possibility that single tubules or parallel arrays of carbon nanotubes might exhibit superconductivity. Because of the small interplanar distance (~3.4 Å) between these cylindrical tubules, it is not expected that dopants can easily be introduced between the cylindrical layers of multilayered tubules, because of steric considerations. One possibility for the future observation of superconductivity in tubules might involve the introduction of a charge transfer dopant into the hollow core of the cylindrical tubule or of a superconducting metal into the core or in the interstitial space between tubules in a tubule array. It has already been demonstrated that capillary action can cause guest species to be introduced into the hollow core of a carbon nanotube [19.27, 66].

Electronic structure calculations of carbon nanotubes show that, because of periodic boundary conditions in the direction perpendicular to the tubule axis, the energy bands of a two-dimensional graphene sheet can be related to one-dimensional energy bands of carbon nanotubes using zone-folding techniques, where the various 1D bands are specified by a small number of k values arising from the periodic boundary conditions. The density of states for the antibonding π band shows many singularities at the band edges of the one-dimensional energy bands which reflect van Hove singularities for the density of states. The energy position of these singularities is much closer to the Fermi energy compared with the case of the two-dimensional van Hove singularity of a graphene layer. In the case of graphite intercalation compounds, we cannot shift the Fermi energy to that of the two-dimensional van Hove singularity by simply doping the graphite with alkali metals, since this singularity is far (about 3 eV $=\gamma_0$) above the Fermi energy. On the other hand, in the case of carbon nanotubes, the energy position of the singularity depends on the tubule diameter and its axial chirality, so that we can much more easily adjust the Fermi energy near the singularity. It may be possible to observe superconductivity in a solid based on arrays of parallel carbon nanotubes, if a high density of states at the Fermi level could be achieved. The observation of superconductivity in a suitably doped single tubule or an array of tubules presents an exciting possibility for future exploration.

References

[19.1] S. Iijima. *Nature (London)*, **354**, 56 (1991).

[19.2] T. W. Ebbesen and P. M. Ajayan. *Nature (London)*, **358**, 220 (1992).

[19.3] T. W. Ebbesen, H. Hiura, J. Fujita, Y. Ochiai, S. Matsui, and K. Tanigaki. *Chem. Phys. Lett.*, **209**, 83 (1993).

[19.4] M. S. Dresselhaus, G. Dresselhaus, and R. Saito. *Phys. Rev. B*, **45**, 6234 (1992).

[19.5] M. S. Dresselhaus, G. Dresselhaus, and R. Saito. *Carbon*, **33**, 883 (1995).

[19.6] R. Saito, M. Fujita, G. Dresselhaus, and M. S. Dresselhaus. *Appl. Phys. Lett.*, **60**, 2204 (1992).

[19.7] M. Endo, H. Fujiwara, and E. Fukunaga. *Meeting of Japanese Carbon Society*, pp. 34–35 (1991). (unpublished).

[19.8] M. Endo, H. Fujiwara, and E. Fukunaga. *Second C_{60} Symposium in Japan*, pp. 101–104 (1992). (unpublished).

[19.9] T. W. Ebbesen. *Annu. Rev. Mater. Sci.*, **24**, 235 (1994).

[19.10] J. C. Charlier and J. P. Michenaud. *Phys. Rev. Lett.*, **70**, 1858 (1993).

[19.11] J. C. Charlier. *Carbon Nanotubes and Fullerenes*. Ph.D. thesis, Catholic University of Louvain, Louvain-la-Neuve, Belgium (1994). Department of Physics of Materials.

[19.12] M. Ge and K. Sattler. *Science*, **260**, 515 (1993).

[19.13] M. J. Gallagher, D. Chen, B. P. Jakobsen, D. Sarid, L. D. Lamb, F. A. Tinker, J. Jiao, D. R. Huffman, S. Seraphin, and D. Zhou. *Surf. Sci. Lett*, **281**, L335 (1993).

[19.14] Z. Zhang and C. M. Lieber. *Appl. Phys. Lett.*, **62**, 2792 (1993).

[19.15] K. Sattler. *Carbon*, **33**, 915 (1995).

[19.16] R. Hoeper, R. K. Workman, D. Chen, D. Sarid, T. Yadav, J. C. Withers, and R. O. Loutfy. *Surf. Sci.*, **311**, L371 (1994).

[19.17] M. Endo, K. Takeuchi, S. Igarashi, K. Kobori, M. Shiraishi, and H. W. Kroto. *J. Phys. Chem. Solids*, **54**, 1841 (1993).

[19.18] M. Endo. Ph.D. thesis, University of Orleans, Orleans, France (1975). (in French).

[19.19] M. Endo. Ph.D. thesis, Nagoya University, Japan (1978). (in Japanese).

[19.20] A. Oberlin, M. Endo, and T. Koyama. *Carbon*, **14**, 133 (1976).

[19.21] A. Oberlin, M. Endo, and T. Koyama. *J. Cryst. Growth*, **32**, 335 (1976).

[19.22] H. W. Kroto, J. R. Heath, S. C. O'Brien, R. F. Curl, and R. E. Smalley. *Nature (London)*, **318**, 162 (1985).

[19.23] W. Krätschmer, L. D. Lamb, K. Fostiropoulos, and D. R. Huffman. *Nature (London)*, **347**, 354 (1990).

[19.24] Z. Y. Kosakovskaya, L. A. Chernozatonskii, and E. A. Fedorov. *JETP Lett. (Pis'ma Zh. Eksp. Teor.)*, **56**, 26 (1992).

[19.25] E. G. Gal'pern, I. V. Stankevich, A. L. Christyakov, and L. A. Chernozatonskii. *JETP Lett. (Pis'ma Zh. Eksp. Teor.)*, **55**, 483 (1992).

[19.26] S. Wang and P. R. Buseck. *Chem. Phys. Lett.*, **182**, 1 (1991).

[19.27] P. M. Ajayan and S. Iijima. *Nature (London)*, **358**, 23 (1992).
[19.28] M. S. Dresselhaus. *Nature (London)*, **358**, 195 (1992).
[19.29] R. Bacon. *J. Appl. Phys.*, **31**, 283 (1960).
[19.30] D. Ugarte. *Nature (London)*, **359**, 707 (1992). see also: ibid H. W. Kroto, p.670.
[19.31] D. S. Bethune, C. H. Kiang, M. S. de Vries, G. Gorman, R. Savoy, J. Vazquez, and R. Beyers. *Nature (London)*, **363**, 605 (1993).
[19.32] M. Endo. *CHEMTECH*, **18**, 568 (1988).
[19.33] S. Iijima and T. Ichihashi. *Nature (London)*, **363**, 603 (1993).
[19.34] M. Fujita, R. Saito, G. Dresselhaus, and M. S. Dresselhaus. *Phys. Rev. B*, **45**, 13834 (1992).
[19.35] R. Saito, M. Fujita, G. Dresselhaus, and M. S. Dresselhaus. *Materials Science and Engineering*, **B19**, 185 (1993).
[19.36] P. W. Fowler, D. E. Manolopoulos, and R. P. Ryan. *Carbon*, **30**, 1235 (1992).
[19.37] S. Iijima. *Mater. Sci. Eng.*, **B19**, 172 (1993).
[19.38] V. P. Dravid, X. Lin, Y. Wang, X. K. Wang, A. Yee, J. B. Ketterson, and R. P. H. Chang. *Science*, **259**, 1601 (1993).
[19.39] M. Endo, K. Takeuchi, K. Kobori, and S. Igarashi. *Hyomen*, **32**, 277 (1994). (In Japanese.)
[19.40] B. I. Dunlap. *Phys. Rev. B*, **46**, 1938 (1992).
[19.41] X. F. Zhang, X. B. Zhang, G. Van Tendeloo, S. Amelinckx, M. Op de Beeck, and J. Van Landuyt. *J. Cryst. Growth*, **130**, 368 (1993).
[19.42] S. Amelinckx, D. Bernaerts, X. B. Zhang, G. V. Tendeloo, and J. V. Landuyt. *Science*, **267**, 1334 (1995).
[19.43] J. Xhie, K. Sattler, N. Venkatasmaran, and M. Ge. *Phys. Rev. B*, **47**, 8917 (1994).
[19.44] J. Xhie, K. Sattler, N. Venkatasmaran, and M. Ge. *Phys. Rev. B*, **47**, 15835 (1993).
[19.45] D. Tománek, S. G. Louie, H. J. Mamin, D. W. Abraham, R. E. Thomson, E. Ganz, and J. Clarke. *Phys. Rev. B*, **35**, 7790 (1987).
[19.46] R. Saito, M. Fujita, G. Dresselhaus, and M. S. Dresselhaus. *Phys. Rev. B*, **46**, 1804 (1992).
[19.47] N. Hamada, S.-I. Sawada, and A. Oshiyama. *Phys. Rev. Lett.*, **68**, 1579 (1992).
[19.48] M. Liu and J. M. Cowley. *Proceedings of the Meeting of the Microscopy Society of America*, p. 752, San Francisco Press (1993).
[19.49] T. C. Chieu, G. Timp, M. S. Dresselhaus, M. Endo, and A. W. Moore. *Phys. Rev. B*, **27**, 3686 (1983).
[19.50] G. Overney, W. Zhong, and D. Tománek. *Z. Phys. D*, **27**, 93 (1993).
[19.51] S. Iijima, T. Ichihashi, and Y. Ando. *Nature (London)*, **356**, 776 (1992).
[19.52] M. Ge and K. Sattler. *Chem. Phys. Lett.*, **230**, 792 (1994).
[19.53] M. Ge and K. Sattler. In C. L. Renchler (ed.), *Novel Forms of Carbon II, MRS Symposia Proceedings*, vol. 349, pp. 313–318, Pittsburgh, PA (1994). Materials Research Society Press.
[19.54] J. B. Howard, D. Chowdhary, and J. B. Vander Sande. *Nature (London)* **370**, 603 (1994).
[19.55] J. Ahrens, M. Bachmann, T. Baum, and J. Griesheimer. *International Journal of Mass Spectrometry and Ion Processes*, **138**, 133 (1994).
[19.56] S. Seraphin, D. Zhou, J. Jiao, J. C. Withers, and R. Loufty. *Carbon*, **31**, 685 (1993).
[19.57] C. H. Kiang, W. A. Goddard, III, R. Beyers, J. R. Salem, and D. S. Bethune. *J. Phys. Chem.*, **98**, 6612 (1994).
[19.58] S. Seraphin and D. Zhou. *Appl. Phys. Lett.*, **64**, 2087 (1994).
[19.59] C. H. Kiang, W. A. Goddard III, R. Beyers, and D. S. Bethune. *Carbon*, **33**, 903 (1995).

[19.60] M. S. Dresselhaus, G. Dresselhaus, K. Sugihara, I. L. Spain, and H. A. Goldberg. *Graphite Fibers and Filaments*, vol. 5 of *Springer Series in Materials Science*. Springer-Verlag, Berlin (1988).
[19.61] J. J. Cuomo and J. M. E. Harper. *IBM Tech. Disclosure Bulletin*, **20**, 775 (1977).
[19.62] J. A. Floro, S. M. Rossnagel, and R. S. Robinson. *J. Vac. Sci. Technol.*, **A1**, 1398 (1983).
[19.63] S. C. Tsang, P. J. F. Harris, and M. L. H. Green. *Nature (London)*, **362**, 520 (1993).
[19.64] P. M. Ajayan, T. W. Ebbesen, T. Ichihashi, S. Iijima, K. Tanigaki, and H. Hiura. *Nature (London)*, **362**, 522 (1993).
[19.65] T. W. Ebbesen, P. M. Ajayan, H. Hiura, and K. Tanigaki. *Nature (London)*, **367**, 519 (1994).
[19.66] P. M. Ajayan and S. Iijima. *Nature (London)*, **361**, 333 (1993).
[19.67] R. E. Smalley. *Mater. Sci. Eng.*, **B19**, 1 (1993).
[19.68] G. B. Adams, O. F. Sankey, J. B. Page, M. O'Keefe, and D. A. Drabold. *Science*, **256**, 1792 (1992).
[19.69] D. H. Robertson, D. W. Brenner, and J. W. Mintmire. *Phys. Rev. B*, **45**, 12592 (1992).
[19.70] D. H. Robertson, D. W. Brenner, and C. T. White. *J. Phys. Chem.*, **96**, 6133 (1992).
[19.71] S. I. Sawada and N. Hamada. *Solid State Commun.*, **83**, 917 (1992).
[19.72] A. A. Lucas, P. Lambin, and R. E. Smalley. *J. Phys. Chem. Solids*, **54**, 587 (1993).
[19.73] P. M. Ajayan, O. Stephan, C. Colliex, and D. Trauth. *Science*, **265**, 1212 (1994).
[19.74] W. A. de Heer, W. S. Bacsa, A. Chatelain, T. Gerfin, R. Humphrey-Baker, L. Forro, and D. Ugarte. *Science*, **268**, 845 (1995).
[19.75] M. Endo and H. W. Kroto. *J. Phys. Chem.*, **96**, 6941 (1992).
[19.76] R. E. Smalley. *Acc. Chem. Res.*, **25**, 98 (1992).
[19.77] S. Iijima. private communication.
[19.78] Y. Yosida. *Fullerene Sci. Tech.*, **1**, 55 (1993).
[19.79] M. Tomita, Y. Saito, and T. Hayashi. *Jpn. J. Appl. Phys.*, **32**, L280 (1993).
[19.80] X. X. Bi, B. Ganguly, G. P. Huffman, F. E. Huggins, M. Endo, and P. C. Eklund. *J. Mater. Res.*, **8**, 1666 (1993).
[19.81] R. S. Ruoff, D. C. Lorents, B. Chan, R. Malhotra, and S. Subramoney. *Science*, **259**, 346 (1993).
[19.82] R. A. Jishi, M. S. Dresselhaus, and G. Dresselhaus. *Phys. Rev. B*, **47**, 16671 (1993).
[19.83] R. A. Jishi, L. Venkataraman, M. S. Dresselhaus, and G. Dresselhaus. *Phys. Rev. B*, **51**, 11176 (1995).
[19.84] R. A. Jishi, L. Venkataraman, M. S. Dresselhaus, and G. Dresselhaus. *Chem. Phys. Lett.*, **209**, 77 (1993).
[19.85] M. Tinkham. *Group Theory and Quantum Mechanics*. McGraw-Hill, New York (1964).
[19.86] C. T. White, D. H. Roberston, and J. W. Mintmire. *Phys. Rev. B*, **47**, 5485 (1993).
[19.87] R. A. Jishi, D. Inomata, K. Nakao, M. S. Dresselhaus, and G. Dresselhaus. *J. Phys. Soc. Jpn.*, **63**, 2252 (1994).
[19.88] R. Saito, G. Dresselhaus, and M. S. Dresselhaus. *Chem. Phys. Lett.*, **195**, 537 (1992).
[19.89] J. W. Mintmire, B. I. Dunlap, and C. T. White. *Phys. Rev. Lett.*, **68**, 631 (1992).
[19.90] K. Harigaya. *Chem. Phys. Lett.*, **189**, 79 (1992).
[19.91] K. Tanaka, A. A. Zakhidov, K. Yoshizawa, K. Okahara, T. Yamabe, K. Yakushi, K. Kikuchi, S. Suzuki, I. Ikemoto, and Y. Achiba. *Phys. Lett.*, **A164**, 221 (1992).
[19.92] J. W. Mintmire, D. H. Robertson, and C. T. White. *J. Phys. Chem. Solids*, **54**, 1835 (1993).
[19.93] P. W. Fowler. *J. Phys. Chem. Solids*, **54**, 1825 (1993).
[19.94] P. R. Wallace. *Phys. Rev.*, **71**, 622 (1947).

[19.95] M. S. Dresselhaus and G. Dresselhaus. *Advances in Phys.*, **30**, 139 (1981).

[19.96] M. S. Dresselhaus, G. Dresselhaus, R. Saito, and P. C. Eklund. In J. L. Birman, C. Sébenne, and R. F. Wallis (eds), *Elementary Excitations in Solids*, chap. 18, p. 387. Elsevier Science Publishers, B.V. 1992, New York (1992).

[19.97] R. Saito, M. Fujita, G. Dresselhaus, and M. S. Dresselhaus. In L. Y. Chiang, A. F. Garito, and D. J. Sandman (eds.), *Electrical, Optical and Magnetic Properties of Organic Solid State Materials, MRS Symposia Proceedings, Boston*, vol. 247, p. 333, Pittsburgh, PA (1992). Materials Research Society Press.

[19.98] R. Saito, G. Dresselhaus, and M. S. Dresselhaus. *J. Appl. Phys.*, **73**, 494 (1993).

[19.99] G. S. Painter and D. E. Ellis. *Phys. Rev. B*, **1**, 4747 (1970).

[19.100] M. S. Dresselhaus, G. Dresselhaus, and R. Saito. *Solid State Commun.*, **84**, 201 (1992).

[19.101] M. S. Dresselhaus, R. A. Jishi, G. Dresselhaus, D. Inomata, K. Nakao, and R. Saito. *Molecular Materials*, **4**, 27 (1994).

[19.102] J. W. Mintmire and C. T. White. *Carbon*, **33**, 893 (1995).

[19.103] M. L. Elert, C. T. White, and J. W. Mintmire. *Mol. Cryst. Liq. Cryst.*, **125**, 329 (1985).

[19.104] C. T. White, D. H. Roberston, and J. W. Mintmire. *Phys. Rev. B*, **47**, 5489 (1993).

[19.105] X. Blase, L. X. Benedict, E. L. Shirley, and S. G. Louie. *Phys. Rev. Lett.*, **72**, 1878 (1994).

[19.106] J. C. Charlier, X. Gonze, and J. P. Michenaud. *Europhys. Lett.*, **29**, 43 (1995).

[19.107] J. C. Charlier, X. Gonze, and J. P. Michenaud. *Carbon*, **32**, 289 (1994).

[19.108] S. G. Louie. In H. Kuzmany, J. Fink, M. Mehring, and S. Roth (eds.), *Progress in Fullerene Research: International Winter School on Electronic Properties of Novel Materials*, p. 303 (1994). Kirchberg Winter School, World Scientific Publishing, Singapore.

[19.109] X. Blase, A. Rubio, S. G. Louie, and M. L. Cohen. *Phys. Rev. B*, **51**, 6868 (1995).

[19.110] X. Blase, A. Rubio, S. G. Louie and M. L. Cohen. *Europhys. Lett.*, **28**, 335 (1994).

[19.111] Z. Weng-Sieh, K. Cherrey, N. G. Chopra, X. Blase, Y. Miyamoto, A. Rubio, M. L. Cohen, S. G. Louie, A. Zettl, and R. Gronsky. *Phys. Rev. B*, **51**, 11229 (1995).

[19.112] P. Lambin, L. Philippe, J. C. Charlier, and J. P. Michenaud. *Computational Mater. Sci.*, **2**, 350 (1994).

[19.113] P. Lambin, J. C. Charlier, and J. P. Michenaud. In H. Kuzmany, J. Fink, M. Mehring, and S. Roth (eds.), *Proceedings of the Winter School on Fullerenes*, pp. 130–134 (1994). Kirchberg Winter School, World Scientific Publishing, Singapore.

[19.114] M. S. Dresselhaus, G. Dresselhaus, and R. Saito. *Mater. Sci. Eng.*, **B19**, 122 (1993).

[19.115] K. Harigaya. *Phys. Rev. B*, **45**, 12071 (1992).

[19.116] J. Y. Yi and J. Bernholc. *Phys. Rev. B*, **47**, 1708 (1993).

[19.117] H. Fröhlich. *Phys. Rev.*, **79**, 845 (1950).

[19.118] H. Fröhlich. *Proc. Roy. Soc. London*, **A215**, 291 (1952).

[19.119] H. Ajiki and T. Ando. *J. Phys. Soc. Jpn.*, **62**, 2470 (1993), Erratum, ibid p. 4267.

[19.120] R. Saito, G. Dresselhaus, and M. S. Dresselhaus. *Phys. Rev. B*, **50**, 14698 (1994).

[19.121] L. Langer, L. Stockman, J. P. Heremans, V. Bayot, C. H. Olk, C. Van Haesendonck, Y. Bruynseraede, and J. P. Issi. *J. Mat. Res.*, **9**, 927 (1994).

[19.122] J. Heremans, C. H. Olk, and D. T. Morelli. *Phys. Rev. B*, **49**, 15122 (1994).

[19.123] X. K. Wang, R. P. H. Chang, A. Patashinski, and J. B. Ketterson. *J. Mater. Res.*, **9**, 1578 (1994).

[19.124] J.-P. Issi, L. Langer, J. Heremans, and C. H. Olk. *Carbon*, **33**, 941 (1995).

[19.125] D. R. Hofstadter. *Phys. Rev. B*, **14**, 2239 (1976).

[19.126] F. A. Butler and E. Brown. *Phys. Rev.*, **166**, 630 (1968).

[19.127] M. Kosaka, T. W. Ebbesen, H. Hiura, and K. Tankgaki. *Chem. Phys. Lett.*, **225**, 161 (1994).
[19.128] S. Wang and D. Zhou. *Chem. Phys. Lett.*, **225**, 165 (1994).
[19.129] C. H. Olk and J. P. Heremans. *J. Mater. Res.*, **9**, 259 (1994).
[19.130] C. A. Klein. *J. Appl. Phys.*, **33**, 3388 (1962).
[19.131] C. A. Klein. *J. Appl. Phys.*, **35**, 2947 (1964).
[19.132] C. A. Klein. In P. L. Walker, Jr. (ed.), *Chemistry and Physics of Carbon*, vol. 2, p. 217. Marcel Dekker, Inc., New York (1966).
[19.133] L. Langer, L. Stockman, J. P. Heremans, V. Bayot, C. H. Olk, C. Van Haesendonck, Y. Bruynseraede, and J. P. Issi. (private communcation).
[19.134] S. N. Song, X. K. Wang, R. P. H. Chang, and J. B. Ketterson. *Phys. Rev. Lett.*, **72**, 697 (1994).
[19.135] P. A. Lee and T. V. Ramakrishnan. *Rev. Mod. Phys.*, **57**, 287 (1985).
[19.136] V. Bayot, L. Piraux, J. P. Michenaud, and J. P. Issi. *Phys. Rev. B*, **40**, 3514 (1989).
[19.137] L. Langer, V. Bayot, J. P. Issi, L. Stockman, C. Van Haesendonck, Y. Bruynseraede, J. P. Heremans, and C. H. Olk. *Extended Abstract of the 22nd Biennial Carbon Conference*, San Diego, CA (1995), p. 348.
[19.138] A. D. Stone. *Phys. Rev. Lett.*, **54**, 2692 (1985).
[19.139] B. L. Altshuler and A. G. Aronov. In A. L. Efros and M. Pollack (eds.), *Electron-electron interactions in disordered systems*, North Holland, Amsterdam (1985).
[19.140] J. W. McClure. *Phys. Rev.*, **119**, 606 (1960).
[19.141] N. Ganguli and K. S. Krishnan. *Proc. Roy. Soc. (London)*, **A177**, 168 (1941).
[19.142] O. Chauvet, L. Forro, W. Bacsa, D. Ugarte, B. Doudin, and W. A. de Heer. *Phys. Rev. B*, **52**, R6963 (1995).
[19.143] P. M. Ajayan, S. Iijima, and T. Ichihashi. *Phys. Rev. B*, **47**, 6859 (1993).
[19.144] K. K. Mani and R. Ramani. *Phys. Status Solidi B*, **61**, 659 (1974).
[19.145] R. Al-Jishi and G. Dresselhaus. *Phys. Rev. B*, **26**, 4514 (1982).
[19.146] P. Lespade, R. Al-Jishi, and M. S. Dresselhaus. *Carbon*, **20**, 427 (1982).
[19.147] R. Nicklow, N. Wakabayashi, and H. G. Smith. *Phys. Rev. B*, **5**, 4951 (1972).
[19.148] R. A. Jishi. (private communication).
[19.149] J. M. Holden, P. Zhou, X.-X. Bi, P. C. Eklund, S. Bandow, R. A. Jishi, K. D. Chowdhury, G. Dresselhaus, and M. S. Dresselhaus. *Chem. Phys. Lett.*, **220**, 186 (1994).
[19.150] H. Hiura, T. W. Ebbesen, K. Tanigaki, and H. Takahashi. *Chem. Phys. Lett.*, **202**, 509 (1993).
[19.151] N. Chandrabhas, A. K. Sood, D. Sundararaman, S. Raju, V. S. Raghunathan, G. V. N. Rao, V. S. Satry, T. S. Radhakrishnan, Y. Hariharan, A. Bharathi, and C. S. Sundar. *PRAMA-Journal of Phys.*, **42**, 375 (1994).
[19.152] M. V. Ellacott, L. S. K. Pang, L. Prochazka, M. A. Wilson, J. D. F. Gerald, and G. H. Taylor. *Carbon*, **32**, 542 (1994).
[19.153] X.-X. Bi, M. Jagtoyen, F. J. Derbyshire, P. C. Eklund, M. Endo, K. D. Chowdhury, and M. S. Dresselhaus. *J. Mat. Res.* (1995) (in press).
[19.154] J. Kastner, T. Pichler, H. Kuzmany, S. Curran, W. Blau, D. N. Weldon, M. Dlamesiere, S. Draper, and H. Zandbergen. *Chem. Phys. Lett.*, **221**, 53 (1994).
[19.155] J. Tersoff and R. S. Ruoff. *Phys. Rev. Lett.*, **73**, 676 (1994).
[19.156] R. S. Ruoff and D. C. Lorents. *Carbon*, **33**, 925 (1995).
[19.157] B. T. Kelly. *Physics of Graphite*. Applied Science (London) (1981).
[19.158] J. Tersoff. *Phys. Rev. B*, **46**, 15546 (1992).
[19.159] R. S. Ruoff, J. Tersoff, D. C. Lorents, S. Subramoney, and B. Chan. *Nature (London)*, **364**, 514 (1993).
[19.160] D. S. Bethune (private communication).

[19.161] K. Harigaya and S. Abe. *Chem. Phys. Lett.* (in press).

[19.162] M. R. Pederson and J. Q. Broughton. *Phys. Rev. Lett.*, **69**, 2689 (1992).

[19.163] T. W. Ebbesen. In P. Jena, S. N. Khanna, B. K. Rao, *et al.* (eds.), *Physics and Chemistry of Finite Systems: From Clusters to Crystals*, vol. 374, Dordrecht (1994). Kluwer Academic Publishers. NATO ASI Series C: Mathematical and Physical Sciences.

[19.164] S. Seraphin, D. Zhou, J. Jiao, J. C. Withers, and R. Loufty. *Nature (London)*, **362**, 503 (1993).

[19.165] P. M. Ajayan, C. Colliex, J. M. Lambert, P. Bernier, and L. Barbedette. *Phys. Rev. Lett.*, **72**, 1722 (1994).

[19.166] S. Subramoney, R. S. Ruoff, D. C. Lorents, B. Chan, R. Malhotra, M. J. Dyer, and K. Parvin. *Carbon*, **32**, 507 (1994).

[19.167] C. Guerret-Piecourt, Y. L. Bouar, A. Loiseau, and H. Pascard. *Nature (London)*, **372**, 159 (1994).

[19.168] S. C. Tsang, Y. K. Chen, P. J. F. Harris, and M. L. H. Green. *Nature (London)*, **372**, 159 (1994).

[19.169] J. S. Speck. *J. Appl. Phys.*, **67**, 495 (1990).

[19.170] E. A. Kmetko. In S. Mrozowski and L. W. Phillips (eds.), *Proceedings of the First and Second Conference on Carbon*, p. 21. Waverly Press, Buffalo, New York (1956).

[19.171] R. D. Heidenreich, W. M. Hess, and L. L. Ban. *J. Appl. Crystallogr.*, **1**, 1 (1968).

[19.172] D. Ugarte. *Chem. Phys. Lett.*, **207**, 473 (1993).

[19.173] D. Ugarte. *Carbon*, **33**, 989 (1995).

[19.174] H. Kroto. *Nature (London)*, **359**, 670 (1992).

[19.175] D. Ugarte. *Europhys. Lett.*, **22**, 45 (1993).

[19.176] K. Yamada, H. Kunishige, and A. B. Sowaoka. *Naturwissenschaften*, **78**, 450 (1991).

[19.177] N. Hatta and K. Murata. *Chem. Phys. Lett.*, **217**, 398 (1994).

[19.178] L. S. Weathers and W. A. Basset. *Phys. Chem. Minerals*, **15**, 105 (1987).

[19.179] V. L. Kuznetsov, A. L. Chuvilin, Y. V. Butenko, I. Y. Mal'kov, and V. M. Titov. *Chem. Phys. Lett.*, **222**, 343 (1994).

[19.180] A. Maiti, J. Bravbec, and J. Bernholc. *Phys. Rev. Lett.*, **70**, 3023 (1993).

[19.181] D. Tománek, W. Zhong, and E. Krastev. *Phys. Rev. B*, **48**, 15461 (1993).

[19.182] D. York, J. P. Lu, and W. Yang. *Phys. Rev. B*, **49**, 8526 (1994).

[19.183] Y. Saito, T. Yoshikawa, M. Okuda, M. Ohkohchi, Y. Ando, A. Kasuya, and Y. Nishina. *Chem. Phys. Lett.*, **209**, 72 (1994).

[19.184] Y. Saito, T. Yoshikawa, M. Okuda, N. Fujimoto, S. Yamamuro, K. Wakoh, K. Sumiyama, K. Suzuki, A. Kasuya, and Y. Nishina. *Chem. Phys. Lett.*, **212**, 379 (1993).

[19.185] Y. Saito. *Carbon*, **33**, 979 (1995).

[19.186] Y. Saito, T. Yoshikawa, M. Okuda, N. Fujimoto, S. Yamamuro, K. Wakoh, K. Sumiyama, K. Suzuki, A. Kasuya, and Y. Nishina. *Chem. Phys. Lett.*, **212**, 379 (1993).

[19.187] L. Wang and J. M. Cowley. *Ultramicroscopy*, **55**, 228 (1994).

CHAPTER 20

Applications of Carbon Nanostructures

Although research on solid C_{60} and related materials is still at an early stage, these materials are already beginning to show many exceptional properties, some of which may lead to practical applications. Small-scale applications will come first; then targeted applications utilizing the special properties of fullerenes are expected to follow [20.1, 2].

Many of the fullerene applications that have been identified thus far have been related to the unusual properties of fullerenes discussed earlier in this book, such as the nonlinear processes associated with the excited molecular triplet states (§13.7), the charge transfer between certain polymers and C_{60} (§13.5), phototransformation properties (§13.3.4), and the strong bonding between C_{60} and metal (or Si) surfaces (§17.14). This chapter first covers applications associated with optical excitations (§20.1) because they are so pervasive, followed by applications to electronics (§20.2), materials and electrochemical applications (§20.3 and §20.4), and finally miscellaneous other applications including nanotechnology, coatings, free-standing membranes, tribology, separations, and sensors (§20.5). The concluding section (§20.6) gives a summary of patents and efforts to commercialize the applications previously discussed in the chapter.

20.1. Optical Applications

Several of the unusual optical properties exhibited by fullerenes show promise for applications. The nonlinear properties of optical absorption in the excited triplet state give rise to optical limiting devices (§20.1.1) and photorefractive devices (§20.1.3), while photoinduced charge transfer between polymer and C_{60} constituents in C_{60}–polymer composites gives rise to photoconductivity applications (§20.1.2), photodiodes (§20.2.2), and photovoltaic devices (§20.2.4).

20.1.1. Optical Limiter

One promising application for fullerenes is as an optical limiter because of the large magnitude of the nonlinear third-order electric susceptibility and nonlinear optical response, which in turn arise from basic optical properties associated with the interplay between allowed and forbidden transitions and between singlet and triplet state transitions [20.3] (see §13.7). Optical limiters are used to protect materials from damage by intense incident light pulses via a saturation of the transmitted light intensity with increasing exposure time and/or incident light intensity.

Outstanding performance for C_{60} relative to presently used optical limiting materials has been observed at 5320 Å for 8 ns pulses (see Fig. 20.1) using solutions of C_{60} in toluene and in chloroform (CH_3Cl) [20.5]. Although C_{70} in similar solutions also shows optical limiting action, the performance of C_{60} was found to be superior. It was found, in fact, that more than 100 photons per C_{60} molecule could be absorbed repetitively in a single nanosecond optical pulse (at a wavelength of 532 nm) [20.6].

The proposed mechanism for the optical limiting is that C_{60} and C_{70} are more absorptive in the infrared (IR) and visible regions of the spectrum

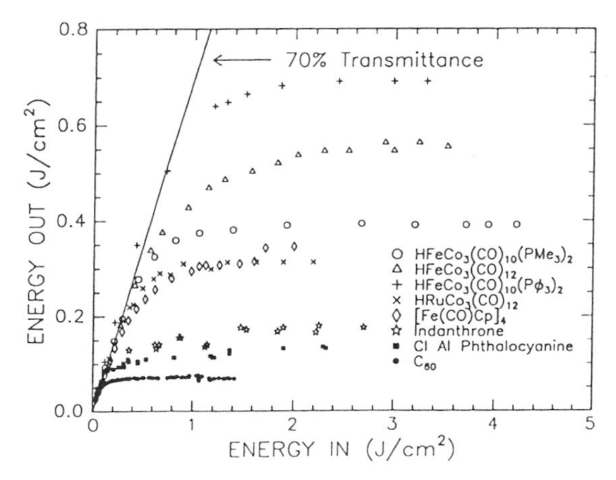

Fig. 20.1. Optical limiting response of a variety of reverse saturation absorption active materials to light pulses at a wavelength of 532 nm [20.4].

when these molecules are excited from the triplet excited state than when excited from the singlet ground state (see §13.2.3) [20.6]. The higher absorption associated with the triplet state (relative to the ground state) can be simply understood in terms of an increased transition probability through dipole-allowed transitions and the larger number of available final states. The larger availability of final triplet states arises because the irreducible representations associated with the triplet states Γ_T are found by taking the direct product $F_{1g} \otimes \Gamma_S$, where F_{1g} denotes the transformation properties of the $S = 1$ spin state and Γ_S denotes the irreducible representations of the orbital singlet states (see Table 12.4 in §12.5).

Fullerene molecules reach the metastable triplet state through the following process, as shown in Fig. 13.7. The absorption of a photon takes an electron from the singlet S_0 ground state to an excited singlet state S_i, which decays rapidly to the lowest excited singlet state S_1. This is followed by an intersystem crossing ($S_1 \rightarrow T_1$) to the lowest triplet state, T_1, although other $S_i \rightarrow T_{i'}$ intersystem crossings also may occur from higher-lying S_i states before the $S_i \rightarrow S_1$ transitions takes place. Because electrons in the triplet state T_1 can be excited via dipole-allowed processes, while electrons in the ground state can be excited only by dipole-forbidden processes near the optical absorption edge, the nonlinear absorption coefficient of C_{60} near 2 eV can be increased by populating the metastable triplet state. Even though the decay time in the triplet state is usually in the μs range in a solid film, the population of the triplet state increases rapidly as the light intensity for optical excitation increases, and consequently the absorption coefficient also increases. These physical properties allow use of C_{60} as an optical limiter [20.5]. For the optical limiting of fast pulses (on a nanosecond time scale) a single C_{60} molecule was found to be able to transfer 200–300 eV (i.e., ~100 photons of energy in the visible range) of energy per pulse from the C_{60} molecules to vibrational excitations of the fullerene and/or solvent molecules without dissociation of either the fullerene or the solvent molecules [20.6].

Figure 20.2 shows that the nonlinear absorption of the triplet state gives rise to a very large increase in absorption at infrared frequencies, while Fig. 20.1 shows excellent optical limiting properties of C_{60} in comparison with other materials which have either been considered or used for optical limiting applications [20.4]. It is suggested by Fig. 20.2 that C_{60} would have even better performance as an optical limiter near ~0.8 eV (pumped by two photon absorption followed by intersystem crossing) than at 532 nm (2.33 eV). Because of the long lifetime of the triplet exciton in frozen solution (50 ms) [20.8, 9], it has been suggested that C_{60} would have superior performance characteristics as an optical limiter if used in a solid matrix solution [20.10]. Although C_{60} and C_{70} offer attractive characteristics for use

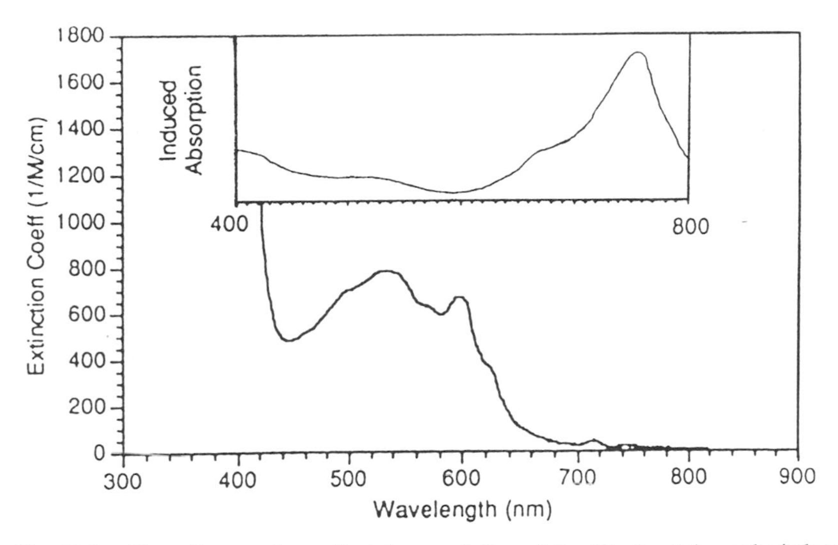

Fig. 20.2. Absorption spectrum of a toluene solution of C_{60}. The inset shows the induced absorption of C_{60} associated with optical transitions from excited triplet states. Note that the wavelength scale of the inset lines up with that for the main figure [20.7] .

as an optical limiter material, damage effects such as are induced by high-intensity ($\sim 2\times 10^{11}$ W/cm^2) light pulses of very short duration (e.g., 300 fs duration at a wavelength of 612 nm) [20.11] require further investigation.

20.1.2. Photoexcited C_{60}–Polymer Composites

As discussed §13.5, experiments on C_{60}–polymer composites [20.12–16] have shown that a very fast (subpicosecond) photoinduced electron transfer occurs from the polymer into the nearby C_{60} molecules, thereby forming a metastable C_{60}^- anion and a mobile hole on the polymer backbone. The mobile hole is responsible for the high photoconductivity of the composite [20.12, 16, 17]. A polymer–C_{60} heterojunction, therefore, would be expected to exhibit a photovoltaic response with a charge separation photoinduced across the interface. These C_{60}–polymer composites and heterojunctions have been investigated for their potential as photoconductors (§20.1.2), rectifying diodes (§20.2.2), for their photorefractive effects (§20.1.3), and for photovoltaic applications (§20.2.4) [20.18–23].

Studies on the conducting polymer PHT [poly(3-hexylthiophene)] [20.24] showed that the addition of 10–30 mol % of C_{60} to form a polymer C_{60}–composite suppresses the interband optical absorption edge of the pristine PHT at ~2 eV and also quenches the PHT photoluminescence, thus indicating that C_{60} acts as an acceptor dopant (i.e., a C_{60}^- anion is formed). Consistent with this view is the appearance at high C_{60} doping levels (~30%) of a broad optical absorption band at 1.1 eV (see §13.4.1), shown previously to be associated with the presence of the C_{60}^- anion [20.25, 26]. However, also present is the characteristic strong absorption band of C_{60} (neutral molecules), which suggests that it is the C_{60} molecules with the shortest distances to the polymer chain that are predominantly involved in the photoinduced steady-state charge transfer, and the more distant C_{60} molecules remain neutral. Several groups showed that even the neutral C_{60} molecules might be involved in the suppression of the PHT photoluminescence via a two-step process involving photoexcitation of the polymer and subsequent electron transfer to the C_{60} to form the C_{60}^- anion with some carrier hopping from a C_{60}^- anion to a neutral C_{60} molecule [20.12, 16, 24].

Photoconducting devices are another interesting application of C_{60}–polymer composites. Polyvinylcarbazole (PVK) doped with a mixture of C_{60} and C_{70} was the first fullerene–polymer system reported to exhibit exceptionally good photoconductive properties and a high potential for xerographic applications [20.18, 27]. The striking enhancement of the photoinduced discharge curve of PVK by doping with a few percent of C_{60} is shown in Fig. 13.34 [20.18]. The charge generation efficiency ϕ is increased by more than a factor of 50 at a wavelength of 500 nm and by a factor of ~4 at 340 nm by doping PVK with as little as ~2.7% C_{60} (by weight). The performance of the fullerene–PVK composite in these experiments was found to compare favorably with one of the best commercial polymer photoconductors (thiapyrylium-doped polycarbonate).

Materials useful for commercial xerography applications should exhibit low dark conductivity, large charge generation efficiency ϕ, and a fast, complete discharge of the surface charge. For light with a wavelength $\lambda >$ 350 nm, efficient photoinduced discharge occurs in C_{60}–polymer composites for both positive and negative charging, but for the strongly absorbing region ($\lambda <$ 350 nm), the discharge occurs only for positive charging [20.18, 28]. The charge generation efficiency ϕ is enhanced by an electric field, and the observed electric field dependence of ϕ is explained by a model due to Onsager [20.29], which assumes that upon absorption the incident photon creates a bound electron–hole pair, where the electron and hole are separated by a distance r_0. The thermalized electron–hole pairs either recombine or are separated into free electrons and holes which are collected in the photocurrent. The fraction of absorbed photons giving rise to free or unbound electron–hole pairs (denoted by ϕ_0) determines the efficiency of the material as a photoconductor. The fitting of the data for the PVK–C_{60} composite to the Onsager model yields values for $r_0 = 19$ Å and $\phi_0 = 0.90$.

The use of fullerenes to enhance the photoconductivity and charge generation efficiency of polymers has been extended to other polymer systems, seeking both enhanced performance and a better understanding of the science behind the device performance. In C_{60}–conjugated-polymer composites, at long wavelengths the fullerenes serve both as light absorbers and as sensitizers for the photoconductivity. But at shorter wavelengths (higher photon energies) light is primarily absorbed by the polymer close to the surface, and in this case an electron in the polymer is excited to a state lying above the lowest excited state for C_{60}, thereby inducing charge transfer of the electron to C_{60} as shown in Figs. 13.35 and 13.37. This charge transfer separates the electron and the hole, thus enhancing hole conduction along the polymer chain. Although high charge generation efficiency was first found for the C_{60}–PVK composite (see Fig. 13.34), later work showed that the charge generation efficiency is even higher for C_{60}–PMPS (phenylmethylpolysilane) than for C_{60}–PVK at low electric fields, as shown in Figs. 13.34 and 20.3.

While photoinduced electron transfer with high quantum efficiency was observed for conjugated polymers (see §13.5), such as poly(*para*-phenylene-vinylene) and its derivatives (e.g., MEH-PPV) [20.16], poor

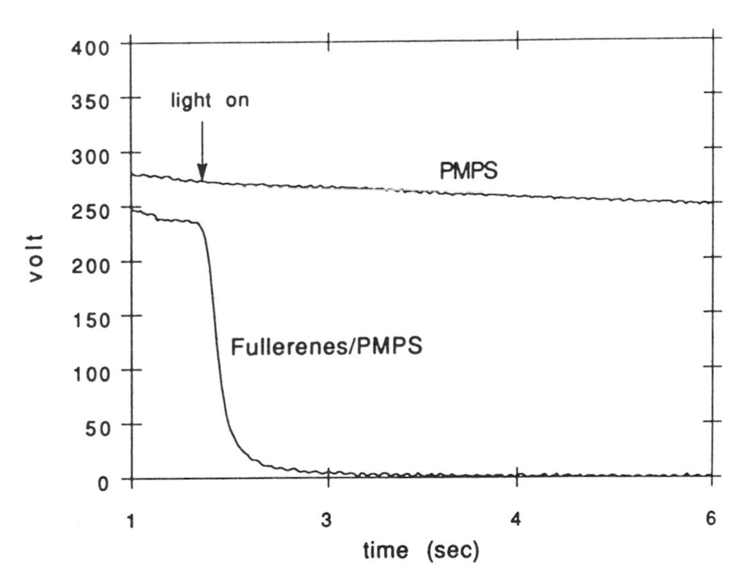

Fig. 20.3. Comparison of photoinduced discharge curves of a 3.2-μm-thick film of undoped PMPS (phenylmethylpolysilane) and a similar fullerene-doped film under illumination by a 50 mW/cm^2 tungsten lamp light source [20.28]. See Fig. 13.34 for similar results for C_{60}–PVK [20.18].

performance was observed for several other polydiacetylene (PDA) derivatives [20.16]. This was attributed to the strong, excitonic binding energy between electrons and holes in those polymers, in contrast to an exciton interaction that is apparently well screened in PPV [20.16]. Further evidence for this proposed explanation is needed. Enhanced photoconductivity by C_{60} doping has also been reported in another conducting polymer poly(3-octyldithiophene) or PODT [20.30] near 1.9 eV and 3.5 eV for C_{60} doping in the range 1–10 mol %. A related polymer P3OT [poly(3-octyldithiophene)] also shows enhanced photoconductivity upon addition of C_{60} [20.16]. Although research progress in this field is strongly dependent on the close collaboration between physicists and chemists, it is the chemical synthesis and polymer processing that become especially important in the ultimate commercialization of this class of materials in practical devices.

20.1.3. Photorefractivity in C_{60}–Polymer Composites

The enhancement of the photoconductivity of polymeric films by the addition of a few weight percent of C_{60} has led to the development of highly efficient photorefractive polymeric films [20.19–21, 31]. We first briefly review the principles underlying photorefractive devices in polymer composites [20.21].

A photorefractive device acts like a transmission grating, by diffracting, or bending, a transmitted light beam away from the forward direction. One coherent light source is used to create (or *write*) a grating in the photorefractive medium, and a second source, incident from the other side of the device, is used to *read* the result of the writing source. The effect of the grating is to diffract the reading beam, and this photorefracted beam is monitored with a detector.

In the photorefractive medium, the "grating" is created by the superposition of two interfering, coherent beams derived from the same laser. A sinusoidal, spatial modulation of light intensity from these beams bathes the sample, and photogenerates carriers in proportion to the local light intensity. These carriers migrate by diffusion (in an inorganic, photorefractive medium) or by the action of an applied electric field (in polymeric, photorefractive materials) away from the location where the carriers were generated to occupy deep trap sites, thereby setting up a steady-state, spatially modulated space charge field. This "field grating" is converted by the *linear* electro-optic effect into an "index grating," or a spatial modulation of the refractive index coefficient, which can be used to diffract a third, "read" beam incident from the other side of the device into a detector.

The experimental apparatus used to study the photorefractive effect is shown schematically in Fig. 20.4. The presence of the C_{60} additive in the photorefractive film stimulates the photogeneration of mobile holes in the polymer. These mobile holes are soon trapped and then set up a modulated space charge field (i.e., the grating). A beam from an InGa–AlP semiconductor diode laser (687 nm wavelength) is incident from one side of the sample and is used to *read* the index grating *written* by the interference grating induced by the He–Ne laser from the other side of the film (see Fig. 20.4). The response due to these three beams is a fourth, photodiffracted *signal* beam monitored by the detector D_1, and D_2 monitors the incident intensity of the "reading" beam. The elements P_j, M_j, and BS_j ($j = 1, 2, \ldots$) in Fig. 20.4 represent optical polarizers, mirrors, and beam splitters, respectively, and the optical wavelength is 687 nm.

The initial studies of the photorefractive effect were carried out in the C_{60}–polymer composite PVK:C_{60}:DEANST [20.22], where the DEANST, or diethyl aminonitrostyrene, was added as a second-order, nonlinear optical molecular constituent to the polyvinylcarbazole (PVK)–C_{60} composite. The photorefractive performance of this PVK–polymer composite was found to be very promising. However, substantial improvements in per-

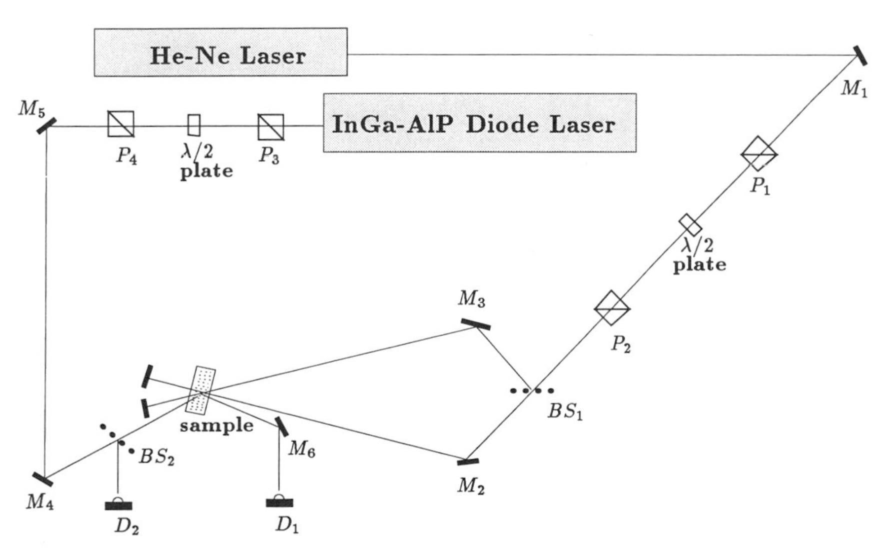

Fig. 20.4. Schematic diagram of the apparatus used for writing a pattern on a 100-μm-thick film of the PVK-TCP:C_{60}:DEANST polymeric composite for photorefractive studies. The He–Ne laser *writes* the grating, which is then *read* by the InGa–AlP semiconductor diode laser [20.21].

formance were made by lowering the glass transition temperature of the polymer through the addition of a plasticizer TCP (tricresyl phosphate), which improves the applied field-induced, noncentrosymmetric alignment of the DEANST molecules. The actual photorefractive device is a three-layer structure with the 100-μm-thick photorefractive PVK-TCP:C_{60}:DEANST film sandwiched between two transparent indium–tin oxide (ITO) electrodes. These electrodes are used to *pole* the device (align the DEANST molecules by an applied electric field), as well as to induce a drift of the photogenerated holes [20.20]. In Fig. 20.5 we display time-resolved diffraction intensity data for PVK:C_{60}:DEANST (the plasticizer was not present

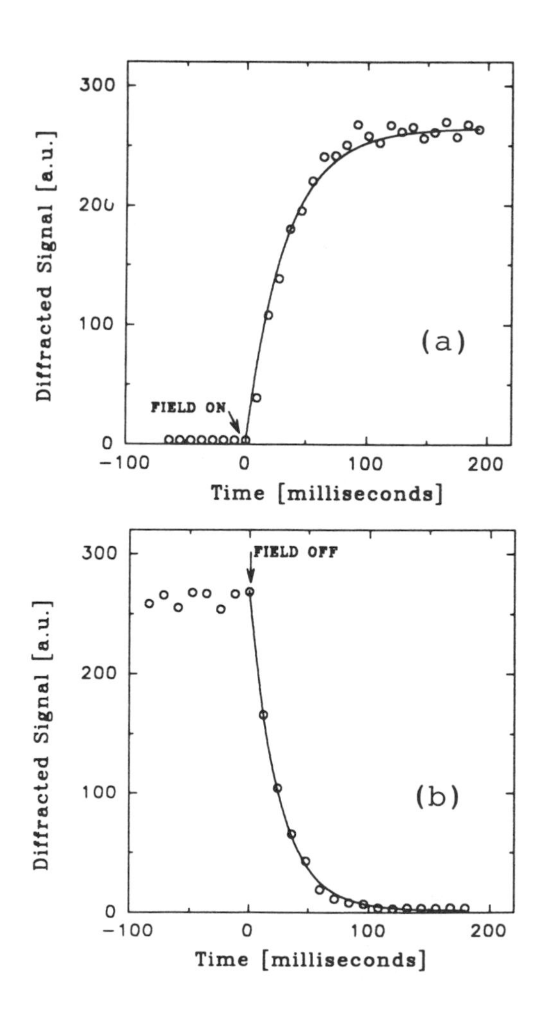

Fig. 20.5. Photorefractive, nondegenerate four-wave mixing time response for a 100-μm-thick film of the PVK:C_{60}:DEANST polymeric composite as the field is turned on (a) and turned off (b). From (a) a rise time of $\tau_r = 62$ ms is obtained and from (b) a decay time of $\tau_d = 46$ ns is obtained for a dc electric field of 41 V/μm. The data for the rise time is fit to a functional form $[1 - \exp(-t/\tau_r)]^2$ and for the decay time to a form $\exp(-2t/\tau_d)$ [20.21].

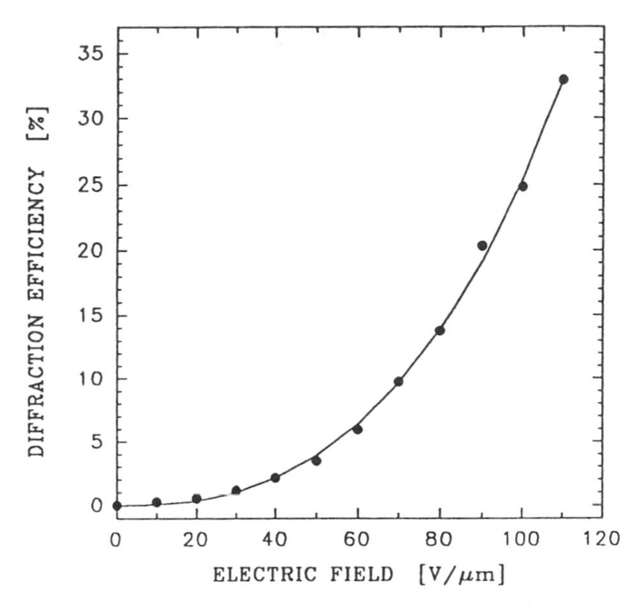

Fig. 20.6. Electric field dependence of the diffraction efficiency measured on a 100-μm-thick film of the PVK-TCP:C_{60}:DEANST polymeric composite between ITO film electrodes [20.20].

in this sample) for an applied field of 41 V/μm [20.21]. The rise and decay times in Fig. 20.5 for the diffracted signal based on a single carrier model were found to be $\tau_r = 62$ ms and $\tau_d = 46$ ms, respectively.

The diffraction efficiency for a 100-μm-thick PVK-TCP:C_{60}:DEANST photorefractive device is shown in Fig. 20.6 as a function of the dc electric field produced by a bias voltage V applied between the transparent ITO electrodes. High efficiencies, greater than 30% and greater than those of any known inorganic photorefractive media, were observed at electric fields of ~100 V/μm [20.21]. To achieve these large E-fields in the device requires the application of a high voltage (~10 kV). In Fig. 20.7 [20.21] we display the photorefractive device performance parameters (diffraction efficiency and response speed) for the best photorefractive inorganic materials [e.g., $BaTiO_3$, $Sr_xBa_{1-x}NbO_3$ (SBN), $Bi_{12}SiO_{20}$ (BSO)] [20.32], for promising photorefractive polymer composite materials (e.g., bisA-NPDA:DEH[1,2] [20.22, 33, 34], PVK:C_{60}:DEANST [20.35, 36], PVK:FDEANST:TNF [20.37], PVK:TPY:DEANST [20.38],

[1]Bisphenol-A-diglycidylether (bisA) and 4-nitro-1,2-phenylenediamine (NPDA).
[2]Diethylamino-benzaldehyde diphenylhydrazone (DEH).

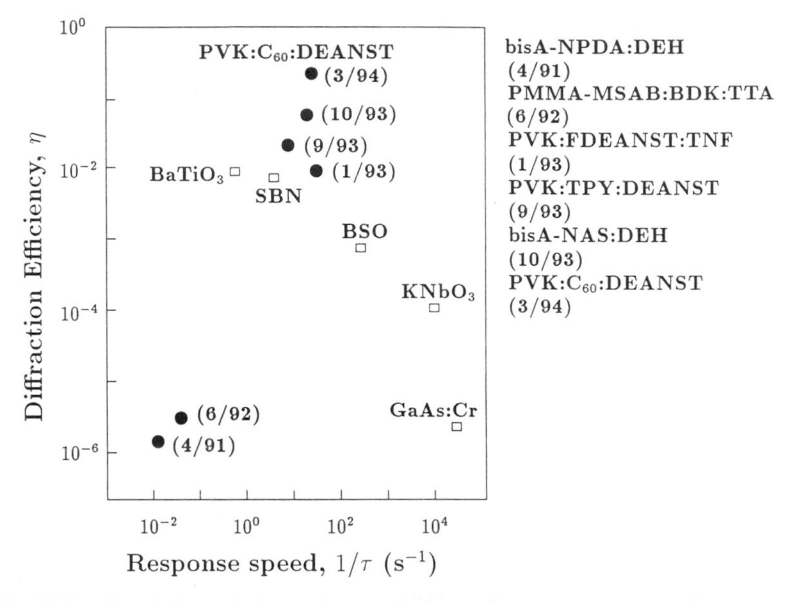

Fig. 20.7. Comparison of photorefractive device performance parameters for many materials and composites including a C_{60}–polymer composite. All diffraction efficiencies are scaled to a sample thickness of 100 μm. Polymer composites are referenced in the figure by the date of the first report. The spectral sensitizers are borondiketonate (BDK), trinitrofluorenone (TNF), and thiapyrylium (TPY). The optically active ingredient to assist charge transport is tri-*p*-tolylamine (TTA) and diethylaminonitrostyrene (DEANST). The best performance to date is with the polyvinylcarbazole (PVK) polymer:C_{60} composite [20.21].

bisA-NAS:DEH[2] [20.23], PMMA-MSAB:BDK:TTA[3] [20.33, 34, 39]) and for PVK:C_{60}:DEANST, a C_{60}–polymer composite with outstanding photorefractive performance. Although the C_{60}–polymer composite PVK:C_{60}:DEANST exhibits a slower response time than several of the fast inorganic materials, the diffraction efficiency of this C_{60}–polymer composite is more than one order of magnitude larger than that of the best inorganic photorefractive materials.

20.2. Electronics Applications

To date, a variety of M/C_{60}/M (M = metal) rectifying diodes, field effect transistors (FETs), as well as photovoltaic and photorefractive devices have

[3]Polymethylmethacrylate (PMMA) and 4'-[(6-(methacroyloxy)hexyl]methylamino]-4-(methyl-sulfonyl) azobenzene (MSAB).

been proposed and measurements have been carried out on a number of actual electronic devices. A few research groups have concentrated on applications based on the strong bonding at the C_{60}–Si interface (see §17.9.3) [20.40–42] and the incorporation of C_{60} into existing microfabrication processes. It has been demonstrated that C_{60} films can be patterned on silicon using conventional photoresist and liftoff techniques [20.43] achieving linewidths between 3 and 5 μm. Moreover, the C_{60} film itself can serve as a negative photoresist (decreasing solubility upon exposure to light) [20.44], yielding line resolution of better than 1 μm as shown below.

While a number of devices based on pristine C_{60} films have been investigated, the most promising electronic applications proposed to date utilize C_{60} in conjunction with selected conducting polymer composites (see §20.1). As discussed in §13.5, experiments on C_{60}–polymer composites [20.12–16] have shown that a very fast (subpicosecond) photoinduced electron transfer from the polymer to a nearby C_{60} molecule occurs, thereby forming a mobile hole on the polymer backbone and a metastable C_{60}^- anion. A polymer–C_{60} heterojunction, therefore, would be expected to exhibit a photovoltaic response, and such devices have been investigated (see §20.2.4). In this section, the structure and properties of C_{60}-based transistors (§20.2.1) and rectifying diodes (§20.2.2) are discussed, and C_{60}–semiconducting polymer heterojunctions are described for both rectifying diodes (§20.2.3) and photovoltaic (§20.2.4) applications. This is followed by a variety of other electronics applications based on fullerenes.

It should be noted that most C_{60} devices are unstable in air due to the diffusion and photodiffusion of dioxygen [20.45, 46] into the large interstitial sites of the fullerene solid. Therefore, if C_{60}–polymer devices are to be commercialized, they need to be packaged in oxygen-resistant coatings. While such packaging is done routinely in the semiconductor industry, fullerenes are unique in that they themselves can be used to make oxygen-resistant seals [20.47].

20.2.1. C_{60} Transistors

In this subsection, field effect transistors based on C_{60} are described and are compared to their conventional silicon-based counterparts. The conventional silicon-based metal–oxide–semiconductor field effect transistor [MOSFET; see Fig. 20.8(a)] consists of two p–n junctions placed immediately adjacent to the region of the semiconductor controlled by the MOS gate. The carriers enter the structure through the source (S), leave through the drain (D), and are subject to the control by the bias voltage on the gate (G). The voltage applied to the gate relative to ground is V_G, while the drain voltage relative to ground is V_D. In this particular device geom-

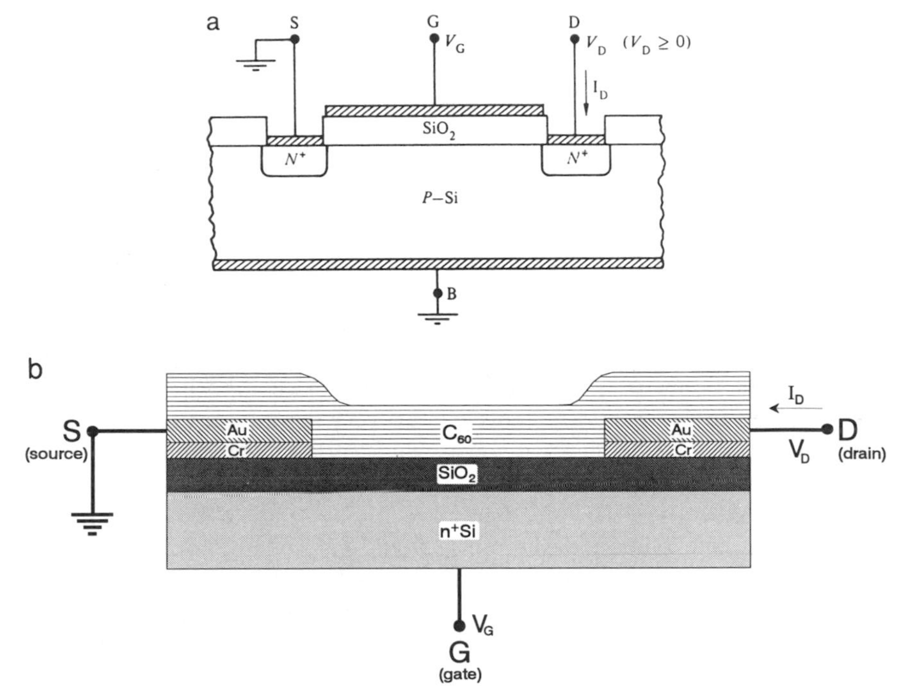

Fig. 20.8. Cross-sectional view of field effect transistor (FET) structures: (a) the terminal designations and biasing conditions for conventional Si-based MOSFETs [20.48]. G, D, B, and S, respectively, denote the ground, drain, base, and source. (b) The corresponding structure for the fullerene (C_{60}) FET device [20.49].

etry the source (S) and base (B) are grounded. As the gate voltage V_G is increased $V_G > V_T$, where V_T is the depletion-inversion transition-point voltage, an inversion layer containing mobile carriers is formed adjacent to the Si surface, creating a source-to-drain channel. Now, by keeping V_G fixed and varying V_D, the current–voltage characteristic I_D *vs.* V_D of the transistor can be determined for various gate voltages.

Fullerene-based *n*-channel FETs have been constructed [20.49–52] and an example of such a device is shown in Fig. 20.8(b). The corresponding current–voltage characteristics for this fullerene FET are shown in Fig. 20.9. In the fullerene FET shown in Fig. 20.8(b), a highly doped *n*-type silicon wafer takes the place of the gate metal, a ~30–300 nm thick layer of SiO_2 serves as the oxide, and the fullerene film is the semiconductor. When an appropriate positive gate voltage V_G is applied (see Fig. 20.9), the drain current I_D increases, which indicates that a conduction channel is formed near the fullerene–insulator interface. The sign of the response indicates

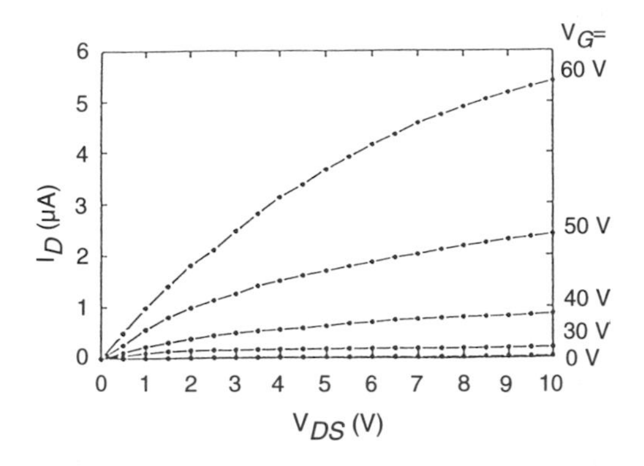

Fig. 20.9. Room temperature drain current (I_D) *vs.* drain–source voltage (V_{DS}) characteristics at various gate voltages V_G (0, 30, 40, 50, 60 V) for a fullerene-derived field effect transistor with conduction in an *n*-type C_{60}/C_{70} channel induced at the SiO_2–C_{60} interface [20.49]. A schematic diagram for the device is shown in Fig. 20.8(b).

that the fullerene transistors are *n*-channel devices, consistent with C_{60} anion formation. Field effect mobilities of up to 0.08 $cm^2/V{\cdot}s$ were obtained from the I_D–V_{DS} characteristics, which compares well with organic FETs, but is considerably lower than the carrier mobilities (0.5–1.0 $cm^2/V{\cdot}s$) observed in C_{60} films by the Dember effect [20.53] and by the time-of-flight mobility measurements on C_{60} single crystals [20.54, 55]. This suggests that localized states at the SiO_2–C_{60} interface may play an important role in the electronic conduction.

Because C_{60}-based FETs operate as *n*-channel devices, they are subject to problems that are not experienced by the majority of organic materials investigated to date, which tend to be *p*-type. The field effect mobilities of 0.08 $cm^2/V{\cdot}s$ and the on–off ratios of $\sim 10^6$ are rapidly degraded by exposure to oxygen, presumably because charge transport in the C_{60} FETs depends on electron transport rather than on hole transport as is common in most organic materials-based FETs [20.52]. The fact that the device performance is not degraded by exposure to nitrogen gas [20.52] suggests that the oxygen molecules act as electron traps in the lattice of the C_{60} molecular solid.

It was also found [20.52] that the treatment of the SiO_2 surface with TDAE (tetrakis-dimethylamino-ethylene) prior to deposition of the C_{60} film reduces the threshold voltage of the device, increases the mobility of the carriers in the C_{60} conduction channel, narrows the barrier width, and increases the band bending at the C_{60}–metal interface.

Overall, the electronic properties of C_{60} as a FET channel material show strong resemblance to amorphous semiconductors. The mobility of fullerene FETs may perhaps be improved by epitaxial growth of C_{60} on the surface or the use of Langmuir–Blodgett techniques, which could reduce the concentration of defect-induced traps.

20.2.2. C_{60}-Based Heterojunction Diodes

Of importance to many potential applications of fullerenes is the nature of the interface between fullerenes and various semiconductors and metallic substrates (see §17.9). It has been shown by several authors [20.56, 57] that Nb/C_{60}/*p*-Si and Ti/C_{60}/*p*-Si heterojunctions (Fig. 20.10) are strongly rectifying. Rectifying diodes have been demonstrated both for C_{60}–metal interfaces as described in this subsection and for C_{60}–polymer heterostructures as reviewed in §20.2.3.

Direct rectification between solid C_{60} and *p*-type crystalline Si has been demonstrated in Nb/C_{60}/*p*-Si and Ti/C_{60}/*p*-Si heterojunctions (as shown in Fig. 20.10) [20.56, 57]. Since the potential barriers at the Nb–C_{60} and Ti–C_{60} interfaces are close to zero [20.56, 58], it has been concluded that it is the C_{60}/*p*-Si interface that is responsible for the strong rectifying properties of this heterostructure [20.56, 57]. Since the rectification occurs between the C_{60} and Si interfaces, and not between the C_{60}–metal interfaces, it is technically not proper to refer to these devices as Schottky diodes. Nevertheless, these heterojunctions were found to be strongly rectifying (by a factor of $\sim 10^4$ between +2 V and −2 V), and, as shown in Fig. 20.11, the sign of rectification is inverted when *n*-Si is used in place of *p*-Si. While rectifying behavior was observed in Al/AlO_x/C_{60}/Au devices [20.59], a different mechanism is responsible for rectification in this case, namely the tunneling of electrons through the AlO_x oxide layer, which is characteristic of the rectification behavior in metal–insulator–semiconductor (MIS) tunnel diodes. MIS devices with fullerenes as the active semiconductor contact were also studied by Pichler *et al.* [20.60].

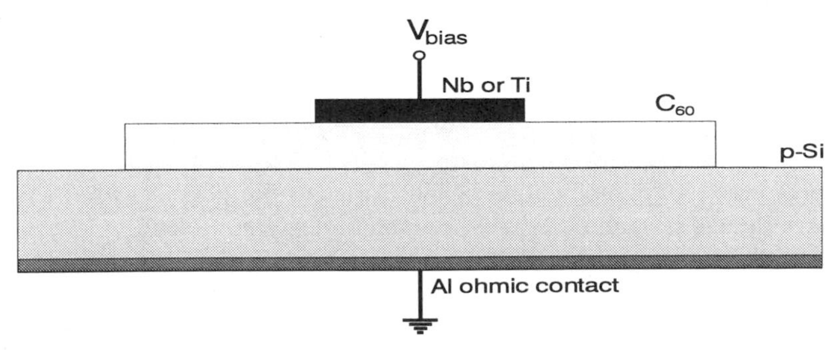

Fig. 20.10. Schematic cross section of an Nb/C_{60}/*p*-Si structure used as a heterojunction diode. Rectification occurs at the C_{60}–Si interface [20.56].

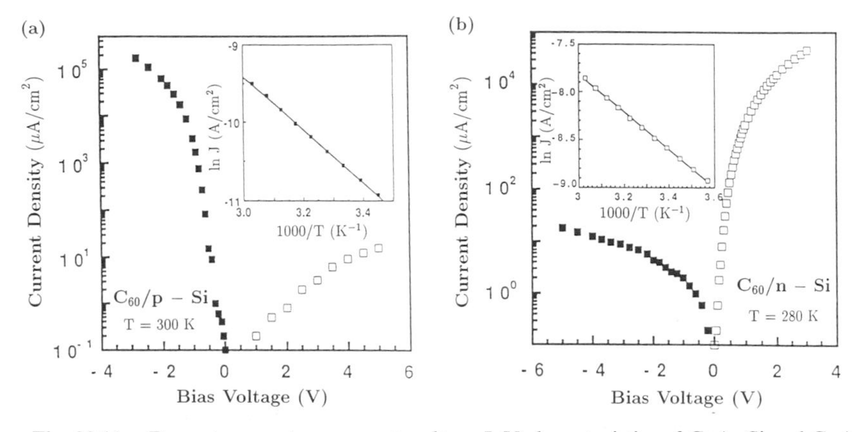

Fig. 20.11. Room temperature current–voltage I–V characteristics of C_{60}/p-Si and C_{60}/n-Si junctions with the filled squares for negative bias voltage and open squares for positive V. The complementary behavior between the junctions with n-Si and p-Si further supports the interpretation that it is the C_{60}–Si interfaces which yield efficient rectification and diode behavior in (Nb or Ti)/C_{60}/Si heterostructures. The insets, showing Arrhenius plots for each junction, yield activation energies (interpreted as barrier heights) of 0.30 eV for C_{60}/n-Si and 0.48 eV for C_{60}/p-Si [20.57].

20.2.3. *C_{60}–Polymer Composite Heterojunction Rectifying Diode*

Rectification behavior has also been demonstrated in C_{60}–polymer heterostructures, such as shown in Fig. 20.12 [20.12, 17]. Because of other optical experiments to be performed on these structures, a thin film transparent electrode, such as indium–tin oxide (ITO), is vacuum-deposited onto a glass substrate to form the lower transparent electrode. Next a thin film of polymer, such as a 1000 Å layer of methyl-ethyl-hydroxyl-polypropylvinyl (MEH-PPV) is spin-coated over the ITO, followed by a vacuum-deposited C_{60} film of similar thickness. The active areas ($\sim$0.1 cm^2) of the device are

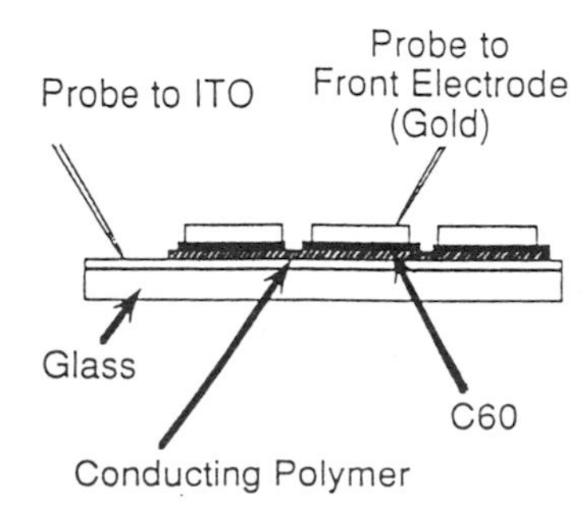

Fig. 20.12. ITO/polymer/C_{60}/Au heterojunction fabricated as a layered structure [20.17]. ITO denotes the optically transparent and electrically conducting indium–tin oxide. An example of a conducting polymer used in heterojunction applications is MEH–PPV. Light enters through the glass substrate and enters the device at the ITO electrode.

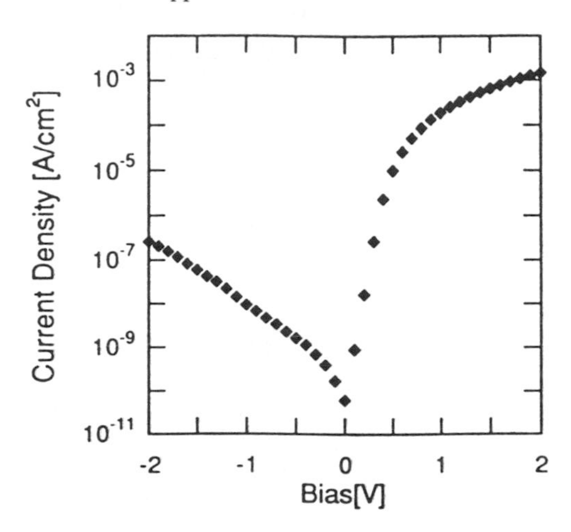

Fig. 20.13. Dark current density *vs.* bias voltage showing rectification behavior in an ITO/MEH-PPV/C_{60}/Au photovoltaic device at room temperature [20.12].

defined by the Au or Al pads that are vacuum-deposited onto the upper C_{60} surface.

The I–V characteristics for such a C_{60}–polymer heterostructure diode are shown in Fig. 20.13 for a C_{60}/MEH-PPV heterojunction (see §20.1.2) measured in the absence of light. Note that the current density is displayed on a log scale and that at a bias voltage of ± 2 V the rectification ratio is about four orders of magnitude, similar to the rectification behavior discussed in §20.2.2 for the Nb/C_{60}/Si rectifying diodes that do not employ a C_{60}–polymer heterojunction. A positive voltage V implies that the ITO electrode is more positive than the metal electrode, so that the MEH-PPV polymer is positively biased with respect to the C_{60}. The forward current in Fig. 20.13 can be seen to exhibit a nearly exponential rise with applied bias voltage. Similar I–V measurements were made on Au/C_{60}/Au, Au/C_{60}/ITO, Au/MEH-PPV/Au, and Au/MEH-PPV/ITO layered structures, and all of these structures were found to exhibit ohmic (linear) I–V characteristics. It must therefore be concluded that the rectification occurs at the MEH-PPV/C_{60} interface [20.12].

20.2.4. C_{60}–Polymer Composite Heterojunction Photovoltaic Devices

C_{60}–polymer composite heterojunctions, such as the ITO/MEH-PPV/C_{60}/Al heterojunction shown in Fig. 20.12, have also been used to demonstrate photovoltaic behavior. In this context we show in Fig. 20.14 the short-circuit current density (closed circles with no applied bias voltage), and the photocurrent density (closed circles with -1 V bias) as a function of the incident light flux on a log–log plot [20.12]. The open circuit (zero-current)

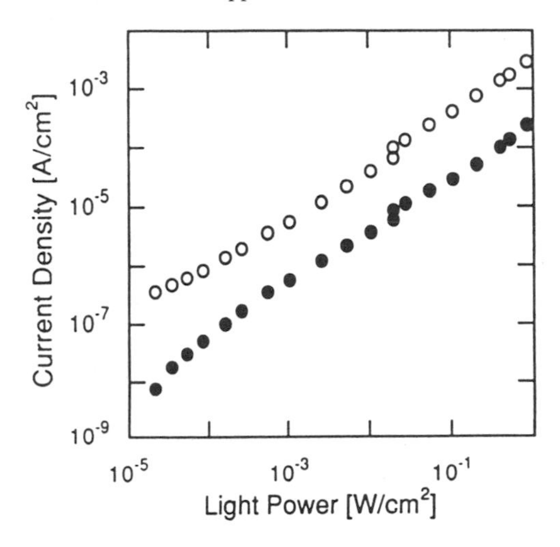

Fig. 20.14. Short-circuit current (closed circles) and photocurrent at −1 V bias (open circles) as a function of light intensity for the ITO/MEH-PPV/C_{60}/Au photovoltaic device [20.12].

voltage of the device under illumination was found to be ~0.5 V. Furthermore, this voltage was identified with the MEH-PPV/C_{60} junction, since such bias voltages applied to other junctions (e.g., C_{60}/Au and C_{60}/Al) did not affect the photoresponse. Both the short-circuit photocurrent and the biased-device photocurrent are reasonably linear over five decades of light intensity in the plot in Fig. 20.14, up to power densities approximately an order of magnitude greater than the terrestrial solar intensity level. The spectral dependence of the photocurrent (taken at −1 V bias) for an ITO/MEH-PPV/C_{60}/Au photovoltaic device is plotted in Fig. 20.15. The onset of the photocurrent near 1.7 eV matches the absorption edge for the MEH-PPV polymer, and the photoconductivity minimum near

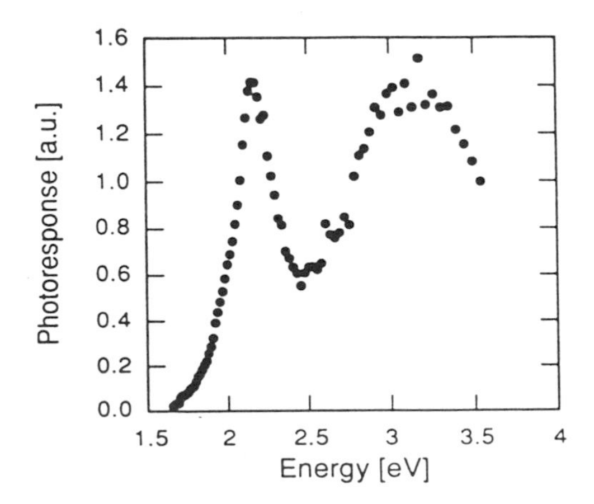

Fig. 20.15. Spectral response of the photocurrent in an ITO/MEH-PPV/C_{60}/Au photodiode in arbitrary units at (reverse) −1 V bias [20.12].

2.5 eV matches the absorption maximum in the polymer. Since the light was incident from the polymer side of the polymer–C_{60} heterojunction (see Fig. 20.12), it was concluded that the electron injection into the C_{60} must be from the polymer chains located very close to the heterojunction. Consistent with this experimental result, and using a charge carrier mobility of 10^{-4} cm^2/V-s for this device, an applied E-field of 10^5 V/cm, and a photocarrier lifetime of $\sim$1 ns [20.61], the distance a carrier could drift before recombination was estimated to be only a few angstroms.

Use of C_{60}–polymer heterojunctions (see Fig. 20.12) for photovoltaic energy conversion device applications [20.12, 16] requires that the polymer be optimized for optical absorption at the peak of the solar spectrum, and the device must present a large effective area to the sunlight. C_{60}–polymer heterostructures appear promising for photovoltaic applications by further improving the materials and device design.

20.2.5. Microelectronic Fabrication and Photoresists

A number of advances have been made toward integrating fullerenes into the microfabrication processes for microelectronics. For instance, the use of fullerene films for patterning, as surfaces of uniform electric potential for electronics applications [20.62], for the passivation of surfaces [20.63], and as adhesives for direct bonding of the silicon wafers [20.42] have been suggested and are further discussed in this subsection.

The basis for patterning in microfabrication processes is the photochemical transformation that is induced by visible and ultraviolet (UV) light. In the absence of diatomic oxygen, solid C_{60} undergoes a photochemical transformation when exposed to visible or ultraviolet light with photon energies greater than the optical absorption edge ($\sim$1.7 eV) [20.64–66]. Phototransformation renders the C_{60} films insoluble in fullerene monomer solvents such as toluene, a property which may be exploitable in photoresist applications. Because the optical penetration depth in crystalline C_{60} in this spectral region is on the order of $\sim$0.1–1 μm, small particle size powders or thin films can undergo a significant fractional volume decrease upon phototransformation. The photochemical process is fairly efficient; a loss of 63% of the signature of monomeric units (i.e., van der Waals–bonded C_{60} molecules) is observed by Raman spectroscopy after the C_{60} film is exposed for $\sim$20 min to a modest light flux (20 mW/mm^2). Heating the transformed films to $\sim$100°C has been reported to return the system to the pristine monomeric state [20.67].

The patterning of micro-meter-sized features on electronically compatible substrates has already been demonstrated using C_{60} on Si (100) [20.43].

The conventional photoresist and liftoff technique yielded lines of C_{60} with widths between 3 and 5 μm and heights of 0.4–0.8 μm [20.43]. This finding provides the opportunity to form many different structures using C_{60} films for a wide variety of applications, such as microelectronic devices, superconducting interconnects, and optical switching devices.

The C_{60} film itself can also serve as a negative photoresist (less soluble) [20.68] upon irradiation with ultraviolet light in vacuum. Such irradiation photo-transforms the fullerene solid and promotes dimer and oligomer formation. The decreased solubility and lower vapor pressure of the phototransformed material enable wet (organic solvent) or dry (thermal or photoinduced sublimation) development of photodefined negative images, as shown on Fig. 20.16, where the sequence of steps in the photolithography process is shown schematically. The negative developing step shows that the C_{60} is retained on the Si substrate where it has been exposed to UV light. This photolithography process yields line resolution of better than 1 μm with good edge definition in the original Si substrate [20.68]. The efficiency of the photolithography process can be greatly increased by exposing the C_{60} film to UV light in the presence of NO gas at ambient pressures [20.69], as discussed below.

Two methods, Raman scattering (see §11.8) and chromatography, are typically used to monitor the photosensitivity of C_{60} as a photoresist, which is related to the degree of cross-linking of the C_{60} film. Raman scattering characterization shows a factor of 10^3 increase in photodimerization through introduction of NO gas at ambient pressures [20.69]. The enhanced photosensitivity of C_{60} in the presence of NO gas was confirmed by high-resolution chromatography characterization. The finding, that amorphous carbon coatings can provide some protection for C_{60} films against the diffusion of oxygen and other species [20.47], could also be useful for specific photolithography applications.

In addition to photon beams, electron beams can be used to write patterns on C_{60} surfaces, either very fine lines on a nanometer scale using an STM tip operating at ~3 eV, or thicker lines using an electron gun operating at ~1500 eV [20.63] (see §7.5.2). This technology also appears promising for specific device applications.

20.2.6. Silicon Wafer Bonding

Utilizing the strong bonding between C_{60} and silicon at the C_{60}–Si interface (see §17.9.3), a thin layer (~600 Å) of C_{60} has been used as an adhesive to bond a 3 inch silicon wafer to another Si wafer at room temperature [20.42]. The thin C_{60} film between the two Si wafers was polycrystalline with grain sizes of ~60 Å, while the Si wafer surface had a (100) orien-

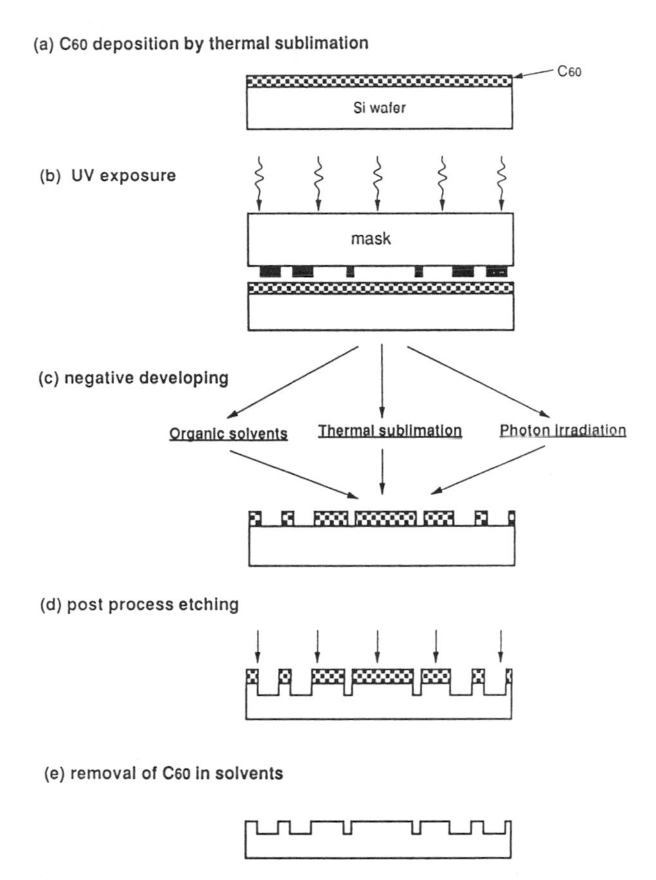

Fig. 20.16. Sequence of steps (deposition, exposure, development, and pattern transfer) used in fullerene photolithography in which the C_{60} acts as a negative photoresist [20.68].

tation and was prepared as a hydrophilic surface (see §9.2.1). Some pressure was needed to bond the pair of wafers together. Interface energies of $\sim$40 erg/cm^2 and $\sim$20 erg/cm^2 were reported for the C_{60}–SiO_2 and C_{60}–Si interfaces and were measured by the crack length generated by inserting a razor blade at the bonding interface [20.42]. Examination of the debonded surfaces did not show adhering C_{60} layers, indicating a lower cohesive energy at the interface than between C_{60} molecules. It was also found that two C_{60}-covered Si wafers can easily be bonded to each other [20.42]. It was further speculated that a thin fullerene film could be used to bond selected smooth surfaces to one another or to bond an Si wafer to certain other smooth surfaces [20.70]. Since the strongest bonding is at the C_{60}–Si interface, a single C_{60} monolayer would be expected to show bonding action to Si.

20.2.7. Passivation of Reactive Surfaces

Using Auger electron spectroscopy (see §17.3) and temperature-programmed desorption (see §17.5), it has been demonstrated that a monolayer of C_{60} is sufficient to passivate a clean aluminum surface from room temperature up to 600 K against oxidation by water vapor, as shown in Fig. 20.17. In this figure, clear Auger lines are seen for aluminum (68 eV) and for carbon (272 eV), but essentially none are seen for oxygen (503 eV) after exposure of Al–C_{60} to about 340 monolayers of water vapor [20.71]. The amount of desorption of C_{60} multilayers to eventually yield monolayer coverage is determined by the intensity of the carbon to aluminum Auger lines in Fig. 20.17. Above ~700 K, the C_{60} molecules were found to become dissolved in the near-surface region of the aluminum substrate. Even in the dissolved phase, some surface passivation of the Al surface by the presence of dissolved C_{60} was reported [20.71]. In addition to inhibiting oxidation of the highly reactive Al surface, a monolayer of C_{60} has been shown to protect the highly reactive Si $(111)7 \times 7$ and Si $(100)2 \times 1$ surfaces against oxidation [20.71–73].

20.2.8. Fullerenes Used for Uniform Electric Potential Surfaces

For some electronics applications, surfaces with a uniform electrical potential are required. In the past, graphite and gold surfaces have been used to provide equipotential surfaces. It has been found that by applying a thin (~10 Å) deposit of germanium to a substrate (such as amorphous carbon, γ-irradiated NaCl or polished phosphorus bronze) and subsequently depositing an overlayer of ~100 Å of fullerenes, a good equipotential surface is obtained [20.62]. The dangling bonds of the thin Ge interface layer would be expected to facilitate bonding to both the substrate and the fullerenes (see §17.9.3). Amorphous fullerene growth and uniform surface coverage of C_{60} on the Ge interface layer would be expected at room temperature, because of the low mobility and high sticking coefficient of the fullerenes to the Ge dangling bonds. The demonstration of this fullerene application was made with 85% C_{60} and 15% C_{70}. Fullerenes are useful for providing equipotential surfaces on metals because fullerenes are chemically stable and easy to deposit in vacuum [20.62].

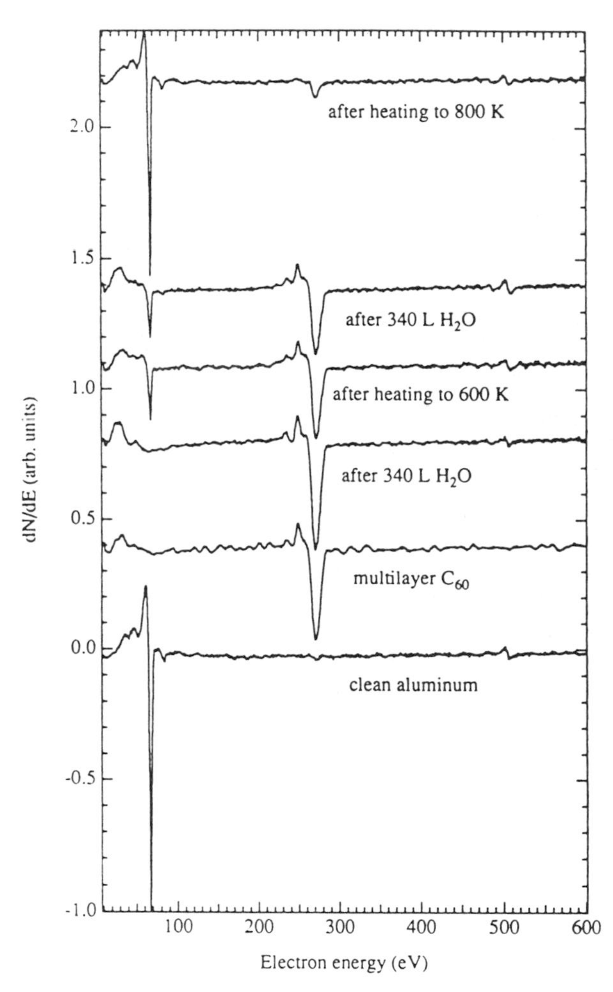

Fig. 20.17. A series of Auger electron spectra showing the passivation of aluminum by multilayer and single layer coverage of C_{60}. The traces are displayed in chronological order from the bottom to the top. The bottom trace is for the Al surface after it was sputter cleaned and annealed at 800 K. The next trace up is for the clean surface exposed to C_{60} for a sufficient time to form a multilayer coverage. The Langmuir (L) denotes a unit of exposure with 1L $= 10^{-6}$ torr sec, so that with a sticking coefficient of unity, 1L corresponds to 1 monolayer coverage. Each succeeding trace is for the sample receiving the indicated treatment and all previous treatments [20.71].

20.3. Materials Applications

In this section we present several current examples of situations where fullerenes offer promise for the synthesis of new materials and for better preparation methods for well-known materials.

20.3.1. Enhanced Diamond Synthesis

The possible use of C_{60} for the fabrication of industrial diamonds offers another area for possible fullerene applications. It is found that when non-hydrostatic pressures (in the range of 20 GPa) are applied rapidly at room temperature to C_{60}, the material is quickly transformed into bulk polycrystalline diamond at high efficiency [20.74]. It is believed that the presence of pentagons in the C_{60} structure promotes the formation of sp^3 bonds during the application of high anisotropic stress.

Normally, for diamond films to grow from a mixture of gaseous CH_4 and H_2, the substrate surface (usually Si) must be pretreated with diamond grit polish or must contain small diamond seeds. Enhanced nucleation of a high density of diamond crystallites on an Si substrate has been reported through the deposition of a 500–1000 Å C_{70} film, activation of the film by positive ion bombardment, followed by chemical vapor deposition (CVD) growth in a microwave plasma reactor. A base layer of ion-activated C_{70} was found to be more effective in promoting sp^3 bonding [20.75, 76] than a similarly treated C_{60} buffer layer.

One possible explanation for the enhancement of the diamond nucleation process by the presence of fullerenes comes from the STM studies of C_{60} on Si (111) surfaces (see §17.9.3) [20.77], which show the C_{60} cages to burst above 1020 K. The microwave discharge described above may serve to clean the Si surface, thereby allowing strong bonding of the C_{60} to the Si surface. With this strong Si–C bond, the surface containing C_{60} can be heated to a high enough temperature to evaporate all the C_{60} except for the surface monolayer, which opens up above 1000 K, leaving active carbon sites for diamond nucleation [20.78].

20.3.2. Enhanced SiC Film Growth and Patterning

Several groups have studied the interaction of C_{60} with Si surfaces at high temperatures [20.77–85], and several groups have investigated the deposition of C_{60} on hot Si substrates as a method for enhancing the growth of SiC films [20.77–81, 84]. Film growth occurs by first nucleating the SiC growth at $\sim$1250 K and then by Si diffusion to the SiC interface to react with the C_{60} molecules that arrive at the free surface in a continuous flux. The Si diffusion coefficient through the SiC film is estimated to be between

10^{-13} and 10^{-12} cm^2/s at a substrate temperature between 1100 and 1170 K [20.71]. SiC growth by this method on Si (previously cleaned at 1300 K) has been demonstrated on both Si (111) and Si (100) substrates heated to temperatures in the 950–1250 K range. Epitaxial SiC (100) 1×1 films up to 1 μm in thickness can be grown in the cubic phase, as shown by x-ray diffraction and low-energy electron diffraction (LEED) patterns, while Auger spectroscopy was used to verify the SiC stoichiometry [20.86].

By exploiting the strong C_{60}–Si interaction on a clean Si surface and the weak C_{60}–Si interaction on a SiO_2 surface in comparison to the C_{60}–C_{60} bonding itself (see Table 17.2), it is possible to pattern the SiC film which grows readily on the clean Si surface but does not grow on the SiO_2 surface. To prepare a patterned SiC film, the SiC growth is carried out on an Si surface that was previously patterned by standard lithographic techniques with contrast provided by the regions of clean Si and clean SiO_2, and patterned SiC film growth to $\sim$1 μm thickness is possible. If Si rather than SiO_2 is desired in the regions where SiC growth has not occurred, the SiO_2 material can be removed with an HF etch. If a free-standing patterned SiC film is desired, liftoff of the SiC film from the substrate can be accomplished using an atomic force microscope (AFM) tip with a force of $\sim 10^{-8}$ N between the tip and the surface, utilizing the weak adhesion and the large interface strain of thick ($\sim$1 μm) SiC films on Si due to the $\sim$20% lattice mismatch between SiC and Si. The AFM tip can also be used to manipulate and position the SiC films on a substrate. The SiC films produced through C_{60} precursors have a density of $\sim$83% of the ideal density for SiC. Patterned SiC films may find use in high-temperature electronics and in micromechanical systems (MEMS) applications utilizing their high strength modulus (310 GPa), hardness (26 GPa), high thermal conductivity, and low coefficient of friction (one half to one third that of Si) [20.87].

20.3.3. Catalytic Properties of C_{60}

A few reports have appeared in the literature referring to the catalytic activity of C_{60} [20.88–91]. Most of these reports have shown that C_{60} and C_{70} exhibit, at best, moderate catalytic activity. The choice of model compound reactions to evaluate their catalytic reactivity has been motivated by the tendency of fullerenes to form anions and hydrogenated adducts. Thus C_{60} has been tested as a catalyst for intermolecular hydrocarbon coupling (i.e., the formation of higher molecular weight derivatives), –C=C–; –C–C– bond cleavage, hydrogenation–dehydrogenation reactions, methane activation, and oxidation of organic solvents. These reactions are important in hydrocarbon refining, where an active catalyst is needed to improve the efficiency by which hydrogen is utilized. Malhotra *et al.* [20.89] investigated the properties of C_{60} and C_{70} as catalysts in the coupling, bond-cleavage, and dehydrogenation reactions of various organic solvents, such as 1,2-dinaphthylmethane ($C_{21}H_{16}$) and mesitylene [$C_6H_3(CH_3)_3$]. The results showed that C_{60} is relatively active in catalyzing these reactions. However, poor selectivity in the reaction products was observed, which implies that a wide range of products was obtained.

As an example of a hydrocarbon coupling reaction, fullerenes were extracted with organic solvents like mesitylene [$C_6H_3(CH_3)_3$] and 1,4-diethylbenzene and the products showed formation of dimers, trimers, and higher oligomers of these solvents, even at room temperature, while part of the C_{60} (10%) reacted with the solvent to form addition products.

As an example of a bond cleavage reaction, 1,2-dinaphthylmethane ($C_{21}H_{16}$) was reacted in the presence of aromatic solvents and C_{60}. The results showed that addition of C_{60} to the reaction medium increased the reaction rate constant by an amount that varied between 30% and 145%, depending on the aromatic solvent used for the reaction. In addition, C_{60} was also found to increase the fraction of dehydrogenated products from 1.3 to 33 [20.89].

Methane activation (formation of higher hydrocarbons from methane) using carbon soot containing fullerenes as a catalyst has also been investigated [20.88]. The results showed that the fullerene soot needed a lower onset temperature for the activation reaction compared to that of a high-surface-area activated carbon. However, it was not clear what role, if any, the nonfullerene components of the soot (e.g., carbon black) played in this reaction, since no reactions were carried out with a pure fullerene catalyst.

Fullerene-catalyzed oxidation and polymerization of organic solvents such as α-pinene ($C_{10}H_{16}$), 4-methyl-1-cyclohexene (C_7H_{12}), and pyridine (C_5H_5N) by C_{60} and C_{70} have been observed [20.90] in the presence of oxygen. Other forms of carbons (graphite and diamond) did not show any activity for the same oxidation reactions. It was found that only the solvents that possessed electron donor ability (such as the presence of carbon–carbon double bonds or a nitrogen lone pair) were converted by C_{60}. Contrary to typical catalytic reactions in which the catalysts are not consumed, in these oxidation reactions spectrometric evidence showed that the fullerenes were consumed as they were broken down into fragments by the attack from hydrocarbon radicals formed during the reaction.

C_{60} has also been found to be a more active catalyst for the conversion of H_2S into sulfur than neutral alumina, and charcoal mixed with alumina [20.92]. This reaction constitutes the main process for the industrial production of sulfur. Proton NMR experiments on the products formed from the reaction of H_2S bubbled through a solution of C_{60} in toluene showed evidence that C_{60} was also hydrogenated during this oxidation reaction.

Fullerenes can also be used as a catalyst support to increase the degree of dispersion of the active material. Nagashima *et al.* [20.93, 94] synthesized a palladium-carrying polymer (see §10.10.2) in which the metal is highly dispersed in the C_{60} matrix so that its usage as a catalyst is enhanced. This C_{60}–Pd polymer was found to catalyze the hydrogenation of diphenyl acetylene at standard temperature and pressure. Normally, a high loading of active metal is needed to catalyze such reactions.

20.3.4. Self-Assembled Monolayers

A rather popular research topic and possible future applications area for fullerenes concerns self-assembled monolayers [20.95, 96] and a fullerene literature in this subject is now beginning to appear [20.97–100]. We give here one simple example of a work already reported.

Layer-by-layer self-assembly has been demonstrated for a polycation monolayer such as PAH [poly(allylamine hydrochloride)] or PPV [poly(phenylene vinylene)] followed by a polyanion monolayer such as sulfonated C_{60} [20.101]. For the sulfonated C_{60}, several [$-(CH_2)_4-SO_3H$] sulfonic acid groups attach to each C_{60} molecule, and the layer-by-layer self-assembly is carried out by dipping a substrate into a solution of the polycation followed by dipping into the polyanion solution, and repeating this process, until the desired number of polycation–polyanion bilayers are assembled. The layer-by-layer assembly of bilayers is demonstrated in Fig. 20.18, where optical absorbance measurements at a wavelength of 420 nm are shown *vs* the number of C_{60}–PAH bilayers. Since only the C_{60} layers are absorbing at this optical wavelength, Fig. 20.18 shows that the absorbance is proportional to the number of bilayers, which is independently measured by a profilometer [20.101]. In the case of the PPV–C_{60} bilayers, both the anion and cation layers are absorbing near the absorption peak (~420 nm), as shown in Fig. 20.19. In contrast, for the C_{60}–PAH

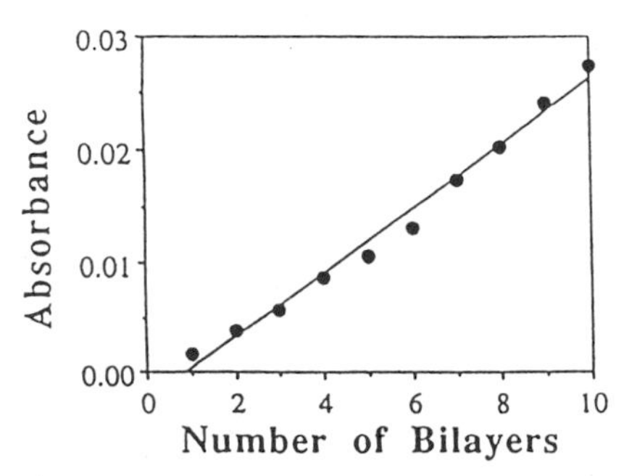

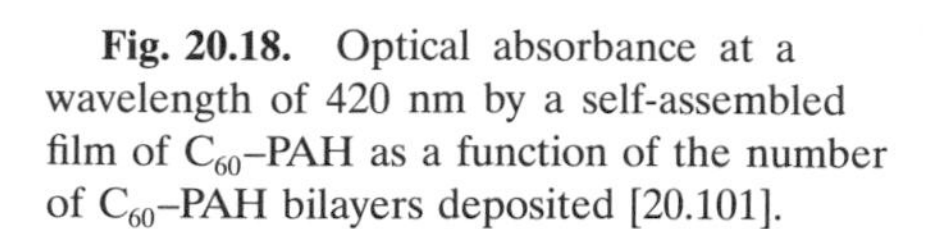
Fig. 20.18. Optical absorbance at a wavelength of 420 nm by a self-assembled film of C_{60}–PAH as a function of the number of C_{60}–PAH bilayers deposited [20.101].

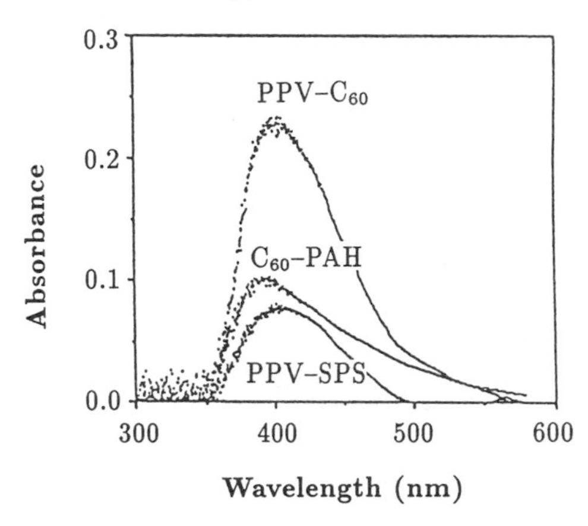

Fig. 20.19. Absorbance spectra in the visible part of the spectral range for films of C_{60}–PAH and PPV–C_{60} containing 15 bilayers. Also shown for comparison is the corresponding spectrum for a sample of 15 bilayers of PPV–SPS (sulfonated polystyrene) [20.101].

bilayers it is only the C_{60} monolayers that are absorbing in this frequency range. On the other hand, the photoluminescence of the PPV monolayers is quenched by the adjacent sulfonated C_{60} in PPV–C_{60} bilayers, as a result of charge transfer from the PPV to the C_{60}. This charge transfer process separates the electron and hole spatially, thereby suppressing electron–hole recombination.

20.3.5. New Chemicals

The synthesis of new fullerene-based chemicals with specific properties has been an active research field, with promise for specific applications to polymer science and technology, biology, and medicine. With regard to pharmaceutical applications, fullerene derivatives of the CAr_2 carbon bridge type have been discussed for practical applications to control the binding of drugs. The water-soluble CAr_2 fulleroid is shown in Fig. 10.12.

Recently, Wudl and co-workers have synthesized a water-soluble derivative of fullerene which appears to inhibit the activity of the HIV virus and has therefore attracted attention for possible use in the treatment of AIDS [20.102–104]. One fullerene-based compound, the bis(monosuccinimide) derivative of p, p'-bis(2-aminoethyl)diphenylC_{60}, has been shown to have selective activity against HIV-1 protease [20.102]. Another example of the many fullerene derivatives which are of medical interest is the water-solubilized fullerene–peptide conjugate which covalently links to the alpha-amino group of the hydrophilic 4–8 sequence of peptide T, known to display potent human monocyte chemotaxis [20.105]. Characterization of this water-soluble fullerene product by several techniques, including UV spectroscopy, shows high chemotactic potency, comparable to that of the parent pentapeptide [20.105].

An example of a fullerene derivative of interest for biological applications is the C_{60}–poly(vinylpyrrolidone) derivative, which shows a significant increase in cell differentiation of the rat limb bud cell (by a factor of 3.2), while the poly(vinylpyrrolidone) by itself inhibits cell differentiation [20.106].

Recently, an all-carbon ladder polymer (i.e., no heteroatoms, such as O or S are included in the polymer backbone) has been synthesized [20.107]. A precursor polymer was first obtained by a Diels–Alder polyaddition of a monomer [reaction (i), Fig. 2.21 in §2.18] and then converted (polymer analogous conversion) by dehydrogenation [reaction (ii), Fig. 2.21] into the fully unsaturated ladder polymer, which is planar. The resulting planar molecule resembles the structure found along the equatorial belt of a fullerene molecule after cutting the belt normal to its length and laying the cut belt onto a plane. It was proposed [20.107] that this class of polymers might combine favorable electronic properties with processability. Possible uses of these polymer-derived carbons may be for electroluminescence and photovoltaics applications. Interestingly, the polymer was found to be stable in oxygen. After exposure to air for several weeks, no apparent change in the UV–visible absorption spectrum was noticed [20.107].

20.4. Electrochemical Applications of C_{60}

As described in §10.4, C_{60} can be significantly hydrogenated (for example, forming $C_{60}H_x$, where $2 \leq x \leq 60$) by means of chemical, electrochemical, and catalytic methods. The upper limit for x (60) has not been confirmed in independent studies. Because of the high density of hydrogen storage per gram of $C_{60}H_x$ that is theoretically possible, this molecule offers the possibility of practical applications similar to those of metal hydrides. These applications include hydrogen storage, hydrogen fuel cells, and electrodes for primary and secondary batteries [20.108, 109].

The term battery refers to an array of electrochemical cells; for example, the common Pb–acid battery is an array of six 2-V cells connected in series to yield a 12-V battery. This battery is an example of a "secondary" battery (as is the common Ni–Cd battery) which is designed to be recharged. Secondary batteries contain lower-energy chemicals which are converted to higher-energy chemicals in the charging process; they return to the lower energy state as they are discharged. Primary batteries, on the other hand, are assembled in the high-energy chemical state and they are ready to convert chemical to electrical energy. Primary batteries are not intended to be recharged; the irreversibility is tied to the chemical reactions which occur during discharge that lead either to a loss of chemicals from the battery or to an irreversible chemical change which prevents full recharging. The

Leclanche cell, developed in the 1880s, is an example of a primary battery and used an amalgamated Zn anode, a MnO_2 cathode and an ammonium chloride electrolyte. Present-day Leclanche cells use an improved solid electrolyte and have an open-circuit voltage of 1.6 V when fresh; in cylindrical form, these batteries are known as the common "D" cell. "Fuel cells" are a special class of primary batteries in which the electrical energy is generated by continuously feeding high-energy reactants into the battery, while the electrical energy and low-energy reactants are continuously removed.

20.4.1. *Hydrogen Storage and Primary Batteries*

The possibility of using C_{60} for hydrogen storage has been explored by investigating the electrochemistry of C_{60} films on Ag electrodes in an aqueous electrolyte for the reduction of hydrogen [20.110]. C_{60} films of 0.5–2.0 μm in thickness were prepared by vacuum evaporation of the fullerenes on silver electrodes. A 5.0 M KOH electrolyte was used. Cyclic voltammetry measurements were done at a scan rate of 0.025 V/s between 0 and 2.5 V, and the results showed that C_{60} films under these conditions can be reduced reversibly. Calculations of the net charge transferred to C_{60}, together with laser ablation mass spectroscopy studies (which gave a molecular mass of 770 ± 5 amu), provided evidence for the hydrogenation of C_{60} to about a stoichiometry of $C_{60}H_{56}$.

The half-reactions proposed for the electrochemical processes at the cathode and anode are:

$$C_{60} + xH_2O + xe^- \rightleftharpoons C_{60}H_x + xOH^- \qquad \text{(cathode)}$$

$$xOH^- \rightleftharpoons \tfrac{x}{2}H_2O + \tfrac{x}{4}O_2 + xe^- \qquad \text{(anode)}$$

Although these electrodes are not fully rechargeable, they might be useful not only in energy storage cells but also in high energy density primary batteries. Before these electrodes are useful for practical applications, difficulties with battery reversibility and with the physical stability of the electrode will have to be overcome. Other studies report the production of $C_{60}H_{36}$ by Birch reduction (see §10.4) and of $C_{60}H_{60}$ by catalytic hydrogenation under high pressure [20.110].

20.4.2. *C_{60} Electrodes for Secondary Batteries*

Important factors in the performance of C_{60} and C_{60} composites as electrode materials for secondary batteries are the charging and discharging capacities of the battery at a constant current density. Figure 20.20 shows the results of the charging and discharging capacity of a C_{60}–Ag electrode (15% C_{60} by volume) at several charge–discharge current densities. During

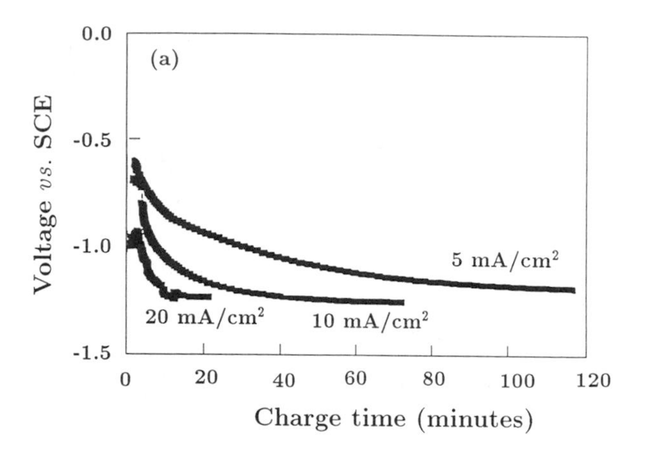

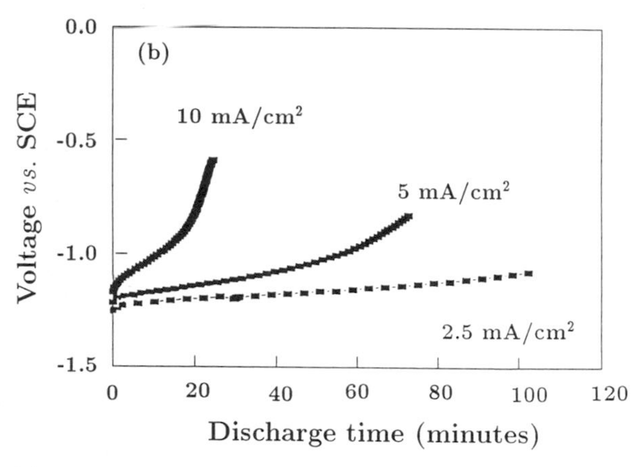

Fig. 20.20. (a) Charge and (b) discharge capacity of a C_{60}–Ag electrode at several constant current densities [20.111].

charging, C_{60} is hydrogenated to $C_{60}H_x$ and the reverse reaction (dehydrogenation) occurs during discharge. For both charging and discharging cycles, the final voltage is dependent on the current density. The capacity of the cell (calculated from the voltage–current curve and the current density), as well as its efficiency, was found to decrease with increasing current density: from 92% efficient for 5 mA cm^{-2} to 60% for 20 mA cm^{-2}. These efficiencies compare favorably with those obtained for an Ni–Cd cell under similar conditions: 71% efficiency for a sealed Ni–Cd cell and 67–80% efficiency for vented cells such as LiH or $LiAlH_4$, which incorporate a lower density of hydrogen atoms [20.110]. Due to the higher hydrogen storage capacity of C_{60} relative to metal hydrides, it is expected that a fullerene-based battery should be lighter per unit of stored charge than the present Ni–Cd batteries and that they should have a longer lifetime.

Also related to battery applications are the results of studies of the intercalation of Li in C_{60} (see §10.3.2). These investigations have shown that the maximum Li-to-C_{60} ratio obtained is 1 to 12 (see §2.21), which is within a factor of two of the Li:C ratio in first stage graphite intercalation compounds (LiC_6) of 1:6 [20.112]. However, the viable Li:C ratio for lithium battery applications is only 1:3, because above this composition the intercalation is not reversible [20.113]. This 1:3 ratio is judged to be too low for commercial applications [20.114]. So far, the success in the electrochemical intercalation of C_{60} with Li and other atoms does not appear to offer significant advantage in terms of specific battery capacity over standard graphite and other carbons. Furthermore, cyclic voltammetry studies and constant-current charge–discharge experiments on C_{60} and C_{70} have shown poor stability of the highly reduced material and poor electrochemical reversibility. These difficulties need to be overcome before C_{60} can be used in commercial battery applications [20.113, 114].

20.5. Other Applications

In this section we review several other areas where fullerenes have potential for applications, including nanotechnology, fullerene coatings, free-standing fullerene membranes, tribology, and lubrication, among others.

20.5.1. Nanotechnology

Nanotechnology is a new, potentially important field at the crossroads of physics, engineering, and material science, which searches for ways to manipulate atoms and molecules in efforts to build new materials and molecular devices. Nanotechnology can be traced back to Richard Feynman's landmark 1959 address entitled "There's Plenty of Room at the Bottom" [20.115] at an American Physical Society meeting, where he suggested the possibility of creating new materials by mechanically assembling molecules, *an atom at a time.* Concurrently, Von Hippel [20.116] predicted dramatic material science possibilities if advances in "molecular designing" and "molecular engineering" of materials could be achieved. Following their lead, other scientists delved into the field. Nanoscale fabrication principles were advanced and popularized later by Drexler [20.95, 96, 117] and many others.

Fullerenes, particularly C_{60}, are perhaps ideal for materials manipulations on a molecular scale, including the forming and tailoring of structures of molecular size. The C_{60} molecules are quite inert, nearly spherical, and nonpolar due to symmetry, their size is very large, the van der Waals bonds between the individual molecules are weak, and thin fullerene films can be grown easily. A C_{60} molecule can be compared to a huge atom, and it is, therefore, possible to manipulate such molecules using the same techniques that have been developed for the manipulation of noble gas atoms using STM probe tips [20.118]. Moreover, some other manipulation methods may be possible for C_{60} that are not possible for any other species. For instance, it has been proposed [20.119] that a C_{60} molecule can be "rolled" (diffused) along the surface of a suitable ionic substrate by a rotating external electric field, utilizing the large size and polarizability of C_{60} as illustrated in Fig. 20.21. A recent tribological study [20.120] suggested that nanocrystalline C_{60} islands grown on certain types of substrates [e.g., NaCl (011)] could be used as transport devices for fabrication processes of nanometer-sized machines. These 0.5–3 ML C_{60} islands might play the role of a tiny "nanosled" transporting larger molecules (e.g., biomolecules) to a desired location [20.120].

Since fullerenes are all-carbon molecules and have been shown to be of interest for biotechnology applications (see §20.3.5), fullerene-based nanodevices may be capable of biological interfacing. Nanoscale fabrication of logic information devices capable of such interfacing should enable construction of valuable nanoscale robots, sensors, and machines.

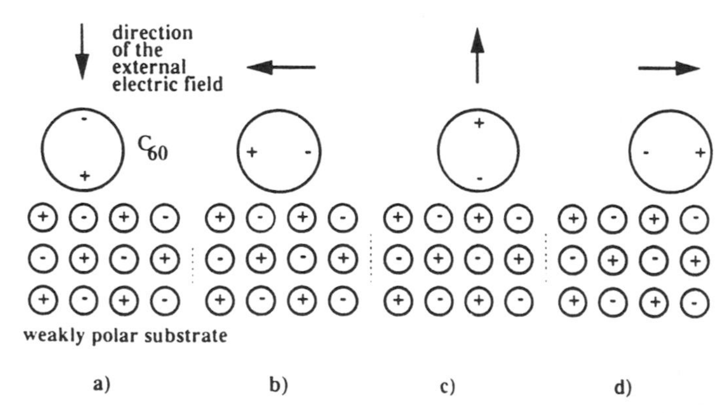

Fig. 20.21. The principle of moving C_{60} molecules by a rotating external electric field. An isolated molecule is assumed to be adsorbed on an ideally flat, weakly polar substrate. A rotating external electric field induces a dipole moment that exceeds the effect of the substrate dipoles on the molecule; the successive directions of the field are shown in (a) through (d). The large diameter of the molecule should allow coupling to successive poles, thereby generating lateral movement [20.119].

To interconnect these nanoscale devices, it may someday be possible to utilize carbon nanotubes for nanowire applications. As discussed in §19.5, by varying the chirality and diameter of carbon nanotubes, it should be possible to alter their properties from metallic to semiconducting to insulating. Such nanotubes could, in their own right, lead to nanoscopic electronic devices based on concentric semiconducting and metallic carbon tubules as shown in Fig. 20.22. The significance of carbon nanotubes as electronic materials is the demonstration of quasi-1D cylindrical wires with a large aspect (length-to-diameter) ratio, thereby opening a new field of 1D physics for the study of 1D cylindrical wires or hollow tubes. There is also the possibility of the application of nanotubes filled with appropriate guest species with regard to low-dimensional transport, magnetism, and superconductivity. The

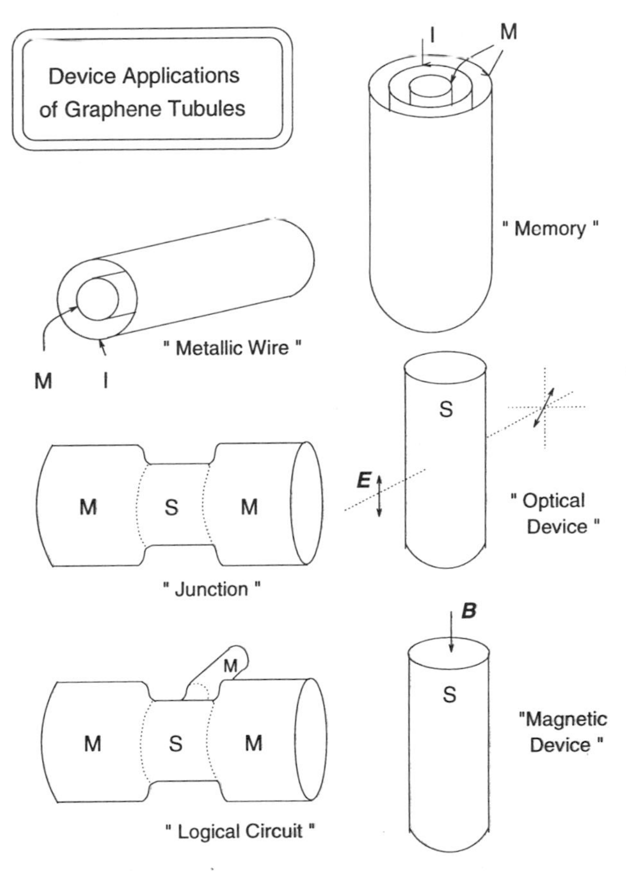

Fig. 20.22. Schematic presentation of several proposed electronic device applications for carbon nanotubes [20.123].

heterostructures shown in Fig. 20.22 could function as an insulation-clad metallic wire, a tunnel junction, a capacitor in memory devices, or a transistor in switching circuitry [20.121, 122]. Because the tubes do not require any doping by impurities, as conventional semiconductors do, but acquire their electronic properties from their geometry, the resulting devices should be highly thermally stable and have high intrinsic mobility. Schematics for several devices have been suggested (see Fig. 20.22).

To facilitate such applications, the mechanical and electronic properties of a free-standing carbon nanowire have been measured [20.124]. By employing the low-energy electron projection microscope for *in situ* visual control, it was possible to attach or detach a fine manipulating tip to or from a carbon fiber network. In this way it was possible to apply an electrical potential to a single nanometer-sized fiber while observing the electron projection images at TV sampling rates. The maximum current passed through individual free-standing wires of diameter ~ 10 nm and lengths $\sim 1\ \mu m$ exceedcd a current density of 10^6 A cm^{-2} at room temperature [20.124].

An alternative method of constructing an ultrathin wire involves filling the hollow cavity of the nanotubes with metallic atoms arranged much like "peas in a pod" [20.125, 126]. Although the lengths of the filled nanowires thus far produced are impractically short [20.126], the use of carbon nanotubes as 1D quantum wires merits further attention.

While the field of nanotechnology is still about halfway between science and science fiction, it is a very active fast-growing field whose promise of nanoscale molecular devices and orders of magnitude increase in the integration density of electronic components is too appealing to be ignored. The discovery of fullerenes has given a significant boost to the field of nanotechnology by providing an abundance of stable, highly symmetric, nonreactive, and relatively large molecules that can, in principle, be manipulated one at a time. Fullerenes at semiconductor interfaces can be utilized to modify electronic device behavior on nanometer length scales. In such applications, C_{60} anions alone represent an almost ideal spherical capacitor that can store charge. Together with other carbon structures, such as tubules (serving as nanowires), a new carbon-based nanotechnology can be envisioned.

20.5.2. Fullerene Coatings for STM Tips

A significant improvement in the atomic resolution of STM images obtained from a graphite surface was found when using metal tunneling tips (such as platinum or platinum–rhodium) containing adsorbed fullerenes on the tip surface. In this case, the tunneling current is directed through these fullerene molecules on the tip. The improvement in STM image quality, forming sharper and more symmetrical images, is attributed both to a reduction in the area of closest approach between the tip and the surface (because of the curvature of the fullerene) and to the increased stabilization of the spatial gap between the tip and the surface, thereby reducing the noise associated with mechanical vibrations. The fullerene-bearing tips are prepared by direct vacuum evaporation or by repetitive scanning of the metal tip over a graphite surface upon which several monolayers of C_{60} molecules had previously been deposited [20.127]. The strong bonding of C_{60} to Pt or Pt–Rh stabilizes the C_{60} on the tip surface. The adsorbed fullerenes remain attached to the tip for several scans. Fullerenes do not possess any of the problems normally associated with semiconductors when considered as candidates for tip materials. C_{60} has no surface states, the local tip morphology is known and is always the same, and variations in faceting and band structure effects are not important [20.127]. For these reasons C_{60}-coated tips are especially attractive as a semiconducting tip for scanning tunneling spectroscopy (STS) applications.

20.5.3. Fullerene Membranes

Free-standing membranes of C_{60} films 2000–6000 Å thick with areas as large as 6×6 mm^2 have been fabricated [20.128] and supported on an Si frame (see §9.2.2). Despite the weak van der Waals bonding between C_{60} molecules, these membranes were found to be robust and have already proven valuable for measurement of the mechanical properties of C_{60} [20.128, 129]. Other potential scientific uses might exploit the low density and low atomic number of the free-standing membrane for use as a sample holder for measurements such as electron energy loss spectroscopy and transmission electron microscopy [20.128]. Fullerene membranes might also have potential for practical applications as a semitransparent material for soft x-rays and as a micropore filter, allowing small molecules to pass through the membrane, but not larger molecules. The sequence of deposition and etching steps necessary to fabricate these C_{60} membranes is shown in Fig. 9.7.

20.5.4. Tribology—Lubricants

A number of experiments have also been done assessing the lubricating properties of pristine fullerene films [20.130–133]. The resilience, load-bearing capacity, low surface energy, chemical stability, and spherical shape of C_{60} molecules and their weak intermolecular bonding offer significant potential for selected mechanical and tribological applications [20.131]. Fullerenes thus may be of interest for higher-temperature applications, for rust prevention, wear protection, and antioxidizing coatings, although it is believed that specific functionalization of fullerenes will be needed for specific applications. It has been speculated that fullerenes can roll like tiny ball bearings when placed between two sliding surfaces and thus may offer the best lubrication performance achieved thus far. In practice, however, fullerenes in some cases tend to clump and compress into high shear strength islands and layers [20.130], while in other cases [20.132] they may bond strongly to the substrate and not slide, or they readily form an undesirable transfer film on the mating slider surface. Therefore, a way of achieving wider fullerene particle dispersion, combined with the proper amount of adhesion of the fullerene molecules to the substrate (perhaps by choosing an appropriate substrate), must be discovered or developed before pristine fullerenes can start being useful as lubricants.

Nanotribology studies [20.134–137] using scanning, atomic, and frictional force microscopies also demonstrate the importance of both the substrate and the film surface morphology for the lubricating properties of fullerene films. For instance, it has been shown that well-ordered epitaxially grown C_{60} films can have significantly better lubricating properties than a layered material (such as graphite, a commonly used solid lubricant) [20.137]. On the other hand, frictional measurements carried out to monitor the buckling of an AFM cantilever [20.135] show increased friction for C_{60}-covered surfaces relative to clean silicon and mica substrates where a fullerene film consisting of disordered 40–60 nm microcrystallites was used.

The most potentially important application of fullerenes in lubrication engineering to date is their use as additives in oils and greases to reduce friction and wear under boundary and mixed-film lubrication conditions [20.138]. Under such conditions, the liquid lubricant film thickness is on the order of the surface roughness of the sliding surfaces, and material-to-material contact may occur between the two surfaces. Solid lubricant additives added to oils and greases tend to prevent or delay seizure at the contact regions [20.138]. The friction and wear performance of C_{60}-rich powder as an additive to oil and grease was found to be comparable to that of other well-known solid lubricants, such as graphite and MoS_2. These improvements have been attributed [20.138] to the formation of a transfer film at the contact region and to the presence of C_{60} molecules acting as tiny ball bearings to facilitate sliding.

Ever since C_{60} was discovered [20.139], the possibility of using fluorinated derivatives such as $C_{60}F_{60}$ as lubricants was considered. It has, however, been shown that highly fluorinated C_{60} is unstable in the presence of trace amounts of H_2O [20.140] and therefore is probably unsuitable for application as a lubricant.

20.5.5. *Separations*

C_{60}, C_{60}–polymer, and the residue remaining after a toluene extraction of C_{60} and C_{70} from arc-derived carbon soot have been investigated as stationary phases for gas chromatography and for the liquid chromatography separation of PAHs (polyaromatic hydrocarbons) and PCBs (polychlorinated biphenyls). Jinno and co-workers [20.141] investigated the use of pure C_{60} as the stationary phase for the separation of PAHs. Due to the high solubility of C_{60} in most common chromatographic mobile phases (e.g., toluene), it is doubtful that this approach will find widespread application. However, the higher fullerenes C_{n_C} ($n_C > 70$) are much less soluble and therefore could overcome this objection. Of course, the expense to separate and accumulate sufficient amounts of higher fullerenes would seem to preclude their use in large chromatography columns.

One approach to generating a relatively inexpensive column material containing higher fullerenes is to use the residue remaining after a toluene extraction of C_{60} and C_{70}. This has been investigated by Stalling *et al.* [20.142]. They found better performance for the separation of coplanar PCBs using this residue than with a commercial column. However, given the small amount of higher fullerenes in the residue ($<1\%$) , it is not clear whether the improved chromatography was derived from the presence of higher fullerenes in the residue or from the special nature of the arc-derived carbon black in the residue.

Finally, Stalling *et al.* [20.143] have investigated the performance of polystyrene divinylbenzene (PSDVE) covalently bonded to C_{60} and C_{70} for the high-pressure chromatographic separation of coplanar PCBs from other PCB isomers. In this study, hexane was found to be the best mobile phase. The separation mechanism was reported to be the "electron donor–acceptor" (EDA) mechanism in which the retention time is governed by the formation of charge transfer complexes of the molecules to be separated with the stationary phase. They found that PSDVE-C_{60}/C_{70} provided enhanced performance over commercial columns for the chromatographic enrichment of coplanar PCBs, PCDDs (polychlorinated dibenzodioxins), and PCDFs (polychlorinated dibenzofuran).

20.5.6. *Sensors*

The possible use of C_{60} films as sensors for volatile polar gases such as NH_3 and tetrachloroethylene has been considered, exploiting the decrease in film resistance with gas adsorption and the resulting charge transfer. A linear dependence of the film resistance on gas concentration has been demonstrated as well as reversible and reproducible behavior on time scales of hours [20.144]. Sensitivity levels of a few ppm NH_3 in air have already been achieved, though many challenges remain before giving serious consideration to this potential application, including selectivity to the detection of specific gas species, response time (now on the order of seconds), sensitivity of the film calibration to humidity levels, stability of the sensor over longer times (days), and cost/performance comparison with alternative sensors for the same gas.

Other sensor applications for fullerenes and fullerene-derived materials have been considered. A good source for these possible applications can be found in the patent literature (see §20.6).

20.6. Commercialization and Patents

Ironically, fullerenes may have been commercialized even before they were discovered [20.145]. In 1984, Bill Burch, a biomedical researcher at the Australian National University in Canberra, developed a new material now known by the trade name Technegas, which is used in the diagnosis of a life-threatening condition called pulmonary embolism, to which postoperative patients are particularly susceptible.

Since 1986, when Technegas was first commercialized by Tetley Manufacturing, a Sydney-based firm, it has been a very successful product, and clinical trials for a new application of this product in the United States are now under way. Technegas is a technetium-bearing aerosol that is inhaled with lead-bearing smog and offers penetration in the lungs superior to that with conventional techniques [20.146, 147]. In his discovery, Burch unwittingly may have become the first person to successfully utilize fullerenes [20.147]. It was not until a year later that these molecular cages were discovered at Rice University [20.139]. Although fullerenes were produced under the same conditions as Burch used to make Technegas, it is not yet known whether the beneficial properties of Technegas are directly connected to fullerenes. Some structural evidence has been presented [20.147] to show that the structure of Technegas is compatible with a model where the radioactive ^{99}Tc isotope is trapped by a fullerene.

It was not until after 1990, when fullerenes were first synthesized in macroscopic quantities [20.148], that the unique physical, chemical, and biological properties of fullerenes were fully appreciated, and fullerenes were used to build some prototype devices. Fullerenes, therefore, now appear ripe for commercial exploitation. Indeed, suggested fullerene uses have become a hot topic at the U.S. Patent Office [20.149].

We summarize in Table 20.1 a wide range of possible practical fullerene applications for which a patent has been granted or a patent application has been disclosed. As suggested by the great number of polymer-related patents, it seems likely that the use of C_{60} and other fullerenes as build-

Table 20.1

Some uses of fullerenes suggested by patents and patent applications.

Use	Patent[a]
Optical applications	
Electrophotographic imaging	WO 9308509 A1, US 5215841, JP 06222575 A2, JP 05289381 A2, JP 05088387 A2, US 5215841, US 5178980, JP 6-298532 A, JP 7-1546 A
Sensitizer of a photorefractive polymer	US 5361148
Solar cells from fullerenes	JP 06244440 A2
Electroluminescent cell	JP 05331458 A2
Magneto-optical recording	JP 06002106 A2
Optical limiters (flash goggles)	US 5172278, US 5325227
Nonlinear optical thin films	JP 5-142597
Photoelectronic devices	WO 9405045, US 5171373, US 5331183
Superconductors	US 5294600, US 5325002, US 5332723, US 5380703, US 5356872, US 5348936
Electronic applications	
Carbon films	JP 06184738
Electrophoretic display	JP 06148693 A2
Bilayer resist-coated substrate	JP 06061138 A2
Tunnel diode	JP 06029556 A2
Photolithography, photopolymer	US 08-037056 A, US 08-073533 A
Double layer capacitor and storage device	US 5319518
Atomic scale electronic switch	US 5323376
Conductive compounds	US 5391323, US 5380595
Materials applications	
Oriented diamond film	WO 9426953 A1
Nucleating diamond growth	WO 9305207 A1
Diamond-like carbon coating	US 5393572
Conversion of fullerenes to diamond	US 5209916, US 5273788, JP 05305227 A2, US 5308661, US 5328676, US 5360477, US 5370855
Physical and chemical separations	
Chromatography materials	WO 9324200 A1, US 5308481
Product separation in diamond fabrication	FR 2674090 A1
Absorbent for gases	JP 06063396 A2
Photoinduced physisorption of oxygen	US 08-073533 A
Sensors	
NO_x detector	JP 06186195 A2
Sensor for analyses in fluid media	DE 4241438 A1, US 5334351
Gas, temperature, particle sensor	JP 06118042, JP 5-099759, DE 4142959
Detection of organic vapors	US 5395589

continues

Table 20.1 *continued*

Use	Patent[a]
Biological and medical applications	
Fullerene-coated surfaces and cell culture	WO 9400552 A1
Diagnostic and/or therapeutic agents	WO 9315768 A1
Biologically active material	JP 3-94692
Magnetic resonance imaging	WO 9303771 A1
Chemicals	
Selective functionalization of fullerenes on molecular sieves	US 5364993
Synthesis of derivatized fullerenes	EP 546718 A2
Lubricants	JP 05179269 A2, US 5374463
Lubricants and adhesives	US 5382718, US 5382719, US 5354926
Charge transfer complexes	WO 9302012 A1
Polysubstituted fullerenes	US 5166248
Liquid ink compositions	US 5114477, US 5322756
Surface coating	US 5120669, US 5368890
Cosmetics	JP 5-17327, JP 5-17328
Catalyst	US 5336828, US 5347024
H_2O_2 production	US 5376353
Polymer chemistry and technology	
Fullerene–diamine adducts	DE 4229979 A1
Fullerene-grafted additives to lube oil	US 5292444
Grafted polymers	US 5292813
Anionically polymerized C_{60}–polymers	US 5270394
Modified viscoelastic polymer properties	EP 544513 A1
Others	US 5276085, US 5278239, US 5281653, US 5294732, US 5296536, US 5296543, US 5302681, US 5367051, US 535079A, DE 4221279 A1
Electrochemical applications	
Secondary battery	JP 05101850 A2
Li secondary batteries	JP 07006764 A2, JP 05275078 A2, US 5302474, JP 05082132 A2, JP 05275083 A2
Nonaqueous batteries	JP 05325974 A2, JP 0531477 A2
Fuel cell electrodes	WO 9422176 A1, US 5277996, JP 05229966 A2, JP 527801 A2
Other applications	
Storage of nuclear materials	US 5350569
Solid-state rocket fuel	US 5341639
Ballistic missile shield	US 4922827
Metallurgy (additives to steel alloys)	US 5288342
Identification tracer in fuels	US 5234475
Electromagnetic shielding resin	JP 5124894

[a]US = United States patent; JP = Japanese; FR = French; DE = German; WO = International patent application; EP = European patent application.

ing blocks for new chemicals will be a very important future application. Rechargeable batteries may turn out to be another promising application, due to the fact that reversible attachment of hydrogen to C_{60} could provide a charge storage per unit mass that is better than current metal hydride battery technology [20.150]. Somewhat similar considerations apply to hydrogen storage, perhaps for vehicle propulsion. Use of fullerenes for conversion to other solid products, such as diamond or SiC, may be of value in the future. As a component of solid fuel rocket propellant, fullerenes or fullerene soot may have practical advantages. Fullerenes as optical limiters and as a component in chromatography columns appear to be small-scale applications already under consideration [20.150].

One of the largest impediments to the large-scale commercialization of fullerenes is their relatively high cost. Although the price has come down from about $1000–$2500 per gram for mixed fullerenes in 1990 to less than $100 per gram for 99% pure C_{60} today, this is still prohibitively expensive for most commercial applications (e.g., batteries). This cost is bound to decrease further, however, since new methods of synthesis, production, purification, separation, processing, and deposition of various fullerenes and their numerous compounds are constantly being discovered.[4] Recently, a joint venture company, Fullerene Technologies Inc., was formed by Mitsubishi and MER (Tucson, Arizona) to explore the commercialization of fullerene-based batteries and hydrogen-storage devices. This joint venture announced plans for a plant having a fullerene production capacity of a few tons per day [20.151]. Other companies known to be actively working and/or publishing on fullerene applications include such giants as Exxon (fullerene precursors), duPont (fullerene-based compounds), Xerox (fullerene inks, toners, developers, and photoreceptors), SRI (fullerene catalysts, electronics, and fibers), NEC (fullerene-based polymers), and AT&T (fullerene photoconductors, superconductors, and photolithography) [20.152, 153].

References

[20.1] J. E. Fischer. *Science*, **264**, 1548 (1994).

[20.2] H. U. ter Meer. In H. Kuzmany, J. Fink, M. Mehring, and S. Roth (eds.), *Progress in Fullerene Research: International Winter School on Electronic Properties of Novel Materials*, p. 535 (1994). Kirchberg Winter School, World Scientific Publishing Co., Ltd., Singapore.

[20.3] S. A. Jenekhe, S. K. Lo, and S. R. Flom. *Appl. Phys. Lett.*, **54**, 2524 (1989).

[20.4] L. W. Tutt and T. F. Boggess. *Progress in Quantum Electronics*, **17**, 299 (1993).

[20.5] L. W. Tutt and A. Kost. *Nature (London)*, **356**, 225 (1992).

[4]Some examples of such patents: U.S. Pat. No. 5,275,705, 5,281,406, 5,300,203, 5,304,366, 5,310,532, 5,316,636, 5,324,495, 5,338,529, 5,338,571, 5,346,683 (bucky tubes synthesis method), 5,348,936, 5,354,926, 5,364,993.

[20.6] J. E. Wray, K. C. Liu, C. H. Chen, W. R. Garrett, M. G. Payne, R. Goedert, and D. Templeton. *Appl. Phys. Lett.*, **64**, 2785 (1994).

[20.7] N. M. Dimitrijevic and P. V. Kamat. *J. Phys. Chem.*, **96**, 4811 (1992).

[20.8] M. R. Wasielewski, M. P. O'Neil, K. R. Lykke, M. J. Pellin, and D. M. Gruen. *J. Am. Chem. Soc.*, **113**, 2774 (1991).

[20.9] Y. Zeng, L. Biczok, and H. Linschitz. *J. Phys. Chem.*, **96**, 5237 (1992).

[20.10] T. W. Ebbesen, Y. Mochizuki, K. Tanigaki, and H. Hiura. *Europhys. Lett.*, **25**, 503 (1994).

[20.11] I. V. Bezel, S. V. Chekalin, Y. A. Matveets, A. G. Stepanov, A. F. Yartsev, and V. S. Letokhev. *Chem. Phys. Lett.*, **218**, 475 (1994). Erratum: ibid vol. 221, p. 332, 1994.

[20.12] N. S. Sariciftci, D. Braun, C. Zhang, V. I. Srdanov, A. J. Heeger, C. Stucky, and F. Wudl. *Appl. Phys. Lett.*, **62**, 585 (1993).

[20.13] N. S. Sariciftci, L. Smilowitz, C. Zhang, V. I. Srdanov, A. J. Heeger, and F. Wudl. *Proc. SPIE - Int. Soc. Opt. Eng.*, **1852**, 297 (1993).

[20.14] B. Kraabel, C. H. Lee, D. McBranch, D. Moses, N. S. Sariciftci, and A. J. Heeger. *Chem. Phys. Lett.*, **213**, 389 (1993).

[20.15] K. Lee, R. A. J. Janssen, N. S. Sariciftci, and A. J. Heeger. *Phys. Rev. B*, **49**, 5781 (1994).

[20.16] N. S. Sariciftci and A. J. Heeger. *Int. J. Mod. Phys. B*, **8**, 237 (1994).

[20.17] N. S. Sariciftci, A. J. Heeger, and F. Wudl. *J. Appl. Phys., Jpn. Solid State Devices and Materials, Intl. Conf. on Solid State Devices and Materials*, pp. 781–784 (1993). 25th International Conference, Chiba, Japan.

[20.18] Y. Wang. *Nature (London)*, **356**, 585 (1992).

[20.19] N. S. Sariciftci, L. Smilowitz, A. J. Heeger, and F. Wudl. *Science*, **258**, 1474 (1992).

[20.20] M. E. Orczyk and P. N. Prasad. *Photonics Science News*, **1**, 3 (1994).

[20.21] M. E. Orczyk, B. Swedek, J. Zieba, and P. N. Prasad. In G. R. Möhlmann (ed.), *Proceedings of the International Society for Optical Engineering (SPIE). "Nonlinear optical properties of organic materials VII"; Proc. vol. 2285*, pp. 166–177, Bellingham, WA (1994). SPIE Optical Engineering Press. San Diego, CA, July 24-29, 1994.

[20.22] S. Ducharme, J. C. Scott, R. J. Twieg, and W. E. Moener. *Phys. Rev. Lett.*, **66**, 1846 (1991).

[20.23] S. Ducharme, A. Gooneskera, B. Jones, J. M. Takacs, and L. Zhang. *Opt. Soc. Am. Tech. Digest*, **17**, 232 (1993).

[20.24] S. Morita, A. A. Zakhidov, and K. Yoshino. *Solid State Commun.*, **82**, 249 (1992).

[20.25] T. Kato, T. Kodama, T. Shida, T. Nakagawa, Y. Matsui, S. Suzuki, H. Shiromaru, K. Yamauchi, and Y. Achiba. *Chem. Phys. Lett.*, **180**, 446 (1991).

[20.26] T. Kato, T. Kodama, M. Oyama, S. Okasaki, T. Shida, T. Nakagawa, Y. Matsui, S. Suzuki, H. Shiromaru, K. Yamauchi, and Y. Achiba. *Chem. Phys. Lett.*, **186**, 35 (1991).

[20.27] A. Hirao, H. Nishizawa, H. Miyamoto, M. Sugiuchi, and M. Hosoya. In Z. H. Kafafi (ed.), *Proceedings of the International Society for Optical Engineering (SPIE)*. Bellingham, WA (1995). SPIE Optical Engineering Press. San Diego, CA, July 13, 1995.

[20.28] Y. Wang, R. West, and C. H. Yuan. *J. Am. Chem. Soc.*, **115**, 3844 (1993).

[20.29] R. G. Kepler, J. M. Zeigler, L. A. Harrah, and S. R. Kurtz. *Phys. Rev. B*, **35**, 2818 (1987).

[20.30] K. Yoshino, X. H. Yin, S. Morita, T. Kawai, and A. A. Zakhidov. *Solid State Commun.*, **85**, 85 (1993).

[20.31] S. M. Silence, C. A. Walsh, J. C. Scott, and W. E. Moerner. *Appl. Phys. Lett.*, **61**, 2967 (1992).

[20.32] G. C. Valley and M. B. Klein. *Opt. Eng.*, **22**, 704 (1983).

[20.33] W. E. Moerner and S. M. Silence. *Chem. Rev.*, **94**, 127 (1994).

[20.34] Y. Cui, Y. Zhang, J. S. Prasad, J. S. Schildkraut, and D. J. Williams. *Appl. Phys. Lett.*, **61**, 2132 (1992).

[20.35] P. N. Prasad, M. E. Orczyk, B. Swedek, and J. Zieba. Opt. Soc. Amer./Amer. Chem. Soc., Washington, DC (1994).

[20.36] Y. Zhang, Y. P. Cui, and P. N. Prasad. *Phys. Rev. B*, **46**, 9900 (1992).

[20.37] M. C. J. M. Donkers, S. M. Silence, C. A. Walsh, F. Hache, D. M. Burland, W. E. Moemer, and R. J. Twieg. *Optics Lett.*, **18**, 1044 (1993).

[20.38] Y. Zhang, S. Ghosal, M. K. Casstevens, and R. Burzynski. *Appl. Phys. Lett.*, **66**, 256 (1995).

[20.39] A. Kost, L. Tutt, M. B. Klein, T. K. Dougherty, and W. E. Elias. *Optics Lett.*, **18**, 334 (1993).

[20.40] A. F. Hebard, O. Zhou, Q. Zhong, R. M. Fleming, and R. C. Haddon. *Thin Solid Films*, **257**, 147 (1995).

[20.41] C. J. Wen, T. Aida, I. Honma, and H. Komiyama. *DENKI KAGAKU*, **62**, 264 (1994).

[20.42] Q.-Y. Tong, C. B. Eom, U. Gösele, and A. F. Hebard. *J. Electrochem. Soc.*, **141**, L137 (1994).

[20.43] A. V. Hamza, M. Balooch, R. J. Tench, M. A. Schildbach, R. A. Hawley-Fedder, H. W. H. Lee, and C. McConaghy. *J. Vacuum Sci. Tech. B*, **11**, 763 (1993).

[20.44] A. F. Hebard. *Annual Rev. Mater. Sci.*, **23**, 159 (1993).

[20.45] P. Zhou, A. M. Rao, K. A. Wang, J. D. Robertson, C. Eloi, M. S. Meier, S. L. Ren, X. X. Bi, and P. C. Eklund. *Appl. Phys. Lett.*, **60**, 2871 (1992).

[20.46] C. Eloi, J. D. Robertson, A. M. Rao, P. Zhou, K. A. Wang, and P. C. Eklund. *J. Mater. Res.*, **8**, 3085 (1993).

[20.47] H. G. Busmann, H. Gaber, H. Strasser, and I. V. Hertel. *Appl. Phys. Lett.*, **64**, 43 (1994).

[20.48] R. F. Pierret. *Field Effect Devices*. Addison–Wesley, Reading, MA (1983). Modular Series on Solid State Devices; Vol. 4.

[20.49] J. Kastner, J. Paloheimo, and H. Kuzmany. In *Proc. of the Int. Winter School on Electronic Properties of High Temperature Superconductors (IWEPS'92)*, vol. 113, pp. 512–515, New York (1993). H. Kuzmany, M. Mehring, and J. Fink (eds.). Springer-Verlag.

[20.50] J. Paloheimo, H. Isotalo, J. Kastner, and H. Kuzmany. *Synth. Metals*, **56**, 3185 (1993).

[20.51] K. Hoshimono, S. Fujimori, and S. Fujita. *Jpn. J. Appl. Phys. Part 2, Lett.*, **32**, L1070 (1993).

[20.52] R. C. Haddon, A. S. Perel, R. C. Morris, T. T. M. Palstra, A. F. Hebard, and R. M. Fleming. *Appl. Phys. Lett.* **67**, 121 (1995).

[20.53] D. Sarkar and N. J. Halas. *Solid State Commun.*, **90**, 261 (1994).

[20.54] E. Frankevich, Y. Maruyama, H. O. Y. Achiba, and K. Kikuchi. *Solid State Commun.*, **88**, 177 (1993).

[20.55] E. Frankevich, Y. Maruyama, and H. Ogata. *Chem. Phys. Lett.*, **214**, 39 (1993).

[20.56] K. M. Chen, Y. Q. Jia, K. Wu, X. D. Zhang, W. B. Zhao, C. Y. Li, and Z. N. Gu. *J. Phys. Condensed Matter*, **6**, L367 (1994).

[20.57] K. M. Chen, Y. Q. Jia, S. X. Jin, K. Wu, and C. Y. Li. MRS Symposium H, paper H7.3 (April 1995).

[20.58] K. M. Chen, Y. Q. Jia, S. X. Jin, K. Wu, X. D. Zhang, W. B. Zhao, C. Y. Li, and Z. N. Gu. (unpublished).

[20.59] H. Yonehara and C. Pac. *Appl. Phys. Lett.*, **61**, 575 (1992).
[20.60] K. Pichler, M. G. Harrison, R. H. Friend, and S. Pekker. *Synth. Metals*, **56**, 3229 (1993).
[20.61] L. Smilowitz, A. W. Hays, G. Wang, A. J. Heeger, and J. E. Bowers. (unpublished).
[20.62] J. B. Camp and R. B. Schwarz. *Appl. Phys. Lett.*, **63**, 445 (1993).
[20.63] Y. B. Zhao, D. M. Poirier, R. J. Pechman, and J. H. Weaver. *Appl. Phys. Lett.*, **64**, 577 (1994).
[20.64] A. M. Rao, P. Zhou, K.-A. Wang, G. T. Hager, J. M. Holden, Y. Wang, W. T. Lee, X.-X. Bi, P. C. Eklund, D. S. Cornett, M. A. Duncan, and I. J. Amster. *Science*, **259**, 955 (1993).
[20.65] P. Zhou, A. M. Rao, K. A. Wang, J. D. Robertson, C. Eloi, M. S. Meier, S. L. Ren, X. X. Bi, P. C. Eklund, and M. S. Dresselhaus. *Appl. Phys. Lett.*, **60**, 2871 (1992).
[20.66] P. C. Eklund, A. M. Rao, P. Zhou, Y. Wang, and J. M. Holden. *Thin Solid Films*, **257**, 185 (1995).
[20.67] Y. Wang, J. M. Holden, X. X. Bi, and P. C. Eklund. *Chem. Phys. Lett.*, **217**, 413 (1994).
[20.68] A. F. Hebard, C. B. Eom, R. M. Fleming, Y. J. Chabal, A. J. Muller, S. H. Glarum, G. J. Pietsch, R. C. Haddon, A. M. Mujsce, M. A. Paczkowski, and G. P. Kochanski. *Appl. Phys. A*, **57**, 299 (1993).
[20.69] K. B. Lyons, A. F. Hebard, D. Inniss, R. L. Opila, Jr., H. L. Carter, Jr., and R. C. Haddon. (unpublished).
[20.70] R. S. Ruoff. *Interface*, **3**, 30 (1994).
[20.71] A. V. Hamza, J. Dykes, W. D. Mosley, L. Dinh, and M. Balooch. *Surf. Sci.*, **318**, 368 (1994).
[20.72] H. Hong, W. E. McMahon, P. Zschack, D. S. Lin, R. D. Aburano, H. Chen, and T. C. Chiang. *Appl. Phys. Lett.*, **61**, 3127 (1992).
[20.73] H. Hong, R. D. Aburano, E. S. Hirshorn, P. Schack, H. Chen, and T. C. Chiang. *Phys, Rev. B*, **47**, 6450 (1993).
[20.74] M. N. Regueiro, P. Monceau, and J.-L. Hodeau. *Nature (London)*, **355**, 237 (1992).
[20.75] R. Meilunas, R. P. H. Chang, S. Z. Lu, and M. M. Kappes. *Appl. Phys. Lett.*, **59**, 3461 (1991).
[20.76] R. Meilunas and R. P. H. Chang. *J. Mater. Res.*, **9**, 61 (1994).
[20.77] M. Balooch and A. V. Hamza. *Appl. Phys. Lett.*, **63**, 150 (1993).
[20.78] A. V. Hamza and M. Balooch. *Chem. Phys. Lett.*, **201**, 404 (1993).
[20.79] T. Hashizume, X. D. Wang, Y. Nishina, H. Shinohara, Y. Saito, Y. Kuk, and T. Sakurai. *Jpn. J. Appl. Phys.*, **31**, L880 (1992).
[20.80] T. Sakurai, X. D. Wang, T. Hashizume, Y. Nishina, H. Shinohara, and Y. Saito. *Appl. Surf. Sci.*, **67**, 281 (1993).
[20.81] X.-D. Wang, T. Hashizume, H. Shinohara, Y. Saito, Y. Nishina, and T. Sakurai. *Phys. Rev. B*, **47**, 15923 (1993).
[20.82] Y. Z. Li, M. Chander, J. C. Patrin, J. H. Weaver, L. P. F. Chibante, and R. E. Smalley. *Phys. Rev. B*, **45**, 13837 (1992).
[20.83] H. Xu, D. M. Chen, and W. N. Creager. *Phys. Rev. Lett.*, **70**, 1850 (1993).
[20.84] D. Chen and D. Sarid. *Phys. Rev. B*, **49**, 7612 (1994).
[20.85] M. Moalem, M. Balooch, A. V. Hamza, W. J. Siekhaus, and D. R. Olander. *J. Chem. Phys.*, **99**, 4855 (1993).
[20.86] A. V. Hamza, M. Balooch, and M. Moalem. *Surf. Sci.*, **317**, L1129 (1994).
[20.87] M. Balooch and A. V. Hamza. *J. Vac. Sci. Technol.*, **B12**, 3218 (1994).
[20.88] H. J. Wu, A. S. Hirschon, R. Malhotra, and R. B. Wilson. *Preprints of Papers, ACS, Fuel Chemistry Div.* **39**, 1233 (1994).

[20.89] R. Malhotra, D. McMillen, D. S. Tse, D. C. Lorents, R. Ruoff, and D. Keegan. *Energy & Fuels*, **7**, 685 (1993).

[20.90] J. Cohen, N. M. Lawandy, and E. M. Suuberg. *Energy & Fuels*, **8**, 810 (1994).

[20.91] A. Darwish, H. W. Kroto, R. Taylor, and D. Walton. *Fullerene Science & Technology*, **1**, 571 (1993).

[20.92] K. Shigematsu, K. Abe, M. Mitani, and K. Tanaka. *Fullerene Science & Technology*, **1**, 309 (1993).

[20.93] H. Nagashima, A. Nakaoka, Y. Saito, M. Kato, T. Kawanishi, and K. Itoh. *J. Chem. Soc. Chem. Commun.*, pp. 377–380 (1992).

[20.94] J. M. Cowley, M.-Q. Liu, B. L. Ramakrishna, T. S. Peace, A. K. Wertsching, and M. R. Pena. *Carbon*, **32**, 746 (1994).

[20.95] K. E. Drexler. *Nanosystems: Molecular Machinery, Manufacturing, and Computation*. John Wiley & Sons, New York (1992).

[20.96] A. Ulman. *An Introduction to Ultrathin Organic Films: from Langmuir–Blodgett to Self-Assembly*. Academic Press, New York (1991).

[20.97] R. E. Smalley. *Accounts Chem. Res.*, **25**, 98 (1992).

[20.98] D. Q. Li and B. I. Swanson. *Langmuir: ACS J. Surf. Colloids*, **9**, 3341 (1993).

[20.99] K. Chen, B. Caldwell, and C. Mirkin. *J. Am. Chem. Soc.*, **115**, 1193 (1993).

[20.100] J. K. Gimzewski, S. Modesti, and R. R. Schlittler. *Phys. Rev. Lett.*, **72**, 1036 (1994).

[20.101] M. Ferreira, M. F. Rubner, and B. R. Hsieh. In A. F. Garito, A. K. Y. Jen, C. Y. C. Lee, and L. R. Dalton (eds.), *Electrical, Optical, and Magnetic Properties of Organic Solid State Materials, MRS Symposia Proceedings*, vol. 328, p. 119, Pittsburgh, PA (1994). Materials Research Socicty Press.

[20.102] R. F. Schinazi, R. Sijbesma, G. Srdanov, C. L. Hill, and F. Wudl. *Antimicrobial Agents and Chemotherapy*, **37**, 1707 (1993).

[20.103] S. H. Friedman, D. L. DeCamp, R. P. Sijbesma, F. Wudl, and G. L. Kenyon. *J. Am. Chem. Soc.*, **115**, 6506 (1993).

[20.104] R. Sijbesma, G. Srdanov, F. Wudl, J. A. Castoro, C. L. Wilkins, S. H. Friedman, D. L. DeCamp, and G. L. Kenyon. *J. Am. Chem. Soc.*, **115**, 6510 (1993).

[20.105] C. Toniolo, A. Bianco, M. Maggini, G. Scorrano, M. Prato, M. Marastoni, R. Tomatis, S. Spisani, G. Palu, and E. D. Blair. *J. Med. Chem.*, **37**, 4558 (1994).

[20.106] T. Tsuchiya, Y. N. Yamakoshi, and N. Miyata. *Biochem. Biophys. Res. Commun.*, **206**, 885 (1995).

[20.107] A. D. Schluter, M. Loffler, and V. Enkelmann. *Nature (London)*, **368**, 831 (1994).

[20.108] P. R. Gifford, in Kirk-Othmer. *Encyclopedia of Chemical Technology*, Ed., M. Howe-Grant. vol. 3, pp. 963–1121. John Wiley & Sons. 4th edition. (1992).

[20.109] K. Kinoshita and E. J. Cairns, in Kirk-Othmer. *Encyclopedia of Chemical Technology*, Vol. 11, pp. pp. 1098–1121. John Wiley & Sons, 4th edition (1992).

[20.110] D. Koruga, S. Hameroff, J. Withers, R. Loutfy, and M. Sundareshan. *Fullerene C_{60}: History, Physics, Nanobiology, Nanotechnology*. North-Holland, Amsterdam (1993). Chapter 10.

[20.111] A. I. Sokolov, Y. A. Kufaev, and E. B. Sonin. *Physica C*, **212**, 19 (1993).

[20.112] M. S. Dresselhaus and G. Dresselhaus. *Adv. Phys.*, **30**, 139 (1981).

[20.113] J. Chlistunoff, D. Cliffel, and A. J. Bard. *Thin Solid Films*, **257**, 166 (1995).

[20.114] L. Seger, L. Q. Wen, and J. B. Schlenoff. *J. Electrochem. Soc.*, **138** (1991).

[20.115] R. P. Feynman. *There's plenty of room at the bottom*. John Wiley & Sons, New York (1961).

[20.116] A. R. von Hippel. *Science*, **138**, 91 (1962).

[20.117] K. E. Drexler. *Engines of Creation: The Coming Era of Nanotechnology*. Anchor Press/Doubleday, New York (1986).

[20.118] J. A. Stroscio and D. M. Eigler. *Science*, **254**, 1319 (1991).

[20.119] J. Viitanen. *J. Vacuum Sci. Tech. B: Microelectronics Processing, Phenomena*, **11**, 115 (1993).

[20.120] R. Lüthi, E. Meyer, H. Haefke, L. Howald, W. Gutmannsbauer, and H.-J. Güntherodt. *Science*, **266**, 1979 (1994).

[20.121] S. Saito, S. I. Sawada, N. Hamada, and A. Oshiyama. *Mater. Sci. Eng.*, **B19**, 105 (1993).

[20.122] R. Saito, M. Fujita, G. Dresselhaus, and M. S. Dresselhaus. *Mat. Sci. Eng.*, **B19**, 185 (1993).

[20.123] R. Saito (1994). private communication.

[20.124] H. Schmid and H. W. Fink. *Nanotechnology*, **5**, 26 (1994).

[20.125] C. M. Jin, T. Guo, Y. Chai, A. Lee, and R. E. Smalley. In C. Taliani, G. Ruani, and R. Zamboni (eds.), *Proceedings of the First Italian Workshop on Fullerenes: Status and Perspectives*, vol. 2 of *World Scientific Advanced Series in Fullerenes*, pp. 21–29, World Scientific, Singapore (1992).

[20.126] R. E. Smalley and P. Norlander. *private communication* (1993).

[20.127] J. Resh, D. Sarkar, J. Kulik, J. Brueck, A. Ignatiev, and N. J. Halas. *Surf. Sci.*, **316**, L1061 (1994).

[20.128] C. B. Eom, A. F. Hebard, L. E. Trimble, G. K. Celler, and R. C. Haddon. *Science*, **259**, 1887 (1993).

[20.129] V. I. Orlov, V. I. Nikitenko, R. K. Nikolaev, I. N. Kremenskaya, and Y. A. Ossipyan. *JETP Lett.*, **59**, 667 (1994).

[20.130] P. J. Blau and C. E. Haberlin. *Thin Solid Films*, **219**, 129 (1992).

[20.131] B. Bhushan, B. K. Gupta, G. W. Vancleef, C. Capp, and J. V. Coe. *Appl. Phys. Lett.*, **62**, 3253 (1993).

[20.132] B. Bhushan and B. K. Gupta. *J. Appl. Phys.*, **75**, 6156 (1994).

[20.133] B. K. Gupta, B. Bhushan, C. Capp, and J. V. Coe. *J. Mater. Res.*, **9**, 2823 (1994).

[20.134] J. Ruan and B. Bhushan. *J. Mater. Res.*, **8**, 3019 (1993).

[20.135] T. Thundat, R. J. Warmack, D. Ding, and R. N. Compton. *Appl. Phys. Lett.*, **63**, 891 (1993).

[20.136] C. M. Mate. *Wear*, **168**, 17 (1993).

[20.137] W. Allers, U. D. Schwartz, G. Gensterblum, and R. Wiesendanger. *Appl. Phys. A*, **59**, 11 (1994).

[20.138] B. K. Gupta and B. Bhushan. *Lubrication Engineering*, **50**, 524 (1994).

[20.139] H. W. Kroto, J. R. Heath, S. C. O'Brien, R. F. Curl, and R. E. Smalley. *Nature (London)*, **318**, 162 (1985).

[20.140] R. Taylor, A. G. Avent, T. J. Dennis, J. P. Hare, H. W. Kroto, D. R. M. Walton, J. H. Holloway, E. G. Hope, and G. J. Langley. *Nature (London)*, **355**, 27 (1992).

[20.141] K. Jinno, K. Yamamoto, J. C. Fetzer, and W. R. Biggs. *J. Microcolumn Separations*, **4**, 187 (1992).

[20.142] D. L. Stalling, C. Y. Guo, K. C. Kuo, and S. Saim. (unpublished).

[20.143] D. L. Stalling, C. Y. Guo, and S. Saim. *J. Chromatographic Sci.*, **31**, 265 (1993).

[20.144] A. W. Synowczyk and J. Heinze. vol. 117, pp. 73–77, Berlin (1993). Springer-Verlag. Springer Series in Solid-State Sciences. H. Kuzmany, J. Fink, M. Mehring and S. Roth (eds.).

[20.145] B. Johnstone. *Far Eastern Economic Review*, **155**, 78 (1992).

[20.146] D. W. J. Mackey, G. D. Willett, and K. J. Fisher. *Journal of Nuclear Medicine*, **34**, P232 (1993).

[20.147] D. W. J. Mackey, W. M. Burch, I. G. Dance, *et al.* *Nuclear Medicine Commun.*, **15**, 430 (1994).
[20.148] W. Krätschmer, L. D. Lamb, K. Fostiropoulos, and D. R. Huffman. *Nature (London)*, **347**, 354 (1990).
[20.149] T. W. Ebbesen. *Nature*, **361**, 218 (1993).
[20.150] D. R. Huffman. *Mater. Lett.*, **21**, 127 (1994).
[20.151] *Chemical Marketing Reporter*, **245**, 3 (1994).
[20.152] R. M. Baum. *Chemical and Engineering News*, **71**, 8 (1993).
[20.153] S. Moore. *Chemical Engineering*, **99**, 41 (1992).

Index

A

A15 structure, 254–256, 568, 619, 641
Ablation (*see* Laser ablation)
Absorption (*see* Optical properties)
Acceptor doping, 37, 39, 41, 231, 236, 256, 258, 288, 416, 533, 557, **572–573**,[1] 664, 748, 874, 907
Acoustic modes (*see* Vibrational modes)
Activated carbons (*see* Carbon materials)
Activation energy, 157, 191, 194, 388, 400, 541, 564, **567–569**, 593–595, 597, 607, 660, 716, 717, 749, 885
Adamantane (*see* Cage structure)
Addition reactions (*see* Chemical reactions)
Adhesion, 697, 710, 888, **889–891**, 894, 905, 906, 910
Adsorption of fullerenes, 697, 700, 712–714, **719–733**, 905
  layer-by-layer, 710
  of gases, 907
Aerogels, 35, 36, 212
Ag surfaces (*see* Substrates)
Aharonov–Bohm effect (*see also* Carbon nanotubes), 818, 819, 823
Aliphatic chains, 48
Alkali metal ions (*see also* Alkali metal fullerides), 226, 235, **240**, 241, 244, 249–253, 303, 323, 390, 445, 517, 560, 563, 582, 621, 646, 696, 748, 863
  ionic radii, 238, **240**, 244, 250–252, 255, 389, 563, 617, 642, 657, 696
Alkali metal fullerides (M_xC_{60}), 38, 225–228, 234, 235, **238–253**, 285, 286, 299, 301, 303, 316, 330, 360, 378–380, 386, 387, 423, 424, **442–447**, 455, 511, **518–532**, 544, 556, **557–566**, 598, 600, 617, 623–625, 662, 665, 676, 679, 690, 702, 707, 729, 747
  binary compounds, 235, 253
  Cs_xC_{60}, 238, 379, 559, 560, 563, 621, 696, **745–748**
    Cs_1C_{60}, 235, 242, 247, 663, 664, **745–748**
    Cs_3C_{60}, 235, 240, 243, 253, 292, 559, 563, 617, 619, 620
    Cs_4C_{60}, 235, 243, 247, 559, 747, 748
    Cs_6C_{60}, 176, 235, 243, 247, 249, 250, 255, 376, 377, 380, 396, 444, 518–521, 523, 702
  intermetallic $M_xM'_{3-x}C_{60}$ compounds, 235, 243, 252, 253, 285–287, 299, 388–390, 445, 563, 617–621, 623, 625, 627
    Ca_2CsC_{60}, 449
    Na_2CsC_{60}, 243, 445, 446, 620, 621
    Na_2KC_{60}, 243, 253, 563
    Na_2RbC_{60}, 243, 251, 252
    $RbCs_2C_{60}$, 243, 617, 619, 620, 625, 657, 664, 665

[1]Note that major sections on a given subject are designated by bold-face.

Alkali metal fullerides (*continued*)
Rb_2CsC_{60}, 243, 252, 621, 625, 637, 657, 664, 665, 667, 668
K_xC_{60}, 238, 239, 297, 379, 380, 382–387, 444, 445, 521, 557–559, 563, 662, 663, 691, 692, 695, 696, 730
K_1C_{60}, 235, 239, 242, **247**, 322, 323, 376, 384, 386, 446, 558, 564, 663, 664, 749
K_2C_{60}, 446, 662
K_3C_{60}, 235, 239, **241–247**, 253, 287, 292, 330, 367, 368, 375–378, 380, 384, 388–390, 405, 406, 428, 442–447, 455, 456, 521, **523–531**, 544, 557–563, 565, 566, 573, 574, 580–585, 587, 598, 601, 602, 604–608, 616, 617, 619, 621, 622, 624, 625, 627–635, 637, 639, 641–643, 645, 646, 656, 657, 662–668, 676, 678, 683, 684, 690, 696, 697, 702, 708, 712, 718, 724, 730
K_3C_{60} (properties), 560
K_4C_{60}, 235, 238, 239, 242, **247–248**, 384, 445, 558
K_6C_{60}, 176, 235, 238, 239, 242, 247, **249–250**, 255, 314, 376–378, 380, 382, 384, 396, 405, 406, 442–445, 455, **517–523**, 558, 623, 662, 663, 690, 696, 702, 724
Li_xC_{60}, 235, 238, 619
metallic phases, 380
Na_xC_{60}, 235, 238, 240, 242, 244, **249–252**, 379, 388–390, 445, 518, 559, 560, 563, 564, 619–621, 696
Na_1C_{60}, **248**, 386
Na_2C_{60}, **242**, 244, 250, 251, 559, 563
Na_3C_{60}, **251**
Na_6C_{60}, **250–251**, 376, 377
$Na_{11}C_{60}$, **251**
Rb_xC_{60}, 379, 382, 385–387, 444, 543, 557–560, 563, 696, 707, 708
Rb_1C_{60}, 235, 242, **247–248**, 380–382, 386, **531–532**, 563, 663, 664, 749
Rb_3C_{60}, 235, **241–247**, 253, 287, 292, 330, 367, 368, 375–378, 380, 382, 383, 386–388, 390, 428, 442–444, **523–531**, 544, 557, 560, 561, 563, 565, 573, 574, 583–585, 587, 602, 604–607, 617, 619, 622, 625, 627, 629, 631–635, 637, 639, 641–643, 645, 657, 662, 664–668, 676, 678, 683, 684, 702, 707, 708, 712, 718
Rb_3C_{60} (properties), 560
Rb_4C_{60}, 235, 242, 247, 248, 250, 382, 387, 663
Rb_6C_{60}, 176, 235, 242, 247, **249–250**, 255, 376, 377, 380, 382, 383, 387, 396, 444, **517–523**, 623, 702, 707, 708
synthesis, 235, 283, **285–287**
Alkali metal fullerides (M_xC_{70}), 235, **262–263**, 393–395, 424, 557, 604, 605, 616, 696, 707
Cs_6C_{70}, 395, 397
K_xC_{70}, 746
K_1C_{70}, 569, 570, 696, 747
K_4C_{70}, 395, 398, 569, 570, 580, 581, 583, 604, 608, 696, 746, 747
K_6C_{70}, 395–397, 696, 708
Rb_xC_{70}, 707, 708
Rb_3C_{70}, 707
Rb_6C_{70}, 395, 397, 707
Alkaline earths
ionic radii, 236, **240**, 254, 255, 566, 568, 621, 622, 696
Alkaline earth C_{70} (M_xC_{70}), **569–570**
Alkaline earth fullerides (M_xC_{60}), 38, 225, 234, 235, 240–242, **253–256**, 299, 330, 424, 445, **447–450**, 556, 557, **566–570**, 621–623, 696
Ba_xC_{60}, 235, **254–256**, 287, **567–569**, 621
Ba_3C_{60}, 242, **255**, 256, 287
$Ba_3Sr_3C_{60}$, 450, 567
Ba_5C_{60}, 568
Ba_6C_{60}, 242, **254–256**, 287, **447–449**, **568**, 621, 623, 696
Ca_xC_{60}, 235, **254–255**, 287, **567–568**, 621, 696
CaC_{60} cluster, 448
Ca_3C_{60}, **447**, 567
Ca_5C_{60}, 242, **254**, **447–448**, 622, 623, 696
charge transfer, 236, 253–256, 567, 568, 623
Sr_xC_{60}, 235, **254–255**, 287, **567–568**, 621
Sr_3C_{60}, 242, **254**, **567**
Sr_4C_{60}, **567**
Sr_6C_{60}, 242, **255**, **447–449**, 621, 623
structure, **254–256**
superconductors, 241, 253, 287, 299, 621, 622, 646

synthesis, 235, 285, **287–288**
Alkenes, 295, 304
polyalkene, 293
Alkyls, 293, 304, 306
Allylic radical, 315
poly(allylamine hydrochloride) (PAH), 896
Amine, 296, 306, 314, 323, 879, 880, 896, 910
aminopolymer, 323
Ammonia (NH_3) addition, 226, 227, 243, **253**, 286, 287, 563, 619–621
$(NH_3)K_3C_{60}$, **253**, 287, 621
$(NH_3)_4Na_2CsC_{60}$, **253**, 287, 620, 621
Amorphous materials, **33–34**, 207, 212, 403, 277, 280, 403, 597, 604, 765, 784, 851, 862, 883, 889, 891
carbon, 23, **33–34**, 207, 212, 765, 784
Angular momentum quantum numbers, 341, 417, 424–429, 465, 470, 472
Anions, 38, 155–157, 225, 227, 241, 242, 245, 246, 249, **511–518**, 531, 534–537, 557, 559, 560, 563, 580, 620, 635, 636, 645, 674, 678, 715, 721, 726, **744–745**, 747–749, 751, 894, 896, 897, 904
C_{60}^{-}, 425, 427, 428, **511–518**, 532, 580, 673–676, 744, 873, 874, 881
C_{60}^{2-}, 425, 427, 428, **511–516**, 673–676, 717, 744
C_{60}^{3-}, 390, 424, 425, 427, 428, 430–433, **511–516**, 560, 566, 577, 621, 648, 657, 662, 664, 673, 674, 676–678, 744, 745
C_{60}^{4-}, 425, 428, **511–516**, 744, 748
C_{60}^{5-}, 428, 673, 744
C_{60}^{6-}, 251, 300, 424, 518, 520, 523, 662, 744
C_{60}^{n-}, 298–300, 330, 331, 376, 384, **424–429**, **511–513**, 516, 518, 580, 646, **744**, 747, 748
C_{70}^{n-}, 298, 300, **432–434**
dopant interaction, 243, 244, 247, 251–253, 255, 257, 289, 317–320, 378, 379, 384, 390, 451, 535, 536, 563, 572, 619, 671, 673
Anisotropy, 200, 261
Annulene structure, 259, 308
Anode (*see* Electrode)
Applications, 13, 21, 34, 40, 46, 47, 124, 209, 274, 281, 304, 310, 317, 487, 488, 532, 533, 588, 689, 785, 857, 863, **870–911**
battery (*see* Batteries)
coatings for STM tips, 901, **904–905**, 909
electrochemical, **898–901**, 909
electronic (*see also* Electronics applications), **880–892**, 909
materials (*see* Materials applications),
membranes, 901, 905, 909
nanotechnology, **901–904**, 909
nanotube-related devices, 814, 817, **903–904**
new chemicals, **897–898**
optical (*see* Optical applications)
other applications, 908–911
patents, **908–911**
sensors, **907–908**, 909
separations, 907, 910
tribology-lubricants, 863, 901, **905–906**, 910
Arc discharge (*see also* Carbon arc), 30, 49, **111–115**, 131, 761, 764, 765
Architecture, **7–8**
Armchair tubule (*see* Carbon nanotubes)
Aromaticity, 295, 308, 312, 325
Arrhenius law, 400, 541, 572, 587, 636, 666, 715, 717, 885
Arts and crafts, 8, 9
Astronomical observations, 3, 5–7
Atomic force microscope (AFM), **714–715**, 761, 762, 776, 777, 894, 906
Au surfaces (*see* Substrates)
Auger electron spectroscopy (AES), 437, 438, 455, 457, 689, 696, **708–712**, 713, 726, 729, 891, 892, 894
KVV spectra, 455, **710–711**
lineshapes, 455, 696, **710**
many-body effects, 455, 457, 689, 696, **708**, **710–711**
scanning, 708
stoichiometric dependence, 689, 708, 894
surface composition, 689, 708, 710, 891, 892, 894
surface studies, 689, **708–709**
temperature dependence, 709, 891, 892
Axisymmetric modes, 336, 337

B

Bacteriophage, 9, 10, 12
Bands (*see* Electronic structure)
Basis functions (*see* Group theory)
Batteries (*see also* Electrochemistry), 32, 302, 304, **898–901**, 910, 911
capacity, 900, 901

Batteries (*continued*)
  charge time, 900
  charging, 300, 303, 898–901
  current density, 899, 900
  discharge time, 900
  discharging, 300, 303, 898–901
  efficiency, 900
  electrodes (*see* Electrodes)
  electrolyte, 300, 302, 899
  energy conversion, 898, 899
  energy storage, 899
  fuel cells, 899, 910
  irreversible, 898, 901
  Leclanche cell, 899
  light weight, 901, 911
  lithium battery, 901, 910
  open circuit voltage, 300, 302, 898, 899
  primary, 898, 899
  recharge, 898, 899, 911
  reversible, 300, 898, 899, 901
  secondary, 898, **899–901**, 910
  solid electrolyte, 899
  stability, 901
  voltage–current curve, 900
BCS theory (*see* Superconductivity)
Belt modes, 97, 101–104, 391, 392, 395
Benzene, 546, 741, 742
Binding energy per C atom, 63, 68, 153
Biological ties, **9–12**, 898, 902, 910
Birch reaction (*see* Chemical reactions)
Body-centered cubic (bcc) structure, 176, **226**, 240, 241, **242**, 249, **250–251**, 254–256, 262, 442, 443, 445, 447, 448, 518, 563, 568, 620, 623, 662, 663, 730
Body-centered tetragonal (bct) structure, **226**, 239, 240, **248**, 262, 619
Bohr magneton (*see* Magnetic properties)
Bond angle, 758
Bond angle disorder, 43
Bond bending, 338, 391
Bond cleavage, 894, 895
Bond lengths, 67, 82, 237
  a_5, **60–63**, 65, 81, 174, 208, 209, 227, 261, 414, 416, 440, 656
  a_6, **60–61**, **63**, 65, 81, 174, 208, 209, 227, 261, 414, 416, 440, 656
  C–C, 16–19, 40, **60–63**, 69, 70, 233, 247, 255, 258, 378, 385, 403, 516, 517, 565, 573, 587, 643, 656, 760, 792
  C–M, 250, 390, 436, 437, 450, 560
  C_{60}–C_{60}, 171, **172**, **248**, **250–252**, 255, 258, 261, 276, 292, 308, 403, 439, 440, 442, 532, 556, 560, 564, 566, 623, 647, 656, 699, 700, 722, 723, 727, 749
  C_{60}–polymer, 874
  C_{70}, 67–69, 391
  C_{70}–C_{70}, 258, 700
  C_{84}–C_{84}, 731
  C_{n_C}–C_{n_C}, 731, 732
  I–I, 257
  intermolecular C–C, 573, 587, 643, 647, 699, 700, 723, 749
  M–C_{60}, 240, 248
  M–M, 250, 450
Bond strength (*see also* Surface–fullerene interaction)
  C_{60}–C_{60}, 689, 894, 905
Bond stretching, 338, 391, 401
Bonding (*see also* van der Waals)
  bridge, 297, 308, 310–313, 897
  carbon bridge, 310, 322
  covalent, 40, 210, 211, 228, 234, 258, 296, 310, 312, 316, 320, 398, 399, 508, 897, 907
  double, 60, 61, 65, 99, 174, **185–189**, 192, 193, 197, 205, 211, 216, 227, 243, 260, **292–297**, 304, **306–309**, 311, 313, 323, 324, 353, 354, 390, 414, 441, 541, 597, 603, 656, 725, 816, 895
  ionic, 228, 316, 519
  Kekulé (*see also* Bond alternation), 65, 579
  metal bridge, 310–313
  oxygen bridge, 310
  π, 37, 62, 85, 91–95, 148, 295, 296, 298, 309, 312, 313, 319, 378, 414–418, 421, 427, 436, 450, 453, 474, 476–478, 544, 576, 593, 636, 645, 646, 691, 694, 703, 705, 707, 708, 752, 807, 863
  single, 60, 61, 65, 294, 295, 305, 309, 314, 414, 656
  sp^1, 18, 32
  sp^2 (trigonal), 5, 15, 18, **32–35**, 37, 41–43, 60, 61, 294, 414, 416, 417, 453, 756, 851, 853
  sp^3 (tetrahedral), 15, 18, **32–34**, 37, 40–44, 61, 294, 414, 851, 893
  σ, 91, 92, 144, 148, 414, 416, 417, 453, 523, 694, 703, 705, 707, 708
  trigonal, 15, 60, 61, 414

Born–von Kárman model (*see* Vibrational modes)
Boron doping, 41, 233, 234, 292, 818
  BC_{59}, 233
Boron nitride (BN), 279, 814
Bravais lattice, 85, 96, 760
Breathing ($A_g(1)$) mode (*see also* Raman scattering), 333, 336, 337, 372
Breit–Wigner–Fano lineshape (*see also* Raman scattering), 340, 378
Brillouin scattering, 211, 212, 347, 397
Brillouin zone, **19**, 28, 180, 183, 335, 341, 373, 445, **801–803**, 806, 807
  zone folding, 178, 183, 340, 341, 372, 373, 792, **802–805**, **809**, 840, 841, 844, 845, 848, 849, 863
Bromine (*see* Halogens)
Buckminster Fuller, 7
Buckminsterfullerene, 7, 44, 45
  bucky ball, 7
  "Bucky Clutcher", 124, 135
  bucky dumbbell, 215
  bucky onion (*see* Carbon onion)
  bucky tube (*see* Carbon nanotubes)
Bulk modulus, 17, 22, 23, **172**, 199, 204, 205, 207, 440, 560, 765

C

C_2 pairs, **143–151**, **153–154**, 167, **210**, 786, 788
  C_2 absorption, **147–150**, 154, 167, 168, **210**, 785–788
  C_2 emission, **150**, **153–154**, **157–163**, 168, 210, 228, 229, 231
C_3 trimers, 786, 788
C_{20}, 49, 69, 70, 95
C_{28}, 451
C_{50}, 163
$C_{60}F_{36}$, 307
$C_{60}F_{60}$, 307, 907
C_{62}, 155
C_{70}, 64, 66–69, 72, 73, **96–104**, 112, 113, 118, 121, 122, 127, 131, 143, 144, 150, 155, 163, **197–202**, 207, **390–395**, 703, 705, 731, 732, 740–742, 746, 747, 751, 753, 871, 872, 874, 883, 891, 893–895, 901, 907
  abundance, 64, 66
  applications, 871, 872, 874
  atomic sites, 67
  band filling, 569
  belt hexagons, 67, **99**, **101–104**, 391, 394, 395, 742
  binding energy per C atom, 66, 199, 432
  bond lengths, 67, 68, 199, 433
  carrier localization, 563
  catalytic properties, 894, 895
  chemistry, 293, 297, 298, 300, 306
  core level spectroscopy, 703, 705
  D_{5h} symmetry, **96–104**
  Debye temperature, 598
  diamond nucleation application, 893
  dielectric constant, 199, 538, 705
  dielectric function, 537, 538, 705
  doped, 145, 258, 262, 263, 395, **569–570**, 580, 581, 587, 593, 594, 605, 608
  dynamics of excited states, 483–487, 541
  electrochemistry, 300, 301, 901
  electron spectroscopy, 699, **702–705**
  electronic structure (*see also* Electronic structure), **418–419**, **432–434**, 730
  enthalphy change, 598
  EPR studies, 486, 569, **679–680**, 753
  equivalence transformation, 101
  excited state lifetimes, 484
  exciton state energies (*see also* Electronic structure), 484, 538, 539, 705
  extraction, 111, 116
  growth, 150, 151, 730
  Hall effect, 580
  high temperature phase, 201
  intermolecular interaction, 539
  inverse photoemission, 693, 706
  ionization potential, 199
  isolated pentagon rule, 64
  lattice constant, 199, 262, 699, 700, 731
  lattice modes, **390–395**
  lifetime of triplet state, 485–487
  low temperature phase, 201
  magnetic susceptibility, 740–742, 746, 751, 753
  metal-coated, 53
  modulus, 199
  molar volume, 207
  molecular alignment, 198–201, 258
  molecular diameter, 67, 199
  molecular solid, 539
  non-linear optics, 544–547
  NMR, 654, 655, 657–661

C_{70} (*continued*)
optical limiting, 871, 872
optical properties, 464, **485–491**, 498, **537–540**, 544
phase transitions, 198–201, 207, 594, 598
pressure dependent resistivity, 586, 587
photoconductivity, 587, 593
photoemission, 693, 694, 696, 706, 718
polymer composites, 874
purification, 111, 127
Raman spectra, 207, **390–392**, 397, 718
rhombohedral phases, **198–200**, 206, 207
separations, 907
shape, 67, 96, 104, 655, 659, 679, 700, 714
size, 714
specific heat, 594, 595, 598
structure, 47, 64, **66–74**, **197–203**, 699, 700
sublimation, 199
surface binding, 710, 730
surface-enhanced Raman effect, 718
symmetry, 71, **96–104**, 537, 541
synthesis, 110–113, 115
TDAE–C_{70} (*see* TDAE), 751
thermal conductivity, 605
thermopower, 581
transient absorption, 485, 486
transport properties, 569, 570
xerography application, 593
C_{74}, 127
C_{76}, 66, 67, **72–73**, 122, 124, 143, **202–203**, **452–453**, 655, 693, **699–700**, **703–706**, 720, 731, 732
C_{78}, 67, **71–73**, 122, 127, 143, 720, 731, 732
C_{80}, 69, **70**, **72**, 75, 76, 90, 101, **202–203**, **452**
C_{82}, 67, **72**, **202–203**, **452–453**, 693, 720, 731, 732
C_{84}, 67, **71–73**, 121, 143, 155, **202–203**, **452–453**, 693, 694, **699–700**, **703**, **705**, 706, 720, 731, 732, 753
C_{140}, **70**, **72**, 75, 76, 90, 772
higher mass fullerenes, 67, 70, 72, 73, 90, 102–105, 110, 112, 116, 128, **153–154**, 202, 331, 390, 753
Cage structure, 1, 5, 6, 7, 8, **44–46**, **48–50**, **60–65**, 74, 75, 143, 158, 228, 294, 417, 419, 777, 860, 908
adamantane, 44, 45
buckminsterfullerene, 44
cage opening, 294, 710, 716, 729, 730, 893
carbon-free fullerenes, **50–52**
cubane, 44
diamantane, 44, 45
dodecahedrane, 44, 46
metallo-carbohedrenes, **48–50**
tetramantane, 44, 45
triamantane, 44, 45
twistane, 44, 46
Calcium (*see also* Alkaline earths), 447
Capacitance, 493, 506, 904
Carbide particle, **29–31**, 782
Carbolites, **32–33**
Carbon arc, 10, 21, 29, 30, 32, 110, **111–115**, 131–134, 138, 155, 156, 761, 778, 779, 782, 786, 789–791, 850–853, 859, 860, 862
inert gas cooling, 21, 111–113, 115, 132, 778, 779, 782, 789, 850, 862
Carbon-carbon bonding (*see* Bonding)
Carbon-carbon distance (*see* Bond lengths)
Carbon materials, 11, **15–58**
activated carbons, 34, 35, 895
aerogels, 34–36
all-carbon polymers, **46–48**
amorphous carbon, 23, **32–34**, 207, 212, 403, 765, 784, 851, 891
arc deposit, 779, **780**, 781, 829, 833, 834, 837, 838, 850–852
cage hydrocarbons, 31, **44–46**
carbynes, 16, 22
carbon blacks, 18, **25–29**, 112, 116, 853, 860, 895, 907
carbon clusters, 3–6, 153
carbon coated particles, **29–31**, 761, 780–782, 790, 791, 829, 850, **862–863**
carbon fibers, 16, **21–24**, 763–765, 832, 854, 857, 911
activated, 34, 35
heat treatment, 765, 775, 782
mesophase pitch, 22, 23
polyacrylonitrile (PAN), 22, 23
precursors, 21
preferred orientation, 21
vapor growth, **22–24**, 763, 765, 775, 779, 782, 783, 786, 791, 854, 857
carbon-free fullerenes, 34, **50–51**
clusters, 5, 153
CVD diamond films, **41–43**
density, 17, 34, 35, 40
diamond (*see* Diamond)
disordered carbons, 18, 19, 24, 25, 28, 30, 35, 830, 850

glassy carbon, **24–26**, 35, 36, 780, 851–853
graphite (*see* Graphite)
graphite-related materials, 15, 851
hard carbons, 24, 25
hexagonal diamond, 40
HOPG (*see* Highly oriented pyrolytic graphite)
intercalation compounds (*see* Intercalation)
interplanar distance, 17–19
isotropic pitch, 34
kish graphite, 20
linear chain, 4
liquid carbon, 16, 25
metal-coated fullerenes, **51–54**
metallo-carbohedrenes (Met-Cars), **48–50**
nanocapsules, 30, 31
nanoparticles, 30, 825
polygonal particles, 861
porous carbons, 24, 25, **34–35**
precursors, 21, 34, 46, 47
pyrolytic carbon (*see also* Pyrolytic carbon), 20, 27, 29–31, 763, 832
shock-quenched phases, 16, 31
whiskers, 21, 765, 783
Carbon nanocones, 772, 774, **777–778**, 779, 783
cone angle, 777, 778
opening angle, 777–779
Carbon nanotubes (*see also* Nanotubes), 1, 13, 17, 21, 24, 29, 65, 76, 80, 90, 96, 104, 111, 144, 145, 147, 168, 272, 712, 740–742, **756–864**, **903–904**, 911
Aharonov–Bohm effect (**H** ∥ **T**), 818, 819, 823
alignment, **785**, 857
applications, 814, 817, **903–904**
interconnect, 903, 904
junction, 903, 904
logic circuit, 903
magnetic device, 903
memory, 903, 904
optical device, 903
switching, 904
armchair, 757–761, 769, 786, 788, 791, 792, 794, **795–797**, 800–803, **804–805**, 806, 807, 809, 810, 812, 814, 815, 841, 842, **843–844**, 845–847, 849, 856
edge, 786
arrays, **815–817**, 818, 863, 864
aspect ratio, 760, **763–765**, 786, 795, 903
bamboo structure, 782, 783
band gap, 803, 807, **808**, 809, **811–813**, 816, 825, 836
diameter dependence, **808**, **827**
barrelenes, 764
basic 1D symmetry operations, 758
basic modulus, 765
bending, 783, 784
bundles, 761, 763–767, 780, 785, 815, 818, 825, 827, 828, 830, 833, 834, 837, 838, 853, 858
microbundle, 780, 829
caps (*see also* Terminations), 756–758, 760, 761, **769–776**, 777, 782–789, 795, 800, 815, 842, 857, 858
carrier density (*see* Carrier density)
chiral angle, 111, **757–760**, 762, 767, 768, 770, 773–775, 792, 797, 802, 803, 809, 811, 819, 825, 828, 838, 841, 843, 844, 846, 849, 854–856, 858, 864, 903
chiral tubes, 96, 757–760, 768–771, 773–775, 786–789, 791–795, **797–800**, 801, 803, **809–814**, 825, 841, 844–846, **848–849**, 864, 903
chiral vector, **757**, 758–760, 770, 791, 793, **794**, **801**, 803, 809, 816, 841
circumferential direction, 797, 801, 803–807, 812, 815, 818, 823
closed-end, 757, 771, 772, 782, 785, 789
coefficient of thermal expansion, 857
cohesive energy, 815
composite, 782
conductance, 826
cones (*see* Carbon nanocones)
curvature, 765, 774, 775, 784, 788, 803, 812, 813, 815, 834, 836, 838, 840, 841, 855, 856, 858
cylindrical shape, 756
defects, 803, 807, 815, 857
1D density of states, 762, 776, 777, 806, 807, **808**, **810**, 818, 825, **826**, 829, 863, 864
diameter, 757, 758, 761, 765, 767, 768, **769–770**, 772, 773, 775–777, 783, 784, 790, 793, **794**, 795, 800, 802, 809, 811–814, 816, 819–822, 825–828, 830 831, 833, 834, 836, 838, 841, 842, 844–850, 853, 854, 856, 864, 903

Carbon nanotubes (*continued*)
diameter distribution, **769–770**, 833, 853, 854
minimum diameter, 769, 784, 795, 813, 816, 823
dimerization, 858
disclination, 773, 774
doping, 624, 803, 807, 817, 818, 863
elastic properties (*see also* Elastic properties), 757, 762, 813, 814, **854–858**
electron energy loss spectroscopy (*see also* EELS), **838–839**
electronic properties, 795, 904
electronic structure (*see also* Electronic structure), 756, 761, 762, 792, **802–839**, 855
ellipsometry, 785
ESR, 825, 836, 837
faceted, 775
Fermi energy, **823**
Fermi function, 836
filled nanotubes, 784, 817, 818, **858–860**, 903, 904
carbides, 859, 860
metals, 859, 860
nanocapillarity, 858, 859
nanowires, 859
oxides, 859, 860
oxygen-enhanced, 858
surface tension, 858
tube opening, 858, 859
wetting, 858
fullerene impurities, 850
fullerene–tubule connection, 756, **757–761**, 763, 775–777, 784, 815, 856
g-factors, **836–837**
geometry, 757, 849, 904
group theory considerations, **791–802**
growth mechanism, 756, 764, 765, 775, 782, **785–791**
active sites, 785
catalysts, 782, 784, 786, 790, 791
dimer addition, 782, 785–788
epitaxial, 786
nucleation, 785, 790
open tube, 768, 788
Stone–Wales, 786
highest common divisor, **791**, 792, 793, **794**, 798–801, 810, 811, 844
hollow tubes, 761, 765, 818, 858, 859, 863, 903
infrared spectra (*see* IR spectrum)
interlayer correlation, 774, 775, 815, 816, 838
interlayer distance, 761, 762, 803, 814, 815, 829, 855, 863
interplanar geometry, 815
intertubule separation, 815
1D lattice vector, 759, **791–795**, **801**
magnetic properties (*see also* Magnetic properties, Magnetic susceptibility), **818–825**, 858
magnetoresistance (*see* Magnetoresistance)
mechanical properties, 765, 775, 854, 857, 904
metallic, 759, 790, 802, 805–811, 813, 815, 821, 822, 826, 829, 834, 848
modulus, 775
multiwall, 756, **761–765**, 768, 773–775, 784, 789, 790, 803, 811, **814–818**, 826, 829, 845, 850
nesting of tubules, 760, 816, 858
open-end, 757, 771, 784–786, 788–790
oxidation resistance, 857
oxygen burning technique, 781, 784, 825, 859
Peierls distortion (*see* Electronic structure)
phonons (*see also* Phonons), **839–854**
projection mapping (*see also* Projection mapping), 76, 77, 770–772
quantum effects, 850
quasi 1D hollow wires, 903, 904
Raman spectra (*see* Raman spectroscopy)
reciprocal lattice vector, **800–802**, 804, 841
1D reciprocal space unit cell, 801–805, 841, 848
resistivity (*see* Transport properties)
screw axis, 757, 773, 775
scroll morphology, 765
semiconducting, 759, 802, 809, 811–814, 817, 820–822, 825–827, 829, 834
single wall, 29, 756–758, 761, **765–769**, 772, 773, 775, 782, 784, 790, 802, **803–814**, 816, 825, 850, 856–858, 863
stability, 784, 815
STM (*see* Scanning tunneling microscopy)
structure, 815
superconductivity, 757, 818, 858, **863–864**

symmetry (*see also* Point groups, Space groups), 813, 823, 840, 849
synthesis, 756, 761, 763, **778–784**
carbon arc, 761, 764, 765, 778, 779, 782, 786, 789–791, 850–853, 859
conditions, 778, 790
electric field, 790
flames, 778
He pressure, 779, 788–790
ion bombardment, 778, **783**
metal-packed anodes, 859
purification, **781**, 857
separation, **780**, 857
temperature, 779, 782, 783, 788–790
transition metal-catalyzed, 763, **766**, 769, 770, 782, 790, 791, 851–853, 859
vapor growth, 762, 763, 778, 779, 782, 783
tensile strength, 765, 775
terminations
bill-like, 772, 774
polyhedral, 771, 772
semitoroidal, 772, 774, 775, 788
thermal conductivity, 857
transition junction, 774
1D translation vector, 757, **791**, 792, 793, **794**, 797, 800, 803, 810, 841
transmission electron microscopy (*see* TEM)
transport measurements (*see* Transport properties)
tubule axis, 760, 762, 765, 775–777, 783, 785, 786, 794, 797, 802, 804, 810, 815, 816, 818–824, 830, 831, 833–838, 841, 842, 845, 863
turbostratic stacking, 760, 761, 815, 829, 862
two coaxial tubules, **814–815**, 816
1D unit cell, 757, **759**, **791–792**, 793, **794**, 799–804, 823, 841–845, 848
wave vector, 802–807, 809–812, 823, 841, 863
allowed, **804**, 807, 812
crossing point, 805–807, 812
zigzag, 757–760, 773, 775–777, 786, 788, 791–793, **795–797**, 800–804, **805–806**, 807–810, 812, 814, 824, 827, 841, **844**, 845, 846, **847**, 849, 856
edge, 786
zone folding (*see* Zone folding)

Carbon onion, 27, 50, 74, 144, 756, 757, 765, 784, **815**, **860–863**
by electron beam irradiation, 860, 861
by laser melting, 860
by shock waves, 860
concentric stacking, 756, 757, 765, 860
diameter, 860
double layer C_{60}@C_{240}, 860, 862
filled onions, 862
from nanodiamonds, 860
fullerene-onion connection, 860, 862
hollow core, 860, 861
inner diameter, 860
interlayer separation, 815, 862
internal oscillations, 862
multilayer, 74, 756, 757, 765, 860
purification, 862
spherical shells, 756, 757, 765, 860
stabilization energy, 862
strain energy, 860
synthesis, 862
Carbynes, 16, **31–32**
Carcinogens, 115, 138
Carrier density, 238, 440, 443, 532, 560, 566, 573, 574, 580, 581, 583–585, 588–594, 602–604, 607, 608, 683
carbon nanotubes, 829, 838
on surface, 574
Carrier generation, **588–591**
Carrier localization (*see also* Localized charge), 51, 456, 540, 563, 565, 566, 592, 608, 621, 676, 739, 747, 749, 831, 838
Carrier mobility, 589, 882, 883, 888, 904
Carrier recombination, 238, 480, 488, 588, 589, 591–593, 888, 897
positron, 238
Carrier scattering, 556, 564, 565, 582
by boundaries, 632
by defects, 561
by impurities, 561
by merohedral disorder, 245, 561, 565, 566
by phonons, 565, 573, 574
mechanism, 557, 561
Carrier sign, 581, 608
Carrier trapping, 541, 588, 589
Carrier velocity, 574, 575
Catalytic agent (*see also* Chemical reactions), 41, 766, 782, 784, 786, 790, 850–852, **894–896**, 898, 910, 911
Cathode (*see* Electrode)

Cations (*see also* Endohedral fullerenes), **154–164**, 236, 256, 303, 426, **428–429**, 702, 744, 745, 896, 897
C_{60}^{+}, **154–164**, 256, 303, 304, 306, 426, 674
C_{60}^{2+}, 426, 429
C_{60}^{n+}, 426, 428, 429, 744, 748
metal, 226, 241, 243, 247, 250, 252, 254–256, 330, 331, 376, 386
$(NH_3)_4Na$, 621
radical, 314
Cavity (*see* Voids)
Cementite (FeC_3), **29–30**, 790, 863
Character table (*see* Group theory)
Charge density wave (CDW), 646, 647
Charge distribution, 694, 725
Charge generation, 533, 534, 556, 557
Charge transfer, 37–39, 365, 442, 443, 447, 449–451, 491, 506, 507, 863
adsorbed layers, 697, 702, 706, 717–722, 724–726, 728
battery (*see also* Batteries), 300, 304, 423, 899
doped C_{60}, **224–227**, **234–236**, **242–244**, 261, 262, 288, 330, 405, 570, 572, 643, 695, 697, 718, 730, 739, 748, 749
alkali metal doped C_{60}, **226**, 227, **235**, 236, **238–253**, 376, 378–380, 382, 384, 442, 443, 518, 520, 557, 560, 563, 580, 619, **623**, 663, 664
alkali metal doped C_{70}, **262–263**, **569–570**
alkaline earth doped C_{60}, **253–256**, **567–568**, **623**
electrolytic, 512
endohedral, 230–233, 435, 672, 733, 744, 745
exciton, 491, 500, 504, 505, 520, 691
photo-induced, 534–536, 871, 873–875, 881
polymers, 533–537, 870, 871, **873–876**, 881, 897, 907
salts, 271, 300, 316, 319, 320, 907, 910
substitutional doping, **233–234**
Charge transfer complex, 316, 319, 907, 910
Charged phonons (*see* IR spectrum)
Chemical reactions, 179, **292–328**
addition reactions, 296, 315
alkoxylation, 304
alkylation, 187, 296, 297, 299, **304–306**, 315, 895
amination, 297, 306
Birch reaction, 304, 305, 899
bond cleavage, 894, 895
bridging, 192, 297, **307–312**
catalytic, **894–896**, 898, 899
cycloaddition, 187, 297, 311, **312–314**
complexation, 318, 319
dehydrogenation, 894, 895, 898, 900
dispersive agent, 896
electrochemical (*see* Electrochemistry)
gas phase, 304
general characteristics, 180, **294–297**
halogenation, 190, 296, 297, **306–307**
high pressure, 304
hydrogenation, 187, 296, 297, 299, 304, 306, 315, 894–896, 898–900
irreversible, 300, 898
oxidation, 186, 189, 297, 298, 300, 303, 304, 895
phenylation, 307
polymerization, **320–325**, 895
reactions with free radicals, 191, 297, **315–316**, 895
redox, 300
reduction, 182, 183, **297–304**
reversible, 898
scission reactions, 296, 894, 895
selectivity, 895
stress catalyzed, 858
substitution reactions, 186, 305, 307, 314
sulfur extraction, 895
Chemical stability (*see also* Stability), 6, **153**, **155–156**, 905
Chemical vapor deposition (CVD), 42–44, 893
Chemisorption, 721
Chemotactic potency, 897
Chiral (*see also* Carbon nanotubes)
fullerenes, 96, 104, 122, 124, 656
structure of C_{76}, 655
Chromatography, 72, 117, **120–125**, 128, **135–137**, 229, 272, 293, 537, 889, 907, 909, 911
fullerene residue, 907
liquid, 72, 272, 537
mobile phases, 121, 122, 124, 125, 907
retention time, 122, 907
stationary phase, 121, 122, 128, 907
Classification of doped fullerenes, 137, 224, 225

Clathrates, 39, 224, 225, 227, **236–237**, 257, 271, **288–289**, 297, 303, 316, 317
$C_{60}C_3H_6Br_2$, 288
$C_{60}C_4H_{10}O$, 288
$C_{60}H_2$, 288
$C_{60}O_2$, 236, 288
$C_{60}S_8$, 236, 288
C_{60}-*n*-pentane, 236, 289
$C_{60}(S_8)_2CS_2$, 236, **237**, 289
C_{70}-based, 288, 289
Clausius–Clapeyron equation, 600
Clausius–Mossotti relation (*see* Optical properties)
Clusters, 3, 37, 44–46, 48–54
CO, 313
Coatings, **904–905**, 909, 910
for STM tips, 904, 905
Coaxial tubules (*see also* Carbon nanotubes), 761, 765, 789, 814–816, 857, 860
Cobalt, 30
Coherence (*see also* Superconductivity)
length, 564, 626, 627, 632, 633
peak, 636, 684
Cohesive energy, **63**, 234, 438, 440, 443, 560, 716, 720, 732, 815
intermolecular, 63, 172, 438, 440, 443, 716, 815
intramolecular, **149**, 150, 152, 153, 415
Collisions with fullerenes, **159–165**
energy transfer, 163
ions, 159, 161
molecules, 160, 161
ions, 159, 161
surfaces, **163–165**
Columnar growth, 779
Combination modes (*see* IR, Raman spectra)
Compatibility relations (*see* Group theory)
Composites
C_{60}–polymer, 302, 304, **532–537**, 871, **873–880**, **881–883**, **885–888**, 896, 897, 899, 907–911
metal–graphite, 131–135, 782, 862, 863
nanotube–polymer, 785, 857
Compressibility, 65, 171, **172**, **175**, **203–204**, 402, 602, 641, 642
Concave surface, 9, 65, 145
Concentric layers, 27, 54, 756, 765, 860, 862
Conductivity (*see* Transport properties)
Convex surface, 65, 145
Corannulene (*see also* Annulene structure), 61, 62, **148–150**, 167
Core level spectroscopy, 694, 695, 697, 698, **703–704**, 705, 706, 708, 711, 719
C 1*s* binding energy, 694, 697, 698, 703, 705, 708
Correlation energy, 421, 456, 712
Corrugation potential, 715
Cost considerations, 908, 911
Coulomb interaction (*see also* Electronic structure), 197, 244, 252, 254, 255, 369, 384, 395, 437, 438, 449, 451, 453–455, 457, 472, 504, 523, 561, 580, 625, 645–647, 689, 710, 711
Coverage, 276, 279, 280, 697, 698
Langmuir, 892
monolayer, 212, 697, 698, 701, 702, 706, 710, 711, 713, 715, 717–719, 721, 723–727, 730, 732, 733, 891–893, 896, 897, 902, 905
multilayer, 212, 697, 698, 701, 702, 713, 715, 717, 721, 724, 726, 727, 730, 732, 891, 892, 896, 897, 902, 905
submonolayer, 698, 715–717, 728, 729, 732, 902
two layer (bilayers), 697, 698, 724, 725, 727, 896, 897
Cross-linking (*see also* Polymerization), 207, 210, 211, 213, 247, 398, 400, 401, 508, 889
Crystal growth (*see also* Single crystal), 148, 271, **272–274**
C_{60}, **272–275**
C_{70}, 273
crystallite dimension, 712
Crystal structure
alkali metal doped C_{60}, 146, **238–253**, 285, 380
alkaline earth doped C_{60}, 151, **253–256**, 285
ambient C_{60} structure, 111
C_{60}, 111, **171–197**, 689
$C_{60}I_2$, 257
$C_{60}I_4$, 256, 257
$C_{60}O$, 256, 308
$C_{60}(S_8)_2CS_2$, 236
C_{60}-*n*-pentane, 237
$C_{61}H_2$, 259
C_{70}, **203–208**
$C_{70}I_2$, 258

Crystal structure (*continued*)
effect of pressure, 125, **203–208**
effect of temperature, 126, **208–209**
halogen doped phases, 152
polymerized C_{60}, 126
photopolymerized C_{70} films, 128
quenched state of M_1C_{60}, 380, 381, 531, 532, 664
rare earth R_xC_{60}, 256
TDAE–C_{60}, **259–261**, 749
Cs_xC_{60} (*see* Alkali metal fullerides)
Cu surfaces (*see* Substrates)
"Cup-shaped" molecules, **317–319**
Curie paramagnetism, 739, 740, **745–747**, 751
Curie temperature, 260, 669
Curie–Weiss temperature, 748
Curie–Weiss paramagnetism, 681, 745–748, 750, 751
Curvature, 74, 144, 294, 704, 758, 905
convex, 145
Cyanopolyynes, 3
Cycloaddition, 211, 212, 215, 248, 311, **312–314**
"2+2", 211, 212, 247, 248, **312–314**, 322–324, 508
"4+2", **312–313**
Cyclopentadiene, 313–315
Cyclopentadienyl radical, 298, 315
Cyclovoltammetry, 228, **300–303**, 325, 899, 901

D

Da Vinci, 1, 2
Dangling bonds (*see also* Passivation), 33, 35, 438, 713, 720, 726, 771, 784, 788, 790, 857, 860
David model, 188
DEANST (*see* Diethyl aminonitrostyrene)
Debye frequency, 349
Debye relaxation, 191
Debye temperature, 172, **594–600**, 603, 617
Debye velocity of sound, 597
Defects, 561, 730, 883
carbon nanotubes, 854
dopant position, 557, 563
dopant vacancies, 563, 582, 583
grain boundaries, 497, 498, 564, 730
heptagonal, 783
impurities, 40, 41, 506, 540, 561, 586, 775, 807, 817, 850, 904
lattice, 586, 588, 663, 714
pentagonal (*see also* Pentagon), 9, 74, 75
screw dislocations, 20, 728
stacking faults, 586
surface, 720
twinning, 20
vacancy, 728
Degenerate four-wave mixing (DFWM), 544–547, 877
Dehydrogenation, 23, 25, 31, 46, 49, 894, 895
Delocalized charge (*see also* Localized charge), 225, 227, 229, 232, 235, 295, 298, 314, 427, 435, 544, 739, 745
delocalized exciton, 490, 491, 500, 504, 505, 673, 680
Dember effect, 588, 589, 883
Degrees of freedom (*see* Vibrational modes)
Density, 17, 32, 35, 40, 171, 172
Density of states (DOS) (*see also* Electronic structure), 451, 452
electronic, 73, 413, 432, 438, 440–442, **454–455**, 495, **500–501**, 503, **504–505**, 510, 586, 587, 590, 689–693, 711, 719, 725, 728, 729, 733
broadening, 636, 693
lattice constant dependence, 585
doped C_{60}, **442–446**, 449, 451, **452**, 556, 560, 570, 574, 576, **583–585**, 586, 587, 599, 607, 616, 617, 619, 621, 622, 624, 636, 637, 644, 665, 684, 695, 696, 745–747
doped C_{70} solid, 747
carbon nanotubes, 762, 776, 777, 806, 807, **808**, 810, 818, **825–827**, 829, 852, 854, 863, 864
energy dependence, 808
per unit length, **810**
singularity, **808**, **826**, 863, 864
2D graphene sheet, 807, **808**, 863
joint, 500
phonon, 28, 330, 331, 354, 367, 394, 606, 852, 854
Desorption (*see also* Temperature-programmed desorption), 709, 710, 713, **715–717**, 720
energy, 713, 715, 716
fullerene mass dependence, 732
rate, 716
temperature, 708, 713, **715–717**, 732
temperature dependence, 708, 709, 713, **715–717**, 720, 721

Diamagnetism (*see also* Magnetic properties), 516, 631, 667, 745, 834
Diamantane (*see* Cage structure)
Diameter (*see also* Carbon nanotubes, Carbon onions), 714, 732, 733
C_{60}, 62, 63, 69, 647, 656, 714, 902
icosahedron, 69, 70, 902
Diamond, 15, 17, 28, 37, **39–41**, 602, 604, 710, 740, 741, 851, 893, 895, 909, 911
hardness, 17
lattice constant, 17
Raman scattering, 42
Diamond-like film, **41–44**
Dielectric constant (*see also* C_{70}, Optical properties), 17, 120, 172, 199, 479, 705–707
various systems, 17, 120
Dielectric function (*see also* Optical properties), 386, 456, 507, 699, 704, 705, 707
$\epsilon_1(\omega)$, 507, 704, 705, 707
$\epsilon_2(\omega)$, 507, 704, 705, 707
Dielectric loss, 456, 507
Dielectric properties, **505–507**
Dielectric relaxation, 190
Diels–Alder reaction, 46, 313, 898
Diethyl aminonitrostyrene (DEANST) (*see also* Composites), **877–880**
Differential scanning calorimeter (DSC), 184, 199, 200, 207, 237, 386, 557, **600–601**
Diffusion, 709, 710, 715, 720, 732, 881, 893
carrier, 833, 876
constant, 188, 507
rate, 28
temperature dependence, 715, 893
Dilatometry, 190, 200, 274
Dimer, 149, 211, 212, 215–217, 248, 465, 532, 728, 889, 895
C_2, **147–148**, 149–151, 167, 210
$(C_{60})_2$, 313, 381, 386, 398
Diodes, 884, **885–886**, 887, 909
Diphenylfulleroid, 308, 309
Dipole moment, 231, 232, 495, 506, 726
Dipole scattering mechanism, 369, 370
Disclination, 771, 772
Disorder (*see* Defects)
Dispersion relations
electron (*see also* Electronic structure)
2D graphene, **803–809**
1D nanotube, 756, 795, **802–825**
3D solid, 181, 337, **437–453**, 699, 730
phonon (*see also* Vibrational modes)
2D graphene, 276, 802, 840, 841
1D nanotube, 756, 795, 802, **839–845**, 853
3D solid, 181, 194, 332, 341, **347–354**, 365, 367, 372, 373, 388, **840–845**, 851–853
Dissipation factor, 506
Disproportionation, 572
Distillation flask, 123
Divalent ions (*see* Alkaline earths)
Donor doping, **234–236**, 240
Doped fullerenes (*see* Alkali metal and Alkaline earth fullerides)
Doping effects
endohedral doping, 140, 224, 225, **228–233**, **435–437**, **670–673**, **733**, 739, **742–743**, 744, 745, 818
exohedral doping, 144, 224, 225, 231, **234–238**, 283, 285–287, **442–452**, **517–532**, 556–575, 579–587, 593, 594, 598, 599, 601, 602, 604–608, 616–625, 627–743, 645–648, 673–680, 683, 684, 708, 728, 730, 739, 742, 743, 745, 746, 818
intercalation (*see* Intercalation compounds)
nanotubes (*see* Carbon nanotubes)
polymers by fullerenes, 874, 876
substitutional doping, 143, 224, 225, **233–234**, 451
Doping techniques, 271, 283, **285–288**, 384, 386, 518, 531, 561, 563
two-temperature method, 285
doping-anneal cycles, 286, 287
Double bonds (*see* Bonding)
Drug applications, 309, 310, 317, 897
Dürer, 1, 2
Dynamical matrix (*see also* Vibrational modes), 88, 104, 335, 341, 391, 841

E

Edges, 60, 64, 80, 148
Einstein modes, 595
Einstein temperature, 597, 603
Einstein term, 598, 599

Effective mass, 441, 528, 560
Elastic constants, 185, 190–192, 608
  carbon nanotubes, **854–858**
Elastic properties
  carbon nanotube, 757, 762, **854–858**
    bend, 854, 855
    compression, 856
    continuum elastic theory, 855–857
    distortion of cross section, 857
    elastic modulus, 855, **856**
    flexibility, 854, 857
    fracture, 854
    hardness, 857
    kink, 854
    density functional theory, **856**
    twist, 854, 855
    stiffness, 856, 857
    straightness, 857
    strain energy, **855–856**
    stretching, 856
  deformation, 336
  energy, 163
Electrical conductivity (*see also* Transport properties)
  carbon nanotubes (*see* Carbon nanotubes)
  contact preparation, 286
Electrochemical applications, **898–901**
Electrochemistry, 225, 228, 297, 298, **299–303**, 423, **511–518**, 674, 675, 870, **898–901**, 910, 911
  batteries (*see also* Batteries), 301–303, **898–901**, 910, 911
  charging, 898–901
  discharging, 898–901
  electrolyte, 300, 302, 512, 513, 899
  energy conversion, 898–900
  hydrogen storage, 898–900
  intercalation, 303, 901
  irreversible, 898, 899, 901
  oxidation, 297, 298, 300, 302
  primary batteries, 898, 899
  recharge, 898, 899
  reduction, **297–304**
  reduction potential, 300
  reversible, 898, 899, 901
  secondary batteries, 898–900
  voltammetry, 228, 298–303, 325, 423, 899, 901
Electrode, 10, 21, 30, 50, 110–115, 131–133, 300–304, 506, 507, 512, 563, 588, 589, 593, 763, 778, 779, 782, 790, 791, 814, 850, 859, 862, 878, 879, 885, 886, 898, **899–901**, 910
  anode, 134, 300, 302, 779, 782, 833, 834, 851, 852, 859, 899
  attachment, 563
  battery (*see also* Batteries), 301
  carbon nanotube synthesis, 763, 778
  cathode, 301, 779, 782, 785, 850, 853, 899
  comb, 506, 507
  counter, 300, 512
  standard, 300, 572
  working, 303, 304, 512
Electrolyte (*see* Electrochemistry)
Electron affinity, 63, 68, 73, 155, 231, 236, 298, 416, 457, 534
Electron beam “writing,” 889
Electron charge density, 238, 440, 443
Electron diffraction (*see also* LEED), 185, 202, 203, 278, 279, 697, **699–700**, 767, 773, 859
Electronegativity, 39, 256
Electron–electron correlation, 96, 229, 421, 456, 479, 503, 528, 556, 558, 561, 584, 621, 648, 711, 712, 746
Electron–electron interaction, 423, 453, 477, 488, 522, 523, 528, 556, 561, 566, 574, 582, 583, 608, 616, 621, 625, 638, 644–647, 665
Electron energy loss spectroscopy (EELS) (*see also* Surface science), 202, 203, 406, 439, 488, 492, 493, 504, 689, 691, 693, 694, **697–708**, 717, 724, **838–839**, 859, 905
  broadening effects, 703
  carbon nanotubes (*see also* Carbon nanotubes), **838–839**
  core level (*see also* Core level spectroscopy), 694, 695, 697, 698, **703–704**, 724, 839
  dielectric function (*see also* Dielectric function), 524, 699, **704–705**, 707
  dispersion, 699, 707
  dopant concentration dependence, 707
  doped fullerenes, 702, 706, **707–708**
  electron energy loss function, 492, 493, 499, 523, 699, **704–707**
  endohedral fullerenes, 229
  high resolution (*see* HREELS)
  higher mass fullerenes, **699–700**, **702–706**

inelastic, 697, 699
interband transitions, 439, 498, **499**, **520**, **522–523**, 699–702, 705, 707
line intensities, 701
longitudinal excitations, 704
optical conductivity, **382**, 501, **524–527**, **530**, **637–638**, 702, **704**, 705, 707, 749
peak energies, 439, 498, **499**, **520**, **522–523**, 702–705, 707, 838, 839
plasmons, 528, 689, 694, 696, 699, 701, 702, **704–706**, 838, 839
polarization effects, 838, 839
selection rules, 364, 365, 369, 370, 697, 702, 704
spin-flipping transitions, 701
surface studies, 697, **699–708**
vibrational modes, 331, 332, 338, 357, 359, 360, **369–370**, **388**
Electron–hole interaction (*see also* Exciton), 468, 497, 522, 539
Electron paramagnetic resonance (*see* EPR)
Electron penetration depth, 697
Electron–phonon coupling (*see also* Superconductivity), 232, 367, 380, 382, 388, 395, 405, 406, 471, 477, 528, 530, 535, 565, **573–580**, 584, 599, 607, 608, 616, 617, 623–626, 633, 638, 641, **643–646**, 648
connection to Jahn–Teller effect, 579, 580, 644
from specific phonons, 576–579, 644, 645
magnitude of coupling constant, **575–578**, 644
renormalization, 575
symmetry considerations, **578**
vertex correction, 575, 644
Electron–plasmon coupling, 584
Electron–positron recombination, 238
Electronics applications, **880–892**, 909
diode rectifiers, 871, 873, 880, 881, **884–888**
electroluminescence, 47, 898, 909
equipotential surface, 888, **891**
field effect transistors, 880, **881–883**
microfabrication, **888–892**, 909
nanotechnology, **901–904**, 909
patterning, 1, 209, **888–890**, 909
photovoltaic devices, 47, 533, 880, 881, **886–888**, 898
polymer–C_{60} heterojunction, 871, 873, 881, **884–886**
silicon wafer bonding, 888, **889–891**
surface passivation, 888, 891, 892
tunnel diodes, 884
Electronic density of states (*see* Density of states, Electronic structure)
Electronic polarizability, 382, 456, 457, 479
Electronic structure (*see also* Hamiltonian), 225, 257, **413–458**, 528, 689, 693, 699
BC_{59}, 451
BN nanotubes, **814**
C_{28}, 451
C_{76}, 452, 453
C_{80}, 452
C_{82}, 452, 453
C_{84}, 452, 453
C_{60} solid, 275, 277, 413, 416, 423, **437–442**, **453–457**, 498, 503, 689
band gap, **416–418**, 438–441, 453, 457, 501, 693
bandwidth, 38, 331, 413, 416, 437–439, **442**, **454–457**, 495, 501, 518, 556, 561, 574, 583, 584, 605, 607, 608, 648, 691, 711, 823, 840, 863, 905
Coulomb repulsion, 453–457, 504
density of states, 438, 440–442, 454, 501, 503–505, 509
dispersion relations, 439, 440, 442, 454, 498, 503, 693
effective mass, 441
electron density contours, 440, 442
half filled band, 556, 558, 561
LDA calculation, 68, 69, 153, 338, 416, 418, 432, **438–442**, 445, **452–456**, 499–503, 516, 520, 574, 577, **583–585**, 645, **811–814**, 855–857
level filling, 442, 455
many-body effects, 413, **421–423**, 438, **453–458**, 692, 696, 708, 710
molecular solid, 437, 491, 574, 619, 719
one-electron model, 183, 413, **414–421**, 422, 438–453, 469, 476–480, 492, 501, 556, 561, 566, 569, 692
with alkaline earths, 282, **447–450**
X-point extrema, 441, 501, 503
C_{60} molecules, 257, 413, **414–423**
ab initio calculations, 414, 420, 421, 438, 439, 453, 503, 585, 863
bandwidth, 416, 495
CNDO/S method, 477
excited states, 414, 418, 422, **424–428**, 466, 469, 472, 478

Electronic structure (*continued*)
filled shell configurations, 93, 95, 417, 421
ground state, 341, 342, 414, 422, 424, 469, 472, 478
HOMO–LUMO gap, 418, 425, 476, 504, 537, 539, 540, 593, 691, 693, 694, 701, 706
Hückel model, 3, 5, 94, 95, **414–416**, 418, 419, 422, 429, 432, 803
intermolecular interactions, 416, 417, 423, 468, 495, 673
level filling, 94, 95, 260, **417–419**, 420–423, 455
local density approximation (LDA), 415, 418
many-electron states, 264, 417, 418, **421–423**
MINDO/3 method, 579
MNDO method, 68, 69, 456, 580
molecular orbitals, 258, 414–420, 423
MOPAC method, 646
spherical approximation, 417–420, 422, 424–426, 429
symmetry-based models, 181, 262, 414, 417, 418, **419–421**, 422, 424
C_{70} molecules (*see also* C_{70}), **418–419**, 420, **432–434**, 570, 730
bandwidth, 432
density of states, 432, 730
excited states, 433, 434, **483–491**
ground state, 434
HOMO–LUMO gap, 432
Hückel calculations, 432
spherical approximation, 432
symmetry lowering, 432
calculations, 277, **413–458**, **468–476**, **478–480**
GW quasiparticle, 453–455
Hartree–Fock (HF), 69, 505, 522, 579
local density functional (LDA), 68, 69, 153, 338, 416, 418, 432, 438–442, 445, 451–456, 499–503, 516, 520, 574, 577, 583–585, 645, 811–814, 855–857
many-body approach (*see also* Hubbard model), 285, 413, 438, **453–457**, 504, 558, 561
molecular approach, 413, 437, 468
one-electron, 437, 449, 468, 469, 476, 478–480, 491, 501, 561, 566, 570
pseudopotential, 361, 439, 440
quantum chemistry, 456, 468, 474, 482
random phase approximation, 479
tight binding model, 69, 166, 215, 414, 438–441, 444, 452, 576, 803, 804, 810, 812–815, 818–823, 827
carbon nanotubes, 756, 761, 762, 792, **802–839**, 840, 863
allowed wave vectors, 804
arrays, **814–817**, 818
band gap, 803, 807, **808**, 809, **811–813**, 816, 825, **827**, 836
band overlap, 827–829, 836
calculation, 812, 813
conduction bands, 804–807, 810, 813
1D density of states, 806, 807, **808**, **810**, 825, 829, 863, 864
degeneracy point, 805, 807, 814, 817, 823, 829
degenerate bands, 804–807, 810, 811, 814
dispersion, 792, 795, 800, 802, 803, **804**, 805, **806**, 807, **809–811**, **817**, **824**
double-wall, **814–815**
effective mass approximation, 820
electron concentration, 827, 829, 830, 832, 838
energy $E(k)$, 803, **804**, **806**, **809–811**, **817**
Fermi level, 828, 829, 832, 863, 864
four-wall, 814
fourfold degeneracy, 806, 810, 811
hole concentration, 827, 829, 830, 832, 838
interlayer interaction, 803, 814, 829, 839
intertubule interaction, 803, 857
Jahn–Teller splitting, 814
$\mathbf{k} \cdot \mathbf{p}$ perturbation theory, 818, 820, 823
K-point degeneracy (*see also* degeneracy point), 807, 814
LDA calculation, 811–814
magnetic field dependence, 803, **818–825**
metallic, 802, 805–811, 813–815, 817, 819, 821, 822, 825, 826, 829, 834, 903
multiwall (*see* Carbon nanotubes)
nearest-neighbor overlap integral, 803, 804, 808, 810–813, 822
non-degenerate bands, 804, 807

non-symmorphic, 803, **809–814**
Peierls distortion (*see also* Peierls distortion), 803, 806, 811, 813, 814, 816
Peierls gap, 806, 811, 812
periodic boundary conditions, 803, **804**, **806**, 807, **809**, 818, 819, 863
periodic wave function, 818
semiconducting, 807–809, 811–814, 817, 819–822, 825–827, 829, 834, 903
semimetal, 828, 830
single-wall (*see* Carbon nanotubes)
strain energy, 816
symmorphic, **803–809**
tight binding model, 803, 804, 810, **812**, 813–815, 818–820, 822, 823, 827
two-band model, **827**, 828
valence bands, 804–807, 810
zero field, **802–818**
zone boundary, 805
zone folding, 792, 801–811, 863
charged polarons, 488
configuration mixing, 468
doped C_{60}, 413, 423, 437, 439, **442–451**, 522, 643
alkali metals, 280, 413, **442–447**, 449
alkaline earths, 282, **447–450**
band filling, 445, 556, 558–561, 563, 567–569, 623, 695
band gap, 225, 444, 531
band hybridization, 236, 413, 445, 447–449, 568, 648, 697, 707
bandwidth, **442–445**, 451, 556, 561, 573, 583, 607, 608, 617, 619, 648, 708
charge transfer, 442, 443, 447, 449, 506, 507, 518, 520, 557, 559, 560, 563, 567
charge contour maps, 443
cohesive energy, 443, 445, 446
electron velocity, 574
density of states, 443–446, 449, 574, 576, 617, 619, 696
dispersion, **444–453**, 606
Fermi level, 442, **444–452**, 574
Fermi surface, 445, 447, 449
Fermi velocity, 574
Fermi wavevector, 574
lattice constant (*see* Lattice constant)
LDA calculation, 442, 444–448
many-body effects, 696
molecular nature, 696
nearly free electron model, 566, 574, 606, 608
screening effects, 574
semimetallic behavior, 448, 449
total energy, 446
wave function overlap, 749, 752
doped C_{70}, **569–570**, 580, 581, 605, 608
endohedral fullerenes (*see* Endohedral fullerenes)
excitonic states for C_{60}, 270, 294, 302, 304, 307, 414, **429–432**, 438, 456, 457, 464, 468, 472–477, 480–483, 503–505, 542, 692, 701, 706, 870
binding energy (*see* Exciton)
delocalized exciton, 489–491, 504, 505
exchange energy (*see* Exciton)
localized exciton (*see also* Exciton), 489–491
quasi-electron, 503, 504
quasi-hole, 503, 504
screening effects, 876
singlet, 429, 465–467, 472, 474, 480, 481, 483, 484, 487–489, 504, 541, 542, 547, 693, 701, 702, 871, 872
singlet–triplet decay, 541
symmetry, 504
triplet, 429, 465–467, 472, 474, 481, 483, 484, 486–491, 504, 515, 541, 543, 547, 678, 693, 701, 702, 870–873
triplet state splitting, 488, 489
triplet–triplet annihilation, 541
excitonic states for C_{70}, **483–488**, **537–539**
halogen doped C_{60}, 451
higher-mass fullerenes, 271, 275, 283, 414, **421–423**, 439, **452–453**
hybridization
band, 236, 445, 449, 568, 623, 646
orbitals, 18, 148, 229, 413, 414, 436, 447, 449, 519, 520, 697, 707, 813
intercalation compounds, **450–451**
$C_{60}C_{32}$, 450
$K_4C_{60}C_{32}$, 450, 451
multiplet states for ions
C_{60}^{n-} ions, **424–425**, **427–428**, 514–518, 559, 621, 646, 748
C_{60}^{n+} ions, **428–429**
$C_{60}^{n\pm}$ ions, 432–434
excited states for C_{60}^{n-}, 423–428, 431, **512–517**

Electronic structure (*continued*)
ground states for $C_{60}^{n\pm}$, 153, 265, 413, **423–429**, 430, 516, 518, 557, 559, 744, 748
Pauli-allowed states, 424–433, 744, 745, 748
polarons, 488, 491
surface, 722, 728
symmetry, **90–96**, 341, 414
vibronic states, 438, 464–466, 468, 470, 473, 474–476
singlet, 473, 475
triplet, 473–475
Electronic transitions (*see also* Optical properties), 382, 384, 464, 465, 472, 495, 500, 537, 587, **700–702**
energies, 493, 500, 501, 509, 510, 512–518, **520–522**, 542
identifications, 469, 470, 478, 500, 501, 509, **520–522**
$g_g \rightarrow t_{1u}$, 470, 478, 500, 522
$h_g \rightarrow t_{1g}$, 380, 500, 520
$h_u \rightarrow h_g$, 478, 500, 522
$h_u \rightarrow t_{1u}$, 500, 522
$h_u \rightarrow t_{2g}$, 470
$h_u \rightarrow t_{1g}$, 380, 470, 478, 522
$t_{1u} \rightarrow h_g$, 500, 520–522
$t_{1u} \rightarrow t_{1g}$, 382, 384, 478, 500, 518, 520–522, 524, 532, 542, 708
$t_{1u} \rightarrow t_{2u}$, 500, 520, 522
Electrophilic properties, 292, 296, 297, 304, 314
Electrostatic, 250, 380, 451, 563
attraction, 39, 241, 243, 255, 380, 619
repulsion, 244, 251
Ellipsometry (*see* Optical properties)
Enantiomers, 124, 125
Encapsulation, 30, 316, 862
Endohedral fullerenes (*see also* Metallofullerenes), 6, 51, 54, 125, 126, 128, 129, **131–137**, 157, 224, 225, **228–233**, 272, 283, **435–437**, 443, 635, 654, 669, **670–673**, 683, 689, **733**, **742–743**, 744, 745
2^+ cations, 131, 135, 229, 231
3^+ cations, 131, 132, 229–232
adsorption, 733
alkali metal, 228, 231, 233, 435
calculations, 435, 436
cation energy levels, 436
cation location, 230, 231, 435–437, 672, 733
charge transfer, 435, 436, 672, 733
composite negative electrode, 132
crystal structure, 233
density of states, 733
diameter, 232, 733
electron diffraction, 232
electronic structure, **435–437**, 733
EPR studies (*see* EPR)
g values, 673
ground state configurations, 744
HOMO–LUMO gap, 435
interaction with surfaces, 733
island growth, 733
isomers, 229, 435
listing, 132, 133, 231, 672, 673
muon dopant, 232, **635–638**, 683, 684
purification, 135–137, 228, 231, 233, 435, 669
shape, 230, 231, 733
STM studies, 230, 232, 733
synthesis, 110, 111, **131–137**, 228, 231, 233
synthesis conditions, 133, 231
TEM, 232
yield, 134–137, 231
Endohedral targets, 131
Energy gap (*see* Electronic structure)
Enthalpy, 196, 197, 594
Entropy, 130, 167, 195–197, 354, 594–596
Epoxide, **258–259**, 303, 307, 308
EPR, 72, 133, 137, 394, 593, **669–683**
angular dependence, 679
C_{70}, 679, 680
depolarization effects, 753
doped C_{60}, 673, 676, **678–679**, 683, 748, 749
doped C_{70}, 569, **746**
endohedral fullerenes, 133, 229, 230, 427, 435, 436, 543, 654, **670–673**
excited fullerenes, 427, 486, 491, 673, **678–679**
fullerene anions, 515, 516, 531, 580, 647, 654, 670, **673–677**
g factor (*see* g-factor)
gyromagnetic ratio, 670
hyperfine interaction, **669–672**
intensity, 680, 749
Jahn–Teller effect, 675, 676, 679
^{139}La, 670, 671

lifetime, 679
lineshape, 680
linewidth, 673, 675, 676, 678, 680, 682, 683, 749, 753
localized spin, 673
magnetic system, 654, **680–683**
motional narrowing, 671, 680
neutral C_{60}, 673, 676
ODMR experiments, 680
phase transition, 680, 683
pulse–echo experiments, 679
solid state, 670
solution spectra, 670–673, 675–677, 680
spectra, 670–675, 677, 678, 681
spin–echo, 753
spin–lattice relaxation, 676, 680, 838
spin susceptibility, 683
TDAE–fullerenes, **680–683**, 748–750, 753
temperature dependence, 673–680, 683
time resolved EPR, **678–679**, 680
triplet decay, 679, 680
unpaired spins, 673, 676
Equilateral triangle, 74, 75, 230
Equipotential surface, 888, **891**
Equivalence transformation, 85–88, 98, 99, 101, 102, 180, 181, 184, 341
ESR, 323, 386, 584, 670, 682, 740, 745, 749, 750, 753, 825, 836, 837
carbon nanotubes, 825, **836–837**
Etching, 889, 890, 894, 905
Ethylene, 26, 29, 260, 323, 853
Euler's theorem, 5, 7, 8, 60, **64**, 143, 145–148, 153, 167, 294, 771, 777
EXAFS, 229–231
Exchange energy (*see* Exciton)
Excited state spectroscopy, **375–376**, 464, 465, 467–469, 472, 473, 476–480, 483–495, 498–503, 508, 526, 538, 539, **542–544**, 870–873
Exciton (*see also* Electronic structure), 414, **429–432**, 434, 438, 439, 456, 497, 693, 701, 706, 872, 876
binding energy, 503, 504, 533, 593, **691**, **693**, 706, 876
charge transfer, 500, 504, 505, 520
exchange energy (*see also* Singlet-triplet splitting), 421, 422, 488, 504, **693**, **702**
Frenkel, 489–491, 500, 504, 505, 522, 523, 542, 691, 692
singlet, 542
triplet, 873
ground state degeneracy, 429, 430, 515–518
lifetime, 873
Exohedral doping (*see also* Alkali metal fullerides, Alkaline earth fullerides, Charge transfer, Doping effects, Doping techniques, Octahedral, Rare earth dopants, Tetrahedral)
empirical criterion, 234
Extraction of fullerenes, **116–121**, 272, 911
filtration, 116
solvent methods, **116–119**
Soxhlet extraction, 116–118, 122, 123
sublimation methods, **118–121**

F

F_{1u}-derived mode (*see also* IR spectrum), **333–335**, **382–388**
Faces of polygons, 60, 64, 80
Face-centered cubic (fcc) structure, **172–173**, 176, 183, 193, 197–199, 202, 203, **226**, 233, **239**, **241–244**, 250, **252–254**, 257, 259, 262, 263, 273, 276, 279, 280, 286, 288, 316, 338, 348, 388, 443, 447, 456, 492, 501, 518, 563, 568, 618–620, 622, 623, 699, 700, 714, 727, 728, 730, 732
pressure induced fcc(pC_{60}), **206–207**, 403, 404
Faceted surface, 27, 273–275
Fermi level, 225, 415, 417, 446, 451, 557, 606–608, 616, 617, 621, 622, 624, 648, 684, 690, 691, 693, 695, 696, 699, **700–702**, 703, 708, 721, 722, 726, 728, 729, 777, 791, 805, 806, 808, 810–813, 817, 818, 828, 829, 832, 863, 864
Fermi liquid theory, 575
Fermi surface, 19, 629, 791
doped fullerenes, **445**, **447**, 449, 456, 528, 574, 580
Fermi velocity, 574
Ferrocene, 319, 320
Ferromagnetism (*see also* TDAE–C_{60}), 260, 683, 745, **748–752**
Fibers (*see* Carbon materials)
Field effect transistor (FET), 880, **881–883**
conduction channel, 882, 883
drain-source voltage, 882, 883
I–V characteristic, 883

Field effect transistor (*continued*)
  n channel device, 883
  structure, 882
Film growth, 271, **274–283**, 713, 720, 721, 723, **893–894**
  at edges, 721, 725, 727, 733
  at steps, 721, 722, 724, 725, 727, 730–732
  commensurate, 722–724
  coverage (*see* Coverage)
  crystallinity, 274, 276–281, 732
  epitaxial, 23, 271, 274, **276–281**, 700, 712, 713, 720, 723, 724, 782, 786, 860, 883, 906
  free-standing (*see* Membranes)
  higher mass fullerenes, 693, 694, 699, 700, 703–706, 731
  hot wall method, 277, 279
  incommensurate, 713, 720, 731
  intercalation, 285
  island, 710, 713, 716, 724–727, 731–733, 902
  Langmuir–Blodgett films, 883
  layer-by-layer, 710, 724, 726
  microstructure, 274, 276, 730
  ordered structures, 722, 724, 726–729
  SiC, **893–894**
  spiral, 714
  sublimation growth, 274
  substrates (*see* Substrates)
  superlattice, 722
  temperature dependence, 271–280, 713, 715–717, 720, 721, 723, 731, 732
  terraced, 714
  vapor growth, **274–276**
Fivefold axes (*see* Symmetry)
Flames, 11, 29, 110, 112, 778
Fluctuations, 174, 522, 563, 564, 566, 570, 583, 599, 608, 621, 663, 812, 832, 833
Fluorinate (*see also* Halogens), 306, 314, 906
Flux quantum, 627, 818, 819, 821–823
Force constant, 233, 335, 338, 348, 349, 363, 374, 391, 576, **840–841**
  bending, 840
  carbon nanotubes, 840
  radial, 840, 841
  stretching, 840
  tangential, 840, 841
Four-member rings, **211–212**, **314**, **322–324**, 401, 402, 508
Fourier transform infrared spectroscopy (FTIR) (*see also* IR spectrum), 493
Fragmentation, 6, 12, 48, 132, 143, 150–152, 155, **156–166**, 167, 168, 173, 292, 465, 710, 895
  collisions with ions, **159–163**, **165–166**
  collisions with surfaces, **163–165**
  contraction, **156–157**
  deformation on impact, 163, 164
  photofragmentation, 6, **157–159**
Free energy, 195, 732, 775
Free radicals, 315, 320, 323
Free-standing C_{60} films (*see* Film growth)
Frenkel exciton (*see* Exciton)
Friction, 714, 715, 894, 906
Frozen matrix, 488–491, 872
Fulgurite, 10, 110
Fulleride (*see also* Alkali metal fullerides, Alkaline earth fullerides), 224
  other fullerides, 572, 573, 664, 682
Fullerite, 455, 556, 690
Fulleroid (C_{61}), 308, 309, 324, 325, 897
  bifulleroids, 325
  $C_{60}C(C_6H_4X)_2$, 309, 310
  Diphenylfulleroid, 308, 309
  $(p\text{-}BrC_6H_4)_2C_{61}$, 309
Fullerols, 314
Functional groups, 296, 306, 310, 311, 314, 315, 317, 906, 910

G

g-factor, 670, 674–676, 680, 743, 753, **836, 838**
  anisotropy, 675, 676, 753, 836, 838
g-value, 670, 673–678, 680–682, 750, 753, 836, 838
  carbon nanotubes, **836, 838**
GaAs substrate (*see* Substrates)
GaSe substrate (*see* Substrates)
GeS substrate (*see* Substrates)
Gas phase purification (*see* Purification)
Geodesic dome, 7, 8
Geological connections (*see also* Fulgurite, Shungite), 1, 9–11, 110
Glass transition (*see also* Magnetic properties, spin glass), 192, 878
Glassy carbon (*see* Carbon materials)
Grain boundaries, 497, 498, 632, 732

Grain sizes, 203, 565, 629, 720, 889
Granular conductors, 565, 629, 632
Graphene planes, 16, 21, 27, 37, 60, 537, 756–758, 760, 776, 777, 782, 784, 791–793, 801, 802, 804–808, 813, 819, 822, 824, 840, 841, 845, 847, 849
  2D Brillouin zone, 841, 845, 848, 849
  2D cosine band, 823
  2D density of states, 807, **808**, 863
  2D electronic dispersion, **802–809**, 813, 819, 822
  2D lattice constant, 75, 804, 810, 820, 822, 824
  2D phonon dispersion, 840, 841, 845
  dynamical matrix, 841
Graphite (*see also* Carbon materials), 3–7, **15–21**, 815, 829, 833, 837
  a-axis alignment, 20
  anisotropy, **16–19**, 22, 822, 833, 836
  c-axis alignment, 20
  band overlap, 17, 19, 829
  bandwidth, 432
  basal plane, 15
  battery applications (*see also* Batteries), 901
  binding energy, 17, 63
  carrier density, 17, 829, 834, 838
  carrier mobility, 17
  compressibility, 17, 175
  conductivity, 25, 26
  core level spectra, 704–706, 711
  Debye temperature, 17, 603
  defects, 20
  density of states, 17, 807, 808
  diamagnetism, 822
  dispersion relations, 19, 851
  elastic constants, 17, 856
  elastic modulus, 16, 17, 21, 23, 856
  electron energy loss spectra, 838, 839
  electronic structure, 19, 416, 802, 806, 807, 829
  exfoliated, 34, 35
  Fermi surface, 19
  force constants, 840, 841
  g-factor, 836, 838
  HOPG (*see* Highly oriented pyrolytic graphite)
  infrared spectrum, 847
  in-plane bonding, 816
  intercalation compounds (*see* Intercalation)
  interlayer 3D stacking, 16, 760, 815
  interlayer separation, 17–19, 822
  Kish graphite, 20
  lattice constants, 17, 18
  lubrication, 906
  magnetic susceptibility, 741–743, **822**, 833, 834, 836–838
    anisotropy, 833
    $H \parallel c$-axis, 822, 833, 834
    $H \perp c$-axis, 822, 833, 834
  melting point, 17, 36, 37
  overlap integral, 803, 804, 813
  phonon dispersion relations, **840**, 841
  π-bands, 807, 813
  plasma frequency, 704, 705
  precursors, 21, 34, 46, 47
  pressure effects, 15, 17
  Raman spectrum, 17, 24, 26–28, 30, 851–853
  reactivity, 895
  resistivity, 17, 829, 838
  ribbons, 35
  single crystal, 16, 18, 20, 21
  sound velocity, 17
  specific heat, 17, 595
  strength, 23
  stress, 20
  structure, 16, 17, 822, 860
  surfaces, 163, 713, 714
  thermal conductivity, 17, 602
  thermal expansion, 17
  thin film, 21
  turbostratic (*see* Turbostratic graphite)
  vermicular, 35
  whiskers, 21, 765, 783
Graphitization, 28, 29, 789
Group theory (*see also* Point groups and Space groups), **80–108**, **175–184**, **791–802**, **843–845**
  Abelian, **797–800**, 844
    phase factors, 798, 799
  angular momentum decomposition, 82, **84**, **100**, **341–347**
    for I, **84**
  basis functions, **82–84**, 91, 92, 98, **108**, 795, **796**, 799
    for D_{5d} and D_{5h}, **98**
    for I, **83**, 92
  character tables, **82**, 85, **86**, **91**, **97–99**, **101–102**, **108**, **795–800**

Group theory (*continued*)
class, 81, 86, 91, 99, 177, 795, 797, 799
compatibility relations, **97–99**, 179, 795
direct product, 86, 88, **89**, 91, 103, 181, 182, 184, 341, 349, 357–359, 361, 372, 384, 427–429, 433, 470–475, 578, 795, 797, 799, 872
equivalence transformation, 85–88, 99, **101–102**, 104, 177, **178**, **180–182**, **184**, 341
inversion (*see also* Parity), 62, 70, 88, 94, 96, 97, 248, 260, 332, 380, 386, 795, 797
irreducible representations, **81–92**, 98–105, 177–181, 332, 333, 340, 343–345, 361, 372, 392, 417, 418, 420, 428, 475, 517, 578, 744, 797–799, 805, 872
isomers, 71–73
subgroups, 97, 99, 244, 342, 795
symmetry axes, 9, 202, 351, 797
fivefold, 1, 9, 10, 12, 62, 66, 69, 80, 81, 82, 85, 88, 96, 97, 99, 102, 103, 105, 186, 194, 200, 393, 420, 598, 659, 757, 775
N-fold, 798
threefold, 62, 81, 85, 88, 178, 181, 184, 186, 187, 190, 194, 199, 311, 659, 723–725, 757
twofold, 9, 12, 48, 62, 81, 85, 88, 176–179, 185–187, 190, 193, 244, 245, 261, 419, 447, 725, 797
symmetry elements, 175, 176, 797, 799, 800
time reversal symmetry, 798
Growth (*see also* Carbon nanotubes, Film growth), **143–152**
C_2 absorption or emission (*see* C_2 pairs)
carbon nanotubes (*see* Carbon nanotubes)
contraction, 132, 143, **156–166**
epitaxial (*see* Film growth)
fragmentation (*see* Fragmentation)
from a corannulene cluster, **148–150**
fullerene growth models, **143–144**
mass spectroscopy characterization, **152–153**
molecular dynamics models for growth, 143, **166–168**
stability issues, **153–156**
Stone–Wales model, **144–147**
transition from C_{60} to C_{70}, **150–152**, 452
Grüneisen parameter, 403

H

h_g-derived band (*see also* Electronic structure), 521, 522
H_g-derived mode (*see also* Vibrational modes), 378, 643
h_u-derived band (*see also* Electronic structure), 437, 441, 448, 451, 453, 456, 501, 518, 522
Hagen–Rubens extrapolation, 529, 530
"Hairy ball," 296
Hall effect (*see also* Transport properties), **580–581**, 608
connection to Seebeck coefficient, 581
sign change, 580, 581
temperature dependence, 580
Halogens, **256–258**, **306–307**, 451
bromine, 258, 451
$C_{60}Br_6$, 305, 307
$C_{60}Br_8$, 307
$C_{60}Br_{24}$, 296, 307
chlorine, 451
$C_{60}Cl_n$, 307
fluorine, 314, 906
$C_{60}F_{36}$, 307
$C_{60}F_{60}$, 296, 307, 906
iodine, 236, 256–258, 664, 665
$C_{60}I_2$, 257, 664, 665
$C_{60}I_4$, 256, 257, 586
$C_{60}I_x$, 236
$C_{70}I_2$, 258
Hamiltonian (*see also* Hubbard model), **419–424**, 456, 471, 646, 669, 701, 740
hyperfine interaction, 669
icosahedral perturbation, 419
many-electron, 421, 422
phenomenological, 421, 422
ring currents, 740
Health and safety issues, 111, 114, 115, 117, 138
Hardness (*see* Mechanical hardness)
Heat capacity (*see* Specific heat)
Heat of formation, 63, 68, 560
Heat treatment temperature, 20, 24, 25, 27, 287, 531, 592, 765, 782, 851
Hebel–Slichter peak (*see also* NMR), 636, 637, 666, 684
Heptagon-pentagon (7,5) pair, **144–147**, 775, 783, 788
Heptagons, 65, **144–147**, 772–774, 788

Herzberg–Teller vibronic states (*see* Optical properties)
Heterojunctions
diodes, 871, 873, 880, 881, 884, 885
Nb/C_{60}/*p*-Si, 884–886
Ti/C_{60}/*p*-Si, 884, 885
photovoltaic devices, **886–888**
polymer–C_{60} composite rectifiers, **884–886**
Hexagonal close packed (hcp) structure, 50, 85, 198–202, 233, 598, 714
High pressure, 15, 16, 31, 175, 403, 587
catalytic reactions, 899
chromatography, 907
High temperature, 15, 16, 31, 36, 37, 175
Highly oriented pyrolytic graphite (HOPG), 20, 21, 157, 164, 713, 714, 851, 853, 854
Hildebrand solubility parameter, 120, 121
Hindered rotation, 259, 329, 657, 749
History, **1–14**, 49, 80, **111–112**, 122, 228
astronomical observations, 3
architectural analogs, 4
biological and geological examples, 5
carbon cluster studies, 3, 111
carbon nanotubes, **761–764**
early history, 1
physical interest, 7
production techniques, 6, 111, 112
recent, 6, 111, 112
superconducting properties, 617
HOMO (highest occupied molecular orbital), 3, **93–95**, 229, 233, 413, **416–420**, **425–428**, 432–435, 438–441, 451, 453, 469, 470, 472, 473, 476–478, 480, 501, 518, 693, 721, 728, 744, 745
C_{60} on surfaces, 721, 728
$C_{60}^{n\pm}$ ions, **425–428**, 518, **744–745**
doped C_{60}, 584
higher mass fullerenes, 432–435, 536–540
HOMO–LUMO gap, 73, **418**, 425, **432**, **435**, **438–439**, 441, 453, 457, 476, 504, 537, 539, 540, 593, 691, 693, 694, 701, 706
HOMO–1, 508–510
molecular C_{60}, 300, 430, 438–440, **453–455**, 457, 469, 470, 473, 476–478, 480, 593, 691, 693, 694, 697
phototransformed C_{60}, 508–510
Honeycomb lattice, 74, 76, 757–760, 774, 776, 777, 792, 803, 806
Hopping, 25, 541, 563, 579, 586, 592, 593, 608, 749
Host-guest complex, **316–320**
C_{60}-hydroquinone, 316, 317
C_{60}-γ cyclodextrin, 317, 318
C_{60}-calixarenes, 317–319
C_{60}-ferrocene, 319, 320
C_{70}-calixarenes, 317–319
HREELS (high resolution electron energy loss spectroscopy) (*see also* EELS), 279, 357, 359, 360, **369–370**, 387, 388, 499, 702, 706, 717, 724, 726
Hubbard model, **456–457**, 644, 646–648
Hermitian conjugate, 646
Hubbard *U*, 438, **455–457**, 556, 558, 561, **646–647**, 689, 691, 693, 696, 708, 711
Hückel theory (*see* Electronic structure)
Hund's rule, **94–96**, 417, **424–428**, **434**, 739, **742–745**, 748
Hydrogen, 296
hydrides, 898, 900, 911
impurities, 43
storage, 898, **899–901**, 911
Hydrogenation (*see* Chemical reactions)
Hydrophilic, 281, 282, 889, 890, 897, 900
Hydrophobic, 281–283
Hyperfine structure (*see also* Rotational spectra, EPR), 137, 229, 230, 344, 345, 424, 428, 663–665, **669–672**, 745
Hysteresis (*see also* Phase transitions), 197, 597, 750–753

I

Icosahedral fullerenes, 60, 61, 66, **70–71**, **74–77**
Icosahedral harmonics, 420
Icosahedral symmetry, **80–96**, 329, 340, 414, 415, 417–419, 744
Icosahedron, **1–12**
truncated, **1–12**, 60, 61, 294
Icosahedron truncum, 2
Impurities, 40, 41, 43, 540, 561, 586, 606, 659, 673, 676, 683, 709, 807, 817, 850, 904
substitutional, 775
Inclusion complex, **316–319**
Indium-tin oxide (ITO), 878, 879, 885
Inelastic neutron scattering (*see* Neutron inelastic scattering)
Inequivalent sites, 179, 181, 246, 255, 258, 259, 391

Infrared spectrum (*see* IR spectrum)
Insolubility, 209, 215
Intercalation compounds
carbon nanotubes, 775
fullerene, 7, 224, 225, **234–236**, 256, **285–288**, 294, 297, 301, 316, 365, 506, 561, 570, 720, 726, 901
graphite, 26, 35, **37–39**, 235, 240, 241, 257, 285, 287, 294, 301, 378, 379, 450, 572, 580, 616, 617, 621, 743, 744, 863, 901
graphite/C_{60}, **450–451**
$C_{60}C_{32}$, 450
$K_4C_{60}C_{32}$, 450, 451
MoS_2, 235
staging, 39
transition metal dichalcogenides, 241, 301, 743
Interfacial strain, 277
Intergalactic medium, 3
Intermolecular charge transport, **441–443**, 556–573
Intermolecular distance (*see also* Bond lengths), 199, 203, 205, 233
Intermolecular hopping integral, 504, 561
Intermolecular modes (*see* Vibrational modes)
Internal friction, 192, 608
Interstellar dust, 6
Interstitial sites, 235, 236, 450, 506, 881
Interstitial voids, 234, 236, 301
Intersystem crossing, 464, **467**, 472, **473**, 481, **484–485**, **487–488**, 680, 872
quantum efficiency, **483–485**, 488, 547
Inverse photoemission spectroscopy (IPES) (*see* Photoemission)
Inversion layer, 882
Inversion symmetry (*see also* Parity), 62, 70, 72, 81, 85–88, 91, 97, 248, 260, 380, 332, 386, 795
Iodine (*see* Halogens)
Ionic size (*see also* Alkali metals, Alkaline earths), 236, 238, 243
Ions (*see* Anions, Cations)
Ion implantation, **165–166**, 893
Ionization potential, 63, 68, 163, 166, 232, 416, 417, 457, 708, 719
IR spectrum, 3, 6, 7, 88, 89, 108, 215, 229, 248, 258, 296, 331, 336–339, 355, 357, 365, 367, **372–374**, **381–386**, 464, 492, **511–513**, 531, 565, 627, 641, 699, 871, 872
anions, **511–518**
C_{60}^{1-}, 512–516
C_{60}^{2-}, 512–514, 516
C_{60}^{3-}, 512–516
C_{60}^{4-}, 513–516
C_{60}^{n-}, 512, 516
$^{12}C_{59}{}^{13}C$, 343
$^{12}C_{60}$, 343, 344, 346
$^{13}C_{60}$, 343, 344, 346
C_{70}, 356, 390–392
carbon nanotubes, 756, 797, 799, 839, 840, 842–845, **845–850**
charged phonon model, **382–385**, 386
combination modes, 360, 361, 363, 367, 374, 494
crystal field effects, 356, 363, 372–374
dispersion effects, 373
dopant dependence, 382, 383, 386
doped C_{60}, **381–387**, 523, 524
doping-induced broadening, 383
doping-induced shifts, 381–387, 749
electron–molecular vibration (EMV) coupling, 382–384, 465, 468, 470, 498
F_{1u} (T_{1u}) mode, 88, 90, 213, **333–335**, 356, 357, 372, 373, **381–387**, 403, 478, 483, 493, 494, 518, 524, 526
vibronic, 475, 476
far IR, 524, 530, 627, 629, 637, 638
first-order, 88, 89, 331, 338, 356–358, 361, 362, 365, 481, 494
FTIR spectrum, 493
gas phase, 342, 356
higher order, 88–90, 331, 332, **361–363**, 365
intermolecular vibrations, 347, 352, 494
isotope effect, 108, **342–347**, 363, 374
lack of overtones, 360, 361
Lorentz oscillator, 386, 494
mid-IR features, 524, 526–528
mode displacements, **335**, 337, 494
oscillator strength, 382–386
phase transition, 372, 374, 386, 494
phototransformed C_{60}, 397, 399, 402
phototransformed C_{70}, 402
pressure-induced broadening, 403, 404
pressure-induced phase, 207, 403, 404
pressure-induced shifts, 403, 404
selection rules, 88, 89, 332, 342, 344, 345, 356, 361, 706, 797, 799
solution spectra, 356

splitting, 373, 374
temperature dependence, 365, 372, 373, 385
TDAE-C_{60}, 406, 749
transmission, 531, 532
unrenormalized mode frequency, 382
Irreducible representations (*see* Group theory)
Isolated pentagon rule, 62, 64, 66, 67, 69, 143, 145, 147, 149, 294, 295, 731, 769, 771
Isomers, 907
fullerene, 66, 69, **71–73**, 90, 101, 102, 104, 110, 111, 122, 143–145, 202, 229, 310, 331, 390, 432, 436, 437, 452, 453, 655, 671, 672, 694, 703, 731, 771
other molecules, 44–46, 215, 257, 259, 310
Isotopes
abundance, 80, 104, **106–108**, 114, 153, 354, 654
$^{12}C_{60}$, 104, **106–108**, 341, 639, 641
$^{13}C_{60}$, 104, **106–108**, 639, 641, 654
$K_3{}^{12}C_{60}$, 639, 641
$K_3{}^{13}C_{60}$, 639, 641
$^{87}Rb_3C_{60}$, 641
$Rb_3({}^{13}C_{1-x}{}^{12}C_x)_{60}$, 641
$Rb_3[({}^{13}C_{60})_{1-x}({}^{12}C_{60})_x]$, 641
Isotopic distribution, **106–108**, 153, 155–157
Isotopic enrichment, 106, 107, 639–641
Isotopic shifts, 108, 639
Isotopic symmetry lowering, 106, 108, **341–347**

J

Jahn–Teller effect, 378, 422–424, 429, 430, 488, 489, 512, 515, 516, 518, 557, 559, **576–580**, 643, 647, 648, 675, 676, 679, 744, 745, 814
dynamic, 578, 579, 648, 676, 679
splittings, 423, 489, 512
static, 579, 679
superconductivity, 577, 643, **647–648**
Joint density of states, 500

K

K_xC_{60} (*see* Alkali metal fullerides)
Keating potential (*see* Phonons)
Kekulé bonding (*see* Bonding)
Kish graphite, 20
Korringa relation, 664, 665
Knight shift, 559, 637, 638, **664**, 666–668
Knudsen cell, 73, 120, 129, 130, 274, 279
Knudsen effusion equation, 129
Kramers–Kronig relation (*see also* Optical properties), 493, 506, 523, 524, 526, 528, 529, 637, 638, 704

L

L_a, 27, 28
L_c, 27
Ladder polymer, 46–48, 898
Landau level quantization, 821, 823, 829, 836
Landé g-factor, 743
Langmuir, **892**
Langmuir–Blodgett techniques, 883, 896, 897
Lanthanum (La), 30, 228–231
La^{3+}, 230, 231, 671, 672, 744
LaC_2 particles, 791, 862
$La@C_{44}$, 228
$La@C_{60}$, 131, 228
$La_2@C_{60}$, 131
$La@C_{74}$, 127
$La@C_{76}$, 671
$La_2@C_{80}$, 229
$La@C_{82}$, 228–230, 435–437, 669, 671
$La@C_{84}$, 671
$La_3@C_{82}$, 229
$La@C_{82}$, 127, 135–137, 228–231
$La_2@C_{82}$, 231
$La_3@C_{82}$, 229, 230
Laser ablation, 3, 4, 26, 37, 110–112, 153, 155, 158, 899
Laser desorption, 126, 152, 158, 159, 210, 214–217
Laser pyrolysis, 26, 29, 30, 790, 791, 851, 853
Laser vaporization, 3, 4, 112, 131, 133
Latent heat, 166, 172, 199, 596, 600, 601
Lattice constant, 193, 208, 731, 821–823
alkali metal doped C_{60}, **242–243**, **247–267**, 386, 442, 444–446, 560, 564, 585, **617–620**, **627**, 629, 643, 657
alkaline earth doped C_{60}, **242**, **253–256**, 566
C_{60}, 171, 172, 175, 184, 206, 208, 276, 492, 699, 700, 722, 723
pressure-induced, 206
$C_{60}(S_8)_2CS_2$, 236, 237

Lattice constant (*continued*)
  C_{60}-n-pentane, 237
  C_{60}-ferrocene, 320
  C_{70}, 199, 262, 699, 700, 731
  carbolite, 32
  carbynes, 32
  other doped C_{60}, 242, 243, 253, 256–259
  doped C_{70}, 262, 263
  epoxide $C_{60}O$, 258, 308
  fullerene mass dependence, 699, 700
  higher fullerenes, 199, 202, 203, 731
  pressure dependence, 587
  Rb_9C_{70}, 263
  TDAE-C_{60}, 260, 261
Lattice distortion, 237, 253, 445, 621, 657, 745, 814
  Peierls, 803, 805, 807, 811–813, 816
Lattice dynamics (*see also* Dynamical matrix), 347, 348
Lattice expansion (*see also* Thermal expansion), 238, 240, 259, 286, 557, 560, 621, 622, 665, 857
Lattice image (*see also* TEM), 27, 29, 233, 276, 278, 784, 855, 858, 859
Lattice mismatch, 276, 277, 281, 537, 713, 720, 722, 724, 731, 894
Lattice modes (*see* Vibrational modes)
LEED (low energy electron diffraction), 213, 279, 280, 369, 697, 700, 894
Lennard–Jones potential, 242, 244, 353
Lewis acid, 256, 258, 288, 307, 572, 573
LDA (*see* Electronic structure)
Libration (*see* Vibrational modes)
Librational energy, 388, 389
Lightening strikes, 10, 11
Liquid chromatography (*see* Chromatography)
Lithium (*see also* Alkali metals), 53, 301–304, 619, 901
  battery (*see also* Batteries), 301, 901
Lithography (*see* Photolithography)
Localization length, 587
Localized carriers (*see also* Transport properties), 838
  charge (*see also* Delocalized charge), 33, 304, 314, 315, 436, 566, 645, 673, 683, 739, 745–749, 823, 836, 883
  excitons (*see also* Excitons), 489–491, 500, 504, 505, 522, 680, 692
  strong, 456, 566, 593
  weak, 570, 583, 831, 832
Localized spins, 747, 749
London penetration depth (*see also* Superconductivity), 530, 626, 629, 668, 683, 684
Lorenz number, 602
Lubrication, 890, 905, 906, 910
  additive to oils, 906, 910
  ball bearings, 906
Luminescence (*see also* Optical properties)
  electroluminescence, 47, 898, 909
LUMO (lowest unoccupied molecular orbital), 3, 236, 414, 415, **417–419**, 425, 432–435, 438–441, 444, 451, 495, 501, 509, 510, 512, 513, 515, 518, 536, 537, 539, 540, 542, 543, 556, 561, **577–579**, 593, 691–694, 696, 701, 706, 744–747
  $C_{60}^{n\pm}$ ions, **512–513**, **515–516**, 518, **744–745**
  C_{60} doped polymers, **536–537**
  C_{60} on surfaces, 721, 726, 728
  C_{60} solid, 495, 501
  doped C_{60}, 236, 300, 444, 451, 556, 561, 563, **577–579**, 584, 606, 617, 619, 648, 691, 692, 696, 701
  endofullerenes, 137, 229, 435, 436, 671
  half-filled band, 556
  higher mass fullerenes, 418, 419, 432–434, 536, 539, 540, 693, 694, 706
  LUMO+1, 236, 254, 470, 508–510, 514, 516, 542, 543, 696, 747
  LUMO+2, 543
  molecular C_{60}, 94, **417–419**, 438, 439, 453, 455, 457, 469, 470, 472, 473, 476, 478, 480, 495, 691–693, 701
  phototransformed C_{60}, 509, 510

M

Magnetic flux quantum (*see* Flux quantum)
Magnetic properties, 260, 261, 668, **680–683**, **739–753**, **818–825**
  anisotropy, 261, 741, 742, 749
  Bohr magneton, 670, 746, 748
  carbon nanotubes, **818–825**
    anisotropy, 818
    diamagnetic behavior, 820, 821, 833, 834
    diameter dependence, **820**, **824**
    field dependence, 819, 821, 822, 824, 835, 836

field orientation, 819–821, 823, 833
Landau levels, 821, 823, 829
Landau quantization (perpendicular), 821, 823
Landau radius (magnetic length), 821, 823, 824, 834, 836
magnetic energy levels, **818–825**
magnetic moment (parallel), 819, 820, 835
magnetization, 818, 821, 833
Pauli susceptibility, 825
periodic magnetic properties, 818, 819, 821, 823, 824
periodic wave function, 818
phase factor, 818
temperature dependence, 835, 836
circular dichroism, 474
coercive field, 750
core diamagnetism, 745–748, 750
diamagnetic behavior, 413, 739, **740–742**, 745
dopants, 230, 416, 739, **742–743**
d shell, 742, 747, 748
f shell, 742, 743, 747, 748
ferromagnetism, 260, 683, 745, **748–752**
fullerene ions, 418, **744–745**
g-value (*see* *g*-value)
hysteresis, 752, 753
itinerant ferromagnetism, 752
level configuration, 742, 743
magnetic moment, 742–750, 752
magnetic phase transition, 684, 739, 748, 751
magnetic specific heat, 752
magnetization, 682, 746, 750, 751
ordered phases, 739, 749, 752, 753
paramagnetism, 739–742, 752, 753
Curie, 739, 740, **745–747**, 751
Curie–Weiss, 745–748, 750, 751
p-level, 418, **747–753**
Pauli, 415, 569, 584, 667, 739, 744, **745–747**, 748, 825
relaxation, 752
remanent magnetization, 682, 750, 752
ring currents, **739–742**, 745
hexagonal, 740, 741
London theory, 740
pentagonal, 740, 741
saturation magnetization, 750, 752
spin glass, 669, 682–684, 752, 753
structural phase transition, 686, 740, 749–752
superparamagnetism, 752
TDAE–C_{60}, 259, 299, 406, 668, 669, 680–683, **748–753**
temperature dependence, 681, 741, 750, 752
transition temperature, 680–683, 748, 750, 752, 753
field dependence, 750, 752
pressure dependence, 750, 752
Magnetic susceptibility, 398, 564, 569, 622, 648, 681, 683, 740, 741, **745–753**, 756
carbon nanotubes, 741, 742, 756, 818–821, 823, 824, **833–837**
arrays, 824
carrier scattering effects, 836
diamagnetic, 833, 834
field dependence, **833–835**, 836
field independence, 833, 836, 837
field misalignment, 822, 823, 833
flux dependence, **821–822**
parallel, 819–823, 833, 834, 837
periodic behavior, 821
perpendicular, 821–824, 833, 834, 837
singularities, 821
temperature dependence, 835–837
$\chi'(T)$, 749, 750, 752, 753
$\chi''(T)$, 749, 750, 753
Curie–Weiss, 745–748, 750, 751
density of states determination, 746
field cooled, 622, 751, 752
higher fullerenes, 753
parallel, 741, 742
perpendicular, 741, 742
temperature dependence, 741, 747–749, 752
zero field cooled, 622, 751
Magnetoresistance, 557, **581–583**, 608, 768, 818, 823, **829–833**
carbon nanotubes, 768, 818, 823, **829–833**
field dependence, **830–832**
negative, 829, **830**, 831
positive, 830, 831
temperature dependence, 829, **830–832**
transverse, 830
universal conductance fluctuations, 830, **832–833**
magnetic field dependence, 581, 582
temperature dependence, 581, 582
transport mechanism, 581, 582

Many-body interactions (*see also* Hubbard model), 422, 438, **453–456**, 468, 556, 558, 561, 584, 585, 599, 648, 692, 696, 708, 710
many-electron, 90, 413, 418, **421–434**, **468–474**, 503, 504
Mass density
of C_{60}, 171, 172, 331
of doped C_{60}, 250
Mass spectrometry, **152–153**
mass resolution, 152, 156, 157
spectra, **4–6**, 31, 48, 49, 52, 53, 64, 67, 121, 127–129, 130, 134, 136, 143, **152–166**, 210, 214–217, 258, 305, 307, 364, 899
time-of-flight, 5, 48, 49, 52, 53, 118, 126, 135, 136, 152–165
Materials applications, **893–898**, 909
catalytic properties, **894–896**
diamond synthesis, 893, 909
membranes, **888–890**, 905
new chemicals, **897–898**
self assembled monolayers, **896–897**
SiC film growth, **893–894**
Mean free path, 129, 172, 524, 564–566, 602, 632, 633, 696
Mechanical properties (*see also* Carbon nanotubes), 281, 894, 905
hardness, 17, 33, 894
modulus, 172
Medical applications, 897, 910
Membranes, **281–284**, **888–890**, 905
Merohedral disorder, 118, 176, 182, 185, 186, 189, 192, **193**, 245–248, 341, 438, 441–444, 454–456, 557, 561, 564–566, 574, 580, 582, 583, 597, 599, 603, 607, 632, 665, 751, 752
Metal–coated fullerenes (*see also* Carbon materials)
Alkali metal coated fullerenes, **51–54**
Alkaline earth coated fullerenes, **52–54**
Metal–insulator transition, 557, 558, 561
Metallic fullerene phases, 380, 445, 521, 523, 531, 532, 556, 561
Metallo–carbohedrenes (met-cars), **48–50**
Metallofullerenes (*see also* Endohedral fullerenes), **435–436**, **733**
Methyl-ethyl-hydroxyl-polypropylvinyl (MEH-PPV), 535, 536, 885–887
Methyl groups, 305, 314, 315
Methylene, 119, 308
Mica substrate (*see* Substrates)
Microdensitometer, 506, 507
Microfabrication, 881, **888–892**
Microstructure, 9, 24, 25, 35, 36, 274, 570, 629, 660, 850
Migdal theorem
criterion, 574, 575, 647
violation, 574, 575
Mobile hole, 534, 535, 873, 877, 881
Mobility, 815, 830, 883, 888, 891, 904
carrier, 17, 37, 589, 882, 883, 888
surface, 710, 713, 714, 721, 722, 891
Molecular alignment (*see also* Orientational alignment), 185–193, 199–202, 242–246, 252, 732
Molecular density, 172, 479
Molecular distortion (*see also* Jahn–Teller effect), 65, 163, 345, 377, 403, 406, 422–424, 430, 436, 443, 489, 512–516, 518, 542, 557, 559, 576–579, 644, 648, 674–676, 679, 744
Molecular dynamics, 69, 143, 164, **166–168**, 194, 209, 215, 237, 390, 394, 398, 654, **657–662**
simulation, 166–168, 231, 237
Molecular manipulation, 712, 715, 889, 894, 901, 902, 904
by STM tip (*see* STM)
by rotating electric field, 902
nanosled, 902
Molecular orbitals (*see also* Electronic structure), **415**
Molecular photophysics, **465–468**
Molecular reorientation, 173, 175, **189–193**, 199, 202, 209, 237, 238, 245, 330, 373, 603, 658, 659
energy barrier, 189, 190, 192, 193, 603, 607, 608
Molecular solid, 60, 85, 90, 257, 260, 329, 376, 378, 413, 423, 437, 493, 495, 539, 544, 883
Molecular volume, 63, 172, 560
Moment of inertia, 63, 68, 189, 329, 331, 342, 347, 394, 732
Monoclinic (*see also* Crystal structure), 200, 201, 237, 260, 598
MoS_2 (*see also* Substrates), 235, 274, 280, 720, 906
Mössbauer spectroscopy, 229, 257

Mott insulator (*see also* Hubbard model), 645
Multilayer structures, 697, 698, 701, 702, 710, 713–717, 721, 726, 730, 732, 756, 760, 775, 816, 826, 859, 863, 891, 892
metal-C_{60}, 261, **570–572**
metal coated fullerenes, **51–54**
multiwall tubule (*see* Carbon nanotubes)
onions (*see* Carbon onions)
transport properties, **570–572**
Muon spin resonance (μSR), 174, 200, 232, 397, 627, 629, 633, 635–638, 654, 668, **683–684**, 750, 753
coherence peak, 636, 684
Hebel–Slichter theory, 636, 637, 684
motional narrowing, 684
muon, 232, **683–684**
muonium, 635
Mu@C_{60}, 635
phase transition, 684
spin relaxation, 636, 637, 684, 753
temperature dependence, 636

N

Na_xC_{60} (*see* Alkali metal fullerides)
Nanotechnology, 870, **901–904**
interconnects, 903
molecular manipulation, 901, 902, 904
nanotubes, 903, 904
Nanotubes
BN, 814
carbon (*see* Carbon nanotubes)
Neutron scattering, 189, 190, 200
neutron diffraction, 20, 34, 60, 65, 174, 180, 186, 189, 190, 200, 202, 245, 246, 252
neutron inelastic scattering (NIS), 69, 88, 174, 184, 186, 331, 332, 338, 347–355, 357, 359, 360, 365, 366, **367–368**, 388–390, 393–395, 597, 644, 840
New materials synthesis, **897–898**, 901, 909–911
Nitrogen doping, 41, 233
Nitrogen lone pair, 895
NO gas, 889
Nonlinear processes, 375, 464, 540, **544–548**, 870–872, 877, 909
Nonlinear optics, 464, **544–548**, **871–873**, 909
absorption, 546, 547, 871–873
applications, 871–873, 909
C_{70}, 544–547, 871, 872
$\chi^{(3)}_{xyyx}$, 545, 547
$\chi^{(3)}_{xxxx}$, 544–547, 871
damage threshold, 548
degenerate four wave mixing (DFWM), 544–547, 877
dependence on light intensity, 545, 871
electric field-induced second harmonic generation, 544, 548
excited state grating, 546
films, 544, 546
Im($\chi^{(3)}_{xxxx}$), 545, 547
interference grating, 544
optical limiting (*see* Optical properties)
polarizability, 544
polarization vector, 544, 545
resonant enhancement, 546
second harmonic generation, 548
solutions, 544, 546, 871
third harmonic generation, 544, 547
third order electric susceptibility (*see* $\chi^{(3)}_{xxxx}$)
two photon process, 546, 872
Nonradiative transition, 480
Nuclear magnetic resonance (NMR), 6, 12, 34, 48, 60–63, 72, 73, 104, 108, 121, 168, 198–200, 207, 246, 257, 258, 296, 305, 307, 308, 314, 324, 452, **635–638**, **654–669**, 684, 748–750, 752, 753, 895
bond lengths, 440, 654, 656
^{13}C, 246, 305, 637, 654, 655, 667–669
chemical shift, 246, 656, 659–662, 749
tensor, 661, 667, 669
C_{70}, 654, 655, 657–661
C_{76}, 655
^{133}Cs, 637, 655, 657, 663, 667, 668
conduction electron effects, 664, 667
contact term, 664
density of states determination, 585, 665
determination of molecular diameter, 62, 63
dipolar coupling, 656, 662, 664, 665
dopant locations, 246, 258, 654–658
electric quadrupole moment, 654
Fermi contact term, 585
gas phase, 658
gyromagnetic ratio, 660, 665
^{1}H, 305, 668, 669, 753, 895
Hebel–Slichter peak, **636**, 637, 666, 684

Nuclear magnetic resonance (*continued*)
higher mass fullerenes, 655
hyperfine interaction, 664, 665
isotope effects, 108, 654
^{39}K, 559, 655, 656, 664
Knight shift, 560, 637, 638, 664, 666–668
Korringa relation, 664, 665
Larmor frequency, 660
line intensity, 656, 657, 662
lineshape, 657–659, 661, 668
anisotropy, 658, 659, 661, 662
linewidth, 657, 658, 661–663, 669
magic angle, 198, 657, 663, 664
magnetic field dependence, 666
molecular chirality, 655
molecular dynamics, 296, 654, 657–661, 669
molecular reorientation, 173, 174, 189, 199, 246, 659
molecular rotation axis, 660
molecular shape, 655
motional narrowing, 657, 662, 669
phase transition, 184, 188, 189, 198, 200, 658–661
paramagnetic behavior, 669
^{87}Rb, 246, 637, 655, 657, 663, 667, 668
site location, 246, 258, **655–657**, 663
solid, 658
solution, 655, 658
spin diffusion, 666
spin fluctuations, 663
spin-lattice relaxation, 188, 585, **635–638**, 657–660, 663–666, 753
spin susceptibility, 637
stoichiometry of dopants, 305, 307, 654, **662–663**
superconducting gap, 627, 633, 635–638, 654, 665, 666
superconducting phase, 654, 664–666
temperature dependence, 635–637, 657–659, 661, 664, 665
tetramethylsilane reference, 655, 659, 662, 667
Nuclear spin, 104, 106, 229, 354, 654, 670–672
Nucleophilic, 297, 299, 300, 304, 306, 314
Number n_C in C_{n_C}, 69–72, 94, 144
Number of isomers, 71
Number of valence electrons, 94, 95

O

Octahedral sites, 172, **226**, 234, 236, 238, **240–241**, **243**, 245, 247, **250–254**, 256, 259, 263, 286, 303, 316, 442, 443, 445, 447, 449, 451, 563, 568, 582, 588, 621, 622, 646, 656, 657, 696, 730, 743, 747, 748
interstitial site radius, 172
ODMR (optically detected magnetic resonance), **488–491**, 593
absorption (ADMR), 489, 491, 543
photoluminescence (PLDMR), 489, 491, 543
Oligomers, 210, 211, 324, 398, 399, 465, 508, 511, 889, 895
Optical applications, **870–880**, 909
limiters, 487, 488, **871–873**, 909
photoconductors, 1, 871, **873–876**
photorefractive devices, 871, 873, **876–880**, 909
photovoltaic device, 871, 873, 909
xerography, 533, 587, 593, 874
Optical properties, 291, 414, 457, **464–555**, 561, 638, 648, 697, 701, 702, 704, 726, 749, 870
absorption, 6, 7, 41, 283, 284, 289, 307, 438, 439, 452, 453, 456, 464–468, 470, 474, 476, 478, 480–482, 484, 489, 491, 492, **495–505**, 689–691, 694, 871–875, 888, 896, 897
coefficient, 476, 479, 480, 484, 485, 492, 493, 546, 547, 872
cross section, 484, 542, 872
edge, 63, 68, 172, 283, 439, 453, 456, 464, 466, **476–477**, 492, **495–496**, 501–505, 534, 537, 539, 541, 544, 546, 569, 590, 592, 690, 691, 693, 701, 706, 873, 887, 888
peaks, 508, 509, 513–515, 520, 521, 896, 897
anions, **511–518**, 531, 534–537, 675, 676
applications (*see* Optical applications)
broadening effects, 479, 494–496, 498, 505, 508–511, 519, 520, 528, 538, 539
C_{70}, 199, 483–489, 491, 498, **537–540**, 541, 569
calculations
ab initio, 495, 503, 519
intermolecular hopping integral, 499

many electron corrections, 503
transitions, **468–469**, 474–480, 482, **499–505**, 515–523
color, 437, 537
Clausius–Mossotti relation, 479, 492, 494, 495
conductivity, 382, 501, 502, 524, 525, 530, 637, 638, 704, 749
conduction electrons, 527, 532, 637
core dielectric constant, 494, 527, 538
decay times, 484, 485, 540, 541, 872
delay times, 485, 486
dielectric constant, 172, 493, 505, 506, 523, 538
dielectric function, 299, 479, 492–495, 499–501, 505, 518, 522, 524, 527, 528, 537, 538, 689
diffraction, 876, 878–880
diffraction efficiency, 879, 880
doped fullerenes, 307, 498, 500, **518–532**, 541, 543, 637, 638
doping dependence, 521, 522, 524
Drude absorption, 749
Drude edge, 524–527
dynamical studies, 302, 307, 476, **483–488**, **488–491**, **540–544**
electric dipole matrix elements, 471, 472, 474–478, 481
electric dipole transitions, 464, **468–476**, 477, 478, 487, 518, 520
allowed, 464, **468–469**, 470, 471, 476, 478–480, 491, 498–501, 508, 510, 514–518, 520–522, 540–543, 872
forbidden, 464, 468, **469–476**, 477, 480, 483, 491, 492, 514, 540, 548, 872
electronic transitions, 414, 424, 440, **467**, 489, **495–505**, **508–509**, **514–515**, **520–524**, 527, 532, 537, 689
electro-optic effect, 876
ellipsometry, 274, 493, 499, **537–539**, 785
emission, 456, 465, 467, 484, 541
energy loss function [Im$(-1/\epsilon(\omega))$], 502, 523
$\epsilon_1(\omega)$, 492–494, 501, 502, 506, 521, 538, 704
$\epsilon_2(\omega)$, 493–496, 501, 502, 506, 518–521, 538, 704
excitations, 425, 429, 465–468, 481, 534, 535, 556, 870
excited state absorption, 484, 485, 487, 488, 503, 520–522, 540–544, 870
excited state decay, 541, 542
excited state relaxation, 540
extinction coefficient, 492, 513, 537, 873
fluorescence, 467, 468
free carrier contribution, 524, 527
gas phase, 464, 465, 479, 498
grating, 876, 877
Herzberg-Teller coupling, **469–476**, 481, 483, 495, 497, 504
higher-mass fullerenes, 296, 307, 537, 540
indium tin oxide (ITO) electrode, 878, 879, 885–887
induced photoabsorption, 484–491, **542–544**, 873
insensitivity to dopant species, 520
intensity dependence, 871–873, 876
internal conversion, 467
intersystem crossing (*see* Intersystem crossing)
joint density of states, 500
Kramers–Kronig analysis, 493, 506, 523–526, 528, 529, 637, 638, 704
lifetime, 484
for excited state, 490, 491, 541
for nonradiative transitions, 483
for radiative transitions, 467, 480, 483, 485, 488
for singlet states, 483–485, 487, 541
for triplet states, 483–489, 872
limiting, 1, 307, 487, 488, 870, **871–873**, 909, 911
Lorentz oscillator, 494, 508, 519–521, 538
luminescence, 88, 307, 332, 355, 357, 359, 360, 365, 366, 453, 464, 468, 474–476, 478, **480–483**, 484, 489–491, 495–497, 508, 510, 533, 535, 536, 541–543, 593, 689, 690, 873, 874, 897
normal state, 523–526, 529, 530, 637
nonlinear effects (*see* Nonlinear optics)
non-exponential decay, 540, 541
optical constants [*see* $\epsilon_1(\omega)$, $\epsilon_2(\omega)$]
optical density, 476, 499, 521, 522, 539, 548
optical pumping, 542
oscillator strength, 468, **469**, 477–480, 494, 500, 501, 503, 505, 518–520, 527–529, 538
phosphorescence, 464, 467, 473–475, 480, 481, 483, 489, 490
branching ratio, 483, 484

Optical properties (*continued*)
lifetime for C_{60}, 481, 487
lifetime for C_{70}, 486, 487
photoconductivity (*see* Photoconductivity)
photorefractive effect, 873, **876–880**
inorganics, 879, 880
polymers, **876–880**
phototransformation (*see* Phototransformation)
$\pi \rightarrow \pi^*$ absorption, 476, 477
plasma frequency (*see* Plasma frequency)
polymer composites, **532–537**
pulse width, 546, 548, 871, 873
pump-probe experiments, 484, 485, 489, 540, 541, 543
reflectivity, 493–495, 518, 519, 521, 524–526, 528–530, 537, 538, 540, 633, 637
refractive index, 492, 538, 547, 876
relaxation time, 542, 544
rotation, 125
$\sigma_1(\omega)$, 524–527, 529, 530
$\sigma_2(\omega)$, 524
saturation effects, 871
screening effects, 478–480, 503, 544
selection rules, 293, 470, 472, 475, 487, 501, 508, 541
singlet–triplet transitions, 473, 481, 485, 547, 871
skin depth, 544, 888
solid state spectra, **491–505**, 537, 872
solid state interaction effects, 497, 501, 504
solution spectra, 464, 465, 469, **476–480**, 481, 486, 487, 490, 491, 496–499, 512–514, 518, 537, 539–541, 871–873
solvent shift, 497
spectral shifts, 496–498, 509–511
spectroscopy (*see also* UV-visible spectroscopy), 281–284, **477**, 481, **482**, **486**, **490**, 492, **493–494**, **496**, **499**, **501–502**, **505**, **509–510**, **512–516**, **519–523**, **525–532**, **536–540**, **543**, **873**, **887**, 896, **897**
spin conservation, 424
spin-flip transitions, 473, 481
superconducting state, 524, **528–530**, 637
surface sensitivity, 544
temperature dependence, 482, 510, 528, 531, 543
transient absorption, 483–485, 541, 542
transient luminescence, 483, 541
transition probability, 872
transmission, 476, 489, 493–495, 499, 508, 518–524, 531, 532, 537–541
transmission grating, 876
traps, 498, 588, 589, 876, 883
triplet–triplet transitions, 543, 547, 871
vibronic transitions, 297, 307, 466, **471–476**, 481, 483, 487, 491, 495, 497, 498, 501, 510, 514, 693, 719, 872
group theory, 299, 307, 471, 475
X–traps, 498
Order parameter, 174, 195–197
Orientational alignment (*see also* Molecular alignment), **185–193**, 237, 238, **244–252**, 255, 370, 371, 374, 388, 587, 683
standard orientation, **185–187**
Orientational disorder (*see also* Merohedral disorder), 173, 174, 193, 227, 237, 252, 254, 259, 340, 341, 349, 354, 441, 495, 563, 574, 586, 603, 607
Orientational potential, **189–190**, **242–244**, 250, 252, 348, 353, 373, 389, 390, 394, 541, 752
Orthorhombic structures
for alkali metal doped C_{60}, 247, 248, 253, 386
for C_{60}, 175
for $C_{60}I_2$, 257
for C_{60}-n-pentane, 237
for $C_{60}(S_8)CS_2$, 236
for C_{70}, 201
for $K_3(NH_3)C_{60}$, 243, 253
for M_1C_{60} phases, 247, 248, 386, 531
for $Na_{172}In_{197}Z_2$, 51
Osmylation, 125, 311, 312
Other applications, **901–911**
Overtones (*see also* Raman spectra), 89, **358–361**, 478
Oxidation (*see also* Chemical reactions), 296, **303–304**, 781, 784, 891, 892, 894
Oxido–annulene bridge, 308
Oxygen, 211, 217, 238, 259, 292, 294, 307, 308, 312, 314, 365, 366, 481, 485, 486, 488, 506, 507, 524, 540, 545, 557, 586, 740, 781, 858, 881, 883, 889, 891, 892, 895
Oxygen-resistant coatings, 881, 891

P

Pa$\bar{3}$ (*see* Space groups)
PAH (*see* Polyaromatic hydrocarbons)
Palladium C_{60}–Pd complex, 322, 896
Parity (*see also* Group theory), 90, 178, 340, 346, 354, 357, 359, 361, 415, 430, 469, 470, 474, 475, 478, 487, 504, 647
Passivation, 281, **888–891**
Patents, 870, **908–910**
Patterning, 213, 881, **888–890**, 893, 894
Pauli paramagnetism, 569, 584, 667, 739, **744–746**, 748, 750, 825
  Cs concentration dependence, 745, 746
  many-body corrections, 584
Pauli principle, 91, 92, 343, 430, 431, 472
  Pauli-allowed states, 95, 343, 345, 421, 424, 425, 427–429, 433, 472, 744
Peierls distortion (*see also* Lattice distortion), 803, 807, 811–813, 816
Pentagon, 2, 5, 7–10, 48, 49, 53, 54, 62, 63, 74, 144–149, 246, 294, 295, 298, 299, 309, 313–315, 333, 441, 489, 541, 723, 725, 769–771, 774, 777, 778, 785–788, 893
Pentagon–heptagon pair (*see* Heptagon-pentagon pair)
Pentagonal pinch $A_g(2)$ mode (*see also* Raman scattering), 211, 333, **336**, 356, 364, 365, 374, **379–382**, 385, 386, 398, 400, 718, 749
Permutation group, 342
Pharmaceutical applications, 897
Phase coexistence, 208, 239, 254, 285, 723
Phase diagram
  alkali metal doped C_{60}, 238, 239, 662
  carbon, 15, 16, 36, 37, 403
  C_{60}, 112, 171, 173, 175, 184, 192, 402, 403, 608
  C_{70}, 208
  Landau phase transition theory, 194
Phase separation, 239, 386, 563, 662, 664
Phase transition, 594, 595
  at T_{01} in C_{60}, 119, 172, 175, 180, 183, **184–190**, **194–197**, 204, 205, 208, 211, 238, 239, 296, 340, 346, 348, 349, **370–374**, 441, 506, 587, 591, 595, 600–602, 608, 658–661, 740, 749
  at T_{01} in doped C_{60}, 252, 254
  at T_{01} in epoxide $C_{60}O$, 258, 259, 308
  at T_{01} in higher mass fullerenes, 732
  at T_{01} in TDAE–C_{60}, 680, 683
  at T_{02} in C_{60}, **190–193**, 593, 596, 597, 601
  enthalpy change, 594–597, 600
  entropy change, 594–596
  first order, 197, 204, 594, 595, 597, 600
  fcc to bcc, 254, 730
  fcc to orthorhombic, 531
  higher order, 594
  hysteresis (*see also* Hysteresis), 197, 597
  in C_{60}-n-pentane, 237
  in other fullerenes, 198–203, 207, 208, 393, 531, 598, 600
  latent heat, 600
  pressure dependence, 587, 600
  $\partial T_{01}/\partial p$, 172, 600
  volume expansion, 600
Phenyl group, 308–310, 315, 324
Phonons (*see also* Vibrational modes, density of states, dispersion relations)
  carbon nanotubes, **839–854**
    acoustic branches, 834, 845
    atomic displacements, 840
    bending mode, 857
    branches, 842, 848, 851, 853
    curvature-dependent effects, 841
    density of states, 852, 854
    diameter dependence, **846–847**
    dispersion, 792, 795, 800, 802, 803, 839, **840–845**, 851, 853
    first principles calculations, 840
    force constant models, 840
    infrared activity, **841–850**
    infrared intensity, 850
    Keating potential, 857
    mode frequencies, 841, 842, **843**, 845, **846–849**
    mode symmetries (*see also* Point groups), 842, **843–845**, 846
    modes, 842, 845, 846, 857
    out-of-plane vibrations, 850
    Raman activity (*see* Raman spectroscopy)
    Raman intensities, 850
    rotations, 842, 845
    silent modes, **842–845**
    spectra, 794, 795, **843**, 845
    translations, 842

Phonons (*continued*)
vibrational degrees of freedom, 834, 844
zone folding, 840, **841**, 844, 845, 848
graphite, 840, 841
Phonon-defect scattering, 603, 604, 606
Phonon-drag, 604, 606–608
Phonon mean free path, 602
Phonon-phonon scattering, 603, 606, 607
scattering rate, 606
Phosphorescence (*see also* Optical properties)
lifetime for C_{60}, 481, 487
lifetime for C_{70}, 486, 487
Photochemical effects, 96, 210, 211, 271, 303, 324, 356, 400, 465, 508, 888, 909
Photoconductivity, 491, 532–534, 557, **587–594**, 871, 873–876, 886, 887, 909, 911
at T_{01} phase transition, 591, 592
at T_{02} phase transition, 592
bimolecular recombination, 598, 591
carrier generation rate, 589, 593, 877
carrier lifetime, 593, 888
carrier localization, 592
carrier trapping, 588, 589, 876
charge generation efficiency, 874, 875
contact preparation, 588, 593
dark conductivity, 533, 588–594, 874, 886
decay time, 589, 590, 878, 879
dependence on oxygen, 588–592
discharge of surface charge, 874
electron-hole recombination, 588, 591, 592, 874, 888
exciton binding energy, 593
light frequency dependence, 590, 592, 593, 887
light intensity dependence, 590, 887
magnitude, 588
monomolecular recombination, 588, 591–593
Onsager model, 874
persistent, 588, 590, 591, 593
photoexcited carriers, 588, 878, 887
photoinduced discharge, 533, 534, 874, 875
polaron self-trapping, 593
quantum efficiency (*see also* Quantum efficiency), 588, 589, 593, 874, 875
recombination, 588, 589, 591–593, 874
relaxation time, 588–590
rise time, 589, 878, 879
sensitivity to oxygen, 588, 590–593
sensitizer, 875, 880, 909
temperature dependence, 590–594
threshold, 592
transient, 589, 590
variable range hopping, 592
Photoemission (PES), 213, 229, 230, 257, 413, 414, 432, 436–438, 441, 448, 452–457, 503, 504, 508, 510, 689, **690–697**, 711, 717, 718, 721, 726
angle-resolved photoemission, 441, 453, 454, 693
doped C_{60}, 559, 561, 564, 568, 583, 584, **695–696**
higher mass fullerenes, 73
inverse photoemission spectroscopy (IPES), 213, 403, 413, 437, 438, 441, 452–454, 456, 457, 503, 504, 583, 689, **690–697**, 706, 726
linewidths, 692, 693, 694
ultraviolet photoelectron spectroscopy (UPS), 559, 690, **691–694**, 696, 718, 726
x-ray photoelectron spectroscopy (XPS), 230, 455, **694–698**, 718, 719, 729
Photodiodes, 871, 873, 909
Photofragmentation, 6, **157–159**
Photoinduced discharge curves, 533, 534
Photoinduced oxygen diffusion, 881
Photoionization, 52, 53, 159, 719
delayed photoionization, 159
Photolithography, 282, 827, 889, 890, 894, 909, 911
Photoluminescence (*see* Optical properties)
Photorefractive device, 533, 870, 873, **876–880**, 909
Photoresist applications, 209, 881, **888–889**, 890, 909
Phototransformation (*see also* Polymerization), **209–217**, 307, **395–402**, 465, 491, 492, **508–511**, 541, 547, 548, 588, 717, 718, 870, 873, 888, 889
Photovoltaic device, 533, 871, 880, 881, **886–888**, 909
open circuit voltage, 886, 887
photocurrent density, 886, 887
short circuit current density, 886, 887
Plasma frequency, **527–528**, 648, 699, **705**
dependence on fullerene mass, 705
Plasmons, 689, 694, 696, 699, 701, 702, **704–706**, 708

dispersion relations, 699
energies, 705
π plasmon, 705, 708
$\pi + \sigma$ plasmon, 705, 708
Point groups (*see also* Group theory)
C_{1h}, 97, 100, 108
C_2, 72, 73, 229, 453
C_{2v}, 71, 72, 229, 352
C_3, 72, 352
C_{3v}, 72, 229, 230
C_m, 201
$C_{N/\Omega}$, 798
D_2, 71–73, 124, 452, 453, 731
D_{2d}, 73, 452, 453
D_{2h}, 352
D_3, 71, 72, 122
D_{3d}, 50, 580
D_{3h}, 50, 71, 72
D_{4h}, 815, 816
D_5, 97, 98, 419
D_{5d}, 66, 71, 72, 97–100, 102–105, 516–518, 795, 797
D_{5h}, 66, 68, 71, 72, 97–105, 355, 391, 392, 418–420, 433, 434, 569, 746, 795
D_{6d}, 73, 97, 100, 102, 103, 105, 391, 795
D_{6h}, 16, 18, 73, 797, 815–817
D_{nd}, 795–797, 805, 842–844
$D_{(2n)d}$, 796, 797
$D_{(2n+1)d}$, 796
D_{nh}, 795–797, 842, 844
$D_{\infty h}$, 795
I, 70, 71, **80–97**, 99, 419
compatibility relations, 97, 99
direct products, 88, 89
double group, 84, 91
double group basis functions, 92
multiplication table, 89
symmetry, 70
I_h, 70, 71, **80–97**, 99, 332, 333, 580, 795
angular momentum states, 84
basis functions, 83
character table, 82
compatibility relations, 97, 99
degeneracy, 333
double group, 91, 92
eigenvectors, 333
electronic states, **90–96**
equivalent atomic sites, 86, 87
irreducible representations, 332, 333
isomers, 71, 72
molecular diameters, 70
symmetry, 62, 70–72, 86, 87, 176, 178, 185, 244, 340, 419, 420, 516, 517, 580
vibrational modes, **86–90**, 340
P_2, 201
P_m, 201
$Pbnm$, 175
S_6 ($\bar{3}$), 181
character table, 182
T_d, 73, 451, 452
T_h, 100, 352
basis functions, 177
character table, 177
equivalence transformation, 341
symmetry, 97, 100, 176, 185, 244, 249, 340
rotation group (*see* Rotation group)
Poisson ratio, 336
Polarizability
electronic, 382, 456, 457, 479
molecular, 479, 492–494, 902
Polarization effects, 506, 507, 544
Polarization energy, 457
Polaron, 488, 491, 528, 535–537, 541, 543, 593
Polyalkenes, 292
Polyaromatic hydrocarbons (PAH), 115–117, 122, 138, 907
Polychlorinated biphenyls (PCB), 907
Polyhedron, 27, 29, 64, 76
Polymerization of fullerenes (*see also* Phototransformation), **209–217**, 248, 293, 294, 297, 315, **316–324**, 329, 374, **395–402**, 403, 491, 664, 712, 895
activation energy, 400
chemical, 294, **320–325**
cross-linking, 207, 210, 211, 213, 247, 398, 400, 401
doping-induced, 247, 248, 380, 381, **531–532**, 664
effect of oxygen, 211, 217, 238, 294
electron beam-induced, 209, **213–214**
infrared, 213, 217, 395, 399
laser desorption mass spectroscopy (LDMS), 210, 214–217
optical studies, 214, 217, 541
photo-transformation, **209–212**, **215–217**, 238, **399–402**, 541
plasma-induced, **214–215**

Polymerization of fullerenes (*continued*)
pressure-induced, 209, **213–214**, 403
Raman (*see* Raman scattering)
rate constant, 400, 401
Polymers, 24, 31, 43, **46–48**, 153, 171, **209–217**, 297, **532–537**, 871, **873–881**, 910
aminopolymers, 323
backbone of polymers, 873, 898
block copolymers, 320, 321, 323
C_{60}–polymer composites (*see* Composites)
cation–anion bilayers, 896, 897
chain length, 544, 874
charge transfer (*see* Charge transfer)
conducting, 885
copolymer, 320, 321, 323, 324
cross-linking, 320–323
doped with fullerenes, 533–536, 587, 588, 871, 873–875, **876–880**, **881–883**, **885–888**
glass transition, 878
graft copolymer, 321
homopolymer, 321
ladder polymer, 46–48, 898
mobile holes, 534, 535, 876, 877, 881
optical properties, **532–537**, 873, 874
π–conjugated, 544
phenylmethylpolysilane (PMPS), 875
photoexcitation, 874
plasticizer, 878
polyanion chains, 322, 896, 897
polycations, 896, 897
poly(allylamine hydrochloride) (PAH), 896, 897
polydiacetylene (PDA), 876
poly-hexylthiophene (PHT), 533, 873, 874
poly(2-methoxy, 5–12′-ethyl-hexyloxyl)-*p*-phenylene vinylene (MEH-PPV), 535, 536, 875, 885–888
poly(3-octyldithiophene) (PODT), 876
poly(phenylene vinylene) (PPV), 876, 896, 897
polyvinylcarbazole (PVK), 533, 534, 587, 874, 875, 877–880
PVK:C_{60}:DEANST, **877–880**
“pendant,” 321–324
“pearl necklace,” 321–323
processibility, 898
side chain, 320, 321
thiapyrylium-doped polycarbonate, 874
Pores (*see also* Micropores, Nanopores), 24, 25, 34–36
Positron, 238
lifetime, 238
probe, 234, 238
trapping, 238
Potassium (*see also* Alkali metal), 239, 286, 314, 330, 450, 559, 568, 646, 664, 702
Pressure, 258, 893
crystal structure dependence, **203–207**, 586, 587
density of states dependence, 587
electrical resistivity dependence, 586
transition temperature T_{01} dependence, 172, 199, 204, 208, 586
thermal activation energy dependence, 587
vibrational spectra dependence, **402–404**
Pressure-induced
phases, 175, 203–208, 240, 243, 402, 403, 563
polymerization, **213–214**
Projection mapping, 51, 69, **74–77**, 770–772, 787
Purification of fullerenes, **121–131**, 911
filtration, 123, 124
gas phase methods, **128–129**
solvent methods, **121–125**
sublimation, **125–128**
Pyrolytic carbon (*see* Carbon materials)
Pyracylene motif, 47, **61–62**, **144–147**, 292, 293, **295–299**, 308, 310, 313, 314

Q

Quantum efficiency (*see also* Photoconductivity), 483–485, 488, 588, 589, 593, 874, 875
Quantum transport (*see also* Transport properties)
1D wires, 858, 904
Mott formula, 586
universal conductance fluctuations (*see* Transport properties)
variable range hopping, 25, 165, 586, 592
weak localization, 570, 583, 831, 832
Quartz crystal balance, 567, 569
Quenched state of M_1C_{60} (*see* Crystal structure)

R

Radiative lifetime (*see* Optical properties)
Raman scattering, 296, 324, **336–339**, **355–361**, 363, 370, **376–381**, 382, 395–397, **398–402**, 485, 511, 643, 699, **717–719**, 749, 850, 888, 889
- alkali metal species dependence, 376–380, 396, 397
- Breit–Wigner–Fano lineshape, 340, 378, 380
- broadening by doping, 377, 378, 380, 406, 643, 718
- C_{60}–metal vibration, 718
- C_{70}, 390–392, 397, 718
- carbon arc material, 851–854
- carbon nanotubes, 756, 768, 797, 799, 839, **841–854**
- combination modes, 358, 370, 478, 852
- cross sections, 846
- crystal field effects, 183, **339–341**, 378
- disorder-induced band, 852–854
- doped C_{60}, **376–381**, 385, 643, 718
- doped C_{70}, 395, 397, 398
- doping-induced mode shifts, 286, 374, 376, **378–380**, 385, 395–398, 718, 749
- endohedral fullerenes, 229
- excited state spectra, **374–375**
- first-order, 88, 89, 334, 338, **355–358**, 365, 481, 850–854
- glassy carbon, 25, 851–854
- graphite, 17, 42, 851, 853, 854
- harmonics, 853
- higher order, 88, 89, 331, 332, 334, **357–361**, 365, 478, 850–854
- intensity dependence, 375, 395, 397, 406
- intermolecular modes, 352, 401, 402
- isotope effect, 108, 334, 347, **363–364**
- Lorentzian lineshape, 372, 375, 378, 400, 401, 852
- metal–C_{60} multilayers, 570
- microprobe, 850
- mode displacements, 335–337, 379, 396
- oxygen effects, 374
- overtones, 358–361, 478
- phase transition, 185, 190, 370, 371, 374, 375
- phototransformed
  - C_{60} mode frequencies, 398–402
  - C_{70}, 402
  - spectra, 210, 211, 217, 356, **397–402** 406, 888, 889
- polarization effect, 88, 104, 179, 180, **355–356**, 378, 392
- pressure dependence, 402, 403
- pressure-induced phase, 207
- Raman active, 211, 215, 332, 338, 849
  - $A_g(1)$ (breathing) mode, 336–339, **355–357**, 371, 376–380, 396
  - $A_g(2)$ (pentagonal pinch) mode, 330, **333–336**, 338, 339, 356, 357, 364, 365, 371, 374–380, 385, 386, 396, 398, 399, 403, 406, 718, 749
  - H_g modes, 330, **333–335**, 338, 339, 356, 357, 359, 370, 371, 376–380, 406, 643
- resonant enhancement, 380
- selection rules, 88, 89, 104, 179, 180, 182, 332, 357, 359, 360, 363, 400, 706, 797, 799
- solution spectrum, 363, 364
- splittings, 340, 341, 370, 372, 378
- surface-induced shifts, 718
- symmetry lowering at surface, 718
- TDAE-C_{60}, 405, 406, 749
- TDAE-C_{70}, 406
- temperature dependence, 364, 370–372, 374, 398, 400–402, 406

Rare earth dopants, 231, 232, 234, 241, 256, **742–743**
Rare gas, 232
Ratchet motion (*see* Rotation of fullerenes)
Rb_xC_{60} (*see* Alkali metal fullerides)
Reactions (*see* Chemical reactions)
Reactive site, 292
Reactivity, 294
Rectification, 884–886
- I–V characteristics, 884–886
- sign, 884

Recombination (*see also* Photoconductivity), 897
Relaxation
- between potential minima, 597
- excited state, 541
- magnetic, 752
- molecular, 189–192
- rates, 674
- spin (*see* Spin–lattice relaxation)
- structural, 229, 535, 597

Reorientation time, 658, 659, 723

Resistivity (*see also* Transport properties), 247, 248, 261, 286, 385, 498, 521, 528, 531, **557–573**
carbon nanotubes (*see* Transport properties)
maximum, 563
minimum, 286, 385, 395, 445, 558
residual, 245, 246, 561
sign of $\partial\rho/\partial T$, 560–562, 565
Rhombohedral structure
$C_{60}S_8CS_2$, 236, 237
pressure induced rh(pC_{60}), 206, 207, 403, 404
Ring currents, 739–742, 745
Ring of carbon atoms, 4, 44–46, 144, 149, 151, 167
Rock salt structure, 226, 242, 247, 248, 380, 531
Rotation of fullerenes (*see also* Phase transitions), 173, 597, 683, 692, 712, 714, 723, 731, 732, 749, 750
axis, 185, 200, 202, 660, 661
freezing of motion, 185, 188, 189, 191–193, 669
hindered, 184, 189, 259, 329, 354, 657, 749
isotropic, 184, 185, 661
period, 173
ratchet-type, 174, 185, 188, 189, 192, 193, 200, 202, 258, 354, 371, 373, 597, 658, 659, 662
uniaxial, 184, 661
Rotational energy levels, 341–343, 594
Rotation group
basis functions, 82–84
spherical harmonics, 82, 83, 100
Rotational potential, 388
Rotational spectra, **342–347**, 348
fine structure, 343–346
hyperfine structure, 344, 345
superfine structure, 344–346
Rubidium (*see also* Alkali metal fullerides), 559
Rutherford back-scattering (RBS), 558, 567, 569

S

Scandium compounds
Sc^{3+}, 230, 671, 744
$Sc@C_{74}$, 230, 232, 233, 733
$Sc_2@C_{74}$, 230, 232, 233, 733
$Sc@C_{82}$, 671
$Sc_2@C_{82}$, 135, 435, 672
$Sc_3@C_{82}$, 229, 230, 671, 672
$Sc_2@C_{84}$, 233, 672
Scanning tunneling microscope (STM), 12, 72, 212, 213, 608, **712–715**, 724–733, 768, 771, 775, 777, **825–827**, 889, 893, 902, **904–905**, 906
carbon nanotubes, 407, 712, 756, 761, 762, 765, 771, 773, 774, 776, 777, 785, **825–827**
barrier height, 826
I–V measurements, 825, **826**, 827
topographic study, 826
tunneling current, 826
endohedral fullerenes, 229, 230, 232
field ion STM, 725
fullerene-surface interaction, **712–714**, 721
higher mass fullerenes, 202, 203, 452
image quality, 904, 905
imaging individual C atoms, 712, 714, 725
inducing polymerization, 712
moire pattern, 776, 777
molecular manipulation by tip, 712, 715, 889, 894, 902
superconducting gap, 627, **633–638**, 645
surface science, 689, 697, 710, **712–715**, 718, 720, 724–733
tip–fullerene interaction, 714, 889, 894, 905
tunneling, 712, 904
use for electron irradiation, 209, 712
use of C_{60} on STM tip, 712, **904–905**
Scanning tunneling spectroscopy (STS), 633–635, 724, 729, 730, **825–827**, 905
Schlegel diagram, 299
Schottky anomaly, 208, 596
Schottky diode, 884
Screening, 387, 455, 456, 479, 480, 503, 544, 573, 574, 577, 625, 647, 667, 690, 691, 740, 790, 838
Thomas–Fermi, 573, 574
Screw axis, 773
Scroll morphology, 21, 765
Seebeck coefficient, 581, **603–608**
Self-assembly, 9, 11, 151, 152, **896–897**
Semiconductor, 39, 225, 518
carbon nanotubes (*see* Carbon nanotubes)
Semimetal, 15, 17, 19, 254

Sensors, 890, 902, **907–908**, 909
  response time, 908
  selectivity, 908
  sensitivity, 907
Separations, 890, 907, 909
Shape (*see also* C_{70}), 331
Shell model, 336, 337
Shock-quenched phases, 16, 31
Shungite, 1, 11, 110
Si (100) substrate (*see* Substrates)
Si (111) substrate (*see* Substrates)
SiC, 709, 710, 713, 716, 730, 732, **893–894**, 911
Silent modes (*see also* Vibrational modes), 90, 332, 334, 359, 363, 365, 699, 706
Simple cubic (sc) structure (*see also* $Pa\bar{3}$, Space group), 172, 184, 205, 242, 248, 258, 370
Single bonds (*see* Bonding)
Single crystal
  doped fullerenes, 323, 524, 566, 570, 583, 629
  fullerene growth, 271, **272–274**
    crystal twinning, 273, 274
    solution growth, 271, 272
    sublimation growth, 271, **272–274**
    temperature gradient method, 273
  fullerenes, 175, **272–274**, 541, 632, 659, 740, 883
Singlet states (*see* Electronic structure)
Singlet–triplet splitting, 422, 467, 473, **504**, **693**, **702**
Site location for dopants (*see also* Octahedral site, Tetrahedral site), 246, 258, 263, 563, 655–657
Site ordering, 252, 563
Site selectivity, 252
Size
  atomic, 251, 261
  ions, 236, 238, 240, 256, 563, 568
  vacancies, 238, 240, 252, 563, 568
Sodium compounds (*see* Alkali metal)
Solubility, 118, 119, **120–121**, 294, 325, 888, 889
Solvation, 294
Solvents, 116–125, 227, 294, 300, 316, 322, 323, 894, 895
  alkanes, 119
  polar, 119
Soot, 64, 110–118, 124, 129, 765, 766, 838, 850–852, 857, 860, 895, 907, 911
Sound velocity, 172, 185, 575
Soxhlet extraction, **116–118**, 122, 123
Space group, 16–18, 40, 80, 85, **175–188**, 193, 199–201, **236–237**, **241–242**, **245**, **248–250**, 253, 254, **256–258**, **260–261**, 332, 349, 445, 560, 595, 623
  carbon nanotubes, **792–793**, 840, 844
    basic symmetry operations, 759, 791, **792–795**, **797**, 798, 809
    chiral nanotubes, **798–799**, 809
    hexagonal $P6/mcc$ (D_{6h}^2), 815–817
    hexagonal $P6/mmm$ (D_{6h}^1), 815, 816
    level degeneracy, 798
    non-symmorphic, 795, **797–800**, 803, 840, 841, 844, 845, 848
    rotation, **792–795**, 797, 809
    symmorphic, **795–797**, 803–809, 840, 841, 845, 848
    tetragonal $P4_2/mmc$ (D_{4h}^9), 815, 816
    translation, **792–795**, 797, 799
  D_{4h}^9 ($P4_2/mmc$), 815, 816
  D_{6h}^1 ($P6/mmm$), 815, 816
  D_{6h}^2 ($P6/mcc$), 815, 816
  D_{6h}^4 ($P6_3/mmc$), 16, 18, 40
  $Fm3$, 176, 177, 180, 193
  $Fm\bar{3}$, 253, 254, 441, 442, 454, 455
  $Fm\bar{3}m$ (O_h^5), 173, 175, 184, 185, 193, 242, 245, 254, 389, 390, 595, 623, 657
    pressure-induced phase, 206, 207
  $I4/mmm$, 248
  $Im\bar{3}$, 242, 249, 255, 256, 623
  $P1$, 320
  $P2$, 260
  $Pa\bar{3}$, 174, **180–184**, 242, 254, 258, 259, 332, 349, 351, 388–390, 445, 447, 454, 455, 595, 623, 658
    site symmetries, **180–181**
  $Pm\bar{3}n$, 242, 254, 255
  $R\bar{3}m$ pressure-induced phase, 206
  symmetry sites, 176, 178, 179
Specific heat, 17, 184, 189, 190, 193, 196, 197, 346, 350, 354, 557, 584, **594–599**, 600, 603, 606, 617, 629, 633, 648, 746, 752
  anomalies, 595–599
  Debye contribution, 594, 597–599
  Debye temperature (*see* Debye temperature)
  degrees of freedom, 594, 595, 597, 598
  Einstein contribution, 595, 597–599

Specific heat (*continued*)
hysteresis, 597
low temperature, 597
magnetic dependence, 752
many-body effects, 585, 599
on cooling, 596
on heating, 596
strong coupling enhancement, 585, 599
structural disorder contribution, 597, 598
superconductivity jump, 598–600
temperature dependence, **594–600**
Spherical approximation, **336–338**, 417
Spherical harmonics, **82–84**, **92–96**, 417
Spherules (*see* Multilayer structures, Carbon onions)
Spin concentration, 569
Spin-glass (*see* Magnetic properties), 669, 752, 753
Spin–lattice relaxation (*see also* EPR, NMR, Muon spin resonance), 188, **635–638**, 657–660, 663–666, 674, 676, 680, 683, 684, 838
electron, 676, 680, 838
muon, 636, 637, 684, 753
nuclear, 188, 585, **635–638**, 657–660, 663–666, 753
Spin–orbit interaction, 467, 472, 481
carbon, 63, 424, 473
fullerenes, 427–429
Spin susceptibility *see also* Magnetic susceptibility, 527, 637
temperature dependence, 637
Sr_xC_{60} (*see* Alkaline earth fullerides)
Stability, 3, 5, 6, 15, 111, 112, 143, 144, **153–156**, 207, 418, 445, 664, 691, 696, 784, 860, 862, 899, 905, 908
Stacking arrangement, 19, 20, 24, 32, 33, 201, 202, 237, 257
Stacking disorder, 23, 198
Standard electrode, 300
Standard orientation, 244–246
Sticking coefficient, 716
Stone–Wales model, **144–147**, 786
Strain, 151, 226, 720, 815, 894
Stress, 62, 161, 166, 282, 309, 724, 893
anisotropic, 893
carbon nanotubes, 775, 858
Structure of fullerenes (*see also* Crystal structure), 43, **60–78**, 234
doped C_{60}, 234, 235, 236
projection method (*see also* Projection mapping), 69, **74–77**
Sublimation, 31, 116, **118–120**, 121, **125–128**, 129–131, 172, 175, 199, 201, 203, 214, 228, 271, **272–274**, 279, 440, 506, 537, 689, 715, 716, 720, 726, 889, 890
crystal growth, 175, 271, **272–274**
heat of sublimation, 121, 172
temperature, 172
Substrate corrugation, 713
Substrates, 41, 153, 214, 274, 276, 277, 699, 700, **719–733**, 857, 884, 894, 902, 906
alkali halides, 720
Al_2O_3, 713, 718, 720
aluminum, 710–713, 720, 891
C_{60}, 697, 712, 713, 716, 717
CaF_2 (111), 720
CsI, 392
diamond, 715
GaAs (110), 274, 713, 720, 721, 730–732
GaSe (001), 274, 279–281, 369, 370
germanium, 891
GeS (001), 274, 279, 280, 369, 717, 720
glass, 214, 588
graphite (HOPG), 713, 714
H-terminated Si (111), 720
heated, 163, 700, 710, 711, 714, 716, 728, 893, 894
In, 365
KBr, 274, 356, 537
mica, 274, 276–280, 714, 715, 720, 906
MoS_2, 274, 280, 720
NaCl, 274, 277, 720
NaCl (011), 902
noble metals, 697, 698, 717–719, **721–725**, 726, 730
Ag (111), 365, 712, 713, **720–725**
Au (001), 712, 720
Au (110), 701, 702, 717, 724
Au (111), 712, 713, 720, **721–725**
Cu (100), 697, 698, 700, 724
Cu (110), 700, 724
Cu (111), 369, 700, 712, **720–725**
oxides, 712
polar materials, 902
potassium, 701
Pyrex, 274
quartz, 41, 274, 495, 501, 509, 510, 518, 519, 537, 539
sapphire (Al_2O_3), 274, 572, 713, 716, 720

Sb, 274
semiconductors, 713, 731
Si, 384, 537, 721, **726–730**, 881, 884, 888–891, 893, 894, 905, 906
Si (100), 274, 279, 281–283, 369, 392, 538, 702, 708, 709, 713, 716, 717, 720, 725, **726–729**, 730–733, 888, 894
Si (111), 281, 708–710, 713–715, 720, **726–730**, 893, 894
Si (110), 281
SiO_2, 713, 716, 720, 890, 894
steel, 713
Suprasil, 274, 400, 495, 496, 498, 499, 508, 509, 518, 538, 539
transition metals, 697, **725–726**, 884
Nb, 884
Pt, 904, 905
Pt–Rh, 904, 905
Rh (111), 369, 697, 702, 710, 717, 725, 726
Ta (110), 717
Ti, 884
W (100), 697, 725, 726
yttrium-stabilized zirconia (YSZ), 570, 571
Sulfonated C_{60}, 896, 897
Sum rules, 529, 704, 705
f-sum, 470
Superconductivity, 7, 226, 235, 241, 243, 252, 253, 255, 287, 299, 423, 428, 447, 524, 564, 600, **616–653**, 663, **863–864**, 909, 911
alkaline earth doped, 241, 253, 287, 299, 621, 622, 646
applications, 909
average phonon frequency, 617, 619, 625, 626, 638, 644, 645
Bardeen–Cooper–Schrieffer (BCS) theory, 255, 299, 529, 575, 576, 616, **617**, 619, **625**, 628, **633–635**, 637–640, 644, 666, 668, 684, 691
carbon nanotubes, 757, 818
clean limit, 633
coherence length, 564, 626, **627**, 629, 630, 632, 633
Pippard length, 633
Coulomb repulsion μ^*, 449, 451, 625, 626, 645–647
critical current, 627
critical field
field dependence, 629
$\partial H_{c2}/\partial T$, 599, 627–629
lower (H_{c1}), 626, 627, 630, 631
thermodynamic, 632
upper (H_{c2}), 599, 626–629, 631, 632
critical temperature, 227, 232, **243**, 285, 367, 378, 388, 445, 449, 451, 524, 528, 561, 562, 575, 616–623, **624–626**, 627, 629, 631, 635–637, 639–641, 643, 644, 663, 665–667
alloy concentration dependence, 620, 621, 644
density of states dependence, 616, 617, 619, 621, 622, 643, 644
lattice constant dependence, 241–243, 617–621, 623, 643, 644
pressure dependence, 641, 642
cuprate superconductors, 528, 564, 617, 627, 630, 639–641
de Almeida–Thouless relation, 631, 632
density of states, 637
diamagnetic shielding, 524
diamagnetism, 631, 667
dirty limit, 632
dopant species sensitivity, 623, 643
electron–phonon coupling, 375, 376, 451, **573–580**, 617, 624, 625, 641, 643–645
intermolecular, 573, 638, 643
intramolecular, 573, 619, 626, 638, 643, 644, 645, 646
librational modes, 638
Eliashberg equations, 625, 638, 644, 645
excited quasiparticles, 637
Fermi surface anisotropy, 629
fluctuations, 566, 583, 621
flux penetration, 624, 627, 630
gap, 232, 365, 528, 616, 627, **633–638**, 641, 645, 664–666, 668
field dependence, 635
temperature dependence, 616, **633–638**, 666, 712
Ginzburg–Landau theory, 626, 628, 630, 633
graphite intercalation compounds, 616, 617
ground state, 528
heavy Fermion, 645
Hubbard model, 644–647
isotope effect, 575, 616, **638–641**, 643, 648

Superconductivity (*continued*)
Jahn–Teller mechanism, 643, 647, 648
Knight shift, 664, 667
London penetration depth, 530, 626, 627, 629, 630, 637, 667, 668, 683, 684
temperature dependence, 629, 637
Ginzburg–Landau, 630
magnetic field effects, 600, 616, **626–633**
magnetic susceptibility, 622, 631
field cooled, 622, 624, 630, 631, 633
temperature dependence, 630, 641
zero field cooled, 622, 624, 630, 631, 633
magnetization, 624, 633, 639
Mattis–Bardeen theory, 638
McMillan's formula, 449, 577, 625, 626
mean free path, 627, 629, 632, 633
Meissner fraction, 631, 633
Meissner phase, 630, 631, 633, 667
microwave surface impedance, 624, 629, 630, 638
organic superconductors, 617, 621, 641
optical properties, 524, **528–530**
pair breaking, 645
pairing interaction energy, 617, 619, 626
pairing mechanism, 573, 578, 616, 617, 619, 624, 626, 638, 641, **643–648**
electron–electron, 616, 625, 638, 644–648
electron–phonon, 616, 624, 626, 638, 641, 643–646, 648
electron–plasmon, 647
Pauli-limiting spin effects, 629
perfect diamagnet, 624
phase diagram, 627, 630–632
pressure dependent effects, 575, 587, 616–619, 627, 633, **641–643**, 645, 648
quasiparticles, 456, 664
shielding fraction, 624, 625, 631
specific heat jump, 598–600, 629
spectral function, 638, 644, 645
spectral weight, 644, 645
spin lattice relaxation, 636, 665–668
Hebel–Slichter peak, 636, 637, 666, 684
strong coupling, 625, 634, 638, 645
temperature dependence of resistivity, 624, 627, **628**
transition metal superconductors, 639
transition temperature width, 561, 562, 565, 625, 627, 640
trapped flux, 636
type II, 632, 633
vertex corrections, 575, 644
vortex fluid phase, 630–632
vortex glass phase, 631–633
vortex pinning, 632
vortices, 626, 663, 666
wave function, 627
weak coupling BCS, 529, 625, 633, 637, 638, 668
Werthamer, Helfand, and Hohenberg (WHH) formula, 628, 629
Supercooling, 35, 199, 200
Superlattice, 39, 193, **570–572**, 722, 723, 896, 897
Surface science, **689–733**
Surface–fullerene interaction, 276, 262, 271, **276–281**, 408, 689, 697, 709, 710, **712–714**, 715–718, 720, 721, 725–728, 730–732, 870, 881, 889–891, 893, 894, 904, 906
binding energy, 697, 712, 713, 714, 715, 716, 726, 890
Surface-enhanced Raman spectroscopy (SERS), 365, **717–719**
Surfaces
area, 34, 35, 116, 288, 784, 895
coverage (*see* Coverage)
density, 727
desorption (*see* Desorption)
energy, 700, 724, 905
reconstruction, 439, 689, 690, 721, 724
stability, 905
states, 438, 690, 725, 905
steps, 724
structure, 697, 712, 725, 905
Symmetry of fullerenes, **80–109**, 177
carbon nanotubes (*see also* Space group), 757, 758, 775, **791–802**, 805
electronic state symmetry, 65, 342, 725
fivefold axes, 1, 9, 10, 12, 62, 63, 66, 67, 69, 80, 81, 82, 85, 88, 96, 97, 99, 102, 103, 105, 186, 194, 200, 393, 394, 420, 598, 659, 757, 775
icosahedral operations, 80, 81
isotopic considerations, 77, **104–108**, **342–347**
lowering symmetry, **96–104**, 177–179, 296, 310, **339–341**, 346, 390–392, 395, 403, 417, 419, 420, 432, 433, 470, 498, 508, 516, 541, 542, 548, 576, 579, 580, 703, 720, 745, 791, 795

threefold axes, 62, 81, 85, 88, 178, 181, 184, 186, 187, 190, 194, 199, 311, 578, 659, 723–725, 757
twofold axes, 9, 12, 48, 62, 81, 85, 88, 176–179, 185–187, 190, 193, 244, 245, 261, 419, 447, 725, 797
vibrational mode symmetry (*see* Vibrational modes)
Synchrotron light source, 158, 260, 363
Synthesis of fullerenes, 3, 12, **110–115**, 911
carbon nanotubes (*see* Carbon nanotubes)
electrodes, 113–115
extraction (*see* Extraction of fullerenes)
gas phase synthesis, 3, 4, 111
helium gas pressure, 110–115
historical perspective, **111–112**
model reactor, 113–115
natural occurrence, 1, 10, 11, 110
purification (*see* Purification of fullerenes)
resistive heating, 110, 111, 115
yield, 112–115

T

t_{1g}-derived band (*see also* LUMO+1), 236, 254, 382, 424, 444, 445, 447, 449, 453, 500, 567, 568, 623, 624, 646, 695, 696
half filled, 567, 623
t_{1u}-derived band (*see also* LUMO), 236, 249, 254, 298, 300, 382, 395, 415, 423, 424, 427, 437, 441, 444, 445, 447–451, 453, 456, 500, 501, 518, 558, 563, 567, 574, 579, 580, 583, 617, 623, 624, 643, 645, 646, 692, 695, 702, 708, 749
half filled, 556, 558, 561, 617, 623
TDAE–C_{60}, **259–260**, 299, **404–406**, **668–669**, **680–684**, **747–753**, 883
Curie temperature, 260, 748, 750, 752, 753
EPR (*see* EPR)
infrared spectra, 406, 749
lattice constants, 260, 261, 750
NMR, 749, 750
magnetic properties (*see also* Magnetic properties), **748–753**
μSR, 750
optical properties, 748, 749
orientational ordering temperature T_{01}, 680, 683, 749, 750, 752
Raman, 405, 406, 749
specific heat, 752
spin concentration, 750
structure, **259–261**, 748, 749
susceptibility, 749
transport, 749
vibrational spectra, **404–406**, 748, 749
TDAE–C_{70}, 262, 404, 406, 751, 753
magnetic properties, 751, 753
structure, **404–406**
Technigas, 908
Temperature-programmed desorption (TPD), 689, 697, 710, 713, **715–717**, 720, 726, 732, 891
heat of desorption, 713, 715, 716, 732
Tetrahedral site, 172, **226**, 236, 238, **240–241**, **243**, 245, **248–254**, 256, 263, 303, 389, 390, 442, 443, 445, 449–451, 518, 563, 567, 568, 582, 621, 646, 656, 657, 730, 743
interstitial site radius, 172
Tetrahedron, 39, 251, 252, 277
Tetrakis-dimethylamino-ethylene (*see* TDAE–C_{60}, TDAE–C_{70})
Thermal bandgap, 572
Thermal conductivity, 35, 36, 42, 172, 185, 189, 190, 193, 200, 286, 557, **602–604**, 856, 857, 894
anomalies, 602
electron contribution, 602
Einstein model, 603
lattice contribution, 602
maximum, 602
temperature dependence, 602–604
Thermal expansion, 172, 185, 208, 238, 506, **601–602**
at T_{01}, 238, 601
coefficient, 172, 557, 560, 601, 602
K_3C_{60}, 560, 601
Thermionic emission, 159
Thermopower (*see also* Seebeck coefficient), 557, 581, 584, **603–608**
diffusion, 604, 606
Fermi level, 606, 607
negative sign, 605, 608
phonon drag, 604, 606–608
temperature dependence, 581, 603, 604, 605, 606, 607, 608
Thin films (*see also* Film growth), 11, 41, 203, 210, 261, **274–283**, 285, 288, 324, 356, 396, 398, 496, 500, 522, 532, 565, 566, 580, 581, 591, 694, 700, 706, 707, 719, 730, 731, 885, 888, 890, 902, 909

Thomas–Fermi screening, 573
Threefold axis (*see* Symmetry)
Tight binding models (*see* Electronic structure, Carbon nanotubes)
Transition metal dopants, 232
Transfer integral (*see* Transport properties)
Transmission electron microscopy (TEM), 9, 26, 27, 29, 147, 190, 202, 229, 232, 233, 276–278, 281, 700, 775, 777, 780, 782–784, 789, 825, 849, 854, 905
  carbon nanotubes, 756, 761, 762, 764–768, 771–774, 825, 849, 854, 855, 858–860
    moire pattern, 774
  carbon onions, 860, 861
Transport properties, 175, 311, 448, **524–528**, **556–594**, **603–608**, 616, 627, 628, 648, 880
  arc deposits, 829
  carbon nanotubes, 768, 785, 814, 826, 829, 830, 832, 858
    carrier concentration, 827, 829, 830, 832, 838
    carrier mobility, 830
    conductance, 826
    conductivity, 830, 831, 838
    electrical contacts, **828**
    four probe measurements, 829–832
    Hall effect, 830, 831
    low T anomalies, 828
    magnetoresistance (*see also* Magnetoresistance), **829–833**
    resistance, 756, 768, 785, **827–829**, 838
    temperature dependence, **827–828**, 830, **831–832**
    two-probe measurements, 830
    2D weak localization, **831**, 832
    universal conductance fluctuations, 830, **832–833**
  carrier density, 238, 440, 443, 560, 566, 573, 574, 646
  carrier generation, 556
  density of states (*see also* Density of states), 556, 560, 570, **583–585**
  doping, 556
    alkali metal, **557–566**
    alkaline earth, **566–570**
  electrical conductivity, 16, 17, 21, 25, 26, 33, 37, 172, 184, 235, 286, 447, **557–573**, 594, 605, 606, 608, 624, 645, 707, 745, 748, 749, 752, 768, 785, 814, 829, 830, 832
    acceptor doping, **572–573**
    activation energy, 558, 564, 567–569, 587, 594, 749, 884, 885
    Boltzmann model, 566
    C_{60}-multilayers, **570–572**
    dopant dependence, 558–560, 566–568, 572
    doping dependence, **557–564**
    doped C_{70}, 556, 557, **569–570**, 580, 594, 608, 746
    Drude model, 527, 528, 565, 566, 749
    effect of merohedral disorder, 583
    elementary estimate, 583
    fluctuation-induced tunneling, 566, 570, 583, 608
    frequency dependence, 557, 565
    hopping, 563, 565, 566, 749, 874
    I–V curves, 825, 826, 885, 886
    linear T dependence, 564
    low temperature, 557, 561, 565
    maximum, 557–561, 564, 567–570, 580, 608
    metallic, 560, 561, 564, 568, 572, 608, 623, 664, 748
    microwave, 565, 622, 624, 629, 630, 749
    minimum, 558, 563, 567–569
    n-type, 818, 882, 883
    p-type, 41, 818, 883, 884
    quadratic T dependence, 564
    sample quality dependence, 561, 565, 568
    sensors, 907, 908
    sign of $\partial\sigma/\partial T$, 560–562, 565
    temperature dependence, **564–570**, 572, 746
    theoretical models, 565
    thermally-activated, 564, 567–569, 572, 573, 587, 593
    undoped C_{60}, 557
    undoped C_{70}, 569
  electron–electron interaction, 582, 583, 645, 646
  granular fullerides, 565, 629, 632
  Hall coefficient, 557, **580–581**, 607, 608, 830
  hole conduction, 572, 573, 875
  localization effects, 565, 566, 570, 573, 574, 582, 583, 592, 608, 645, 749, 831, 832
  magnetoresistance (*see also* Magnetoresistance), 557, **581–583**, 608

mean free path, 524, 564–566, 627
normal state properties, 556, 564, 574, 581, 645
photoconductivity, 557, **587–594**
pressure dependence, 587
rectification (*see* Rectification)
resistivity (*see* Resistivity)
scattering, 556, 557, 561, 565, 581, 582, 838
screening effects, 573, 574, 577, 647
sheet resistance, 571
temperature dependence, 557, 564–570, 572, 580–582, 644, 648, 746, 749
thermal conductivity, 17, 22, 35, 36, 42, 172, 185, 189, 193, 200, 557, **602–604**, 856, 857, 894
carbon nanotube, 856, 857
thermopower, 581, 584, **603–608**
transfer integral, 557, 645–647
Van der Pauw four-terminal, 570, 572
variable range hopping, 586, 592
wave function overlap, 556, 557, 559, 563, 577, 620, 643
Traps, 588, 589, 883
Tree-ring morphology, 23
Tribology, 902, 905, 906
Triclinic structure, 320
Tricresyl phosphate (TCP), 878
Triphenylphosphine (PPh_3), 310
Triple point for C_{60}, 173, 209
Triplet state, 156, 211, 366, 374, 375, 422, 427, 429, 430, 433, 465–467, 472–475, 480, 481, **483–491**, 504, 515, 516, 541, 543, 547, 673, 676, **678–680**, 693, 701, 702, 870–873
Truncated icosahedron, 1–3, 5, 8, 60–62, 75, 80–82, 174, 294, 345, 414, 441, 655
Truss structure, 8
Tunneling, 541, 634, 884
Turbostratic graphite, 19, 20, 24, 760, 761, 815, 829, 862
Twinning, 20, 124, 125, 198, 273, 275
Two phase coexistence, 239

U

U@C_{28} (*see* Endohedral fullerenes)
Ultrahigh vacuum (UHV), 279, 287, 558, 567, 710
Ultrasonic attenuation, 185, 190
Universal conductance fluctuations (*see also* Transport properties), 830, **832–833**
Unpaired electrons, 137, 298, 315, 427, 436, 516, 676, 681
Unpaired spins, 215, 315, 673, 676, 740, 749, 751
UV–visible spectroscopy, 47, 210, 214–217, 229, 284, 319, 324, 464, 468, 476, **492–493**, **498**, **501–502**, **513**, **897**, 898

V

Vacuum level, 717, 720
Valence band, 19, 33, 41, 432, 437, 441, 442, 449, 451, 501, 503, 504, 518, 521, 693, 711, 719, 804–807, 809, 810, 817, 829
carbon nanotubes, 719, 804–807, 809, 810, 817, 829
van der Waals interaction, 37, 175, 197, 203, 210, 225, 227, 242, 251, 277, 281, 316, 318, 320, 338, 398, 451, 713, 714, 857, 862, 888
bond, 888, 902, 905
force, 37, 236, 242, 714, 857
radius, 250
van Hove singularity, 863
Vapor deposition (*see also* Chemical vapor deposition), 41, 778, 893
Vapor grown carbon fibers (VGCF) (*see also* Carbon materials), **22–24**, 763, 765, 775, 779, 782, 783, 786, 791, 854, 857
Vapor growth method
C_{60} and C_{70}, 271, **272–274**
carbon nanotubes, 763, 782, 788, 789
thin films (*see* Film growth)
Vapor phase, 4, 21, 129, 133, 161, 173, 763, 782, 788
Vapor pressure, 36, **129–131**, 436, 716, 889
temperature dependence, **129–131**, 716
Vapor transport, 129, 272, 273, 277
Vaporization, 36, 129, 131
Variable angle spectroscopic ellipsometry (VASE), 493, 538, 539
Variable-range hopping (*see also* Quantum transport), 25, 165, 586, 592
Velocity of sound, 17, 172, 185, 190, 192, 575, 597, 602
Vertices, 60, 64, 80, 81, 148
Vibrational modes, **329–406**, 437, 493, 556, 574, 699, 706, 872
calculations, 354, 361, 391, 393, 394
ab initio, 338, 361

Vibrational modes (*continued*)
Born–von Kárman model, 349
force constant model, 338, 361, 393, 394
point charge model, 352, 353
C_{60} intermolecular modes, 183, 184, 329–331, **347–355**, 367, 376, 476, 573, 594, 595, 597, 607, 638
acoustic, 329, 330, 348, 351, 375
crystal field effects, 178
density of states, 330, 331
dispersion relations, 348–354
doping dependence, **388–390**
doping-induced mode broadening, 388
doping-induced mode shifts, 388
Γ-point, 348–353
librations, 108, 183, 189, 194, 329–331, 340, 346, **347–354**, 375, 388, 389, 574, 594, 595, 598, 638
metal–C_{60} vibrations, 330, 331, 376
optic modes, 329, 330, 352, 375
symmetry, 175, 178, 181, 183, **332–339**, 349, 352, 354
temperature dependence, 388, 389
translational modes, 348–350, 352–354
C_{60} intramolecular modes, 183, 184, 330, **332–347**, **355–375**, 465, 475, 478, 482, 498, 530, 573, 594, 595, 607, 638, 644
average phonon frequency, 577, 578
degrees of freedom, 333, 335
doping-induced mode shifts, 726
dynamical matrix, 335, 341, 391
eigenvectors, 332–335
gas phase, 338
high frequency, 574, 578, 645
isotope effects, **341–347**, 359, 639, 641
low frequency, 574, 578, 598, 645
mode degeneracies, 333, 498
mode frequencies, 334, 338, 339, 357, 359–361, 392–394, 466, 481, 483, 493, 496, 504, 519, 565, 595, 644
phase transitions, 215
radial motion, 330, 331, 333
silent modes, 90, 332, 334, 359, 363, 365–367, 369, 699, 706
solid state effects, 332, 338, **339–341**, 644
solution phase, 338
symmetry considerations, **86–90**, **97–104**, **183–184**, **332–339**, 366
tangential frequencies, 334
tangential motion, 330, 331, 333, 335
vibration-rotation spectra, **341–347**
C_{70} intermolecular modes, 390, **393–395**
C_{70} intramolecular modes, **390–394**
degrees of freedom, 86, 355, 390, 392, 507, 594, 595, 619, 843–845
intermolecular, 359, 594, 597, 598, 619
intramolecular, 594
rotational, 350, 594, 597, 712, 720
doped C_{60}, **375–390**, 519, **574–580**
H_g mode–electron coupling, 577
mode shifts, 376–388
doped C_{70}, 393–395
Einstein modes, 595
energy bandwidth, 413, 574, 648
experimental techniques, **331–332**
higher mass fullerenes, 215
mode classifications, 200, 329–332
pressure dependence, **402–404**
TDAE-C_{60}, **404–406**, 749
TDAE-C_{70}, **404–406**
unit conversion, 331
Vibronic states (*see* Electronic structure)
Virus, 9, 11, 12, 309
Viscosity, 323
Voids, 240, 241
Voltammogram (*see* Electrochemistry)
Volume of molecule, 68
Volume of unit cell, 248, 250
Volume expansion, 172, 238, 259, 600, 601, 621

W

W (100) surface (*see* Substrates)
Weak localization, 566, 582, 583, 608, 831, 832
Wear protection, 906
Wiedemann–Franz law, 602
Work function, 172, 234, 717, 718, 724, 726

X

Xe, **165–166**
XeF_2, 304
X-ray absorption, 257
X-ray diffraction (XRD), 26, 28, 29, 31, 186, 200, 229, 237, 247, 250, 252, 255, 256, 260, 261, 274, 276, 279, 308, 311, 319, 440, 570, 664, 679, 707, 854, 860, 894

θ–2θ scans, 282, 283
phase transitions at T_{01}, 184, 252
X-ray photoelectron spectroscopy (XPS) (*see* Photoemission)

Y

Young's modulus, 23, 172, 281, 336
Ytterbium compounds
structure, 256
Yb^{3+}, 743
Yb_3C_{60}, 242
Yb_xC_{60}, 234, 256
Yttrium compounds
carbide (YC_2), 859, 862
Y^{3+}, 230, 743, 744
^{89}Y, 672
$Y_2@C_{60}$, 132
$Y@C_{82}$, 133, 137, 230, 671, 672
$Y_2@C_{82}$, 131, 133, 230
$Y_m@C_{n_C}$, 136
Y_2O_3, 132, 133, 862
Yttrium-stabilized zirconia (YSZ) (*see* Substrates)

Z

Zero-field cooled (ZFC), 622, 624, 630, 633, 751
Zero-gap semiconductor, 19, 33, 537, 805, 807
Zigzag tubule (*see* Carbon nanotubes)
Zirconium compounds (*see also* Metallo-carbohedrenes)
Zr_mC_n, 50
$Zr@C_{28}$, 132, 231
Zone-folding (*see* Brillouin zone)
carbon nanotubes, 792, **801–811**, 840, **841**, 844, 845, 848, 849, 863